A Worldwide Timeline

Native Americans in California used a wide variety of numeration systems

George Pólya

FOUR COLORS SUFFICE

University of Illinois honors proof of the four-color theorem with a postmark

Grace Hopper

Across the World

Pioneering work with computers

Mayan calendric numbering

Quipu knotted cord numerical recording

Aztec stone calendar

For a map of birthplaces of notable mathematicians see: http://www.math.bme.hu

The Global

Isaac Newton

René Descartes

Gottfried von Leibniz

• Nikolay Lobachevsky

Charles Babbage

Leibniz's reckoning machine

I727YL789
Oldest written numerals in Europe

da Vinci's polyhedron

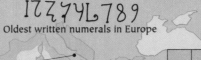
The golden rectangle

Subrahmanyan
Chandrasekhar •

Archimedes

_ΞΥΓ67S
Hindu numerals

Rhind papyrus

Hypatia

Srinivasa
Ramanujan

Euclid

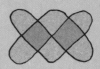
al-Khowârizmî, Algebra manuscript

Page from first printed edition
of Euclid's *Elements*

Bushoong sand drawing

Moslem manuscript of 1258
showing Pythagorean theorem

Cloud patterns analyzed using fractal geometry

Nature of Mathematics

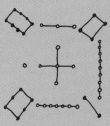

Magic Square — China

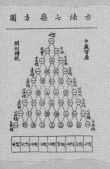

Zhū Shijiē — Pascal's triangle

The *Chou-Pei* text

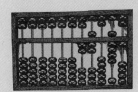

Takakazu Seki Kōwa

Abacus — China
(suan-pan)

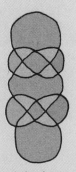

Malekula of Oceania
sand drawings

William Youden
(Youden squares)

$$\begin{bmatrix} x_1 & x_2 & x_3 \\ x_4 & x_5 & x_6 \\ x_7 & x_8 & x_9 \end{bmatrix}$$

Maori game of logic
and strategy

Alexander Aitken

THE NATURE OF MATHEMATICS

Eighth Edition

OTHER BROOKS/COLE TITLES BY KARL J. SMITH

Mathematics: Its Power and Utility, Fifth Edition
Trigonometry for College Students, Seventh Edition
Essentials of Trigonometry, Third Edition

The Nature of Mathematics

Eighth Edition

Karl J. Smith

BROOKS/COLE PUBLISHING COMPANY

I(T)P® An International Thomson Publishing Company

Pacific Grove ■ Albany ■ Belmont ■ Bonn ■ Boston ■ Cincinnati ■ Detroit ■ Johannesburg ■ London
Madrid ■ Melbourne ■ Mexico City ■ New York ■ Paris ■ Singapore ■ Tokyo ■ Toronto ■ Washington

Sponsoring Editor: *Robert W. Pirtle*
Marketing Team: *Jennifer Huber, Margaret Parks*
Editorial Assistant: *Melissa Duge*
Production Editor: *Tessa A. McGlasson*
Production: *Susan L. Reiland*
Manuscript Editor: *Susan L. Reiland*
Permissions Editor: *Lillian Campobasso*
Interior Design: *Nancy Benedict, Kathi Townes*

Interior Illustration: *TECH•arts*
Historical Note Portraits: *Steve Nau*
Cover Design: *Vernon T. Boes*
Cover Photo: *The Stock Connection*
Photo Researcher: *Sue C. Howard*
Typesetting: *Progressive Information Technologies*
Cover Printing: *Phoenix Color Corp.*
Printing and Binding: *World Color/Taunton*

For more information, contact:

BROOKS/COLE PUBLISHING COMPANY
511 Forest Lodge Road
Pacific Grove, CA 93950
USA

International Thomson Publishing Europe
Berkshire House 168-173
High Holborn
London WC1V 7AA
England

Thomas Nelson Australia
102 Dodds Street
South Melbourne, 3205
Victoria, Australia

Nelson Canada
1120 Birchmount Road
Scarborough, Ontario
Canada M1K 5G4

International Thomson Editores
Seneca 53
Col. Polanco
México, D. F., México C.P. 11560

International Thomson Publishing GmbH
Königswinterer Strasse 418
53227 Bonn
Germany

International Thomson Publishing Asia
221 Henderson Road
#05-10 Henderson Building
Singapore 0315

International Thomson Publishing Japan
Hirakawacho Kyowa Building, 3F
2-2-1 Hirakawacho
Chiyoda-ku, Tokyo 102
Japan

Printed in the United States of America

10   9   8   7   6   5   4

LIBRARY OF CONGRESS CATALOGING-IN-PUBLICATION DATA
Smith, Karl J.
  The nature of mathematics / Karl J. Smith.—8th ed.
  Includes index.
  ISBN 0-534-34988-9
  1. Mathematics.   I. Title.
QA39.2.S599   1998
510—dc21   97-40528
                                              CIP

I dedicate this book with love to
Melissa and Benjamin.

Preface

This book was written for those students who need a mathematics course to satisfy the general university competency requirement in mathematics. Because of the university requirement, many students enrolling in a course that uses my book have postponed taking this course as long as possible, are dreading the experience, and are coming with a great deal of anxiety. I wrote this book with one overriding goal: to create a positive attitude toward mathematics. Rather than simply presenting the technical details needed to proceed to the next course, I have attempted to give insight into what mathematics is, what it accomplishes, and how it is pursued as a human enterprise. However, at the same time, I have included in this eighth edition a great deal of material to help students estimate, calculate, and solve problems *outside* the classroom or textbook setting.

I frequently encounter people who tell me about their unpleasant experiences with mathematics. I have a true sympathy for those people, and I recall one of my elementary school teachers who assigned additional arithmetic problems as punishment. This can only create negative attitudes toward mathematics, which is indeed unfortunate. If elementary school teachers and parents have positive attitudes toward mathematics, their children cannot help but see some of the beauty of the subject. I want students to come away from this course with the feeling that mathematics can be pleasant, useful, and practical—and enjoyed for its own sake.

The prerequisites for this course vary considerably, as do the backgrounds of students. Some schools have no prerequisites, but other schools have an intermediate algebra prerequisite. The students, as well, have heterogeneous backgrounds. Some have little or no mathematics skills; others have had a great deal of mathematics. Even though the usual prerequisite for using this book is intermediate algebra, a careful selection of topics and chapters would allow a class with a beginning algebra prerequisite to study effectively from this book.

This book was written to meet the needs of all of these students and schools. How did I accomplish that goal? First, the chapters are almost independent of one another, and can be covered in any order appropriate to a particular audience. Second, the problems are designed to be the core of the course. There are problems that every student will find easy and will provide the opportunity for success; there are also problems that are very challenging. Much interesting material appears in the problems, and students should get into the habit of reading (not necessarily working) all the problems whether they are assigned or not.

A Problems: mechanical or drill problems

B Problems: require understanding of the concepts

Problem Solving Problems: require problem-solving skills or original thinking

Individual Research: requires research or library work

Group Research: requires not only research or library work, but also group participation (See the index for a list of group projects.)

The major themes of this book are problem solving and estimation in the context of presenting the great ideas in the history of mathematics. I believe that *learning to solve problems is the principal reason for studying mathematics.* Problem solving is the process of applying previously acquired knowledge to new and unfamiliar situations. Solving word problems in most textbooks is one form of problem solving, but students also should be faced with non–text-type problems. In the first section of this edition I introduce students to Polya's problem-solving techniques, and these techniques are used throughout the book to solve non–text-type problems. These problem-solving examples are found throughout the book (marked as **Polya's Method** examples). Also new to this edition are problems in *each* section that require Polya's method for problem solving.

Students should learn the language and notation of mathematics. Most students who have trouble with mathematics do not realize that mathematics *does require hard work.* The usual pattern for most mathematics students is to open the book to the assigned page of problems, and begin working. Only after getting "stuck" is an attempt made to "find it in the book." The final resort is reading the text. In this book the students are asked not only to "do math problems," but also to "experience mathematics." This means it is necessary to become involved with the **concepts** being presented, not "just get answers." In fact, the slogan "Mathematics Is Not a Spectator Sport" is not just an advertising slogan, but an invitation which suggests that the only way to succeed in mathematics is to become involved with it. Students will learn to receive mathematical ideas through listening, reading, and visualizing. They are expected to present mathematical ideas by speaking, writing, drawing pictures and graphs, and demonstrating with concrete models. The problems in each section that

are designated **In Your Own Words** provide practice in communication skills.

A Personal Note

Writing a mathematics textbook is both enjoyable and challenging. To make mathematics come alive, I have included many items not usually found in a textbook. For example, I've included cartoons and quotations, and have used the margins for news clippings and historical notes. The historical notes are not strictly biographical reports, but instead focus on the people to convey some of the humanness of mathematics. Nearly every major mathematician (and many minor ones) has some part of his or her life to tell on the pages of this book. At the end of each chapter there is an interview of a living mathematician. I sat down and made a list of those persons who are the most famous, or those whom I greatly respect. I did not know how my request for an interview would be received, but to my surprise each of these persons was most gracious in providing me not only with biographical information, but with personal details of their lives so that I could share some of their humanness with users of this book. I treasure the correspondence I had with these people.

A Note for Instructors

Feel free to arrange the material in a different order from that presented in the text. I have written the chapters to be as independent of one another as possible. There is much more material in this book than could be covered in a single course. This book can be used in classes designed for liberal arts, teacher training, finite mathematics, college algebra, or a combination of these.

I have written an extensive *Instructor's Manual* to accompany this book. It includes the complete solutions to all the problems (including the "Problem Solving" problems) as well as teaching suggestions and transparency masters. For those who wish to integrate the computer into the entire course, there are computer problems in both BASIC and LOGO to accompany each chapter.

Also available are sample tests, not only in hard copy form, but also in electronic form for both IBM and Macintosh formats.

Some of the significant features of the book are shown on the following pages.

Each chapter opens with **IN THE REAL WORLD**. These sections ask some question or raise some issue that requires the development of some material in this chapter. The end of the chapter provides a commentary on this introduction.

Designed for your success, this **PREVIEW** on the chapter opener helps to anticipate what follows in this chapter.

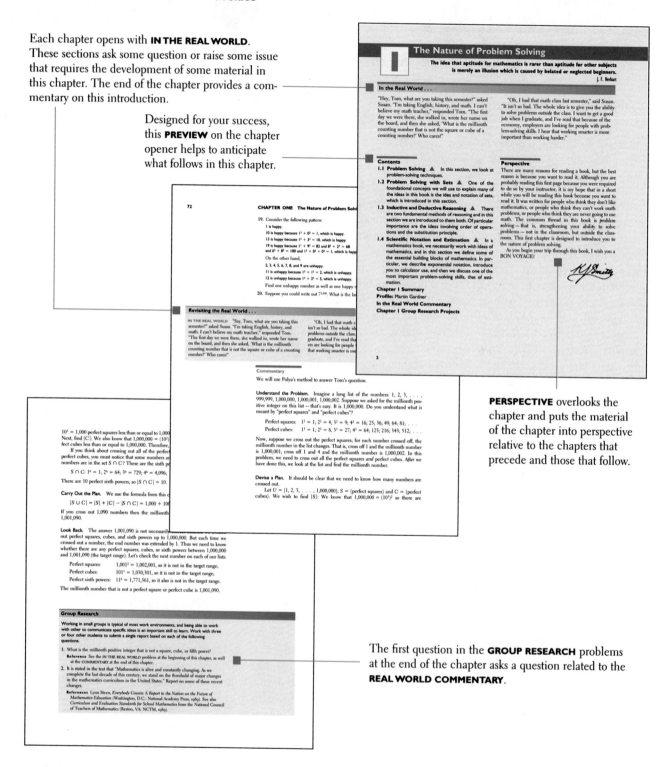

PERSPECTIVE overlooks the chapter and puts the material of the chapter into perspective relative to the chapters that precede and those that follow.

The first question in the **GROUP RESEARCH** problems at the end of the chapter asks a question related to the **REAL WORLD COMMENTARY**.

It is important to be able to communicate your ideas. This book gives you many such opportunities.

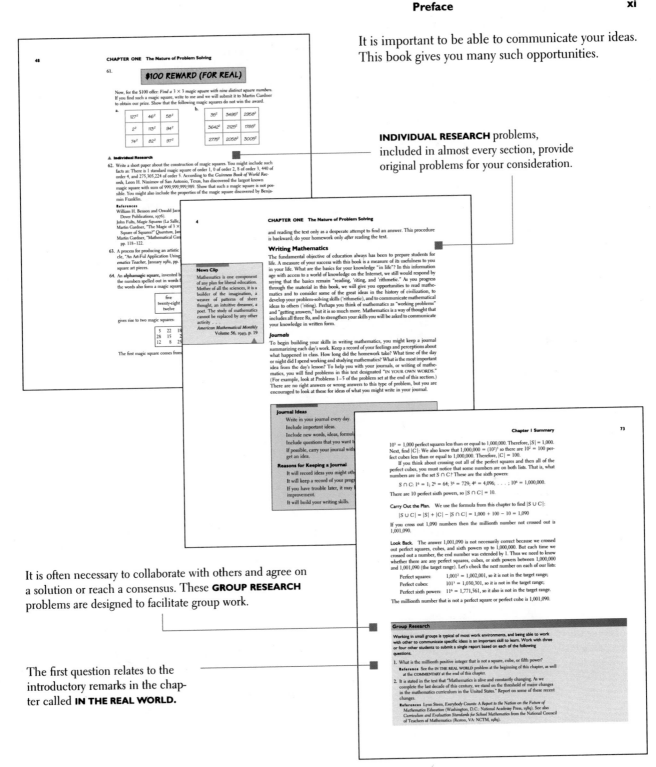

INDIVIDUAL RESEARCH problems, included in almost every section, provide original problems for your consideration.

It is often necessary to collaborate with others and agree on a solution or reach a consensus. These **GROUP RESEARCH** problems are designed to facilitate group work.

The first question relates to the introductory remarks in the chapter called **IN THE REAL WORLD.**

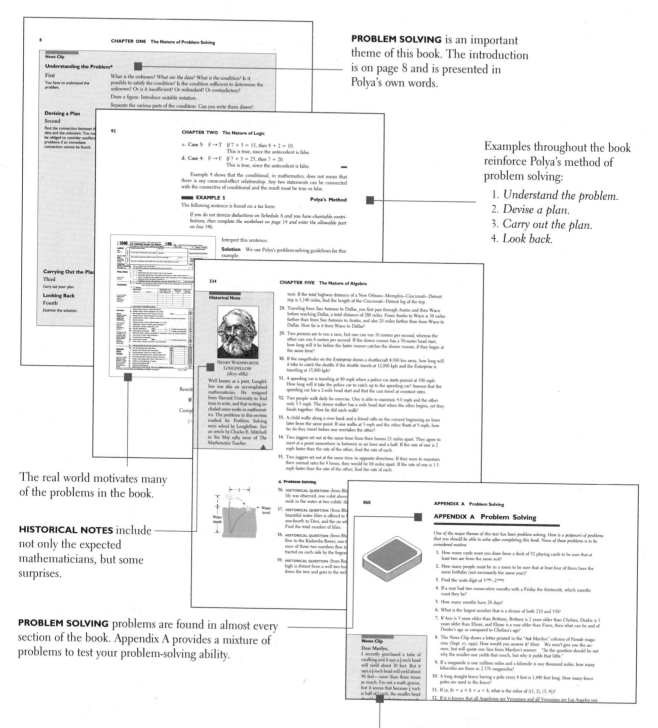

PROBLEM SOLVING is an important theme of this book. The introduction is on page 8 and is presented in Polya's own words.

Examples throughout the book reinforce Polya's method of problem solving:

1. *Understand the problem.*
2. *Devise a plan.*
3. *Carry out the plan.*
4. *Look back.*

The real world motivates many of the problems in the book.

HISTORICAL NOTES include not only the expected mathematicians, but some surprises.

PROBLEM SOLVING problems are found in almost every section of the book. Appendix A provides a mixture of problems to test your problem-solving ability.

NEWS CLIPS motivate many of the problems. Here is a "Dear Marilyn" problem.

Problems, Problems, Problems, Problems....

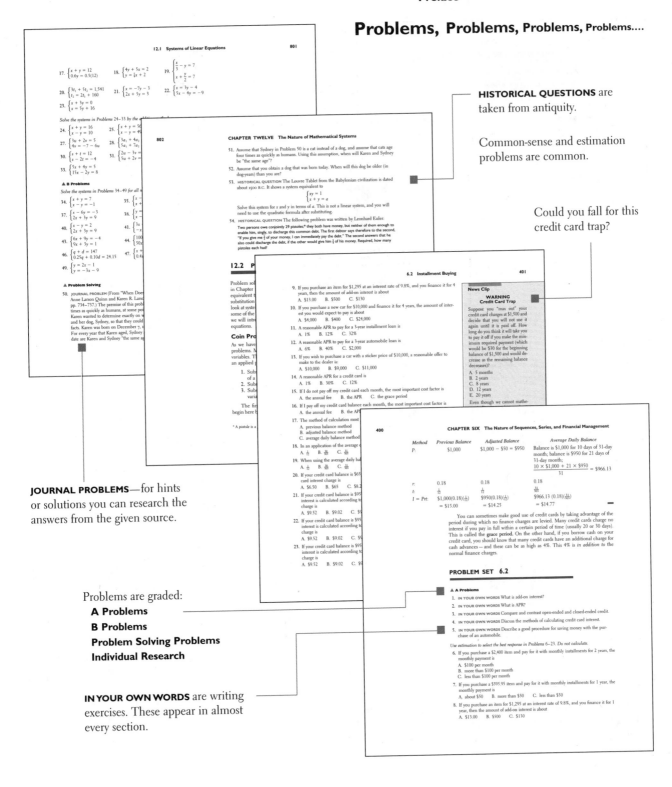

HISTORICAL QUESTIONS are taken from antiquity.

Common-sense and estimation problems are common.

Could you fall for this credit card trap?

JOURNAL PROBLEMS—for hints or solutions you can research the answers from the given source.

Problems are graded:
A Problems
B Problems
Problem Solving Problems
Individual Research

IN YOUR OWN WORDS are writing exercises. These appear in almost every section.

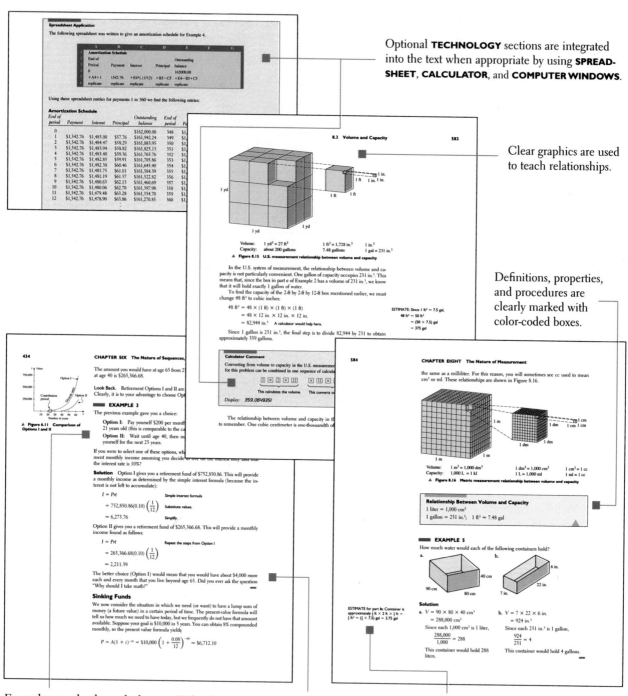

Optional **TECHNOLOGY** sections are integrated into the text when appropriate by using **SPREAD-SHEET**, **CALCULATOR**, and **COMPUTER WINDOWS**.

Clear graphics are used to teach relationships.

Definitions, properties, and procedures are clearly marked with color-coded boxes.

Examples are clearly marked; author's notes are provided to explain steps.

"Why should I take math?" or, in terms of this question, "Would I rather have $6,273.76/mo or $2,211.39/mo? Here is how!

Author's notes are frequent.

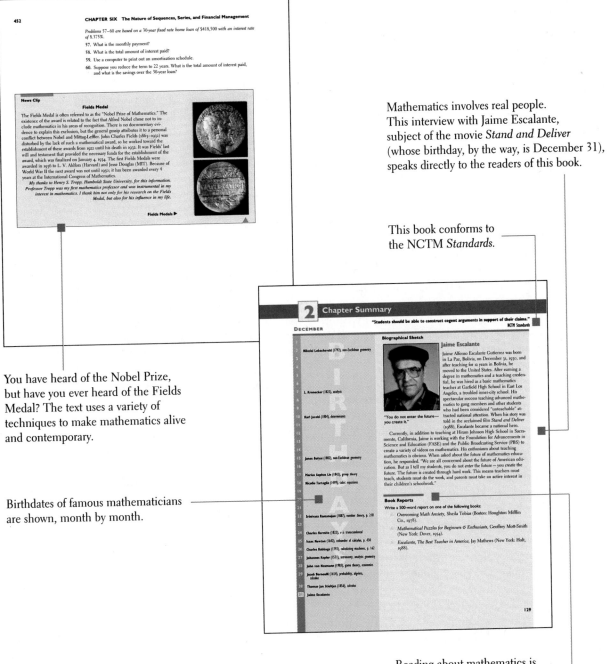

Mathematics involves real people. This interview with Jaime Escalante, subject of the movie *Stand and Deliver* (whose birthday, by the way, is December 31), speaks directly to the readers of this book.

This book conforms to the NCTM *Standards*.

You have heard of the Nobel Prize, but have you ever heard of the Fields Medal? The text uses a variety of techniques to make mathematics alive and contemporary.

Birthdates of famous mathematicians are shown, month by month.

Reading about mathematics is encouraged with suggested book reports at the end of each chapter.

Changes from the Previous Edition

You will find this edition substantially unchanged from the seventh edition. I have added group research projects at the end of each chapter, included z-scores in statistics, and you will find many new and interesting problems, as well as a glossary of terms. I have retained all of the familiar features that were in the previous edition.

Acknowledgments

I would like to thank Diana Gerardi for her valuable suggestions in improving my book. I also appreciate the suggestions of the reviewers of this edition: Brenda Allen, Georgia College and State University; Nancy Angle, Cerritos College; V. Sagar Bakhshi, Virginia State University; Daniel C. Biles, Western Kentucky University; Barry Brenin, Hofstra University; Robert Cicenia, Pace University Pleasantville—Briarville Campus; Mickle Duggan, East Central University; King Jamison, Middle Tennessee State University; Valerie Melvin, Cape Fear Community College; Barbara Ostrick, Hofstra University; Mary Anne C. Petruska, Pensacola Junior College; Joan Raines, Middle Tennessee State University; and Jean Woody, Tulsa Community College.

One of the nicest things about writing a successful book is all of the letters and suggestions I've received. I would like to thank the following people who gave suggestions for previous editions of this book: Jeffrey Allbritten, Peter R. Atwood, John August, Charles Baker, Jerald T. Ball, Carol Bauer, George Berzsenyi, Jan Boal, Kolman Brand, Chris C. Braunschweiger, T. A. Bronikowski, Charles M. Bundrick, T. W. Buquoi, Eugene Callahan, Michael W. Carroll, Joseph M. Cavanaugh, James R. Choike, Mark Christie, Gerald Church, Wil Clarke, Lynn Cleaveland, Penelope Ann Coe, Thomas C. Craven, Gladys C. Cummings, Ralph De Marr, Maureen Dion, Charles Downey, Mickle Duggan, Samuel L. Dunn, Beva Eastman, William J. Eccles, Gentil Estevez, Ernest Fandreyer, Loyal Farmer, Gregory N. Fiore, Robert Fleiss, Richard Freitag, Gerald E. Gannon, Ralph Gellar, Gary Gislason, Mark Greenhalgh, Martin Haines, Abdul Rahim Halabieh, John J. Hanevy, Michael Helinger, Robert L. Hoburg, Caroline Hollingsworth, Scott Holm, Libby W. Holt, Peter Hovanec, M. Kay Hudspeth, Carol M. Hurwitz, James J. Jackson, Vernon H. Jantz, Charles E. Johnson, Nancy J. Johnson, Michael Jones, Martha C. Jordan, Judy D. Kennedy, Linda H. Kodama, Daniel Koral, Helen Kriegsman, C. Deborah Laughton, William Leahey, John LeDuc, William A. Leonard, Adolf Mader, John Martin, Cherry F. May, George McNulty, Carol McVey, Charles C. Miles, Allen D. Miller, John Mullen, Charles W. Nelson, John Palumbo, Gary Peterson, Michael Petricig, Michael Pinter, James V. Rauff, Richard Rempel, Paul M. Riggs, Jane Rood, Peter Ross, O. Sassian, Mickey G. Settle, James R. Smart, Andrew Simoson, Glen T. Smith, Donald G. Spencer, Gustavo Valadez-Ortiz, John Vangor, Arnold Villone, Clifford H. Wagner, James Walters, Barbara Williams, Stephen S. Willoughby, and Bruce Yoshiwara.

Nancy Angle and Jean Woody did a superb job of checking all of the examples and checking the accuracy of the answers. I would especially like to thank Robert J. Wisner of New Mexico State for his countless suggestions and ideas over the many editions of this book; Tessa McGlasson, Craig Barth, Jeremy Hayhurst, Paula Heighton, Gary Ostedt, and Bob Pirtle of Brooks/Cole; as well as Jack Thornton, for the sterling leadership and inspiration he has been to me from the inception of this book to the present.

The production of this book was a true team effort, and I especially appreciate Susan Reiland for her help in countless ways, including editing, accuracy checking, and giving me tireless support and help (she is a real miracle worker). I would also like to thank the photo researcher, Sue C. Howard; the permissions researcher, Lillian Campobasso; and Kathi Townes, Brian Betsill, and Stephanie Kuhns at TECH·arts for the long hours, superb work, and for "going the extra mile" for me in putting this book together.

Finally, my thanks go to my wife, Linda, who has always been there for me. Without her, this book would exist only in my dreams, and I would have never embarked as an author.

Karl J. Smith
Sebastopol, CA
email: smithkjs@wco.com

Contents

*Optional sections

*Optional sections

*Optional sections

* Optional sections

*Optional sections

12 | The Nature of Mathematical Systems   796

Appendices

*Optional sections

To the Student

A Fable

Once upon a time, two young ladies, Shelley and Cindy, came to a town called Mathematics. People had warned them that this is a particularly confusing town. Many people who arrived in Mathematics were very enthusiastic, but could not find their way around, became frustrated, gave up, and left town.

Shelley was strongly determined to succeed. She was going to learn her way through the town. For example, to learn how to go from her dorm to class, she concentrated on memorizing this clearly essential information: she had to walk 325 steps south, then 253 steps west, then 129 steps in a diagonal (southwest), and finally 86 steps north. It was not easy to remember all of that, but fortunately she had a very good instructor who helped her to walk this same path 50 times. To stick to the strictly necessary information, she ignored much of the beauty along the route, such as the color of the adjacent buildings or the existence of trees, bushes, and nearby flowers. She always walked blindfolded. After repeated exercising, she succeeded in learning her way to class and also to the cafeteria. But she could not learn the way to the grocery store, the bus station, or a nice restaurant; there were just too many routes to memorize. It was so overwhelming! Finally she gave up and left town; Mathematics was too complicated for her.

Cindy, on the other hand, was of a much less serious nature. To the dismay of her instructor, she did not even intend to memorize the number of steps of her walks. Neither did she use the standard blindfold that students need for learning. She was always curious, looking at the different buildings, trees, bushes, and nearby flowers or anything else not necessarily related to her walk. Sometimes she walked down dead-end alleys to find out where they were leading, even if this was obviously superfluous. Curiously, Cindy succeeded in learning how to walk from one place to another. She even found it easy and enjoyed the scenery. She eventually built a building on a vacant lot in the city of Mathematics.[*]

[*]My thanks to Emilio Roxin of the University of Rhode Island for the idea for this fable.

THE NATURE OF MATHEMATICS

Eighth Edition

The Nature of Problem Solving

The idea that aptitude for mathematics is rarer than aptitude for other subjects is merely an illusion which is caused by belated or neglected beginners.

J. F. Herbart

In the Real World . . .

"Hey, Tom, what are you taking this semester?" asked Susan. "I'm taking English, history, and math. I can't believe my math teacher," responded Tom. "The first day we were there, she walked in, wrote her name on the board, and then she asked, 'What is the millionth counting number that is not the square or cube of a counting number?' Who cares!"

"Oh, I had that math class last semester," said Susan. "It isn't so bad. The whole idea is to give you the ability to solve problems *outside* the class. I want to get a good job when I graduate, and I've read that because of the economy, employers are looking for people with problem-solving skills. I hear that working smarter is more important than working harder."

Contents

Perspective

There are many reasons for reading a book, but the best reason is because you want to read it. Although you are probably reading this first page because you were required to do so by your instructor, it is my hope that in a short while you will be reading this book because you *want* to read it. It was written for people who think they don't like mathematics, or people who think they can't work math problems, or people who think they are never going to use math. The common thread in this book is *problem solving* — that is, strengthening your ability to solve problems — not in the classroom, but outside the classroom. This first chapter is designed to introduce you to the nature of problem solving.

As you begin your trip through this book, I wish you a BON VOYAGE!

1.1 PROBLEM SOLVING

A Word of Encouragement

Do you think of mathematics as a difficult, foreboding subject that was invented hundreds of years ago? Do you think that you will never be able (or even want) to use mathematics? If you answered "yes" to either of these questions, then I want you to know that I have written this book for you. I have tried to give you some insight into how mathematics is developed, and to introduce you to some of the people behind the mathematics. In this book, I will present some of the great ideas of mathematics, and then we will look at how these ideas can be used in an everyday setting to build your problem-solving abilities. *The most important prerequisite for this course is an openness to try out new ideas — a willingness to experience the suggested activities rather than to sit on the sideline as a spectator.* I have attempted to make this material interesting by putting it together differently from the way you might have had mathematics presented in the past. You will find this book difficult if you wait for the book or the teacher to give you answers — instead *be willing to guess, experiment, estimate, and manipulate,* and try out problems *without fear of being wrong!*

There is a common belief that mathematics is to be pursued only in a clear-cut logical fashion. This belief is perpetuated by the way mathematics is presented in most textbooks. Often it is reduced to a series of definitions, methods to solve various types of problems, and theorems. These theorems are justified by means of proofs and deductive reasoning. I do not mean to minimize the importance of proof in mathematics, for it is the very thing that gives mathematics its strength. But the power of the imagination is every bit as important as the power of deductive reasoning. As the mathematician Augustus De Morgan once said, "The power of mathematical invention is not reasoning but imagination."

Hints for Success

Mathematics is different from other subjects. One topic builds upon another, and you need to make sure that you understand *each* topic before progressing to the next one.

You must make a commitment to attend each class. Obviously, unforeseen circumstances can come up, but you must plan to attend class regularly. Pay attention to what your teacher says and does, and take notes. If you must miss class, write an outline of the text corresponding to the missed material, including working out each text example on your notebook paper.

You must make a commitment to daily work. Do not expect to save up and do your mathematics work once or twice a week. It will take a daily commitment on your part, and you will find mathematics difficult if you try to "get it done" in spurts. You could not expect to become proficient in tennis, soccer, or playing the piano by practicing once a week, and the same is true of mathematics. Try to schedule a regular time to study mathematics each day.

Read the text carefully. Many students expect to get through a mathematics course by beginning with the homework problems, then reading some examples, and reading the text only as a desperate attempt to find an answer. This procedure is backward; do your homework only *after* reading the text.

Writing Mathematics

The fundamental objective of education always has been to prepare students for life. A measure of your success with this book is a measure of its usefulness to you in your life. What are the basics for your knowledge "in life"? In this information age with access to a world of knowledge on the Internet, we still would respond by saying that the basics remain "reading, 'riting, and 'rithmetic." As you progress through the material in this book, we will give you opportunities to read mathematics and to consider some of the great ideas in the history of civilization, to develop your problem-solving skills ('rithmetic), and to communicate mathematical ideas to others ('riting). Perhaps you think of mathematics as "working problems" and "getting answers," but it is so much more. Mathematics is a way of thought that includes all three Rs, and to strengthen your skills you will be asked to communicate your knowledge in written form.

Journals

To begin building your skills in writing mathematics, you might keep a journal summarizing each day's work. Keep a record of your feelings and perceptions about what happened in class. How long did the homework take? What time of the day or night did I spend working and studying mathematics? What is the most important idea from the day's lesson? To help you with your journals, or writing of mathematics, you will find problems in this text designated "IN YOUR OWN WORDS." (For example, look at Problems 1–5 of the problem set at the end of this section.) There are no right answers or wrong answers to this type of problem, but you are encouraged to look at these for ideas of what you might write in your journal.

News Clip

Mathematics is one component of any plan for liberal education. Mother of all the sciences, it is a builder of the imagination, a weaver of patterns of sheer thought, an intuitive dreamer, a poet. The study of mathematics cannot be replaced by any other activity . . .
American Mathematical Monthly
Volume 56, 1949, p. 19

Journal Ideas

Write in your journal every day.

Include important ideas.

Include new words, ideas, formulas, or concepts.

Include questions that you want to ask later.

If possible, carry your journal with you so you can write in it anytime you get an idea.

Reasons for Keeping a Journal

It will record ideas you might otherwise forget.

It will keep a record of your progress.

If you have trouble later, it may help you diagnose areas for change or improvement.

It will build your writing skills.

Individual Research

At the end of many sections you will find problems requiring some library research. I hope that as you progress through the course you will find one or more topics that interest you so much that you will want to do additional reading on that topic, even if it is not assigned. Your instructor may assign one or more of these as term papers. One of the best ways for you to become aware of all the books and periodicals that are available is to log onto the Internet, or visit the library to research specific topics.

Preparing a mathematics paper or project can give you interesting and worthwhile experiences. In preparing a paper or project, you will get experience in using resources to find information, in doing independent work, in organizing your presentation, and in communicating ideas orally, in writing, and in visual demonstrations. You will broaden your background in mathematics and encounter new mathematical topics that you never before knew existed. In setting up an exhibit, you will experience the satisfaction of demonstrating what you have accomplished. It may be a way of satisfying your curiosity and your desire to be creative. It is an opportunity for developing originality, craftsmanship, and new mathematical understandings. If you are requested to do some individual research problems, here are some suggestions.

1. *Select a topic that has interest potential.* Do not do a project on a topic that does not interest you. Suggestions are given throughout this book under the heading "Individual Research."
2. *Find as much information about the topic as possible.* Many of the Individual Research problems have one or two references to get you started. In addition, check the following sources:

   PERIODICALS: *The Mathematics Teacher, Teaching Children Mathematics* (formerly *Arithmetic Teacher*), and *Scientific American*; each of these has its own cumulative index; also check the *Reader's Guide.*

   SOURCE BOOKS: *The World of Mathematics* by Newman (Simon & Schuster, New York, 1956, 4 vols.) is a gold mine of ideas. *Mathematics*, a Time–Life book by David Bergimini, may provide you with many ideas. *Encyclopedias* can be consulted after you have some project ideas; however, I do not have in mind that the term project necessarily be a term paper.

   INTERNET: Use one or more search engines on the Internet for information on a particular topic. The more specific you can be in describing what you are looking for, the better the engine will be able to find material on your topic.

3. *Prepare and organize your material into a concise, interesting report.* Include drawings in color, pictures, applications, and examples to get the reader's attention and add meaning to your report.
4. *Build an exhibit that will tell the story of your topic.* Remember the science projects in high school? That type of presentation might be appropriate. Use models, applications, and charts that lend variety. Give your paper or exhibit a catchy, descriptive title.
5. A **term** *project cannot be done in one or two evenings.*

Group Research

Working in small groups is typical of most work environments, and being able to work with others to communicate specific ideas is an important skill to learn. At the end of each chapter is a list of suggested group projects, and you are encouraged to work with three or four others to submit a single report.

Problem Solving

We begin this study of **problem solving** by looking at the *process* of problem solving. As a mathematics teacher, I often hear the comment, "I can do mathematics, but I can't solve word problems." There *is* a great fear and avoidance of "real-life" problems because they do not fit into the same mold as the "examples in the book." Few practical problems from everyday life come in the same form as those you study in school.

To compound the problem, learning to solve problems takes time. All too often, the mathematics curriculum is so packed with content that the real process of problem solving is slighted, and because of time limitations, becomes an exercise in mimicking the instructor's steps instead of developing into an approach that can be used long after the final examination is over.

Before we build problem-solving skills, it is necessary to build certain prerequisite skills necessary for problem solving. It is my goal to develop your skills in the mechanics of mathematics, in understanding the important concepts, and finally in applying those skills to solve a new type of problem. I have segregated the problems in this book to help you build these different skills:

In Your Own Words	This type of problem asks you to discuss or rephrase main ideas or procedures using your own words.
A Problems	These are mechanical and drill problems.
B Problems	These problems require an understanding of the concepts.
Problem Solving	These require problem-solving skills or original thinking.
Problems for Research	These problems require research or library work. Most are intended for individual research but a few at the end of chapters are group research projects.

The model for problem solving that we will use was first published in 1945 by the great, charismatic mathematician George Polya. His book *How to Solve It* (Princeton University Press, 1973) has become a classic. In Polya's book you will find this problem-solving model as well as a treasure trove of strategy, know-how rules of thumb, good advice, anecdotes, history, and problems at all levels of mathematics. His problem-solving model is as follows.

Guidelines for Problem Solving

First   You have to *understand the problem.*
Second   *Devise a plan.* Find the connection between the data and the
           unknown. Look for patterns, relate to a previously solved
           problem or a known formula, or simplify the given information
           to give you an easier problem.
Third   *Carry out the plan.*
Fourth   *Look back.* Examine the solution obtained.

Polya's original statement of this procedure is reprinted in the box on the next page. These steps can be amplified by some individual strategies that might be used at appropriate moments:

1. Sort out the irrelevant information.
2. If you cannot solve the proposed problem, look around for an appropriate related problem (a simpler one, if possible).
3. Work backward.
4. Work forward.
5. Narrow the condition.
6. Widen the condition.
7. Seek a counterexample.
8. Guess and test.
9. Divide and conquer (that is, break the problem into simpler parts).
10. Change the conceptual mode; make it simpler, if possible.
11. Work with the "opposite" situation or question.

Let's apply this procedure for problem solving to the map shown in Figure 1.1; we refer to this problem as the **street problem.** Melissa lives at the YWCA (point

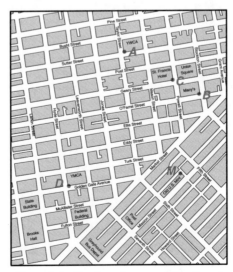

▲ **Figure 1.1   Portion of a map of San Francisco**

News Clip

Understanding the Problem*

First

You have to *understand* the problem.

What is the unknown? What are the data? What is the condition? Is it possible to satisfy the condition? Is the condition sufficient to determine the unknown? Or is it insufficient? Or redundant? Or contradictory?

Draw a figure. Introduce suitable notation.

Separate the various parts of the condition. Can you write them down?

Devising a Plan

Second

Find the connection between the data and the unknown. You may be obliged to consider auxiliary problems if an immediate connection cannot be found.

Have you seen it before? Or have you seen the same problem in a slightly different form?

Do you know a related problem? Do you know a theorem that could be useful?

Look at the unknown! And try to think of a familiar problem having the same or a similar unknown.

Is the problem related to one you have solved before? Could you use it?

Could you use its result? Could you use its method? Should you introduce some auxiliary element in order to make its use possible?

Could you restate the problem? Could you restate it still differently? Go back to definitions.

If you cannot solve the proposed problem try to solve first some related problem. Could you imagine a more accessible related problem? A more general problem? A more special problem? An analogous problem? Could you solve a part of the problem? Keep only a part of the condition, drop the other part; how far is the unknown then determined, how can it vary? Could you derive something useful from the data? Could you think of other data appropriate to determine the unknown? Could you change the unknown or the data, or both if necessary, so that the new unknown and the new data are nearer to each other? Did you use all the data? Did you see the whole condition? Have you taken into account all essential notions involved in the problem?

Carrying Out the Plan

Third

Carry out your plan.

Carrying out your plan of the solution, *check each step.* Can you see clearly that the step is correct? Can you prove that it is correct?

Looking Back

Fourth

Examine the solution.

Can you *check the result?* Can you check the argument?

Can you derive the result differently? Can you see it at a glance?

Can you use the result or the method for some other problem?

* This is taken word for word as it was written by Polya in 1941. It was printed in *How to Solve It* (Princeton, N.J.: Princeton University Press, 1973).

A) and works at Macy's (point *B*). She usually walks to work. How many different routes can Melissa take?

Where would you begin with this problem?

Step 1. **Understand the Problem.** Can you restate it in your own words? Can you trace out one or two possible paths? What assumptions are reasonable? We assume that she will not do any backtracking — that is, she always travels toward her destination. We also assume that she travels along the city streets — she cannot cut diagonally across a lot or a block.

Step 2. **Devise a Plan.** Simplify the question asked. Consider the simplified drawing shown in Figure 1.2.

Step 3. **Carry Out the Plan.** Count the number of ways it is possible to arrive at each point, or as it is sometimes called, a *vertex*.

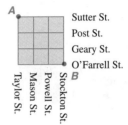

▲ **Figure 1.2 Simplified portion of Figure 1.1**

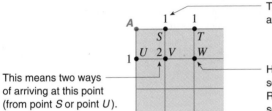

This means one way to arrive at this point.

This means two ways of arriving at this point (from point *S* or point *U*).

How many ways to this point? Do you see why the answer is 2 + 1 = 3 ways? Remember, no backtracking is allowed, so you must get here from point *V* or from point *T*.

Now fill in all the possibilities on Figure 1.3, as shown by the above procedure.

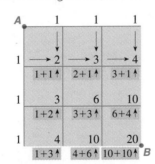

▲ **Figure 1.3 Map with solution**

Step 4. **Look Back.** Does the answer 20 different routes make sense? Do you think you could fill in all of them?

▬▬▬ EXAMPLE 1

In how many different ways could Melissa get from the YWCA (point *A*) to the St. Francis Hotel (point *C* in Figure 1.1), using the method of Figure 1.3?

Solution Draw a simplified version of Figure 1.3, as shown in the margin.

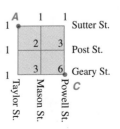

There are 6 different paths.

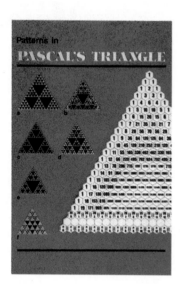

Problem Solving by Patterns

Let's formulate a general solution. Consider a map with a starting point A:

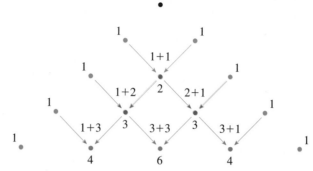

Do you see the pattern for building this figure? Each new row is found by adding the two previous numbers, as shown by the arrows. This pattern is known as **Pascal's triangle.** In Figure 1.4 the rows and diagonals are numbered for easy reference.

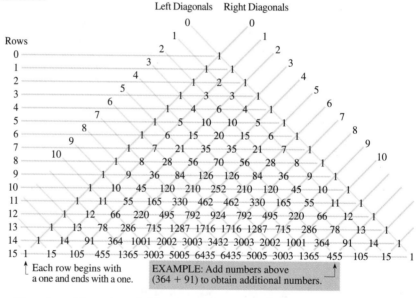

▲ **Figure 1.4   Pascal's triangle**

How does this apply to Melissa's trip from the YWCA to Macy's? It is 3 blocks down and 3 blocks over. Look at Figure 1.4, and count out these blocks as shown in Figure 1.5.

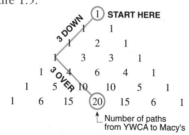

Remember that Melissa must walk on the streets and cannot "cut through the middle of a block."

▲ **Figure 1.5   Using Pascal's triangle to solve the street problem**

EXAMPLE 2

In how many different ways could Melissa get from the YWCA (point A in Figure 1.1) to the YMCA (point D)?

Solution Look at Figure 1.1; from point A to point D is 7 blocks down and 3 blocks left. Use Figure 1.4 as follows:

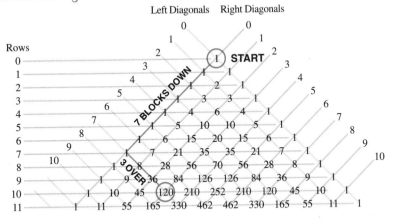

We see that there are 120 paths.

Pascal's triangle applies to the street problem only if the streets are rectangular. If the map shows irregularities (for example, diagonal streets or obstructions), then you must revert back to numbering the vertices.

EXAMPLE 3

In how many different ways could Melissa get from the YWCA (point A) to the Old U.S. Mint (point M)?

Solution If the streets are irregular or if there are obstructions, you cannot use Pascal's triangle, but you can still count the blocks in the same fashion, as shown at the right. There are 52 paths from point A to point M (if, as usual, we do not allow backtracking).

Problem solving is a difficult task to master, and you are not expected to master it after one section of this book (or even after several chapters of this book). However, you must make building your problem-solving skills an ongoing process. One of

Eyn Newe

Vnnd wolgegründte

underwegsung aller Kauffmanß Rech:
nung in dreyen Büchern/mit schönen Re:
geln vñ fragstucken begriffen . Sunder:
lich was sorel vnnd behendigkait in der
Welsche Practica vñ Tollern gebraucht
wirdt / des gleychen fürmals weder in
Teütscher noch in Welscher sprach nie
gedrückt. durch Petrum Apianu
von Leyßnick/ð Astronomei
zü Ingolstat Ordina:
riü / verfertiger.

The title page of an arithmetic book by Petrus Apianus in 1527 is reproduced above. It was the first time Pascal's triangle appeared in print.

the most important aspects of problem solving is to relate new problems to old problems. The problem-solving techniques outlined here should be applied when you are faced with a new problem. When you are faced with a problem similar to one you have already worked, you can apply previously developed techniques (as we did in Examples 2 and 3). Now, because Example 4 seems to be a new type of problem, we again apply the guidelines for problem solving.

■■■ EXAMPLE 4   Polya's Method

A jokester tells you that he has a group of cows and chickens and that he counted 13 heads and 36 feet. How many cows and chickens does he have?

Solution   Let's use Polya's problem-solving guidelines.

Understand the Problem.   A good way to make sure you understand a problem is to attempt to phrase it in a simpler setting:

one chicken and one cow:	2 heads and 6 feet (chickens have two feet; cows have four)
two chickens and one cow:	3 heads and 8 feet
one chicken and two cows:	3 heads and 10 feet

Devise a Plan.   How you organize the material is often important in problem solving. Let's organize the information into a table:

No. of chickens	No. of cows	No. of heads	No. of feet
0	13	13	52

Do you see why we started here? The problem says we must have 13 heads. There are other possible starting places (13 chickens and 0 cows, for example), but an important aspect of problem solving is to start with *some* plan.

No. of chickens	No. of cows	No. of heads	No. of feet
1	12	13	50
2	11	13	48
3	10	13	46
4	9	13	44

Carry Out the Plan.   Now, look for patterns. Do you see that as the number of cows decreases by one and the number of chickens increases by one, the number of feet must decrease by two? *Does this make sense to you?* Remember, step 1 requires that you not just push numbers around, but that you understand what you are doing. Since we need 36 feet for the solution to this problem, we see

$$44 - 36 = 8$$

so the number of chickens must increase by an additional four. The answer is 8 chickens and 5 cows.

Look Back.

No. of chickens	No. of cows	No. of heads	No. of feet
8	5	13	36

Check: 8 chickens have 16 feet and 5 cows have 20 feet, so the total number of heads is $8 + 5 = 13$, and the number of feet is 36. ▬

▬ **EXAMPLE 5** **Polya's Method**

If a family has 5 children, in how many different orders could the parents have a 3-boy, 2-girl family?

Solution

Understand the Problem. Part of understanding the problem might involve estimation. For example, if a family has 1 child, there are 2 possible orders (B or G). If a family has 2 children, there are 4 orders (BB, BG, GB, GG); for 3 children, 8 orders; for 4 children, 16 orders; and for 5 children, a total of 32 orders. This means that an answer of 140 possible orders is an unreasonable answer.

Devise a Plan. You might begin by enumeration:

BBBGG, BBGBG, BBGGB, . . .

This would seem to be too tedious. Instead, rewrite this as a simpler problem and look for a pattern:

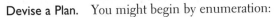

1 child: B ← one way   2 children: BB ← one way
           G ← one way   BG ⎫
                                            GB ⎭ two ways
3 children: BBB ← one way   GG ← one way
              BBG ⎫
              BGB ⎬ three ways
              GBB ⎭
              BGG ⎫
              GBG ⎬ three ways
              GGB ⎭
              GGG ← one way

Look at the possibilities:

1 child   1B   1G
2 children   1BB   2   1GG
                                ↑
                    ways for 1 boy and 1 girl

3 children   1BBB   3   3   1GGG   Look familiar?
                        ↑   ↑
                               ways for 1 boy and 2 girls
              ways for 2 boys and 1 girl

Look at Pascal's triangle in Figure 1.4; for 5 children, look at row 5.

Carry Out the Plan.

row 5:	1	5	10	10	5	1
	↑	↑	↑	↑	↑	↑
	5 boys	4 boys 1 girl	3 boys 2 girls	2 boys 3 girls	1 boy 4 girls	5 girls

They could have 3 boys and 2 girls in a total of 10 ways.

Look Back. We estimated that there are a total of 32 ways a family could have 5 children; let's sum the number of possibilities we found in carrying out the plan to see if it totals 32:

$$1 + 5 + 10 + 10 + 5 + 1 = 32$$

The following example illustrates the necessity of carefully reading the question.

EXAMPLE 6

Nick and Marsha are driving from Santa Rosa, CA, to Los Angeles, a distance of 460 miles. They leave at 11:00 A.M. and average 50 mph. On the other hand, Mary and Dan leave at 1:00 P.M. in Dan's sports car. Who is closer to Los Angeles when they meet for dinner in San Luis Obispo at 5:00 P.M.?

Solution

Understand the Problem. If they are sitting in the same restaurant, then they are all the same distance from Los Angeles.

The last example of this section illustrates that problem solving may require that you change the conceptual mode.

EXAMPLE 7

If you have been reading the information in the margins, you may have noticed that Blaise Pascal was born in 1623 and died in 1662. You may also have noticed that the first time Pascal's triangle appeared in print was in 1527. How can this be?

Solution It was a reviewer of this book who brought this apparent discrepancy to my attention. The facts are all correct. How could Pascal's triangle have been in print almost 100 years before he was born? The fact is, the number pattern we call Pascal's triangle is *named after* Pascal, but was not *discovered* by Pascal. This number pattern seems to have been discovered several times, by Johann Scheubel in the 16th century, by the Chinese mathematician Nakone Genjun, and recent research has traced the triangle pattern as far back as Omar Khayyám.

"Wait!" you exclaim. "How was I to answer the question in Example 7 — I don't know all those facts about the triangle." You are not expected to know these facts, but you are expected to begin to think critically about the information you are given, and the assumptions you are making. It was never stated that Blaise Pascal was the first to think of or publish Pascal's triangle!

PROBLEM SET 1.1

▲ A Problems

1. **IN YOUR OWN WORDS** In the text it was stated that "the most important prerequisite for this course is an openness to try out new ideas — a willingness to experience the suggested activities rather than to sit on the sideline as a spectator." Do you agree or disagree that this is the *most* important prerequisite? Discuss.

2. **IN YOUR OWN WORDS** What do you think the primary goal of mathematics education should be? What do you think it is in the United States? Discuss the differences between what it is and what you think it should be.

3. **IN YOUR OWN WORDS** In the chapter perspective (did you read it?) it was pointed out that this book was written for people who think they don't like mathematics, or people who think they can't work math problems, or people who think they are never going to use math. Do any of those descriptions apply to you, or to someone you know? Discuss.

4. **IN YOUR OWN WORDS** Discuss Polya's problem-solving model.

5. **IN YOUR OWN WORDS** At the beginning of this section, three hints for success were listed. Discuss each of these from your perspective. Are there any other hints that you might add to this list?

6. Given Pascal's triangle, as shown in Figure 1.4, write down the next two rows.

7. Describe the location of the numbers 1, 2, 3, 4, 5, . . . in Pascal's triangle.

8. Describe the location of the numbers 1, 4, 10, 20, 35, . . . in Pascal's triangle.

9. **IN YOUR OWN WORDS** In Example 2, the solution was found by going 7 blocks down and 3 blocks over. Could the solution also have been obtained by going 3 blocks over and 7 blocks down? Would this Pascal's triangle solution end up in a different location? Describe a property of Pascal's triangle that is relevant to an answer for this question.

10. If a family has 5 children, in how many ways could the parents have 2 boys and 3 girls as children?

11. If a family has 6 children, in how many ways could the parents have 3 boys and 3 girls as children?

12. If a family has 7 children, in how many ways could the parents have 4 boys and 3 girls as children?

13. If a family has 8 children, in how many ways could the parents have 3 boys and 5 girls as children?

In Problems 14–17, what is the number of direct routes from point A to point B?

14.

15.

16.

						B

A

17.

						B

A

Use the map in Figure 1.6 to determine the number of different paths from point A to the point indicated in Problems 18–21. Remember, no backtracking is allowed.

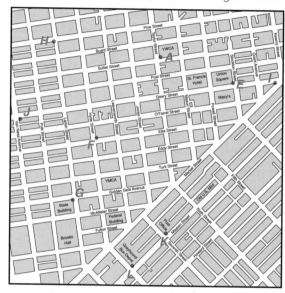

▲ **Figure 1.6   Map of a portion of San Francisco**

18. *E*   19. *F*   20. *G*   ´21. *H*

▲ B Problems

22. If an island's only residents are penguins and bears, and if there are 16 heads and 34 feet on the island, how many penguins and how many bears are on the island?

23. Below are listed three problems. Do not solve these problems; simply tell which one is most like Problem 22.

   A. A penguin in a tub weighs 8 lb, and a bear in a tub weighs 800 lb. If the penguin and the bear together weigh 802 lb, how much does the tub weigh?

   B. A bottle and a cork cost $1.10, and the bottle is a dollar more than the cork. How much does the cork cost?

   C. Bob has 15 roses and 22 carnations. Carol has twice as many roses and half as many carnations. How many flowers does Carol have?

24. Ten full crates of walnuts weigh 410 pounds, whereas an empty crate weighs 10 pounds. How much do the walnuts alone weigh?

25. There are three separate, equal-size boxes, and inside each box there are two separate small boxes, and inside each of the small boxes there are three even smaller boxes. How many boxes are there all together?

26. If you expect to get 50,000 miles on each tire from a set of five tires (four and one spare), how should you rotate the tires so that each tire gets the same amount of wear, and how far can you drive before buying a new set of tires?

27. **a.** What is the sum of the numbers in row 1 of Pascal's triangle?
    b. What is the sum of the numbers in row 2 of Pascal's triangle?
    c. What is the sum of the numbers in row 3 of Pascal's triangle?
    d. What is the sum of the numbers in row 4 of Pascal's triangle?

28. What is the sum of the numbers in row n of Pascal's triangle?

Use the map in Figure 1.6 to determine the number of different paths from point A to the point indicated in Problems 29–32. Remember, no backtracking is allowed.

29. *I*   30. *J*   31. *K*   32. *L*

33. **IN YOUR OWN WORDS** Suppose you have a long list of numbers to add, and you have misplaced your calculator. Discuss the different approaches that could be used for adding this column of numbers.

34. **IN YOUR OWN WORDS** You are faced with a long division problem, and you have misplaced your calculator. You do not remember how to do long division. Discuss your alternatives to come up with the answer to your problem.

35. **IN YOUR OWN WORDS** You have 10 items in your grocery cart. Six people are waiting in the express lane (10 items or less); one person is waiting in the first checkout stand and two people are waiting in another checkout stand. The other checkout stands are closed. What additional information do you need in order to decide which lane to enter?

36. **IN YOUR OWN WORDS** You drive up to your bank and see five cars in front of you waiting for two lanes of the drive-through banking services. What additional information do you need in order to decide whether to drive through or park your car and enter the bank to do your banking?

37. A boy cyclist and a girl cyclist are 10 miles apart and pedaling toward each other. The boy's rate is 6 miles per hour, and the girl's rate is 4 miles per hour. There is also a friendly fly zooming continuously back and forth from one bike to the other. If the fly's rate is 20 miles per hour, by the time the cyclists reach each other, how far does the fly fly?

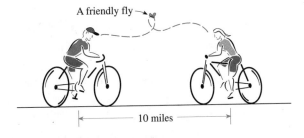

A friendly fly

10 miles

38. Two volumes of Newman's *The World of Mathematics* stand side by side, in order, on a shelf. A bookworm starts at page i of Volume I and bores its way in a straight line to the last page of Volume II. Each cover is 2 mm thick, and the first volume is $\frac{17}{19}$ as thick as the second volume. The first volume is 38 mm thick without its cover. How far does the bookworm travel?

39. Alex, Beverly, and Cal live on the same straight road. Alex lives 10 miles from Beverly and Cal lives 2 miles from Beverly. How far does Alex live from Cal?

40. In a different language, *liro cas* means "red tomato." The meaning of *dum cas dan* is "big red barn" and *xer dan* means "big horse." What do you think the words for "red barn" are in this language?

41. Write down a three-digit number. Write the number in reverse order. Subtract the smaller of the two numbers from the larger to obtain a new number. Write down the new number. Reverse the digits again, but add the numbers this time. Do this process for another three-digit number. Do you get a pattern, and does your pattern work for all three-digit numbers?

42. Start with a common fraction between 0 and 1. Form a new fraction, using the following rules:

    New denominator: Add the numerator and denominator of the original fraction.

    New numerator: Add the new denominator to the original numerator.

    Write the new fraction and use a calculator to find a decimal equivalent to four decimal places. Repeat these steps again, this time calling the new fraction the original. Continue the process until a pattern appears about the decimal equivalent. What is the decimal equivalent?

43. The number 6 has four divisors — namely 1, 2, 3, and 6. List all numbers less than 20 that have exactly four divisors.

Problems 44–57 are not typical math problems but are problems that require only common sense (and sometimes creative thinking).

44. How many 3-cent stamps are there in a dozen?

45. Which is more, six dozen dozen or half a dozen dozen?

46. Which weighs more — a ton of coal or a ton of feathers?

47. If you take 7 cards from a deck of 52 cards, how many cards do you have?

48. At six o'clock the grandfather clock struck 6 times. If it was 30 seconds between the first and last strikes, how long will it take the same clock to strike noon?

49. Oak Park cemetery in Oak Park, New Jersey, will not bury anyone living west of the Mississippi. Why?

50. Two U.S. coins total $.30, yet one of these coins is not a nickel. What are the coins?

51. Two girls were born on the same day of the same month of the same year to the same parents, but they are not twins. Explain how this is possible.

52. How many outs are there in a baseball game that lasts the full 9 innings?

53. If posts are spaced 10 feet apart, how many posts are needed for 100 feet of straight-line fence?

54. A farmer has to get a fox, a goose, and a bag of corn across a river in a boat that is large enough only for him and one of these three items. If he leaves the fox alone with the goose, the fox will eat the goose. If he leaves the goose alone with the corn, the goose will eat the corn. How does he get all the items across the river?

55. Can you place ten lumps of sugar in three empty cups so that there is an odd number of lumps in each cup?

56. Six glasses are standing in a row. The first three are empty, and the last three are full of water. By handling and moving only one glass, it is possible to change this arrangement so that no empty glass is next to another empty one and no full glass is next to another full glass. How can this be done?

57. Suppose you are chasing someone with a 10-mile head start, and that you are running at a rate that is one mile per hour faster than the person you are chasing. Also suppose that a fly is flying back and forth between the two of you at the rate of 20 miles per hour. How far would the fly fly (if the fly could fly nonstop all that time) by the time you reached the other person?

▲ **Problem Solving**

Each section of the book has one or more problems designated by PROBLEM SOLVING. *These problems may require additional insight, information, or effort to solve. True problem-solving ability comes from solving problems that "are not like the examples" but rather require independent thinking. I hope you will make it a habit to read these problems and attempt to work those that interest you, even though they may not be part of your regular class assignment.*

58. Consider the routes from *A* to *B* and notice that there is now a barricade blocking the path. Work out a general solution for the number of paths with a blockade, and then illustrate your general solution by giving the number of paths for each of the following street patterns.

a. 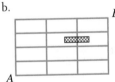 b. 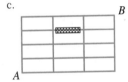 c.

59. **HISTORICAL QUESTION** Thoth, an ancient Egyptian god of wisdom and learning, has abducted Ahmes, a famous Egyptian scribe, in order to assess his intellectual prowess. Thoth places Ahmes before a large funnel set in the ground (see Figure 1.7). It has a

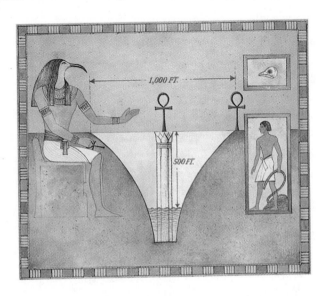

▲ **Figure 1.7 Ahmes' dilemma. Note that there are two ankh-shaped towers. One stands on a cylindrical platform in the center of the funnel. The platform's surface is at ground level. The distance from the platform surface to the liquid is 500 ft. The other ankh tower is on land, at the edge of the funnel.**

circular opening 1,000 ft in diameter, and its walls are quite slippery. If Ahmes attempts to enter the funnel, he will slip down the wall. At the bottom of the funnel is a sleep-inducing liquid that will instantly put Ahmes to sleep for eight hours if he touches it.* Thoth hands Ahmes two objects: a rope 1,006.28 ft in length and the skull of a chicken. Thoth says to Ahmes, "If you are able to get to the central tower and touch it, we will live in harmony for the next millennium. If not, I will detain you for further testing. Please note that with each passing hour, I will decrease the rope's length by a foot." How can Ahmes reach the central ankh tower and touch it?

60. A very magical teacher had a student select a two-digit number between 50 and 100 and write it on the board out of view of the instructor. Next, the student was asked to add 76 to the number, producing a three-digit sum. If the digit in the hundreds place is added to the remaining two-digit number and this result is subtracted from the original number, the answer is 23, which was predicted by the instructor. How did the instructor know the answer would be 23? *Note:*   This problem is dedicated to my friend Bill Leonard of Cal State, Fullerton. His favorite number is 23.

61. A magician divides a deck of cards into two equal piles, counts down from the top of the first pile to the seventh card, and shows it to the audience without looking at it herself. These seven cards are replaced faced down in the same order on top of the pile. She then picks up the other pile and deals the top three cards up in a row in front of her. She forms three columns of cards, by making them add up to 10 (the jack, queen, and king count as 10, and others count at face value). That is, if the card in the first column is a 4, she counts out six more cards; if the card is a queen, no additional cards are needed. The remainder of this pile is placed on top of the first pile. Next the magician adds the values of the three face-up cards, and this number of cards down in the deck is the card that was originally shown to the audience. Explain why this trick works.

▲ **Individual Research**

At the end of many sections you will find problems requiring some library research. One or two references to many of these problems have been provided to get you started.

62. What do the following people have in common?

> **Ralph Abernathy,** civil rights leader
>
> **Harry Blackmun,** Associate Justice of the U.S. Supreme Court
>
> **David Dinkins,** Mayor of New York City
>
> **Art Garfunkel,** folk-rock singer
>
> **Alexander Solzhenitsyn,** Nobel prize-winning novelist
>
> **J. P. Morgan,** banking, steel, and railroad magnate
>
> **Michael Jordan,** basketball superstar

63. Find some puzzles, tricks, or magic stunts that are based on mathematics. Write a paper describing the tricks and also indicate why they work.

> **References** William Schaaf, *A Bibliography of Recreational Mathematics* (Washington, D.C.: National Council of Teachers of Mathematics, 1970). See also the *Journal of Recreational Mathematics.*

* From "The Thoth Maneuver," by Clifford A. Pickover, *Discover*, March 1996, p. 108. © 1996 by Walt Disney Publishing Group, Inc.

1.2 PROBLEM SOLVING WITH SETS

The following problem appeared on the 1987 Examination of the California Assessment Program as an open-ended problem:

> James knows that half the students from his school are accepted at the public university nearby. Also, half are accepted at the local private college. James thinks that this adds up to 100%, so he will surely be accepted at one or the other institution. Explain why James may be wrong. If possible, use a diagram in your explanation.

In this section, we will introduce some terminology and notation that will be useful not only in answering this examination question, but will also form the basis for many problem-solving techniques.

Terminology of Sets

One of the most fundamental concepts in mathematics — and in life, for that matter — is the sorting of objects into certain similar groupings. Every language has an abundance of words that mean "a collection" or "a grouping." For example, we speak of a *herd* of cattle, a *flock* of birds, a *school* of fish, a track *team*, a stamp *collection*, and a *set* of dishes. All these grouping words serve the same purpose, and in mathematics we use the word **set** to refer to any collection of objects.

A useful way to depict sets is to draw a circle or an oval as a representation for the set. The objects in the set, called its **members** or **elements,** are depicted inside the circle, and objects not in the set are shown outside the circle. The **universal set** contains all the elements under consideration in a given discussion. This representation of a set is called a **Venn diagram,** after John Venn (1834–1923). The Swiss mathematician Leonhard Euler (1707–1783) also used circles to illustrate principles of logic, so sometimes these diagrams are called **Euler circles.** However, Venn was the first to use them in a general way.

▬▬ EXAMPLE 1

Let the universal set be all of the cards in a deck of cards. Draw a Venn diagram for the set of hearts.

Solution   It is customary to represent the universal set as a rectangle (labeled U) and the set of hearts (labeled H) as a circle. The sets involved are too large to list all of the elements in either H or U, but we can say that the two of hearts (labeled $\heartsuit 2$) is a member of H, whereas the two of diamonds (labeled $\diamondsuit 2$) is not a member of H. We write $\heartsuit 2 \in H$, whereas $\diamondsuit 2 \notin H$.

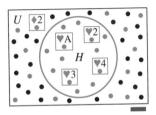

The customary notation for listing the elements in a set is to use braces. For Example 1,

$$H = \{\heartsuit A, \heartsuit 2, \heartsuit 3, \heartsuit 4, \heartsuit 5, \heartsuit 6, \heartsuit 7, \heartsuit 8, \heartsuit 9, \heartsuit 10, \heartsuit J, \heartsuit Q, \heartsuit K\}$$

A flock of birds is an everyday example of the mathematical concept of set.

▲ **Figure 1.10   Relationships between two sets A and B**

The elements that are not in H are referred to as the **complement** of H, and this is written using an overbar: $\overline{H} = \{$spades, diamonds, clubs$\}$. These relationships are shown in Figure 1.8.

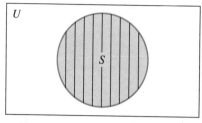

a. To represent a set S, shade the interior of S; the answer is everything shaded (shown as a color screen).

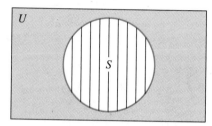

b. To represent a set $\overline{S}$, shade S; the answer is everything not shaded (shown as a color screen).

▲ **Figure 1.8   Venn diagrams for a set and for its complement**

Notice that any set S divides the universe into two regions, as shown in Figure 1.9.

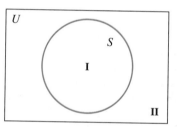

▲ **Figure 1.9   General representation of a set S**

The **cardinality** of a set is the number of elements in the set. If you are familiar with a deck of cards, you know the cardinality of U in Example 1 is 52, and the cardinality of H is 13. We symbolize the cardinality of a set as follows:

$$|H| = 13 \quad \text{and} \quad |U| = 52$$

🚫 Note that $H \neq |H|$. In words, a set is *not* the same as its cardinality. 🚫

Most applications will involve more than one set, so we begin by considering the relationships between two sets A and B. The various possible relationships are shown in Figure 1.10. We say that A is a **subset** of B, written $A \subseteq B$, if every element

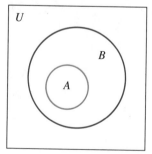

a. $A \subseteq B$

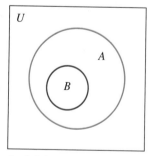

b. $B \subseteq A$

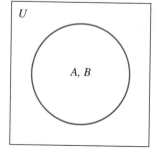

c. $A = B$

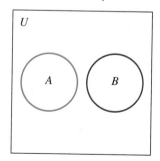

d. A and B are disjoint

of A is also an element of B (see Figure 1.10a). Similarly, $B \subseteq A$ if every element of B is also an element of A (Figure 1.10b). Two sets A and B are equal, written $A = B$, if the sets have exactly the same elements (Figure 1.10c). Finally, A and B are **disjoint** if they have no elements in common (Figure 1.10d).

Sometimes we are given two sets X and Y, and we know nothing about the way they are related. In this situation, we draw a general figure, such as the one shown in Figure 1.11.

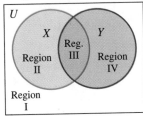

▲ **Figure 1.11 General Venn diagram for two sets**

X is regions II and III.

Y is regions III and IV.

$\overline{X}$ is regions I and IV.

$\overline{Y}$ is regions I and II.

If $X \subseteq Y$, then region II is empty.

If $Y \subseteq X$, then region IV is empty.

If $X = Y$, then regions II and IV are empty.

If X and Y are disjoint, then region III is empty.

We can generalize for more sets. The general Venn diagram for three sets divides the universe into eight regions, as shown in Figure 1.12.

■■■ EXAMPLE 2

Name the regions in Figure 1.12 described by each of the following.

a. A **b.** C **c.** $\overline{A}$ **d.** $\overline{B}$ **e.** $A \subseteq B$ **f.** A and C are disjoint

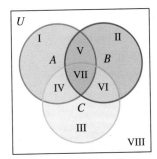

▲ **Figure 1.12 General Venn diagram for three sets**

Solution

a. A is regions I, IV, V, and VII.

b. C is regions III, IV, VI, and VII.

c. $\overline{A}$ is regions II, III, VI, and VIII.

d. $\overline{B}$ is regions I, III, IV, and VIII.

e. $A \subseteq B$ means that regions I and IV are empty.

f. A and C are disjoint means that regions IV and VII are empty. ▬

Operations With Sets

Suppose we consider two general sets, B and Y, as shown in Figure 1.13. If we show the set Y using a yellow highlighter, and the set B using a blue highlighter, it is easy

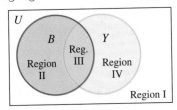

▲ **Figure 1.13 Venn diagram showing intersection and union. If we color one set blue and the other set yellow, then the intersection is green and the union is any colored region.**

to visualize two operations. The **intersection** of the sets is the region shown in green (the parts that are *both* yellow and blue). We see this is region III, and we describe this using the word "**and**." The **union** of the sets is the part shown in *any* color (the parts that are yellow or blue or green). We see this is regions II, III, and IV, and we describe this using the word "**or**."

⊘ Intersection (∩) is translated as *and*, and union (∪) as *or*. ⊘

Operations on Sets: Intersection and Union

The **intersection** of sets A and B, denoted by $A \cap B$, is the set consisting of all elements common to A and B.

The **union** of sets A and B, denoted by $A \cup B$, is the set consisting of all elements of A or B or both.

■■■■ **EXAMPLE 3**

Let $U = \{1, 2, 3, 4, 5, 6, 7, 8, 9\}$, $A = \{2, 4, 6, 8\}$, $B = \{1, 3, 5, 7\}$, and $C = \{5, 7\}$. Draw a Venn diagram showing these sets, and find:

a. $A \cup C$ **b.** $B \cup C$ **c.** $B \cap C$ **d.** $A \cap C$

Solution   The Venn diagram is shown at the right. Begin with the general Venn diagram, and then place each element of the universe in one of the eight regions, as required by the definitions of A, B, and C.

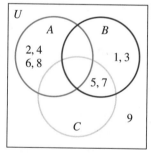

a. $A \cup C = \{2, 4, 6, 8\} \cup \{5, 7\}$
$= \{2, 4, 5, 6, 7, 8\}$
Notice that the union consists of all elements in A or in C or in both. Also note that the order in which the elements are listed is not important.

b. $B \cup C = \{1, 3, 5, 7\} \cup \{5, 7\}$
$= \{1, 3, 5, 7\}$
Notice that, even though the elements 5 and 7 appear in both sets, they are listed only once. That is, the sets $\{1, 3, 5, 7\}$ and $\{1, 3, 5, 5, 7, 7\}$ are equal (exactly the same).

c. $B \cap C = \{1, 3, 5, 7\} \cap \{5, 7\} = \{5, 7\}$
The intersection contains the elements common to both sets. Notice that the resulting set has a name (it is called C), so we write
$$B \cap C = C$$

d. $A \cap C = \{2, 4, 6, 8\} \cap \{5, 7\} = \{\ \ \}$
These sets have no elements in common, so we write $\{\ \ \}$. This set, consisting of no elements, is more commonly called the **empty set** and is denoted by $\varnothing$.

■■

Suppose we consider the cardinality of the various sets in Example 3:

$|U| = 9, \quad |A| = 4, \quad |B| = 4, \quad \text{and} \quad |C| = 2$

$|A \cup C| = 6 \quad (\text{part } \mathbf{a})$

$|B \cup C| = 4 \quad (\text{part } \mathbf{b})$

$|B \cap C| = 2 \quad (\text{part } \mathbf{c})$

$|A \cap C| = 0 \quad (\text{part } \mathbf{d})$

The cardinality of an intersection is found by looking at the number of elements in the intersection. For sets with small cardinalities, we can find the cardinality of the union by direct counting, but if the sets have large cardinalities, it might not be easy to find the union and then the cardinality by direct counting. Some students might want to find $|B \cup C|$ by adding $|B|$ and $|C|$, but you can see from Example 3 that $|B \cup C| \neq |B| + |C|$. However, if you look at the Venn diagram for the number of elements in the union of two sets, the situation becomes quite clear, as shown in Figure 1.14.

$|X| + |Y|$ adds this region twice

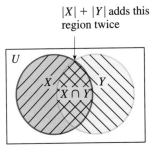

▲ Figure 1.14 Venn diagram for the number of elements in the union of two sets

Formula for the Cardinality of the Union of Two Sets

For any two sets X and Y,

$$|X \cup Y| = |X| + |Y| - |X \cap Y|$$

The elements in the intersection are counted twice.

This corrects for the "error" introduced by counting those elements in the intersection twice.

■ EXAMPLE 4

Suppose a survey indicates that 45 students are taking mathematics and 41 are taking English. How many students are taking math or English?

Solution At first, it might seem that all you do is add 41 and 45, but such is not the case. Let $M = \{\text{persons taking math}\}$ and $E = \{\text{persons taking English}\}$.

To find out how many students are taking math and English, we need to know the number in this intersection.

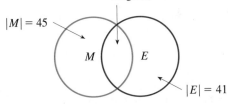

$|M| = 45$

$M \quad E$

$|E| = 41$

As you can see, you need further information. Problem solving requires that you not only recognize what known information is needed when answering a question, but also recognize when additional information is needed. Suppose 12 students are taking both math and English. In this case we see that:

By Formula:

$$|M \cup E| = |M| + |E| - |M \cap E|$$
$$= 45 + 41 - 12$$

$$= 74$$

By Diagram:

First, fill in 12 in $M \cap E$. Then,
$|M| = 45$ $|E| = 41$
Fill in 33   Fill in 29
$(45 - 12 = 33)$. $(41 - 12 = 29)$.

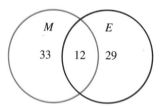

The total number is $33 + 12 + 29 = 74$.

Example 4 looks very much like the open-ended examination question we posed at the beginning of this section. You will find that open-ended question in the problem set.

For three sets, the situation is a little more involved. There is a formula for the number of elements, but it is easier to use Venn diagrams, as illustrated by Example 5. Remember, the overall procedure is to fill in the number in the innermost region first and work your way outward through the Venn diagram using subtraction.

■■■■ EXAMPLE 5   Polya's Method

A survey of 100 randomly selected students gave the following information:

> 45 students are taking mathematics.
> 41 students are taking English.
> 40 students are taking history.
> 15 students are taking math and English.
> 18 students are taking math and history.
> 17 students are taking English and history.
> 7 students are taking all three.

a. How many are taking only mathematics?

b. How many are taking only English?

c. How many are taking only history?

d. How many are not taking any of these courses?

Solution We use Polya's problem-solving guidelines for this example.

Understand the Problem. We are considering students who are members of one or more of three sets. If U represents the universe, then $|U| = 100$. We also define

the three sets:

M = {students taking mathematics}

E = {students taking English}

H = {students taking history}

Devise a Plan.   The plan is to draw a Venn diagram, and then to fill in the various regions. We fill in the innermost region first, and then work our way outward (using subtraction) until the number of elements of the eight regions formed by the three sets is known.

Carry Out the Plan.

Step 1. We note $|M \cap E \cap H| = 7$.

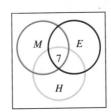

Step 2. Fill in the other inner portions.

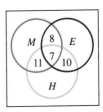

$|E \cap H| = 17$, but 7 have previously been accounted for, so an additional 10 ($17 - 7 = 10$) are added to the Venn diagram.

$|M \cap H| = 18$; fill in $18 - 7 = 11$.

$|M \cap E| = 15$; fill in $15 - 7 = 8$.

Step 3. Fill in the other regions.

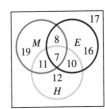

$|H| = 40$, but 28 have previously been accounted for in the set H, so there are an additional 12 members ($40 - 11 - 7 - 10 = 12$).

$|E| = 41$; fill in $41 - 8 - 7 - 10 = 16$.

$|M| = 45$; fill in $45 - 11 - 7 - 8 = 19$.

Step 4. Add all the numbers listed in the sets of the Venn diagram to see that 83 students have been accounted for. Since 100 students were surveyed, we see that 17 are not taking any of the three courses. We now have the answers to the questions directly from the Venn diagram:

a. 19   **b.** 16   **c.** 12   **d.** 17

Look Back.   Does our answer make sense? Add all the numbers in the Venn diagram as a check to see that we have accounted for the 100 students.   ▬

Combined Operations With Sets

We conclude this section by looking at some combined operations of several sets. For sets, we perform operations from left to right; however, if there are parentheses, operations within them are performed first.

▬▬▬ **EXAMPLE 6**

Illustrate using Venn diagrams: **a.** $\bar{A} \cup \bar{B}$ **b.** $\overline{A \cup B}$

Solution

a. This is a combined operation that should be read from left to right. Find the complements of A and B and *then* find the union.

Step 1.   Shade $\bar{A}$ (vertical lines). Then shade $\bar{B}$ (horizontal lines).

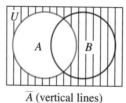   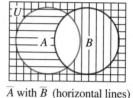

$\bar{A}$ (vertical lines) $\bar{A}$ with $\bar{B}$ (horizontal lines)

Step 2.

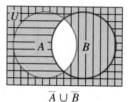

$\bar{A} \cup \bar{B}$

$\bar{A} \cup \bar{B}$ is every portion that is shaded. We show that here using a color highlighter.

b. This is a combined operation that should be interpreted to mean $\overline{(A \cup B)}$, which is the complement of the union. *First* find $A \cup B$ (vertical lines), and *then* find the complement (color highlighter). Compare this with part **a.**

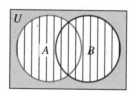

Notice that $\overline{A \cup B} \neq \bar{A} \cup \bar{B}$. If they were equal, the final highlighted color portions of the Venn diagrams from Example 6 would be the same.   ▬

▬▬▬ **EXAMPLE 7**   **Polya's Method**

Prove $\overline{A \cup B} = \bar{A} \cap \bar{B}$.

Solution   We use Polya's problem-solving guidelines for this example.

Understand the Problem.   We wish to prove the given statement is true *for all* sets A and B, so we cannot work with a *particular* example.

Devise a Plan.   The procedure is to draw separate Venn diagrams for the left and the right sides, and then to compare them to see if they are identical.

Carry Out the Plan.

Step 1. Draw a diagram for the expression on the left side of the equal sign. The final result is shown with color highlighter. (See Example 6 for details.)

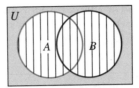

$\overline{A \cup B}$
(see Example 6 for details)

Step 2. Draw a diagram for the expression on the right side of the equal sign.

$\overline{A}$ (vertical lines)

$\overline{B}$ (horizontal lines)

The final result, $\overline{A} \cap \overline{B}$, is the part with both vertical and horizontal lines (as shown with the color highlighter).

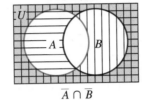

$\overline{A} \cap \overline{B}$

Step 3. Compare the portions shaded by the color highlighter in the two Venn diagrams. They are the same, so we have proved

$$\overline{A \cup B} = \overline{A} \cap \overline{B}$$

The result proved in Example 7 is called **De Morgan's law.** In the problem set you are asked to prove the second part of De Morgan's law.

De Morgan's Laws

For any sets X and Y:

$$\overline{X \cup Y} = \overline{X} \cap \overline{Y} \qquad \overline{X \cap Y} = \overline{X} \cup \overline{Y}$$

We conclude this section by finding the regions described by some combined operations for three sets.

■■■ EXAMPLE 8

Using the eight regions labeled at the right, describe each of the following sets:

a. $A \cup B$ **b.** $A \cap C$

c. $B \cap C$ **d.** $\overline{A}$

e. $\overline{A \cup B}$ **f.** $A \cap B \cap C$

g. $A \cup (B \cap C)$ **h.** $\overline{A \cup B} \cap C$

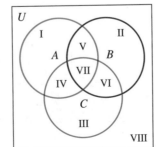

Solution

a. A (vertical lines); B (horizontal lines); $A \cup B$ is everything shaded and is highlighted.

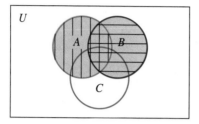

Regions I, II, IV, V, VI, and VII

b. A (vertical lines); C (horizontal lines); $A \cap C$ is everything shaded twice and is highlighted.

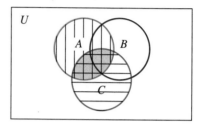

Regions IV and VII

c. B (vertical lines); C (horizontal lines); $B \cap C$ is everything shaded twice as highlighted.

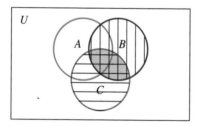

Regions VI and VII

d. A (vertical lines); $\overline{A}$ is everything not shaded, as highlighted.

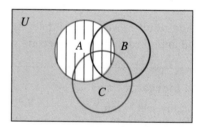

Regions II, III, VI, and VIII

e. $A \cup B$ (vertical lines); $\overline{A \cup B}$ is everything not shaded, as highlighted.

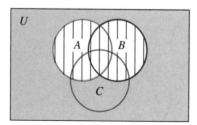

Regions III and VIII

f. $A \cap B \cap C$ is $(A \cap B) \cap C$, so we show $A \cap B$ (vertical lines) and C (horizontal lines); highlight regions shaded twice.

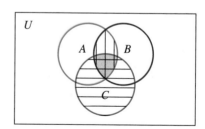

Region VII

g. Parentheses first, $B \cap C$ (vertical lines); A (horizontal lines); $A \cup (B \cap C)$ is everything shaded, as highlighted.

h. $\overline{A \cup B}$ (vertical lines); C (horizontal lines); everything shaded twice is highlighted.

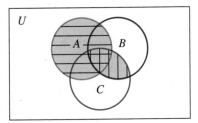

Regions I, IV, V, VI, and VII

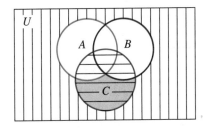

Region III ▬

PROBLEM SET 1.2

▲ A Problems

1. **IN YOUR OWN WORDS** What do we mean by the operations of *union, intersection,* and *complementation?*

2. **IN YOUR OWN WORDS** Consider the sets X and Y. Write each of the following:
   a. a union of complements
   b. a complement of a union
   c. a complement of an intersection
   d. an intersection of complements

3. **IN YOUR OWN WORDS** What do we mean by *De Morgan's laws?*

4. **IN YOUR OWN WORDS** Give an example of a set with cardinality 0.

5. **IN YOUR OWN WORDS** Give an example of a set with cardinality greater than 1 million.

6. **IN YOUR OWN WORDS** This section began with an open-ended question:

   James knows that half the students from his school are accepted at the public university nearby. Also, half are accepted at the local private college. James thinks that this adds up to 100%, so he will surely be accepted at one or the other institution. Explain why James may be wrong. If possible, use a diagram in your explanation.

Perform the given set operations in Problems 7–14. Let
$U = \{1, 2, 3, 4, 5, 6, 7, 8, 9, 10\}$.

7. $\{2, 6, 8\} \cup \{6, 8, 10\}$

8. $\{2, 6, 8\} \cap \{6, 8, 10\}$

9. $\{1, 2, 3, 4, 5\} \cap \{3, 4, 5, 6, 7\}$

10. $\{1, 2, 3, 4, 5\} \cup \{3, 4, 5, 6, 7\}$

11. $\{2, 5, 8\} \cup \{3, 6, 9\}$

12. $\{2, 5, 8\} \cap \{3, 6, 9\}$

13. $\overline{\{2, 8, 9\}}$

14. $\overline{\{1, 2, 5, 7, 9\}}$

Let $U = \{1, 2, 3, 4, 5, 6, 7\}, A = \{1, 2, 3, 4\}, B = \{1, 2, 5, 6\},$ *and* $C = \{3, 5, 7\}$. *List all the members of each of the sets in Problems 15–20.*

15. $A \cup B$

16. $A \cup C$

17. $B \cap C$

18. $A \cap C$

19. $\overline{A}$

20. $\overline{C}$

Draw Venn diagrams for each of the relationships in Problems 21–32.

21. $X \cup Y$

22. $X \cap Z$

23. $\overline{Y}$

24. $\overline{A} \cup B$

25. $A \cap \overline{B}$

26. $\overline{A \cap C}$

27. $A \cap (B \cup C)$

28. $A \cup (B \cup C)$

29. $\overline{(A \cup B) \cup C}$

30. $(A \cap B) \cup (A \cap C)$

31. $A \cap \overline{B \cup C}$

32. $\overline{A} \cup \overline{B} \cup C$

▲ B Problems

33. Draw a Venn diagram showing the relationship among cats, dogs, and animals.

34. Draw a Venn diagram showing the relationship among trucks, buses, and cars.

35. Draw a Venn diagram showing people who are over 30, people who are 30 or under, and people who drive a car.

36. Draw a Venn diagram showing people in a classroom wearing some black, people wearing some blue, and people wearing some brown.

37. Draw a Venn diagram showing that all Chevrolets are automobiles.

38. Draw a Venn diagram showing birds, bees, and living creatures.

39. Santa Rosa Junior College enrolled 29,000 students in the fall of 1996. It was reported that of that number, 58% were female. In addition, 62% were over the age of 25. How many students were there in each category if 40% of those over the age of 25 were male? Draw a Venn diagram showing these relationships.

40. In 1995 the United States population was approximately 263 million. It was reported that of that number, 79% were white, 12% were black, and 9% were Hispanic. If $\frac{1}{2}\%$ have one black and one white parent, 2% have one black and one Hispanic parent, and 1% have one white and one Hispanic parent, how many people are there in each category? Draw a Venn diagram showing these relationships.

In Problems 41–44, use set notation to identify the shaded region.

41.

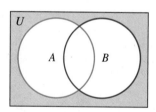

42.

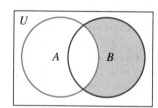

43.

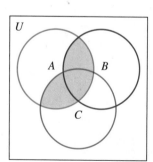

44.

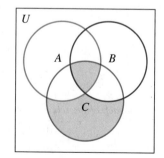

In Problems 45–50, use Venn diagrams to prove or disprove each expression. Remember to draw a diagram for the left side of the equation and another for the right side. If the final shaded portions are the same, then you have proved the result. If the final shaded portions are not identical, then you have disproved the result.

45. $\overline{A \cup B} = \overline{A} \cup \overline{B}$

46. $\overline{A} \cap \overline{B} = A \cup B$

47. $(A \cup B) \cup C = A \cup (B \cup C)$

48. $A \cup (B \cap C) = (A \cup B) \cup (A \cup C)$

49. $A \cap (B \cup C) = (A \cap B) \cap (A \cap C)$

50. De Morgan's law: $\overline{X \cap Y} = \overline{X} \cup \overline{Y}$

51. Listed below are five female and five male Wimbledon tennis champions, along with their country of citizenship and handedness.

Female	*Male*

Female

Steffi Graf, Germany, right
Martina Navratilova, U.S., left
Chris Evert Lloyd, U.S., right
Evonne Goolagong, Australia, right
Virginia Wade, Britain, right

Male

Michael Stich, Germany, right
Stefan Edberg, Sweden, right
Boris Becker, Germany, right
Pat Cash, Australia, right
John McEnroe, U.S., left

Using Figure 1.15, indicate in which region each of the above individuals would be placed.

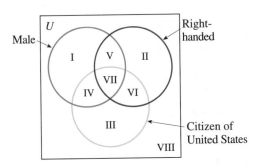

▲ **Figure 1.15   Problem 51**

52. Montgomery College has a 50-piece band and a 36-piece orchestra. If 14 people are members of both the band and the orchestra, can the band and orchestra travel in two 40-passenger buses?

53. From a survey of 100 college students, a marketing research company found that 75 students owned stereos, 45 owned cars, and 35 owned both cars and stereos.
  a. How many students owned either a car or a stereo (but not both)?
  b. How many students do not own either a car or a stereo?

54. In a survey of a TriDelt chapter with 50 members, 18 were taking mathematics, 35 were taking English, and 6 were taking both. How many were not taking either of these subjects?

55. In a senior class at Rancho Cotati High School, there were 25 football players and 16 basketball players. If 7 persons played both sports, how many different people played in these sports?

56. The fire department wants to send booklets on fire hazards to all teachers and home-owners in town. How many booklets does it need, using these statistics?

    50,000 homeowners
    4,000 teachers
    3,000 teachers who own their own homes

57. In a recent survey of 100 persons, the following information was gathered:

    59 use shampoo A.
    51 use shampoo B.
    35 use shampoo C.
    24 use shampoos A and B.
    19 use shampoos A and C.
    13 use shampoos B and C.
    11 use all three.

    Let A = {persons who use shampoo A}; B = {persons who use shampoo B}; and C = {persons who use shampoo C}. Use a Venn diagram to show how many persons are in each of the eight possible categories.

58. Matt E. Matic was applying for a job. To determine whether he could handle the job, the personnel manager sent him out to poll 100 people about their favorite types of TV shows. His data were as follows:

    59 preferred comedies.
    38 preferred variety shows.
    42 preferred serious drama.
    18 preferred comedies and variety programs.
    12 preferred variety and serious drama.
    16 preferred comedies and serious drama.
    7 preferred all types.
    2 did not like any of these TV show types.

    If you were the personnel manager, would you hire Matt on the basis of this survey?

59. A poll was taken of 100 students at a commuter campus to find out how they got to campus. The results were:

    42 said they drove alone.
    28 rode in a carpool.
    31 rode public transportation.
    9 used both carpools and public transportation.
    10 used both a carpool and sometimes their own cars.
    6 used buses as well as their own cars.
    4 used all three methods.

    How many used none of the above-mentioned means of transportation?

60. In an interview of 50 students,

    12 liked Proposition 8 and Proposition 13.
    18 liked Proposition 8, but not Proposition 5.
    4 liked Proposition 8, Proposition 13, and Proposition 5.
    25 liked Proposition 8.
    15 liked Proposition 13.
    10 liked Proposition 5, but not Proposition 8 or Proposition 13.
    1 liked Proposition 13 and Proposition 5, but not Proposition 8.

    a. Show the completed Venn diagram.
    b. Of those surveyed, how many did not like any of the three propositions?
    c. How many like Proposition 8 and Proposition 5?

▲ **Problem Solving**

61. Draw a Venn diagram for four sets, and label the 16 regions.

62. The Venn diagram in Figure 1.16 shows five sets. It was drawn by Allen J. Schwenk of the U.S. Naval Academy. As you can see, there are 32 separate regions. Describe the following regions:

   a. 1   **b.** 11   **c.** 21   **d.** 31   **e.** 16

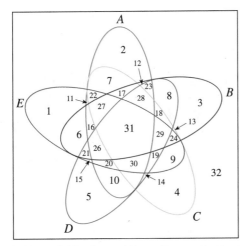

▲ **Figure 1.16   General Venn diagram for 5 sets**

63. Using the Venn diagram in Figure 1.16, specify which region is described by:

   a. $\overline{A \cup B \cup C \cup D \cup E}$ **b.** $A \cap \overline{B \cup C \cup D \cup E}$

64. Human blood is typed Rh⁺ (positive blood) or Rh⁻ (negative blood). This Rh factor is called an *antigen*. There are two other antigens known as A and B types. Blood lacking both A and B types is called type O. Sketch a Venn diagram showing the three types of antigens A, B, and Rh, and label each of the eight regions. For example, O⁺ is inside the Rh set (anything in Rh is positive), but outside both A and B. On the other hand, O⁻ is that region outside all three circles.

65. Each of the circles in Figure 1.17 is identified by a letter, each having a number value from 1 to 9. Where the circles overlap, the number is the sum of the values of the letters in the overlapping circles. What is the number value for each letter?*

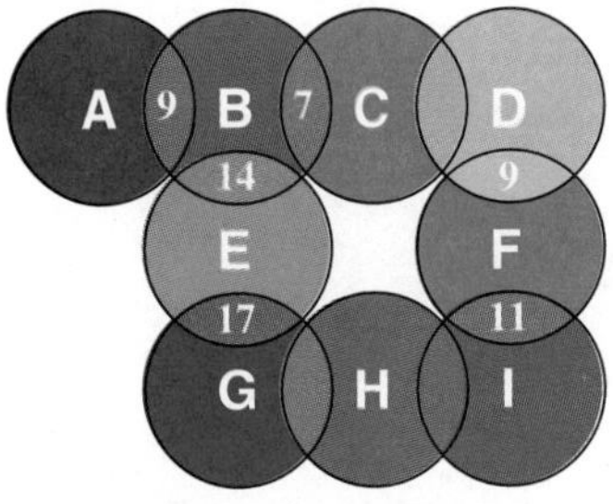

66. On the *NBC Nightly News* on Thursday, May 25, 1995, Tom Brokaw read a brief report on computer use in the United States. The story compared computer users by ethnic background, and Brokaw reported that 14% of blacks and 13% of Hispanics use computers, but 27% of whites use computers. Brokaw then commented that computer use by whites was equal to that of blacks and Hispanics combined.

   a. Draw a Venn diagram showing the white, black, Hispanic, Asian, and native American populations of the United States, along with U.S. computer users.

   b. Use the Venn diagram from part **a** to show that percentages cannot be added as was done by Brokaw.

▲ **Figure 1.17   Circle intersection puzzle**

* From "Perception Puzzles," by Jean Moyer, *Sky*, January 1995, p. 120.

▲ **Individual Research**

67. See how many of the 32 regions of the Venn diagram in Figure 1.16 you can describe using set notation similar to that shown in Problem 63. Answers are not unique.

    Reference B. Grunbaum, "The Construction of Venn Diagrams," *College Mathematics Journal*, 1984, pp. 238–247. Also, see Shoeleh Mutameni, "Venn Diagrams Finale," *The Mathematics Teacher*, December 1994, pp. 744–746.

68. Venn diagrams can be used to illustrate both additive and subtractive color mixing. Write a report discussing the creation of colors in the Venn diagrams shown in Figure 1.18.

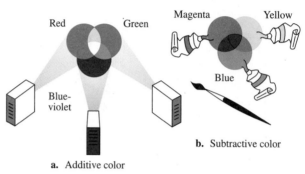

a. Additive color

b. Subtractive color

▲ **Figure 1.18   Color mixing**

69. Suppose in the small California town of Ferndale it is the practice of many of the men to be shaved by the barber. Now, the barber has a rule that has come to be known as the barber's rule: *He shaves those men and only those men who do not shave themselves.* The question is: Does the barber shave himself? If he does shave himself, then according to the barber's rule, he does not shave himself. On the other hand, if he does not shave himself, then, according to the barber's rule, he shaves himself. We can only conclude that there can be no such barber's rule. But why not?

1.3   INDUCTIVE AND DEDUCTIVE REASONING

Studying numerical patterns is one frequently used technique of problem solving. Let's begin with some simple patterns.

A Pattern of Nines

A very familiar pattern is found in the ordinary "times" tables. By pointing out patterns, teachers can make it easier for children to learn some of their multiplication tables. For example, consider the multiplication table for 9s:

$$1 \times 9 = 9 \qquad\qquad 6 \times 9 = 54$$
$$2 \times 9 = 18 \qquad\qquad 7 \times 9 = 63$$
$$3 \times 9 = 27 \qquad\qquad 8 \times 9 = 72$$
$$4 \times 9 = 36 \qquad\qquad 9 \times 9 = 81$$
$$5 \times 9 = 45 \qquad\qquad 10 \times 9 = 90$$

What patterns do you notice? You should be able to see many number relationships by looking at the totals. For example, notice that the sum of the digits to the right of the equality is 9 in all the examples ($1 + 8 = 9$, $2 + 7 = 9$, $3 + 6 = 9$, and so on). Will this always be the case for multiplication by 9? (Consider $11 \times 9 = 99$. The sum of the digits is 18. However, notice the result if you add the digits of 18.) Do you see any other patterns? Can you explain why they "work"? This pattern of adding after multiplying by 9 generates a sequence of numbers: 9, 9, 9, ... We call this the *nine pattern*. The two number tricks described in the News Clip in the margin use this nine pattern.

■■■■■ **EXAMPLE 1**   **Polya's Method**

Find the *eight pattern*.

Solution   We use Polya's problem-solving guidelines for this example.

Understand the Problem.   What do we mean by *eight pattern*? Do you understand the example for the *nine pattern*?

Devise a Plan.   We will carry out the multiplications of successive counting numbers by 8 and if there is more than a single-digit answer, we add the digits. What we are looking for is a pattern for these single-digit numerals.

Carry Out the Plan.

$$1 \times 8 = 8, \quad 2 \times 8 = 16, \quad 3 \times 8 = 24, \quad 4 \times 8 = 32, \quad 5 \times 8 = 40, \ldots$$

Single digits: 8 $1 + 6 = 7$ $2 + 4 = 6$ $3 + 2 = 5$ $4 + 0 = 4$

Continue with some additional terms:

$6 \times 8 = 48$ and $4 + 8 = 12$ and $1 + 2 = 3$
$7 \times 8 = 56$ and $5 + 6 = 11$ and $1 + 1 = 2$
$8 \times 8 = 64$ and $6 + 4 = 10$ and $1 + 0 = 1$
$9 \times 8 = 72$ and $7 + 2 = 9$
$10 \times 8 = 80$ and $8 + 0 = 8$

We now see the eight pattern: 8, 7, 6, 5, 4, 3, 2, 1, 9, 8, 7, 6, ...

Look Back.   Let's do more than check the arithmetic, since this pattern seems clear. The problem seems to be asking whether we understand the concept of a *nine pattern* or an *eight pattern*. Verify that the *seven pattern* is 7, 5, 3, 1, 8, 6, 4, 2, 9, 7, 5, 3, 1, ...   ■

Order of Operations

Complicated arithmetic problems can sometimes be solved by using patterns. Given a difficult problem, a problem solver will often try to *solve a simpler, but similar, problem*. The second suggestion for solving using Polya's problem-solving procedure stated, "If you cannot solve the proposed problem, look around for an appropriate related problem (a simpler one, if possible)." For example, suppose we wish to compute the following number:

$$10 + 123{,}456{,}789 \times 9$$

Instead of doing a lot of arithmetic, let's study the following pattern:

$$2 + 1 \times 9$$
$$3 + 12 \times 9$$
$$4 + 123 \times 9$$

Do you see the next entry in this pattern? Do you see that if we continue the pattern we will eventually reach the desired expression of $10 + 123{,}456{,}789 \times 9$? Using Polya's strategy, we begin by working these easier problems. Thus, we begin with $2 + 1 \times 9$. There is a possibility of ambiguity in calculating this number:

Left-to-Right	*Multiplication First*
$2 + 1 \times 9 = 3 \times 9 = 27$	$2 + 1 \times 9 = 2 + 9 = 11$

Although either of these might be acceptable in certain situations, it is not acceptable to get two different answers to the same problem. We therefore all agree to do a problem like this by multiplying first. If we wish to change this order, we use parentheses, as in $(2 + 1) \times 9 = 27$.

Order of Operations

1. First, perform any operations enclosed in parentheses.
2. Next, perform multiplications and divisions as they occur by working from left to right.
3. Finally, perform additions and subtractions as they occur by working from left to right.

Thus, the correct result for $2 + 1 \times 9$ is 11. Also,

$$3 + 12 \times 9 = 3 + 108 = 111$$
$$4 + 123 \times 9 = 4 + 1{,}107 = 1{,}111$$
$$5 + 1{,}234 \times 9 = 5 + 11{,}106 = 11{,}111$$

Do you see a pattern? If so, then make a prediction about the desired result. If you do not see a pattern, continue with this pattern to see more terms, or go back and try another pattern. For this example, we predict

$$10 + 123{,}456{,}789 \times 9 = 1{,}111{,}111{,}111$$

The most difficult part of this type of problem solving is coming up with a correct pattern. For this example, you might guess that

$2 + 1 \times 1$
$3 + 12 \times 2$
$4 + 123 \times 3$
$5 + 1,234 \times 4$
$\vdots$

leads to $10 + (123,456,789 \times 9)$. Calculating, we find

$2 + 1 \times 1 = 2 + 1 = 3$
$3 + 12 \times 2 = 3 + 24 = 27$
$4 + 123 \times 3 = 4 + 369 = 373$
$5 + 1,234 \times 4 = 5 + 4,936 = 4,941$

If you begin a pattern and it does not lead to a pattern of answers, then you need to remember that part of Polya's problem-solving procedure is to work both backward and forward. Be willing to give up one pattern and begin another.

We also point out that the patterns you find are not unique. One last time, we try a pattern for $10 + 123,456,789 \times 9$:

$10 + 1 \times 9 = 10 + 9 = 19$
$10 + 12 \times 9 = 10 + 108 = 118$
$10 + 123 \times 9 = 10 + 1,107 = 1,117$
$10 + 1,234 \times 9 = 10 + 11,106 = 11,116$
$\vdots \qquad\qquad \vdots$

We do see a pattern here (although not quite as easily as the one we found with the first pattern for this example):

$10 + 123,456,789 \times 9 = 1,111,111,111$

Inductive Reasoning

The type of reasoning used here and in the first sections of this book — first observing patterns and then predicting answers for more complicated problems — is called **inductive reasoning.** It is a very important method of thought and is sometimes called the *scientific method.* It involves reasoning from particular facts or individual cases to a general **conjecture** — a statement you think may be true. That is, a generalization is made on the basis of some observed occurrences. The more individual occurrences we observe, the better able we are to make a correct generalization. Peter in the *B.C.* cartoon makes the mistake of generalizing on the basis of a single observation.

■■■■ **EXAMPLE 2**　　　　　　　　　　　　　　　　　　**Polya's Method**

What is the sum of the first 100 consecutive odd numbers?

Solution　We use Polya's problem-solving guidelines for this example.

Understand the Question.　Do you know what the terms mean? *Odd numbers* are 1, 3, 5, . . . , and *sum* means to add:

$$1 + 3 + 5 + \cdots + ?$$
　　　　　　　　　　↑

The first thing you need to understand is what the last term will be, so you will know when you have reached 100 consecutive odd numbers.

1 + 3 is two terms.
1 + 3 + 5 is three terms.
1 + 3 + 5 + 7 is four terms.

It seems as if the last term is always one less than twice the number of terms. Thus, the sum of the first 100 consecutive odd numbers is

$$1 + 3 + 5 + \cdots + 195 + 197 + 199$$
　　　　　　　　　　　　　　↑
　　　　　　This is one less than 2(100).

Devise a Plan.　The plan we will use is to look for a pattern:

$$1 = 1 \quad \text{one term}$$
$$1 + 3 = 4 \quad \text{sum of two terms}$$
$$1 + 3 + 5 = 9 \quad \text{sum of three terms}$$

Do you see a pattern yet? If not, continue:

$$1 + 3 + 5 + 7 = 16$$
$$1 + 3 + 5 + 7 + 9 = 25$$

Carry Out the Plan.　It appears that the sum of 2 terms is 2^2; of 3 terms, 3^2; of 4 terms, 4^2; and so on. The sum of the first 100 consecutive odd numbers is therefore 100^2.

Looking Back.　Does $100^2 = 10,000$ seem correct?　　　　　　　　　　—

Deductive Reasoning

Another method of reasoning used in mathematics is called **deductive reasoning.** This method of reasoning produces results that are *certain* within the logical system being developed. That is, deductive reasoning involves reaching a conclusion by using a formal structure based on a set of **undefined terms** and a set of accepted unproved **axioms** or **premises.** The conclusions are said to be proved and are called **theorems.** We consider deductive reasoning at greater length in Chapter 2.

The most useful axiom in problem solving is the principle of substitution: *If two quantities are equal, one may be substituted for the other without changing the truth or falsity of the statement.*

Substitution Property

If $a = b$, then a may be **substituted** for b in any mathematical statement without affecting the truth or falsity of the given mathematical statement.

The simplest way to illustrate the substitution property is to use it in evaluating a formula.

■ EXAMPLE 3

A billiard table is 4 ft by 8 ft. Find the perimeter.

Solution   Use an appropriate formula,

$$P = 2\ell + 2w \quad P = \text{PERIMETER}, \ell = \text{LENGTH}, \text{and } w = \text{WIDTH}$$

Substitute the known values into the formula:

$$\ell = 8 \quad w = 4$$
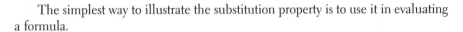
These arrows mean substitution.

$$P = 2(8) + 2(4)$$
$$= 16 + 8$$
$$= 24$$

"For a minute I thought we had him stymied!"

The perimeter is 24 ft.   ▬

Problem solving depends not only on the substitution property, but also on translating statements from English into mathematical symbols. On a much more advanced level, this process is called **mathematical modeling** (we will consider some mathematical modeling later in this book). However, for now, we will simply call it **translating**. Example 4 reviews much of the terminology from your previous mathematics courses. These terms include **sum** to indicate the result obtained from addition, **difference** for the result from subtraction, **product** for the result of a multiplication, and **quotient** for the result of a division.

■ EXAMPLE 4

Write the word statement or phrase in symbols. Do not simplify the translated expressions.

a. The sum of seven and a number
b. The difference of a number subtracted from seven
c. The quotient of two numbers
d. The product of two consecutive numbers
e. The difference of the squares of a number and two
f. The square of the difference of two from a number

News Clip

Mathematics, of course, is not the *only* cornerstone of opportunity in today's world. Reading is even more fundamental as a basis for learning and for life. What is different today is the great increase in the importance of mathematics to so many areas of education, citizenship, and careers.

Everybody Counts, p. 3

g. The sum of two times a number and six is equal to two times the sum of a number and three.

Solution

a. Since *sum* indicates addition, this can be rewritten as

   (SEVEN) + (A NUMBER)

   Now select some variable — say, s = A NUMBER. The symbolic statement is $7 + s$.

b.   (SEVEN) − (A NUMBER)

   Let d = A NUMBER; then the expression is $7 - d$.

c.   A NUMBER
   ―――――――――
        ANOTHER NUMBER

   If there is more than one unknown in a problem and no relationship between those unknowns is given, more than one variable may be needed. Let m = A NUMBER and n = ANOTHER NUMBER. Then the expression is $\dfrac{m}{n}$.

d.   (A NUMBER)(NEXT CONSECUTIVE NUMBER)

   If there is more than one unknown in a problem but a given relationship exists between those unknowns, *do not choose more variables than you need* for the problem. In this problem, a consecutive number means one more than the first number:

   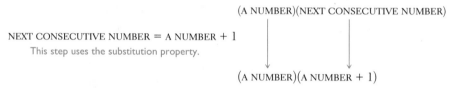

   NEXT CONSECUTIVE NUMBER = A NUMBER + 1
   This step uses the substitution property.

   (A NUMBER)(A NUMBER + 1)

   Let x = A NUMBER; then the variable expression is

   $x(x + 1)$

e.   (A NUMBER)^2 − 2^2

   Let x = A NUMBER; then

   $x^2 - 2^2$

   Recall that this is called a *difference of squares*.

f.   (A NUMBER − 2)^2

   Let x = A NUMBER; then

   $(x - 2)^2$

g. If "a number" is referred to more than once in a problem, it is assumed to be the same number.

   $2(\text{A NUMBER}) + 6 = 2(\text{A NUMBER} + 3)$

   Let n = A NUMBER;

   $2n + 6 = 2(n + 3)$

PROBLEM SET 1.3

▲ A Problems

1. **IN YOUR OWN WORDS** Discuss the nature of *inductive* and *deductive reasoning*.
2. **IN YOUR OWN WORDS** Explain what is meant by the *seven pattern*.
3. **IN YOUR OWN WORDS** What do we mean by *order of operations*?
4. **IN YOUR OWN WORDS** What is the scientific method?
5. **IN YOUR OWN WORDS** What is the substitution principle?

Perform the operations in Problems 6–19.

6. **a.** $5 + 2 \times 6$ **b.** $7 + 3 \times 2$
7. **a.** $14 + 6 \times 3$ **b.** $30 \div 5 \times 2$
8. **a.** $3 \times 8 + 3 \times 7$ **b.** $3(8 + 7)$
9. **a.** $1 + 3 \times 2 + 4 + 3 \times 6$ **b.** $3 + 6 \times 2 + 8 + 4 \times 3$
10. **a.** $10 + 5 \times 2 + 6 \times 3$ **b.** $4 + 3 \times 8 + 6 + 4 \times 5$
11. **a.** $(5 + 6) \div 2$ **b.** $5 + 6 \div 2$
12. **a.** $8 + 2(3 + 12) - 5 \times 3$ **b.** $25 - 4(12 - 2 \times 6) + 3$
13. **a.** $12 + 6/3$ **b.** $(12 + 6)/3$
14. **a.** $450 + 550/10$ **b.** $\dfrac{450 + 550}{10}$
15. **a.** $20/2 \cdot 5$ **b.** $20/(2 \cdot 5)$
16. $3 + 9 - 3 \times 2 + 2 \times 6 \div 3$ 17. $[(3 + 9) \div 3] \times 2 + [(2 \times 6) \div 3]$
18. $3 + [(9 \div 3) \times 2] + [(2 \times 6) \div 3]$ 19. $[(3 + 9) \div (3 \times 2)] + [(2 \times 6) \div 3]$

Problems 20–26 are modeled after Example 1. Find the requested pattern.

20. three pattern   **21.** four pattern   **22.** five pattern   **23.** six pattern
24. What is the sum of the first 25 consecutive odd numbers?
25. What is the sum of the first 50 consecutive odd numbers?
26. What is the sum of the first 1,000 consecutive odd numbers?

Rewrite the word statement or phrase in Problems 27–37, using symbols. Do not simplify; simply translate.

27. **a.** The sum of three and the product of two and four
    b. The product of three and the sum of two and four
28. **a.** The quotient of three divided by the sum of two and four
    b. Ten times the sum of four and three
29. **a.** Eight times nine plus ten
    b. Eight times the sum of nine and ten
30. **a.** The sum of three squared and five squared
    b. The square of the sum of three and five

31. **a.** The square of three increased by the cube of two
    b. The cube of three decreased by the square of two

32. **a.** The square of a number plus five
    b. Six times the cube of a number

33. **a.** The sum of the squares of 4 and 9
    b. The square of the sum of 4 and 9

34. **a.** The sum of a number and one is equal to one added to the number.
    b. Two added to a number is equal to the sum of the number and two.

35. **a.** Three times the sum of a number and four is equal to sixteen.
    b. Five times the sum of a number and one is equal to the sum of five times the number and five.

36. **a.** The sum of six times a number and twelve is equal to six times the sum of the number and two.
    b. A number times the sum of seven and the number is equal to zero.

37. **a.** The sum of a number squared and eight times the number is equal to six.
    b. The square of a number plus eight times the number is equal to the number times the sum of eight and the number.

▲ B Problems

In Problems 38–46, write a formula to express the given relation.

38. The area A of a parallelogram is the product of the base b and the height h.

39. The area A of a triangle is one-half the product of the base b and the height h.

40. The area A of a rhombus is one-half the product of the diagonals p and q.

41. The area A of a trapezoid is the product of one-half the height h and the sum of the bases a and b.

42. The volume V of a cube is the cube of the length s of an edge.

43. The volume V of a rectangular solid is the product of the length ℓ, the width w, and the height h.

44. The volume V of a cone is one-third the product of pi, the square of the radius r, and the height h.

45. The volume V of a circular cylinder is the product of pi, the square of the radius r, and the height h.

46. The volume V of a sphere is the product of four-thirds pi times the cube of the radius r.

47. Consider the following pattern:

$$9 \times 1 - 1 = 8$$
$$9 \times 21 - 1 = 188$$
$$9 \times 321 - 1 = 2,888$$
$$9 \times 4,321 - 1 = 38,888$$

Use this pattern and inductive reasoning to find the next problem and the next answer in the sequence.

48. Use Problem 47 to predict the answer to $9 \times 987,654,321 - 1$.

49. Use Problem 47 to predict the answer to $9 \times 10,987,654,321 - 1$.

50. Consider the following pattern:

$$123,456,789 \times 9 = 1,111,111,101$$
$$123,456,789 \times 18 = 2,222,222,202$$
$$123,456,789 \times 27 = 3,333,333,303$$

Use this pattern and inductive reasoning to find the next problem and the next answer in the sequence.

51. Use Problem 50 to predict the answer to $123,456,789 \times 63$.

52. Use Problem 50 to predict the answer to $123,456,789 \times 81$.

▲ Problem Solving

53. How many squares are there in Figure 1.19?

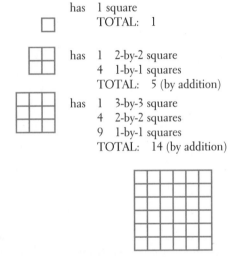

Hint: has   1 square
              TOTAL: 1

              has   1   2-by-2 square
                    4   1-by-1 squares
              TOTAL: 5 (by addition)

              has   1   3-by-3 square
                    4   2-by-2 squares
                    9   1-by-1 squares
              TOTAL: 14 (by addition)

▲ **Figure 1.19   How many squares?**

54. How many triangles are there in Figure 1.20?

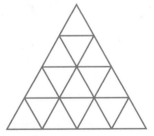

▲ **Figure 1.20   How many triangles?**

55. A four-inch cube is painted red on all sides. It is then cut into one-inch cubes. What fraction of all the one-inch cubes are painted on exactly one side?

56. Problem 55 gives rise to several patterns. Suppose you consider a set of painted cubes, each of which is made up of several smaller cubes. Use patterns to fill in the blanks in the following table. The last entries (for a cube with length of edge 10 in.) have been filled in so that you can check the patterns you obtain.

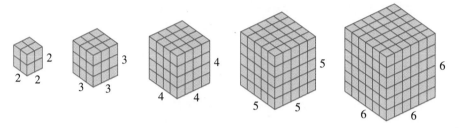

			Number of smaller cubes with the indicated number of painted faces		
Length of edge	*Total cubes*	3	2	1	0
a.	2	8				
b.	3	27				
c.	4					
d.	5					
e.	6					
f.	7					
g.	8					
h.	9					
	10	1,000	8	96	384	512

57. You have 9 coins, but you are told that one of the coins is counterfeit and weighs just a little more than an authentic coin. How can you determine the counterfeit with 2 weighings on a two-pan balance scale?

58. A **magic square** is an arrangement of numbers in the shape of a square, with the sum of each vertical column, each horizontal row, and each diagonal all equal. The numbers of rows and columns are the same, and this number is called the **order** of the magic square. A **standard magic square** is made up of the consecutive counting numbers starting with 1. The first known example of a magic square comes from China.

Legend tells us that around the year 200 B.C. the emperor Yu of the Shang dynasty received the following magic square etched on the back of a tortoise's shell:

The incident supposedly took place along the Lo River, so this magic square has come to be known as the Lo-shu magic square. The even numbers are black (female numbers) and the odd numbers are white (male numbers). Translate this magic square into modern symbols.

59. Rearrange the cards in the formulation below so that each horizontal, vertical, and diagonal line of three adds up to 15.

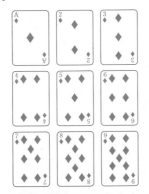

60.

$100 REWARD

In 1987 Martin Gardner (see the profile at the end of this chapter) offered $100 to anyone who could find a 3 × 3 magic square made with consecutive primes. The prize was won by Harry Nelson of Lawrence Livermore Laboratories. He produced the following simplest such square:

1,480,028,201	1,480,028,129	1,480,028,183
1,480,028,153	1,480,028,171	1,480,028,189
1,480,028,159	1,480,028,213	1,480,028,141

Prove that this is a magic square.

61.

> # $100 REWARD (FOR REAL)

Now, for the $100 offer: *Find a* 3×3 *magic square with nine distinct square numbers.* If you find such a magic square, write to me and we will submit it to Martin Gardner to obtain our prize. Show that the following magic squares do not win the award.

a.

127^2	46^2	58^2
2^2	113^2	94^2
74^2	82^2	97^2

b.

35^2	3495^2	2958^2
3642^2	2125^2	1785^2
2775^2	2058^2	3005^2

▲ **Individual Research**

62. Write a short paper about the construction of magic squares. You might include such facts as: There is 1 standard magic square of order 1, 0 of order 2, 8 of order 3, 440 of order 4, and 275,305,224 of order 5. According to the *Guinness Book of World Records*, Leon H. Nissimov of San Antonio, Texas, has discovered the largest known magic square with sum of 999,999,999,989. Show that such a magic square is not possible. You might also include the properties of the magic square discovered by Benjamin Franklin.

References

William H. Benson and Oswald Jacoby, *New Recreations with Magic Squares* (New York: Dover Publications, 1976).

John Fults, *Magic Squares* (La Salle, IL: Open Court, 1974).

Martin Gardner, "The Magic of 3×3 — The $100 Question: Can You Make a Magic Square of Squares?" *Quantum*, January/February 1996, pp. 24–26.

Martin Gardner, "Mathematical Games Department," *Scientific American*, January 1976, pp. 118–122.

63. A process for producing an artistic pattern using magic squares is described in an article, "An Art-Ful Application Using Magic Squares" by Margaret J. Kenney (*The Mathematics Teacher*, January 1982, pp. 83–89). Read the article and design some magic square art pieces.

64. An **alphamagic square**, invented by Lee Sallows, is a magic square so that not only do the numbers spelled out in words form a magic square, but the numbers of letters of the words also form a magic square. For example,

five	twenty-two	eighteen
twenty-eight	fifteen	two
twelve	eight	twenty-five

gives rise to two magic squares:

5	22	18
28	15	2
12	8	25

and

4	9	8
11	7	3
6	5	10

The first magic square comes from the numbers represented by the words in the alphamagic square, and the second magic square comes from the numbers of letters in the words of the alphamagic square. Find another alphamagic square.

1.4 SCIENTIFIC NOTATION AND ESTIMATION

"How Big Is the Cosmos?" The size of the known cosmos can be seen by looking at the cubes in Figure 1.21.

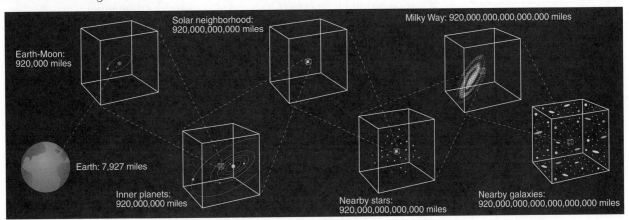

▲ **Figure 1.21   Size of the universe.* Each successive cube is a thousand times as wide and a billion times as voluminous as the one before it. The contents and width of each one are described in the caption beneath it.**

How can the human mind comprehend such numbers? Scientists often work with very large numbers. Distances such as those in Figure 1.21 are measured in terms of the distance that light, moving at 186,000 miles per second, travels in a year. In this section, we turn to patterns to see if there is an easy way to deal with very large and very small numbers. We will also discuss estimation as a problem-solving technique.

Exponential Notation

We often encounter expressions that comprise multiplication of the same numbers. For example,

$$10 \cdot 10 \cdot 10 \quad \text{or} \quad 6 \cdot 6 \cdot 6 \cdot 6 \cdot 6 \cdot 6 \cdot 6 \quad \text{or}$$
$$15 \cdot 15 \cdot 15 \cdot 15 \cdot 15 \cdot 15 \cdot 15 \cdot 15 \cdot 15 \cdot 15 \cdot 15 \cdot 15 \cdot 15 \cdot 15$$

These numbers can be written more simply using what is called **exponential notation:**

$$\underbrace{10 \cdot 10 \cdot 10}_{3 \text{ factors}} = 10^3 \qquad \underbrace{6 \cdot 6 \cdot 6 \cdot 6 \cdot 6 \cdot 6 \cdot 6}_{7 \text{ factors}} = 6^7$$

$$\underbrace{15 \cdot 15 \cdot 15 \cdot 15 \cdot 15 \cdot 15 \cdot 15 \cdot 15 \cdot 15 \cdot 15 \cdot 15 \cdot 15 \cdot 15 \cdot 15}_{14 \text{ factors}} = 15^{14}$$

* Illustration is adapted from *The Universe*, the Life Nature Library.

Exponential Notation

For any number b and any counting number* n,

$$b^n = \underbrace{b \cdot b \cdot b \cdot \cdots \cdot b}_{n \text{ factors}}, \quad b^0 = 1, \quad b^{-n} = \frac{1}{b^n}$$

The number b is called the **base**, the number n in b^n is called the **exponent**, and the number b^n is called a **power** or **exponential**.

EXAMPLE 1

Write without exponents.
a. 10^5 **b.** 6^2 **c.** 7^5 **d.** 2^{63} **e.** 3^{-2} **f.** 8.9^0

Solution

a. $10^5 = 10 \cdot 10 \cdot 10 \cdot 10 \cdot 10 = 100,000$

b. $6^2 = 6 \cdot 6 = 36$

⊘ *Note:* 10^5 is not five multiplications, but rather five factors of 10. ⊘

c. $7^5 = 7 \cdot 7 \cdot 7 \cdot 7 \cdot 7$ or $16,807$

d. $2^{63} = \underbrace{2 \cdot 2 \cdot 2 \cdot \cdots \cdot 2 \cdot 2}_{63 \text{ factors}}$ or $9,223,372,036,854,775,808$

Note: You are not expected to find the form at the right; the factored form is acceptable.

e. $3^{-2} = \dfrac{1}{3^2} = \dfrac{1}{9}$ **f.** $8.9^0 = 1$

Scientific Notation

There is a similar pattern for multiplications of any number by a power of 10. Consider the following examples, and notice what happens to the decimal point.

$9.42 \times 10^1 = 94.2$ We find these answers by direct multiplication.
$9.42 \times 10^2 = 942.$
$9.42 \times 10^3 = 9,420.$
$9.42 \times 10^4 = 94,200.$

Do you see the pattern? Look at the decimal point (which is included for emphasis). If we multiply 9.42×10^5, how many places to the right will the decimal point be moved?

$$9.42 \times 10^5 = 9\,42,000.$$

↓   ↑
5 places to the right

---

* Counting numbers are the numbers we use for counting—namely 1, 2, 3, 4, ... Sometimes they are also called **natural numbers**. The integers are the counting numbers, their opposites, and 0, namely . . . , -3, -2, -1, 0, 1, 2, 3, ... We consider the properties of those numbers in Chapter 4.

Using this pattern, can you multiply the following *without direct calculation?*

9.42×10^{12} This answer is found by observing the
$= 9,420,000,000,000$ pattern, not by direct multiplication.

The pattern also extends to smaller numbers:

$9.42 \times 10^{-1} = 0.942$ ⎫   These numbers are found by direct multiplication.
$9.42 \times 10^{-2} = 0.0942$ ⎬   For example,
$9.42 \times 10^{-3} = 0.00942$ ⎭ $9.42 \times 10^{-2} = 9.42 \times \dfrac{1}{100}$
$\phantom{9.42 \times 10^{-3} = 0.00942}$ $= 9.42 \times 0.01 = 0.0942$

Do you see that the same pattern also holds for multiplying by 10 with a negative exponent? Can you multiply the following *without direct calculation?*

$$9.42 \times 10^{-6} = 0.000009\,42$$
$\uparrow\downarrow$
Moved six places to the left

These patterns lead to a useful way for writing large and small numbers, called **scientific notation**.

> ### Scientific Notation
> The **scientific notation** for a number is that number written as a power of 10 times another number x, such that x is between 1 and 10; that is, $1 \le x < 10$.

■ EXAMPLE 2
Write the given numbers in scientific notation.
a. 123,400   **b.** 0.000035   **c.** 1,000,000,000,000   **d.** 7.35

Solution
a. $123,400 = 1.234 \times 10^5$
b. $0.000035 = 3.5 \times 10^{-5}$
c. $1,000,000,000,000 = 10^{12}$ Technically, this is 1×10^{12} with the 1 understood.
d. 7.35 (or 7.35×10^0) is in scientific notation.

■ EXAMPLE 3
Assuming that light travels at 186,000 miles per second, what is the distance (in miles) that light travels in 1 year? Give your answer in scientific notation.

Solution   One year is 365.25 days $= 365.25 \times 24$ hours
$= 365.25 \times 24 \times 60$ minutes
$= 365.25 \times 24 \times 60 \times 60$ seconds
$= 31,557,600$ seconds

Since light travels 186,000 miles each second and there are 31,557,600 seconds in 1 year, we have

$$186{,}000 \times 31{,}557{,}600 = 5{,}869{,}713{,}600{,}000 \approx 5.87 \times 10^{12}$$

Thus, light travels 5.87×10^{12} miles in 1 year.   ▬

Calculators

Throughout the book we will include calculator comments for those of you who have (or expect to have) a calculator. Calculators are classified according to their ability to do different types of problems, as well as by the type of logic they use to do the calculations. The problem of selecting a calculator is further complicated by the multiplicity of brands from which to choose. Therefore, choosing a calculator and learning to use it require some sort of instruction.

For most nonscientific purposes, a four-function calculator with memory is sufficient for everyday use. If you anticipate taking several mathematics and/or science courses, you will find that a scientific calculator is a worthwhile investment. These calculators use essentially three types of logic: **arithmetic, algebraic,** and **RPN.** In the previous section, we discussed the correct order of operations, according to which the correct value for

$$2 + 3 \times 4$$

is 14 (multiply first). An algebraic calculator will "know" this and will give the correct answer, whereas an arithmetic calculator will simply work from left to right and obtain the incorrect answer, 20. Therefore, if you have an arithmetic-logic calculator, you will need to be careful about the order of operations. Some arithmetic-logic calculators provide parentheses, $[($ $)]$, so that operations can be grouped, as in

$$\boxed{2}\ \boxed{+}\ \boxed{(}\ \boxed{3}\ \boxed{\times}\ \boxed{4}\ \boxed{)}\ \boxed{=}$$

but then you must remember to insert the parentheses.

The last type of logic is RPN. A calculator using this logic is characterized by $\boxed{\text{ENTER}}$ or $\boxed{\text{SAVE}}$ keys and does not have an equal key $\boxed{=}$. With an RPN calculator, the operation symbol is entered after the numbers have been entered.

These three types of logic can be illustrated by the problem $2 + 3 \times 4$:

Arithmetic Logic:	$\boxed{3}\ \boxed{\times}\ \boxed{4}\ \boxed{=}\ \boxed{+}\ \boxed{2}\ \boxed{=}$	Input to match order of operations
Algebraic Logic:	$\boxed{2}\ \boxed{+}\ \boxed{3}\ \boxed{\times}\ \boxed{4}\ \boxed{=}$	Input is the same as the problem
RPN Logic:	$\boxed{2}\ \boxed{\text{ENTER}}\ \boxed{3}\ \boxed{\text{ENTER}}\ \boxed{4}\ \boxed{\times}\ \boxed{+}$	Operations input last

In this book, we will illustrate the examples using algebraic logic. If you have a calculator with RPN logic, you can use your owner's manual to change the examples to RPN. We will also indicate the keys to be pushed by drawing boxes around the numerals and operational signs as shown.

Calculator Comment

The different levels of calculators are distinguished primarily by their price.

1. **Four-function calculator with memory (under $10).**
   These calculators have a keyboard consisting of the numerals, the four arithmetic operations or functions (addition $+$, subtraction $-$, multiplication $\times$, and division $\div$) and a memory register indicated by a key marked $\boxed{M}$, $\boxed{STO}$, or $\boxed{M+}$.

2. **Scientific calculators ($8–$40).** These calculators include additional mathematical functions, such as square root $\boxed{\sqrt{}}$; trigonometric keyboard, $\boxed{SIN}$, $\boxed{COS}$, and $\boxed{TAN}$; logarithmic $\boxed{LOG}$; and exponential $\boxed{EXP}$. Depending on the particular brand, a scientific model may have other keys as well. Most calculators sold today are scientific calculators.

3. **Programmable/graphics calculators ($50–$600).** These calculators "remember" the sequence of steps for complex calculations. Graphics calculators can display a graph of an input equation.

4. **Special-purpose calculators ($40–$400).** Special-use calculators for business, statistics, surveying, medicine, and even gambling and chess are available.

■■■ EXAMPLE 4

Show the calculator steps for $14 + 38$.

Solution   Be sure to turn your calculator on, or clear the machine if it is already on. A clear button is designated by $\boxed{C}$, and the display will show 0 after the clear button is pushed. You will need to check these steps every time you use your calculator, but after a while it becomes automatic. We will not remind you of this on each example.

Press	Display	
$\boxed{1}$	1	Here we show each numeral in a single box, which means you key in one numeral at a time, as shown. But from now on, this will be shown as $\boxed{14}$.
$\boxed{4}$	14	
$\boxed{+}$	14	
$\boxed{3}\,\boxed{8}$	38	
$\boxed{=}$	52	

Notice that every time you push $\boxed{+}$, a subtotal is shown in the display. After completing Example 4, you can either continue with the same problem or start a new problem. If the next button pressed is an operation button, the result 52 will be carried over to the new problem. If the next button pressed is a numeral, the 52 will be lost and a new problem started. For this reason, it is not necessary to press $\boxed{C}$ to clear between problems. The button $\boxed{CE}$ is called the *clear entry* key and is used if you make a mistake keying in a number and do not want to start over with the problem. For example, if you want $2 + 3$ and accidentally push

$\boxed{2}\,\boxed{+}\,\boxed{4}$ you can then push $\boxed{CE}\,\boxed{3}\,\boxed{=}$

to obtain the correct answer. This is especially helpful if you are in the middle of a long problem.

■■■ EXAMPLE 5

Show the calculator steps and display for $4 + 3 \times 5 - 7$.

Solution

Press:	$\boxed{4}$	$\boxed{+}$	$\boxed{3}$	$\boxed{\times}$	$\boxed{5}$	$\boxed{-}$	$\boxed{7}$	$\boxed{=}$
Display:	4	4	3	3	5	19	7	12

If you have an algebraic-logic calculator, your machine will perform the correct order of operations. If it is an arithmetic-logic calculator, it will give the incorrect answer 28 unless you input the numbers using the order-of-operations agreement.

■■■ EXAMPLE 6

Repeat Example 3 using a calculator; that is, find

$$365.25 \times 24 \times 60 \times 60 \times 186{,}000$$

Solution When you press these calculator keys and then press $=$ the display will probably show something that looks like:

5.86971 12 or *5.86971 + 12* or *5.86971E12*

This display is a form of scientific notation. The 12 or + 12 at the right (separated by one or two blank spaces) is the exponent on the 10 when the number in the display is written in scientific notation. That is,

5.86971 12 means 5.86971×10^{12} ▬

Suppose you have a particularly large number that you wish to input into a calculator, say 920,000,000,000,000,000,000 miles divided by 7,927 miles (from Figure 1.21). You can input 7,927 but if you attempt to input the larger number you will be stuck when you fill up the display (9 or 12 digits). Instead, you will need to write

$$920,000,000,000,000,000,000 = 9.2 \times 10^{20}$$

This may be entered by pressing an $\boxed{\text{EE}}$, $\boxed{\text{EEx}}$, or $\boxed{\text{EXP}}$ key:

$\boxed{9.2}$ $\boxed{\text{EE}}$ $\boxed{20}$ $\boxed{\div}$ $\boxed{7927}$ $\boxed{=}$ *Display:* *1.160590387E17*

> 🚫 Do not confuse the scientific notation keys on your calculator with the exponent key. Exponent keys are labeled $\boxed{y^x}$ or $\boxed{10^x}$. 🚫

This means that the last cube in Figure 1.21 is about 1.16×10^{17} times larger than the earth.

Scientific notation is represented in a slightly different form on many calculators and computers, and this new form is sometimes called **floating-point form**. When representing very large or very small answers, most calculators will automatically output the answers in floating-point notation. The following example compares the different forms for large and small numbers with which you should be familiar.

▬▬ **EXAMPLE 7**

Write each given number in scientific notation and in calculator notation.
a. 745   **b.** 1,230,000,000   **c.** 0.00573   **d.** 0.00000 06239

Solution The form given in this example is sometimes called **fixed-point form** or **decimal notation** to distinguish it from the other forms.

Decimal or Fixed-Point Notation	Scientific Notation	Floating-Point	
a. 745	7.45×10^2	7.45	02
b. 1,230,000,000	1.23×10^9	1.23	09
c. 0.00573	5.73×10^{-3}	5.73	-03
d. 0.00000 06239	6.239×10^{-7}	6.239	-07 ▬

Estimation

Part of problem solving is using common sense about the answers you obtain. This is even more important when using a calculator, because there is a misconception that if a calculator or computer displays an answer, "it must be correct." Reading and understanding the problem are parts of the process of problem solving.

When problem-solving, you must ask whether the answer you have found is reasonable. If I ask for the amount of rent you must pay for an apartment, and you do a calculation and arrive at an answer of $16.25, you know that you have made a mistake. Likewise, an answer of $135,000 would not be reasonable. As we progress through this course you will be using a calculator for many of your calculations, and with a calculator you can easily press the wrong button and come up with an outrageous answer. One aspect of *looking back* is using common sense to make sure the answer is reasonable. The ability to recognize the difference between reasonable answers and unreasonable ones is important not only in mathematics, but whenever you are problem-solving. This ability is even more important when you use a calculator, because pressing the incorrect key can often cause very unreasonable answers.

Whenever you try to find an answer, you should ask yourself whether the answer is reasonable. How do you decide whether an answer is reasonable? One way is to **estimate** an answer. Webster's *New World Dictionary* tells us that as a verb, to *estimate* means "to form an opinion or a judgment about" or to calculate "approximately." In the 1986 *Yearbook* of the National Council of Teachers of Mathematics, we find:

> The broad *mathematical context* for an estimate is usually one of the following types:
>
> A.   An exact value is known but for some reason an estimate is used.
>
> B.   An exact value is possible but is not known and an estimate is used.
>
> C.   An exact value is impossible.

We will work on building your estimation skills throughout this book.

▬▬ EXAMPLE 8

If your salary is $9.75 per hour, your annual salary is approximately

A. $5,000   B. $10,000   C. $15,000   D. $20,000   E. $25,000

Solution   Problem solving often requires some assumptions about the problem. For this problem, we are not told how many hours per week you work, or how many weeks per year you are paid. We assume a 40-hour week, and we also assume that you are paid for 52 weeks per year.

> *Estimate:*   Your hourly salary is about $10 per hour.
>   A 40-hour week gives us $40 \times \$10 = \$400$ per week.
>
>   For the estimate, we calculate the wages for 50 weeks instead of 52:
>   50 weeks yields $50 \times \$400 = \$20,000$.

The answer is **D**.   ▬

▬▬ EXAMPLE 9

Estimate the distance from Orlando International Airport to Disney World.

Solution   Note that the scale is 20 miles to 1 in. Looking at the map, you will note that it is approximately $\frac{3}{4}$ in. from the airport to Disney World. This means that we estimate the distance to be 15 miles.   ▬

There are two important reasons for estimation: (1) to form a reasonable opinion, or (2) to check the reasonableness of an answer. If reason (1) is our motive, we should not think it necessary to follow an estimation by direct calculation. To do so would defeat the purpose of the estimation. On the other hand, if we are using the estimate for reason (2) — to see whether an answer is reasonable — we might perform the estimate as a check on the calculated answer for Example 3 (the problem about the speed of light):

$$186,000 \text{ miles per second} \approx 2 \times 10^5 \text{ miles per second}$$

and

$$\text{one year} \approx 4 \times 10^2 \text{ days} \approx \underbrace{4 \times 10^2}_{\text{days}} \times \underbrace{2 \times 10}_{\text{hr per day}} \times \underbrace{6 \times 10}_{\text{min per hr}} \times \underbrace{6 \times 10}_{\text{sec per hr}}$$

$$= \underbrace{(4 \times 2 \times 6 \times 6)}_{\text{seconds per year}} \times 10^5 \approx (10 \times 36) \times 10^5 \approx 3.6 \times 10^7 \text{ seconds}$$

Thus, one light year is about $(2 \times 10^5)(3.6 \times 10^7) \approx 7.2 \times 10^{12}$ miles. This estimate seems to confirm the reasonableness of the answer we obtained in Example 3.

Laws of Exponents

In working out the previous estimation for Example 3, we used some properties of exponents that we can derive by, once again, turning to some patterns. Consider

$$2^3 \times 2^4 = 8 \times 16 = 128 \quad \text{or} \quad 2^7$$
$$3^2 \times 3^3 = 9 \times 27 = 243 \quad \text{or} \quad 3^5$$
$$10^2 \times 10^5 = 100 \times 100,000 = 10,000,000 \quad \text{or} \quad 10^7$$

When we *multiply powers* of the same base, we *add* exponents. This is called the **addition law of exponents.**

Suppose we wish to raise a power to a power. We can apply the addition law of exponents. Consider

$$(2^3)^2 = 2^3 \cdot 2^3 = 2^{3+3} = 2^6$$
$$(10^2)^3 = 10^2 \cdot 10^2 \cdot 10^2 = 10^{2+2+2} = 10^6$$

When we *raise a power to a power*, we *multiply* the exponents. This is called the **multiplication law of exponents.**

A third law is needed to raise products to powers. Consider

$$(2 \cdot 3)^2 = 6^2 = 36 \quad \text{and} \quad 36 = 4 \cdot 9 = 2^2 \cdot 3^2$$

Thus, $(2 \cdot 3)^2 = 2^2 \cdot 3^2$.

$$(3 \times 10^4)^2 = 3^2 \times (10^4)^2 = 3^2 \times 10^8 = 9 \times 10^8$$

This result, called the **distributive law of exponents,** says that to *raise a product to a power,* raise each factor to that power and then multiply.

Similar patterns can be observed for quotients. We now summarize the five laws of exponents.

Laws of Exponents

Let a and b be any nonzero real numbers, and let m and n be any integers.

Addition law: $b^m \cdot b^n = b^{m+n}$

Multiplication law: $(b^n)^m = b^{mn}$

Subtraction law: $\dfrac{b^m}{b^n} = b^{m-n}$

Distributive laws: $(ab)^m = a^m b^m$

$\qquad\qquad\qquad\quad \left(\dfrac{a}{b}\right)^m = \dfrac{a^m}{b^m}$

■ **EXAMPLE 10**   **Polya's Method**

Under $\frac{3}{4}$ impulse power the *Starship Enterprise* will travel 1 million kilometers (km) in 3 minutes.* Compare full impulse power with the speed of light, which is approximately 1.08×10^9 kilometers per hour (km/hr).

Solution   We use Polya's problem-solving guidelines for this example.

Understand the Problem.   You might say, "I don't know anything about Star Trek," but with most problem solving in the real world, the problems you are asked to solve are often about situations with which you are unfamiliar. Finding the necessary information to understand the question is part of the process. We assume that full impulse is the same as 1 impulse power, so that if we multiply $\frac{3}{4}$ impulse power by $\frac{4}{3}$ we will obtain ($\frac{3}{4} \times \frac{4}{3} = 1$) full impulse power.

Devise a Plan.   We will calculate the distance traveled (in kilometers) in one hour under $\frac{3}{4}$ power, and then will multiply that result by $\frac{4}{3}$ to obtain the distance in kilometers per hour under full impulse power.

Carry Out the Plan.

$$\frac{3}{4}\text{ impulse power} = \frac{1{,}000{,}000 \text{ km}}{3 \text{ min}} \qquad \text{Given}$$

$$= \frac{10^6 \text{ km}}{3 \text{ min}} \cdot \frac{20}{20} \qquad \text{Multiply by } \frac{20}{20} \text{ to change 3 minutes}$$

$$= \frac{10^6 \times 2 \times 10 \text{ km}}{60 \text{ min}} \qquad \begin{array}{l}\text{to 60 minutes so that we can} \\ \text{convert to hours.}\end{array}$$

$$= \frac{2 \times 10^7 \text{ km}}{1 \text{ hr}}$$

$$= 2 \times 10^7 \text{ km/hr}$$

* STARTREK, The Next Generation (episode that first aired the week of May 15, 1993).

We now multiply both sides by $\frac{4}{3}$ to find the distance under full impulse.

$$\frac{4}{3}\left(\frac{3}{4}\text{ impulse power}\right) = \frac{4}{3} \times 2 \times 10^7 \text{ km/hr}$$

$$\text{full impulse power} = \frac{8}{3} \times 10^7 \text{ km/hr}$$

$$\approx 2.666666667 \times 10^7 \text{ km/hr}$$

Comparing this to the speed of light, we see

$$\frac{2.666666667 \times 10^7}{1.08 \times 10^9} = \frac{2.666666667}{1.08} \times 10^{7-9} \approx 0.025$$

Look Back. We see that full impulse power is about 2.5% of the speed of light.

Comprehending Large Numbers

We began this section by looking at the size of the cosmos. But just how large is large? Most of us are accustomed to hearing about millions, billions (budgets or costs of disasters), or even trillions (the national debt is about $5 trillion), but how do we really understand the magnitude of these numbers? Would you do a better job than Dennis' parents in the cartoon at explaining "How much is a million?"

More than a million balloons were released at one time on December 5, 1985, at Disneyland by the city of Anaheim, California.

A **million** is a fairly modest number, 10^6. Yet if we were to count one number per second, nonstop, it would take us about 278 hours or approximately $11\frac{1}{2}$ days to count to a million. Not a million days have elapsed since the birth of Christ (a million days is about 2,700 years). A large book of about 700 pages contains about a million letters. A million balloons (see photograph in the margin) were released at a single time. How large a room would it take to hold 1,000,000 inflated balloons?

But the age in which we live has been called the age of billions. How large is a **billion?** How long would it take you to count to a billion? Go ahead — make a guess. To get some idea about how large a billion is, let's compare it to some familiar units:

- If you gave away $1,000 *per day*, it would take you more than 2,700 *years* to give away a billion dollars.
- A stack of a billion $1 bills would be more than 59 miles high.
- At 8% interest, a billion dollars would earn you $219,178.08 interest *per day!*
- A billion seconds ago, John F. Kennedy was assassinated.
- A billion minutes ago was about 100 A.D.
- A billion hours ago, people had not yet appeared on earth.

But a billion is only 10^9, a mere nothing when compared with the real giants. Keep these magnitudes (sizes) in mind. Earlier in this section we noticed that a cube containing our solar neighborhood is 9.2×10^{11} miles. (See Figure 1.21.) This is less than a billion times the size of the earth. (Actually, it is $9.2 \times 10^{11} \div 7,927 \approx 1.2 \times 10^8$.) Now we are moving beyond that comparison to some real giants.

There is an old story of a king who, being under obligation to one of his subjects, offered to reward him in any way the subject desired. Being of mathematical mind and modest tastes, the subject simply asked for a chessboard with one grain of wheat on the first square, two on the second, four on the third, and so forth. The old king was delighted with this modest request! Alas, the king was soon sorry he granted the request.

▬▬ **EXAMPLE 11**   **Polya's Method**

Estimate the magnitude of the grains of wheat on the last square of a chessboard.

Solution   We use Polya's problem-solving guidelines for this example.

Understand the Problem.   Each square on the chessboard has grains of wheat placed on it. To answer this question you need to know that a chessboard has 64 squares. The first square has $1 = 2^0$ grains, the next has $2 = 2^1$ grains, the next $4 = 2^2$, and so on. Thus, he needed 2^{63} grains of wheat for the last square alone. We showed this number in Example 1d.

Devise a Plan.   We know (from Example 1) that $2^{63} \approx 9.22337 \times 10^{18}$. We need to find the size of a grain of wheat, and then convert 2^{63} grains into bushels. Finally, we need to state this answer in terms we can understand.

Carry Out the Plan.   I went to a health food store, purchased some raw wheat, and found that there are about 250 grains per cubic inch (in.³). I also went to a dictionary and found that a bushel is 2,150 in.³. Thus, the number of grains of wheat in a bushel is

$$2{,}150 \times 250 = 537{,}500 = 5.375 \times 10^5$$

To find the number of bushels in 2^{63} grains we need to divide:

$$(9.22337 \times 10^{18}) \div (5.375 \times 10^5) = \frac{9.22337}{5.375} \times 10^{18-5}$$
$$\approx 1.72 \times 10^{13}$$

Look Back.   This answer does not mean a thing without looking back and putting it in terms we can understand. I went to a *World Almanac* and found that in 1990 the U.S. wheat production was 2,738,594,000 bushels. To find the number of years it would take the United States to produce the necessary wheat for the last square of the chessboard, we need to divide the production into the amount needed:

$$\frac{1.72 \times 10^{13}}{2.74 \times 10^9} = \frac{1.72}{2.74} \times 10^{13-9} \approx 0.63 \times 10^4$$

This is 6,300 years!

What is the name of the largest number you know? Recently we have heard about the national debt, which exceeds $5 **trillion**. Table 1.1 shows some large numbers.

TABLE 1.1	Some Large Numbers	
1	*one:*	1
10	*ten:*	10
10^2	*hundred:*	100
10^3	*thousand:*	1,000
10^6	*million:*	1,000,000 ← Dennis the Menace cartoon
10^9	*billion:*	1,000,000,000 ← Discussed in text
10^{12}	*trillion:*	1,000,000,000,000 ← National debt $5 trillion
10^{15}	*quadrillion:*	1,000,000,000,000,000 ← Number of all words ever printed
10^{18}	*quintillion:*	1,000,000,000,000,000,000
10^{21}	*sextillion:*	1,000,000,000,000,000,000,000
		⋮
10^{63}	*vigintillion:*	1 followed by 63 zeros
		⋮
10^{100}	*googol:*	10,000,000,000,000,000,000,000,000,000,000,000,000,000,000,000,-000,000,000,000,000,000,000,000,000,000,000,000,000,000,000,-000,000,000

News Clip

Do Things Really Change?

"Students today can't prepare bark to calculate their problems. They depend upon their slates which are more expensive. What will they do when their slate is dropped and it breaks? They will be unable to write!"

Teacher's Conference, 1703

"Students depend upon paper too much. They don't know how to write on slate without chalk dust all over themselves. They can't clean a slate properly. What will they do when they run out of paper?"

Principal's Association, 1815

"Students today depend too much upon ink. They don't know how to use a pen knife to sharpen a pencil. Pen and ink will never replace the pencil."

National Association of Teachers, 1907

"Students today depend upon store bought ink. They don't know how to make their own. When they run out of ink they will be unable to write words or ciphers until their next trip to the settlement. This is a sad commentary on modern education."

The Rural American Teacher, 1929

"Students today depend upon these expensive fountain pens. They can no longer write with a straight pen and nib (not to mention sharpening their own quills). We parents must not allow them to wallow in such luxury to the detriment of learning how to cope in the real business world, which is not so extravagant."

PTA Gazette, 1941

"Ball point pens will be the ruin of education in our country. Students use these devices and then throw them away. The American virtues of thrift and frugality are being discarded. Business and banks will never allow such expensive luxuries."

Federal Teacher, 1950

"Students today depend too much on hand-held calculators."

Anonymous, 1995

PROBLEM SET 1.4

▲ **A Problems**

1. **IN YOUR OWN WORDS** What do we mean by *exponent?*

2. **IN YOUR OWN WORDS** Define *scientific notation* and discuss why it is useful.

3. **IN YOUR OWN WORDS** Do you plan to use a calculator for working the problems in this book? If so, what type of logic does it use?

4. **IN YOUR OWN WORDS** How many classrooms would be necessary to hold 1,000,000 inflated balloons?

5. **IN YOUR OWN WORDS** What is the largest number whose name you know? Describe the size of this number.

6. **IN YOUR OWN WORDS** What is a *trillion?* Do not simply define this number, but discuss its magnitude (size) in terms that are easy to understand.

Write each of the numbers in Problems 7–10 in scientific notation and in floating-point notation (as on a calculator).

7. **a.** 3,200   **b.** 0.0004   **c.** 64,000,000,000

8. a. 23.79   b. 35,000,000,000   c. 0.000001
9. a. 63,000,000 b. 5,629   c. 0.00000 00000 0034
10. a. 0.00000 123 b. googol   c. 1,200,300,000

Write each of the numbers in Problems 11–14 in fixed-point notation.

11. a. 7^2 b. 7.2×10^{10} c. $4.56 \quad +3$
12. a. 2^6 b. 2.1×10^{-3} c. $4.07 \quad +4$
13. a. 6^3 b. 4.1×10^{-7} c. $4.8 \quad -7$
14. a. $3.217 \quad -7$ b. 6.81×10^0 c. $8.89 \quad -11$

15. Indicate the calculator keys you would press to evaluate each of the following numbers. Also state the brand and model number you are using.

   a. 5^6 b. 8.56×10^6 c. 45,000,000,000,000,000

16. **IN YOUR OWN WORDS** State the brand and model number of your calculator and then explain the keys you press to evaluate exponents and the keys you press to use scientific notation.

Write each of the numbers in Problems 17–20 in scientific notation.

17. The velocity of light in a vacuum is about 30,000,000,000 cm/sec.
18. The wavelength of the orange-red line of krypton 86 is about 6,100 Å.
19. The distance between Earth and Mars (220,000,000 miles) when drawn to scale is 0.00000 25 in.
20. Our gross domestic product is approaching $6 trillion. It has been proposed that this number be used to define a new monetary unit, a *light buck*. That is, one light buck is the amount necessary to generate domestic goods and services at the rate of $186,000 per second.

Write each of the numbers in Problems 21–24 in fixed-point notation.

21. A kilowatt-hour is about 3.6×10^6 joules.
22. A ton is about 9.06×10^2 kilograms.
23. The volume of a typical neuron is about 3×10^{-8} cm^3.
24. If the sun were a light bulb, it would be rated at 3.8×10^{25} watts.
25. Estimate the distance from Los Angeles International Airport to Disneyland.

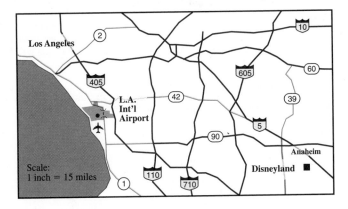

26. Estimate the distance from Fish Camp to Yosemite Village in Yosemite National Park.

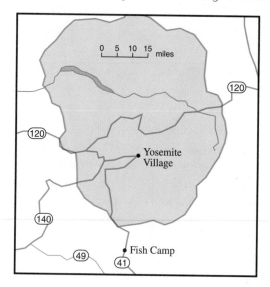

In Problems 27–36, first estimate your answer and then calculate the exact answer.

27. How many hours are there in 360 days?

28. How many pages are necessary to make 1,850 copies of a manuscript that is 487 pages long? (Print on one side only.)

29. In 1995 the Internal Revenue Service allowed a $3,900 deduction for each dependent. If a family has four dependents, what is the allowed deduction?

31. If your monthly salary is $1,543 per month, what is your annual salary?

32. If you are paid $6.25 per hour, what is your annual salary?

33. If you are paid $31,200 per hear, what is your hourly salary?

34. If your car gets 23 miles per gallon, how far can you go on 15 gallons of gas?

35. Estimate the total price of the following grocery items.

lettuce, $.59; sliced turkey, $2.50; chips, $1.89; paper towels, $.99; $\frac{1}{2}$ gal orange juice, $1.97; 1 lb cheddar cheese, $1.99; 3 cucumbers, $.99; 1 lb frozen corn, $1.09; 2 bars soap, $1.79; 1 tube toothpaste, $1.89; 1 lb peaches, $.79; 1 lb elbow macaroni, $1.39; 3 yogurt, $1.61.

36. Estimate the total cost of the following plumbing items.

120 1$\frac{1}{2}$-in. PVC pipe, $76.80; 1$\frac{1}{4}$-in. ball valve, $17.89; 1$\frac{1}{4}$-in. PVC fittings, $24.28; 1-in. PVC fittings, $14.16; $\frac{3}{4}$-in. PVC fittings, $6.48; 1-in. ball valve, $12.84.

▲ **B Problems**

Compute the results in Problems 37–42. Leave your answers in scientific notation.

37. a. $(6 \times 10^5)(2 \times 10^3)$ **b.** $\dfrac{6 \times 10^5}{2 \times 10^3}$

38. a. $\dfrac{(5 \times 10^4)(8 \times 10^5)}{4 \times 10^6}$ b. $\dfrac{(6 \times 10^{-3})(7 \times 10^8)}{3 \times 10^7}$

39. a. $\dfrac{(6 \times 10^7)(4.8 \times 10^{-6})}{2.4 \times 10^5}$ b. $\dfrac{(2.5 \times 10^3)(6.6 \times 10^8)}{8.25 \times 10^4}$

40. a. $\dfrac{(2xy^{-2})(2^{-1}x^{-1}y^4)}{x^{-2}y^2}$ b. $\dfrac{x^2y(2x^3y^{-5})}{2^{-2}x^4y^{-8}}$

41. a. $\dfrac{0.00016 \times 500}{2,000,000}$ b. $\dfrac{15,000 \times 0.0000004}{0.005}$

42. a. $\dfrac{4,500,000,000,000 \times 0.00001}{50 \times 0.0003}$ b. $\dfrac{0.0348 \times 0.00000\ 00000\ 00002}{0.000058 \times 0.03}$

Estimate the number of items in each photograph in Problems 43–46.

43.

44.

45.

46.

47. The Associated Press recently reported that W. M. Keck Observatory on the island of Hawaii took infrared pictures of galaxy 4C41.17 at a distance of 72 trillion billion miles from earth. Write this number in scientific notation.

48. The United Press International article shown in the margin says that our atmosphere weighs 5 quadrillion, 157 trillion tons. Write this number in scientific notation.

49. A box of oranges contains approximately 96 oranges. If the U.S. annual production of oranges is 186,075,000 boxes, estimate the number of oranges produced each year in the United States.

50. Approximately how high would a stack of 1 million $1 bills be? (Assume there are 233 new $1 bills per inch.)

51. Estimate how many pennies it would take to make a stack 1 in. high. Approximately how high would a stack of 1 million pennies be?

52. There are 2,260,000 grains in a pound of sugar. If the U.S. annual production of sugar is 30,000,000 tons, estimate the number of grains of sugar produced in a year in the United States. Use scientific notation. (*Note:* There are 2,000 lb per ton.)

53. **a.** What is the largest number you can represent on your calculator?
    b. What is the largest number you can think of using only three digits?
    c. Use your calculator to convert this number into scientific notation.

54. Zerah Colburn (1804–1840) toured America when he was 6 years old to display his calculating ability. He could instantaneously give the square and cube roots of large numbers. It is reported that it took him only a few seconds to find 8^{16}. Use your calculator to *help you* find this number exactly (not in scientific notation).

55. Jedidiah Buxion (1707–1772) never learned to write, but given any distance he could tell you the number of inches, and given any length of time he could tell you the number of seconds. If he listened to a speech or a sermon, he could tell the number of words or syllables in it. It reportedly took him only a few moments to mentally calculate the number of cubic inches in a right-angle block of stone 23,451,789 yards long, 5,642,732 yards wide, and 54,465 yards thick. Estimate this answer on your calculator. You will use scientific notation because of the limitations of your calculator, but remember that Jedidiah gave the *exact* answer by working the problem in his head.

56. George Bidder (1806–1878) not only possessed exceptional power at calculations but also went on to obtain a good education. He could give immediate answers to problems of compound interest and annuities. One question he was asked was: "If the moon is 238,000 miles from the earth and sound travels at the rate of 4 miles per minute, how long would it be before the inhabitants of the moon could hear the Battle of Waterloo?" By calculating *mentally*, he gave the answer in less than one minute! First make an estimate, and then use your calculator to give the answer in days, hours, and minutes, to the nearest minute.

57. **IN YOUR OWN WORDS** Enter your favorite counting number from 1 to 9 into the display of a calculator. Multiply by 259; then multiply this result by 429. What is your answer? Try this for three different choices. Explain.

58. **IN YOUR OWN WORDS** Pick any three-digit number, and repeat the digits to make a six-digit number. Divide by 7, then 11, then 13. What is the result? Repeat for three different numbers. Explain.

59. When my daughter was 2 years old, she had a toy that consisted of colored rings of different sizes, as shown in Figure 1.22. Suppose we wish to move the "tower" from stand A to stand C. To make this interesting, let's agree to the following rules: (1) move only one ring at a time; (2) at no time may a larger ring be placed on a smaller ring. According to these rules, we can move the tower from stand A to C in seven moves (try it).

▲ **Figure 1.22   Tower problem**

　　a. If we add another ring, the tower can be moved in 15 moves. Explain how.
　　b. How many moves would be required for five rings?

60. A sheet of notebook paper is approximately 0.003 in. thick. Tear the sheet in half so that there are 2 sheets. Repeat so that there are 4 sheets. If we repeat again, there will be a pile of 8 sheets. Continue in this fashion until the paper has been halved 50 times. If it were possible to complete the process, how high would you guess the final pile would be? After you have guessed, *compute* the height.

▲ **Problem Solving**

61. If it takes one second to write down each digit, how long will it take to write down all the numbers from 1 to 1,000?

62. If it takes one second to write down each digit, how long will it take to write down all the numbers from 1 to 1,000,000?

63. Imagine that you have written down the numbers from 1 to 1,000. What is the total number of zeros you have recorded?

64. Imagine that you have written down the numbers from 1 to 1,000,000. What is the total number of zeros you have recorded?

65. Estimate the number of bricks required to build a solid wall 100 ft by 10 ft by 1 ft.

66. Estimate the number of bricks in one solid tower shown in Figure 1.23a. The height of a tower is 70 ft.

　　a. Interior view of church　　　　b. Exterior view of church

▲ **Figure 1.23   Grundtvig's Church on Bispebjerg in Copenhagen is made of bricks.**

67. Estimate the number of bricks needed to build the church shown in Figure 1.23b. The church has a height of 95 ft, length 250 ft, and width 120 ft. The tower is 156 feet high.

California

158,600
square miles

World population
about
5,500,000,000

How much space would each person have (make a guess)? See Problem 68.

68. If the entire population of the world moved to California and each person were given an equal amount of area, how much space would you *guess* that each person would have (multiple choice)?

    A. 8 in.^{2}
    B. 8 ft^2
    C. 80 ft^2
    D. 800 ft^2
    E. 1 mi^2

69. The total population of the earth is about 5.5×10^9.

    a. If each person has the room of a prison cell (50 ft^2) and if there are about 2.8×10^7 ft^2 in 1 mi^2, how many people could fit into a square mile?
    b. How many square miles would be required to accommodate the entire population of the earth?
    c. If the total land area of the earth is about 5.2×10^7 mi^2, and if all the land area were divided equally, how many acres of land would each person be allocated?

70. It is known that a person's body has about one gallon of blood in it, and that a cubic foot will hold about 7.5 gallons of liquid. It is also known that Central Park in New York has an area of 840 acres or about 1.3 mi^2. If walls were built around the park, how tall would those walls need to be to contain the blood of all 5,500,000,000 people in the world?

▲ **Individual Research**

71. Answer the question posed in Problem 68 for your own state. If you live in California, then use Florida.

    Reference Check an almanac to find the area of your state.

72. Problem 59 is an example of a famous problem called the *Tower of Hanoi.* The ancient Brahman priests were to move a pile of 64 such disks of decreasing size, after which the world would end. This task would require $2^{64} - 1$ moves. Try to estimate how long this would take at the rate of one move per second. Do some research on this problem and write a paper.

    Reference
    Frederick Schuh, *The Masterbook of Mathematical Recreations* (New York: Dover Publications, 1968).
    Michael Schwager, "Another Look at the Tower of Hanoi," *The Mathematics Teacher*, September 1977, pp. 528–533.

"Numeracy is the ability to cope confidently with the mathematical demands of adult life."

Mathematics Counts

OCTOBER

1
2
3
4
5   **Bernhard Bolzano** (1781), logic, function theory
6   **Julius Dedekind** (1831), number theory
7
8
9
10
11
12
13
14
15   **Evangelista Torricelli** (1608), analytic geometry
16
17
18
19
20
21   **Martin Gardner**
22
23   **Hippocrates** (450 B.C.), first paid math teacher
24
25   **Evariste Galois** (1811), modern algebra, p. 291
26
27
28
29
30
31   **Karl Weierstrass** (1815), function theory

Biographical Sketch

"Progress in science and technology is unthinkable without progress in mathematics."

Martin Gardner

Martin Gardner was born in Tulsa, Oklahoma, on October 21, 1914, and is well known as the author of the "Mathematical Games" column in *Scientific American* from 1957 to 1980. He has written some fifty books, most of them about mathematics, science, and philosophy. His *Annotated Alice* and *More Annotated Alice* are in-depth commentaries on mathematician Lewis Carroll's two immortal fantasies.

"I am not a professional mathematician," Gardner confesses, "and my few discoveries are elementary, but I love mathematics with passion, and there are times I regret not having chosen it for a career."

"Progress in science and technology is unthinkable without progress in mathematics," he adds. "No persons today can call themselves educated if they are unfamiliar with the basics of mathematics. It is more essential now to a liberal education than, say, a knowledge of Shakespeare or an appreciation of Mozart or Michelangelo. Our nation faces a bleak future if it does not find better ways to teach mathematics to children, and to overcome what has been called its deplorable mathematical innumeracy."

Book Reports

Write a 500-word report on one of these books:

▲ *Mathematical Magic Show*, Martin Gardner (New York: Alfred A. Knopf, 1977).

▲ *How to Solve It: A New Aspect of Mathematical Method*, George Polya (New Jersey: Princeton University Press, 1945, 1973).

Important Terms

And [1.2]	Element [1.2]	Intersection [1.2]	Scientific notation [1.4]
Axiom [1.3]	Empty set [1.2]	Laws of exponents [1.4]	Set [1.2]
Base [1.4]	Estimate [1.4]	Member [1.2]	Street problem [1.1]
Billion [1.4]	Euler circles [1.2]	Million [1.4]	Subset [1.2]
Cardinality [1.2]	Exponent [1.4]	Or [1.2]	Substitution property [1.3]
Complement [1.2]	Exponential notation [1.4]	Order of operations [1.3]	Theorem [1.3]
Conjecture [1.3]	Fixed-point form [1.4]	Pascal's triangle [1.1]	Trillion [1.4]
Deductive reasoning [1.3]	Floating-point form [1.4]	Power [1.4]	Union [1.2]
De Morgan's laws [1.2]	Googol [1.4]	Premise [1.3]	Universal set [1.2]
Disjoint [1.2]	Inductive reasoning [1.3]	Problem solving [1.1]	Venn diagram [1.2]

Important Ideas

Guidelines for problem solving [1.1]
Complement [1.2]
Operations on sets [1.2]
   Intersection [1.2]
   Union [1.2]
Formula for the cardinality of the union of two sets [1.2]

De Morgan's laws [1.2]
Order of operations [1.3]
Substitution property [1.3]
Mathematical modeling [1.3]
Exponential notation [1.4]
Five laws of exponents [1.4]

Types of Problems

Use Polya's method to solve a problem [1.1]
Use Pascal's triangle as an aid to problem solving [1.1]
Draw Venn diagram for complement, union, and intersection [1.2]
Answer questions involving surveys and cardinality of sets [1.2]
Carry out combined operations with sets [1.2]
Prove statements involving sets, including De Morgan's laws [1.2]
Write out formulas for given word statements [1.3]
Use inductive or deductive reasoning to answer a question [1.3]

Distinguish inductive from deductive reasoning [1.3]
Translate math statements (distinguish sums, differences, products, and quotients) [1.3]
Write out exponential numbers without using exponents [1.4]
Write a large or small number in scientific notation [1.4]
Use a calculator to answer numerical questions [1.4]
Estimate answers to numerical questions [1.4]
Simplify numerical problems by using the laws of exponents [1.4]
Explain the relative size of large and small numbers [1.4]

CHAPTER 1   Review Questions

▲ **Figure 1.24   Chessboard**

1. In your own words describe Polya's problem-solving model.

2. In how many ways can a person walk 5 blocks north and 4 blocks west, if the streets are arranged in a standard rectangular arrangement?

3. A chessboard consists of 64 squares, as shown in Figure 1.24. The rook can move one or more squares horizontally or vertically. Suppose a rook is in the upper left-hand corner of a chessboard. Tell how many ways the rook can reach the point marked "X." Assume that the rook always moves toward its destination.

4. Compute 111,111,111 × 111,111,111. Do not use direct multiplication; show all your work.

5. What is meant by "order of operations"?

6. Let $U = \{1, 2, 3, 4, 5, 6, 7, 8, 9, 10\}$, $A = \{1, 3, 5, 7, 9\}$, $B = \{2, 4, 6, 9, 10\}$. Find:
   a. $A \cup B$ b. $A \cap B$ c. $\bar{B}$ d. $\bar{A} \cap (B \cup A)$

7. Draw Venn diagrams for
   a. $X \cap Y$ b. $X \cup Y$ c. $\bar{X}$ d. $\bar{X} \cap Y$

8. Prove or disprove:
   a. $A \cup (B \cap C) = (A \cup B) \cap C$ b. $(A \cup B) \cap C = (A \cap C) \cup (B \cap C)$

9. Does the story in the News Clip illustrate inductive or deductive reasoning?

10. Show the calculator keys you would press as well as the calculator display for the result. Also state the brand and model of the calculator you are using.
    a. 2^{63} b. $\dfrac{9.22 \times 10^{18}}{6.34 \times 10^6}$

11. Assume that there is a \$281.9 billion budget "windfall." Which of the following choices would come closest to liquidating this windfall?
    A. Buy the entire U.S. population a steak dinner.
    B. Burn one dollar per second for the next 1,000 years.
    C. Give \$80,000 to every resident of San Francisco and use the remainder of the money to buy a \$200 CD player for every resident of China.

12. One method for estimating the capacity of a boat is to divide the product of the length and width of the boat by 15. Write a formula to represent this idea.

13. What is wrong, if anything, with the following "Great Tapes" advertisement.

> Instead of reading the 100 greatest books of all time, buy these beautifully transcribed books on tape. If you listen to only one 45-minute tape per day you will complete the greatest books of all time in only one year.

14. A survey of 70 college students showed the following data:

    42 had a car; 50 had a TV; 30 had a bicycle; 17 had a car and a bicycle; 35 had a car and a TV; 25 had a TV and a bicycle; 15 had all three

    How many students had none of the 3 items?

15. What is scientific notation?

16. In 1995, it was reported that an iceberg separated from Antarctica. The size of this iceberg was reported equal to 7×10^{16} ice cubes. Convert this size to meaningful units.

17. The national debt in 1997 was about \$5,200,000,000,000. Write this number in scientific notation. Suppose that in 1997 there were 266,500,000 people in the United States. If the debt is divided equally among the people, how much is each person's share?

18. Assume that your classroom is 20 ft $\times$ 30 ft $\times$ 10 ft. If you fill this room with dollar bills (assume that a stack of 233 dollar bills is 1 in. tall), how many classrooms would it take to contain the national debt of \$5,200,000,000,000?

News Clip

Q: What has 18 legs and catches flies?
A: I don't know, what?
Q: A baseball team. What has 36 legs and catches flies?
A: I don't know that, either.
Q: Two baseball teams. If the United States has 100 senators, and each state has 2 senators, what does . . .
A: I know this one!
Q: Good. What does each state have?
A: Three baseball teams!

19. Consider the following pattern:

    I is happy.

    10 is happy because $1^2 + 0^2 = 1$, which is happy.

    13 is happy because $1^2 + 3^2 = 10$, which is happy.

    19 is happy because $1^1 + 9^2 = 82$ and $8^2 + 2^2 = 68$
    and $6^2 + 8^2 = 100$ and $1^2 + 0^2 + 0^2 = 1$, which is happy.

    On the other hand,

    2, 3, 4, 5, 6, 7, 8, and 9 are unhappy.

    11 is unhappy because $1^2 + 1^2 = 2$, which is unhappy.

    12 is unhappy because $1^2 + 2^2 = 5$, which is unhappy.

    Find one unhappy number as well as one happy number.

20. Suppose you could write out $7^{1,000}$. What is the last digit?

Revisiting the Real World . . .

IN THE REAL WORLD   "Say, Tom, what are you taking this semester?" asked Susan. "I'm taking English, history, and math. I can't believe my math teacher," responded Tom. "The first day we were there, she walked in, wrote her name on the board, and then she asked, 'What is the millionth counting number that is not the square or cube of a counting number?' Who cares!"

"Oh, I had that math class last semester," said Susan. "It isn't so bad. The whole idea is to give you the ability to solve problems *outside* the class. I want to get a good job when I graduate, and I've read that because of the economy, employers are looking for people with problem-solving skills. I hear that working smarter is more important than working harder."

Commentary

We will use Polya's method to answer Tom's question.

Understand the Problem.   Imagine a long list of the numbers: 1, 2, 3, . . . , 999,999, 1,000,000, 1,000,001, 1,000,002. Suppose we asked for the millionth positive integer on this list — that's easy. It is 1,000,000. Do you understand what is meant by "perfect squares" and "perfect cubes"?

Perfect squares: $1^2 = 1$; $2^2 = 4$, $3^2 = 9$; $4^2 = 16$; 25; 36; 49; 64; 81; . . .

Perfect cubes: $1^3 = 1$; $2^3 = 8$, $3^3 = 27$; $4^3 = 64$; 125; 216; 343; 512; . . .

Now, suppose we cross out the perfect squares; for each number crossed off, the millionth number in the list changes. That is, cross off 1 and the millionth number is 1,000,001; cross off 1 and 4 and the millionth number is 1,000,002. In this problem, we need to cross out all the perfect squares *and* perfect cubes. *After* we have done this, we look at the list and find the millionth number.

Devise a Plan.   It should be clear that we need to know how many numbers are crossed out.

Let $U = \{1, 2, 3, . . . , 1,000,000\}$; $S = \{$perfect squares$\}$ and $C = \{$perfect cubes$\}$. We wish to find $|S|$: We know that $1,000,000 = (10^3)^2$ so there are

$10^3 = 1,000$ perfect squares less than or equal to $1,000,000$. Therefore, $|S| = 1,000$. Next, find $|C|$: We also know that $1,000,000 = (10^2)^3$ so there are $10^2 = 100$ perfect cubes less than or equal to $1,000,000$. Therefore, $|C| = 100$.

If you think about crossing out all of the perfect squares and then all of the perfect cubes, you must notice that some numbers are on both lists. That is, what numbers are in the set $S \cap C$? These are the sixth powers:

$$S \cap C: 1^6 = 1; \ 2^6 = 64; \ 3^6 = 729; \ 4^6 = 4,096; \ \ldots \ ; \ 10^6 = 1,000,000.$$

There are 10 perfect sixth powers, so $|S \cap C| = 10$.

Carry Out the Plan. We use the formula from this chapter to find $|S \cup C|$:

$$|S \cup C| = |S| + |C| - |S \cap C| = 1,000 + 100 - 10 = 1,090$$

If you cross out 1,090 numbers then the millionth number not crossed out is 1,001,090.

Look Back. The answer $1,001,090$ is not necessarily correct because we crossed out perfect squares, cubes, and sixth powers up to $1,000,000$. But each time we crossed out a number, the end number was extended by 1. Thus we need to know whether there are any perfect squares, cubes, or sixth powers between $1,000,000$ and $1,001,090$ (the target range). Let's check the next number on each of our lists:

Perfect squares: $1,001^2 = 1,002,001$, so it is not in the target range;

Perfect cubes: $101^3 = 1,030,301$, so it is not in the target range;

Perfect sixth powers: $11^6 = 1,771,561$, so it also is not in the target range.

The millionth number that is not a perfect square or perfect cube is $1,001,090$.

Group Research

Working in small groups is typical of most work environments, and being able to work with other to communicate specific ideas is an important skill to learn. Work with three or four other students to submit a single report based on each of the following questions.

1. What is the millionth positive integer that is not a square, cube, or fifth power?

   Reference See the IN THE REAL WORLD problem at the beginning of this chapter, as well at the COMMENTARY at the end of this chapter.

2. It is stated in the text that "Mathematics is alive and constantly changing. As we complete the last decade of this century, we stand on the threshold of major changes in the mathematics curriculum in the United States." Report on some of these recent changes.

   References Lynn Steen, *Everybody Counts: A Report to the Nation on the Future of Mathematics Education* (Washington, D.C.: National Academy Press, 1989). See also *Curriculum and Evaluation Standards for School Mathematics* from the National Council of Teachers of Mathematics (Reston, VA: NCTM, 1989).

3. A teacher assigned five problems, A, B, C, D, and E. Not all students turned in answers to all of the problems. Here is a tally of the percentage of students turning in each problem:

A: 46%	A, B: 25%	A, B, C: 13%	A, B, C, D: 7%	A, B, C, D, E: 4%
B: 40%	B, C: 26%	A, B, E: 19%	A, B, C, E: 8%	
C: 43%	C, D: 26%	A, D, E: 16%	A, B, D, E: 9%	
D: 38%	D, E: 22%	B, C, D: 12%	A, C, D, E: 11%	
E: 41%	A, E: 30%	C, D, E: 14%	B, C, D, E: 6%	

What percentage of the students did not turn in any problems? Assume that no students turned in combinations not listed.

4. Do some research on Pascal's triangle, and see how many properties you can discover. You might begin by answering these questions:

  a. What are the successive powers of 11?
  b. Where are the natural numbers found in Pascal's triangle?
  c. What are triangular numbers and how are they found in Pascal's triangle?
  d. What are the tetrahedral numbers and how are they found in Pascal's triangle?
  e. What relationships do the patterns in Figure 1.25 have to Pascal's triangle?

Multiples of 2

Multiples of 3

▲ **Figure 1.25**   **Patterns in Pascal's triangle**

References

James N. Boyd, "Pascal's Triangle," *Mathematics Teacher*, November 1983, pp. 559–560.

Dale Seymour, *Visual Patterns in Pascal's Triangle* (Palo Alto, CA: Dale Seymour Publications, 1986).

Karl J. Smith, "Pascal's Triangle," *Two-Year College Mathematics Journal*, Volume 4 (Winter 1973).

5. A famous mathematician, Bertrand Russell, created a whole series of paradoxes by considering situations such as the barber's rule (see Problem 69, Problem Set 1.2). Write a paper explaining what is meant by a *paradox*. Use the Historical Note below for some suggestions about mathematicians who have done work in this area.

Historical Note

About the time that Cantor's work began to gain acceptance, certain inconsistencies began to appear. One of these inconsistencies, called Russell's paradox, is what Problem 5 is about. Other famous paradoxes in set theory have been studied by many famous mathematicians, including Zermelo, Frankel, von Neumann, Bernays, and Poincaré. These studies have given rise to three main schools of thought concerning the foundation of mathematics.

---

2 The Nature of Logic

> . . . the two great components of the critical movement, though distinct in origin and following separate paths, are found to converge at last in the thesis: Symbolic Logic is Mathematics, Mathematics is Symbolic Logic, the twain are one.
>
> C. J. Keyser

In the Real World . . .

"What's up?" asked Shannon. "I'm going nuts filling out these applications!" exclaimed Missy. "They must make these forms obscure just to test us. Look at this question on the residency questionnaire for the California State University system:"

> Complete and submit this residency questionnaire only if your response to item 34a of Part A was other than 'California' or your response to item 37 was 'no.'

"Maybe I can help," added Shannon. "What are the questions in items 34a and 37?" Missy responded, "Well, for what it's worth, here they are:"

> Question 34a: Were you born in California? *Answer yes or no.*
> Question 37: . . .

"I quit!" shouted Missy. "Wait a minute," calmed Shannon. "We know that Question 37 can be answered by a yes or no response, so we can figure this out. Let's look at the possibilities."

Contents

Perspective

In the first chapter we studied patterns that allow us to form conjectures using inductive reasoning. Remember that inductive reasoning is the process of forming conjectures that are based on a number of observations of specific instances. The conjectures may be probable, but they are not necessarily true. For example, based on our observations of planetary motion, we may predict that tomorrow the sun will rise at 6:13 A.M. Although it is very probable that this prediction is correct, it is not absolutely certain.

There is, however, a type of reasoning that produces *certain* results. It is based on accepting certain *premises* and then *deducing* certain inescapable conclusions. Not only much of mathematics, but also much of the reasoning we do in the world, are based on the principles of logic introduced in this chapter.

We begin with *simple statements*, which are either true or false, and then we combine those statements using certain connectives, such as *and, or, not, because, either . . . or, neither . . . nor,* and *unless*. These connected statements are called *compound statements,* and we discover the truth or falsity of compound statements using truth tables. The final step in our journey through this chapter is to begin with an argument, translate it into symbols, derive logical conclusions, and then translate back into English to complete a proof of a logical argument.

2.1 DEDUCTIVE REASONING

Terminology

There is a type of reasoning that produces results based on certain laws of logic. For example, consider the following argument.

1. If you read the *Times*, then you are well informed.
2. You read the *Times*.
3. Therefore, you are well informed.

Statements 1, 2, and 3 are called an **argument.** If you accept statements 1 and 2 of the argument as true, then you *must* accept statement 3 as true. Statements 1 and 2 are called the **hypotheses** or **premises** of the argument, and statement 3 is called the **conclusion.** Such reasoning is called **deductive reasoning** and, if the conclusion follows from the hypotheses, the reasoning is said to be **valid.**

The purpose of this chapter is to build a logical foundation to aid you, not only in your study of mathematics and other subjects, but also in your day-to-day contact with others.

Logic is a method of reasoning that accepts only those conclusions that are inescapable. This is possible because of the strict way in which every concept is defined. That is, everything must be defined in a way that leaves no doubt or vagueness in meaning. Nothing can be taken for granted, and dictionary definitions are not usually sufficient. For example, in English one often defines a sentence as "a word or group of words stating, asking, commanding, requesting, or exclaiming something; conventional unit of connected speech or writing, usually containing a subject and predicate, beginning with a capital letter, and ending with an end mark." In symbolic logic, we use the word *statement* to refer to a declarative sentence.

> **Statement**
>
> A **statement** is a declarative sentence that is either true or false, but not both true and false.

All of the following are statements, since they are either true or false:

1. School starts on Friday the 13th.
2. $5 + 6 = 16$
3. Fish swim.
4. Mickey Mouse is president.

If the sentence is a question or a command, or if it is vague or nonsensical, then it cannot be classified as true or false; thus, we would not call it a statement.

SATORU ISAKA
(1959–)

Satoru Isaka is a researcher of fuzzy logic, the most recent advance in the study of logical systems. Dr. Isaka specializes in control systems and flexible machine intelligence at OMRON Advanced Systems in Santa Clara, California. He is interested in applying machine learning concepts to biomedical and factory automation systems.

For example:

    5. Go away!
    6. What are you doing?
    7. This sentence is false.

These are not statements by our definition, since they cannot possibly be either true or false.

Difficulty in simplifying arguments may arise because of their length, the vagueness of the words used, the literary style, or the possible emotional impact of the words used. Consider the following two arguments:

  I. If you use heroin, then you first used marijuana.
     Therefore, if you use marijuana, then you will use heroin.
 II. If George Washington was assassinated, then he is dead.
     Therefore, if he is dead, he was assassinated.

Logically, these two arguments are exactly the same, and both are **invalid** forms of reasoning. Nearly everyone would agree that the second is invalid, but many people see the first as valid, because of the emotional appeal of the words used.

To avoid these difficulties, and to be able to simplify a complicated logical argument, we set up an *artificial symbolic language*. This procedure was first suggested by Leibniz in his search to unify all of mathematics. What we will do is invent a notational shorthand. We denote simple statements with letters such as p, $q, r, s, \ldots$, and then define certain connectives. The problem, then, is to *translate the English statements into symbolic form, simplify the symbolic form, and then translate the symbolic form back into English statements.*

The key to simplifying complicated logical arguments is to consider only **simple statements** connected by certain operators (connectives), each of which is defined precisely. Some of these operators are *not, and, or, neither . . . nor, if . . . then, unless,* and *because.*

A **compound statement** is formed by combining simple statements with one or more of these operators. Because of our basic definition of a statement, we see that the **truth value** of any statement is either true (T) or (F), but we should point out that the newest chapter of logic is presently being written. This new logic, called **fuzzy logic,** defines gradations of "T" or "F" to describe real-world concepts that are, by nature, often vague. The difference between the logic we study in this chapter and fuzzy logic is something Aristotle called the **law of the excluded middle.** As we saw in Chapter 1, an object either does or does not belong to a set. There is no middle ground: The number 5 belongs fully to the set of counting numbers, and not to its complement set. This law of the excluded middle says that an object cannot belong both to a set and to its complement set, and thus avoids the contradiction of an object that both is and is not the same object at the same time. Sets that are fuzzy break the law of the excluded middle — to some degree. The study of fuzzy logic requires concepts of probability (Chapter 9), as well as an understanding of the logic we introduce in this chapter. For purposes of this course, we accept the law of the excluded middle as an axiom, and work within a framework that forces every statement to be either true or false.

In Eastern culture, the yin-yang symbol shows the duality of nature: male–female, light–dark, good–evil.

Truth Value

The **truth value** of a *simple statement* is either true (T) or false (F).

The **truth value** of a *compound statement* is true or false and depends only on the truth values of its simple component parts. It is determined by using the rules for connecting those parts with well-defined operators.

It is not sufficient to assume that we know the meanings of the operators *and, or, not,* and so on, even though they may seem obvious and simple. The strength of logic is that it does not leave any meanings to chance or to individual interpretation. In defining the truth values of these words, we will, however, try to conform to common usage.

Conjunction

If *p* and *q* represent two simple statements, then "*p* and *q*" is the compound statement using the operator called **conjunction.** The word *and* is symbolized by $\wedge$. For example,

*I have a penny **and** a quarter in my pocket.*

When is this compound statement true? There are four logical possibilities (not necessarily all true):

1. I have a penny. I have a quarter.
2. I have a penny. I do not have a quarter.
3. I do not have a penny. I have a quarter.
4. I do not have a penny. I do not have a quarter.

Let *p* and *q* represent two simple statements as follows:

p: I have a penny.   *q*: I have a quarter.

The four possibilities can be shown in table form:

	p	*q*
1.	T	T
2.	T	F
3.	F	T
4.	F	F

If *p* and *q* are both true, we would certainly say that the compound statement is true; otherwise, we say it is false. Thus, we define "*p* $\wedge$ *q*" according to Table 2.1. The common usage of the word *and*, used when we discussed the intersection of sets in Chapter 1, conforms to the technical definition of conjunction.

It is worth noting that statements *p* and *q* need not be related. For example,

Fish swim and there are 100 U.S. senators.

is a true compound statement.

Historical Note

GOTTFRIED LEIBNIZ
(1646–1716)

The second great period for logic came with the use of symbols to simplify complicated logical arguments. This was first done when the great German mathematician Gottfried Leibniz, at the age of 14, attempted to reform Aristotelian logic. He called his logic the *universal characteristic* and wrote, in 1666, that he wanted to create a general method in which truths of reason would be reduced to a calculation so that errors of thought would appear as computational errors.

TABLE 2.1   Definition of Conjunction

p	*q*	*p* $\wedge$ *q*
T	T	T
T	F	F
F	T	F
F	F	F

Historical Note

GEORGE BOOLE
(1815–1864)

Gottfried Leibniz advanced Aristotelian logic with his universal characteristic. However, the world took little notice of Leibniz' logic until George Boole completed his book *An Investigation of the Laws of Thought.* Boole considered various mathematical operators by separating them from the other commonly used symbols. This idea was popularized by Bertrand Russell (1872–1970) and Alfred North Whitehead (1861–1947) in their monumental *Principia Mathematica.* In this work, they began with a few assumptions and three undefined terms, and built a system of symbolic logic. From this, they then formally developed the theory of arithmetic and mathematics.

Disjunction

The operator *or*, denoted by $\vee$, is called **disjunction.** The meaning of this simple word is ambiguous, as we can see by considering the following examples:

   I. *I have a penny or a quarter in my pocket.*
   II. *The president is speaking in New York or in California at* 7:00 P.M. *tonight.*

Let's represent four statements as follows:

   p: I have a penny in my pocket.

   q: I have a quarter in my pocket.

   n: The president is speaking in New York at 7:00 P.M. tonight.

   c: The president is speaking in California at 7:00 P.M. tonight.

In everyday terms, what do we mean by each of these examples?

Statement I may mean:

I have a penny in my pocket. I have a quarter in my pocket. I have both a penny and a quarter in my pocket.

Statement II may mean:

The president is speaking in New York at 7:00 P.M. *tonight.*
The president is speaking in California at 7:00 P.M. *tonight.*
It does *not* mean that he will do both.

These statements illustrate different usages of the word *or.* Now we are forced to select a single meaning for the operator, so we choose a definition that conforms to statement I. That is, "$p \vee q$" means "p or q, perhaps both." Thus, we define "$p \vee q$" according to Table 2.2.

TABLE 2.2   **Definition of Disjunction**

p	q	$p \vee q$
T	T	T
T	F	T
F	T	T
F	F	F

In logic, the statement defined in Table 2.2 is called the **inclusive *or*.** The second meaning of the word *or* ("*p* or *q*, but not both") is called the **exclusive *or*.** In this book, we will translate the exclusive *or* by using the operators **either . . . or.** Thus, in this book, we would translate the two examples as:

   I. *I have a penny or a quarter in my pocket.*
   II *The president is speaking either in New York or in California at* 7:00 P.M. *tonight.*

The common meaning of the word *or*, as used when we discussed the union of sets in Chapter 1, conforms to the technical definition of disjunction.

Negation

The operator **not,** denoted by ~, is called **negation.** Table 2.3 serves as a straightforward definition of negation. The negation of a true statement is false, and the negation of a false statement is true.

TABLE 2.3   Definition of Negation

p	$\sim p$
T	F
F	T

◼◼◼◼ EXAMPLE 1

For the statement *t*: Otto is telling the truth, translate ~*t*.

Solution ~*t*:   Otto is not telling the truth.

We also can translate ~*t* as "It is not the case that Otto is telling the truth." You must be careful when negating statements containing the words *all, none,* or *some.* For example, write the negation of

   All students have pencils.

Go ahead—write it down before reading on. Did you write "No students have pencils" or "All students do not have pencils"? Remember that if a statement is false, then its negation must be true. The correct negations are:

   Not all students have pencils.

   At least one student does not have a pencil.

   It is not the case that all students have pencils.

   Some students do not have pencils.

In mathematics, the word **some** is used to mean "at least one." Table 2.4 gives some of the common negations.

TABLE 2.4   Negation of *All, Some,* and *No*

Statement	Negation
All	Some . . . not
Some	No
Some . . . not	All
No	Some

■■■■ **EXAMPLE 2**

Write the negation of each statement.

a. All people have compassion.

b. Some animals are dirty.

c. Some students do not take Math 10.

d. No students are enthusiastic.

Solution

a. Some people do not have compassion.

b. No animal is dirty.

c. All students take Math 10.

d. Some students are enthusiastic.

We consider the negation of compound statements in the next section.

Order of Operations

Parentheses are used to indicate the order of operations. Thus,

$\sim (n \wedge c)$ means the negation of the statement "n and c,"

and is read as "it is not the case that n and c." On the other hand,

$\sim n \wedge c$ means "the negation of n and the statement c,"

which is read "not n and c."

■■■■ **EXAMPLE 3**

Translate the given statements into words.

a. $p \wedge q$ **b.** $\sim p$ **c.** $\sim (p \wedge q)$ **d.** $\sim p \wedge q$ **e.** $\sim (\sim q)$

Let p: I eat spinach; q: I am strong.

Solution

a. A translation of $p \wedge q$ is "I eat spinach, and I am strong."

b. A translation of $\sim p$ is "I do not eat spinach."

c. A translation for $\sim (p \wedge q)$ is "It is not the case that I eat spinach and am strong."

d. A translation for $\sim p \wedge q$ is "I do not eat spinach, and I am strong."

e. A translation of $\sim (\sim q)$ is "I am not not strong" or, if we assume that "not strong" is the same as "weak," then we can translate this as "I am not weak."

We can now consider the truth value of a compound statement using parentheses.

■■■■ **EXAMPLE 4**

Suppose n is a true statement, and c is a false statement. What is the truth value of the compound statement $(n \vee c) \wedge \sim (n \wedge c)$?

Solution Begin with the symbolic statement, fill in the truth values given in the problem, and then simplify using the correct order of operations and the definitions in Tables 2.1–2.3.

$$(n \vee c) \wedge \sim (n \wedge c)$$
$$(T \vee F) \wedge \sim (T \wedge F)$$
$$\quad T \quad \wedge \quad \sim F$$
$$\quad T \quad \wedge \quad T$$
$$\quad\quad T$$

Thus, the compound statement is true. This result does not depend on the particular statements n and c. As long as n is true and c is false, the result of the compound statement is the same — namely, true. ▬

We conclude this section by finding the truth values for the statements in Example 3.

▬▬ **EXAMPLE 5**

Suppose that p is T and q is F. Find the truth values for the statements in Example 3.

a. $p \wedge q$ **b.** $\sim p$ **c.** $\sim (p \wedge q)$ **d.** $\sim p \wedge q$ **e.** $\sim (\sim q)$

Let p: I eat spinach; q: I am strong.

Solution

a. $p \wedge q$

$\quad T \wedge F$ Substitute truth values.

$\quad\quad F$ Thus, the statement "I eat spinach, and I am strong" is false.

b. $\sim p$

$\quad \sim T$ Substitute.

$\quad\quad F$ The statement "I do not eat spinach" is false.

c. $\sim (p \wedge q)$

$\quad \sim (T \wedge F)$ Substitute.

$\quad\quad \sim F$ Definition of conjunction

$\quad\quad\quad T$ Definition of negation

The statement "It is not the case that I eat spinach and am strong" is true.

d. $\sim p \wedge q$

$\quad \sim T \wedge F$ Substitute.

$\quad F \wedge F$ Definition of negation

$\quad\quad F$ Definition of conjunction

The statement "I do not eat spinach, and I am strong" is false.

e. $\sim(\sim q)$

$\quad \sim(\sim F)$ Substitute.

$\quad \sim(T)$ Definition of negation

$\qquad F$ Definition of negation again

The statement "I am not weak" is false.

PROBLEM SET 2.1

▲ A Problems

1. **IN YOUR OWN WORDS** What is an operator?

2. **IN YOUR OWN WORDS** What do we mean by *conjunction?* Include as part of your answer the definition.

3. **IN YOUR OWN WORDS** What do we mean by *disjunction?* Include as part of your answer the definition.

4. **IN YOUR OWN WORDS** What do we mean by *negation?* Include as part of your answer the definition.

5. **IN YOUR OWN WORDS** Describe the procedure for finding the truth value of a compound statement.

According to the definition, which of the examples in Problems 6–9 are statements?

6. **a.** Hickory, Dickory, Dock, the mouse ran up the clock.
   b. $3 + 5 = 9$
   c. Is John ugly?
   d. John has a wart on the end of his nose.

7. **a.** March 18, 1999, is a Monday.
   b. Division by zero is impossible.
   c. Logic is difficult.
   d. $4 - 6 = 2$

8. **a.** $6 + 9 \neq 7 + 8$
   b. Thomas Jefferson was the 23rd president.
   c. Sit down and be quiet!
   d. If wages continue to rise, then prices will also rise.

9. **a.** Dan and Mary were married on August 3, 1979.
   b. $6 + 12 \neq 10 + 8$
   c. Do not read this sentence.
   d. Do you have a cold?

Write the negation of each statement in Problems 10–19.

10. All mathematicians are ogres.

11. All dogs have fleas.

12. Some integers are negative.

13. Some people do not pay taxes.

14. No even integers are divisible by 5.

15. No triangles are squares.

16. All squares are rectangles.

17. All counting numbers are divisible by 1.

18. Some apples are rotten.

19. Some integers are not odd.

20. Let p: Prices will rise; q: Taxes will rise. Translate each of the following statements into symbols.
    a. Prices will rise, or taxes will not rise.
    b. Prices will rise, and taxes will not rise.
    c. Prices will rise, and taxes will rise.
    d. Prices will not rise, and taxes will rise.

21. Assume that prices rise and taxes also rise. Under these assumptions, which of the statements in Problem 20 are true?

22. Assume that prices rise and taxes do not rise. Under these assumptions, which of the statements in Problem 20 are true?

23. Let p: Prices will rise; q: Taxes will rise. Translate each of the following statements into words.
    a. $p \vee q$ **b.** $\sim p \wedge q$ **c.** $p \vee \sim q$ **d.** $\sim p \vee \sim q$

24. Let p: Paul is peculiar; q: Paul likes to read mathematics textbooks. Translate each of the following statements into symbols.
    a. Paul is peculiar and he likes to read mathematics textbooks.
    b. Paul likes to read mathematics textbooks or he is peculiar.
    c. Paul likes to read mathematics textbooks and he is not peculiar.
    d. Paul does not like to read mathematics textbooks and Paul is not peculiar.

25. Assume that Paul likes to read mathematics textbooks, and also that he is not peculiar. Under these assumptions, which of the statements in Problem 24 are true?

26. Assume that Paul does not like to read mathematics textbooks, and also that he is not peculiar. Under these assumptions, which of the statements in Problem 24 are true?

27. Let p: Paul is peculiar; q: Paul likes to read mathematics textbooks. Translate each of the following statements into words.
    a. $p \wedge q$ **b.** $\sim p \wedge q$ **c.** $p \vee \sim q$ **d.** $\sim p \vee \sim q$

28. Let p: Today is Friday; q: There is homework tonight. Translate each of the following statements into words.
    a. $p \wedge q$ **b.** $\sim p \wedge q$ **c.** $p \vee \sim q$ **d.** $\sim p \vee \sim q$

29. Assume p is T and q is T. Under these assumptions, which of the statements in Problem 27 are true?

30. Assume p is F and q is T. Under these assumptions, which of the statements in Problem 28 are true?

Find the truth value for each of the compound statements in Problems 31–38.

31. Assume r is F, s is T, and t is T.
    a. $(r \vee s) \vee t$ **b.** $(r \wedge s) \wedge \sim t$

32. Assume r is T, s is T, and t is T.
    a. $r \wedge (s \vee t)$ b. $(r \wedge s) \vee (r \wedge t)$

33. Assume p is T, q is T, and r is F.
    a. $(p \vee q) \wedge r$ b. $(p \wedge q) \wedge {\sim}r$

34. Assume p is T, q is T, and r is T.
    a. $p \wedge (q \vee r)$ b. $(p \wedge q) \vee (p \wedge r)$

35. Assume p is T, q is T, and r is T.
    a. $(p \vee q) \vee (r \wedge {\sim}q)$ b. ${\sim}({\sim}p) \vee (p \wedge q)$

36. Assume p is T, q is F, and r is T.
    a. $(p \wedge q) \vee (p \wedge {\sim}r)$ b. $({\sim}p \vee q) \wedge {\sim}p$

37. Assume p is T, and q is F.
    a. ${\sim}(p \wedge q)$ b. ${\sim}p \wedge q$

38. Assume p is T and q is T.
    a. $({\sim}p \vee q) \wedge {\sim}p$ b. $(p \wedge {\sim}q) \vee (p \wedge q)$

39. Does the News Clip in the margin illustrate inductive or deductive reasoning? Explain your answer.

40. Does the *B.C.* cartoon illustrate inductive or deductive reasoning? Explain your answer.

41. Prove the law of double negation. That is, prove that for any proposition p, ${\sim}({\sim}p)$ has the same truth value as p.

▲ **B Problems**

Translate the statements in Problems 42–50 into symbols. For each simple statement, be sure to indicate the meanings of the symbols you use. Answers are not unique.

42. W. C. Fields is eating, drinking, and having a good time.

43. Sam will not seek and will not accept the nomination.

44. Jack will not go tonight, and Rosamond will not go tomorrow.

45. Fat Albert lives to eat and does not eat to live.

46. The decision will depend on judgment or intuition, and not on who paid the most.

47. The successful applicant for the job will have a B.A. degree in liberal arts or psychology.

48. The winner must have an A.A. degree in drafting or three years of professional experience.

49. **a.** Dinner includes soup and salad, or the vegetable of the day.
    b. Dinner includes soup, and salad or the vegetable of the day.

50. **a.** Marsha finished the sign and table, or a pair of chairs.
    b. Marsha finished the sign, and the table or a pair of chairs.

In Problems 51–58, find the truth value when p is T, *q is* F, *and r is* F.

51. $(p \wedge q) \wedge r$
52. $p \vee (q \wedge r)$
53. $(p \vee q) \wedge \sim(p \vee \sim q)$
54. $(p \wedge \sim q) \vee (r \wedge \sim q)$
55. $\sim(\sim p) \wedge (q \vee p)$
56. $(r \wedge p) \vee (q \wedge r)$
57. $\sim(r \wedge q) \wedge (q \vee \sim q)$
58. $(q \vee \sim q) \wedge [(p \wedge \sim q) \vee (\sim r \vee r)]$

▲ **Problem Solving**

59. Smith received the following note from Melissa: "Dr. Smith, I wish to explain that I was really joking when I told you that I didn't mean what I said about reconsidering my decision not to change my mind." Did Melissa change her mind or didn't she?

60. Their are three errers in this item. See if you can find all three.

▲ **Individual Research**

61. What do the following people have in common?

   Ira Glasser, executive director of the ACLU

   Bram Stoker, author of *Dracula*

   David Robinson, basketball star

   Ed Thorpe, inventor of programmed-trading on Wall Street

   Clifford Brown, 1950s jazz trumpeter

2.2   TRUTH TABLES AND THE CONDITIONAL

Truth Tables

The connectives *and, or,* and *not* introduced in the previous section are called the **fundamental operators.** To go beyond these and consider additional operators, we need a device known in logic as a **truth table.** A truth table shows how the truth values of compound statements depend on the fundamental operators. Tables 2.1, 2.2, and 2.3 from the previous section should be memorized. They are summarized in Table 2.5 on page 88.

TABLE 2.5   Truth Table of the Fundamental Operators

p	q	Conjunction $p \wedge q$	Disjunction $p \vee q$	Negation	
				$\sim p$	$\sim q$
T	T	T	T	F	F
T	F	F	T	F	T
F	T	F	T	T	F
F	F	F	F	T	T

▬▬ EXAMPLE I

Construct a truth table for the compound statement:

Alfie did not come last night and did not pick up his money.

Solution   First identify the operators: *not . . . and . . . not*. Next, choose variables to represent the simple statements. Let

p: Alfie came last night. q: Alfie picked up his money.

Translate the English statement into symbols: $\sim p \wedge \sim q$.
     Finally, construct a truth table. List all possible combinations of truth values for the simple statements (see columns A and B).

A	B
p	q
T	T
T	F
F	T
F	F

Insert the truth values for $\sim p$ and $\sim q$ (see columns C and D).

A	B	C	D
p	q	$\sim p$	$\sim q$
T	T	F	F
T	F	F	T
F	T	T	F
F	F	T	T

Finally, insert the truth values for $\sim p \wedge \sim q$ (see column E).

A	B	C	D	E
p	q	$\sim p$	$\sim q$	$\sim p \wedge \sim q$
T	T	F	F	F
T	F	F	T	F
F	T	T	F	F
F	F	T	T	T

The only time the compound statement is true is when *both p and q are false.* ▬

▬ EXAMPLE 2

Construct a truth table for $\sim(\sim p)$.

Solution

p	$\sim p$	$\sim(\sim p)$
T	F	T
F	T	F

▬

Notice from Example 2 that $\sim(\sim p)$ and p have the same truth values. If two statements have the same truth values, one can replace the other in any logical expression. This means that the double negation of a statement is the same as the original statement.

> **Law of Double Negation**
> $\sim(\sim p)$ may be replaced by p in any logical expression. ▲

▬ EXAMPLE 3

Construct a truth table to determine when the following statement is true.

$$\sim(p \wedge q) \wedge [(p \vee q) \wedge q]$$

Solution   We begin as before and move from left to right with parentheses taking precedence, focusing our attention on no more than two columns at a time (refer to Table 2.5 to find the correct entries).

A	B	C	D	E	F	G
p	q	$p \wedge q$	$\sim(p \wedge q)$	$p \vee q$	$(p \vee q) \wedge q$	$\sim(p \wedge q) \wedge [(p \vee q) \wedge q]$
T	T	T	F	T	T	F
T	F	F	T	T	F	F
F	T	F	T	T	T	T
F	F	F	T	F	F	F

The compound statement is true only when p is false and q is true. ▬

THERE'S TOO MUCH UNCERTAINTY ABOUT OUR RELATIONSHIP?

if

Conditional

We can now use truth tables to prove certain useful results and to introduce some additional operators. The first one we'll consider is called the **conditional**. The statement **"if p, then q"** is called a *conditional statement*. It is symbolized by $p \rightarrow q$, where p is called the **antecedent** and q is called the **consequent**. There are several ways of using a conditional, as illustrated by the following examples.

1. We use "if–then" to indicate a *logical* relationship — one in which the consequent follows logically from the antecedent:

   If $\sim(\sim p)$ has the same truth value as p, then p can replace $\sim(\sim p)$.

2. We can use "if–then" to indicate a causal relationship.

   If John drops that rock, then it will land on my foot.

3. We can use "if–then" to report a decision on the part of the speaker.

   If John drops that rock, then I will hit him.

4. We can use "if–then" when the consequent follows from the antecedent by the very definition of the words used.

   If John drives a Geo, then John drives a car.

5. Finally, we can use "if–then" to make a *material implication*. There is no logical, causal, or definitional relationship between the antecedent and consequent; we use the expression simply to convey humor or emphasis.

   If John gets an A on that test, then I'm a monkey's uncle.

   The consequent is obviously false, and the speaker wishes to emphasize that the antecedent is also false.

Our task is to state a definition of the conditional that applies to all of these *if–then* statements. We will approach the problem by asking under what circumstances a given conditional would be false. Let's consider another example.

Suppose I make you a promise: "If I receive my check tomorrow, then I will pay you the $10 that I owe you." If I keep my promise, let's agree to say the statement is true; if I don't, then it is false. Let

p: I receive my check tomorrow.

q: I will pay you the $10 that I owe you.

We symbolize the promise by $p \rightarrow q$. There are four possibilities:

	p	q	
Case 1:	T	T	*I receive my check tomorrow, and I pay you the $10. In this case, the promise, or conditional, is true.*
Case 2:	T	F	*I receive my check tomorrow, and I do not pay you the $10. In this case, the conditional is false, since I did not fulfill my promise.*
Case 3:	F	T	*I do not receive my check tomorrow, but I pay you the $10. In this case, I certainly didn't break my promise, so the conditional is true.*

Case 4:   F   F   *I do not receive my check tomorrow, and I do not pay you the* $10. Here, again, I did not break my promise, so the conditional is true.

Actually, the promise was not tested for cases 3 and 4, since I didn't receive my check. Assume the principle of "innocent until proven guilty." The only time that I will have broken my promise is in case 2. The test for a conditional is to determine when it is false. In symbols,

$p \rightarrow q$ is false whenever $p \wedge \sim q$ is true (case 2)

or

$p \rightarrow q$ is true whenever $\sim(p \wedge \sim q)$ is true.

Construct a truth table for $\sim(p \wedge \sim q)$.

A	B	C	D	E
p	q	$\sim q$	$p \wedge \sim q$	$\sim(p \wedge \sim q)$
T	T	F	F	T
T	F	T	T	F
F	T	F	F	T
F	F	T	F	T

We use this truth table to define the conditional $p \rightarrow q$ as shown in Table 2.6. The following examples illustrate the definition of the conditional.

TABLE 2.6   Definition of Conditional

p	q	$p \rightarrow q$
T	T	T
T	F	F
F	T	T
F	F	T

■■■ EXAMPLE 4

Make up an example illustrating each of the four cases for the conditional.

Solution   Examples can vary.

a. Case 1:   T → T   *If* $7 < 14$; *then* $7 + 2 < 14 + 2$.
This is true, since both component parts are true.

b. Case 2:   T → F   *If* $7 + 5 = 12$, *then* $7 + 10 = 15$.
This is false, since the antecedent is true, but the consequent is false.

c. Case 3:   F → T   *If 7 + 5 = 15, then 8 + 2 = 10.*
This is true, since the antecedent is false.

d. Case 4:   F → F   *If 7 + 5 = 25, then 7 = 20.*
This is true, since the antecedent is false.   ▬

Example 4 shows that the conditional, in mathematics, does not mean that there is any cause-and-effect relationship. *Any* two statements can be connected with the connective of conditional and the result must be true or false.

▬▬▬ **EXAMPLE 5**   **Polya's Method**

The following sentence is found on a tax form:

If you do not itemize deductions on Schedule A and you have charitable contributions, then complete the worksheet on page 14 and enter the allowable part on line 34b.

Interpret this sentence.

Solution   We use Polya's problem-solving guidelines for this example.

Understand the Problem.   We need to rephrase, or analyze, this statement so that we can determine an action to take under all possible circumstances.

Devise a Plan.   We will translate the statement into symbolic form. It could then be analyzed using a truth table.

Carry Out the Plan.   First identify the operators:

If you do **not** itemize deductions on Schedule A **and** you have charitable contributions, **then** complete the worksheet on page 14 **and** enter the allowable part on line 34b.

Next, assign variables to simple statements:

d: You itemize deductions on Schedule A.
c: You have charitable contributions.
w: You complete the worksheet on page 14.
b: You enter the allowable part on line 34b.

Rewrite the sentence, making substitutions for the variables:

If not *d* **and** *c*, **then** (*w* **and** *b*).

Complete the translation into symbols:

$$(\sim d \wedge c) \rightarrow (w \wedge b)$$

Look Back.   We check to see whether the statement is true if d, c, w, and b are all true:

$$(\sim d \wedge c) \rightarrow (w \wedge b)$$
$$(\sim T \wedge T) \rightarrow (T \wedge T)$$
$$(F \wedge T) \rightarrow T$$
$$F \rightarrow T$$
$$T$$

The given statement is true.   ▬

Translations for the Conditional

The *if* part of a conditional need not be stated first. All of the following statements have the same meaning:

Conditional translation	Example
If p, then q.	If you are 18, then you can vote.
q, if p.	You can vote, if you are 18.
p, only if q.	You are 18 only if you can vote.
$P \subseteq Q$.	P is a subset of Q.
All p are q.	All 18-year-olds can vote.

In addition to these translations for the conditional, there are related statements.

Converse; Inverse; Contrapositive

Given the conditional $p \rightarrow q$, we define:

the **converse** is $q \rightarrow p$;
the **inverse** is $\sim p \rightarrow \sim q$;
the **contrapositive** is $\sim q \rightarrow \sim p$.

▬▬ **EXAMPLE 6**

Write the converse, inverse, and contrapositive of the statement: *If it is a 300ZX, then it is a car.*

Solution   Let p: It is a 300ZX; q: It is a car. The given statement is symbolized as $p \rightarrow q$.

Converse: $q \rightarrow p$	If it is a car, then it is a 300ZX.	
Inverse: $\sim p \rightarrow \sim q$	If it is not a 300ZX, then it is not a car.	
Contrapositive: $\sim q \rightarrow \sim p$	If it is not a car, then it is not a 300ZX. ▬	

As you can see from Example 6, not all these statements are equivalent in meaning. The contrapositive and the original statement always have the same truth values, as do the converse and the inverse. We see this in Table 2.7, and summarize it in the following box.

Law of Contraposition

A conditional may always be replaced by its contrapositive without having its truth value affected.

TABLE 2.7   Truth Table for Variations of the Conditional

p	q	$\sim p$	$\sim q$	Statement $p \rightarrow q$	Converse $q \rightarrow p$	Inverse $\sim p \rightarrow \sim q$	Contrapositive $\sim q \rightarrow \sim p$
T	T	F	F	T	T	T	T
T	F	F	T	F	T	T	F
F	T	T	F	T	F	F	T
F	F	T	T	T	T	T	T

▬▬ EXAMPLE 7

Assume that the following statement is true:

$p \rightarrow \sim q$ If you obey the law, then you will not go to jail.

Write the converse, inverse, and contrapositive, and tell which are true.

Solution   We note that p: you obey the law; q: you go to jail.

Converse: $\sim q \rightarrow p$ If you do not go to jail, then you obey the law. *False*

Inverse: $\sim p \rightarrow q$ *Note:* $\sim(\sim q)$ is replaced by q (double negation).
                              If you do not obey the law, then you will go to jail. *False*

Contrapositive: $q \rightarrow \sim p$ If you go to jail, then you did not obey the law. *True*   ▬

PROBLEM SET   2.2

▲ A Problems

1. **IN YOUR OWN WORDS** What is a truth table?
2. **IN YOUR OWN WORDS** What is a conditional? Discuss.
3. **IN YOUR OWN WORDS** What is the law of double negation?
4. **IN YOUR OWN WORDS** What is the law of contraposition?

Construct a truth table for the statements given in Problems 5–24.

5. $\sim p \vee q$ 6. $\sim p \wedge \sim q$ 7. $\sim(p \wedge q)$

8. $\sim r \vee \sim s$

9. $\sim(\sim r)$

10. $(r \wedge s) \vee \sim s$

11. $p \wedge \sim q$

12. $\sim p \vee \sim q$

13. $(\sim p \wedge q) \vee \sim q$

14. $(p \wedge \sim q) \wedge p$

15. $p \vee (p \rightarrow q)$

16. $(p \wedge q) \rightarrow p$

17. $(p \rightarrow \sim q) \rightarrow (q \rightarrow \sim p)$

18. $(p \rightarrow q) \rightarrow (\sim q \rightarrow \sim p)$

19. $[p \wedge (p \vee q)] \rightarrow p$

20. $[p \vee (p \wedge q)] \rightarrow p$

21. $(p \vee q) \vee r$

22. $(p \wedge q) \wedge \sim r$

23. $[(p \vee q) \wedge \sim r] \wedge r$

24. $[p \wedge (q \vee \sim p)] \vee r$

Write the converse, inverse, and contrapositive of the statements in Problems 25–30.

25. $\sim p \rightarrow \sim q$

26. $\sim r \rightarrow t$

27. $\sim t \rightarrow \sim s$

28. If you break the law, then you will go to jail.

29. I will go Saturday if I get paid.

30. If you brush your teeth with Smiles toothpaste, then you will have fewer cavities.

▲ B Problems

Translate the sentences in Problems 31–38 into if–then form.

31. All triangles are polygons.

32. All prime numbers greater than 2 are odd numbers.

33. All good people go to heaven.

34. Everything happens to everybody sooner or later if there is time enough. (G. B. Shaw)

35. We are not weak if we make a proper use of these means which the God of Nature has placed in our power. (Patrick Henry)

36. All useless life is an early death. (Goethe)

37. All work is noble. (Thomas Carlyle)

38. Everything's got a moral if only you can find it. (Lewis Carroll)

First decide whether each simple statement in Problems 39–42 is true or false. Then state whether the given compound statement is true or false.

39. If $5 + 10 = 16$, then $15 - 10 = 3$.

40. The moon is made of green cheese only if Mickey Mouse is president.

41. If $1 + 1 = 10$, then the moon is made of green cheese.

42. $3 \cdot 2 = 6$ only if water runs uphill.

In Problems 43–48, fill in the blanks with a symbolic statement that follows from the given statement.

43. The applicant for the position must have a two-year college degree in drafting or five years of experience in the field.

    Let a: You are an applicant for the position.
          q: You are qualified for the position.
          e: You have a two-year college degree in drafting.
          f: You have five years of experience in the field.

    a. $a \rightarrow$ _____   **b.** $(a \wedge e) \rightarrow$ _____   **c.** $(a \wedge f) \rightarrow$ _____

44. To qualify for a loan, the applicant must have a gross income of at least $35,000 if single or combined income of $50,000 if married.

    Let q: You qualify for a loan.
         m: You are married.
         i: You have an income of at least $35,000.
         b: Your spouse has an income of at least $35,000.

    a. $(m \wedge \sim i) \rightarrow$ _____
    b. $(\sim m \wedge i) \rightarrow$ _____
    c. $[m \wedge (i \wedge b)] \rightarrow$ _____

45. To qualify for the special fare, you must fly on Monday, Tuesday, Wednesday, or Thursday, and you must stay over a Saturday evening.

    Let q: You qualify for the special fare.
         m: You fly on Monday.
         t: You fly on Tuesday.
         w: You fly on Wednesday.
         h: You fly on Thursday.
         s: You stay over a Saturday evening.

    _____ $\rightarrow q$

46. This contract is noncancellable by tenant for 60 days.
    Let t: You are a tenant.
         d: It is within 60 days.
         c: This contract can be cancelled.

    $(t \wedge d) \rightarrow$ _____

47. The tenant agrees to lease the premises for 12 months beginning on September 1 and at a monthly rental charge of $800.
    Let t: You are a tenant.
         m: You lease the premises for 12 months.
         s: You will begin on September 1.
         p: You will pay $800 per month.

    $t \rightarrow$ _____

48. Utilities, except for water and garbage, are paid by the tenant.
    Let t: You are a tenant.
         w: You pay water.
         g: You pay garbage.
         u: You pay the other utilities.

    $t \rightarrow$ _____

49. Repeat Example 5 except this time assume that you do not complete the worksheet on page 14, but all other statements are true.

50. Under what conditions will the statement in Example 5 be true?

Translate the statements (taken from tax forms) in Problems 51–56 into symbolic form.

51. If the qualifying person is a child and not your dependent, enter this child's name.

52. If the amount of line 31 is less than $26,673 and a child lives with you, turn to page 27.

53. If line 32 is $86,025 or less, multiply $2,500 by the total number of exemptions claimed on line 6e.

54. If married and filing a joint return, enter your spouse's earned income.

55. If you are a student or disabled, see line 6 of instructions.

56. If the income on line 1 was reported to you on Form W-2 and the "Statutory employee" box on that form was checked, see instructions for Schedule C, line 1, and check here.

Tell which of the statements in Problems 57–60 are true.

57. Let p: $2 + 3 = 5$; q: $12 - 7 = 5$.
    a. $\sim p \vee q$ b. $\sim p \wedge \sim q$ c. $\sim(p \wedge q)$

58. Let p: 2 is prime; q: 1 is prime.
    a. $\sim p \vee \sim q$ b. $\sim(\sim p)$ c. $(p \wedge q) \vee \sim q$

59. Let p: $5 + 8 = 10$; q: $4 + 4 = 8$.
    a. $(p \wedge \sim q) \wedge p$ b. $p \vee (p \rightarrow q)$ c. $(p \wedge q) \rightarrow p$

60. Let p: $1 + 1 = 2$; q: $9 - 3 = 5$.
    a. $(p \rightarrow \sim q) \rightarrow (q \rightarrow \sim p)$ b. $(p \rightarrow q) \rightarrow (\sim q \rightarrow \sim p)$

▲ **Problem Solving**

61. If Apollo can do anything, could he make an object that he could not lift?

62. Decide about the truth or falsity of the following statement:

    If wishes were horses, then beggars could ride.

▲ **Individual Research**

63. Do some research to explain the differences between the words *necessary* and *sufficient*.

64. Sometimes statements p and q are described as *contradictory*, *contrary*, or *inconsistent*. Consult a logic text, and then define these terms using truth tables.

2.3 OPERATORS AND LAWS OF LOGIC

Biconditional, Implication, and Logical Equivalence

In the previous section we took great care to point out that a statement $p \rightarrow q$ and its converse $q \rightarrow p$ do not have the same truth values. However, it may be the case that $p \rightarrow q$ *and also* $q \rightarrow p$. In this case, we write

$$p \leftrightarrow q$$

and call this operator the **biconditional.** To determine the truth values of the biconditional, we construct a truth table for $(p \rightarrow q) \wedge (q \rightarrow p)$:

p	q	$p \rightarrow q$	$q \rightarrow p$	$(p \rightarrow q) \wedge (q \rightarrow p)$
T	T	T	T	T
T	F	F	T	F
F	T	T	F	F
F	F	T	T	T

Signpost in New York City

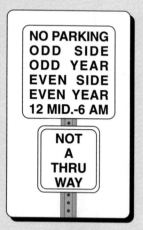

Many drivers must throw up their hands in utter confusion when they come upon this wacky sign in Medford, Massachusetts. Local police say the sign is really easy to decipher: You can't park on the odd side of the street — the side in which the house numbers are odd — from midnight to 6 A.M. in an odd-numbered year such as 1995. Likewise, parking on the even side is taboo in an even year. The reason for this zany arrangement is to keep one side clear of snow for emergency vehicle use.

This leads us to define the biconditional so that it is true only when both p and q are true or when both p and q are false (that is, whenever they have the same truth values). This definition is shown in Table 2.8.

TABLE 2.8   Definition of Biconditional

p	q	$p \leftrightarrow q$
T	T	T
T	F	F
F	T	F
F	F	T

In mathematics and logic, $p \leftrightarrow q$ is translated in several ways, all of which have the same meaning:

Biconditional translation

1. p if and only if q.
2. q if and only if p.
3. If p then q, and conversely.
4. If q then p, and conversely.

▬▬ EXAMPLE 1

Rewrite the following as one statement:

1. *If a polygon has three sides, then it is a triangle.*
2. *If a polygon is a triangle, then it has three sides.*

Solution   A polygon is a triangle if and only if it has three sides.   ▬

The set of logical possibilities for which a given statement is true is called its **truth set.** If its truth set is identical to its universal set, then the statement is a logically true statement and is called a **tautology.** This means that a compound statement is a tautology if you obtain only Ts on a truth table.

▬▬ EXAMPLE 2

Is $(p \vee q) \to (\sim q \to p)$ a tautology?

Solution

p	q	$p \vee q$	$\sim q$	$\sim q \to p$	$(p \vee q) \to (\sim q \to p)$
T	T	T	F	T	T
T	F	T	T	T	T
F	T	T	F	T	T
F	F	F	T	F	T

Thus, the given statement is a tautology.   ▬

If a conditional is a tautology, as in Example 2, then it is called an **implication** and is symbolized by $\Rightarrow$. That is, Example 2 can be written

$$(p \vee q) \Rightarrow (\sim q \rightarrow p)$$

The implication symbol $p \Rightarrow q$ is pronounced "p implies q."

A biconditional statement $p \leftrightarrow q$ that is also a tautology (that is, always true) is a **logical equivalence**, written $p \Leftrightarrow q$ and read "p is logically equivalent to q."

▬▬ EXAMPLE 3

Show that $(p \rightarrow q) \Leftrightarrow \sim (p \wedge \sim q)$.

Solution   We begin by constructing a truth table for $(p \rightarrow q) \leftrightarrow \sim (p \wedge \sim q)$.

p	q	$p \rightarrow q$	$\sim q$	$p \wedge \sim q$	$\sim (p \wedge \sim q)$	$(p \rightarrow q) \rightarrow \sim (p \wedge \sim q)$
T	T	T	F	F	T	T
T	F	F	T	T	F	T
F	T	T	F	F	T	T
F	F	T	T	F	T	T

Since all possibilities are true (it is a tautology), we see it is a logical equivalence and we write

$$(p \rightarrow q) \Leftrightarrow \sim (p \wedge \sim q)$$

▬

Notice the difference between the symbols $\rightarrow$ and $\Rightarrow$ as well as $\leftrightarrow$ and $\Leftrightarrow$. The conditional ($\rightarrow$) and biconditional ($\leftrightarrow$) are used as logical connectives and the truth table for these may be true or false. If, because of the statements being connected, the truth table gives *only true values*, then we use the symbols $\Rightarrow$ and $\Leftrightarrow$ in place of $\rightarrow$ and $\leftrightarrow$ and call them implication ($\Rightarrow$) and logical equivalence ($\Leftrightarrow$).

Laws of Logic

We can use the idea of logical equivalence to write two previously stated laws:

Law of double negation: $\sim(\sim p) \Leftrightarrow p$

Law of contraposition: $(\sim q \rightarrow \sim p) \Leftrightarrow (p \rightarrow q)$

You will be asked to prove these laws in the problem set. Other important laws are developed in the next section. Two additional laws, called **De Morgan's laws**, will be considered in this section.

De Morgan's Laws

$$\sim(p \vee q) \Leftrightarrow \sim p \wedge \sim q$$
$$\sim(p \wedge q) \Leftrightarrow \sim p \vee \sim q$$

There is a strong tie between set theory and symbolic logic, so it is no coincidence that there are De Morgan's laws both in set theory and in symbolic logic. We prove the second one here and leave the first for the problem set. The proof, by truth table, is shown:

p	q	$p \wedge q$	$\sim(p \wedge q)$	$\sim p$	$\sim q$	$\sim p \vee \sim q$	$\sim(p \wedge q) \leftrightarrow \sim p \vee \sim q$
T	T	T	F	F	F	F	T
T	F	F	T	F	T	T	T
F	T	F	T	T	F	T	T
F	F	F	T	T	T	T	T

Since the last column is all Ts, we can write $\sim(p \wedge q) \Leftrightarrow \sim p \vee \sim q$.

Negation of a Compound Statement

De Morgan's laws can be used to write the negation of a compound statement using conjunction and disjunction.

■ EXAMPLE 4

Write the negation of the compound statements.

a. John went to work or he went to bed.

b. Alfie didn't come last night and didn't pick up his money.

Solution   Begin by writing the statement in symbolic form, then find the negation, and finally use the rules of logic to simplify before translating back into English.

a. Let w: John went to work; b: John went to bed.

The symbolic form is:	$w \vee b$
Negation:	$\sim(w \vee b)$
Simplify (De Morgan's law):	$\sim w \wedge \sim b$
Translate back into English:	

       John did not go to work and he did not go to bed.

b. Let p: Alfie came last night; q: Alfie picked up his money.

Symbolic form:	$\sim p \wedge \sim q$
Negation:	$\sim(\sim p \wedge \sim q)$
Simplify (De Morgan's law):	$\sim(\sim p) \vee \sim(\sim q)$
Simplify (law of double negation):	$p \vee q$
Translate:	Alfie came last night or he picked up his money.

    Sometimes it is necessary to find the negation of a conditional, $p \rightarrow q$. It can be shown (see Problem 38) that

$$(p \rightarrow q) \Leftrightarrow (\sim p \vee q)$$

Therefore, the negation of $p \rightarrow q$ is equivalent to the negation of $\sim p \vee q$:

$$\sim (p \rightarrow q) \Leftrightarrow \sim (\sim p \vee \sim q)$$
$$\Leftrightarrow \sim (\sim p) \wedge \sim q \qquad \text{De Morgan's law}$$
$$\Leftrightarrow p \wedge \sim q \qquad \text{Law of double negation}$$

Negation of a Conditional
The negation of $p \rightarrow q$ is $p \wedge \sim q$.

■■■ **EXAMPLE 5** **Polya's Method**

The officer said to the detective, "If John was at the scene of the crime, then he knows that Jean could not have done it." "No, Colombo," answered the detective, "that is not correct. John was at the scene of the crime and Jean did it." Was the detective's statement a correct negation of Colombo's statement?

Solution We use Polya's problem-solving guidelines for this example.

Understand the Problem. We are asking for the negation of a given statement, which we then wish to compare with the proposed negation.

Devise a Plan. Translate the given statement into symbolic form, find the negation, simplify, and then translate back into English.

Carry Out the Plan. Let p: John was at the scene of the crime; q: Jean committed the crime. Then $p \rightarrow \sim q$ is a symbolic statement of the given statement, "If John was at the scene of the crime, then he knows that Jean could not have done it." We now form the negation and then simplify:

$$\sim (p \rightarrow \sim q) \Leftrightarrow p \wedge \sim (\sim q) \qquad \text{Negation of a conditional}$$
$$\Leftrightarrow p \wedge q \qquad \text{Law of double negation}$$

Translate back into English: John was at the scene of the crime and Jean committed the crime.

Look Back. The detective's negation was correct. ■

Miscellaneous Operators

Occasionally, we encounter other operators, and it is necessary to formulate precise definitions of these additional operators. As Table 2.9 (on page 102) shows, they are all defined in terms of our previous operators.

■■■ **EXAMPLE 6**

Show that the statement "p because q" as defined in Table 2.9 is equivalent to conjunction.

TABLE 2.9   Additional Operators

p	q	Either p or q $(p \vee q) \wedge \sim(p \wedge q)$	Neither p nor q $\sim(p \vee q)$	p unless q $\sim q \rightarrow p$	p because q $(p \wedge q) \wedge (q \rightarrow p)$	No p is q $p \rightarrow \sim q$
T	T	F	F	T	T	F
T	F	T	F	T	F	T
F	T	T	F	T	F	T
F	F	F	T	F	F	T

Solution   According to Table 2.9, "p because q" means $(p \wedge q) \wedge (q \rightarrow p)$ and conjunction means $p \wedge q$.

p	q	$(p \wedge q) \wedge (q \rightarrow p)$	$p \wedge q$	$(p \wedge q) \wedge (q \rightarrow p) \leftrightarrow p \wedge q$
T	T	T	T	T
T	F	F	F	T
F	T	F	F	T
F	F	F	F	T
		↑ From Table 2.9	↑ From Table 2.5	

The last column is all Ts, so we can write $(p \wedge q) \wedge (q \rightarrow p) \Leftrightarrow p \wedge q$.   ▬

PROBLEM SET   2.3

▲ A Problems

1. **IN YOUR OWN WORDS** Discuss the difference between the conditional and the biconditional.

2. **IN YOUR OWN WORDS** Discuss the procedure for finding the negation of compound statements.

3. **IN YOUR OWN WORDS** Discuss when you use the symbols $\leftrightarrow$ and $\Leftrightarrow$.

4. **IN YOUR OWN WORDS** Discuss when you use the symbols $\rightarrow$ and $\Rightarrow$.

Use truth tables in Problems 5–12 to determine whether the given compound statement is a tautology.

5. $(p \wedge q) \vee (p \rightarrow \sim q)$ 6. $(p \vee q) \wedge (q \rightarrow \sim p)$

7. $(\sim p \rightarrow q) \rightarrow p$ 8. $(\sim q \rightarrow p) \rightarrow q$

9. $(p \wedge q) \leftrightarrow (p \vee q)$ 10. $(p \rightarrow q) \leftrightarrow (\sim p \vee q)$

11. $(p \vee q) \leftrightarrow (p \wedge q)$ 12. $(p \vee \sim q) \leftrightarrow (\sim p \vee q)$

Verify the indicated definition in Problems 13–16 from Table 2.9 using a truth table.

13. either p or q 14. neither p nor q

15. p unless q 16. no p is q

Translate the statements in Problems 17–26 into symbols. For each simple statement, indicate the meanings of the symbols you use.

17. Neither smoking nor drinking is good for your health.

18. I will not buy a new house unless all provisions of the sale are clearly understood.

19. To obtain the loan, I must have an income of $85,000 per year.

20. I am obligated to pay the rent because I signed the contract.

21. I cannot go with you because I have a previous engagement.

22. No man is an island.

23. Either I will invest my money in stocks or I will put it in a savings account.

24. Be nice to people on your way up 'cause you'll meet 'em on your way down. (Jimmy Durante)

25. No person who has once heartily and wholly laughed can be altogether irreclaimably bad. (Thomas Carlyle)

26. If by the mere force of numbers a majority should deprive a minority of any clearly written constitutional right, it might, in a moral point of view, justify revolution. (Abraham Lincoln)

Use the tautology $(p \rightarrow q) \Leftrightarrow (\sim p \vee q)$ to write each statement in Problems 27–32 in an equivalent form.

27. If I go, then I paid $100.

28. If the cherries have turned red, then they are ready to be picked.

29. We will not visit New York or visit the Statue of Liberty.

30. Melissa will not watch Jay Leno or the NBC late night orchestra.

31. The sun is shining or I will not go to the park.

32. The money is available or I will not take my vacation.

▲ B Problems

33. Show that the definition for *neither p nor q* could also be $\sim p \wedge \sim q$.

34. Prove the law of double negation by using a truth table.

35. Prove the law of contraposition by using a truth table.

36. Prove De Morgan's law: $\sim (p \vee q) \Leftrightarrow \sim p \wedge \sim q$

37. In the text we used laws of logic to prove that $\sim (p \rightarrow q) \Leftrightarrow (p \wedge \sim q)$. Use a truth table to prove this result.

38. Prove $(p \rightarrow q) \Leftrightarrow (\sim p \vee q)$.

Write the negation of the compound statements in Problems 39–52.

39. $p \rightarrow q$ 40. $p \rightarrow \sim q$

41. $\sim p \rightarrow q$ 42. $\sim p \rightarrow \sim q$

43. John went to Macy's or Sears.

44. Jane went to the basketball game or to the soccer game.

45. Tim is not here and he is not at home.

46. Sally is not on time and she missed the boat.

47. If I can't go with you then I'll go with Bill.

48. If you're out of Schlitz, you're out of beer.

49. If $x + 2 = 5$, then $x = 3$.

50. If $x - 5 = 4$, then $x = 1$.

51. If $x = -5$, then $x^2 = 25$.

52. $2x + 3y = 8$ if $x = 1$ and $y = 2$.

▲ Problem Solving

53. To qualify for a loan of $200,000, an applicant must have a gross income of $72,000 if single, or $100,000 combined income if married. It is also necessary to have assets of at least $50,000. Write these statements symbolically, and decide whether Liz, who is single and has assets of $125,000, satisfies the conditions for obtaining a loan. She has an income of $58,000.

54. An airline advertisement states, "OBTAIN 40% OFF REGULAR FARE." Read the fine print:

    You must purchase your tickets between January 5 and February 15 and fly round trip between February 20 and May 3. You must also depart on a Monday, Tuesday, or Wednesday, and return on a Tuesday, Wednesday, or Thursday. You must also stay over a Saturday night.

    Write out these conditions symbolically.

55. The contract states, "No alterations, redecorating, tacks, or nails may be made in the building, unless written permission is obtained." Write this statement symbolically.

56. The contract states, "The tenant shall not let or sublet the whole or any portion of the premises to anyone for any purpose whatsoever, unless written permission from the landlord is obtained." Write this statement symbolically.

57. One day in a foreign country I met three politicians. Now, all of the politicians of this country belonged to one of two political parties. The first was the Veracious party, consisting of persons who could tell only the truth. The other party, called the Deceit Party, consisted of persons who were chronic liars. I asked these politicians to which party they belonged. The first said something I did not hear. The second remarked, "He said he belonged to the Veracious party." The third said, "You're a liar!" To which party did the third politician belong?

58. Translate into symbolic form: *Either Alfie is not afraid to go, or Bogie and Clyde will have lied.*

59. A man is about to be electrocuted but is given a chance to save his life. In the execution chamber are two chairs, labeled 1 and 2, and a jailer. One chair is electrified; the other is not. The prisoner must sit on one of the chairs, but before doing so, he may ask the jailer one question, to which the jailer must answer yes or no. The jailer is a consistent liar or else a consistent truth teller, but the prisoner does not know which. Knowing that the jailer either deliberately lies or faithfully tells the truth, what question should the prisoner ask?

▲ Individual Research

60. Between now and the end of the course, look for logical arguments in newspapers, periodicals, and books. Translate these arguments into symbolic form. Turn in as many of them as you find. Be sure to indicate where you found each argument.

2.4 THE NATURE OF PROOF

One of the greatest strengths of mathematics is its concern with the logical proof of its propositions. Any logical system must start with some undefined terms, definitions, and postulates or axioms. We have seen several examples of each of these. From here, other assertions can be made. These assertions are called theorems, and they must be proved using the rules of logic. In this section, we are concerned not so much with "proving mathematics" as with investigating the nature of proof in mathematics since the idea of proof does occupy a great portion of a mathematician's time.

In Chapter 1 we discussed inductive reasoning. Experimentation, guessing, and looking for patterns are all part of inductive reasoning. After a conjecture or generalization has been made, it needs to be proved using the rules of logic. After the conjecture is proved, it is called a **theorem**. Often several years will pass from the conjectural stage to the final proved form. Certain definitions must be made, and certain axioms or postulates must be accepted. Perhaps the proof of the conjecture will require the results of some previous theorems.

A **syllogism** is a form of reasoning in which two statements or premises are made and a logical conclusion is drawn from them. In this book, we will consider three types of syllogisms: direct reasoning, indirect reasoning, and transitive reasoning.

"I think you should be more explicit here in step two."

Direct Reasoning

The simplest type of syllogism is **direct reasoning**. This type of argument consists of two *premises*, or *hypotheses*, and a *conclusion*. For example,

$p \rightarrow q$	If you receive an A on the final, then you will pass the course.
p	You receive an A on the final.
$\therefore q$	Therefore, you pass the course.

The three-dot symbol $\therefore$ is used to symbolize the word *therefore*, which is used to separate the conclusion from the premises. We can use a truth table to prove direct reasoning. We begin by noting that the argument form can be rewritten as

$$[(p \rightarrow q) \wedge p] \rightarrow q$$

p	q	$p \rightarrow q$	$(p \rightarrow q) \wedge p$	$[(p \rightarrow q) \wedge p] \rightarrow q$
T	T	T	T	T
T	F	F	F	T
F	T	T	F	T
F	F	T	F	T

Since the argument is always true, we can write $[(p \rightarrow q) \wedge p] \Rightarrow q$, which proves the reasoning form called direct reasoning. It is also called **modus ponens, law of detachment,** or **assuming the antecedent.**

Direct Reasoning

Major premise: $p \rightarrow q$

Minor premise: p

Conclusion: $\therefore q$

■ **EXAMPLE 1**

Use direct reasoning to formulate a conclusion for each of the given arguments.

a. If you play chess, then you are intelligent.
   You play chess.

b. If $x + 2 = 3$, then $x = 1$.
   $x + 2 = 3$.

c. If you are a logical person, then you will understand this example.
   You are a logical person.

Solution

a. You are intelligent.

b. $x = 1$.

c. You understand this example.   —

Indirect Reasoning

The following syllogism illustrates what we call **indirect reasoning.**

$p \rightarrow q$ If you receive an A on the final, then you will pass the course.

$\sim q$ You did not pass the course.

$\therefore \sim p$ Therefore, you did not receive an A on the final.

We can prove this is valid by using a truth table (see Problem 6) or by using direct reasoning as follows:

$[(\sim q \rightarrow \sim p) \wedge \sim q] \Rightarrow \sim p$ Direct reasoning

$[(p \rightarrow q) \wedge \sim q] \Rightarrow \sim p$ Law of contraposition

This type of reasoning is also called **modus tollens** or **denying the consequent.**

Indirect Reasoning

Major premise: $p \rightarrow q$

Minor premise: $\sim q$

Conclusion: $\therefore \sim p$

■ **EXAMPLE 2**

Formulate a conclusion for each statement by using indirect reasoning.

a. If the cat takes the rat, then the rat will take the cheese.
The rat does not take the cheese.

b. If x is an even number, then $3x$ is an even number.
$3x$ is not an even number.

c. If you received an A on the test, then I am Napoleon.
I am not Napoleon.

Solution

a. The cat does not take the rat.

b. x is not an even number.

c. You did not receive an A on the test.

■

Transitive Reasoning

Sometimes we must consider some extended arguments. Transitivity allows us to reason through several premises to some conclusion. The argument form is given in the box.

Transitive Reasoning

Premise:	$p \rightarrow q$
Premise:	$q \rightarrow r$
Conclusion:	$\therefore p \rightarrow r$

We can prove transitivity by using a truth table. Notice that for three statements, we need a truth table with eight possibilities.

p	q	r	$p \rightarrow q$	$q \rightarrow r$	$p \rightarrow r$	$(p \rightarrow q) \wedge (q \rightarrow r)$	$[(p \rightarrow q) \wedge (q \rightarrow r)] \rightarrow (p \rightarrow r)$
T	T	T	T	T	T	T	T
T	T	F	T	F	F	F	T
T	F	T	F	T	T	F	T
T	F	F	F	T	F	F	T
F	T	T	T	T	T	T	T
F	T	F	T	F	T	F	T
F	F	T	T	T	T	T	T
F	F	F	T	T	T	T	T

↑
All Ts

Since transitivity is always true, we may write

$$[(p \rightarrow q) \wedge (q \rightarrow r)] \Rightarrow (p \rightarrow r)$$

■■■■■ **EXAMPLE 3**

Formulate a conclusion for each argument.

a. If you attend class, then you will pass the course.
   If you pass the course, then you will graduate.
b. If you graduate, then you will get a good job.
   If you get a good job, then you will meet the right people.
   If you meet the right people, then you will become well known.
c. If $x + 2x + 3 = 9$ then $3x + 3 = 9$.
   If $3x + 3 = 9$, then $3x = 6$.
   If $3x = 6$, then $x = 2$.

Solution

a. If you attend class, then you will graduate.
b. The transitive law can be extended to several premises: If you graduate, then you will become well known.
       Notice that we can apply the transitive law to both parts **a** and **b**: If you attend class, then you will become well known.
c. If $x + 2x + 3 = 9$, then $x = 2$.   ▬

Logical Proof

The point of studying these various argument forms is to acquire the ability to apply them for longer and more involved arguments.

■■■■■ **EXAMPLE 4**

Form a valid conclusion using all these statements. We number the premises for easy reference.

1. If I receive a check for $500, then we will go on vacation.
2. If the car breaks down, then we will not go on vacation.
3. The car breaks down.

Solution   Begin by changing the argument into symbolic form.

1. $c \rightarrow v$ where c: I receive a $500 check.
                               v: We will go on vacation.
2. $b \rightarrow \sim v$ b: The car breaks down.
3. b

Next, simplify the argument. For this example, we rearrange the premises. Notice the numbers to help you keep track of the premises.

2. $b \rightarrow \sim v$		Given
1′. $\underline{\sim v \rightarrow \sim c}$		Contrapositive of the first premise
$\therefore\ b \rightarrow \sim c$		Transitive
3. $\underline{b}$		Given
$\therefore\ \sim c$		Direct reasoning

Finally, we translate the conclusion back into words: *I did not receive a check for* $500. ▬

■■■ **EXAMPLE 5**   Polya's Method

Form a valid conclusion using all the statements.*

1. All unripe fruit is unwholesome.
2. All these apples are wholesome.
3. No fruit grown in the shade is ripe.

Solution   We use Polya's problem-solving guidelines for this example.

Understand the Problem.   The forms in this argument are not exactly like those we are used to seeing. We recall that the statement *all p is q* can be translated as *if p then q*.

Devise a Plan.   The procedure we will use is to (Step 1) translate into symbols, (Step 2) simplify using logical arguments, and finally (Step 3), translate the symbolic form back into English.

Carry Out the Plan.

Step 1:

1. $\sim r \rightarrow (\sim w)$ where r: This fruit is ripe.
2. $a \rightarrow w$ w: This fruit is wholesome.
3. $s \rightarrow (\sim r)$ a: This fruit is an apple.
                                     s: This fruit is grown in the shade.

Note: The sentence "No p is q" is translated as $p \rightarrow (\sim q)$.

Step 2:

Let $(1')$, $w \rightarrow r$, replace (1) by the law of contraposition.
Let $(3')$, $r \rightarrow \sim s$, replace (3) by the law of contraposition.
Rearranging the premises, we have:

2. $a \rightarrow w$
$1'.$ $w \rightarrow r$
――――――――――
      $\therefore a \rightarrow r$ Transitive
$3'.$ $r \rightarrow (\sim s)$
――――――――――
      $a \rightarrow (\sim s)$ Transitive

Step 3:

   Conclusion: *All these apples were not grown in the shade.* If we assume that $\sim s$ = grown in the sun, then the conclusion can be stated more simply: *All these apples were grown in the sun.*

Look Back.   Does this conclusion seem reasonable?

* From Lewis Carroll, *Symbolic Logic and The Game of Logic* (New York: Dover Publications, 1958).

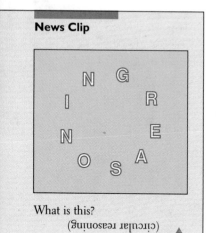
Fallacies

Sometimes *invalid arguments* are given. The remainder of this section is devoted to some of the more common **logical fallacies**.

Fallacy of the Converse

▬▬ EXAMPLE 6

Show that the following argument is not valid.

> If a person reads the *Times*, then she is well informed.
> This person is well informed.
> Therefore, this person reads the *Times*.

Solution   This argument has the following form:

$$p \rightarrow q$$
$$\underline{q}$$
$$\therefore p$$

By considering the associated truth table, we can test the validity of our argument.

p	q	$p \rightarrow q$	$(p \rightarrow q) \wedge q$	$[(p \rightarrow q) \wedge q] \rightarrow p$
T	T	T	T	T
T	F	F	F	T
F	T	T	T	F
F	F	T	F	T

We see that the result is not always true; thus the argument is invalid.   ▬

If $p \rightarrow q$ were replaced by $q \rightarrow p$, the argument in the preceding example would become valid. That is, the argument would be valid if the direct statement and the converse had the same truth values, which in general they do not. For this reason the argument is sometimes called the **fallacy of the converse**.

We can often show that a given argument is invalid by finding a **counterexample**. In the preceding example we found a counterexample by looking at the truth table. The entry in the third row is false, so the argument can be shown to be false in the case in which p is false and q is true. In terms of this example, a person could never see the *Times* (p false) and still be well informed (q true).

Fallacy of the Inverse

Consider the following argument:

> If a person reads the *Times*, then he is well informed.
> This person does not read the *Times*.
> Therefore, this person is not well informed.

As we have seen, a person who reads the *Tribune* might also be well informed. This line of reasoning is called the **fallacy of the inverse** (sometimes also called the *fallacy of denying the antecedent*). A truth table for

$$[(p \rightarrow q) \wedge (\sim p)] \rightarrow (\sim q)$$

shows that the fallacy of the inverse is not valid. (The truth table is left as a problem.)

▬▬ EXAMPLE 7

Test the validity of the following argument.

> If a person goes to college, he will make a lot of money.
>
> You do not go to college.
>
> Therefore, you will not make a lot of money.

Solution   This argument has the following form:

$$p \rightarrow q$$
$$\underline{\sim p}$$
$$\therefore \sim q$$

We recognize this reasoning as the fallacy of the inverse (the fallacy of denying the antecedent). ▬

False Chain Pattern

In certain areas of this country, it is thought that thunderstorms cause milk to sour. This belief is an example of the fallacy we will call the **false chain pattern**. It can be shown as follows:

> Hot, humid weather favors thunderstorms.
>
> Hot, humid weather favors bacterial growth, which causes milk to sour.
>
> Therefore, thunderstorms cause milk to sour.

The false chain pattern is illustrated by

$$p \rightarrow q$$
$$\underline{p \rightarrow r}$$
$$\therefore q \rightarrow r$$

and can be proved invalid by construction of an appropriate truth table.

The three types of common fallacies that we have discussed can be summarized as follows:

Fallacy of the Converse (assuming the consequent)	*Fallacy of the Inverse* (denying the antecedent)	*False Chain Pattern*
$p \rightarrow q$	$p \rightarrow q$	$p \rightarrow q$
$\underline{q}$	$\underline{\sim p}$	$\underline{p \rightarrow r}$
$\therefore p$	$\therefore \sim q$	$\therefore q \rightarrow r$

Study these fallacies, and notice how they differ from direct reasoning, indirect reasoning, and transitivity.

PROBLEM SET 2.4

▲ **A Problems**

1. **IN YOUR OWN WORDS** Explain what we mean by direct reasoning.
2. **IN YOUR OWN WORDS** Explain what we mean by indirect reasoning.
3. **IN YOUR OWN WORDS** Explain what we mean by transitive reasoning.
4. **IN YOUR OWN WORDS** What do we mean by logical fallacies?
5. **IN YOUR OWN WORDS** What is a syllogism?
6. Prove $[(p \rightarrow q) \wedge \sim q] \rightarrow \sim p$. What is the name we give to this type of reasoning?
7. By constructing a truth table, show that
$$[(p \rightarrow q) \wedge (p \rightarrow r)] \rightarrow (q \rightarrow r)$$
   is an invalid argument. What is the name of this fallacy?
8. Prove that $[(p \rightarrow q) \wedge \sim p] \rightarrow \sim q$ is an invalid argument. What is the name of this fallacy?
9. **IN YOUR OWN WORDS** There are similarities between Polya's problem-solving method and the steps a detective will go through in solving a case. Rewrite each of these problem-solving steps using the language that a detective solving a case might use. (See Problems 61–62 below for examples of detective-type problems.)

Understand the Case.

> What are you looking for?

Investigate the Case.

> Have you solved a similar case?
> What are the facts?

Analyze the Facts/Data.

> What information is important?
> What information is not important?
> What pieces of information do not seem to fit together logically?
> Which data are inconsistent with the given information?

Reexamine the Facts.

> Do the facts support the solution?
> Can we obtain a conviction?

Determine whether each argument in Problems 10–18 is valid or invalid. Give reasons for your answer.

10. $p \rightarrow q$
$\sim q$
$\therefore \sim p$

11. $p \rightarrow q$
q
$\therefore p$

12. $p \vee q$
$\sim p$
$\therefore q$

13. $p \rightarrow q$
$\sim p$
$\therefore \sim q$

14. $p \rightarrow q$
p
$\therefore q$

15. $p \lor q$
$\underline{\sim q}$
$\therefore p$

16. $p \to \sim q$
$\underline{q\qquad}$
$\therefore \sim p$

17. $\sim p \to q$
$\underline{\sim p}$
$\therefore q$

18. $p \to q$
$\underline{q \to r}$
$\therefore p \to r$

Determine whether each argument in Problems 19–41 is valid or invalid. If valid, name the type of reasoning and if invalid, determine the error in reasoning.

19. If I inherit $1,000, I will buy you a cookie.
    I inherit $1,000.
    Therefore, I will buy you a cookie.

20. If a^2 is even, then a must be even.
    a is odd.
    Therefore, a^2 is odd.

21. All snarks are fribbles.
    All fribbles are ugly.
    Therefore, all snarks are ugly.

22. All cats are animals.
    This is not an animal.
    Therefore, this is not a cat.

23. If I don't get a raise in pay, I will quit.
    I don't get a raise in pay.
    Therefore, I quit.

24. If Fermat's Last Theorem is ever proved, then my life is complete.
    My life is not complete.
    Therefore, Fermat's Last Theorem is not proved.

25. If Congress appropriates the money, the project can be completed.
    Congress appropriates the money.
    Therefore, the project can be completed.

26. If Alice drinks from the bottle marked "poison," she will become sick.
    Alice does not drink from a bottle that is marked "poison."
    Therefore, she does not become sick.

27. Blue-chip stocks are safe investments.
    Stocks that pay a high rate of interest are safe investments.
    Therefore, blue-chip stocks pay a high rate of interest.

28. If Al understands logic, then he enjoys this sort of problem.
    Al does not understand logic.
    Therefore, Al does not enjoy this sort of problem.

29. If Al understands a problem, it is easy.
    This problem is not easy.
    Therefore, Al does not understand this problem.

30. If Mary does not have a little lamb, then she has a big bear.
    Mary does not have a big bear.
    Therefore, Mary has a little lamb.

31. If $2x - 4 = 0$, then $x = 2$.
    $x \neq 2$
    Therefore, $2x - 4 \neq 0$.

32. If Todd eats Krinkles cereal, then he has "extra energy."
    Todd has "extra energy."
    Therefore, Todd eats Krinkles cereal.

33. If Ron uses Slippery oil, then his car is in good running condition.
    Ron's car is in good running condition.
    Therefore, Ron uses Slippery oil.

34. If you get a fill-up of gas, you will get a free car wash.
    You get a fill-up of gas.
    Therefore, you will get a free car wash.

35. If the San Francisco 49ers lose, then the Dallas Cowboys win.
    If the Dallas Cowboys win, then they will go to the Super Bowl.
    Therefore, if the San Francisco 49ers lose, then the Dallas Cowboys will go to the
      Super Bowl.

36. If Missy uses Smiles toothpaste, then she has fewer cavities.
    Therefore, if Missy has fewer cavities, then she uses Smiles toothpaste.

37. All mathematicians are eccentrics.
    All eccentrics are rich.
    Therefore, all mathematicians are rich.

38. (Let x be an integer and y a nonzero integer for this argument.)
    If Q is a rational number, then $Q = \frac{x}{y}$, where $\frac{x}{y}$ is a reduced fraction.
    $Q \neq \frac{x}{y}$, where $\frac{x}{y}$ is a reduced fraction.
    Therefore, Q is not a rational number.

39. If you like beer, you'll like Bud.   40. No students are enthusiastic.
    You don't like Bud.   You are enthusiastic.
    Therefore, you don't like beer.   Therefore, you are not a student.

41. If the crime occurred after 4:00 A.M., then Smith could not have done it.
    If the crime occurred at or before 4:00 A.M., then Jones could not have done it.
    The crime involved two persons, if Jones did not commit the crime.
    Therefore, if Smith committed the crime, it involved two persons.

▲ **B Problems**

In Problems 42–60, form a valid conclusion, using all the statements for each argument. Give reasons.

42. If you learn mathematics, then you are intelligent.
    If you are intelligent, then you understand human nature.

43. If I am idle, then I become lazy.
    I am idle.

44. If we go to the concert, then we are enlightened.
    We are not enlightened.

45. If you climb the highest mountain, then you feel great.
    If you feel great, then you are happy.

46. All Trebbles are Frebbles.
    All Frebbles are expensive.

47. $a = 0$ or $b = 0$ 48. If $a \cdot b = 0$, then $a = 0$ or $b = 0$.
    $a \neq 0$ $a \cdot b = 0$

49. If we interfere with the publication of false information, we are guilty of suppressing
      the freedom of others.
    We are not guilty of suppressing the freedom of others.

50. If 2 divides a positive integer N and if N is greater than 2, then N is not a prime
      number.
    N is a prime number.

51. If I eat that piece of pie, I will get fat.
    I will not get fat.

52. If a nail is lost, then a shoe is lost.
    If a shoe is lost, then a horse is lost.
    If a horse is lost, then a rider is lost.
    If a rider is lost, then a battle is lost.
    If a battle is lost, then a kingdom is lost.

53. If we win first prize, we will go to Europe.
    If we are ingenious, we will win first prize.
    We are ingenious.

54. If I am tired, then I cannot finish my homework.
    If I understand the material, then I can finish my homework.

55. Babies are illogical.
    Nobody is despised who can manage
      a crocodile.
    Illogical persons are despised.

56. All hummingbirds are richly colored.
    No large birds live on honey.
    Birds that do not live on honey are dull
      in color.

57. No ducks waltz.
    No officers ever decline to waltz.
    All my poultry are ducks.

58. Everyone who is sane can do logic.
    No lunatics are fit to serve on a jury.
    None of your sons can do logic.

59. If the government awards the contract to Airfirst Aircraft Company, then Senator
      Firstair stands to earn a great deal of money.
    If Airsecond Aircraft Company does not suffer financial setbacks, then Senator Firstair
      does not stand to earn a great deal of money.
    The government awards the contract to Airfirst Aircraft Company.

60. If you go to college, then you get a good job.
    If you get a good job, then you make a lot of money.
    If you do not obey the law, then you do not make a lot of money.
    You go to college.

▲ **Problem Solving**

61. "The Case of the Dead Professor"*

    The detective, Columbo, had just arrived at the scene of the crime and found that the professor
    had been at the lab working for hours. He seemed to have electrocuted himself and ended up
    blowing the fuses for the whole building. Later in the night, the janitor came to clean up the lab
    and found the professor's body. Columbo suspected the janitor, but when he was questioned, he
    vehemently denied killing the professor. He said, "I came to work late. When I got off the eleva-
    tor, I found the professor dead with his head on the lab table. I wish I could tell you more, but
    that is all I know."
        Then Columbo said, "I think you can tell us more at headquarters." Downtown under gruel-
    ing interrogation, the janitor confessed. What made Columbo suspect the janitor?

62. "The Case of the Tumbled Tower"*

    Dwayne got up at 6:00 A.M. and was watching the sun rise from his bedroom window. After the
    sun came up, he started working on his toothpick tower in his room. The tower was very fragile.
    While he was working on his tower, his little brother came into the room bugging him. Dwayne's
    brother wanted to be more like Dwayne and he wanted to build something out of toothpicks,
    too. He was very jealous of Dwayne.

* Problems 61 and 62 were adapted from "Solving the Mystery," by Frances R. Curicio and J. Lewis
McNeece, *The Mathematics Teacher*, November 1993, pp 682–685.

In the afternoon, Dwayne went out to buy candy at the candy store and he left his little brother home (even though he was supposed to be babysitting). On his way out of the house, he heard on the radio that there was going to be a slight westerly wind coming. He then left the house, forgetting that he had left his bedroom window open. When he returned and saw that he had left the window open, he thought that the wind had blown the tower over. But then he remembered something and said that his little brother must have knocked over the tower. What did he remember?

Form a valid conclusion using all of the statements in Problems 63–65.

63. No kitten that loves fish is unteachable.
    No kitten with a tail will play with a gorilla.
    Kittens with whiskers always love fish.
    No teachable kitten has green eyes.
    No kittens without whiskers have tails.

64. When I work a logic problem without grumbling, you may be sure it is one that I can understand.
    These problems are not arranged in regular order, like the problems I am used to.
    No easy problem ever makes my head ache.
    I can't understand problems that are not arranged in regular order, like those I am used to.
    I never grumble at a problem unless it gives me a headache.

65. Every idea of mine that cannot be expressed as a syllogism is really ridiculous.
    None of my ideas about rock stars is worth writing down.
    No idea of mine that fails to come true can be expressed as a syllogism.
    I never have any really ridiculous idea that I do not at once refer to my lawyer.
    All my dreams are about rock stars.
    I never refer any idea of mine to my lawyer unless it is worth writing down.

2.5   PROBLEM SOLVING USING LOGIC

In Chapter 1 we laid the foundation for problem solving. The key to building problem-solving skills is to encounter problem solving in a variety of contexts. We will now apply another method of proof to solving some logic puzzles that have been around for a long time but that nevertheless continue to challenge the lay reader.

■■■■ EXAMPLE I   Polya's Method

Odd Ball Problem   You are given nine steel balls of the same size and color. One of the nine balls is slightly heavier in weight; the others all weigh the same. Using a two-pan balance, what is the minimum number of weighings necessary to find the ball of different weight?

Solution   We use Polya's problem-solving guidelines for this example.

Understand the Problem.   Do you understand the terminology? Are you familiar with a two-pan balance?

Devise a Plan.   One of the techniques for Polya's method is to guess and test. That is the method we will use here. We will find a solution by trial and error (that is, guess), and then test it. Next we will see whether we can find a solution with fewer weighings. We will accomplish this by first solving some simpler problems.

Carry Out the Plan.   Suppose we had two steel balls. Then one weighing would suffice. What about three steel balls? One weighing would still suffice. How? Put one ball on each pan. If it doesn't balance, then you have found the heavier one. If it balances, then the one not weighed is the heavier one. Does this give you an idea for the nine balls?

Let's try two weighings:

1. Divide the 9 steel balls into 3 groups of 3. First weighing: Balance 3 balls against 3 balls.
   a. The weighing either balances or doesn't balance (law of excluded middle).
   b. If it balances, then the heavier one is in the group not weighed. If it doesn't balance, then take the group with the heavier ball.
2. Divide the 3 steel balls from the heavier group. Second weighing: Balance 1 ball against 1 ball.
   a. The weighing either balances or doesn't balance. (Why?)
   b. If it doesn't balance, then the heavier ball is the one that tips the scale. If it does balance, then the heavier ball is the one not weighed.

Look Back.   Two weighings is the solution.   ▬

▬▬■ EXAMPLE 2   **Polya's Method**

Liar Problem   On the South Side, a member of the mob had just knocked off a store. Since the boss had told them all to lay low, he was a bit mad. He decided to have a talk with the boys.

From the boys' comments, can you help the Boss figure out who committed the crime? Assume that the Boss is telling the truth.

Solution   We use Polya's problem-solving guidelines for this example.

Understand the Problem.

Let a: Alfie did it.

b: Bogie did it.

c: Clyde did it.

d: Dirty Dave did it.

f: Fingers did it.

Translate the sentences into symbolic form:

Alfie said:	$b \vee c$
Bogie said:	$\sim(f \vee b)$
Clyde said:	$[\sim(b \vee c)] \wedge \sim[\sim(f \vee b)]$
Dirty Dave said:	$[(b \vee c) \wedge (f \vee b)] \vee \{\sim(b \vee c) \wedge [\sim(f \vee b)]\}$
Fingers said:	$\sim([(b \vee c) \wedge (f \vee b)] \vee \{\sim(b \vee c) \wedge [\sim(f \vee b)]\})$

Assume that a is true, and check all these statements. Next, assume that b is true, and check all the statements. Do the same for c, d, and f. Since we assume the boss is telling the truth, we look for those cases (if any) in which three are truthful and two are lying.

Devise a Plan.

We can summarize our analysis by using a matrix (a rectangular array) with vacant cells for all possible pairings of the elements in each set.

	a	b	c	d	f
a					
b					
c					
d					
f					

Let us assume that the a at the left means that we are making the assumption that Alfie did it. Then we place a 1 or a 0 in each position to indicate the truth or falsity, respectively, of each witness' statement with the assumption that person x (as listed at the left) is guilty.

Carry Out the Plan.

Conclusion:		Witness statements:				
		a	b	c	d	f
Alfie did it:	a	0	1	0	1	0
Bogie did it:	b	1	0	0	1	0
Clyde did it:	c	1	1	0	0	1
Dave did it:	d	0	1	0	1	0
Fingers did it:	f	0	0	1	0	1

Next we assume that Bogie did it, and we fill in the second row of the matrix. We continue until the matrix is complete, as shown. Now if we take the boss's statement as the major premise, we see that in only one case are there three truthful mobsters and two liars. Thus we can say that Clyde knocked off the store.

Look Back. Assume that Clyde knocked off the store and check each of the mob's comments to see that in this case there are three truthful mobsters and two liars.   ▬

▬▬▬ **EXAMPLE 3**   **Polya's Method**

Prisoner Problem   During an ancient war three prisoners were brought into a room. In the room was a large box containing three white hats and two black hats. Each man was blindfolded, and one of the hats was placed on his head. The men were lined up, one behind the other, facing the wall. The blindfold of the man farthest from the wall was removed, and he was permitted to look at the hats of the two men in front of him. If he knew (not guessed) the color of the hat on his head, he would be freed. However, he was unable to tell. The blindfold was then taken from the head of the next man, who could see only the hat of the one man in front of him. This man had the same chance for freedom, but he, too, was unable to tell the color of his hat. The remaining man then told the guards the color of the hat he was wearing and was released. What color hat was he wearing, and how did he know?

Solution   We use Polya's problem-solving guidelines for this example.

Understand the Problem.   The situation looks like this:

A   B   C

Devise a Plan.   We consider what the last man in line could see, and then decide which of these possibilities fit the conditions of what the middle man in the line saw.

Carry Out the Plan.

1. What did man C see? There are four possibilities:

	I	II	III	IV
A	white	white	black	black
B	white	black	white	black

If he had seen possibility IV, he would have known that he had on a white hat. (Why?) He did not know, so it must be possibility I, II, or III.

2. What did man B see? If he had seen a black hat, then he would have known that his hat was white. (Why?) This rules out possibility III.
3. Since both of the remaining possibilities, I and II, have a white hat on the front man, man A *knew* he had a white hat.

Look Back.   Reread the question to see whether all information of the problem is satisfied.

■ EXAMPLE 4   Polya's Method

Reporter's Error Problem   Emor D. Nilap, a news reporter who was a little backward, was sent to cover a billiards tournament. Since he wanted to do a good job, he rounded up some human-interest facts to make his story a little more interesting. He gathered the following information:

1. The men competing were Pat, Milt, Dick, Joe, and Karl.
2. Milt had once beaten the winner at tennis.
3. Pat and Karl frequently played cards together.
4. Joe's finish ahead of Karl was totally unexpected.
5. The man who finished fourth left the tournament after his match and did not see the last two games.
6. The winner had not met the man who came in fifth prior to the day of the tournament.
7. The winner and the runner-up had never met until Joe introduced them just before the final game.

When Emor returned to his office, he found that he had forgotten the order in which the men had finished but he was able to figure it out from the information provided. Can you figure it out?

Solution   We use Polya's problem-solving guidelines for this example.

Understand the Problem.   We need to find the order in which the men finished the tournament.

Devise a Plan.   The procedure we will use is to form a matrix of all possibilities and then use the rules of logic to eliminate the parts that are impossible; what remains is the solution.

Carry Out the Plan.   From (1) we are able to make the following matrix of the possible solutions:

Pat	Milt	Dick	Joe	Karl
1	1	1	1	1
2	2	2	2	2
3	3	3	3	3
4	4	4	4	4
5	5	5	5	5

The steps are listed with the "work" shown after the last step.

  a. We are given seven premises. We begin by the process of elimination; that is, we cross out those positions that it would be impossible for the men to occupy. The order in which we work the problem is not unique.
  b. From (7), Joe is not the winner or the runner-up. Thus we cross out 1 and 2 under Joe's name and label those places b, as shown below.
  c. By (5), Joe did not finish fourth; cross out and label c.
  d. Joe finished ahead of Karl, by (4); therefore Karl was not first, second, or third. By the same premise, Joe could not have been last. (Why?)
  e. Therefore Joe finished third, which means that Pat, Milt, and Dick are not in third place.
  f. By (2), Milt was not the winner.
  g. By (7), Milt was not the runner-up.
  h. By (6), Milt was not fifth.
  i. Therefore Milt finished fourth, which means that Pat, Dick, and Karl are not in fourth place.
  j. Therefore Karl finished fifth, which means that the others did not finish fifth.
  k. By (6) and the fact that Karl finished last, along with (3), we see that Pat could not have finished first.
  ℓ. Therefore Pat is second, which means that Dick was not second.
  m. Therefore Dick finished first, and the problem is finished.

Pat	Milt	Dick	Joe	Karl
X^k	X^f	$\boxed{1}^m$	X^b	X^d
$\boxed{2}^\ell$	Z^g	Z^e	Z^b	Z^d
3^e	3^e	3^e	$\boxed{3}^e$	3^d
4^i	$\boxed{4}^i$	4^i	4^c	4^i
5^i	5^h	5^i	5^d	$\boxed{5}^i$

Look Back. We have found that Dick finished first, Pat second, Joe third, Milt fourth, and Karl last. Reread each of the statements to make sure this solution does not cause any inconsistencies. ▬

▬▬▬ **EXAMPLE 5**   **Polya's Method**

Turkey Problem   A man I know once owned a number of turkeys. One day, one of his gobblers flew over the man's fence and laid an egg on a neighbor's property. To whom did the egg belong — to the man who owned the gobbler, to the gobbler, or to the neighbor?

Solution   We use Polya's problem-solving guidelines for this example.

Understand the Problem.   To understand the question, you must understand the terminology. Since a turkey gobbler is a male turkey, it could not lay an egg. You must be careful to use common sense along with the rules of logic when answering puzzle problems. ▬

PROBLEM SET 2.5

▲ **Problem Solving**

1. **The Hat Game**   Harry, Larry, and Moe are sitting in a circle so that they see each other. The game is played by a judge placing either a black or a white hat on each person's head so that each person can see the others' hats, but not the hat on his or her own head. Now when the judge says "GO," all players who see a black hat must raise their hands. The first player to deduce (not guess) the color of his own hat wins the game. Now, the judge put a black hat on each player's head, and then said, "GO." All players raised their hands and after a few moments Moe said he knew his hat was black. How did he deduce this?

2. **Broken Window Problem**   Three children were playing baseball; their names were Alice, Bob, and Cole. One of them hit a home run and broke your expensive plate glass window, and you went out to question the children. Each child made two statements.

> *Alice:*   "Cole didn't do it. Bob did it."
>
> *Bob:*   "I didn't do it. Cole did it."
>
> *Cole:*   "I didn't do it. Alice did it."

Now if you know that one of the three always tells the truth, one always lies, and the other tells the truth half the time, can you point to the guilty party?

3. **The Marble Players**   Four boys were playing marbles; their names were Gary, Harry, Iggy, and Jack. One of the boys had 9 marbles, another had 15, and each of the other two had 12. Their ages were 3, 10, 17, and 18, but not respectively. Gary shot before Harry and Jack. Jack was older than the boy with 15 marbles. Harry had fewer than 15 marbles. Jack shot before Harry. Iggy shot after Harry. If Iggy was 10 years old, he did not have 15 marbles. Gary and Jack together had an even number of marbles. The youngest boy was not the one with 15 marbles. If Harry had 12 marbles, he wasn't the youngest. The 10-year-old shot after the 17-year-old. In what order did they shoot, and how old was each boy?

4. **Bear Problem**   A fox, hunting for a morsel of food, spotted a huge bear about 100 yards due east of him. Before the hunter could become the hunted, the crafty fox ran due north for 100 yards but then realized the bear had not noticed him. Thus he stopped and remained hidden. At this point the bear was due south of the fox. What was the color of the bear?

5. **Flower Garden**   I visited a beautiful flower garden yesterday and counted exactly 50 flowers. Each flower was either red or yellow, and the flowers were not all the same color. My friend made the following observation: No matter which two flowers you might have picked, at least one was bound to be red. From this can you determine how many were red and how many were yellow?

6. **Sock Problem**   I have a habit of getting up before the sun rises. My socks are all mixed up in the drawer, which contains 10 black and 20 blue socks. I reach into the drawer and grab some socks in the dark. How many socks do I need to take from the drawer to be sure that I have a matched pair?

7. **Whodunit?**   Daniel Kilraine was killed on a lonely road, two miles from Pontiac, at 3:30 A.M. on March 3, 1997. Otto, Curly, Slim, Mickey, and The Kid were arrested a week later in Detroit and questioned. Each of the five made four statements, three of

which were true and one of which was false. One of these men killed Kilraine. Who-dunit? Their statements were:*

   Otto: "I was in Chicago when Kilraine was murdered. I never killed anyone. The Kid is the guilty man. Mickey and I are pals."

   Curly: "I did not kill Kilraine. I never owned a revolver in my life. The Kid knows me. I was in Detroit the night of March 3rd."

   Slim: "Curly lied when he said he never owned a revolver. The murder was committed on March 3rd. One of us is guilty. Otto was in Chicago at the time."

   Mickey: "I did not kill Kilraine. The Kid has never been in Pontiac. I never saw Otto before. Curly was in Detroit with me on the night of March 3rd."

   The Kid: "I did not kill Kilraine. I have never been in Pontiac. I never saw Curly before. Otto lied when he said I am guilty."

8. **Coin Problem**   Suppose you are given 12 coins, one of which is counterfeit and weighs a little more or less than the real coins. Using a two-pan balance scale, what is the minimum number of weighings necessary to find the counterfeit coin? (If you are interested in a general solution to this type of problem, see T. H. O'Beirne's book, *Puzzles and Paradoxes*, Oxford University Press, New York, 1965, Chapters 2 and 3.)

9. **Mixed Bag Problem**   Suppose I have three bags, one with two peaches, another with two plums, and a third mixed bag with one peach and one plum. Now I give the bags (in mixed-up order) to Alice, Betty, and Connie. I tell the three to look into their bags and that I want each to make a statement about the contents of her bag, but I want them to lie. Here is what they say:

   Alice: I have two peaches.

   Betty: I have two plums.

   Connie: I have one peach and one plum.

Now here is the game. I want you to develop a strategy of asking one of the three to reach into her bag and pull out one fruit and show it to you. The fruit is then returned to the bag and you ask another to do the same thing. Continue until you can deduce which bag is the mixed bag. What is the minimum possible number of necessary moves? Explain.

10. **Rectangle Tangle**   Fill in each blank with a digit so that every statement is true.†

> In this rectangle (including the smaller rectangles), the number of occurrences of the digit 1 is __a__, of 2 is __b__, and of 7 is __c__.
>
> > In this rectangle (including the smaller rectangles), the number of occurrences of the digit 1 is __d__, of 2 is __e__, and of 7 is __f__.
> >
> > > In this rectangle (including the smaller rectangles), the number of occurrences of the digit 3 is __g__, of 4 is __h__, of 5 is __i__, and of 6 is __j__.
> > >
> > > > In this rectangle (including the smaller rectangle), the number of occurrences of the digit 1 is __k__, of 2 is __ℓ__, of 3 is __m__, and of 5 is __n__.
> > > >
> > > > > In this rectangle, the number of occurrences of the digit 1 is __p__, of 2 is __q__, of 3 is __r__, and of 4 is __s__.

* Reprinted with permission of The Macmillan Company from *Introduction to Logic* (2nd ed.), by Irving Copi. Copyright © 1961 by The Macmillan Company.
† By Guney Mentes in *Games*, August 1996, p. 43.

11. **Baseball Problem**   Nine men play the positions on a baseball team. Their names are Brown, White, Adams, Miller, Green, Hunter, Knight, Smith, and Jones. Determine from the following information the position played by each man.
   a. Brown and Smith each won $10 playing poker with the pitcher.
   b. Hunter is taller than Knight and shorter than White, but all three weigh more than the first baseman.
   c. The third baseman lives across the corridor from Jones in the same apartment house.
   d. Miller and the outfielders play bridge in their spare time.
   e. White, Miller, Brown, the right fielder, and the center fielder are bachelors, and the rest are married.
   f. Of Adams and Knight, one plays an outfield position.
   g. The right fielder is shorter than the center fielder.
   h. The third baseman is a brother of the pitcher's wife.
   i. Green is taller than the infielders, the pitcher, and the catcher except for Jones, Smith, and Adams.
   j. The second baseman beat Jones, Brown, Hunter, and the catcher at cards.
   k. The third baseman, the shortstop, and Hunter each made $150 speculating in General Motors stock.
   ℓ. The second baseman is engaged to Miller's sister.
   m. Adams lives in the same apartment house as his own sister but dislikes the catcher.
   n. Adams, Brown, and the shortstop each lost $200 speculating in grain.
   o. The catcher has three daughters, the third baseman has two sons, and Green is being sued for divorce.

2.6   LOGIC CIRCUITS

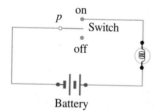

▲ **Figure 2.1   Schematic diagram showing how a light might be connected to a switch**

One of the applications of symbolic logic that is easiest to understand is that of an electrical circuit (see Figure 2.1). Consider the following symbols that we use for the parts of a circuit:

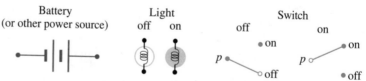

There are two types of circuits for connecting two switches together; the first is called a **series** circuit, and the other is called a **parallel** circuit.

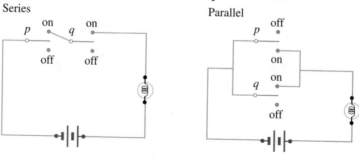

▬▬ EXAMPLE 1

How can symbolic logic be used to describe circuits?

Solution   We use Polya's problem-solving guidelines for this example.

Understand the Problem.   When two switches are connected in *series*, current will flow only when both switches are closed. When they are connected in *parallel*, current will flow if either switch is closed.

Devise a Plan.   Let switch p be considered as a logical proposition that is either true or false. If p is true, then the switch will be considered closed (current flows), and if p is false, then the switch will be considered open (current does not flow).

Carry Out the Plan.   A series circuit of two switches p and q can be specified as a logical statement $p \wedge q$. A parallel circuit can be specified as a logical statement $p \vee q$.

Look Back.   Each circuit has four possible states, as shown.

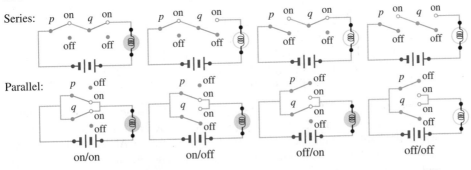

The three fundamental operators are conjunction, disjunction, and negation. In Example 1 we saw that a series circuit can be represented as a conjunction, and a parallel circuit as a disjunction. The circuit for negation is relatively easy, as is shown in Figure 2.2.

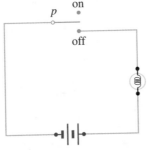

▲ **Figure 2.2 Negation circuit**

It is often necessary to combine circuits. We symbolically represent these as **gates.** The series circuit is called an **AND-gate,** the parallel circuit is called an **OR-**

gate, and the negation circuit is called a **NOT-gate.** The notation for each of these is shown in Figure 2.3.

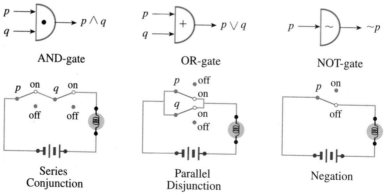

▲ **Figure 2.3   Circuits and gates**

p	q	$p \wedge q$	$p \vee q$	$\sim p$
T; on	T; on	T; light on	T; light on	F; light off
T; on	F; off	F; light off	T; light on	F; light off
F; off	T; on	F; light off	T; light on	T; light on
F; off	F; off	F; light off	F; light off	T; light on

Notice that the NOT-gate has a single input (proposition p) and a single output (proposition $\sim p$). That is, current will flow out of the NOT-gate in those cases in which the light is on for the negation circuit. Similarly, the AND-gate and OR-gate have two switches, p and q. These are symbolized as two inputs, p and q. The single output stands for the light.

We can construct logic circuits to simulate logical truth tables. For example, suppose we wish to find the truth values for $\sim(p \vee q)$. We should design the following circuit (remember that parentheses indicate the operation to be performed first):

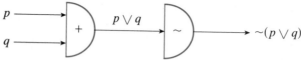

EXAMPLE 2

Design a circuit for $\sim p \wedge (\sim q)$.

Solution

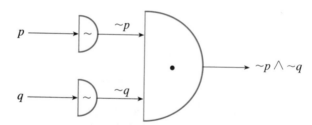

■■■■ EXAMPLE 3

Design a circuit that will find the truth values for $(p \vee q) \wedge q$.

Solution We use gates to symbolize $(p \vee q) \wedge q$:

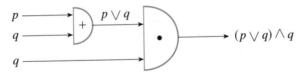

Next we must have some way of determining when the output values of $(p \vee q) \wedge q$ are true and when they are false. We can do this by again connecting a light to the circuit. When $(p \vee q) \wedge q$ is true, the light should be on; when it is false, the light should be off. We represent this connection as follows:

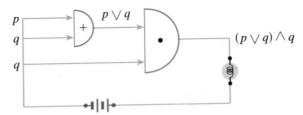

You may have noticed the relationship among sets, logic, and circuits. These relationships are summarized in Table 2.10.

TABLE 2.10 Sets, Logic, and Circuits

Sets	Logic	Circuits
Elements: A, B, C, . . .	Propositions: $p, q, r, \ldots$	Switches
Intersection: ∩	Conjunction: $\wedge$	Series circuit
Union: ∪	Disjunction: $\vee$	Parallel circuit
Complement: $\overline{A}$	Negation: $\sim p$	Negation circuit

PROBLEM SET 2.6

▲ A Problems

1. **IN YOUR OWN WORDS** Explain when the light on a circuit is on and when it is off.

2. **IN YOUR OWN WORDS** Compare De Morgan's laws for sets and for logic propositions.

a.

b.

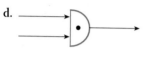

c.

p •——————   • on

   ○ off

d.

e.

3. IN YOUR OWN WORDS What do the circuit symbols in the margin mean?

Using switches, design a circuit that would find the truth values for the statements in Problems 4–12 (answers are not unique).

4. $p \wedge q$ **5.** $p \vee q$ **6.** $\sim p \wedge q$ **7.** $p \wedge \sim q$ **8.** $\sim(p \vee q)$

9. $\sim(p \wedge q)$ **10.** $p \rightarrow q$ **11.** $q \rightarrow p$ **12.** $p \rightarrow \sim q$

▲ B Problems

Using AND-gates, OR-gates, and NOT-gates, design circuits that would find the truth values for the sentences in Problems 13–18.

13. $p \wedge \sim q$ **14.** $(p \wedge q) \vee r$ **15.** $(p \wedge q) \vee (p \wedge r)$

16. $\sim(\sim p \vee q)$ **17.** $\sim[(p \wedge q) \vee r]$ **18.** $(\sim p \wedge q) \vee (p \vee \sim q)$

19. The conditional $p \rightarrow q$ was defined to be equivalent to $\sim(p \wedge \sim q)$. Use gates to represent this conditional.

20. The cost of the circuit is often a factor in work with switching circuits. The circuit you constructed in Problem 19, by using the definition of the conditional, required three gates. Construct a truth table for $p \rightarrow q$ and $\sim p \vee q$. What do you notice about the truth values of the result? Design circuits for $p \rightarrow q$ by using only one OR-gate and one NOT-gate.

Using the results of Problems 19 and 20, design circuits for the conditional statements given in Problems 21–23.

21. $\sim p \rightarrow q$ **22.** $p \rightarrow \sim q$ **23.** $\sim q \rightarrow \sim p$

▲ Problem Solving

24. a. Suppose an engineer designs the following circuit:
$$\sim\{\sim[(p \vee q) \wedge (\sim p)] \wedge (\sim p)\}$$
What will the circuit look like?

b. If each gate costs 2¢ and a company is going to manufacture one million items using this circuit, how much will the circuits cost?

c. Use truth tables to find the following.
   i. $\sim\{\sim[(p \vee q) \wedge \sim p] \wedge (\sim p)\}$
   ii. $p \vee q$
   iii. What do you notice about the final column of each of these truth tables?

d. Draw a simpler circuit that will output the same values as those in part **a**.

e. If the company is going to manufacture one million items using this simpler circuit, how much will the circuits cost? How much will the company save by using this simpler circuit rather than the more complicated one?

25. Alfie, Bogie, and Clyde are the members of a Senate committee. Design a circuit that will output *yea* or *nay* depending on the way the majority of the committee members voted.

26. Suppose that the Senate committee of Problem 25 has five members. Design a circuit that will output the result of the majority of the committee's vote.

"Students should be able to construct cogent arguments in support of their claims."

NCTM Standards

DECEMBER

1

2 **Nikolai Lobachevski** (1792), non-Euclidean geometry

3

4

5

6

7 **L. Kronecker** (1823), analysis

8

9

10 **Karl Jacobi** (1804), determinants

11

12

13

14

15 **Janos Bolyai** (1802), non-Euclidean geometry

16

17 **Marius Sophus Lie** (1842), group theory

18 **Nicollo Tartaglia** (1499), cubic equations

19

20

21

22 **Srinivasa Ramanujan** (1887), number theory, p. 218

23

24 **Charles Hermite** (1822), *e* is transcendental

25 **Isaac Newton** (1642), cofounder of calculus, p. 430

26 **Charles Babbage** (1792), calculating machines, p. 162

27 **Johannes Kepler** (1571), astronomy, analytic geometry

28 **John von Neumann** (1903), game theory, economics

29 **Jacob Bernoulli** (1654), probability, algebra, calculus

30 **Thomas Jan Stieltjes** (1856), calculus

31 **Jaime Escalante**

Biographical Sketch

"You do not enter the future— you create it."

Jaime Escalante

Jaime Alfonso Escalante Gutierrez was born in La Paz, Bolivia, on December 31, 1930, and after teaching for 12 years in Bolivia, he moved to the United States. After earning a degree in mathematics and a teaching credential, he was hired as a basic mathematics teacher at Garfield High School in East Los Angeles, a troubled inner-city school. His spectacular success teaching advanced mathematics to gang members and other students who had been considered "unteachable" attracted national attention. When his story was told in the acclaimed film *Stand and Deliver* (1988), Escalante became a national hero.

Currently, in addition to teaching at Hiram Johnson High School in Sacramento, California, Jaime is working with the Foundation for Advancements in Science and Education (FASE) and the Public Broadcasting Service (PBS) to create a variety of videos on mathematics. His enthusiasm about teaching mathematics is obvious. When asked about the future of mathematics education, he responded. "We are all concerned about the future of American education. But as I tell my students, you do not *enter* the future — you *create* the future. The future is created through hard work. This means teachers must teach, students must do the work, and parents must take an active interest in their children's schoolwork."

Book Reports

Write a 500-word report on one of the following books:

▲ *Overcoming Math Anxiety*, Sheila Tobias (Boston: Houghton Mifflin Co., 1978).

▲ *Mathematical Puzzles for Beginners & Enthusiasts*, Geoffrey Mott-Smith (New York: Dover, 1954).

▲ *Escalante, The Best Teacher in America*, Jay Mathews (New York: Holt, 1988).

Important Terms

And [2.1]	De Morgan's laws [2.3]	Inclusive *or* [2.1]	Or [2.1]
AND-gate [2.6]	Denying the consequent	Indirect reasoning [2.4]	OR-gate [2.6]
Antecedent [2.2]	[2.4]	Invalid argument [2.1]	Parallel circuit [2.6]
Argument [2.1]	Direct reasoning [2.4]	Inverse [2.2]	Premise [2.1]
Assuming the antecedent	Disjunction [2.1]	Law of contraposition [2.2]	Series circuit [2.6]
[2.4]	Either . . . or [2.1]	Law of detachment [2.4]	Simple statement [2.1]
Because [2.3]	Exclusive *or* [2.1]	Law of double negation [2.2]	Some [2.1]
Biconditional [2.3]	Fallacy [2.4]	Law of the excluded middle	Statement [2.1]
Compound statement [2.1]	Fallacy of the converse [2.4]	[2.1]	Syllogism [2.4]
Conclusion [2.1]	Fallacy of the inverse [2.4]	Logical equivalence [2.3]	Tautology [2.3]
Conditional [2.2]	False chain pattern [2.4]	Modus ponens [2.4]	Theorem [2.4]
Conjunction [2.1]	Fundamental operators [2.2]	Modus tollens [2.4]	Transitive reasoning [2.4]
Consequent [2.2]	Fuzzy logic [2.1]	Negation [2.1]	Truth set [2.3]
Contrapositive [2.2]	Gates [2.6]	Neither . . . nor [2.3]	Truth value [2.1; 2.2]
Converse [2.2]	Hypotheses [2.1]	No *p* is *q* [2.3]	Unless [2.3]
Counterexample [2.4]	If . . . then [2.2]	Not [2.1]	Valid argument [2.1]
Deductive reasoning [2.1]	Implication [2.3]	NOT-gate [2.6]	

Important Ideas

What is truth value? [2.1]

Know the definition of fundamental operators [2.1]

Law of double negation [2.2, 2.3]

Definition of conditional [2.2]

Converse, inverse, and contrapositive [2.2]

Law of contraposition [2.2, 2.3]

Definition of biconditional [2.3]

De Morgan's laws [2.3]

Find the negation of a compound statement [2.3]

Negation of a conditional [2.3]

Additional operators (either, neither, unless, because, and no *p* is *q*) [2.3]

Direct reasoning [2.4]

Indirect reasoning [2.4]

Transitive reasoning [2.4]

Fallacies [2.4]

Gates [2.6]

Circuits [2.6]

Relationship between sets, logic, and circuits [2.6]

Types of Problems

Determine whether a sentence is a statement [2.1]

Write the negation of *all*, *some*, and *not* [2.1]

Truth value of simple and compound statements [2.1]

Translate statements into symbolic form [2.1–2.3]

Construct a truth table for a given symbolic form [2.2]

Write the converse, inverse, and contrapositive for a given statement [2.2]

Determine whether a given symbolic statement is true or false [2.2]

Decide whether a given statement is a tautology [2.3]

Write the negation of a compound statement [2.3]

Given certain real-life premises, reach conclusions [2.3, 2.4]

Prove logical statements [2.3, 2.4]

Determine whether a given argument is valid or invalid [2.4]

Find a valid conclusion for a given argument [2.4]

Solve logical puzzles [2.5]

Design a circuit to simulate truth values of a given logical statement [2.6]

Use gates to design a circuit [2.6]

CHAPTER 2   Review Questions

1. Discuss the difference between definitions and theorems.

2. Complete the following truth table.

p	q	$\sim p$	$p \wedge q$	$p \vee q$	$p \rightarrow q$	$p \leftrightarrow q$

Construct truth tables for the statements in Problems 3–6.

3. $\sim(p \wedge q)$

4. $[(p \vee \sim q) \wedge \sim p] \rightarrow \sim q$

5. $[(p \wedge q) \wedge r] \rightarrow p$

6. $\sim(p \wedge q) \Leftrightarrow (\sim p \vee \sim q)$

7. State and prove the principle of direct reasoning.

8. Give an example of indirect reasoning.

9. Give an example of a common fallacy of logic, and show why it is a fallacy.

10. Is the following valid?

$$p \rightarrow q$$
$$\underline{\sim q}$$
$$\therefore \sim p$$

Can you support your answer?

11. Write the negation of each of the following statements.
   a. All birds have feathers.
   b. Some apples are rotten.
   c. No car has two wheels.
   d. Not all smart people attend college.
   e. If you go on Tuesday, then you cannot win the lottery.

12. Which of the following statements are true?
   a. If $111 + 1 = 1,000$, then I'm a monkey's uncle.
   b. If $6 + 2 = 10$ and $7 + 2 = 9$, then $5 + 2 = 7$.
   c. If $6 + 2 = 10$ or $5 + 2 = 7$, then $7 + 2 = 9$.
   d. If $5 + 2 = 7$ and $7 + 2 \neq 9$, then $6 + 2 = 10$.
   e. If the moon is made of green cheese, then $5 + 7 = 57$.

13. Let p: P is a prime number; q: $P + 2$ is a prime number. Translate the following statements into verbal form.
   a. $p \rightarrow q$ b. $(p \vee q) \wedge [\sim(p \wedge q)]$

14. Consider this statement: "All computers are incapable of self-direction."
   a. Translate this statement into symbolic form.
   b. Write the contrapositive of the statement.

15. Translate into symbols and identify the argument.

    If there are a finite number of primes, then there is some natural number, greater than 1, that is not divisible by any prime.

    Every natural number greater than 1 is divisible by a prime number.

    Therefore, there are infinitely many primes.

16. **a.** Using switches, design a circuit that would find the truth values for the following statement: $q \rightarrow \sim p$
    b. Using AND-gates, OR-gates, or NOT-gates, design a circuit that would find the truth values for $[(p \wedge q) \wedge p] \rightarrow q$.

Form a valid conclusion using all the statements in Problems 17–19.

17. If I attend to my duties, I am rewarded.
    If I am lazy, I am not rewarded.
    I am lazy.

18. All organic food is healthy.
    All artificial sweeteners are unhealthy.
    No prune is nonorganic.

19. All squares are rectangles.
    All rectangles are quadrilaterals.
    All quadrilaterals are polygons.

20. **Table Puzzle**

    The mathematics department of a very famous two-year college consists of four people: Josie, who is department chairman, Maureen, Terry, and Warren. For department meetings they always sit in the same seats around a square table. Their hobbies are (in alphabetical order) baking, gardening, hiking, and surfing. Josie, whose hobby is gardening, sits on Terry's left. Maureen sits at the hiker's right. Warren, who faces Terry, is not the baker.

    Make a drawing showing who sits where, and state the hobby of each person.

Revisiting the Real World . . .

IN THE REAL WORLD   "What's up?" asked Shannon. "I'm going nuts filling out these applications!" exclaimed Missy. "They must make these forms obscure just to test us. Look at this question on the residency questionnaire for the California State University system:"

   Complete and submit this residency questionnaire only if your response to item 34a of Part A was other than 'California' or your response to item 37 was 'no.'

"Maybe I can help," added Shannon. "What are the questions in items 34a and 37?" Missy responded, "Well, for what it's worth, here they are."

   Question 34a: Were you born in California? *Answer yes or no.*
   Question 37:   . . .

"I quit!" shouted Missy. "Wait a minute," calmed Shannon. "We know that Question 37 can be answered by a yes or no response, so we can figure this out. Let's look at the possibilities."

Commentary

Once again, we use Polya's method to look at the possibilities as suggested by Shannon.

Understand the Problem. Sometimes the wording of a statement in English can be complicated or convoluted, so that to understand it, you might apply some of the techniques of this chapter for simplification. The problem of this situation is to simplify and clarify the statement on the residency questionnaire for the California State University system.

Devise a Plan. We will translate into symbols, simplify the logical symbols, and then translate back into English to see whether we can rewrite the question more simply.

Carry Out the Plan. Let c: You complete this questionnaire; s: You submit this questionnaire; p: Your response to item 34a was other than "California"; q: Your response to item 37 was "yes." The statement can now be written in symbolic form:

$$(c \wedge s) \text{ only if } (p \vee \sim q) \qquad \text{or} \qquad (c \wedge s) \rightarrow (p \vee \sim q)$$

Consider the contrapositive of this statement:

$$\sim(p \vee \sim q) \rightarrow \sim(c \wedge s)$$

which (by De Morgan's laws and double negation) can be rewritten as

$$(\sim p \wedge q) \rightarrow \sim(c \wedge s)$$

This statement is (in English): "If your response to item 34a was 'California' and you answered 'yes' to item 37, then you do not have to complete and submit this residency questionnaire." Consider the following truth table:

p	q	$\sim p$	$\sim p \wedge q$
T	T	F	F
T	F	F	F
F	T	T	T
F	F	T	F

$\rightarrow$ When you are a California resident and answer "yes" to #37, do not complete or submit questionnaire. Otherwise, you should complete and submit the questionnaire.

Look Back. Remember, using your logic skills to solve problems, you can simplify your work.

Group Research

Working in small groups is typical of most work environments, and being able to work with others to communicate specific ideas is an important skill to learn. Work with three or four other students to submit a single report based on each of the following questions.

1. Write a symbolic statement for each of the following verbal statements.
   (1) Either Donna does not like Elmer because Elmer is bald, or Donna likes Frank and George because they are handsome twins. (2) Either neither you nor I am honest, or Hank is a liar because Iggy did not have the money.

2. Three schools have a track meet and enter one person in each of the events. The number of events is unknown, and so is the scoring system — except that the winner of each event scores a certain number of points, second place scores fewer points, and third place scores fewer still. Georgia won with 22, and Alabama and Florida tied with 9 each. Florida won the high jump. Who won the mile run?*

3. Suppose a prisoner must make a choice between two doors. One door leads to freedom and the other door is booby-trapped so that it leads to death. The doors are labeled as follows:

   Door 1: This door leads to freedom and the other door leads to death.

   Door 2: One of these doors leads to freedom and one of these doors leads to death.

   If exactly one of the signs is true, which door should the prisoner choose? Give reasons for your answer.

4. Five cabbies have been called to pick up five fares at the Hilton Towers. On arrival, they find that their passengers are slightly intoxicated. Each man has a different first and last name, a different profession, and a different destination; in addition, each man's wife has a different first name. Unable to determine who's who and who's going where, the cabbies want to know: **Who is the accountant? What is Winston's last name? Who is going to Elm Street?** Use only the following facts to answer these questions:
   1. Sam is married to Donna.
   2. Barbara's husband gets into the third cab.
   3. Ulysses is a banker.
   4. The last cab goes to Camp St.
   5. Alice lives on Denver Street.
   6. The teacher gets into the fourth cab.
   7. Tom gets into the second cab.
   8. Eve is married to the stock broker.
   9. Mr. Brown lives on Denver St.
   10. Mr. Camp gets into the cab in front of Connie's husband.
   11. Mr. Adams gets into the first cab.
   12. Mr. Duncan lives on Bourbon St.
   13. The lawyer lives on Anchor St.
   14. Mr. Duncan gets into the cab in front of Mr. Evans.
   15. The accountant is three cabs in front of Victor.
   16. Mr. Camp is in the cab in front of the teacher.

---

*From "Ask Marilyn" by Marilyn Vos Savant, *Parade Magazine*, October 31, 1993.

5. Consider the apparatus shown in Figure 2.4. Notice that there are 12 chutes (numbered 1 to 12), and if you drop a ball into the chute it will slide down the tube until it reaches an AND-gate or an OR-gate. If two balls reach an AND-gate, then one ball will pass through, but if only one ball reaches an AND-gate, it will not pass through. If one or two balls reach an OR-gate, then one ball will pass through. The object is to obtain a reward by having a ball reach the location called REWARD. What is the fewest number of balls that can be released to gain the reward?

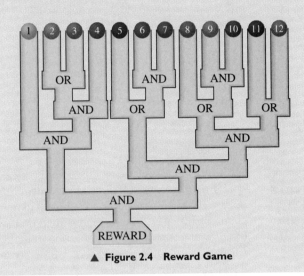

▲ **Figure 2.4   Reward Game**

3 The Nature of Calculation

To err is human, but to really foul things up requires a computer.

In the Real World

"It's just not fair!" exclaimed José. "Half my class have computers at home. I just went to the student store and the cheapest computer and printer I can find is over $1,000. Forget that!"

"You're right, I gave up on getting my own computer," said Sheri. "But I use the library computer all the time. There's some network, I forget its name, that allows me to access every main library in the world! It is unbelievable. Last week I was able to access some information at the Stanford library!"

"Do you think that library computer will help me with my paper? I need to *invent* my own numeration system," said José. "How am I supposed to do what took centuries for others to do? As I've said before, it's just not fair!"

Contents

Perspective

It is nearly impossible to graduate from college today and not have some computer skills. We are all affected by computers, and even though we do not need to understand all about what makes them operate any more than we need to know about an internal combustion engine to drive a car, it is worthwhile to learn just a little about how they do what they do.

We study numeration systems in this chapter so that we can better understand our own numeration system, as well as gain some insight into the internal workings of a computer. In addition to looking at computers today, we take a peek back into the history of computers.

Some knowledge about computers and how they work may help you to understand how better to deal with computers in your day-to-day life.

3.1 EARLY NUMERATION SYSTEMS

A **numeration system** consists of a set of basic symbols and some rules for making other symbols from them, the purpose being the identification of numbers. The invention of a precise and "workable" numeration system is one of the greatest inventions of humanity. It is certainly equal to the invention of the alphabet, which enabled humans to carry the knowledge of one generation to the next. It is simple for us to use the symbol 17 to represent this many objects:

● ● ● ● ● ● ● ● ● ● ● ● ● ● ● ● ●

However, this is not the first and probably will not be the last numeration system to be developed; it took centuries to arrive at this stage of symbolic representation. Here are some of the ways that 17 has been written:

Tally: |||| |||| |||| || *Egyptian:* ∩ | | | | | | | *Roman:* XVII

Mayan: $\overset{..}{=}$ *Linguistic:* seventeen, seibzehn, dix-sept

The concept represented by each of these symbols is the same, but the symbols differ. The concept or idea of "seventeenness" is called a **number;** the symbol used to represent the concept is called a **numeral.** The difference between *number* and *numeral* is analogous to the difference between a person and his or her name.

Three of the earliest civilizations known to use numerals were the Egyptian, Babylonian, and Roman. We shall examine these systems for two reasons. First, it will help us to more fully understand our own numeration (decimal) system; second, it will help us to see how the ideas of these other systems have been incorporated into our system.

Egyptian Numeration System

Perhaps the earliest type of written numeration system developed was a **simple grouping system.** The Egyptians used such a system by the time of the first dynasty, around 2850 B.C. The symbols of the Egyptian system were part of their hieroglyphics and are shown in Table 3.1.

> **News Clip**
>
> There was a woman who decided she wanted to write to a prisoner in a nearby state penitentiary. However, she was puzzled over how to address him, since she knew him only by a string of numerals. She solved her dilemma by beginning her letter: "Dear 5944930, may I call you 594?"

TABLE 3.1	Egyptian Hieroglyphic Numerals	
Our numeral (decimal)	*Egyptian numeral*	*Descriptive name*
1	|	Stroke
10	∩	Heel bone
100	ᕀ	Scroll
1,000	⚘	Lotus flower
10,000	⌒	Pointing finger
100,000	⌒	Polliwog
1,000,000	⚥	Astonished man

Any number is expressed by using these symbols **additively;** each symbol is repeated the required number of times, but with no more than nine repetitions.

▄▄▄ EXAMPLE 1

Write 12,345 using Egyptian numerals.

Solution

$$12,345 = \ \text{𝄞 𝄢 𝄢 } {}^{9\,9}_{\ 9}\ {}^{\cap\cap}_{\cap\cap}\ ||||| $$

The position of the individual symbols is not important. For example,

$$\text{𝄞 𝄢 𝄢 } {}^{9\,9}_{\ 9}\ {}^{\cap\cap}_{\cap\cap}\ ||||| = \text{𝄢 𝄢 } {}^{9\,9}_{\ 9}\ \text{𝄞 } {}^{\cap\cap}_{\cap\cap}\ |||||$$
$$= |||||\,\text{𝄞 𝄢 } {}^{\ \ 9\,9}_{\cap\cap\cap\cap}\ \text{𝄢 } 9$$

The Egyptians had a simple **repetitive-type arithmetic.** Addition and subtraction were performed by repeating the symbols and regrouping.

▄▄▄ EXAMPLE 2

Use Egyptian arithmetic to find

a. $245 + 457$ **b.** $142 - 67$

Solution

a.

$$\begin{array}{r} 9\,9 \ {}^{\cap\cap}_{\cap\cap}\ ||||| \\ 9\,9\ \cap\cap\cap\ ||||| \\ +\,9\,9\ \cap\cap\ || \\ \hline 9\,9\,9\ \cap\cap\cap\cap \qquad \\ 9\,9\,9\ \cap\cap\cap\cap\cap\ ||||||||||\ || \end{array}$$

Regroup:
$$\begin{array}{r} 9\,9\,9\ \cap\cap\cap\cap\boxed{\cap} \\ 9\,9\,9\ \cap\cap\cap\cap\cap\ || \end{array}$$

Regroup again:
$$\begin{array}{r} 9\,9\,9\ \boxed{9}\ || \\ 9\,9\,9 \end{array}$$

b.

$$\boxed{9}\ {}^{\cap\cap}_{\cap\cap}\ ||$$
$$-\ {}^{\cap\cap\cap}_{\cap\cap\cap}\ ||||||||$$

Regroup:
$$\boxed{\cap\cap\cap\cap\ \cap\cap}\ |||||||||\ ||$$

$$-\ {}^{\cap\cap\cap}_{\cap\cap\cap}\ |||||$$

$$\begin{array}{r} \cap\cap\cap\cap \\ \cap\cap\cap\ ||||| \end{array}$$

In Egyptian arithmetic, multiplication and division were performed by successions of additions that did not require the memorization of a multiplication table and could easily be done on an abacus-type device (see Problem 62).

The Egyptians also used unit fractions (that is, fractions of the form $1/n$, for some counting number n). These were indicated by placing an oval over the numeral for the denominator. Thus,

$$\frac{\text{⬭}}{|||} = \frac{1}{3} \qquad \frac{\text{⬭}}{||||} = \frac{1}{4} \qquad \frac{\text{⬭}}{\cap} = \frac{1}{10} \qquad \frac{\text{⬭}}{{}^{\cap\cap\cap}_{|||}} = \frac{1}{33}$$

The fractions $\frac{1}{2}$ and $\frac{2}{3}$ were exceptions:

$$\frac{\text{⬭}}{||} \ \text{or}\ \sqsubset = \frac{1}{2} \qquad \text{⬭}\ \text{or}\ \text{♇} = \frac{2}{3}$$

"*No, no!* 𒀭 *as in* 𒀭°𒆳𒈨𒐊."

Fractions that were not unit fractions (with the exception of $\frac{2}{3}$ just mentioned) were represented as sums of unit fractions. For example,

$$\frac{2}{7} \quad \text{could be expressed as} \quad \frac{1}{4} + \frac{1}{28}$$

Repetitions were not allowed, so $\frac{2}{7}$ would not be written as $\frac{1}{7} + \frac{1}{7}$. Why so difficult? We don't know, but the Rhind papyrus (see Figure 3.1) includes a table for all such decompositions for odd denominators from 5 to 101.

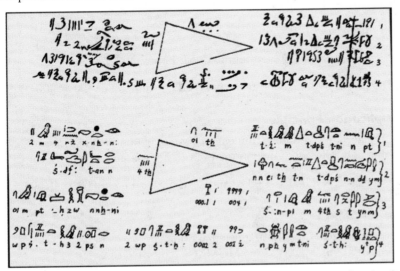

▲ **Figure 3.1   This is a portion of the Rhind papyrus showing the area of a triangle.**

Roman Numeration System

The Roman numeration system was used in Europe until the 18th century, and is still used today in outlining, numbering certain pages in books, and for certain copyright dates. The Roman numerals were selected letters from the Roman alphabet, as shown in Table 3.2.

TABLE 3.2	Roman Numerals						
Decimal numeral	1	5	10	50	100	500	1,000
Roman numeral	I	V	X	L	C	D	M

The Roman numeration system is *repetitive*, like the Egyptian system, but has two additional properties, called the **subtraction principle** and the **multiplication principle**.

Subtraction Principle

One property of the Roman numeration system is based on the **subtraction principle.** Reading from the left, we add the value of each numeral unless its value is

smaller than the value of the numeral to its right, in which case we subtract it from that number. Only the numbers 1, 10, 100, . . . , can be subtracted, and only from the next two higher numbers. For example,

I, II, III, IIII or IV, V, VI, VII, VIII, VIIII or IX, X, XI, . . .

are the successive counting numbers. Instead of IIII we write IV, which (using the subtraction principle) means $5 - 1 = 4$. The I can be subtracted from V or X, and the X can be subtracted from L or C; thus

XL means $50 - 10 = 40$

The C can be subtracted only from D or M. For example,

DC means $500 + 100 = 600$ but   CD means $500 - 100 = 400$

Multiplication Principle

A bar is placed over a Roman numeral or a group of numerals to mean 1,000 times the value of the Roman numerals themselves. For example,

$\overline{V}$ means $5 \times 1,000 = 5,000$ and $\overline{XL}$ means $40 \times 1,000 = 40,000$

▬▬ EXAMPLE 3

Write each Roman numeral as a decimal numeral.
a. CCLXIII   **b.** MXXIV   **c.** MCMXC   **d.** $\overline{CD}$CLIX

Solution

a. CCLXIII $= 100 + 100 + 50 + 10 + 1 + 1 + 1 = 263$
b. MXXIV $= 1,000 + 10 + 10 + (5 - 1) = 1,024$
c. MCMXC $= 1,000 + (1,000 - 100) + (100 - 10) = 1,990$
d. $\overline{CD}$CLIX $= (500 - 100) \times 1,000 + 100 + 50 + (10 - 1) = 400,159$ ▬

▬▬ EXAMPLE 4

Write each decimal numeral as a Roman numeral.
a. 49   **b.** 379   **c.** 345,123   **d.** 1,997

Solution

a. $49 = (50 - 10) + (10 - 1) = $ XLIX
b. $379 = 100 + 100 + 100 + 50 + 20 + (10 - 1) = $ CCCLXXIX
c. $345,123 = 345 \times 1,000 + 123$

$$= [100 + 100 + 100 + (50 - 10) + 5] \times 1,000 + (100 + 20 + 3)$$
$$= \overline{CCCXLV}CXXIII$$

d. $1,997 = 1,000 + (1,000 - 100) + (100 - 10) + 7$
$$= MCMXCVII$$ ▬

Babylonian Numeration System

The Babylonian numeration system differed from the Egyptian in several respects. Whereas the Egyptian system was a simple grouping system, the Babylonians employed a much more useful **positional system.** Since they lacked papyrus, they used mostly clay as a writing medium, and thus the Babylonian cuneiform was much less pictorial than the Egyptian system. They employed only two wedge-shaped characters, which date from 2000 B.C. and are shown in Table 3.3.

TABLE 3.3	**Babylonian Cuneiform Numerals**
Our numeral (decimal)	Babylonian numeral
1	▼
2	▼▼
9	▼▼▼▼▼▼▼▼▼
10	◁
59	◁◁◁ ▼▼▼▼▼ / ◁◁ ▼▼▼▼

Historical Note

There are many clay tablets available to us (more than 50,000), and the work of interpreting them is continuing. Most of our knowledge of the tablets is less than 100 years old and has been provided by Professors Otto Neugebauer and F. Thureau-Dangin.

Notice from the table that, for numbers 1 through 59, the system is *repetitive.* However, unlike in the Egyptian system, the position of the symbols is important. The ◁ symbol *must* appear to the left of any ▼s to represent numbers smaller than 60. For numbers larger than 60, the symbols ◁ and ▼ are written to the left of ◁ (as in a positional system), and now take on a new value that is 60 times larger than the original value. That is,

$$\text{▼ ◁◁◁ ▼▼▼} \quad \text{means} \quad (1 \times 60) + 35$$

Study the following example.

■ EXAMPLE 5

Write each of the Babylonian numerals in decimal form.

a. ▼▼▼ ◁ ◁◁◁ ▼▼▼▼▼▼▼▼▼

b. ▼▼▼▼▼ ◁ ▼

c. ◁◁ ▼▼▼ ◁ ▼▼▼▼▼▼▼▼▼

Solution

a. ▼▼▼ ◁ ◁◁◁ ▼▼▼▼▼▼▼▼▼ $= (3 \times 60) + 59 = 239$

b. ▼▼▼▼▼ ◁ ▼ $= (5 \times 60) + 11 = 311$

c. ◁◁ ▼▼▼ ◁ ▼▼▼▼▼▼▼▼▼ $= (23 \times 60) + 16 = 1,396$

This system is called a *sexagesimal system* and uses the principle of position. However, the system is not fully positional, because only numbers larger than 60 use the position principle; numbers within each basic 60-group are written by a simple grouping system. A true positional sexagesimal system would require 60 different symbols.

The Babylonians carried their positional system a step further. If any numerals were to the left of the second 60-group, they had the value of 60×60 or 60^2. Thus,

$$▼▼\ ⊲⊲\ ▼▼▼\ ⊲⊲\ ▼▼ = (2 \times 60^2) + (45 \times 60) + 24$$
$$= 7{,}200 + 2{,}700 + 24$$
$$= 9{,}924$$

The Babylonians also made use of a *subtractive symbol*, $\ulcorner$. That is, 38 could be written

$$⊲⊲⊲\ ▼▼▼▼ \quad \text{or} \quad ⊲⊲⊲\ \overline{▼▼}$$

◼◼◼ EXAMPLE 6

Change each Babylonian numeral to decimal form.

a. ▼▼▼⊲⊲|▼   **b.** ▼▼ ⊲⊲ ▼▼▼⊲ ▼▼▼   **c.** ⊲⊲ ⊲⊲⊲ ▼▼▼▼

Solution

a. $(3 \times 60) + 20 - 1 = 199$

b. $(2 \times 60^2) + (23 \times 60) + 16 = 7{,}200 + 1{,}380 + 16 = 8{,}596$

c. Note the ambiguity; this might be 54 or it might be

$$(20 \times 60) + 34 = 1{,}234$$
◼

◼◼◼ EXAMPLE 7

Write 4,571 as a Babylonian numeral.

Solution

$$
\begin{aligned}
1 \times 60^2 &= 3{,}600 \\
16 \times 60 &= 960 \\
11 \times 1 &= \underline{11} \\
& \ 4{,}571
\end{aligned}
$$

Thus, ▼ ⊲ ▼▼▼ ⊲ ▼ is the Babylonian numeral.   ◼

Although this positional numeration system was in many ways superior to the Egyptian system, it suffered from the lack of a zero or place-holder symbol. For example, how is the number 60 represented? Does ▼▼ mean 2 or 61? In Example 6c the value of the number had to be found from the context. (Scholars tell us that such ambiguity can be resolved only by a careful study of the context.) However, in later Babylon, around 300 B.C., records show that there is a zero symbol, an idea that was later used by the Hindus.

Arithmetic with the Babylonian numerals is quite simple, since there are only two symbols. A study of the Babylonian arithmetic will be left for the reader (see Problem 6).

Other Historical Systems

One of the richest areas in the history of mathematics is the study of historical numeration systems. In this book, we discuss the Egyptian numeration system because it illustrates a simple grouping system, the Babylonian numeration system because it illustrates a positional numeration system, and the Roman system because it is still in limited use today.

Other numeration systems of interest from a historical point of view are the Mayan, Greek, and Chinese numeration systems. These are summarized in Table 3.4.

TABLE 3.4   Historical Numeration Systems

Numeration system	Classification	Selected numerals							
		1	2	5	10	50	100	500	1,000
Decimal system	positional (10s)	1	2	5	10	50	100	500	1,000
Egyptian system	grouping	|	||	|||||	∩	∩∩∩ ∩∩	9	999 99	𖤓
Roman system	grouping	I·	II	V	X	L	C	D	M
Babylonian system (also called Sumerian)	positional (60s)	▼	▼▼	▼▼▼ ▼▼	◁	◁◁◁ ◁◁	▼◁◁ ◁◁	▼▼▼▼◁ ▼▼▼▼◁	◁▼▼▼◁◁ ▼▼▼◁◁
Greek	grouping	α	β	ε	ι	ν	ρ	φ	α′
Mayan	positional (vertical)	·	··	—	=	·· =	⊖	⊙	⊙̈
Chinese	positional (vertical)	╱	╱╱	𠄌	十	𠄡	丙	𠄡百	千

PROBLEM SET 3.1

▲ **A Problems**

1. **IN YOUR OWN WORDS** Explain the difference between *number* and *numeral*. Give examples of each.

2. **IN YOUR OWN WORDS** Discuss the similarities and differences between a simple grouping system and a positional system. Give examples of each.

3. **IN YOUR OWN WORDS** What do you regard as the shortcomings and contributions of the Egyptian numeration system?

4. **IN YOUR OWN WORDS** What do you regard as the shortcomings and contributions of the Roman numeration system?

5. **IN YOUR OWN WORDS** What do you regard as the shortcomings and contributions of the Babylonian numeration system?

6. **IN YOUR OWN WORDS** Discuss addition and subtraction for Babylonian numerals. Show examples.

7. Tell which of the named properties apply to the Egyptian, Roman, and Babylonian numeration systems.
   a. grouping system   **b.** positional system
   c. repetitive system   **d.** additive system
   e. subtractive system   **f.** multiplicative system

Write each numeral in Problems 8–35 as a decimal numeral. By decimal numeral, we mean the system we use in everyday arithmetic.

8. 𒀭 99 ∩∩∩ ∩∩ |||||| 9. ⬯ 99   10. 𒀮   11. ⬭ ∩

12. 9 ∩∩|||⊏   13. ⬯ ||∩   14. ∩∩||||⊏   15. ⬯ 9

16. 𝒶𒀭 99∩∩∩ 9 ∩∩ || 17. 𐌰| 18. 9 𒀭

19. 𒀭 99999 99999 ∩∩∩∩∩ ∩∩∩∩ |||||| 20. ◁◁ ▼▼▼▼   21. ▼▼▼◁◁▼

22. ▼◁▼   23. ◁▼◁▼   24. ▼◁◁◁ |▼ ◁

25. ◁◁◁ |▼▼   26. ◁◁◁ ◁◁ |▼▼▼   27. ◁◁ ▼▼▼ ▼▼

28. XLVIII   29. DCCIX   30. MCMXCVII   31. MMI

32. $\overline{\text{DL}}$ 33. $\overline{\text{CD}}$ 34. $\overline{\text{V}}$MMDC   35. $\overline{\text{IX}}$DCCXII

The Rhind papyrus contains many problems. Two are symbolized on the following simulated papyrus. The one in Problem 37 is an 18th century Mother Goose rhyme. Answer the question posed in each problem.

36.

In each of 7 houses are 7 cats.

Each cat kills 7 mice.

Each mouse would have eaten 7 spelt (wheat).

Each ear of spelt would have produced 7 kehat of grain.

Query: How much grain is saved by the 7 houses' cats?

37.

As I was going to St. Ives

I met a man with seven wives.

Every wife had seven sacks,

Every sack had seven cats,

Every cat had seven kits.

Kits, cats, sacks, and wives,

How many were there going to St. Ives?

▲ **B Problems**

Write each of the numerals in Problems 38–43 in the following systems: **a.** *Egyptian*   **b.** *Roman*   **c.** *Babylonian*

38. 47   39. 75   40. 258   41. 521   42. 852   43. 1,998

44. Kathryn had a dream in which she was selling roses in an international market. She started out with 143 roses. Then her first customer, an Egyptian, bought ∩||||| of them. Soon afterward, a Babylonian asked to buy ◁◁◁ ▼▼▼▼▼▼ roses; later XIV roses were bought by (you guessed it) a Roman. How many roses did she have left to sell at the end of her dream?

45. In a strange mix-up, several people (none of whom spoke the same language) needed to pool their resources to survive the scorching desert heat. Eric had 45 bottles of water; Dimetrius had ∩∩∩ |||| bottles; Sparticus had ʊ ʊ ʊ ʊ ʊ ▼▼▼▼ bottles. How many bottles of water did they have all together?

Perform the indicated operations in Problems 46–55.

46.　　𒀸 ∩∩ ||
　　　　 ∩∩
　　+ 99 ∩∩∩∩ ||||||
　　　　 ∩∩∩

47.　　99 ∩ |
　　 − ∩∩ ||||

48.　　99 ∩∩ ||
　　+ 9 ∩∩∩∩ |||||
　　　　 ∩∩∩∩

49.　　99 ∩∩ ||
　　 − 9 ∩∩∩∩ |||||
　　　　 ∩∩∩∩

50.　　ʊʊʊ ▼▼▼▼
　　+ ʊ ▼▼▼
　　　　 ▼▼▼

51.　　ʊʊ ▼▼▼▼
　　 − ʊ ▼▼▼
　　　　 ▼▼▼

52.　　▼ʊʊ▼▼▼▼
　　+ ʊʊʊ ▼▼▼▼
　　　 ʊ　 ▼▼▼▼

53.　　▼ʊ▼▼
　　 − ʊʊʊ ▼▼▼▼
　　　 ʊ　 ▼▼▼

54.　　▼ʊʊ▼▼▼
　　 − ʊʊ ▼▼▼
　　　 ʊʊ ▼▼

55.　　▼ʊʊ▼▼
　　 − ʊʊ ▼▼▼▼
　　　 ʊʊ

▲ **Problem Solving**

56. The Yale tablet from the Babylonian collection, written between 1900 B.C. and 1600 B.C., contains the following algebra problem:

The length of a rectangle exceeds the width by 7, and the area is 60. What are the dimensions of the rectangle?

57. Another question from the Yale tablet (see Problem 56) is the following algebra problem:

The length of a rectangle exceeds the width by 10 and the area is 600. What are the dimensions of the rectangle?

58. **IN YOUR OWN WORDS** The people from the Long Lost Land had only the following four symbols: □, △, ▽, and ○. Write out the first 20 numbers. Find □ + ▽ and △ × △. (Use your imagination to invent a system to answer these questions.)

59. **IN YOUR OWN WORDS** In Chapter 1 we introduced Pascal's triangle. The reproduction in Figure 3.2 is from a 14th-century Chinese manuscript. Even though we have not discussed these ancient Chinese numerals, see if you can reconstruct the basics of their numeration system. (Use your imagination.)

60. **IN YOUR OWN WORDS** The Ionic Greek numeration system (approximately 3000 B.C.) counts as follows:

$$\alpha, \beta, \gamma, \delta, \epsilon, \mathcal{V}, \zeta, \eta, \theta, \iota, \iota\alpha, \iota\beta, \iota\gamma, \iota\epsilon, \iota\mathcal{V}, \iota\zeta, \theta\eta, \iota\theta, \kappa, \kappa\alpha, \ldots$$

Other numbers are: λ (for 30), μ (for 40), ν (for 50), ξ (for 60), o (for 70), π (for 80), $\mathcal{Q}$ (for 90), ρ (for 100), σ (for 200), τ (for 300), υ (for 400), ϕ (for 500), χ (for 600), ψ (for 700), ω (for 800), and Π (for 900). Try to reconstruct the basics of their numeration system. (Use your imagination.)

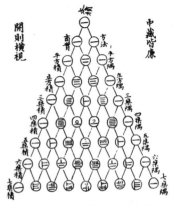

▲ Figure 3.2 Chinese triangle

▲ **Individual Research**

61. What do the following people have in common?

> **Eamon de Valera,** Prime minister and President of the Republic of Ireland
>
> **Tom Lehrer,** songwriter-parodist
>
> **Edmund Husserl,** the "Father of Phenomenology"
>
> **Frank Ryan,** quarterback for the Cleveland Browns

62. Write a paper discussing the Egyptian method of multiplication.

References

James Newman, *The World of Mathematics*, Vol. I (New York: Simon and Schuster, 1956), pp. 170–178.

Howard Eves, *Introduction to the History of Mathematics*, 3rd ed. (New York: Holt, Rinehart, and Winston, 1969).

3.2   HINDU–ARABIC NUMERATION SYSTEM

The numeration system in common use today (the one we have been calling the decimal system) has ten symbols — namely, 0, 1, 2, 3, 4, 5, 6, 7, 8, and 9. The selection of ten digits was no doubt a result of our having ten fingers (digits).

The symbols originated in India in about 300 B.C. However, because the early specimens do not contain a zero or use a positional system, this numeration system offered no advantage over other systems then in use in India.

The date of the invention of the zero symbol is not known. The symbol did not originate in India but probably came from the late Babylonian period via the Greek world.

By the year 750 A.D. the zero symbol and the idea of a positional system had been brought to Baghdad and translated into Arabic. We are not certain how these numerals were introduced into Europe, but it is likely that they came via Spain in the 8th century. Gerbert, who later became Pope Sylvester II, studied in Spain and was the first European scholar known to have taught these numerals. Because of their origins, these numerals are called the **Hindu–Arabic numerals.** Since ten basic symbols are used, the Hindu–Arabic numeration system is also called the *decimal numeration system,* from the Latin word *decem,* meaning "ten."

Although we now know that the decimal system is very efficient, its introduction met with considerable controversy. Two opposing factions, the "algorists" and the "abacists," arose. Those favoring the Hindu–Arabic system were called algorists, since the symbols were introduced into Europe in a book called (in Latin) *Liber Algorismi de Numero Indorum,* by the Arab mathematician al-Khowârizmî. The word *algorismi* is the origin of our word *algorithm.* The abacists favored the status quo — using Roman numerals and doing arithmetic on an abacus. The battle between the abacists and the algorists lasted for 400 years. The Roman Catholic church exerted great influence in commerce, science, and theology. The church criticized those using the "heathen" Hindu–Arabic numerals and consequently kept the world using Roman numerals until 1500. Roman numerals were easy to write and learn, and addition and subtraction with them were easier than with the "new" Hindu–Arabic numerals. It seems incredible that our decimal system has been in general use only since about the year 1500.

Let's examine the Hindu–Arabic or decimal numeration system a little more closely:

1. It uses ten symbols, called digits.
2. Larger numbers are expressed in terms of powers of 10.
3. It is positional.

Consider how we count objects:

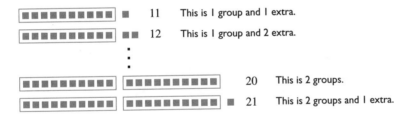

At this point we could invent another symbol as the Egyptians did (you might suggest 0, but remember that 0 represents no objects), or we could reuse the digit symbols by repeating them or by altering their positions. That is, we agree to use 10 to mean 1 group of

■■■■■■■■■■

The symbol 0 was invented as a place-holder to show that the 1 here is in a different position from the 1 representing ■. We continue to count:

■■■■■■■■■■ ■   11   This is 1 group and 1 extra.
■■■■■■■■■■ ■■ 12   This is 1 group and 2 extra.
⋮
■■■■■■■■■■ ■■■■■■■■■■   20   This is 2 groups.
■■■■■■■■■■ ■■■■■■■■■■ ■   21   This is 2 groups and 1 extra.

We continue in the same fashion until we have 9 groups and 9 extra. What's next? It is 10 groups or a group of groups:

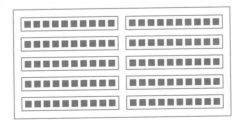

We call this group of groups a 10 · 10 or 10^2 or a **hundred**. We again use position and repeat the symbol 1 with still different meaning: 100.

▰▰ EXAMPLE 1

What does 134 mean?

Solution   134 means that we have 1 group of 100, 3 groups of 10, and 4 extra. In symbols,

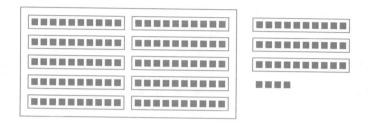

We denote this more simply by writing:

These represent the number in each group.

$$(1 \times 10^2) - (3 \times 10) + 4$$

These are the names of the groups.

This leads us to the meaning *one hundred, three tens, four ones.*   ▬

The representation, or the meaning, of the number 134 in Example 1 is called **expanded notation.**

▰▰ EXAMPLE 2

Write 52,613 in expanded form.

Solution

$$52,613 = 50,000 + 2,000 + 600 + 10 + 3$$
$$= 5 \times 10^4 + 2 \times 10^3 + 6 \times 10^2 + 1 \times 10 + 3$$ ▬

▰▰ EXAMPLE 3

Write $4 \times 10^8 + 9 \times 10^7 + 6 \times 10^4 + 3 \times 10 + 7$ in decimal form.

Solution   You can use the order of operations and multiply out the digits, but you should be able to go directly to decimal form if you remember what place-value means:

$$49\ 00600\ 37 = 490,060,037$$

Notice that there were no powers of 10^6, 10^5, 10^3, or 10^2.   ▬

A period, called a **decimal point** in the decimal system, is used to separate the fractional parts from the whole parts. The positions to the right of the decimal point are fractions:

$$\frac{1}{10} = 10^{-1}, \quad \frac{1}{100} = 10^{-2}, \quad \frac{1}{1,000} = 10^{-3}$$

To complete the pattern, we also sometimes write $10 = 10^1$ and $1 = 10^0$.

▄▄▄ EXAMPLE 4

Write 479.352 using expanded notation.

Solution

$$479.352 = 400 + 70 + 9 + 0.3 + 0.05 + 0.002$$

$$= 400 + 70 + 9 + \frac{3}{10} + \frac{5}{100} + \frac{2}{1,000}$$

$$= 4 \times 10^2 + 7 \times 10^1 + 9 \times 10^0 + 3 \times 10^{-1} + 5 \times 10^{-2} + 2 \times 10^{-3}$$ ▬

PROBLEM SET 3.2

▲ A Problems

1. **IN YOUR OWN WORDS** Illustrate the meaning of 123.

2. **IN YOUR OWN WORDS** Illustrate the meaning of 145.

3. **IN YOUR OWN WORDS** Illustrate the meaning of 1,134 by showing the appropriate groupings.

4. **IN YOUR OWN WORDS** What is expanded notation?

5. **IN YOUR OWN WORDS** Define b^n for b any nonzero number and n any counting number.

6. **IN YOUR OWN WORDS** Define b^{-n} for b any nonzero number and n any counting number.

Give the meaning of the numeral 5 in each of the numbers in Problems 7–10.

7. 805   8. 508   9. 0.00567   10. 58,000,000

Write the numbers in Problems 11–27 in decimal notation.

11. a. 10^5 b. 10^3 12. a. 10^6 b. 10^4

13. a. 10^{-4} b. 10^{-3} 14. a. 10^{-2} b. 10^{-6}

15. a. 5×10^3 b. 5×10^2 16. a. 8×10^{-4} b. 7×10^{-3}

17. a. 6×10^{-2} b. 9×10^{-5} 18. a. 5×10^{-6} b. 2×10^{-9}

19. $1 \times 10^4 + 0 \times 10^3 + 2 \times 10^2 + 3 \times 10^1 + 4 \times 10^0$

20. $6 \times 10^1 + 5 \times 10^0 + 0 \times 10^{-1} + 8 \times 10^{-2} + 9 \times 10^{-3}$

21. $5 \times 10^5 + 2 \times 10^4 + 1 \times 10^3 + 6 \times 10^2 + 5 \times 10^1 + 8 \times 10^0$

22. $6 \times 10^7 + 4 \times 10^3 + 1 \times 10^0$

23. $7 \times 10^6 + 3 \times 10^{-2}$ 24. $6 \times 10^9 + 2 \times 10^{-3}$

25. $5 \times 10^5 + 4 \times 10^2 + 5 \times 10^1 + 7 \times 10^0 + 3 \times 10^{-1} + 4 \times 10^{-2}$

26. $3 \times 10^3 + 2 \times 10^1 + 8 \times 10^0 + 5 \times 10^{-1} + 4 \times 10^{-2} + 2 \times 10^{-4}$

27. $2 \times 10^4 + 6 \times 10^2 + 4 \times 10^{-1} + 7 \times 10^{-3} + 6 \times 10^{-4} + 9 \times 10^{-5}$

Write each of the numbers in Problems 28–43 in expanded notation.

28. 741   29. 728,407   30. 0.096421   31. 27.572

32. 47.00215   33. 521   34. 6,245   35. 2,305,681

36. 428.31   37. 5,245.5   38. 0.00000 527   39. 100,000.001

40. 893.0001   41. 8.00005   42. 678,000.01   43. 57,285.9361

▲ B Problems

One of the oldest devices used for calculation is the abacus, as shown in Figure 3.3.

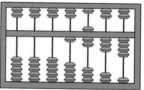

$10^6\ 10^5\ 10^4\ 10^3\ 10^2\ 10^1\ 10^0$

▲ **Figure 3.3   Abacus**

Each rod names one of the positions we use in counting. Each bead on the bottom represents one unit in that column and each bead on the top represents 5 units in that column. The number illustrated in Figure 3.3 is 1,734. What number is illustrated by the drawings in Problems 44–49?

44.
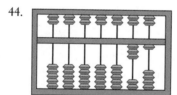

45.
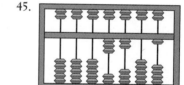

46.
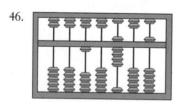

47.
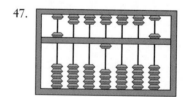

48.
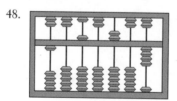

49.
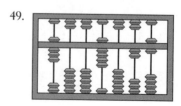

Sketch an abacus to show the numbers given in Problems 50–57.

50. 132   **51.** 849   **52.** 3,214   **53.** 9,387

54. 1,998   **55.** 2,001   **56.** 3,000,400   **57.** 8,007,009

▲ **Problem Solving**

58. Can you find a pattern?

$$0, 1, 2, 10, 11, 12, 20, 21, 22, 100, \ldots$$

59. Can you find a pattern?

$$0, 1, 2, 3, 4, 10, 11, 12, 13, 14, 20, 21, \ldots$$

60. Notice the following pattern for multiplication by 11:

$$14 \times 11 = 154 \qquad 51 \times 11 = 561$$
$$\uparrow \qquad\qquad\qquad\qquad \uparrow$$

Insert the sum of the original two digits between those digits.

Use expanded notation to show why this pattern "works."

▲ **Individual Research**

61. What are some of the significant events in the development of mathematics? Who are some of the famous people who have contributed to mathematical knowledge?

References
Howard Eves, *In Mathematical Circles*, Vols. 1 and 2 (Boston: Prindle, Weber, & Schmidt, 1969), *Mathematical Circles Revisited* (1971), *Mathematical Circles Adieu* (1977).
Virginia Newell et al., *Black Mathematicians and Their Works* (Ardmore, PA: Dorrence & Company, 1980).
Mona Fabricant, Sylvia Svitak, and Patricia Clark Kenschaft, "Why Women Succeed in Mathematics," *Mathematics Teacher*, February 1990, pp. 150–154 (with references).

▲ **Group Research**

62. Construct a history chart and time line showing the significant developments in mathematics. [*Hint:* You can begin with the inside cover flaps in this book.]

3.3 DIFFERENT NUMERATION SYSTEMS

In the previous section we discussed the Hindu–Arabic numeration system and grouping by tens. However, we could group by twos, fives, twelves, or any other counting number. In this section, we summarize numeration systems with bases other than ten. This not only will help you understand our own numeration system, but will give you insight into the numeration systems used with computers, namely base 2 (**binary**), base 8 (**octal**), and base 16 (**hexadecimal**).

Number of Symbols

The number of symbols used in a particular base depends on the method of grouping for that base. For example, in base ten the grouping is by tens, and in base five the grouping is by fives. Suppose we wish to count ■ ■ ■ ■ ■ ■ ■ ■ ■ ■ ■ in various

Historical Note

A study of the Native Americans of California yields a wide variety of number bases. Several of these bases are discussed by Barnabus Hughes in an article entitled "California Indian Arithmetic" in *The Bulletin of the California Mathematics Council* (Winter 1971/1972). According to Professor Hughes, the Yukis, who live north of Willits, have both the quaternary (base four) and the octonary (base eight) systems. Instead of counting on their fingers, these Native Americans enumerated the spaces between the fingers.

The sketch shows a Karok woman; there are no known photographs of the Yukis because they were nearly eliminated at the beginning of the 20th century. The Native Americans in the western corner of California use the decimal system, but those in the rest of the state use the quinary (base five) system for small numbers, but switch to a vigesimal (base twenty) system for large numbers.

bases. Let's look for patterns in Table 3.5. Note the use of the subscript following the numeral to keep track of the base in which we are working.

TABLE 3.5   Grouping in Various Bases

Base	Symbols	Method of grouping	Notation
two	0, 1		1011_{two}
three	0, 1, 2		102_{three}
four	0, 1, 2, 3		23_{four}
five	0, 1, 2, 3, 4		21_{five}
six	0, 1, 2, 3, 4, 5		15_{six}
seven	0, 1, 2, 3, 4, 5, 6		14_{seven}
eight	0, 1, 2, 3, 4, 5, 6, 7		13_{eight}
nine	0, 1, 2, 3, 4, 5, 6, 7, 8		12_{nine}
ten	0, 1, 2, 3, 4, 5, 6, 7, 8, 9		11_{ten}

Do you see any patterns? Suppose we wish to continue this pattern. Can we group by elevens or twelves? We can, provided new symbols are "invented." For base eleven (or higher bases), we use the symbol T to represent ■■■■■■■■■■. For base twelve (or higher bases), we use E to stand for ■■■■■■■■■■■. For bases larger than twelve, other symbols can be invented.

For example, $2T_{twelve}$ means that there are two groupings of twelve and T (ten) extra:

We continue with the pattern from Table 3.5 by continuing beyond base ten in Table 3.6.

TABLE 3.6   Grouping in Various Bases

Base	Symbols	Method of grouping	Notation
eleven	$0, 1, 2, \ldots, 8, 9, T$		10_{eleven}
twelve	$0, 1, 2, \ldots, 8, 9, T, E$		E_{twelve}
thirteen	$0, 1, 2, \ldots, 8, 9, T, E, U$		$E_{thirteen}$
fourteen	$0, 1, 2, \ldots, 9, T, E, U, V$		$E_{fourteen}$

Do you see more patterns? Can you determine the number of symbols in the base b system? Remember that b stands for some counting number greater than 1. Look for a pattern:

$$\textbf{base } b, b \textbf{ symbols} \left\{ \begin{array}{l} \text{base two, two symbols} \\ \text{base three, three symbols} \\ \text{base four, four symbols} \\ \text{base five, five symbols} \\ \qquad \vdots \\ \text{base ten, ten symbols} \\ \text{base eleven, eleven symbols} \\ \qquad \vdots \\ \text{base twenty, twenty symbols} \\ \qquad \vdots \end{array} \right.$$

Change from Base b to Base Ten

To change from base b to base ten, we write the numerals in expanded notation. The resulting number is in base ten.

■ EXAMPLE 1

Change each number to base ten.

a. 1011.01_{two} **b.** 1011.01_{four} **c.** 1011.01_{five}

Solution

a. $1011.01_{two} = 1 \times 2^3 + 0 \times 2^2 + 1 \times 2^1 + 1 \times 2^0 + 0 \times 2^{-1} + 1 \times 2^{-2}$
$= 8 + 0 + 2 + 1 + 0 + 0.25 = 11.25$

b. $1011.01_{four} = 1 \times 4^3 + 0 \times 4^2 + 1 \times 4^1 + 1 \times 4^0 + 0 \times 4^{-1} + 1 \times 4^{-2}$
$= 64 + 0 + 4 + 1 + 0 + 0.0625 = 69.0625$

c. $1011.01_{five} = 1 \times 5^3 + 0 \times 5^2 + 1 \times 5^1 + 1 \times 5^0 + 0 \times 5^{-1} + 1 \times 5^{-2}$
$= 125 + 0 + 5 + 1 + 0 + 0.04 = 131.04$

Change from Base Ten to Base b

To see how to change from base ten to any other valid base, let's again look for a pattern:

To change from base ten to base two, group by twos.

To change from base ten to base three, group by threes.

To change from base ten to base four, group by fours.

To change from base ten to base five, group by fives.

$\vdots$

The groupings from this pattern are summarized in Table 3.7.

TABLE 3.7   Place-value Chart

Base			Place value			
2	$2^5 = 32$	$2^4 = 16$	$2^3 = 8$	$2^2 = 4$	$2^1 = 2$	$2^0 = 1$
3	$3^5 = 243$	$3^4 = 81$	$3^3 = 27$	$3^2 = 9$	$3^1 = 3$	$3^0 = 1$
4	$4^5 = 1,024$	$4^4 = 256$	$4^3 = 64$	$4^2 = 16$	$4^1 = 4$	$4^0 = 1$
5	$5^5 = 3,125$	$5^4 = 625$	$5^3 = 125$	$5^2 = 25$	$5^1 = 5$	$5^0 = 1$
7	$7^5 = 16,807$	$7^4 = 2,401$	$7^3 = 343$	$7^2 = 49$	$7^1 = 7$	$7^0 = 1$
8	$8^5 = 32,768$	$8^4 = 4,096$	$8^3 = 512$	$8^2 = 64$	$8^1 = 8$	$8^0 = 1$
10	$10^5 = 100,000$	$10^4 = 10,000$	$10^3 = 1,000$	$10^2 = 100$	$10^1 = 10$	$10^0 = 1$
12	$12^5 = 248,832$	$12^4 = 20,736$	$12^3 = 1,728$	$12^2 = 144$	$12^1 = 12$	$12^0 = 1$

The next example shows how we can interpret this grouping process in terms of a simple division.

■■■ EXAMPLE 2   **Polya's Method**

Convert 42 to base two.

Solution   We use Polya's problem-solving guidelines for this example.

Understand the Problem.   Using Table 3.7, we see that the largest power of two smaller than 42 is 2^5, so we begin with $2^5 = 32$:

$$42 = 1 \times 2^5 + 10$$
$$10 = 0 \times 2^4 + 10$$
$$10 = 1 \times 2^3 + 2$$
$$2 = 0 \times 2^2 + 2$$
$$2 = 1 \times 2^1 + 0$$

We could now write out 42 in expanded notation.

Devise a Plan.   Instead of carrying out the steps by using Table 3.7, we will begin with 42 and carry out repeated division, saving each remainder as we go.

Carry Out the Plan.   We are changing to base 2, so we do repeated division by 2:

$$\begin{array}{r} 21 \quad \text{r. } 0 \leftarrow \text{save remainder} \\ 2\overline{)42} \end{array}$$

Next we need to divide 21 by 2, but instead of rewriting our work we work our way up:

$$\begin{array}{r} 10 \quad \text{r. } 1 \leftarrow \text{save all remainders} \\ 2\overline{)21} \quad \text{r. } 0 \leftarrow \text{save remainder} \\ 2\overline{)42} \end{array}$$

Continue by doing repeated division.

Stop when you get a zero here

↓

0   r. 1 | Answer is found by reading down
2)$\overline{1}$ r. 0
2)$\overline{2}$ r. 1
2)$\overline{5}$ r. 0
2)$\overline{10}$ r. 1
2)$\overline{21}$ r. 0 ↓
2)$\overline{42}$

Thus, $42 = 101010_{two}$.

Look Back. $101010_{two} = 1 \times 2^5 + 1 \times 2^3 + 1 \times 2 = 32 + 8 + 2 = 42.$ ▬

▬▬ **EXAMPLE 3**

Write 42 in   **a.** base three   **b.** base four

Solution

a. Begin with $3^3 = 27$ (from Table 3.7):

$42 = 1 \times 3^3 + 15$ 0   r. 1
$15 = 1 \times 3^2 + 6$ 3)$\overline{1}$ r. 1
$6 = 2 \times 3^1 + 0$ 3)$\overline{4}$ r. 2
$0 = 0 \times 3^0$ 3)$\overline{14}$ r. 0
                              3)$\overline{42}$

Thus, $42 = 1120_{three}$.

b. Begin with $4^2 = 16$ (from Table 3.7):

$42 = 2 \times 4^2 + 10$ 0   r. 2
$10 = 2 \times 4^1 + 2$ 4)$\overline{2}$ r. 2
                              4)$\overline{10}$ r. 2
                              4)$\overline{42}$

Thus, $42 = 222_{four}$.   ▬

How would you change from base ten to base seven? To base eight? To base b? We see from the above pattern that we group by b's or perform repeated division by b.

■■■■ **EXAMPLE 4**   **Polya's Method**

Suppose you need to purchase 1,000 name tags and can buy them by the gross (144), the dozen (12), or individually. The name tags cost $.50 each, $4.80 per dozen, and $50.40 per gross. How should you order to minimize the cost?

Solution   We use Polya's problem-solving guidelines for this example.

Understand the Problem.   If you purchase 1,000 tags individually, the cost is $.50 × 1,000 = $500. This is not the least cost possible, because of the bulk discounts. We need to find the maximum number of gross, the number of dozens, and then purchase the remainder individually.

Devise a Plan.   We will proceed by repeated division by 12, which we recognize as equivalent to changing the number to base twelve.

Carry Out the Plan.   Change 1,000 to base 12:

$$
\begin{array}{r}
0 \\ \hline
12)\overline{6} \\
12)\overline{83} \\
12)\overline{1,000}
\end{array}
\quad
\begin{array}{l}
\text{r. } 6 \\
\text{r. } 11, \text{ or } E \text{ in base twelve} \\
\text{r. } 4
\end{array}
$$

Thus, $1,000 = 6E4_{twelve}$ so you must purchase 6 gross, 11 dozen, and 4 individual tags.

Look Back.   The cost is 6 × $50.40 + 11 × $4.80 + 4 × $.50 = $357.20. As you can see, this is considerably less expensive than purchasing the individual name tags.   ■

PROBLEM SET   3.3

▲ **A Problems**

1. **IN YOUR OWN WORDS** Explain how to change from base eight to base ten.

2. **IN YOUR OWN WORDS** Explain how to change from base sixteen to base ten.

3. **IN YOUR OWN WORDS** Explain how to change from base ten to base eight.

4. **IN YOUR OWN WORDS** Explain how to change from base ten to base sixteen.

5. **IN YOUR OWN WORDS** Explain how to change from base two to base eight.

6. **IN YOUR OWN WORDS** Discuss the binary (base two) numeration system.

7. Count the number of people in the indicated base.

    a. base ten   **b.** base five   **c.** base thirteen

    d. base eight   **e.** base two   **f.** base twelve

8. Count the number of people in the indicated base.

    a. base ten   **b.** base five   **c.** base three

    d. base eight   **e.** base two   **f.** base nine

In Problems 9–13, write the numbers in expanded notation.

  9. a. 643_{eight} **b.** 5387.9_{twelve}

10. a. 110111.1001_{two} **b.** 5411.1023_{six}

11. a. 64200051_{eight} **b.** 1021.221_{three}

12. a. 323000.2_{four} **b.** 234000_{five}

13. a. 3.40231_{five} **b.** 2033.1_{four}

Change the numbers in Problems 14–28 to base ten.

14. 527_{eight} **15.** 527_{twelve} **16.** 1101.11_{two}

17. $25\ TE_{twelve}$ **18.** 431_{five} **19.** 65_{eight}

20. 1011.101_{two} **21.** 11101000110_{two} **22.** 573_{twelve}

23. 4312_{eight} **24.** 2110_{three} **25.** 4312_{five}

26. 537.1_{eight} **27.** 3721_{eight} **28.** 5742_{eight}

29. Change 628 to base four.   **30.** Change 724 to base five.

31. Change 427 to base twelve.   **32.** Change 256 to base two.

33. Change 615 to base eight.   **34.** Change 412 to base five.

35. Change 615 to base two.

37. Change 795 to base seven.

39. Change 4,731 to base twelve.

41. Change 76 to base four.

36. Change 5,133 to base twelve.

38. Change 512 to base two.

40. Change 52 to base three.

42. Change 602 to base eight.

▲ B Problems

Use number bases to answer the questions given in Problems 43–55.

43. Change 52 days to weeks and days.

44. Change 158 hours to days and hours.

45. Change 55 inches to feet and inches.

46. Change 39 ounces to pounds and ounces.

47. Change 500 into gross, dozens, and units.

48. Change $4.59 into quarters, nickels, and pennies.

49. Using only quarters, nickels, and pennies, what is the minimum number of coins needed to make $.84?

50. Suppose you have two quarters, four nickels, and two pennies. Use base five to write a numeral to indicate your financial status.

51. A bookstore ordered 9 gross, 5 dozen, and 4 pencils. Write this number in base twelve and in base ten.

52. Change $8.34 into the smallest number of coins consisting of quarters, nickels, and pennies.

53. Change 44 days to weeks and days.

54. Change 54 months to years and months.

55. Change 29 hours to days and hours.

▲ Problem Solving

56. The *duodecimal numeration system* refers to the base twelve system, which uses the symbols 0, 1, 2, 3, 4, 5, 6, 7, 8, 9, T, E. Historically, a numeration system based on 12 is not new. There were 12 tribes in Israel and 12 Apostles of Christ. In Babylon, 12 was used as a base for the numeration system before it was replaced by 60. In the 18th and 19th centuries, Charles XII of Sweden and Georg Buffon (1707–1788) advocated the adoption of the base twelve system. Even today there is a Duodecimal Society of America that advocates the adoption of this system. According to the Society's literature, no one "who thought long enough — three to 17 minutes — to grasp the central idea of the duodecimal system ever failed to concede its superiority." Study the duodecimal system from 3 to 17 minutes, and comment on whether you agree with the Society's statement.

▲ Individual Research

57. Is it possible to have a numeration system with a base that is negative? Before you answer, see "Numeration Systems with Unusual Bases," by David Ballew, in *The Mathematics Teacher*, May 1974, pp. 413–414. Study the topic of negative bases, and present a report to the class.

3.4   HISTORY OF CALCULATING DEVICES

One of the biggest surprises of our time has been the impact of the personal computer on everyone's life. Few people foresaw that computers would jump the boundaries of scientific and engineering communities and create a revolution in our way of life in the last half of the 20th century. No one expected to see a computer sitting on a desk, much less on desks in every type of business, large and small, and even in our homes. Today no one is ready to face the world without some knowledge of computers. What has created this change in our lives is not only the advances in technology that have made computers small and affordable, but the tremendous imagination shown in developing ways to use them.

"It says, 'don't fold, spindle, or mutilate.'"

In this section we will consider the historical achievements leading to the easy availability of calculators and computers.

First Calculating Tool: Finger Calculating

The first "device" for arithmetic computations is finger counting. It has the advantages of low cost and instant availability. You are familiar with addition and subtraction on your fingers, but here is a method for multiplication by 9. Place both hands as shown in the margin. To multiply 4×9, simply bend the fourth finger from the left as shown:

The answer is read as 36, the bent finger serving to distinguish between the first and second digits in the answer. What about 36×9? Separate the third and fourth fingers from the left (as shown in the margin), since 36 has 3 tens. Next, bend the sixth finger from the left, since 36 has 6 units. Now the answer can be read directly from the fingers:

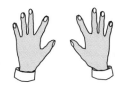

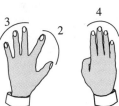

The answer is 324.

Aristophanes devised a complicated finger-calculating system in about 500 B.C., but it was very difficult to learn. Figure 3.4 (page 160) shows an illustration from a manual published about two thousand years later, in 1520.

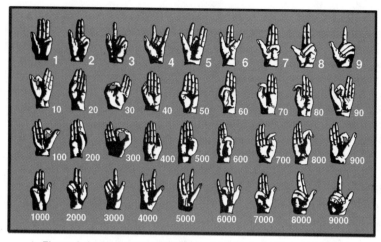

▲ **Figure 3.4   Finger calculation was important in the Middle Ages.**

Other Early Calculating Devices

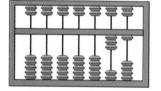

▲ **Figure 3.5   Abacus: The number shown is 31.**

Numbers can be represented by stones, slip knots, or beads on a string. These devices evolved into the abacus, as shown in Figure 3.5. Abaci (plural of abacus) were used thousands of years ago, and are still used today. In the hands of an expert, they even rival calculators for speed in performing certain calculations. An abacus consists of rods that contain sliding beads, four or five in the lower section, and two in the upper that equal one in the lower section of the next higher denomination. The abacus is useful today for teaching mathematics to youngsters. One can actually see the "carry" in addition and the "borrowing" in subtraction.

In 1624, John Napier invented a device similar to a multiplication table with movable parts, as shown in Figure 3.6. These rods are known as **Napier's rods** or **Napier's bones.**

▲ **Figure 3.6   Napier's rods (1617)**

A device used for many years was a **slide rule** (see Figure 3.7), which was also invented by Napier. The answers given by a slide rule are only visual approximations and do not have the precision that is often required.

▲ **Figure 3.7 Slide rule**

Mechanical Calculators

The 17th century saw the beginnings of calculating machines. When Pascal was 19, he began to develop a machine to add long columns of figures (see Figure 3.8). He built several versions, and since all proved to be unreliable, he considered this project a failure; but the machine introduced basic principles that are used in modern calculators.

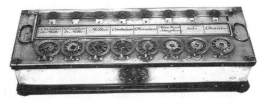

▲ **Figure 3.8 Pascal's calculator (1642)**

The next advance in mechanical calculators came from Germany in about 1672, when the great mathematician Gottfried Leibniz studied Pascal's calculators, improved them, and drew up plans for a mechanical calculator. In 1695 a machine was finally built (see Figure 3.9), but this calculator also proved to be unreliable.

▲ **Figure 3.9 Leibniz' reckoning machine (1695)**

In the 19th century, an eccentric Englishman, Charles Babbage, developed plans for a grandiose calculating machine, called a "difference engine," with thousands of gears, ratchets, and counters (see Figure 3.10).

▲ **Figure 3.10 Babbage's calculating machines**

CHARLES BABBAGE
(1792–1871)

Babbage continued for 40 years trying to build his "engines." He was constantly seeking financial help to subsidize his projects. However, his eccentricities often got him into trouble. For example, he hated organ-grinders and other street musicians, and for this reason people ridiculed him. He was a no-nonsense man and would not tolerate inaccuracy. Had he been willing to compromise his ideas of perfection with the technical skill available, he might have developed a workable machine.

Four years later, in 1826, even though Babbage had still not built his difference engine, he began an even more elaborate project — the building of what he called an "analytic engine." This machine was capable of an accuracy to 20 decimal places, but it, too, could not be built because the technical knowledge to build it was not far enough advanced. Much later, IBM built both the difference and analytic engines based on Babbage's design, but using modern technology, and they work perfectly.

Hand-Held Calculators

In the past few years pocket calculators have been one of the fastest growing items in the United States. There are probably two reasons for this increase in popularity. Most people (including mathematicians!) don't like to do arithmetic, and a good calculator is very inexpensive. It is assumed that you have access to a calculator to use with this book, and the basics of their use were introduced in Chapter 1.

First Computers

The devices discussed thus far in this section would all be classified as calculators. With some of the new programmable calculators, the distinction between a calculator and a computer is less well defined than it was in the past. Stimulated by the need to make ballistics calculations and to break secret codes during World War II, researchers made great advances in calculating machines. During the late 1930s and early 1940s John Vincent Atanasoff, a physics professor at Iowa State University, and his graduate student Clifford E. Berry built an electronic digital computer. However, the first working electronic computers were invented by J. Presper Eckert and John W. Mauchly at the University of Pennsylvania. All computers now in use derive from the original work they did between 1942 and 1946. They built the first fully electronic digital computer called ENIAC (Electronic Numerical Integrator and Calculator). The ENIAC filled a space 30 × 50 feet, weighed 30 tons, had 18,000 vacuum tubes, cost $487,000 to build, and used enough electricity to operate three 150-kilowatt radio stations. Vacuum tubes generated a lot of heat and the room where computers were installed had to be kept in a carefully controlled atmosphere; still, the tubes had a substantial failure rate, and to "program" the computer, many switches and wires had to be adjusted.

There is some controversy over who actually invented the first computer. In the early 1970s there was a lengthy court case over a patent dispute between Sperry Rand (who had acquired the ENIAC patent) and Honeywell, who represented those who originally worked on the ENIAC project. The judge in that case ruled that "between 1937 and 1942, Atanasoff . . . developed and built an automatic electronic digital computer for solving large systems of linear equations. . . . Eckert and Mauchly did not themselves invent the automatic electronic digital computer, but instead derived that subject matter from one Dr. John Vincent Atanasoff." On the other hand, others believe that a court of law is not the place for deciding questions about the history of science and give Eckert and Mauchly the honor of inventing the first computer. In 1980, the Association for Computing Machinery honored them as founders of the computer industry.

The UNIVAC I, built in 1951 by the builders of the ENIAC, became the first commercially available computer. Unlike the ENIAC, it could handle alphabetic data as well as numeric data. The invention of the transistor in 1947 and solid-state devices in the 1950s provided the technology for smaller, faster, more reliable machines.

Present-Day Computers

In 1958, Seymour Cray developed the first **supercomputer,** a computer that can handle at least 10 million instructions per second, to be used in major scientific research and military defense installations. The newest version of this computer, the C-90, is illustrated in Figure 3.11.

▲ **Figure 3.11   The Cray C-90 supercomputer**

Throughout the 1960s and early 1970s, computers continued to become faster and more powerful; they became a part of the business world. The "user" often never saw the machines. A **job** would be submitted to be "batch processed" by someone trained to run the computer. The notion of "time sharing" developed, so that the computer could handle more than one job, apparently simultaneously, by switching quickly from one job to another. Using a computer at this time was often frustrating; an incomplete job would be returned because of an error and the user would have to resubmit the job. It often took days to complete the project.

The large computers, known as *mainframes*, were followed by the **minicomputers,** which took up less than 3 cubic feet of space and no longer required a controlled atmosphere. These were still used by many people at the same time, though often directly through the use of terminals.

In 1976, Steven Jobs and Stephen Wozniak built the Apple, the first personal computer to be commercially successful. This computer proved extremely popular. Small businesses could afford to purchase these machines that, with a printer attached, took up only about twice the space of a typewriter. Technology "buffs" could purchase their own computers. This computer was designed so that innovations that would improve or enlarge the scope of performance of the machine could be added with relatively little difficulty. Owners could open up their machines and install new devices. This increased contact between the user and the machine pro-

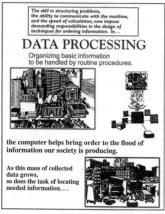

The skill in structuring problems, the ability to communicate with the machine, and the speed of calculation, now impose demanding responsibilities in the design of techniques for ordering information. In...

DATA PROCESSING

Organizing basic information to be handled by routine procedures.

the computer helps bring order to the flood of information our society is producing.

As this mass of collected data grows, so does the task of locating needed information...

A program for

INFORMATION RETRIEVAL

Locating and displaying specific material from a description of its content.

defines a pattern of words and values...   that the machine can seek out.

The care with which a pattern is defined is important.

The problem is—not to get too little...

or too much.

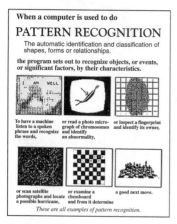

When a computer is used to do

PATTERN RECOGNITION

The automatic identification and classification of shapes, forms or relationships.

the program sets out to recognize objects, or events, or significant factors, by their characteristics.

To have a machine listen to a spoken phrase and recognize the words,   or read a photo micrograph of chromosomes and identify an abnormality,   or inspect a fingerprint and identify its owner,

or scan satellite photographs and locate a possible hurricane,   or examine a chessboard and from it determine   a good next move.

These are all examples of pattern recognition.

▲ **Figure 3.12   Common computer uses**

duced an atmosphere of tremendous creativity and innovation. In many cases, individual owners brought their own machines to work to introduce their superiors to the potential usefulness of the personal computer.

In the fast-moving technology of the computer world, the invention of language to describe it changes as fast as teenage slang. Today there are battery-powered **laptops,** weighing about 10 pounds, which are more powerful than the huge ENIAC. In 1989 a pocket computer weighing only 1 pound was introduced. The most recent trend has been to link together several personal computers so that they can share software and different users can easily access the same documents. These are called **LANs,** or local area **networks.** The most widely known networks are the **Internet,** which was first described by Paul Baran of the RAND Corp. in 1962, and the **World Wide Web** (www), developed by Tim Berners based on Baran's ideas and released in 1992. The growth of the Internet and the World Wide Web could not have been predicted. The figure on the facing page shows some of the significant events in the growth of the Internet from 1962 to the present. In 1962 there were 4 hosts, and in 1992 there were 727,000. By January 1, 1996, there were almost 10 million. In addition to educational access through most schools, you can access the Internet through services such as Compuserve, America Online, or Prodigy.

Uses for a Computer

The machine, or computer itself, is referred to as **hardware.** A personal computer might sell for under $1,000, but many extras, called **peripherals,** can increase the price by several thousand dollars.

Communication with computers has become more sophisticated, and when a computer is purchased it often comes with several *programs* that allow the user to communicate with the computer. A **computer program** is a set of step-by-step directions that instructs a computer how to carry out a certain task. Programs that allow the user to carry out particular tasks on a computer are referred to as **software.**

The business world has welcomed the advances made possible by the use of personal computers and powerful programs, known as **software packages,** that enable the computer to do many diverse tasks. The most important computer applications are **word processing, database managers,** and **spreadsheets.**

Today, computers are used for a variety of purposes (see Figure 3.12). **Data processing** involves large collections of data on which relatively few calculations need to be made. For example, a credit card company could use the system to keep track of the balances of its customers' accounts. **Information retrieval** involves locating and displaying material from a description of its content. For example, a police department may want information about a suspect, or a realtor may want listings that meet the criteria of a client. **Pattern recognition** is the identification and classification of shapes, forms, or relationships. For example, a chess playing program will examine the chessboard carefully to determine a good next move. The computer carries out these tasks by accepting information, performing mathematical and logical operations with the information, and then supplying the results of the operations as new information. Finally, a computer can be used for **simulation** of a real situation. For example, a prototype of a new airplane can be tested under various circumstances to determine its limitations.

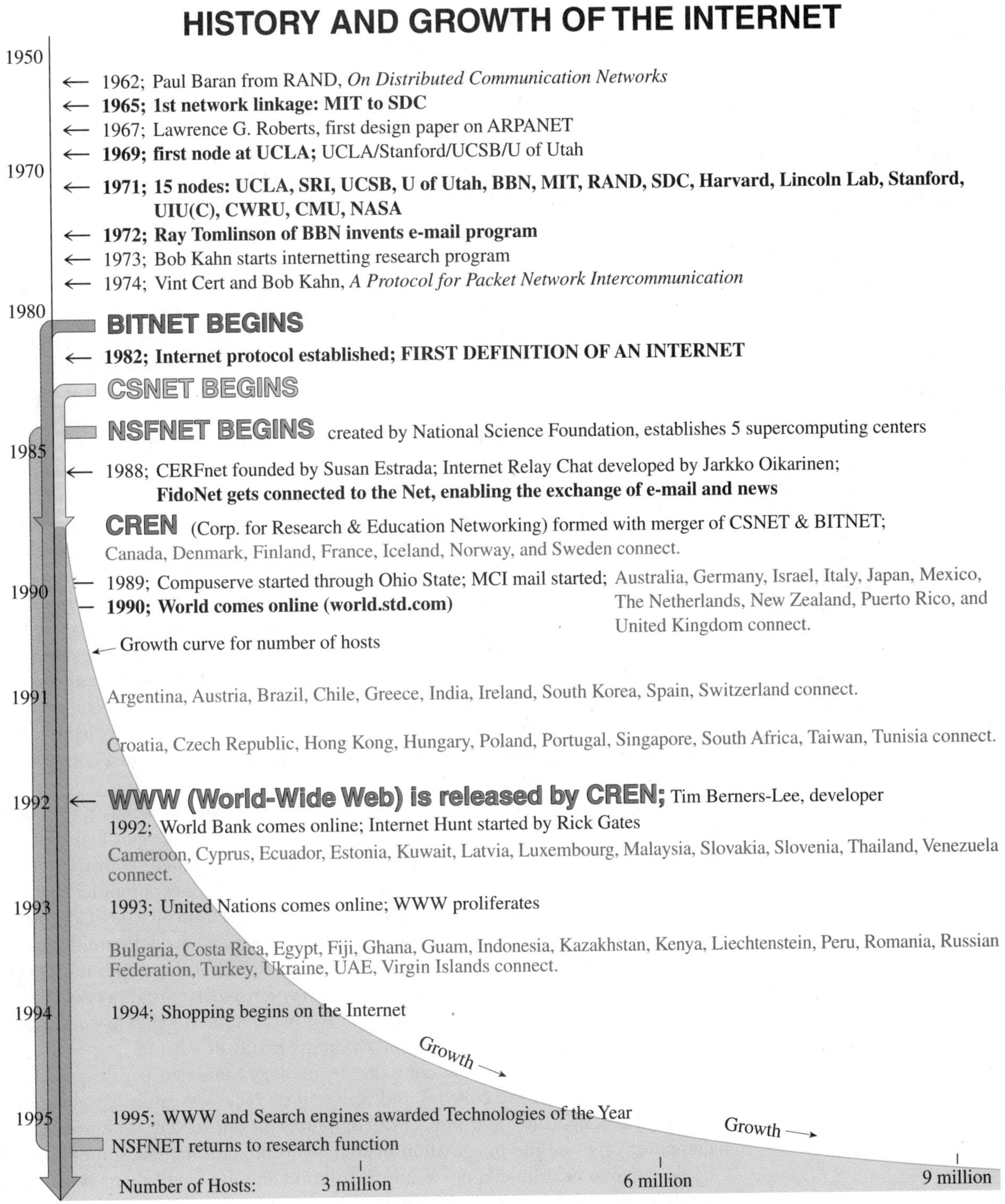

HISTORY AND GROWTH OF THE INTERNET

1950

← 1962; Paul Baran from RAND, *On Distributed Communication Networks*

← **1965; 1st network linkage: MIT to SDC**

← 1967; Lawrence G. Roberts, first design paper on ARPANET

← **1969; first node at UCLA;** UCLA/Stanford/UCSB/U of Utah

1970

← **1971; 15 nodes: UCLA, SRI, UCSB, U of Utah, BBN, MIT, RAND, SDC, Harvard, Lincoln Lab, Stanford, UIU(C), CWRU, CMU, NASA**

← **1972; Ray Tomlinson of BBN invents e-mail program**

← 1973; Bob Kahn starts internetting research program

← 1974; Vint Cert and Bob Kahn, *A Protocol for Packet Network Intercommunication*

1980

BITNET BEGINS

← **1982; Internet protocol established; FIRST DEFINITION OF AN INTERNET**

CSNET BEGINS

NSFNET BEGINS created by National Science Foundation, establishes 5 supercomputing centers

1985

← 1988; CERFnet founded by Susan Estrada; Internet Relay Chat developed by Jarkko Oikarinen; **FidoNet gets connected to the Net, enabling the exchange of e-mail and news**

CREN (Corp. for Research & Education Networking) formed with merger of CSNET & BITNET; Canada, Denmark, Finland, France, Iceland, Norway, and Sweden connect.

1990

← 1989; Compuserve started through Ohio State; MCI mail started; Australia, Germany, Israel, Italy, Japan, Mexico,
1990; World comes online (world.std.com) The Netherlands, New Zealand, Puerto Rico, and United Kingdom connect.

← Growth curve for number of hosts

1991

Argentina, Austria, Brazil, Chile, Greece, India, Ireland, South Korea, Spain, Switzerland connect.

Croatia, Czech Republic, Hong Kong, Hungary, Poland, Portugal, Singapore, South Africa, Taiwan, Tunisia connect.

1992 ← **WWW (World-Wide Web) is released by CREN;** Tim Berners-Lee, developer

1992; World Bank comes online; Internet Hunt started by Rick Gates

Cameroon, Cyprus, Ecuador, Estonia, Kuwait, Latvia, Luxembourg, Malaysia, Slovakia, Slovenia, Thailand, Venezuela connect.

1993

1993; United Nations comes online; WWW proliferates

Bulgaria, Costa Rica, Egypt, Fiji, Ghana, Guam, Indonesia, Kazakhstan, Kenya, Liechtenstein, Peru, Romania, Russian Federation, Turkey, Ukraine, UAE, Virgin Islands connect.

1994

1994; Shopping begins on the Internet

Growth →

1995

1995; WWW and Search engines awarded Technologies of the Year

Growth →

NSFNET returns to research function

| Number of Hosts: | 3 million | 6 million | 9 million |

▲ **Figure 3.13 The growth and history of the Internet**

STEPHEN HAWKING
(1942–   )

Stephen Hawking, a physicist known the world over for his study of the origins of the universe, has had a degenerative disease for many years. It has left him unable to speak, although his mind remains as brilliant as ever. He communicates with his colleagues and family and even gives speeches using a computer with a voice synthesizer. There has recently been a movie featuring his life and his accomplishments, and in 1992 he even appeared in an episode of *Star Trek: The Next Generation.* He is one of the great thinkers of our time.

By using a **modem,** installed internally or attached as a peripheral, and a program called a **communications package,** a computer can transfer information to and from other computers directly or over a phone line. If you send software or data from your computer to another, this is called **uploading;** if you get software or data from another computer, this is called **downloading.** A person or company may set up a computer specifically to be accessed by modem. The phone number is then made available to a private list of people or to the public. Some companies specialize in making databases available commercially. There is a charge to use these services based on the time you are connected, or the time you are **online.** Networks have been designed specifically to enable people to send messages to each other via computer. Such networks, known as **electronic mail** or **e-mail,** have become commonplace among large universities and companies, and are even available commercially to the public. Individuals or groups set up computers accessible by modem to share information about a common interest. These are called **bulletin boards.** Often, though not always, these are free and the only charge to the user is the cost of the telephone call.

Most companies and bulletin boards require a **password** or special procedures to be able to go online with their computers. A major problem for certain computer installations has been to make sure that it is impossible for the wrong person to get into their computer. There have been many instances where people have "broken into" computers just as a safecracker breaks into a safe. They have been clever enough to discover how to access the system. A few of these people have been caught and prosecuted. A security break of this nature can be very serious. By transferring funds in a bank computer, someone can steal money. By breaking into a military installation, a spy could steal secrets or commit sabotage. By breaking into a university's computers, one could change grades or records. **Computer abuse** falls into two categories: illegal (as described above, but also including illegal copying of programs), and abuse caused by ignorance. Ignorance of computers leads to the assumption that the output data are always correct or that the use of a computer is beyond one's comprehension.

With the tremendous advances in computer technology, financial investors and analysts can simulate various possible outcomes and select strategies that best suit their goals. Designers and engineers use computer-aided design, known as CAD. Using computer graphics, sculptors can model their work on a computer that will rotate the image so that the proposed work can be viewed from all sides. Computers have been used to help the handicapped with learning certain skills and by taking over certain functions. Computers can be used to "talk" for those without speech. For those without sight, there are artificial vision systems that translate a visual image to a tactile "image" that can be "seen" on their back. A person who has never seen a candle burn can learn the shape and dynamic quality of a flame.

Along with the advances in computer technology came new courses, and even new departments, in the colleges and universities. New specialties developed involving the design, use, and impact of computers on our society. The feats of the computer captured the imagination of some who thought that a computer could be designed to be as intelligent as a human mind and the field known as **artificial intelligence** was born. Many researchers worked on natural language translation with very high hopes initially, but the progress has been slow. Progress continues in "expert systems," programs that involve the sophisticated strategy of pushing the computer ever closer to human-like capabilities. The quest has produced powerful uses for computers.

What does the future hold? Earlier editions of this book attempted to make predictions of the future, and in every case those predictions became outdated during the life of the edition. Instead of a prediction, we invite you to visit a computer store.

PROBLEM SET 3.4

▲ A Problems

1. **IN YOUR OWN WORDS** Describe some of the computing devices that were used before the invention of the electronic computer.

2. **IN YOUR OWN WORDS** Distinguish between hardware and software.

3. **IN YOUR OWN WORDS** Give at least one example of each of the indicated computer functions: data processing, information retrieval, pattern recognition, and simulation.

4. **IN YOUR OWN WORDS** Name three places where you have seen computers in operation. Were these computers personal computers, were they connected to a large computer housed outside the building, or were they connected to other computers at the same site?

5. **IN YOUR OWN WORDS** What are five common uses for a computer in the office? Describe the purpose of each.

6. **IN YOUR OWN WORDS** It has been said that "computers influence our lives increasingly every year, and the trend will continue." Do you see this as a benefit or a detriment to humanity? Explain your reasons.

7. **IN YOUR OWN WORDS** A heated controversy rages about the possibility of a computer actually thinking. Do you believe that is possible? Do you think a computer can eventually be taught to be truly creative?

8. Use finger multiplication to do the following calculations:
   a. 3×9 **b.** 7×9 **c.** 27×9 **d.** 48×9 **e.** 56×9

9. Use finger multiplication to do the following calculations:
   a. 5×9 **b.** 8×9 **c.** 35×9 **d.** 47×9 **e.** 68×9

Problems 10–21 list a specific task. Decide whether a computer should be used, and if so, is it because of the computer's speed, complicated computations, repetition, or is there some other reason?

10. Guiding a missile to its target

11. Controlling the docking of a spacecraft

12. Tabulating averages for a teacher's classes

13. Identifying the marital status indicated on tax returns reporting income more than $100,000

14. Sorting the mail according to zip code

15. Assembling automobiles on an assembly line

16. Turning on and off the lights in a public office building

17. Typing and printing a term paper

18. Teaching a person to solve quadratic equations

19. Teaching a person to play the piano

20. Teaching a person how to repair a flat tire

21. Compiling a weather report and forecast

Problems 22–26 are multiple-choice questions designed to get you to begin thinking about computer use and abuse.

22. As part of a patient monitoring system in a local hospital, the computer (a) constantly monitors the patient's pulse and respiration rate; (b) sounds an alarm if the rates fall outside a preset range; and (c) notifies the nearest nursing station if either or both of its systems fail to function properly. The computer's capabilities are suited to these tasks because the tasks call for:
    A. Speed
    B. Repetition
    C. Program modification
    D. Simulation
    E. Data processing

23. Computers are capable of organizing large amounts of data. Which of the following situations would call for a computer having that capability?
    A. Printing the names and addresses of all employees
    B. Turning lights on and off at specified times
    C. Running a statistical analysis of family incomes as taken from the 1990 census
    D. Monitoring the temperature and humidity in a large office building
    E. All of the above

24. Karlin Publishing purchased a program for its computer that would enable it to streamline the printing process and decrease the time it takes to print each page. The program was too fast for Karlin's printer, so the company hired Shannon Sovndal to eliminate the problem. The human function required by Shannon for this job was:
    A. Performing calculations
    B. Modifying the hardware
    C. Interpreting output
    D. Modifying the software
    E. None of the above

25. Melissa knew that if she flunked mathematics she would be ineligible to play in her team's last five basketball games. When she found out that one of her classmates had been able to change someone else's grade on the school computer, she had this same person change her math grade from F to C. This example of computer abuse was:
    A. Invasion of privacy
    B. Stealing funds
    C. Pirating software
    D. Falsifying information
    E. Stealing information

26. One example of computer abuse is categorized as assuming that computers are largely incomprehensible. Which of the following situations would indicate this abuse?

   A. A secretary refuses promotion because he will have to interact with the company's computer.
   B. A clerk inputs the wrong information into the computer to defraud the company.
   C. A student changes the grade of a classmate.
   D. A hacker (computer hobbyist) copies a copyrighted program for three of her friends.
   E. A credit card customer wants to receive his bill on the 25th of the month instead of on the 10th, but when he calls to have it changed he is told that there is nothing that can be done since that is the way the computer was programmed.

▲ **B Problems**

27. **IN YOUR OWN WORDS** There is a great deal of concern today about **invasion of privacy** by computer. With more and more information about all of us being kept in computerized databases, there is the increasing possibility that a computer may be used to invade our privacy. Discuss this issue.

28. **IN YOUR OWN WORDS** Have you ever been told that something cannot be changed because "that is the way the computer does it"? Discuss this issue.

29. **IN YOUR OWN WORDS** Have you ever been told that something is right because the computer did it? For example, suppose I have my computer print, THE VALUE OF π IS 3.141592. Is this statement necessarily correct? What confidence do you place in computer results? What confidence do you think any of us should place on computer results? Discuss.

30. **IN YOUR OWN WORDS** Definitions for "thinking" are given. *According to each of the given definitions,* answer the question, "Can computers think?"

   a. To remember
   b. To subject to the process of logical thought
   c. To form a mental picture of
   d. To perceive or recognize
   e. To have feeling or consideration for
   f. To create or devise
   g. To have the ability to learn

31. What do you mean by "thinking" and "reasoning"? Try to formulate these ideas as clearly as possible, and then discuss the following question: "Can computers think or reason?"

▲ **Individual Research**

32. **IN YOUR OWN WORDS** One of the things that can go wrong when we use a computer is due to **machine error.** Such errors can be caused by an electric surge or a power failure. Talk to some users of computers and gather some anecdotal information about computer problems due to machine errors.

33. **IN YOUR OWN WORDS** It is possible that a computer can make a mistake due to a **programming error.** There are three types of programming errors: *syntax errors, run-time errors,* and *logic errors.* Do some research and write a paragraph about each of these types of programming errors.

Historical Note

ALAN M. TURING
(1912–1954)

Turing played a major role in the development of early computers. He was concerned with answering the question, "Can computers think?" (See Problems 30 and 31.) Today, computer memory has become so inexpensive that we may soon have computers that can pass the so-called turing test.

34. **IN YOUR OWN WORDS** One category of possible computer errors includes **data errors** or **input errors**. What is meant by this type of error?

35. In his book *Future Shock*, Alvin Toffler divided humanity's time on earth into 800 lifetimes. The 800th lifetime, in which we now live, has produced more knowledge than the previous 799 combined, and this has been made possible by computers. How have computers made this possible? What impact has this had on our lives? Do people actually know more today because they have ready access to vast amounts of knowledge?

36. In his article "Toward an Intelligence Beyond Man's" (*Time*, February 20, 1978), Robert Jastrow claimed that by the 1990s computer intelligence would match that of the human brain. He quoted Dartmouth President Emeritus John Kemeny as saying we would "see the ultimate relation between man and computer as a symbiotic union of two living species, each completely dependent on the other for survival." He called the computer a new form of life. Now, with the advantage of a historical perspective, do you think his predictions came to pass? Now that we have reached the 1990s, do you agree or disagree with the hypothesis that a computer could be a "new form of life"?

37. Software bugs can have devastating effects. For example, during the Persian Gulf War, a bug prevented a Patriot missile from firing at an incoming Iraqi Scud missile, which crashed into an Army barracks, killing 28 people. Another example involves Ashton-Tate, a company that never recovered its reputation after shipping bug-filled accounting software to its customers. Do some research to find recent (within the last 5 years) examples of major problems caused by software bugs.

38. Build a working model of Napier's rods.

39. "I became operational at the HAL Plant in Urbana, Ill., on January 12, 1997," the computer HAL declares in Arthur C. Clarke's 1968 novel, *2001: A Space Odyssey*. Now that time has passed and many advances have been made in computer technology between 1968 and today. Write a paper showing the similarities and differences between HAL and the computers of today.

40. Write a paper and prepare a classroom demonstration on the use of an abacus. Build your own device as a project.

41. Write a paper regarding the invention of the first electronic computer.

    References:
    Allen R. Mackintosh, "The First Electronic Computer," *Physics Today*, March 1987, pp. 25–32. Includes additional references.
    Bryan Bunch and Alexander Hellemans, eds., *The Timetables of Technology*, Touchstone Books, 1993.

42. Write a history of the hand-held calculator.

    References:
    W. D. Smith, "Hand-Held Calculators: Tool or Toy?" *New York Times*, August 20, 1972, p. 7.
    L. Buckwalter, "Now It's Pocket Calculators," *Mechanics Illustrated*, February 1973, pp. 108–109.
    Kathy B. Hamrick, "History of the Hand-Held Electronic Calculator," *MAA Monthly*, October 1996, pp. 633–639.

3.5 COMPUTERS AND THE BINARY NUMERATION SYSTEM

Almost everyone has had, or someday will have, firsthand experience with a computer. In this book, when we refer to a computer, we are referring to a personal computer (PC) because this is the type you are more likely to have an opportunity to use. To operate a computer, you don't *need* to know anything about how it works. We will keep our discussion general, so that what we say should apply equally to Macintosh, IBM format computers, Radio Shack, or other personal computers.

The Hardware

In this section we will familiarize you with the various components of a typical personal computer system that you might find in a home, classroom, or office. We will also discuss the link between the electronic signals of the computer and the human mind that enables us to harness the power and speed of the computer.

The basic parts of a computer are shown in Figure 3.14.

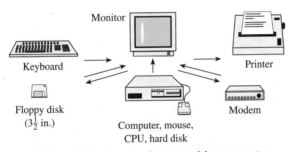

Monitor

Keyboard

Floppy disk
($3\frac{1}{2}$ in.)

Computer, mouse,
CPU, hard disk

Printer

Modem

▲ **Figure 3.14 Input and output with a computer**

We will focus our attention in this section on the terminology you will need to know to purchase a computer. The central unit of a computer contains a small microchip, called the *central processing unit*, where all processing takes place, and the computer memory, where the information is stored while the computer is completing a given task. The microchip and the unit that contains this chip are collectively referred to as the CPU. In 1972, Intel developed the first 8-bit microprocessor called the 8008, replaced successively by the 8080, 80386, 80486, Pentium, 80686, and Pentium 2 chips. In 1979, Motorola introduced the 68000 chip, followed by the 68020 chip, both used by the Macintosh computer. The Intel MMX and the Motorola 68030 are the current chips at the time of publication.

Although you do not need to understand electronic circuitry, a simple conceptual model of the inside of a computer can be very useful. The memory of a computer can be thought of as rows and rows of little storage boxes, each one with an address — yes, an address, just like your address. These boxes are of two types: ROM (read-only memory) and RAM (random-access memory). ROM is very special: you can never change what is in it and it is not erased when the computer is turned off.

It is programmed by the manufacturer to contain routines for certain processes that the computer must always use. The CPU can fetch information from it, but cannot send new results back. It is well protected because it is at the heart of the machine's capabilities. On the other hand, programs and data are stored in RAM while the computer is completing a task. If the power to the computer is interrupted for any time, however brief, everything in RAM is lost forever.

Your computer will also have memory. When referring to memory, it is convenient to talk in terms of a thousand bytes, called a kilobyte (KB or K), or in terms of a thousand kilobytes, called a megabyte (MB or MEG). Your computer will come with 2, 4, or more MEG of RAM. You will also need a **hard drive,** which is used as additional memory to store the programs you purchase. Hard disks in sizes of 40, 80, 120, or more megabytes are commonly installed in microcomputers.

To use a computer, you will need **input** and **output** devices. A **keyboard** and a **mouse** are the most commonly used input devices. The most common output devices are a **monitor** and a **printer.** One of the monitor's main functions is to enable you to keep track of what is going on in the computer. The letters or pictures seen on a monitor are displayed by little dots called **pixels,** just as they are on a television. As the number of dots is increased, we say that the **resolution** of the monitor increases. The better the resolution, the easier on your eyes. Most monitors today are color, and we list them in order of increasing resolution: RGB (red, green, blue), CGA, EGA, VGA, and Super VGA. Most monitors purchased today are VGA or Super VGA. Much of the expense in using a computer is associated with the purchase of a printer. Most printers today are *dot-matrix, ink-jet,* or *laser printers.*

Binary Numeration System

For a computer to be of use in calculating, the human mind had to invent a way of representing numbers and other symbols in terms of electronic devices. In fact, it is the simplest of electronic devices, the switch, that is at the heart of communication between machines and human beings. A switch can be only "on" or "off," as illustrated in Figure 3.15.

▲ **Figure 3.15   Two-state device. Light bulbs serve as a good example of two-state devices. The 1 is symbolized by "on," and the 0 is symbolized by "off."**

Decimal		Binary
0	↔	0
1	↔	1
2	↔	10
3	↔	11
4	↔	100
5	↔	101
6	↔	110
7	↔	111
8	↔	1000
9	↔	1001
10	↔	1010
11	↔	1011
12	↔	1100
13	↔	1101
14	↔	1110
15	↔	1111
16	↔	10000
17	↔	10001
.		.
.		.
.		.
31	↔	11111
32	↔	100000
.		.
.		.
.		.

▲ **Figure 3.16   Binary and decimal numeration. Can you see the pattern?**

We can see that it would be easy to represent the numbers "one" and "zero" by "on" and "off," but that won't get us very far. The idea that is key to allowing us to represent every symbol that we need is that of the **binary numeration system.** Binary numerals represent the numbers 1, 2, 3, . . . using only the symbols 1 and 0, as shown in Figure 3.16. Recall that expanded notation provides the link between the binary numeration and decimal numeration systems, and that for a binary numeration system the grouping is by twos.

EXAMPLE 1

What does 1110_{two} mean?

Solution

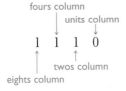

We see this means: $1 \times 8 + 1 \times 4 + 1 \times 2 + 0 = 8 + 4 + 2 = 14$ ▬

EXAMPLE 2

Change 10111_{two} to base 10.

Solution $10111_{two} = 1 \times 16 + 0 \times 8 + 1 \times 4 + 1 \times 2 + 1 \times 1$
$$= 16 + 4 + 2 + 1 = 23$$ ▬

To change a number from decimal to binary, it is necessary to find how many units (1 if odd, 0 if even), how many twos, how many fours, and so on are contained in that given number. This can be accomplished by repeated division by 2.

EXAMPLE 3

Change 47 to a binary numeral.

Solution

$$
\begin{array}{ll}
0 \quad\text{r. } 1 & \text{There is one thirty-two.} \\
2\overline{)\ 1} \quad\text{r. }0 & \text{There are zero sixteens.} \\
2\overline{)\ 2} \quad\text{r. } 1 & \text{There is one eight.} \\
2\overline{)\ 5} \quad\text{r. } 1 & \text{There is one four.} \\
2\overline{)11} \quad\text{r. } 1 & \text{There is one two.} \\
2\overline{)23} \quad\text{r. } 1 & \text{There is one unit.} \\
2\overline{)47} &
\end{array}
$$

If you read down the remainders, you obtain the binary numeral 101111. *Check:* 101111 means

$$1 \times 32 + 0 \times 16 + 1 \times 8 + 1 \times 4 + 1 \times 2 + 1 \times 1 = 47$$ ▬

ASCII Code

In the computer, each switch, called a **bit** (from <u>bi</u>nary dig<u>it</u>), represents a 1 or 0, depending on whether it is on or off. A group of eight of these switches is lined up to form a **byte**, which can represent any number from 0 to $11111111_{two} = 255_{ten}$.

Of course, when we use computers we need to represent more than numerals. There must be a representation for every letter and symbol that we wish to use. For example, we will need to represent expressions such as

$$x = 3.45 - y$$

The code that has been most commonly used is the American Standard Code for Information Interchange, developed in 1964 and called ASCII (pronounced "ask-key") code. A partial list of this code is shown in Table 3.8.

TABLE 3.8   Partial Listing of ASCII Code. The entire ASCII code has 127 symbols.

ASCII code	Symbol	ASCII code	Symbol	ASCII code	Symbol
32	Space	56	8	91	[
33	!	57	9	92	\
34	"	58	:	93	]
35	#	59	;	94	^
⋮	⋮	⋮	⋮	95	-
48	0	65	A	96	`
49	1	66	B	97	a
50	2	67	C	98	b
51	3	⋮	⋮	⋮	⋮
52	4	87	W	120	x
53	5	88	X	121	y
54	6	89	Y	122	z
55	7	90	Z	⋮	⋮

TODAY ONLY

MARKED DOWN TO

SCHOCHET

From this table you can see that the word *cat* would be represented by the ASCII code 99 97 116. In the machine, this would be represented by binary numerals, each one in a different byte. Thus, *cat* is represented as shown:

Symbol	ASCII Code	Binary	Byte
c	99	01100011	off on on off off off on on
a	97	01100001	off on on off off off off on
t	116	01110100	off on on on off on off off

The computer is able to store large amounts of information by having a large number of bytes, each one with an address:

ADDRESS **1**	INFORMATION
ADDRESS **2**	INFORMATION
ADDRESS **3**	IS
ADDRESS **4**	STORED
ADDRESS **5**	IN
ADDRESS **6**	EACH
ADDRESS **7**	NUMBERED
ADDRESS **8**	LOCATION
ADDRESS **9**	INFORMATION
ADDRESS **10**	INFORMATION

A computer has billions of such addresses. For example, suppose we want to store the information, 3 *cats*. We need six locations or addresses:

Address	*Information*	
1000	00110011	← Code for 3
1001	00100000	← Code for space
1010	01100011	← Code for *c*
1011	01100001	← Code for *a*
1100	01110100	← Code for *t*
1101	01110011	← Code for *s*

During the process of solving a problem, the storage unit will contain not only the set of instructions that specify the sequence of operations required to complete the problem, but also the input data. Notice that an address number has nothing to do with the contents of that address. The early computers stored these individual bits with switches, vacuum tubes, or cathode ray tubes, but computers today use much more sophisticated means for storing bits of information. These advances in storing information have enabled computers to use codes that use many more than eight bits. For example, a location might look like the following:

Address 00101   0001111100011011011000000001101

PROBLEM SET 3.5

▲ A Problems

1. **IN YOUR OWN WORDS** What does the CPU in a computer do?
2. **IN YOUR OWN WORDS** What are input and output devices?
3. **IN YOUR OWN WORDS** What is the difference between ROM and RAM?
4. **IN YOUR OWN WORDS** Describe Figure 3.15 on page 172.

What decimal number is represented by the light bulbs shown in Problems 5–8?

5.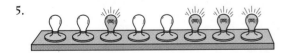

6.

7.

8.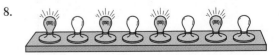

Write each number given in Problems 9–20 as a decimal numeral.

9. 1101_{two} 10. 1001_{two} 11. 1011_{two} 12. 1111_{two}

13. 11101_{two} 14. 10111_{two} 15. 11011_{two} 16. 11111_{two}

17. 1100011_{two} 18. 1110111_{two} 19. 10111000_{two} 20. 11111111_{two}

Write each number given in Problems 21–32 as a binary numeral.

21. 13   22. 15   23. 35   24. 46   25. 51   26. 63

27. 64   28. 256   29. 128   30. 615   31. 795   32. 803

▲ B Problems

Write the ASCII codes for the capital letters in Problems 33–36.

33. DO   34. PRINT   35. END   36. SAVE

Write the words for the ASCII codes given in Problems 37–40.

37. 72 65 86 69   38. 70 85 78   39. 83 84 85 68 89   40. 72 65 82 68

The binary numeration system has only two symbols, 0 and 1, which makes counting in the system simple. Arithmetic is particularly easy in the binary system, since the only arithmetic "facts" one needs are the following:

Addition

+	0	1
0	0	1
1	1	10

Multiplication

×	0	1
0	0	0
1	0	1

Perform the indicated operations in Problems 41–46.

41. 11_{two}
    $+ 10_{two}$

42. 1011_{two}
    $+ 111_{two}$

43. 110_{two}
    $+ 111_{two}$

44. 1101_{two}
    $+ 1100_{two}$

45. 11011_{two}
    $- 10110_{two}$

46. 10110_{two}
    $\times\ 101_{two}$

47. **IN YOUR OWN WORDS** There is a slogan among computer operators: *GIGO (garbage in, garbage out)*. This phrase reflects the fact that a computer will do exactly what it is told to do. Consider the following humorous story.

    A very large computer system has been created for the military. It was built and staffed by the best computer people in the country. "The system is now ready to answer questions," said the spokesperson for the project. A four-star general bit off the end of a cigar, looked whimsically at his comrades and said, "Ask the machine if there will be war or peace." The machine replied: YES. "Yes *what?*" bellowed the general. The operator typed in this question, and the machine answered: YES SIR!

    In terms of what you know about logic from Chapter 2, why did the computer answer the first question YES correctly?

48. **IN YOUR OWN WORDS** The **octal numeration system** refers to the base eight system, which uses the symbols 0, 1, 2, 3, 4, 5, 6, 7. Describe a process for converting octal numbers to decimal and decimal to octal.

TABLE 3.9   Octal–Binary Equivalents

Octal	Binary
0	000
1	001
2	010
3	011
4	100
5	101
6	110
7	111

Computer programmers often use octal numerals (see Problem 48) instead of binary numerals because it is easy to convert from binary to octal directly. Table 3.9 shows this conversion. A binary number 1101110111 is separated into three-digit groupings by starting with the right end of the number and supplying leading zeros at the left if necessary: 001 101 110 111. The binary groups are then replaced by their octal equivalents:

$$001_{two} = 1_{eight} \qquad 101_{two} = 5_{eight} \qquad 110_{two} = 6_{eight} \qquad 111_{two} = 7_{eight}$$

and the binary number is converted to its octal equivalent: 1567. *Conversely, an octal number can be expanded to a binary number using the same table of equivalents:*

$$5307_{eight} = 101\ 011\ 000\ 111_{two}$$

Use this information for Problems 49–58.

Convert the numbers in Problems 49–52 to the binary system.

49. a. 5_{eight} **b.** 6_{eight} **50. a.** 14_{eight} **b.** 624_{eight}

51. a. 7045_{eight} **b.** 3062_{eight} **52. a.** 5700_{eight} **b.** 04320_{eight}

Convert the numbers in Problems 53–58 to the octal system.

53. a. 101_{two} **b.** 100_{two} **54. a.** 011_{two} **b.** 001_{two}

55. $000\ 000\ 111\ 111\ 101\ 000_{two}$ **56.** $100\ 000\ 000\ 101\ 110\ 111_{two}$

57. $111\ 111\ 011\ 101\ 010\ 001_{two}$ **58.** $100\ 101\ 011\ 001\ 010\ 111_{two}$

▲ **Problem Solving**

59. You will need to use a pair of scissors for this problem. Construct the five cards shown below.

Card 1

Card 2

Card 3

Card 4

Card 5

Back of Card 5

"YES" is on back of this corner

Cut out those parts of the cards that are white. Ask someone to select a number between 1 and 31. Hold the cards face up on top of one another, and ask the person if his or her number is on the card. If the answer is yes, place the card on the table face up with the word *yes* at the upper left corner. If the answer is no, place the card face up with the word *no* at the upper left corner. Repeat this procedure with each card, placing the cards on top of one another on the table. After you have gone through the five cards, turn the entire stack over. The chosen number will be seen. Explain why this trick works.

▲ **Individual Research**

60. Visit a computer store, talk to a salesperson about the available computers, and then write a paper on your experiences.

61. Find out what local, state, and federal governments have stored in their computers about you and your family. Find out what you can see and what others can see. This will provide you with an interesting intellectual journey, if you wish to take it.

"The availability of low-cost calculators, computers, and related new technology has already dramatically changed the nature of business, industry, government, sciences, and social sciences."

NCTM Standards

JANUARY

Biographical Sketch

"I think of mathematics as a vast musical instrument."

Donald Knuth

Donald Knuth was born in Milwaukee, Wisconsin, on January 10, 1938. His monumental work, *The Art of Computer Programming*, demonstrates the significant interaction between mathematics and computer science. He is also well known as the designer of the T_EX typesetting system. In addition to being one of the world's leading scholars of computer science, he is an accomplished organist, composer, and novelist. The fact that his first publication was for *Mad Magazine* tells a little about the diversity of his interests. At the age of 41 he was awarded the National Medal of Science by President Carter.

"Computer programming is an art, an aesthetic experience like composition of poetry or music. I like to think of mathematics as a vast musical instrument on which one can play a great variety of beautiful melodies. The best programs are works of literature — addressed to human beings, not to machines," proclaims Dr. Knuth from his Stanford University office.

Professor Knuth says that what he enjoys most about mathematics is discovering and exploring patterns. He likes those questions that arise naturally when he tries to get a quantitative understanding of basic computer methods, which have both mathematical beauty and practical importance. "I can have fun while making other people happy with the things I happen to turn up."

Book Report

Write a 500-word report on the following book:

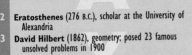   *How Computers Work*, Ron White (Emeryville, CA: Ziff-Davis Press, 1993).

Important Terms

Artificial intelligence [3.4]
ASCII code [3.5]
Binary numeration system [3.5]
Bit [3.5]
Bulletin boards [3.4]
Communications package [3.4]
Computer [3.4]
Computer abuse [3.4]
Data processing [3.4]
Database manager [3.4]
Decimal point [3.2]
Downloading [3.4]

e-mail [3.4]
Expanded notation [3.2]
Hard drive [3.5]
Hardware [3.4]
Hindu–Arabic numerals [3.2]
Information retrieval [3.4]
Input device [3.5]
Internet [3.4]
Keyboard [3.5]
Laptops [3.4]
Minicomputers [3.4]
Modem [3.4]
Monitor [3.5]

Mouse [3.5]
Multiplication principle [3.1]
Network [3.4]
Number [3.1]
Numeral [3.1]
Numeration system [3.1]
Online [3.4]
Output device [3.5]
Password [3.4]
Pattern recognition [3.4]
Peripheral [3.4]
Pixel [3.5]
Positional system [3.1]

Printer [3.5]
Program [3.4]
RAM [3.5]
Resolution [3.5]
ROM [3.5]
Simple grouping system [3.1]
Simulation [3.4]
Software [3.4]
Software packages [3.4]
Spreadsheet [3.4]
Subtraction principle [3.1]
Upload [3.4]
Word processing [3.4]
World Wide Web [3.4]

Important Ideas

The difference between number and numeral [3.1]
Properties of numeration systems [3.1]
Properties of the Hindu–Arabic numeration system [3.2]
The number of symbols in a base b system [3.3]

The history of calculating devices [3.4]
What do we mean by computer hardware? [3.4]
What do we mean by computer software? [3.4]
How can "data" be stored by a computer? [3.5]

Types of Problems

Write decimal numerals for numbers written in the Egyptian, Babylonian, or Roman numerations systems [3.1]
Write decimal numerals in the Egyptian, Babylonian, and Roman numeration systems [3.1]
Perform addition and subtraction in the Egyptian and Babylonian numeration systems [3.1]
Give the meaning of a particular numeral in the Hindu–Arabic numeration system [3.2]
Write the decimal representation for a number written in expanded notation [3.2]
Write a decimal numeral in expanded notation [3.2]
Use an abacus to illustrate the meaning of a decimal number [3.2]

Count objects in various number bases [3.3]
Change numbers from base b to base 10 [3.3]
Change numbers from base 10 to base b [3.3]
Solve applied problems by using number bases [3.3]
Know the principal uses for a computer [3.4]
Know the principal computer abuses [3.4]
Use the binary numeration system to represent a number [3.5]
Use the binary numeration system and the ASCII code to represent a word [3.5]
Convert from binary to octal and from octal to binary numeration systems [3.5]

CHAPTER 3   Review Questions

1. **IN YOUR OWN WORDS** What do we mean when we say that a numeration system is positional? Give examples.

2. **IN YOUR OWN WORDS** Is addition easier in a positional system or in a grouping system? Discuss and show examples.

3. **IN YOUR OWN WORDS** What are some of the characteristics of the Hindu–Arabic numeration system?

4. **IN YOUR OWN WORDS** Briefly discuss some of the events leading up to the invention of the computer.

5. **IN YOUR OWN WORDS** Discuss some computer abuses.

6. **IN YOUR OWN WORDS** Briefly describe each of the given computer terms.

    a. hardware   **b.** software   **c.** word processing   **d.** modem

    e. e-mail   **f.** RAM   **g.** computer bulletin board

Write the numbers given in Problems 7–10 in expanded notation.

7. one billion   8. 436.20001   9. 523_{eight} 10. 1001110_{two}

11. Write $4 \times 10^6 + 2 \times 10^4 + 5 \times 10^0 + 6 \times 10^{-1} + 2 \times 10^{-2}$ in decimal notation.

Write the numbers in Problems 12–15 in base ten.

12. 11101_{two} 13. 1111011_{two} 14. 122_{three} 15. 821_{twelve}

Write the numbers in Problems 16–19 in base two.

16. 12   17. 52   18. 1997   19. one million

20. **a.** Write 1,331 in base twelve.   **b.** Write 100 in base five.

Revisiting the Real World . . .

IN THE REAL WORLD   "It's just not fair!" exclaimed José. "Half my class have computers at home. I just went to the student store and the cheapest computer and printer I can find is over $1,000. Forget that!"

    "You're right, I gave up on getting my own computer," said Sheri. "But I use the library computer all the time. There's some network, I forget its name, that allows me to access every main library in the world! It is unbelievable. Last week I was able to access some information at the Stanford library!"

    "Do you think that library computer will help me with my paper? I need to *invent* my own numeration system," said José. "How am I supposed to do what took centuries for others to do? As I've said before, it's just not fair!"

Commentary

We will present an outline only, and Problem 1 of the Group Research below will ask you to fill in the details.

    Suppose we want to count the students in a classroom. The first thing we will need to do to invent a numeration system is to associate a given symbol and sound with a specific number of elements. Consider the groupings in Figure 3.17.

 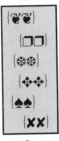  

    a.   **b.**   **c.**   **d.**

▲ **Figure 3.17 The objects in each of the groupings above have some common property. Can you tell what it is?**

We suggest the following names for the common properties:

Figure 3.17a: *fe*

Figure 3.17b: *fi*

Figure 3.17c: *fo*

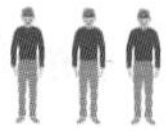

Figure 3.17d: *fum*

We can now use these names for larger sets. We can create an additive system, a grouping system, or a positional system. If we continue counting by forming a group called a *fiddle*, we could count:

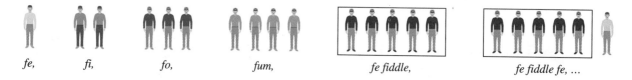

fe,   *fi,*   *fo,*   *fum,*   *fe fiddle,*   *fe fiddle fe,* ...

Can you continue counting? *fe fiddle fi, fe fiddle fo, fe fiddle fum, fi fiddle, fi fiddle fe, fi fiddle fi, fi fiddle fo, fi fiddle fum, fo fiddle, fo fiddle fe, fo fiddle fi, fo fiddle fo, fo fiddle fum, fum fiddle, fum fiddle fe,* . . .

Group Research

Working in small groups is typical of most work environments, and being able to work with others to communicate specific ideas is an important skill to learn. Work with three or four other students to submit a single report based on each of the following questions.

1. Invent an original numeration system.

2. Organize a debate. One side represents the algorists and the other side the abacists. The year is 1400. Debate the merits of the Roman numeration system and the Hindu–Arabic numeration system.

   Reference Barbara E. Reynolds, "The Algorists vs. The Abacists: An Ancient Controversy on the Use of Calculators," *The College Mathematics Journal*, Vol. 24, No. 3, May 1993, pp. 218–223. Includes additional references.

3. Organize a debate. The issue: Can computers think?

4. In a now famous paper, Alan Turing asked, "What would we ask a computer to do before we would say that it could think?" In the 1950s, Turing devised a test for "thinking" that is now known as the **turing test**. Dr. Hugh Loebner, a New York philanthropist, has offered $100,000 for the first machine that fools a judge into thinking it is a person. In 1991, the Computer Museum in Boston held a contest in

which 10 judges at the museum held conversations on terminals with eight respondents around the world, including six computers and two humans. The conversations of about 15 minutes each were limited to particular subjects, such as wine, fishing, clothing, and Shakespeare, but in a true turing test, the questions could involve any topic. Work as a group to decide the questions you would ask. Do you think a computer will ever be able to pass the test?

References

Betsy Carpenter, "Will Machines Ever Think?" *U.S. News & World Report*, October 17, 1988, pp. 64–65.

Stanley Wellborn, "Machines That Think," *U.S. News & World Report*, December 5, 1983, pp. 59–62.

5. Construct an exhibit on ancient computing methods. Some suggestions for your exhibit are charts of sample computations by ancient methods, pebbles, tally sticks, tally marks in sand, Roman numeral computations, abaci, Napier's bones, and old computing devices. You should consider answering the following questions as part of your exhibit: How do you multiply with Roman numerals? What is the scratch system? What is the lattice method of computation? What changes in our methods of long multiplication and long division have taken place over the years? How did the old computing machines work? Who invented the slide rule?

4 The Nature of Numbers

> In most sciences one generation tears down what another has built, and what one has established another undoes. In mathematics alone each generation builds a new story to the old structure.
>
> Herman Hankel

In the Real World . . .

"I'm never going to have children!" exclaimed Shelly. "Sometimes my little brother drives me nuts!" "I know what you mean," added Mary. "Just yesterday my brother asked me to explain what 5 means. The best I could do was to show him my fist."

"Stop! That's beginning to sound too much like my math class — just yesterday Ms. Jones asked us whether $\sqrt{2}$ is rational or irrational, **and** we need to justify our answer," Shelly said with a bit of anger. "I don't have the faintest; the only radical I know is my brother. I don't know why we should know about weird numbers like $\sqrt{2}$."

Contents

Perspective

Before we can do mathematical work, we need to have some building blocks for our journey. Those building blocks are sets of numbers. We assume that you know about some of these sets of numbers, but to form a common basis for the rest of the textbook, we will discuss some sets of numbers and their properties in this section.

The first set of numbers we encounter as children is the set of counting numbers. As we build more and more complex sets of numbers in this chapter, we move from the counting numbers to the integers (which include the counting numbers, zero, and their opposites), to the fractions, which we characterize as numbers whose decimal representations terminate or repeat. Even though this is a very useful set of numbers, there are certain applications (finding the length of a diagonal of a square, for example) that require numbers that cannot be written as terminating or repeating decimals. This set is called the set of irrational numbers. We complete our discussion of sets of numbers in this chapter by considering the union of the rationals and irrationals to form the set of real numbers.

4.1 NATURAL NUMBERS

The most basic set of numbers used by any society is the set of numbers used for counting:

$$\mathbb{N} = \{1, 2, 3, 4, 5, 6, 7, 8, 9, 10, 11, \ldots\}$$

This set of numbers is called the set of counting numbers or **natural numbers.** Let's assume that you understand what the numbers in this set represent, and you understand the operation of addition, $+$.

There are a few self-evident properties of addition for this set of natural numbers. They are called "self-evident" because they almost seem too obvious to be stated explicitly. For example, if you jump into the air, you expect to come back down. That assumption is well founded in experience and is also based on an assumption that jumping has certain undeniable properties. But astronauts have found that some very basic assumptions are valid on earth and false in space. Recognizing these assumptions (properties, axioms, laws, or postulates) is important.

Closure Property

When we add or multiply any two natural numbers, we know that we obtain a natural number. This "knowing" is an assumption based on experiences (inductive reasoning), but we have actually experienced only a small number of cases for all the possible sums and products of numbers. The scientist — and the mathematician in particular — is very skeptical about making assumptions too quickly. The assumption that the sum or product of two natural numbers is a natural number is given the name *closure*. The property is phrased in terms of sets and operations. Think of a set as a "box"; there is a label on the box — say, addition. If *all* additions of numbers in the box have answers that are already *in* the box, then we say the set is **closed** for addition. If there is at least one answer that is *not* contained in the box, then the set is said to be **not closed** for that operation.

> **Closure for + in $\mathbb{N}$; Closure for · in $\mathbb{N}$**
>
> Let $\mathbb{N}$ be the set of natural (or counting) numbers. Let a and b be any natural numbers.
>
> 1. $a + b$ is a natural number. We say $\mathbb{N}$ is **closed for addition.**
> 2. ab is a natural number. We say $\mathbb{N}$ is **closed for multiplication.**

EXAMPLE 1

Are the given sets closed for the given operations?

a. $A = \{1, 2, 3, 4, 5, 6, 7, 8, 9, 10\}$ for $+$

b. $B = \{0, 1\}$ for $\times$

Historical Note

The word *add* comes from the Latin word *adhere*, which means "to put to." Widman first used "+" and "−" signs in 1489 when he stated, "What is −, that is minus, what is +, that is more." The symbol "+" is believed to be a derivation of the Latin *et* ("and").

A closed box

Closure for + in $\mathbb{N}$
Closure for · in $\mathbb{N}$

Solution

a. The set A is *not closed* for addition because

$$5 + 7 = 12 \quad \text{and} \quad 12 \notin A$$

Note: The fact that $5 + 3 = 8$ is in the set does not change the fact that A is not closed for addition.

You need find only one *counterexample* to show that a property does not hold.

$5 + 7 = 12$; 12 is *not* in the box; the box is not closed.

$A = \{1, 2, 3, 4, 5, 6, 7, 8, 9, 10\}$

A box that is not closed

b. The set B is *closed* for the operation of multiplication, because all possible products are in B:

$$0 \times 0 = 0 \qquad 0 \times 1 = 0 \qquad 1 \times 0 = 0 \qquad 1 \times 1 = 1$$

Commutative and Associative Properties

The word *commute* can mean to travel back and forth from home to work; this back-and-forth idea can help you remember that the *commutative property* applies if you read from left to right or from right to left.

Another self-evident property of the natural numbers concerns the order in which they are added. It is called the **commutative property for addition** and states that the *order* in which two numbers are added makes no difference; that is (if we read from left to right),

$$a + b = b + a$$

for any two natural numbers a and b. The commutative property allows us to re-arrange numbers; it is called the *property of order*. Together with another property, called the *associative property*, it is used in calculation and simplification.

The word *associate* can mean connect, join, or unite; with this property you associate two of the added numbers.

The **associative property for addition** allows us to group numbers for addition. Suppose you wish to add three numbers — say, 2, 3, and 8:

$$2 + 3 + 8$$

To add these numbers you must first add two of them and then add this sum to the third. The associative property tells us that, no matter which two numbers are added first, the final result is the same. If parentheses are used to indicate the numbers to be added first, then this property can be symbolized by

$$(2 + 3) + 8 = 2 + (3 + 8)$$

The parentheses indicate the numbers to be added first. This associative property for addition holds for *any* three or more natural numbers.

Add the column of numbers in the margin. How long does it take? Five seconds is long enough if you use the associative and commutative properties for addition:

$$(9 + 1) + (8 + 2) + (7 + 3) + (6 + 4) + 5 = 10 + 10 + 10 + 10 + 5 = 45$$

However, it takes much longer if you don't rearrange (commutative) and regroup

9
8
7
6
5
4
3
2
+ 1

(associative) the numbers:

$$
\begin{aligned}
(9 + 8) + (7 + 6 + 5 + 4 + 3 + 2 + 1) &= (17 + 7) + (6 + 5 + 4 + 3 + 2 + 1) \\
&= (24 + 6) + (5 + 4 + 3 + 2 + 1) \\
&= (30 + 5) + (4 + 3 + 2 + 1) \\
&= (35 + 4) + (3 + 2 + 1) \\
&= (39 + 3) + (2 + 1) \\
&= (42 + 2) + 1 \\
&= 44 + 1 \\
&= 45
\end{aligned}
$$

The properties of associativity and commutativity are not restricted to the operation of addition. For example, these properties also hold for $\mathbb{N}$ and multiplication, as we will now discuss.

Multiplication is defined as repeated addition.

Multiplication

For $b \neq 0$, **multiplication** is defined as follows:

$$
a \times b \quad \text{means} \quad \underbrace{b + b + b + \cdots + b}_{a \text{ addends}}
$$

If $a = 0$, then $0 \times b = 0$.

We can now check commutativity and associativity for multiplication in the set $\mathbb{N}$ of natural numbers:

Commutativity: $\quad 2 \times 3 \overset{?}{=} 3 \times 2$

Associativity: $\quad (2 \times 3) \times 4 \overset{?}{=} 2 \times (3 \times 4)$

The question mark above the equal sign signifies that we should not assume the conclusion (namely, that the expressions on both sides are equal) until we check the arithmetic. Even though we can check these properties for particular natural numbers, it is impossible to check them for *all* natural numbers, so we accept the following axioms.

Commutative and Associative Properties

For any natural numbers a, b, and c:

Commutative Properties

Addition: $\quad a + b = b + a$

Multiplication: $\quad ab = ba$

Associative Properties

Addition: $\quad (a + b) + c = a + (b + c)$

Multiplication: $\quad (ab)c = a(bc)$

Historical Note

The word *multiply* comes from the Latin word *multiplicare*, which means "having many folds." Oughtred introduced the symbol $\times$ for multiplication in 1631, and Harriot introduced the dot in the same year. In 1698 Leibniz wrote to Bernoulli, saying, "I do not like the times sign $\times$ as a symbol for multiplication, as it is easily confounded with x, ... Often I simply relate two quantities by an interposed dot." Today, multiplication is denoted by a cross (usually in arithmetic) as in $3 \times a$; a dot, as in $3 \cdot a$; parentheses, as in $3(a)$; or juxtaposition, as in $3a$.

To distinguish between the commutative and associative properties, remember the following:

1. When the *commutative property* is used, the *order* in which the elements appear from left to right is changed, but the grouping is not changed.
2. When the *associative property* is used, the elements are *grouped* differently, but the order in which they appear is not changed.
3. If *both* the order and the grouping have been changed, then both the commutative and associative properties have been used.

We are not confined to $\mathbb{N}$ when discussing the associative and commutative properties (or any of the properties, for that matter). Nor are we restricted to addition and multiplication for our operations. Indeed, it is often fun to form your own "group" of numbers and see whether the properties hold for your group under your designated operation.

■■■ EXAMPLE 2   Polya's Method

Is the operation of union for sets an associative operation?

Solution   We use Polya's problem-solving guidelines for this example.

Understand the Problem.   Let $U = \{1, 2, 3, 4, 5, 6, 7, 8, 9, 10\}$, $P = \{1, 4, 7\}, Q = \{2, 4, 9, 10\}$, and $R = \{6, 7, 8, 9\}$. Does

$$(P \cup Q) \cup R = P \cup (Q \cup R)?$$
$$(P \cup Q) \cup R = \{1, 2, 4, 7, 9, 10\} \cup \{6, 7, 8, 9\} = \{1, 2, 4, 6, 7, 8, 9, 10\}$$
$$P \cup (Q \cup R) = \{1, 4, 7\} \cup \{2, 4, 6, 7, 8, 9, 10\} = \{1, 2, 4, 6, 7, 8, 9, 10\}$$

For this example, the operation of union for sets is associative. If we had seen

$$(P \cup Q) \cup R \neq P \cup (Q \cup R)$$

then we would have had a counterexample. Although they are equal, we cannot say that the property is true for all possibilities. However, all is not lost because it did help us to understand the question.

Devise a Plan.   Use Venn diagrams.

Carry Out the Plan.   Recall that the union is the entire shaded area.

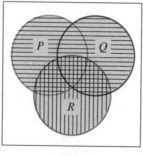 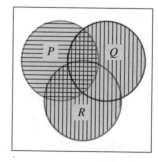

$(P \cup Q) \cup R$ $P \cup (Q \cup R)$

Look Back.   The operation of union for sets is an associative operation since the parts shaded in yellow are the same for both diagrams. ▬

Distributive Property

Are there properties in $\mathbb{N}$ that involve both operations? Consider an example.

▬▬ **EXAMPLE 3**

Suppose you are selling tickets for a raffle, and the tickets cost $2 each. You sell 3 tickets on Monday and 4 tickets on Tuesday. How much money did you collect?

Solution I   You sold a total of $3 + 4 = 7$ tickets, which cost $2 each, so you collected $2 \times 7 = 14$ dollars. That is,

$$2 \times (3 + 4) = 14$$

Solution II   You collected $2 \times 3 = 6$ dollars on Monday and $2 \times 4 = 8$ dollars on Tuesday for a total of $6 + 8 = 14$ dollars. That is,

$$(2 \times 3) + (2 \times 4) = 14$$

Since these solutions are equal, we see

$$2 \times (3 + 4) = (2 \times 3) + (2 \times 4)$$ ▬

Do you suppose this would be true if the tickets cost a dollars and you sold b tickets on Monday and c tickets on Tuesday? Then the equation would be

$$a \times (b + c) = (a \times b) + (a \times c)$$

or simply $a(b + c) = ab + ac$

This example illustrates the **distributive property.**

Distributive Property for Multiplication over Addition
   $a(b + c) = ab + ac$

In the set $\mathbb{N}$ of counting numbers, is addition distributive over multiplication? We wish to check

$$3 + (4 \times 5) \stackrel{?}{=} (3 + 4) \times (3 + 5)$$

Checking:

$$3 + (4 \times 5) = 3 + 20 = 23 \quad \text{and} \quad (3 + 4) \times (3 + 5) = 7 \times 8 = 56$$

Thus addition is not distributive over multiplication in the set of counting numbers.

The distributive property can also help to simplify arithmetic. Suppose you wish to multiply 9 by 71. You can use the distributive property to do the following mental multiplication:

$$9 \times 71 = 9 \times (70 + 1) = (9 \times 70) + (9 \times 1) = 630 + 9 = 639$$

This allows you to do the problem quickly and simply in your head.

Since these properties hold for the operations of addition and multiplication, we might reasonably ask whether they hold for other operations. **Subtraction** is defined as the opposite of addition.

Subtraction

The operation of **subtraction** is defined in terms of addition:

$$a - b = x \quad \text{means} \quad a = b + x$$

To verify the commutative property for subtraction, we check a particular example:

$$3 - 2 \overset{?}{=} 2 - 3$$

Now, $3 - 2 = 1$, but $2 - 3$ doesn't even exist in the set of natural numbers. Therefore, the commutative property does not hold for subtraction in $\mathbb{N}$.

Furthermore, to provide the result of the operation of subtraction for $2 - 3$, we must find a number that when added to 3 gives the result 2. But there is *no such natural number*. Thus, the set of natural numbers is *closed* for addition and multiplication, but is *not closed* for subtraction. In Section 4.3, we will add elements to the set of natural numbers to create the set of *integers*, which will be closed for subtraction as well as for addition and multiplication.

To make sure you understand the properties discussed in this section, this set of problems focuses on the properties of closure, commutativity, associativity, and distributivity rather than on the set of natural numbers and the operations of addition, multiplication, and subtraction. Since you are so familiar with the set of natural numbers and with these operations, you could probably answer questions about them without much reflection on the concepts involved. Therefore, in the problem set we work with some operations other than addition, multiplication, and subtraction. When we refer to a table, rows are horizontal and columns are vertical.

> From the definition of subtraction, $2 - 3 = \Box$ means $2 = \Box + 3$ so we need to find a number that, when added to 3, gives the result 2.

PROBLEM SET 4.1

▲ A Problems

IN YOUR OWN WORDS *Explain each of the words or concepts in Problems 1–8.*

1. Natural number   2. Multiplication   3. Subtraction

4. Closure   5. Commutativity   6. Associativity

7. Distributivity   8. Contrast commutativity and associativity

Use the definition of multiplication to show what each expression in Problems 9–14 means.

9. **a.** $3 \cdot 4$ **b.** $4 \cdot 3$ 10. **a.** $5 \cdot 2$ **b.** $2 \cdot 5$

11. **a.** $2 \cdot 1{,}845$ **b.** $1{,}845 \cdot 2$ 12. **a.** $3 \cdot 14{,}500$ **b.** $14{,}500 \cdot 3$

13. **a.** ab **b.** ba 14. **a.** xy **b.** yx

In Problems 15–24, *classify each as an example of the commutative property, the associative property, or both.*

15. $3 + 5 = 5 + 3$

16. $2 + 3 + 5 = 2 + 5 + 3$

17. $6 + (2 + 3) = (6 + 2) + 3$

18. $6 + (2 + 3) = (6 + 3) + 2$

19. $6 + (2 + 3) = 6 + (3 + 2)$

20. $6 + (2 + 3) = (2 + 3) + 6$

21. $(4 + 5)(6 + 9) = (4 + 5)(9 + 6)$

22. $(4 + 5)(6 + 9) = (6 + 9)(4 + 5)$

23. $(3 + 5) + (2 + 4) = (3 + 5) + (4 + 2)$

24. $(3 + 5) + (2 + 4) = (3 + 4) + (5 + 2)$

25. "Isn't this one just too sweet, dear?" asked the wife as she tried on a beautiful diamond ring. "No," the husband replied. "It's just too dear, sweet." Does this story remind you of the associative or the commutative property?

26. Is the operation of putting on your shoes and socks commutative?

27. In the English language, the meanings of certain phrases can be very different depending on the association of the words. For example,

(MAN EATING) TIGER

is not the same as

MAN (EATING TIGER)

Decide whether each of the following groups of words is associative.

a. HIGH SCHOOL STUDENT   **b.** SLOW CURVE SIGN

c. BARE FACTS PERSON.   **d.** RED FIRE ENGINE

e. TRAVELING SALESMAN JOKE   **f.** BROWN SMOKING JACKET

28. Think of three nonassociative word triples as shown in Problem 27.

"*How thrilling—what time does it start?*"

▲ **B Problems**

29. Consider the set $A = \{1, 4, 7, 9\}$ with an operation $*$ defined by the table.

$*$	1	4	7	9
1	9	7	1	4
4	7	9	4	1
7	1	4	7	9
9	4	1	9	7

$a * b$ means find the entry in row a and column b; for example, $7 * 9 = 9$ (the entry in row 7 and column 9). Find each of the following.

a. $7 * 4$ **b.** $9 * 1$ **c.** $1 * 7$ **d.** $9 * 9$ **e.** $4 * 7$

30. Consider the set $F = \{1, -1, i, -i\}$ with an operation $\times$ defined by the table.

$\times$	1	-1	i	$-i$
1	1	-1	i	$-i$
-1	-1	1	$-i$	i
i	i	$-i$	-1	1
$-i$	$-i$	i	1	-1

$a \times b$ means find the entry in row a and column b; for example, $-1 \times (-i) = i$ (the entry in row -1 and column $-i$). Find each of the following.

a. $-1 \times i$ b. $i \times i$ c. $-i \times i$ d. $-i \times 1$ e. $1 \times i$

31. Consider the set A and the operation $*$ from Problem 29. Does the set A satisfy the given property for the operation of $*$? Give reasons for your answers.

a. Closure   b. Associative   c. Commutative

32. Consider the set F and the operation of $\times$ from Problem 30. Does the set F satisfy the given property for the operation of $\times$? Give reasons for your answers.

a. Closure   b. Associative   c. Commutative

33. Let a be the process of putting on a shirt; let b be the process of putting on a pair of socks; and let c be the process of putting on a pair of shoes. Let $\bigstar$ be the operation of "followed by."

a. Is $\bigstar$ commutative for $\{a, b, c\}$?   b. Is $\bigstar$ associative for $\{a, b, c\}$?

34. Consider the set $\mathbb{N}$ of counting numbers and an operation $\leftarrow$ which means "select the first of the two." That is,

$$4 \leftarrow 3 = 4; \qquad 3 \leftarrow 4 = 3; \qquad 5 \leftarrow 7 = 5; \qquad 6 \leftarrow 6 = 6$$

a. Is the set $\mathbb{N}$ closed for the operation of $\leftarrow$? Give reasons.
b. Is $\leftarrow$ associative for $\mathbb{N}$? Give reasons.
c. Is $\leftarrow$ commutative for $\mathbb{N}$? Give reasons.

35. Consider the operation $\bullet$ defined by the table in the margin.
a. Find $\square \bullet \triangle$.   b. Find $\triangle \bullet \circ$.
c. Does $\circ \bullet \square = \square \bullet \circ$? Is the set commutative for $\bullet$?
d. Does $(\circ \bullet \triangle) \bullet \triangle = \circ \bullet (\triangle \bullet \triangle)$?

36. Let $\downarrow$ mean "select the smaller number" and $\rightarrow$ mean "select the second of the two." Is $\rightarrow$ distributive over $\downarrow$ in the set of counting numbers, $\mathbb{N}$?

37. Do the following problems mentally using the distributive property.
a. 6×82 b. 8×41 c. 7×49 d. 5×99 e. 4×88

38. Is the operation of union for sets a commutative operation? Give reasons.

39. Is the operation of intersection for sets a commutative operation? Give reasons.

40. Is the operation of intersection for sets an associative operation? Give reasons.

41. Is the set of even natural numbers closed for the operation of addition? Give reasons.

42. Is the set of even natural numbers closed for the operation of multiplication? Give reasons.

43. Is the set of odd natural numbers closed for the operation of multiplication? Give reasons.

44. Is the set of odd natural numbers closed for the operation of addition? Give reasons.

45. Let $S = \{1, 2, 3, \ldots, 99, 100\}$. Define an operation $\oplus$ as $a \oplus b = 2a + b$. Check the commutative and associative properties.

46. Let $S = \{1, 2, 3, \ldots, 999, 1000\}$. Define an operation $\otimes$ as $a \otimes b = 2(a + b)$. Check the commutative and associative properties.

▲ **Problem Solving**

47. Consider a soldier facing in a given direction (say north). Let us denote "left face" by ℓ, "right face" by r, "about face" by a, and "stand fast" by f. (The element f means

"don't move from your present position." It does not mean "return to your original position.") Then we define $H = \{\ell, r, a, f\}$ and an operation $\star$ meaning "followed by." Thus, $\ell \star \ell = a$ means "left face" followed by "left face" and is the same as the single command "about face." Complete the table in the margin.

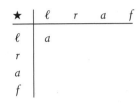

$\star$	ℓ	r	a	f
ℓ	a			
r				
a				
f				

48. Does the set H and operation $\star$ from Problem 47 satisfy the following properties? Give reasons.

   a. Closure   **b.** Associative   **c.** Commutative

49. Mensa is an association for people with high IQs. An advertisement for the organization offers the following challenge: "Take this instant test to see if you're a genius."

   > Put the appropriate plus or minus signs between the numbers, in the correct places, so that the sum total will equal 1.
   >
   > 0 1 2 3 4 5 6 7 8 9 = 1

   The advertisement also states, "This problem stumps 45% of the Mensa members who try it. And they *all* have IQs in the top 2% nationwide. If you can solve it, you might have what it takes to join us. To find out, . . . write to American Mensa, 2626 East 14th St., Brooklyn, NY 11235."

50. **The Vanishing Leprechaun Puzzle** The puzzle shown in Figure 4.1 consists of three pieces and was originally published by W. A. Elliott Company, 212 Adelaide St. W., Toronto, Canada M5H 1W7.

▲ **Figure 4.1 Leprechaun Puzzle: How many leprechauns?**

If we place the pieces together as shown at the top, we see 15 leprechauns. However, if we *commute* pieces A and B as shown at the bottom, we count 14 leprechauns! Clearly, from Figure 4.1 we see

$$AB \neq BA$$

Can you explain where the vanishing leprechaun went?

▲ **Individual Research**

51. What do the following people have in common?

> **Corazon Aquino,** former President of the Philippines
>
> **Leon Trotsky,** revolutionary
>
> **Carole King,** singer–songwriter
>
> **Heloise (Poncé Cruse Evans),** columnist, *Hints from Heloise*
>
> **Florence Nightingale,** pioneer in professional nursing

4.2   PRIME NUMBERS

Divisibility

A set of numbers that is important, not only in algebra but in all of mathematics, is the set of prime numbers. To understand prime numbers, you must first understand the idea of divisibility, along with some new terminology and notation.

The natural number 10 is divisible by 2, since there is a counting number 5 so that $10 = 2 \cdot 5$; it is not divisible by 3, since there is no counting number k such that $10 = 3 \cdot k$. This leads us to the following definition.

Divisibility

If m and d are counting numbers, and if there is a counting number k so that $m = d \cdot k$, we say that d **is a divisor of** m, d **is a factor of** m, d **divides** m, and m **is a multiple of** d. We denote this relationship by $d \mid m$.

⊘ Do not confuse this notation with the notation sometimes used for fractions: 5/30 means 5 divided by 30, which is the fraction $\frac{1}{6}$, and 5|30 means "5 divides 30." ⊘

That is, $5 \mid 30$ is read "5 divides 30" and means that there exists some counting number k — namely, 6 — such that $30 = 5 \cdot k$.

▬▬ EXAMPLE I

Tell whether each of the following is true or false, and give the meaning of each.

a. $7 \mid 63$ **b.** $8 \mid 104$ **c.** $14 \mid 2$ **d.** $6 \mid 15$

Solution

a. $7 \mid 63$ is read "7 divides 63" and is true since we can find a counting number k — namely, 9 — such that $63 = 7 \cdot k$.

b. $8 \mid 104$ is true, since $104 = 8 \cdot 13$.

c. $14|2$ is false because we can find no counting number k so that $2 = 14 \cdot k$. We write $14 \nmid 2$ to say that 14 does not divide 2.

d. $6|15$ is false, because we can find no counting number k so that $15 = 6 \cdot k$.

It is easy to see that 1 divides every counting number m, since $m = 1 \cdot m$. Also, by the commutative property of multiplication, $m = m \cdot 1$; thus every counting number m divides itself. We have proved the following theorem.

Least Number of Divisors

Every counting number greater than 1 has at least two distinct divisors: itself and 1.

Consider the number 341,592. Is this number divisible by 2? By 3? By 4? By 5? You may know some ways of answering these questions without the necessity of actually doing the division.

■ EXAMPLE 2   **Polya's Method**

Find a rule for the divisibility of any number N by 2.

Solution   We use Polya's problem-solving guidelines for this example.

Understand the Problem.   You might already know the rule for divisibility by 2. It says that *if the last digit of the number is even,* then *the number is divisible by 2.* That is, if the number N ends in 0, 2, 4, 6, or 8, it is divisible by 2. This example asks us to show why this rule "works."

Devise a Plan.   We will begin with a simpler example, say, 341,592. We write the number in expanded notation:

$$341{,}592 = 3 \times 10^5 + 4 \times 10^4 + 1 \times 10^3 + 5 \times 10^2 + 9 \times 10^1 + 2$$

The question to answer is "When will this number be divisible by 2?" Associate all the digits except the last one:

$$341{,}592 = \underbrace{3 \times 10^5 + 4 \times 10^4 + 1 \times 10^3 + 5 \times 10^2 + 9 \times 10^1}_{\text{This is divisible by 2.}} + 2$$

The associated part is *always* divisible by 2, since $2|10$, $2|10^2$, $2|10^3$, $2|10^4$, and $2|10^5$.

Carry Out the Plan.   For the number M we see that since $2|10$, $2|10^2$, . . ., $2|10^n$, the divisibility of M by 2 depends solely on whether 2 divides the last digit.

Look Back.   We see that $2|341{,}592$ since $2|2$. Also $2|838$ since $2|8$, and $2 \nmid 839$ since $2 \nmid 9$.

Similar rules apply for divisibility by 4 or 8. You might even expect to try the same type of rule for divisibility by 3, but this situation is not quite so easy, as shown by the following example.

■■■■■■ **EXAMPLE 3**   **Polya's Method**

Find a rule for divisibility by 3.

Solution   We use Polya's problem-solving guidelines for this example.

Understand the Problem.   Try a simple example. We see that $3|84$ since we can find a counting number — namely, 28 — so that $84 = 3 \cdot k$. We also note that $3 \nmid 4$, so the same type of rule that worked for divisibility by 2 will not work for 3.

Devise a Plan.   We will once again look at the expanded notation. Consider a simpler problem — say, the divisibility of 341,592 by 3:

$$341,592 = 3 \times 10^5 + 4 \times 10^4 + 1 \times 10^3 + 5 \times 10^2 + 9 \times 10^1 + 2$$

The plan is to make each product divisible by 3. We do this by adding and subtracting 1 from each term containing 10^b, where b is a natural number.

Carry Out the Plan.

$$341,592 = 3 \times (10^5 - 1 + 1) + 4 \times (10^4 - 1 + 1) + 1 \times (10^3 - 1 + 1)$$
$$+ 5 \times (10^2 - 1 + 1) + 9 \times (10^1 - 1 + 1) + 2$$

We now use the distributive and associative properties to rewrite this expression:

$$341,592 = [3 \times (10^5 - 1) + 3] + [4 \times (10^4 - 1) + 4] + [1 \times (10^3 - 1) + 1]$$
$$+ [5 \times (10^2 - 1) + 5] + [9 \times (10^1 - 1) + 9] + 2$$
$$= \underbrace{[3(10^5 - 1) + 4(10^4 - 1) + 1(10^3 - 1) + 5(10^2 - 1) + 9(10^1 - 1)]}$$

This is divisible by 3.

$$+ \underbrace{[3 + 4 + 1 + 5 + 9 + 2]}$$

This is the sum of the digits.

Notice what we have done: $10^1 - 1 = 9$; $10^2 - 1 = 99$; $10^3 - 1 = 999$; $10^4 - 1 = 9,999$; $10^5 - 1 = 99,999$. These are all divisible by 3, and hence we see that if $3|(3 + 4 + 1 + 5 + 9 + 2)$, then $3|341,592$. Checking, we see that $3|24$ (sum of digits), so $3|341,592$. Furthermore, since $3|(10^n - 1)$ for any counting number n, we see that the divisibility of a number N by 3 depends on whether 3 divides the sum of the digits.

Look Back.   If the sum of the digits of a number N is divisible by 3, then N is divisible by 3.   ■

Some of the more common **rules of divisibility** are shown in Table 4.1.

TABLE 4.1   Rules of Divisibility for a Counting Number _N_

N is divisible by	Test
1	all N
2	if the last digit is divisible by 2.
3	if the sum of the digits is divisible by 3.
4	if the number formed by the last two digits is divisible by 4.
5	if the last digit is 0 or 5.
6	if the number is divisible by 2 and by 3.
8	if the number formed by the last three digits is divisible by 8.
9	if the sum of the digits is divisible by 9.
10	if the last digit is 0.
12	if the number is divisible by 3 and by 4.

Prime Numbers

Since every counting number greater than 1 has at least two divisors, can any number have more than two?

Checking: 2 has exactly two divisors: 1, 2
3 has exactly two divisors: 1, 3
4 has more than two divisors: 1, 2, and 4

Thus, some numbers (such as 2 and 3) have exactly two divisors, and some (such as 4 and 6) have more than two divisors. Do any counting numbers have fewer than two divisors?

We now state a definition that classifies each counting number according to the number of divisors it has.

Prime Number

A **prime number** is a counting number that has exactly two divisors. A counting number that has more than two divisors is called a **composite number.**

We see that 2 is prime, 3 is prime, 4 is composite (since it is divisible by three counting numbers), 5 is prime, 6 is composite (since it is divisible by 1, 2, 3, and 6). Note that every counting number greater than 1 is either prime or composite. The number 1 is neither prime nor composite.

One method for finding primes smaller than some given number was first used by a Greek mathematician, Eratosthenes, more than 2,000 years ago. The technique

Historical Note

In 1978, Hugh C. Williams of the University of Manitoba discovered a prime number:

11,111,111,111,111,111,111,
111,111,111,111,111,111,111,
111,111,111,111,111,111,111,
111,111,111,111,111,111,111,
111,111,111,111,111,111,111,
111,111,111,111,111,111,111,
111,111,111,111,111,111,111,
111,111,111,111,111,111,111,
111,111,111,111,111,111,111,
111,111,111,111,111,111,111,
111,111,111,111,111,111,111,
111,111,111,111,111,111,111,
111,111,111,111,111,111,111,
111,111,111,111,111,111,111,
111,111,111,111,111,111,111,
111

He called this number R_{317} for short. Do some research and see if you can find some newly discovered prime numbers.

Historical Note

Eratosthenes was a highly talented and versatile person: athlete, astronomer, geographer, historian, poet, and mathematician. In addition to his sieve for prime numbers, he invented a mechanical "mean finder" for duplicating the cube, and he made some remarkably accurate measurements of the size of the earth. Unfortunately, most of his mathematical contributions are lost. It is said that he committed suicide by starving himself to death upon discovering that he had ophthalmia.

is known as the **sieve of Eratosthenes.** Suppose we wish to find the primes less than 100. We prepare a table of counting numbers 1–100, as shown in Table 4.2.

TABLE 4.2   Finding Primes Using the Sieve of Eratosthenes

~~1~~	②	③	~~4~~	⑤	~~6~~	⑦	~~8~~	~~9~~	~~10~~
⑪	~~12~~	⑬	~~14~~	~~15~~	~~16~~	⑰	~~18~~	⑲	~~20~~
~~21~~	~~22~~	㉓	~~24~~	25	~~26~~	~~27~~	~~28~~	㉙	~~30~~
㉛	~~32~~	~~33~~	~~34~~	35	~~36~~	㊲	~~38~~	~~39~~	~~40~~
㊶	~~42~~	㊸	~~44~~	~~45~~	~~46~~	㊼	~~48~~	~~49~~	~~50~~
~~51~~	~~52~~	㊾	~~54~~	55	~~56~~	57	~~58~~	㊾	~~60~~
㊶	~~62~~	~~63~~	~~64~~	65	~~66~~	㊻	~~68~~	~~69~~	~~70~~
㊼	~~72~~	㊽	~~74~~	~~75~~	~~76~~	~~77~~	~~78~~	㊾	~~80~~
~~81~~	~~82~~	㊿	~~84~~	85	~~86~~	~~87~~	~~88~~	㊿	~~90~~
~~91~~	~~92~~	~~93~~	~~94~~	95	~~96~~	㊼	~~98~~	~~99~~	100

Cross out 1, since it is not classified as a prime number.

Draw a circle around 2, the smallest prime number. Then cross out every following multiple of 2, since each is divisible by 2 and thus is not prime.

Draw a circle around 3, the next prime number. Then cross out each succeeding multiple of 3. Some of these numbers, such as 6 and 12, will already have been crossed out because they are also multiples of 2.

Circle the next open prime, 5, and cross out all subsequent multiples of 5.

The next prime number is 7; circle 7 and cross out multiples of 7. Since 7 is the largest prime less than $\sqrt{100} = 10$, we now know that all the remaining numbers are prime.

The process is a simple one, since you do not have to cross out the multiples of 3 (for example) by checking for divisibility by 3 but can simply cross out every third number. Thus, anyone who can count can find primes by this method. Also, notice that in finding the primes under 100, we had crossed out all the composite numbers by the time we crossed out the multiples of 7. That is, to find all primes less than 100: (1) Find the largest prime smaller than or equal to $\sqrt{100} = 10$ (7 in this case); (2) cross out multiples of primes up to and including 7; and (3) all the remaining numbers in the chart are primes.

This result generalizes. If you wish to find all primes smaller than n:

1. Find the *largest* prime less than or equal to $\sqrt{n}$.
2. Cross out the multiples of primes less than or equal to $\sqrt{n}$.
3. All the remaining numbers in the chart are primes.

Phrasing this another way, if n is composite, then one of its factors must be less than or equal to $\sqrt{n}$. That is, if $n = ab$, then it can't be true that *both a and b* are greater than $\sqrt{n}$ (otherwise $ab > \sqrt{n}\,\sqrt{n} = n = ab$, so $ab > ab$ is a contradiction). Thus, one of the factors must be less than or equal to $\sqrt{n}$.

Prime Factorization

Prime numbers are fundamental to many mathematical processes. In particular, we use prime numbers in working with rational numbers later in this chapter. You will need to understand *prime factorization, greatest common factor,* and *least common multiple.*

   The operation of **factoring** is the reverse of the operation of multiplying. For example, multiplying 3 by 6 yields $3 \cdot 6 = 18$ and this answer is unique (only one answer is possible). In the reverse process, called factoring, you are given the number 18 and asked for numbers that can be multiplied together to give 18. This process is *not* unique; we list several different factorizations of 18:

$$18 = 1 \cdot 18 = 18 \cdot 1 = 2 \cdot 9 = 9 \cdot 2 = 1 \cdot 1 \cdot 2 \cdot 9 = 3 \cdot 6 = 2 \cdot 3 \cdot 3 = \cdots$$

There are, in fact, infinitely many possibilities. We make some agreements so the process gives a unique answer:

1. We will not consider the order in which the factors are listed as important. That is, $2 \cdot 9$ and $9 \cdot 2$ are considered the same factorization.
2. We will not consider 1 as a factor when writing out any factorizations. That is, prime numbers do not have factorizations.
3. Recall that we are working in the set of counting numbers; thus, $18 = 36 \cdot \frac{1}{2}$ and $18 = (-2)(-9)$ are *not* considered factorizations of 18.

With these agreements, we have greatly reduced the possibilities:

$$18 = 2 \cdot 9 = 3 \cdot 6 = 2 \cdot 3^2$$

These are the only three possible factorizations. Notice that the last factorization contains only prime factors; thus it is called the **prime factorization** of 18.

   It should be clear that, if a number is composite, it can be factored into two counting numbers greater than 1. Each of these two numbers will be prime or composite. If both are prime, then we have a prime factorization. If one or more is composite, we repeat the process, and continue until we have written the original number as a product of primes. It is also true that this representation is unique. This is one of the most important results in arithmetic, and we state it formally.

Fundamental Theorem of Arithmetic

Every counting number greater than 1 is either a prime or a product of primes, and the prime factorization is unique (except for the order in which the factors appear).

EXAMPLE 4

Find the prime factorizations: **a.** 385   **b.** 1,400

Solution

a. One of the easiest ways to find the prime factors of a number is to try division by each of the prime numbers in order: 2, 3, 5, 7, ... The rules of divisibility in Table 4.1 may help. For this example, we need to check primes up to $\sqrt{385} \approx 19$. If none of the primes up to 19 is a factor of 385, then 385 is prime. We see by inspection that 385 is not divisible by 2 or 3. It is divisible by 5, so

$$385 = 5 \cdot 77$$

Since 77 is composite ($77 = 7 \cdot 11$), we write

$$385 = 5 \cdot 7 \cdot 11$$

Many people prefer to write these using a **factor tree**:

$$385$$
$$5 \cdot 77$$
$$5 \cdot 7 \cdot 11$$

b. Using a factor tree, we may begin with *any* factors of 1,400:

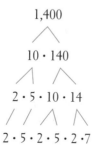

$$1,400$$
$$10 \cdot 140$$
$$2 \cdot 5 \cdot 10 \cdot 14$$
$$2 \cdot 5 \cdot 2 \cdot 5 \cdot 2 \cdot 7$$

We write the prime factorization using exponents:

$$1,400 = 2^3 \cdot 5^2 \cdot 7$$

The answer to Example 4 is written in what is called **canonical form.** The *canonical representation* of a number is the representation of that number as a product of primes using exponential notation with the factors arranged in order of increasing magnitude.

EXAMPLE 5

Find the canonical representation of 3,465.

Solution   We use a factor tree.

3,465

5 · 693

5 · 3 · 231

5 · 3 · 3 · 77

5 · 3 · 3 · 7 · 11

The prime factorization in canonical form is $3^2 \cdot 5 \cdot 7 \cdot 11$.   —

Greatest Common Factor

Suppose we look at the set of factors common to a given set of numbers:

Factors of 18: $\{1, 2, 3, 6, 9, 18\}$

Common factors: $\{1, 2, 3, 6\}$

Factors of 12: $\{1, 2, 3, 4, 6, 12\}$

The *greatest common factor* is the largest number in the set of common factors.

Greatest Common Factor

The **greatest common factor (g.c.f.)** of a set of numbers is the largest number that divides evenly into each of the numbers in the given set.

The procedure for finding the greatest common factor involves the canonical form of the given numbers. For example, suppose we want to find the greatest common factor of 24 and 30. We could, of course, find all factors, then the intersection (common factors), and finally the greatest one in that set. Instead, find the canonical representation of each:

Factors of:
24: $\{1, 2, 3, 4, 6, 8, 12, 24\}$
30: $\{1, 2, 3, 5, 6, 10, 15, 30\}$
Common factors: $\{1, 2, 3, 6\}$
    g.c.f. $= 6$

$$24 = 2^3 \cdot 3 \quad = 2^3 \cdot 3^1 \cdot 5^0$$
$$30 = 2 \cdot 3 \cdot 5 = 2^1 \cdot 3^1 \cdot 5^1$$

Select one representative from each of the columns in the factorization. The representative we select when finding the g.c.f. is the one with the smallest exponent. The g.c.f. is the product of these representatives:

$$\text{g.c.f.} = 2^1 \cdot 3^1 \cdot 5^0$$

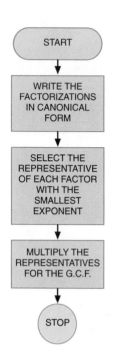

▲ **Figure 4.2   Procedure for finding the g.c.f.**

The procedure for finding the g.c.f. is summarized in Figure 4.2.

■■■ EXAMPLE 6

Find the greatest common factor of the given sets of numbers.

a. 300, 144   **b.** 15, 28

Solution

a.
$$300 = 2^2 \cdot 3^1 \cdot 5^2$$
$$144 = 2^4 \cdot 3^2 \cdot 5^0$$
$$\text{g.c.f.} = 2^2 \cdot 3^1 \cdot 5^0$$
$$= 4 \cdot 3 \cdot 1 = 12$$

b.
$$15 = 2^0 \cdot 3^1 \cdot 5^1 \cdot 7^0$$
$$28 = 2^2 \cdot 3^0 \cdot 5^0 \cdot 7^1$$
$$\text{g.c.f.} = 2^0 \cdot 3^0 \cdot 5^0 \cdot 7^0$$
$$= 1 \cdot 1 \cdot 1 \cdot 1 = 1$$

           ■

If the greatest common factor of two numbers is 1, we say that the numbers are **relatively prime.** Notice that 15 and 28 are relatively prime, but they themselves are not prime. It is possible for relatively prime numbers to be composite numbers.

Least Common Multiple

The greatest common factor is the largest number in the intersection of the factors of a set of given numbers. On the other hand, the *least common multiple* is the smallest number in the intersection of the multiples of a set of given numbers.

Multiples of:
24: {24, 48, 72, 96, 120, 144, . . .}
30: {30, 60, 90, 120, 150, . . .}
Common multiples: {120, 240, . . .}
    l.c.m. = 120

Least Common Multiple

The **least common multiple (l.c.m.)** of a set of numbers is the smallest number that each of the numbers in the set divides into evenly.

    An algorithm for finding the least common multiple is very much like the one for the g.c.f. The process, as before, begins by finding the canonical representations of the numbers involved. For example, to obtain the l.c.m. of the numbers 24 and 30, write each in canonical form:

$$24 = 2^3 \cdot 3^1 \cdot 5^0 \qquad 30 = 2^1 \cdot 3^1 \cdot 5^1$$

For the l.c.m., we choose the representative of each factor with the largest exponent. The l.c.m. is the product of these representatives:

$$\text{l.c.m.} = 2^3 \cdot 3^1 \cdot 5^1 = 120$$

    The procedure for finding the least common multiple is shown in Figure 4.3.

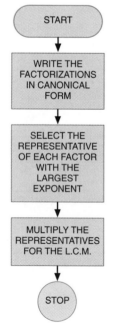

▲ **Figure 4.3   Procedure for finding the l.c.m.**

■■■ EXAMPLE 7

Find the least common multiple of 300, 144, and 108.

Solution

$$300 = 2^2 \cdot 3^1 \cdot 5^2$$
$$144 = 2^4 \cdot 3^2 \cdot 5^0$$
$$108 = 2^2 \cdot 3^3 \cdot 5^0$$
$$\text{l.c.m.} = 2^4 \cdot 3^3 \cdot 5^2 = 16 \cdot 27 \cdot 25 = 10,800$$

This means that the *smallest* number that 300, 144, and 108 *all* divide into evenly is 10,800.

In Pursuit of the Largest Prime Number

The method of Eratosthenes gives a finite list of primes, but it is not very satisfactory to use if we wish to determine whether a given number n is a prime. For centuries, mathematicians have tried to find a formula that would yield *every* prime. Let's try to find a formula that results in giving only primes. A possible candidate is

$$n^2 - n + 41$$

If we try this formula for $n = 1$, we obtain $1^2 - 1 + 41 = 41$.

For $n = 2$: $\quad 2^2 - 2 + 41 = 43,\quad$ a prime
For $n = 3$: $\quad 3^2 - 3 + 41 = 47,\quad$ a prime

So far, so good. That is, we are obtaining only primes. Continuing, we keep finding only primes for n up to 40:

For $n = 40$: $\quad 40^2 - 40 + 41 = 1,601,\quad$ a prime

Inductively, we might conclude that the formula yields only primes, but the next value provides a counterexample:

For $n = 41$: $\quad 41^2 - 41 + 41 = 41^2 = 1,681,\quad$ not a prime!

A more serious attempt to find a prime number formula was made by Pierre de Fermat, who tried the formula

$$2^{2^n} + 1$$

For $n = 1$: $\quad 2^{2^1} + 1 = 5,\quad$ a prime
For $n = 2$: $\quad 2^{2^2} + 1 = 2^4 + 1 = 17,\quad$ a prime
For $n = 3$: $\quad 2^{2^3} + 1 = 257,\quad$ a prime
For $n = 4$: $\quad 2^{2^4} + 1 = 2^{16} + 1 = 65,537,\quad$ a prime
For $n = 5$: $\quad 2^{2^5} + 1 = 4,294,967,297$

Is 4,294,967,297 a prime? The answer is not easy. See the Historical Note in the margin. It turns out this number is not prime! It is divisible by 641. Whether this formula generates any other primes is unknown.

In 1644, the French priest and number theorist Marin Mersenne (1588–1648) stated without proof that the number

$$2^{251} - 1$$

Historical Note

The mathematician Pierre de Fermat (see Historical Note on page 315) was not able to show that 4,294,967,297 was not a prime. Leonhard Euler (see page 798) later discovered it is divisible by 641. In the 1800s there was a young American named Colburn who had a great capacity to do things in his head. He was shown this number and asked if it was a prime. He said, "No, because 641 divides it." When asked how he knew that, he replied, "Oh, I just felt it." He was unable to explain his great gift—he just had it.

Historical Note

DAVID HILBERT
(1862–1943)

In 1900, David Hilbert delivered an address before the International Congress of Mathematicians in Paris. Instead of solving a problem, Hilbert presented a list of 23 problems. The mathematical world went to work on these problems and, even today, is still working on some of them. In the process of solving these problems, entire new frontiers of mathematics were opened. Hilbert's 10th problem was solved in 1970 by Yuri Matyasievich, who built upon the work of Martin Davis, Hilary Putnam, and Julia Robinson. A consequence of Hilbert's 10th problem is the fact that there must exist a polynomial with integer coefficients that will produce only prime numbers.

is composite. In the 19th century, mathematicians finally proved Mersenne correct when they discovered that this number was divisible by both 503 and 54,217. Mersenne did discover, however, that

$$2^{257} - 1$$

is a prime number.

In 1970, a young Russian named Matyasievich discovered several explicit polynomials of this sort that generate only prime numbers, but all of those he discovered are too complicated to reproduce here. The largest known prime number at that time was

$$2^{11,213} - 1$$

which was discovered at the University of Illinois through the use of number theory and computers. The mathematicians were so proud of this discovery that the following postmark was used on the university's postage meter:

Some other large prime numbers and the dates of their discovery are shown in Table 4.3.

TABLE 4.3 Some Large Primes		
Prime Number	*Date of Discovery*	*Source*
$2^{257} - 1$	1644	Marin Mersenne
$2^{11,213} - 1$	December 1970	University of Illinois
$2^{19,937} - 1$	March 1971	Bryant Tuckerman
$2^{21,701} - 1$	October 1978	Laura Nickel and Curt Knoll
$2^{23,209} - 1$	January 1979	Curt Knoll
$2^{86,243} - 1$	May 1983	David Slowinski
$2^{216,091} - 1$	November 1985	Amdahl Benchmark Center
$2^{756,839} - 1$	February 1992	David Slowinski and Paul Gage
$2^{858,433} - 1$	April 1994	David Slowinski and Paul Gage

Infinitude of the Primes

Table 4.3 shows some very large primes. Is there a largest prime? If there is no largest prime, then there must be infinitely many primes.

■■■ **EXAMPLE 8**   **Polya's Method**

Show that there is no largest prime.

Solution We use Polya's problem-solving guidelines for this example.

Understand the Problem. Is there a prime larger than the largest known prime shown in Table 4.3? If we find one, then we are finished. If we can't find one, is there a way we can still show it is not the largest prime?

Devise a Plan. We will consider a simpler problem. Suppose we believe that 19 is the largest prime. The task is to show that there exists a prime larger than 19. There are two ways to proceed. We could simply find a larger prime — say, 23. But what if we can't find a larger one? We will proceed by showing that 19 can't be the largest prime without actually finding it. Consider the number $M = (2 \cdot 3 \cdot 5 \cdot 7 \cdot 11 \cdot 13 \cdot 17 \cdot 19) + 1$. This number is larger than 19. Is it a prime?

Carry Out the Plan. According to our assumption, M must be composite, since it is larger than 19. But if it is composite, it has a prime divisor. Check all the primes:

2 does not divide M since $2|(2 \cdot 3 \cdot 5 \cdot 7 \cdot 11 \cdot 13 \cdot 17 \cdot 19)$, and thus 2 does not divide 1 more than this number.

3 does not divide M for the same reason.

Repeat for every known prime.

Thus, if M is not divisible by any known prime, then either it must be prime or there is a prime divisor larger than 19. In either case, we have found a prime larger than 19.

Look Back. If *anyone* claims to be in possession of the largest prime, we need only carry out an argument like the above to find a larger prime. Thus, we are saying that there are infinitely many primes, since it is impossible to have a largest prime.

Historical Note

The only standing ovation ever given at a meeting of the American Mathematical Association was given to a mathematician named Cole in 1903. It was commonly believed that $2^{67} - 1$ was a prime. Cole first multiplied out 2^{67} and then subtracted 1. Moving to another board, he wrote

$$761{,}838{,}257{,}287$$
$$\times \quad 193{,}707{,}721$$

He multiplied it out and came up with the same result as on the first blackboard. He had factored a number thought to be a prime! Here is what it looks like using a computer:

$2^{67} - 1$
$= 147573952589676412927$

$761838257287 \times 193707721$
$= 147573952589676412927$

PROBLEM SET 4.2

▲ A Problems

1. **IN YOUR OWN WORDS** What is a prime number?

2. **IN YOUR OWN WORDS** Describe a process for finding a prime factorization.

3. **IN YOUR OWN WORDS** What is the canonical representation of a number?

4. **IN YOUR OWN WORDS** What does g.c.f. mean, and what is the procedure for finding the g.c.f. of a set of numbers?

5. **IN YOUR OWN WORDS** What does l.c.m. mean, and what is the procedure for finding the l.c.m. of a set of numbers?

6. **IN YOUR OWN WORDS** Compare and contrast finding the g.c.f. and l.c.m. of a set of numbers.

7. **IN YOUR OWN WORDS** Are there infinitely many primes? Prove that there is no largest prime number.

Which of the numbers in Problems 8–11 are prime?

8. a. 59   b. 57   c. 1   d. 1,995

9. a. 63   b. 73   c. 79   d. 1,997

10. a. 43   b. 97   c. 171   d. 1,999

11. a. 91   b. 87   c. 111   d. 2,001

Are the statements in Problems 12–15 true or false?

12. a. $6|48$ b. $7|48$ c. $8|48$ d. $9|48$

13. a. $6\!\!\not|39$ b. $5\!\!\not|30$ c. $16\!\!\not|576$ d. $10\!\!\not|148{,}729{,}320$

14. a. $15|5$ b. $5|83{,}410$ c. $2|628{,}174$ d. $10|148{,}729{,}320$

15. a. $15|4{,}814$ b. $17|255$ c. $3|7{,}823$ d. $9|7{,}823{,}572$

16. Find all prime numbers less than or equal to 300.

17. What is the largest prime you need to consider to be sure that you have excluded, in the sieve of Eratosthenes, all primes less than or equal to:

     a. 200   b. 500   c. 1,000   d. 1,000,000

Write the prime factorization for each of the numbers in Problems 18–38. If the number is prime, so state.

18. a. 24   b. 30   c. 300   d. 144

19. a. 28   b. 76   c. 215   d. 125

20. a. 108   b. 740   c. 699   d. 123

21. a. 120   b. 90   c. 75   d. 975

22. a. 490   b. 4,752   c. 143   d. 51

23. 83   24. 97   25. 127   26. 113   27. 377   28. 151

29. 105   30. 187   31. 67   32. 229   33. 315   34. 111

35. 567   36. 568   37. 2,869   38. 793

Find the g.c.f. and l.c.m. of the sets of numbers in Problems 39–47.

39. {60, 72}   40. {95, 1425}   41. {12, 54, 171}

42. {11, 13, 23}   43. {12, 20, 36}   44. {6, 8, 10}

45. {9, 12, 14}   46. {3, 6, 15, 54}   47. {90, 75, 120}

▲ **B Problems**

48. Bill and Sue both work at night. Bill has every sixth night off and Sue has every eighth night off. If they are both off tonight, how many nights will it be before they are both off again at the same time?

49. Two movie theaters, UAI and UAII, start their movies at 7:00 P.M. The movie at UAI takes 75 minutes and the movie at UAII takes 90 minutes. If the shows run continuously, when will they again start at the same time?

50. We used a sieve of Eratosthenes in Table 4.2 by arranging the first 100 numbers into 10 rows and 10 columns. Repeat the sieve process for the first 100 numbers by arranging the numbers in the following patterns.

a. By 6:

1	2	3	4	5	6
7	8	9	10	11	12
13	14	15	. . .		

b. By 7:

1	2	3	4	5	6	7
8	9	10	. . .			

c. By 21:   1   2   3   . . .

As you are using these sieves, look for patterns. Describe some of the patterns you notice. Do you think that any of these sieves are better than the one shown in Table 4.2? Why or why not?

51. Use the sieve in Problem 50a to make a conjecture about primes and multiples of 6.

52. **Lucky Numbers** Set up a sieve similar to the one illustrated in Table 4.4.

TABLE 4.4   Lucky Numbers

1	2̸	3	4̸	5̸	6̸
7	8̸	9	10	1̸1̸	12
13	14	15	16	1̸7̸	18
19	20	21	22	2̸3̸	24
25	26	27	28	29	30
31	32	33	34	3̸5̸	36
37	38	39	40	4̸1̸	42
43	44	45	46	4̸7̸	48

Start counting with 1 *each time.*

Cross out every second number (shown as /).

The next uncrossed number is 3, so cross out every 3rd number that remains (shown as ×).

The next uncrossed number is 7, so cross out every 7th number that remains (shown as −).

Continue in the same fashion. The numbers that are not crossed out are called **lucky numbers.**

What are the lucky numbers less than 100?

53. Pairs of consecutive odd numbers that are primes are called *prime twins*. For example, 3 and 5, 11 and 13, and 41 and 43 are prime twins. Can you find any others?

54. Three consecutive odd numbers that are primes are called *prime triplets*. It is easy to show that 3, 5, and 7 are the only prime triplets. Can you explain why this is true?

55. **Goldbach's Conjecture** In a letter to Leonhard Euler in 1742, Christian Goldbach observed that every even number except 2 seemed representable as the sum of two primes. Euler was unable to prove or disprove this conjecture, and it remains unsolved to this day. Write the following numbers as the sum of two primes (the first three are worked for you):

$$4 = 2 + 2 \qquad 6 = 3 + 3 \qquad 8 = 5 + 3 \qquad 10 =$$
$$12 = \qquad 14 = \qquad 16 = \qquad 18 =$$
$$20 = \qquad 40 = \qquad 80 = \qquad 100 =$$

Historical Note

In 1742 the mathematician Christian Goldbach observed that every even number (except 2) seemed representable as the sum of two primes (see Problem 55). To date, this conjecture has not been proved, but the Russian mathematician L. Schnirelmann (1905–1938) proved that every positive integer can be represented as the sum of not more than 300,000 primes. That may seem like a long way off from Goldbach's conjecture, but at least 300,000 is a finite number! Later, another mathematician, I. M. Vinogradoff, proved that there exists a number N such that all numbers larger than N can be written as the sum of, at most, four primes.

56. Let $S = \{1, 2, 3, 5, 6, 10, 15, 30\}$, and define an operation $\mathcal{M}$ meaning *least common multiple*. For example,

$$5 \ \mathcal{M} \ 10 = 10, \quad 5 \ \mathcal{M} \ 6 = 30, \quad 10 \ \mathcal{M} \ 30 = 30$$

   a. Is S closed for the operation of $\mathcal{M}$?   b. Is S associative for $\mathcal{M}$?
   c. Is S commutative for $\mathcal{M}$?

57. Use an argument similar to the one in the text to show that 23 is not the largest prime.

▲ **Problem Solving**

58. In the text, we showed that 19 is not the largest prime by considering

$$M = 2 \cdot 3 \cdot 5 \cdot 7 \cdot 11 \cdot 13 \cdot 17 \cdot 19 + 1$$

   Now, M is either prime or composite. If it is prime, then since it is larger than 19, we have a prime larger than 19. If it is composite, it has a prime divisor larger than 19. In either case, we find a prime larger than 19. Show that this number M does not always generate primes. That is,

$$2 + 1 = 3, \quad \text{a prime}$$
$$2 \cdot 3 + 1 = 7, \quad \text{a prime}$$
$$2 \cdot 3 \cdot 5 + 1 = 31, \quad \text{a prime}$$

   Find an example in which the product of consecutive primes plus 1 does not yield a prime.

59. Find the one composite number in the following set:

$$31$$
$$331$$
$$3331$$
$$33331$$
$$333331$$
$$3333331$$
$$33333331$$
$$333333331$$

60. In the text we tried some formulas that might have generated only primes, but, alas, they failed. Below are some other formulas. Show that these, too, do not generate only primes.

   a. $n^2 + n + 41$ b. $n^2 - 79n + 1{,}601$ c. $2n^2 + 29$
   d. $9n^2 - 489n + 6{,}683$ e. $n^2 + 1$, n an even integer

61. For what values of n is $11 \cdot 14^n + 1$ a prime? *Hint:*   Consider n even, and then consider n odd.

62. What is the smallest natural number that is divisible by the first 20 counting numbers?

63. Some primes are 1 more than a square. For example, $5 = 2^2 + 1$. Can you find any other primes p so that $p = n^2 + 1$?

64. Some primes are 1 less than a square. For example, $3 = 2^2 - 1$. Can you find any other primes p so that $p = n^2 - 1$?

65. A formula that generates all prime numbers is given by David Dunlop and Thomas Sigmund in their book *Problem Solving with the Programmable Calculator* (Englewood Cliffs, N.J.: Prentice-Hall, 1983). The authors claim that the formula $\sqrt{1 + 24n}$

produces every prime number except 2 and 3, but give no proof or reference to a proof. Create a table, and give an argument to support or find a counterexample to disprove their claim.

66. Consider the product

$$1 \cdot 2 \cdot 3 \cdot \cdots \cdot 1{,}995 \cdot 1{,}996 \cdot 1{,}997$$

What is the last digit?

67. Consider the product

$$1 \cdot 2 \cdot 3 \cdot \cdots \cdot 1{,}995 \cdot 1{,}996 \cdot 1{,}997$$

From the product, cross out all even factors as well as all multiples of 5. What is the last digit?

68. Consider the product

$$1 \cdot 2 \cdot 3 \cdot \cdots \cdot 1{,}995 \cdot 1{,}996 \cdot 1{,}997$$

From the product, cross out all even factors as well as all multiples of 5. This problem was invented because of the year 1997. Write up a solution for any ending number.

69. **HISTORICAL QUESTION** The Pythagoreans studied numbers to find certain mystical properties in them. Certain numbers they studied were called *perfect numbers*. A **perfect number** is a counting number that is equal to the sum of all its divisors that are less than the number itself. A divisor that is less than the number itself is called a **proper divisor.** The proper divisors of 6 are {1, 2, 3} and $1 + 2 + 3 = 6$, so 6 is a perfect number. It is not hard to show that 6 is the smallest perfect number. On the other hand, 24 is not perfect, since its proper divisors are {1, 2, 3, 4, 6, 8, 12}, which have the sum $1 + 2 + 3 + 4 + 6 + 8 + 12 = 36$. The Pythagoreans discovered the first four perfect numbers. Fourteen centuries later the fifth perfect number was discovered. There are only 33 known perfect numbers. The 32nd perfect number is

$$2^{N-1}(2^N - 1)$$

where N is the largest known prime (see Table 4.3). It is known that all even perfect numbers are of the form shown for the 33rd perfect number. Show that if $N = 5$, the resulting number is perfect.

70. **HISTORICAL QUESTION** The Pythagoreans studied numbers that they called *amicable* or *friendly*. A pair of numbers is **friendly** if each number is the sum of the proper divisors of the other (a proper divisor includes the number 1, but not the number itself). The Pythagoreans discovered that 220 and 284 are friendly. The proper divisors of 220 are {1, 2, 4, 5, 10, 11, 20, 22, 44, 55, 110}, and $1 + 2 + 4 + 5 + 10 + 11 + 20 + 22 + 44 + 55 + 110 = 284$. Also, the proper divisors of 284 are {1, 2, 4, 71, 142}, and $1 + 2 + 4 + 71 + 142 = 220$. The next pair of friendly numbers was found by Pierre de Fermat: 17,296 and 18,416. In 1638 the French mathematician René Descartes found a third pair, and the Swiss mathematician Leonhard Euler found more than 60 pairs. Another strange development was the discovery in 1866 of another pair of friendly numbers, 1,184 and 1,210. The discoverer, Nicolo Pagonini, was a 16-year-old Italian schoolboy. Evidently the great mathematicians had overlooked this simple pair. Today there are more than 600 known pairs of friendly numbers. Show that 1,184 and 1,210 are friendly.

▲ **Individual Research**

71. The largest known prime, $2^{858,433} - 1$, is a number that has 258,716 digits. A number this large is hard to comprehend. Remember the chessboard problem in Example 11,

Section 1.4, when we looked at the enormous size of the number 2^{64}. Write a paper making the size of this number meaningful to a nonmathematical reader.

72. Investigate some of the properties of primes not discussed in the text. Why are primes important to mathematicians? Why are primes important in mathematics? What are some of the important theorems concerning primes?

Reference Martin Gardner, "The Remarkable Lore of Prime Numbers," *Scientific American*, March 1964.

4.3   INTEGERS

Historically, an agricultural-type society would need only natural numbers, but what about a subtraction such as

$$5 - 5 = ?$$

Certainly society would have a need for a number representing $5 - 5$, so a new number, called **zero**, was invented, so that $5 = 5 + 0$ (remember the definition of subtraction). If this new number is annexed to the set of natural numbers, a set called the set of **whole numbers** is formed:

$$\mathbb{W} = \{0, 1, 2, 3, 4, \ldots\}$$

This one annexation to the existing numbers satisfied society's needs for several thousand years.

However, as society evolved, the need for bookkeeping advanced, and eventually the need to answer this:

Can we annex new numbers to the set $\mathbb{W}$ so that it is possible to carry out *all* subtractions?

The numbers that need to be annexed are the opposites of the natural numbers. The opposite of 3, which is denoted by -3, is the number that when added to 3 gives 0. If we add these opposites to the set $\mathbb{W}$ we have the following set:

$$\mathbb{Z} = \{\ldots, -3, -2, -1, 0, 1, 2, 3, \ldots\}$$

This set is known as the set of **integers.** It is customary to refer to certain subsets of $\mathbb{Z}$ as follows:

1. Positive integers: $\{1, 2, 3, 4, \ldots\}$
2. Zero: $\{0\}$
3. Negative integers: $\{-1, -2, -3, \ldots\}$

Now with this new enlarged set of numbers, are we able to carry out all possible additions, subtractions, and multiplications? Before we answer this question, let's review the process by which we operate within the set of integers. It is assumed that you have had an algebra course, so the following summary is intended only as a review.

You might recall that the process for describing the operations with integers requires the notion of *absolute value*, which represents the distance of a number from the origin when plotted on a number line. We give an algebraic definition.

Absolute Value

The **absolute value** of x, denoted by $|x|$, is defined as

$$|x| = \begin{cases} x, & \text{if } x \geq 0 \\ -x, & \text{if } x < 0 \end{cases}$$

▰▰▰ EXAMPLE 1

Find the absolute value of the numbers:

a. $|5|$ **b.** $|-5|$ **c.** $|-(-3)|$

Solution

a. $|5| = 5$,   since $5 \geq 0$

b. $|-5| = 5$,   since $-5 < 0$ and $-(-5) = 5$

c. $|-(-3)| = |3| = 3$,   since $3 \geq 0$

Addition of Integers

If one (or both) of the integers is 0, then we use the identity property to write $x + 0 = 0 + x = x$, for all x. We could introduce the addition of nonzero integers in terms of number lines, and we note that if the numbers we're adding have the same sign, the result is the same as the sum of the absolute values, except for a plus or a minus sign (since their directions on a number line are the same). If we're adding numbers with different signs, their directions are opposite, so the net result is the difference of the absolute values with a sign to indicate final position. This is summarized by the following procedure for adding integers.

Addition of Integers

To add nonzero integers x and y, look at the signs of x and y:

1. If the signs are the same:

$$\begin{array}{l} \text{POSITIVE} + \text{POSITIVE} = \text{POSITIVE} \\ \text{NEGATIVE} + \text{NEGATIVE} = \text{NEGATIVE} \end{array} \quad \left\{ |x| + |y| \right.$$

$\underbrace{\hspace{3cm}}_{\text{Sign part}}$ $\underbrace{\hspace{3cm}}_{\text{Whole-number part}}$

2. If the signs are different:

$$\left.\begin{array}{l} \text{POSITIVE} + \text{NEGATIVE} \\ \text{NEGATIVE} + \text{POSITIVE} \end{array}\right\} = \left\{\begin{array}{l}\text{Sign of the larger} \\ \text{absolute value}\end{array}\right. \left\{\begin{array}{l}\text{Subtract smaller} \\ \text{absolute value from} \\ \text{larger absolute value}\end{array}\right.$$

$\underbrace{\hspace{3cm}}_{\text{Sign part}}$ $\underbrace{\hspace{4cm}}_{\text{Whole-number part}}$

Notice that all integers consist of two parts: a sign part and a whole-number part.

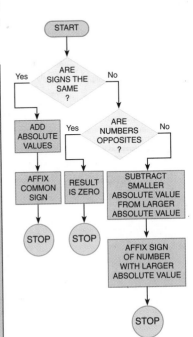

▬▬▬ **EXAMPLE 2**

Add the integers:

a. $41 + 13$　　**b.** $-41 + (-13)$　　**c.** $41 + (-13)$　　**d.** $-41 + 13$

Solution

a. POSITIVE + POSITIVE:　$41 + 13 = 54$ Add absolute values.
b. NEGATIVE + NEGATIVE:　$-41 + (-13) = -54$ Add absolute values.
c. POSITIVE + NEGATIVE:　$41 + (-13) = 28$ Subtract absolute values.
d. NEGATIVE + POSITIVE:　$-41 + 13 = -28$ Subtract absolute values. ▬▬

Calculator Comment

When using a calculator, you must distinguish between a negative sign, as in -5, and a subtraction sign, as in $8 - 5$. For entering negative numbers into a calculator you'll find a key marked

$\boxed{+/-}$　or　$\boxed{\text{CHS}}$

These keys change the sign of a number. For example, $5 + (-6)$ is entered as

$\boxed{5}\ \boxed{+}\ \boxed{6}\ \boxed{+/-}\ \boxed{=}$

▬▬▬ **EXAMPLE 3**

Indicate the sequence of keys to enter $(-8) + (-5)$ into a calculator.

Solution

$\boxed{8}\ \boxed{+/-}\ \boxed{+}\ \boxed{5}\ \boxed{+/-}\ \boxed{=}$
    ↑   ↑
  Change sign   Change sign
  or negative key or negative key

Notice that a calculator has separate keys for subtraction $\boxed{-}$ and opposite $\boxed{+/-}$. The $\boxed{+/-}$ key changes the sign of a number to the opposite of its present sign. For example, $5 - (-2)$ would be entered as

$\boxed{5}\ \boxed{-}\ \boxed{2}\ \boxed{+/-}\ \boxed{=}$

Because the calculator assumes that all numbers entered are positive, a negative number is obtained by taking the opposite of a positive.

Multiplication of Integers

For whole numbers, multiplication is defined as repeated addition, since we say that $5 \cdot 4$ means

$$\underbrace{4 + 4 + 4 + 4 + 4}_{5 \text{ addends}}$$

However, we cannot do this for the integers, since $(-5) \cdot 4$ or

$$\underbrace{4 + 4 + 4 + \cdots + 4}_{-5 \text{ addends does not make sense}}$$

Even though you may remember how to multiply integers, we consider four patterns.

POSITIVE · POSITIVE   We know how to multiply positive numbers since these are counting numbers. *The product of two positive numbers is a positive number.*

POSITIVE · NEGATIVE   Consider, for example, $3 \cdot (-4)$. We look at the pattern:

$3 \cdot 4 = 12$
$3 \cdot 3 = 9$
$3 \cdot 2 = 6$
$3 \cdot 1 = 3$
$3 \cdot 0 = 0$

What comes next? **Answer this question before reading further.**

$3 \cdot (-1) = -3$
$3 \cdot (-2) = -6$
$3 \cdot (-3) = -9$
$3 \cdot (-4) = -12$

Do you know how to continue? Try building a few more such patterns using different numbers. What did you discover about the product of a positive number and a negative number? *The product of a positive number and a negative number is a negative number.*

NEGATIVE · POSITIVE   Since we now know how to multiply a positive by a negative, we know the result here must be the same because of the commutative property. *The product of a negative number and a positive number is a negative number.*

NEGATIVE · NEGATIVE   Consider the example $-3 \cdot (-4)$. Let's build another pattern:

$-3 \cdot 4 = -12$
$-3 \cdot 3 = -9$
$-3 \cdot 2 = -6$
$-3 \cdot 1 = -3$
$-3 \cdot 0 = 0$

What comes next? **Answer this question before reading further.**

$-3 \cdot (-1) = 3$
$-3 \cdot (-2) = 6$
$-3 \cdot (-3) = 9$
$-3 \cdot (-4) = 12$

Thus, as the pattern indicates: *The product of two negative numbers is a positive number.* We summarize our discussion in the following box.

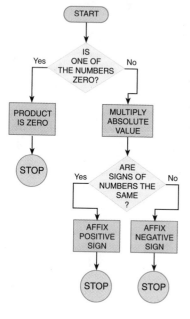

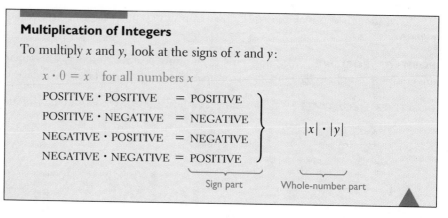

Multiplication of Integers

To multiply x and y, look at the signs of x and y:

$x \cdot 0 = x$ for all numbers x

$$\left.\begin{array}{lcl}\text{POSITIVE} \cdot \text{POSITIVE} & = & \text{POSITIVE} \\ \text{POSITIVE} \cdot \text{NEGATIVE} & = & \text{NEGATIVE} \\ \text{NEGATIVE} \cdot \text{POSITIVE} & = & \text{NEGATIVE} \\ \text{NEGATIVE} \cdot \text{NEGATIVE} & = & \text{POSITIVE}\end{array}\right\} \quad |x| \cdot |y|$$

Sign part   Whole-number part

▪ EXAMPLE 4

Multiply the given integers:

a. $(41)(13)$ **b.** $(-41)(-13)$ **c.** $(41)(-13)$ **d.** $(-41)(13)$

Solution

a. POSITIVE · POSITIVE: $(41)(13) = 533$ positive

b. NEGATIVE · NEGATIVE: $(-41)(-13) = 533$ positive

c. POSITIVE · NEGATIVE: $(41)(-13) = -533$ negative

d. NEGATIVE · POSITIVE: $(-41)(13) = -533$ negative

Subtraction of Integers

What about subtracting negative numbers? Negative already indicates "going back." Does subtraction of a negative indicate "going ahead?" Consider the following pattern:

$$4 - 4 = 0$$
$$4 - 3 = 1$$
$$4 - 2 = 2$$
$$4 - 1 = 3$$
$$4 - 0 = 4$$

Stop and look for patterns:

$$4 - (-1) = 5$$
$$4 - (-2) = 6$$
$$4 - (-3) = 7$$

Guided by these results, we make the following procedure for subtraction of integers.

> **Subtraction**
>
> $$x - y = x + (-y)$$
>
> To subtract, add the opposite (of the number being subtracted).

■ EXAMPLE 5

Subtract the given integers:

a. $41 - 13$ **b.** $41 - (-13)$ **c.** $-41 - 13$ **d.** $-41 - (-13)$

Solution

a. POSITIVE − POSITIVE: $41 - 13 = 41 + (-13) = 28$

subtraction   addition   opposite   complete the addition

b. POSITIVE − NEGATIVE: $41 - (-13) = 41 + 13 = 54$

c. NEGATIVE − POSITIVE: $-41 - 13 = -41 + (-13) = -54$

d. NEGATIVE − NEGATIVE: $-41 - (-13) = -41 + 13 = -28$ ▬

Division of Integers

Let's take an overview of what has been done in this chapter. We began with the *natural numbers*, which are closed for addition and multiplication. Next, we defined subtraction and created a situation where it was impossible to subtract some numbers from others. After looking at the prime numbers and factorization, we then "created" another set (called the *integers*) that includes not only the natural numbers, but also zero and the opposite of each of its members.

Since the subtraction of integers is defined in terms of addition, we can easily show that the integers are closed for subtraction. You are asked to do this in Problem 58. We now will define division and then ask the question, "Is the set of integers closed for division?"

Division is defined as the opposite operation of multiplication.

> **Division**
>
> If a, b, and z are integers, where $b \neq 0$, then **division** $a \div b$ is written as $\dfrac{a}{b}$ and is defined in terms of multiplication:
>
> $$\frac{a}{b} = z \quad \text{means} \quad a = bz$$

Since division is defined in terms of multiplication, the rules for dividing integers are identical to those for multiplication. We summarize the procedure for

$x \div y$, but first we must make sure $y \neq 0$, because division by zero is not defined.

$$\left.\begin{array}{l} \text{POSITIVE} \div \text{POSITIVE} \quad = \text{POSITIVE} \\ \text{POSITIVE} \div \text{NEGATIVE} \quad = \text{NEGATIVE} \\ \text{NEGATIVE} \div \text{POSITIVE} \quad = \text{NEGATIVE} \\ \text{NEGATIVE} \div \text{NEGATIVE} = \text{POSITIVE} \end{array}\right\} \qquad \dfrac{|x|}{|y|}$$

$$\underbrace{}_{\text{Sign part}} \qquad \underbrace{}_{\text{Whole-number part}}$$

▰▰▰ EXAMPLE 6

Divide the given integers.

a. $12 \div 6$ **b.** $-18 \div 2$ **c.** $10 \div (-2)$ **d.** $-65 \div (-13)$

Solution

a. POSITIVE ÷ POSITIVE: $\dfrac{12}{6} = 2$

b. NEGATIVE ÷ POSITIVE: $\dfrac{-18}{2} = -9$

c. POSITIVE ÷ NEGATIVE: $\dfrac{10}{-2} = -5$

d. NEGATIVE ÷ NEGATIVE: $\dfrac{-65}{-13} = 5$

   Notice that for $\frac{a}{b}$, we require $b \neq 0$. Why do we not allow division by zero? We consider two possibilities.

1. Division of a nonzero number by zero:

$$a \div 0 \quad \text{or} \quad \frac{a}{0} = x$$

What does this mean? Is there such a number x so that this makes sense? We see that any number x would have to be such that $a = 0 \cdot x$. But $0 \cdot x = 0$ for all x, and since $a \neq 0$, we see that such a situation is impossible. That is, $a \div 0$ does not exist.

2. Division of zero by zero:

$$0 \div 0 \quad \text{or} \quad \frac{0}{0} = x$$

What does this mean? Is there such a number x? We see that *any* x makes this true, since $0 \cdot x = 0$ for all x. But this leads to certain absurdities, for example:

If $\dfrac{0}{0} = 2$, then this checks since $0 \cdot 2 = 0$; also

if $\dfrac{0}{0} = 5$, then this checks since $0 \cdot 5 = 0$.

News Clip

Abraham Lincoln used a biblical reference (Mark 3:25) to initiate his campaign in 1858. He said, "A house divided against itself cannot stand." I offer a corollary to Lincoln's statement: "A house divided by itself is one (provided, of course, that the house is not zero)."

But since both 2 and 5 are equal to the *same* number, we would conclude that $2 = 5$. This is absurd, so we say that division of zero by zero is excluded (or that it is *indeterminate*).

Is the set of integers closed for division? Certainly we can find many examples in which an integer divided by an integer is an integer. Does this mean that the set of integers is closed? What about $1 \div 2$ or $4 \div 5$? These numbers do not exist in the set of integers; thus, the set is *not* closed for division. Now, as long as society has no need for such division problems, the question of inventing new numbers will not arise. However, as the need to divide 1 into 2 or more parts arises, some new numbers will have to be invented so that the set will be closed for division. We'll do this in the next section. The problem with inventing such new numbers is that it must be done in such a way that the properties of the existing numbers are left unchanged. That is, closure for addition, subtraction, and multiplication must be retained.

PROBLEM SET 4.3

▲ **A Problems**

1. **IN YOUR OWN WORDS** Explain how to add integers.

2. **IN YOUR OWN WORDS** Explain how to subtract integers.

3. **IN YOUR OWN WORDS** Explain how to multiply integers.

4. **IN YOUR OWN WORDS** Explain how to divide integers.

5. **IN YOUR OWN WORDS** Explain the difference between $0 \div 5$ and $5 \div 0$.

6. **IN YOUR OWN WORDS** Why is division by 0 not defined?

7. Suppose you are given the following number line:

a. Explain why a move from *H* to *N* is described by 7.
b. Describe a move from *T* to *F*.

8. Suppose you are given the number line in Problem 7.
a. Explain why a move from *U* to *I* is described by -4.
b. Describe a move from *U* to *A*.

Simplify the expressions in Problems 9–51.

9. a. $5 + 3$ b. $-5 + 3$ 10. a. $4 + (-7)$ b. $-2 + (-4)$

11. a. $7 + 3$ b. $-9 + 5$ 12. a. $-10 + 4$ b. $-8 + (-10)$

13. a. $-15 + 8$ b. $|-14 + 2|$ 14. a. $|8 + (-10)|$ b. $-38 + (-14)$

15. a. $10 - 7$ b. $7 - 10$ 16. a. $6 - (-4)$ b. $0 - (-15)$

17. a. $3(-6)$ b. $-5(4)$

18. a. $\dfrac{-4}{-2}$ b. $\dfrac{-6}{-3}$

19. a. $14(-5)$ b. $-14(-5)$

20. a. $-5(8-12)$ b. $[-5(8)] - 12$

21. a. $\dfrac{-12}{4}$ b. $\dfrac{-63}{-9}$

22. a. $\dfrac{12}{-4}$ b. $(-6)\dfrac{14}{-2}$

23. a. $\dfrac{-528}{-4}$ b. $(-1)^3$

24. a. $(-1)^4$ b. $10\left(\dfrac{-8}{-2}\right)$

25. a. $7(-8)$ b. $-5(15)$

26. a. $-5(-6)$ b. $\dfrac{-42}{3}$

27. a. -2^2 b. $(-2)^2$

28. a. $(-3)^2$ b. -3^2

29. a. $162 + (-12)$ b. $-12 + [(-4) + (-3)]$

30. a. $-5 + [-6 + 10]$ b. $[6 + (-8)] + (6 + 8)$

31. a. $-46 - (-46)$ b. $|7 - (-3)|$

32. a. $|-5 - (-10)|$ b. $|5 - (-5)|$

33. a. $-7 - (-18)$ b. $62 - (-112)$

34. a. $-4 - 8$ b. $31 + (-16)$

35. a. $-14 - 21$ b. $-9 + 16 + (-11)$

36. a. $-3 - [(-6) - 4]$ b. $5 + (-19) + |15|$

37. a. $|-8| + (-8)$ b. $-9 - (4 - 5)$

38. a. $-(23 + 14)$ b. $-7 - (6 - 4)$

39. a. $|-3| - [-(-2)]$ b. $15 - (-7)$

40. a. $-6 - (-6)$ b. $-18 - 5$

41. a. $-8 + 7 + 16$ b. $14 + (-10) - 8 - 11$

42. a. $6 + (-8) - 5$ b. $-2(-3) + (-1)(6)$

43. a. $-5(7) - (-9)$ b. $-2(3) - (-8)$

44. a. $\dfrac{-32}{-8} - 5 - (-7)$ b. $\dfrac{-15}{-5} - 4 - (-8)$

45. a. $6 - (-2)$ b. $-5 - (-3)$

46. a. $-15 - (-6)$ b. $-4 + |6 - 8|$

47. a. $[-54 \div (-9)] \div 3$ b. $-54 \div [(-9) \div 3]$

48. a. $[48 \div (-6)] \div (-2)$ b. $48 \div [(-6) \div (-2)]$

49. a. $15 - (-3) - |4 - 11|$ b. $-12 + (-7) - 10 - 14$

50. a. $-5(2) + (-3)(-4) - 6(-7)$
    b. $-8(3) - 6(-4) - 2(-8)$

51. a. $1(-2)(3)(-4)(5)(-6)(7)(-8)(9)(-10)$
    b. $1 + (-2) + 3 + (-4) + 5 + (-6) + 7 + (-8) + 9 + (-10)$

▲ B Problems

52. Perform the indicated operations. Let k be a natural number.
    a. $(-1)^6$ **b.** $(-1)^{67}$ **c.** $(-1)^{1998}$ **d.** $(-1)^{2k}$ **e.** $(-1)^{2k+1}$

53. Perform the indicated operations. Let k be a natural number.
    a. $(-2)^2$ **b.** $(-2)^3$ **c.** $(-2)^4$ **d.** $(-2)^5$
    e. Is $(-2)^{2k+1}$ positive or negative?

54. Draw a Venn diagram showing $\mathbb{N}$, $\mathbb{W}$, and $\mathbb{Z}$.

55. **a.** What is the intersection of the set of positive integers and the set of negative integers?
    b. Is the union of these two sets the set of integers?

56. **a.** State the commutative property.
    b. Is $\mathbb{Z}$ commutative for addition? Give reasons.
    c. Is $\mathbb{Z}$ commutative for subtraction? Give reasons.
    d. Is $\mathbb{Z}$ commutative for multiplication? Give reasons.
    e. Is $\mathbb{Z}$ commutative for division? Give reasons.

57. **a.** State the associative property.
    b. Is $\mathbb{Z}$ associative for addition? Give reasons.
    c. Is $\mathbb{Z}$ associative for subtraction? Give reasons.
    d. Is $\mathbb{Z}$ associative for multiplication? Give reasons.
    e. Is $\mathbb{Z}$ associative for division? Give reasons.

58. Show that the set $\mathbb{Z}$ is closed for subtraction.

▲ Problem Solving

59. Find a finite subset of $\mathbb{Z}$ that is closed for multiplication.

60. Multiply $1{,}234{,}567 \times 9{,}999{,}999$
    a. using a calculator   **b.** using patterns

61. **Four Fours** B.C. apparently has a mental block against fours, as we can see from the cartoon.

See if you can handle fours by writing the numbers from 1 to 10 using four 4s, operation symbols, or possibly grouping symbols, for each. Here are the first three completed for you:

$$\frac{4}{4} + 4 - 4 = 1 \qquad \frac{4}{4} + \frac{4}{4} = 2 \qquad \frac{4 + 4 + 4}{4} = 3$$

4.4   RATIONAL NUMBERS

Historical Note

The symbol ÷ was adopted by John Wallis (1616–1703) and was used in Great Britain and in the United States [but not on the European continent, where the colon (:) was used]. In 1923, the National Committee on Mathematical Requirements stated: "Since neither ÷ nor : as signs of division play any part in business life, it seems proper to consider only the needs of algebra, and to make more use of the fractional form and (where the meaning is clear) of the symbol "/" and to drop the symbol ÷ in writing algebraic expressions."
—*From* Report of the National Committee on Mathematical Requirements *under the auspices of the Mathematical Association of America, Inc.* (1923), *p. 81.*

Historically, the need for a closed set for division came before the need for closure for subtraction. We need to find some number k so that

$$1 \div 2 = k$$

As we saw in Section 3.1, the ancient Egyptians limited their fractions by requiring the numerators to be 1. The Romans avoided fractions by the use of subunits; feet were divided into inches and pounds into ounces, and a twelfth part of the Roman unit was called an *uncia.*

However, people soon felt the practical need to obtain greater accuracy in measurement and the theoretical need to close the number system with respect to the operation of division. In the set $\mathbb{Z}$ of integers, some divisions are possible:

$$\frac{10}{-2}, \quad \frac{-4}{2}, \quad \frac{-16}{-8}, \quad \cdots$$

However, certain others are not:

$$\frac{1}{2}, \quad \frac{-16}{5}, \quad \frac{5}{12}, \quad \cdots$$

Just as we extended the set of natural numbers by creating the concept of opposites, we can extend the set of integers. That is, the number $\frac{5}{12}$ is defined to be that number obtained when 5 is divided by 12. This new set, consisting of the integers as well as the quotients of integers, is called the set of *rational numbers.*

Rational Number

The set of **rational numbers,** denoted by $\mathbb{Q}$, is the set of all numbers of the form

$$\frac{a}{b}$$

where a and b are integers, and $b \neq 0$.

Historical Note

The Arabic word for fraction is *al-kasr* and is derived from the Latin word *fractus* meaning "to break." The English word *fraction* was first used by Chaucer in 1321 and the fraction bar was used in 1556 by Tartaglia, who said, "We write the numerator above a little bar and the denominator below it."

Notice that a rational number has fractional form. In arithmetic you learned that if a number is written in the form $\frac{a}{b}$ it means $a \div b$ and that a is called the **numerator** and b the **denominator.** Also, if a and b are both positive, $\frac{a}{b}$ is called

a **proper fraction** if $a < b$;

an **improper fraction** if $a > b$; and

a **whole number** if b divides evenly into a.

It is assumed that you know how to perform the basic operations with fractions, but we will spend the next few pages reviewing those operations.

Fundamental Property of Fractions

If the greatest common factor of the numerator and denominator of a given fraction is 1, then we say the fraction is in lowest terms or **reduced.** If the greatest common factor is not 1, then divide both the numerator and denominator by this greatest common factor using the following property.

Fundamental Property of Fractions

If $\frac{a}{b}$ is any rational number and x is any nonzero integer, then

$$\frac{a \cdot x}{b \cdot x} = \frac{x \cdot a}{x \cdot b} = \frac{a}{b}$$

That is, given some fraction that you wish to simplify:

1. Find the g.c.f. of the numerator and denominator (this is x in the fundamental property).
2. Use the fundamental property to simplify the fraction.

The fundamental property works only for *factors* and NOT for terms.

■■■ EXAMPLE 1

Reduce the given fractions.

a. $\dfrac{24}{30}$ b. $\dfrac{300}{144}$

Solution

a. First, find the greatest common factor:

$$24 = 2^3 \cdot 3^1 \cdot 5^0$$
$$30 = 2^1 \cdot 3^1 \cdot 5^1$$
$$\text{g.c.f.} = 2^1 \cdot 3^1 \cdot 5^0 = 6$$

Next, use the fundamental property to simplify the fraction:

$$\frac{24}{30} = \frac{6 \cdot 2^2}{6 \cdot 5} = \frac{2^2}{5} = \frac{4}{5}$$
$$\uparrow$$
$$\text{g.c.f.}$$

b. $300 = 2^2 \cdot 3^1 \cdot 5^2$

$144 = 2^4 \cdot 3^2 \cdot 5^0$

$\text{g.c.f.} = 2^2 \cdot 3^1 \cdot 5^0 = 12$

$$\frac{300}{144} = \frac{12 \cdot 5^2}{12 \cdot 2^2 \cdot 3} = \frac{5^2}{2^2 \cdot 3} = \frac{25}{12}$$

Note that $\frac{25}{12}$ is reduced because the g.c.f. of the numerator and denominator is 1. Even though it is required to reduce fractions, it is not necessary to avoid improper fractions.

In this book, we agree to leave all fractional answers in reduced form.

Operations with Rational Numbers

If $\frac{a}{b}$ and $\frac{c}{d}$ are rational numbers, then

Addition: $$\frac{a}{b} + \frac{c}{d} = \frac{ad}{bd} + \frac{bc}{bd} = \frac{ad + bc}{bd}$$

Subtraction: $$\frac{a}{b} - \frac{c}{d} = \frac{ad}{bd} - \frac{bc}{bd} = \frac{ad - bc}{bd}$$

Multiplication: $$\frac{a}{b} \times \frac{c}{d} = \frac{ac}{bd}$$

Division: $$\frac{a}{b} \div \frac{c}{d} = \frac{ad}{bc} \quad (c \neq 0)$$

You will note that both addition and subtraction require that we first obtain common denominators. That process requires a multiplication of fractions, so we begin with an example reviewing multiplication of rational numbers.

■■■■ EXAMPLE 2

Multiply the given rational numbers.

a. $\frac{1}{3} \times \frac{2}{5}$ **b.** $\frac{2}{3} \times \frac{-4}{7}$ **c.** $-5 \times \frac{-2}{3}$ **d.** $3\frac{1}{2} \times 2\frac{3}{5}$ **e.** $\frac{3}{4} \times \frac{2}{3}$

Solution

a. $\dfrac{1}{3} \times \dfrac{2}{5} = \boxed{\dfrac{1 \times 2}{3 \times 5}} = \dfrac{2}{15}$

This step is often done in your head.

b. $\dfrac{2}{3} \times \dfrac{-4}{7} = \dfrac{-8}{21}$

c. When multiplying a whole number and a fraction, write the whole number as a fraction, and then multiply:

$$-5 \times \frac{-2}{3} = \frac{-5}{1} \times \frac{-2}{3} = \frac{10}{3}$$

d. When multiplying mixed numbers, write the mixed numbers as improper fractions, and *then* multiply:

$$3\frac{1}{2} \times 2\frac{3}{5} = \frac{7}{2} \times \frac{13}{5} = \frac{91}{10}$$

e. For most complicated fractions, the best procedure is to reduce *before* multiplying rather than after, as illustrated here:

$$\frac{\overset{1}{3}}{\underset{2}{\cancel{4}}} \times \frac{\overset{1}{\cancel{8}}}{\underset{1}{3}} = \frac{1 \times 1}{2 \times 1} = \frac{1}{2}$$

■

━━ **EXAMPLE 3**   **Polya's Method**

Justify the rule for division of rational numbers.

Solution   We use Polya's problem-solving guidelines for this example.

Understand the Problem.   Given $\frac{a}{b} \div \frac{c}{d}$ where $c \neq 0$. Note that we don't say $b \neq 0$ and $d \neq 0$, because the definition of rational numbers excludes these possibilities, but does not exclude $c = 0$, so this is the condition that must be stated. The example asks us to show where the rule for division as stated in the previous box comes from. Let

$$\frac{a}{b} \div \frac{c}{d} = \square$$

We are looking for the value of $\square$. To understand what we are doing here, look at a more familiar problem:

$$\frac{2}{3} \div \frac{4}{5} = \boxed{} \quad \text{means} \quad \frac{4}{5} \times \boxed{} = \frac{2}{3}$$

What do we put into the box to get the answer? Do it in two steps. First multiply $\frac{4}{5}$ by $\frac{5}{4}$ to obtain 1, and *then* multiply by $\frac{2}{3}$ to obtain the result that makes the equation true:

$$\frac{4}{5} \times \boxed{\frac{5}{4} \times \frac{2}{3}} = \frac{2}{3}$$

Thus, $\dfrac{2}{3} \div \dfrac{4}{5} = \boxed{\dfrac{5}{4} \times \dfrac{2}{3}}$. This seems to suggest that we "invert" the fraction we are dividing by, and then multiply.

Devise a Plan.   We will use the definition of division, and then the Fundamental Property of Fractions to multiply the numerator and denominator by the same number.

Carry Out the Plan.

$$\frac{a}{b} \div \frac{c}{d} = \frac{\dfrac{a}{b}}{\dfrac{c}{d}} \qquad \text{Write the division using fractional notation.}$$

$$= \frac{\dfrac{a}{b} \times \dfrac{d}{c}}{\dfrac{c}{d} \times \dfrac{d}{c}} \qquad \begin{array}{l}\text{Multiply numerator and denominator by } \frac{d}{c};\\ \text{Fundamental Property of Fractions}\end{array}$$

$$= \frac{\dfrac{ad}{bc}}{\dfrac{cd}{dc}} \qquad \text{Multiplication of the "big" fractions}$$

$$= \frac{\dfrac{ad}{bc}}{1} \qquad \text{Reduce the fraction } \tfrac{cd}{dc}.$$

$$= \frac{ad}{bc}$$

Look Back.   The result here, $\dfrac{a}{b} \div \dfrac{c}{d} = \dfrac{ad}{bc}$, checks with the entry in the box. ▬

▬▬ **EXAMPLE 4**

Divide the given rational numbers.

a. $\dfrac{-3}{4} \div \dfrac{-4}{7}$ b. $\dfrac{4}{3} \div \dfrac{8}{9}$

Solution

a. $\dfrac{-3}{4} \div \dfrac{-4}{7} = \dfrac{-3}{4} \times \dfrac{7}{-4} = \dfrac{-21}{-16} = \dfrac{21}{16}$

b. $\dfrac{4}{3} \div \dfrac{8}{9} = \dfrac{4}{\overset{1}{\cancel{3}}} \times \dfrac{\overset{3}{\cancel{9}}}{8} = \dfrac{1 \times 3}{1 \times 2} = \dfrac{3}{2}$

▬

To carry out addition or subtraction of fractions, you must find the least common denominator. The least common denominator is the same as the least common multiple.

▬▬ **EXAMPLE 5**

Simplify the given expressions.

a. $\dfrac{5}{24} + \dfrac{7}{30}$ b. $\dfrac{19}{300} + \dfrac{55}{144} + \dfrac{25}{108}$ c. $\dfrac{7}{18} - \dfrac{-5}{24}$ d. $\dfrac{-1}{15} - \dfrac{27}{50}$

Solution

a. The denominators are 24 and 30, so we find the l.c.m. of these numbers:

$$24 = 2^3 \cdot 3^1 \cdot 5^0$$
$$30 = 2^1 \cdot 3^1 \cdot 5^1$$
$$\text{l.c.m.} = 2^3 \cdot 3^1 \cdot 5^1 = 120$$

$$\frac{5}{24} = \frac{5}{24} \cdot \frac{5}{5} = \frac{25}{120}$$

We are simply multiplying each fraction by the identity 1, since $\frac{5}{5}$ and $\frac{4}{4}$ are both equal to 1.

$$+\frac{7}{30} = \frac{7}{30} \cdot \frac{4}{4} = \frac{28}{120}$$

$$\frac{53}{120}$$

The answer is in reduced form, since 53 and 120 are relatively prime.

b. If the numbers are complicated, it is easier to work in factored form. In Section 4.2 we found the l.c.m. of 300, 144, and 108 to be $2^4 \cdot 3^3 \cdot 5^2$. Thus,

$$\frac{19}{300} = \frac{19}{2^2 \cdot 3 \cdot 5^2} \cdot \frac{2^2 \cdot 3^2}{2^2 \cdot 3^2} = \frac{19 \cdot 2^2 \cdot 3^2}{2^4 \cdot 3^3 \cdot 5^2}$$

This is the step during which we multiply each fraction by 1.

$$\frac{55}{144} = \frac{5 \cdot 11}{2^4 \cdot 3^2} \cdot \frac{3 \cdot 5^2}{3 \cdot 5^2} = \frac{3 \cdot 5^3 \cdot 11}{2^4 \cdot 3^3 \cdot 5^2}$$

$$+\frac{25}{108} = \frac{5^2}{2^2 \cdot 3^3} \cdot \frac{2^2 \cdot 5^2}{2^2 \cdot 5^2} = \frac{2^2 \cdot 5^4}{2^4 \cdot 3^3 \cdot 5^2}$$

Adding: $\quad \dfrac{19 \cdot 2^2 \cdot 3^2}{2^4 \cdot 3^3 \cdot 5^2} + \dfrac{3 \cdot 5^3 \cdot 11}{2^4 \cdot 3^3 \cdot 5^2} + \dfrac{2^2 \cdot 5^4}{2^4 \cdot 3^3 \cdot 5^2}$

$$= \frac{684}{2^4 \cdot 3^3 \cdot 5^2} + \frac{4,125}{2^4 \cdot 3^3 \cdot 5^2} + \frac{2,500}{2^4 \cdot 3^3 \cdot 5^2} = \frac{7,309}{2^4 \cdot 3^3 \cdot 5^2}$$

Now 7,309 is not divisible by 2, 3, or 5; thus, 7,309 and $2^4 \cdot 3^3 \cdot 5^2$ are relatively prime, and the solution is complete:

$$\frac{7,309}{2^4 \cdot 3^3 \cdot 5^2} \quad \text{or} \quad \frac{7,309}{10,800}$$

c. $\quad 18 = 2^1 \cdot 3^2$

$\quad\quad 24 = 2^3 \cdot 3^1$

$\quad$ l.c.m. $= 2^3 \cdot 3^2 = 72$

$$\frac{7}{18} = \frac{7}{18} \cdot \frac{4}{4} = \frac{28}{72}$$

$$-\frac{5}{24} = \frac{5}{24} \cdot \frac{3}{3} = \frac{15}{72}$$

$$\frac{43}{72}$$

d. $\quad 15 = \quad\quad 3^1 \cdot 5^1$

$\quad\quad 50 = 2^1 \cdot \quad\quad 5^2$

$\quad$ l.c.m. $= 2^1 \cdot 3^1 \cdot 5^2 = 150$

$$\frac{-1}{15} = \frac{-1}{15} \cdot \frac{10}{10} = \frac{-10}{150}$$

$$-\frac{27}{50} = \frac{-27}{50} \cdot \frac{3}{3} = \frac{-81}{150}$$

$$\frac{-91}{150}$$

Notice with subtraction it is usually easier to add the opposite.

Calculator Comment

Even though there are some special-purpose calculators that work in fractional form, the commonly accepted procedure when working with rational expressions is to work in decimal form. You will use the definition of fractions to translate the fractional bar into the division key. You must be careful about the order of operations and interpret the fractional bar in algebra using parentheses or the equal key on a calculator. Consider the following examples.

For $\dfrac{2+3}{5}$, press: $\boxed{2}\,\boxed{+}\,\boxed{3}\,\boxed{=}\,\boxed{\div}\,\boxed{5}\,\boxed{=}$ Display: *1*

Note: You must press the equal key to group together the numbers above the fractional bar before doing the division. Contrast this with the next sequence:

For $2+\dfrac{3}{5}$, press: $\boxed{2}\,\boxed{+}\,\boxed{3}\,\boxed{\div}\,\boxed{5}\,\boxed{=}$ Display: *2.6**

For $\dfrac{5}{24}+\dfrac{7}{30}$ (Example 5a), press: $\boxed{5}\,\boxed{\div}\,\boxed{24}\,\boxed{+}\,\boxed{7}\,\boxed{\div}\,\boxed{30}\,\boxed{=}$ Display: *.4416666667*

Note: The decimal representation on a calculator is often a decimal approximation of the exact answer. For example,

$\dfrac{2}{3}$, press: $\boxed{2}\,\boxed{\div}\,\boxed{3}\,\boxed{=}$ Display: *.6666666667*

We know that $\frac{2}{3}$ does not *equal* .6666666667, but rather the calculator shows a rounded answer. This example, however, will tell you a little about your calculator. It shows the maximum number of decimal places your calculator uses and it also shows you whether your calculator *rounds* (last digit is a 7) or *truncates* (last digit is a 6).

The set $\mathbb{Q}$ is closed for the operations of addition, subtraction, multiplication, and nonzero division. As an example, we will show that the rationals are closed for addition. We need to show that, given any two elements of $\mathbb{Q}$, their sum is also an element of $\mathbb{Q}$. Suppose

$$\frac{x}{y} \quad \text{and} \quad \frac{w}{z} \quad \text{are any two rational numbers.}$$

By definition of addition,

$$\frac{x}{y} + \frac{w}{z} = \frac{xz + wy}{yz}$$

We now need to show that $\dfrac{xz + wy}{yz}$ is a rational number. Since x and w are integers, and y and z are nonzero integers, we know from closure of the integers for multi-

* We are assuming here that you are using an algebraic calculator. If your calculator display shows 1 for both of these steps, then you have an arithmetic calculator.

plication that *xz* and *wy* are also integers. Since the set of integers is closed for addition, we know that $xz + wy$ is also an integer. This means that, since *yz* is a nonzero integer,

$$\frac{xz + wy}{yz} \quad \text{is a rational number.}$$

Thus, the rational numbers are closed for addition.

We conclude this section by considering a historical application.*

◼◼◼ EXAMPLE 6   Polya's Method

Article 1, Section 2, of the United States Constitution introduces what we will call the **apportionment problem.** There are four historically significant methods of apportioning seats in the U.S. House of Representatives. These are the methods of Thomas Jefferson, John Quincy Adams, Daniel Webster, and Alexander Hamilton. For simplicity, we will apply their methods to fictional examples rather than for the allocation of seats from the 50 states. Consider College Town with district populations as follows:

North: 8,600   South: 5,400   East: 7,200   West: 3,800

Suppose each council member is to represent 2,400 citizens. Decide on a fair representation.

Solution

Understand the Problem. Let's begin by doing some simple arithmetic.

North: $\dfrac{8,600}{2,400} = 3.5833\ldots$ South: $\dfrac{5,400}{2,400} = 2.25$

East: $\dfrac{7,200}{2,400} = 3$ West: $\dfrac{3,800}{2,400} = 1.58333\ldots$

Only one district (East) can be awarded the exact number of council members to which it is entitled. The quotients for the three other districts need to be rounded in some way.

Devise a Plan. Historically, there were four plans devised.

Adams' plan: Any quotient with a decimal portion must be rounded up to the next whole number.

Jefferson's plan: Any quotient with a decimal portion must be rounded down to the previous whole number.

Webster's plan: Any quotient with a decimal portion must be rounded to the nearest whole number.

Hamilton's plan: This plan is a variation of Jefferson's plan. Such a plan is necessary in situations where the size of the representative body is specified by

---

Historical Note

Article 1, Section 2
United States Constitution

Representation and direct taxes shall be apportioned among the several states which may be included in the Union, according to the respective numbers .... The number of Representatives shall not exceed one for every thirty thousand, but each state shall have at least one representative.

---

* This problem and solution are from an article "Decimals, Rounding, and Apportionment," by Kay I. Meeks, *The Mathematics Teacher*, October 1992, pp. 523–525.

law or regulation, as is the case with the U.S. House of Representatives. Jefferson's plan will not assign all the representatives available (since all quotients are rounded down). Hamilton's plan allocates the remainder, one at a time, to districts on the basis of the decreasing order of the decimal portion of the quotients. This method is sometimes called the *method of the largest fractions*. It will be easier to illustrate this method after we carry out the other plans.

Carry Out the Plan.   The results of these four plans are shown in the following table. The population is divided by the specified number of council members (2,400) to find the quotient.

Results of City Council Apportionment

District	Population	Quotient	Adams	Jefferson	Webster	Hamilton
North	8,600	3.5833 . . .	4	3	4	3
South	5,400	2.25	3	2	2	2
East	7,200	3.0	3	3	3	3
West	3,800	1.5833 . . .	2	1	2	2
TOTAL	25,000		12	9	11	10

The methods of Adams, Jefferson, and Webster are straightforward.

Some explanation of Hamilton's method is in order. First, notice that with Adams', Jefferson's, and Webster's methods, we do not have a preconceived target number of representatives that will result. In some situations (as with the U.S. House of Representatives) there is a target number of representatives. Assume there must be 10 representatives for College Town; notice that none of the first three plans gives 10 representatives. Hamilton's method takes care of this situation as follows. Determine the total population (in this case, 25,000). Form the fraction for the district population divided by the total population and multiply by the quota (in this case, 10):

North: $\dfrac{8,600}{25,000} \cdot 10 = 3.44$ Decimal rank 3

South: $\dfrac{5,400}{25,000} \cdot 10 = 2.16$ Decimal rank 4

East: $\dfrac{7,200}{25,000} \cdot 10 = 2.88$ Decimal rank 1

West: $\dfrac{3,800}{25,000} \cdot 10 = 1.52$ Decimal rank 2

Hamilton's plan first rounds down (as does Jefferson's plan). This will create a total that does not exceed the target. If there are any remaining seats available, a rank will be determined by looking at the decimal part of each ratio only, and the remaining seats will be awarded one-by-one (until they are all assigned) according to the magnitude (size) of the decimal part, which determines their rank. Thus, Hamilton's plan adds one seat to the East (highest decimal, so it is rank 1) and one to the West (second highest decimal, so it is rank 2).

Look Back. What are the advantages and disadvantages of each of these apportionment methods? What happens if the decimal part (as determined by Hamilton's plan) turns out to be the same for two or more districts? ▬

PROBLEM SET 4.4

▲ **A Problems**

1. **IN YOUR OWN WORDS** What does it mean for a fraction to be reduced? Describe a process for reducing a fraction.

2. **IN YOUR OWN WORDS** Describe the process for multiplying fractions.

3. **IN YOUR OWN WORDS** Describe the process for dividing fractions.

4. **IN YOUR OWN WORDS** Describe the process for adding fractions.

5. **IN YOUR OWN WORDS** Use algebra to show where the formula for subtracting fractions comes from.

Completely reduce the fractions in Problems 6–17.

6. a. $\dfrac{3}{9}$ b. $\dfrac{6}{9}$ 7. a. $\dfrac{2}{10}$ b. $\dfrac{3}{12}$

8. a. $\dfrac{4}{12}$ b. $\dfrac{6}{12}$ 9. a. $\dfrac{14}{7}$ b. $\dfrac{38}{19}$

10. a. $\dfrac{92}{20}$ b. $\dfrac{72}{15}$ 11. a. $\dfrac{42}{14}$ b. $\dfrac{16}{24}$

12. a. $\dfrac{18}{30}$ b. $\dfrac{70}{105}$ 13. a. $\dfrac{50}{400}$ b. $\dfrac{140}{420}$

14. a. $\dfrac{150}{1,000}$ b. $\dfrac{2,500}{10,000}$ 15. a. $\dfrac{78}{455}$ b. $\dfrac{75}{500}$

16. a. $\dfrac{240}{672}$ b. $\dfrac{5,670}{12,150}$ 17. a. $\dfrac{2,431}{3,003}$ b. $\dfrac{47,957}{54,808}$

Perform the indicated operations in Problems 18–42. (Recall that negative exponents are sometimes used to denote fractions. For example, $\frac{1}{7} = 7^{-1}$.)

18. a. $\dfrac{2}{3} + \dfrac{7}{9}$ b. $\dfrac{-5}{7} + \dfrac{4}{3}$ c. $\dfrac{-12}{35} - \dfrac{8}{15}$

19. a. $3 + 3^{-1}$ b. $2 + 2^{-1}$ c. $2^{-1} + 3^{-1}$

20. a. $3^{-1} + 5^{-1}$ b. $2^{-1} + 5^{-1}$ c. $2^{-1} + 3^{-1} + 5^{-1}$

21. a. $\dfrac{7}{9} - \dfrac{2}{3}$ b. $\dfrac{4}{7} - \dfrac{-5}{9}$ c. $\dfrac{-3}{5} - \dfrac{-6}{9}$

22. a. $\dfrac{2}{3} \times \dfrac{5}{7}$ b. $\dfrac{-1}{8} \times \dfrac{2}{-5}$ c. $5 \times \dfrac{6}{7}$

23. a. $\dfrac{4}{9} \div \dfrac{2}{3}$ b. $\dfrac{105}{-11} \div \dfrac{-15}{33}$ c. $\dfrac{47}{-5} \times \dfrac{-15}{26}$

24. a. $\dfrac{5}{3} \div \dfrac{7}{12}$ b. $\dfrac{-2}{9} \div \dfrac{6}{7}$ c. $\dfrac{6}{7} \div \dfrac{-3}{7}$

25. a. 7×7^{-1} b. -14×14^{-1} c. -12×12^{-1}

26. a. $6 \div 6^{-1}$ b. $-5 \div 5^{-1}$ c. $-4 \div 4^{-1}$

27. a. $\dfrac{1}{10} \cdot \dfrac{-2}{5}$ b. $\left(\dfrac{3}{7} \cdot \dfrac{3}{5}\right) \div \dfrac{1}{2}$ c. $\dfrac{2}{-3} \cdot \dfrac{-2}{15}$

28. a. $\dfrac{-3}{4} \cdot \dfrac{119}{200} + \dfrac{-3}{4} \cdot \dfrac{81}{200}$ b. $\dfrac{-3}{4}\left(\dfrac{119}{200} + \dfrac{81}{200}\right)$

29. a. $\dfrac{4}{5}\left(\dfrac{17}{95}\right) + \dfrac{4}{5}\left(\dfrac{78}{95}\right)$ b. $\dfrac{4}{5}\left(\dfrac{17}{95} + \dfrac{78}{95}\right)$

30. a. $\dfrac{2}{3} - \dfrac{7}{12}$ b. $\dfrac{-7}{24} + \dfrac{-13}{16}$

31. a. $\dfrac{28}{9} - \dfrac{4}{27}$ b. $\dfrac{1}{-8} + \dfrac{1}{-7}$

32. a. $-0.3 + 0.16 + 0.476$ b. $5^{-1} + 5^{-2} + 5$

33. a. $-5 + (-5)^2$ b. $-3 + (-3)^2$

34. a. $\dfrac{-2}{15} + \dfrac{3}{5} + \dfrac{7}{12}$ b. $-2\tfrac{3}{5} + 4\tfrac{1}{8} - 7\tfrac{1}{10}$

35. a. $\dfrac{\frac{1}{2} + \frac{-2}{3}}{\frac{5}{6} - \frac{-3}{5}}$ b. $\dfrac{\frac{1}{3} - \frac{-1}{4}}{\frac{7}{8} - \frac{3}{16}}$

36. a. $\dfrac{2^{-1} + 3^{-2}}{2^{-1} + 3^{-1}}$ b. $\dfrac{6 + 2^{-1}}{\frac{1}{2} + \frac{1}{3}}$

37. $\dfrac{11}{144} + \dfrac{17}{300} + \dfrac{7}{108}$ 38. $\dfrac{11}{108} - \dfrac{7}{144} + \dfrac{23}{300}$ 39. $\dfrac{7}{60} - \dfrac{19}{90} + \dfrac{21}{51}$

40. $\dfrac{14}{90} + \dfrac{7}{60} - \dfrac{11}{50}$ 41. $\dfrac{143}{210} + \dfrac{15}{124} + \dfrac{11}{1,085}$ 42. $\dfrac{15}{484} - \dfrac{5}{234} + \dfrac{27}{200}$

▲ **B Problems**

43. Consider the following apportionment problem for College Town:

        North: 8,700   South: 5,600   East: 7,200   West: 3,500

Suppose each council member is to represent 2,400 citizens. What is the representation for the plans of Adams, Jefferson, Webster, and Hamilton? For Hamilton's plan, assume that there must be 11 representatives.

44. Consider the following apportionment problem:

        North: 18,200   South: 12,900   East: 17,600   West: 13,300

Suppose each council member is to represent 2,400 citizens. What is the representation for the plans of Adams, Jefferson, Webster, and Hamilton? For Hamilton's plan, assume that there must be 25 representatives.

45. Consider the following apportionment problem:

     North: 18,200   South: 12,900   East: 17,600   West: 13,300

    Suppose each council member is to represent 5,000 citizens. What is the representation for the plans of Adams, Jefferson, Webster, and Hamilton? For Hamilton's plan, assume that there must be 12 representatives.

46. Consider the following apportionment problem:

North:	1,820,000	Northeast:	2,950,000
East:	1,760,000	Southeast:	1,980,000
South:	1,200,000	Southwest:	2,480,000
West:	3,300,000	Northwest:	1,140,000

    Suppose each house member is to represent 35,000 citizens. What is the representation for the plans of Adams, Jefferson, Webster, and Hamilton? For Hamilton's plan, assume that there must be 475 representatives.

47. Show that the set $\mathbb{Q}$ of rationals is closed for subtraction.

48. Show that the set $\mathbb{Q}$ of rationals is closed for nonzero division.

49. Is the set $\mathbb{Z}$ of integers closed for addition, subtraction, multiplication, and nonzero division?

50. Is $\mathbb{Q}$, the set of rationals, associative for addition?

51. Is $\mathbb{Q}$, the set of rationals, commutative for addition?

52. Is $\mathbb{Q}$, the set of rationals, associative and/or commutative for multiplication?

▲ Problem Solving

53. An elderly rancher died and left her estate to her three children. She bequeathed her 17 prize horses, to be divided in the following proportions: 1/2 to the eldest, 1/3 to the second child, and 1/9 to the youngest. How would you divide this estate?

54. The children (see Problem 53) decided to call in a very wise judge to help in the distribution of the rancher's estate. The judge arrived with a horse of his own. He put his horse in with the 17 belonging to the estate, and then told each child to pick from among the 18 in the proportions stipulated by the will (but be careful, he warned, not to pick *his* horse). The first child took nine horses, the second child took six, and the third child, two. The 17 horses were thus divided among the children. The wise judge took his horse from the corral, took a fair sum for his services, and rode off into the sunset.

    The youngest son complained. The oldest son received 9 horses (but was entitled to only $17/2 = 8.5$ horses). The judge was asked about this, and he faxed the children the following message: "You all received more than you deserved. The eldest received 1/2 of an 'extra' horse, the middle child received 1/3 more; and the youngest, 1/9 of a horse 'extra.'" Apportion the horses according to Adams', Jefferson's, and Webster's plans. Which plan gives the appropriate distribution of horses?

55. The children (see Problem 53) decided to call in a very wise judge to help in the distribution of the rancher's estate. They informed the judge that the 17 horses were not of equal value. The children agreed on a ranking of the 17 horses (#1 being the best and #17 being a real dog of a horse). They asked the judge to divide the estate

fairly so that each child would receive not only the correct number of horses but horses whose average rank would also be the same. For example, if a child received horses 1 and 17, the number of horses is two and the average value is $\dfrac{1 + 17}{2} = 9$. How did the judge apportion the horses?

A unit fraction (a fraction with a numerator of 1) is sometimes called an Egyptian fraction. On page 138, we said that the Egyptians expressed their fractions as sums of distinct (different) unit fractions. How might the Egyptians have written the fractions in Problems 56–59?

56. $\frac{3}{4}$ **57.** $\frac{47}{60}$ **58.** $\frac{42}{120}$ **59.** $\frac{7}{17}$

60.

A quantity and its two-thirds and its half and its one-seventh together make 33. Find the quantity.

Papyrus

answer: $14 + \dfrac{1}{4} + \dfrac{1}{56} + \dfrac{1}{97} + \dfrac{1}{194} + \dfrac{1}{388} + \dfrac{1}{679} + \dfrac{1}{776}$

Recall that the Egyptians used only unit fractions to represent numbers. Is the answer given on the papyrus correct?

61. The sum and difference of the same two squares may be primes, as in this example:

$9 - 4 = 5$ and $9 + 4 = 13$

Can the sum and difference of the same two primes be squares? Can you find more than one example?

▲ **Individual Research**

62. We mentioned that the Egyptians wrote their fractions as sums of unit fractions. Show that every positive fraction less than 1 can be written as a sum of unit fractions.

Reference Bernhardt Wohlgemuth, "Egyptian Fractions," *Journal of Recreational Math,* Vol. 5, No. 1 (1972), pp. 55–58.

63. The Egyptians had a very elaborate and well developed system for working with fractions. Write a paper on Egyptian fractions.

References

George Berzsenyi, "Egyptian Fractions," *Quantum,* November/December 1994, p. 45. Follow-up comment in *The College Mathematics Journal,* March 1995, p. 165.
Richard Gillings, *Mathematics in the Time of the Pharaohs* (New York: Dover Publications, 1982).
Spencer Hurd, "Egyptian Fractions: Ahmes to Fibonacci to Today," *Mathematics Teacher,* October 1991, pp. 561–568.

4.5   IRRATIONAL NUMBERS

Pythagorean Theorem

We have been considering numbers as they relate to practical problems. However, numbers can be appreciated for their beauty and interrelationships. The Pythagoreans, to our knowledge, were among the first to investigate numbers for their own sake.

Much of the Pythagoreans' lifestyle was embodied in their beliefs about numbers. They considered the number 1 the essence of reason; the number 2 was identified with opinion; and 4 was associated with justice because it is the first number that is the product of equals (the first perfect squared number, other than 1). Of the numbers greater than 1, odd numbers were masculine and even numbers were feminine; thus, 5 represented marriage, since it was the union of the first masculine and feminine numbers (2 + 3 = 5).

The Pythagoreans were also interested in special types of numbers that had mystical meanings: perfect numbers, friendly numbers, deficient numbers, abundant numbers, prime numbers, triangular numbers, square numbers, and pentagonal numbers. Other than the prime numbers we have already considered, the *perfect square* numbers are probably the most interesting. They are called **perfect squares** because they can be arranged into squares (see Figure 4.4). They are found by squaring the counting numbers.

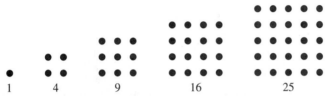

▲ **Figure 4.4   Square numbers 1, 4, 9, 16, and 25. Other square numbers (not pictured) are 36, 49, 64, 81, 100, 121, 144, 169, ...**

The Pythagoreans discovered the famous property of square numbers that today bears Pythagoras' name. They found that if they constructed any right triangle and then constructed squares on each of the legs of the triangle, the area of the larger square was equal to the sum of the areas of the smaller squares (see Figure 4.5).

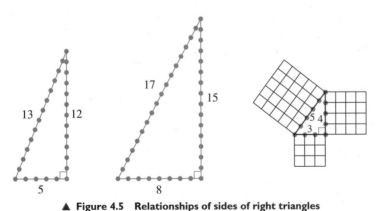

▲ **Figure 4.5   Relationships of sides of right triangles**

Today we state the Pythagorean theorem algebraically by saying that if a and b are the lengths of the **legs** (or sides) of a right triangle, and c is the length of the **hypotenuse** (the longest side), then the square of the length of the hypotenuse is equal to the sum of the squares of the lengths of the other two sides.

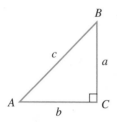

> **Pythagorean Theorem**
>
> For a right triangle ABC, with sides of length a, b, and hypotenuse c,
>
> $$a^2 + b^2 = c^2$$
>
> Also, if $a^2 + b^2 = c^2$ for a triangle with sides a, b, and c, then $\triangle ABC$ is a right triangle.

Square Roots

The Pythagoreans were overjoyed with this discovery, but it led to a revolutionary idea in mathematics — one that caused the Pythagoreans many problems.

Legend tells us that one day while the Pythagoreans were at sea, one of their group came up with the following argument. Suppose each leg of a right triangle is 1; then

$$a^2 + b^2 = c^2$$
$$1^2 + 1^2 = c^2$$
$$2 = c^2$$

If we denote the number whose square is 2 by $\sqrt{2}$, we have $\sqrt{2} = c$. The symbol $\sqrt{2}$ is read "square root of two." This means that $\sqrt{2}$ is that number such that, if multiplied by itself, is exactly equal to 2; i.e.,

$$\sqrt{2} \times \sqrt{2} = 2$$

> **Square Root**
>
> If $n \geq 0$, then the **positive square root of n**, denoted by $\sqrt{n}$, is defined as that number for which
>
> $$\sqrt{n}\,\sqrt{n} = n$$

■ EXAMPLE 1

Use the definition of square root to find the indicated products.

a. $\sqrt{3} \times \sqrt{3}$ **b.** $\sqrt{4} \times \sqrt{4}$ **c.** $\sqrt{5} \times \sqrt{5}$ **d.** $\sqrt{15} \times \sqrt{15}$
e. $\sqrt{16} \times \sqrt{16}$ **f.** $\sqrt{144} \times \sqrt{144}$ **g.** $\sqrt{200} \times \sqrt{200}$

Solution

a. 3   **b.** 4   **c.** 5   **d.** 15   **e.** 16   **f.** 144   **g.** 200   ▬

Some square roots are rational. For example,

$$\sqrt{4} \times \sqrt{4} = 4$$

and we also know $2 \times 2 = 4$, so it seems that $\sqrt{4} = 2$. But wait! We also know $(-2) \times (-2) = 4$, so isn't it just as reasonable to say $\sqrt{4} = -2$? Mathematicians have agreed that the square root symbol may be used only to denote positive numbers, so that $\sqrt{4} = 2$ and NOT -2.

Irrational Numbers

What about square roots of numbers that are not perfect squares? Is $\sqrt{2}$, for example, a rational number? Remember, if $\sqrt{2} = a/b$, where a/b is some fraction so that

$$\frac{a}{b} \cdot \frac{a}{b} = 2$$

then it is *rational*. The Pythagoreans were among the first to investigate this question. Now remember that, for the Pythagoreans, mathematics and religion were one; they asserted that all natural phenomena could be expressed by whole numbers or ratios of whole numbers. Thus, they believed that $\sqrt{2}$ must be some whole number or fraction (ratio of two whole numbers). Suppose we try to find such a rational number:

$$\frac{7}{5} \times \frac{7}{5} = \frac{49}{25} = 1.96 \quad \text{or, try again:} \quad \frac{707}{500} \times \frac{707}{500} = \frac{499{,}849}{250{,}000} = 1.999396$$

We are "getting closer" to 2, but we are still not quite there, so we really get down to business and use a calculator:

$\boxed{2}\ \boxed{\sqrt{}}$ *Display:* $\boldsymbol{1.414213562}$

If you square this number, do you obtain 2? Notice that the last digit of this multiplication will be 4; what should it be if it were the square root of 2? Even if we use a computer to find the following possibility for $\sqrt{2}$, we see that its square is still not 2:

1.41421356237309504880168872420969807856967187537694807317667973799007324784

Can you give a brief argument showing why that can't be $\sqrt{2}$? The real-world situation for this chapter asks for a *proof* that $\sqrt{2}$ is not rational. Such a number is called an *irrational number*. The set of **irrational numbers** is the set of numbers whose decimal representations do not terminate nor do they repeat.

It can be shown that not only $\sqrt{2}$ is irrational, but also $\sqrt{3}, \sqrt{5}, \sqrt{6}, \sqrt{7}, \sqrt{8},$ $\sqrt{10}$; in fact, the square root of any whole number that is not a perfect square is irrational. Also the cube root of any whole number that is not a perfect cube, and so on, is an irrational number. The number π, which is the ratio of the circumference of any circle to its diameter, is also not rational. In everyday work, you will use irrational numbers when finding the circumference and area of a circle, which we consider in Chapter 7.

We also use irrational numbers when applying the Pythagorean theorem. Since the Pythagorean theorem asserts $a^2 + b^2 = c^2$, then

$$c = \sqrt{a^2 + b^2}$$

Also, if you wish to find the length of one of the legs of a right triangle, say a, when

News Clip

One of the most famous irrational numbers is π. You are, of course, familiar with this number used to find the area or circumference of a circle. Technically, it is defined as the ratio of the circumference of a circle to its diameter. You may remember its decimal approximation of 3.1416; as an irrational number, it cannot be written as a terminating or a repeating decimal. This number was recently featured in Leslie Nielsen's spoof *Spy Hard*, where Nicollette Sheridan plays Russian Agent 3.14.

Historical Note

The Egyptians and the Chinese knew of the Pythagorean theorem before the Pythagoreans (but they didn't call it the Pythagorean theorem, of course). An early example is shown above. It is attributed to Chou Pei, who was probably a contemporary of Pythagoras. Below we have shown a copy from a Moslem manuscript, written in 1258, which shows the Pythagorean theorem.

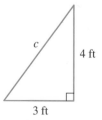

you know both b and c, you can use the formula

$$a = \sqrt{c^2 - b^2}$$

■ EXAMPLE 2

If a 13-ft ladder is placed against a building so that the base of the ladder is 5 ft away from the building, how high up does the ladder reach?

Solution   Consider Figure 4.6. Since h is one of the legs, use the formula $a = \sqrt{c^2 - b^2}$.

$$h = \sqrt{13 - 5^2} \quad \text{Substitute the lengths of the hypotenuse and the known leg.}$$

unknown

$$= \sqrt{169 - 25} = \sqrt{144} = 12$$

Thus, the ladder reaches 12 ft up the side of the building.

▲ **Figure 4.6   Ladder problem**

The Pythagorean theorem is of value only when dealing with a right triangle. Carpenters often make use of this property when they want to construct a right angle. That is, if $a^2 + b^2 = c^2$, then an angle of the triangle must be a right angle.

■ EXAMPLE 3

A carpenter wants to make sure that the corner of a room is square (is a right angle). If she measures out sides (legs) of 3 ft and 4 ft, how long should she make the diagonal (hypotenuse)?

Solution   The triangle (corner of the room) is shown in Figure 4.7. The hypotenuse is the unknown, so use the formula

$$c = \sqrt{a^2 + b^2}$$

▲ **Figure 4.7   Building a right angle**

Thus,

$$c = \sqrt{3^2 + 4^2} = \sqrt{9 + 16} = \sqrt{25} = 5$$

She should make the diagonal 5 ft long. ▬

Your answers to both Examples 2 and 3 are rational. Suppose the result is irrational. You can either leave the result in **radical form** or approximate the result, as shown in Example 4.

▬▬ EXAMPLE 4

Suppose you need to attach several guy wires to your TV antenna, as shown in Figure 4.8. If one guy wire is attached 20 ft away from the base of an antenna, and the height of the wire on the antenna is 30 ft, what is the exact length of one guy wire, and what is the length to the nearest foot?

Solution   The length of the guy wire is the length of the hypotenuse of a right triangle. Thus,

$$c = \sqrt{a^2 + b^2} = \sqrt{20^2 + 30^2} = \sqrt{400 + 900} = \sqrt{1,300}$$

The exact length of the guy wire is $\sqrt{1,300}$ and is irrational since 1,300 is not a perfect square. What is this length as an approximate rational number? Because

$$30^2 = 900 \quad \text{and} \quad 40^2 = 1,600$$

and 1,300 is between 900 and 1,600, we know $\sqrt{1,300}$ is between 30 and 40. For a better approximation, we can use a calculator:

$\boxed{1300}$ $\boxed{\sqrt{}}$ *Display:*   **36.05551275**

The guy wire is 36 ft (to the nearest ft). If the application will not allow any length less than $\sqrt{1,300}$, then, instead of rounding, take the *next larger foot*. ▬

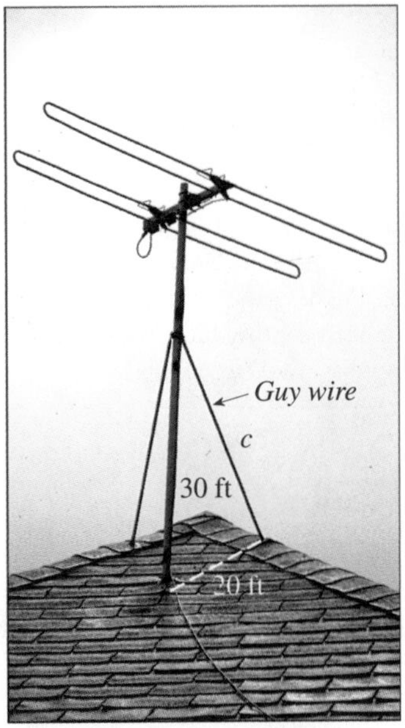

▲ **Figure 4.8   TV antenna**

Operations with Square Roots

There are times when a square root is irrational and yet we do not want a rational approximation. In such cases, we will need to know certain laws of square roots and when a square root is simplified.

Laws of Square Roots

Let a and b be positive numbers. Then:

1. $\sqrt{0} = 0$ 2. $\sqrt{a^2} = a$ 3. $\sqrt{ab} = \sqrt{a}\sqrt{b}$ 4. $\sqrt{\dfrac{a}{b}} = \dfrac{\sqrt{a}}{\sqrt{b}}$

5. A square root is **simplified** if:

The **radicand** (the number under the radical sign) has no factor with an exponent larger than 1 when it is written in factored form.

The radicand is not written as a fraction or by using negative exponents.

There are no square root symbols used in the denominators of fractions.

EXAMPLE 5

Simplify $\sqrt{8}$.

Solution

Step 1.   Factor the radicand: $\sqrt{8} = \sqrt{2^3}$

Step 2.   Write the radicand as a product of as many factors with exponents of 2 as possible; if there is a remaining factor, it will have an exponent of 1:

$$\sqrt{2^3} = \sqrt{2^2 \cdot 2^1}$$

Step 3.   Use Law 3 for square roots: $\sqrt{2^2 \cdot 2^1} = \sqrt{2^2} \cdot \sqrt{2^1}$

Step 4.   Use Law 2 for square roots: $\sqrt{2^2} \cdot \sqrt{2^1} = 2\sqrt{2}$ ▬

Notice that the simplified form in Example 5 still contains a radical, so $\sqrt{8}$ is an irrational number. We call $2\sqrt{2}$ the **exact** simplified representation for $\sqrt{8}$ and the calculator representation ($\boxed{8}$ $\boxed{\sqrt{}}$) 2.828427125 is an **approximation**. The whole process of simplifying radicals depends on factoring the radicand and separating out the square factors, and is usually condensed as shown by the following example.

EXAMPLE 6

Simplify the radical expression and assume that the variables are positive. If the expression is simplified, so state.

a. $\sqrt{441}$ b. $\sqrt{2{,}100}$ c. $\sqrt{(x + y)^2}$ d. $\sqrt{x^2 + y^2}$ e. $\dfrac{7\sqrt{2}}{\sqrt{6}}$ f. $\sqrt{0.1}$

g. $\sqrt{\dfrac{5x^2}{27y}}$ h. $\sqrt{2^2 - 4(1)(-5)}$ i. $\sqrt{10^2 - 4(5)(-15)}$ j. $3 + \sqrt{5}$

k. $6 + 3\sqrt{5}$ l. $\dfrac{3 + \sqrt{5}}{3}$ m. $\dfrac{6 + 3\sqrt{5}}{3}$ n. $\dfrac{-(-2) + \sqrt{(-2)^2 - 4(2)(-1)}}{2(2)}$

Solution   There are many parts to this example, and you should review them carefully because they lay the groundwork for future sections in this book.

a. If you do not see any factors that are square numbers, you can use a factor tree:

$$\sqrt{441} = \sqrt{3^2 \cdot 7^2} = 3 \cdot 7 = 21$$

b. $\sqrt{2{,}100} = \sqrt{10^2 \cdot 3 \cdot 7} = 10\sqrt{21}$

c. $\sqrt{(x + y)^2} = x + y$

d. $\sqrt{x^2 + y^2}$ is simplified; remember that we are looking for square factors, not square terms.

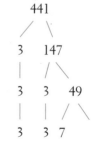

e. This example came from a *Peanuts* cartoon, and the solution shown in the cartoon is correct.

PEANUTS reprinted by permission of UFS, Inc.

f. $\sqrt{0.1} = \sqrt{\dfrac{1}{10}} = \dfrac{\sqrt{1}}{\sqrt{10}} \cdot \dfrac{\sqrt{10}}{\sqrt{10}} = \dfrac{\sqrt{10}}{10}$

This is also written as $\frac{1}{10}\sqrt{10}$ or $0.1\sqrt{10}$.

g. $\sqrt{\dfrac{5x^2}{27y}} = \sqrt{\dfrac{5x^2}{3^2 \cdot 3y}} = \dfrac{x\sqrt{5}}{3\sqrt{3y}} \cdot \dfrac{\sqrt{3y}}{\sqrt{3y}} = \dfrac{x\sqrt{5 \cdot 3y}}{3(3y)} = \dfrac{x\sqrt{15y}}{9y}$

h. $\sqrt{2^2 - 4(1)(-5)} = \sqrt{24} = 2\sqrt{6}$

i. $\sqrt{10^2 - 4(5)(-15)} = \sqrt{100 + 300} = \sqrt{400} = 20$

j. $3 + \sqrt{5}$ is simplified.

k. $6 + 3\sqrt{5}$ is simplified, but sometimes is written in factored form (using the distributive property) as $3(2 + \sqrt{5})$.

l. $\dfrac{3 + \sqrt{5}}{3}$ is simplified.

m. $\dfrac{6 + 3\sqrt{5}}{3} = \dfrac{3(2 + \sqrt{5})}{3} = 2 + \sqrt{5}$

n. $\dfrac{-(-2) + \sqrt{(-2)^2 - 4(2)(-1)}}{2(2)} = \dfrac{2 + \sqrt{4 + 4(2)}}{4} = \dfrac{2 + \sqrt{12}}{4}$

$= \dfrac{2 + 2\sqrt{3}}{4} = \dfrac{2(1 + \sqrt{3})}{4} = \dfrac{1 + \sqrt{3}}{2}$

PROBLEM SET 4.5

▲ A Problems

1. **IN YOUR OWN WORDS** What is the Pythagorean theorem?

2. **IN YOUR OWN WORDS** Explain the two meanings of the square root symbol — as an operation and as a number.

3. **IN YOUR OWN WORDS** Discuss the exact and decimal approximations for an irrational number.

4. **IN YOUR OWN WORDS** What does it mean for a square root to be simplified?

5. **IN YOUR OWN WORDS** A computer approximation for $\sqrt{2}$ is

$$1.41421356237309504880168872420969807856967187537694$$

Give a brief argument showing why that can't be $\sqrt{2}$.

6. **IN YOUR OWN WORDS** What do you think is meant by a triangular number?

7. Write the limerick at the left in symbols, and then tell whether it is true or false.*

A dozen, a gross, and a score
Plus three times the square root
  of four
Divided by seven
Plus five times eleven
Is nine squared and not a bit more.

Use the definition of square root to find the indicated products in Problems 8–11.

8. a. $\sqrt{6} \times \sqrt{6}$ b. $\sqrt{7} \times \sqrt{7}$ c. $\sqrt{9} \times \sqrt{9}$

9. a. $\sqrt{14} \times \sqrt{14}$ b. $\sqrt{30} \times \sqrt{30}$ c. $\sqrt{36} \times \sqrt{36}$

10. a. $\sqrt{807} \times \sqrt{807}$ b. $\sqrt{169} \times \sqrt{169}$ c. $\sqrt{400} \times \sqrt{400}$

11. a. $\sqrt{2.5} \times \sqrt{2.5}$ b. $\sqrt{2.4} \times \sqrt{2.4}$ c. $\sqrt{0.25} \times \sqrt{0.25}$

Classify each number in Problems 12–17 as rational or irrational. If it is rational, write it without a square root symbol. If it is irrational, approximate it with a rational number correct to the nearest thousandth.

12. a. $\sqrt{9}$ b. $\sqrt{25}$ 13. a. $\sqrt{10}$ b. $\sqrt{30}$

14. a. $\sqrt{36}$ b. $\sqrt{50}$ 15. a. $\sqrt{169}$ b. $\sqrt{400}$

16. a. $\sqrt{500}$ b. $\sqrt{1,000}$ 17. a. $\sqrt{1,024}$ b. $\sqrt{1,936}$

Simplify the expressions in Problems 18–37. Assume that the variables are positive.

18. a. $-\sqrt{16}$ b. $-\sqrt{144}$ c. $\sqrt{125}$ d. $\sqrt{96}$

19. a. $\sqrt{1,000}$ b. $\sqrt{2,800}$ c. $\sqrt{2,240}$ d. $\sqrt{4,410}$

20. a. $3\sqrt{75}$ b. $2\sqrt{90}$ c. $5\sqrt{48}$ d. $3\sqrt{96}$

21. a. $\sqrt{\frac{1}{2}}$ b. $\sqrt{\frac{1}{3}}$ c. $\sqrt{\frac{3}{5}}$ d. $\sqrt{\frac{3}{7}}$

22. a. $-\sqrt{0.1}$ b. $-\sqrt{0.4}$ c. $\sqrt{0.75}$ d. $\sqrt{0.05}$

23. a. $\frac{1}{\sqrt{2}}$ b. $\frac{-1}{\sqrt{3}}$ c. $\frac{2}{\sqrt{5}}$ d. $\frac{5}{\sqrt{10}}$

24. a. $\sqrt{5^2 - 4(3)(2)}$ b. $\sqrt{7^2 - 4(5)(2)}$

25. a. $\sqrt{3^2 - 4(5)(-3)}$ b. $\sqrt{8^2 - 4(1)(-1)}$

26. a. $\sqrt{10^2 - 4(5)(-5)}$ b. $\sqrt{12^2 - 4(3)(12)}$

27. a. $\sqrt{6^2 - 4(3)(-2)}$ b. $\sqrt{2^2 - 4(1)(-1)}$

28. a. $\frac{6 + 2\sqrt{5}}{2}$ b. $\frac{8 - 4\sqrt{3}}{4}$ 29. a. $\frac{12 - 3\sqrt{2}}{6}$ b. $\frac{6 - 2\sqrt{5}}{4}$

30. a. $\frac{4 + 6\sqrt{3}}{-2}$ b. $\frac{6 - 8\sqrt{3}}{-2}$ 31. a. $\frac{3 - 9\sqrt{x}}{3}$ b. $\frac{9 + 3\sqrt{x}}{-3}$

* From *Omni*, March 1995, "Games" department, p. 104.

32. a. $\dfrac{3}{\sqrt{x}}$ b. $\dfrac{-7}{\sqrt{y}}$ 33. a. $\sqrt{\dfrac{4x^2}{25y}}$ b. $\sqrt{\dfrac{5y}{16x}}$

34. $\dfrac{-7 + \sqrt{7^2 - 4(2)(3)}}{2(2)}$

35. $\dfrac{-(-2) - \sqrt{(-2)^2 - 4(6)(-3)}}{2(6)}$

36. $\dfrac{-10 - \sqrt{10^2 - 4(3)(6)}}{2(3)}$

37. $\dfrac{-(-12) + \sqrt{(-12)^2 - 4(1)(-1)}}{2(1)}$

▲ B Problems

38. How far from the base of a building must a 26-ft ladder be placed so that it reaches 10 ft up the wall?

39. How high up on a wall does a 26-ft ladder reach if the bottom of the ladder is placed 10 ft from the base of the building?

40. If a carpenter wants to make sure that the corner of a room is square and measures out 5 ft and 12 ft along the walls, how long should he make the diagonal?

41. If a carpenter wants to be sure that the corner of a building is square and measures out 6 ft and 8 ft along the sides, how long should she make the diagonal?

42. What is the exact length of the hypotenuse if the legs of a right triangle are 2 in. each?

43. What is the exact length of the hypotenuse if the legs of a right triangle are 3 ft each?

44. An empty lot is 400 ft by 300 ft. How many feet would you save by walking diagonally across the lot instead of walking the length and width?

45. A television antenna is to be erected and held by guy wires. If the guy wires are 15 ft from the base of the antenna and the antenna is 10 ft high, what is the exact length of each guy wire? What is the length of each guy wire to the nearest foot? If three guy wires are to be attached, how many feet of wire should be purchased if it can't be bought by a fraction of a foot?

46. A diagonal brace is to be placed in the wall of a room. The height of the wall is 8 ft and the wall is 20 ft long. What is the exact length of the brace? What is the length of the brace to the nearest foot?

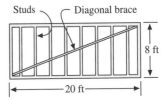

47. A balloon rises at a rate of 4 ft per second when the wind is blowing horizontally at a rate of 3 ft per second. After three seconds, how far away from the starting point, in a direct line, is the balloon?

48. Consider a square inch as shown in Figure 4.9a.

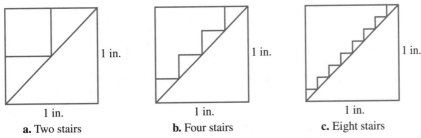

a. Two stairs   **b.** Four stairs   **c.** Eight stairs

▲ **Figure 4.9   Find the length of steps**

Find the total length of the segments making up the stairs in each part of Figure 4.9.

49. Find an irrational number between 1 and 3.

50. Find an irrational number between 0.53 and 0.54.

51. Find an irrational number between $\frac{1}{11}$ and $\frac{1}{10}$.

52. Without using a radical symbol, write an irrational number using only 2s and 3s.

53. Suppose three squares are made of gold plate of uniform thickness and you are offered one of either the large square or the two smaller ones. Which choices would you make for each of the squares having sides whose lengths are given below?

   a. 1 in. and 1 in. or 2 in.   b. 3 in. and 4 in. or 5 in.
   c. 4 in. and 5 in. or 7 in.   d. 5 in. and 7 in. or 9 in.
   e. 10 in. and 11 in. or 15 in.

54. You may have seen a *geoboard*, which is simply a board with pegs labeled A, B, C, ... Two simple geoboards are shown in Figure 4.10. Suppose that the horizontal and vertical distance between pegs is 1 inch. We can then calculate the distance from A to D in Figure 4.10a to be $\sqrt{2}$ (by the Pythagorean theorem). If we adopt the agreement that the distance from A to A (or from any point to itself) is 0, we can find the distance from any peg to any other peg.

   Use the 3 × 3 pegboard shown in Figure 4.10b to fill in the following table. Note that some of the entries have been filled in for you.

	A	B	C	D	E	F	G	H	I
A	0	1	2		$\sqrt{2}$				
B		0							
C			0						
D				0					
E					0				
F						0			
G							0		
H								0	
I									0

a. 2 × 2 geoboard

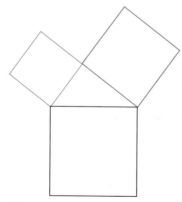

b. 3 × 3 geoboard

▲ **Figure 4.10**   **Distances on a geoboard**

▲ **Problem Solving**

55. The Pythagorean theorem tells us that the sum of the squares of the lengths of the legs of a right triangle is equal to the square of the length of the hypotenuse. Verify that the theorem is true by tracing the squares and fitting them onto the pattern of the upper two squares shown in Figure 4.11. Next, cut the squares along the lines and rearrange the pieces so that the pieces all fit into the large square in Figure 4.11.

▲ **Figure 4.11**   **Pythagorean theorem squares**

56. Repeat Problem 55 for the following squares.

57. A man wants to board a plane with a 5-ft long steel rod, but airline regulations say that the maximum length of any object or parcel checked on board is 4 ft. Without bending or cutting the rod, or altering it in any way, how did the man check it through without violating the rule?

58. HISTORICAL QUESTION The Historical Note in this section introduced the great mathematician Karl Gauss. Gauss kept a scientific diary containing 146 entries, some of which were independently discovered and published by others. On July 10, 1796, he wrote

What do you think this meant? Illustrate with some numerical examples.

59. What mathematical property is illustrated in Figure 4.12?

▲ **Figure 4.12   Can you discover this mathematical property?**

60. How are the square numbers embedded in Pascal's triangle?

61. How are the triangular numbers embedded in Pascal's triangle?

▲ **Individual Research**

62. Write a paper or prepare an exhibit illustrating the Pythagorean theorem. Here are some questions you might consider:

What is the history of the Pythagorean theorem?

What are some unusual proofs of the Pythagorean theorem?

What are some of the unusual relationships that exist among Pythagorean numbers?

What models can be made to visualize the Pythagorean theorem?

4.6   REAL NUMBERS

Definition of Real Numbers

You are familiar with the rational numbers (fractions, for example) and the irrational numbers (π or square roots of certain numbers, for example), and now we wish to consider the most general set of numbers to be used in elementary mathematics. This set consists of the annexation of the irrational numbers to the set of rational numbers, and is called the set of *real numbers*.

> **Real Numbers**
>
> The set of **real numbers**, denoted by $\mathbb{R}$, is defined as the union of the set of rationals and the set of irrationals.

Historical Note

History of the Decimal Point

Write the decimal 24.375:

Author	Time	Notation
Unknown		$24\frac{375}{1,000}$
Simon Stevin	1585	24 3(1)7(2)5(3)
François Viète	1600	24|375
Johannes Kepler	1616	24(375
John Napier	1616	24:3 7 5
Henry Biggs	1624	24375
William Oughtred	1631	24|375
Balam	1653	24:375
Ozanam	1691	(1)(2)(3) 24·3 7 5

Decimal Representation of Real Numbers

Let's consider the decimal representation of a real number. If a number is *rational*, then its decimal representation is either **terminating** or **repeating**.

▪ EXAMPLE 1

Find the decimal representation of each of the given rational numbers.

a. $\dfrac{1}{4}$ b. $\dfrac{5}{8}$ c. $\dfrac{58}{10}$ d. $\dfrac{2}{3}$ e. $\dfrac{1}{6}$ f. $\dfrac{5}{11}$ g. $\dfrac{1}{7}$

Solution

a. $\dfrac{1}{4} = 0.25$ [1] ÷ [4]   This is a terminating decimal.

b. $\dfrac{5}{8} = 0.625$ [5] ÷ [8]   This is a terminating decimal.

c. $\dfrac{58}{10} = 5.8$ [58] ÷ [10]   This is a terminating decimal.

d. $\dfrac{2}{3} = 0.666\ldots$ [2] ÷ [3]   *Display:* .6666666667
This is a repeating decimal.

Notice that calculators always represent decimals as terminating decimals, so you need to *interpret* the calculator display as a repeating decimal. If you look at the long division, you can see that the division never terminates:

```
     .666 . . .
3)2.000 . . .
  1.8
  ───
   20
   18
   ──
    2 . . .
```

e. $\dfrac{1}{6} = 0.166\ldots$ $\boxed{1}\ \boxed{\div}\ \boxed{6}$ This is a repeating decimal.

f. $\dfrac{5}{11} = 0.4545\ldots$ $\boxed{5}\ \boxed{\div}\ \boxed{11}$ This is a repeating decimal.

g. $\dfrac{1}{7} \approx 0.143$ $\boxed{1}\ \boxed{\div}\ \boxed{7}$ *Display:*   *.1428571429*

It may happen that you do not recognize a pattern by looking at the calculator display. For most of our work, an approximation of the result will suffice. Can you use long division to show that the decimal representation *must* terminate (have a 0 remainder) or *must* repeat? For $\frac{1}{7}$ it repeats after six digits.

When a decimal repeats, we sometimes use an overbar to indicate the numerals that repeat. For Example 1,

One digit repeats: $\frac{2}{3} = 0.\overline{6},\ \ \frac{1}{6} = 0.1\overline{6},$
Two digits repeat: $\frac{5}{11} = 0.\overline{45},$
Six digits repeat: $\frac{1}{7} = 0.\overline{142857}$

Real numbers that are *irrational* have decimal representations that are *nonterminating* and *nonrepeating*:

$$\sqrt{2} = 1.414213\ldots \qquad \pi = 3.141592\ldots$$

In each of these examples, the numbers exhibit no repeating pattern and are irrational. Other decimals that do not terminate or repeat are also irrational:

$$0.12345678910111213\ldots \qquad 0.10110111011110111110\ldots$$

We now have some different ways to classify real numbers:

1. Positive, negative, or zero
2. A rational number or an irrational number
   a. If it terminates, it is rational.
   b. If it eventually repeats, it is rational.
   c. If it has a nonterminating and nonrepeating decimal, it is irrational.

We have illustrated the procedure for changing from a fraction to a decimal: Divide the numerator by the denominator. To reverse the procedure and to change from a terminating decimal representation of a rational number to a fractional representation, use expanded notation. Recall that $10^{-1} = \frac{1}{10}$, $10^{-2} = \frac{1}{100}$, $10^{-3} = \frac{1}{1,000}$, ..., $10^{-n} = \frac{1}{10^n}$. Thus 0.5 means $5 \times 10^{-1} = 5 \cdot \frac{1}{10} = \frac{5}{10} = \frac{1}{2}$.

▬▬▬ **EXAMPLE 2**

Change the terminating decimals to fractional form.

a. 0.123   **b.** 56.28   **c.** 0.3479

WE'LL NEVER BE EQUALS.

3 π

β.

Solution   Write each in expanded notation.

a. $0.123 = 1 \times 10^{-1} + 2 \times 10^{-2} + 3 \times 10^{-3}$

$$= \frac{1}{10} + \frac{2}{100} + \frac{3}{1,000} = \frac{123}{1,000}$$

b. $56.28 = \dfrac{5,628}{100} = \dfrac{1,407}{25}$ The steps shown in part *a* can often be done mentally.

c. $0.3479 = \dfrac{3,479}{10,000}$ Make sure the fraction is reduced.

Real Number Line

If we once again consider a number line and associate points on the number line with rational numbers, it appears that the number line is just about "filled up" by the rationals. The reason for this feeling of "fullness" is that the rationals form what is termed a **dense set**. That is, between every two rationals we can find another rational (see Figure 4.13).

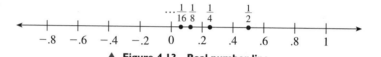

▲ **Figure 4.13   Real number line**

If we plot *all* the points of this dense set called the rationals, are there still any "holes"? In other words, is there any room left for any of the irrationals? We have shown that $\sqrt{2}$ is irrational. We can show that there is a place on the number line representing this length by using the Pythagorean theorem, as shown in Figure 4.14.

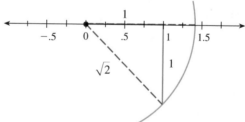

▲ **Figure 4.14   Finding an irrational "hole" on a real number line**

We could show that other irrationals have their places on the number line (see Question 1 of the Group Research at the end of the chapter as well as Figure 4.15).

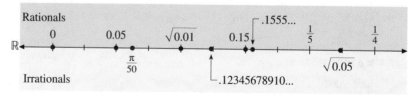

▲ **Figure 4.15   Real number line showing some rationals and some irrationals**

The relationships among the various sets of numbers we have been discussing are shown in Figure 4.16.

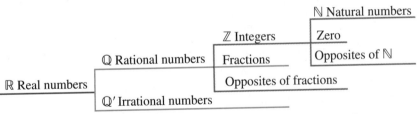

▲ **Figure 4.16 Classifications within the set of real numbers, $\mathbb{R}$**

Properties of Real Numbers

At the beginning of this chapter, we stated some properties of natural numbers that apply to sets of real numbers as well. We repeat them here for easy reference. Let a, b, and c be real numbers; we denote this by a, b, $c \in \mathbb{R}$. Then,

	Addition	*Multiplication*
Closure:	$(a + b) \in \mathbb{R}$	$ab \in \mathbb{R}$
Associative:	$(a + b) + c = a + (b + c)$	$(ab)c = a(bc)$
Commutative:	$a + b = b + a$	$ab = ba$

Distributive for multiplication over addition: $a(b + c) = ab + ac$

There are two additional properties that are important in the set of real numbers: the identity and inverse properties.

In Section 3.1 we mentioned that the development of the concept of zero and a number for it did not take place at the same time as the development of the counting numbers. The Greeks were using the letter "oh" for zero as early as 150 A.D., but the predominant system in Europe was the Roman numeration system, which did not include zero. It was not until the 15th century, when the Hindu–Arabic numeration system finally replaced the Roman system, that the zero symbol came into common usage.

The number 0 (zero) has a special property for addition that allows it to be added to any real number without changing the value of that number. This property is called the **identity property for addition** of real numbers.

Identity for Addition

There exists in $\mathbb{R}$ a number 0, called **zero,** so that

$$0 + a = a + 0 = a$$

for any $a \in \mathbb{R}$.

Remember that when you are studying algebra, you are studying ideas and not just rules about specific numbers. A mathematician would attempt to isolate the *concept* of an identity. First, does an identity property apply for other operations?

Multiplication	*Subtraction*	*Division*
$\square \times a = a \times \square = a$	$\triangle - a = a - \triangle = a$	$\triangledown \div a = a \div \triangledown = a$
↑ same number ↑	↑ same number ↑	↑ same number ↑

Is there a real number that will satisfy any of the blanks for multiplication, subtraction, or division?

The second property involves another special, important number in $\mathbb{R}$, namely, the number 1 (one). This number has the property that it can multiply any real number without changing the value of that number. This property is called the **identity property for multiplication** of real numbers.

Identity For Multiplication

There exists in $\mathbb{R}$ a number 1, called **one**, so that

$$1 \times a = a \times 1 = a$$

for any $a \in \mathbb{R}$.

Notice that there is no real number that satisfies the identity property for subtraction or division. There may be identities for other operations or for sets other than the set of real numbers, as indicated in the next example.

▬▬ EXAMPLE 3

Is there an identity set for the operation of union of sets?

Solution   We wish to find a single set $\square$ so that

$$\square \cup X = X \cup \square = X$$

holds for every set X. We see that the empty set satisfies this condition; that is,

$$\varnothing \cup X = X \cup \varnothing = X$$

Thus, $\varnothing$ is the identity set for the operation of union.   ▬

In Section 4.3 we spoke of opposites when we were adding and subtracting integers. Recall the property of opposites:

$$5 + (-5) = 0 \quad -128 + 128 = 0 \quad a + (-a) = 0$$

When opposites are added, the result is zero, the identity for addition. This idea, which can be generalized, is called the **inverse property for addition**.

Inverse Property for Addition

For each $a \in \mathbb{R}$, there is a unique number $(-a) \in \mathbb{R}$, called the **opposite** (or **additive inverse**) of a, so that

$$a + (-a) = -a + a = 0$$

Historical Note

The Egyptians did not have a zero symbol, but their numeration system did not require such a symbol. On the other hand, the Babylonians, with their positional system, had a need for a zero symbol but did not really use one until around 150 A.D. The Mayan Indians' numeration system was one of the first to use a zero symbol, not only as a place-holder but as a number zero. The exact time of its use is not known, but the symbol was noted by the early 16th century Spanish expeditions into Yucatan. Evidently, the Mayans were using the zero long before Columbus arrived in America.

Recall that the product of a number and its reciprocal is one, the identity for multiplication. The reciprocal of a number, then, is the multiplicative inverse of the number, as we will now show.

Inverse for multiplication   Inverse for multiplication

$$5 \times \square = \square \times 5 = 1 \qquad -128 \times \triangle = \triangle \times (-128) = 1$$

$$5 \times \frac{1}{5} = \frac{1}{5} \times 5 = 1 \qquad -128 \times \frac{1}{-128} = \frac{1}{-128} \times (-128) = 1$$

Since $\frac{1}{5} \in \mathbb{R}$, $\frac{1}{5}$ is an inverse   Since $\frac{1}{-128} \in \mathbb{R}$, $\frac{1}{-128}$ is an inverse
of 5 for multiplication.   of -128 for multiplication.

To show this inverse property for multiplication in a general way, we seek a replacement for the box for each and every real number a:

$$a \times \square = \square \times a = 1$$

Does the inverse property for multiplication hold for every real number a? No, because if $a = 0$, then

$$0 \times \square = \square \times 0 = 1$$

does *not* have a replacement for the box in $\mathbb{R}$. However, the inverse property for multiplication holds for all *nonzero* replacements of a, and we adopt this condition as part of the inverse property for multiplication of real numbers.

Inverse Property for Multiplication

For *each* number $a \in \mathbb{R}$, $a \neq 0$, there exists a number $a^{-1} \in \mathbb{R}$, called the **reciprocal** (or **multiplicative inverse**) of a, so that

$$a \times a^{-1} = a^{-1} \times a = 1$$

EXAMPLE 4

Given the set $A = \{-1, 0, 1\}$. Does this set A have an element that satisfies the inverse property for multiplication?

Solution   Before we can talk about the inverse property, we need to find the identity for the operation — in this case, multiplication. Without an identity for multiplication, we cannot have an inverse for multiplication. The operation of multiplication has the identity 1. Check to see that *each* element of A has an inverse:

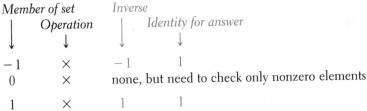

Member of set	Operation	Inverse	Identity for answer
-1	$\times$	-1	1
0	$\times$	none, but need to check only nonzero elements	
1	$\times$	1	1

Since every nonzero element in the set has an inverse, we say the inverse property for multiplication is satisfied.

We have now introduced several properties that we can apply to a given set with a given operation. If a set satisfies the closure, associative, identity, and inverse properties for a particular operation, then we call that set a *group*. We use the symbol ∘ to stand for any operation. This operation might be $+$, $\times$, or any other *given* operation.

Group

Let $\mathbb{S}$ be any set, let ∘ be any operation, and let a, b, and c be any elements of $\mathbb{S}$. We say that $\mathbb{S}$ is a **group** for the operation of ∘ if the following properties are satisfied:

1. The set $\mathbb{S}$ is *closed* for ∘: $(a \circ b) \in \mathbb{S}$
2. The set $\mathbb{S}$ is *associative* for ∘: $(a \circ b) \circ c = a \circ (b \circ c)$
3. The set $\mathbb{S}$ satisfies the *identity* for ∘:   There exists a number $I \in \mathbb{S}$ so that $x \circ I = I \circ x = x$ for *every* $x \in \mathbb{S}$.
4. The set $\mathbb{S}$ satisfies the *inverse* for ∘:   For *each* $x \in \mathbb{S}$, there exists a corresponding $x^{-1} \in \mathbb{S}$ so that $x \circ x^{-1} = x^{-1} \circ x = I$, where I is the identity element in $\mathbb{S}$.

Furthermore, $\mathbb{S}$ is called a **commutative group** (or **Abelian group**) if the following property is satisfied:

5. The set $\mathbb{S}$ is *commutative* for ∘: $a \circ b = b \circ a$

A commutative group is called Abelian to honor the mathematician Niels Abel (see Historical Note on page 282).

▬▬ EXAMPLE 5

Is the set $\mathbb{N}$ of natural numbers a group for multiplication?

Solution   We have already studied the properties of $\mathbb{N}$, so we can form hasty conclusions:

1. Closure: The product of any two natural numbers is a natural number.
2. Associative: Yes
3. Identity: Yes, namely 1
4. Inverse: No, since there is no number $\square$ in $\mathbb{N}$ so that

$$3 \times \square = \square \times 3 = 1$$

All we need to do to show that a property doesn't hold is to come up with one counterexample. We might also note that an inverse exists, namely $\frac{1}{3}$, but $\frac{1}{3} \notin \mathbb{N}$.

Therefore, $\mathbb{N}$ does not form a group for multiplication.   ▬

■■■ **EXAMPLE 6** Polya's Method

Let us partition the set of even and odd natural numbers into two sets, E (even) and O (odd). Consider the operations of addition (+) and multiplication (×) in the set $\{E, O\}$. Does the set form a group for either or both of these operations?

Solution We use Polya's problem-solving guidelines for this example.

Understand the Problem. In this example, we are considering a set that consists of two elements, E and O (never mind that each of these elements is also a set). How could we possibly define addition? $E + O$ means that we take *any* even number and add to it *any* odd number, and then ask whether the result is even, odd, or something else; in this case we conclude that the answer must be an odd number, so we write $E + O = O$. Consider the operations of addition and multiplication:

Addition	Multiplication
even + even = even	even × even = even
even + odd = odd	even × odd = even
odd + even = odd	odd × even = even
odd + odd = even	odd × odd = odd

These operations can be summarized in table format:

+	E	O		×	E	O
E	E	O		E	E	E
O	O	E		O	E	O

Devise a Plan. We will check the properties for each of these operations one at a time. If and when we find a counterexample for one of the group properties, we will have the conclusion that it is not a group for that operation.

Carry Out the Plan.

	Addition	Multiplication
1. Closure:	Yes	Yes

Every entry in the tables is either an E or O, both of which are in the set.

	Addition	Multiplication
2. Associative:	Yes	Yes

We prove this by checking all possibilities (there are several). We show a few here:

$$(E + E) + E = E + (E + E) \qquad (E \times E) \times E = E \times (E \times E)$$
$$(E + E) + O = E + (E + O) \qquad (E \times E) \times O = E \times (E \times O)$$
$$(E + O) + E = E + (O + E) \qquad (E \times O) \times E = E \times (O \times E)$$
$$\vdots \qquad\qquad\qquad\qquad \vdots$$

	Addition	Multiplication
3. Identity:	Yes, it is E.	Yes, it is O.

	Addition	Multiplication
4. Inverse:	Yes	No

Inverse of E is E E does not have an inverse
     since $E + E = E$ because $E \times ? = O$
Inverse of O is O
     since $O + O = E$

Conclusion. The set $\{E, O\}$ is a *group* for $+$, but not for $\times$.

5. Commutative: Yes   Yes

     The tables are symmetric with respect to the principal diagonal. That is, $E + O = O + E$ and $E \times O = O \times E$.

Look Back. The set $\{E, O\}$ is a commutative group for addition.   ▬

     Example 6 showed that a given set was a commutative group for one operation but not for another. If a set is a commutative group for two operations, and *also* satisfies the distributive property, it is called a *field*. We define a field in terms of the set $\mathbb{R}$ and the operations of $+$ and $\times$.

Field

A **field** is a set $\mathbb{R}$, with two operations $+$ and $\times$ satisfying the following properties for any elements $a, b, c \in \mathbb{R}$:

	Addition, $+$	Multiplication, $\times$
Closure:	1. $(a + b) \in \mathbb{R}$	2. $ab \in \mathbb{R}$
Associative:	3. $(a + b) + c$ $= a + (b + c)$	4. $(a \times b) \times c$ $= a \times (b \times c)$
Identity:	5. There exists $0 \in \mathbb{R}$ so that $0 + a = a + 0 = a$ for every element a in $\mathbb{R}$.	6. There exists $1 \in \mathbb{R}$ so that $1 \times a = a \times 1 = a$ for every element a in $\mathbb{R}$.
Inverse:	7. For each $a \in \mathbb{R}$, there is a unique number $(-a) \in \mathbb{R}$ so that $a + (-a) = (-a) + a = 0$	8. For each $a \in \mathbb{R}$, $a \neq 0$, there is a unique number $\frac{1}{a} \in \mathbb{R}$ so that $a \times \frac{1}{a} = \frac{1}{a} \times a = 1$
Commutative:	9. $a + b = b + a$	10. $ab = ba$

Distributive for multiplication over addition:

$$11.\ a \times (b + c) = a \times b + a \times c$$

▲

     The set of real numbers is a field, but there are other fields. In our definition of a field we used $\mathbb{R}$ and the operations of addition and multiplication for the sake of understanding, but for the mathematician, a field is defined as a set with *any two* operations satisfying the 11 stated properties.

PROBLEM SET   4.6

▲ A Problems

1. **IN YOUR OWN WORDS** What is the distinguishing characteristic between the rational and irrational numbers?

2. **IN YOUR OWN WORDS** Draw a Venn diagram showing the natural numbers, integers, rationals, and irrationals where the universe is the set of reals.

3. **IN YOUR OWN WORDS** Explain the identity property.

4. **IN YOUR OWN WORDS** Explain the inverse property.

5. What is a group?

6. What is a field?

7. Tell whether each number is an element of $\mathbb{N}$ (a natural number), $\mathbb{Z}$ (an integer), $\mathbb{Q}$ (a rational number), $\mathbb{Q}'$ (an irrational number), or $\mathbb{R}$ (a real number). Since these sets are not all disjoint, you may need to list more than one set for each answer.
   a. 7   **b.** 4.93   **c.** 0.656656665 . . .   **d.** $\sqrt{2}$
   e. 3.14159   **f.** $\frac{17}{43}$ **g.** $0.00\overline{27}$ **h.** $\sqrt{9}$

8. Tell whether each number is an element of $\mathbb{N}$ (a natural number), $\mathbb{Z}$ (an integer), $\mathbb{Q}$ (a rational number), $\mathbb{Q}'$ (an irrational number), or $\mathbb{R}$ (a real number). Since these sets are not all disjoint, you may need to list more than one set for each answer.
   a. 19   **b.** 6.48   **c.** 1.868686868 . . .   **d.** $\sqrt{8}$
   e. π **f.** $0.00\overline{12}$ **g.** $\sqrt{16}$ **h.** $\sqrt{1,000}$

Express each of the numbers in Problems 9–20 as a decimal.

9. **a.** $\dfrac{3}{2}$ **b.** $\dfrac{7}{10}$ 10. **a.** $\dfrac{5}{2}$ **b.** $\dfrac{9}{10}$

11. **a.** $\dfrac{2}{5}$ **b.** $\dfrac{47}{100}$ 12. **a.** $\dfrac{3}{5}$ **b.** $\dfrac{27}{15}$

13. **a.** $\dfrac{5}{6}$ **b.** $\dfrac{2}{7}$ 14. **a.** $\dfrac{7}{8}$ **b.** $\dfrac{3}{7}$

15. **a.** $\dfrac{3}{25}$ **b.** $2\frac{1}{6}$ 16. **a.** $14\frac{2}{11}$ **b.** $2\frac{4}{9}$

17. **a.** $\dfrac{2}{3}$ **b.** $2\frac{2}{13}$ 18. **a.** $\dfrac{15}{3}$ **b.** $\dfrac{12}{11}$

19. **a.** $\dfrac{-4}{5}$ **b.** $-\dfrac{2}{3}$ 20. **a.** $-\dfrac{17}{6}$ **b.** $\dfrac{-14}{5}$

Express each of the numbers in Problems 21–30 as the quotient of two integers.

21. **a.** 0.5   **b.** 0.8   22. **a.** 0.25   **b.** 0.75
23. **a.** 0.08   **b.** 0.12   24. **a.** 0.65   **b.** 2.3
25. **a.** 0.45   **b.** 0.234   26. **a.** 0.111   **b.** 0.52
27. **a.** 98.7   **b.** 0.63   28. **a.** 0.24   **b.** 16.45
29. **a.** 15.3   **b.** 6.95   30. **a.** 0.64   **b.** 6.98

Identify each of the properties illustrated in Problems 31–40.

31. $5 + 7 = 7 + 5$ **32.** $3(4 + 8) = 3(4) + 3(8)$

33. $5 \cdot 1 = 1 \cdot 5$ **34.** $5 \cdot \frac{1}{5} = 1$

35. $a + (10 + b) = (a + 10) + b$ **36.** $a + (10 + b) = (10 + b) + a$

37. mustard + catsup = catsup + mustard

38. (red + blue) + yellow = red + (blue + yellow)

39. $15 + [a + (-a)] = 15 + 0$ **40.** $A \cap (B \cup C) = (A \cap B) \cup (A \cap C)$

▲ B Problems

Check whether each of the sets and operations in Problems 41–50 form a group.

41. $\mathbb{N}$ for +   **42.** $\mathbb{N}$ for −   **43.** $\mathbb{W}$ for +   **44.** $\mathbb{W}$ for ×   **45.** $\mathbb{Z}$ for +

46. $\mathbb{Z}$ for ×   **46.** $\mathbb{Q}$ for +   **48.** $\mathbb{Q}$ for −   **49.** $\mathbb{Q}$ for ×   **50.** $\mathbb{Q}$ for ÷

51. a. Given the set $\{1, 2, 3, 4\}$ and the operation ×, construct a multiplication table showing all possible answers for the numbers in the set.

  b. Given the set $\{1, 2, 3, 4\}$ and the operation ∗ defined by $a \ast b = 2a$, construct a table for ∗ showing all possible answers for numbers in the set.

  c. Verify as many of the field properties as possible for the operations of × and ∗.

52. List the rationals between 0 and 1 by roster.

Let ∘ be an arbitrary operation in Problems 53–60. Describe the operation ∘ for each problem.

53. $5 \circ 3 = 8; 7 \circ 2 = 9; 9 \circ 1 = 10; 8 \circ 2 = 10; \ldots$

54. $5 \circ 3 = 15; 7 \circ 2 = 14; 9 \circ 1 = 9; 8 \circ 2 = 16; \ldots$

55. $5 \circ 3 = 2; 7 \circ 2 = 5; 9 \circ 1 = 8; 8 \circ 2 = 6; \ldots$

56. $1 \circ 9 = 11; 2 \circ 7 = 10; 9 \circ 0 = 10; 9 \circ 8 = 18; \ldots$

57. $8 \circ 0 = 1; 5 \circ 4 = 21; 1 \circ 0 = 1; 5 \circ 6 = 31; \ldots$

58. $4 \circ 6 = 20; 8 \circ 2 = 20; 7 \circ 9 = 32; 6 \circ 8 = 28; \ldots$

59. $4 \circ 7 = 1; 4 \circ 5 = 3; 7 \circ 3 = 11; 12 \circ 9 = 15; \ldots$

60. $4 \circ 7 = 17; 5 \circ 6 = 26; 6 \circ 4 = 37; 2 \circ 8 = 5; \ldots$

▲ Problem Solving

61. How can you tell whether a proper fraction will represent a terminating decimal without actually performing the division?

62. Symmetries of a Square   Cut out a small square and label it as shown in Figure 4.17.

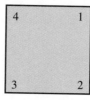
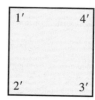

Front   Back

▲ **Figure 4.17   Square construction**

Be sure that 1 is in front of 1′, 2 is in front of 2′, 3 is in front of 3′, and 4 is in front of 4′. We will study certain *symmetries* of this square — that is, the results that are obtained when a square is moved around according to certain rules that we will establish.

Hold the square with the front facing you and the 1 in the top right-hand corner as shown in the margin. This is called the *basic position*.

Now rotate the square 90° clockwise so that 1 moves into the position formerly held by 2 and so that 4 and 3 end up on top. We use the letter A to denote this rotation of the square. That is, A indicates a clockwise rotation of the square through 90°. Other symmetries can be obtained similarly according to Table 4.5. You should be able to tell how each of the results in the table was found. Do this before continuing with the problem.

Basic position

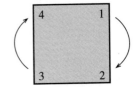

Apply *A*

Result

TABLE 4.5 Symmetries of a square

Element	Description	Result
A	90° clockwise rotation	3 4 / 2 1
B	180° clockwise rotation	2 3 / 1 4
C	270° clockwise rotation	1 2 / 4 3
D	360° clockwise rotation	4 1 / 3 2
E	Flip about a **horizontal** line through the middle of the square	3′ 2′ / 4′ 1′
F	Flip about a **vertical** line through the middle of the square	1′ 4′ / 2′ 3′
G	Flip along a line drawn from **upper** left to lower right	4′ 3′ / 1′ 2′
H	Flip along a line drawn from **lower** left to upper right	2′ 1′ / 3′ 4′

We now have a set of elements: {A, B, C, D, E, F, G, H}. We must define an operation that combines a pair of these symmetries. Define an operation ★ which means "followed by." Consider, for example, A ★ B: Start with the basic position, apply A, *followed by* B (without returning to basic position) to obtain the result shown:

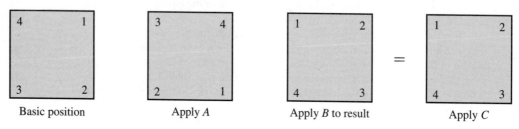

Basic position   Apply A   Apply B to result   Apply C

Thus, we say A ★ B = C, meaning "A followed by B is the same as applying the single element C."

a. Complete a table for the operation ★ and the set

$$\{A, B, C, D, E, F, G, H\}$$

b. Is the set closed for ★?
c. Is the set associative for ★?
d. Is the set commutative for ★?
e. Does the set have an identity for ★?
f. Does the inverse property hold for the set and the operation ★?
g. Is the set of symmetries of a square with the operation ★ a group?

4.7   A FINITE ALGEBRA

When you hear the word *algebra*, you probably think of the algebra you encountered in high school. The dictionary says that **algebra** is a generalization of arithmetic. Most people speak of algebra in the singular, but to a mathematician **algebra** refers to a structure, a set of symbols that operate according to certain agreed-upon properties. A mathematical dictionary reveals that we can speak of *an algebra with a unit element, a simple algebra, an algebra of subsets, a commutative algebra,* or *an algebra over a field.* In the next chapter we consider the algebra studied in high school, but in this section we will look at a different algebra that focuses on the definition of operations and on the properties we have studied in this chapter.

Clock Arithmetic

We now consider a mathematical system based on a 12-hour clock (see Figure 4.18). We'll need to define some operations for this set of numbers, and we'll use the way we tell time as a guide to our definitions. For example, if you have an appointment at 4:00 P.M. and you are two hours late, you arrive at 6:00 P.M. On the other hand, if your appointment is at 11:00 A.M. and you're two hours late, you arrive at 1:00 P.M. That is, on a 12-hour clock,

$$4 + 2 = 6 \quad \text{and} \quad 11 + 2 = 1$$

▲ **Figure 4.18   A 12-hour clock**

Using the clock as a guide, we define an operation called "clock addition," which is different from ordinary addition because it consists of a *finite* set, {1, 2, 3, 4, 5, 6, 7, 8, 9, 10, 11, 12}, and is *closed* for addition, as you can see by looking at Table 4.6.

TABLE 4.6   Addition on a 12-Hour Clock

+	1	2	3	4	5	6	7	8	9	10	11	12
1	2	3	4	5	6	7	8	9	10	11	12	1
2	3	4	5	6	7	8	9	10	11	12	1	2
3	4	5	6	7	8	9	10	11	12	1	2	3
4	5	6	7	8	9	10	11	12	1	2	3	4
5	6	7	8	9	10	11	12	1	2	3	4	5
6	7	8	9	10	11	12	1	2	3	4	5	6
7	8	9	10	11	12	1	2	3	4	5	6	7
8	9	10	11	12	1	2	3	4	5	6	7	8
9	10	11	12	1	2	3	4	5	6	7	8	9
10	11	12	1	2	3	4	5	6	7	8	9	10
11	12	1	2	3	4	5	6	7	8	9	10	11
12	1	2	3	4	5	6	7	8	9	10	11	12

EXAMPLE 1

Use a 12-hour clock to find the following sums.

a. $7 + 5$ **b.** $9 + 5$ **c.** $4 + 11$ **d.** $12 + 3$

Solution

a. $7 + 5 = 12$ **b.** $9 + 5 = 2$ **c.** $4 + 11 = 3$ **d.** $12 + 3 = 3$ ▬

Subtraction, multiplication, and division may be defined as they are in ordinary arithmetic.

Elementary Operations

Subtraction: $a - b = x$ means $a = b + x$

Multiplication: $a \times b = ab$ means $\underbrace{b + b + \cdots + b}_{a \text{ addends}}$

Zero multiplication: If $a = 0$, then $a \times b = 0 \times b = 0$.

Division: $a \div b = \dfrac{a}{b} = x$ means $a = bx$ provided b has an inverse for multiplication.

EXAMPLE 2

Carry out the given operations on a 12-hour clock by using the definitions for the elementary operations.

a. $4 - 9$ **b.** 4×9 **c.** $4 \div 7$ **d.** $\dfrac{4}{9}$

Solution

a. $4 - 9 = x$ means $4 = 9 + x$.

That is, what number when added to 9 produces 4? From Table 4.6, we see $x = 7$.

b. 4×9 means $9 + 9 + 9 + 9 = 12$ (from Table 4.6).

c. $4 \div 7 = t$ means $4 = 7t$.

That is, what number can be multiplied by 7 to obtain the result 4? We can proceed by trial and error:

$$7 \times 1 = 7; \quad 7 \times 2 = 2; \quad 7 \times 3 = 9; \quad 7 \times 4 = 4; \quad \ldots$$

We see $t = 4$.

d. $\dfrac{4}{9}$ means $4 \div 9 = s$ or $4 = 9s$.

We need to find the number that, when multiplied by 9, produces 4. Once again, proceed by checking each possibility:

$$9 \times 1 = 9; \quad 9 \times 2 = 6; \quad 9 \times 3 = 3; \quad 9 \times 4 = 12;$$
$$9 \times 5 = 9; \quad 9 \times 6 = 6; \quad 9 \times 7 = 3; \quad 9 \times 8 = 12;$$
$$9 \times 9 = 9; \quad 9 \times 10 = 6; \quad 9 \times 11 = 3; \quad 9 \times 12 = 12$$

Even though this process is tedious, it is complete because we are dealing with a finite set, and since we have checked all possibilities we see there is no such number. We say $\frac{4}{9}$ does not exist.   ▬

We see from Example 2 that it might be worthwhile to construct a multiplication table. However, since the addition and multiplication tables for a 12-hour clock are rather large, we shorten our arithmetic system by considering a clock with fewer than 12 hours.

Modulo Five Arithmetic

Consider a mathematical system based on a 5-hour clock, numbered 0, 1, 2, 3, and 4, as shown in Figure 4.19. Clock addition on this 5-hour clock is the same as on an ordinary clock, except that the only numbers in this set are {0, 1, 2, 3, 4}.

We define the operations of addition, subtraction, multiplication, and division just as we did for a 12-hour clock. Since subtraction and division are defined in terms of addition and multiplication, respectively, we need only the two operation tables that are shown in Table 4.7.

▲ **Figure 4.19   A 5-hour clock**

TABLE 4.7   Addition and Multiplication Tables

+	0	1	2	3	4		×	0	1	2	3	4
0	0	1	2	3	4		0	0	0	0	0	0
1	1	2	3	4	0		1	0	1	2	3	4
2	2	3	4	0	1		2	0	2	4	1	3
3	3	4	0	1	2		3	0	3	1	4	2
4	4	0	1	2	3		4	0	4	3	2	1

Instead of speaking about "arithmetic on the 5-hour clock," mathematicians usually speak of "modulo 5 arithmetic." The set $\{0, 1, 2, 3, 4\}$, together with the operations defined in Table 4.7, is called a **modulo 5** or **mod 5** system. Suppose it is 4 o'clock on a 5-hour clock. What time will the clock show 9 hours later? We could write

$$4 + 9 = 3 \quad \text{and} \quad 2 + 1 = 3,$$

thus, $4 + 9 = 2 + 1$. Since we do not wish to confuse this with ordinary arithmetic in which

$$4 + 9 \neq 2 + 1$$

we use the following notation:

$$4 + 9 \equiv 2 + 1, (\text{mod } 5)$$

which is read "$4 + 9$ is congruent to $2 + 1$, mod 5." We define *congruence* as follows.

Congruence Mod *m*

The real numbers a and b are **congruent modulo *m***, written $a \equiv b$, (mod m), if a and b differ by a multiple of m.

▄▄▄ EXAMPLE 3

Decide whether each statement is true or false.

a. $3 \equiv 8$, (mod 5)   **b.** $3 \equiv 53$, (mod 5)   **c.** $3 \equiv 19$, (mod 5)

Solution

a. $3 \equiv 8$, (mod 5) because $8 - 3 = 5$, and 5 is a multiple of 5.

b. $3 \equiv 53$, (mod 5) because $53 - 3 = 50$, and 50 is a multiple of 5.

c. $3 \not\equiv 19$, (mod 5) because $19 - 3 = 16$, and 16 is not a multiple of 5.   ▬

Another way to determine whether two numbers are congruent mod m is to divide each by m and check the remainders. If the remainders are the same, then the numbers are congruent mod m. For example, $3 \div 5$ gives a remainder 3, and $53 \div 5$ gives a remainder of 3, so $3 \equiv 53$, (mod 5).

▬▬ EXAMPLE 4

Solve each equation for x. In other words, find a replacement for x that makes each equation true.

a. $4 + 9 \equiv x$, (mod 5)   **b.** $15 + 92 \equiv x$, (mod 5)

c. $2 + 4 \equiv x$, (mod 5)   **d.** $2 - 4 \equiv x$, (mod 5)

e. $7 \times 5 \equiv x$, (mod 7)   **f.** $3 - 5 \equiv x$, (mod 12)

Solution

a. $4 + 9 = 13 \equiv 3$, (mod 5)

b. $15 + 92 = 107 \equiv 2$, (mod 5)

c. $2 + 4 = 6 \equiv 1$, (mod 5)

d. $2 - 4 \equiv 7 - 4$, (mod 5)   Since $2 \equiv 7$, (mod 5), we can replace 2 by 7.
    $\equiv 3$, (mod 5)

e. $7 \times 5 = 35 \equiv 0$, (mod 7)

f. $3 - 5 \equiv 15 - 5$, (mod 12)   Since $3 \equiv 15$, (mod 12)
    $\equiv 10$, (mod 12)

Notice that every whole number is congruent modulo 5 to exactly one element in the set $I = \{0, 1, 2, 3, 4\}$, which contains all possible remainders when dividing by 5. Since the number of elements in I is rather small, we can easily solve equations by trying the numbers 0, 1, 2, 3, and 4. In modulo 7 try numbers in the set $\{0, 1, 2, \ldots, 6\}$ and in modulo 10 try numbers in the set $\{0, 1, 2, \ldots, 9\}$. Consider the following example.

▬▬ EXAMPLE 5

The manager of a TV station hired a college student, U. R. Stuck, as an election-eve runner. The request for reimbursement turned in is shown in the margin. The station manager refused to pay the bill, since the mileage was not honestly recorded. How did he know? *Note:*   The answer has nothing to do with the illegible digits in the mileage report, and you should also assume that the mileage is rounded to the nearest mile.

REQUEST FOR REIMBURSEMENT FOR MILEAGE
NAME _____ U.R. Stuck _____
ADDRESS ____ 1234 Fifth St. ____
S.S. NO. _____ 576-38-4459 _____
ENDING MILEAGE ____ 14,2̸,8 ____
BEGINNING MILEAGE ___ 14,8̸93 ___
NO. OF TRIPS____ 8 ____

Solution   Let $x =$ the number of miles in one round trip. For the reimbursement request, we see there are 8 trips, so the total mileage must be $8x$. Since the last digits are legible, we see the difference is 5, (mod 10). Thus,

$$8x \equiv 5, \text{ (mod 10)}$$

We solve this by checking each possible value:

$x = 0$: $8(0) \equiv 0$, (mod 10)

$x = 1$: $8(1) \equiv 8$, (mod 10)

$x = 2$: $8(2) = 16 \equiv 6$, (mod 10)

$x = 3$: $8(3) = 24 \equiv 4$, (mod 10)

$x = 4$: $8(4) = 32 \equiv 2$, (mod 10)

$x = 5$: $8(5) = 40 \equiv 0$, (mod 10)

$x = 6$: $8(6) = 48 \equiv 8$, (mod 10)

$x = 7$: $8(7) = 56 \equiv 6$, (mod 10)

$x = 8$: $8(8) = 64 \equiv 4$, (mod 10)

$x = 9$: $8(9) = 72 \equiv 2$, (mod 10)

These are the only possible values for x in modulo 10, and none gives a mileage reading for 5, (mod 10) in the units digit. Therefore, Stuck did not report the mileage honestly. ▬

▬ **EXAMPLE 6**   **Polya's Method**

Suppose you are planning to buy paper to cover some shelves. You need to cover 1 shelf 100 inches long, and to minimize the amount of waste you need to cover at least 30 11-inch shelves, but not more than 50 11-inch shelves. The paper can be purchased in multiples of 36 inches. How much paper should you buy to minimize the waste?

Solution   We use Polya's problem-solving guidelines for this example.

Understand the Problem.   Suppose I need to cover 30 shelves; then I will need $30 \times 11 + 100$ inches of paper. Since $30 \times 11 + 100 = 430$ inches and the paper comes in 36-inch lengths, I can find $430 \div 36 = 11.9\overline{4}$. This means the *minimum* purchase is $12 \times 36 = 432$ inches of paper. On the other hand, if I need to cover 50 shelves, then I will need $50 \times 11 + 100 = 650$ inches; $650 \div 36 = 18.0\overline{5}$. This requires 19 units of length 36 inches: $36 \times 19 = 684$ inches.

Devise a Plan.   We know the minimum purchase and we also know the maximum purchase. We need to find a solution that specifies a general solution (for any number of shelves between 30 and 50). Since the paper comes in multiples of 36 inches, we must buy $36x$ inches of paper, where x represents the number of multiples we buy. Suppose we need to cover k shelves at 11 inches and an extra shelf at 100 inches for a total of $11k + 100$ inches. We will write this as a congruence modulo 11 and then solve for x.

Carry Out the Plan.   The amount of shelf paper required is $11k + 100$, which means

$$36x = 11k + 100$$
$$36x \equiv 100, \text{(mod 11)}$$
$$3x \equiv 1, \text{(mod 11)}\qquad \text{Note } 36 \equiv 3, \text{(mod 11) and } 100 \equiv 1, \text{(mod 11).}$$

Since we are working in modulo 11, and if we assume no waste, we can check all possibilities (since the set of possibilities is *finite*, namely, 0, 1, 2, 3, 4, 5, 6, 7, 8, 9, and 10; all these congruences are modulo 11):

$$x = 0: 3(0) = 0 \not\equiv 1 \qquad x = 1: 3(1) = 3 \not\equiv 1 \qquad x = 2: 3(2) = 6 \not\equiv 1$$
$$x = 3: 3(3) = 9 \not\equiv 1 \qquad x = 4: 3(4) = 12 \equiv 1 \qquad x = 5: 3(5) = 15 \not\equiv 1$$
$$x = 6: 3(6) = 18 \not\equiv 1 \qquad x = 7: 3(7) = 21 \not\equiv 1 \qquad x = 8: 3(8) = 24 \not\equiv 1$$
$$x = 9: 3(9) = 27 \not\equiv 1 \qquad x = 10: 3(10) = 30 \not\equiv 1$$

The solution is $x \equiv 4$, (mod 11).

Look Back. We know that the minimum is 12 multiples of the 36-inch paper, and the maximum is 19 multiples. To *minimize the waste* we see that $x \equiv 4$, (mod 11) must be $x = 4, 15, 26, \ldots$. This answer means that to minimize the waste we should buy 15 multiples of the 36-inch paper:

$$36(15) = 540$$

inches of paper. This amount allows us to cover the 100-inch board and 40 11-inch boards. ▄

Group Properties for a Modulo System

It appears that some modulo systems "behave" like ordinary algebra and others do not. The distinction is found by determining which systems form a field. Let's first explore the group properties for the set $I = \{0, 1, 2, 3, 4\}$ and the operation of addition, modulo 5.

Closure for +: The set I of elements modulo 5 is closed with respect to addition. That is, for any pair of elements, there is a unique element that represents their sum, and that is also a member of the original set.

Associative for +: Addition of elements modulo 5 satisfies the associative property. That is,

$$(a + b) + c \equiv a + (b + c)$$

for all elements a, b, and $c \in I$. As a specific example, we evaluate $2 + 3 + 4$ in two ways:

$$(2 + 3) + 4 = 9 \equiv 4, \text{ (mod 5)}$$
$$2 + (3 + 4) = 9 \equiv 4, \text{ (mod 5)}$$

Identity for +: The set I of elements modulo 5 includes an identity element for addition. That is, the set contains an element 0 such that the sum of any given element and zero is the given element. In modulo 5,

$$0 + 0 \equiv 0, \quad 1 + 0 \equiv 1, \quad 2 + 0 \equiv 2, \quad 3 + 0 \equiv 3, \quad 4 + 0 \equiv 4$$

Inverse for +: Each element in arithmetic modulo 5 has an inverse with respect to addition. That is, for each element $a \in I$, there exists a unique element $a' \in I$ such that $a + a' \equiv a' + a \equiv 0$. The element a' is said to be the *inverse* of a. Specifically (in mod 5):

elements in the set   identity
$$\downarrow \qquad\qquad \downarrow$$

The inverse of 0 is 0: $0 + 0 \quad\equiv 0$
The inverse of 1 is 4: $1 + 4 \quad\equiv 0$
The inverse of 2 is 3: $2 + 3 \quad\equiv 0$
The inverse of 3 is 2: $3 + 2 \quad\equiv 0$
The inverse of 4 is 1: $4 + 1 \quad\equiv 0$
$$\uparrow$$
inverses

Since the closure, associative, identity, and inverse properties are satisfied for addition modulo 5, we conclude that it is a *group*.

■■■ EXAMPLE 7

Is the set {0, 1, 2, 3, 4, 5} a group for multiplication modulo 6?

Solution   This set is closed, associative, and has an identity element 1 for multiplication. However, the inverse property is not satisfied (in mod 6):

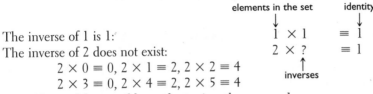

elements in the set   identity
$$\downarrow \qquad\qquad \downarrow$$

The inverse of 1 is 1: $1 \times 1 \quad\equiv 1$
The inverse of 2 does not exist: $2 \times\ ? \quad\equiv 1$
$2 \times 0 \equiv 0, 2 \times 1 \equiv 2, 2 \times 2 \equiv 4$ $\uparrow$
$2 \times 3 \equiv 0, 2 \times 4 \equiv 2, 2 \times 5 \equiv 4$ inverses
None of the possible products gives the answer 1.

The set is *not* a group for multiplication modulo 6.   ▬

■■■ EXAMPLE 8

Is the set $I = \{0, 1, 2, 3, 4\}$ a field for + and × modulo 5?

Solution   We have shown (on the previous page) that four of the 11 field properties are satisfied.

Commutative for +:   Addition in arithmetic modulo 5 satisfies the commutative property. That is,

$$a + b \equiv b + a$$

where a and b are any elements in I. Specifically, we see that the entries in Table 4.7 are symmetric with respect to the principal diagonal.

We can now say that I is a commutative group for addition. We continue by verifying that it is also a commutative group for multiplication.

Closure for ×:   The set I is closed for multiplication, as we can see from Table 4.7.

Associative for ×:   This property is satisfied, and the details are left for you to verify.

Identity for ×: The identity element for multiplication is 1, since (in mod 5):

$$0 \times 1 \equiv 0, \quad 1 \times 1 \equiv 1, \quad 2 \times 1 \equiv 2,$$
$$3 \times 1 \equiv 3, \quad 4 \times 1 \equiv 4$$

Inverse for ×: The inverse for multiplication can be checked by finding the inverse of each element (mod 5).

	elements in the set	identity
	↓	↓
There is no inverse of 0:	$0 \times ?$	$\equiv 1$
The inverse of 1 is 1:	1×1	$\equiv 1$
The inverse of 2 is 3:	2×3	$\equiv 1$
The inverse of 3 is 2:	3×2	$\equiv 1$
The inverse of 4 is 4:	4×4	$\equiv 1$
	↑	
	inverses	

Since every nonzero element has an inverse for multiplication, we say the inverse property for multiplication is satisfied.

The set I is a group for multiplication.

Commutative for ×: The set is commutative for multiplication, as we can see by looking at Table 4.7.

The set I is a commutative group for multiplication.

Distributive for × over +: We need to check $a(b + c) \equiv ab + ac$. Check some particular examples:

$$2(3 + 4) = 2(7) = 14 \equiv 4, \text{ (mod 5)}$$
$$2(3) + 2(4) = 6 + 8 = 14 \equiv 4, \text{ (mod 5)}$$

Thus, $2(3 + 4) \equiv 2(3) + 2(4)$. Also,

$$4(1 + 2) \equiv 4(1) + 4(2)$$

and

$$3(2 + 4) \equiv 3(2) + 3(4)$$

These examples seem to imply that the distributive property holds.

The set I is a field for the operations of addition and multiplication, modulo 5.

PROBLEM SET 4.7

▲ **A Problems**

1. **IN YOUR OWN WORDS** What do we mean by clock arithmetic?

2. **IN YOUR OWN WORDS** How can the operations of addition and multiplication be defined for a 24-hour clock?

3. **IN YOUR OWN WORDS** Discuss the meaning of the definition of congruence modulo m.

4. Define precisely the concept of congruence modulo m.

Perform the indicated operations in Problems 5–14 using arithmetic for a 12-hour clock.

5. **a.** $9 + 6$ **b.** $5 - 7$ 6. **a.** 5×3 **b.** 2×7

7. **a.** $7 + 10$ **b.** $7 - 9$ 8. **a.** 6×7 **b.** $1 \div 5$

9. **a.** $5 + 7$ **b.** $4 - 8$ 10. **a.** $2 - 6$ **b.** 9×3

11. **a.** 4×8 **b.** 2×3 12. **a.** $1 \div 12$ **b.** $10 + 6$

13. **a.** $3 \times 5 - 7$ **b.** $7 + 3 \times 2$

14. **a.** $5 \times 2 - 11$ **b.** $5 \times 8 + 5 \times 4$

Which of the statements in Problems 15–20 are true?

15. **a.** $5 + 8 \equiv 1, \pmod 6$ **b.** $4 + 5 \equiv 1, \pmod 7$

16. **a.** $5 \equiv 53, \pmod 8$ **b.** $102 \equiv 1, \pmod 2$

17. **a.** $47 \equiv 2, \pmod 5$ **b.** $108 \equiv 12, \pmod 8$

18. **a.** $5{,}670 \equiv 270, \pmod{365}$ **b.** $2{,}001 \equiv 39, \pmod{73}$

19. **a.** $1{,}996 \equiv 0, \pmod{1{,}996}$ **b.** $246 \equiv 150, \pmod 6$

20. **a.** $126 \equiv 1, \pmod 7$ **b.** $144 \equiv 12, \pmod{144}$

Perform the indicated operations in Problems 21–28.

21. **a.** $9 + 6, \pmod 5$ **b.** $7 - 11, \pmod{12}$

22. **a.** $4 \times 3, \pmod 5$ **b.** $1 \div 2, \pmod 5$

23. **a.** $5 + 2, \pmod 4$ **b.** $2 - 4, \pmod 5$

24. **a.** $6 \times 6, \pmod 8$ **b.** $5 \div 7, \pmod 9$

25. **a.** $4 + 3, \pmod 5$ **b.** $6 - 12, \pmod 8$

26. **a.** $121 \times 47, \pmod{121}$ **b.** $2 \div 3, \pmod 7$

27. **a.** $7 + 41, \pmod 5$ **b.** $4 - 5, \pmod{11}$

28. **a.** $62 \times 4, \pmod 2$ **b.** $7 \div 12, \pmod{13}$

Solve each equation for x in Problems 29–37. Assume that k is any counting number.

29. **a.** $x + 3 \equiv 0, \pmod 7$ **b.** $4x \equiv 1, \pmod 5$

30. **a.** $x + 5 \equiv 2, \pmod 9$ **b.** $4x \equiv 1, \pmod 6$

31. **a.** $x - 2 \equiv 3, \pmod 6$ **b.** $3x \equiv 2, \pmod 7$

32. **a.** $5x \equiv 2, \pmod 7$ **b.** $7x + 1 \equiv 3, \pmod{11}$

33. **a.** $x^2 \equiv 1, \pmod 4$ **b.** $x \div 4 \equiv 5, \pmod 9$

34. **a.** $x^2 \equiv 1, \pmod 5$ **b.** $4 \div 6 \equiv x, \pmod{13}$

35. **a.** $4k \equiv x, \pmod 4$ **b.** $4k + 2 \equiv x, \pmod 4$

36. $2x^2 - 1 \equiv 3, \pmod 7$ 37. $5x^3 - 3x^2 + 70 \equiv 0, \pmod 2$

▲ **B Problems**

38. Assume that today is Monday (day 2). Determine the day of the week it will be at the end of each of the following periods. (Assume no leap years.)
    a. 24 days   **b.** 155 days   **c.** 365 days   **d.** 2 years

39. Assume that today is Friday (day 6). Determine the day of the week it will be at the end of each of the following periods. (Assume no leap years.)
    a. 30 days   **b.** 195 days   **c.** 390 days   **d.** 3 years

40. Your doctor tells you to take a certain medication every 8 hours. If you begin at 8:00 A.M., show that you will not have to take the medication between midnight and 7:00 A.M.

41. Suppose you make six round trips to visit a sick aunt and wish to record your mileage to the nearest mile. You forget the original odometer reading, but you do remember that the units digit has increased by 8 miles. What are the possible distances between your house and your aunt's house?

42. Suppose you are planning to purchase some rope. You need between 15 and 20 pieces that are 7 inches long and one piece that is 80 inches long. The rope can be purchased in multiples of 12 inches. How much rope should you buy to minimize waste?

43. If you know that your aunt in Problem 41 lives somewhere between 10 and 15 miles from your house, how far exactly is her house, given the information in Problem 41?

44. Is the set {0, 1, 2, 3, 4, 5} a group for addition modulo 6?

45. Is the set {0, 1, 2, 3, 4, 5, 6} a group for addition modulo 7?

46. Is the set {0, 1, 2, 3} a field for addition and multiplication modulo 4?

Problems 47–52 involve the set {0, 1, 2, 3, 4, 5, 6, 7, 8, 9, 10} and addition and multiplication modulo 11.

47. Make a table for addition and multiplication.

48. Is the set a group for addition?

49. Is the set a group for multiplication?

50. Is the set a commutative group for addition?

51. Is the set a commutative group for multiplication?

52. Is the set a field for the operations of addition and multiplication?

▲ **Problem Solving**

An International Standard Book Number (ISBN) is used to identify books. The ISBN for this book is 0-534-34988-9.

The first digit, 0, identifies the book as one published in an English-speaking country. The next three digits, 534, identify the publisher (Brooks/Cole) and the next five digits identify the particular book (this one). The last digit is a check digit, which is used as follows. Multiply the first digit by 10, the next digity by 9, and so on:

$$10(0) + 9(5) + 8(3) + 7(4) + 6(3) + 5(4) + 4(9) + 3(8) + 2(8) + 1(9) = 220$$

This number is congruent to 0, (mod 11). This is not by chance. The check digit, 9, is chosen so that this sum is 0, (mod 11). In other words, the check digit x for the book with ISBN 0-534-34015-x is found by considering

$$10(0) + 9(5) + 8(3) + 7(4) + 6(3) + 5(4) + 4(0) + 3(1) + 2(5) = 148 \equiv 5$$

This means that for check digit x, $5 + x \equiv 0$, (mod 11). The one-digit solution to this equation is $x = 6$, so the check digit is 6. What is the check digit for a book with the given ISBN in Problems 53–56?

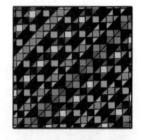

53. 0-534-13728-*x*   **54.** 0-691-02356-*x*

55. 0-8028-1430-*x*   **56.** 9-68-7270-82-*x*

57. What are the possible check digits for ISBNs? What is the check digit for the book with ISBN 0-13-330325-X? What do you think the X stands for in this ISBN? (*Note:* X *is the check digit, and is not a variable.*)

58. If it is now 2 P.M., what time will it be 99,999,999,999 hours from now?

59. Write a schedule for 12 teams so that each team will play every other team once and no team will be idle.

60. **One Hundred Fowl Problem** (an old Chinese puzzle)   A man buys 100 birds for $100. A rooster is worth $10, a hen is worth $3, and chicks are worth $1 a pair. How many roosters, hens, and chicks did he buy if he bought at least one of each type?

▲ **Individual Research**

61. What is a Diophantine equation?

   References Warren J. Himmelberger, "Puzzle Problems and Diophantine Equations," *The Mathematics Teacher*, February 1973, pp. 136–138.
   For a more complete reference, see any number theory textbook.

62. **Modular Art**   Many interesting designs such as those shown in the margin can be created using patterns based on modular arithmetic. Prepare a report for class presentation based on the article "Using Mathematical Structures to Generate Artistic Designs," by Sonia Forseth and Andrea Price Troutman, *The Mathematics Teacher*, May 1974, pp. 393–398. Another source is "Mod Art: The Art of Mathematics" by Susan Morris, *Technology Review*, March/April 1979.

4.8   CRYPTOGRAPHY

Cryptography is the art of writing or deciphering messages in code. Work in cryptography combines theoretical foundations with practical applications. It involves, at some level, ways to generate random-looking sequences and to detect and evaluate nonrandom effects. You might think of governments, covert operations, and spies when you think of cryptography, but there are many applications outside government. Cable television companies, whose signals are easily intercepted from satellite relays, use encryption schemes to prevent unauthorized use of their transmissions. Banks encrypt financial transactions and records in their computers for purposes of authenticity and integrity — that is, to make sure that the sender and contents are really as they appear — as well as for privacy. Cryptography figured prominently in a recent movie on the war against the drug cartel in Colombia. A cryptologist needs

RAE WA XLHH XDL
WAYLTSAT XDNX VM DL
XALZS'X PAIIEXL IR
LGLPEXVAS. V FVHH SLYLT
YAXL MAT DVI NWNVS!

Games magazine, December 1985

skills in communications, engineering, speech research, signals processing, and the design of specialized computers.

Simple Codes

Simple codes can be formed by replacing one letter by another. You may have seen this type of code in a children's magazine, or as a puzzle problem in a newspaper. The television show *Wheel of Fortune* uses a variation of this simple code-breaking skill. The cartoon at the left is from a popular game magazine. Codes such as this can easily be broken by using some logic and a knowledge of our language. Here is an analysis of how this code could be broken:

RAE WA XLHH XDL WAYLTSAT XDNX VM DL KALZS'X
PAIIEXL IR LGLPEXVAS, V FVHH SLYLT YAXL MAT DVI
NWNVS!

We have color-coded the solution.

1. The hint for this puzzle was "A four-letter word that starts and ends with the same letter is often THAT." From this, we conclude that the coded word XDNX is *that*, so that X is T, D is H, N is A. Fill in these letters above the letters of the puzzle:

         T   TH   THAT   H   T
    RAE WA XLHH XDL WAYLTSAT XDNX VM DL KALZS'X

         T   T   T   H
    PAIIEXL IR LGLPEXVAS, V FVHH SLYLT YAXL MAT DVI

    A A
    NWNVS!   TH

2. The three-letter word XDL must be the word *the*, so fill in E for L:

         TE   THE   E   THAT   HE   E   T
    RAE WA XLHH XDL WAYLTSAT XDNX VM DL KALZS'X

         TE   E E T   E E   TE   H
    PAIIEXL IR LGLPEXVAS, V FVHH SLYLT YAXL MAT DVI

    A A
    NWNVS!

3. The single letter V is not A since N is A, so we guess that it is I:

         TE   THE   E   THAT I HE   E   T
    RAE WA XLHH XDL WAYLTSAT XDNX VM DL KALZS'X

         TE   E E TI   I I   E E   TE   HI
    PAIIEXL IR LGLPEXVAS, V FVHH SLYLT YAXL MAT DVI

    A AI
    NWNVS!

4. The coded word XLHH begins with TE_ _; since the last two letters are the same, we assume that H is the letter L:

TELL THE   E   THAT I HE E   T
RAE WA XLHH XDL WAYLTSAT XDNX VM DL KALZS'X

      TE   E E TI   I I LL E E   TE   HI
PAIIEXL IR LGLPEXVAS, V FVHH SLYLT YAXL MAT DVI

A AI
NWNVS!

5. The coded word FVHH ends with _ILL; assume F is W. Also we see the coded word VM, which is I_, so we also assume that the coded M is decoded as F:

TELL THE   E   THAT I F HE   E   T
RAE WA XLHH XDL WAYLTSAT XDNX VM DL KALZS'X

      TE   E E TI   I WILL E E   TE F   HI
PAIIEXL IR LGLPEXVAS, V F VHH SLYLT YAXL MAT DVI

A AI
NWNVS!

6. Now, observing the context of the cartoon and making some assumptions about what a prisoner might be saying, we make some trial-and-error guesses: He wants a pardon from the governor, so we count letters and fill in the blanks for the word *governor*:

O GO TELL THE GOVERNOR THAT I F HE OE NT
RAE WA XLHH XDL WA YLT S AT XDNX VM DL KALZS'X

   O   TE   E E TION I WILL NEVER VOTE F OR HI
PAIIEXL IR LGLPEXVA S, V FVHH SLYLT YAXL MAT DVI

AGAIN
NWNVS!

7. We can now fill in the completed deciphered message. Note that spaces translate as spaces:

YOU GO TELL THE GOVERNOR THAT IF HE DOESN'T COMMUTE MY EXECUTION, I WILL NEVER VOTE FOR HIM AGAIN!

Modular Codes*

Rather than simply substituting one letter for another (which is an easily broken code), more sophisticated codes can be developed. To **encrypt** a message means to scramble it by something called an **encoding key.** The result is a secret or coded message, called **cyphertext.** The coded message is then unscrambled using a secret **decoding key.** This coding procedure requires that both the sender and the receiver know, and conceal, the encoding and decoding keys. Suppose, for example, that

* This code requires Section 4.7.

News Clip

European mathematicians Pomerance, Rumely, and Adleman, building on recent work by theorists in the United States, have solved a 3,000-year-old problem in mathematics, an achievement that raises questions about the security of recently developed secret codes.

The problem, one of the oldest in mathematics, is how to determine whether a number is prime — that is, whether it can be divided by any number other than itself and 1. The smallest prime numbers are 2, 3, 5, 7, and 11.

Using the new method, a test on a 97-digit number, which previously could have taken as long as 100 years with the fastest computer, was done in 77 seconds.

It's possible that the method may help determine the divisors of a large number that is not prime. One way to tell whether a number is prime is to try to divide it by other numbers. If the number is very large, trying to divide it by other numbers is hopeless.

In 1640, Pierre de Fermat invented a test for determining primes without doing division. In the last decade, mathematicians decided that they could run Fermat's test on a suspected prime number many times, and if it consistently passed, there was a high probability that the number was prime.

News Clip

On October 11, 1988, Mark S. Manasse of Digital Corporation's Systems Research Systems and Arjen K. Lenstra of the University of Chicago linked over a dozen users of some 400 computers on three continents to find the factors of a 100-digit number.

Several of the most secure cipher systems invented in the past decade are based on the fact that large numbers are extremely difficult to factor, even using the most powerful computers for long periods of time. The accomplishment of factoring a 100-digit number "is likely to prompt cryptographers to reconsider their assumptions about cipher security," Lenstra said in a telephone interview. His colleague Manasse added, "What this shows is that a cryptographer should avoid basing a cipher on any factorable number smaller than about 200 digits. The cipher system still works, but we have upped the ante." Using larger numbers makes the work of cryptographers more cumbersome.

The Mathematics Teacher,
Jan. 1990, p. 70.

we wish to send the following secret message:

THE FBI HAS BED BUGS.

Suppose we select some encoding key — say, multiply by 2. Then we code the message according to some given modular number, say 29, as shown in Figure 4.20. Notice that we use 29 instead of 0; this is still considered a modulo 29 system since $29 \equiv 0$, (mod 29).

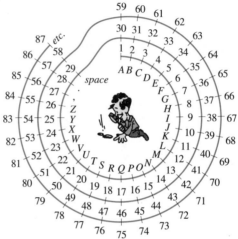

▲ **Figure 4.20   Modular 29 code**

We look up the numerical value for each letter, space, and punctuation mark according to Figure 4.20.

Message: 20 8 5 29 6 2 9 29 8 1 19 29 2 5 4 29 2 21 7 19 28
T H E   F B I   H A S   B E D   B U G S .

Note that the "code" 20-8-5-29-6-2-9-29-8-1-19-29-2-5-4-29-2-21-7-19-28 would be "easy" to break. Why?

Now we encode by multiplying each of these numbers by the encoding key; then we *modulate*, or write each of these answers modulo 29:

	20 8 5 29 6 2 9 29 8 1 19 29 2 5 4 29 2 21 7 19 28
Message:	T H E F B I H A S B E D B U G S .
Numerical value:	20- 8 - 5 -29- 6 - 2 - 9 -29- 8 - 1 -19-29- 2 - 5 - 4 -29- 2 -21- 7 -19-28
Encode (encoding key 2):	40-16-10-58-12- 4 -18-58-16- 2 -38-58- 4 -10- 8 -58- 4 -42-14-38-56
Modulate:	11-16-10-29-12- 4 -18-29-16- 2 - 9 -29- 4 -10- 8 -29- 4 -13-14- 9 -27
Coded message:	K P J L D R P B I D J H D M N I ,

The coded message is KPJ LDR PBI DJH DMNI,

The decoding key is the inverse of the encoding key. Since the encoding key is to multiply by 2, the decoding key is to divide by 2.

■■■ EXAMPLE 1

Decode the following message for which the encoding key is to multiply by 2:   NBXXMC RI B CAXX FBK,

Solution First use Figure 4.20 to find the numerical value:

14-2-24-24-13-3-29-18-9-29-2-29-3-1-24-24-29-6-2-11-27

The decoding key is to divide by 2, so we look at the code and write the odd numbers so that they are evenly divisible by 2, (mod 29):

14- 2 -24-24- 42-32-58- 18- 38-58- 2 - 58-32-30- 24-24- 58- 6 - 2 - 40-56

Decoding key (divide by 2):   7 - 1 -12-12-21-16-29- 9 -19-29- 1 -29-16-15-12-12-29- 3 - 1 -20-28

Decoded message (Figure 4.20):   G-A- L -L -U- P - - I - S - -A- - P -O- L - L - -C-A-T- .

The decoded message is: GALLUP IS A POLL CAT. ▬

Unbreakable Codes

In 1970, mathematicians Whitfield Diffie and Martin Hellman showed a way to make the keys public. Suppose a code has two keys, an encoding key and a decoding key, and also suppose it is impossible to compute one key from the other in the sense that no person or computer would be able to do it; then this would constitute an unbreakable code. Here is the way such a code works. Everyone owns a unique pair of keys, one of which remains private, but the other is public in the sense that it is listed in a readily available directory. To send a message, you look up the public key for the person to whom the message is to be sent. You use the public key to scramble the message. The receiver then uses their private key to decode the message with total secrecy and privacy.

Three mathematicians, Rivest, Shamir, and Adleman, created a public key algorithm known as RSA, from their initials. This method depends on ideas of prime numbers and factoring studied in this chapter. Consider two prime numbers, say, 11 and 7. Now if I hand you their product, 77, this product can be made public, while the factors 11 and 7 remain secret. To *encode* you need the number 77, but to *decode* you need the factors. Now, *if* the factors are large enough (say, 200 digits each), then the product is so large that it can never be factored, and the result is an unbreakable code.*

Why would we need an unbreakable public-key code? Some applications include protecting money transfers from tampering, shielding sensitive business data from competitors, and protecting computer software from viruses. The government has published a *digital signature algorithm* (DSA), which depends on a single very large prime number.

PROBLEM SET 4.8

▲ **A Problems**

Number the letters of the alphabet from 1 to 26; code a comma as 27, period as 28, and space as 29. Encode the messages in Problems 1–4.

1. NEVER SAY NEVER.   2. THE EAGLE HAS LANDED.

3. YOU BET YOUR LIFE.   4. MY BANK BALANCE IS NEGATIVE.

* RSA Laboratories, Redwood City, CA, tests it ability to create difficult ciphers by establishing a series of cryptographic contests. Its first challenge required 140 days to solve. The coded message was "Strong cryptography makes the world a better place." If you are interested in this type of challenge, check out www.rsa.com for additional contests.

Number the letters of the alphabet from 1 to 26 and code a blank as 29. Decode the messages in Problems 5–8.

5. 1-18-5-29-23-5-29-8-1-22-9-14-7-29-6-21-14-29-25-5-20

6. 9-29-12-15-22-5-29-13-1-20-8-5-13-1-20-9-3-19

7. 6-1-9-12-21-18-5-29-20-5-1-3-8-5-19-29-19-21-3-3-5-19-19

8. 19-17-21-1-18-5-29-13-5-1-12-19-29-13-1-11-5-29-18-15-21-14-4-29-16-5-15-16-12-5

Give the decoding key for the encoding keys in Problems 9–14.

9. Multiply by 8.   10. Multiply by 20 and add 2.

11. Divide by 6.   12. Divide by 4 and subtract 3.

13. Multiply by 4 and add 2, then double the result.

14. Multiply by 3 and subtract 3, then divide the result by 2.

▲ B Problems

Use Figure 4.20 to encode or decode the messages in Problems 15–23.

15. Encoding key: Multiply by 3.

   NEVER SAY NEVER.

16. Encoding key: Multiply by 5.

   THE EAGLE HAS LANDED.

17. Encoding key: Multiply by 2 and add 5.

   YOU BET YOUR LIFE.

18. Encoding key: Multiply by 4 and subtract 10.

   MY BANK BALANCE IS NEGATIVE.

19. Encoding key: Multiply by 3.

   XEJSBQ LEJSBQ,. C RCGG UEQZ

20. Encoding key: Multiply by 2 and add 10.

   LJMQKVTSSKQJ.SJKITJ,ZKJULECSJ.IJSKGTKITJTESTSJSETTM

21. Encoding key: Multiply by 2 and subtract 7.

   SKAXBVIXAVXVGKQDV.WSYQCLT

22. Encoding key: Multiply by 2 and subtract 11.

   ALL PERSONS BY NATURE DESIRE KNOWLEDGE.

23. Encoding key: Multiply by 3 and add 5.

   SHOW ME A DROPOUT FROM A DATA PROCESSING SCHOOL AND I WILL SHOW YOU A NINCOMPUTER.

▲ Problem Solving

The ciphers in Problems 24–27 are taken from Games Magazine, *December 1985.*

24. L FPHWDJ QJPQFJ TDP MJJQ BPKR. WDJU HVJ IPTHVBR TDP DHXJG'W KPW WDJ KSWR WP ALWJ QJPQFJ WDJNRJFXJR. (*Hint:* Ciphertext pattern QJPQFJ often represents PEOPLE.)

25. GHJWHY NDW LOGKL TGRLBK, UBLRGXV, GHV XYOZLD WH DZL DWR VWS ZL RXBOJ G UGH MWX GOO LYGLWHZHSL. (*Hint:* A three-letter word after a series of words set off by commas is often AND.)

26. GRAPE MA QLMVFGCB, BLLCRQU XRUX GPMRQUB, TFQBMPQMEK BTXLYNEL DLBM BXFVB PUPRQBM LPTX FMXLG. (*Hint:* Ciphertext B represents S. Note its high frequency as a first and last letter. *Bonus hint:* The fifth word is *not* THAT.)

27. MZDGBWV-MVCWZ WTZ HR ZHMAD XHKEBWTEG NZHLTUCG TUCWV CELTZHEXCEB RHZ PCWBSCZ GBWBTHE. (*Hint:* The five vowels, A to U, are represented by C, H, K, T, and W, but not necessarily in that order.)

Cryptic arithmetic is a type of mathematics puzzle in which letters have been replaced by the digits of numbers. Replace each letter by a digit (the same digit for the same letter throughout; different digits for different letters), and the arithmetic will be performed correctly. Break the codes in Problems 28–30.

28.
```
  SEND
+ MORE
------
 MONEY
```

29.
```
  THIS
    IS
+ VERY
------
  EASY
```

30.
```
   DAD
  SEND
+ MORE
------
 MONEY
```

▲ Individual Research

31. Prepare an exhibit on cryptography. Include devices or charts for writing and deciphering codes, coded messages, and illustrations of famous codes from history. For example, codes are found in literature in *Before the Curtain Falls*, by J. Rives Childs; *The Gold Bug*, by Edgar Allan Poe, *Voyage to the Center of Earth*, by Jules Verne.

References

Richard V. Andree, "Cryptography as a Branch of Mathematics," *The Mathematics Teacher*, November 1952.

Martin Gardner, "Mathematical Games Department," *Scientific American*, August 1972, pp. 114–118.

Dennis Shasta, *Codes, Puzzles, and Conspiracy*, Menlo Park, CA: Dale Seymour Publications, 1993.

News Clip

Under increasing pressure from Congress and the computer industry, the Clinton administration on Friday proposed a series of new data-scrambling initiatives that it said would permit U.S. companies to more effectively compete internationally — including liberalizing export controls on software used to encrypt data and developing an international system for managing the mathematical keys used to scramble data.

Press Democrat news services, July 8, 1996

"More new jobs will require more postsecondary mathematics education."

A Challenge of Numbers

Biographical Sketch

"I didn't study mathematics because of its practical uses, but because it was fun!"

Mina Rees

Mina Rees was born in Cleveland, Ohio, on August 2, 1902. She is not only a distinguished mathematician, but also a pioneer as a woman in mathematics.

In 1946, Dr. Rees was invited by the U.S. Navy to establish the prestigious mathematics research program in the newly created Office of Naval Research. By the time she left the department in 1953, she was deputy science director and had done considerable work in the field of cryptography, and much of her research is still classified. She says, "We can't be too specific, except we can say that there was some very exciting mathematics, much of which could not be published in the form in which it was originally done because of the context."

She has an enormously impressive list of awards, including the Award for Distinguished Service to Mathematics, a President's Certificate of Merit, and the Public Welfare Medal given by the National Academy of Sciences.

Her advice for young women contemplating a mathematical career: "I want them to be mathematicians if they like mathematics!" Now, as always, Mina's love for mathematics is clearly evident.

Book Reports

Write a 500-word report on one of the following books:

▲ *Estimation and Mental Computation*, Harold L. Schoen and Marilyn J. Zweng (Reston, VA: Yearbook of the National Council of Teachers of Mathematics, 1986).

▲ *The Man Who Knew Infinity*, S. Kanigel (New York: Charles Scribners, 1991).

Important terms

Abelian group [4.6]
Absolute value [4.3]
Additive inverse [4.6]
Algebra [4.7]
Apportionment problem
　[4.4]
Associative property [4.1]
Canonical form [4.2]
Closed set [4.1]
Closure [4.1]
Commutative group [4.6]
Commutative property [4.1]
Composite number [4.2]
Congruent modulo m [4.7]
Counting numbers [4.1]
Cryptography [4.8]
Cyphertext [4.8]
Decoding key [4.8]
Denominator [4.4]
Dense set [4.6]

Distributive property [4.1]
Divisibility [4.2]
Division [4.3]
Division by zero [4.3]
Divisor [4.2]
Encoding key [4.8]
Encrypt [4.8]
Factoring [4.2]
Field [4.6]
Fundamental property of
　fractions [4.4]
Fundamental theorem of
　arithmetic [4.2]
g.c.f. [4.2]
Greatest common factor
　[4.2]
Group [4.6]
Hypotenuse [4.5]
Identity [4.6]
Improper fraction [4.4]

Integer [4.3]
Inverse property [4.6]
Irrational number [4.5]
Law of square roots [4.5]
l.c.m. [4.2]
Least common multiple [4.2]
Leg [4.5]
Modular codes [4.8]
Modulo 5 [4.7]
Multiple [4.2]
Multiplication [4.1]
Multiplicative inverse [4.6]
Natural numbers [4.1]
Number line [4.6]
Numerator [4.4]
Opposite [4.6]
Perfect number [4.2]
Perfect square [4.5]
Prime factorization [4.2]
Prime number [4.2]

Proper divisor [4.2]
Proper fraction [4.4]
Pythagorean theorem [4.5]
Radical form [4.5]
Radicand [4.5]
Rational number [4.4]
Real number line [4.6]
Real numbers [4.6]
Reciprocal [4.6]
Reduced fraction [4.4]
Relatively prime [4.2]
Repeating decimal [4.6]
Rules of divisibility [4.2]
Sieve of Eratosthenes [4.2]
Square number [4.5]
Square root [4.5]
Subtraction [4.1]
Terminating decimal [4.6]
Whole number [4.3; 4.4]
Zero [4.3; 4.6]

Important Ideas

Properties of numbers:
　commutative property [4.1]
　associative property [4.1]
　distributive property [4.1]
　identity [4.6]
　inverse [4.7]
Relationships among sets of numbers:
　natural numbers [4.1]
　whole numbers [4.3]
　integers [4.3]
　rational numbers [4.4]
　irrational numbers [4.5]
　real numbers [4.6]

Operations: Integers [4.3]; Rationals [4.4]; Irrationals [4.5];
　Reals [4.6]
Multiplication as repeated addition [4.1]
Subtraction as an addition property [4.1]
Divisibility [4.2]
Rules of divisibility [4.2]
Factoring [4.2]
There is no largest prime number [4.2]
Pythagorean theorem [4.5]
Know the difference between a square root and an irrational
　number [4.5]
Simplify a radical expression [4.5]
Group and field [4.6]
Congruence, mod m [4.7]

Types of Problems

Determine whether a given set with a given operation is
　closed [4.1]
Recognize and distinguish the commutative and associative
　properties [4.1]
Apply the distributive property with a variety of operations
　[4.1]
Tell whether one number divides another number [4.2]
Determine whether a given number is prime or not [4.2]

Find the prime factorization and write the answer in
　canonical form [4.2]
Find the least common multiple of a set of numbers [4.2]
Find the greatest common factor of a set of numbers [4.2]
Show that there is no largest prime number [4.2]
Use problem-solving techniques to solve applied problems
　[4.2–4.8]

Find the absolute value of a number [4.3]

Carry out operations with integers [4.3]

Reduce fractions using the fundamental property of fractions [4.4]

Carry out operations with fractions [4.4]

Use rational numbers to solve an apportionment problem [4.4]

Use the definition of square root to simplify radical expressions [4.5]

Classify numbers as rational or irrational [4.5]

Carry out operations with square roots [4.5]

Find the decimal representation of a fraction [4.6]

Be able to classify real numbers [4.6]

Change terminating decimals to fractional form [4.6]

Find the identity given some set and some operation [4.6]

Find the opposite of a number and/or verify the inverse property for addition [4.6]

Find the reciprocal of a number and/or verify the inverse property for multiplication [4.6]

Carry out operations in modular arithmetic [4.7]

Solve modular equations [4.7]

Given a set and an operation, determine whether they form a group [4.6, 4.7]

Given a set and two operations, determine whether they form a field [4.6, 4.7]

Be able to break some simple codes [4.8]

CHAPTER 4   Review Questions

Perform the indicated operations in Problems 1–8.

1. $-4 + 5(-3)$

2. $\dfrac{4}{7} + \dfrac{5}{9}$

3. $\dfrac{7}{30} + \dfrac{5}{42} + \dfrac{5}{99}$

4. $-\sqrt{10} \cdot \sqrt{10}$

5. $\left(\dfrac{11}{12} + 2\right) + \dfrac{-11}{12}$

6. $\dfrac{3^{-1} + 4^{-1}}{6}$

7. $\dfrac{-7}{9} \cdot \dfrac{99}{174} + \dfrac{-7}{9} \cdot \dfrac{75}{174}$

8. $\dfrac{-3 + \sqrt{3^2 + 4(2)(3)}}{2(2)}$

Reduce each fraction in Problems 9–13. If it is reduced, so state.

9. $\dfrac{8}{3}$ 10. $\dfrac{16}{18}$ 11. $\dfrac{100}{825}$ 12. $\dfrac{184}{207}$ 13. $\dfrac{1,209}{2,821}$

Find the decimal representation for each number in Problems 14–18, and classify it as rational or irrational.

14. $\dfrac{3}{8}$ 15. $\dfrac{7}{3}$ 16. $\dfrac{3}{7}$ 17. $\dfrac{125}{20}$ 18. $\dfrac{3}{13}$

Find the prime factorization of each number given in Problems 19–23. If it is prime, so state.

19. 89   20. 101   21. 349   22. 1,001   23. 6,825

Solve the equations in Problems 24–26 for x.

24. $\frac{x}{5} \equiv 2$, (mod 8)   25. $2x \equiv 3$, (mod 7)   26. $2x^2 + 7x + 1 \equiv 0$, (mod 2)

27. Find the greatest common factor for the set of numbers {49, 1001, 2401}.

28. Find the least common multiple for the set of numbers {49, 1001, 2401}.

29. If $a \odot b = a \times b + a + b$, what is the value of $(1 \odot 2) \odot (3 \odot 4)$?

30. If $a \uparrow b$ stands for the larger number of the pair and $a \downarrow b$ stands for the lesser number of the pair, what is the value of $(1 \downarrow 2) \uparrow (2 \downarrow 3)$?

Define each of the operations given in Problems 31–33.

31. multiplication   32. subtraction   33. division

34. **IN YOUR OWN WORDS** Explain why we cannot divide by zero.

35. Find an irrational number between 34 and 35.

36. What is a field? Describe each property.

37. Consider the set $\mathbb{N}$ and let $a, b \in \mathbb{N}$. Let Ø be an operation defined by a Ø b = g.c.f. of a and b. Is $\mathbb{N}$ a commutative group for the operation of Ø?

38. The news clip is from the column "Ask Marilyn," *Parade Magazine*, April 15, 1990. Answer the question asked in the article.

39. Suppose you are building a stairway (such as shown in Figure 4.21). Assume that the vertical rise is 8 ft, the horizontal run is 12 ft, and the maximum rise for each step is 8 inches. How many steps are necessary? What is the total length of the segments making up the stairs, and what is the length of the diagonal?

News Clip

There are 1,000 tenants and 1,000 apartments. The first tenant opens every door. The second closes every other door. The third tenant goes to every third door, opening it if it is closed and closing it if it is open. The fourth tenant goes to every fourth door, closing it if it is open and opening it if it is closed. This continues with each tenant until the 1,000th tenant closes the 1,000th door. Which doors are open?

Anita Mueller, Arnold, MD

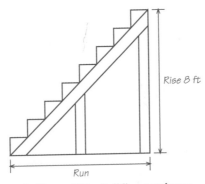

Rise 8 ft

Run

▲ **Figure 4.21   Building a stairway**

40. Consider the following apportionment problem:

North: 2,400,000   Northeast: 4,000,000

East: 2,700,000   Southeast: 2,500,000

South: 2,800,000   Southwest: 3,500,000

West: 4,100,000   Northwest: 2,600,000

Suppose each House member is to represent 250,000 citizens. What is the representation according to the plans of Adams, Jefferson, Webster, and Hamilton? For Hamilton's plan assume that there must be 100 representatives.

Revisiting the Real World . . .

IN THE REAL WORLD   "I'm never going to have children!" exclaimed Shelly. "Sometimes my little brother drives me nuts!" "I know what you mean," added Mary. "Just yesterday my brother asked me to explain what 5 means. The best I could do was to show him my fist."

"Stop! That's beginning to sound too much like my math class — just yesterday Ms. Jones asked us whether $\sqrt{2}$ is rational or irrational, **and** we need to justify our answer," Shelly said with a bit of anger. "I don't have the faintest; the only radical I know is my brother. I don't know why we should know about weird numbers like $\sqrt{2}$."

Commentary: Polya's method will lead the way.

Understand the Problem.   The number $\sqrt{2}$ must be either rational or irrational. If it is rational, then it can be written in the form of a common fraction or in decimal form as a terminating or a repeating decimal.

Devise a Plan.   We will assume that $\sqrt{2}$ is rational, and then we will show that this leads to a contradiction, which will in turn force us to conclude that $\sqrt{2}$ is not rational — that is, that $\sqrt{2}$ is irrational.

Carry Out the Plan.   Suppose $\sqrt{2} = \frac{a}{b}$, where a and b are whole numbers such that $\frac{a}{b}$ is a reduced fraction. Square both sides:

$$2 = \frac{a^2}{b^2}$$

$$2b^2 = a^2 \qquad \text{Multiply both sides by } b^2.$$

The number on the left-hand side is even because it contains a factor of 2. Hence the number on the right-hand side must also be even. But if a^2 is even, then a must be even. If a is even, then it must have 2 as a factor, so that $a = 2d$ for some integer d. We substitute this value into our equation:

$$2b^2 = (2d)^2 \qquad \text{Substitution}$$
$$2b^2 = 4d^2 \qquad \text{Simplify.}$$
$$b^2 = 2d^2 \qquad \text{Divide both sides by 2.}$$

We now see that b^2 is an even number, so b must be even. Since *both* a and b are even numbers, then we see that $\frac{a}{b}$ is not a reduced fraction. This contradicts the result that $\frac{a}{b}$ is reduced.

Look Back.   The only possible mistake in our reasoning is that our assumption that $\sqrt{2}$ equals a fraction is incorrect. In other words, $\sqrt{2}$ cannot be a ratio of two whole numbers. This means that $\sqrt{2}$ is irrational.

Group Research

Working in small groups is typical of most work environments, and being able to work with others to communicate specific ideas is an important skill to learn. Work with three or four other students to submit a single report based on each of the following questions.

1. With only a straightedge and compass, use a number line and the Pythagorean theorem to construct a segment whose length is $\sqrt{2}$. Measure the segment as accurately as possible, and write your answer in decimal form. *Do not use a calculator or any tables.* Now, continue your work to construct segments whose lengths are $\sqrt{3}, \sqrt{4}, \sqrt{5}, \ldots$.

2. **Four Fours**   Write the numbers from 1 to 100 (inclusive) using exactly four fours. See Problem 61 in Problem Set 4.3 to help you get started.

3. **Pythagorean Theorem**   Write out three different proofs of the Pythagorean theorem.

4. **Symmetries of a Cube**   Consider a cube labeled as shown in Figure 4.22. List all the possible symmetries of this cube. See Problem 63 in Problem Set 4.6 to help you get started.

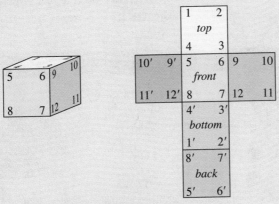

▲ **Figure 4.22   Symmetries of a cube**

5. The Babylonians estimated square roots using the following formula:

$$\text{If } n = a^2 + b, \text{ then } \sqrt{n} \approx a + \frac{b}{2a}.$$

For example, if $n = 11$, then $n = 9 + 2$, so that $n = 11$, $a = 3$, and $b = 2$.

perfect square

This Babylonian approximation for $\sqrt{11}$ is found by

$$\sqrt{11} \approx 3 + \frac{2}{2(3)} \approx 3.3333 \ldots$$

With a calculator, we obtain $\sqrt{11} \approx 3.31662479$. Write a paper about this approximation method. Here are some questions you might consider:

a. Can b be negative?

b. Consider the following possibilities:

$$|b| < a^2 \qquad |b| = a^2 \qquad |b| > a^2$$

Can you formulate any conclusions about the appropriate hypotheses for the Babylonian approximation?

c. Put this formula into a historical context.

The Nature of Algebra

The fact that algebra has its origin in arithmetic, however widely it may in the end differ from that science, led Sir Isaac Newton to designate it "Universal Arithmetic," a designation which, vague as it is, indicates its character better than any other.

George E. Chrystal

In the Real World . . .

Rhonda, a very small person, needs to move a very large, heavy object. She does not know what to do, so she asks her friend, Jane, for help. Jane tells Rhonda, "Let's get a big board. I remember from school that you can move a big object with a lever."

"I get it," Rhonda cried loudly. "I have just the right stuff out back. Let's see . . . . Look at how I've put this together."

"What have you placed on the other end of the lever?" asked Jane.
"Well, it is a can of Coke. Will that not work?" asked Rhonda.

"That is really dumb, Rhonda," said Jane sarcastically.

Rhonda quickly responded, "Not if the weights of the can of Coke and the trunk are the same. You do not know how much each of these containers weighs, and I'm prepared to show you that they weigh the same!"

Contents

5.1 Polynomials
5.2 Factoring
5.3 Evaluation, Applications, and Spreadsheets
5.4 Equations
5.5 Inequalities
5.6 Algebra in Problem Solving
5.7 Ratios, Proportions, and Problem Solving
5.8 Modeling Uncategorized Problems
5.9 Introduction to Computer BASIC
Chapter 5 Summary
Profile: Shirley Frye
In the Real World Commentary
Chapter 5 Group Research Projects

Perspective

It has been said that algebra is the greatest labor-saving device ever invented. Even though you have probably had algebra sometime in your past, the essential algebraic ideas you will need for the rest of this text are covered or reviewed in this chapter. We begin with the basic building blocks of algebra — namely, polynomials and operations with polynomials. We continue with algebraic manipulations when we reverse the operation of multiplication and consider factoring in the second section. Some applications of these ideas are introduced in the next three sections, where we evaluate expressions, look at an application from genetics, take a peek at computer spreadsheets, and then solve equations and inequalities. We conclude this chapter with further applications of problem solving.

5.1 POLYNOMIALS

Many people think of algebra as simply a high school mathematics course in which variables are manipulated. This chapter reviews many of these procedures; however, the word *algebra* refers to a structure, or a set of axioms that forms the basis for what is accepted and what is not. For example, in the previous chapter we defined a field, which involves a set, two operations, and 11 specified properties. If you studied Section 4.7, you investigated an algebra with a finite number of elements. As you study the ordinary algebra presented in this chapter, you should remember this is only one of many possible algebras. Additional algebras are often studied in more advanced mathematics courses.

In this chapter we will review much of the algebra you have previously studied. Recall that a **term** is a number, a variable, or the product of numbers and variables. Thus, $10x$ is one term, but $10 + x$ is not (because the terms 10 and x are connected by addition and not by multiplication). A fundamental notion in algebra is that of a **polynomial**, which is a term or the sum of terms. We classify polynomials by the number of terms and by degree:

A polynomial with one term is called a **monomial.**

A polynomial with two terms is called a **binomial.**

A polynomial with three terms is called a **trinomial.**

There are other words that could be used for polynomials with more than three terms, but this classification is sufficient. To classify by degree, we recall that the **degree of a term** is the number of *variable* factors in that term. Thus, $3x$ is first-degree, $5xy$ is second-degree, 10 is zero-degree, $2x^2$ is second-degree, and $9x^2y^3$ is fifth-degree. The **degree of a polynomial** is the largest degree of any of its terms. A first-degree term is sometimes called **linear** and a second-degree term is sometimes referred to as **quadratic.**

■■■ EXAMPLE 1

Classify each polynomial by number of terms and by degree. If the expression is not a polynomial, so state.

a. x b. $x^2 + 5x - 7$ c. $x^3 - \dfrac{2}{x}$ d. $x^2y^3 - xy^2$ e. 5

Solution

a. Monomial, degree 1

b. Trinomial, degree 2

c. Not a polynomial because it involves division by a variable. Notice that $x^3 - \dfrac{x}{2}$

   is a polynomial because it can be written as $x^3 - \frac{1}{2}x$ and fractional coefficients are permitted.

d. Binomial, degree 5

e. Monomial, degree 0

Historical Note

Algebra and algebraic ideas date back 4,000 years to the Babylonians and the Egyptians; the Hindus and the Greeks also solved algebraic problems. The title of Arab mathematician al-Khowârizmî's text (about 825 A.D.), *Hisâb al-jabr w'almugâbalah,* is the origin of the word *algebra.* The book was widely known in Europe through Latin translations. The word *al-jabr* or *al-gebra* became synonymous with equation solving. Interestingly enough, the Arabic word *al-jabr* was also used in connection with medieval barbers. The barber, who also set bones and let blood in those times, was known as an *algebrista.* In the 19th century, there was a significant change in the prevalent attitude toward mathematics: Up to that time, mathematics was expected to have immediate and direct applications, but now mathematicians became concerned with the structure of their subject and with the *validity,* rather than the *practicality,* of their conclusions. Thus, there was a move toward *pure mathematics* and away from *applied mathematics.* ▲

When writing polynomials, it is customary to arrange the terms from the highest-degree to the lowest-degree term. If terms have the same degree they are usually listed in alphabetical order.

Simplification of Polynomials

When working with polynomials, it is necessary to simplify algebraic expressions. The key ideas of simplification are *similar terms* and the *distributive property*. Terms that differ only in the numerical coefficients are called **like terms** or **similar terms**.

▧▧▧ EXAMPLE 2

Simplify the given algebraic expressions.

a. $-12x - (-5)x$ 　　　　　　　 b. $-3x - 6x + 2x$

c. $2x + 3y + 5x - 2y$ 　　　　 d. $5xy^2 - xy^2 - 4x^2y$

e. $(4x - 5) + (5x^2 + 2x - 3)$ 　　 f. $(4x - 5) - (5x^2 + 2x - 3)$

Solution

a. $-12x - (-5x) = -12x + 5x = -7x$ 　　 Add the opposite of $(-5x)$.

b. $-3x - 6x + 2x = -7x$

c. $2x + 3y + 5x - 2y = 7x + y$ 　　 Note the similar terms.

d. $5xy^2 - xy^2 - 4x^2y = 4xy^2 - 4x^2y$

e. $(4x - 5) + (5x^2 + 2x - 3) = 5x^2 + (4x + 2x) + (-5 - 3)$

$\qquad\qquad\qquad\qquad\qquad\quad = 5x^2 + 6x - 8$

f. Recall that to subtract a polynomial, you subtract *each* term of that polynomial:

$$(4x - 5) - (5x^2 + 2x - 3) = 4x - 5 - 5x^2 - 2x - (-3)$$

$$= -5x^2 + 2x - 2$$ ▬

Remember from beginning algebra that $-x = (-1)x$, so to subtract a polynomial you can do it as shown in Example 2**f** — by subtracting *each* term — or you can think of it as an application of the distributive property:

$$(4x - 5) - (5x^2 + 2x - 3) = 4x - 5 + (-1)(5x^2 + 2x - 3)$$

$$= 4x - 5 + (-1)(5x^2) + (-1)(2x) + (-1)(-3)$$

$$= -5x^2 + 2x - 2$$

The distributive property is also important in multiplying polynomials.

▧▧▧ EXAMPLE 3

Simplify the given algebraic expressions.

a. $3x(x^2 - 1)$ 　　 b. $(2x - 3)(x + 1)$ 　　 c. $(x + 2)(x^2 + 5x - 2)$

d. $(2x + 1)^3$

Solution 　 In each case, we will distribute the expression on the left: $3x$, $(2x - 3)$, $(x + 2)$, and $(2x + 1)$, respectively. You will find your work easier to read if you work down with the equal signs aligned rather than across your paper. And finally,

we use the laws of exponents (Section 1.4, p. 58) to simplify expressions such as
$x(x^2) = x^1x^2 = x^{1+2} = x^3$.

a. $3x(x^2 - 1) = 3x(x^2) + 3x(-1)$ ← This step is usually done mentally, and is not written down.

$= 3x^3 - 3x$

b. $(2x - 3)(x + 1) = (2x - 3)(x) + (2x - 3)(1)$ Mental step

$= 2x^2 - 3x + 2x - 3$ Distributive property

$= 2x^2 - x - 3$ Add similar terms.

c. $(x + 2)(x^2 + 5x - 2) = (x + 2)(x^2) + (x + 2)(5x) + (x + 2)(-2)$

$= x^3 + 2x^2 + 5x^2 + 10x - 2x - 4$

$= x^3 + 7x^2 + 8x - 4$

d. $(2x + 1)^3 = (2x + 1)(2x + 1)(2x + 1)$ Definition of cube

$= (2x + 1)[(2x + 1)(2x) + (2x + 1)(1)]$

$= (2x + 1)[4x^2 + 2x + 2x + 1]$

$= (2x + 1)(4x^2 + 4x + 1)$

$= (2x + 1)(4x^2) + (2x + 1)(4x) + (2x + 1)(1)$

$= 8x^3 + 4x^2 + 8x^2 + 4x + 2x + 1$

$= 8x^3 + 12x^2 + 6x + 1$

$\bigcirc$ Note that $(2x)^3 + (1)^3 = 8x^3 + 1 \neq 8x^3 + 12x^2 + 6x + 1$. $\bigcirc$ ▬

Shortcuts with Products

It is frequently necessary to multiply binomials, and even though we use the distributive property, we want to be able to carry out the process quickly and efficiently in our heads. Consider the following example, which leads us from multiplication using the distributive property to an efficient process that is usually called FOIL.

▬▬▬ **EXAMPLE 4**

Simplify

a. $(2x + 3)(4x - 5)$ **b.** $(5x - 3)(2x + 3)$ **c.** $(4x - 3)(3x - 2)$

Solution

a. $(2x + 3)(4x - 5) = (2x + 3)(4x) + (2x + 3)(-5)$ Distributive property

$= \underbrace{8x^2}_{\substack{\uparrow \\ \text{Product} \\ \text{of lines} \\ \text{terms}}} + \underbrace{12x + (-10x)}_{\substack{\uparrow \\ \text{Sum of products} \\ \text{of inner terms} \\ \text{and outer terms}}} + \underbrace{(-15)}_{\substack{\uparrow \\ \text{Product} \\ \text{of last} \\ \text{terms}}}$

$= 8x^2 + 2x - 15$

b. $(5x - 3)(2x + 3) = \underbrace{10x^2}_{} + \underbrace{(15x - 6x)}_{} + \underbrace{(-9)}_{}$

Mentally: FIRST   OUTER terms   LAST
                    terms   and INNER terms   terms

$$= 10x^2 + 9x - 9$$

c. $(4x - 3)(3x - 2) = 12x^2 - 17x + 6$ Mentally

You are encouraged to carry out the binomial multiplication mentally, as shown in Example 4c. To help you remember the process, we sometimes call this binomial multiplication FOIL to remind you:

First terms + **O**uter terms + **I**nner terms + **L**ast terms

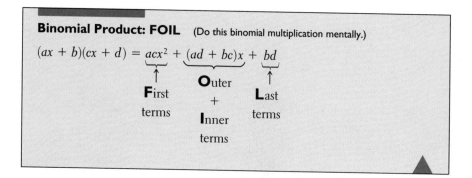

Binomial Product: FOIL   (Do this binomial multiplication mentally.)

$$(ax + b)(cx + d) = \underbrace{acx^2}_{} + \underbrace{(ad + bc)x}_{} + \underbrace{bd}_{}$$

First   **O**uter   **L**ast
terms   +   terms
          Inner
          terms

EXAMPLE 5

Simplify (mentally):

a. $(2x - 3)(x + 3)$ **b.** $(x + 3)(3x - 4)$ **c.** $(5x - 2)(3x + 4)$

Solution

a. $(2x - 3)(x + 3) = \underbrace{2x^2}_{F} + \underbrace{3x}_{O+I} - \underbrace{9}_{L}$

b. $(x + 3)(3x - 4) = 3x^2 + 5x - 12$

c. $(5x - 2)(3x + 4) = 15x^2 + 14x - 8$

A second shortcut involves raising a binomial to an integral power — for example, $(2x + 1)^3$ or $(a + b)^8$. We look for a pattern with the following example.

EXAMPLE 6   **Polya's Method**

Expand (multiply out) the expression $(a + b)^8$.

Solution   We use Polya's problem-solving guidelines for this example.

Understand the Problem. We could begin by using the definition of 8th power and the distributive property:

$$(a + b)^8 = \underbrace{(a + b)(a + b) \cdots (a + b)}_{\text{8 factors of } a + b}$$

However, we soon see that this is too lengthy to complete directly.

Devise a Plan. We will consider a pattern of successive powers; that is, consider $(a + b)^n$ for $n = 0, 1, 2, \ldots$.

Carry Out the Plan. We begin by actually doing the multiplications:

$n = 0$:	$(a + b)^0 =$	1
$n = 1$:	$(a + b)^1 =$	$1 \cdot a + 1 \cdot b$
$n = 2$:	$(a + b)^2 =$	$1 \cdot a^2 + 2 \cdot ab + 1 \cdot b^2$
$n = 3$:	$(a + b)^3 =$	$1 \cdot a^3 + 3 \cdot a^2b + 3 \cdot ab^2 + 1 \cdot b^3$
$n = 4$:	$(a + b)^4 =$	$1 \cdot a^4 + 4 \cdot a^3b + 6 \cdot a^2b^2 + 4 \cdot ab^3 + 1 \cdot b^4$

$$\vdots \qquad\qquad \vdots$$

First, ignore the coefficients (shown in color) and focus on the variables:

$(a + b)^1$:	a	b			
$(a + b)^2$:	a^2	ab	b^2		
$(a + b)^3$:	a^3	a^2b	ab^2	b^3	
$(a + b)^4$:	a^4	a^3b	a^2b^2	ab^3	b^4

$$\vdots \qquad\qquad\qquad \vdots$$

Do you see a pattern? As you read from left to right, the powers of a decrease and the powers of b increase. Note that the sum of the exponents for each term is the same as the original exponent:

$$(a + b)^n: \quad a^nb^0 \quad a^{n-1}b^1 \quad a^{n-2}b^2 \cdots a^{n-r}b^r \cdots a^2b^{n-2} \quad a^1b^{n-1} \quad a^0b^n$$

Next, consider the numerical coefficients (shown in color):

$(a + b)^0$:	1
$(a + b)^1$:	1 1
$(a + b)^2$:	1 2 1
$(a + b)^3$:	1 3 3 1
$(a + b)^4$:	1 4 6 4 1

$$\vdots \qquad\qquad \vdots$$

Do you see the pattern? Recall Pascal's triangle from Section 1.1 (see Figure 5.1).

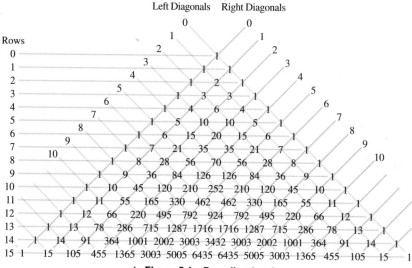

▲ Figure 5.1   Pascal's triangle

To find $(a + b)^8$, we look at the 8th row of Pascal's triangle for the coefficients to complete the product:

$$(a + b)^8 = a^8 + 8a^7b + 28a^6b^2 + 56a^5b^3 + 70a^4b^4 + 56a^3b^5 + 28a^2b^6 + 8ab^7 + b^8$$

Look Back.   Does this pattern seem correct? We have verified the pattern directly for $n = 0, 1, 2, 3$, and 4. With a great deal of algebraic work, you can verify by direct multiplication that the pattern checks for $(a + b)^5$.   ▬

Binomial Theorem

The pattern we discovered for $(a + b)^8$ is a very important theorem in mathematics. It is called the **binomial theorem.** The difficulty in stating this theorem is in relating the coefficients of the expansion to Pascal's triangle. We write $\binom{4}{2}$ to represent the number in row 4, diagonal 2 of Pascal's triangle (see Figure 5.1). We see

$$\binom{4}{2} = 6 \qquad \binom{6}{4} = 15 \qquad \binom{5}{3} = 10 \qquad \binom{15}{7} = 6{,}435$$

Binomial Theorem

For any positive integer n,

$(a + b)^n$

$= \binom{n}{0}a^n + \binom{n}{1}a^{n-1}b + \binom{n}{2}a^{n-2}b^2 + \cdots + \binom{n}{n-1}ab^{n-1} + \binom{n}{n}b^n$

where $\binom{n}{r}$ is the number in the nth row, rth diagonal of Pascal's triangle.

Pascal's triangle is efficient for finding the numerical coefficients for exponents that are relatively small, as shown in Figure 5.1. However, for larger exponents we will need some additional work, which will be presented in Chapter 9.

▬▬ EXAMPLE 7

Expand $(x - 2y)^4$.

Solution   In this example, let $a = x$ and $b = -2y$, and look at row 4 of Pascal's triangle for the coefficients.

$$(a + b)^4 = a^4 + 4a^3b + 6a^2b^2 + 4ab^3 + b^4 \qquad \text{Binomial theorem, } n = 4$$
$$(x - 2y)^4 = x^4 + 4x^3(-2y) + 6x^2(-2y)^2 + 4x(-2y)^3 + (-2y)^4$$
$$= x^4 - 8x^3y + 24x^2y^2 - 32xy^3 + 16y^4 \qquad\qquad \text{▬}$$

Polynomials and Areas

Although we will discuss area in Section 8.2, we can assume that you know the area formulas for squares and rectangles:

> ⃠ This is EASY if you cut out these figures and use them as manipulatives. Rearrangement of strips is hard to show in a book, but if you try it with cutout pieces, you will like it! ⃠

$$\text{Area of a square:} \quad A = s^2 \qquad \text{Area of rectangle:} \quad A = \ell w$$

Areas of squares and rectangles are often represented as trinomials. Consider the area represented by the trinomial $x^2 + 3x + 2$. This expression is made up of three terms: x^2, $3x$, and 2. Translate each of these terms into an area:

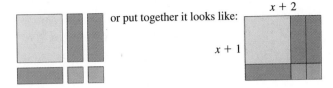

Now you must rearrange these pieces into a rectangle. Think of them as cutouts and move them around. There is only one way (not counting order) to fit them into a rectangle:

Note: The rectangle with dimensions $x + 2$ by $x + 1$ has the same area as one with dimensions $x + 1$ by $x + 2$.

Thus, $(x + 2)(x + 1) = x^2 + 3x + 2$. These observations provide another way (besides the distributive property) for verifying the shortcut method of FOIL.

▬▬ EXAMPLE 8

Find the product $(x + 3)(x + 1)$ both algebraically and geometrically.

Solution

Algebraic: $(x + 3)(x + 1) = x^2 + 4x + 3$

Geometric: Draw a rectangle with sides that measure
$x + 3$ and $x + 1$:

There is one square: x^2
four rectangles: $4x$
three units: 3

Thus, $(x + 3)(x + 1) = x^2 + 4x + 3$

PROBLEM SET 5.I

▲ A Problems

1. **IN YOUR OWN WORDS** What is a polynomial?

2. **IN YOUR OWN WORDS** What is the degree of a polynomial?

3. **IN YOUR OWN WORDS** Discuss the process of adding and subtracting polynomials.

4. **IN YOUR OWN WORDS** Discuss the process of multiplying polynomials using the distributive property.

5. **IN YOUR OWN WORDS** Discuss the process of multiplying binomials using FOIL.

6. **IN YOUR OWN WORDS** Discuss the process of multiplying binomials using areas.

7. **IN YOUR OWN WORDS** What is the binomial theorem?

Simplify each expression in Problems 8–22.

8. $(x + 3) + (5x - 7)$

9. $(2x - 4) - (3x + 4)$

10. $(x + y + 2z) + (2x + 5y - 4z)$

11. $(x - y - z) + (2x - 5y - 3z)$

12. $(5x^2 + 2x - 5) + (3x^2 - 5x + 7)$

13. $(2x^2 - 5x + 4) + (3x^2 - 2x - 11)$

14. $(x + 2y - 3z) - (x - 5y + 4z)$

15. $(x^2 + 4x - 3) - (2x^2 + 9x - 6)$

16. $(3x - x^2) - 5(2 - x) - (2x + 3)$

17. $3(x - 5) - 2(x + 8)$

18. $6(x + 1) - 2(x + 1)$

19. $3(2x^2 + 5x - 5) + 2(5x^2 - 3x + 6)$

20. $2(4x^2 - 3x + 2) - 3(x^2 - 5x - 8)$

21. $2(x + 3) - 3(x^2 - 3x + 1) + 4(x - 5)$

22. $3(x - 1) - 2(x^2 + 4x + 5) - 5(x + 8)$

In Problems 23–29 multiply mentally.

23. **a.** $(x + 3)(x + 2)$ **b.** $(y + 1)(y + 5)$ **c.** $(z - 2)(z + 6)$
    d. $(s + 5)(s - 4)$

24. **a.** $(x + 1)(x - 2)$ **b.** $(y - 3)(y + 2)$ **c.** $(a - 5)(a - 3)$
    d. $(b + 3)(b - 4)$

25. **a.** $(c + 1)(c - 7)$ **b.** $(z - 3)(z + 5)$ **c.** $(2x + 1)(x - 1)$
    d. $(2x - 3)(x - 1)$

26. **a.** $(x + 1)(3x + 1)$ **b.** $(x + 1)(3x + 2)$ **c.** $(2a + 3)(3a - 2)$
    d. $(2a + 3)(3a + 2)$

27. **a.** $(x + y)(x + y)$ **b.** $(x - y)(x - y)$ **c.** $(x + y)(x - y)$
    d. $(a + b)(a - b)$

28. **a.** $(5x - 4)(5x + 4)$ **b.** $(3y - 2)(3y + 2)$ **c.** $(a + 2)^2$ **d.** $(b - 2)^2$

29. **a.** $(x + 4)^2$ **b.** $(y - 3)^2$ **c.** $(s + t)^2$ **d.** $(u - v)^2$

▲ **B Problems**

Simplify the expressions in Problems 30–37.

30. $(5x + 1)(3x^2 - 5x + 2)$ 31. $(2x - 1)(3x^2 + 2x - 5)$

32. $(3x - 1)(x^2 + 3x - 2)$ 33. $(5x + 1)(x^3 - 2x^2 + 3x)$

34. $(5x + 1)(3x^2 - 5x + 2) - (x^3 - 4x^2 + x - 4)$

35. $3(3x^2 - 5x + 2) - 4(x^3 - 4x^2 + x - 4)$

36. $(x - 2)(2x - 3) - (x + 1)(x - 5)$ 37. $(2x - 3)(3x + 2) + (x + 2)(x + 3)$

Find each product in Problems 38–43 both algebraically and geometrically.

38. $(x + 2)(x + 4)$ 39. $(x + 1)(x + 4)$ 40. $(x + 2)(x + 5)$

41. $(x + 3)(x + 4)$ 42. $(2x + 3)(3x + 2)$ 43. $(2x + 1)(2x + 3)$

Use the binomial theorem to expand each binomial given in Problems 44–51.

44. $(x + 1)^3$ 45. $(x - 1)^3$ 46. $(x + y)^5$ 47. $(x + y)^6$

48. $(x - y)^7$ 49. $(x - y)^8$ 50. $(5x - 2y)^3$ 51. $(2x - 3y)^4$

52. Write out the first three terms in the expansion $(x + y)^{12}$.

53. Write out the last three terms in the expansion $(x + y)^{14}$.

54. The number of desks in one row is $5d + 2$. How many desks are there in a room of $2d - 1$ rows if they are arranged in a rectangular arrangement?

55. An auditorium has $6x + 2$ seats in each row and $51x - 7$ rows. If the chairs are in a rectangular arrangement, how many seats are there in the auditorium?

56. Each apartment in a building rents for $800 - d$ dollars per month. What is the monthly income from $6d + 12$ units?

57. If a boat is traveling at a rate of $6b + 15$ miles per hour for a time of $10 - 2b$ hours, what is the distance traveled by the boat?

▲ **Problem Solving**

58. Binomial products can be used to do mental calculations. For example, to multiply a pair of two-digit numbers whose tens digit is the same and whose units digits add up to 10, mentally multiply the units digit and then multiply the tens digit of the first number by one more than the tens digit of the second. For example,

$$8 \times 9 = 72$$
$$84 \times 86 = 72 \; 24$$
$$4 \times 6 = 24$$

Mentally multiply the given numbers.

a. 62×68　　**b.** 57×53　　**c.** 63×67　　**d.** 95^2　　**e.** 75^2

59. Suppose the given numbers for a mental calculation (see Problem 58) are $10x + y$ and $10x + z$. Notice that these two numbers have the same tens digit. Also assume that $y + z = 10$, which says that the units digits of the two numbers sum to 10. Algebracially show why the mental calculation described in Problem 58 "works."

60. Devise a procedure for mentally multiplying three-digit numbers with the same first two digits and units digits that add up to 10.

▲ **Individual Research**

61. What do the following people have in common?

> **Carl T. Rowan,** columnist for the *Washington Post*
>
> **Lewis Carroll (Charles Dodgson),** author of *Alice in Wonderland*
>
> **Christopher Wren,** architect of St. Paul's Cathedral in London
>
> **Virginia Wade,** tennis player, Wimbledon champion
>
> **Lawrence Leighton Smith,** conductor and pianist

5.2 FACTORING

We have called numbers that are multiplied *factors*. In Section 4.2 we defined a *prime number* and the *prime factorization* of numbers. The same process can be applied to algebraic expressions. The goal here is completed in Section 5.4 when we use factoring to solve quadratic equations. Factoring is also used extensively in algebra, so to understand some of the algebraic processes, it is necessary to understand factoring.

The approach we take in this section is different from that you will normally see in an algebra course. So often algebra is learned by brute force and memorization or symbol manipulation, but our development uses a geometric visualization that may help in your understanding not only of algebraic processes, but of geometric ones as well.

Using Areas to Factor

In the previous section, we showed how areas can be used to understand multiplication of binomials. We now use areas to factor polynomials.

▬▬ EXAMPLE 1

Factor $2x^2 + 7x + 6$ using areas.

Solution　First draw the areas for the terms:

Rearrange to form a rectangle:

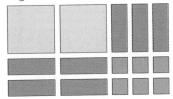

Push these pieces together to form a single rectangle:

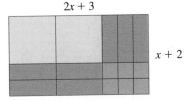

Thus, $2x^2 + 7x + 6 = (2x + 3)(x + 2)$

The following example shows how to use areas when factoring a polynomial with a subtraction, as well as with a polynomial that is not factorable.

▬▬ **EXAMPLE 2**

Factor the following polynomials, if possible, by using areas.

a. $x^2 + 2x - 3$ **b.** $x^2 + 5x + 8$

Solution

a. First draw the rectangles for the terms:

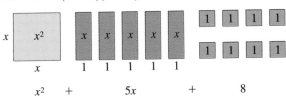

From these rectangles we must subtract three squares; we will indicate these squares to be subtracted by

Arrange the positive pieces and then place the negative pieces (the gray ones) on top of the positive pieces:

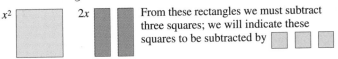

This is $x - 1$; move this piece over here

Thus, $x^2 + 2x - 3 = (x + 3)(x - 1)$.

b.

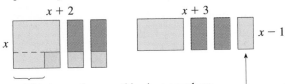

$x^2 \quad + \quad 5x \quad + \quad 8$

There is no way of arranging all the pieces to form a rectangle.

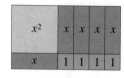 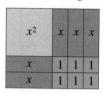

We see that this trinomial is not factorable.   ▬

Using Algebra to Factor

By thinking through the steps for factoring trinomials by using areas, we can develop a process for algebraic factoring. Consider $x^2 + 3x + 2$ and compare the process below with the discussion at the beginning of this section. We consider the first term and the last term of the given trinomial.

First term: Area of the large square with side x.
                     ↓   ↓   ↓
$$x^2 + 3x + 2 = (x \quad 2)(x \quad 1)$$
                     ↑   ↑   ↑
Last term: Area of two squares with sides of length 1.

There may be several ways of rearranging the areas for these first two steps (first term and last term). What you want to do is to rearrange them so that they form a rectangle. By looking at the binomial product and recalling the process we called FOIL, you see that the sum of the outer product and the inner product, which gives the middle term of the trinomial, will be the factorization that gives a rectangular area:

Outer + inner   ┌─outer product─┐
      ↓   │ inner product │
                     ↓   ↓ ↓   ↓
$$x^2 + 3x + 2 = (x + 2)(x + 1)$$

This method of factorization is called FOIL.

▬▬▬ EXAMPLE 3

Factor $2x^2 + 7x + 6$ using FOIL.

Solution   This is the same as Example 1, but now we use FOIL.

First term: $2x^2 + 7x + 6 = (2x \quad)(x \quad)$

Last term: There are several possibilities:

$$(2x + 6)(x + 1)$$
$$(2x + 1)(x + 6)$$
$$(2x + 2)(x + 3)$$
$$(2x + 3)(x + 2)$$

These are equivalent to the different arrangements of "pieces" to form the rectangle in the area method shown in Example 1. Only one will form a rectangle and that is precisely the one that will give the middle term of the trinomial, namely, $7x$:

Middle term: $2x^2 + 7x + 6 = (2x + 3)(x + 2)$

This factorization is unique (except for the order of the factors).

Procedure for Factoring Trinomials

To factor a trinomial:

1. Find the factors of the second-degree term, and set up the binomials.
2. Find the factors of the constant term, and consider all possible binomials (mentally). Think of the factors that will form a rectangle.
3. Determine the factors that yield the correct middle term. If no pair of factors produces the correct full product, then the trinomial is not factorable using integers.

This type of factoring is called FOIL.

EXAMPLE 4

Factor, if possible.

a. $2x^2 + 11x + 12$ **b.** $x^2 + 3x + 5$

Solution

a. $2x^2 + 11x + 12 = (2x\quad)(x\quad)$
Try (mentally): 1, 12; 12, 1; 2, 6; 6, 2; 3, 4; and 4, 3. Use the pair that gives the middle term, $11x$:

$$2x^2 + 11x + 12 = (2x + 3)(x + 4)$$

b. $x^2 + 3x + 5 = (x\quad)(x\quad)$
Try (mentally): 1, 5; and 5, 1. Since neither of these gives the middle term, $3x$, we say that the trinomial is not factorable.

Common Factoring

If several terms share a factor, then that factor is called a **common factor**. For example, the binomial $5x^2 + 10x$ has three common factors: 5, x, and $5x$. To factor this sum of two terms, we must change it to a product, and this can be done several ways:

$$5x^2 + 10x = 5(x^2 + 2x)$$
$$5x^2 + 10x = x(5x + 10)$$
$$5x^2 + 10x = 5x(x + 2)$$

The last of these possibilities has the greatest common factor as a factor, and is said to be **completely factored.**

When combining common factoring with trinomial factoring, the procedure will be easiest if you look for common factors first.

■■■ EXAMPLE 5

Completely factor $6x^3 - 21x^2 - 12x$.

Solution

$$6x^3 - 21x^2 - 12x = 3x(2x^2 - 7x - 4) \quad \text{Common factor first}$$
$$= 3x(2x \quad)(x \quad) \quad \text{Now use FOIL, first terms}$$
$$= 3x(2x + 1)(x - 4) \quad \text{Last terms}$$ ▬

Difference of Squares

The last type of factorization we will consider is called a **difference of squares.** Suppose we start with one square, a^2:

From this square, we wish to subtract another square, b^2:

This gray square should be smaller than the first ($a^2 > b^2$). Place this square (since it is gray) on top of the larger square:

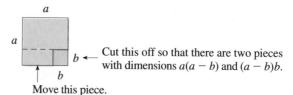

Cut this off so that there are two pieces with dimensions $a(a - b)$ and $(a - b)b$.

Move this piece.

Rearrange these two pieces by moving the smaller one from the bottom and positioning it vertically at the right:

$a - b$ [ ] [ ]   ← Moved piece

The dimension of this new arrangement is $(a - b)$ by $(a + b)$. Thus,

$$a^2 - b^2 = (a - b)(a + b)$$

Difference of Squares

$$a^2 - b^2 = (a - b)(a + b)$$

▬▬ EXAMPLE 6

Completely factor each expression, if possible.

a. $x^2 - 36$ **b.** $8x^2 - 50$ **c.** $6x^4 - 18x^2 - 24$

Solution

a. $x^2 - 36 = (x - 6)(x + 6)$ Difference of squares

b. $8x^2 - 50 = 2(4x^2 - 25)$ Common factor first

$\qquad\qquad = 2(2x - 5)(2x + 5)$ Difference of squares

c. $6x^4 - 18x^2 - 24 = 6(x^4 - 3x^2 - 4)$ Common factor

$\qquad\qquad\qquad = 6(x^2 - 4)(x^2 + 1)$ FOIL

$\qquad\qquad\qquad = 6(x - 2)(x + 2)(x^2 + 1)$ Difference of squares   ▬

PROBLEM SET 5.2

▲ A Problems

1. **IN YOUR OWN WORDS** Outline a procedure for factoring.

2. **IN YOUR OWN WORDS** What is a difference of squares?

If possible, completely factor the expressions in Problems 3–36.

3. $10xy - 6x$ 4. $5x + 5$ 5. $8xy - 6x$

6. $6x - 2$ 7. $x^2 - 4x + 3$ 8. $x^2 - 2x - 3$

9. $x^2 - 5x + 6$ 10. $x^2 - x - 6$ 11. $x^2 - 7x + 12$

12. $x^2 - 7x - 8$ 13. $x^2 - x - 30$ 14. $x^2 + 9x + 14$

15. $x^2 - 2x - 35$ 16. $x^2 - 6x - 16$ 17. $3x^2 + 7x - 10$

18. $2x^2 + 7x - 15$ 19. $2x^2 - 7x + 3$ 20. $3x^2 - 10x + 3$

21. $3x^2 - 5x - 2$ 22. $6y^2 - 7y + 1$ 23. $2x^2 + 9x + 4$

24. $7x^2 + 4x - 3$ 25. $3x^2 + x - 2$ 26. $3x^2 + 7x + 2$

27. $5x^3 + 7x^2 - 6x$ 28. $8x^3 + 12x^2 + 4x$ 29. $7x^4 + 11x^3 - 6x^2$

30. $3x^4 + 3x^3 - 36x^2$ 31. $x^2 - 64$ 32. $x^2 - 169$

33. $25x^2 + 50$ 34. $16x^2 - 25$ 35. $x^4 - 1$

36. $x^8 - 1$

▲ B Problems

Factor each expression in Problems 37–48 by using areas.

37. $x^2 + 5x + 6$ 38. $x^2 + 7x + 12$ 39. $x^2 + 4x + 3$ 40. $x^2 + 5x + 4$

41. $x^2 + 6x + 8$ 42. $x^2 + 2x + 2$ 43. $x^2 - 1$ 44. $x^2 - 4$

45. $x^2 + x - 2$ 46. $x^2 + x - 6$ 47. $x^2 - x - 2$ 48. $x^2 - x - 6$

49. The area of a rectangle is $x^2 - 2x - 143$ square feet. What are the dimensions of the figure?

50. What are the dimensions of a rectangle whose area is $x^2 - 4x - 165$ square feet?

51. What is the time needed to travel a distance of $6x^2 + 5x - 4$ miles if the rate is $3x + 4$ mph?

52. If an auditorium has $x^2 - 50x - 600$ seats arranged in a rectangular fashion, what is the number of rows and how many seats are there in each row?

Factor each expression in Problems 53–58, if possible.

53. $(x + 2)(x + 4) + (5x + 6)(x - 1)$ 54. $(2x + 1)(3x + 2) + (3x + 5)(x - 2)$

55. $x^6 - 13x^4 + 36x^2$ 56. $x^6 - 26x^4 + 25x^2$

57. $20x^2y^2 + 17x^2yz - 10x^2z^2$ 58. $12x^2y^2 + 10x^2yz - 12x^2z^2$

▲ Problem Solving

59. Pick any three consecutive integers — for example, 4, 5, and 6. The square of the middle term is 1 more than the product of the first and third; for example, 25 is 1 more than $4(6) = 24$. Prove this is true for any three consecutive integers.

60. Illustrate the property in Problem 59 geometrically.

▲ Individual Research

61. Write a paper on the relationship between geometric areas and algebraic expressions.
    References
    Albert B. Bennett, Jr., "Visual Thinking and Number Relationships," *The Mathematics Teacher*, April 1988.
    Robert L. Kimball, "Sharing Teaching Ideas: Using Pattern Analysis to Determine the Squares of Three Consecutive Integers," *The Mathematics Teacher*," January 1986.

5.3   EVALUATION, APPLICATIONS, AND SPREADSHEETS

If $x = a$, then x and a name the same number; x may then be replaced by a in any expression, and the value of the expression will remain unchanged. When you replace variables by given numerical values and then simplify the resulting numerical expression, the process is called **evaluating an expression.**

▇▇▇▇ EXAMPLE 1

Evaluate $a + cb$, where $a = 2$, $b = 11$, and $c = 3$.

Solution $a + cb$ *Remember: cb means c **times** b.*

Step 1.   Replace each variable with the corresponding numerical value. You may need additional parentheses to make sure you don't change the order of operations.

$$a + c \cdot b$$
$$\downarrow \ \downarrow \ \downarrow \ \downarrow \ \downarrow$$
$$2 + 3 \ (11) \qquad \leftarrow \text{Parentheses are necessary so that the product } cb$$
$$\text{is not changed to 311.}$$

Step 2.   Simplify:

$$2 + 3(11) = 2 + 33 \quad \text{Multiplication before addition}$$
$$= 35$$

▬

■■■■ **EXAMPLE 2**

Evaluate the following where $a = 3$ and $b = 4$.

a. $a^2 + b^2$ **b.** $(a + b)^2$

Solution

a. $a^2 + b^2 = 3^2 + 4^2$
$$= 9 + 16$$
$$= 25 \quad \text{Remember the order of operations:}$$
Multiplication comes first, and $3^2 = 3 \cdot 3$,
$4^2 = 4 \cdot 4$, which is multiplication.

b. $(a + b)^2 = (3 + 4)^2$
$$= 7^2$$
$$= 49 \quad \text{Order of operations; parentheses first.}$$

Notice that $a^2 + b^2 \neq (a + b)^2$.   ▬

Remember that a particular variable is replaced by a single value when an expression is evaluated. You should also be careful to write capital letters differently from lowercase letters, because they often represent different values. This means that you should not assume that $A = 3$ just because $a = 3$. On the other hand, it is possible that other variables *might* have the value 3. For example, just because $a = 3$, do not assume that another variable — say, t — cannot also have the value $t = 3$.

■■■■ **EXAMPLE 3**

Let $a = 1$, $b = 3$, $c = 2$, and $d = 4$. Find the value of the given capital letters.

a. $G = bc - a$ **b.** $H = 3c + 2d$ **c.** $I = 3a + 2b$ **d.** $R = a^2 + b^2d$

e. $S = \dfrac{2(b + d)}{2c}$ **f.** $T = \dfrac{3a + bc + b}{c}$

After you have found the value of a capital letter, write it in the box that corresponds to its numerical value. This exercise will help you check your work.

	37	9	5	14	6	

Solution

a. $G = bc - a$ **b.** $H = 3c + 2d$ **c.** $I = 3a + 2b$
  $= 3(2) - 1$ $= 3(2) + 2(4)$ $= 3(1) + 2(3)$
  $= 6 - 1$ $= 6 + 8$ $= 3 + 6$
  $= 5$ $= 14$ $= 9$

d. $R = a^2 + b^2 d$
$\quad = 1^2 + 3^2(4)$
$\quad = 1 + 9(4)$
$\quad = 37$

e. $S = \dfrac{2(b + d)}{2c}$
$\quad = \dfrac{2(3 + 4)}{2(2)}$
$\quad = \dfrac{2(7)}{4}$
$\quad = \dfrac{7}{2}$

f. $T = \dfrac{3a + bc + b}{c}$
$\quad = \dfrac{3(1) + 3(2) + 3}{2}$
$\quad = \dfrac{3 + 6 + 3}{2}$
$\quad = \dfrac{12}{2}$
$\quad = 6$

After you have filled in the appropriate boxes, the result is

37	9	5	14	6	
R	I	G	H	T	

In algebra, variables are usually represented by either lowercase or capital letters. However, in other disciplines, variables are often represented by other symbols or combinations of letters. For example, I recently took a flight on Delta Airlines and a formula $VM = \sqrt{A} \times 3.56$ was given as an approximation for the distance you can see from a Delta jet (or presumably any other plane). The article defined VM as the distance you can view in miles when flying at an altitude of A feet. For this example VM is interpreted as a single variable, and not as V *times* M as it normally would be in algebra.

An Application from Genetics

This application is based on the work of Gregor Mendel (1822–1884), an Austrian monk, who formulated the laws of heredity and genetics. Mendel's work was later amplified and explained by a mathematician, G. H. Hardy (1877–1947), and a physician, Wilhelm Weinberg (1862–1937). For years Mendel taught science without any teaching credentials because he had failed the biology portion of the licensing examination! His work, however, laid the foundation for the very important branch of biology known today as genetic science.

Assume that traits are determined by *genes,* which are passed from parents to their offspring. Each parent has a pair of genes, and the basic assumption is that each offspring inherits one gene from each parent to form the offspring's own pair. The genes are selected in a random, independent way. In our examples, we will assume that the researcher is studying a trait that is both easily identifiable (such as color of a rat's fur) and determined by a pair of genes consisting of a *dominant* gene, denoted by A, and a *recessive* gene, denoted by *a*. The possible pairings are called *genotypes:*

AA is called *dominant,* or homozygous.

A*a* is called *hybrid,* or heterozygous; genetically, the genotype *a*A is the same as A*a.*

aa is called *recessive.*

The physical appearance is called the *phenotype:*

Genotype *AA* has phenotype *A.*

Genotype *Aa* has phenotype *A* (since *A* is dominant).

Genotype *aA* has phenotype *A.*

Genotype *aa* has phenotype *a.*

In genetics, a square called a *Punnett square* is used to display genotype. For example, suppose two individuals with genotypes *Aa* are mated, as represented by the following Punnett square:

	Parent 2	
	A	*a*
A	*AA*	*Aa*
a	*aA*	*aa*

Parent 1

We see the result is *AA* + *Aa* + *aA* + *aa* = *AA* + 2*Aa* + *aa*. This reminds us of the binomial product

$$(p + q)^2 = p^2 + 2pq + q^2$$

Let's use binomial multiplication to find the genotypes and phenotypes of a particular example. In population genetics, we are interested in the percent, or frequency, of genes of a certain type in the entire population under study. In other words, imagine taking two genes from each person in the population and putting them into an imaginary pot. This pot is called the *gene pool* for the population. Geneticists study the gene pool to draw conclusions about the population.

▬▬▬ EXAMPLE 4

Suppose a certain population has two eye color genes: *B* (brown eyes, dominant) and *b* (blue eyes, recessive). Suppose we have an isolated population in which 70% of the genes in the gene pool are dominant *B*, and the other 30% are recessive *b*. What fraction of the population has each genotype? What percent of the population has each phenotype?

Solution   Let *p* = 0.7 and *q* = 0.3. Since *p* and *q* give us 100% of all the genes in the gene pool, we see that *p* + *q* = 1. Since

$$(p + q)^2 = p^2 + 2pq + q^2$$

We can find the percents:

genotype *BB*: $p^2 = (0.7)^2 = 0.49$,   so 49% have *BB* genotype
genotype *bB* or *Bb*: $2pq = 2(0.7)(0.3) = 0.42$,   so 42% have this genotype
genotype *bb*: $q^2 = (0.3)^2 = 0.09$,   so 9% have *bb* genotype
Check genotypes: $0.49 + 0.42 + 0.09 = 1.00$

the devices themselves, the electronics and mechanics are
referred to as hardware, but...

the directions that make the hardware perform operations
are known as

SOFTWARE

A computer's programs, plus the procedure for their use.

$\left(\begin{array}{c} \text{PROGRAM} \\ \text{A set of instructions for performing} \\ \text{computer operations.} \end{array} \right)$

As for the phenotypes, we look only at outward appearances, and since brown is dominant, *BB*, *bB*, and *Bb* all have brown eyes; this accounts for 91%, leaving 9% with blue eyes.

Spreadsheets

In Section 3.4 we discussed the notion of computer software and mentioned that one of the most important computer applications is in using something called a *spreadsheet*. A **spreadsheet** is a computer program used to manipulate data and carry out calculations, or chains of calculations. If you have access to a computer and software such as Excel, Lotus 1-2-3, or Quattro-Pro, you might use that software in conjunction with this section. However, it is not necessary to have this software (or even access to a computer) to be able to study variables and the evaluation of formulas using the ideas of a spreadsheet. In fact, your first inclination when reading this might be to skip over this and say to yourself, "I don't know anything about a spreadsheet, so I will not read this. Besides, my instructor is not requiring this anyway." However, regardless of whether this is assigned, chances are that sooner or later you will be using a spreadsheet.

One of the most interesting, and important, new ways of representing variables is as a **cell**, or a "box."

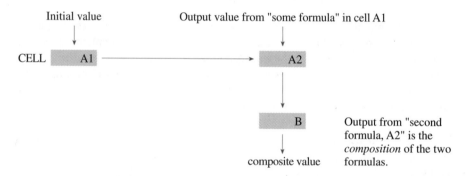

The information in a spreadsheet is stored in a rectangular array of **cells.** The content of each cell can be a number, text, or a formula. The power of a spreadsheet is that a cell's numeric value can be linked to the contents of another cell. For example, it is possible to define the content of one cell as the sum of the contents of two other cells. Furthermore, if the value of a cell is changed anywhere in the spreadsheet, all values dependent on it are recalculated and the new values are displayed immediately.

Instead of designating variables as letters (such as x, y, z, . . .) as we do in algebra, a spreadsheet designates variables as cells (such as B2, A5, Z146, . . .). If you type something into a cell, the spreadsheet program will recognize it as text if it begins with a letter, and as a number if it begins with a numeral. It also recognizes the usual mathematical symbols of $+$, $-$, $*$ (for $\times$), $/$ (for $\div$), and $\wedge$ (for raising to a power). Parentheses are used in the usual fashion as grouping symbols. To enter a formula, you must begin with $+$ or @. Compare some algebraic and spreadsheet evaluations:

Algebra	*Comment*	*Spreadsheet*	*Comment*
$3(x + y)$	Variables x and y	$3*(A1 + A2)$	Variables are in contents of cells A1 and A2.
$x^2 + 2x - 5$	Variable is x	$+B3\wedge2+2*B3-5$	Variable is the content of cell B3. Begins with + to indicate that it is a formula.
$\dfrac{A + B}{2}$	Formula for the average of variables A and B	$+(A3 + A4)/2$	Variables are the contents of cells A3 and A4.

▬▬ EXAMPLE 5

Translate each formula into spreadsheet notation.

a. $x + \dfrac{y}{2}$ **b.** $\dfrac{6x^2 + y}{2x}$ **c.** $-3^2 + (-5)^2 + z^3 + (z + 3)^2$

Solution

a. $+A1 + A2/2$ where the value of x is the content of cell A1 and the value of y is the content of cell A2

b. $+(6A1\wedge2 + A2)/(2A1)$ where the value of x is in cell A1 and the value of y is in cell A2

c. $+(-3\wedge2) + (-5)\wedge2 + A3\wedge3 + (A3 + 3)\wedge2$ where the value of z is the content of cell A3   ▬

For our purposes, we will assume that a spreadsheet program has an almost unlimited number of rows and columns. We will represent a typical spreadsheet as follows:

	A	B	C	D	E	F	G
1							
2							
3							
4							
5							

As an example, we will consider the way in which a spreadsheet program could be used to set up an electronic checkbook. We might fill in the spreadsheet as follows:

	A	B	C	D	E	F	G
1	DESCRIPTION	DEBIT	DEPOSIT	BALANCE			
2	Beginning balance						
3				$+D2-B3+C3$			
4				$+D3-B4+C4$			
5				$+D4-B5+C5$			

After some entries are filled in, the spreadsheet might look like the following:

	A	B	C	D	E	F	G
1	DESCRIPTION	DEBIT	DEPOSIT	BALANCE			
2	Beginning balance			1000.00			
3	School bookstore	250.00		750.00			
4	Paper route		100.00	850.00			
5	Ski trip	300.00		550.00			

The power of a spreadsheet derives from the way variables are referenced by cells. For example, if you go back to the spreadsheet and enter a beginning balance of $2,500 in cell D2, *all* the other entries *automatically* change:

	A	B	C	D	E	F	G
1	DESCRIPTION	DEBIT	DEPOSIT	BALANCE			
2	Beginning balance			2500.00			
3	School bookstore	250.00		2250.00			
4	Paper route		100.00	2350.00			
5	Ski trip	300.00		2050.00			

Once the spreadsheet has been set up, the user will enter information in column A and, depending on whether a check has been written or a deposit made, make an entry in either column B or column C. The entries in column D (beginning with cell D3) are automatically calculated by the spreadsheet program. Empty cells are assumed to have the value 0.

It should be clear that each cell from D3 downward needs to contain a different formula. In this example, the column letters and operations of the formula remain unchanged, but each row number is increased by 1 from the cell above. If each of these formulas had to be entered by hand, one by one, it is obvious that setting up a spreadsheet would be very time-consuming. This is not the case, however, and a typical spreadsheet program allows the user to copy the formula from one cell into another cell and at the same time *automatically* change its formula references. Thus, with a single command, each cell in column D is given the correct formula. We will call this command **replicating** a formula or cell. Formulas replicated down a column have their row numbers incremented, and formulas replicated across a row have the column letters incremented.

■ EXAMPLE 6

Consider the following spreadsheet with the indicated formulas.

	A	B	C	D	E	F	G
1	x	$y = 10x$	$x + y$	x^2			
2		+10*A2	+A2+B2	+A2^2			
3	+A2+1	+10*A3	+A3+B3	+A3^2			
4							
5							

a. Describe what the spreadsheet would show if cells A3 . . . D3 were replicated in rows 4 and 5.

b. Describe what the spreadsheet would show if column D were replicated in column E.

Solution

a.

	A	B	C	D	E	F	G
1	x	$y = 10x$	$x + y$	x^2			
2		+10*A2	+A2+B2	+A2^2			
3	+A2+1	+10*A3	+A3+B3	+A3^2			
4	+A3+1	+10*A4	+A4+B4	+A4^2			
5	+A4+1	+10*A5	+A5+B5	+A5^2			

b.

	A	B	C	D	E	F	G
1	x	$y = 10x$	$x + y$	x^2	x^2		
2		+10*A2	+A2+B2	+A2^2	+B2^2		
3	+A2+1	+10*A3	+A3+B3	+A3^2	+B3^2		
4							
5							

▬▬▬ EXAMPLE 7

Given that cell A2 has the value 12, show what the spreadsheet shown in the solution for Example 6a would look like.

Solution

	A	B	C	D	E	F	G
1	x	$y = 10x$	$x + y$	x^2			
2	12	120	132	144			
3	13	130	143	169			
4	14	140	154	196			
5	15	150	165	225			

We might once again remind you of the power of a computer spreadsheet. Note that the *entire* answer shown in Example 7 would *immediately* be filled in as soon as you fill in the number 12 in cell A2. If you now go back and reenter another number into cell A2, the entire spreadsheet would *immediately* change because every cell is ultimately defined in terms of the content of cell A2 in this example spreadsheet.

	A	B
1	1	
2	1	
3	2	
4	3	
5	5	
6	8	
7	13	
8	21	
9	34	
10	55	
11		

▬▬▬ EXAMPLE 8

Suppose we look at a spreadsheet and use the entries 1, 1, 2, 3, 5, 8, 13, 21, 34, 55 in successive entries in column A. Suppose that 1 is entered into cell A1 and 1 into cell A2. What is the formula to be placed into cell A3?

Solution   Cell A3 should contain the formula +A1 + A2. Note that if this formula were replicated down column A through cell A10, the given numbers would be shown.

The real power of a spreadsheet program lies not in its ability to perform calculations, but rather in its ability to answer "what if" types of questions. For instance, for the set of numbers in Example 8, what if the number in cell A1 is divided by A2, and then the number in cell A2 is divided by A3, and so on for 50 such numbers? In the problem set you are asked to use a spreadsheet to detect a pattern for these quotients.

■■■ EXAMPLE 9

If $100 is deposited in an account that pays 5% interest compounded yearly, then at the end of the first year, the account will contain $100 + (0.05)*100 = \$105$. At the end of two years, the account will contain $105 + (0.05)*105 = \$110.25$. Suppose cell A1 contains the value 100. Then what formula must be placed in cell A2 if it is to contain the amount in the account at the end of the first year?

Solution   Cell A2 should contain the formula $A1 + (0.05)*A1$. Note that if this formula were replicated down column A to cell A11, the column would show the amount in the account at the end of each year through 10 years.   ■

To take advantage of the "what if" power of a spreadsheet, the previous example could be set up to allow for a variable interest rate. This can be done in the following way:

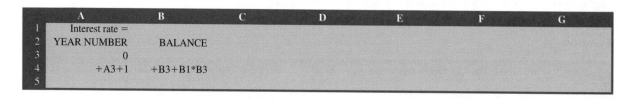

	A	B	C	D	E	F	G
1	Interest rate =						
2	YEAR NUMBER	BALANCE					
3	0						
4	+A3+1	+B3+B1*B3					
5							

Note that when a number is inserted into cell B1, that number will act as the interest rate, and the number inserted into cell B3 will act as the amount of deposit. Therefore, we see that B1 is the variable representing the interest rate and B3 is the variable representing the beginning balance. If row 4 is replicated into rows 5 to 10, we will then obtain the data for the next 7 years. The problem with this replication, however, is that the reference to cell B1 will change as the replication takes place down column B. This difficulty is overcome by using a special character that holds a column or a row constant. The symbol we will use for this purpose is #. (Many spreadsheets use the symbol $ for this purpose.) Thus, we would change the formula in cell B4 above to

$$+B3 + \#B\#1*B3$$

to mean that we want not only column B to remain constant, but also we want row 1 to remain unchanged when this entry is replicated. Note that # applies only to the character directly following its placement. We show the first 7 rows (4 years) of such a spreadsheet for which the rate is 4% and the initial deposit is $1,000.

	A	B	C	D	E	F	G
1	Interest rate =	0.04					
2	YEAR NUMBER	BALANCE					
3	0	1000.00					
4	1	1040.00					
5	2	1081.60					
6	3	1124.86					
7	4	1169.86					

PROBLEM SET   5.3

▲ A Problems

1. **IN YOUR OWN WORDS** What is a variable?

2. **IN YOUR OWN WORDS** What is a spreadsheet?

3. **IN YOUR OWN WORDS** A colleague of mine speculated over lunch a few years ago:

   "Someday there will be just one programming language. IBM and Macintosh formats will merge. The trend is just the same as it is with the languages we speak in the world. Someday the entire world will speak English."

   Comment on this statement.

4. **IN YOUR OWN WORDS** Although the terms *program* and *software* can often be interchanged, there is a subtle difference in their usage. Discuss. What is meant when the term *user-friendly* is used to describe software?

In Problems 5–12, write each expression in spreadsheet notation. Let x be in cell A1, y in A2, and z in A3.

5. **a.** $\frac{2}{3}x^2$ **b.** $5x^2 - 6y^2$ 6. **a.** $3x^2 - 17$ **b.** $14y^2 + 12x^2$

7. **a.** $12(x^2 + 4)$ **b.** $\dfrac{15x + 7}{2}$ 8. **a.** $\dfrac{3x + 1}{12}$ **b.** $3x + \frac{1}{12}$

9. **a.** $(5 - x)(x + 3)^2$ **b.** $6(x + 3)(2x - 7)^2$

10. **a.** $(x + 1)(2x - 3)(x^2 + 4)$ **b.** $(2x - 3)(3x^2 + 1)$

11. **a.** $\frac{1}{4}x^2 - \frac{1}{2}x + 12$ **b.** $\frac{2}{3}x^2 + \frac{1}{3}x - 17$

12. **a.** $1 - \dfrac{x}{yz}$ **b.** $\dfrac{1 - x}{yz}$

In Problems 13–18, write each spreadsheet expression in ordinary algebraic notation. Let cell A1 represent the variable x, A2 the variable y, A3 the variable z; B1 is a, B2 is b, and B3 is c.

13. **a.** 4*A1+3   **b.** 5*A1^2−3*A1+4

14. **a.** 36*A3^2−13*A3+2   **b.** 13*A2^2+(15/2)

15. **a.** +(5/4)*A1+14^2   **b.** +(5/4*A1+14)^2

16. **a.** 5*A1^2−3*A2+4   **b.** 4*(A1^2+5)*(3*A1^2−3)^2

17. **a.** +A1/A2*A3   **b.** +A1/(A2*A3)

18. **a.** $+B2 + B1*\#B\#3$ **b.** $+A1 + \#B2^{\wedge}2$

In Problems 19–42, let $w = 2$, $x = 1$, $y = 2$, *and* $z = 4$. *Find the values of the given capital letters.*

19. $A = x + z + 8$

20. $B = 5x + y - z$

21. $C = 10 - w$

22. $D = 3z$

23. $E = 25 - y^2$

24. $F = w(y - x + wz)$

25. $G = 5x + 3z + 2$

26. $H = 3x + 2w$

27. $I = 5y - 2z$

28. $J = 2w - z$

29. $K = wxy$

30. $L = x + y^2$

31. $M = (x + y)^2$

32. $N = x^2 + 2xy + y^2 + 1$

33. $P = y^2 + z^2$

34. $Q = w(x + y)$

35. $R = z^2 - y^2 - x^2$

36. $S = (x + y + z)^2$

37. $T = x^2 + y^2 z$

38. $U = \dfrac{w + y}{z}$

39. $V = \dfrac{3wyz}{x}$

40. $W = \dfrac{3w + 6z}{xy}$

41. $X = (x^2 z + x)^2 z$

42. $Y = (wy)^2 + w^2 y + 3x$

▲ **B Problems**

43. This problem will help you check your work in Problems 19–42. Fill in the capital letters from Problems 19–42 to correspond with their numerical values (the letter O has been filled in for you). Some letters may not appear in the boxes. When you are finished, darken all the blank spaces to separate the words in the secret message. Notice that some of the blank spaces have also been filled in to help you.

13	5	19	21	3	11	13	14	2	49	22	50
17	7	21	23	19	11	21	13	17	21	49	17
5	13	3	O	11		49	13	48	2	10	19
12	21	48	2	8	21	26	21	48	21	11	22
2	10	48	21	10	17	21	12				

44. A certain population has two eye color genes: B (brown eyes, dominant) and b (blue eyes, recessive). Suppose we have an isolated population in which 75% of the genes in the gene pool are dominant B, and the other 25% are recessive b. What percent of the population has each genotype? What percent of the population has each phenotype?

45. A certain population has two fur color genes: B (black, dominant) and b (brown, recessive). Suppose we have an isolated population in which 65% of the genes in the gene pool are dominant B, and the other 35% are recessive b. What percent of the population has each genotype? What percent of the population has each phenotype?

46. A population of self-pollinating pea plants has two genes: T (tall, dominant) and t (short, recessive). Suppose we have an isolated population in which 50% of the genes

in the gene pool are dominant T, and the other 50% are recessive t. What percent of the population has each genotype? What percent of the population has each phenotype?

47. When flowers known as "four-o'clocks" are crossed, there is an incomplete dominance. A four-o'clock is red if the genotype is rr and white if it is ww. If rr is crossed with ww, a hybrid rw results. Suppose a population of 20% red four-o'clocks is mixed with a population of 80% white four-o'clocks. What percent of the population has each genotype? What percent of the population has each phenotype?

Suppose that a spreadsheet contains the following values:

	A	B	C	D	E
1	6	−4	2	3	
2					
3					
4					
5					

Determine the value of cell E1 *if it contains the formula given in Problems* 48–53.

48. +A1 + B1   49. 2A1 + 3*B1   50. +C1*(A1 + 3*B1)

51. +A1 − B1/C1   52. +(A1 − B1)/C1   53. +A1/C1*D1

Draw a spreadsheet like the one shown and fill in the values of the missing cells assuming that the formula given in Problems 54–59 *has been entered in cell* C1 *and replicated across row* 1.

	A	B	C	D	E
1	1	3			
2					
3					
4					
5					

54. +A1 + B1   55. +A1 − B1   56. +B1 − A1

57. +#A#1 + B1   58. +#A#1*B1   59. +A1*#B#1

In Problems 60–63, *consider the following spreadsheet, in which the formula for each cell except* B1 *is displayed. Determine the value of each cell given the value in cell* B1.

	A	B	C	D
1	+C2+1		+B2+1	
2	+C3+1	+A3+1	+C1+1	
3	+A2+1	+A1+1	+B1+1	

60. 1   61. −10   62. 0   63. 100

▲ **Problem Solving**

64. What do you notice about the nine cells of the spreadsheet given in Problems 60–63?

65. A certain population has two fur color genes: B (black, dominant) and b (brown, recessive). Suppose you look at a population that is 25% brown and 75% black. Estimate the percentages of the gene pool that are B and b.

66. A population of self-pollinating pea plants has two genes: T (tall, dominant) and t (short, recessive). Suppose you look at a population that is 36% short. Estimate the percentages of the gene pool that are T and t.

67. Earlobes can be characterized as attached or free hanging. Free hanging (F) are dominant, and attached (f) are recessive. Survey your class to determine the percentage that are attached; let q^2 be this number between 0 and 1; this represents the ff genotype. Estimate the gene pool for your class.

68. There are three genes in the gene pool for blood: A, B, and O. Two of these three are present in a person's blood; A and B dominate O, whereas A and B are codominant. This gives the following possibilities:

genotype	phenotype
AA	type A blood
AO	type A blood
AB	type AB blood
BO	type B blood
BB	type B blood
OO	type O blood

Let p, q, and r represent the frequencies of the genes A, B, and O in the blood gene pool. Suppose a certain population has 20% type A, 30% type B, and 50% type O. Construct a Punnett square and find $(p + q + r)^2$ to answer the following questions. What percent of the population has each genotype? What percent of the population has each phenotype?

69. The owner of an auto dealership would like to create a spreadsheet in which sales performance information regarding each salesperson can be tabulated. She would like to use the following format:

	A	B	C	D	E
1	NAME	SALES	COST	PROFIT	COMMISSION
2	John Adams	100,000	80,000	20,000	1,600
3					
4					
5					

Column A is to contain the name of each salesperson; column B, the gross sales; column C, the cost; column D, the profit; and column E, the commission, calculated at 8% of the profit. Construct a spreadsheet for 20 employees.

70. A math instructor calculates semester grades by doubling the sum of three midterm exams, multiplying the homework score by 1.5, multiplying the final exam score by 3.5, and adding these three results. Set up a spreadsheet for doing these calculations using the following format:

	A	B	C	D	E	F	G	H
1	NAME	EXAM 1	EXAM 2	EXAM 3	TOT EXAM	HW	FINAL	SEM TOT
2	Debra Bailey	80	75	85	240	80	90	915
3								
4								
5								

Construct a spreadsheet for 30 students.

71. Write a spreadsheet program to calculate the quotients of consecutive Fibonacci numbers.

▲ **Individual Research**

72. Write a history of computer programming. Include information about the Jacquard loom, Ada Byron, who is also known as Lady Lovelace, John von Neumann, Grace Hopper, as well as computer languages known as BASIC, LOGO, PASCAL, COBOL, ADA, PROLOG, C, FORTRAN, and JAVA.

5.4 EQUATIONS

Even though there are many aspects of algebra that are important to the scientist and mathematician, the ability to solve simple equations is important to the lay person and can be used in a variety of everyday applications.

An **equation** is a statement of equality that may be true or false or may depend on the value of a variable. If it is always true, as in

$$2 + 3 = 5$$

then it is called an *identity*. If it is always false, as in

$$2 + 3 = 15$$

then it is called a *contradiction*. If it depends on the value of a variable, as in

$$2 + x = 15$$

then it is a *conditional equation*. The values that make the conditional equation true are said to **satisfy** the equation and are called the **solutions** or **roots** of the equation. Generally, when we speak of equations we mean conditional equations. Our concern when solving equations is to find the numbers that satisfy a given equation, so we look for things to do to equations to make the solutions or roots more obvious. Two equations with the same solutions are called **equivalent equations.** An equivalent equation may be easier to solve than the original equation, so we try to get successively simpler equivalent equations until the solution is obvious. There are certain procedures you can use to create equivalent equations. In this section, we will discuss solving the two most common types of equations you will encounter: *linear* and *quadratic*.

Linear equations:	$ax + b = 0$	$(a \neq 0)$
Quadratic equations:	$ax^2 + bx + c = 0$	$(a \neq 0)$

Linear Equations

To solve a linear equation, you can use one or more of the following equation properties.

Equation Properties

Addition property	Adding the same number to both sides of an equation results in an equivalent equation.
Subtraction property	Subtracting the same number from both sides of an equation results in an equivalent equation.
Multiplication property	Multiplying both sides of a given equation by the same nonzero number results in an equivalent equation.
Division property	Dividing both sides of a given equation by the same nonzero number results in an equivalent equation.

When these properties are used to obtain equivalent equations, **the goal is to isolate the variable on one side of the equation,** as illustrated in Example 1. You can always check the solution to see whether it is correct; substituting the solution into the original equation will verify that it satisfies the equation. Notice how the field properties are used when solving these equations.

■■■ EXAMPLE 1

Solve the given equations.

a. $x + 15 = 25$ **b.** $x - 36 = 42$ **c.** $3x = 75$ **d.** $\dfrac{x}{5} = -12$

e. $15 - x = 0$

⊘ The goal is to isolate the variable on one side of the equal sign. ⊘

Solution

a.
$$x + 15 = 25 \quad \text{Given equation}$$
$$x + 15 - 15 = 25 - 15 \quad \text{Subtract 15 from both sides.}$$
$$x = 10 \quad \text{Carry out the simplification.}$$

The root (solution) of this simpler equivalent equation is now obvious (it is 10). We often display the answer in the form of this simpler equation, $x = 10$, with the variable isolated on one side.

⊘ Perform the opposite operation to find a simpler equivalent equation. ⊘

b.
$$x - 36 = 42 \quad \text{Given equation}$$
$$x - 36 + 36 = 42 + 36 \quad \text{Add 36 to both sides.}$$
$$x = 78 \quad \text{Simplify.}$$

c.
$$3x = 75 \quad \text{Given}$$
$$\frac{3x}{3} = \frac{75}{3} \quad \text{Divide both sides by 3.}$$
$$x = 25 \quad \text{Simplify.}$$

d.
$$\frac{x}{5} = -12 \qquad \text{Given}$$

$$5\left(\frac{x}{5}\right) = 5(-12) \qquad \text{Multiply both sides by 5.}$$

$$x = -60 \qquad \text{Simplify.}$$

e.
$$15 - x = 0 \qquad \text{Given}$$

$$15 - x + x = 0 + x \qquad \text{Add } x \text{ to both sides.}$$

$$15 = x \qquad \text{Simplify.}$$

The equation $15 = x$ is the same as $x = 15$. This is a general property of equality called the **symmetric property of equality:** If $a = b$, then $b = a$.

While Example 1 illustrates the basic properties of equations, you will need to solve more complicated linear equations. In the following examples, some of the steps are left for mental calculations.

■ EXAMPLE 2

Find the root of the given equations.

a. $5x + 3 - 4x = 6 + 9$ **b.** $6x + 3 - 5x - 7 = 11(-2)$

c. $5x + 2 = 4x - 7$ **d.** $4x + x = 20$

e. $3(m + 4) + 5 = 5(m - 1) - 2$

Solution

a.
$$5x + 3 - 4x = 6 + 9 \qquad \text{Given}$$
$$x + 3 = 15 \qquad \text{Simplify.}$$
$$x = 12 \qquad \text{Subtract 3 from both sides.}$$

b.
$$6x + 3 - 5x - 7 = 11(-2) \qquad \text{Given}$$
$$x - 4 = -22 \qquad \text{Simplify.}$$
$$x = -18 \qquad \text{Add 4 to both sides.}$$

c.
$$5x + 2 = 4x - 7 \qquad \text{Given}$$
$$x + 2 = -7 \qquad \text{Subtract } 4x \text{ from both sides.}$$
$$x = -9 \qquad \text{Subtract 2 from both sides.}$$

d.
$$4x + x = 20 \qquad \text{Given}$$
$$5x = 20 \qquad \text{Simplify.}$$
$$x = 4 \qquad \text{Divide both sides by 5.}$$

e.
$$3(m + 4) + 5 = 5(m - 1) - 2 \qquad \text{Given}$$
$$3m + 12 + 5 = 5m - 5 - 2 \qquad \text{Simplify (distributive property).}$$
$$3m + 17 = 5m - 7 \qquad \text{Simplify.}$$
$$17 = 2m - 7 \qquad \text{Subtract } 3m \text{ from both sides.}$$
$$24 = 2m \qquad \text{Add 7 to both sides.}$$
$$12 = m \qquad \text{Divide both sides by 2.}$$

Quadratic Equations

To solve quadratic equations, you must first use the equation properties to write the equation in the form

$$ax^2 + bx + c = 0$$

There are two commonly used methods for solving quadratic equations. The first uses factoring, and the second uses the quadratic formula. Both of these methods require that you apply the linear equation properties to obtain a 0 on one side. Next, look to see whether the polynomial is factorable. If so, use the zero-product rule to set each factor equal to 0, and then solve each of those equations. If the polynomial is not factorable, then use the quadratic formula.

Zero-Product Rule

If $A \cdot B = 0$, then $A = 0$ or $B = 0$, or $A = B = 0$.

If the product of two numbers is 0, then at least one of the factors must be 0.

⊘ With a quadratic equation, first obtain a 0 on one side. ⊘

◼◼◼◼ EXAMPLE 3

Solve each equation.

a. $x^2 = x$ **b.** $x(x - 8) = 4(x - 9)$

Solution

a.

$x^2 = x$	Given
$x^2 - x = 0$	Subtract x from both sides.
$x(x - 1) = 0$	Factor, if possible.
$x = 0, \quad x - 1 = 0$	Zero-product rule; set each factor equal to 0.
$x = 1$	Solve each of the resulting equations.

The equation has two roots, $x = 0$ and $x = 1$. Usually you will set each factor equal to 0 and solve mentally.

b.

$x(x - 8) = 4(x - 9)$	Given
$x^2 - 8x = 4x - 36$	Simplify.
$x^2 - 12x + 36 = 0$	Subtract $4x$ from both sides; add 36 to both sides.
$(x - 6)(x - 6) = 0$	Factor.
$x = 6$	Set each factor equal to 0 and mentally solve. Since the factors are the same, there is one root, 6. In such a case we say the root of 6 has *multiplicity two*. ◼

If the quadratic expression is not easily factorable (after you obtain a 0 on one side), then you can use the quadratic formula, which is derived in most high school algebra books.

Quadratic Formula

If $ax^2 + bx + c = 0$, $a \neq 0$, then

$$x = \frac{-b \pm \sqrt{b^2 - 4ac}}{2a}$$

Historical Note

The convention followed in this section of using the last letters of the alphabet for variables and the first letters for constants is credited to René Descartes (1596–1650). Many symbols had been used before his time. In the earliest forms of algebra, statements were written out in words. "A certain number" gave way to the names of colors that were used by some Arabian mathematicians. *Cosa, censo, cubo* were abbreviated by *co, ce, cu*, then by Q, S, C before Descartes used x, x^2, and x^3. The quadratic formula, as you see it today, is the result of a slow and tedious evolution of ideas and symbols and is not the work of a single person.

▬▬ EXAMPLE 4

Solve the given equations.

a. $2x^2 + 4x + 1 = 0$ **b.** $x^2 = 6x - 13$

Solution

a. Note that $2x^2 + 4x + 1 = 0$ has a 0 on one side and also that the left-hand side does not easily factor, so we will use the quadratic formula. We begin by (mentally) identifying $a = 2$, $b = 4$, and $c = 1$.

$$x = \frac{-(4) \pm \sqrt{(4)^2 - 4(2)(1)}}{2(2)}$$ Substitute for *a*, *b*, and *c* in the quadratic formula.

$$= \frac{-4 \pm \sqrt{8}}{4}$$ Simplify under the square root.

$$= \frac{-4 \pm 2\sqrt{2}}{4}$$ Simplify radical.

$$= \frac{2(-2 \pm \sqrt{2})}{4}$$ Factor a 2 out of the numerator so that we can reduce the fraction. This step is usually done mentally.

$$= \frac{-2 \pm \sqrt{2}}{2}$$

b.

$$x^2 = 6x - 13$$ Given

$$x^2 - 6x + 13 = 0$$ Obtain a 0 on one side.

$$x = \frac{-(-6) \pm \sqrt{(-6)^2 - 4(1)(13)}}{2(1)}$$

$$= \frac{6 \pm \sqrt{-16}}{2}$$

The square root of a negative number is not defined in the set of real numbers. Thus, since we are working in the set of real numbers, we say there is no solution. ▬

Since most of us have access to a calculator, we often estimate the roots of quadratic equations with radicals as rational (decimal) approximations. We illustrate this with the next calculator example.

▬▬ EXAMPLE 5

Solve $5x^2 + 2x - 2 = 0$ and approximate the roots to the nearest hundredth.

Solution　From the quadratic formula, where $a = 5$, $b = 2$, and $c = -2$:

$$x = \frac{-b \pm \sqrt{b^2 - 4ac}}{2a}$$

$$= \frac{-2 \pm \sqrt{2^2 - 4(5)(-2)}}{2(5)} \approx 0.46, -0.86$$

Calculator Comment

We can use calculators to help us solve quadratic equations. Since there are many types of calculators, we can only offer some suggestions, and you will need to check your owner's manual.

Solve $5x^2 + 2x - 2 = 0$ using an **algebraic calculator.** To approximate the roots:

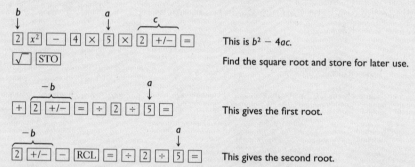

This is $b^2 - 4ac$.

Find the square root and store for later use.

This gives the first root.

This gives the second root.

Some of the steps shown here could be combined because these are simple numbers. These steps give the numerical approximation for a quadratic equation with real roots. For this quadratic equation the roots are (to four decimal places) 0.4633 and -0.8633.

Since you will have occasion to use the quadratic formula over and over again, and since many calculators have programming capabilities, this is a good time to consider writing a simple program to give the real roots for a quadratic equation. First write the equation in the form $ax^2 + bx + c = 0$, input the a, b, and c values into the calculator as A, B, and C. The program will then output the two real values (if they exist). Each brand of graphing calculator is somewhat different, but it is instructive to illustrate the general process. Press the PRGM key. You will then be asked to name the program; we call our program QUAD. Next, input the formula for the two roots (from the quadratic formula). Finally, display the answer:

:($-$B$+\sqrt{\ }$(B^2$-$4AC))/(2A)

:Disp Ans

:($-$B$-\sqrt{\ }$(B^2$-$4AC))/(2A)

:Disp Ans

For Example 5, input the A, B, and C values as follows:

5 STO→ A　2 STO→ B　-2 STO→ C　PRGM QUAD

Then run the program for the DISPLAY:　.4633249581

$-$.8633249581

News Clip

PIERRE DE FERMAT
(1601–1665)

Fermat's Last Theorem

Pierre de Fermat was a lawyer by profession, but he was an amateur mathematician in his spare time. He became Europe's finest mathematician, and he wrote well over 3,000 mathematical papers and notes. However, he published only one, because he did them just for fun. Every theorem that Fermat said he proved has subsequently been verified — with one defying solution until 1993! This problem, known as Fermat's Last Theorem (see Problems 73 and 74) was written by Fermat in the margin of a book:

> To divide a cube into two cubes, a fourth power, or in general any power whatever above the second, into powers of the same denomination, is impossible, and I have assuredly found an admirable proof of this, but the margin is too narrow to contain it.

Many of the most prominent mathematicians since his time have tried to prove or disprove this conjecture, and on June 23, 1993, during the third of a series of lectures at a conference held at the Newton Institute in Cambridge, it was reported that British mathematician Andrew Wiles of Princeton had proved a theorem for which Fermat's Last Theorem is a corollary.

Fermat's Last Theorem has been mentioned in literary works ranging from *Sherlock Holmes* to *Star Trek: The Next Generation*. In 1983, a 23-year-old German mathematician, Gerd Faltings, proved that the number of possible exceptions is finite, and in 1988 there were reports in the *Los Angeles Times* and *The Chronicle of Higher Education* that a Japanese mathematician, Yoichi Miyaoka, proved the famous theorem, but alas, there were errors. In 1989, *Discover* proclaimed, "Fermat Still Has the Last Laugh," and *Time* reported the error in an article (April 18, 1988) entitled "The Joy of Math, or Fermat's Revenge." History repeated itself once again when it was reported that Wiles had proved this elusive theorem (*Focus*, the newsletter of the Mathematical Association of America, August 1993), but by 1994 the editor of *Focus*, Keith Devlin, reported that the earlier report was premature. He proclaims, "For most mathematicians, the only thing to do is wait and see" (*Focus*, April 1994, p. 3).

PROBLEM SET 5.4

▲ A Problems

1. **IN YOUR OWN WORDS** Describe a procedure for solving first-degree equations.

2. **IN YOUR OWN WORDS** Describe a procedure for solving second-degree equations.

Solve the equations in Problems 3–22.

3. **a.** $x - 5 = 10$ **b.** $x - 8 = 14$ **c.** $6 = x - 2$

4. **a.** $-13 = x - 49$ **b.** $x + 2 = 7$ **c.** $x + 8 = 2$

5. **a.** $8 + x = 4$ **b.** $12 + x = 15$ **c.** $18 + x = 10$

6. **a.** $\dfrac{x}{4} = 8$ **b.** $\dfrac{x}{2} = 18$ **c.** $\dfrac{x}{-4} = 11$

7. **a.** $7 = \dfrac{x}{-8}$ **b.** $-4 = \dfrac{x}{-10}$ **c.** $12 = \dfrac{x}{-9}$

8. **a.** $4x = 12$ **b.** $-8x = -96$ **c.** $-x = 5$

9. a. $13x = 0$ b. $-19x = 0$ c. $-\frac{1}{2}x = 0$

10. a. $A + 13 = 18$ b. $5 = 3 + B$ c. $-5X = -1$

11. a. $6 = C - 4$ b. $2D + 2 = 10$ c. $15E - 5 = 0$

12. a. $16F - 5 = 11$ b. $6 = 5G - 24$ c. $4(H + 1) = 4$

13. a. $5(I - 7) = 0$ b. $\dfrac{I}{5} = 3$ c. $\dfrac{2K}{3} = 6$

14. a. $\dfrac{3L}{4} = 5$ b. $1 = \dfrac{2M}{3} + 7$ c. $7 = \dfrac{2N}{3} + 11$

15. a. $-5 = \dfrac{2P + 1}{3}$ b. $\dfrac{2 - 5Q}{3} = 4$ c. $\dfrac{5R - 1}{2} = 5$

16. $5(6S - 81) = -3(15 + 5S)$ 17. $3T + 3(T + 2) + (T + 4) + 11 = 0$

18. $3(U - 3) - 2(U - 12) = 18$ 19. $6(V - 2) - 4(V + 3) = 10 - 42$

20. $5(W + 3) - 6(W + 5) = 0$ 21. $6(Y + 2) = 4 + 5(Y - 4)$

22. $5(Z - 2) - 3(Z + 3) = 9$

▲ B Problems

23. This problem should help you check your work in Problems 10–22. Fill in the capital letters from Problems 10–22 to correspond with their numerical values (the letter O has been filled in for you). Some letters may not appear in the boxes. When you are finished, darken all the blank spaces to separate the words in the secret message. Notice that one of the blank spaces has also been filled in to help you.

12	−3	0	7	8	−7	−8	$\frac{11}{5}$	O	2	$\frac{20}{3}$	$\frac{1}{3}$	−9	$\frac{4}{5}$		11
−15	7	$\frac{20}{3}$	$\frac{20}{3}$	−1	3	−6	4	O	3	2	−3	$\frac{1}{3}$	4	$\frac{20}{3}$	−28
−1	5	8	8	7	8	−3	−5	−28	O	3	$\frac{1}{6}$	$\frac{1}{6}$	7	−6	12
1	7	−6	4	7	−6	6	$\frac{1}{4}$	$\frac{1}{3}$	$\frac{11}{5}$	$\frac{11}{5}$	O	$\frac{11}{5}$	8	!	!

Solve the equations in Problems 24–45.

24. $x^2 = 10x$ 25. $x^2 = 14x$ 26. $5x + 66 = x^2$

27. $15x^2 + 4x = 4$ 28. $x^3 = 4x$ 29. $x^3 = x$

30. $4x(x - 9) = 9(1 - 4x)$ 31. $4(9x - 1) = 9x(4 - x)$ 32. $x^2 + 7x + 2 = 0$

33. $x^2 - 3x + 1 = 0$ 34. $x^2 - 5x - 3 = 0$ 35. $x^2 - 6x + 9 = 0$

36. $x^2 - 6x + 7 = 0$ 37. $x^2 - 6x + 6 = 0$ 38. $3x^2 + 5x - 4 = 0$

39. $2x^2 - x + 3 = 0$ 40. $4x^2 + 2x = -5$ 41. $6x^2 = 13x - 6$

42. $3x^2 = 11x + 4$ 43. $6x^2 = 17x + 3$ 44. $6x^2 = 5x$

45. $9x^2 = 2x$

Use a calculator to obtain solutions correct to the nearest hundredth in Problems 46–53.

46. $x^2 + 4 = 3\sqrt{2}\,x$

47. $x^2 - 4\sqrt{3}\,x + 9 = 0$

48. $\sqrt{2}\,x^2 + 2x - 3 = 0$

49. $4x^2 - 2\sqrt{5}\,x - 2 = 0$

50. $0.02x^2 + 0.831x + 0.0069 = 0$

51. $68.38x^2 - 4.12x - 198.41 = 0$

52. $x^2 - 11.001x + 24.098 = 0$

53. $x^2 + 4.09x = 0.078$

54. Young's Rule for calculating a child's dosage for medication is

$$\text{CHILD'S DOSE} = \frac{\text{AGE OF CHILD}}{\text{AGE OF CHILD} + 12} \times \text{ADULT DOSE}$$

   a. If an adult's dose of a particular medication is 100 mg, what is the dose for a 10-year-old child?
   b. If a 12-year-old child's dose of a particular medication is 10 mg, what is the adult's dose?

55. Fried's Rule for calculating an infant's dosage for medication is

$$\text{INFANT'S DOSE} = \frac{\text{AGE OF INFANT IN MONTHS}}{150} \times \text{ADULT DOSE}$$

   a. If an adult's dose of a particular medication is 50 mg, what is the dosage for a 10-month old infant?
   b. If a 15-month old infant is to receive 7.5 mg of a medication, what is the equivalent adult dose?

56. Use the equation $PV = k$ to find P where $V = 4.8$ and $k = 14{,}400$.

57. Use the equation $PV = P'V'$ to find V' where $P = 740$, $V = 450$, and $P' = 750$.

▲ **Problem Solving**

58. An approximation for π can be obtained from

$$\frac{\pi^2}{6} = 1 + \frac{1}{2^2} + \frac{1}{3^2} + \frac{1}{4^2} + \cdots$$

   a. Solve for π.
   b. Find an approximation for π using the first 20 terms.
   c. Compare your answer to part **b** with an approximation of π you probably used in grade school, namely $\frac{22}{7}$.

▲ **Individual Research**

59. Look in a high school algebra book and write out a derivation of the quadratic formula.

60. Find any replacements for x, y, and z such that $x^n + y^n = z^n$, where n is greater than 2 and where x, y, and z are counting numbers. Write a history of this problem, known as Fermat's Last Theorem.

5.5 INEQUALITIES

Comparison Property

The techniques of the previous section can also be applied to quantities that are not equal. If we are given any two numbers x and y,

then obviously either

$$x = y \quad \text{or} \quad x \neq y$$

If $x \neq y$, then either

$$x < y \quad \text{or} \quad x > y$$

Comparison Property[*]

For any two numbers x and y, exactly one of the following is true:

1. $x = y$ x is equal to y (the same as)
2. $x > y$ x is greater than y (larger than)
3. $x < y$ x is less than y (smaller than)

This means that if two quantities are not exactly equal, we can relate them with a greater-than or a less-than symbol (called an **inequality symbol**). The solution of

$$x < 3$$

has more than one value, and it becomes very impractical to write "The answers are 2, 1, -110, 0, $2\frac{1}{2}$, 2.99, ..." Instead, we relate the answer to a number line, as shown in Figure 5.2. The fact that 3 is not included (3 is not less than 3) in the

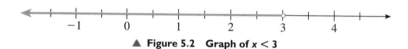

▲ **Figure 5.2 Graph of x < 3**

solution set is indicated by an open circle at the point 3.

If we want to include the endpoint $x = 3$ with the inequality $x < 3$, we write $x \leq 3$ and say "x is less than or equal to 3." We define two additional inequality symbols:

$$x \geq y \quad \text{means } x > y \text{ or } x = y$$
$$x \leq y \quad \text{means } x < y \text{ or } x = y$$

■ EXAMPLE 1

Graph the solution sets for the given inequalities.

a. $x \leq 3$ **b.** $x > 5$ **c.** $-2 \leq x$

Solution

a. Notice that in this example the endpoint is included because with $x \leq 3$, it is possible that $x = 3$. This is shown as a solid dot on the number line:

[*] Sometimes this is called the trichotomy property.

b. $x > 5$

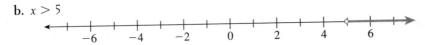

c. $-2 \le x$

Although you can graph this directly, you will have less chance of making a mistake when working with inequalities if you rewrite them so that the variable is on the left; that is, reverse the inequality to read $x \ge -2$. Notice that the direction of the arrow has been changed, because the symbol requires that the arrow always point to the smaller number. When you change the direction of the arrow, we say that you have *changed the order of the inequality*. For example, if $-2 \le x$, then $x \ge -2$, and we say the order has been changed. The graph of $x \ge -2$ is shown on the following number line:

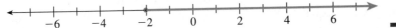

On a number line, if $x < y$, then x is to the left of y. Suppose the coordinates x and y are plotted as shown in Figure 5.3.

▲ **Figure 5.3   Number line showing two coordinates, x and y**

If you add 2 to both x and y, you obtain $x + 2$ and $y + 2$. From Figure 5.4, you see that $x + 2 < y + 2$.

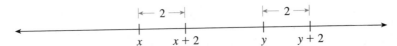

▲ **Figure 5.4   Number line with 2 added to both x and y**

If you add some number c, there are two possibilities, as shown in Figure 5.5.

$c > 0$ ($c > 0$ is read "c is positive")
$c < 0$ ($c < 0$ is read "c is negative")

c positive:

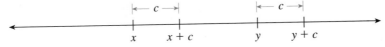

c negative:

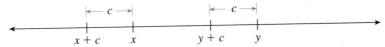

▲ **Figure 5.5   Adding positive and negative values to x and y**

If $c > 0$, then $x + c$ is still to the left of $y + c$. If $c < 0$, then $x + c$ is still to the left of $y + c$, as shown in Figure 5.5. In both cases, $x < y$, which justifies the following property.

Addition Property of Inequality

If $x < y$, then

$$x + c < y + c$$

Also,   if $x \leq y$, then $x + c \leq y + c$
        if $x > y$, then $x + c > y + c$
        if $x \geq y$, then $x + c \geq y + c$

Because this **addition property of inequality** is essentially the same as the addition property of equality, you might expect that there is also a multiplication property of inequality. We would hope that we could multiply both sides of an inequality by some number c without upsetting the inequality. Consider some examples. Let $x = 5$ and $y = 10$, so that $5 < 10$.

Let $c = 2$: $5 \cdot 2 < 10 \cdot 2$
               $10 < 20$ True
Let $c = 0$: $5 \cdot 0 < 10 \cdot 0$
               $0 < 0$ False
Let $c = -2$: $5(-2) < 10(-2)$
               $-10 < -20$ False

You can see that you cannot multiply both sides of an inequality by a constant and be sure that the result is still true. However, if you restrict c to a positive value, then you can multiply both sides of an inequality by c. On the other hand, if c is a negative number, then the order of the inequality should be reversed. This is summarized by the **multiplication property of inequality.**

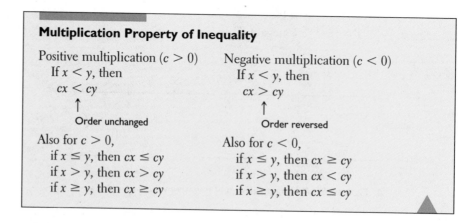

Multiplication Property of Inequality

Positive multiplication $(c > 0)$ Negative multiplication $(c < 0)$
  If $x < y$, then   If $x < y$, then
    $cx < cy$ $cx > cy$
    ↑   ↑
    Order unchanged   Order reversed

Also for $c > 0$,   Also for $c < 0$,
  if $x \leq y$, then $cx \leq cy$ if $x \leq y$, then $cx \geq cy$
  if $x > y$, then $cx > cy$ if $x > y$, then $cx < cy$
  if $x \geq y$, then $cx \geq cy$ if $x \geq y$, then $cx \leq cy$

The same properties hold for positive and negative division. We can summarize with the following statement.

Solution of Inequalities

The procedure for solving inequalities is the same as the procedure for solving equations except that, if you multiply or divide by a negative number, you reverse the order of the inequality.

In summary, given $x < y$, $x \leq y$, $x > y$, or $x \geq y$:

⊘The **inequality symbols are the** *same* if you

1. Add the same number to both sides.
2. Subtract the same number from both sides.
3. Multiply both sides by a positive number.
4. Divide both sides by a positive number.

} This works the same as with equations.

⊘The **inequality symbols are** *reversed* if you

1. Multiply both sides by a negative number.
2. Divide both sides by a negative number.
3. Interchange the x and the y.

} This is where inequalities differ from equations.

EXAMPLE 2

Solve: **a.** $-x \geq 2$ **b.** $\dfrac{x}{-3} < 1$ **c.** $5x - 3 \geq 7$

Solution

a. $-x \geq 2$

  $x \leq -2$ Multiply both sides by -1 and remember to reverse the order of the inequality.

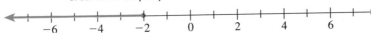

b. $\dfrac{x}{-3} < 1$

  $x > -3$ Multiply both sides by -3 and reverse the order of the inequality.

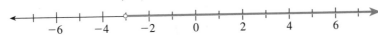

c. $5x - 3 \geq 7$

  $5x - 3 + 3 \geq 7 + 3$

  $5x \geq 10$

  $\dfrac{5x}{5} \geq \dfrac{10}{5}$

  $x \geq 2$

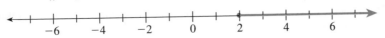

PROBLEM SET   5.5

▲ **A Problems**

1. **IN YOUR OWN WORDS** What is the comparison property?
2. **IN YOUR OWN WORDS** Describe a procedure for solving a first-degree inequality.

Graph the solution sets in Problems 3–20.

3. $x < 5$	4. $x \geq 6$	5. $x \geq -3$
6. $x \leq -2$	7. $4 \geq x$	8. $-1 < x$
9. $-5 \leq x$	10. $1 > x$	11. $-3 > x$
12. $\dfrac{x}{2} > 3$	13. $4 < \dfrac{x}{2}$	14. $-2 > \dfrac{x}{-4}$
15. $x < 50$	16. $x \geq 100$	17. $x \geq -125$
18. $x \leq -75$	19. $45 \geq x$	20. $-40 < x$

Solve the inequalities in Problems 21–44.

21. $x + 7 \geq 3$	22. $x - 2 \leq 5$	23. $x - 2 \geq -4$
24. $10 < 5 + y$	25. $-4 < 2 + y$	26. $-3 < 5 + y$
27. $2 > -s$	28. $-t \leq -3$	29. $-m > -5$
30. $5 \leq 4 - y$	31. $3 > 2 - x$	32. $5 \geq 1 - w$
33. $2x + 6 \leq 8$	34. $3y - 6 \geq 9$	35. $3 > s + 9$
36. $2 < 2s + 8$	37. $4 \leq a + 2$	38. $3 > 2b - 13$
39. $3s + 2 > 8$	40. $5t - 7 \geq 8$	41. $7u - 5 \leq 9$
42. $9 - 2v < 5$	43. $5 - 3w > 8$	44. $2 - x \geq 3x + 10$

▲ **B Problems**

Solve the inequalities in Problems 45–56.

45. $7 - 5A < 2A + 7$	46. $B > 3(1 + B)$
47. $3C > C + 19$	48. $2(D + 7) > 2 - D$
49. $5E - 4 < 3E - 6$	50. $3(3F - 2) > 4F - 3$
51. $4G - 1 > 3(G + 2)$	52. $5(4 + H) < 3(H + 1)$
53. $2 - 3I < 7(1 - I)$	54. $7(J - 2) + 5 \leq 3(2 + J)$
55. $-3(K - 3) - 2 \leq K + 3(K + 4)$	56. $2(L + 2) - 5 \geq L - 2(L + 1)$

57. Suppose that seven times a number is added to 35 and the result is positive. What are the possible numbers?

58. If the opposite of a number must be greater than twice the number, what are the possible numbers?

59. Suppose that three times a number is added to 12 and the result is negative. What are the possible numbers?

60. If the opposite of a number must be less than 5, what are the possible numbers satisfying this condition?

▲ **Problem Solving**

61. If a number is four more than its opposite, what are the possible numbers?

62. If a number is six less than twice its opposite, what are the possible numbers?

63. If a number is less than four more than its opposite, what are the possible numbers?

64. If a number is less than six minus twice its opposite, what are the possible numbers?

65. Current postal regulations state that no package may be sent if its combined length, width, and height exceed 72 in. What are the possible dimensions of a box to be mailed with equal height and width if the length is four times the height?

5.6   ALGEBRA IN PROBLEM SOLVING

One of the goals of problem solving is to be able to apply techniques that you learn in the classroom to situations outside the classroom. However, a first step is to learn to solve contrived textbook-type word problems to develop the problem-solving skills you will need outside the classroom.

We will now rephrase Polya's problem-solving guidelines in a setting that is appropriate to solving word problems. This procedure is summarized in the following box.

Procedure for Problem Solving in Algebra

First: You have to *understand the problem.*   This means you must read the problem and note what it is all about. Focus on processes rather than numbers. You cannot work a problem you do not understand. A sketch may help in understanding the problem.

Second: *Devise a plan.*   **Write down a verbal description of the problem using operation signs and an equal or inequality sign.** Note the following common translations.

Symbol	Verbal Description
$=$	is equal to; equals; are equal to; is the same as; is; was; becomes; will be; results in
$+$	plus; the sum of; added to; more than; greater than; increased by
$-$	minus; the difference of; the difference between; is subtracted from; less than; smaller than; decreased by; is diminished by
$\times$	times; product; is multiplied by; twice ($2\times$); triple ($3\times$)
$\div$	divided by; quotient of

(continued)

Third: *Carry out the plan.*   In the context of word problems, we need to proceed deductively by carrying out the following steps.
Choose a variable.   If there is a single unknown, choose a variable. If there are several unknowns, you can use the substitution property to reduce the number of unknowns to a single variable. Later we will consider word problems with more than one unknown.
Substitute.   Replace the verbal phrase for the unknown with the variable.
Solve the equation.   This is generally the easiest step. Translate the symbolic statement (such as $x = 3$) into a verbal statement. Probably no variables were given as part of the word problem, so $x = 3$ is not an answer. Generally, word problems require an answer stated in words. Pay attention to units of measure and other details of the problem.

Fourth: *Look back.*   Be sure your answer makes sense by checking it with the original question in the problem. **Remember to answer the question that was asked.**

In this section, we will focus on common types of word problems that are found in most textbooks. You might say, "I want to learn how to become a problem solver, and textbook problems are not what I have in mind; I want to do *real* problem solving." But to become a problem solver, you must first learn the basics, and there is good reason why word problems are part of a textbook. We start with these problems *to build a problem-solving **procedure** that can be expanded to apply to problem solving in general.*

Number Relationships

The first type of word problem we consider involves number relationships. These are designed to allow you to begin thinking about the *procedure* to use when solving word problems.

▬▬ EXAMPLE 1

If you add 10 to twice a number, the result is 22. What is the number?

Solution   Read the problem carefully. Make sure you know what is given and what is wanted. Next, write a verbal description (without using variables), using operation signs and an equal sign, but still using the key words. This is called *translating* the problem:

$$10 + 2(\text{A NUMBER}) = 22$$

When there is a single unknown, choose a variable.   ⃠ IMPORTANT: Do not *begin* by choosing a variable; choose a variable only *after* you have translated the problem. 🚫   With more complicated problems you will not know at the start what the variable should be.

Let n = A NUMBER.

Use the substitution property:

$$10 + 2(\text{A NUMBER}) = 22$$
$$\downarrow$$
$$10 + 2n \qquad = 22$$

Solve the equation and check the solution in the original problem to see if it makes sense:

$$10 + 2n = 22$$
$$2n = 12$$
$$n = 6$$

Check: Add 10 to twice 6 and the result is 22. State the solution to a word problem in words: The number is 6. ▬

The second type of number problem involves **consecutive integers.** If n = AN INTEGER, then

THE SECOND CONSECUTIVE INTEGER = $n + 1$,   and

THE THIRD CONSECUTIVE INTEGER = $n + 2$

Also, if E = AN EVEN INTEGER, then

$E + 2$ = THE NEXT CONSECUTIVE EVEN INTEGER

or if F = AN ODD INTEGER, then

$F + 2$ = THE NEXT CONSECUTIVE ODD INTEGER

🚫 Notice that you add 2 when you are writing down consecutive evens or consecutive odds. 🚫

▬▬▬ **EXAMPLE 2**

Find three consecutive integers whose sum is 42.

Solution   Read the problem. Write down a verbal description of the problem, using operation signs and an equal sign:

INTEGER + NEXT INTEGER + THIRD INTEGER = 42

This problem has three variables, but they are related. If

x = INTEGER, then

$x + 1$ = NEXT INTEGER

$x + 2$ = THIRD INTEGER

Substitute the variables into the verbal equation:

INTEGER + NEXT INTEGER + THIRD INTEGER = 42
$$\downarrow \qquad\qquad \downarrow \qquad\qquad\quad \downarrow$$
$$x \quad + \quad x + 1 \quad + \quad x + 2 \quad = 42$$

Solve the equation:

$$x + x + 1 + x + 2 = 42$$
$$3x + 3 = 42$$
$$3x = 39$$
$$x = 13$$

Check: $13 + 14 + 15 = 42$

The integers are 13, 14, and 15.   ▬

Distance Relationships

The first example of a problem with several variables that we will consider involves distances. The relationships may seem complicated, but if you draw a figure and remember that the total distance is the sum of the separate parts, you will easily be able to analyze this type of problem.

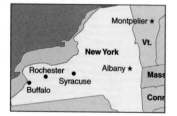

▬▬ EXAMPLE 3   Polya's Method

The drive from Buffalo to Albany is 210 miles across northern New York State. On this route, you pass Rochester and then Syracuse before reaching the capital city of Albany. It is 20 miles less from Buffalo to Rochester than from Syracuse to Albany, and 10 miles farther from Rochester to Syracuse than from Buffalo to Rochester. How far is it from Rochester to Syracuse?

Solution   First, you should begin by examining the problem. Remember that you cannot solve a problem you do not understand. It must make sense before mathematics can be applied to it. All too often poor problem solvers try to begin solving the problem too soon. Take your time when trying to understand the problem. Start at the beginning of the problem, and make a sketch of the situation, as shown in Figure 5.6. The cities are located in the order sketched on the line.

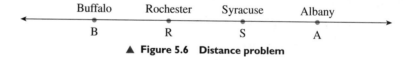

▲ **Figure 5.6   Distance problem**

Second, devise a plan. The cities are located in the order sketched in Figure 5.6, but how can that fact give us an equation? To have an equation, you must find an equality. Which quantities are equal? The distances from Buffalo to Rochester, from Rochester to Syracuse, and from Syracuse to Albany must add up to the distance from Buffalo to Albany. That is, the sum of the parts must equal the whole distance — so start there. This is what we mean when we say *translate*.

Translate.

(DIST. B TO R) + (DIST. R TO S) + (DIST. S TO A) = (DIST. B TO A)

Notice that there appear to be four variables. We now use the substitution property to *evolve*.

Evolve.   With this step we ask whether we know the value of any quantity in the equation. The first sentence of the problem tells us that the total distance is 210 miles, so

(DIST. B TO R) + (DIST. R TO S) + (DIST. S TO A) = (DIST. B TO A)

$\qquad\qquad\qquad\qquad\qquad\qquad\qquad\qquad\quad$↓ Buffalo to Albany = 210

(DIST. B TO R) + (DIST. R TO S) + (DIST. S TO A) = 210

Also, part of evolving the equation is to use substitution for the relationships that are given as part of the problem. We now translate the other pieces of given information by adding to the smaller distance in each case:

(DIST. B TO R) + 20 = (DIST. S TO A)

(DIST. R TO S) = (DIST. B TO R) + 10

We now use substitution to let the equation evolve into one with a single unknown.

Polya's method now tells us to carry out the plan. We use substitution on the above equations:

(DIST. B TO R) + (DIST. R TO S) + (DIST. S TO A)$\qquad\qquad$= 210

$\qquad\qquad\qquad\quad$↑$\qquad\qquad\qquad\qquad\quad$↑

$\qquad\qquad$(DIST. B TO R + 10)$\qquad$(DIST. B TO R) + 20

(DIST. B TO R) + [(DIST. B TO R) + 10] + [(DIST. B TO R) + 20] = 210

This equation now has a single variable, so we now let

d = DIST. B TO R

and substitute into the equation:

$d + [d + 10] + [d + 20] = 210$

Solve:

$$3d + 30 = 210$$
$$3d = 180$$
$$d = 60$$

The equation is solved. Does that mean that the answer to the problem is "$d = 60$"? No, the question asks for the distance from Rochester to Syracuse, which is $d + 10$. So now interpret the solution and **answer** the question: The distance from Rochester to Syracuse is 70 miles.

Notice that the steps we used above can be summarized as **translate, evolve, solve,** and **answer.** Polya's procedure requires that we look back. To be certain that the answer makes sense in the original problem, you should always check the solution:

$$60 + 70 + 80 \stackrel{?}{=} 210$$
$$210 = 210 \quad ✔$$

▬▬ EXAMPLE 4

Polya's Method

Once upon a time (about 450 B.C.) a Greek named Zeno made up several word problems that became known as Zeno's paradoxes. This problem is not a paradox, but was inspired by one of Zeno's problems. Consider a race between Achilles and a tortoise. The tortoise has a 100-meter head start. Achilles runs at a rate of 10 meters per second, whereas the tortoise runs 1 meter per second (it is an extraordinarily swift tortoise). How long does it take Achilles to catch up with the tortoise?

Solution   We use Polya's problem-solving guidelines for this example.

Understand the Problem.   Before you begin, make sure you understand the problem. It is often helpful to draw a figure or diagram to help you understand the problem. The situation for this problem is shown in Figure 5.7.

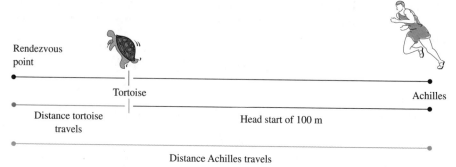

▲ **Figure 5.7   Achilles and tortoise problem**

Devise a Plan.   The plan we will use is *translate, evolve, and solve*.

Carry Out the Plan.
Translate.

$$(\text{ACHILLES' DISTANCE TO RENDEZVOUS}) = (\text{TORTOISE'S DISTANCE TO RENDEZVOUS}) + (\text{HEAD START})$$

Evolve.   The equation we wrote has three unknowns. Our goal is to evolve this equation into one with a single unknown so that we can choose that as our variable. The evolution of the equation requires that we use the substitution property to replace the unknowns with known numbers or with other expressions that, in turn, will lead to an equation with one unknown. We will begin this problem by substituting the known number.

$$(\text{ACHILLES' DISTANCE TO RENDEZVOUS}) = (\text{TORTOISE'S DISTANCE TO RENDEZVOUS}) + (\text{HEAD START})$$
$$\downarrow$$
$$(\text{ACHILLES' DISTANCE TO RENDEZVOUS}) = (\text{TORTOISE'S DISTANCE TO RENDEZVOUS}) + 100$$

There are still two unknowns, which we can change by using the following distance–rate–time formulas:

$$\text{ACHILLES' DISTANCE} = (\text{ACHILLES' RATE})(\text{TIME TO RENDEZVOUS})$$

$$\text{TORTOISE'S DISTANCE} = (\text{TORTOISE'S RATE})(\text{TIME TO RENDEZVOUS})$$

These values are now substituted into the equation:

(ACHILLES' DISTANCE TO RENDEZVOUS) = (TORTOISE'S DISTANCE TO RENDEZVOUS) + 100

(A's RATE)(TIME TO RENDEZVOUS) = (T's RATE)(TIME TO RENDEZVOUS) + 100

There are now three unknowns, but the values for two of *these* unknowns are given in the problem.

(A's RATE)(TIME TO RENDEZVOUS) = (T's RATE)(TIME TO RENDEZVOUS) + 100

$\quad\downarrow\qquad\qquad\qquad\qquad\qquad\qquad\downarrow$

10 (TIME TO RENDEZVOUS) = 1 (TIME TO RENDEZVOUS) + 100

The equation now has a single unknown, so let

t = TIME TO RENDEZVOUS

The last step in the evolution of this equation is to substitute the variable:

$10t = t + 100$

Solve.

$9t = 100$

$t = \dfrac{100}{9}$

Look Back. It takes Achilles $11\frac{1}{9}$ seconds to catch up with the tortoise.   ▬

Pythagorean Theorem

Many word problems are concerned with relationships involving a right triangle. If two sides of a right triangle are known, the third can be found by using the Pythagorean theorem. Remember, if a right triangle has sides a and b and hypotenuse c, then

$a^2 + b^2 = c^2$

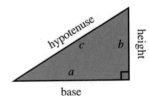

▬▬▬ **EXAMPLE 5**   **Polya's Method**

If the area of a right triangle is one square unit, and the height is two units longer than the base, find the lengths of the sides of the triangle to the nearest thousandth.

Solution We use Polya's problem-solving guidelines for this example. First, understand the problem. Draw a picture to help you understand the relationships.

Translate.

$\text{AREA} = \dfrac{1}{2}(\text{BASE})(\text{HEIGHT})$

Evolve. The area is known, so begin by substituting 1 for AREA,

$\text{AREA} = \dfrac{1}{2}(\text{BASE})(\text{HEIGHT})$

$\quad\downarrow$

$1 \;\;= \dfrac{1}{2}(\text{BASE})(\text{HEIGHT})$

Next, the problem tells us that HEIGHT = BASE + 2. Therefore,

$$1 = \frac{1}{2}(\text{BASE})(\text{HEIGHT})$$
$$\downarrow$$
$$1 = \frac{1}{2}(\text{BASE})(\text{BASE} + 2)$$

There is now a single unknown, so we choose a variable for the unknown. Let b = BASE. Then

$$1 = \frac{1}{2}b(b + 2)$$
$$2 = b(b + 2)$$
$$2 = b^2 + 2b$$
$$0 = b^2 + 2b - 2$$

Apply the quadratic formula:

$$b = \frac{-2 \pm \sqrt{2^2 - 4(1)(-2)}}{2(1)}$$
$$= \frac{-2 \pm \sqrt{12}}{2}$$
$$= \frac{-2 \pm 2\sqrt{3}}{2}$$
$$= -1 \pm \sqrt{3}$$

Since the base cannot be negative, we have

$$\text{BASE} \quad = -1 + \sqrt{3} \approx 0.7320508$$
$$\text{HEIGHT} = \text{BASE} + 2 = -1 + \sqrt{3} + 2 = 1 + \sqrt{3} \approx 2.7320508$$

The Pythagorean theorem is necessary for finding the third side. When finding the hypotenuse, be sure not to work with the approximate values. **You should round only once in a problem, and that is when you are finding your answer.**

$$(\text{HYPOTENUSE})^2 = (-1 + \sqrt{3})^2 + (1 + \sqrt{3})^2$$
$$= 1 - 2\sqrt{3} + 3 + 1 + 2\sqrt{3} + 3$$
$$= 8$$

From the square root property,

$$\text{HYPOTENUSE} = \pm \sqrt{8} = 2\sqrt{2} \approx 2.8284271$$

Positive value only, since hypotenuse represents a distance

Answer.   State the answer to the correct number of decimal places. The sides of the triangle are 0.732, 2.732, and 2.828.   ▬

Have you ever wondered why sidewalks, pipes, or tracks have expansion joints every few feet? The next example may help you to understand why; it considers the unlikely situation in which 1-mile sections of pipe are connected together.

■■■ **EXAMPLE 6**   **Polya's Method**

A 1-mile-long pipeline connects two pumping stations. Special joints must be used along the line to provide for expansion and contraction due to changes in temperature. However, if the pipeline were actually one continuous length of pipe fixed at each end by the stations, then expansion would cause the pipe to bow. Approximately how high would the middle of the pipe rise if the expansion were just 1 inch over the mile?

Solution   We use Polya's problem-solving guidelines for this example.

Understand the Problem.   First, understand the problem. Draw a picture as shown in Figure 5.8.

▲ **Figure 5.8   Pipeline problem**

Before beginning the solution to this example, try to guess the answer. Consider the following choices for the rise in pipe:

A. 1 inch   **B.** 1 foot   **C.** 1 yard   **D.** 5 yards   **E.** 1 mile

Go ahead, choose one of these. This is one problem for which the answer was not intuitive for the author. I guessed incorrectly when I first considered this problem.

Devise a Plan.   For purposes of solution, notice from Figure 5.8 that we are assuming that the pipe bows in a circular arc. A triangle would produce a reasonable approximation since the distance x should be quite small compared to the total length. Since a right triangle is used to model this situation, the Pythagorean theorem may be used. The method we will use is to *translate, evolve, and solve.*

Carry Out the Plan.
   Translate.

$$(\text{SIDE})^2 + (\text{HEIGHT})^2 = (\text{HYPOTENUSE})^2$$

   Evolve.   The side is one-half the length of pipe, so it is 0.5 mile. Also, since the expansion is 1 inch, one-half the arc would have length 0.5 mile + 0.5 inch.

$$(\text{SIDE})^2 + (\text{HEIGHT})^2 = (\text{HYPOTENUSE})^2$$
$$\downarrow \qquad\qquad\qquad\qquad \downarrow$$
$$(0.5 \text{ mile})^2 + (\text{HEIGHT})^2 = (0.5 \text{ mile} + 0.5 \text{ in.})^2$$

There is a single unknown, so let h = HEIGHT. Also, notice that there is a mixture of units. Let us convert all measurements to inches. We know that

$$1 \text{ mile} = 5{,}280 \text{ ft} = 5{,}280(12 \text{ in.}) = 63{,}360 \text{ in.}$$

$$\frac{1}{2} \text{ mile} = 31{,}680 \text{ in.}$$

$$(0.5 \text{ mile})^2 + (\text{HEIGHT})^2 = (0.5 \text{ mile} + 0.5 \text{ in.})^2$$
$$\qquad \downarrow \qquad\qquad\qquad\qquad\qquad\qquad \downarrow$$
$$(31{,}680)^2 + \qquad h^2 \qquad = \qquad (31{,}680.5)^2$$

Solve.

$$h^2 = (31{,}680.5)^2 - (31{,}680)^2$$
$$h = \sqrt{(31{,}680.5)^2 - (31{,}680)^2}$$
$$\approx 177.99$$

Look Back.   The solution, 177.99 in., is approximately 14.8 ft. This is an extraordinary result if you consider that the pipe expanded only 1 *inch*. The pipe would bow approximately 14.8 ft at the middle.   ▬

PROBLEM SET 5.6

▲ A Problem

1. **IN YOUR OWN WORDS** Outline a procedure for solving word problems in algebra.

2. **IN YOUR OWN WORDS** Explain what is meant by *translate, evolve,* and *solve.*

Solve Problems 3–40. Because you are practicing a **procedure,** *you must show all of your work.*

3. If you add seven to twice a number, the result is seventeen. What is the number?

4. If you subtract twelve from twice a number, the result is six. What is the number?

5. If you add fifteen to twice a number, the result is seven. What is the number?

6. If you multiply a number by five and then subtract negative ten, the difference is negative thirty. What is the number?

7. If 12 is subtracted from twice a number, the difference is four times the number. What is the number?

8. If 6 is subtracted from three times a number, the difference is twice the number. What is the number?

9. Find two consecutive integers whose sum is 117.

10. Find two consecutive even integers whose sum is 94.

11. Find three consecutive integers whose sum is 54.

12. The sum of three consecutive integers is 105. What are the integers?

13. The sum of four consecutive integers is 74. What are the integers?

14. The sum of two consecutive even integers is 30. What is the larger number?

15. The sum of two consecutive odd integers is 48. What is the smaller integer?

16. Find two consecutive odd integers whose sum is 424.

▲ B Problems

17. A house and a lot are appraised at $212,400. If the value of the house is five times the value of the lot, how much is the house worth?

18. A cabinet shop produces two types of custom-made cabinets. If the cost of one type of cabinet is four times the cost of the other, and the total price for one of each type of cabinet is $4,150, how much does each cabinet cost?

19. To stimulate his daughter in the pursuit of problem solving, a math professor offered to pay her $8 for every equation correctly solved and to fine her $5 for every incorrect solution. At the end of the first 26 problems of this problem set, neither owed any money to the other. How many problems did the daughter solve correctly?

20. A professional gambler reported that at the end of the first race at the track he had doubled his money. He bet $30 on the second race and tripled the money he came with. He bet $54 on the third race and quadrupled his original bankroll. He bet $72 on the fourth race and lost it, but still had $48 left. With how much money did he start?

21. The hypotenuse of a right triangle is 13 in., and one leg is 6 in. shorter than the other. Find the dimensions of the figure.

22. A 10-ft pole is to be erected and held in the ground by four guy wires attached at the top. The guy wires are attached to the ground at a distance of 15 ft from the base of the pole. What is the exact length of each guy wire? How much wire should be purchased if it cannot be purchased in fractions of a foot?

23. A diagonal brace is to be placed on the wall of a room. The height of the wall is 8 ft, and the wall is 14 ft long. What is the exact length of the brace, and what is the length of the brace to the nearest foot?

24. A college and the local movie theater are ten blocks apart. On his way to the college from the movie, Joe passes a sports arena first and then his dorm. It is a block farther from the college to the arena than from the arena to the dorm, and it is three blocks farther from his dorm to the theater than from the arena to the dorm. How far is it from the arena to his dorm?

25. My car pool begins at my house, and I pick up (in order) John, Terry, and Amber before reaching work, a distance of 39 miles. It is one mile farther from my house to John's house than it is from John's to Terry's, and it is one mile farther from Terry's house to Amber's house than from Amber's to work. How far is it from my house to John's house if Amber lives the same distance from work as I live from John's house?

26. In traveling from Jacksonville to Miami you pass through Orlando and then through Palm Beach. It is 10 miles farther from Orlando to Palm Beach than it is from Jacksonville to Orlando. The distance between Jacksonville and Orlando is 90 miles more than the distance from Palm Beach to Miami. If it is 370 miles from Jacksonville to Miami, how far is it from Jacksonville to Orlando?

27. The drive from New Orleans to Memphis is 90 miles shorter than the drive from Memphis to Cincinnati, but 150 miles farther than the drive from Cincinnati to De-

troit. If the total highway distance of a New Orleans–Memphis–Cincinnati–Detroit trip is 1,140 miles, find the length of the Cincinnati–Detroit leg of the trip.

28. Traveling from San Antonio to Dallas, you first pass through Austin and then Waco before reaching Dallas, a total distance of 280 miles. From Austin to Waco is 30 miles farther than from San Antonio to Austin, and also 20 miles farther than from Waco to Dallas. How far is it from Waco to Dallas?

29. Two persons are to run a race, but one can run 10 meters per second, whereas the other can run 6 meters per second. If the slower runner has a 50-meter head start, how long will it be before the faster runner catches the slower runner, if they begin at the same time?

30. If the rangefinder on the *Enterprise* shows a shuttlecraft 4,500 km away, how long will it take to catch the shuttle if the shuttle travels at 12,000 kph and the *Enterprise* is traveling at 15,000 kph?

31. A speeding car is traveling at 80 mph when a police car starts pursuit at 100 mph. How long will it take the police car to catch up to the speeding car? Assume that the speeding car has a 2-mile head start and that the cars travel at constant rates.

32. Two people walk daily for exercise. One is able to maintain 4.0 mph and the other only 3.5 mph. The slower walker has a mile head start when the other begins, yet they finish together. How far did each walk?

33. A child walks along a river bank and a friend rafts on the current beginning an hour later from the same point. If one walks at 3 mph and the other floats at 9 mph, how far do they travel before one overtakes the other?

34. Two joggers set out at the same time from their homes 21 miles apart. They agree to meet at a point somewhere in between in an hour and a half. If the rate of one is 2 mph faster than the rate of the other, find the rate of each.

35. Two joggers set out at the same time in opposite directions. If they were to maintain their normal rates for 4 hours, they would be 68 miles apart. If the rate of one is 1.5 mph faster than the rate of the other, find the rate of each.

▲ **Problem Solving**

36. **HISTORICAL QUESTION** (from Bhaskara, ca. 1120 A.D.) "In a lake the bud of a water lily was observed, one cubit above the water, and when moved by the gentle breeze, it sunk in the water at two cubits' distance." Find the depth of the water.

37. **HISTORICAL QUESTION** (from Bhaskara, ca. 1120 A.D.) "One third of a collection of beautiful water lilies is offered to Mahadev, one-fifth to Huri, one-sixth to the Sun, one-fourth to Devi, and the six which remain are presented to the spiritual teacher." Find the total number of lilies.

38. **HISTORICAL QUESTION** (from Bhaskara, ca. 1120 A.D.) "One-fifth of a hive of bees flew to the Kadamba flower; one-third flew to the Silandhara; three times the differ-ence of these two numbers flew to an arbor, and one bee continued flying about, at-tracted on each side by the fragrant Keteki and the Malati." Find the number of bees.

39. **HISTORICAL QUESTION** (from Brahmagupta, ca. 630 A.D.) "A tree one hundred cubits high is distant from a well two hundred cubits; from this tree one monkey climbs down the tree and goes to the well, but the other leaps in the air and descends by the

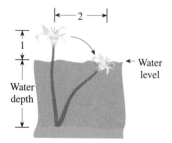

hypotenuse from the high point of the leap, and both pass over an equal space." Find the height of the leap.

40. **HISTORICAL QUESTION** "Ten times the square root of a flock of geese, seeing the clouds collect, flew to the Manus lake; one-eighth of the whole flew from the edge of the water amongst a multitude of water lilies; and three couples were observed playing in the water." Find the number of geese.

5.7   RATIOS, PROPORTIONS, AND PROBLEM SOLVING

Ratios

Ratios and proportions are powerful problem-solving tools in algebra. Ratios are a way of comparing two numbers or quantities — for example, the compression ratio of a car, the gear ratio of a transmission, the pitch of a roof, the steepness of a road, or a player's batting average. A **ratio** expresses a size relationship between two sets and is defined as the quotient of two numbers. It is written using the word *to*, a colon, or a fraction; that is, if the ratio of men to women is **5 to 4,** this could also be written as $5:4$ or $\frac{5}{4}$.

Ratio

$\frac{a}{b}$ is called the **ratio** of a to b. The two parts a and b are called its terms.

We will emphasize the idea that a ratio can be written as a fraction (or as a quotient of two numbers). Since a fraction can be reduced, a ratio can also be reduced.

▰▰▰ **EXAMPLE 1**

Reduce the given ratios to lowest terms.

a. 4 to 52   **b.** 15 to 3   **c.** $1\frac{1}{2}$ to 2   **d.** $1\frac{2}{3}$ to $3\frac{3}{4}$

Solution

a. A ratio of 4 to 52

$$\frac{4}{52} = \frac{1}{13}$$

A ratio of 1 to 13

b. A ratio of 15 to 3

$$\frac{15}{3} = 5$$

Write this as $\frac{5}{1}$ because a ratio compares two numbers.

A ratio of 5 to 1

c. A ratio of $1\frac{1}{2}$ to 2

$$\frac{1\frac{1}{2}}{2} = 1\frac{1}{2} \div 2$$

$$= \frac{3}{2} \times \frac{1}{2}$$

$$= \frac{3}{4}$$

A ratio of 3 to 4

d. A ratio of $1\frac{2}{3}$ to $3\frac{3}{4}$

$$\frac{1\frac{2}{3}}{3\frac{3}{4}} = 1\frac{2}{3} \div 3\frac{3}{4}$$

$$= \frac{5}{3} \div \frac{15}{4}$$

$$= \frac{5}{3} \times \frac{4}{15}$$

$$= \frac{4}{9}$$

A ratio of 4 to 9

Proportions

A **proportion** is a statement of equality between ratios. In symbols,

$$\frac{a}{b} \underset{\uparrow}{=} \frac{c}{d}$$
$$\underset{\uparrow}{} \quad \underset{\uparrow}{}$$

"*a* is to *b*" "as" "*c* is to *d*"

The notation used in some books is $a:b::c:d$. Even though we won't use this notation, we will use words associated with this notation to name the terms:

$$a : b :: c : d$$

In the more common fractional notation, we have

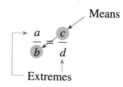

EXAMPLE 2

Read each proportion, and name the means and the extremes.

a. $\dfrac{2}{3} = \dfrac{10}{15}$ **b.** $\dfrac{m}{5} = \dfrac{3}{8}$

Solution

a. $\dfrac{2}{3} = \dfrac{10}{15}$ *Read:*   Two is to three as ten is to fifteen.

Means: 3 and 10   *Extremes:*   2 and 15

b. $\dfrac{m}{5} = \dfrac{3}{8}$ *Read:*   *m* is to five as three is to eight.

   Means: 5 and 3   *Extremes:*   *m* and 8   ▬

   The following property is fundamental to our study of proportions and percents.

Property of Proportions

If the product of the means equals the product of the extremes, then the ratios form a proportion.

Also, if the ratios form a proportion, then the product of the means equals the product of the extremes.

In symbols,

$$\frac{a}{b} = \frac{c}{d}$$

$$\underbrace{b \times c}_{\uparrow} = \underbrace{a \times d}_{\uparrow}$$

product of means $=$ product of extremes

Calculator Comment

Since ratios can be written as fractions, they can be written in decimal form. This is a particularly useful representation of a ratio when a calculator is used.

▬ **EXAMPLE 3**

Write the decimal forms for the given ratios.

a. 3 to 5   b. 2 to 3

Solution

a. 3 to 5 $\boxed{3}\,\boxed{\div}\,\boxed{5}\,\boxed{=}$ .6
b. 2 to 3 $\boxed{2}\,\boxed{\div}\,\boxed{3}\,\boxed{=}$ .666666667   ▬

▬ **EXAMPLE 4**

Tell whether each pair of ratios forms a proportion.

a. $\dfrac{3}{4}, \dfrac{36}{48}$ b. $\dfrac{5}{16}, \dfrac{7}{22}$

Solution

a. *Means*   *Extremes*   b. *Means*   *Extremes*

4×36 3×48 16×7 5×22

$144 = 144$ $112 \neq 110$

Thus,   Thus,

$$\frac{3}{4} = \frac{36}{48}$$ $$\frac{5}{16} \neq \frac{7}{22}$$

They form a proportion.   They do not form a proportion.   ▬

You can use the cross-product method not only to see if two fractions form a proportion, but also to compare the sizes of two fractions. For example, if a and c are whole numbers, and b and d are counting numbers, then the following property can be used to compare the sizes of two fractions:

If $ad = bc$, then $\dfrac{a}{b} = \dfrac{c}{d}$.

If $ad < bc$, then $\dfrac{a}{b} < \dfrac{c}{d}$.

If $ad > bc$, then $\dfrac{a}{b} > \dfrac{c}{d}$.

▬ **EXAMPLE 5**

Insert $=$, $<$, or $>$ as appropriate.

a. $\dfrac{3}{4} \underline{\hspace{1cm}} \dfrac{18}{24}$ b. $\dfrac{1}{2} \underline{\hspace{1cm}} \dfrac{1}{3}$ c. $\dfrac{2}{3} \underline{\hspace{1cm}} \dfrac{7}{8}$ d. $\dfrac{3}{5} \underline{\hspace{1cm}} \dfrac{4}{7}$

Solution

a. $\dfrac{3}{4} \underline{\hspace{1cm}} \dfrac{18}{24}$ b. $\dfrac{1}{2} \underline{\hspace{1cm}} \dfrac{1}{3}$ c. $\dfrac{2}{3} \underline{\hspace{1cm}} \dfrac{7}{8}$ d. $\dfrac{3}{5} \underline{\hspace{1cm}} \dfrac{4}{7}$

$3(24) ? 4(18)$ $1(3) ? 2(1)$ $2(8) ? 3(7)$ $3(7) ? 5(4)$

$72 ? 72$ $3 ? 2$ $16 ? 21$ $21 ? 20$

$=$ $>$ $<$ $>$ ▬

Comparing decimal numbers can sometimes be confusing; for example, which is larger,

0.6   or   0.58921?

The larger number is 0.6. To see this, simply write each decimal with the same number of places by affixing trailing zeros, and then compare. That is, write

0.60000
0.58921

Now, it is easy to see that $0.60000 > 0.58921$ because 60,000 hundred thousandths is larger than 58,921 hundred thousandths.

■■■■ EXAMPLE 6

Insert $<$ or $>$ as appropriate.

a. 0.28 _____ 0.3 b. 0.001 _____ 0.01

c. 0.005 _____ 0.00482 d. 3 _____ 0.98712

Solution

a. 0.28 _____ 0.3
   $0.28 < 0.30$ since $28 < 30$

b. 0.001 _____ 0.01
   $0.001 < 0.010$ since $1 < 10$

c. 0.005 _____ 0.00482
   $0.00500 > 0.00482$

d. 3 _____ 0.98712
   $3.00000 > 0.98712$ ■

Solving Proportions

The usual setting for a proportion problem is that three of the terms of the proportion are known and one of the terms is unknown. It is always possible to find the
missing term by solving an equation. However, when given a proportion, we first
use the property of proportions (which is equivalent to multiplying both sides of the
equation by the same number).

■■■■ EXAMPLE 7

Find the missing term of each proportion.

a. $\dfrac{3}{4} = \dfrac{w}{20}$ b. $\dfrac{3}{4} = \dfrac{27}{y}$ c. $\dfrac{2}{x} = \dfrac{8}{9}$ d. $\dfrac{t}{15} = \dfrac{3}{5}$

Solution

a. $\dfrac{3}{4} = \dfrac{w}{20}$

PRODUCT OF MEANS = PRODUCT OF EXTREMES

$$4w = 3(20)$$

$$w = \dfrac{3(\overset{5}{\cancel{20}})}{\underset{1}{\cancel{4}}}$$

Solve the equation by dividing both
sides by 4. Notice that 4 is the
number opposite the unknown:

$$\dfrac{3}{4} = \dfrac{w}{20}$$

$$w = 15$$

b. $\dfrac{3}{4} = \dfrac{27}{y}$

PRODUCT OF MEANS = PRODUCT OF EXTREMES

$$4(27) = 3y$$

$$\dfrac{4(\overset{9}{\cancel{27}})}{\underset{1}{\cancel{3}}} = y$$

Divide both sides by 3; notice that
3 is the number opposite the
unknown:

$$\dfrac{3}{4} = \dfrac{27}{y}$$

$$36 = y$$

c. $\dfrac{2}{x} = \dfrac{8}{9}$

PRODUCT OF MEANS = PRODUCT OF EXTREMES

$$8x = 2(9)$$

$$x = \dfrac{\overset{1}{2(9)}}{\underset{4}{8}}$$

Divide both sides by 8; notice that 8 is the number opposite the unknown:

$$\dfrac{2}{x} = \dfrac{8}{9}$$

$$x = \dfrac{9}{4}$$

d. $\dfrac{t}{15} = \dfrac{3}{5}$

PRODUCT OF MEANS = PRODUCT OF EXTREMES

$$3(15) = 5t$$

$$\dfrac{\overset{3}{3(\cancel{15})}}{\underset{1}{\cancel{5}}} = t$$

Divide both sides by 5; notice that 5 is the number opposite the unknown:

$$\dfrac{t}{15} = \dfrac{3}{5}$$

$$9 = t$$

    Notice that the unknown term can be in any one of four positions, as illustrated by the four parts of Example 7. But even though you can find the missing term of a proportion (called **solving the proportion**) by the technique used in Example 7, it is easier to think in terms of **the cross-product divided by the number opposite the unknown.** This method is easier than actually solving the equation because it can be done quickly using a calculator, as shown in the following examples.

Procedure for Solving Proportions

1. Find the product of the means or the product of the extremes, whichever does not contain the unknown term.
2. Divide this product by the number that is opposite the unknown term.

▬▬▬ EXAMPLE 8

Solve the proportion for the unknown term.

a. $\dfrac{5}{6} = \dfrac{55}{y}$

b. $\dfrac{5}{b} = \dfrac{3}{4}$

c. $\dfrac{2\frac{1}{2}}{5} = \dfrac{a}{8}$

Solution

a. $\dfrac{5}{6} = \dfrac{55}{y}$

$y = \dfrac{6 \times 55}{5}$ ← Product of the extremes
                ← Number opposite the unknown

$\quad = \dfrac{6 \times \overset{11}{\cancel{55}}}{\underset{1}{\cancel{5}}}$ You can cancel to simplify many of these problems.

$\quad = 66$ $\boxed{6}\ \boxed{\times}\ \boxed{55}\ \boxed{\div}\ \boxed{5}\ \boxed{=}$ **66**

b. $\dfrac{5}{b} = \dfrac{3}{4}$

$b = \dfrac{5 \times 4}{3}$ ← Product of the means
              ← Number opposite the unknown

$\quad = \dfrac{20}{3}$ or $6\frac{2}{3}$ $\boxed{5}\ \boxed{\times}\ \boxed{4}\ \boxed{\div}\ \boxed{3}\ \boxed{=}$

Display: **6.66666667**   Interpret this as $6\frac{2}{3}$.

Notice that your answers don't have to be whole numbers. This means that the correct proportion is

$$\frac{5}{6\frac{2}{3}} = \frac{3}{4}$$

c. $\dfrac{2\frac{1}{2}}{5} = \dfrac{a}{8}$

$a = \dfrac{2\frac{1}{2} \times 8}{5}$ ← Product of the extremes
              ← Number opposite the unknown

$\quad = \dfrac{\frac{5}{2} \times \frac{8}{1}}{5}$

$\quad = \dfrac{20}{5}$

$\quad = 4$ $\boxed{2.5}\ \boxed{\times}\ \boxed{8}\ \boxed{\div}\ \boxed{5}\ \boxed{=}$ **4**

> **Calculator Comment**
>
> Proportions are quite easy to solve if you have a calculator. You can multiply the cross-terms and divide by the number opposite the variable, all in one calculator sequence.

Many applied problems can be solved using a proportion. Whenever you are working an applied problem, you should estimate an answer so that you will know whether the result you obtain is reasonable.

When setting up a proportion with units, be sure that like units occupy corresponding positions, as illustrated in Examples 9–12. The proportion is obtained by applying the sentence "*a* is to *b* as *c* is to *d*" to the quantities in the problem.

■■■■ EXAMPLE 9   Polya's Method

If 4 cans of cola sell for $1.89, how much will 6 cans cost?

Solution We use Polya's problem-solving guidelines for this example.

Understand the Problem.   We see that 4 cans sell for $1.89 and 8 cans sell for 2($1.89) = $3.78. We need to find the cost for 6 cans, which must be somewhere between $1.89 and $3.78.

Devise a Plan.   There are many possible plans for solving this problem. We will form a proportion.

Carry Out the Plan.   Solve "4 cans is to $1.89 as 6 cans is to what?"

$$\frac{4 \text{ cans}}{1.89 \text{ dollars}} = \frac{6 \text{ cans}}{x \text{ dollars}}$$

Product of the means

$$x = \frac{1.89 \times 6}{4}$$

$$\boxed{1.89}\ \boxed{\times}\ \boxed{6}\ \boxed{\div}\ \boxed{4}\ \boxed{=}$$

Number opposite the unknown

$$= 2.835$$

Look Back.   We see that 6 cans will cost $2.84.   ▬

▬▬ EXAMPLE 10

If a 120-mile trip took $8\frac{1}{2}$ gallons of gas, how much gas is needed for a 240-mile trip?

Solution   "120 miles is to $8\frac{1}{2}$ gallons as 240 miles is to how many gallons?"

miles   miles
↓   ↓

$$\frac{120}{8\frac{1}{2}} = \frac{240}{x}$$

↑   ↑
gallons   gallons

$$x = \frac{8\frac{1}{2} \times \overset{2}{\cancel{240}}}{\underset{1}{\cancel{120}}} = 17 \qquad \boxed{8.5}\ \boxed{\times}\ \boxed{240}\ \boxed{\div}\ \boxed{120}\ \boxed{=}$$

The trip will require 17 gallons.   ▬

▬▬ EXAMPLE 11

If the property tax on a $65,000 home is $416, what is the tax on an $85,000 home?

Solution   "$65,000 is to $416 as $85,000 is to what?"

value   value
↓   ↓

$$\frac{65,000}{416} = \frac{85,000}{x}$$

↑   ↑
tax   tax

$$x = \frac{416 \times 85,000}{65,000}$$

You can do the arithmetic by canceling or on a calculator:

$$x = \frac{416 \times \overset{\overset{17}{\cancel{85}}}{\cancel{85{,}000}}}{\underset{\underset{13}{\cancel{65}}}{\cancel{65{,}000}}}$$ $\boxed{416}\ \boxed{\times}\ \boxed{85000}\ \boxed{\div}\ \boxed{65000}\ \boxed{=}$

$= 544$

The tax is \$544.

■ EXAMPLE 12

In a can of mixed nuts the ratio of cashews to peanuts is 1 to 6. If a given machine releases 46 cashews into a can, how many peanuts should be released?

Solution   First, understand the problem. This problem is comparing two quantities, cashews and peanuts, so consider writing a proportion.

   Translate.

This is the given ratio.
$$\frac{\text{NUMBER OF CASHEWS}}{\text{NUMBER OF PEANUTS}} = \frac{1}{6}$$

Evolve.   The NUMBER OF CASHEWS = 46, so by substitution

$$\frac{46}{\text{NUMBER OF PEANUTS}} = \frac{1}{6}$$

Let p = NUMBER OF PEANUTS: then the equation is

$$\frac{46}{p} = \frac{1}{6}$$

   Solve.

$6(46) = p$
$276 = p$

Answer.   Thus, 276 peanuts should be released.

PROBLEM SET   5.7

▲ A Problems

1. **IN YOUR OWN WORDS** What do we mean by ratios and proportions?

2. **IN YOUR OWN WORDS** How does the property of proportions relate to solving equations?

3. **IN YOUR OWN WORDS** How does the procedure for solving proportions relate to solving equations?

4. **IN YOUR OWN WORDS** Describe a process for setting up a proportion, given an applied problem.

Write the ratios given in Problems 5–16 as reduced fractions.

5. A cement mixture calls for 60 pounds of cement for 3 gallons of water. What is the ratio of cement to water?

6. What is the ratio of water to cement in Problem 5?

7. About 106 baby boys are born for every 100 baby girls. Write this as a simplified ratio of males to females.

8. Use Problem 7 to write a simplified ratio of females to males.

9. If you drive 119 miles on $8\frac{1}{2}$ gallons of gas, what is the simplified ratio of miles to gallons?

10. If you drive 180 miles on $7\frac{1}{2}$ gallons of gas, what is the simplified ratio of miles to gallons?

11. If you drive 279 miles on $15\frac{1}{2}$ gallons of gas, what is the simplified ratio of miles to gallons?

12. If you drive 151.7 miles on 8.2 gallons of gas, what is the simplified ratio of miles to gallons?

13. Find the ratio of $2\frac{3}{4}$ to $6\frac{1}{4}$.

14. Find the ratio of $6\frac{1}{4}$ to $2\frac{3}{4}$.

15. Find the ratio of $5\frac{2}{3}$ to $2\frac{1}{3}$.

16. Find the ratio of $6\frac{1}{3}$ to $4\frac{1}{3}$.

Tell whether each pair of ratios in Problems 17–21 forms a proportion.

17. a. $\dfrac{7}{1}, \dfrac{21}{3}$ b. $\dfrac{6}{8}, \dfrac{9}{12}$ c. $\dfrac{3}{6}, \dfrac{5}{10}$

18. a. $\dfrac{7}{8}, \dfrac{6}{7}$ b. $\dfrac{9}{2}, \dfrac{10}{3}$ c. $\dfrac{6}{5}, \dfrac{42}{35}$

19. a. $\dfrac{85}{18}, \dfrac{42}{9}$ b. $\dfrac{403}{341}, \dfrac{13}{11}$ c. $\dfrac{20}{70}, \dfrac{4}{14}$

20. a. $\dfrac{3}{4}, \dfrac{75}{100}$ b. $\dfrac{2}{3}, \dfrac{67}{100}$ c. $\dfrac{5}{3}, \dfrac{7\frac{1}{3}}{4}$

21. a. $\dfrac{3}{2}, \dfrac{5}{3\frac{1}{2}}$ b. $\dfrac{5\frac{1}{5}}{7}, \dfrac{4}{5}$ c. $\dfrac{1}{3}, \dfrac{33\frac{1}{3}}{100}$

Insert =, <, or > as appropriate in Problems 22–27.

22. a. $\dfrac{1}{6}$ ____ $\dfrac{1}{8}$ b. $\dfrac{1}{4}$ ____ $\dfrac{1}{3}$ c. $\dfrac{1}{5}$ ____ $\dfrac{1}{8}$

23. a. $\dfrac{4}{5}$ ____ $\dfrac{7}{8}$ b. $\dfrac{3}{8}$ ____ $\dfrac{2}{5}$ c. $\dfrac{3}{5}$ ____ $\dfrac{2}{3}$

24. a. $\dfrac{25}{5}$ ____ $\dfrac{10}{2}$ b. $\dfrac{14}{15}$ ____ $\dfrac{42}{45}$ c. $\dfrac{11}{16}$ ____ $\dfrac{7}{12}$

25. a. 0.8 ____ 0.8001   b. 0.8 ____ 0.7999   c. 8 ____ 2.81

26. **a.** 2.8 _____ 2.88 **b.** 2 _____ 0.9875 **c.** 0.4958 _____ 0.49

27. **a.** π _____ 3.1416 **b.** $\sqrt{2}$ _____ 1.4142 **c.** $\sqrt{3}$ _____ 1.7320508

Solve each proportion in Problems 28–51 for the item represented by a letter.

28. $\dfrac{5}{1} = \dfrac{A}{6}$ 29. $\dfrac{1}{9} = \dfrac{4}{B}$ 30. $\dfrac{C}{2} = \dfrac{5}{1}$ 31. $\dfrac{7}{D} = \dfrac{1}{8}$ 32. $\dfrac{12}{18} = \dfrac{E}{12}$

33. $\dfrac{12}{15} = \dfrac{20}{F}$ 34. $\dfrac{G}{24} = \dfrac{14}{16}$ 35. $\dfrac{4}{H} = \dfrac{3}{15}$ 36. $\dfrac{2}{3} = \dfrac{I}{24}$ 37. $\dfrac{4}{5} = \dfrac{3}{J}$

38. $\dfrac{3}{K} = \dfrac{2}{5}$ 39. $\dfrac{L}{18} = \dfrac{5}{6}$ 40. $\dfrac{7\frac{1}{5}}{9} = \dfrac{M}{5}$ 41. $\dfrac{4}{2\frac{2}{3}} = \dfrac{3}{N}$ 42. $\dfrac{P}{4} = \dfrac{4\frac{1}{2}}{6}$

43. $\dfrac{5}{2} = \dfrac{Q}{12\frac{3}{5}}$ 44. $\dfrac{5}{R} = \dfrac{7}{12\frac{3}{5}}$ 45. $\dfrac{1\frac{1}{3}}{\frac{1}{9}} = \dfrac{S}{2\frac{2}{3}}$ 46. $\dfrac{33}{2\frac{1}{5}} = \dfrac{3\frac{3}{4}}{T}$ 47. $\dfrac{U}{1\frac{1}{2}} = \dfrac{\frac{1}{2}}{\frac{3}{4}}$

48. $\dfrac{\frac{1}{2}}{\frac{2}{3}} = \dfrac{\frac{3}{4}}{V}$ 49. $\dfrac{\frac{3}{2}}{\frac{1}{2}} = \dfrac{X}{\frac{2}{3}}$ 50. $\dfrac{9}{Y} = \dfrac{1\frac{1}{2}}{3\frac{2}{3}}$ 51. $\dfrac{Z}{2\frac{1}{3}} = \dfrac{1\frac{1}{2}}{4\frac{1}{5}}$

▲ B Problems

52. This problem will help you check your work in Problems 28–51. Fill in the capital letters from Problems 28–51 to correspond with their numerical values in the boxes. For example, if

$$\frac{W}{7} = \frac{10}{14}$$

then

$$W = \frac{7 \times 10}{14} = 5$$

Now find the box or boxes with number 5 in the corner and fill in the letter W. This has already been done for you. (The letter O has also been filled in for you.) Some letters may not appear in the boxes. When you are finished filling in the letters, darken all the blank spaces to separate the words in the secret message. Notice that one of the blank spaces has also been filled in to help you.

24	5	$\frac{1}{4}$	20	16	32		3	9	*O*	36	15	8	4	$\frac{2}{3}$	7
5 W	16	15	15	12	1	2	56		1 *O*	36	$\frac{1}{4}$	8	56	15	22
13	30	32	32	16	32	$\frac{1}{4}$	7	22	1 *O*	12	$\frac{1}{2}$	16	2	24	
25	16	2	56	16	2	21	6	8	9	9	*O*	9	32	!	!

53. If 4 melons sell for \$.52, how much would 7 melons cost?

54. If a 184-mile trip took $11\frac{1}{2}$ gallons of gas, how much gas is needed for a 160-mile trip?

55. If a 121-mile trip took $5\frac{1}{2}$ gallons of gas, how many miles can be driven with a full tank of 13 gallons?

56. If a family uses $3\frac{1}{2}$ gallons of milk per week, how much milk will this family need for four days?

57. If 2 gallons of paint are needed for 75 ft of fence, how many gallons are needed for 900 ft of fence?

58. If Jack jogs 3 miles in 40 minutes, how long will it take him (to the nearest minute) to jog 2 miles at the same rate?

59. If Jill jogs 2 miles in 15 minutes, how long will it take her (to the nearest minute) to jog 5 miles at the same rate?

60. A moderately active 140-pound person will use 2,100 calories per day to maintain that body weight. How many calories per day are necessary to maintain a moderately active 165-pound person?

61. If $\dfrac{V}{T} = \dfrac{V'}{T'}$, find V' when $V = 175$, $T = 300$, and $T' = 273$.

62. If $\dfrac{PV}{T} = \dfrac{P'V'}{T'}$, find V' when $V = 12$, $P = 2$, $T = 300$, $P' = 8$, and $T' = 400$.

63. The *pitch* of a roof is the ratio of the rise to the half-span. If a roof has a rise of 8 feet and a span of 24 feet, what is the pitch?

64. What is the pitch of a roof (see Problem 63) with a 3-foot rise and a span of 12 feet?

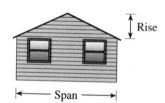

Rise

Span

65. You've probably seen advertisements for posters that can be made from any photograph. If the finished poster will be 2 ft by 3 ft, it's likely that part of your original snapshot will be cut off. Suppose that you send in a photo that measures 3 in. by 5 in. If the shorter side of the enlargement will be 2 ft, what size should the longer side of the enlargement be so that the entire snapshot is shown in the poster?

66. Suppose you wish to make a scale drawing of your living room, which measures 18 ft by 25 ft. If the shorter side of the drawing is 6 in., how long is the longer side of the scale drawing?

67. If the property tax on a $180,000 home is $1,080, what is the tax on a $130,000 home?

▲ **Problem Solving**

68. To dilute a medication you can use the following formula:

$$\frac{\text{PERCENT OF DILUTE SOLUTION}}{\text{PERCENT OF ORIGINAL SOLUTION}} = \frac{\text{AMOUNT OF STRONG SOLUTION NEEDED}}{\text{AMOUNT OF DILUTED SOLUTION WANTED}}$$

How many units of a 10% solution are needed to prepare 500 units of a 2% solution?

69. At a certain hamburger stand, the owner sold soft drinks out of two 16-gallon barrels. At the end of the first day, she wished to increase her profit, so she filled the soft-drink barrels with water, thus diluting the drink served. She repeated the procedure at the end of the second and third days. At the end of the fourth day, she had 10 gallons remaining in the barrels, but they contained only 1 pint of pure soft drink. How much pure soft drink was served in the four days?

70. Answer the question in the following *Peanuts* cartoon strip.

PEANUTS reprinted by permission of UFS, Inc.

5.8 MODELING UNCATEGORIZED PROBLEMS

One major criticism of studying the common types of word problems presented in most textbooks is that students can fall into the habit of solving problems by using a template or pattern and be successful in class without ever developing independent problem-solving abilities. Even though problem solving may require algebraic skills, it also requires many other skills; a goal of this book is to help you develop a general problem-solving ability so that when presented with a problem that does not fit a template, you can apply techniques that will lead to a solution.

Most problems that do not come from textbooks are presented with several variables, along with one or more relationships among those variables. In some instances, insufficient information is given; in others there may be superfluous information or inconsistent information. A common mistake when working with these problems is to assume that you know which variable to choose as the unknown in the equation *at the beginning of the problem*. This leads to trying to take too much into your memory at the start, and as a result it is easy to get confused.

Developing a Strategy

Now you are in a better position to practice Polya's problem-solving techniques because you have had some experience in carrying out the second step of the process, which we repeat here for convenience.

Guideline for Problem Solving

First: You have to *understand the problem.*

Second: *Devise a plan.* Find the connection between the data and the unknown. Look for patterns, relate to a previously solved problem or to a known formula, or simplify the given information to give you an easier problem.

Third: *Carry out the plan.*

Fourth: *Look back.* Examine the solution obtained.

Remember that the key to becoming a problem solver is to develop the skill to solve problems *that are not like the examples shown in the text*. For this reason, the problems in this section will require that you go beyond copying the techniques developed in the first part of this chapter.

■■■■ EXAMPLE 1   Polya's Method

In California, state funding of education is based on the average daily attendance (ADA). Develop a formula for determining the ADA at your school.

Solution   We use Polya's problem-solving guidelines for this example. Most attempts at mathematical modeling come from a real problem that needs to be solved. There is no "answer book" to tell you when you are correct. There is often the need to do additional research to find needed information, and the need *not* to use certain information that you have to arrive at a solution.

Understand the Problem.   What is meant by ADA? The first step might be a call to your school registrar to find out how ADA is calculated. For example, in California it is based on weekly student contact hours (WSCH), and 1 ADA = 525 WSCH. Thus, you might begin by writing the formula

$$\frac{WSCH}{525} = TOTAL\ ADA$$

This means that 525 WSCH is 525/525 = 1 ADA or 1,050 WSCH is

$$\frac{1,050}{525} = 2\ ADA$$

In other words, we divide the WSCH by 525 to find the ADA.

Devise a Plan.   Mathematical modeling requires that we check this formula to see whether it properly models the TOTAL ADA at your school. How is the WSCH determined? In California, the roll sheets are examined at two census dates, and WSCH is calculated as the average of the census numbers. In mathematical modeling, you often need formulas that are not specified as part of the problem. In this case, we need a formula for calculating an average. We use the **mean:**

$$WSCH = \frac{Census\ 1 + Census\ 2}{2}$$

We now have a second attempt at a formula for TOTAL ADA:

$$\frac{Census\ 1 + Census\ 2}{2} \div 525 = TOTAL\ ADA$$

Carry Out the Plan.   Does this properly model the TOTAL ADA? Perhaps, but suppose a taxpayer or congressperson brings up the argument that funding should take into account absent people, because not all classes have 100% attendance each day. You might include an *absence factor* as part of the formula. In California, the agreed absence factor for funding purposes is 0.911. How would we incorporate this into the TOTAL ADA calculation?

$$\frac{Census\ 1 + Census\ 2}{2} \cdot \frac{Absence\ factor}{525} = TOTAL\ ADA$$

This formula could be simplified to

$$\text{TOTAL ADA} = 0.0008676190476(c_1 + c_2)$$

where c_1 and c_2 are the numbers of students on the roll sheets on the first and second census dates, respectively.

Look Back. In fact, a Senate bill in California states this formula as

$$\frac{\text{Census 1 WSCH} + \text{Census 2 WSCH}}{2} \cdot \frac{0.911 \text{ Absence factor}}{525} = \text{TOTAL ADA}$$

This is equivalent to the formula we derived. ▬

▬▬ EXAMPLE 2   Polya's Method

Suppose you wish to record a 3-hour movie on your VCR and want to obtain the best quality recording on a single videotape. Explain how you could do this. There are three modes for recording on your VCR: SP (standard play; tape will record 2 hours), LP (long play; tape will record 4 hours), and SLP (super long play; tape will record 6 hours). The best quality recording is on SP, medium quality on LP, and least quality on SLP.*

Solution   We use Polya's problem-solving guidelines for this example.

Understand the Problem.   If we record the movie at the SP speed, we will need more than one videotape. If we record the movie at the LP speed, we will be able to record the movie, but an empty 1-hour space would be left at the end, and we would be sacrificing quality by recording the entire movie at the LP speed. It seems that the solution is to begin in the LP mode and at some "crucial moment" switch to the SP mode with the goal that the 3-hour movie should just fill the tape.

Devise a Plan.   We begin with some assumptions. In reality, the actual lengths of VCR tapes are not exact. We will assume, however, that the times specified in the problem are exact. That is, a tape will play for exactly 2, 4, or 6 hours depending on the chosen speed. We also assume a linear relationship between recording time and the amount of tape used. Some experimentation will show us that the number of counts-per-hour on a built-in counter varies and is not linear, so we cannot use this meter on our VCR to determine when to make the switch.

Let us begin by drawing a picture of our problem:

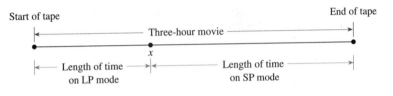

---
* The idea for this example is from an article by Gregory N. Fiore in *The Mathematics Teacher*, October 1988.

Let x be the length of time that we record on the LP mode. Then $3 - x$ is the length of time that we record on the SP mode.

How do we represent the fraction of the tape that is recorded in the LP mode?

If $x = 3$, then $\frac{3}{4}$ of the tape is used; remember that 4 hours of movies can be recorded in the LP mode.

If $x = 2$, then $\frac{2}{4}$ of the tape is used.

If $x = 1$, then $\frac{1}{4}$ of the tape is used.

Thus, the model seems to be

$$\frac{\text{TIME RECORDING IN THE LP MODE}}{4} = \frac{x}{4}$$

How do we represent the fraction of the tape that is recorded in the SP mode? Since this mode fills the tape in 2 hours, we see that the fraction of the tape recorded in the SP mode is

$$\frac{\text{TIME RECORDING IN THE SP MODE}}{2} = \frac{3 - x}{2}$$

Since we want to fill the entire tape with the movie, both fractions must add up to 1:

$$\frac{x}{4} + \frac{3 - x}{2} = 1$$

Carry Out the Plan.   Simplify this equation by multiplying both sides by 4:

$$x + 2(3 - x) = 4$$
$$x + 6 - 2x = 4$$
$$x = 2$$

The mode should be changed after 2 hours.

Look Back.   If we record for 2 hours on the LP mode, we have used up $\frac{1}{2}$ of the tape. We then switch to the SP mode and record for 1 hour, which uses up the other half of the tape.   ▬

▬ EXAMPLE 3   Polya's Method

An inlet pipe on a swimming pool can be used to fill the pool in 24 hours. The drain pipe can be used to empty the pool in 30 hours. If the pool is half-filled and then the drain pipe is accidentally opened, how long will it take to fill the pool?

Solution   We use Polya's problem-solving guidelines for this example.

Understand the Problem.   Draw a picture, if necessary. Water comes into the pool until it is half full. Then water begins draining out of the pool at the same time it is coming in. Is there a solution? What if the drain pipe alone emptied the pool in

Swimming pool
(cross-section)
                      Fill
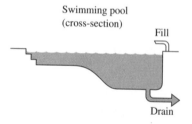
         Drain

20 hours? In that case, the pool would never be more than half full, and instead we could ask when it would be empty. Estimate an answer. Since it takes less time to fill than to drain, the inlet pipe is working faster than the drain, so it will fill. How long? Well, you might guess that it must be longer than 24 hours.

Devise a Plan. Let us denote the amount of water drained as the amount of work done. Then we see

WORK DONE BY INLET PIPE − WORK DONE BY DRAIN = ONE JOB COMPLETED

But wait, is this correct? The drain works only after the pool is half full. One of Polya's suggestions is to work a simpler problem. How long does it take to fill the first half of the pool? That's easy. Since the pool fills in 24 hours, it must be half full in 12 hours. Now, we can focus on filling the second half of the pool, so we come back to the guide we wrote down above.

Another step we can take when devising a plan is to ask whether there is an appropriate formula. In this case,

WORK DONE BY INLET PIPE = (RATE OF INLET PIPE)(TIME TO FILL)

$$= \frac{1}{24} \text{ (TIME TO FILL)}$$

WORK DONE BY DRAIN = (RATE OF DRAIN)(TIME TO FILL)

$$= \frac{1}{30} \text{ (TIME TO FILL)}$$

Carry Out the Plan. Substitute these formulas into the guide (for the second half of the pool):

WORK DONE BY INLET PIPE − WORK DONE BY DRAIN = ONE-HALF JOB COMPLETED

$$\uparrow \qquad\qquad \uparrow \qquad\qquad \uparrow$$

$$\frac{1}{24} \text{ (TIME TO FILL)} \quad - \quad \frac{1}{30} \text{ (TIME TO FILL)} \quad = \quad \frac{1}{2}$$

Let t = TIME TO FILL

> Remember that this is the time required to fill the second half of the pool.

$$\tfrac{1}{24}t - \tfrac{1}{30}t = \tfrac{1}{2}$$
$$(120)\tfrac{1}{24}t - (120)\tfrac{1}{30}t = (120)\tfrac{1}{2} \quad \text{Multiply both sides by 120.}$$
$$5t - 4t = 60$$
$$t = 60$$

It will take 12 hours to fill the first half of the pool, but it will take 60 more hours to fill the second half. Thus, it will take 72 hours to fill the pool.

Look Back. Does this answer make sense? In 72 hours, $\frac{72}{24} = 3$ swimming pools could be filled. The drain is working for 60 hours, so this is the equivalent of $\frac{60}{30} = 2$ swimming pools drained. Let's see: $3 - 2 = 1$ swimming pool, so it checks.

∎

We have refrained from working so-called age problems because they seem to be a rather useless type of word problem — one that occurs only in algebra books. They do, however, give us a chance to practice our problem-solving skills so they are worth considering. By the way, the precedent for age problems in algebra books goes back over 2,000 years (see Problem 34 about Diophantus). The age problem given in the next example was found in the first three frames of a *Peanuts* cartoon:

PEANUTS reprinted by permission of UFS, Inc.

███ **EXAMPLE 4**   **Polya's Method**

A man has a daughter and a son. The son is three years older than the daughter. In one year the man will be six times as old as the daughter is now, and in ten years he will be fourteen years older than the combined ages of his children. What is the man's present age?

Solution   We use Polya's problem-solving guidelines for this example.

Understand the Problem.   There are three people: a man, his daughter, and his son. We are concerned with their ages now, in one year, and in ten years.

Devise a Plan.   Does this look like any of the problems we have solved? We need to distinguish among the persons' ages now and some time in the future. Let us begin by translating the known information:

Ages now:

MAN'S AGE NOW
DAUGHTER'S AGE NOW
SON'S AGE NOW

We are given that

SON'S AGE NOW = DAUGHTER'S AGE NOW + 3   First equation

Ages in one year:

MAN'S AGE NOW + 1
DAUGHTER'S AGE NOW + 1
SON'S AGE NOW + 1

We are given that

MAN'S AGE NOW + 1 = 6(DAUGHTER'S AGE NOW)

This is the same as

MAN'S AGE NOW = 6(DAUGHTER'S AGE NOW) − 1   Second equation

Ages in ten years:

MAN'S AGE NOW + 10
DAUGHTER'S AGE NOW + 10
SON'S AGE NOW + 10

We are given that

MAN'S AGE NOW + 10 = (SON'S AGE NOW + 10) + (DAUGHTER'S AGE NOW + 10) + 14

14 years older

Subtract 10 from both sides:

MAN'S AGE NOW = SON'S AGE NOW + DAUGHTER'S AGE NOW + 24   Third equation

The plan is to begin with the third equation and evolve this into an equation we can solve.

Carry Out the Plan.

MAN'S AGE NOW = SON'S AGE NOW + DAUGHTER'S AGE NOW + 24

SON'S AGE NOW = (DAUGHTER'S AGE NOW + 3)   First equation

MAN'S AGE NOW = 6(DAUGHTER'S AGE NOW) − 1   Second equation

Substitute (as shown by the arrows):

6(DAUGHTER'S AGE NOW) − 1 = (DAUGHTER'S AGE NOW + 3) + DAUGHTER'S AGE NOW + 24

There is now one unknown, so let d = DAUGHTER'S AGE NOW. Then

$$6d - 1 = d + 3 + d + 24$$
$$6d - 1 = 2d + 27$$
$$4d = 28$$
$$d = 7$$

From the second equation, MAN'S AGE NOW = 6(7) − 1 = 41. The man's present age is 41.

Look Back.   From the first equation, SON'S AGE NOW = 7 + 3 = 10. In one year the man will be 42, and this is 6 times the daughter's present age. In ten years the man will be 51 and the children will be 20 and 17; the man will be 14 years older than 20 + 17 = 37.

▬▬▬ **EXAMPLE 5**   **Polya's Method**

To number the pages of a bulky volume, the printer used 4,001 digits. It was estimated that the volume has 1,000 pages. Is this a good estimate? How many actual
pages has the volume?

Solution   We use Polya's problem-solving guidelines for this example.

Understand the Problem.   Page 1 uses one digit; page 49 uses two digits; page 125
uses three digits. Assume that the first page is page 1 and that the pages are numbered consecutively.

Devise a Plan.   This does not seem to fit any of the problem types we have considered in this book. Can you reduce this to a simpler problem? If the book has exactly
9 numbered pages, how many digits does the printer use? If the book has exactly
99 numbered pages, how many digits does the printer use? The plan is to count
the number of digits on single-digit pages, then those on double-digit pages, then
on triple-digit pages.

Carry Out the Plan.

Single-digit pages:	9 digits total	
Double-digit pages:	$2(90) = 180$ digits total	Do you see why this is not 2(99)?
		Remember, you are counting 90
		pages (namely, from page 10 to
Triple-digit pages:	$3(900) = 2,700$ digits total	page 99).

We have now accounted for 2,889 digits. Since we are looking for 4,001 digits and
since we can estimate four-digit pages to be $4(9,000) = 36,000$, we need to know
how many pages past 999 are necessary. Let x be the total number of pages. Then
the number of digits on four-digit pages is

Four-digit pages: $4(x - 999)$

Now we can solve for x by using the following equation:

$$9 + 180 + 2,700 + 4(x - 999) = 4,001$$
$$4x + 2,889 - 3,996 = 4,001$$
$$4x - 1,107 = 4,001$$
$$4x = 5,108$$
$$x = 1,277$$

Look Back.   The book has 1,277 pages.   ▬

PROBLEM SET 5.8

▬▬▬▬▬▬▬▬▬▬▬▬▬▬▬▬▬▬▬▬▬▬▬▬▬▬▬▬▬

▲ **A Problems**

1. **IN YOUR OWN WORDS** Discuss the difference between solving word problems in textbooks and problem solving outside the classroom.

2. **IN YOUR OWN WORDS** In the first problem set of this book, we asked you to discuss Polya's problem-solving model. Now that you have spent some time reading this book, we ask the same question again to see whether your perspective has changed at all. Discuss Polya's problem-solving model.

3. The product of two consecutive positive even numbers is 440. What are the numbers?

4. The product of two consecutive positive odd numbers is 255. What are the numbers?

5. The sum of an integer and its reciprocal is $\frac{5}{2}$. What is the integer?

6. The sum of a number and its reciprocal is $2\frac{1}{6}$. What is the number?

7. The sum of a number and twice its reciprocal is 3. What is the number?

8. The sum of a number and three times its reciprocal is 4. What is the number?

9. Sonoma State University has a 60-piece band and a 32-piece orchestra. If 19 people are members of both the band and the orchestra, can the band and orchestra travel in two 40-passenger buses?

10. In a sample of defective tires, 72 have defects in materials, 89 have defects in workmanship, and 17 have defects of both types. How many tires are in the sample of defective tires? Assume that the sample consists of only tires with defects in materials or workmanship.

11. The Standard Oil and the Sears buildings have a combined height of 2,590 ft. The Sears Building is 318 ft taller. What is the height of each of these Chicago towers?

12. The combined area of New York and California is 204,192 square miles. The area of California is 108,530 square miles more than that of New York. Find the land area of each state.

13. What is the sum of the first 100 consecutive even numbers?

▲ **B Problems**

14. An inlet pipe on a swimming pool can be used to fill the pool in 24 hours. The drain pipe can be used to empty the pool in 30 hours. If the pool is two-thirds filled and then the drain pipe is accidentally opened, how long will it take to fill the pool?

15. An inlet pipe on a swimming pool can be used to fill the pool in 36 hours. The drain pipe can be used to empty the pool in 40 hours. If the pool is half filled and then the drain pipe is accidentally opened, how long will it take to fill the pool?

16. A survey of 100 persons at Better Widgets finds that 40 jog, 25 swim, 16 cycle, 15 both swim and jog, 10 swim and cycle, 8 jog and cycle, and 3 persons do all three. How many people at Better Widgets do not take part in any of these three exercise programs?

17. A survey of executives of Fortune 500 companies finds that 520 have MBA degrees, 650 have business degrees, and 450 have both degrees. How many executives with MBAs have nonbusiness degrees?

18. Solve the videotape example for taping a movie of length ℓ where $2 < \ell \le 4$. (If $\ell \le 2$, then the solution is obvious: just use the SP mode).

19. Reconsider Problem 18 if you are recording the movie and are eliminating the commercials from the taping process. Suppose that the movie is ℓ hours long and that there are c minutes of commercials each hour.

20. A swimming pool is 15 ft by 30 ft and an average of 5 ft in depth. It takes 25 minutes longer to fill than to drain the pool. If it can be drained at a rate of 15 ft³/min faster than it can be filled, what is the drainage rate?

▲ **Problem Solving**

The following problems do not match the examples in this book, but all the problem-solving techniques necessary to answer these questions have been covered in this book. These problems are designed to test your problem-solving abilities.

21. After paying a real estate agent a commission of 6% on the selling price, $2,525 in other costs, and $65,250 on the mortgage, you receive a cash settlement of $142,785. What was the selling price of your house?

22. An investor has $100,000 to invest and can choose between insured savings at 8.5% interest or annuities paying 12.25%. If the investor wants to earn an annual income of $9,850, how much should be invested in each?

23. The longest rod that will just fit inside a rectangular box, if placed diagonally top to bottom, is 17 inches. The box is 1 inch shorter and 3 inches longer than it is wide. How much must you cut off the rod so that it will lie flat in the bottom of the container? What are the dimensions of the box?

24. A canoeist rows downstream in 1.5 hours and back upstream in 3 hours. What is the rate of the current if the distance rowed is 9 miles in each direction?

25. A plane makes a 630-mile flight with the wind in 2.5 hours; the return flight against the wind takes 3 hours. Find the wind speed.

26. A plane makes an 870-mile flight in $3\frac{1}{3}$ hours against a strong head wind, but returns in 50 minutes less with the wind. What is the plane's speed without the wind?

27. A plane with a light tail wind makes its 945-mile flight in 3 hours. The return flight against the wind takes a half hour longer. What is the wind speed?

28. A tour agency is booking a tour and has 100 people signed up. The price of a ticket is $2,500 per person. The agency has booked a plane seating 150 people at a cost of $240,500. Additional costs to the agency are incidental fees of $500 per person. For each $5 that the price is lowered, a new person will sign up. How much should be charged for a ticket so that the income and expenses will be the same? The lower price will be charged to all participants, even those already signed up.

29. Foley's Village Inn offers the following menu in its restaurant:

Main course	Dessert	Beverage
Prime rib	Ice cream	Coffee
Steak	Sherbet	Tea
Chicken	Cheesecake	Milk
Ham	Chocolate mousse	Sanka
Shrimp		Soft drink
Hamburger		

In how many different ways can someone order a meal consisting of one choice from each category?

30. Imagine that you have written down the numbers from 1 to 1,000,000. What is the total number of zeros you have written down?

31. Suppose you have a large bottle with a canary inside. The bottle is sealed, and it's on a scale. The canary is standing on the bottom of the bottle. Then the canary starts to fly around inside of it. Does the reading on the scale change?

32. Suppose it were possible to count all of the individual strands of hair on your head. Also suppose that it is possible to do that for any number of people. For example, if you have 4,890 hairs on your head and I have 1,596 hairs, then the product of the numbers of hairs on both our heads in 7,804,440. How many hairs are there in the product for all persons in New York City at midnight on New Year's Eve January 1, 1998?

33. An ancient society called a number *sacred* if it could be written as the sum of the squares of two counting numbers.
    a. List five sacred numbers.
    b. Is the product of any two sacred numbers sacred?

34. **HISTORICAL QUESTION** Diophantus wrote the following puzzle about himself in *Anthologia Palatina*:

> *Here lies Diophantus. The wonder behold—*
> *Through art algebraic, the stone tells how old:*
> *"God gave him his boyhood one-sixth of this life,*
> *One-twelfth more as youth while whiskers grew rife;*
> *And then yet one-seventh their marriage begun;*
> *In five years there came a bouncing new son.*
> *Alas, the dear child of master and sage*
> *Met fate at just half his dad's final age.*
> *Four years yet his studies gave solace from grief;*
> *Then leaving scenes earthly he, too, found relief."*

What is Diophantus' final age?

5.9   INTRODUCTION TO COMPUTER BASIC

Earlier in this chapter we discussed an important category of computer programs called spreadsheets. Computer programs must be written in a computer language. One of the most widely used languages was developed in 1965 by John Kemeny to teach introductory computer programming at Dartmouth College. It is known as BASIC, and today has many different variations. In this section, we'll discuss a small number of the many BASIC commands and show how they can be used to write programs that may be useful to you in understanding and exploring some of the concepts covered in this book.* Even if you do not have occasion to do much programming, the techniques you learn in this section will make you more com-

---

*If you are interested in doing more programming, there are numerous books available to help you learn BASIC.

fortable in using the common applications packages, as well as strengthen your algebraic skills.

There are three categories of BASIC computer commands: *system commands, editor commands,* and *program commands* (see Figure 5.9). We will discuss the system commands you will need, but you must find out how to use them for the particular computer you will be using. The newer versions of BASIC provide a work environment of menus that can be made to appear on the screen as **pull-down windows** anytime you need them. Editor commands allow you to make corrections, and the program commands instruct BASIC to perform certain tasks.

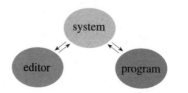

▲ **Figure 5.9 Types of commands in BASIC**

System Commands

When you run an existing program, you must LOAD it into the computer memory. To **load** a program means to copy it from a floppy disk or a hard disk into the computer memory, RAM. The original program remains unaltered.

If you wish to enter a new program or alter an old one, you must leave the system and enter the editor. Find out, before you begin, how to enter and how to exit the editor. Otherwise, you run the risk of sitting in front of the computer feeling frustrated and silly, or worse, of losing everything you have just typed. It is important to learn how to SAVE your work. You must learn how to give your program (that is, your work) a name so that you can keep both the original and a new version. Many versions of BASIC will automatically keep track of this for you. Many programs are similar to each other and you can often save a lot of time by editing a copy of an existing program.

Be sure to save your program *before* you run it. If it has a tiny mistake, the machine can get "hung up" so that the only thing to do is to start over. If the program is only in machine memory (that is, you have not saved it first), all your work will be lost. If you have a copy of it, you can LOAD again.

If you want a copy of your program on paper, called a hard copy, you will need to find out how to PRINT out a copy. Occasionally, there is an error in a program that causes the computer either to do the same thing over and over, or to refuse to do anything at all. To find out what program you are working on, you can use the system command called LIST. This command tells the computer to list the program in its memory. You will then see it on the monitor, or you can PRINT to obtain a hard copy.

Editor Commands

The function of the editor is to allow you to enter your program and make changes. Find the **cursor** and experiment by moving the cursor around the monitor. You can either type over a mistake or insert something new into the text. See if you can find the INSERT key. Usually you will want to set this key so that it will automatically insert. You also need to find out how to delete single letters, words, or lines. All these can vary according to your equipment and version of BASIC.

Checklist: What you need to know about your system before you get started:

> How to enter and exit the editor
>
> How to move the cursor, insert and delete text
>
> How to save a program with a given name (chosen by you)
>
> How to run a program (RUN command)
>
> How to list a program (LIST command)
>
> How to reboot (restart) when a program is "hung up"

Program Commands

A BASIC program is made up of a sequence of statements called **commands.** Different commands instruct the computer to carry out different processes. In the newer forms of BASIC, such as structured BASIC, line numbers are not used. However, since many of you may still be using versions of BASIC that require line numbers, and since structured BASIC allows line numbers, we will use them in this book. If your version of BASIC does not require line numbers, simply omit them from what is shown in this book. The commands we use in this book will work for both structured BASIC and for older forms of BASIC. If line numbers are required, the computer will execute the commands in ascending order of line numbers. We usually use the line numbers 10, 20, 30 (rather than 1, 2, 3) so that we can insert new statements to debug or modify the program at some later time. If line numbers are not required, the computer will execute the commands in the order in which they are typed in the program.

In this book we will discuss how to include comments in the program, input and output commands, and then how to execute three important programming concepts: **assignment, loop,** and **branch.**

Comments or Remarks in a Program

One way to make a program as easy as possible to read is to include **comments** (or remarks) within the program. If informative comments are included in a well-organized program, then someone (you or your instructor) reading the program sometime after it was written will have no difficulty understanding what the program does. It is a good idea to include the name of the file (the program you are writing) and your name on the first line. To include a remark in the program, start the line with the command REM. If the remark requires more than one line, the command REM must be placed at the beginning of each line as in the examples below. Any information on a line so marked will be ignored when the program is running.

Print Command

Before we start telling the computer what to do, we will discuss how to input and output commands. It is important for a program to be able to send messages, called **prompts,** to the screen or a printer. To send a message to the screen (the monitor),

Historical Note

JOHN KEMENY
(1927–1992)

John Kemeny was born in Budapest, Hungary, and is famous for inventing BASIC. He was a professor of mathematics and computer science as well as President of Dartmouth College. The most enjoyable part of mathematics for Kemeny was the thrill of discovering new theorems. After he came to the United States, he worked on the Los Alamos Project and was research assistant to Albert Einstein. In 1979, President Carter named Kemeny to head a 12-member commission that investigated the accident at the Three Mile Island nuclear power plant. Tom Kurtz and Kemeny wanted to open up computing to all Dartmouth students, so they wrote the BASIC computer language for their students. Kemeny spent his lifetime working for students, and when he died was working on a new version of BASIC called TruBASIC. ▲

we use the command PRINT. To send output to a printer, we use the command LPRINT.

Suppose you wish to have the computer type the phrase "I love you." To do this, you will use a command PRINT.

```
This is the line number.
    ↓
   ┌─┐
   10    PRINT ''I love you.''
                └─────┘
This is the command.
   20  END
```

A set of directions to the computer is called a **program.** The box above shows a very simple program. The line number is your (the programmer's) choice. It specifies the order in which the commands of a program will be carried out. The line number is followed by a command, and the command is followed by some specific directions. Note that whatever is enclosed in quotation marks after the command PRINT will be printed by the computer. The end of the program is designated by the END command. When you type RUN, the program will be executed (carried out).

■■■ EXAMPLE I

Tell what each of the programs will do when given a RUN command.

Program A

```
10    PRINT ''I love you.''
20    END
```

Program B

```
10    PRINT ''I am Robbie the Robot.''
20    PRINT ''I love you.''
30    PRINT ''Goodbye''
40    END
```

Solution

a. RUN: I love you.

b. RUN: I am Robbie the Robot.
         I love you.
         Goodbye

The following example does some numerical calculations. BASIC programs recognize the order of operations of algebra and use the operation symbols we introduced for spreadsheets in Section 5.3.

■■■ EXAMPLE 2

Write the following BASIC programs.

a. Calculate $5 \times 3 - 6$.

b. Calculate the area of a circle with radius 6.3. Use the formula $A = \pi r^2$ and approximate π by 3.1416.

Solution

a. 10 PRINT 5*3−6   b. 10 PRINT 3.1416*6.3^2
    20 END   20 END

Several computations can be done with one program. Each new line in the program will require that the answer be on a new line. Consider the following example.

```
10  PRINT 55 + 11
20  PRINT 55*11
30  PRINT 55 − 11, 55/11
50  END
RUN
66
605
44        5
```

A single PRINT statement can contain more than one expression, as shown above; notice that in such a case the result is printed on the same line (up to the permissible width of the page). If the output for one line is too long, the end of it automatically wraps to the next line. If a PRINT statement contains more than one expression, those expressions are separated by commas or semicolons.

A PRINT command can also be used for spaces in the output. For example, the command PRINT followed by no variables or numerals will simply generate a line feed — that is, skip a line.

```
10  PRINT 1 + 2
20  PRINT 2 + 3
30  PRINT
40  PRINT
50  PRINT 4 + 5
60  END

RUN
 3
 5

 9
```

Remember, the PRINT command has several functions. The first is to perform arithmetic calculations (as in lines 10, 20, and 50), the second is to type out text (as illustrated in the following example), and the third is to skip lines (as in lines 30 and 40).

```
10 PRINT ''4 + 7 ='', 4 + 7
20 END

RUN
4 + 7 = 11
```

Input Command

To send information into the computer from the keyboard, we must name a variable, such as X or Y and use the command INPUT:

```
INPUT X   or   INPUT X, Y
```

After the command INPUT X, the user must type in a value for X and then press RETURN. After the command INPUT X, Y, the user must type in two numbers, separated by a comma, and then press RETURN.

■■■■ EXAMPLE 3

Write a program to evaluate $x^3 - 3x^2 + 2x - 8$ for different values of x.

Solution

```
10 PRINT ''I WILL EVALUATE X^3 - 3X^2 + 2X - 8.''
20 PRINT ''WHAT IS THE VALUE OF X'';
30 INPUT X
40 PRINT ''IF X ='',X,''THEN THE VALUE IS '',X^3-3*X^2+2*X-8
50 END
```

Notice the semicolon at the end of line 20. This causes the computer **not** to do a line feed when it moves to line 30. The INPUT command by itself prints out "?", so if the line in front of the INPUT is written as a question (without question mark, but with a semicolon), the user will have a proper prompt for the question mark. Let us begin by taking a closer look at line 40:

Material in quotation marks is printed out exactly as shown within the quotes:

```
40 PRINT ''IF X = '' ,X, '', THEN THE VALUE IS '' , X^3-3*X^2+2*X-8
```

Material without quotation marks will print out values; in this example, X without quotes will print out *the value of x*, and X^3-3*X^2+2*X-8 will print out the *value* of the expression $x^3 - 3x^2 + 2x - 8$ for the input value of x.

```
RUN

I WILL EVALUATE X^3 - 3X^2 + 2X - 8.
WHAT IS THE VALUE OF X? 1    ← This is the value of x (user's choice).
IF X = 1, THEN THE VALUE IS -8

RUN

WHAT IS THE VALUE OF X? 5.25
IF X = 5.25, THEN THE VALUE IS 64.515625
```

In algebra, variables are single letters, but in programming, variables can contain up to eight letters.* For example, in algebra you might write $I = Prt$, but a programmer could define the variables by using words:

```
INTEREST = PRNCPL*RATE*TIME
```

If you want to include input that is not just a number (for example, a name or other symbol), you need to follow the name of the variable by a dollar sign.

■ EXAMPLE 4

Follow the directions of the given program to tell what the output will be.

```
10 PRINT ''WHAT IS YOUR NAME'';
20 INPUT NAME$
30 PRINT ''WHAT IS YOUR AGE'';
40 INPUT AGE
50 END
```

Solution

```
RUN
WHAT IS YOUR NAME? KARL    ← Fill in your own name here.
WHAT IS YOUR AGE? 39
```

If you wanted to add a question about your phone number to Example 4, you would use a dollar sign because the telephone number does not consist only of numerals (it contains a hyphen).

Assignment: Let Command

We assign values to locations by using a LET command.† For example, we might say

```
10 LET A = 2
20 LET B = 5
30 LET PI = 3.1416
40 PRINT A, B
50 PRINT PI
60 PRINT A + B
70 PRINT
80 PRINT PI*B^2
100 END
```

If you want to change the value of A from 2 to 8 in the above program, there are two ways to do this:

```
LET A = 8    LET A = A + 6
```

The second way of changing the value A to the value 8 illustrates that the equal sign is not used in the same way as in algebra.

---

* There are certain combinations of letters that are reserved for special purposes, but except for those limitations, you can choose any combination of eight (or fewer) symbols to be a variable.

† In the newer versions of BASIC, the "LET" may be omitted. A statement could be written X = 6. We will continue the practice of using LET, for instructional purposes.

The exact meaning of

LET A = A + 6

is

LET *the new* A = *the old* A + 6

Whatever is on the left of the equal sign is new; whatever is on the right is old.

Also, in algebra you are used to a symmetric property of equality which says that $A = 2$ is the same as $2 = A$. This is not true in programming. The command

90 LET A = B

inserted in the above program will cause the value of A to be changed from 2 to 5, and B will remain unchanged. On the other hand, if we instead used

90 LET B = A

then B will change in value from 5 to 2 and A will be unchanged.

▬▬ EXAMPLE 5

Tell what the output will be with the following program.

```
10 LET X = 3
20 LET Y = 7
30 LET X = X + Y
40 LET Y = Y - X
50 PRINT X
60 PRINT Y
70 END
```

Solution   Two things may have caught your eye when you tried to run this program. The first is that it would be better to indicate in the output which number is X and which is Y. Replace lines 50 and 60 with:

```
50 PRINT ''X = '';X
60 PRINT ''Y = '';Y
```

We are using a semicolon on these lines to combine two commands on one line. By two commands, we mean a print command to print text (the part with the quotes) and the part where a numerical value for the variable is printed.

The second thing to note is that the value of Y may be a little surprising. Did you expect it to be 4 since $7 - 3 = 4$? Let's see why that is not the case.

Make a table of successive values for X and Y to trace out the values for the variable on each line, as well as the eventual output. Remember that if you were using a computer, the only thing you would see is what is listed in the "Output" column of the following table.

	X	Y	*Output*
10 LET X = 3	3		
20 LET Y = 7	3	7	
30 LET X = X + Y	10	7	
40 LET Y = Y − X	10	− 3	
50 PRINT ''X = ''; X			$X = 10$
60 PRINT ''Y = ''; Y			$Y = - 3$
70 END			

Suppose you want to have the output for lines 50 and 60 on the same line. The semicolon is also used at the **end** of a line with a print command to instruct the computer not to give a line feed when moving to the next line. For example, if line 50 were changed to:

```
50 PRINT ''X = '';X;
```

the output would be:

```
X = 10   Y = -3
```

Loops

If you want to do the same thing over and over again, and do not want to type in the instructions repeatedly, you can use the idea of a **loop.** A loop executes a sequence of commands and then goes back to the start of that sequence and does it again and again. This command is called the FOR-NEXT loop. Consider the following program. Notice the spacing in the program; the indentation for the PRINT command in the loop can help you to follow through the loop.

```
10 FOR J = 1 to 4
20     PRINT ''Happy Birthday''
30 NEXT J
40 END
```

The program sets the value of J = 1 the first time it encounters the command and every time it passes the NEXT J command it raises the value by 1 and returns to the FOR statement. As long as J is no larger than 4, the command indented (PRINT "Happy Birthday") is executed again.

```
RUN
Happy Birthday
Happy Birthday
Happy Birthday
Happy Birthday
```

The counting variable (J in this example) can also be part of the calculation, as shown in Example 6.

■■■ EXAMPLE 6

' What is the output for the following program?

```
10 LET A = 0
20 FOR K = 1 TO 5
30      LET A = A + K
40 NEXT K
50 PRINT ''THE SUM OF THE FIRST 5 COUNTING NUMBERS IS'', A
60 END
```

Solution Trace the values of A and K:

Line number	Value of A	Value of K	Output
10	0		
20	0	1	
30	1	1	
40	1	2	

Transfers back to line 20

20	Check to see if the value of K is 5 (the last no. shown)		
30	3	2	
40	3	3	

Transfers back to line 20

20	Check to see if the value of K is 5		
30	6	3	
40	6	4	

Transfers back to line 20

20	Check to see if the value of K is 5		
30	10	4	
40	10	5	

Transfers back to line 20

20	Check to see if the value of K is 5; it is, so this is the last time through the loop; this time skip over step 40 (to end the loop)		
30	15	5	
50	15	5	

THE SUM OF THE FIRST 5 COUNTING NUMBERS IS 15   ■

We can easily modify this program so that it will find the sum of the first N counting numbers:

```
 5 PRINT
10 LET A = 0
15 PRINT ''I will find the sum of the first N counting numbers.''
16 PRINT
17 PRINT ''What is the value of N'';
18 INPUT N
20 FOR K = 1 TO N
30    LET A = A + K
40 NEXT K
50 PRINT ''THE SUM OF THE FIRST''; N; ''COUNTING NUMBERS IS''; A
60 END
```

Branch

Sometimes we want to transfer from one point of the program to another only if certain conditions are met. The IF-THEN and IF-ELSE commands give the computer its basic decision-making ability. The structure of these commands allows us to force the computer to take one of two paths depending on the truth or falsity of a logical expression. The command has the following two forms:

Form 1:
```
IF...THEN
   ...
END IF
```
Example:
```
IF X > 2 THEN
         PRINT X
END IF
```

Form 2:
```
IF...THEN
   ...
ELSE
   ...
END IF
```
Example:
```
IF X > 2 THEN
         Y = X^2 + 3
ELSE
         Y = 3*X + 1
END IF
```

TABLE 5.I BASIC symbols	
Math Symbol	*BASIC*
=	=
>	>
≥	>=
<	<
≤	<=
≠	<>

These are two versions of the "IF" statement. In the first example, the statement PRINT X will be executed (carried out) only if $X > 2$; otherwise it will be skipped. In the second example, the formula $Y = X^2 + 3$ will be used to calculate a value for Y if $X > 2$ and the formula $Y = 3X + 1$ will be used to calculate Y if X is less than or equal to 2. Sometimes you may want other relation symbols as shown in Table 5.1.

▬▬▬ EXAMPLE 7

What would the following program output for you?

```
10 PRINT ''What is your age'';
20 INPUT AGE
30 IF AGE < 18 THEN
40         PRINT ''You are still too young to vote.''
50 ELSE
60         PRINT ''Be sure to register and VOTE in the next election.''
70 END IF
80 END
```

Solution   The answer to this example depends on your age.   ▬

PROBLEM SET 5.9

▲ **A Problems**

1. **IN YOUR OWN WORDS** What is a computer program?

2. **IN YOUR OWN WORDS** What is a line number?

3. **IN YOUR OWN WORDS** Distinguish editor, system, and program commands.

In Problems 4–9, assume that the given program is in a computer. Without using a computer, tell what a computer would output if you give a RUN command.

4.
```
10 PRINT 6+5
20 PRINT 6-5
30 PRINT 6*5, 6/3
40 END
```

5.
```
10 PRINT ''6+5='', 6+5
20 PRINT ''6-5='', 6-5
30 PRINT ''6*5='', 6*5
35 PRINT ''6/3='', 6/3
40 END
```

6.
```
10 LET A=2
20 PRINT ''Working''
30 LET B=3
40 PRINT A+B
50 END
```

7.
```
10 PRINT 4+5
20 PRINT ''Hello, '';
30 PRINT ''I LIKE YOU!''
40 PRINT ''4+5 = '';4+5
50 END
```

8.
```
 10 PRINT ''T'';
 20 PRINT ''H'';
 30 FOR J = 1 TO 2
 40    PRINT ''IS '';
 50 NEXT J
 60 PRINT ''C'';
 70 PRINT ''O'';
 80 FOR J = 1 TO 2
 90    PRINT ''R'';
100 NEXT J
110 PRINT ''ECT.''
120 END
```

9.
```
 10 PRINT ''THE '';
 20 PRINT ''M'';
 30 FOR K = 1 to 2
 40    PRINT ''ISS'';
 50 NEXT K
 60 PRINT ''IPPI '';
 70 PRINT ''IS '';
 80 PRINT ''W'';
 90 PRINT ''ET'';
100 PRINT ''.'';
110 END
120 PRINT ''HIPPY''
```

10. The program in Problem 8 is inefficient from a programmer's standpoint, because the same output can be generated by using only two commands. Refine the program in Problem 8 by writing a new program that is more efficient. By "refine" we mean write a better program that generates the same output.

11. The program in Problem 9 is inefficient from a programmer's standpoint, because the same output can be generated by using only two commands. Refine the program in Problem 9 by writing a new program that is more efficient. By "refine" we mean write a better program that generates the same output.

12. What is the output for the following program if you input a mental age of 12 and a chronological age of 10?

```
10 PRINT ''I will calculate IQ.''
20 PRINT ''What is mental age'';
30 INPUT M
40 PRINT ''What is chronological age'';
50 INPUT C
60 PRINT
70 PRINT ''IQ is '';
80 PRINT (M/C)*100
90 END
```

13. What is the output for the following program if you input a length of 5 and a width of 4?

```
10 PRINT ''I will calculate the perimeter '';
20 PRINT ''and area of a rectangle.''
30 PRINT ''What is the length'';
40 INPUT L
50 PRINT ''What is the width'';
60 INPUT W
70 PRINT ''The perimeter is '';
80 PRINT 2*(L+W);
90 PRINT ''AREA is '';
100 PRINT L*W
110 END
```

In Problems 14–21, write a BASIC program to perform the indicated calculations. If a variable is used, include an INPUT statement to request a particular value of the variable.

14. $5,000(1 + 0.06)^2$

15. $12,850(1 + \frac{0.09}{12})^{360}$

16. $\dfrac{23.5^2 - 5(61.1)}{2}$

17. $\dfrac{2(5.8) - 8.9^2}{5}$

18. $5x^2 + 17x - 128$

19. $12.8x^2 + 14.76x + 6\frac{1}{3}$

20. $5,000\left(1 + \dfrac{R}{12}\right)^{12n}$

21. $P\left(1 + \dfrac{R}{12}\right)^{360}$

Rewrite the programs in Problems 22–23 in four steps so that the output is unchanged.

22.

```
10 LET X = 5
20 LET Y = 9
30 LET A = X*Y
40 PRINT ''The area is ''; A
50 END
```

23.

```
10 LET R=2
20 LET PI=3.1416
30 LET A = PI*R^2
40 PRINT ''A = ''; A
50 END
```

24. Make three different choices for *x*, *y*, and *z*, and show the output.

```
10 INPUT X, Y, Z
20 LET M = (X+Y+Z)/3
30 PRINT ''X = ''; X, ''Y = ''; Y, ''Z = ''; Z, ''M = ''; M
40 PRINT "Try this again for some other values."
50 END
```

▲ B Problems

25. Suppose you are considering a job with a company as a sales representative. You are offered $100 per week plus 5% commission on sales. Write a program that will compute your weekly salary if you input the total amount of sales.

26. A manufacturer of sticky widgets wishes to determine a monthly cost of manufacturing on an item that costs $2.68 per item for materials and $14 per item for labor. The company also has fixed expenses of $6,000 per month (this includes advertising, taxes, plant facilities, and so on). Write a program that will compute monthly cost if you input the number of items manufactured.

In Problems 27–33, assume that the given program is active. Without using a computer, tell what a computer would output if you give a RUN command.

27.

```
10 FOR N = 1 TO 10
20    LET A = 2*N + 3
30    PRINT N, A
40 NEXT N
50 END
```

28.

```
10 LET A = 0
20 FOR J = 1 TO 5
30    LET A = A + J
40 NEXT J
50 PRINT A
60 END
```

29.

```
10 LET A = 1
20 FOR K = 1 TO 6
30    LET A = A*K
40 NEXT K
50 PRINT A
60 END
```

30.

```
10 LET A = 2
20 FOR K = 1 TO 4
30    LET A = A*K
40    PRINT A
50 NEXT K
60 END
```

31.
```
10 PRINT ''NUMBER'', ''NUMBER SQUARED''
20 FOR N = 1 TO 10
30    LET S = N^2
40     PRINT N, S
50 NEXT N
60 END
```

32.
```
10 PRINT ''N''
20 FOR N = 1 TO 10
30    LET V = 3 * N^2 - 20
40    IF V > 100 THEN
50           PRINT N
60    END IF
70 NEXT N
80 PRINT ''Values to N = 10 checked.''
90 END
```

33.
```
10 PRINT ''EVALUATE F''
20 FOR N = 1 TO 10
30    IF N < = 5 THEN
40           PRINT N
50    ELSE
60           PRINT N^2 - 36
70    END IF
80 NEXT N
90 PRINT ''The values to F have been checked from 1 to 10.''
100 END
```

Write a BASIC program to perform the tasks named in Problems 34–41.

34. Compare two numbers and tell which number is the larger number.

35. Calculate the sum of the first 100 numbers.

36. Calculate the square of the sum of the first 100 numbers.

37. Calculate the sum of the squares of the first 100 numbers.

38. Print out the first 100 square numbers.

39. Print the first one hundred numbers backwards.

40. Print out the first one hundred even numbers.

41. Print out the average of your test scores in a certain class.

42. The old song "100 Bottles of Beer on the Wall" is quite tedious because of the lyrics. Write a computer program that will generate an output:

100 bottles of beer on the wall.

99 bottles of beer on the wall.

98 bottles of beer on the wall.

⋮

2 bottles of beer on the wall.

1 bottle of beer on the wall.

That's all folks!

▲ **Problem Solving**

43. Write a BASIC program to find the roots of a linear equation $ax + b = 0$, $a \neq 0$.

44. Write a BASIC program to find the roots of a quadratic equation $ax^2 + bx + c = 0$, $a \neq 0$. In BASIC the command PRINT SQR(2) is used to find the square root of the number 2.

45. Think of a number. Add 9. Square. Subtract the square of the original number. Subtract 61. Multiply by 2. Add 24. Subtract 36 times the original number. Take the square root. What is the result?

  a. Show algebraically why everyone who works the problem correctly gets the same answer.

  b. Write a BASIC program to solve the puzzle. In BASIC, the square root of a number N is found by giving the command SQR(N).

46. This problem uses a function called RND that is part of the BASIC language. It generates a list of "random" numbers between 0 and 1. A list of random numbers jumps around from number to number with no pattern; each number is just as likely to be the next on the list as any other. Of course, since a program is used to make the list, the numbers are not really random, but are reasonably close to random numbers. The list of numbers depends on a "seed." If you use the same seed, you will get the same list. Report the results of running the following program 5 times on a computer where you choose 5 random numbers for each run.

```
10  REM RANDOM NUMBER GENERATOR
20  PRINT ''HOW MANY RANDOM NUMBERS DO YOU WANT'';
30  INPUT N
40  PRINT ''YOU MUST INPUT A POSITIVE NUMBER TO BEGIN THE RANDOM'';
45  PRINT ''NUMBER GENERATOR. WHAT IS YOUR NUMBER'';
50  INPUT SEED
60  PRINT
70  LET R = RND(-SEED)
80  FOR J = 1 TO N
90    LET R = RND
100   PRINT R
110 NEXT J
120 END
```

"Algebra is generous, she often gives more than is asked of her."

D'Alembert

NOVEMBER

1

2 **George Boole** (1815), logic, p. 80

3

4 **Johann Bernoulli** (1744), probability

5

6

7 **James Stirling** (1692), differential calculus

8 **Felix Hausdorff** (1869), set theory, topology

9 **Benjamin Banneker** (1731), surveyor; early American black mathematician, p. 491

10

11 **Bonaventura Cavalieri** (1598), geometry, astronomy

12

13 **Shirley Frye**

14

15

16 **Jean d'Alembert** (1717), differential equations, physics
August Möbius (1790), one-sided surface

17

18

19

20 **Benoit Mandelbrot** (1924), fractal geometry, p. 519

21

22

23 **John Wallis** (1616), logic, cryptography

24

25

26 **Norbert Wiener** (1894), communication theory, cybernetics

27

28

29 **Johann Doppler** (1803), acoustics, optics

30

Biographical Sketch

"I appreciate the structure and beauty of mathematics . . ."

Shirley Frye

Shirley Frye was born in Chicago, Illinois, on November 13, 1929, and has actively taught mathematics to students and teachers of all levels during 5 decades. She is past president of both the National Council of Supervisors of Mathematics (NCSM) and the National Council of Teachers of Mathematics (NCTM). She served as a key leader in the development and introduction of the NCTM's *Standards* documents, which provided direction for the dramatic reform in mathematics education. Shirley has received numerous awards, including the Louise Hay Award for Contributions to Mathematics Education.

"I appreciate the structure and beauty of mathematics that are continually being discovered and revealed in the world around us. I am excited about helping students and teachers learn school mathematics and developing a 'friendly feeling' toward numbers!" Shirley exclaimed with excitement in her voice.

Sexual stereotypes are hard to break. One evening, Shirley was standing at the door of her classroom meeting parents who were attending an open house. As she greeted the parents of students in her calculus class a father said, "You can't be the calculus teacher, you are not a gray-haired old man!" Shirley adds, "I then determined that changing the image of mathematics teachers was a necessity."

Book Reports

Write a 500-word report on the following book:

▲ *Hypatia's Heritage*, Margaret Alic (Boston: Beacon Press, 1986).

This book is a history of women in science from antiquity through the 19th century. Since most of recorded history has been male-dominated, history books reflect this male bias and have ignored the history of women. This book is a rediscovery of women in science.

Important Terms

Addition property [5.4]
Addition property of
   inequality [5.5]
Assignment [5.9]
BASIC [5.9]
Binomial [5.1]
Binomial theorem [5.1]
Branch [5.9]
Cell [5.3]
Command [5.9]
Common factor [5.2]
Comparison property [5.5]
Cursor [5.9]
Degree [5.1]
Difference of squares [5.2]

Division property [5.4]
Equation [5.4]
Equation properties [5.4]
Equivalent equations [5.4]
Evaluate an expression [5.3]
Factor [5.2]
FOIL [5.1]
FOR-NEXT [5.9]
IF-THEN [5.9]
Inequality [5.5]
INPUT [5.9]
LET [5.9]
Like terms [5.1]
Linear [5.1]
Linear equation [5.4]

LOAD [5.9]
Loop [5.9]
Monomial [5.1]
Multiplication property [5.4]
Multiplication property of
   inequality [5.5]
Multiplicity [5.4]
Polynomial [5.1]
PRINT [5.9]
Program [5.9]
Prompt [5.9]
Proportion [5.7]
Pull-down windows [5.9]
Quadratic [5.1]
Quadratic equation [5.4]

Quadratic formula [5.4]
Ratio [5.7]
Replication [5.3]
Root [5.4]
Satisfy [5.4]
SAVE [5.9]
Similar terms [5.1]
Solution [5.4]
Spreadsheet [5.3]
Subtraction property [5.4]
Symmetric property of
   equality [5.4]
Term [5.1]
Trinomial [5.1]
Zero-product rule [5.4]

Important Ideas

Know the basic terminology of polynomials [5.1]
Classify polynomials by degree and by number of terms [5.1]
Multiply binomials algebraically and geometrically [5.1]
Factor polynomials algebraically and geometrically [5.2]
Procedure for factoring trinomials [5.2]
Factor by looking for a common factor or a difference of
   squares [5.2]
Linear equations (use equation properties to get the variable
   on one side) [5.4]
Quadratic equations (use zero-product rule or the quadratic
   formula) [5.4]
Comparison property and number lines [5.5]
Linear inequalities (inequality properties; when inequalities
   are reversed) [5.5]

Solving proportions (know the property of proportions) [5.7]
BASIC programming [5.9]
Procedure for problem solving [5.6]
   Polya's method
   translate, evolve, solve, and answer
   develop a strategy
Problem-solving examples
   genetics [5.3]
   linear and quadratic equations [5.4]
   linear inequalities [5.6]
   number relationships [5.6]
   distance relationships [5.6]
   Pythagorean theorem applications [5.6]
   proportion problems [5.7]

Types of Problems

Simplify an algebraic expression [5.1]
Multiply binomials mentally [5.1]
Expand a binomial using the binomial theorem [5.1]
Factor a polynomial [5.2]
Evaluate an algebraic expression [5.3]
Use a spreadsheet to evaluate (including the replicate
   command) [5.3]
Solving equations (both linear and quadratic) [5.4]

Use a calculator to solve an equation [5.4]
Solving linear inequalities [5.5]
Solving proportion problems [5.7]
Use problem-solving methods to answer questions whose
   solutions are not obvious [5.8]
Solving proportion problems [5.7]
Read and follow BASIC programs [5.9]
Write BASIC programs [5.9]

CHAPTER 5   Review Questions

1. Simplify $(x - 1)(x^2 + 2x + 8)$.

2. Simplify $x^2(x^2 - y) - xy(x^2 - 1)$.

Completely factor the expressions in Problems 3–6.

3. $a^2b + ab^2$

4. $x^2 - 5x - 6$

5. $3x^2 - 27$

6. $x^2 + 5x + 6$ (use areas)

Solve the equations in Problems 7–18.

7. $2x + 5 = 13$

8. $3x + 1 = 7x$

9. $\dfrac{2x}{3} = 6$

10. $2x - 7 = 5x$

11. $\dfrac{P}{100} = \dfrac{3}{20}$

12. $\dfrac{25}{W} = \dfrac{80}{12}$

13. $x^2 = 4x + 5$

14. $4x^2 + 1 = 6x$

15. $3 < -x$

16. $2 - x \geq 4$

17. $3x + 2 \leq x + 6$

18. $14 > 5x - 1$

19. Fill in $=$, $>$, or $<$:

    a. $\dfrac{2}{3}$ ——— $\dfrac{67}{100}$

    b. $\dfrac{3}{4}$ ——— $\dfrac{75}{100}$

    c. $\dfrac{23}{27}$ ——— $\dfrac{92}{107}$

    d. 0.05 ——— 0.1

    e. 0.99 ——— 0.909

20. In a recipe for bread, flour and water are to be mixed in a ratio of 1 part water to 5 parts flour. If $2\frac{1}{2}$ cups of water is called for, how much flour (to the nearest tenth cup) is needed?

21. The sum of four consecutive even integers is 100. Find the integers.

22. A population of self-pollinating pea plants has two genes: T (tall, dominant) and t (short, recessive). Suppose we have an isolated population in which 52% of the genes in the gene pool are dominant T, and the other 48% are recessive t. What percent of the population has each genotype? What percent of the population has each phenotype?

23. The airline distance from Chicago to San Francisco is 1,140 miles farther than the flight from New York to Chicago, and 540 miles shorter than the distance from San Francisco to Honolulu. What is the length of each leg of the 4,980-mile New York–Chicago–San Francisco–Honolulu flight?

24. Write a BASIC language program to ask the value of x and then compute $18x^2 + 10$.

25. Use a spreadsheet to ask the value of x and then compute $18x^2 + 10$.

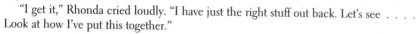

Revisiting the Real World . . .

IN THE REAL WORLD   Rhonda, a very small person, needs to move a very large, heavy object. She does not know what to do, so she asks her friend, Jane, for help. Jane tells Rhonda, "Let's get a big board. I remember from school that you can move a big object with a lever."

   "I get it," Rhonda cried loudly. "I have just the right stuff out back. Let's see . . . . Look at how I've put this together."

"What have you placed on the other end of the lever?" asked Jane.
"Well, it is a can of Coke. Will that not work?" asked Rhonda.

   "That is really dumb, Rhonda," said Jane sarcastically.
   Rhonda quickly responded, "Not if the weights of the can of Coke and the trunk are the same. You do not know how much each of these containers weighs, and I'm prepared to show you that they weigh the same!"

Commentary

Polya's method leads the way.

Understand the Problem.   In some sort of a silly game, Rhonda wants to show Jane that the weight of the can of Coke is the same as the weight of the heavy trunk.

Devise a Plan.   Rhonda asserts: "There must be some weight w (probably very large) so that

$$T = c + w$$

where T is the weight of trunk, and c is the weight of the can of Coke. The plan is to use the elementary properties of equations presented in this chapter to conclude that $T = c$.

Carry Out the Plan.

$T = c + w$	Given
$T(T - c) = (c + w)(T - c)$	Multiply both sides by $T - c$.
$T^2 - Tc = cT + wT - c^2 - wc$	Distributive property
$T^2 - Tc - wT = cT - c^2 - wc$	Subtract wT from both sides.
$T(T - c - w) = c(T - c - w)$	Use the distributive property to factor both sides.
$T = c$	Divide both sides by $T - c - w$.

"Thus," says Rhonda, "The weight of the trunk is the same as the weight of the can of Coke! Since their weights are the same, the two objects will balance on the lever."

Group Research

Working in small groups is typical of most work environments, and being able to work with others to communicate specific ideas is an important skill to learn. Work with three or four other students to submit a single report based on each of the following questions.

1. One step was left off of Polya's method in the above commentary; what was that? Sometimes the most important part of problem solving is to look back and see whether the answer makes sense. It does not! But what is wrong? Where is the error in the reasoning for this situation?

2. **HISTORICAL QUESTION** The Babylonians solved the quadratic equation $x^2 + px = q$ ($q > 0$) without the benefit of algebraic notation. Tablet 6967 at Yale University finds a positive solution to this equation to be

$$x = \sqrt{\left(\frac{p}{2}\right)^2 + q} - \frac{p}{2}$$

Using modern algebraic notation, show that this result is correct.

3. **HISTORICAL QUESTION** In 1907, the University of Göttingen offered the Wolfskehl Prize of 100,000 marks to anyone who could prove Fermat's Last Theorem, which seeks any replacements for x, y, and z such that $x^n + y^n = z^n$ (where n is greater than 2 and x, y, and z are counting numbers). In 1937, the mathematician Samuel Drieger announced that 1324, 731, and 1961 solved the equation. He would not reveal n — the power — but said that it was less than 20. That is,

$$1,324^n + 731^n = 1,961^n$$

However, it is easy to show that this cannot be a solution *for any n*. See if you can explain why by investigating some patterns for powers of numbers ending in 4 and 1.

4. Prove that there are infinitely many integers such that the sum of the digits in their square equals the sum of the digits in their cube.

5. **JOURNAL PROBLEM** (From *Journal of Recreational Mathematics*, Vol. II, #2) Translate the following message: Wx utgtuz f pbkz tswx wlx xwozm pbkzr, f exbmwo cxlzm xm ts jzszmfi fsv cxlzm lofwzgzm tswx wlx cxlzmr xe woz rfnz uzsxntsfwtxs fkxgz woz rzpxsu tr tncxrrtkiz, fsu T ofgz frrbmzuiv exbsu fs funtmfkiz cmxxe xe wotr kbw woz nfmjts tr wxx sfmmxl wx pxswftsu tw. *Ctzmmz Uz Ezmnfw*

6

The Nature of Sequences, Series, and Financial Management

To get money is difficult, to keep it is more difficult, but to spend it wisely is most difficult of all.

Anonymous

In the Real World . . .

Ron and Lorraine purchased a home some time ago, and they still owe $20,000. Their payments are $195 per month, and they have 10 years left to pay. They want to free themselves of the monthly payments by paying off the loan. They go to the bank and are told that, to pay off the home loan, they must pay $20,000 plus a "prepayment penalty" of 3%. (Older home loans often had prepayment clauses ranging from 1% to 5% of the amount to be paid off. Today, most lenders will waive this penalty. However, you must still request that the bank do this.) This means that, to own their home outright, Ron and Lorraine would have to pay $20,600 ($20,000 + 3% of $20,000). They want to know whether this is a wise financial move.

Contents

Perspective

The stated goal of this course is to strengthen your ability to solve problems — not the classroom type of problems, but those problems that you may encounter as an employee, a manager, or in everyday living. You can apply your problem-solving ability in your financial life. A goal of this chapter might well be to put some money into your bank account that you would not have had if you had not read this chapter. As a preview to this chapter, consider the question asked in Example 3 of Section 6.5:

Suppose you are 21 years old and will make monthly deposits to a bank account paying 10% annual interest compounded monthly. Which is the better option?

Option I: Pay yourself $200 per month for 5 years and then leave the balance in the bank until age 65. (Total amount of deposits is $200 × 5 × 12 = $12,000.)

Option II: Wait until you are 40 years old (the age most of us start thinking seriously about retirement) and then deposit $200 per month until age 65. (Total amount of deposits is $200 × 25 × 12 = $60,000.)

Warning!! The wrong answer to this question could cost you $4,000/mo for the rest of your life!

The real-world situation described above is a true story of the author and two of his friends, Ron and Lorraine. The purpose of this chapter is to help you with similar financial decisions.

6.1 INTEREST

Amount of Simple Interest

Certain arithmetic skills enable us to make intelligent decisions about how we spend the money we earn. One of the most fundamental mathematical concepts that consumers, as well as business people, must understand is *interest.* Simply stated, **interest** is money paid for the use of money. We receive interest when we let others use our money (when we deposit money in a savings account, for example), and we pay interest when we use the money of others (for example, when we borrow from a bank).

The amount of the deposit or loan is called the **principal** or **present value,** and the interest is stated as a percent of the principal, called the **interest rate.** The **time** is the length of time for which the money is borrowed or lent. The interest rate is usually an *annual interest rate,* and the time is stated in years unless otherwise given.

Simple Interest Formula

$$\text{INTEREST} = \text{PRESENT VALUE} \times \text{RATE} \times \text{TIME}$$

$$I = Prt$$

$I = $ AMOUNT OF INTEREST
$P = $ PRESENT VALUE (or PRINCIPAL)
$r = $ ANNUAL INTEREST RATE
$t = $ TIME (in years)

Suppose you save 20¢ per day, but only for a year. At the end of a year you will have saved $73. If you then put the money into a savings account paying 3.5% interest, how much interest will the bank pay you after one year? The present value (P) is $73, the rate ($r$) is 3.5% = 0.035, and the time (t, in years) is 1. Therefore,

$$I = Prt$$
$$= 73(0.035)(1)$$
$$= 2.555$$

You can do this computation on a calculator: $\boxed{73}$ $\boxed{\times}$ $\boxed{.035}$ $\boxed{=}$

Round money answers to the nearest cent: After one year, the interest is $2.56.

■■■ EXAMPLE 1

How much interest will you earn in three years with an initial deposit of $73?

Solution $I = Prt = 73(0.035)(3) = 7.665$. After three years, the interest is $7.67. ■

Future Value

There is a difference between asking for the amount of interest, as illustrated in Example 1, and asking for the **future value.** The future amount is the amount you will have after the interest is added to the principal, or present value. Let A = FUTURE VALUE. Then

$$A = P + I$$

■■■■ EXAMPLE 2

Suppose you see a car with a price of $12,436 that is advertised at $290 per month for 5 years. What is the amount of interest paid?

Solution   The present value is $12,436. The future value is the total amount of all the payments:

Monthly payment ⏜ Number of years ↓

$$\underbrace{\$290}_{} \times \underbrace{12}_{} \times 5 = \$17,400$$

Number of payments per year

Therefore, the amount of interest is:

$$
\begin{aligned}
I &= A - P \\
&= 17{,}400 - 12{,}436 \\
&= 4{,}964
\end{aligned}
$$

The amount of interest is $4,964.   ■

Calculator Comment

You can carry out all these calculations in one step:

290 × 12 × 5

– 12436 =

Display: *4964*

Interest for Part of a Year

The numbers in Example 2 were constructed to give a "nice" answer, but the length of time for an investment is not always a whole number of years. There are two ways to convert a number of days into a year:

> **Exact interest:**   365 days per year
>
> **Ordinary interest:**   360 days per year

Most applications and businesses use ordinary interest. So in this book, unless it is otherwise stated, assume ordinary interest; that is, use 360 for the number of days in a year:

Time in Ordinary Interest Formula

$$t = \frac{\text{ACTUAL NUMBER OF DAYS}}{360}$$

Calculating the time using ordinary interest often requires a calculator, as illustrated by Example 3 and the following Calculator Comment.

Calculator Comment

■■■ EXAMPLE 3

Suppose you want to save $3,650, and put $3,000 in the bank at 8% simple interest. How long must you wait?

$$I = Prt$$
$$650 = 3,000(0.08)t \qquad I = 3,650 - 3,000 = 650; P = 3,000; r = 0.08$$
$$650 = 240t$$
$$\frac{650}{240} = t$$

This is 2 years plus some part of a year. To change the fractional part to days, do not clear your calculator, but subtract 2 (the number of years); then multiply the fractional part by 360:

$\boxed{650}\ \boxed{\div}\ \boxed{240}\ \boxed{-}\ \boxed{2}\ \boxed{=}\ \boxed{\times}\ \boxed{360}\ \boxed{=}$　Display: **255**

The time is 2 years 255 days.

■■■ EXAMPLE 4

Suppose that you borrow $1,200 on March 25 at 21% simple interest. How much interest accrues to September 15 (174 days later)? What is the total amount that must be repaid?

Solution　We are given $P = 1,200$, $r = 0.21$ and

$$t = \frac{174}{360} \quad \begin{array}{l} \leftarrow \text{Actual number of days} \\ \leftarrow \text{Assume ordinary interest} \end{array}$$

$$I = Prt$$
$$= 1,200(0.21)\left(\frac{174}{360}\right)$$
$$= 121.8$$

> **Calculator Comment**
>
> Notice that a calculator is almost essential when working with a part of a year: $\boxed{1200}\ \boxed{\times}\ \boxed{.21}\ \boxed{\times}$ $\boxed{174}\ \boxed{\div}\ \boxed{360}\ \boxed{=}$
>
> Display: **121.8**

The amount of interest is $121.80. To find the amount that must be repaid, find the future value:

$$A = P + I = 1,200 + 121.80 = 1,321.80$$

The amount that must be repaid is $1,321.80.

It is worthwhile to derive a formula for future value because sometimes we will not calculate the interest separately as we did in Example 4.

$$\text{FUTURE VALUE} = \text{PRESENT VALUE} + \text{INTEREST}$$

$$A = P + I$$

$$= P + Prt \qquad \text{Substitute } I = Prt.$$

$$= P(1 + rt) \qquad \text{Distributive property}$$

Future Value Formula (Simple Interest)

$$A = P(1 + rt)$$

EXAMPLE 5

If $10,000 is deposited in an account earning $5\frac{3}{4}\%$ simple interest, what is the future value in 5 years?

Solution   We identify $P = 10,000$, $r = 0.0575$, and $t = 5$.

$$A = P(1 + rt)$$

$$= 10,000(1 + 0.0575 \times 5)$$

$$= 10,000(1 + 0.2875) \qquad \text{Don't forget order of}$$
$$\qquad\qquad\qquad\qquad\qquad \text{operations; multiplication first.}$$

$$= 10,000(1.2875)$$

$$= 12,875$$

The future value in 5 years is $12,875.

EXAMPLE 6

Suppose you have decided that you will need $4,000 per month on which to live in retirement. If the rate of interest is 8%, how much must you have in the bank when you retire so that you can live on interest only?

Solution   We are given $I = 4,000$, $r = 0.08$, and $t = \frac{1}{12}$ (one month $= \frac{1}{12}$ year):

$$I = Prt$$

$$4,000 = P(0.08)\left(\frac{1}{12}\right) \qquad \text{Substitute.}$$

$$48,000 = (0.08)P \qquad \text{Multiply both sides by 12.}$$

$$600,000 = P \qquad \text{Divide both sides by 0.08.}$$

You must have $600,000 on deposit to earn $4,000 per month at 8%.

Annual Compounding

Most banks do not pay interest according to the simple interest formula; instead, after some period of time, they add the interest to the principal and then pay interest on this new, larger amount. When this is done, it is called **compound interest.**

▬▬ EXAMPLE 7

Compare simple and compound interest for a $1,000 deposit at 8% interest for 3 years.

Solution   First, calculate the future value using simple interest:

$$A = P(1 + rt)$$
$$= 1,000(1 + 0.08 \times 3)$$
$$= 1,000(1.24) \qquad \text{Order of operations; multiplication first}$$
$$= 1,240$$

With simple interest, the future value in 3 years is $1,240.

Next, assume that the interest is **compounded annually.** This means that the interest is added to the principal after 1 year has passed. This new amount then becomes the principal for the following year. Since the time period for each calculation is 1 year, we let $t = 1$ for each calculation.

First year $(t = 1)$: $A = P(1 + r)$
$$= 1,000(1 + 0.08)$$
$$= 1,080$$
$$\downarrow$$

Second year $(t = 1)$: $A = P(1 + r)$ One year's principal is previous year's balance.
$$= 1,080(1 + 0.08)$$
$$= 1,166.40$$
$$\downarrow$$

Third year $(t = 1)$: $A = P(1 + r)$
$$= 1,166.40(1 + 0.08)$$
$$= 1,259.71$$

With interest compounded annually, the future value in 3 years is $1,259.71. The earnings from compounding are $19.71 more than from simple interest.   ▬

The problem with compound interest relates to the difficulty of calculating it. Notice that, to simplify the calculations in Example 7, the variable representing time t was given the value 1, and the process was repeated three times. Also notice that, after the future value was found, it was used as the principal in the next step. What if we wanted to compound annually for 20 years instead of for 3 years? Look at Example 7 to discover the following pattern:

Simple interest
(20 *years*)

$$A = P(1 + rt)$$
$$= 1{,}000(1 + 0.08 \times 20)$$
$$= 1{,}000(1 + 1.6)$$
$$= 1{,}000(2.6)$$
$$= 2{,}600$$

Annual compounding
(20 *years*)

$$A = \underbrace{P(1 + r)} \qquad \text{First year}$$
$$\downarrow$$
$$= \overbrace{P(1 + r)(1 + r)} \qquad \text{Second year}$$
$$= \underbrace{P(1 + r)^2} \qquad \text{Second year simplified}$$
$$\downarrow$$
$$= \overbrace{P(1 + r)^2(1 + r)} \qquad \text{Third year}$$
$$= P(1 + r)^3 \qquad \text{Third year simplified}$$
$$\vdots$$
$$= P(1 + r)^{20} \qquad \text{Twentieth year}$$

For a period of 20 years, starting with \$1,000 at 8% compounded annually, we have

$$A = 1{,}000(1.08)^{20}$$

The difficulty lies in calculating this number. For years we relied on extensive tables for obtaining numbers such as this, but the availability of calculators has made such calculations accessible to all. The calculation is shown in the Calculator Comment box in the margin. Round money answers to the nearest cent: \$4,660.96 is the future value of \$1,000 compounded annually at 8% for 20 years. This compounding yields \$2,060.96 *more* than simple interest.

Compounding Periods

Most banks compound interest more frequently than once a year. For instance, a bank may pay interest as follows:

Semiannually: twice a year or every 180 days

Quarterly: 4 times a year or every 90 days

Monthly: 12 times a year or every 30 days

Daily: 360 times a year

To write a formula for various compounding periods, we need to introduce three new variables. First, let

$n = $ NUMBER OF TIMES INTEREST IS CALCULATED EACH YEAR

Calculator Comment

You will need an exponent key. These are labeled in different ways, depending on the brand. It might be $\boxed{y^x}$ or $\boxed{x^y}$ or $\boxed{\wedge}$. In this book we will show exponents by using $\boxed{y^x}$, but you should press the appropriate key on your own brand of calculator.

$\boxed{1000}$ $\boxed{\times}$ $\boxed{1.08}$ $\boxed{y^x}$ $\boxed{20}$ $\boxed{=}$

Display: *4660.957144*

That is,

$n = 1$ for *annual* compounding

$n = 2$ for *semiannual* compounding

$n = 4$ for *quarterly* compounding

$n = 12$ for *monthly* compounding

$n = 360$ for *daily* compounding

Second, let

N = NUMBER OF COMPOUNDING PERIODS

That is,

$N = nt$

Third, let

i = RATE PER PERIOD

That is,

$$i = \frac{r}{n}$$

We can now summarize the variables we use for interest.

Variables Used with Interest Formulas

A = FUTURE VALUE   This is the principal plus interest.

P = PRESENT VALUE   This is the same as the principal.

r = INTEREST RATE   This is the *annual* interest rate.

t = TIME   This is the time in *years*.

n = NUMBER OF COMPOUNDING PERIODS EACH YEAR

$N = nt$ This is the number of compounding periods.

$i = \dfrac{r}{n}$ This is the rate per period.

■■■■ **EXAMPLE 8**

Fill in the blanks, given an annual interest rate of 12% and a time period of 3 years for the following compounding periods.

a. annual   **b.** semiannual   **c.** quarterly   **d.** monthly   **e.** daily

News Clip

Money makes money, and the money that money makes makes more money.

Benjamin Franklin

Solution

		n	r	t	$N = nt$	$i = \dfrac{r}{n}$
	Compounding Period		Yearly Rate	Time	Number of Periods	Rate per Period
a.	Annual	1	12%	3 yr	3	12%
b.	Semiannual	2	12%	3 yr	6	6%
c.	Quarterly	4	12%	3 yr	12	3%
d.	Monthly	12	12%	3 yr	36	1%
e.	Daily	360	12%	3 yr	1,080	0.03%

To find the number of periods, use $N = nt$; for part **e**:
$N = 360(3) = 1,080$

To find the rate per period use $i = \dfrac{r}{n}$;

for part **e**: $i = \dfrac{0.12}{360} = 0.000333\ldots.$

As a percent, $i \approx 0.03\%$.

We are now ready to state the future value formula for compound interest, which is sometimes called the **compound interest formula.** For these calculations you will need access to a calculator with an exponent key.

Future Value Formula (Compound Interest)

$$A = P(1 + i)^N$$

Procedure for Using the Future Value Formula

Identify the present value, P.
Identify the rate, r.
Identify the time, t.
Identify the number of times interest is compounded each year, n.
Calculate $N = nt$.

Calculate $i = \dfrac{r}{n}$.

Use the future value formula with the calculated values of i and N.

◼◼◼ EXAMPLE 9

Find the future value of $1,000 invested for 10 years at 8% interest

a. compounded annually.   **b.** compounded semiannually.

c. compounded quarterly.   **d.** compounded daily.

Solution Identify the variables: $P = 1,000$, $r = 0.08$, $t = 10$.

a. $n = 1$: Calculate $N = nt = 1(10) = 10$ and $i = \dfrac{r}{n} = \dfrac{0.08}{1} = 0.08$

$\boxed{1000}\ \boxed{\times}\ \boxed{(}\ \boxed{1}\ \boxed{+}\ \boxed{.08}\ \boxed{)}\ \boxed{y^x}\ \boxed{10}\ \boxed{=}$ Display: **2158.924997**

The amount (to the nearest cent) is $2,158.92.

b. $n = 2$: Calculate $N = nt = 2(10) = 20$ and $i = \dfrac{r}{n} = \dfrac{0.08}{2} = 0.04$

$\boxed{1000}\ \boxed{\times}\ \boxed{(}\ \boxed{1}\ \boxed{+}\ \boxed{.04}\ \boxed{)}\ \boxed{y^x}\ \boxed{20}\ \boxed{=}$ Display: **2191.123143**

The amount is $2,191.12.

c. $n = 4$: Calculate $N = nt = 4(10) = 40$ and $i = \dfrac{r}{n} = \dfrac{0.08}{4} = 0.02$

$\boxed{1000}\ \boxed{\times}\ \boxed{(}\ \boxed{1}\ \boxed{+}\ \boxed{.02}\ \boxed{)}\ \boxed{y^x}\ \boxed{40}\ \boxed{=}$ Display: **2208.039664**

The amount is $2,208.04.

d. $n = 360$: Calculate $N = nt = 360(10) = 3,600$ and $i = \dfrac{r}{n} = \dfrac{0.08}{360}$

$\boxed{1000}\ \boxed{\times}\ \boxed{(}\ \boxed{1}\ \boxed{+}\ \boxed{.08}\ \boxed{\div}\ \boxed{360}\ \boxed{)}\ \boxed{y^x}\ \boxed{3600}\ \boxed{=}$ **2225.343141**

The amount is $2,225.34.

Inflation

Any discussion of compound interest is incomplete without a discussion of **inflation.** The same procedure we used to calculate compound interest can be used to calculate the effects of inflation. The government releases reports of monthly and annual inflation rates. In 1981 the inflation rate was nearly 9%, but in 1996 it was less than 2%. Keep in mind that inflation rates can vary tremendously and that the best we can do in this section is to assume different constant inflation rates. For our purposes in this book, we will assume $n = 1$ (annual compounding) when working inflation problems.

■ EXAMPLE 10

Richard is earning $30,000 per year and would like to know what salary he could expect in 20 years if inflation continues at an average rate of 9%.

Solution $P = 30,000$, $r = 0.09$, and $t = 20$. In this book, assume $n = 1$ for inflation problems. This means that $N = 20$ and $i = 0.09$. Find

$$A = P(1 + i)^N \qquad \text{Future value formula}$$
$$= 30,000(1 + 0.09)^{20} \qquad \text{Substitute variables.}$$
$$\approx 168,132.32 \qquad \text{Use a calculator.}$$

The answer means that, if inflation continues at a constant 9% rate, an annual salary of $168,000 will have about the same purchasing power in 20 years as a salary of $30,000 today. ■

Present Value

Sometimes we know the future value of an investment and wish to know its present value. Such a problem is called a **present value** problem. The formula follows directly from the future value formula (by division).

Present Value Formula

$$P = A \div (1 + i)^N = A(1 + i)^{-N}$$

▬▬ EXAMPLE 11

Suppose that you want to take a trip to Tahiti in 5 years and you decide that you will need $5,000. To have that much money set aside in 5 years, how much money should you deposit now into a bank account paying 6% compounded quarterly?

Solution   In this problem, P is unknown and A is given: $A = 5,000$. We also have $r = 0.06$, $t = 5$, and $n = 4$. Calculate:

$$N = nt = 4(5) = 20 \quad \text{and} \quad i = \frac{r}{n} = \frac{0.06}{4}$$

$$P = A(1 + i)^{-N} = 5,000 \left(1 + \frac{0.06}{4}\right)^{-20}$$

$$\boxed{5000}\;\boxed{\times}\;\boxed{(}\;\boxed{(}\;\boxed{1}\;\boxed{+}\;\boxed{.06}\;\boxed{\div}\;\boxed{4}\;\boxed{)}\;\boxed{y^x}\;\boxed{20}\;\boxed{+/-}\;\boxed{=} \quad \mathit{3712.352091}$$

$\uparrow$

Negative exponent

You should deposit $3,712.35.

▬▬ EXAMPLE 12

An insurance agent wishes to sell you a policy that will pay you $100,000 in 30 years. What is the value of this policy in today's dollars, if we assume a 9% inflation rate?

Solution   This is a present value problem for which $A = 100,000$, $r = 0.09$, $n = 1$, and $t = 30$. We calculate

$$N = nt = 1(30) = 30 \quad \text{and} \quad i = \frac{r}{n} = \frac{0.09}{1} = 0.09$$

To find the present value, calculate

$$P = 100,000(1 + 0.09)^{-30} \approx \$7,537.11$$

This means that the agent is offering you an amount comparable to $7,537.11 in terms of today's dollars.

■■■■ **EXAMPLE 13**   **Polya's Method**

Your first child has just been born. You want to give her 1 million dollars when she retires at age 65. If you invest your money on the day of her birth, how much do you need to invest so that she will have $1,000,000 on her 65th birthday?

Solution We use Polya's problem-solving guidelines for this example.

Understand the Problem. You want to make a single deposit and let it be compounded for 65 years, so that at the end of that time there will be 1 million dollars. Neither the rate of return nor the compounding period is specified. We assume daily compounding, a constant rate of return over the time period, and need to determine whether we can find an investment to meet our goals.

Devise a Plan. With daily compounding, $N = nt = 360(65) = 23,400$. We will use the present value formula, and experiment with different interest rates:

$$P = 1,000,000 \left(1 + \frac{r}{360}\right)^{-23,400}$$

Carry Out the Plan.

Interest Rate	Formula	Value of P
2%	$1,000,000 \left(1 + \dfrac{0.02}{360}\right)^{-23,400}$	$\approx 272,541.63$
5%	$1,000,000 \left(1 + \dfrac{0.05}{360}\right)^{-23,400}$	$\approx 38,782.96$
8%	$1,000,000 \left(1 + \dfrac{0.08}{360}\right)^{-23,400}$	$\approx 5,519.75$
12%	$1,000,000 \left(1 + \dfrac{0.12}{360}\right)^{-23,400}$	≈ 410.27
20%	$1,000,000 \left(1 + \dfrac{0.20}{360}\right)^{-23,400}$	≈ 2.27

Look Back. Look down the list of interest rates and compare the rates with the amount of deposit necessary. Generally, the greater the risk of an investment, the higher the rate. Insured savings accounts may pay lower rates, bonds may pay higher rates for long-term investments, and other investments in stamps, coins, or real estate may pay the highest rates. The amount of the investment necessary to build an estate of 1 million dollars is dramatic!   ■■

PROBLEM SET 6.1

▲ **A Problems**

1. **IN YOUR OWN WORDS** What is interest?

2. **IN YOUR OWN WORDS** Contrast amount of interest and interest rate.

3. **IN YOUR OWN WORDS** Compare and contrast simple and compound interest.

4. **IN YOUR OWN WORDS** Compare and contrast present value and future value.

Use estimation to select the best response in Problems 5–14. Do not calculate.

5. If you deposit $100 in a bank account for a year, then the amount of interest is likely to be
   A. $1   B. $5   C. $105   D. impossible to estimate

6. If you deposit $100 in a bank account for a year, then the future value is likely to be
   A. $1   B. $5   C. $105   D. impossible to estimate

7. If you purchase a new automobile and finance it for four years, the amount of interest you might pay is
   A. $400   B. $100   C. $4,000   D. impossible to estimate

8. What is a reasonable monthly income when you retire?
   A. $300   B. $10,000   C. $500,000   D. impossible to estimate

9. To retire and live on the interest only, what is a reasonable amount to have in the bank?
   A. $300   B. $10,000   C. $500,000   D. impossible to estimate

10. If $I = Prt$ and $P = \$49,236.45$, $r = 10.5\%$, and $t = 2$ years, estimate I.
    A. $10,000   B. $600   C. $50,000   D. $120,000

11. If $I = Prt$ and $I = \$398.90$, $r = 9.85\%$, and $t = 1$ year, estimate P.
    A. $400   B. $40   C. $40,000   D. $4,000

12. If $t = 3.52895$, then the time is about 3 years and how many days?
    A. 30   B. 300   C. 52   D. 150

13. If a loan is held for 450 days, then t is about
    A. 450   B. 3   C. $1\frac{1}{4}$ D. 5

14. If a loan is held for 180 days, then t is about
    A. 180   B. $\frac{1}{2}$ C. $\frac{1}{4}$ D. 3

In Problems 15–20 compare the future amounts (A) you would have if the money were invested at simple interest and if it were invested with interest to be compounded annually.

15. $1,000 at 8% for 5 years   16. $5,000 at 10% for 3 years

17. $2,000 at 12% for 3 years   18. $2,000 at 12% for 5 years

19. $5,000 at 12% for 20 years   20. $1,000 at 14% for 30 years

In Problems 21–34, find the future value, using the future value formula and a calculator.

21. $35 at 17.65% compounded annually for 20 years

22. $155 at 21.25% compounded annually for 25 years

23. $835 at 3.5% compounded semiannually for 6 years

24. $9,450 at 7.5% compounded semiannually for 10 years

25. $575 at 5.5% compounded quarterly for 5 years

26. $3,450 at 4.3% compounded quarterly for 8 years

27. $9,730.50 at 7.6% compounded monthly for 7 years

28. $3,560 at 9.2% compounded monthly for 10 years

29. $45.67 at 3.5% compounded daily for 3 years

30. $34,500 at 6.9% compounded daily for 2 years

31. $89,500 at 6.2% compounded monthly for 30 years

32. $119,400 at 7.5% compounded monthly for 30 years

33. $225,500 at 8.65% compounded daily for 30 years

34. $355,000 at 9.5% compounded daily for 30 years

Find the total amount that must be repaid on the notes described in Problems 35–40.

35. $1,500 borrowed at 21% simple interest. What is the total amount to be repaid 55 days later?

36. $2,300 borrowed at 19% simple interest. What is the total amount to be repaid 230 days later?

37. $23,400 borrowed at 14% simple interest. What is the total amount to be repaid 95 days later?

38. $2,100,000 borrowed at 8% simple interest. What is the total amount to be repaid 10 days later?

39. $8,553 borrowed at 16.5% simple interest. What is the total amount to be repaid 3 years 125 days later?

40. $15,225 borrowed at 14.5% simple interest. What is the total amount to be repaid 4 years 230 days later?

41. Find the cost of each item in 5 years, assuming an inflation rate of 9%.
    a. cup of coffee, $1.25   b. Sunday paper, $1.50
    c. Big Mac, $1.95   d. gallon of gas, $1.35

42. Find the cost of each item in 10 years, assuming an inflation rate of 5%.
    a. movie admission, $5.00   b. CD, $14.95
    c. textbook, $40.00   d. electric bill, $65

43. Find the cost of each item in 10 years, assuming an inflation rate of 12%.
    a. phone bill, $45   b. pair of shoes, $65
    c. new suit, $370   d. monthly rent, $600

44. Find the cost of each item in 20 years, assuming an inflation rate of 6%.
    a. TV set, $600   b. small car, $10,000
    c. car, $18,000   d. tuition, $16,000

▲ B Problems

45. How much would you have in 5 years if you purchased a $1,000 5-year savings certificate that paid 4% compounded quarterly?

46. What is the interest on $1,500 for 5 years at 12% compounded monthly?

47. What is the interest on $2,400 for 5 years at 12% compounded monthly?

48. What is the future value after 15 years if you deposit $1,000 for your child's education and the interest is guaranteed at 16% compounded quarterly?

49. Suppose you see a car with an advertised price of $18,490 at $480 per month for 5 years. What is the amount of interest paid?

50. Suppose you see a car with an advertised price of $14,500 at $410.83 per month for 4 years. What is the amount of interest paid?

51. Suppose you see a home and finance $285,000 at $2,293.17 per month for 30 years. What is the amount of interest paid?

52. Suppose you see a home and finance $170,000 at $1,247.40 per month for 30 years. What is the amount of interest paid?

53. Find the cost of a home in 30 years, assuming an annual inflation rate of 10%, if the present value of the house is $125,000.

54. Find the cost of the monthly rent for a two-bedroom apartment in 30 years, assuming an annual inflation rate of 10%, if the current rent is $650.

55. Suppose that an insurance agent offers you a policy that will provide you with a yearly income of $50,000 in 30 years. What is the comparable salary today, assuming an inflation rate of 6%?

56. Suppose that an insurance agent offers you a policy that will provide you with a yearly income of $50,000 in 30 years. What is the comparable salary today, assuming an inflation rate of 10%?

57. Suppose an insurance agent offers you a policy that will provide you with a yearly income of $50,000 in 30 years. What is the comparable salary today, assuming an inflation rate of 4%?

58. If a friend tells you she earned $5,075 interest for the year on a 5-year certificate of deposit paying 5% simple interest, what is the amount of the deposit?

59. If Rita receives $45.33 interest for a deposit earning 3% simple interest for 240 days, what is the amount of her deposit?

60. In 1996, the U.S. national debt was 5 trillion dollars.
    a. If this debt is shared equally by the 300 million citizens, how much would it cost each of us?
    b. If the interest is 6%, what is the interest on the national debt *each second*? Assume a 365-day year.

▲ **Problem Solving**

61. The News Clip in the margin is typical of what you will see in a newspaper. It gives two rates, the **annual yield** or **effective rate** (8.33%) and a **nominal rate** (8%). Since banks pay interest compounded for different periods (quarterly, monthly, daily, for example), they calculate a rate for which annual compounding will yield the same amount at the end of 1 year. That is, for an 8% rate

Nominal Rate	Effective Rate
8%, annual compounding	8%
8%, semiannual compounding	8.16%
8%, quarterly compounding	8.24%
8%, monthly compounding	8.30%
8%, daily compounding	8.33%

To find a formula for effective rate we recall that the compound interest formula is:

$$A = P(1 + i)^N = P\left(1 + \frac{r}{n}\right)^{nt}$$

The effective rate Y is a rate that will equal the amount from the simple interest formula for a period of 1 year:

$$A = P(1 + Y)$$

Find a formula for effective (annual) rate, Y, for which the future value for compound interest is equal to the future value for simple interest at the end of 1 year.

62. Find the effective yield for the following investments (see Problem 61). Round to the nearest hundredth of a percent.

    a. 6%, compounded quarterly   **b.** 6%, compounded monthly
    c. 4%, compounded semiannually   **d.** 4%, compounded daily

63. If John wants to retire with $10,000 per month,* how much principal is necessary to generate this amount of monthly income if the interest rate is 15%?

64. If Melissa wants to retire with $50,000 per month,* how much principal is necessary to generate this amount of monthly income if the interest rate is 12%?

65. If Jack wants to retire with $1,000 per month, how much principal is necessary to generate this amount of monthly income if the interest rate is 6%?

▲ Individual Research

66. What do the following people have in common?

    Paul Painlevé, President of France

    Omar Khayyam, author of *The Rubaiyat*

    Emanual Lasker, world chess champion

    James Moriarty, Sherlock Holmes's nemesis, author of *The Dynamics of an Asteroid*

67. Conduct a survey of banks, savings and loan companies, and credit unions in your area. Prepare a report on the different types of savings accounts available and the interest rates they pay. Include methods of payment as well as interest rates.

68. Do you expect to live long enough to be a millionaire? Suppose that your annual salary today is $19,000. If inflation continues at 12%, how long will it be before $19,000 increases to an annual salary of a million dollars?

69. Consult an almanac or some government source, and then write a report on the current inflation rate. Project some of these results to the year of your own expected retirement.

6.2 INSTALLMENT BUYING

Consumer Loans

Two types of consumer credit allow you to make installment purchases. The first, called **closed-ended,** is the traditional installment loan. An **installment loan** is an

---

* You might think these are exorbitant monthly incomes, but if you assume 10% average inflation for 40 years, a monthly income of $220 today will be equivalent to about $10,000 per month in 40 years.

agreement to pay off a loan or a purchase by making equal payments at regular intervals for some specific period of time. In this book, it is assumed that all install-ment payments are made monthly.

There are two common ways of calculating installment interest. The first method uses simple interest and is called *add-on interest*; the second uses compound interest and is called *amortization*. We discuss the simple interest application in this section, and the compound interest application in Section 6.6.

In addition to closed-ended credit, it is common to obtain a type of consumer credit called **open-ended, revolving credit,** or more commonly, a **credit card** loan. MasterCard, VISA, and Discover cards, as well as those from department stores and oil companies, are examples of open-ended loans. This type of loan allows for pur-chases or cash advances up to a specified maximum **line of credit** and has a flexible repayment schedule.

Add-On Interest

The most common method for calculating interest on installment loans is by a method known as **add-on interest.** It is nothing more than an application of the simple interest formula. It is called *add-on interest* because the interest is *added to* the amount borrowed so that both interest and the amount borrowed are paid for over the length of the loan. You should be familiar with the following variables:

$$P = \text{AMOUNT TO BE FINANCED (present value)}$$

$$r = \text{ADD-ON INTEREST RATE}$$

$$t = \text{TIME (in years) TO REPAY THE LOAN}$$

$$I = \text{AMOUNT OF INTEREST}$$

$$A = \text{AMOUNT TO BE REPAID (future value)}$$

$$m = \text{AMOUNT OF THE MONTHLY PAYMENT}$$

$$N = \text{NUMBER OF PAYMENTS}$$

Installment Loan Formulas

AMOUNT OF INTEREST: $I = Prt$

AMOUNT TO BE REPAID: $A = P + I$ or $A = P(1 + rt)$

NUMBER OF PAYMENTS: $N = 12t$

AMOUNT OF EACH PAYMENT: $m = \dfrac{A}{N}$

■■■■■ **EXAMPLE 1**

You want to purchase a computer that has a price of $1,399, and you decide to pay for it with installments over 3 years. The store tells you that the interest rate is 15%. What is the amount of each monthly payment?

Solution   You ask the clerk how the interest is calculated, and you are told that the store uses add-on interest.

Thus,

$$P = 1,399, \quad r = 0.15, \quad t = 3, \quad N = 36$$

Two-Step Solution	*One-Step Solution*

$$I = Prt = 1,399(0.15)(3) = 629.55 \qquad A = P(1 + rt)$$
$$A = P + I = 1,399 + 629.55 = 2,028.55 \qquad = 1,399(1 + 0.15 \cdot 3) = 2,028.55$$
$$m = \frac{2,028.55}{36} \approx 56.35$$

The amount of each monthly payment is $56.35.   ▬

The most common applications of installment loans are for the purchase of a car or a home. Interest for purchasing a car is determined by the add-on method, but interest for purchasing a home is not. We will, therefore, delay our discussion of home loans until after we have discussed compound interest. The next example shows a calculation for a car loan.

▬▬ **EXAMPLE 2**   **Polya's Method**

Suppose that you have decided to purchase a Saturn SC2 Coupe and want to determine the monthly payment if you pay for the car in 4 years. The value of your trade-in is $4,100.

Solution   We use Polya's problem-solving guidelines for this example.

Understand the Problem.   Not enough information is given, so you need to ask some questions of the car dealer:

> **Sticker price** of the car (as posted on the window): $14,935
>
> Dealer's preparation charges (as posted on the window): $350.00
>
> Total asking (sticker) price: $15,285
>
> Tax rate (determined by the state):   7%
>
> Add-on interest rate: 8%

You need to make an offer.

Devise a Plan.   The plan is to offer the dealer 5% over dealer's cost. Assuming that the dealer accepts that offer, then we will calculate the monthly payment.

Carry Out the Plan.   If you are serious about getting the best price, find out the **dealer's cost** — the price the dealer paid for the car you want to buy. In this book, we will tell you the dealer's cost, but in the real world you will need to do some homework to find it (consult a reporting service, an automobile association, a credit union, or the April issue of *Consumer Reports*). Assume that the dealer's cost for this car is $12,993.45. You decide to offer the dealer 5% *over* this cost. We will call this a **5% offer:**

$$\$12,993.45(1 + 0.05) = \$13,643.12$$

Calculator Comment

Problems such as in Example 1 are best done with a calculator. You can carry out the entire calculation in one step: $\boxed{1399}$ $\boxed{\times}$ $\boxed{(}$ $\boxed{1}$ $\boxed{+}$ $\boxed{.15}$ $\boxed{\times}$ $\boxed{3}$ $\boxed{)}$ $\boxed{=}$ $\boxed{\div}$ $\boxed{36}$ $\boxed{=}$

Display: *56.3486111*   ▲

1997 Saturn SC2 Coupe

You will notice that we ignored the sticker price and the dealer's preparation charges. Our offer is based only on the *dealer's cost*. Most car dealers will accept an offer that is between 5% and 10% over what they actually paid for the car. For this example, we will assume that the dealer accepted a price of $13,600. We also assume that we have a trade-in with a value of $4,100.

Here is a list of calculations shown on the sales contract:

Sale price of Saturn:	$13,600.00
Destination charges:	200.00
Subtotal:	$13,800.00
Tax (7% rate)	966.00
Less trade-in	4,100.00
Amount to be financed:	10,666.00

We now calculate several key amounts:

Interest: $I = Prt = 10{,}666(0.08)(4) = 3{,}413.12$

Amount to be repaid: $A = P + I = 10{,}666 + 3{,}413.12 = 14{,}079.12$

Monthly payment: $m = \dfrac{14{,}079.12}{48} = 293.315$

Look Back.   The monthly payment for the car is $293.32.   ▬

Annual Percentage Rate (APR)

An important aspect of add-on interest is that the actual rate you pay exceeds the quoted add-on interest rate. The reason for this is that you do not keep the entire amount borrowed for the entire time. For the car payments calculated in Example 2, the principal used was $10,666, but you do not *owe* this entire amount for 4 years. After the first payment, you owe *less* than this amount. In fact, after you make 47 payments, you will owe only $293.32; but the calculation shown in Example 2 assumes that the principal remains constant for 4 years.

To see this a little more clearly, consider a simpler example. Suppose you borrow $2,000 for 2 years with 10% add-on interest. The amount of interest is

$$\$2{,}000 \times 0.10 \times 2 = \$400$$

Now if you pay back $2,000 + $400 at the end of two years the annual interest rate is 10%. However, if you make a partial payment of $1,200 at the end of the first year and $1,200 at the end of the second year, your total paid back is still the same ($2,400), but you have now paid a higher annual interest rate. Why? Take a look at Figure 6.1. On the left we see the interest on $2,000 is $400. But if you make a partial payment (figure on the right), we see that $200 for the first year is the correct interest, but the remaining $200 interest piled on the remaining balance of $1,000 is 20% interest (not the stated 10%). Note that since you did not owe $2,000 for 2 years, the interest rate, r, necessary to give $400 interest can be calculated using

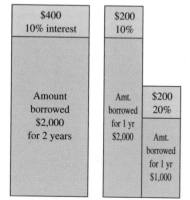

▲ **Figure 6.1   Interest on a $2,000 two-year loan**

$I = Prt$:

$$(2,000)r(1) + (1,000)r(1) = 400$$
$$3,000r = 400$$
$$r = \frac{400}{3,000} = 0.13333 \text{ or } 13.3\%$$

This number, 13.3% is called the **annual percentage rate.** This number is too difficult to calculate, as we have just done here, if the number of months is very large. We will, instead, use the formula given in the following box.

APR Formula

The **annual percentage rate,** or **APR,** is the rate paid on a loan when the rate is based on the actual amount owed for the length of time that it is owed. It can be found for an add-on interest rate, r, with N payments by using the formula

$$\text{APR} = \frac{2Nr}{N + 1}$$

In 1969 a Truth-in-Lending Act was passed by Congress; it requires all lenders to state the true annual interest rate, which is called the **annual percentage rate (APR)** and is based on the actual amount owed. Regardless of the rate quoted, when you ask a salesperson what the APR is, the law requires that you be told this rate. This regulation enables you to compare interest rates *before* you sign a contract, which must state the APR even if you haven't asked for it.

■■■ EXAMPLE 3

In Example 1, we considered the purchase of a computer with a price of $1,399, paid for in installments over 3 years at an add-on rate of 15%. What is the APR (rounded to the nearest tenth of a percent)?

Solution   Knowing the amount of the purchase is not necessary when finding the APR. We need to know only N and r. Since N is the number of payments, we have $N = 12(3) = 36$, and r is given as 0.15:

$$\text{APR} = \frac{2(36)(0.15)}{36 + 1} \approx 0.292$$

The APR is 29.2%.

Calculator Comment

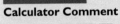

Display: *.2918918919*

■■■ EXAMPLE 4

Consider a Blazer with a price of $18,436 that is advertised at a monthly payment of $384.00 for 60 months. What is the APR (to the nearest tenth of a percent)?

Solution   We are given $P = 18,436$, $m = 384$, and $N = 60$. The APR formula requires that we know the rate r.

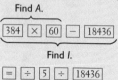
The future value is the total amount to be repaid ($A = P + I$) and the amount of interest is $I = Prt$.

$A = 384(60) = 23,040$, so $I = A - P = 23,040 - 18,436 = 4,604$. Since $N = 12t$, we see that $t = 5$ when $N = 60$.

$$I = Prt$$
$$4,604 = 18,436(r)(5)$$
$$920.8 = 18,436r \qquad \text{Divide both sides by 5.}$$
$$0.0499457583 = r \qquad \text{Divide both sides by 18,436.}$$

Finally, for the APR formula,

$$APR = \frac{2Nr}{N+1}$$
$$= \frac{2(60)(0.0499457583)}{61} \qquad \text{Don't round until the last step.}$$
$$\approx 0.098$$

The APR is 9.8%.

Open-Ended Credit

The most common type of open-ended credit used today involves credit cards issued by VISA, MasterCard, Discover, American Express, department stores, and oil companies. Because you don't have to apply for credit each time you want to charge an item, this type of credit is very convenient.

When comparing the interest rates on loans, you should use the APR. Earlier, we introduced a formula for add-on interest; but for credit cards, the stated interest rate *is* the APR. However, the APR on credit cards is often stated as a daily or a monthly rate. For credit cards, we use a 365-day year rather than a 360-day year.

▰▰▰ EXAMPLE 5

Convert the given credit card rate to APR (rounded to the nearest tenth of a percent).

a. $1\frac{1}{2}\%$ per month   **b.** Daily rate of 0.05753%

Solution

a. Since there are 12 months per year, multiply a monthly rate by 12 to get the APR:

$$1\tfrac{1}{2}\% \times 12 = 18\% \text{ APR}$$

b. Multiply the daily rate by 365 to obtain the APR:

$$0.05753\% \times 365 = 20.99845\%$$

Rounded to the nearest tenth, this is equivalent to 21.0% APR.

Many credit cards charge an annual fee; some charge $1 every billing period the card is used, whereas others are free. These charges affect the APR differently, depending on how much the credit card is used during the year and on the monthly balance. If you always pay your credit card bill in full as soon as you receive it, the card with no yearly fee would obviously be the best for you. On the other hand, if you use your credit card to stretch out your payments, the APR is more important than the flat fee. For our purposes, we won't use the yearly fee in our calculations of APR on credit cards.

Like annual fees, the interest rates or APRs for credit cards vary greatly. Because VISA and MasterCard are issued by many different banks, the terms can vary greatly even in one locality.

Calculating Credit Card Interest

An interest charge added to a consumer account is often called a **finance charge**. The finance charges can vary greatly even on credit cards that show the *same* APR, depending on the way the interest is calculated. There are three generally accepted methods for calculating these charges: *previous balance, adjusted balance,* and *average daily balance.*

Methods of Calculating Interest on Credit Cards

For credit card interest, use the simple interest formula, $I = Prt$.

Previous balance method: Interest is calculated on the previous month's balance. With this method, P = previous balance, r = annual rate, and $t = \frac{1}{12}$.

Adjusted balance method: Interest is calculated on the previous month's balance *less* credits and payments. With this method, P = adjusted balance, r = annual rate, and $t = \frac{1}{12}$.

Average daily balance method: Add the outstanding balances for *each day* in the billing period, and then divide by the number of days in the billing period to find what is called the *average daily balance.* With this method, P = average daily balance, r = annual rate, and t = number of days in the billing period divided by 365.

In Example 6 we compare the finance charges on a $1,000 credit card purchase, using these three different methods.

■■■ EXAMPLE 6

Calculate the interest on a $1,000 credit card bill that shows an 18% APR, assuming that $50 is sent on January 3 and is recorded on January 10. Contrast the three methods for calculating the interest.

Solution The three methods are *previous balance method, adjusted balance method,* and *average daily balance method.* All three methods use the formula $I = Prt$.

Method	Previous Balance	Adjusted Balance	Average Daily Balance
P:	$1,000	$1,000 − $50 = $950	Balance is $1,000 for 10 days of 31-day month; balance is $950 for 21 days of 31-day month; $$\frac{10 \times \$1,000 + 21 \times \$950}{31} = \$966.13$$
r:	0.18	0.18	0.18
t:	$\frac{1}{12}$	$\frac{1}{12}$	$\frac{31}{365}$
$I = Prt$:	$1,000(0.18)($\frac{1}{12}$) = $15.00	$950(0.18)($\frac{1}{12}$) = $14.25	$966.13 (0.18)($\frac{31}{365}$) = $14.77

You can sometimes make good use of credit cards by taking advantage of the period during which no finance charges are levied. Many credit cards charge no interest if you pay in full within a certain period of time (usually 20 or 30 days). This is called the **grace period.** On the other hand, if you borrow cash on your credit card, you should know that many credit cards have an additional charge for cash advances — and these can be as high as 4%. This 4% is *in addition to* the normal finance charges.

PROBLEM SET 6.2

▲ A Problems

1. **IN YOUR OWN WORDS** What is add-on interest?

2. **IN YOUR OWN WORDS** What is APR?

3. **IN YOUR OWN WORDS** Compare and contrast open-ended and closed-ended credit.

4. **IN YOUR OWN WORDS** Discuss the methods of calculating credit card interest.

5. **IN YOUR OWN WORDS** Describe a good procedure for saving money with the purchase of an automobile.

Use estimation to select the best response in Problems 6–23. Do not calculate.

6. If you purchase a $2,400 item and pay for it with monthly installments for 2 years, the monthly payment is
   A. $100 per month
   B. more than $100 per month
   C. less than $100 per month

7. If you purchase a $595.95 item and pay for it with monthly installments for 1 year, the monthly payment is
   A. about $50   B. more than $50   C. less than $50

8. If you purchase an item for $1,295 at an interest rate of 9.8%, and you finance it for 1 year, then the amount of add-on interest is about
   A. $13.00   B. $500   C. $130

9. If you purchase an item for $1,295 at an interest rate of 9.8%, and you finance it for 4 years, then the amount of add-on interest is about
   A. $13.00   B. $500   C. $130

10. If you purchase a new car for $10,000 and finance it for 4 years, the amount of interest you would expect to pay is about
    A. $4,000   B. $400   C. $24,000

11. A reasonable APR to pay for a 3-year installment loan is
    A. 1%   B. 12%   C. 32%

12. A reasonable APR to pay for a 3-year automobile loan is
    A. 6%   B. 40%   C. $2,000

13. If you wish to purchase a car with a sticker price of $10,000, a reasonable offer to make to the dealer is:
    A. $10,000   B. $9,000   C. $11,000

14. A reasonable APR for a credit card is
    A. 1%   B. 30%   C. 12%

15. If I do not pay off my credit card each month, the most important cost factor is
    A. the annual fee   B. the APR   C. the grace period

16. If I pay off my credit card balance each month, the most important cost factor is
    A. the annual fee   B. the APR   C. the grace period

17. The method of calculation most advantageous to the consumer is the
    A. previous balance method
    B. adjusted balance method
    C. average daily balance method

18. In an application of the average daily balance method for the month of August, t is
    A. $\frac{1}{12}$ B. $\frac{30}{365}$ C. $\frac{31}{365}$

19. When using the average daily balance method for the month of September, t is
    A. $\frac{1}{12}$ B. $\frac{30}{365}$ C. $\frac{31}{365}$

20. If your credit card balance is $650 and the interest rate is 12% APR, then the credit card interest charge is
    A. $6.50   B. $65   C. $8.25

21. If your credit card balance is $952, you make a $50 payment, the APR is 12%, and the interest is calculated according to the previous balance method, then the finance charge is
    A. $9.52   B. $9.02   C. $9.06

22. If your credit card balance is $952, you make a $50 payment, the APR is 12%, and the interest is calculated according to the adjusted balance method, then the finance charge is
    A. $9.52   B. $9.02   C. $9.06

23. If your credit card balance is $952, you make a $50 payment, the APR is 12%, and the interest is calculated according to the average daily balance method, then the finance charge is
    A. $9.52   B. $9.02   C. $9.06

News Clip

**WARNING
Credit Card Trap**

Suppose you "max out" your credit card charges at $1,500 and decide that you will not use it again until it is paid off. How long do you think it will take you to pay it off if you make the minimum required payment (which would be $30 for the beginning balance of $1,500 and would decrease as the remaining balance decreases)?

A. 5 months
B. 2 years
C. 8 years
D. 12 years
E. 20 years

Even though we cannot mathematically derive this result, it does seem appropriate to consider this real-life question. Go ahead, make a guess!

   If you keep the payment at a constant $30, the length of time for repayment is 8 years. However, if you make only the required minimum payment, this "maxed" credit card will take 20 years to pay off! The correct answer is E.

Convert each credit card rate in Problems 24–29 to the APR *

24. Oregon, $1\frac{1}{4}$% per month   **25.** Arizona, $1\frac{1}{3}$% per month

26. New York, $1\frac{1}{2}$% per month   **27.** Tennessee, 0.02740% daily rate

28. Ohio, 0.02192% daily rate   **29.** Nebraska, 0.03014% daily rate

Calculate the monthly finance charge for each credit card transaction in Problems 30–35. Assume that it takes 10 days for a payment to be received and recorded, and that the month has 30 days.

	Balance	Rate	Payment	Method
30.	$300	18%	$50	Previous balance
31.	$300	18%	$50	Adjusted balance
32.	$300	18%	$50	Average daily balance
33.	$3,000	15%	$2,500	Previous balance
34.	$3,000	15%	$2,500	Adjusted balance
35.	$3,000	15%	$2,500	Average daily balance

▲ B Problems

Round your answers in Problems 36–41 to the nearest dollar.

36. Make a 6% offer on a Chevrolet Corsica that has a sticker price of $14,385 and a dealer cost of $13,378.

37. Make a 5% offer on a Ford Escort that has a sticker price of $13,205 and a dealer cost of $12,412.70.

38. Make a 10% offer on a Saturn that has a sticker price of $12,895 and a dealer cost of $11,219.

39. Make a 10% offer on a Nissan Pathfinder that has a sticker price of $32,129 and a dealer cost of $28,916.

40. Make a 7% offer on a Dodge Ram that has a sticker price of $20,650 and a dealer cost of $18,172.

41. Make a 5% offer on a BMW that has a sticker price of $62,490 and a dealer cost of $51,242.

Find the amount of interest, the monthly payment, and the APR (rounded to the nearest tenth of a percent) for each of the loans described in Problems 42–51.

42. Purchase a living room set for $3,600 at 12% add-on interest for 3 years.

43. Purchase a stereo for $2,500 at 13% add-on interest for 2 years.

44. A $1,500 loan at 11% add-on rate for 2 years.

45. A $2,400 loan at 15% add-on rate for 2 years.

46. A $4,500 loan at 18% add-on rate for 5 years.

47. A $1,000 loan at 12% add-on rate for 2 years.

48. Purchase an oven for $650 at 11% add-on interest for 2 years.

---

* These rates were the listed finance charges on purchases of less than $500 on a Citibank VISA statement.

49. Purchase a refrigerator for $2,100 at 14% add-on interest for 3 years.

50. Purchase a car for $5,250 at 9.5% add-on rate for 2 years.

51. Purchase a car for $42,700 at 2.9% add-on rate for 5 years.

52. A newspaper advertisement offers a $9,000 car for nothing down and 36 easy monthly payments of $317.50. What is the total amount paid for both car and financing?

53. A newspaper advertisement offers a $4,000 used car for nothing down and 36 easy monthly payments of $141.62. What is the total amount paid for both car and financing?

54. A newspaper advertisement offers a $14,350 car for nothing down and 48 easy monthly payments of $488.40. What is the total amount paid for both car and financing?

55. A car dealer will sell you the $16,450 car of your dreams for $3,290 down and payments of $339.97 per month for 48 months. What is the total amount paid for both car and financing?

56. A car dealer will sell you a used car for $6,798 with $798 down and payments of $168.51 per month for 48 months. What is the total amount paid for both car and financing?

In Problems 57–66 round the percent to the nearest tenth of a percent.

57. A newspaper advertisement offers a $9,000 car for nothing down and 36 easy monthly payments of $317.50. What is the simple interest rate?

58. A newspaper advertisement offers a $4,000 used car for nothing down and 36 easy monthly payments of $141.62. What is the simple interest rate?

59. A newpaper advertisement offers a $14,350 car for nothing down and 48 easy monthly payments of $488.40. What is the simple interest rate?

60. A car dealer will sell you the $16,450 car of your dreams for $3,290 down and payments of $339.97 per month for 48 months. What is the simple interest rate?

61. A car dealer will sell you a used car for $6,798 with $798 down and payments of $168.51 per month for 48 months. What is the simple interest rate?

62. A newspaper advertisement offers a $9,000 car for nothing down and 36 easy monthly payments of $317.50. What is the APR?

63. A newspaper advertisement offers a $4,000 used car for nothing down and 36 easy monthly payments of $141.62. What is the APR?

64. A newpaper advertisement offers a $14,350 car for nothing down and 48 easy monthly payments of $488.40. What is the APR?

65. A car dealer will sell you the $16,450 car of your dreams for $3,290 down and payments of $339.97 per month for 48 months. What is the APR?

66. A car dealer will sell you a used car for $6,798 with $798 down and payments of $168.51 per month for 48 months. What is the APR?

67. A car dealer carries out the calculations shown at the right. What is the annual percentage rate?

List price	$5,368.00
Options	$1,625.00
Destination charges	$ 200.00
Subtotal	$7,193.00
Tax	$ 431.58
Less trade-in	$2,932.00
Amount to be financed	$4,692.58
8% interest for	
48 months	$1,501.63
Total	$6,194.21
MONTHLY PAYMENT	$ 129.05

68. A car dealer carries out the following calculations:

List price	$15,428.00
Options	$ 3,625.00
Destination charges	$ 350.00
Subtotal	$19,403.00
Tax	$ 1,164.18
Less trade-in	$ 7,950.00
Amount to be financed	$12,617.18
5% interest for 60 months	$ 3,154.30
Total	$15,771.48
MONTHLY PAYMENT	$ 262.86

What is the annual percentage rate?

69. A car dealer carries out the following calculations:

List price	$ 9,450.00
Options	$ 1,125.00
Destination charges	$ 300.00
Subtotal	$10,875.00
Tax	$ 652.50
Less trade-in	$ 0.00
Amount to be financed	$11,527.50
11% interest for 48 months	$ 5,072.10
Total	$16,599.60
MONTHLY PAYMENT	$ 345.83

What is the annual percentage rate?

▲ **Problem Solving**

70. The finance charge statement on a Sears Revolving Charge Card statement is shown here. Why do you suppose that the limitation on the 50¢ finance charge is for amounts less than $28.50?

> **SEARS ROEBUCK AND CO.**
> **SEARSCHARGE SECURITY AGREEMENT**
>
> 4. FINANCE CHARGE. If I do not pay the entire New Balance within 30 days (28 days for February statements) of the monthly billing date, a FINANCE CHARGE will be added to the account for the current monthly billing period. **The FINANCE CHARGE** will be either a minimum of 50¢ if the Average Daily Balance is $28.50 or less, or a periodic rate of 1.75% per month **(ANNUAL PER-CENTAGE RATE of 21%)** on the Average Daily Balance.

71. Marsha needs to have a surgical procedure done and does not have the $3,000 cash necessary for the operation. Upon talking to an administrator at the hospital, she finds that it will accept MasterCard, VISA, and Discover credit cards. All of these credit cards have an APR of 18%, so she figures that it does not matter which card she uses, even though she plans to take a year to pay off the loan. Assume that Marsha makes a payment of $300 and then receives a bill. Show the interest from credit cards of 18% APR according to the previous balance, adjusted balance, and average daily balance methods. Assume that the month has 31 days and that it takes 14 days for Marsha's payment to be mailed and recorded.

72. Karen and Wayne need to buy a refrigerator because theirs just broke. Unfortunately, their savings account is depleted, and they will need to borrow money to buy a new one. The bank offers them a personal loan at 21% (APR), and Sears offers them an installment loan at 15% (add-on rate). Suppose that the refrigerator at Sears costs $1,598 plus 5% sales tax, and Karen and Wayne plan to pay for the refrigerator for 3 years. Should they finance it with the bank or with Sears?

73. Karen and Wayne need to buy a refrigerator because theirs just broke. Unfortunately, their savings account is depleted, and they will need to borrow money to buy a new one. Sears offers them an installment loan at 15% (add-on rate). If the refrigerator at Sears costs $1,598 plus 5% sales tax, and Karen and Wayne plan to pay for the refrigerator for 3 years, what is the monthly payment?

74. **Rule of 78** With a typical installment loan, you are asked to sign a contract stating the terms of repayment. If you pay off the loan early, you are entitled to an interest rebate. For example, if you finance $500 and are charged $90 interest (APR 8.64%), the total to be repaid is $590 with 24 monthly payments of $24.59. After 1 year, you decide to pay off the loan, so you figure that the rebate should be $45 (half of the interest for 2 years), but instead you are told the interest rebate is only $23.40. What happened? Look at the fine print on the contract. It says interest will be refunded according to the Rule of 78. The formula for the rebate is as follows:

$$\text{INTEREST REBATE} = \frac{k(k + 1)}{n(n + 1)} \times \text{FINANCE CHARGE}$$

where k is the number of payments remaining and n is the total number of payments. Determine the interest rebate on the following:
   a. $1,026 interest on an 18-month loan; pay off loan after 12 months.
   b. $350 interest on a 2-year loan with 10 payments remaining.
   c. $10,200 borrowed at 11% on a 4-year loan with 36 months remaining.
   d. $51,000 borrowed at 10% on a 5-year loan with 18 payments remaining.

▲ **Individual Research**

75. Karen says that she has heard something about APR rates but doesn't really know what the term means. Wayne says he thinks it has something to do with the prime rate, but he isn't sure what. Write a short paper explaining APR to Karen and Wayne.

76. Some savings and loan companies advertise that they pay interest *continuously*. Do some research to explain what this means.

77. Select a car of your choice, find the list price, and calculate 5% and 10% price offers. Check out available money sources in your community, and prepare a report showing the different costs for the same car. Back up your figures with data.

6.3 SEQUENCES

Sequences or Progressions

Patterns are sometimes used as part of an IQ test. It used to be thought that an IQ test measured "innate intelligence" and that a person's IQ score was fairly constant. Today, it is known that this is not the case. IQ test scores can be significantly changed by studying the types of questions asked. Even if you have never taken an IQ test, you have taken (or will take) tests that ask pattern-type questions.

An example of an IQ test is shown below. It is a so-called "quickie" test, but it illustrates a few mathematical patterns (see Problems 1–6). The purpose of this section is to look at some simple patterns, to become more proficient in recognizing them, and then to apply them to develop some important financial formulas.

Are You A Genius?

Each problem is a series of some sort—that is, a succession of either letters, numbers or drawings—with the last item in the series missing. Each series is arranged according to a different rule and, in order to identify the missing item, you must figure out what that rule is.

Now, it's your turn to play. Give yourself a maximum of 20 minutes to answer the 15 questions. If you haven't finished in that time, stop anyway. In the test problems done with drawings, it is always the top row or group that needs to be completed by choosing one drawing from the bottom row.

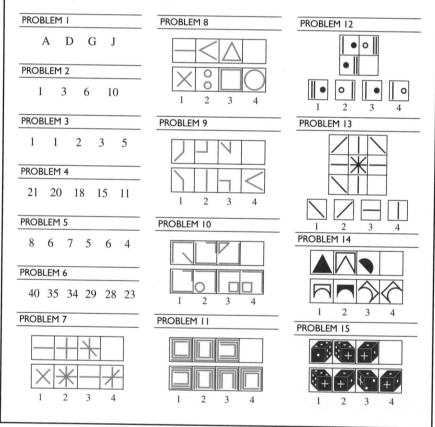

PROBLEM 1

A   D   G   J

PROBLEM 2

1   3   6   10

PROBLEM 3

1   1   2   3   5

PROBLEM 4

21   20   18   15   11

PROBLEM 5

8   6   7   5   6   4

PROBLEM 6

40   35   34   29   28   23

Arithmetic Sequences

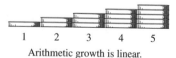

1   2   3   4   5
Arithmetic growth is linear.

Perhaps the simplest pattern is shown by the counting numbers themselves: 1, 2, 3, 4, 5, 6, ... A list of numbers having a first number, a second number, a third number, and so on, is called a **sequence.** The numbers in a sequence are called the **terms** of the sequence. The sequence of counting numbers is formed by adding 1 to each term to obtain the next term. Your math assignments may well have been identified by some sequence: "Do the multiples of 3 from 3 to 30" — that is, 3, 6, 9, . . . , 27, 30. This sequence is formed by adding 3 to each term. Sequences obtained by adding the same number to each term to obtain the next term are called *arithmetic sequences* or *arithmetic progressions*.

Arithmetic Sequence

An **arithmetic sequence** is a sequence whose consecutive terms differ by the same real number, called the **common difference.**

■ EXAMPLE 1

Show that each sequence is arithmetic, and find the missing term.

a. 1, 4, 7, 10, 13, —— , . . .
b. 20, 14, 8, 2, -4, -10, —— , . . .
c. $a_1, a_1 + d, a_1 + 2d, a_1 + 3d, a_1 + 4d,$ —— , . . .

Solution

a. Look for a common difference by subtracting each term from the succeeding term:

$$4 - 1 = 3, \quad 7 - 4 = 3, \quad 10 - 7 = 3, \quad 13 - 10 = 3$$

If the difference between each pair of consecutive terms of the sequence is the same number, then that number is the common difference; in this case it is 3. To find the missing term, simply add the common difference. The next term is

$$13 + 3 = 16$$

b. The common difference is -6. The next term is found by adding the common difference:

$$-10 + (-6) = -16$$

c. The common difference is d, so the next term is $(a_1 + 4d) + d = a_1 + 5d$

⊘ Do not check only the first difference. *All* the differences must be the same. ⊘

Example 1c leads us to a formula for arithmetic sequences:

a_1 is the first term of an arithmetic sequence.

a_2 is the second term of an arithmetic sequence, and
$$a_2 = a_1 + d$$

a_3 is the third term of an arithmetic sequence, and
$$a_3 = a_2 + d = (a_1 + d) + d = a_1 + 2d$$

a_4 is the fourth term of an arithmetic sequence, and
$$a_4 = a_3 + d = (a_1 + 2d) + d = a_1 + 3d$$
$$\vdots$$

a_{43} is the 43rd term of an arithmetic sequence, and
$$a_{43} = a_1 + 42 \, d$$

one less than the term number
$$\vdots$$

This pattern leads to the following definition.

General Term of an Arithmetic Sequence

The **general term** of an arithmetic sequence $a_1, a_2, a_3, \ldots, a_n$, with common difference d is

$$a_n = a_1 + (n - 1)d$$

■ EXAMPLE 2

If $a_n = 26 - 6n$, list the sequence.

Solution

$$a_1 = 26 - 6(1) = 20 \qquad a_1 \text{ means evaluate } 26 - 6n \text{ for } n = 1.$$
$$a_2 = 26 - 6(2) = 14$$
$$a_3 = 26 - 6(3) = 8$$
$$a_4 = 26 - 6(4) = 2$$

The sequence is 20, 14, 8, 2, ...

Geometric Sequences

A second type of sequence is the *geometric sequence* or *geometric progression*. If each term is *multiplied* by the same number (instead of *added* to the same number) to obtain successive terms, a geometric sequence is formed.

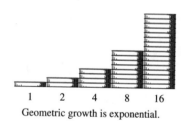

1   2   4   8   16
Geometric growth is exponential.

Geometric Sequence

A **geometric sequence** is a sequence whose consecutive terms have the same quotient, called the **common ratio.**

If the sequence is geometric, the number obtained by dividing any term into the following term of that sequence will be the same nonzero number. No term of a geometric sequence may be zero.

■ EXAMPLE 3

Show that each sequence is geometric, and find the common ratio.

a. 2, 4, 8, 16, 32, _____ , . . .   **b.** 10, 5, $\frac{5}{2}, \frac{5}{4}, \frac{5}{8}$, _____ , . . .

c. $g_1, g_1 r, g_1 r^2, g_1 r^3, g_1 r^4$, . . .

Solution

a. First, verify that there is a common ratio:

$$\frac{4}{2} = 2, \quad \frac{8}{4} = 2, \quad \frac{16}{8} = 2, \quad \frac{32}{16} = 2$$

The common ratio is 2, so to find the next term, multiply the common ratio by the preceding term. The next term is

$$32(2) = 64$$

b. The common ratio is $\frac{1}{2}$ (be sure to check *each* ratio). The next term is found by multiplication:

$$\frac{5}{8}\left(\frac{1}{2}\right) = \frac{5}{16}$$

c. The common ratio is r, and the next term is

$$g_1 r^4(r) = g_1 r^5$$

As with arithmetic sequences, we denote the terms of a geometric sequence by using a special notation. Let $g_1, g_2, g_3, \ldots, g_n, \ldots$ be the terms of a *geometric* sequence. Example 3c leads us to a formula for geometric sequences:

$$g_2 = rg_1$$
$$g_3 = rg_2 = r(rg_1) = r^2 g_1$$
$$g_4 = rg_3 = r(r^2 g_1) = r^3 g_1$$
$$\vdots$$

one less than the term number

$$g_{92} = r^{91} g_1$$
$$\vdots$$

Look for patterns to find the following formula.

General Term of a Geometric Sequence

For a geometric sequence $g_1, g_2, g_3, \ldots, g_n, \ldots$, with common ratio r, the **general term** is

$$g_n = g_1 r^{n-1}$$

EXAMPLE 4

List the sequence generated by $g_n = 50(2)^{n-1}$.

Solution

$$g_1 = 50(2)^{1-1} = 50 \qquad g_2 = 50(2)^{2-1} = 100$$
$$g_3 = 50(2)^{3-1} = 200 \qquad g_4 = 50(2)^{4-1} = 400$$

The sequence is 50, 100, 200, 400, ...

Fibonacci-Type Sequences

Even though our attention is focused on arithmetic and geometric sequences, it is important to realize there can be other types of sequences.

The next type of sequence came about, oddly enough, by looking at the birth patterns of rabbits. In the 13th century, Leonardo Fibonacci wrote a book, *Liber Abaci*, in which he discussed the advantages of the Hindu–Arabic numerals over Roman numerals. In this book, one problem was to find the number of rabbits alive after a given number of generations. Let us consider what he did with this problem.

■■■■ EXAMPLE 5   Polya's Method

Suppose a pair of rabbits will produce a new pair of rabbits in their second month, and thereafter will produce a new pair every month. The new rabbits will do exactly the same. Start with one pair. How many pairs will there be in 10 months?

Solution   We use Polya's problem-solving guidelines for this example, since it does not seem to match the previous problems we have encountered.

Understand the Problem.   We can begin to understand the problem by looking at the following chart:

Number of Months	Number of Pairs	Pairs of Rabbits (the pairs shown in color are ready to reproduce in the next month)
Start	1	
1	1	
2	2	
3	3	
4	5	
5	8	
⋮	⋮	Same pair (rabbits never die)

Devise a Plan.   We look for a pattern with the sequence 1, 1, 2, 3, 5, 8, . . . ; it is not arithmetic and it is not geometric. It looks as if (after the first two months) each new number can be found by adding the two previous terms.

Carry Out the Plan.

$$1 + 1 = 2$$
$$1 + 2 = 3$$
$$2 + 3 = 5$$
$$3 + 5 = 8$$
$$5 + 8 = ?$$

Do you see the pattern? The sequence is 1, 1, 2, 3, 5, 8, 13, 21, 34, 55, 89, ...

Look Back. Using this pattern, Fibonacci was able to compute the number of pairs of rabbits alive after 10 months (it is the tenth term after the first 1): 89. He could also compute the number of pairs of rabbits after the first year or any other interval. Without a pattern, the problem would indeed be a difficult one. ■

General Terms of a Fibonacci-Type Sequence

A **Fibonacci-type sequence** is a sequence in which the **general term** is given by the formula

$$s_n = s_{n-1} + s_{n-2}$$

where s_1 and s_2 are given. *The* **Fibonacci sequence** is that sequence for which $s_1 = s_2 = 1$.

In other words, since s_n represents the nth term, s_{n-1} and s_{n-2} are the two previous terms. A Fibonacci-type sequence is one in which terms are found by adding the two previous terms. The first two terms of the sequence must be given. If it is *the* Fibonacci sequence, then the first two terms are 1.

■ **EXAMPLE 6**

a. If $s_1 = 5$ and $s_2 = 2$, list the first five terms of this Fibonacci-type sequence.

b. If s_1 and s_2 represent any first numbers, list the first eight terms of this Fibonacci-type sequence.

Solution

a. 5, 2, 7, 9, 16

b. s_1 and s_2 are given.

$$s_3 = s_2 + s_1$$
$$s_4 = s_3 + s_2 = (s_1 + s_2) + s_2 = s_1 + 2s_2$$
$$s_5 = s_4 + s_3 = (s_1 + 2s_2) + (s_1 + s_2) = 2s_1 + 3s_2$$
$$s_6 = s_5 + s_4 = (2s_1 + 3s_2) + (s_1 + 2s_2) = 3s_1 + 5s_2$$
$$s_7 = s_6 + s_5 = (3s_1 + 5s_2) + (2s_1 + 3s_2) = 5s_1 + 8s_2$$
$$s_8 = s_7 + s_6 = (5s_1 + 8s_2) + (3s_1 + 5s_2) = 8s_1 + 13s_2$$
$$\vdots$$

Look at the coefficients in the algebraic simplification and notice that the Fibonacci sequence 1, 1, 2, 3, 5, 8, . . . is part of the construction of any Fibonacci-type sequence regardless of the terms s_1 and s_2. ▬

Historically, there has been much interest in the Fibonacci sequence. It is used in botany, zoology, business, economics, statistics, operations research, archeology, architecture, education, and sociology. There is even an official Fibonacci Association. An example of Fibonacci numbers occurring in nature is illustrated by a sunflower. The seeds are arranged in spiral curves as shown in Figure 6.2. If we count the number of clockwise spirals (13 and 21 in this example), they are successive terms in the Fibonacci sequence. This is true of all sunflowers and, indeed, of the seed head of any composite flower such as the daisy or aster.

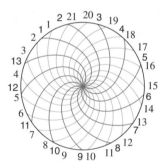

▲ **Figure 6.2   The arrangement of the pods (phyllotaxy) of a sunflower illustrates a Fibonacci sequence.**

▬▬ **EXAMPLE 7**

Classify the given sequences as arithmetic, geometric, Fibonacci-type, or none of the above. Find the next term for each sequence and give its general term if it is arithmetic, geometric, or Fibonacci-type.

a. 15, 30, 60, 120, . . .   **b.** 15, 30, 45, 60, . . .

c. 15, 30, 45, 75, . . .   **d.** 15, 20, 26, 33, . . .

e. 3, 3, 3, 3, . . .   **f.** 15, 30, 90, 360, . . .

Solution

a. 15, 30, 60, 120, . . . does not have a common difference, but a common ratio of 2, so this is a geometric sequence. The next term is $120(2) = 240$. The general term is $g_n = 15(2)^{n-1}$.

b. 15, 30, 45, 60, . . . has a common difference of 15, so this is an arithmetic sequence. The next term of the sequence is $60 + 15 = 75$. The general term is $a_n = 15 + (n - 1)15 = 15 + 15n - 15 = 15n$.

c. 15, 30, 45, 75, . . . does not have a common difference or a common ratio. Next, we check for a Fibonacci-type sequence by adding successive terms: $15 + 30 = 45$, $30 + 45 = 75$, so this is a Fibonacci-type sequence. The next term is $45 + 75 = 120$. The general term is $s_n = s_{n-1} + s_{n-2}$ where $s_1 = 15$ and $s_2 = 30$.

d. 15, 20, 26, 33, . . . does not have a common difference or a common ratio. Since $15 + 20 \neq 26$, we see it is not a Fibonacci-type sequence. We do see a pattern, however, when looking at the differences:

$$20 - 15 = 5; \quad 26 - 20 = 6; \quad 33 - 26 = 7$$

The next difference is 8. Thus, the next term is $33 + 8 = 41$.

e. 3, 3, 3, 3, . . . has a common difference of 0 and the common ratio is 1, so it is both arithmetic and geometric. The next term is 3. The general term is $a_n = 3 + (n - 1)0 = 3$ or $g_n = 3(1)^{n-1} = 3$.

f. 15, 30, 90, 360, . . . does not have a common difference or a common ratio. We see a pattern when looking at the ratios:

$$\frac{30}{15} = 2; \quad \frac{90}{30} = 3; \quad \frac{360}{90} = 4$$

The next ratio is 5. Thus, the next term is $360(5) = 1,800$.   ▬

General terms can be given for sequences that are not arithmetic, geometric, or Fibonacci-type. Consider the following example.

▬▬▬ EXAMPLE 8

Find the first four terms for the sequences with the given general terms. If the general term defines an arithmetic, geometric, or Fibonacci-type sequence, so state.

a. $s_n = n^2$

b. $s_n = (-1)^n - 5n$

c. $s_n = s_{n-1} + s_{n-2}$, where $s_1 = -4$ and $s_2 = 6$

d. $s_n = 2n$

e. $s_n = 2n + (n - 1)(n - 2)(n - 3)(n - 4)$

Solution

a. Since $s_n = n^2$, we find $s_1 = (1)^2 = 1$; $s_2 = (2)^2 = 4$, $s_3 = (3)^2$, ... The sequence is 1, 4, 9, 16, ...

b. Since $s_n = (-1)^n - 5n$, we find

$$s_1 = (-1)^1 - 5(1) = -6$$
$$s_2 = (-1)^2 - 5(2) = -9$$
$$s_3 = (-1)^3 - 5(3) = -16$$
$$s_4 = (-1)^4 - 5(4) = -19$$
$$\vdots$$

The sequence is $-6, -9, -16, -19, \ldots$.

c. $s_n = s_{n-1} + s_{n-2}$ is the form of a Fibonacci-type sequence. Using the two given terms, we find:

$$s_1 = -4 \hspace{3cm} \text{Given}$$
$$s_2 = 6 \hspace{3.2cm} \text{Given}$$
$$s_3 = s_2 + s_1 = 6 + (-4) = 2$$
$$s_4 = s_3 + s_2 = 2 + 6 = 8$$

The sequence is $-4, 6, 2, 8, \ldots$.

d. $s_n = 2n$; $s_1 = 2(1) = 2$; $s_2 = 2(2) = 4$; $s_3 = 2(3) = 6$; $\ldots$. The sequence is $2, 4, 6, 8, \ldots$; this is an arithmetic sequence.

e. $s_n = 2n + (n-1)(n-2)(n-3)(n-4)$

$$s_1 = 2(1) + (1-1)(1-2)(1-3)(1-4) = 2$$
$$s_2 = 2(2) + (2-1)(2-2)(2-3)(2-4) = 4$$
$$s_3 = 2(3) + (3-1)(3-2)(3-3)(3-4) = 6$$
$$s_4 = 2(4) + (4-1)(4-2)(4-3)(4-4) = 8$$

The sequence is $2, 4, 6, 8, \ldots$.

Examples 8d and 8e show that if only a finite number of successive terms are known and no general term is given, then a *unique* general term cannot be given. That is, if we are given the sequence

$$2, \quad 4, \quad 6, \quad 8, \ldots$$

the next term is probably 10 (if we are thinking of the general term of Example 8d), but it *may* be something different. In Example 8e,

$$s_5 = 2(5) + (5-1)(5-2)(5-3)(5-4) = 34$$

This gives the unlikely sequence $2, 4, 6, 8, 34, \ldots$. In general, you are looking for the simplest general term; nevertheless, you must remember that answers are not unique *unless the general term is given.*

Spreadsheet Application

If you have access to a spreadsheet program, it is easy to generate the terms of a sequence. The spreadsheet below shows the first 25 terms for the sequences given in Example 8.

	A	B	C	D	E	F	G
1	Term	a	b	c	d	e	
2	1	1	−6	−4	2	2	
3	2	4	−9	6	4	4	
4	3	9	−16	2	6	6	
5	4	16	−19	8	8	8	
6	5	25	−26	10	10	34	
7	6	36	−29	18	12	132	
8	7	49	−36	28	14	374	
9	8	64	−39	46	16	856	
10	9	81	−46	74	18	1698	
11	10	100	−49	120	20	3044	
12	11	121	−56	194	22	5062	
13	12	144	−59	314	24	7944	
14	13	169	−66	508	26	11906	
15	14	196	−69	822	28	17188	
16	15	225	−76	1330	30	24054	
17	16	256	−79	2152	32	32792	
18	17	289	−86	3482	34	43714	
19	18	324	−89	5634	36	57156	
20	19	361	−96	9116	38	73478	
21	20	400	−99	14750	40	93064	
22	21	441	−106	23866	42	116322	
23	22	484	−109	38616	44	143684	
24	23	529	−116	62482	46	175606	
25	24	576	−119	101098	48	212568	
26	25	625	−126	163580	50	255074	

PROBLEM SET 6.3

▲ **A Problems**

1. **IN YOUR OWN WORDS** What is a sequence?

2. **IN YOUR OWN WORDS** What do we mean by a general term?

3. **IN YOUR OWN WORDS** What is an arithmetic sequence?

4. **IN YOUR OWN WORDS** What is a geometric sequence?

5. **IN YOUR OWN WORDS** What is a Fibonacci sequence?

In Problems 6–31,
a. *Classify the sequences as arithmetic, geometric, Fibonacci, or none of these.*
b. *If arithmetic, give d; if geometric, give r; if Fibonacci, give the first two terms; and if none of these, state a pattern using your own words.*
c. *Supply the next term.*

6. 2, 4, 6, 8, ――― , . . .

7. 2, 4, 8, 16, ――― , . . .

8. 2, 4, 6, 10, ――― , . . .

9. 5, 15, 25, ――― , . . .

10. 5, 15, 45, ――― , . . .

11. 5, 15, 20, ――― , . . .

12. 1, 5, 25, ――― , . . .

13. 25, 5, 1, ――― , . . .

14. 9, 3, 1, ――― , . . .

15. 1, 3, 9, ――― , . . .

16. 21, 20, 18, 15, 11, ――― , . . .

17. 8, 6, 7, 5, 6, 4, ――― , . . .

18. 2, 5, 8, 11, 14, ――― , . . .

19. 3, 6, 12, 24, 48, ――― , . . .

20. 5, -15, 45, -135, 405, ――― , . . .

21. 10, 10, 10, ――― , . . .

22. 2, 5, 7, 12, ――― , . . .

23. 3, 6, 9, 15, ――― , . . .

24. 1, 8, 27, 64, 125, ――― , . . .

25. 8, 12, 18, 27, ――― , . . .

26. 3^2, 3^5, 3^8, 3^{11}, ――― , . . .

27. 4^5, 4^4, 4^3, 4^2, ――― , . . .

28. $\frac{1}{2}$, $\frac{1}{3}$, $\frac{2}{3}$, $\frac{1}{4}$, $\frac{3}{4}$, $\frac{1}{5}$, $\frac{2}{5}$, $\frac{3}{5}$, $\frac{4}{5}$, $\frac{1}{6}$, ――― , . . .

29. $\frac{1}{10}$, $\frac{1}{5}$, $\frac{3}{10}$, $\frac{2}{5}$, $\frac{1}{2}$, ――― , . . .

30. $\frac{4}{3}$, 2, 3, $4\frac{1}{2}$, ――― , . . .

31. $\frac{7}{12}$, $\frac{2}{3}$, $\frac{3}{4}$, $\frac{5}{6}$, ――― , . . .

▲ B Problems

In Problems 32–49,
a. *Find the first three terms of the sequences whose nth terms are given.*
b. *Classify the sequence as arithmetic (give d), geometric (give r), both, or neither.*

32. $s_n = 4n - 3$

33. $s_n = -3 + 3n$

34. $s_n = 10n$

35. $s_n = 2 - n$

36. $s_n = 7 - 3n$

37. $s_n = 10 - 10n$

38. $s_n = \dfrac{2}{n}$

39. $s_n = 1 - \dfrac{1}{n}$

40. $s_n = \dfrac{n-1}{n+1}$

41. $s_n = \dfrac{1}{2}n(n+1)$

42. $s_n = \dfrac{1}{6}n(n+1)(2n+1)$

43. $s_n = \dfrac{1}{4}n^2(n+1)^2$

44. $s_n = (-1)^n$

45. $s_n = (-1)^{n+1}$

46. $s_n = -5$

47. $s_n = \dfrac{2}{3}$

48. $s_n = (-1)^n(n+1)$

49. $s_n = (-1)^{n-1}n$

Find the requested terms in Problems 50–57.

50. Find the 15th term of the sequence $s_n = 4n - 3$.

51. Find the 69th term of the sequence $s_n = 7 - 3n$.

52. Find the 20th term of the sequence $s_n = (-1)^n(n+1)$.

53. Find the 3rd term of the sequence $s_n = (-1)^{n+1}5^{n+1}$.

54. Find the first five terms of the sequence where $s_1 = 2$ and $s_n = 3s_{n-1}$, $n \geq 2$.

55. Find the first five terms of the sequence where $s_1 = 3$ and $s_n = \frac{1}{3}s_{n-1}$, $n \geq 2$.

56. Find the first five terms of the sequence where $s_1 = 1$, $s_2 = 1$, and $s_n = s_{n-1} + s_{n-2}$, $n \geq 3$.

57. Find the first five terms of the sequence where $s_1 = 1$, $s_2 = 2$, and $s_n = s_{n-1} + s_{n-2}$, $n \geq 3$.

▲ Problem Solving

58. Is the following sequence a Fibonacci sequence?

a_n is one more than the nth term of the Fibonacci sequence.

59. Is the following sequence a Fibonacci sequence?

$1, 1, 2, 3, 5, 8, \ldots, a_n$ where a_n is the integer nearest to $\dfrac{1}{\sqrt{5}} \left[\dfrac{1 + \sqrt{5}}{2} \right]^n$

60. Apartment blocks of n floors are to be painted blue and yellow, with the rule that no two adjacent floors can be blue. (They can, however, be yellow.) Let a_n be the number of ways to paint a block with n floors.*

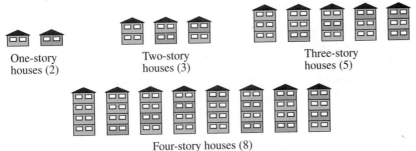

One-story
houses (2)

Two-story
houses (3)

Three-story
houses (5)

Four-story houses (8)

Does the sequence $a_1, a_2, a_3, \ldots$ form a Fibonacci sequence?

61. Consider the following magic trick.

The magician asks the audience for any two numbers (you can limit it to the counting numbers between 1 and 10 to keep the arithmetic manageable). Add these two numbers to obtain a third number. Add the second and third numbers to obtain a fourth. Continue until ten numbers are obtained. Then ask the audience to add the ten numbers, while the magician instantly gives the sum.

* From "Fibonacci Forgeries" by Ian Steward, *Scientific American*, May 1995, p. 104.

First number:	5
Second number:	9
(3) Add:	14
(4)	23
(5)	37
(6)	60
(7)	97
(8)	157
(9)	254
(10)	411
Add column:	1,067

Consider the example shown in the margin. The trick depends on the magician's ability to multiply quickly and mentally by 11. Consider the following pattern of multiplication by 11:

$$11 \times 10 = 110$$
$$11 \times 11 = 121$$
$$11 \times 12 = 132$$
$$11 \times 13 = 1\ 3 \qquad\qquad = 143$$

original two digits

$$11 \times 52 = 5\ 2 \qquad\qquad = 572$$

sum of the two digits

$$11 \times 74 = 7\ 4 \qquad\qquad = 7\ \boxed{11}\ 4 = 814 \qquad \text{Do the usual carry.}$$

Consider 11 times the 7th number in the example in the margin:
$11 \times 97 = 9\ \boxed{16}\ 7 = 1{,}067$. Note that is the sum of the ten numbers. Explain why this magician's trick works, and why it is called Fibonacci's magic trick.

62. Fill in the blanks so that ____ , 8, ____ , ____ , 27, ____ , . . . is
   a. an arithmetic sequence.
   b. a geometric sequence.
   c. a sequence that is neither arithmetic nor geometric, for which you are able to write a general term.

6.4   SERIES

If the terms of a sequence are added, the expression is called a **series.** We first consider a finite sequence along with its associated sum. Note that a capital letter is used to indicate the sum.

⊘ Write a lowercase *s* on your paper; now write a capital *S*. Do they look different? You need to distinguish between the lowercase and capital *S* in your *own* written work. ⊘

> **Finite Series**
>
> The indicated sum of the terms of a finite sequence
>
> $$s_1, s_2, s_3, \ldots, s_n$$
>
> is called a **finite series** and is denoted by
>
> $$S_n = s_1 + s_2 + s_3 + \cdots + s_n$$

▬ **EXAMPLE I**

a. Find S_4 where $s_n = 26 - 6n$.   b. Find S_3 where $s_n = (-1)^n n^2$.

Solution

a. $S_4 = s_1 + s_2 + s_3 + s_4$

$$= \overbrace{[26 - 6(1)]}^{s_1} + \overbrace{[26 - 6(2)]}^{s_2} + \overbrace{[26 - 6(3)]}^{s_3} + \overbrace{[26 - 6(4)]}^{s_4}$$

$$= 20 + 14 + 8 + 2$$

$$= 44$$

b. $S_3 = s_1 + s_2 + s_3$

$$= \overbrace{[(-1)^1(1)^2]}^{s_1} + \overbrace{[(-1)^2(2)^2]}^{s_2} + \overbrace{[(-1)^3(3)^2]}^{s_3}$$

$$= -1 + 4 + (-9)$$

$$= -6$$

The terms of the sequence in Example 1b alternate in sign: $-1, 4, -9,$ $16, \ldots$. A factor of $(-1)^n$ or $(-1)^{n+1}$ in the general term will cause the sign of the terms to alternate, creating a series called an **alternating series.**

Summation Notation

Before we continue to discuss finding the sum of the terms of a sequence, we need a handy notation, called **summation notation.** In Example 1a we wrote

$$S_4 = s_1 + s_2 + s_3 + s_4$$

Using summation notation (or, as it is sometimes called, **sigma notation**), we could write this sum using the Greek letter Σ:

$$S_4 = \sum_{k=1}^{4} s_k = s_1 + s_2 + s_3 + s_4$$

The sigma notation evaluates the expression (s_k) immediately following the sigma (Σ) sign, first for $k = 1$, then for $k = 2$, then for $k = 3$, and finally for $k = 4$, and then adds these numbers. That is, the expression is evaluated for *consecutive counting numbers* starting with the value of k listed at the bottom of the sigma ($k = 1$) and ending with the value of k listed at the top of the sigma ($k = 4$). For example, consider $s_k = 2k$ with $k = 1, 2, 3, 4, 5, 6, 7, 8, 9, 10$. Then

This is the last natural number in the domain. It is called the *upper limit.*

↓

$$\sum_{k=1}^{10} 2k\} \leftarrow \text{This is the function being evaluated. It is called the } \textit{general term.}$$

↑

This is the first natural number in the domain. It is called the *lower limit.*

Thus, $\displaystyle\sum_{k=1}^{10} 2k = 2(1) + 2(2) + 2(3) + 2(4) + 2(5) + 2(6) + 2(7) + 2(8)$
$$+ 2(9) + 2(10) = 110$$

The words **evaluate** and **expand** are both used to mean write out an expression in summation notation, and then sum the resulting terms, if possible.

▬▬▬ EXAMPLE 2

a. Evaluate $\displaystyle\sum_{k=3}^{6} (2k + 1)$ **b.** Expand $\displaystyle\sum_{k=3}^{n} \frac{1}{2^k}$

Solution

$$
\textbf{a.} \ \underbrace{\sum_{k=3}^{6} (2k + 1) = \underbrace{(2 \cdot 3 + 1)}_{\text{Evaluate the expression } 2k + 1 \text{ for } k = 3} + (2 \cdot 4 + 1) + \overbrace{(2 \cdot 5 + 1)}^{k=5} + \overbrace{(2 \cdot 6 + 1)}^{k=6}}
$$

$$
= 7 + 9 + 11 + 13 = 40
$$

$$
\textbf{b.} \ \sum_{k=3}^{n} \frac{1}{2^k} = \frac{1}{2^3} + \frac{1}{2^4} + \frac{1}{2^5} + \frac{1}{2^6} + \cdots + \frac{1}{2^{n-1}} + \frac{1}{2^n}
$$

Arithmetic Series

Row 1:   1
Row 2:   6
Row 3:   11
Row 4:   16

▲ **Figure 6.3   How many blocks?**

An **arithmetic series** is the sum of the terms of an arithmetic sequence. Let us consider a rather simple-minded example. How many blocks are shown in the stack in Figure 6.3?

We can answer this question by simply counting the blocks: there are 34 blocks. Somehow it does not seem like this is what we have in mind with this question. Suppose we ask a better question. How many blocks are in a similar building with n rows? We notice that the number of blocks (counting from the top) in each row forms an arithmetic series:

$$1 + 6 + 11 + 16 + \cdots$$

Look for a pattern:

Denote one row by $A_1 = 1$ block

two rows: $A_2 = 1 + 6 = 7$

three rows: $A_3 = 1 + 6 + 11 = 18$

four rows: $A_4 = 1 + 6 + 11 + 16 = 34$ (shown in Figure 6.3)

$$\vdots$$

What about 10 rows?

$$A_{10} = 1 + 6 + 11 + 16 + 21 + 26 + 31 + 36 + 41 + 46$$

Instead of adding all these numbers directly, let us try an easier way. Write down A_{10} twice, once counting from the top and once counting from the bottom:

$$
\begin{array}{ccccccccccc}
A_{10} = & 1 & + & 6 & + & 11 & + & 16 & + & 21 & + & 26 & + & 31 & + & 36 & + & 41 & + & 46 \\
 & \updownarrow & & \updownarrow & & \updownarrow & & \updownarrow & & \updownarrow & & \updownarrow & & \updownarrow & & \updownarrow & & \updownarrow & & \updownarrow \\
A_{10} = & 46 & + & 41 & + & 36 & + & 31 & + & 26 & + & 21 & + & 16 & + & 11 & + & 6 & + & 1
\end{array}
$$

Add these equations:

$$2A_{10} = 47 + 47 + 47 + 47 + 47 + 47 + 47 + 47 + 47 + 47$$

number of terms

$$2A_{10} = \overbrace{10}^{} (47) = 470$$

sum of 1st and last terms

$$A_{10} = 10 \underbrace{\left(\frac{47}{2}\right)}_{} = 235 \quad \text{Divide both sides by 2.}$$

average of 1st and last terms

This pattern leads us to a formula for n terms. We note that the number of blocks is an arithmetic sequence with $a_1 = 1$, $d = 5$; thus since $a_n = a_1 + (n - 1)d$, we have for the blocks in Figure 6.3 a stack starting with 1 and ending (in the nth row) with

$$a_n = 1 + (n - 1)5 = 1 + 5n - 5 = 5n - 4$$

Thus,

Number of blocks in n rows   Number of rows
$$\downarrow \qquad\qquad\qquad\qquad \downarrow$$
$$A_n \qquad\qquad = n \underbrace{\left[\frac{1 + (5n - 4)}{2}\right]}_{}$$

average of 1st and last terms

$$= n \left[\frac{5n - 3}{2}\right]$$

$$= \tfrac{1}{2}(5n^2 - 3n)$$

This formula can be used for the number of blocks for any number of rows. Looking back, we see

For $n = 1$: $A_1 = \tfrac{1}{2}[5(1)^2 - 3(1)] = 1$

For $n = 4$: $A_4 = \tfrac{1}{2}[5(4)^2 - 3(4)] = 34$ (Figure 6.3)

For $n = 10$: $A_{10} = \tfrac{1}{2}[5(10)^2 - 3(10)] = 235$

If we carry out these same steps for A_n where $a_n = a_1 + (n - 1)d$, we derive the following formula for the sum of the terms of an arithmetic sequence.

Arithmetic Series Formula

The sum of the terms of an arithmetic sequence $a_1, a_2, a_3, \ldots, a_n$ with common difference d is

$$A_n = \sum_{k=1}^{n} a_k = n\left(\frac{a_1 + a_n}{2}\right) \quad \text{or} \quad A_n = \frac{n}{2}[2a_1 + (n - 1)d]$$

In other words, the sum of n terms of an arithmetic sequence is n times the average of the first and last terms.

The last part of the formula for A_n is used when the last term is not explicitly stated or known. To derive this formula we know $a_n = a_1 + (n - 1)d$ so

$$A_n = n\left(\frac{a_1 + a_n}{2}\right)$$

$$= n\left(\frac{a_1 + [a_1 + (n - 1)d]}{2}\right)$$

$$= \frac{n}{2}[a_1 + a_1 + (n - 1)d]$$

$$= \frac{n}{2}[2a_1 + (n - 1)d]$$

▇▇ EXAMPLE 3

In a classroom of 35 students, each student "counts off" by threes (i.e., 3, 6, 9, 12, . . .). What is the sum of the students' numbers?

Solution   We recognize the sequence 3, 6, 9, 12, . . . as an arithmetic sequence with the first term $a_1 = 3$ and the common difference $d = 3$. The sum of these numbers is denoted by A_{35} since there are 35 students "counting off":

$$A_{35} = \frac{35}{2}[2(3) + (35 - 1)3] = 1,890$$

Geometric Series

A **geometric series** is the sum of the terms of a geometric sequence. To motivate a formula for a geometric series, we once again consider an example. Suppose Charlie Brown receives a chain letter, and he is to copy this letter and send it to six of his friends.

You may have heard that chain letters "do not work." Why not? Consider the number of people who could become involved with this chain letter if we assume that everyone carries out their task and does not break the chain. The first mailing would consist of six letters. The second mailing involves 42 letters since the second mailing of 36 letters is added to the total:

$$6 + 36 = 42$$

1st mailing:

2nd mailing:

The number of letters in each successive mailing is a number in the geometric sequence

6,   36,   216,   1296, . . .   or   6, 6^2, 6^3, 6^4, . . .

How many people receive letters with 11 mailings, assuming that no person receives a letter more than once? To answer this question, consider the series associated with a geometric sequence. We begin with a pattern:

Denote one mailing by $G_1 = 6$

two mailings: $G_2 = 6 + 6^2$

three mailings: $G_3 = 6 + 6^2 + 6^3$

$\vdots$

eleven mailings: $G_{11} = 6 + 6^2 + \cdots + 6^{11}$

We could probably use a calculator to find this sum, but we are looking for a formula, so we try something different. In fact, this time we will work out the general formula. Let

$$G_n = g_1 + g_2 + g_3 + \cdots + g_n$$
$$G_n = g_1 + g_1 r + g_1 r^2 + \cdots + g_1 r^{n-1}$$

Multiply both sides of this latter equation by r:

$$rG_n = g_1 r + g_1 r^2 + g_1 r^3 + \cdots + g_1 r^n$$

Notice that, except for the first and last terms, all the terms in the expansions for G_n and rG_n are the same, so that if we subtract one equation from the other, we have

$$G_n - rG_n = g_1 - g_1 r^n$$

We now solve for G_n:

$$(1 - r)G_n = g_1(1 - r^n)$$
$$G_n = \frac{g_1(1 - r^n)}{1 - r} \quad \text{if } r \neq 1$$

For Charlie Brown's chain letter problem, $g_1 = 6$, $n = 11$, and $r = 6$ so we find

$$G_{11} = \frac{6(1 - 6^{11})}{1 - 6} = \frac{6}{5}(6^{11} - 1) = 435{,}356{,}466$$

This is more than the number of people in the United States! The number of letters in only two more mailings would exceed the number of men, women, and children in the whole world.

> **Geometric Series**
>
> The sum of the terms of a geometric sequence $g_1, g_2, g_3, \ldots, g_n$ with common ratio r $(r \neq 1)$ is
>
> $$G_n = \frac{g_1(1 - r^n)}{1 - r}$$

▬▬ EXAMPLE 4

Suppose some eccentric millionaire offered to hire you for a month (say, 31 days) and offered you the following salary choice. She will pay you $0.5 million per day or else will pay you 1¢ for the first day, 2¢ for the second day, 4¢ for the third day, and so on for the 31 days. Which salary should you accept?

Solution   If you are paid $500,000 per day, your salary for the 31 days is

$$\$500,000(31) = \$15,500,000$$

Now, if you are paid using the doubling scheme, your salary is (in cents)

$$\overbrace{}^{\text{31st day}}$$
$$1 + 2 + 4 + 8 + \cdots + \text{last day} \quad \text{or} \quad 2^0 + 2^1 + 2^2 + 2^3 + \cdots + 2^{30}$$

We see that this is the sum of the geometric sequence where $g_1 = 1$ and $r = 2$. We are looking for G_{31}:

$$G_{31} = \frac{1(1 - 2^{31})}{1 - 2} = -(1 - 2^{31}) = 2^{31} - 1$$

Using a calculator, we find this to be 2,147,483,647 cents or $21,474,836.47. You should certainly accept the doubling scheme (starting with 1¢, but do not ask for any days off). ▬

▬▬ EXAMPLE 5   Polya's Method

The NCAA men's basketball tournament has 64 teams. How many games are necessary for the playoffs?

Solution   We use Polya's problem-solving guidelines for this example.

Understand the Problem.   Most tournaments are formed by drawing an elimination schedule similar to the one shown in Figure 6.4. This is sometimes called a *two-team elimination tournament*.

Devise a Plan.   We could obtain the answer to the question by direct counting, but instead we will find a general solution, working backward. We know there will be 1 championship game, and 2 semifinal games; continuing to work backward, there are 4 quarter-final games, . . . :

$$1 + 2 + 2^2 + 2^3 + \cdots$$

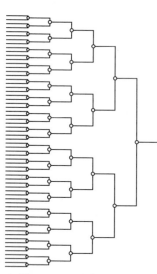

▲ **Figure 6.4   NCAA playoffs**

We recognize this as a geometric series. To reach 64 teams we note that since there are 64 teams, the first round has $64/2 = 32$ games. Also, we know that $2^5 = 32$, so we must find

$$1 + 2 + 2^2 + 2^3 + 2^4 + 2^5$$

Carry Out the Plan. We note that $g_1 = 1$, $r = 2$, and $n = 6$:

$$G_6 = \frac{1(1 - 2^6)}{1 - 2} = 2^6 - 1 = 63$$

Thus, the NCAA playoffs will require 63 games.

Look Back. We not only see that the NCAA tournament has 63 games for a playoff tournament, but, in general, if there are 2^n teams in a tournament, there will be $2^n - 1$ games. However, do not forget to check by estimation or by using common sense. Each game eliminates one team, and so 63 teams must be eliminated to crown a champion. ▬

Infinite Geometric Series

We have just found a formula for the sum of the first n terms of a geometric sequence. Sometimes it is also possible to find the sum of an entire infinite geometric series. Suppose an infinite geometric series

$$g_1 + g_2 + g_3 + g_4 + \cdots$$

is denoted by G. This is an **infinite series**. The **partial sums** are defined by

$$G_1 = g_1; \quad G_2 = g_1 + g_2; \quad G_3 = g_1 + g_2 + g_3; \quad \cdots$$

Consider the partial sums for an infinite geometric series with $g_1 = \frac{1}{2}$ and $r = \frac{1}{2}$. The geometric sequence is $\frac{1}{2}, \frac{1}{4}, \frac{1}{8}, \frac{1}{16}, \ldots$. The first few partial sums can be found as follows:

$$G_1 = \frac{1}{2}; \quad G_2 = \frac{1}{2} + \frac{1}{4} = \frac{3}{4}; \quad G_3 = \frac{1}{2} + \frac{1}{4} + \frac{1}{8} = \frac{7}{8}; \quad \cdots$$

Does this series have a sum if you add *all* its terms? It does seem that as you take more terms of the series, the sum is closer and closer to 1. Analytically, we can check the partial sums on a calculator or spreadsheet (see margin). Geometrically, you can see (Figure 6.5) that if the terms are laid end-to-end as lengths on a number line, each term is half the remaining distance to 1.

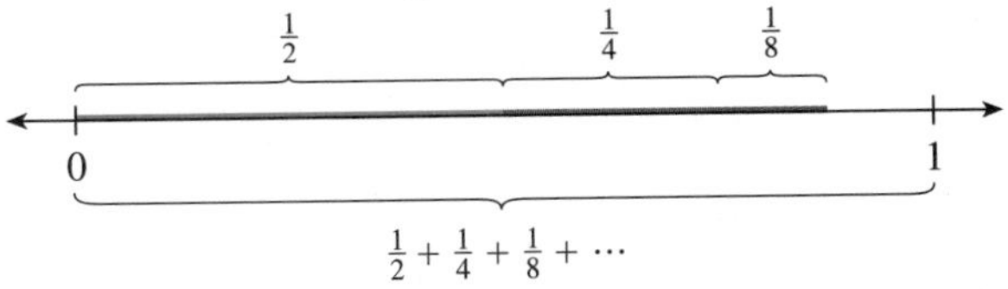

The infinite! No other question has ever moved so profoundly the spirit of man [or woman]; no other idea has so fruitfully stimulated his [her] intellect; yet no other concept stands in greater need of clarification than that of the infinite. . . .

David Hilbert

▲ **Figure 6.5** Series $\frac{1}{2} + \frac{1}{4} + \frac{1}{8} + \cdots$

It appears that the partial sums are getting closer to 1 as n becomes larger. We *can* find the sum of an infinite geometric sequence.

Spreadsheet Application

It is easy to use a spreadsheet (review page 300) to look at the partial sums of a sequence. For the sequence $\frac{1}{2}, \frac{1}{4}, \frac{1}{8}, \ldots$, define the cells as shown.

	A	B	C
1	n	term	partial sum
2	1	.5	+B2
3	+A2+1	+B2*.5	+C2+B3

The output is:

n	term	partial sum
1	.5	.5
2	.25	.75
3	.125	.875
4	.0625	.9375
5	.03125	.96875
6	.015625	.984375
7	.0078125	.9921875
8	.00390625	.99609375
9	.00195313	.99804688
10	.00097656	.99902344
11	.00048828	.99951172
12	.00024414	.99975586
13	.00012207	.99987793
14	.00006104	.99993896
15	.00003052	.99996948
16	.00001526	.99998474
17	.00000763	.99999237
18	.00000381	.99999619
19	.00000191	.99999809
20	.00000095	.99999905

Consider

$$G_n = \frac{g_1(1 - r^n)}{1 - r} = \frac{g_1 - g_1 r^n}{1 - r} = \frac{g_1}{1 - r} - \frac{g_1}{1 - r} r^n$$

Now, g_1, r, and $1 - r$ are fixed numbers. If $|r| < 1$, then r^n approaches 0 as n grows, and thus G_n approaches $\frac{g_1}{1 - r}$.

Infinite Geometric Series

If $g_1, g_2, g_3, \ldots, g_n, \ldots$ is an infinite geometric sequence with a common ratio r such that $|r| < 1$, then its sum is denoted by G and is found by

$$G = \frac{g_1}{1 - r}$$

If $|r| \geq 1$, the infinite geometric series has no sum.

EXAMPLE 6

The path of each swing of a pendulum is 0.85 as long as the path of the previous swing (after the first). If the path of the tip of the first swing is 36 in. long, how far does the tip of the pendulum travel before it eventually comes to rest?

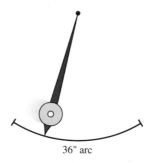

36" arc

Solution

$$
\begin{aligned}
\text{TOTAL DISTANCE} &= 36 + 36(0.85) + 36(0.85)^2 + \cdots \\
&= \frac{36}{1 - 0.85} \\
&= 240
\end{aligned}
$$

Infinite geometric series; $g_1 = 36$; $r = 0.85$

The tip of the pendulum travels 240 in.

Summary of Sequence and Series Formulas

We conclude this section by repeating the important formulas related to sequences and series, as shown in Table 6.1.

TABLE 6.1

Type	Definition	Notation	Formula
Sequences	A list of numbers having a first term, a second term, . . .	s_n	
Arithmetic	Sequence whose terms differ by a constant called the common difference, d	a_n	$a_n = a_1 + (n - 1)d$
Geometric	Sequence whose terms differ by a constant called the common ratio, r	g_n	$g_n = g_1 r^{n-1}$
Fibonacci	Sequence whose first two terms are given, and subsequent terms are found by adding the two previous terms		$s_n = s_{n-1} + s_{n-2}, n \geq 3$

Type	Definition	Notation	Formula
Series	The indicated sum of terms of a sequence	S_n	
Arithmetic	Sum of terms of an arithmetic sequence	A_n	$A_n = \dfrac{n}{2}(a_1 + a_n)$

$$A_n = \sum_{k=1}^{n} a_k = a_1 + a_2 + a_3 + \cdots + a_n$$

Use when first and last terms are known.

$$A_n = \frac{n}{2}[2a_1 + (n - 1)d]$$

Use when first term and common difference are known.

Geometric — Sum of terms of a geometric sequence — G_n — $G_n = \dfrac{g_1(1 - r^n)}{1 - r} \; (r \neq 1)$

$$G_n = \sum_{k=1}^{n} g_k = g_1 + g_2 + g_3 + \cdots + g_n$$

Sum of terms of an infinite geometric sequence — G — $G = \dfrac{g_1}{1 - r} \qquad (|r| < 1)$

$$G = g_1 + g_2 + g_3 + \cdots$$

PROBLEM SET 6.4

▲ **A Problems**

1. **IN YOUR OWN WORDS** Distinguish a sequence and a series.

2. **IN YOUR OWN WORDS** Explain summation notation.

3. **IN YOUR OWN WORDS** What is a partial sum?

4. **IN YOUR OWN WORDS** Distinguish a geometric series and an infinite geometric series.

Evaluate the expressions in Problems 5–16.

5. $\displaystyle\sum_{k=2}^{6} k$
6. $\displaystyle\sum_{k=1}^{4} k$
7. $\displaystyle\sum_{k=3}^{5} k$
8. $\displaystyle\sum_{k=1}^{4} k^2$

9. $\displaystyle\sum_{k=2}^{6} k^2$
10. $\displaystyle\sum_{k=3}^{5} (k^2 - 1)$
11. $\displaystyle\sum_{k=2}^{5} (10 - 2k)$
12. $\displaystyle\sum_{k=2}^{5} (100 - 5k)$

13. $\displaystyle\sum_{k=1}^{10} [1^k + (-1)^k]$
14. $\displaystyle\sum_{k=1}^{5} (-2)^{k-1}$
15. $\displaystyle\sum_{k=0}^{4} 3(-2)^k$
16. $\displaystyle\sum_{k=1}^{3} (-1)^k(k^2 + 1)$

▲ B Problems

If possible, find the sum of the infinite geometric series in Problems 17–24.

17. $1 + \frac{1}{2} + \frac{1}{4} + \cdots$ **18.** $1 + \frac{1}{3} + \frac{1}{9} + \cdots$

19. $1 + \frac{3}{4} + \frac{9}{16} + \cdots$ **20.** $1 + \frac{3}{2} + \frac{9}{4} + \cdots$

21. $1{,}000 + 500 + 250 + \cdots$ **22.** $100 + 50 + 25 + \cdots$

23. $-100 + 50 - 25 + \cdots$ **24.** $-20 + 10 - 5 + \cdots$

25. Find the sum of the first 5 odd positive integers.

26. Find the sum of the first 5 even positive integers.

27. Find the sum of the first 5 positive integers.

28. Find the sum of the first 10 odd positive integers.

29. Find the sum of the first 10 even positive integers.

30. Find the sum of the first 10 positive integers.

31. Find the sum of the first 100 odd positive integers.

32. Find the sum of the first 100 even positive integers.

33. Find the sum of the first 100 positive integers.

34. Find the sum of the first n odd positive integers.

35. Find the sum of the first n even positive integers.

36. Find the sum of the first n positive integers.

37. Find the sum of the first 20 terms of the arithmetic sequence whose first term is 100 and whose common difference is 50.

38. Find the sum of the first 50 terms of the arithmetic sequence whose first term is -15 and whose common difference is 5.

39. Find the sum of the even integers between 41 and 99.

40. Find the sum of the odd integers between 80 and 100.

41. Find the sum of the odd integers between 48 and 136.

The game of pool uses 15 balls numbered from 1 through 15 (see Figure 6.6). In the game of rotation, a player attempts to "sink" a ball in a pocket of the table and receives the number of points on the ball. Answer the questions in Problems 42–45.

42. How many points would a player who "runs the table" receive? (To "run the table" means to sink all the balls.)

43. Suppose Missy sinks balls 1 through 8, and Shannon sinks balls 9 through 15. What are their respective scores?

44. Suppose Missy sinks the even-numbered balls and Shannon sinks the odd-numbered balls. What are their respective scores?

45. Suppose we consider a game of "super pool," which has 30 consecutively numbered balls on the table. How many points would a player receive to "run the table"?

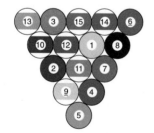

▲ **Figure 6.6   Pool balls**

46. The *Peanuts* cartoon expresses a common feeling regarding chain letters. Consider the total number of letters sent after a particular mailing:

$$\text{1st mailing:} \quad 6$$
$$\text{2nd mailing:} \quad 6 + 36 = 42$$
$$\text{3rd mailing:} \quad 6 + 36 + 216 = 258$$

Determine the total number of letters sent in five mailings of the chain letter.

47. How many blocks would be needed to build a stack like the one shown in Figure 6.7 if the bottom row has 28 blocks?

48. Repeat Problem 47 if the bottom row has 87 blocks.

49. Repeat Problem 47 if the bottom row has 100 blocks.

50. A culture of bacteria increases by 100% every 24 hours. If the original culture contains 1 million bacteria, find the number of bacteria present after 10 days.

51. Use Problem 50 to find a formula for the number of bacteria present after d days.

52. How many games are necessary for a two-team elimination tournament with 32 teams?

53. Games like "Wheel of Fortune" and "Jeopardy" have one winner and two losers. A three-team game tournament is illustrated by Figure 6.8. If "Jeopardy" has a Tournament of Champions consisting of 27 players, what is the necessary number of games?

54. How many games are necessary for a three-team elimination tournament with 729 teams?

55. A pendulum is swung 20 cm and allowed to swing freely until it eventually comes to rest. Each subsequent swing of the bob of the pendulum is 90% as far as the preceding swing. How far will the bob travel before coming to rest?

56. The initial swing of the tip of a pendulum is 25 cm. If each swing of the tip is 75% of the preceding swing, how far does the tip travel before eventually coming to rest?

57. A flywheel is brought to a speed of 375 revolutions per minute (rpm) and allowed to slow and eventually come to rest. If, in slowing, it rotates three-fourths as fast each subsequent minute, how many revolutions will the wheel make before returning to rest?

58. A rotating flywheel is allowed to slow to a stop from a speed of 500 rpm. While slowing, each minute it rotates two-thirds as many times as in the preceding minute. How many revolutions will the wheel make before coming to rest?

59. Advertisements say that a new type of superball will rebound to 9/10 of its original height. If it is dropped from a height of 10 ft, how far, based on the advertisements, will the ball travel before coming to rest?

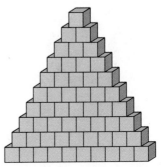

▲ **Figure 6.7 How many blocks?**

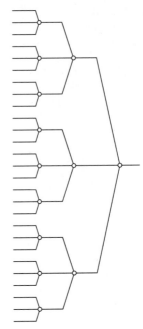

▲ **Figure 6.8 Three-team tournament**

60. A tennis ball is dropped from a height of 10 ft. If the ball rebounds to 2/3 of its height on each bounce, how far will the ball travel before coming to rest?

▲ **Problem Solving**

61. **a.** How many blocks are there in the solid figure shown in Figure 6.9?
    b. How many blocks are there in a similar figure with 50 layers?

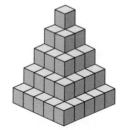

▲ **Figure 6.9   How many blocks?**

6.5   ANNUITIES

Sir Isaac Newton, one of the greatest mathematicians of all time, worked at the mint, but apparently did not like to apply mathematics to money. In a book written by Rev. J. Spence in 1858, it is written, "Sir Isaac Newton, though so deep in algebra and fluxions, could not readily make up a common account: and, when he was Master of the Mint, used to get somebody else to make up his accounts for him." There is an important lesson here: If you do not like to work with money, hire someone to help you. When Albert Einstein was asked what was the most amazing formula he knew, you might think he would have given his famous formula $E = mc^2$, but he did not. He thought the most amazing formula was the compound interest formula.

Indeed, one of the most fundamental mathematical concepts for business people and consumers is the idea of interest. We have considered present- and future-value problems involving a fixed amount, so we refer to such problems as **lump-sum problems.** On the other hand, it is far more common to encounter financial problems based on monthly or other periodic payments; we refer to such problems as **periodic-payment problems.** Consider the situation in which the monthly payment is known and the future value is to be determined. A sequence of payments into or out of an interest-bearing account is called an **annuity.** If the payments are made into an interest-bearing account at the end of each time period, and if the frequency of payments is the same as the frequency of compounding, the annuity is called an *ordinary annuity.* The amount of an annuity is the sum of all payments made plus all accumulated interest. In this book we will assume that all annuities are ordinary annuities.

The best way to understand what we mean by an annuity is to consider an example. Suppose you decide to give up smoking and save the $2 per day you spend on cigarettes. How much will you save in 5 years? If you save the money without earning any interest, you will have

$$\$2 \times 365 \times 5 = \$3,650$$

However, let us assume that you save $2 per day, and at the end of each month you deposit the $60 (assume that all months are 30 days; that is, assume ordinary interest) into a savings account earning 12% interest compounded monthly. Now, how much will you have in 5 years? This is an example of an annuity. We could solve the problem using a spreadsheet to simulate our bank statement. Part of such a spreadsheet is shown in the margin. Notice that even though the interest on $60 for one month is $.60, the total monthly increase in interest is not linear, because the interest is compounded.

To derive a formula for annuities, let us consider this problem for a period of 6 months, and calculate the amounts plus interest for *each* deposit separately. We need a new variable to represent the amount of periodic deposit. Since this is usually a monthly payment, we let m = periodic payment.

We are given $m = 60$, $r = 0.12$, and $n = 12$ (monthly deposit means monthly compounding for an ordinary annuity); we calculate $i = \frac{0.12}{12} = 0.01$. The time, t, varies for each deposit. Remember that the deposit comes at the *end* of the month.

First deposit will earn 5 months' interest:	$60(1 + 0.01)^5 =$	63.06
Second deposit will earn 4 months' interest:	$60(1 + 0.01)^4 =$	62.44
Third deposit will earn 3 months' interest:	$60(1 + 0.01)^3 =$	61.82
Fourth deposit will earn 2 months' interest:	$60(1 + 0.01)^2 =$	61.21
Fifth deposit will earn 1 month's interest:	$60(1 + 0.01)^1 =$	60.60
Sixth deposit will earn no interest:		60.00
TOTAL IN THE ACCOUNT		369.13

This leads us to the following pattern (using variables). The total after 6 months is

$$A = m + m(1 + i)^1 + m(1 + i)^2 + m(1 + i)^3 + m(1 + i)^4 + m(1 + i)^5$$

$$= \sum_{k=1}^{6} m(1 + i)^{k-1}$$

For the total after N periods, we recognize this as a geometric series. That is, it is the sum of the terms of a geometric sequence with $g_1 = m$ and common ratio $(1 + i)$. The sum, G_n, is the future value A. Thus (from the formula for the sum of a geometric sequence), we have

$$A = \frac{m[1 - (1 + i)^N]}{1 - (1 + i)}$$

$$= \frac{m[1 - (1 + i)^N]}{-i}$$

$$= \frac{m[(1 + i)^N - 1]}{i}$$

Spreadsheet Application

Time	Amt saved	Interest	Total in acct
start	$0	$0	$0
1 mo	$60	$0	$60
2 mo	$60	$.60	$120.60
3 mo	$60	$1.21	$181.81
4 mo	$60	$1.82	$243.63
5 mo	$60	$2.44	$306.07
6 mo	$60	$3.06	$369.13
⋮			

We are using the following formula for the sum of the terms of a geometric sequence:

$$G_n = \frac{\overset{m}{\underset{\uparrow}{g_1}}(1 - r^n)}{\underset{\underset{1+i}{\uparrow}}{1 - r}}$$

$\uparrow$
A

Ordinary Annuity Formula

Let $i = \dfrac{r}{n}$ and $N = nt$ with periodic payment m; then the future value is found with the formula

$$A = m\left[\frac{(1 + i)^N - 1}{i}\right]$$

■■■■ EXAMPLE 1

How much do you save in 5 years if you deposit $60 at the end of each month into an account paying 12% compounded monthly?

Solution   The rate is $i = \dfrac{0.12}{12} = 0.01$ and $N = 5(12) = 60$. Thus

$$A = m\left[\frac{(1 + i)^N - 1}{i}\right] = 60\left[\frac{(1 + 0.01)^{60} - 1}{0.01}\right] \approx 4{,}900.180191$$

The future value is $4,900.18. The graph is shown in Figure 6.10.

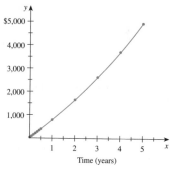

▲ **Figure 6.10   Growth of a bank account with a $60 monthly deposit. The graph shows both deposits and interest compounded monthly at 12% annual rate.**   ━

■■■■ EXAMPLE 2   **Polya's Method**

Suppose you are 21 years old and will make monthly deposits to a bank account paying 10% annual interest compounded monthly.

> **Option I:**   Pay yourself $200 per month for 5 years and then leave the balance in the bank until age 65. (Total amount of deposits is $200 × 5 × 12 = $12,000.)

> **Option II:**   Wait until you are 40 years old (the age most of us start thinking seriously about retirement) and then deposit $200 per month until age 65. (Total amount of deposits is $200 × 25 × 12 = $60,000.)

Compare the amounts you would have from each of these options.

Solution   We use Polya's problem-solving guidelines for this example.

Understand the Problem.   When most of us are 21 years old we do not think about retirement. However, if we do, the results can be dramatic. With this example, we investigate the differences if we save early (for 5 years), or later (for 25 years).

Devise a Plan.   We calculate the value of the annuity for the 5 years of the first option, and then calculate the effect of leaving the value at the end of 5 years (the annuity) in a savings account until retirement. This part of the problem is a future value problem because it becomes a lump-sum problem when deposits are no longer made (after 5 years). For the second option, we calculate the value of the annuity for 25 years.

Carry Out the Plan.
Option I: $200 per month for 5 years at 10% annual interest is an annuity with $m = 200$, $r = 0.10$, $t = 5$, and $n = 12$.

We find $i = \dfrac{r}{n} = \dfrac{0.10}{12}$ and $N = nt = 12(5) = 60$. Then

$$A = 200\left[\frac{\left(1 + \dfrac{0.1}{12}\right)^{60} - 1}{\dfrac{0.1}{12}}\right] \approx 15{,}487.41443$$

At the end of 5 years the amount in the account is $15,487.41. This money is left in the account, so it is now a future-value problem for 39 years (ages 26 to 65); for this part of the problem $P = 15{,}487.41$, $r = 0.1$, $t = 39$, and $n = 12$, so that

$$i = \frac{r}{n} = \frac{0.1}{12} \qquad N = nt = 12(39) = 468$$

$$A = P(1 + i)^N = 15{,}487.41\left(1 + \frac{0.1}{12}\right)^{468} \approx 752{,}850.8644$$

(It is 752,851.0799 if you use the calculator value for A.) The amount you would have at age 65 from 5 years of payments to yourself starting at age 21 is $752,850.86.

Option II: If you wait until you are 40 years old, then it is an annuity problem with $m = 200$, $r = 0.10$, $t = 25$, and $n = 12$;

$$i = \frac{0.10}{12} \quad \text{and} \quad N = 12(25) = 300$$

$$A = 200\left[\frac{\left(1 + \dfrac{0.1}{12}\right)^{300} - 1}{\dfrac{0.1}{12}}\right] \approx 265{,}366.6805$$

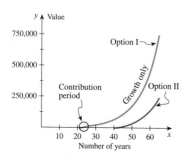

▲ **Figure 6.11   Comparison of Options I and II**

The amount you would have at age 65 from 25 years of payments to yourself starting at age 40 is $265,366.68.

Look Back.   Retirement Options I and II are compared graphically in Figure 6.11. Clearly, it is to your advantage to choose Option I.   ▬

▬ **EXAMPLE 3**

The previous example gave you a choice:

> **Option I:**   Pay yourself $200 per month for 5 years beginning when you are 21 years old (this is comparable to the car payments you might make).
>
> **Option II:**   Wait until age 40, then make payments of $200 per month to yourself for the next 25 years.

If you were to select one of these options, what effect would it have on your retirement monthly income assuming you decide to live on the interest only and that the interest rate is 10%?

Solution   Option I gives you a retirement fund of $752,850.86. This will provide a monthly income as determined by the simple interest formula (because the interest is not left to accumulate):

$$I = Prt \qquad \text{Simple interest formula}$$

$$= 752{,}850.86(0.10)\left(\frac{1}{12}\right) \qquad \text{Substitute values.}$$

$$\approx 6{,}273.76 \qquad \text{Simplify.}$$

Option II gives you a retirement fund of $265,366.68. This will provide a monthly income found as follows:

$$I = Prt \qquad \text{Repeat the steps from Option I}$$

$$= 265{,}366.68(0.10)\left(\frac{1}{12}\right)$$

$$\approx 2{,}211.39$$

The better choice (Option I) would mean that you would have about $4,000 *more* each and every month that you live beyond age 65. Did you ever ask the question "Why should I take math?"   ▬

Sinking Funds

We now consider the situation in which we need (or want) to have a lump sum of money (a future value) in a certain period of time. The present-value formula will tell us how much we need to have today, but we frequently do not have that amount available. Suppose your goal is $10,000 in 5 years. You can obtain 8% compounded monthly, so the present value formula yields

$$P = A(1 + i)^{-N} = \$10{,}000\left(1 + \frac{0.08}{12}\right)^{-60} \approx \$6{,}712.10$$

However, this is more than you can afford to put into the bank now. The next choice is to make a series of small equal investments to accumulate at 8% compounded monthly, so that the end result is the same — namely, $10,000 in 5 years. The account you set up to receive those investments is called a **sinking fund.**

To find a formula for a sinking fund, we begin with the formula for an ordinary annuity, and solve for m (which is the unknown for a sinking fund):

$$A = m \left[\frac{(1 + i)^N - 1}{i} \right]$$

$$Ai = m[(1 + i)^N - 1] \qquad \text{Multiply both sides by } i.$$

$$\frac{Ai}{(1 + i)^N - 1} = m \qquad \text{Divide both sides by } [(1 + i)^N - 1].$$

Sinking Fund Formula

If the future value (A) is known and you wish to find the amount of the monthly payment (m), use the formula

$$m = \frac{Ai}{(1 + i)^N - 1}$$

EXAMPLE 4

Suppose you want to have $10,000 in 5 years and decide to make monthly payments into an account paying 8% compounded monthly. What is the amount of each monthly payment?

Solution You might begin by estimation. If you are paid *no* interest, then since you are making 60 equal deposits, the amount of each deposit would be

$$\frac{10,000}{60} \approx \$166.67$$

Therefore, since you are paid interest, you would expect the amount of each deposit to be somewhat less than this estimation. We identify the known values: $A = \$10,000$ (future value); $r = 0.08$; $t = 5$; and $n = 12$. Calculate $i = \dfrac{0.08}{12}$ and $N = nt = 12(5) = 60$. Finally,

$$m = \frac{Ai}{(1 + i)^N - 1} = \frac{10,000 \left(\dfrac{0.08}{12} \right)}{\left(1 + \dfrac{0.08}{12} \right)^{60} - 1} \approx 136.0972762$$

The necessary monthly payment to the account is $136.10.

Calculator Comment

Most graphing calculators have the ability to accept programs. Modern calculators require no special ability to write a program. There are, however, some necessary considerations. When using a calculator, we have the calculator find the values of i and N, and we input the values of the variables into the memory locations as shown on the following chart:

Variable Input to Storage

$$
\begin{array}{ccc}
P & \boxed{\text{STO}} & \boxed{\text{P}} \\
A & \boxed{\text{STO}} & \boxed{\text{A}} \\
m & \boxed{\text{STO}} & \boxed{\text{M}} \\
r & \boxed{\text{STO}} & \boxed{\text{R}} \\
n & \boxed{\text{STO}} & \boxed{\text{N}} \\
\end{array}
$$

When evaluating the financial formulas, we will substitute $\dfrac{r}{n}$ for i and nt for N. Also, when using a calculator, generally the only difficulty is the proper use of parentheses:

	Formula	*With Substitutions*	*Calculator Notation*
For future value:	$A = P(1 + i)^N$	$A = P\left(1 + \dfrac{r}{n}\right)^{nt}$	$P * (1 + R/N)\verb|^|(N * T)$
For present value:	$P = A(1 + i)^{-N}$	$P = A\left(1 + \dfrac{r}{n}\right)^{-nt}$	$A * (1 + R/N)\verb|^|(-N * T)$
For ordinary annuity:	$A = m\left[\dfrac{(1 + i)^N - 1}{i}\right]$	$A = m\left[\dfrac{\left(1 + \dfrac{r}{n}\right)^{nt} - 1}{\dfrac{r}{n}}\right]$	$M * (((1 + R/N)$ $\verb|^|(N * T) - 1)/(R/N))$
For sinking fund:	$m = \dfrac{Ai}{(1 + i)^N - 1}$	$m = \dfrac{A\left(\dfrac{r}{n}\right)}{\left(1 + \dfrac{r}{n}\right)^{nt} - 1}$	$(A * R/N)/((1 + R/N)$ $\verb|^|(N * T)-1)$

To write a program, press the $\boxed{\text{PRGM}}$ key and then input the formula. Most calculator programs also require some statement that forces the output of an answer. Check with your owner's manual, but this is usually a one-step process. To use your calculator for one of these formulas, you simply need to identify the type of financial problem, input the values for the appropriate variables, and then run the program containing the formula you wish to evaluate.

PROBLEM SET 6.5

▲ A Problems

1. IN YOUR OWN WORDS What do we mean by a lump-sum problem?

2. IN YOUR OWN WORDS Why should we call an annuity a periodic-payment problem?

3. IN YOUR OWN WORDS What is an annuity?

4. IN YOUR OWN WORDS What is a sinking fund?

5. IN YOUR OWN WORDS Distinguish an annuity from a sinking fund.

Find i and N for N = 12, and then use a calculator to evaluate

$$A = m \left[\frac{(1 + i)^N - 1}{i} \right]$$

for m, r, and t (respectively) given in Problems 6–23.

6. $50; 5%; 3 yr   7. $50; 6%; 3 yr   8. $50; 8%; 3 yr

9. $50; 12%; 3 yr   10. $50; 5%; 30 yr   11. $50; 6%; 30 yr

12. $50; 8%; 30 yr   13. $50; 12%; 30 yr   14. $100; 5%; 10 yr

15. $100; 6%; 10 yr   16. $100; 8%; 10 yr   17. $100; 12%; 10 yr

18. $150; 5%; 35 yr   19. $150; 6%; 35 yr   20. $150; 8%; 35 yr

21. $150; 12%; 35 yr   22. $650; 5%; 30 yr   23. $650; 6%; 30 yr

In Problems 24–35, find the value of each annuity at the end of the indicated number of years. Assume that the interest is compounded with the same frequency as the deposits.

	Amount of deposit m	Frequency compounded n	Rate r	Time (in yr) t
24.	$500	annually	8%	30
25.	$500	annually	6%	30
26.	$250	semiannually	8%	30
27.	$600	semiannually	2%	10
28.	$300	quarterly	6%	30
29.	$100	monthly	4%	5
30.	$200	quarterly	8%	20
31.	$400	quarterly	11%	20
32.	$30	monthly	8%	5
33.	$5,000	annually	4%	10
34.	$2,500	semiannually	8.5%	20
35.	$1,250	quarterly	3%	20

Find the amount of monthly payment necessary for each deposit to a sinking fund in Problems 36–47.

	Amount needed A	Frequency compounded n	Rate r	Time (in yr) t
36.	$7,000	annually	8%	5
37.	$25,000	annually	11%	5
38.	$25,000	semiannually	12%	5
39.	$50,000	semiannually	14%	10
40.	$165,000	semiannually	2%	10
41.	$3,000,000	semiannually	3%	20
42.	$500,000	quarterly	8%	10
43.	$55,000	quarterly	10%	5
44.	$100,000	quarterly	8%	8
45.	$35,000	quarterly	8%	12
46.	$45,000	monthly	8%	6
47.	$120,000	monthly	7%	30

▲ **B Problems**

48. Self-employed persons can make contributions for their retirement into a special tax-deferred account called a **Keogh account.** Suppose you are able to contribute $20,000 into this account at the end of each year. How much will you have at the end of 20 years if the account pays 8% annual interest?

49. The owner of Sebastopol Tree Farm deposits $650 at the end of each quarter into an account paying 8% compounded quarterly. What is the value of the account at the end of 5 years?

50. The owner of Oak Hill Squirrel Farm deposits $1,000 at the end of each quarter into an account paying 8% compounded quarterly. What is the value at the end of 5 years, 6 months?

51. Clearlake Optical has a $50,000 note that comes due in 4 years. The owners wish to create a sinking fund to pay this note. If the fund earns 8% compounded semiannually, how much must each semiannual deposit be?

52. A business must raise $70,000 in 5 years. What should be the size of the owners' quarterly payment to a sinking fund paying 8% compounded quarterly?

▲ **Problem Solving**

53. Clearlake Optical has developed a new lens. The owners plan to issue a $4,000,000 30-year bond with a contract rate of 5.5% paid annually to raise capital to market this new lens. This means that Clearlake will be required to pay 5.5% interest each year for 30 years. To pay off the debt, they will also set up a sinking fund paying 8% interest compounded annually. What size annual payment is necessary for interest and sinking fund combined? (Answer to the nearest dollar.)

54. The owners of Bardoza Greeting Cards wish to introduce a new line of cards but need to raise $200,000 to do it. They decide to issue 10-year bonds with a contract rate of 6% paid semiannually. This means Bardoza must make interest payments to the bond holders each 6 months for 10 years. They also set up a sinking fund paying 8% interest compounded semiannually. How much money will they need to make the semiannual interest payments as well as payments to the sinking fund? (Answer to the nearest dollar.)

55. John and Rosamond want to retire in 5 years and can save $150 every three months. They plan to deposit the money at the end of each quarter into an account paying 6.72% compounded quarterly. How much will they have at the end of 5 years?

56. You want to retire at age 65. You decide to make a deposit to yourself at the end of each year into an account paying 13%, compounded annually. Assuming you are now 25 and can spare $1,200 per year, how much will you have when you retire at age 65?

57. In 1997 the maximum Social Security deposit was $3,949.40. Suppose you are 25 and make a deposit of this amount into an account at the end of each year. How much would you have (to the nearest dollar) when you retire if the account pays 8% compounded annually and you retire at age 65?

58. Repeat Problem 56 using your own age.

59. Repeat Problem 57 using your own age.

▲ **Individual Research**

60. Outline a program for your own retirement. In the process of writing this paper, answer the following questions. You will need to state your assumptions about interest and inflation rates.
    a. What monthly amount of money today would provide you a comfortable living?
    b. Using the answer to part **a**, project that amount to your retirement, calculating the effects of inflation. Use your own age and assume that you will retire at age 65.
    c. How much money would you need to have accumulated to provide the amount you found in part **b** if you decide to live on the interest only?
    d. If you set up a sinking fund to provide the amount you found in part **c**, how much would you need to deposit each month?
    e. Offer some alternatives to a sinking fund.
    f. Draw some conclusions about your retirement.

6.6 AMORTIZATION

Present Value of an Annuity

In the previous sections we considered the lump-sum formulas for present value and for future value. We also considered the periodic payment formula, which provided the future value of periodic payments into an interest-bearing account. With the annuity formula, we assumed that we knew the amount of the periodic payment.

In the real world, if you need or want to purchase an item and you do not have the entire amount necessary, you can generally save for the item, or purchase the item and pay for it with monthly payments. Let us consider an example. Suppose you can afford $275 per month for an automobile. One possibility is to put the $275 per month into a bank account until you have enough money to buy a car. To find the future value of this monthly payment, you would need to know the interest rate (say, 5.2%) and the length of time you will make these deposits (say, 48 months). The future value in the account is found using the annuity formula:

$$A = m\left[\frac{(1+i)^N - 1}{i}\right] = 275\left[\frac{(1+\frac{0.052}{12})^{48} - 1}{\frac{0.052}{12}}\right] \approx 14,638.04004$$

This means that if you make these payments to yourself, in 4 years you will have saved $275 \times 48 = $13,200$, but because of the interest paid on this account, the total amount in the account is $14,638.04.

However, it is more common to purchase a car and agree to make monthly payments to pay for it. Given the annuity calculation above, can we conclude that we can look for a car that costs $14,638? No, because if we borrow to purchase the car we need to *pay* interest charges rather than *receive* the interest as we did with the annuity. Well, then, how much can we borrow when we assume that we know the monthly payment, the interest rate, and the length of time we will pay? This type of financial problem is called **present value of an annuity.**

⊘ The financial problem called *present value of an annuity* is not the same as a present-value financial problem. ⊘

Continuing with this car payment example, we need to ask, "What is the present value of 14,638.04004?" This is found using the formula $P = A(1+i)^{-N}$:

$$P = 14,638.04004(1 + \tfrac{0.052}{12})^{-48} \approx 11,894.46293$$

This means we could finance a loan in the amount of $11,894.46 and pay it off in 48 months with payments of $275, provided the interest rate is 5.2% compounded monthly.

We will now derive a formula for the present value of an annuity.

Since $A = P(1+i)^N$ and $A = m\left[\frac{(1+i)^N - 1}{i}\right]$, we see

$$P(1+i)^N = m\left[\frac{(1+i)^N - 1}{i}\right] \qquad \text{Substitution}$$

$$P = \frac{m}{i}\left[\frac{(1+i)^N - 1}{(1+i)^N}\right] \qquad \text{Divide both sides by } (1+i)^N.$$

$$= \frac{m}{i}\left[1 - \frac{1}{(1+i)^N}\right]$$

$$= \frac{m}{i}\left[1 - (1+i)^{-N}\right]$$

Present Value of an Annuity Formula

If the monthly payment is known (m) and you wish to find the present value of those monthly payments, use the **present value of an annuity formula:**

$$P = m\left[\frac{1 - (1+i)^{-N}}{i}\right]$$

EXAMPLE I

You look at your budget and decide that you can afford $250 per month for a car. What is the maximum loan you can afford if the interest rate is 13% and you want to repay the loan in 4 years?

Solution This requires the present value of an annuity formula. First, determine the given variables:

$m = 250, r = 0.13, t = 4, n = 12$; calculate $i = \frac{r}{n} = \frac{0.13}{12}$ and

$N = nt = 12(4) = 48$. Thus,

$$P = m\left[\frac{1 - (1 + i)^{-N}}{i}\right] = 250\left[\frac{1 - (1 + \frac{0.13}{12})^{-48}}{\frac{0.13}{12}}\right] \approx 9{,}318.797438$$

This means that you can afford to finance about $9,319. To keep the payments at $250, a down payment should be made to lower the price of the car to the finance amount.

> **Calculator Comment**
>
> A program for the present value of an annuity would be input as follows:
>
> M*((1−(1+R/N)^(−N*T))/ (R/N))

EXAMPLE 2   **Polya's Method**

Suppose your budget (or your banker) tells you that you can afford $1,575 per month for a house payment. If the current interest rate for a home loan is 10.5% and you will finance the home for 30 years, what is the expected price of the home you can afford? Assume that you have saved the money to make the required 20% down payment.

Solution We use Polya's problem-solving guidelines for this example.

Understand the Problem. Most textbook problems give you a price for a home, and ask for the monthly payment. This problem takes the real-life situation in which you know the monthly payment and wish to know the price of the house you can afford.

Devise a Plan. We first find the amount of the loan, and then find the down payment by subtracting the amount of the loan from the price of the home we wish to purchase.

Carry Out the Plan. We first find the amount of the loan; this is the present value of an annuity. We know $m = 1{,}575, r = 0.105, t = 30, n = 12$; calculate $i = \frac{r}{n} = \frac{0.105}{12}$ and $N = nt = 12(30) = 360$. Thus,

$$P = m\left[\frac{1 - (1 + i)^{-N}}{i}\right] = 1{,}575\left[\frac{1 - (1 + \frac{0.105}{12})^{-360}}{\frac{0.105}{12}}\right] \approx 172{,}180.2058$$

> **Calculator Comment**
>
> If you do not have a programmable calculator, it is a good idea to store the value of i because it is used more than once in these financial formulas.

The loan that you can afford is about $172,180.

What about the down payment? If there is a 20% down payment, then the amount of the loan is 80% of the price of the home. We know the amount of the loan, so the problem is now to determine the home price such that 80% of the price is equal to the amount of the loan. In symbols,

$0.80x = 172{,}180.2058$

which means that you simply divide the calculator output for the amount of the loan by 0.8 to find 215,225.2573.

Look Back.   Given the conditions of the problem, we can look back to check our work (rounded to the nearest dollar for the check):

Purchase price of home: $215,225

Less 20% down payment: $43,045

Amount of loan: $172,180

▬▬ EXAMPLE 3

Theresa's Social Security benefit is $450 per month if she retires at age 62 instead of age 65. What is the present value of an annuity that would pay $450 per month for 3 years if the current interest rate is 10% compounded monthly?

Solution   Here, $m = 450$, $r = 0.10$, $t = 3$, and $n = 12$; calculate $i = \frac{r}{n} = \frac{0.10}{12}$ and $N = nt = 12(3) = 36$. Then

$$P = m\left[\frac{1 - (1 + i)^{-N}}{i}\right] = 450\left[\frac{1 - (1 + \frac{0.10}{12})^{-36}}{\frac{0.10}{12}}\right] \approx 13{,}946.05601$$

The present value of this decision is $13,946.06. This represents the value of the additional 3 years of Social Security payments.

Monthly Payments

The process of paying off a debt by systematically making partial payments until the debt (principal) and the interest are repaid is called **amortization.** If the loan is paid off in regular equal installments, then we use the formula for the present value of an annuity to find the monthly payments by algebraically solving for m:

$$P = m\left[\frac{1 - (1 + i)^{-N}}{i}\right]$$

$$Pi = m[1 - (1 + i)^{-N}] \qquad \text{Multiply both sides by } i.$$

$$\frac{Pi}{1 - (1 + i)^{-N}} = m \qquad \text{Divide both sides by } 1 - (1 + i)^{-N}.$$

Amortization Formula

If the amount of the loan is known (P), and you wish to find the amount of the monthly payment (m), use the formula

$$m = \frac{Pi}{1 - (1 + i)^{-N}}$$

▬▬ EXAMPLE 4

In 1990, the average price of a new home was $162,000 and the interest rate was 11%. If this amount is financed for 30 years at 11% interest, what is the monthly payment?

Solution Given $P = 162{,}000$, $r = 0.11$, $t = 30$, and $n = 12$; calculate $i = \frac{r}{n} = \frac{0.11}{12}$ and $N = nt = 12(30) = 360$. Then

$$m = \frac{Pi}{1 - (1+i)^{-N}} = \frac{162{,}000(\frac{0.11}{12})}{1 - (1 + \frac{0.11}{12})^{-360}} \approx 1{,}542.763901$$

The monthly payment is $1,542.76. This is the payment for interest and principal to pay off a 30-year 11% loan of $162,000.

The schedule of payments on a loan showing how much of each payment goes to pay the principal and how much goes to repay the interest is called an **amortization schedule.** For a home loan this schedule is rather long, so we show the input information for the loan in Example 4 in the following Spreadsheet Application. If you have a computer and spreadsheet program, you can output the entire amortization schedule.

Calculator Comment

To program the installment formula into a calculator input it as

(P*R/N)/(1 − (1+R/N)^(−N*T))

Spreadsheet Application

The following spreadsheet was written to give an amortization schedule for Example 4.

	A	B	C	D	E
1	**Amortization Schedule**				
2	End of				Outstanding
3	Period	Payment	Interest	Principal	balance
4	0				162000.00
5	+A4+1	1542.76	+E4*(.11/12)	+B5−C5	+E4−B5+C5
6	replicate	replicate	replicate	replicate	replicate

Using these spreadsheet entries for payments 1 to 360 we find the following entries:

Amortization Schedule

End of period	Payment	Interest	Principal	Outstanding balance	End of period	Payment	Interest	Principal	Outstanding balance
0				$162,000.00	348	$1,542.76	$172.66	$1,370.10	$17,465.46
1	$1,542.76	$1,485.00	$57.76	$161,942.24	349	$1,542.76	$160.10	$1,382.66	$16,082.80
2	$1,542.76	$1,484.47	$58.29	$161,883.95	350	$1,542.76	$147.43	$1,395.33	$14,687.47
3	$1,542.76	$1,483.94	$58.82	$161,825.13	351	$1,542.76	$134.64	$1,408.12	$13,279.34
4	$1,542.76	$1,483.40	$59.36	$161,765.76	352	$1,542.76	$121.73	$1,421.03	$11,858.31
5	$1,542.76	$1,482.85	$59.91	$161,705.86	353	$1,542.76	$108.70	$1,434.06	$10,424.25
6	$1,542.76	$1,482.30	$60.46	$161,645.40	354	$1,542.76	$95.56	$1,447.20	$8,977.05
7	$1,542.76	$1,481.75	$61.01	$161,584.39	355	$1,542.76	$82.29	$1,460.47	$7,516.58
8	$1,542.76	$1,481.19	$61.57	$161,522.82	356	$1,542.76	$68.90	$1,473.86	$6,042.72
9	$1,542.76	$1,480.63	$62.13	$161,460.69	357	$1,542.76	$55.39	$1,487.37	$4,555.35
10	$1,542.76	$1,480.06	$62.70	$161,397.98	358	$1,542.76	$41.76	$1,501.00	$3,054.35
11	$1,542.76	$1,479.48	$63.28	$161,334.70	359	$1,542.76	$28.00	$1,514.76	$1,539.59
12	$1,542.76	$1,478.90	$63.86	$161,270.85	360	$1,553.70	$14.11	$1,539.59	$0.00

It is noteworthy to see how much interest is paid for the home loan of Example 4. There are 360 payments of $1,542.76, so the total amount repaid is

$$360(1,542.76) = 555,393.60$$

Since the loan was for $162,000, the interest is

$$\$555,393.60 - \$162,000 = \$393,393.60$$

The amount of interest paid can be reduced by making a larger down payment. For Example 4, if a 20% down payment (which is fairly standard) was made, then $162,000(0.80) = \$129,600$ is the amount to be financed. If this amount is amortized over 20 years (instead of 30), the monthly payments are $1,337.72. The total amount paid on *this* loan is

$$\underset{\substack{\uparrow \\ \text{number of payments}}}{240} \overset{\overbrace{\qquad\qquad}^{\text{amount of each payment}}}{(1,337.72)} + \underbrace{32,400}_{\text{down payment}} = 353,452.80$$

The savings yielded over the term of the loan by making a 20% down payment for 20 years rather than financing the entire amount for 30 years is

$$\$555,393.60 - \$353,452.80 = \$201,940.80$$

It is interesting to see how much is saved in Example 4 by lowering the interest rates. In 1997 the fixed interest rate was 6.45%. If $P = \$162,000$, $r = 0.0645$, $t = 30$, $n = 12$, $i = \frac{0.0645}{12}$, and $N = 60$, then

$$m = \frac{Pi}{1 - (1 + i)^{-N}} = \frac{162,000\left(\frac{0.0645}{12}\right)}{1 - \left(1 + \frac{0.0645}{12}\right)^{-360}} \approx 1,018.629046$$

The interest on the 30-year 11% loan in Example 4 was $393,393.60; now at 6.45% interest for 30 years the total interest is

$$360(\$1,018.63) - \$162,000 = \$204,706.80$$

a savings of almost $190,000.

Finally, if we consider *both* the lower 6.45% rate *and* make a 20% down payment, the *savings* (as compared to Example 4) is $358,795.16, more than *double* the price of the house in the first place!

PROBLEM SET   6.6

▲ **A Problems**

1. **IN YOUR OWN WORDS** What does amortization mean?

2. **IN YOUR OWN WORDS** Describe when you would use the present value of an annuity formula.

3. The variables m, n, r, t, i, N, A, and P are used in the various financial formulas. Tell what each of these variables represents.

Find i and N for n = 12, and then use a calculator to evaluate

$$P = m \left[\frac{1 - (1 + i)^{-N}}{i} \right]$$

for the values of the variables m, r, and t (respectively) given in Problems 4–12.

4. $50; 5%; 5 yr   5. $50; 6%; 5 yr   6. $50; 8%; 5 yr

7. $150; 5%; 30 yr   8. $150; 6%; 30 yr   9. $150; 8%; 30 yr

10. $1,050; 5%; 30 yr   11. $1,050; 6%; 30 yr   12. $1,050; 8%; 30 yr

Find i and N for n = 12, and then use a calculator to evaluate

$$m = \frac{Pi}{1 - (1 + i)^{-N}}$$

for the values of the variables P, r, and t (respectively) given in Problems 13–21.

13. $14,000; 5%; 5 yr   14. $14,000; 10%; 5 yr   15. $14,000; 19%; 5 yr

16. $150,000; 8%; 30 yr   17. $150,000; 9%; 30 yr   18. $150,000; 10%; 30 yr

19. $260,000; 12%; 30 yr   20. $260,000; 9%; 30 yr   21. $260,000; 8%; 30 yr

Find the present value of the ordinary annuities in Problems 22–33.

	Amount of deposit m	Frequency compounded n	Rate r	Time (in yr) t
22.	$500	annually	8%	30
23.	$500	annually	6%	30
24.	$250	semiannually	8%	30
25.	$600	semiannually	2%	10
26.	$300	quarterly	6%	30
27.	$100	monthly	4%	5
28.	$200	quarterly	8%	20
29.	$400	quarterly	11%	20
30.	$30	monthly	8%	5
31.	$75	monthly	4%	10
32.	$50	monthly	8.5%	20
33.	$100	quarterly	3%	40

▲ **B Problems**

Find the monthly payment for the loans in Problems 34–45.

34. $500 loan for 12 months at 12%

35. $100 loan for 18 months at 18%

36. $4,560 loan for 20 months at 21%

37. $3,520 loan for 30 months at 19%

38. Used-car financing of $2,300 for 24 months at 15%

39. New-car financing of 2.9% on a 30-month $12,450 loan

40. Furniture financed at $3,456 for 36 months at 23%

41. A refrigerator financed for $985 at 17% for 15 months

42. A $112,000 home bought with a 20% down payment and the balance financed for 30 years at 11.5%

43. A $108,000 condominium bought with a 30% down payment and the balance financed for 30 years at 12.05%

44. Finance $450,000 for a warehouse with a 12.5% 30-year loan

45. Finance $859,000 for an apartment complex with a 13.2% 20-year loan

46. How much interest (to the nearest dollar) would be saved in Problem 42 if the home were financed for 15 rather than 30 years?

47. How much interest (to the nearest dollar) would be saved in Problem 43 if the condominium were financed for 15 rather than 30 years?

48. Melissa agrees to contribute $500 to the alumni fund at the end of each year for the next 5 years. Shannon wants to match Melissa's gift, but he wants to make a lump-sum contribution. If the current interest rate is 12.5% compounded annually, how much should Shannon contribute to equal Melissa's gift?

49. A $1,000,000 lottery prize pays $50,000 per year for the next 20 years. If the current rate of return is 12.25%, what is the present value of this prize?

50. Suppose you have an annuity from an insurance policy and you have the option of being paid $250 per month for 20 years or having a lump-sum payment of $25,000. Which has more value if the current rate of return is 10% compounded monthly?

51. An insurance policy offers you the option of being paid $750 per month for 20 years or a lump sum of $50,000. Which has more value if the current rate of return is 9%, compounded monthly, and you expect to live for 20 years?

52. You look at your budget and decide that you can afford $250 per month for a car. What is the maximum loan you can afford if the interest rate is 13% and you want to repay the loan in 4 years?

▲ Problem Solving

53. I recently found a real-life advertisement in the newspaper, which is shown in the margin (only the phone number has been changed). Suppose that you have won a $20,000,000 lottery, paid in 20 annual installments. How much would be a fair price to be paid today for the assignment of this prize? Assume the interest rate is 5%.

54. The bottom notice (not circled) states that $6,000 is needed for 90 days, and that the advertiser is willing to pay 20% interest. How much would you expect to receive in 90 days if you gave this party the $6,000?

55. Suppose your gross monthly income is $5,500 and your current monthly payments are $625. If the bank will allow you to pay up to 36% of your gross monthly income (less current monthly payments) for a monthly house payment, what is the maximum loan you can obtain if the rate for a 30-year mortgage is 9.65%?

56. Suppose your gross monthly income is $4,550 and your spouse's gross monthly salary is $3,980. Your monthly bills are $1,235. The home you wish to purchase costs $355,000 and the loan is an 11.85% 30-year loan. How much down payment

(rounded to the nearest hundred dollars) is necessary to be able to afford this home? This down payment is what percent of the cost of the home? Assume the bank will allow you to pay up to 36% of your gross monthly income (less current monthly payments) for house payments.

57. Suppose you want to purchase a home for $225,000 with a 30-year mortgage at 10.24% interest. Suppose also that you can put down 25%. What are the monthly payments? What is the total amount paid for principal and interest? What is the amount saved if this home is financed for 15 years instead of for 30 years?

58. The McBertys have $30,000 in savings to use as a down payment on a new home. They also have determined that they can afford between $1,500 and $1,800 per month for mortgage payments. If the mortgage rates are 11% per year compounded monthly, what is the price range for houses they should consider for a 30-year loan?

59. Rework Problem 58 for a 20-year loan instead of a 30-year loan.

60. Rework Problem 58 if interest rates go down to 10.2%.

61. As the interest rate increases, determine whether each of the given amounts increases or decreases. Assume that all other variables remain constant.
    a. future value   **b.** present value

62. As the interest rate increases, determine whether each of the given amounts increases or decreases. Assume that all other variables remain constant.
    a. annuity   **b.** present value of an annuity

63. As the interest rate increases, determine whether each of the given amounts increases or decreases. Assume that all other variables remain constant.
    a. monthly value for a sinking fund
    b. monthly value for amortization

6.7   BUILDING FINANCIAL POWER

For many students, the most difficult part of working with finances (other than the problem of lack of funds) is in determining *which* formula to use. If we wish to free ourselves from the problem of lack of funds, we must use our knowledge in a variety of situations that occur outside the classroom.

Summary of Financial Formulas

We will now outline a procedure for using the financial formulas we have introduced in this chapter. First, review the meaning of the variables, as shown in the margin.

   When classifying financial formulas, the first question to ask is:

Is it a lump-sum problem?

   If it is, then what is the unknown?

   If FUTURE VALUE is the unknown, then it is a *future-value* problem.

$$A = P(1 + i)^N$$

Definitions of variables:
P = present value (principal)
A = future value
I = amount of interest
r = annual interest rate
t = number of years
n = periods; number of times compounded each year

Calculate:
i = rate per period = $\dfrac{r}{n}$
N = number of periods = nt

If PRESENT VALUE is the unknown, then it is a *present-value* problem.

$$P = A(1 + i)^{-N}$$

Is it a periodic payment problem?

If it is, then is the periodic payment known?

PERIODIC PAYMENT Known

Find the future value: ORDINARY ANNUITY

$$A = m \left[\frac{(1 + i)^N - 1}{i} \right]$$

Find the present value: PRESENT VALUE OF AN ANNUITY

$$P = m \left[\frac{1 - (1 + i)^{-N}}{i} \right]$$

PERIODIC PAYMENT Unknown

Future value known: SINKING FUND

$$m = \frac{Ai}{(1 + i)^N - 1}$$

Present value known: AMORTIZATION

$$m = \frac{Pi}{1 - (1 + i)^{-N}}$$

PROBLEM SET   6.7

▲ **A Problems**

1. **IN YOUR OWN WORDS** What is the formula for the present value of an annuity? What is the unknown?

2. **IN YOUR OWN WORDS** What is the amortization formula? What is the unknown?

3. **IN YOUR OWN WORDS** What are a reasonable down payment and monthly payment for a home in your area?

4. **IN YOUR OWN WORDS** What are a reasonable down payment and monthly payment for a new automobile?

5. **IN YOUR OWN WORDS** Outline a procedure for identifying the type of financial formula for a given applied problem.

Classify the type of financial formula for the information given in Problems 6–11.

Lump sum problems

	P	A
6.	known	unknown
7.	unknown	known

Periodic payment problems

	P	A	m
8.	unknown		known
9.		unknown	known
10.		known	unknown
11.	known		unknown

12. Match each formula in Column A with the type of financial problem in Column B.

Column A	Column B
$A = m\left[\dfrac{(1 + i)^N - 1}{i}\right]$	Annuity
$P = m\left[\dfrac{1 - (1 + i)^{-N}}{i}\right]$	Amortization
$m = \dfrac{Ai}{(1 + i)^N - 1}$	Present value of an annuity
$m = \dfrac{Pi}{1 - (1 + i)^{-N}}$	Sinking fund

Classify the financial problems in Problems 13–16, and then answer each question by assuming a 12% interest rate compounded annually.

13. Find the value of a $1,000 certificate in 3 years.

14. Deposit $300 at the end of each year. What is the total in the account in 10 years?

15. An insurance policy pays $10,000 in 5 years. What lump-sum deposit today will yield $10,000 in 5 years?

16. What annual deposit is necessary to give $10,000 in 5 years?

▲ **B Problems**

In Problems 17–52: **a.** *State the type; and* **b.** *Answer the question.*

17. Find the value of a $1,000 certificate in $2\frac{1}{2}$ years if the interest rate is 12% compounded monthly.

18. You deposit $300 at the end of each year into an account paying 12% compounded annually. How much is in the account in 10 years?

19. An insurance policy pays $10,000 in 5 years. What lump-sum deposit today will yield $10,000 in 5 years if it is deposited at 12% compounded quarterly?

20. A 5-year term insurance policy has an annual premium of $300, and at the end of 5 years, all payments and interest are refunded. What lump-sum deposit is necessary to equal this amount if you assume an interest rate of 10% compounded annually?

21. What annual deposit is necessary to have $10,000 in 5 years if all the money is deposited at 9% interest compounded annually?

22. A $5,000,000 apartment complex loan is to be paid off in 10 years by making 10 equal annual payments. How much is each payment if the interest rate is 14% compounded annually?

23. The price of automobiles has increased at 6.25% per year. How much would you expect a $20,000 automobile to cost in 5 years if you assume annual compounding?

24. The amount to be financed on a new car is $9,500. The terms are 7% for 4 years. What is the monthly payment?

25. What deposit today is equal to 33 annual deposits of $500 into an account paying 8% compounded annually?

26. If you can afford $875 for your house payments, what is the loan you can afford if the interest rate is 6.5% and the monthly payments are made for 30 years?

27. What is the monthly payment for a home costing $125,000 with a 20% down payment and the balance financed for 30 years at 12%?

28. Ricon Bowling Alley will need $80,000 in 4 years to resurface the lanes. What lump sum would be necessary today if the owner of the business can deposit it in an account that pays 9% compounded semiannually?

29. Rita wants to save for a trip to Tahiti, so she puts $2.00 per day into a jar. After 1 year she has saved $730 and puts the money into a bank account paying 10% compounded annually. She continues to save in this manner and makes her annual $730 deposit for 15 years. How much does she have at the end of that time period?

30. Karen receives a $12,500 inheritance that she wants to save until she retires in 20 years. If she deposits the money in a fixed 11% account, compounded daily, how much will she have when she retires?

31. You want to give your child a million dollars when he retires at age 65. How much money do you need to deposit into an account paying 9% compounded monthly if your child is now 10 years old?

32. An accounting firm agrees to purchase a computer for $150,000 (cash on delivery) and the delivery date is in 270 days. How much do the owners need to deposit in an account paying 18% compounded quarterly so that they will have $150,000 in 270 days?

33. For 5 years, Thompson Cleaners deposits $900 at the end of each quarter into an account paying 8% compounded quarterly. What is the value of the account at the end of 5 years?

34. In 1980 the inflation rate hit 16%. Suppose that the average cost of a textbook in 1980 was $15. What is the expected cost in the year 2000 if we project this rate of inflation on the cost? (Assume annual compounding.)

35. What is the necessary amount of monthly payments to an account paying 18% compounded monthly in order to have $100,000 in $8\frac{1}{3}$ years if the deposits are made at the end of each month?

36. Thomas' Grocery Store is going to be remodeled in 5 years, and the remodeling will cost $300,000. How much should be deposited now in order to pay for this remodeling if the account pays 12% compounded monthly?

37. If an apartment complex will need painting in $3\frac{1}{2}$ years and the job will cost $45,000, what amount needs to be deposited into an account now in order to have the necessary funds? The account pays 12% interest compounded semiannually.

38. Teal and Associates needs to borrow $45,000. The best loan they can find is one at 12% that must be repaid in monthly installments over the next $3\frac{1}{2}$ years. How much are the monthly payments?

39. Certain Concrete Company deposits $4,000 at the end of each quarter into an account paying 10% interest compounded quarterly. What is the value of the account at the end of $7\frac{1}{2}$ years?

40. Major Magic Corporation deposits $1,000 at the end of each month into an account paying 18% interest compounded monthly. What is the value of the account at the end of $8\frac{1}{3}$ years?

41. What is the future value of $112,000 invested for 5 years at 14% compounded monthly?

42. What is the future value of $800 invested for 1 year at 10% compounded daily?

43. What is the future value of $9,000 invested for 4 years at 20% compounded monthly?

44. If $5,000 is compounded annually at 5.5% for 12 years, what is the future value?

45. If $10,000 is compounded annually at 8% for 18 years, what is the future value?

46. You owe $5,000 due in 3 years, but you would like to pay the debt today. If the present interest rate is compounded annually at 11%, how much should you pay today so that the present value is equivalent to the $5,000 payment in 3 years?

47. Ricon Movie Theater will need $20,000 in 5 years to replace the seats. What deposit should be made today in an account that pays 9% compounded semiannually?

48. The Fair View Market must be remodeled in 3 years. It is estimated that remodeling will cost $200,000. How much should be deposited now (to the nearest dollar) to pay for this remodeling if the account pays 10% compounded monthly?

49. A laundromat will need seven new washing machines in $2\frac{1}{2}$ years for a total cost of $2,900. How much money (to the nearest dollar) should be deposited at the present time to pay for these machines? The interest rate is 11% compounded semiannually.

50. A computerized checkout system is planned for Abel's Grocery Store. The system will be delivered in 18 months at a cost of $560,000. How much should be deposited today (to the nearest dollar) into an account paying 7.5% compounded daily?

51. A lottery offers you a choice of $1,000,000 per year for 5 years or a lump-sum payment. What lump-sum payment (rounded to the nearest dollar) would equal the annual payments if the current interest rate is 14% compounded annually?

52. A contest offers the winner $50,000 now or $10,000 now and $45,000 in one year. Which is the better choice if the current interest rate is 10% compounded monthly and the winner does not intend to use any of the money for one year?

▲ **Problem Solving**

Problems 53–56 are based on a 30-year fixed rate home loan of $185,500 with an interest rate of 7.75%.

53. What is the monthly payment?

54. What is the total amount of interest paid?

55. Use a computer to print out an amortization schedule.

56. Suppose you reduce the term to 20 years. What is the total amount of interest paid, and what is the savings over the 30-year loan?

Problems 57–60 are based on a 30-year fixed rate home loan of $418,500 with an interest rate of 8.375%.

57. What is the monthly payment?

58. What is the total amount of interest paid?

59. Use a computer to print out an amortization schedule.

60. Suppose you reduce the term to 22 years. What is the total amount of interest paid, and what is the savings over the 30-year loan?

News Clip

Fields Medal

The Fields Medal is often referred to as the "Nobel Prize of Mathematics." The existence of the award is related to the fact that Alfred Nobel chose not to include mathematics in his areas of recognition. There is no documentary evidence to explain this exclusion, but the general gossip attributes it to a personal conflict between Nobel and Mittag-Leffler. John Charles Fields (1863–1932) was disturbed by the lack of such a mathematical award, so he worked toward the establishment of these awards from 1922 until his death in 1932. It was Fields' last will and testament that provided the necessary funds for the establishment of the award, which was finalized on January 4, 1934. The first Fields Medals were awarded in 1936 to L. V. Ahlfors (Harvard) and Jesse Douglas (MIT). Because of World War II the next award was not until 1950; it has been awarded every 4 years at the International Congress of Mathematics.

My thanks to Henry S. Tropp, Humboldt State University, for this information. Professor Tropp was my first mathematics professor and was instrumental in my interest in mathematics. I thank him not only for his research on the Fields Medal, but also for his influence in my life.

Fields Medals ▶

"To get money is difficult, to keep it is more difficult, but to spend it wisely is most difficult of all."

MAY   Anonymous

Biographical Sketch

"I'm glad I majored in mathematics because it provided me a foundation for my current business career."

Herman Cain

Herman Cain was born in Memphis, Tennessee, on December 13, 1945, and is the co-owner of Godfather's Pizza, Inc. He has received numerous awards and serves on several prestigious boards, including the Federal Reserve Bank of Kansas City, of which he is vice-chairman.

The story of his career is a story of work, determination, and a unique recipe for success: **focus.** He graduated from Morehouse College with a B.S. in mathematics, and then earned his master's degree in Computer Science from Purdue University. He attributes his success in business to a blending of people skills and analytical abilities allowing him to quickly diagnose a situation and plan and develop strategies for success. When I asked him what he enjoyed most about mathematics he said, "It is everywhere in daily life—in both my personal life and my business life. Mathematics has allowed me to make complex 'things' simple. I call if *focus* which allows me to solve business problems of all types."

"I never dreamed of being president of a pizza company. My formal education, of which mathematics was a major part, has certainly allowed me to be blessed with a *career* and not just a *job*. A *career* is exciting! A job is Just Ordinary and Boring."

Book Report

Write a 500-word report on the following book:

▲ *What are Numbers?*, (Glenview, IL: Scott, Foresman and Company, 1969) by Louis Auslander.

"Suppose a ball has the property that whenever it is dropped, it bounces up 1/2 the distance that it fell. For instance, if we drop it from 6 feet, it bounces up 3 feet. Let us hold the ball 6 feet off the ground and drop it. How far will the ball go before coming to rest?" Include the solution of this problem, which is discussed in this book (p. 112) as part of your report.

Important Terms

Add-on interest [6.2]
Adjusted balance method [6.2]
Alternating series [6.4]
Amortization [6.6]
Amortization schedule [6.6]
Annual compounding [6.1]
Annual percentage rate [6.2]
Annuity [6.5]
APR [6.2]
Arithmetic sequence [6.3]
Arithmetic series [6.4]
Average daily balance method [6.2]
Closed-ended loan [6.2]
Common difference [6.3]
Common ratio [6.3]
Compound interest [6.1]
Credit card [6.2]
Daily compounding [6.1]
Dealer's cost [6.2]
Evaluate a summation [6.4]

Exact interest [6.1]
Expand a summation [6.4]
Fibonacci sequence [6.3]
Finance charge [6.2]
Finite series [6.4]
Five percent offer [6.2]
Future value [6.1]
Geometric sequence [6.3]
Geometric series [6.4]
Infinite series [6.4]
Inflation [6.1]
Installment loan [6.2]
Interest [6.1]
Interest rate [6.1]
Line of credit [6.2]
Lump-sum problem [6.5]
Monthly compounding [6.1]
Open-ended loan [6.2]
Ordinary interest [6.1]
Partial sum [6.4]
Periodic payment problem [6.5]

Present value [6.1]
Present value of an annuity [6.6]
Previous balance method [6.2]
Principal [6.1]
Quarterly compounding [6.1]
Revolving credit [6.2]
Semiannual compounding [6.1]
Sequence [6.3]
Series [6.4]
Sigma notation [6.4]
Simple interest formula [6.1]
Sinking fund [6.5]
Sticker price [6.2]
Summation notation [6.4]
Term of a sequence [6.3]
Time [6.1]

Important Ideas

Find the amount of simple interest [6.1]
Future value formula with simple interest [6.1]
Simple interest formula [6.1]
Distinguish ordinary and exact interest [6.1]
Future value formula for simple interest [6.1]
Variables used with interest formulas [6.1]
Future value formula with compound interest [6.1]
Present value formula with compound interest [6.1]
Distinguish between add-on interest and amortization [6.2; 6.6]
Distinguish between add-on interest and annual percentage rate [6.2]
Distinguish among previous balance, adjusted balance, and average daily balance methods for credit card interest [6.2]
Distinguish among arithmetic, geometric, and Fibonacci sequences [6.3]
General term of an arithmetic sequence [6.3]

General term of a geometric sequence [6.3]
General term of a Fibonacci sequence [6.3]
Distinguish sequences and series [6.3; 6.4]
Be able to use summation notation [6.4]
Distinguish between arithmetic and geometric series [6.4]
Formulas for an arithmetic series [6.4]
Formulas for a geometric series [6.4]
Formula for an infinite geometric series [6.4]
Distinguish among the various financial problems [6.7]; in particular, those involving periodic payments:
  annuity [6.5]
  sinking fund [6.5]
  present value of an annuity [6.6]
  amortization [6.6]
Use a calculator when working with financial formulas [6.1–6.7]

Types of Problems

Be able to estimate reasonable answers to financial problems [6.1–6.6]
Find the amount of interest if you are given the purchase price, length of loan, and the monthly payment [6.1]

Find the total amount to be repaid for a simple interest loan [6.1]
How much must be deposited into a bank account to provide a given monthly income [6.1]

Compare the amount of interest and future value for simple and compound interest [6.1]

Be able to calculate the future value due to inflation [6.1]

Be able to calculate the present value due to inflation [6.1]

Be able to calculate the monthly payment using add-on interest [6.2]

Determine the monthly payment for an automobile using add-on interest [6.2]

Find the APR if you are given the price, length of time, and the add-on interest rate [6.2]

Calculate the APR given the amount financed, the monthly payment, the length of time financed [6.2]

Calculate the APR for credit cards [6.2]

Be able to calculate the amount of credit card interest by using the previous balance method, the adjusted balance method, and the average daily balance method [6.2]

Make an appropriate offer on a new car [6.2]

Find the amount of interest, the monthly payment, and the

APR for consumer transactions [6.2]

Classify a given sequence as arithmetic, geometric, or Fibonacci; be able to find the next term [6.3]

Write out the terms of a sequence when given the general term; find a specific term [6.3]

Evaluate (expand) summation expressions [6.4]

Find sums by classifying and using the series formulas [6.4]

Calculate the amount you can save by making a periodic payment into an interest-bearing account (annuity) [6.5]

Calculate the amount you need to deposit into an interest-bearing account to have a given amount at some time in the future [6.5]

Given a monthly payment, interest rate, and length of time, find the amount of loan [6.6]

Find the present value of an annuity [6.6]

Find the monthly payment for an amortized loan [6.6]

Work applied financial problems [6.1–6.7]

CHAPTER 6   Review Questions

1. In your own words, explain the difference between a sequence and a series. Include in your discussion how to distinguish arithmetic, geometric, and Fibonacci-type sequences, and give the formulas for their general terms. What are the relevant formulas for the sums of arithmetic and geometric series?

2. In your own words, outline a procedure for identifying financial formulas. Include in your discussion future value, present value, ordinary annuity, present value of an annuity, installment, and sinking fund classifications.

3. Classify each sequence as arithmetic, geometric, Fibonacci-type, or none of these. Find an expression for the general term if it is one of these types; otherwise, give the next two terms.

   a. 5, 10, 15, 20, . . .   **b.** 5, 10, 20, 40, . . .   **c.** 5, 10, 15, 25, . . .
   d. 5, 10, 20, 35, . . .   **e.** 5, 50, 500, 5,000, . . .   **f.** 5, 50, 5, 50, . . .

4. **a.** Evaluate $\sum_{k=1}^{3} (k^2 - 2k + 1)$ **b.** Expand $\sum_{k=1}^{4} \dfrac{k-1}{k+1}$

5. Suppose someone tells you she has traced her family tree back 10 generations. What is the minimum number of people on her family tree if there were no intermarriages?

6. A certain bacterium divides into two bacteria every 20 minutes. If there are 1,024 bacteria in the culture now, how many will there be in 24 hours, assuming that no bacteria die? Leave your answer in exponential form.

7. Suppose that the car you wish to purchase has a sticker price of $22,730 with a dealer cost of $18,579. Make a 5% offer for this car (rounded to the nearest hundred dollars).

8. Suppose that the amount to be financed for a car purchase is $13,500 at an add-on interest rate of 2.9% for 2 years. What are the monthly installment and the amount of interest that you will pay?

9. Find the APR for the loan described in Problem 8.

10. If a car with a cash price of $11,450 is offered for nothing down with 48 monthly payments of $353.04, what is the APR?

11. From the consumer's point of view, which method of calculating interest on a credit card is most advantageous? Illustrate the three types of calculating interest for a purchase of $525 with 9% APR for a 31-day month in which it takes 7 days for your $100 payment to be received and recorded.

12. If you purchase a home today for $185,000, what would you expect it to be worth in 30 years if you assume an inflation rate of 8%?

13. Suppose that you want to have $1,000,000 in 50 years. To achieve this goal, how much do you need to deposit today if you can earn 9% interest compounded monthly?

14. What is the monthly payment for a home loan of $154,000 if the rate is 8% and the time is 20 years?

15. Suppose that you expect to receive a $100,000 inheritance when you reach 21 in three years and four months. What is the present value of your inheritance if the current interest rate is 6.4% compounded monthly?

Revisiting the Real World . . .

IN THE REAL WORLD   Ron and Lorraine purchased a home some time ago, and they still owe $20,000. Their payments are $195 per month, and they have 10 years left to pay. They want to free themselves of the monthly payments by paying off the loan. They go to the bank and are told that, to pay off the home loan, they must pay $20,000 plus a "prepayment penalty" of 3%. (Older home loans often had prepayment clauses ranging from 1% to 5% of the amount to be paid off. Today, most lenders will waive this penalty. However, you must still request that the bank do this.) This means that, to own their home outright, Ron and Lorraine would have to pay $20,600 ($20,000 + 3% of $20,000). They want to know whether this is a wise financial move.

Commentary

Consider the situation of Ron and Lorraine described at the beginning of this chapter. They wish to know whether it is a wise financial move to use $20,600 to pay off their home mortgage.

The first question is how much principal do Ron and Lorraine need to generate $195 per month in income (to make their mortgage payments) if they are able to invest it at 15% interest? "Wait! Where can you find 15% interest? My bank pays only 5% interest!" There are investments you could make (not a deposit into a savings account) that could yield a 15% return on your money. Problem-solving often requires that you make certain assumptions to have sufficient information to answer a question. This first question is an application of simple interest, because the interest is withdrawn each month and is not left to accumulate:

I is the amount withdrawn each month.

$$I = Prt \leftarrow t = \tfrac{1}{12} \text{ because it is a monthly income.}$$
$$195 = P(0.15)(\tfrac{1}{12})$$
$$2{,}340 = 0.15P \qquad \text{Multiply both sides by 12.}$$
$$15{,}600 = P \qquad \text{Divide both sides by 0.15.}$$

This means that a deposit of $15,600 will be sufficient to pay off their home loan, with the added advantage that, when their home is eventually paid for, they will still have their $15,600!

Now, since they have $20,600 to invest, they can also allow the remaining $5,000 to grow at 15% interest. This is an example of compound interest because the interest is not withdrawn, but instead accumulates. Since there are 10 years left on the home loan, we have $P = 5,000; r = 0.15, t = 10$, and $n = 1$ (assume annual compounding). Calculate:

$$N = nt = 10 \quad \text{and} \quad i = \frac{r}{n} = \frac{0.15}{1} = 0.15$$

Thus,

$$A = 5,000(1 + 0.15)^{10} \approx 20,227.79$$

This means that Ron and Lorraine have two options:

1. Use their $20,600 to pay off the home loan. They will have the home paid for and will not need to make any further payments.

2. Dispose of their $20,600 as follows: Deposit $15,600 into an account to make their house payments. They will not need to make any further payments. At the end of 10 years they will still have the $15,600. Deposit the excess $5,000 into an account and let it grow for 10 years. The accumulated value will be $20,227.79. Under Option 2 they will have $35,827.79 in 10 years in addition to all the benefits gained under Option 1.

Group Research

Working in small groups is typical of most work environments, and being able to work with others to communicate specific ideas is an important skill to learn. Work with three or four other students to submit a single report based on each of the following questions.

1. Suppose you have just inherited $30,000 and need to decide what to do with the money. Write a paper discussing your options and the financial implications of those options. The paper you turn in should offer several alternatives and then the members of your group should reach a consensus of the best course of action.

2. Write a short paper about Fibonacci numbers. You might check *The Fibonacci Quarterly*, particularly "A Primer on the Fibonacci Sequence," Parts I and II, in the February and April 1963 issues. The articles, written by Verner Hogatt and S. L. Basin, are considered classic articles on the subject. One member of your group should investigate the relationship of the Fibonacci numbers to nature, another the algebraic properties of the sequence, and another the history of the sequence.

3. Write a computer program to calculate Fibonacci numbers.

4. Suppose you were hired for a job paying $21,000 per year and were given the following options:

Option A:	Annual salary increase of $1,200
Option B:	Semiannual salary increase of $300
Option C:	Quarterly salary increase of $75
Option D:	Monthly salary increase of $10

Each person should write the arithmetic series for the total amount of money earned in 10 years under a different option.

Your group should reach a consensus as to which is the best option. Give reasons and show your calculations in the paper that your group submits.

5. It is not uncommon for the owner of a home to receive a letter similar to the one shown below. Write a paper based on this letter. Different members of your group can work on different parts of the question, but you should submit one paper from your group.

   a. What is the letter about?

Dear Customer:

Previously we provided you an enrollment package for our BiWeekly Advantage Plan. We have had an excellent response from our mortgagors, yet we have not received your signed enrollment form and program fee to start your program.

How effective is the BiWeekly Advantage Plan? Reviewing your loan as of the week of this letter we estimate:

       MORTGAGE INTEREST SAVINGS OF $99,287.56
       LOAN TERM REDUCED BY 7 YEARS.

Many financial advisors agree it is a sound investment practice to pay off a mortgage early, particularly in light of growing economic uncertainty.

Act today! Don't let procrastination or indecision deny you and your family the opportunity to obtain mortgage-free homeownership sooner than you ever expected.

Please call us toll-free at 1-800-555-6060 if you would like more information about our BiWeekly Advantage Plan and a free personalized mortgage analysis. Our telephone representatives are available to serve you from 9 am to 5 pm (EST).

          Sincerely,

---

b. A computer printout (shown at the right) was included with the letter. Assuming that these calculations are correct, discuss the advantages or disadvantages of accepting this offer.

c. The plan as described in the letter costs $375 to sign up. I called the company and asked what their plan would do that I could not do myself by simply making 13 payments a year to my mortgage holder. The answer I received was that the plan would do nothing more, but the reason people do sign up is because they do not have the self-discipline to make the midmonthly payments to themselves. Why is a biweekly payment equivalent to 13 annual payments instead of equivalent to a monthly payment?

Interest Rate:	8.3750
Current Balance:	$ 416,640.54
Monthly Payment:	$ 3,180.91

Homeownership: Remaining Terms		
	Years	Months
Current Payment Plan	29	5
BiWeekly Plan	22	5
REDUCTION[3]	7	0

Total Remaining Principal & Interest Payments:	
Current Payment Plan	$ 1,122,849.24
BiWeekly Plan	$ 923,561.68
INTEREST SAVINGS[3]	$ 199,287.56

d. The representative of the company told me that more than 250,000 people have signed up. How much income has the company received from this offer?

e. You calculated the income the company has received from this offer in part **d**, but that is not all it receives. It acts as a bonded and secure "holding company" for your funds (because the mortgage company does not accept a "two-week" payment). This means that the company receives the use (interest value) on your money for two weeks out of every month. This is equivalent to half the year. Let's assume that the average monthly payment is $1,000 and that the company has 250,000 payments that they hold for half the year. If the interest rate is 5% (a secure guaranteed rate), how much potential interest can be received by this company?

The Nature of Geometry

Geometry is the science created to give understanding and mastery of the external relations of things; to make easy the explanation and description of such relations and the transmission of this mastery.

G. B. Halsted

In the Real World

"Carol, where have you been? I haven't seen you in over a year!" said Tom.

"I've been on an archeological dig," Carol responded. "Donald and I have been looking at the Garden Houses of Ostia. They were built in the 2nd century as part of the Roman Empire, and they were excavated in the first part of this century." "I've never heard of Ostia," Tom said with a swing of his arm. "Where or what is Ostia?" asked Tom.

"Ostia was a boom town in the second century," Carol said with enthusiasm. "Take a look at the map. The Tiber River was the lifeblood of the city Ostia, whose population reached 50,000 at its peak. They constructed an artificial harbor on the Tyrrhenian Sea, and the city became a major port of Rome, which is about 25 km away. Many apartment buildings were built, many four stories high, and they included not only living space, but also shops and gardens. We are trying to reconstruct life in the city."

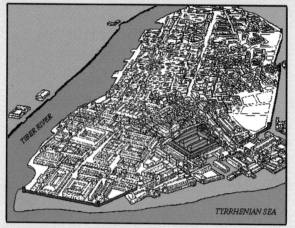

Ostia as it was in the 2nd century. The town of 50,000 required many apartment buildings (red).

"Those apartments look interesting," Tom commented. "It looks like there are 10 doors. With so many doors, I wonder if it is possible to pass through each door only once and stay either outside the apartment building or inside in the shaded hallway?"

Contents

Perspective

Geometry, or "earth measure," was one of the first branches of mathematics. Both the Egyptians and the Babylonians needed geometry for construction, land measurement, and commerce. They both discovered the Pythagorean theorem, although it was not proved until the Greeks developed geometry formally. This formal development utilizes deductive logic (which was briefly discussed in Chapter 2), beginning with certain assumptions, called **axioms** or **postulates**. Historically, the first axioms that were accepted seemed to conform to the physical world. In this chapter, we look at that body of mathematics known as *geometry*.

7.1 GEOMETRY

Euclidean Geometry

A book called Euclid's *Elements* collected all the material known about geometry and organized it into a logical deductive system. We do not know very much about the life of Euclid except that he was the first professor of mathematics at the University of Alexandria (which opened in 300 B.C.). The 13-book *Elements* is the most widely used and studied book in history, with the exception of the Bible.

Geometry involves **points** and sets of points called **lines, planes,** and **surfaces.** Certain concepts in geometry are called **undefined terms.** For example, what is a line? You might say, "I know what a line is!" But try to define a line. Is it a set of points? Any set of points? What is a point?

1. A point is something that has no length, width, or thickness.
2. A point is a location in space.

Certainly these are not satisfactory definitions because they involve other terms that are not defined. We will therefore take the terms *point, line,* and *plane* as undefined.

We often draw physical models or pictures to represent these concepts; however, we must be careful not to try to prove assertions by looking at pictures, since a picture may contain hidden assumptions or ambiguities. For example, consider Figure 7.1. If you look at Figure 7.1a, you might call it a square. If you look at Figure 7.1b, you might say it is something else. But what if we have in mind a cube, as shown in Figure 7.1c?

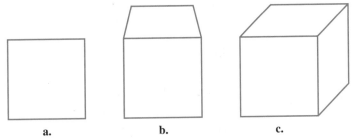

a. b. c.

▲ **Figure 7.1   Three views of the same object. What do you see? A square? A cube? Figures can be ambiguous (not clear, or with hidden meaning).**

Even if you view this object as a cube, do you see the same cube as everyone else? Is the fly in Figure 7.2 inside or outside the cube? Thus, although we may use a figure to help us understand a problem, we cannot prove results by this technique.

Geometry can be separated into two categories:

1. Traditional (which is the geometry of Euclid)
2. Transformational (which is more algebraic than the traditional approach)

When Euclid was formalizing traditional geometry, he based it on five postulates. In mathematics, a result that is proved on the basis of some agreed-upon postulates is called a **theorem.**

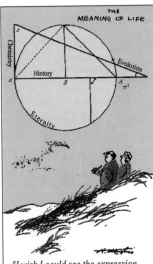

"I wish I could see the expression on the faces of my students who said there was no value in studying geometry."

News Clip

Early civilizations observed from nature certain simple shapes such as triangles, rectangles, and circles. The study of geometry began with the need to measure and understand the properties of these simple shapes.

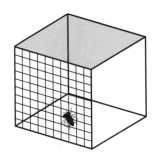

▲ **Figure 7.2   Is the fly on the cube or in the cube? On which face?**

Historical Note

EUCLID
(ca. 300 B.C.)

We know very little about the actual life of Euclid, but we do know that he was the first professor of mathematics at the University of Alexandria. The *Elements* was a 13-volume masterpiece of mathematical thinking. It was described by Augustus De Morgan (1806–1871) with the statement: "The thirteen books of Euclid must have been a tremendous advance, probably even greater than that contained in the *Principia* of Newton."

EUCLID'S POSTULATES

1. A straight line can be drawn from any point to any other point.
2. A straight line extends infinitely far in either direction.
3. A circle can be described with any point as center and with a radius equal to any finite straight line drawn from the center.
4. All right angles are equal to each other.
5. Given a straight line and any point not on this line, there is one and only one line through that point that is parallel to the given line.*

The first four of these postulates were obvious and noncontroversial, but the fifth one was different. This fifth postulate looked more like a theorem than a postulate. It was much more difficult to understand than the other four postulates, and for more than 20 centuries mathematicians tried to derive it from the other postulates or to replace it by a more acceptable equivalent. Two straight lines in the same plane are said to be **parallel** if they do not intersect.

Today we can either accept it as a postulate (without proof) or deny it. If it is denied, it turns out that no contradiction results; in fact, if it is not accepted, other geometries called **non-Euclidean geometries** result. If it is accepted, then the geometry that results is consistent with our everyday experiences and is called **Euclidean geometry.**

Let's look at each of Euclid's postulates. The first one says that a straight line can be drawn from any point to any other point.

To connect two points, you need a device called a **straightedge** (a device that we assume has no markings on it; you will use a ruler, but not to measure, when you are treating it as a straightedge). The portion of the line that connects points *A* and *B* in Figure 7.3 is called a **line segment.**

▲ **Figure 7.3   A line segment**

The second postulate says that we can draw a straight line. This seems straightforward and obvious, but we should point out that we will indicate a line by putting arrows on each end. If the arrow points in only one direction, the figure is called a **ray.** To construct a line segment of length equal to a given line segment, we need a device called a **compass.** Figure 7.4 shows a compass, which is used to mark off and duplicate lengths, but not to measure them. If objects have exactly the same size and shape, they are called **congruent.**

Pointer   Pencil
▲ **Figure 7.4   A compass**

---

* The fifth postulate stated here is the one usually found in high school geometry books. It is sometimes called Playfair's axiom and is equivalent to Euclid's original statement as translated from the original Greek by T. L. Heath: "If a straight line falling on two straight lines makes the interior angle on the same side less than two right angles, the two straight lines, if produced infinitely, meet on that side on which the angles are less than the two right angles."

We can use a straightedge and compass to **construct** a figure so that it meets certain requirements. To *construct a line segment congruent to a given line segment,* copy a segment AB on any line ℓ. First fix the compass so that the pointer is on point A and the pencil is on B, as shown in Figure 7.5a. Then, on line ℓ, choose a point C. Next, without changing the compass setting, place the pointer on C and strike an arc at D, as shown in Figure 7.5b.

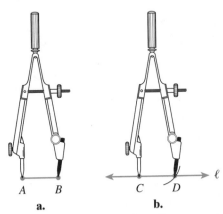

a.   b.

▲ **Figure 7.5** **Constructing a line segment**

Euclid's third postulate leads us to a second construction. The task is to construct a circle, given its center and radius. These steps are summarized in Figure 7.6.

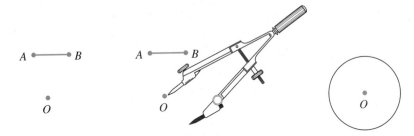

a. Given, a point and a radius of length $\overline{AB}$.

b. Set the legs of the compass on the ends of radius $\overline{AB}$; move the pointer to point O without changing the setting.

c. Hold the pointer at point O and move the pencil end to draw the circle.

▲ **Figure 7.6** **Construction of a circle**

We will demonstrate the fourth postulate in the next section when we consider angles.

The final construction of this section will demonstrate the fifth postulate. The task is to construct a line through a point P parallel to a given line ℓ, as shown in Figure 7.7. First, draw any line through P that intersects ℓ at a point A, as shown in Figure 7.8a on page 464.

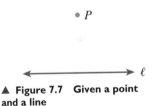

▲ **Figure 7.7** **Given a point and a line**

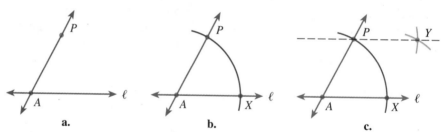

▲ **Figure 7.8   Construction of a line parallel to a given line through a given point**

Now, draw an arc with the pointer at *A* and radius *AP*, and label the point of intersection of the arc and the line *X*, as shown in Figure 7.8b. With the same opening of the compass, draw an arc first with the pointer at *P* and then with the pointer at *X*. Their point of intersection will determine a point *Y*. Draw the line through both *P* and *Y*. This line is parallel to ℓ.

Transformational Geometry

We now turn our attention to the second category of geometry. Transformational geometry is quite different from traditional geometry. It begins with the idea of a **transformation.** For example, one way to transform one geometric figure into another is by a *reflection*. Given a line *L* and a point *P*, as shown in Figure 7.9, we call the point *P'* the **reflection** of *P* about the line *L* if *PP'* is perpendicular to *L* and is also bisected by *L*. Each point in the plane has exactly one reflection point corresponding to a given line *L*. A reflection is called a *reflection transformation*, and the line of reflection is called the **line of symmetry.**

Symmetry is a very useful concept in geometry. There are line symmetries, rotational symmetries, and plane symmetries. The easiest way to describe a line symmetry is to say that if you fold a paper along its line of symmetry, then the figure will fold onto itself to form a perfect match, as shown in Figure 7.10.

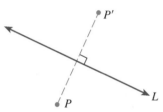

▲ **Figure 7.9   A reflection**

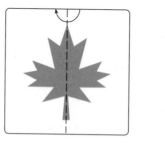

▲ **Figure 7.10   Line symmetry on the Maple Leaf of Canada**

Many everyday objects exhibit a line of symmetry. From snowflakes in nature, to the Taj Mahal in architectural design, to many flags and logos, we see examples of lines of symmetry (see Figure 7.11).

Snowflakes

Taj Mahal

Arizona

Chrysler logo; Arizona flag

▲ **Figure 7.11   Examples of line symmetry**

Other transformations include *translations*, *rotations*, *dilations*, and *contractions*, which are illustrated in Figure 7.12.

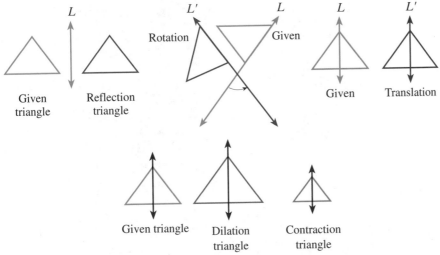

▲ **Figure 7.12   Transformations of a fixed geometric figure**

Geometry is also concerned with the study of the relationships between geometric figures. A primary relationship is that of **congruence**. A second relationship is called **similarity**. Two figures are said to be **similar** if they have the same shape, although not necessarily the same size. These ideas are considered in Sections 7.3 and 7.4.

If we were to develop this text formally, we would need to be very careful about the statement of our postulates. The notion of mathematical proof requires a very precise formulation of all postulates and theorems. For example, Euclid based many proofs about congruence on a postulate that said, "Things that coincide with one another are equal to one another" (Common Notions, Book I, *Elements*). This

News Clip

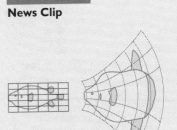

We find an excellent example of a grid transformation in this scheme devised by the Scottish biologist D'Arcy Wentworth Thompson. He used this grid transformation when studying the tissue layers of cell embryo.

Calligraphers sometimes experiment with some inversion transformations.

Wordplay

John Langdon has been creating such word designs for about 15 years, and the two shown here are from his book *Wordplay*. The following inversion was found in the March 1992 issue of *OMNI* magazine.

science

Here is a favorite of mine from Langdon's book:

minimum

statement is not formulated precisely enough to be used as justification for some of the things Euclid did with congruence. Even the explanation we gave for congruence is not precise enough to be used in a formal course. However, since we are not formally developing geometry, we simply accept the general drift of the statement. Remember, though, that you cannot base a mathematical proof on general drift. On the other hand, there are some facts that we *know to be true* that tell us that certain properties are impossible. For example, if someone claims to be able to trisect an angle with a straightedge and compass, we know *without looking at the construction* that the construction is wrong. This can be frustrating to someone who believes that he or she has accomplished the impossible. In fact, it was so frustrating to Daniel Wade Arthur that he was motivated to take out a paid advertisement in the *Los Angeles Times*, which is reproduced in the News Clip.

News Clip

Trisect Angle Using only Compass, Straight Edge and Pencil
Daniel Wade Arthur
Cinema Hotel, 1119 No. McCadden Pl., Hollywood 38, Calif.

1. Given any angle, acute or obtuse, with vertex at A.
2. With any convenient radius, draw an arc of a circle, cutting the sides of the angle at B & C.
3. Draw the straight line BC.
4. Construct the bisector of the angle, intersecting BC at O.
5. With O as center and radius BO or OC, draw a circle intersecting the angle bisector at L, between O and A.
6. Draw a straight line from B to L and from C to L.
7. Make an arc of the right angle L.
8. Trisect the right angle which will be at M.
9. With C as the center and radius of OC or BO, mark off an arc on the semicircle in the angle which falls at Z.
10. Trisect the straight line BC which will be at W and J.
11. Make a straight line from Z to M and from M to J.
12. Bisect the line ZM so that it will make a perpendicular and extend the line on out.
13. Bisect the line MJ so that it will make a perpendicular and make it cross the line that has bisected the line ZM.
14. With S as the center and radius of SZ make an arc through the work that has been done.

I have put within angle A a right angle which is angle L which was obtained by constructing a circle in the angle and so now there is a semicircle within the angle. The arc of any size angle with the width of BC can be put inside the area of the semicircle with the straight line BC the minimum and the semicircle the maximum arc. Within the area of the semicircle there is the arc of the right angle. I trisected each of the arcs the straight line at J, the arc of the right angle at M and the semicircle at Z. Any three points not on a straight line a circle can pass through and so Z, M and J are not in a straight line because perpendicular bisectors between Z and M and M and J will intersect at S. No matter what the size of the angle is if it has the same width of BC its arc will fall inside the area of the semicircle until the area will apparently become full but it never gets full because the angle can be any length and its arc always gets closer to the straight line if the angle gets longer and it never reaches even though it may seem to. The greater the angle gets its arc will get closer to the semicircle and the smaller the angle gets its arc will get closer to the straight line BC. Since the arc of angle A has to be within the area of the semicircle I have found three points within the area of the semicircle and the arc that passes through these points which are trisected areas will trisect the arc angle A, therefore, and arc trisects an arc. The arc of angle S with its radius of SZ will trisect the arc of angle A between M and J on the construction.

PROBLEM SET 7.1

▲ A Problems

1. Is the woman in the figure a young woman or an old woman?

2. **IN YOUR OWN WORDS** Describe a procedure for constructing a line segment congruent to a given segment.

3. **IN YOUR OWN WORDS** Describe a procedure for constructing a circle with a radius congruent to a given segment.

4. **IN YOUR OWN WORDS** Discuss what it means to be an undefined term.

5. **IN YOUR OWN WORDS** Describe line symmetry.

6. **IN YOUR OWN WORDS** What is the difference between an axiom and a theorem?

7. **IN YOUR OWN WORDS** What are the two categories into which geometry is usually separated?

Construct a segment congruent to each segment given in Problems 8–10.

8. A line segment

9. A line segment

10. A line segment

In Problems 11–16, construct a figure that meets the given requirements.

11. Circle with given radius

12. Circle with given radius

13. Circle with given radius

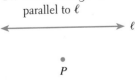

14. Line through P parallel to ℓ

15. Line through Q parallel to m

16. Line through R parallel to n

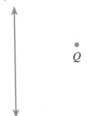

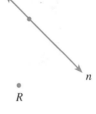

Find at least one line of symmetry for each of the illustrations in Problems 17–22.

17.

The Bell System

18.

The Yellow Pages

19.

20.

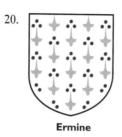

Ermine

21.

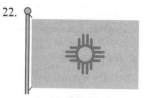

Alabama

22.

New Mexico

▲ **B Problems**

Which of the pictures in Problems 23–29 illustrate at least one line of symmetry? If a picture does illustrate such symmetry, show a line of symmetry; if it does not, tell why.

23.

Great Seal of the United States

24.

Boy Scouts

25.

Tressure with fleurs-de-lis

26.

27.

28.

29.

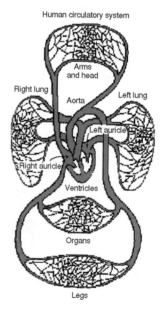

Escher's *Drawing Hands*

Study the patterns shown here. When folded they will form cubes spelling CUBE. Letter each pattern in Problems 30–35. Answers are not unique.

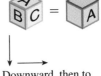

30.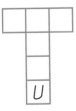

31.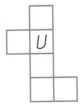

32.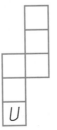

33.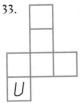

34.

35.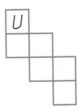

The arrows in Problems 36–40 indicate the direction(s) that the given cube has been rotated 90°. Study the examples shown in the margin. Pick the cube illustrating the appropriate rotations.

To the right

36. CHOICE: A   B   C   D

To the left

37.

CHOICE: A   B   C   D

Upward

38.

Downward

39.

Downward, then to the right

40.

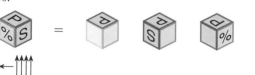

▲ **Problem Solving**

41. The letters of the alphabet can be sorted into the following categories:
(1) FGJLNPQRSZ; (2) BCDEK; (3) AMTUVWY; and (4) HIOX. What defines these
categories?

42. The "Mirror Image" illustration shows a problem condensed from *Games* magazine.
Answer the questions asked in that problem.

Mirror Image
by Diane Dawson

In the Fabulous Kingdom, anything can happen—and usually does.
When we held a looking glass up to this page, we noticed a few incongruities
between it and its "reflection" on the facing page. Can you spot thirty
substantial differences here, with or without the help of a mirror?
(Since both illustrations were hand-drawn there are bound
to be hairline differences—these should be ignored.)

ɘgɒmI ɿoɿɿiM
by Diane Dawson

In the Fabulous Kingdom, anything can happen—and usually does.
When we held a looking glass up to this page, we noticed a few incongruities
between it and its "reflection" on the facing page. Can you spot thirty
substantial differences here, with or without the help of a mirror?
(Since both illustrations were hand-drawn there are bound
to be hairline differences—these should be ignored.)

▲ **Individual Research**

43. What are optical illusions? What is the optical illusion in Problem 28? Find some unusual optical illusions and illustrate with charts, models, advertisements, pictures, or
illusions.

References
Martin Gardner, "Mathematical Games," *Scientific American*, May 1970.
Richard Gregory, "Visual Illusions," *Scientific American*, November 1968.
Lionel Penrose, "Impossible Objects: A Special Type of Visual Illusion," *The British Journal of Psychology*, February 1958.

44. Many curves can be illustrated by using only straight line segments. The basic design is drawn by starting with an angle, as shown. The result is called *aestheometry* and is depicted in Figure 7.13. Make up your own angle design.

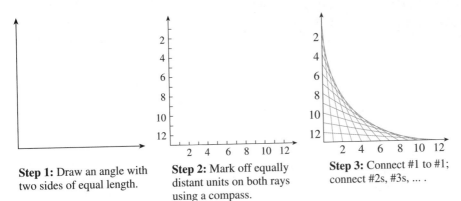

Step 1: Draw an angle with two sides of equal length.

Step 2: Mark off equally distant units on both rays using a compass.

Step 3: Connect #1 to #1; connect #2s, #3s, ...

a. Angle design

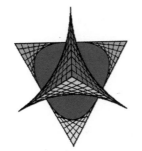

b. Angle design

c. Circle design

▲ **Figure 7.13   Aestheometry designs**

45. A second basic *aestheometric* design (see Problem 44) begins with a circle as shown:

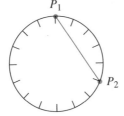

P_1

P_2

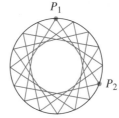

P_1

P_2

Step 1: Draw a circle and mark off equally spaced

Step 2: Choose any two points and connect them.

Step 3: Connect succeeding points around the circle.

Construct various designs using circles or parts of a circle.

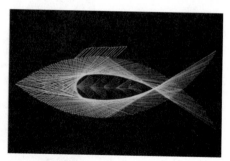

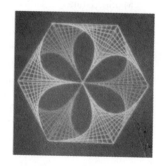

46. In 1928 Frank Ramsey, an English mathematician, showed that there are patterns implicit in any large structure. The accompanying diagram (which resembles the aestheometric design of Problems 44 and 45) typifies the problems that Ramsey theory addresses. "How many people does it take to form a group that always contains either four mutual acquaintances or four mutual strangers?" In the diagram, points represent people. A red edge connects people who are mutual acquaintances, and a blue edge joins people who are mutual strangers. In the group of 17 points shown, there are not four points whose network of edges are either completely red or completely blue. Therefore, it takes more than 17 people to guarantee that there will always be four people who are either acquaintances or strangers. Write a report on Ramsey theory. What is Ramsey Tic-Tac-Toe?

Reference Ronald L. Graham and Joel H. Spencer, "Ramsey Theory," *Scientific American,* July 1990, pp. 112–117.

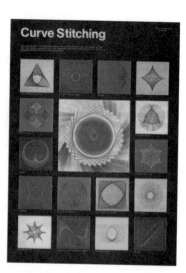

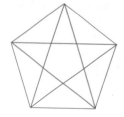

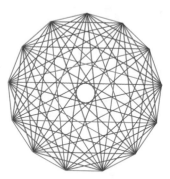

47. Euclid clearly made a distinction between the definition of a figure and the proof that such a figure could be constructed. Two very famous problems in mathematics focus on this distinction:

1. *Trisect an angle* using only a straightedge and compass.
2. *Square a circle:* Using only a straightedge and compass, construct a square with an area equal to the area of a given circle.

These problems have been *proved* to be impossible (as compared with unsolved problems that *might* be possible). Write a paper discussing the nature of an unsolved problem as compared with an impossible problem.

7.2 POLYGONS AND ANGLES

When the Euclidean geometry introduced in the preceding section is studied in high school as an entire course, it is usually presented in a *formal* manner using definitions, axioms, and theorems. The development of this chapter is *informal*, which means that we base our results on observations and intuition. We begin by assuming that you are familiar with the ideas of **point, line,** and **plane.**

If you consider a point on a line, that point separates the line into three parts: two **half-lines** and the point itself. A **ray** is the union of a half-line and its endpoint and is denoted by naming the endpoint as well as any point on the ray, as shown in Figure 7.14.

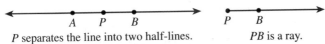

P separates the line into two half-lines.   *PB* is a ray.

▲ **Figure 7.14   Half-lines and rays**

Polygons

A **polygon** is a geometric figure that has three or more straight sides, all of which lie on a flat surface or plane so that the starting point and the ending point are the same. Polygons can be classified according to their number of sides, as shown in Figure 7.15.

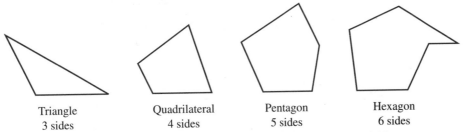

| Triangle | Quadrilateral | Pentagon | Hexagon |
| 3 sides | 4 sides | 5 sides | 6 sides |

▲ **Figure 7.15   Polygons classified according to number of sides**

Polygons (not pictured)
Heptagon: 7 sides
Octagon: 8 sides
Nonagon: 9 sides
Decagon: 10 sides
Dodecagon: 12 sides
n-gon: *n* sides

Angle

A connecting point of two sides is called a **vertex** (plural **vertices**) and is usually designated by a capital letter. An **angle** is composed of two rays or segments with a common endpoint. The angles between the sides of a polygon are sometimes also denoted by a capital letter, but other ways of denoting angles are shown in Figure 7.16.

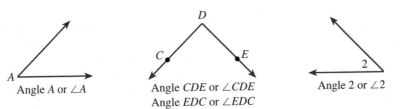

Angle *A* or ∠*A*   Angle *CDE* or ∠*CDE*   Angle 2 or ∠2
Angle *EDC* or ∠*EDC*

▲ **Figure 7.16   Ways of denoting angles**

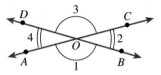

▲ Figure 7.17

■■■■ EXAMPLE 1

Locate each of the following angles in Figure 7.17.

Solution

a. ∠AOB   b. ∠COB   c. ∠BOA

d. ∠DOC   e. ∠3   f. ∠4

―

To construct an angle congruent to a given angle B, first draw a ray from B', as shown in Figure 7.18a. Next, mark off an arc with the pointer at the vertex of the given angle and label the points A and C. Without changing the compass, mark off a similar arc with the pointer at B', as shown in Figure 7.18b. Label the point C' where this arc crosses the ray from B'. Place the pointer at C and set the compass to the distance from C to A. Without changing the compass, put the pointer at C' and strike an arc to make a point of intersection A' with the arc from C', as shown in Figure 7.18c. Finally, draw a ray from B' through A'.

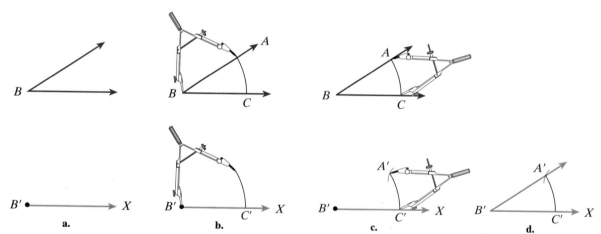

▲ **Figure 7.18   Construction of an angle congruent to a given angle**

Two angles are said to be **equal** if they describe the same angle. If we write m in front of an angle symbol, we mean the measure of the angle rather than the angle itself. Notice in Example 1 that parts **a** and **c** name the *same angle*, so ∠AOB = ∠BOA. Also notice the single and double arcs used to mark angles in Example 1; these are used to denote angles with equal measure, so $m∠COB = m∠AOD$ and $m∠COD = m∠AOB$. Denoting an angle by a single letter is preferred except in the case (as shown by Example 1) where several angles share the same vertex.

Angles are usually measured using a unit called a **degree,** which is defined to be $\frac{1}{360}$ of a full revolution. The symbol ° is used to designate degrees. To measure an angle, you can use a **protractor,** but in this book the angle whose measures we need will be labeled as in Figure 7.19.

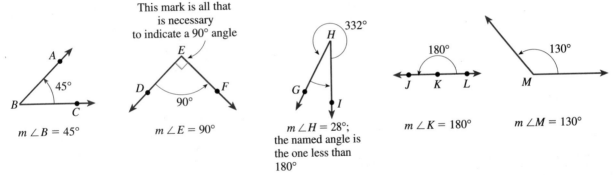

▲ **Figure 7.19 Labeling angles**

Angles are sometimes classified according to their measures, as shown in Table 7.1. It is also easy to see the plausibility of Euclid's fourth postulate that all right angles are equal to one another.

TABLE 7.1 Types of Angles

Angle measure	Classification
Less than 90°	Acute
Equal to 90°	Right
Between 90° and 180°	Obtuse
Equal to 180°	Straight

EXAMPLE 2

Label the angles *B, E, H, K,* and *M* in Figure 7.19 by classification.

Solution

∠*B* is acute; ∠*E* is right; ∠*H* is acute; ∠*K* is straight; ∠*M* is obtuse. ■

Two angles with the same measure are said to be **congruent.** If the sum of the measures of two angles is 90°, they are called **complementary angles,** and if the sum is 180°, they are called **supplementary angles.**

Consider any two distinct (different) intersecting lines in a plane, and let *O* be the point of intersection as shown in Figure 7.20. These lines must form four angles. Angles with a common ray are called **adjacent angles.** We say that ∠1 and ∠2, ∠2 and ∠3, ∠3 and ∠4, as well as ∠4 and ∠1 are pairs of adjacent angles. We also say that ∠1 and ∠3 as well as ∠2 and ∠4 are pairs of **vertical angles.**

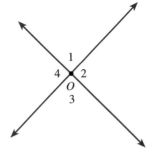

▲ **Figure 7.20 Angles formed by intersecting lines**

EXAMPLE 3

Classify the named angles shown in Figure 7.21.

a. ∠AOB   b. ∠BOC   c. ∠DAO

Classify the named pairs of angles shown in Figure 7.21.

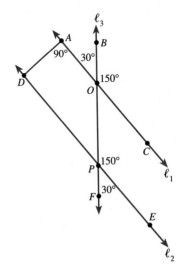

▲ **Figure 7.21 Classifying angles**

d. $\angle AOB$ and $\angle BOC$ **e.** $\angle BOC$ and $\angle OPE$ **f.** $\angle AOB$ and $\angle COP$
g. $\angle FPE$ and $\angle EPO$

Solution   a. Acute   **b.** Obtuse   **c.** Right   **d.** Supplementary or adjacent ·
e. Congruent   **f.** Vertical   **g.** Supplementary or adjacent   ▬

Angles with Parallel Lines

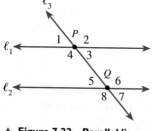

Consider three lines arranged similarly to those shown in Example 3. Suppose that two of the lines, say ℓ_1 and ℓ_2, are **parallel** (that is, they lie in the same plane and never intersect), and also that a third line ℓ_3 intersects the parallel lines at points P and Q, as shown in Figure 7.22. The line ℓ_3 is called a **transversal.**
    We make some observations about angles:

Figure 7.22   Parallel lines cut by a transversal

Vertical angles are congruent.

Alternate interior angles are pairs of angles whose interiors lie between the parallel lines, but on opposite sides of the transversal. Alternate interior angles are congruent.

Alternate exterior angles are pairs of angles that lie outside the parallel lines, but on opposite sides of the transversal. Alternate exterior angles are congruent.

Corresponding angles are two nonadjacent angles whose interiors lie on the same side of the transversal such that one angle lies between the parallel lines and the other does not. Corresponding angles are congruent.

▬▬ EXAMPLE 4

Consider Figure 7.23.

a. Name the vertical angles.   **b.** Name the alternate interior angles.
c. Name the corresponding angles.   **d.** Name the alternate exterior angles.

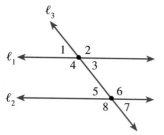

Figure 7.23   Angles formed by parallel lines and a transversal

Solution

a. The vertical angles are: $\angle 1$ and $\angle 3$; $\angle 2$ and $\angle 4$; $\angle 5$ and $\angle 7$; $\angle 6$ and $\angle 8$
b. The alternate interior angles are: $\angle 4$ and $\angle 6$; $\angle 3$ and $\angle 5$
c. The corresponding angles are: $\angle 1$ and $\angle 5$; $\angle 2$ and $\angle 6$; $\angle 3$ and $\angle 7$; $\angle 4$ and $\angle 8$
d. The alternate exterior angles are: $\angle 1$ and $\angle 7$; $\angle 2$ and $\angle 8$ ▬

    To summarize the results from Example 4, notice that the following angles are congruent, written $\simeq$:

$$\angle 1 \simeq \angle 3 \simeq \angle 5 \simeq \angle 7 \quad \text{and} \quad \angle 2 \simeq \angle 4 \simeq \angle 6 \simeq \angle 8$$

Also, all pairs of adjacent angles are supplementary.

▬▬ EXAMPLE 5

Find the measures of the eight numbered angles in Figure 7.23, where ℓ_1 and ℓ_2 are parallel. Assume that $m\angle 5 = 50°$.

Solution

$m\angle 1 = 50°$ Corresponding angles

$m\angle 2 = 180° - 50° = 130°$ $\angle 1$ and $\angle 2$ are supplementary.

$m\angle 3 = 50°$ $\angle 3$ and $\angle 1$ are vertical angles.

$m\angle 4 = 130°$ Vertical angles

$m\angle 5 = 50°$ Given

$m\angle 6 = 130°$ Supplementary angles

$m\angle 7 = 50°$ Vertical angles

$m\angle 8 = 130°$ Supplementary angles

Perpendicular Lines

If two lines intersect so that the adjacent angles are equal, then the lines are **perpendicular.** Simply, lines that intersect to form angles of 90° (right angles) are called perpendicular lines. In Figure 7.24, lines ℓ_3 and ℓ_4 intersect to form a right angle, and therefore they are perpendicular lines.

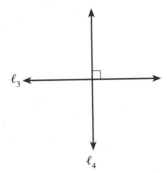

▲ **Figure 7.24   Perpendicular lines**

In a diagram on a printed page, any line that is parallel to the top and bottom edge of the page is considered **horizontal.** Lines that are perpendicular to a horizontal line are considered to be **vertical.** In Figure 7.25, line ℓ_5 is a horizontal line and line ℓ_6 is a vertical line.

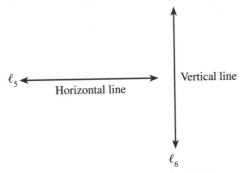

▲ **Figure 7.25   Horizontal and vertical lines**

▰▰ EXAMPLE 6

The diagram in Figure 7.26 shows lines in the same plane. Which of the given statements are true?

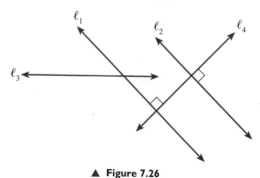

▲ **Figure 7.26**

a. Lines ℓ_1 and ℓ_2 are parallel and horizontal.

b. Lines ℓ_4 and ℓ_2 are intersecting but not perpendicular.

c. Lines ℓ_2 and ℓ_3 are intersecting lines.

Solution

a. Lines ℓ_1 and ℓ_2 are parallel, but they are not horizontal. Thus, the statement (which is a conjunction) is false.

b. Lines ℓ_4 and ℓ_2 are both intersecting and perpendicular. The statement is false.

c. Lines ℓ_2 and ℓ_3 are intersecting. Recall that the arrows on ℓ_3 indicate that it goes on without end in both directions, so it will intersect ℓ_2. This statement is true.

▰

PROBLEM SET   7.2

▲ **A Problems**

1. **IN YOUR OWN WORDS** What is an angle?

2. **IN YOUR OWN WORDS** Distinguish between equal angles and congruent angles.

3. **IN YOUR OWN WORDS** Distinguish between a half-line and a ray.

4. **IN YOUR OWN WORDS** Distinguish between horizontal and vertical lines.

5. **IN YOUR OWN WORDS** Describe right angles, acute angles, and obtuse angles.

6. **IN YOUR OWN WORDS** Describe adjacent, vertical, and corresponding angles.

7. **IN YOUR OWN WORDS** Describe parallel lines.

Name the polygons in Problems 8–19 according to the number of sides.

8.　　　　　　　　　9.　　　　　　　　10.

11. 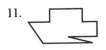   12.    13.

14.   15. 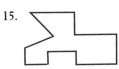   16.

17.   18.    19.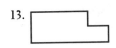

▲ B Problems

Using only a straightedge and a compass, reproduce the angles given in Problems 20–25.

20. 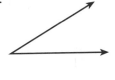   21. 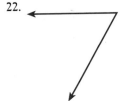   22.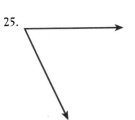

23. 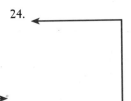   24.   25.

In Problems 26–34 classify the requested angles shown in Figure 7.27, where ℓ_1 and ℓ_2 are parallel.

26. $\angle 3$ if $m\angle 1$ is 30°
27. $\angle 6$ if $m\angle 5$ is 90°
28. $\angle 18$ if $m\angle 17$ is 105°
29. $\angle 19$ if $m\angle 11$ is 70°
30. $\angle 10$ if $m\angle 11$ is 90°
31. $\angle 16$ if $m\angle 15$ is 30°
32. $\angle 6$ if $m\angle 16$ is 120°
33. $\angle 9$ if $m\angle 19$ is 110°
34. $\angle 1$ if $m\angle 2$ is 130°

In Problems 35–43 classify the pairs of angles shown in Figure 7.27.

35. $\angle 2$ and $\angle 4$
36. $\angle 13$ and $\angle 14$
37. $\angle 9$ and $\angle 12$
38. $\angle 9$ and $\angle 10$
39. $\angle 9$ and $\angle 17$
40. $\angle 9$ and $\angle 11$
41. $\angle 12$ and $\angle 18$
42. $\angle 7$ and $\angle 13$
43. $\angle 10$ and $\angle 20$

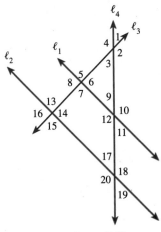

▲ **Figure 7.27**

Each of the four diagrams in Figure 7.28 shows lines in the same plane. Classify each of the statements in Problems 44–49 as true or false.

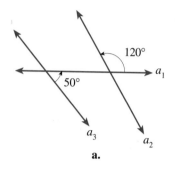
a.

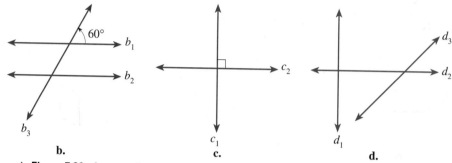

b.   c.   d.

▲ **Figure 7.28   Intersecting lines in the same plane**

44. The lines a_1 and a_3 are both intersecting and parallel.

45. The lines b_1 and b_2 are vertical.

46. The lines c_1 and c_2 are not intersecting.

47. The lines d_1 and d_3 are intersecting.

48. The lines c_2 and d_2 are horizontal.

49. The lines a_2 and a_3 are both vertical and parallel.

In Problems 50–55, find the measures of all the angles in Figure 7.29.

50. Given $m\angle 7 = 110°$ 51. Given $m\angle 2 = 65°$ 52. Given $m\angle 6 = 19°$

53. Given $m\angle 1 = 153°$ 54. Given $m\angle 5 = 120°$ 55. Given $m\angle 3 = 163°$

56. The legs of a picnic table form an isosceles triangle as shown in Figure 7.30. If $m\angle ACB = 90°$, find the measures of angles x and y so that the top of the table will be parallel to the ground.

57. Repeat Problem 56 where $m\angle ACB = 85°$.

58. Repeat Problem 56 where $m\angle ACB = 92°$.

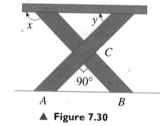

▲ **Figure 7.29 ℓ_1 is parallel to ℓ_2**

▲ **Figure 7.30**

▲ **Problem Solving**

59. The first illustration in the accompanying figure shows a cube with the top cut off. Use solid lines and shading to depict seven other different views of a cube with one side cut off.

60. Allow me to start you on a journey in Golygon City. You can take a similar trip in New York, Tokyo, or almost any large city whose streets form a grid of squares. Here

are your directions. Stroll down a city block, and at the end turn left or right. Walk two more blocks, turn left or right, then walk another three blocks, turn once more, and so on. Each time you turn, you must walk straight one block farther than before. If after a number of turns you arrive at your starting point, you have traced a golygon, as shown in Figure 7.31. A *golygon* consists of straight-line segments that have lengths (measured in miles, meters, or whatever unit you prefer) of one, two, three, and so on units. Draw some golygons. Can you make a conjecture about golygons?*

61. Read the News Clip in the margin. Without the poetry, the 1821 puzzle can be stated as follows: Arrange nine trees so they occur in ten rows of three trees each. If *t* represents the number of trees, and *r* is the number of rows, look at the following pattern:

▲ **Figure 7.31   A sample golygon**

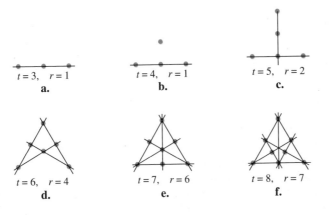

$t = 3,\ r = 1$	$t = 4,\ r = 1$	$t = 5,\ r = 2$
a.	b.	c.
$t = 6,\ r = 4$	$t = 7,\ r = 6$	$t = 8,\ r = 7$
d.	e.	f.

Use this pattern to lead you to a solution of the problem.

News Clip

Your aid I want, nine trees to
   plant
In rows just half a score;
And let there be in each row
   three.
Solve this; I ask no more.

This problem, in verse form, was first published by John Jackson in 1821. Can you answer the question?

7.3   TRIANGLES

One of the most frequently encountered polygons is the triangle. Every triangle has six parts: three sides and three angles. We name the sides by naming the endpoints of the line segments, and we name the angles by identifying the vertex (see Figure 7.32).

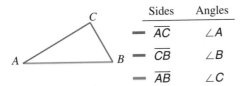

	Sides	Angles
—	$\overline{AC}$	$\angle A$
—	$\overline{CB}$	$\angle B$
—	$\overline{AB}$	$\angle C$

▲ **Figure 7.32   A standard triangle showing the six parts**

* From "Mathematical Recreations," by A. K. Dewdney, *Scientific American*, July 1990, p. 118.

Triangles are classified both by sides and by angles:

By Sides 　　　　　　　*By Angles*

Scalene: no equal sides　　**Acute:** three acute angles

Isosceles: two equal sides　**Right:** one right angle

Equilateral: three equal sides　**Obtuse:** one obtuse angle

We say that two triangles are **congruent** if they have the same size and shape. Suppose that we wish to construct a triangle with vertices *D*, *E*, and *F*, congruent to △*ABC* as shown in Figure 7.32. We would proceed as follows (as shown in Figure 7.33):

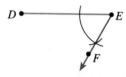

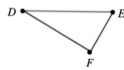

1. Mark off segment $\overline{DE}$ so that it is congruent to $\overline{AB}$. We write this as $\overline{DE} \simeq \overline{AB}$.

2. Construct angle *E* so that it is congruent to angle *B*. We write this as $\angle E \simeq \angle B$.

3. Mark off segment $\overline{EF} \simeq \overline{BC}$.

You can now see that, if you connect points *D* and *F* with a straightedge, the resulting △*DEF* has the same size and shape as △*ABC*. The procedure we used here is called SAS, meaning we constructed two sides and an *included angle* (an angle between two sides) congruent to two sides and an included angle of another triangle. We call these **corresponding parts.** There are other procedures for constructing congruent triangles; some of these are discussed in Problem Set 7.3. For this example, we say △*ABC* ≃ △*DEF*. From this we conclude that all six corresponding parts are congruent.

▲ **Figure 7.33　Constructing congruent triangles**

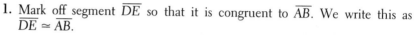

$$\Delta ABC \qquad \simeq \qquad \Delta DEF$$

A corresponds to *D*
B corresponds to *E*
C corresponds to *F*

■■■ **EXAMPLE 1**

Name the corresponding parts of the given triangles.

a. △*ABC* ≃ △*A′B′C′*

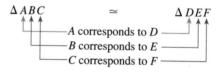

b. △*RST* ≃ △*UST*

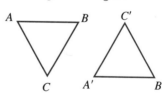

Solution

a. $\overline{AB}$ corresponds to $\overline{A'B'}$
   $\overline{AC}$ corresponds to $\overline{A'C'}$
   $\overline{BC}$ corresponds to $\overline{B'C'}$
   $\angle A$ corresponds to $\angle A'$
   $\angle B$ corresponds to $\angle B'$
   $\angle C$ corresponds to $\angle C'$

b. $\overline{RS}$ corresponds to $\overline{US}$
   $\overline{RT}$ corresponds to $\overline{UT}$
   $\overline{ST}$ corresponds to $\overline{ST}$
   $\angle R$ corresponds to $\angle U$
   $\angle RTS$ corresponds to $\angle UTS$
   $\angle RST$ corresponds to $\angle UST$

—

One of the most basic properties of a triangle involves the sum of the measures of its angles. To discover this property for yourself, place a pencil with an eraser along one side of any triangle as shown in Figure 7.34a.

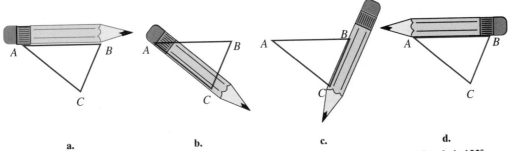

a.   **b.**   **c.**   **d.**

▲ **Figure 7.34** **Demonstration that the sum of the measures of the angles in a triangle is 180°**

Now rotate the pencil to correspond to the size of $\angle A$ as shown in Figure 7.34b. You see your pencil is along side AC. Next, rotate the pencil through $\angle C$, as shown in Figure 7.34c. Finally, rotate the pencil through $\angle B$. Notice that the pencil has been rotated the same amount as the sum of the angles of the triangle. Also notice that the orientation of the pencil is exactly reversed from the starting position. This leads us to the following important property.

Sum of the Measures of Angles in a Triangle

The sum of the measures of the angles in any triangle is 180°.

▬▬ EXAMPLE 2

Find the missing angle measure in the triangle in Figure 7.35.

Solution Let x represent the missing angle measure.

$$65 + 82 + x = 180$$
$$147 + x = 180$$
$$x = 33$$

The missing angle's measure is 33°.

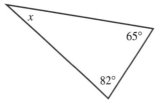

▲ **Figure 7.35** **What is x?** ▬

◼◼◼ EXAMPLE 3

Find the measures of the angles of a triangle if it is known that the measures are x, $2x - 15$, and $3(x + 17)$ degrees.

Solution   Using the theorem for the sum of the measures of angles in a triangle, we have

$$x + (2x - 15) + 3(x + 17) = 180$$
$$x + 2x - 15 + 3x + 51 = 180 \quad \text{Eliminate parentheses.}$$
$$6x + 36 = 180 \quad \text{Combine similar terms.}$$
$$6x = 144 \quad \text{Subtract 36 from both sides.}$$
$$x = 24 \quad \text{Divide both sides by 6.}$$

Now find the angle measures:

$$x = 24$$
$$2x - 15 = 2(24) - 15 = 33$$
$$3(x + 17) = 3(24 + 17) = 123$$

The angles have measures of 24°, 33°, and 123°.   ▬

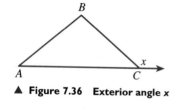

▲ **Figure 7.36   Exterior angle x**

An **exterior angle** of a triangle is the angle on the other side of an extension of one side of the triangle. An example is the angle whose measure is marked as x in Figure 7.36. Notice that the following relationships are true for any $\triangle ABC$ with exterior angle x:

$$m\angle A + m\angle B + m\angle C = 180° \quad \text{and} \quad m\angle C + x = 180°$$

Thus,

$$m\angle A + m\angle B + m\angle C = m\angle C + x$$
$$m\angle A + m\angle B = x \quad \text{Subtract } m\angle C \text{ from both sides.}$$

Exterior Angles of a Triangle

The measure of the exterior angle of a triangle equals the sum of the measures of the two opposite interior angles.

◼◼◼ EXAMPLE 4

Find the value of x in Figure 7.37.

Solution

$$\underbrace{63 + 42}_{\text{sum of interior angles}} = \underset{\underset{x}{\uparrow}}{x} \quad \overset{\text{exterior angle}}{}$$

$$105 = x$$

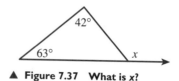

▲ **Figure 7.37   What is x?**

The measure of the exterior angle is 105°.   ▬

PROBLEM SET   7.3

▲ A Problems

1. **IN YOUR OWN WORDS** What is a triangle?

2. **IN YOUR OWN WORDS** What is the sum of the measures of the angles of a triangle?

2. **IN YOUR OWN WORDS** Explain the notation $\triangle ABC \simeq \triangle DEF$.

Name the corresponding parts of the triangles in Problems 4–7. Single and double marks are used to indicate segments of equal length.

4.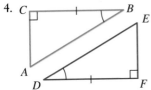

5.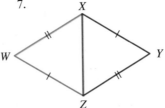

6.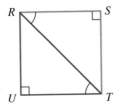

7.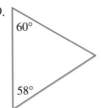

In Problems 8–13, find the measure of the third angle in each triangle.

8.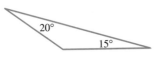

9.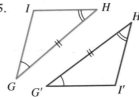

10.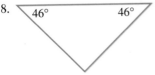

11.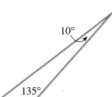

12.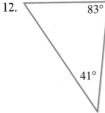

13.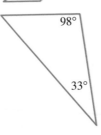

Find the measure of the indicated exterior angle in each of the triangles in Problems 14–19.

14.

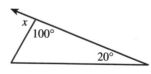

15.

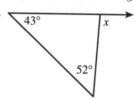

16.

17.

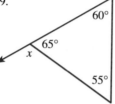

18.

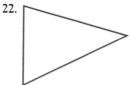

19.

Using only a straightedge and a compass, reproduce the triangles given in Problems 20–25.

20.

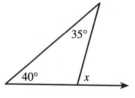

21.

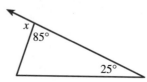

22.

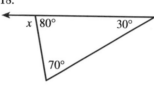

23.

24.

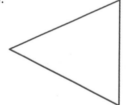

25.

▲ **B Problems**

Use algebra to find the value of x in each of the triangles in Problems 26–31. Notice that the measurement of the angle is not necessarily the same as the value of x.

26.
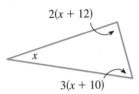

27.
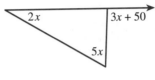

28.
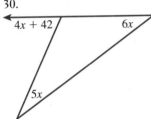

29.
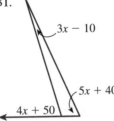

30.
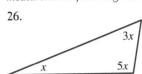

31.
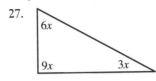

32. Find the measures of the angles of a triangle if it is known that the measures of the angles are x, $14 + 3x$, and $3(x + 25)$.

33. Find the measures of the angles of a triangle if it is known that the measures of the angles are x, $3x - 10$, and $3(55 - x)$.

34. In the text we constructed congruent triangles by using SAS. Reproduce the triangle shown in Problem 22 by using SSS. This means to construct the triangle by using the lengths of the three sides.

35. In the text we constructed congruent triangles by using SAS. Reproduce the triangle shown in Problem 23 by using SSS. This means to construct the triangle by using the lengths of the three sides.

36. In the text we constructed congruent triangles by using SAS. Reproduce the triangle shown in Problem 24 by using ASA. This means to construct the triangle by using a side included between two angles.

37. In the text we constructed congruent triangles by using SAS. Reproduce the triangle shown in Problem 25 by using ASA. This means to construct the triangle by using a side included between two angles.

38. Consider the following definitions:
    A polygon with four sides is a **quadrilateral.**

    A **trapezoid** is a quadrilateral with two parallel sides.

    A **parallelogram** is a quadrilateral with opposite sides parallel.

    A **rhombus** is a parallelogram with adjacent sides equal.

A **rectangle** is a parallelogram that contains a right angle.

A **square** is a rectangle with two adjacent sides of equal length.

Determine whether each sentence is true or false.

   a. Every square is a rectangle.
   b. Every square is a parallelogram.
   c. Every square is a rhombus.
   d. Every rhombus is a square.
   e. Every square is a quadrilateral.

39. Determine whether each sentence is true or false. (See Problem 38 for definitions, if necessary.)

   a. A parallelogram is a rectangle.
   b. A rectangle is a parallelogram.
   c. A trapezoid is a quadrilateral.
   d. A quadrilateral is a trapezoid.
   e. A parallelogram is a trapezoid.

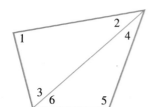

40. Show that *the sum of the measures of the interior angles of any triangle is* 180° by carrying out the following steps:

   a. Draw three triangles: one with all acute angles, one with a right angle, and a third with an obtuse angle.
   b. Tear apart the angles of each triangle you've drawn.
   c. Place the pieces together to form a straight angle.

41. Show that *the sum of the measures of the interior angles of any quadrilateral is* 360° by carrying out the following steps:

   a. Draw any quadrilateral, as illustrated (but draw your quadrilateral so it has a different shape from the one shown here).
   b. Divide the quadrilateral into two triangles by drawing a diagonal (a line segment connecting two nonadjacent vertices). Label the angles of your triangles as shown in the figure.
   c. What is the sum $m\angle 1 + m\angle 2 + m\angle 3$?
   d. What is the sum $m\angle 4 + m\angle 5 + m\angle 6$?
   e. The sum of the measures of the angles of the quadrilateral is

$$(m\angle 1 + m\angle 2 + m\angle 3) + (m\angle 4 + m\angle 5 + m\angle 6)$$

   What is this sum?
   f. Do you think this argument will apply for *any* quadrilateral?

▲ **Problem Solving**

42. Look at Problem 41. What is the sum of the measures of the interior angles of any pentagon?

43. Look at Problem 41. What is the sum of the measures of the interior angles of any octagon?

7.4 SIMILAR TRIANGLES

Congruent figures have exactly the same size and shape. However, it is possible for figures to have exactly the same shape without having the same size. Such figures are called *similar figures*. In this section we will focus on **similar triangles**. If $\triangle ABC$ is similar to $\triangle DEF$, we write

$$\triangle ABC \sim \triangle DEF$$

Similar triangles are shown in Figure 7.38. Since these figures have the same shape, we talk about **corresponding angles** and **corresponding sides**. The corresponding angles of similar triangles are those angles that are equal. The corresponding sides are those sides that are opposite equal angles.

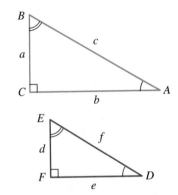

▲ **Figure 7.38   Similar triangles**

$m\angle A = m\angle D$, so these are corresponding angles.

$m\angle B = m\angle E$, so these are corresponding angles.

$m\angle C = m\angle F$, so these are corresponding angles.

Side $\overline{BC}$ is opposite $\angle A$ and side $\overline{EF}$ is opposite $\angle D$ so we say that $\overline{BC}$ corresponds to $\overline{EF}$.

$\overline{AC}$ corresponds to $\overline{DF}$.

$\overline{AB}$ corresponds to $\overline{DE}$.

Even though corresponding angles are equal, corresponding sides do not need to have the same length. If they do have the same length, the triangles are congruent. However, when they are not the same length, we can say they are proportional. From Figure 7.38 we see that the lengths of the sides are labeled a, b, c and d, e, f. When we say the sides are proportional, we mean

Primary Ratios:

$$\frac{a}{b} = \frac{d}{e} \qquad \frac{a}{c} = \frac{d}{f} \qquad \frac{b}{c} = \frac{e}{f}$$

Reciprocals:

$$\frac{b}{a} = \frac{e}{d} \qquad \frac{c}{a} = \frac{f}{d} \qquad \frac{c}{b} = \frac{f}{e}$$

Similar Triangle Theorem

Two triangles are similar if two angles of one triangle are equal to two angles of the other triangle. If the triangles are similar, then their corresponding sides are proportional.

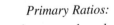 **EXAMPLE 1**

Identify pairs of triangles that are similar in Figure 7.39 on page 490.

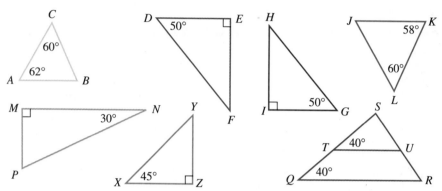

▲ **Figure 7.39**   **Which of these triangles are similar?**

Solution $\triangle ABC \sim \triangle JKL$; $\triangle DEF \sim \triangle GIH$; $\triangle SQR \sim \triangle STU$ ▬

▬▬ **EXAMPLE 2**

Given the similar triangles in Figure 7.40, find the unknown lengths marked b' and c'.

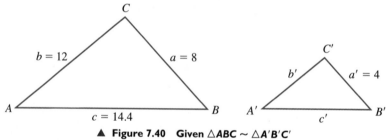

▲ **Figure 7.40**   **Given** $\triangle ABC \sim \triangle A'B'C'$

Solution   Since corresponding sides are proportional (other proportions are possible), we have

$$\frac{a}{b} = \frac{a'}{b'} \qquad\qquad \frac{a}{c} = \frac{a'}{c'}$$

$$\frac{8}{12} = \frac{4}{b'} \qquad\qquad \frac{8}{14.4} = \frac{4}{c'}$$

$$b' = \frac{4 \times 12}{8} \qquad\qquad c' = \frac{14.4 \times 4}{8}$$

$$b' = 6 \qquad\qquad c' = 7.2$$ ▬

Finding similar triangles is simplified even further if we know that the triangles are right triangles, because then the triangles are similar if one of the acute angles of one triangle has the same measure as an acute angle of the other.

▬▬ **EXAMPLE 3**   **Polya's Method**

Find the height of a tree that is difficult to measure directly.

Solution   We use Polya's problem-solving guidelines for this example.

Understand the Problem. We need to find the height of some tree without measuring it directly.

Devise a Plan. We assume that it is a sunny day, and will measure the height of the tree by measuring its shadow on the ground. For reference, we also measure the length of a shadow of an object of known height (say our own height, a meterstick, or a yardstick). We will then use similar triangles and proportions to find the height of the tree.

Carry Out the Plan. Suppose that a tree and a yardstick are casting shadows as shown in Figure 7.41. If the shadow of the yardstick is 3 yards long and the shadow of the tree is 12 yards long, we use similar triangles to estimate h, the height of the tree, if we know that $m\angle S = m\angle S'$.

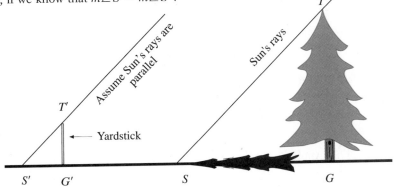

▲ **Figure 7.41**   **Finding the height of a tall object by using similar triangles**

Since $\angle G$ and $\angle G'$ are right angles, and since $m\angle S = m\angle S'$, $\triangle SGT \sim \triangle S'G'T'$. Therefore corresponding sides are proportional.

$$\frac{1}{3} = \frac{h}{12}$$ You solved proportions like this in Chapter 5.

$$h = \frac{1 \times 12}{3}$$

$$h = 4$$

Look Back. The tree is 4 yards, or 12 ft, tall.

There is a relationship between the sizes of the angles of a right triangle and the ratios of the lengths of the sides. In a right triangle, the side opposite the right angle is called the **hypotenuse.** Each of the acute angles of a right triangle has one side that is the hypotenuse; the other side of that angle is called the **adjacent side.** In $\triangle ABC$ with right angle at C, as shown in Figure 7.42,

the hypotenuse is c,

the side adjacent to $\angle A$ is b,

the side adjacent to $\angle B$ is a.

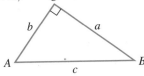

▲ **Figure 7.42**   **A right triangle**

Historical Note

BENJAMIN BANNEKER
(1731–1806)

Geometrical concepts are extremely important in surveying. One of the most perfectly planned cities is Washington, D.C. The center of the city is the Capitol, and the city is divided into four sections: Northwest, Northeast, Southwest, and Southeast. These are separated by North, South, East, and Capitol streets, all of which converge at the Capitol. The initial surveying of this city was done by a group of mathematicians and surveyors that included one of the first Afro-American mathematicians, Benjamin Banneker. He was born in 1731 in Maryland and showed extreme talent in mathematics without the benefit of formal education. His story, as well as those of other prominent black mathematicians, can be found in a book by Virginia K. Newell, Joella H. Gipson, L. Waldo Rich, and Beauregard Stubblefield, entitled *Black Mathematicians and Their Works* (Ardmore, Pennsylvania: Dorrance, 1980).

We also talk about an **opposite side.** The side opposite $\angle A$ is a, and the side opposite $\angle B$ is b.

You might recall the Pythagorean theorem, which was introduced in Chapter 4.

Pythagorean Theorem

For any right triangle with sides a and b and hypotenuse c,

$$a^2 + b^2 = c^2$$

Also, if a, b, and c are sides of a triangle so that

$$a^2 + b^2 = c^2$$

then the triangle is a right triangle.

■ EXAMPLE 4

A carpenter wants to make sure that the corner of a closet is square (a right angle). If she measures out sides of 3 feet and 4 feet, how long should she make the diagonal (hypotenuse)?

Solution   The hypotenuse is the unknown, so use the Pythagorean theorem:

$$
\begin{aligned}
c &= \sqrt{a^2 + b^2} \\
&= \sqrt{3^2 + 4^2} \quad \text{The sides are 3 and 4.} \\
&= \sqrt{9 + 16} \\
&= \sqrt{25} \\
&= 5
\end{aligned}
$$

She should make the diagonal 5 feet long.   ▬

The names that we give to the primary ratios in a right triangle are defined in the following box.

Trigonometric Ratios

In a right triangle ABC with right angle at C,

$\sin A$ (pronounced "sine of A") is the ratio $\dfrac{\text{opposite side of } A}{\text{hypotenuse}}$

$\cos A$ (pronounced "cosine of A") is the ratio $\dfrac{\text{adjacent side of } A}{\text{hypotenuse}}$

$\tan A$ (pronounced "tangent of A") is the ratio $\dfrac{\text{opposite side of } A}{\text{adjacent side of } A}$

■■■■ EXAMPLE 5

Given a right triangle with sides 5 and 12, find the trigonometric ratios for the angles A and B. Show your answers in both common fraction and decimal fraction form, with decimals rounded to four places.

Solution   First use the Pythagorean theorem to find the length of the hypotenuse:

$$c = \sqrt{5^2 + 12^2}$$
$$= \sqrt{25 + 144}$$
$$= \sqrt{169}$$
$$= 13$$

$$\sin A = \frac{5}{13} \approx 0.3846; \quad \cos A = \frac{12}{13} \approx 0.9231; \quad \tan A = \frac{5}{12} \approx 0.4167;$$

$$\sin B = \frac{12}{13} \approx 0.9231; \quad \cos B = \frac{5}{13} \approx 0.3846; \quad \tan B = \frac{12}{5} = 2.4 \quad \blacksquare$$

Tables of ratios for different angles are available (see, for example, the table in Appendix B in the back of the book). Certain calculators have keys for the sine, cosine, and tangent ratios. If you have such a calculator, you should press the angle first and then the key for trigonometric ratio.

■■■■ EXAMPLE 6

Find the trigonometric ratios by using either Appendix B or a calculator.

a. sin 45°   **b.** cos 32°   **c.** tan 19°

Solution

a. sin 45° ≈ 0.7071 from Appendix B; by calculator, press 45 SIN .*

b. cos 32° ≈ 0.8480 from Appendix B; press 32 COS .

c. tan 19° ≈ 0.3443 from Appendix B; press 19 TAN .   ■

Trigonometric ratios are useful in a variety of situations, as illustrated in the next example.

■■■■ EXAMPLE 7

The angle from the ground to the top of the Great Pyramid of Cheops is 52° if a point on the ground directly below the top is 351 ft away. What is the height of the pyramid?

Solution   From Figure 7.43 we see that for height h

$$\tan 52° = \frac{h}{351}$$

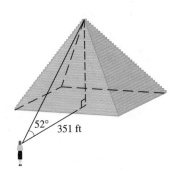

▲ **Figure 7.43   Calculating the height of the Great Pyramid of Cheops**

* Or SIN 45 , depending on brand and model of calculator.

Solving for h by multiplying both sides by 351, we obtain

$$h = 351 \tan 52°$$

Now we need to know the ratio for tan 52°. By table, tan 52° ≈ 1.2799, so $h ≈ 351(1.2799) = 449.2449$ or about 449 ft.

By calculator, press ⌷351⌷ ⌷×⌷ ⌷52⌷ ⌷TAN⌷ ⌷=⌷ .

Display: **449.2595129** or about 449 ft.   ▬

Notice that the calculator and table answers in Example 7 are not identical, because both the table and the calculator give approximations of the exact value of tan 52°.

PROBLEM SET 7.4

▲ **A Problems**

1. **IN YOUR OWN WORDS** Contrast congruent and similar triangles.

2. **IN YOUR OWN WORDS** What does it mean when we say the corresponding sides of two congruent triangles are proportional?

3. **IN YOUR OWN WORDS** What is the Pythagorean theorem?

4. **IN YOUR OWN WORDS** What is a sine?

5. **IN YOUR OWN WORDS** What is a cosine?

6. **IN YOUR OWN WORDS** What is a tangent?

Use the right triangle in Figure 7.44 to answer the questions in Problems 7–16.

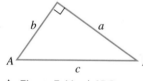

▲ **Figure 7.44** △**ABC**

7. What is the side opposite ∠A?   8. What is the side opposite ∠B?

9. What is the side adjacent ∠A?   10. What is the side adjacent ∠B?

11. What is the hypotenuse?   12. What is sin A?

13. What is sin B?   14. What is cos A?

15. What is cos B?   16. What is tan B?

In Problems 17–22, tell whether it is possible to conclude that the triangles are similar.

17.

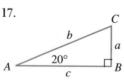 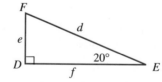

18.

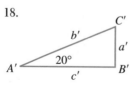 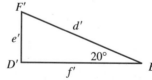

19.

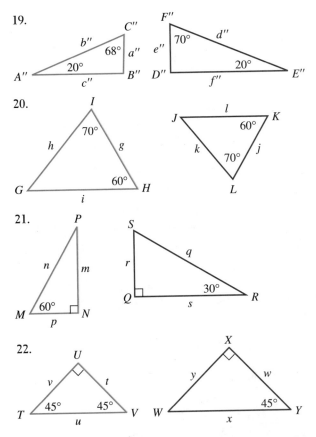

20.

21.

22.

Given two similar triangles, as shown in Figure 7.45, find the unknown lengths in Problems 23–30.

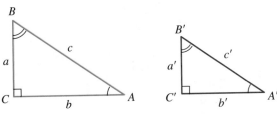

▲ **Figure 7.45** Similar triangles

23. $a = 4$, $b = 8$; find c.

24. $a' = 7$, $b' = 3$; find c'.

25. $a = 4$, $b = 8$, $a' = 2$; find b'.

26. $b = 5$, $c = 15$, $b' = 3$; find c'.

27. $c = 6$, $a = 4$, $c' = 8$; find a'.

28. $a' = 7$, $b' = 3$, $a = 5$; find b.

29. $b' = 8$, $c' = 12$, $c = 4$; find b.

30. $c' = 9$, $a' = 2$, $c = 5$; find a.

Find the trigonometric ratios in Problems 31–42 by using either the table in Appendix B or a calculator. Round your answers to four decimal places.

31. $\sin 56°$ **32.** $\sin 15°$ **33.** $\sin 61°$ **34.** $\sin 18°$

35. cos 54°   **36.** cos 8°   **37.** cos 90°   **38.** cos 34°

39. tan 24°   **40.** tan 52°   **41.** tan 75°   **42.** tan 89°

In Problems 43–52, find the sine, cosine, and tangent for the angle A.

43.

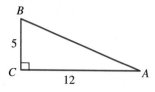

44.

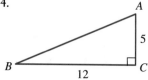

45.

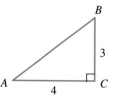

46.

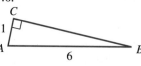

47.

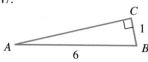

48.

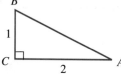

49.

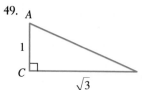

50.

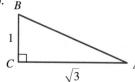

51.

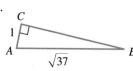

52.

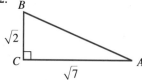

▲ **B Problems**

53. Use similar triangles and a proportion to find the length of the lake shown in Figure 7.46.

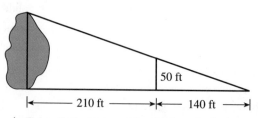

▲ **Figure 7.46   Determining the length of a lake**

54. Use similar triangles and a proportion to find the height of the house shown in Figure 7.47.

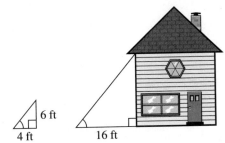

▲ **Figure 7.47   Determining the height of a building**

55. If a tree casts a shadow of 12 ft at the same time that a 6-ft person casts a shadow of $2\frac{1}{2}$ ft, find the height of the tree (to the nearest foot).

56. If the angle from the horizontal to the top of a building is 38° and the horizontal distance from its base is 90 ft, what is the height of the building (to the nearest foot)?

57. If the angle from the horizontal to the top of a tower is 52° and the horizontal distance from its base is 85 ft, what is the height of the tower (to the nearest foot)?

▲ **Problem Solving**

58. What is the radius of the largest circle you can cut from a rectangular poster board with measurements 11 in. by 17 in.? (See Figure 7.48.)

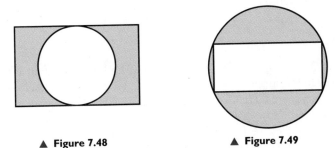

▲ **Figure 7.48**   ▲ **Figure 7.49**

59. What is the width of the largest rectangle with length 16 in. you can cut from a circular piece of cardboard having a radius of 10 in.? (See Figure 7.49.)

60. If the distance from the earth to the sun is 92.9 million miles, and the angle formed between Venus, the earth, and the sun is 47° (as shown in the illustration), find the distance from the sun to Venus (to the nearest hundred thousand miles).

61. Use the information and illustration in Problem 60 to find the distance from the earth to Venus (to the nearest hundred thousand miles).

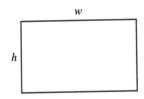

7.5   GOLDEN RECTANGLES

Certain rectangles hold some special interest because of the relationship between the lengths of their height and width. Consider a rectangle with height h and width

Historical Note

The divine proportion has many interesting properties. It was derived by the 15th century mathematician Luca Pacioli. He defined it by dividing a line segment into two parts so that the length of the smaller part is proportional to the length of the larger part as the length of the larger part is to the length of the entire segment. *Would you mind repeating that?* Consider a segment of unit length and divide it into two parts with lengths x and $1 - x$. Then the divine proportion is

$$\frac{1 - x}{x} = \frac{x}{1}$$

If you can solve this equation for x you will find $x = 1/\tau$.

George Markowsky points out in his article, "Misconceptions About the Golden Ratio," that the first time the term *golden section* was seen in print (in German) was in a book by Martin Ohm, and it first appeared in English in the 1875 edition of *Encyclopedia Britannica.* ▲

w, and consider the proportion

$$\frac{h}{w} = \frac{w}{h + w}$$

This relationship is called the *divine proportion.* If $h = 1$ we can solve the resulting equation for w:

$$\frac{1}{w} = \frac{w}{1 + w}$$

$$1 + w = w^2 \qquad \text{Product of means = product of extremes}$$

$$w^2 - w - 1 = 0 \qquad \text{Simplify.}$$

$$w = \frac{1 \pm \sqrt{(-1)^2 - 4(1)(-1)}}{2} = \frac{1 \pm \sqrt{5}}{2}$$

Quadratic formula

Since w is a length, we disregard the negative value to find

$$w = \frac{1 + \sqrt{5}}{2} \approx 1.6180339888$$

This number is called the **golden ratio,** and is denoted by τ (pronounced *tau*).

A rectangle that satisfies this proportion for finding the golden ratio is called a **golden rectangle** and can easily be constructed using a straightedge and a compass. Consider the proportion

$$\frac{h}{w} = \frac{w}{h + w}$$

which means the ratio of the height (h) to the width (w) is the same as the ratio of the width (w) to the sum of its height and width. To draw such a rectangle, we can begin with *any* square $CDHG$, as shown in Figure 7.50. This square is shown with its interior shaded. Now divide the square into two equal parts, as shown by the dashed segment labeled $\overline{AB}$. Set your compass so that it measures the length of $\overline{AC}$, and draw an arc, with center at A and a radius of length $\overline{AC}$, so that it intersects the extension of side $\overline{HD}$; label this point E. Now draw side $\overline{EF}$. The resulting rectangle $EFGH$ is a golden rectangle; $CDEF$ is also a golden rectangle.

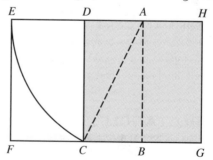

▲ **Figure 7.50   Constructing a golden rectangle *EFGH***

There are many interesting properties associated with the golden ratio τ. In Chapter 6 we considered the Fibonacci sequence

$$1, \quad 1, \quad 2, \quad 3, \quad 5, \quad 8, \quad 13, \quad 21, \quad 34, \quad 55, \quad 89, \quad 144, \ldots$$

Suppose we consider the ratios of the successive terms of the sequence:

$$\frac{1}{1} = 1.0000; \quad \frac{2}{1} = 2.0000; \quad \frac{3}{2} = 1.5000; \quad \frac{5}{3} \approx 1.667; \quad \frac{8}{5} = 1.6000;$$

$$\frac{13}{8} = 1.625; \quad \frac{21}{13.} \approx 1.615; \quad \frac{34}{21} \approx 1.619; \quad \frac{55}{34} \approx 1.618; \quad \frac{89}{55} \approx 1.618$$

If you continue to find these ratios, you will notice that the sequence oscillates about a number approximately equal to 1.618, which is τ.

Suppose we repeat this same procedure with another sequence, this time choosing *any* two nonzero numbers, say 4 and 7, as a starting point. Construct a Fibonacci-type sequence:

$$4, \quad 7, \quad 11, \quad 18, \quad 29, \quad 47, \quad 76, \quad 123, \quad 199, \quad 322, \ldots$$

Next, form the ratios of the successive terms:

$$\frac{7}{4} = 1.750; \quad \frac{11}{7} \approx 1.571; \quad \frac{18}{11} \approx 1.636; \quad \frac{29}{18} \approx 1.611; \quad \frac{47}{29} \approx 1.621; \ldots$$

These ratios are oscillating about the same number, τ!

▬▬ EXAMPLE 1   Polya's Method

Find some numbers with the property that the number and its reciprocal differ by 1.

Solution   We use Polya's problem-solving guidelines for this example.

Understand the Problem.   We know, for example, that if x is such a number, then $\frac{1}{x} + 1$ must be equal to the original number.

Devise a Plan.   We let x be any nonzero number. Then we need to set $\frac{1}{x} + 1$ equal to x, and solve the resulting equation.

Carry Out the Plan.

$$x = \frac{1}{x} + 1$$

$$x^2 = 1 + x \qquad \text{Multiply both sides by } x.$$

$$x^2 - x - 1 = 0$$

$$x = \frac{1 \pm \sqrt{(-1)^2 - 4(1)(-1)}}{2} = \frac{1 \pm \sqrt{5}}{2}$$

▲ **Figure 7.51** A 15-oz box of Kellogg's Sugar Corn Pops is 30 cm by 19 cm, for a ratio of 1.6; a 1-lb box of C&H sugar is 17 cm by 10 cm, for a ratio of 1.7.

Look Back. We recognize the positive value as τ:

$$\tau = \frac{1 + \sqrt{5}}{2} \approx 1.618033989$$

Use your calculator to find the reciprocal $\frac{1}{\tau} \approx 0.618033989$.

It has been said that many everyday rectangular objects have a length-to-width ratio of about $1.6 : 1$, as illustrated in Figure 7.51. Psychologists have tested individuals to determine the rectangles they find most pleasing; the results are those rectangles whose length-to-width ratios are near the golden ratio. George Markowsky, on the other hand, in a recent article "Misconceptions About the Golden Ratio," suggests that with rectangles with greatly different length-to-width ratios, the golden rectangle is the most pleasing; but when confronted with rectangles with ratios "close" to the golden ratio, subjects are unable to select the "best" rectangle.

If we observe many works of art, we can see evidence of the golden rectangles. Whether the artist had such rectangles in mind is open to speculation, but we can see golden rectangles in the work of Albrecht Dürer, Leonardo da Vinci, George Bellows, Pieter Mondriaan, and Georges Seurat (see Figure 7.52).

▲ **Figure 7.52** *La Parade* **by the French impressionist Georges Seurat**

The Parthenon in Athens has been used as an example of a building with a height-to-width ratio that is almost equal to the golden ratio.

Parthenon

■■■ EXAMPLE 2 **Polya's Method**

If the Parthenon is 101 feet wide, what is its height (to the nearest foot) if we assume the dimensions are in a golden ratio?

Solution We use Polya's problem-solving guidelines for this example.

Understand the Problem. First, understand the problem. Since the Parthenon is built to satisfy the golden ratio, the height h and the width w satisfy the following proportion:

$$\frac{h}{w} = \frac{w}{h + w}$$

Devise a Plan.   The width is 101 feet, so

$$\frac{h}{101} = \frac{101}{h + 101}$$

We will solve this equation for h.

Carry Out the Plan.   There is only one unknown, which is written in variable form, so we now solve the equation for h:

$$h(h + 101) = 101^2$$
$$h^2 + 101h - 101^2 = 0$$
$$h = \frac{-101 \pm \sqrt{101^2 - 4(1)(-101^2)}}{2}$$
$$= \frac{-101 \pm 101\sqrt{1 + 4}}{2}$$

≈ 62.4 Disregard the negative solution, since distances are nonnegative.

Look Back.   The Parthenon is about 62 feet high.

Many studies of the human body itself involve the golden ratio (remember $\tau \approx 1.62$). Figure 7.53 shows a drawing of an idealized athlete.

David (1501–1504) **by Michelangelo illustrates many golden ratios.**

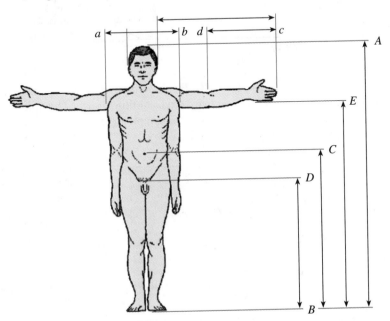

▲ **Figure 7.53   Proportions of the human body**

Let AC = distance from the top of the head to the navel;
     CB = distance from the navel to the floor; and
     AB = height.

Then, $\dfrac{CB}{AC} \approx \tau$ and $\dfrac{AB}{CB} \approx \tau.$

Also, let ab = shoulder width and bc = arm length. Then

$$\frac{bc}{ab} \approx \tau$$

See if you can find other ratios on the human body that approximate τ. Figure 7.54 shows a study of the human face by da Vinci, in which the rectangles approximate golden rectangles.

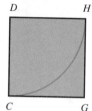

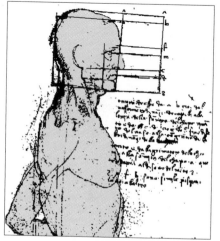

Dynamic symmetry of a human face (Leonardo da Vinci)

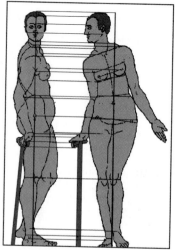

Proportions of the human body by Albrecht Dürer

▲ **Figure 7.54   Golden rectangles used in art**

The last application of the golden rectangle we will consider in this section is related to the manner in which a chambered nautilus grows. The spiral of the chambered nautilus can be seen in the photograph in the margin, and can be constructed by following the steps in Figure 7.55.

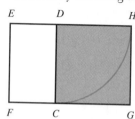

Begin with any square and draw a quarter circle as shown.

Form a golden rectangle.

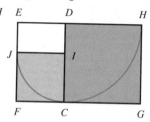

Draw a square within the new rectangle *CDEF*; draw a semicircle as shown.

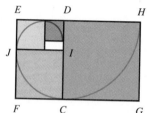

Repeat the process. The resulting curve is called a logarithmic spiral.

▲ **Figure 7.55   A spiral is constructed using a golden rectangle.**

PROBLEM SET 7.5

▲ A Problems

1. **IN YOUR OWN WORDS** What is a golden ratio?

2. **IN YOUR OWN WORDS** What is a golden rectangle?

3. **IN YOUR OWN WORDS** What is the divine proportion?

4. **IN YOUR OWN WORDS** What is τ?

5. **a.** Pick any two nonzero numbers. Construct a Fibonacci-type sequence beginning with these two numbers.
   b. Form the ratios of successive terms, and show that after a while they oscillate around the golden ratio.

6. Repeat Problem 5 for two other nonzero numbers.

7. The Lucas sequence is a Fibonacci-type sequence with $s_1 = 1$ and $s_2 = 3$. Write out the first ten terms of the Lucas sequence and find the ratios of the successive terms. How do these compare with τ?

8. If the Parthenon in Greece is 60 ft tall (at the apex) and 97 ft wide, find the ratio of width to height and compare it with τ.

9. The Great Pyramid of Giza has dimensions as follows: height, $h = 481$ ft; base, $b = 756$ ft; and slant height, $s = 612$ ft. Is the ratio of any of these dimensions related to τ?

10. The American Pyramid of the Sun at Teotihuacán, Mexico, has a height of 216 ft, a base of 700 ft, and a slant height of 411 ft. Does the ratio of any of these measurements approximate τ?

11. How closely does the ratio of the dimensions of a polo field (160 yd by 300 yd) approximate τ?

12. How closely does the ratio of the dimensions of a basketball court (36 ft by 78 ft) approximate τ?

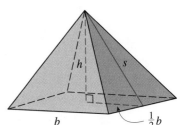

▲ B Problems

13. **a.** What are the length and width of a standard index card?
    b. Write the ratio of length to width as a decimal and compare it with τ.

14. **a.** What are the length and width of a standard brick?
    b. Write the ratio of length to width as a decimal and compare it with τ.

15. **a.** What are the length and width of this textbook?
    b. Write the ratio of length to width as a decimal and compare it with τ.

Use Figure 7.53 to find the requested measurements in Problems 16–21 using your own body as the model.

16. $CB \div AC$
17. $AB \div CB$
18. $AD \div AE$
19. $bc \div ab$
20. $ac \div bc$
21. $dc \div bd$

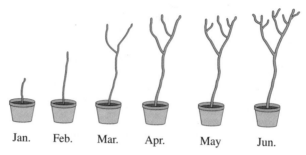

A plant grows for two months and then adds a new branch. Each new branch grows for two months, and then adds another branch. After the second month, each branch adds a new branch every month. Use this information in Problems 22–25, assuming that the growth begins in January.

Jan.   Feb.   Mar.   Apr.   May   Jun.

22. How many branches will there be in March?

23. How many branches will there be in June?

24. How many branches will there be after 12 months?

25. Form a sequence of ratios of successive terms by considering the number of branches for the first 12 months. How do these numbers compare with τ?

Leo Moser studied the effect that two face-to-face panes of glass have on light reflected through the panes. If a ray is unreflected, it has just one path through the glass. If it has one reflection, it can be reflected two ways. For two reflections, it can be reflected three ways. Use this information in Problems 26–29.

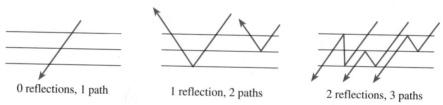

0 reflections, 1 path   1 reflection, 2 paths   2 reflections, 3 paths

26. Show the possible paths for three reflections.

27. Show the possible paths for four reflections.

28. Make a conjecture about the number of paths for n reflections.

29. Form a sequence of ratios of the number of successive paths. How do these numbers compare with τ?

30. If a rectangle is to have sides with lengths in the golden ratio, what is the width if the height (the shorter side) is 2 units?

31. If a window is to be 5 feet wide, how high should it be, to the nearest tenth of a foot, to be a golden rectangle?

32. If a canvas for a painting is 18 inches wide, how high should it be, to the nearest inch, to be in the divine proportion?

33. A photograph is to be printed on a rectangle in the divine proportion. If it is 9 cm high, how wide is it, to the nearest centimeter?

34. If the Parthenon is 60 ft high, what is its width, to the nearest foot, if we assume the building conforms to the golden ratio?

▲ **Problem Solving**

35. "When is an open book a closed book?"* If the length-to-width ratio of a book remains unchanged when that book is opened, then that book is said to be in the librarian's ratio.

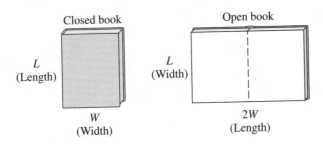

Closed book   Open book

L (Length) L (Width)

W (Width) $2W$ (Length)

(Note that the length, L, of the closed book becomes the width of the opened book, and twice the width of the closed book becomes the length of the opened book.) Find an approximate and an exact representation for constant L/W.

36. Does this book satisfy the librarian's ratio (see Problem 35)?

▲ **Individual Research**

37. Do some research on the length-to-width ratios of the packaging of common household items. Form some conclusions. Find some examples of the golden ratio in art. Do some research on dynamic symmetry.

    References
    Philip J. Davis and Reuben Hersh, *The Mathematical Experience* (Boston: Houghton Mifflin Company, 1981), pp. 168–171.
    H. E. Huntley, *The Divine Proportion: A Study in Mathematical Beauty* (New York: Dover Publications, 1970).

38. In Example 2 we assumed the width of the Parthenon to be 101 ft and found the height to be 62.4 (assuming the golden ratio). If you worked Problem 34, you assumed the height to be 60 ft and found the width using the golden ratio to be 97 ft. Are the numbers from Example 2 and from Problem 34 consistent? Can you draw any conclusions?

    Reference George Markowsky, "Misconceptions About the Golden Ratio," *The College Mathematics Journal*, January 1992.

---

* My thanks to Monte J. Zerger of Friends University in Wichita, Kansas, for this problem. I found it in Vol. 18 of *The Journal of Recreational Mathematics*.

Historical Note

CZARINA CATHERINE
(1729–1796)

In 1783 Catherine II (Catherine the Great) of Russia, originally named Sophie von Anhalt–Zerbst, appointed Princess Daschkoff to the directorship of the Imperial Academy of Sciences in St. Peterburg. Since women were seldom so highly honored in those days, the appointment received wide publicity. The Princess decided to begin her directorship with a short address to the assembled members of the Academy. She invited Leonhard Euler, who at the time was elderly and blind. Since Euler was then the most respected scientist in Russia, he was to have the special seat of honor. However, as the Princess sat down a local professor named Schtelinn maneuvered into Euler's seat. When Princess Daschkoff saw Schtelinn seating himself next to her, she turned to Euler and said, "Please be seated anywhere, and the chair you choose will naturally be the seat of honor." This act charmed Euler and all present — except the arrogant Professor Schtelinn.

7.6   KÖNIGSBERG BRIDGE PROBLEM

In the 18th century, in the German town of Königsberg (now a Russian city), a popular pastime was to walk along the bank of the Pregel River and cross over some of the seven bridges that connected two islands, as shown in Figure 7.56.

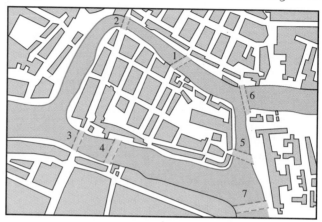

▲ **Figure 7.56   Königsberg bridges**

One day a native asked a neighbor this question, "How can you take a walk so that you cross each of our seven bridges once and only once?" The problem intrigued the neighbor, and soon caught the interest of many other people of Königsberg as well. Whenever people tried it, they ended up either not crossing a bridge at all or else crossing one bridge twice. This problem was brought to the attention of the Swiss mathematician Leonhard Euler, who was serving at the court of the Russian empress Catherine the Great in St. Petersburg. The method of solution we discuss here was first developed by Euler, and led to the development of two major topics in geometry. The first is *networks*, which we discuss in this section, and the second is *topology*, which we discuss in the next section.

Networks

We will use Polya's problem-solving method for the Königsberg bridge problem.

Understand the Problem.   To understand the problem, Euler began by drawing a diagram for the problem, as shown in Figure 7.57a.

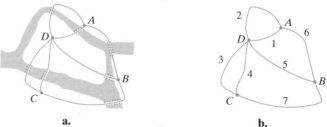

a.   b.

▲ **Figure 7.57   Networks for the Königsberg bridge problem**

Next, Euler used one of the great problem-solving procedures — namely, to change the conceptual mode. That is, he **let the land area be represented as points, and let the bridges be represented by arcs or line segments connecting the given points.** As part of understanding the problem, we can do what Euler did — we can begin by tracing a diagram like the one shown in Figure 7.57b.

Devise a Plan.   To solve the bridge problem, we need to draw the figure without lifting the pencil from the paper. Figures similar to the one in Figure 7.57b are called **networks** or **graphs.** In a network, the points where the line segments meet (or cross) are called **vertices,** and the lines representing bridges are called **edges** or **arcs.** Each separated part of the plane formed by a network is called a **region.**

■■■■ EXAMPLE 1

Complete the table for each of the given networks.

a.   b.   c.

d.   e.   f.

Solution

Graph	Edges (E)	Vertices (V)	Regions (R)	V + R − 2
a.	3	3	2	3
b.	4	3	3	4
c.	5	3	4	5
d.	4	4	2	4
e.	5	4	3	5
f.	6	4	4	6

A network is said to be **traversable** if it can be traced in one sweep without lifting the pencil from the paper and without tracing the same edge more than once. Vertices may be passed through more than once.

■■■■ EXAMPLE 2

Which of the networks of Example 1 are traversable?

Solution a. yes   b. yes   c. yes   d. yes   e. yes   f. no   ■

It is assumed that you worked Example 2 by actually tracing out the networks. However, a more complicated network, such as the Königsberg bridge problem, will require some analysis. We will follow Euler's lead and classify vertices. The number of edges to vertex A in Figure 7.57 is 3, so the vertex A is called an **odd**

vertex. In the same way, D is an odd vertex, because 5 edges connect to it. Euler discovered that there must be a certain number of odd vertices in any network if you are to travel it in one journey without retracing any edge. You may start at any vertex and end at any other vertex, as long as you travel the entire network. Also the network must connect each point (this is called a **closed network**).

Let's examine the network more carefully and look for a pattern, as shown in Table 7.2.

TABLE 7.2   Arrivals and Departures for Networks

Number of arcs at a vertex	Description	Possibilities
1	1 departure (starting point) 1 arrival (ending point)	
2	1 arrival (arrive then depart) and 1 departure (depart then arrive)	
3	1 arrival, 2 departures: (depart-arrive-depart; starting point) 2 arrivals, 1 departure: (arrive-depart-arrive; ending point)	
4	2 arrivals, 2 departures	
5	2 arrivals, 3 departures (starting point) 3 arrivals, 2 departures (ending point)	

We see that, if the vertex is odd, then it must be a starting point or an ending point. What is the largest number of starting and ending points in any network? [*Answer:* Two — one starting point and one ending point.]

Carry Out the Plan: Count the number of odd vertices:

If there are no odd vertices, the network is traversable and any point may be a starting point. The point selected will also be the ending point.

If there is one odd vertex, the network is not traversable. A network cannot have only one starting or ending point without the other.

If there are two odd vertices, the network is traversable; one odd vertex must be a starting point and the other odd vertex must be the ending point.

If there are more than two odd vertices, the network is not traversable. A network cannot have more than one starting point and one ending point.

Carrying out this plan on the Königsberg bridge problem, we classify the vertices; there are four odd vertices, so the network is not traversable.

Look Back. We have solved the problem, but you should note that saying it cannot be done is not the same thing as saying "I can't do the problem." We can do the problem, and the solution is certain.

■ EXAMPLE 3

Are the following networks traversable? Do not answer by trial and error, but by analyzing the number of odd vertices.

a.
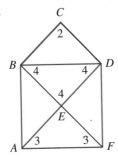

b.
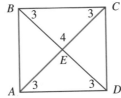

Solution

a. You may remember seeing this network in elementary school. It has four even vertices (*B*, *C*, *D*, and *E*) and two odd vertices (*A* and *F*), and it is therefore traversable. To traverse it, you must start at *A* or *F* (that is, at an odd vertex). The solution is shown.

b. This network has one even vertex and four odd vertices, so it is not traversable.

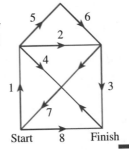

Floor-Plan Problem

This problem, which is related to the Königsberg bridge problem, involves taking a trip through all the rooms and passing through each door only once. Let's label the rooms in Figure 7.58a as A, B, C, D, E, and F. Rooms A, C, E, and F have two doors, and rooms B and D each have three doors; in Figure 7.58b, it looks as if there are five rooms, but since there are doors that lead to the "outside," we must count the outside as a room. So this figure also has six rooms labeled A, B, C, D, E, and F. Rooms A, B, and C each have 5 doors, rooms D and E each have 4 doors, and room F has 9 doors.

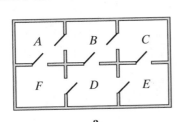

a.

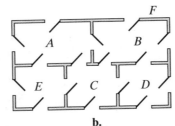

b.

▲ **Figure 7.58** Floor plan problem

Make a conjecture about the solution to this problem, which we call the **floor-plan problem**. We note the following relationships:

Network	Königsberg	Floor-Plan Problem
edge (arc) ↔	bridge ↔	door
vertex ↔	land region ↔	room

Once we have made this connection between the floor-plan and Königsberg problems, the solution to the floor-plan problems shown in Figure 7.58 is easy.

◼◼◼ EXAMPLE 4   **Polya's Method**

Solve the floor-plan problems:

a. 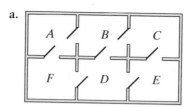   b.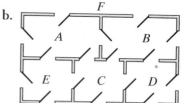

Solution   We use Polya's problem-solving guidelines for this example.

Understand the Problem.   The floor-plan problem asks the questions: "Can we travel into each room and pass through every door once?"

Devise a Plan.   We associate the floor-plan problem to the bridge problem, and classify each room as even or odd, according to the number of doors in that room. It will be possible if there are no rooms with an odd number of doors, or if there are exactly two rooms with an odd number of doors.

Carry Out the Plan.

a. There are six rooms, and rooms B and D are odd, so this floor plan can be traversed. The solution requires that we begin in either room B or room D, and finish in the other.

b. There are six rooms, and rooms A, B, C, and F are odd (with D and E even). Since there are more than two odd rooms, this floor plan cannot be traversed. If one of the doors connecting two of the odd rooms is blocked, then the floor plan could be traversed.

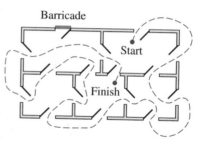

Look Back. We can check our work by actually drawing possible routes, as shown above. ▬

PROBLEM SET 7.6

▲ A Problems

1. **IN YOUR OWN WORDS** Describe the Königsberg bridge problem.
2. **IN YOUR OWN WORDS** Describe the floor-plan problem.
3. **IN YOUR OWN WORDS** Describe the solution to the Königsberg bridge problem.

Which of the networks in Problems 4–15 are traversable? If a network can be traversed, show how.

4.
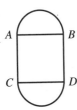

5.
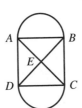

6.

7.

8.
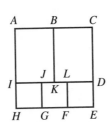

9.

10.
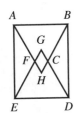

11.
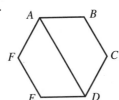

12.
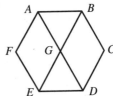

13.
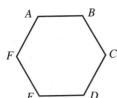

14.
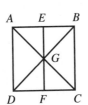

15.
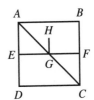

For which of the floor plans in Problems 16–21 can you pass through all the rooms while going through each door only once? If it is possible, show how it might be done.

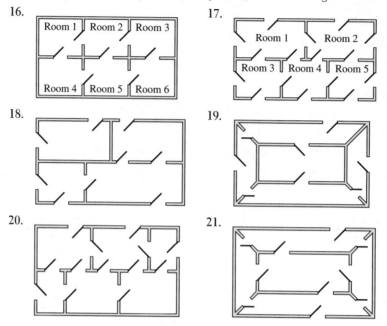

16.

17.

18.

19.

20.

21.

▲ **B Problems**

22. After Euler solved the Königsberg bridge problem, an eighth bridge was built as shown in Figure 7.59. Is this network traversable? If so, show how.

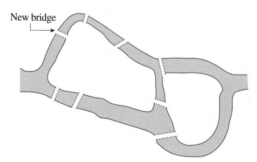

New bridge

▲ **Figure 7.59   Königsberg with eight bridges**

23. On Saturday evenings, a favorite pastime of the high school students in Santa Rosa, California, is to cruise certain streets. The selected routes are shown in Figure 7.60. Is it possible to choose a route so that all of the permitted streets are traveled exactly once?

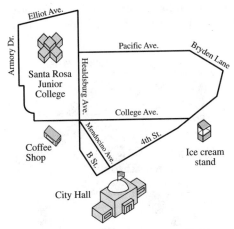

▲ **Figure 7.60   Santa Rosa street problem**

24. A simplified map of New York City, showing the subway connections between Manhattan and The Bronx, Queens, and Brooklyn, is shown in Figure 7.61. Is it possible to travel on the New York subway system and use each subway exactly once? You can visit each borough (The Bronx, Queens, Brooklyn, or Manhattan) as many times as you wish.

▲ **Figure 7.61   New York City subways**

25. A portion of London's Underground transit system is shown in Figure 7.62. Is it possible to travel the entire system and visit each station while taking each route exactly once?

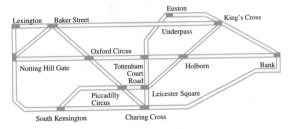

▲ **Figure 7.62   London Underground**

26. In Massachusetts there is a recreation of a New England village called Old Sturbridge Village. A map is shown in Figure 7.63. Is it possible to stroll the streets marked in yellow? Give reasons for your answer.

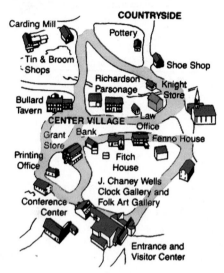

▲ **Figure 7.63**   **Old Sturbridge Village problem**

27. Reconsider the question in Problem 26 if the street in front of the church across from the Knight Store is opened.

28. The edges of a cube form a three-dimensional network. Are the edges of a cube traversable?

▲ **Problem Solving**

29. What is the sum of the measures of the angles of a tetrahedron?
    (*Hint:*   Consider the sum of the measures of the angles of a cube. A cube has six square faces, and since each face has four right angles, the sum of the measures of the angles on each face is 360°; hence, the sum of the measures of the face angles of a cube is 6(360°) = 2160°.)

30. What is the sum of the measures of the angles of a pentagonal prism? (See Figure 7.64.)

▲ **Figure 7.64**   **Pentagonal prism**

31. Count the number of vertices, edges (arcs), and regions for each of Problems 4–15. Let V = number of vertices, E = number of edges, and R = number of regions. Compare $V + R$ with E. Make a conjecture relating V, R, and E. This finding is called *Euler's formula for networks.*

32. On a planet far, far away, Luke finds himself in a strange building with hexagon-shaped rooms as shown in the figure. In his search for the princess, he always moves to an adjacent room and always in a southerly direction.

    a. How many paths are there to room 1? to room 2? to room 3? to room 4?
    b. How many paths are there to room 10?
    c. How many paths are there to room 13?

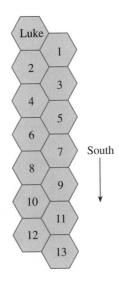

33. How many paths are there to room *n* in Problem 32?

34. Emil Torday told the story of seeing some African children playing with sand:

    > The children were drawing, and I was at once asked to perform certain impossible tasks; great was their joy when the white man failed to accomplish them.*

    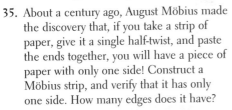

    One task was to trace the figure in the sand with one continuous sweep of the finger.

    a. What is the children's secret for drawing this pattern?
    b. Draw this figure; why is it difficult to do this without knowing something about networks?

35. About a century ago, August Möbius made the discovery that, if you take a strip of paper, give it a single half-twist, and paste the ends together, you will have a piece of paper with only one side! Construct a Möbius strip, and verify that it has only one side. How many edges does it have?

36. Construct a Möbius strip (see Problem 35). Cut the strip in half down the center. Describe the result.

37. Construct a Möbius strip (see Problem 35). Cut the strip in half down the center. Cut it in half again. Describe the result.

38. Construct a Möbius strip (see Problem 35). Cut the strip along a path that is one-third the distance from the edge. Describe the result.

39. Construct a Möbius strip (see Problem 35). Mark a point A on the strip. Draw an arc from A around the strip until you return to the point A. Do you think you could connect *any* two points on the sheet of paper without lifting your pencil?

40. Take a strip of paper 11 in. by 1 in., and give it three half-twists; join the ends together. How many edges and sides does this band have? What happens if you cut down the center of this piece?

▲ **Individual Research**

41. The Garden House of Ostia in the IN THE REAL WORLD problem at the beginning of this chapter is of interest because of the geometry used in its construction. The key to its construction, according to archeologists Donald and Carol Watts, is a "sacred cut." In searching the records of the architect Vitruvius, they found that the basic pattern begins with a square (called the reference square) and its diagonals. Next quarter-circles, centered on the corners of the square are drawn, each with a radius equal to

* Quoted by Claudia Zaslavsky in *Africa Counts* (Boston: Prindle, Weber, & Schmidt, 1973) from *On the Trail of the Bushongo* by Emil Torday.

half of the diagonal. The arcs pass through the center of the square and intersect two adjoining sides; together they cut the sides into three segments, with the central segment being larger than the other two. By connecting the intersection points, you can divide the reference square into nine parts, as described in the article. At the center of the grid is another square that can serve as the foundation for the next sacred cut. Experiment by drawing or quilting some "sacred cut" designs.

Reference "A Roman Apartment Complex," by Donald J. Watts and Carol Martin Watts. *Scientific American*, December 1986, pp. 132–139.

42. What is a Klein bottle? Examine the picture. Can you build or construct a physical model? You can use the limerick as a hint.

Klein bottle

7.7   TOPOLOGY AND FRACTALS

A brief look at the history of geometry illustrates, in a very graphical way, the historical evolution of many mathematical ideas and the nature of changes in mathematical thought. The geometry of the Greeks included very concrete notions of space and geometry. They considered space to be a locus in which objects could move freely about, and their geometry was a geometry of congruence. This geometry is known as Euclidean geometry, which we discussed in Section 7.1. In this section we investigate two very different branches of geometry that question, or alter, the way we think of space and dimension.

Topology

In the 17th century, space came to be conceptualized as a set of points, and, with the non-Euclidean geometries of the 19th century (see Section 7.8), mathematicians gave up the notion that geometry had to describe the physical universe. The existence of multiple geometries was accepted, but space was still thought of as a geometry of congruence. The emphasis shifted to sets, and geometry was studied as a mathematical system. Space could be conceived as a set of points together with an abstract set of relations in which these points are involved. The time was right for geometry to be considered as the theory of such a space, and in 1895 Jules-Henri Poincaré published a book using this notion of space and geometry in a systematic development. This book was called *Vorstudien zur Topologie (Introductory Studies in Topology)*. However, topology was not the invention of any one person, and the names of Cantor, Euler, Fréchet, Hausdorff, Möbius, and Riemann are associated

Historical Note

The set theory of Cantor (see the Historical Note on page 22) provided a basis for topology, which was presented for the first time by Jules-Henri Poincaré (1854–1912) in *Analysis Situs*. A second branch of topology was added in 1914 by Felix Hausdorff (1868–1942) in *Basic Features of Set Theory*. Earlier mathematicians, including Euler, Möbius, and Klein, had touched on some of the ideas we study in topology, but the field was given its major impetus by L. E. J. Brouwer (1882–1966). Today much research is being done in topology, which has practical applications in astronomy, chemistry, economics, and electrical circuitry.

with the origins of **topology.** Today it is a broad and fundamental branch of mathematics.

To obtain an idea about the nature of topology, consider a device called a *geoboard.* You may have used a geoboard in elementary school. Suppose we stretch one rubber band over the pegs to form a square and another to form a triangle, as shown in Figure 7.65.

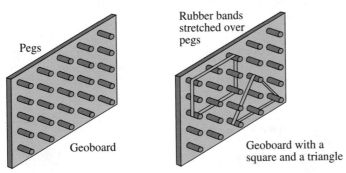

▲ **Figure 7.65   Creating geometric figures with a geoboard**

In the geometries we've been considering, these figures (the square and the triangle) would be seen as different. However, in topology, these figures are viewed as the same object. Topology is concerned with discovering and analyzing the essential similarities and differences between sets and figures. One important idea is called *elastic motion,* which includes bending, stretching, shrinking, or distorting the figure in any way that allows the points to remain distinct. It does not include cutting a figure unless we "sew up" the cut *exactly* as it was before.

Topologically Equivalent Figures

Two geometric figures are said to be **topologically equivalent** if one figure can be elastically twisted, stretched, bent, shrunk, or straightened into the same shape as the other. One can cut the figure, provided at some point the cut edges are "glued" back together again to be exactly the same as before.

The children and their distorted images are topologically equivalent.

Rubber bands can be stretched into a wide variety of shapes. All forms in Figure 7.66 are topologically equivalent. We say that a curve is **planar** if it lies flat in a plane.

▲ **Figure 7.66   Topologically equivalent curves**

All of the curves in Figure 7.66 are *planar simple closed curves*. A curve is **closed** if it divides the plane into three disjoint subsets: the set of points on the curve itself, the set of points *interior* to the curve, and the set of points *exterior* to the curve. It is said to be **simple** if it has only one interior. Sometimes a simple closed curve is called a **Jordan curve**. Notice that, to pass from a point in the interior to a point in the exterior, it is necessary to cross over the given curve an odd number of times. This property remains the same for any distortion, and is therefore called an *invariant* property.

Two-dimensional surfaces in a three-dimensional space are classified according to the number of cuts possible without slicing the object into two pieces. The number of cuts that can be made without cutting the figure into two pieces is called its **genus**. The genus of an object is the number of holes in the object. (See Figure 7.67.)

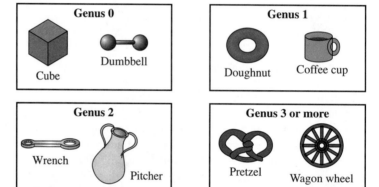

▲ **Figure 7.67   Genus of the surfaces of some everyday objects. Look at the number of holes in the object.**

For example, no closed cut can be made around a sphere without cutting it into two pieces, so its genus is 0. In three dimensions, you can generally classify the genus of an object by looking at the number of holes the object has. A doughnut, for example, has genus 1 since it has 1 hole. In mathematical terms, we say it has genus 1 since only one closed cut can be made without dividing it into two pieces. All figures with the same genus are topologically equivalent. Figure 7.68 shows that a doughnut and a coffee cup are topologically equivalent, and Figure 7.69 shows objects of genus 0, genus 1, and genus 2.

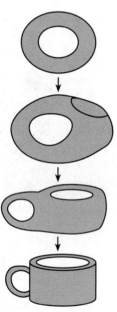

▲ **Figure 7.68   A doughnut is topologically equivalent to a coffee cup.**

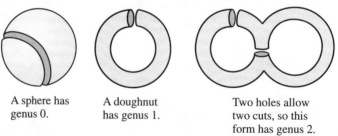

A sphere has genus 0.

A doughnut has genus 1.

Two holes allow two cuts, so this form has genus 2.

▲ **Figure 7.69   Genus of a sphere, a doughnut, and a two-holed doughnut**

Four-Color Problem

One of the earliest and most famous problems in topology is the four-color problem. It was first stated in 1850 by the English mathematician Francis Guthrie. It states that any map on a plane or a sphere can be colored with at most four colors so that any two countries that share a common boundary are colored differently. (See Figure 7.70, for example.) All attempts to prove this conjecture had failed until Kenneth Appel and Wolfgang Haken of the University of Illinois announced their proof in 1976. The university honored their discovery using the illustrated postmark.

FOUR COLORS
SUFFICE

Since the theorem was first stated, many unsuccessful attempts have been made to prove it. The first published "incorrect proof" is due to Kempe, who enumerated four cases and disposed of each. However, in 1990, an error was found in one of those cases, which it turned out was subcategorized as 1,930 different cases. Appel and Haken reduced the map to a graph as Euler did with the Königsberg bridge problem. They reduced each country to a point and used computers to check every possible arrangement of four colors for each case, requiring more than 1,200 hours of computer time to verify the proof.

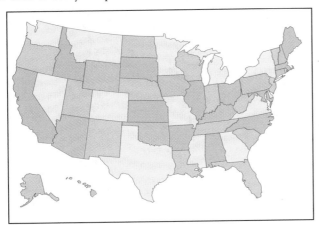

▲ **Figure 7.70   Every map can be colored with four colors.**

Fractal Geometry

One of the newest and most exciting branches of mathematics is called **fractal geometry.** Fractals have been used recently to produce realistic computer images in the movie *Star Trek II*, and the new supercrisp high-definition television (HDTV) uses fractals to squeeze the HDTV signal into existing broadcast channels. In February 1989, Iterated System, Inc., began marketing a $32,500 software package for creating models of biological systems from fractals.

Fractals were invented by Benoit M. Mandelbrot over 30 years ago, but have become important only in the last few years because of computers. Mandelbrot's

first book on fractals appeared in 1975; in it he used computer graphics to illustrate the fractals. The book inspired Richard Voss, a physicist at IBM, to create stunning landscapes, earthly and otherworldly (see Figure 7.71). "Without computer graphics, this work could have been completely disregarded," Mandelbrot acknowledges.

What exactly is a fractal? We are used to describing the dimension of an object without having a precise definition: A point has 0 dimension; a line, 1 dimension; a plane, 2 dimensions; and the world around us, 3 dimensions. We can even stretch our imagination to believe that Einstein used a four-dimensional model. However, what about a dimension of 1.5? Fractals allow us to define objects with noninteger dimension. For example, a jagged line is given a fractional dimension between 1 and 2, and the exact value is determined by the line's "jaggedness."

We will illustrate this concept by constructing the most famous fractal curve, the so-called "snowflake curve." Start with a line segment $\overline{AB}$:

$$A \text{———————} B$$

Divide this segment into thirds, by marking locations C and D:

$$A \quad\quad C \quad\quad D \quad\quad B$$

$N = 3$ segments

$r = \dfrac{1}{3}$ is the length of each segment

Now construct an equilateral triangle CED on the middle segment and then remove the middle segment $\overline{CD}$:

E

$A \quad\quad C \quad\quad D \quad\quad B$

$N = 4$ segments:

$r = \dfrac{1}{3}$ is the length of each segment

Now, repeat the above steps for each segment:

E

$A \quad\quad C \quad\quad D \quad\quad B$

$N = 16$ segments

$r = \dfrac{1}{9}$ is the length of each segment

$N = 3; r = \frac{1}{3}; D = 1$
$N = 4; r = \frac{1}{3}; D \approx 1.26$
$N = 16; r = \frac{1}{9}; D \approx 1.26$
$N = 64; r = \frac{1}{27}; D \approx 1.26$

Again, repeat the process:

E

$A \quad\quad C \quad\quad D \quad\quad B$

If you repeat this process (until you reach any desired level of complexity), you have a fractal curve with dimension between 1 and 2.

The actual description of the dimension is more difficult to understand. Mandelbrot defined the dimension as follows:

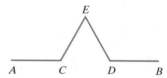

$$D = \frac{\log N}{\log \frac{1}{r}}$$

where N is any integer and *r* is the length of each segment.* For the illustrations above, the dimension is calculated in the margin. Some fractal images are shown in Figure 7.71.

▲ **Figure 7.71   Fractal images of a mountainscape (generated by Richard Voss) and fern**

News Clip

Guest Essay: What Good Are Fractals?

Okay, fractals can make sense out of chaos (see the Guest Essay on page 527), but what can you do with them? It is a question currently being asked by physicists and other scientists at many professional meetings, says Alan Norton, a former associate of Mandelbrot who is now working on computer architectures at IBM. For a young idea still being translated into the dialects of each scientific discipline, the answer, Norton says, is: Quite a bit. The fractal dimension may give scientists a way to describe a complex phenomenon with a single number.

Harold Hastings, professor of mathematics at Hofstra University on Long Island, is enthusiastic about modeling the Oke-fenokee Swamp in Georgia with fractals. From aerial photographs, he has studied vegetation patterns and found that some key tree groups, such as cypress, are patchier and show a larger fractal dimension than others.

Shaun Lovejoy, a meteorologist who works at Météorologie Nationale, the French national weather service in Paris, con-firmed that clouds follow fractal patterns. Again, by analyzing satellite photographs, he found similarities in the shape of many cloud types that formed over the Indian Ocean. From tiny puff-like clouds to an enormous mass that extended from Central Africa to Southern India, all exhibited the same fractal dimension. Prior to Mandelbrot's discovery of fractals, cloud shapes had not been candidates for mathematical analysis and meteorologists who theorize about the origin of weather ignored them. Lovejoy's work suggests that the atmosphere on a small-scale weather pattern near the Earth's surface resembles that on a large-scale pattern extending many miles away, an idea that runs counter to current theories.

The occurrence of earthquakes. The surfaces of metal fractures. The path a computer program takes when it scurries through its memory. The way our own neurons fire when we go searching through *our* memories. The wish list for a fractal description grows. Time will tell if the fractal dimension becomes invaluable to scientists interested in building mathematical models of the world's workings.

From "Geometrical Forms Known as Fractals Find Sense in Chaos," by Jeanne McDermott,
Smithsonian, December 1983, p. 116.

---

* Mandelbrot defined *r* as the ratio $\dfrac{L}{N}$, where L is the sum of the lengths of the N line segments.

Tessellations

The construction of the snowflake curve reminds us of another interesting mathematical construction, called a **tessellation**. By skillfully altering a basic polygon (such as a triangle, rectangle, or hexagon), the artist Escher was able to produce artistic tessellations such as that shown in Figure 7.72. We can describe a procedure for reproducing a simple tessellation based on the Escher print in Figure 7.72.

Step 1 Start with an equilateral triangle *ABC*. Mark off the same curve on sides *AB* and *AC* as shown in Figure 7.73. Mark off another curve on side *BC* that is symmetric about the midpoint *P*. If you choose the curves carefully, as Escher did, an interesting figure suitable for tessellating will be formed.

▲ **Figure 7.72 Escher print:**
Circle Limit I

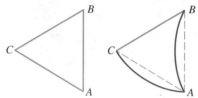 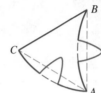 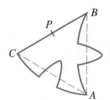 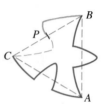 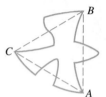

▲ **Figure 7.73 Tessellation pattern**

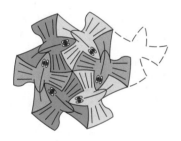

Step 2 Six of these figures accurately fit together around a point forming a hexagonal array. If you trace and cut one of these basic figures, you can continue the tessellation over as large an area as you wish.

PROBLEM SET 7.7

▲ **A Problems**

1. **IN YOUR OWN WORDS** What do we mean by topology?

2. **IN YOUR OWN WORDS** What is the four-color problem?

3. Group the figures into classes so that all the elements within each class are topologically equivalent, and no elements from different classes are topologically equivalent.

A. 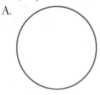   B. 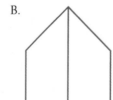   C.

D.    E.   F.

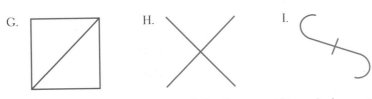

G. H. I.

4. Group the figures into classes so that all the elements within each class are topologically equivalent, and no elements from different classes are topologically equivalent.

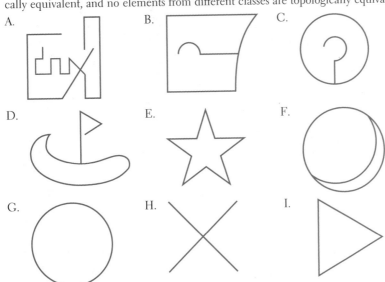

A. B. C.

D. E. F.

G. H. I.

5. Which of the figures in Problem 3 are simple closed figures?

6. Which of the figures in Problem 4 are simple closed figures?

7. Group the objects into classes so that all the elements within each class are topologically equivalent and no elements from different classes are topologically equivalent.

   A. a glass   B. a bowling ball   C. a sheet of typing paper
   D. a sphere   E. a ruler   F. a banana
   G. a sheet of two-ring-binder paper

8. Group the objects into classes so that all the elements within each class are topologically equivalent and no elements from different classes are topologically equivalent.

   A. a bolt   B. a straw   C. a horseshoe   D. a sewing needle
   E. a brick   F. a pencil   G. a funnel with a handle

In Problems 9–13, determine whether each of the points A, B, and C is inside or outside of the simple closed curve.

9. 10. 11.

TESSELLATIONS

Each of these tessellations fills a plane in a pattern without gaps and without overlapping. Which of these examples show one shape tessellating? Which show two or more shapes tessellating? Which examples appear to be three-dimensional?

12.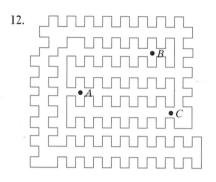

13.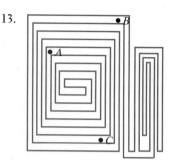

▲ **B Problems**

14. **a.** Let X be a point obviously outside the figure given in Problem 9. Draw $\overline{AX}$. How many times does it cross the curve? Repeat for $\overline{BX}$ and $\overline{CX}$.
    b. Repeat part **a** for the figure given in Problem 10.
    c. Repeat part **a** for the figure given in Problem 11.
    d. Make a conjecture based on parts **a–c**. This conjecture involves a theorem called the *Jordan curve theorem*.

15. One of the simplest map-coloring rules of topology involves a map of "countries" with straight lines as boundaries. How many colors would be necessary if all countries had only straight-line boundaries? Note that a common point is not considered a common boundary.

16. Find the length of the following snowflake curve: Begin with a line segment of length 1 unit.

a. What is the length of the snowflake curve after the first step?

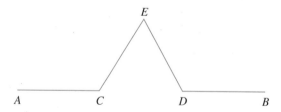

b. What is the length of the snowflake curve after the second step?

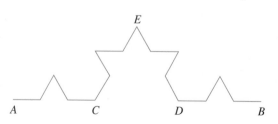

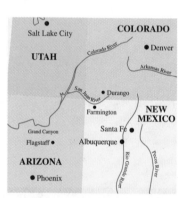

How many colors are needed for the 4-corners area? (see Problem 15)

Snowflake curve

c. What is the length of the snowflake curve after the third step?

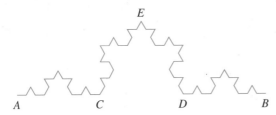

17. Group the letters of the alphabet into classes so that all the elements within each class are topologically equivalent and no elements from different classes are topologically equivalent.

A   B   C   D   E   F   G   H   I   J   K   L   M

N   O   P   Q   R   S   T   U   V   W   X   Y   Z

18. Construct a fractal curve by forming squares (rather than triangles as shown in the text). For example, the first step is:

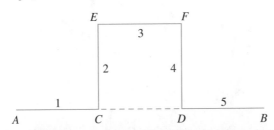

19.

This is an example of an 18th century painting showing two ladies and a servant, serenaded by a musician. The process of painting a distorted image on canvas, which becomes complete when viewed as a reflection, is called *anamorphosis*. What topological relation does the picture on canvas have with the image in the reflecting glass?

20. Construct a tessellation using rectangles.

21. Construct a tessellation using hexagons.

▲ Problem Solving

22. Answer the questions after reading the poem in the News Clip.
   a. If a right-handed mitten is turned inside out, as is suggested in the poem, will it still fit a right hand?
   b. Is a right-handed mitten topologically equivalent to a left-handed mitten?

23. Some mathematicians were reluctant to accept the proof of the four-color problem because of the necessity of computer verification. The proof was not "elegant" in the sense that it required the computer analysis of a large number of cases. Study the map in Figure 7.74 and determine for yourself whether it is the *first five-color* map, providing a counterexample for the computerized "proof."

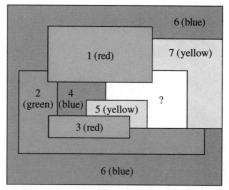

▲ **Figure 7.74   Is this the world's first five-color map?**

News Clip

He killed the noble Mudjokivis.
Of the skin he made him
   mittens,
Made them with the fur side
   inside.
Made them with the skin side
   outside.
He, to get the warm inside,
Put the skin side outside;
He, to get the cold side outside,
Put the warm fur side inside.
That's why he put the fur side
   inside.

Why he put the skin side
   outside,
Why he turned them inside
   outside.

What is the color of the white region? Consider the following table:

Region	Blue	Yellow	Green	Red
1	x	x	x	
2	x	x		x
3	x	x	x	
4		x	x	x
5	x		x	x
6		x	x	x
7	x		x	x
?	x	x	x	x

The x's indicate the colors that share a boundary with the given region. As you can see, the white region is bounded by all four colors, so therefore requires a fifth color.

Example of a tessellation

▲ **Individual Research**

24. Write a paper on the relationship between art and mathematics. You can either do it from a historical perspective, or use examples from present-day artists. An excellent reference (along with a bibliography) is shown in the reference.

    Reference Norman Slawsky, "The Artist as Mathematician," *The Mathematics Teacher,* April 1977.

25. Make drawings of geometric figures on a piece of rubber inner tube. Demonstrate to the class various ways in which these figures can be distorted.

News Clip

Guest Essay: Chaos
JACK WADHAMS, GOLDEN WEST COLLEGE

An exciting new topic in mathematics and an attempt to bring order to the universe has been labeled **chaos theory,** which provides refreshing insight into how very simple beginnings can yield structures of incredible complexity and of enchanting beauty. Let's trace a chaos path to simple beginnings.

Water flowing through a pipe offers one of the simplest physical models of chaos. Pressure is applied to the end of the pipe and the water flows in straight lines. More pressure increases the speed of the laminar flow until the pressure reaches a critical value, and a radically new situation evolves — turbulence. A simple laminar flow suddenly changes to a flow of beautiful complexity consisting of swirls within swirls. Before turbulence, the path of any particle was quite predictable. After a minute change in the pressure, turbulence occurred and predictability was lost. **Chaos** is concerned with systems in which minute changes suddenly transform predictability into unpredictability.

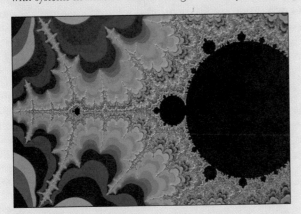 

The simple quadratic equation $z_{n+1} = z_n^2 - c$ (where z is a complex number, a number you considered in algebra, but is not in the domain of real numbers we are assuming in this book) offers a fascinating mathematical example of chaos. With this equation and a computer, a graph of unimaginable complexity and surprising beauty can be constructed. To create a graph we plot a grid of points (often more than a million) from a 4×4 square region R centered at the origin of the complex plane. We let $z_0 = 0$ and $c = a + bi$ be a point in the region R. Substitute z_1 and c into the right-hand side of the equation to obtain a second number z_2. After many such iterations, we obtain the sequence of numbers $z_0, z_1, \ldots, z_i \ldots$, which approaches a fixed value or it does not. If it does approach a fixed value, we paint a black dot on the computer screen at $a + bi$; otherwise, we paint a white dot. This process of selecting a point from region R, done millions of times, produces regions of black and white dots. If the boundary between the convergent and divergent regions is connected, a bumpy curve called the Mandelbrot set is formed. The graph is painted black

(continued)

on the interior of the Mandelbrot set and white on the exterior. We find that points selected far from the Mandelbrot set (the boundary) behave in a very predictable fashion (like the laminar flow in the pipe). However, when we choose points near the Mandelbrot set (the boundary), the behavior is quite unpredictable. We have encountered chaos. Because of the complexity of the Mandelbrot set, it is difficult to determine whether a given point is outside or inside the set.

This path to chaos has led to the Mandelbrot set, the most complicated set in mathematics, and it turns out to be fractal in nature, yet it is generated from a simple quadratic equation. That is, from a very simple quadratic equation, a set of surprising complexity and beauty emerges. Regions of chaotic behavior, under magnification, reveal a self-similar fractal. Our mathematical analogy awakens the possibility that the universe may also originate from a few simple relations that evolve into profoundly complex structures, like our brain, capable of understanding its own origins and still asking about its origins. This chaos path to simple beginnings seems to be fractal.

7.8 PROJECTIVE AND NON-EUCLIDEAN GEOMETRIES

Projective Geometry

As Europe passed out of the Middle Ages and into the Renaissance, artists were at the forefront of the intellectual revolution. No longer satisfied with flat-looking scenes, they wanted to portray people and objects as they looked in real life. The artists' problem was one of dimension, and dimension is related to mathematics, so many of the Renaissance artists had to solve some original mathematics problems. How could a flat surface be made to look three-dimensional?

One of the first (but rather unsuccessful) attempts at portraying depth in a painting is shown in Figure 7.75, Duccio's *Last Supper*. Notice that the figures are in a boxed-in room. This technique is characteristic of the period, and was an attempt to make perspective easier to define.

▲ Figure 7.75 Duccio's *Last Supper* illustrates perspective that is incorrect.

Artists finally solved the problem of perspective by considering the surface of the picture to be a window through which the object was painted. This technique was pioneered by Paolo Uccello (1397–1475), Piero della Francesca (1416–1492),

Leonardo da Vinci (1452–1519), and Albrecht Dürer (1471–1528). As the lines of vision from the object converge at the eye, the picture captures a cross-section of them, as shown in Figure 7.76.

▲ **Figure 7.76** Albrecht Dürer's *Designer of the Lying Woman* shows how the problem of perspective can be overcome. The point in front of the artist's eye fixes the point of viewing the painting. The grid on the window corresponds to the grid on the artist's canvas.

Non-Euclidean Geometry

Euclid's so-called fifth postulate (see Section 7.1) caused problems from the time it was stated. Several different formulations of this postulate are given in the margin. It somehow doesn't seem like the other postulates but rather, like a theorem that should be proved. In fact, this postulate even bothered Euclid himself, since he didn't use it until he had proved his 29th theorem. Many mathematicians tried to find a proof for this postulate.

One of the first serious attempts to prove Euclid's fifth postulate was made by Girolamo Saccheri (1667–1733), an Italian Jesuit. Saccheri's plan was simple. He constructed a quadrilateral, later known as a **Saccheri quadrilateral,** with base angles A and B right angles, and with sides $\overline{AC}$ and $\overline{BD}$ the same length, as shown in Figure 7.77.

▲ **Figure 7.77 A Saccheri quadrilateral**

As you may know from high school geometry, the summit angles C and D are also right angles. However, this result uses Euclid's fifth postulate. Now it is also true that *if* the summit angles are right angles, *then* Euclid's fifth postulate holds. The problem, then, was to establish the fact that angles C and D are right angles. Here is the plan:

1. Assume that the angles are obtuse and deduce a contradiction.
2. Assume that the angles are acute and deduce a contradiction.
3. Therefore, by the first two steps, the angles must be right angles.
4. From step 3, Euclid's fifth postulate can be deduced.

Historical Note

There are many ways of stating Euclid's fifth postulate, and several historical ones are repeated here.

1. **Poseidonius** (ca. 135–51 B.C.): Two parallel lines are equidistant.

2. **Proclus** (410–485 A.D.): If a line intersects one of two parallel lines, then it also intersects the other.

3. **Legendre** (1752–1833): A line through a point in the interior of an angle other than a straight angle intersects at least one of the arms of the angle.

4. **Bolyai** (1802–1860): There is a circle through every set of three noncollinear points.

5. **Playfair** (1748–1819): Given a straight line and any point not on this line, there is one and only one line through that point that is parallel to the given line.

Playfair's statement is the one most often used in high school geometry textbooks. All of these statements are *equivalent,* and, if you accept any one of them, you must accept them all.

▲

GEORG BERNHARD RIEMANN
(1826–1866)

Riemann was director of the Göttingen Observatory, serving from 1859 to 1866, when he died of tuberculosis at the age of 40. During his short career, he made significant contributions to mathematics. Another mathematician, Janos Bolyai (1802–1860), was working on an alternative geometry about the same time as Lobachevski and Riemann. Clearly, the world was nearly ready to accept the idea that a geometry need not be based on our physical experience dictated by Euclid's fifth postulate. Because of his work, hyperbolic geometry is sometimes referred to as **Bolyai–Lobachevski** geometry.

It turned out not to be as easy as he thought, because he was not able to establish a contradiction. He gave up the search because, he said, it "led to results that were repugnant to the nature of a straight line."

Saccheri never realized the significance of what he had started, and his work was forgotten until 1889. However, in the meantime, Johann Lambert (1728–1777) and Adrien-Marie Legendre (1752–1833) similarly investigated the possibility of eliminating Euclid's fifth postulate by proving it from the other postulates.

By the early years of the 19th century, three accomplished mathematicians began to suspect that the parallel postulate was independent and could not be eliminated by deducing it from the others. The great mathematician Karl Gauss, whom we've mentioned before, was the first to reach this conclusion, but since he didn't publish this finding of his, the credit goes to two others. In 1811, an 18-year-old Russian named Nikolai Lobachevski pondered the possibility of a "non-Euclidean" geometry — that is, a geometry that did not assume Euclid's fifth postulate. In 1829, he published his ideas in an article entitled "Geometrical Researches on the Theory of Parallels." The postulate he used was subsequently named after him.

The Lobachevskian Postulate

The summit angles of a Saccheri quadrilateral are acute.

This axiom, in place of Euclid's fifth postulate, leads to a geometry that we call **hyperbolic geometry.** If we use the plane as a model for Euclidean geometry, what model could serve for hyperbolic geometry? A rough model for this geometry can be seen by placing two trumpet bells together as shown in Figure 7.78a. It is called a **pseudosphere** and is generated by a curve called a *tractrix* (as shown in Figure 7.78b). The tractrix is rotated about the line $\overline{AB}$. The pseudosphere has the property that, through a point not on a line, there are many lines parallel to a given line.

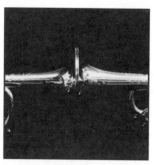

Two trumpets placed end to end serve as a physical model of a pseudosphere.

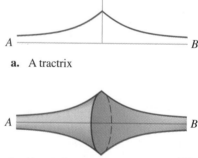

a. A tractrix

b. A tractrix rotated about the line $\overline{AB}$

▲ **Figure 7.78   A tractrix can be used to generate a pseudosphere.**

Georg Riemann (1826–1866), who also worked in this area, pointed out that, although a straight line may be extended indefinitely, it need not have infinite length. It could instead be similar to the arc of a circle, which eventually begins to retrace itself. Such a line is called *re-entrant*. An example of a re-entrant line is found by considering a great circle on a sphere. A **great circle** is a circle on a sphere with a diameter equal to the diameter of the sphere. With this model, a Saccheri quadrilateral is constructed on a sphere with the summit angles obtuse. The resulting geometry is called **elliptic geometry.** The shortest path between any two points on a sphere is an arc of the great circle through those points; these arcs correspond to line segments in Euclidean geometry. In 1854, Riemann showed that, with some other slight adjustments in the remaining postulates, another consistent non-Euclidean geometry can be developed. Notice that the fifth, or parallel, postulate fails to hold because any two great circles on a sphere must intersect at two points (see Figure 7.79).

We have not, by any means, discussed all possible geometries. We have merely shown that the Euclidean geometry that is taught in high school is not the only possible model. A comparison of some of the properties of these geometries is shown in Table 7.3.

▲ **Figure 7.79   A sphere showing the intersection of two great circles**

TABLE 7.3   Comparison of Major Two-Dimensional Geometries

Euclidean geometry	Hyperbolic geometry	Elliptic geometry
Euclid (about 300 B.C.)	Gauss, Bolyai, Lobachevski (ca. 1830)	Riemann (ca. 1850)
Given a point not on a line, there is one and only one line through the point parallel to the given line.	Given a point not on a line, there are an infinite number of lines through the point that do not intersect the given line.	There are no parallels.
Geometry is on a plane:	Geometry is on a pseudosphere:	Geometry is on a sphere:
The sum of the angles of a triangle is 180°.	The sum of the angles of a triangle is less than 180°.	The sum of the angles of a triangle is more than 180°.
$m \angle D = 90°$	$m \angle D < 90°$	$m \angle D > 90°$
Lines are infinitely long.	Lines are infinitely long.	Lines are finite in length.

PROBLEM SET 7.8

▲ **A Problems**

1. **IN YOUR OWN WORDS** What is a non-Euclidean geometry?

2. **IN YOUR OWN WORDS** Compare the major two-dimensional geometries.

3. **IN YOUR OWN WORDS** Why do you think Euclidean geometry remains so prevalent today even though we know there are other valid geometries?

Which of the figures in Problems 4–7 are Saccheri quadrilaterals?

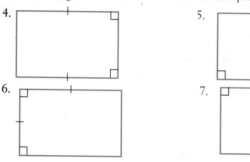

4.

5.

6.

7.

▲ **B Problems**

8. Consider the spheres shown in Figure 7.80.

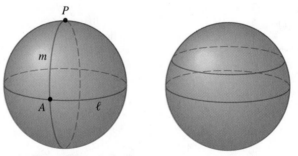

▲ **Figure 7.80   A "line" on a sphere is a great circle.**

a. Why do the lines on a sphere appear parallel?

b. In Table 7.3, we said that there are no parallels on a sphere. What's wrong with our reasoning in part **a**?

9. On a globe, locate San Francisco, Miami, and Detroit. Connect these cities with the shortest paths to form a triangle. Next, use a protractor to measure the angles. What is their sum?

10. On a globe, locate Tokyo, Seattle, and Honolulu. Connect these cities with the shortest paths to form a triangle. Next, use a protractor to measure the angles. What is their sum?

11. Walking north or south on the earth is defined as walking along the meridians, and walking east or west is defined as walking along parallels. There are points on the sur-

face of the earth from which it is possible to walk a mile south, then a mile east, then a mile north, and be right back where you started. Find one such point.

▲ Problem Solving

12. Find two additional points satisfying the conditions stated in Problem 11.

13. Escher's work *Circle Limit III* is a tessellation (see Section 7.7) based on Lobachevskian geometry. Consider the following model. Draw a circle.

Escher: *Circle Limit III*

Now draw a "perpendicular circle," which is a circle whose tangents to the original circle at the points of intersection are perpendicular. The two circles are called **orthogonal circles**.

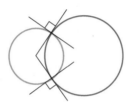

Next, set up another model of Lobachevskian geometry. Points of the plane are points inside the circle. Lines are both diameters of the circle and arcs of orthogonal circles as well as being inside the original circle. On the circle you have drawn, draw several lines for this model.

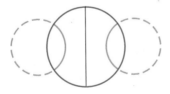

Circle Limit III, the Escher work shown here, is an example that uses this model. The white arcs through the backbones of the fish are lines in this model.

▲ Individual Research

14. The German artist Albrecht Dürer (1471–1528) is not only a Renaissance artist, but also somewhat of a mathematician. Do some research on the mathematics of Dürer.

References
Charles Lenz, "Modeling the Matrix of Albrecht Dürer," *Modeling and Simulation*, 1979, pp. 2149–2161.
Francis Russell, *The World of Dürer*, 1471–1528 (New York: *Time*, 1967).
Karen Walton, "Albrecht Dürer's Renaissance Connections between Mathematics and Art," *The Mathematics Teacher*, April 1994, pp. 278–282.

15. Write a paper on perspective. How are three-dimensional objects represented in two dimensions?

    References

    Morris Kline, *Mathematics, a Cultural Approach* (Reading, MA: Addison-Wesley, 1962), Chapters 10–11.

    C. Stanley Ogilvy, *Excursions in Geometry* (New York: Oxford University Press, 1969), Chapter 7.

16. The discovery and acceptance of non-Euclidean geometries had an impact on all of our thinking about the nature of scientific truth. Can we ever know truth in general? Write a paper on the nature of scientific laws, the nature of an axiomatic system, and the implications of non-Euclidean geometries.

    Reference Philip J. Davis and Reuben Hersh, *The Mathematical Experience* (Boston: Houghton-Mifflin, 1981).

7.9   INTRODUCTION TO COMPUTER LOGO

A computer language called LOGO is very useful in teaching concepts of geometry, especially to children. Every computer system has some type of log-in procedure that will vary from facility to facility. You will need to check with your instructor to find out the log-in procedure for the computer or terminal you will be using. If you have your own computer, you will need a copy of LOGO to carry out the procedures of this section.

PRINT Command

Suppose you wish to have the computer type "I love you." To do this, you will use a command called PRINT:

```
PRINT [I love you.]
```
Command — Whatever is enclosed in brackets will be printed.

Notice that the prompt symbol in LOGO is "?" with a blinking box, called the **cursor.** You can write a program by giving it a name preceded by the word TO.

```
TO EXAMPLE

>PRINT [I am Robbie the robot.]
>PRINT [I love you.]
>PRINT [Goodbye]
```

To tell the computer the program is complete, type END. The computer will respond with the message:

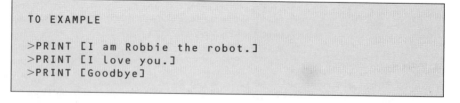

```
EXAMPLE DEFINED
```
This is the name you assigned to the program.

After printing this message, the program will return to the prompt symbol "?". To run the program, you now type the name of the program, EXAMPLE, followed by the ENTER key.

CLEARTEXT and CLEARSCREEN Commands

Two useful commands are CLEARTEXT (abbreviated CT) and CLEARSCREEN (abbreviated CS). CLEARTEXT clears the text from the screen, and CLEAR-SCREEN clears the entire screen (both text and graphics).

REPEAT Command

One very useful command in LOGO enables you to repeat a process a large number of times. This REPEAT command is used as follows:

Number of times to be repeated
↓
REPEAT 5 [EXAMPLE]
Command   Name of program to be repeated.

To interrupt a command, you can type CTRL-G at the same time.

TURTLE Commands

Although you can use LOGO for computations and formula evaluations, the real benefit of LOGO is in working with geometry (graphics). These graphics are done with the aid of what is called a **turtle.** To see the turtle, give the command SHOW-TURTLE (abbreviated ST). Notice that the turtle appears in the middle of the screen as a triangle, and the prompt symbol "?" and the cursor "■" are now at the bottom of the screen. This position (of the turtle in the middle of the screen) is called HOME for the turtle; you can cause the turtle to return home at any time by giving the command HOME. The turtle can be moved around the screen by giving it a distance and a direction.

Forward 50
This is the number of steps you want the turtle to move.

A step is used to denote distance and is approximately the thickness of one line. The possible directions are

FORWARD   abbreviated FD
BACK   abbreviated BK

Directions are specified in terms of degrees:

Right 90
This is the number of degrees you want the turtle to turn.

The possible turns are

RIGHT	abbreviated RT
LEFT	abbreviated LT

Practice moving the turtle around the screen, then CLEARSCREEN and try the following examples.

■■■■ EXAMPLE I

Draw a square.

Solution   The answers may vary. One possibility is shown.

```
FD  30
RT  90
FD  30
RT  90
FD  30
RT  90
FD  30
RT  90
```

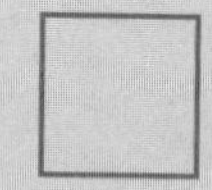

EDIT Command

Suppose you want to write a program to reproduce the square we drew in Example 1 at some later time. You could write a program in the same way we did when we wrote the program called EXAMPLE. However, it is easier to write programs using the EDIT command. Type the word **EDIT**, followed by a space, a quotation mark, and then the name you wish to give the program. Notice that the quotation mark only precedes the name of the program.

EDIT ''SQUARE   Note: Don't forget the quotation mark.
                  This is a very common error. If you do
                  forget, the program will respond:
                  I DON'T KNOW HOW TO SQUARE.

The program's response is

TO SQUARE ■   This is called the title line of the program.
              The cursor will be at the end of the line.
              Press ENTER and type the program you
              wish to call SQUARE. Assume the
              program called SQUARE is the program
              we wrote in Example I.

You can make corrections to a program with the EDIT command by moving the cursor around in the program as follows:

CTRL–F   F for *forward*

CTRL–B   B for *backward*

CTRL–C   C for *complete.* Use this when you have
completed the edit. If you do not type END,
LOGO will insert it for you when you type CTRL-C.

With this introduction, you can build an impressive library of designs.

LOGO Designs

▬▬▬ EXAMPLE 2

Write a program for the given diagram.

Solution

This is the name of this program.

TO SQUARE 4
REPEAT 4 [RT 90 SQUARE]

This means: Turn right, then run program called SQUARE.

Repeat 4 times

END

▬▬▬ EXAMPLE 3

What will be printed by LOGO with the following directions?

```
TO SQUARE 8
REPEAT 8 [RT 45 SQUARE]
END
```

Solution   The line TO SQUARE 8 is the name of the program. The square from Example 1 is repeated 8 times; each time it is drawn it is rotated 45°. The design is shown.

The next thing we might want to do is draw some squares (or other figures) of different size. You could, of course, rewrite the program with different dimensions every time you wanted a different square. This is not very efficient, however; instead we write a routine that accepts inputs such as

This denotes the size of the square.

SQUARE 50

Compare the following program with the program called SQUARE. We call this new program SQUARER, which allows us to input the length of the square, as well as to define a square that turns to the right. The word LENGTH functions as a variable, and the colon tells LOGO that SQUARER requires an input value.

```
TO SQUARER: LENGTH
FD :LENGTH
RT 90
FD :LENGTH
RT 90
FD :LENGTH
RT 90
FD :LENGTH
RT 90
END
```

■■■ EXAMPLE 4

What is the output for the following programs?

a.
```
TO SQUARES

SQUARER 10
SQUARER 20
SQUARER 30
SQUARER 40
SQUARER 50
END
```

b.
```
TO DIAMONDS

RT 45
REPEAT 4 [SQUARES RT 90]
END
```

Solution a.

b.

Once a program is written, you can use one of these commands:

SAVE ''SQUARES   Saves the program whose name follows the quotation marks.

LOAD ''SQUARES	Retrieves the program whose name follows the quotation marks.
ERASE ''SQUARES	Erases the program whose name follows the quotation marks.
ERASEFILE ''SQUARES	Erases the program saved under the name following the quotation marks.

To look at what is stored, you can type CATALOG.

Miscellaneous Commands

The following miscellaneous commands, along with a brief explanation, may be useful to you while you are programming in LOGO.

FULLSCREEN	Gives the entire screen for the turtle.
TEXTSCREEN	Gives the entire screen for the text.
SPLITSCREEN	Gives both the turtle field and four lines of text field.
HIDETURTLE (HT)	Makes the turtle invisible.
SHOWTURTLE (ST)	Makes the turtle visible.
PENUP (PU)	Lets you move the turtle without drawing any line.
PENDOWN (PD)	Puts the pen back to normal position.
PENERASE	The turtle becomes an eraser instead of a drawing instrument.
WAIT 60	LOGO will wait one second (60) before running the next command. WAIT 30 is one-half second.

CIRCLE Commands

LOGO has two commands that allow you to draw circles. One turns the circle to the left, and the other to the right.

CIRCLEL 20	"L" for left; the radius is 20.
CIRCLER 20	"R" for right; the radius is 20.

■■■ EXAMPLE 5

Draw a bull's-eye.

Solution A first attempt might be

```
TO CIRCLES
CIRCLER 10
CIRCLER 20
CIRCLER 30
CIRCLER 40
CIRCLER 50
END
```

This is not correct since you can see it gives a lopsided set of circles. Instead, first draw a circle with the starting point in the center by using the PENUP command for the radius of the circle.

TO CIRCLE :SIZE	The variable SIZE denotes the size of the circle.
PU	Raises pen
FD :SIZE	Moves to the circle itself.
RT 90	This positions the turtle.
PD	This command puts the pen back down.
CIRCLER: SIZE	This draws the circle (to the right).
PU	
LT 90	This aims the turtle toward the center.
BK: SIZE	This command moves the turtle back to the center.
PD	

Now, for the concentric circles, define a program that we'll call BULLSEYE.

```
TO BULLSEYE

CIRCLE 10
CIRCLE 20
CIRCLE 30
CIRCLE 40
CIRCLE 50
END
```

ARC Command

Many designs require only part of a circle, which is called an **arc.** There are two arc commands.

Arc right   Degrees of the arc to be drawn

ARCR 30 90

Radius of the circle

The other arc command is ARCL, which is similar, except that it draws the arc to the left instead of the right.

■■■■ EXAMPLE 6

Write a program to draw a flower.

Solution

```
TO PETAL :SIZE
ARCR :SIZE 90 RT 90
ARCR :SIZE 90 RT 90
END
```

This draws one petal. For several petals, try different possibilities, such as:

```
TO PETALS
REPEAT 8 [PETAL 30 RT 45]
END
```

PROBLEM SET 7.9

▲ A Problems

1. **IN YOUR OWN WORDS** What is LOGO?

2. **IN YOUR OWN WORDS** Tell what each of the following LOGO commands means.
   a. CTRL-F   b. CTRL-C   c. CTRL-B

3. **IN YOUR OWN WORDS** How can you see the steps of a LOGO program that is not printed on the screen?

In Problems 4–15, assume that the given program is in a computer. Without using a computer, show what the turtle would draw. Notice that each program is named by problem number so that the routine in one problem can be used in another and you will be able to find it easily.

4.
```
TO PROBLEM4
FD 60
LT 90
FD 40
LT 90
FD 60
LT 90
FD 40
LT 90
END
```

5.
```
TO PROBLEM5
FD 50
LT 120
FD 50
LT 120
FD 50
LT 120
END
```

6.
```
TO PROBLEM6
FD 40
PROBLEM4
END
```

7.
```
TO PROBLEM7
FD 40
PROBLEM4
FD 10
PROBLEM5
END
```

8.
```
TO PROBLEM8
REPEAT 2 [PROBLEM6 FD 60]
END
```

9.
```
TO PROBLEM9
REPEAT 2 [PROBLEM7 FD 60]
END
```

10.
```
TO PROBLEM10
REPEAT 4 [PROBLEM6 RT 90]
END
```

11.
```
TO PROBLEM11
REPEAT 4 [PROBLEM7 RT 360]
END
```

12.
```
TO PROBLEM12
PROBLEM4
RT 90
FD 80
LT 90
PROBLEM4
LT 90
FD 40
LT 90
FD 40
PROBLEM4
RT 90
FD 80
LT 90
PROBLEM4
LT 90
FD 40
LT 90
FD 40
END
```

13.
```
TO PROBLEM13 :SIDE
FD :SIDE
RT 72
FD :SIDE
RT 72
FD :SIDE
RT 72
FD :SIDE
RT 72
FD :SIDE
RT 72
END
```

14.
```
TO PROBLEM14
FD 50
BK 100
FD 50
PU
RT 90
FD 10
LT 90
PD
END
```

15.
```
TO PROBLEM15
PU
LT 90
FD 50
RT 90
PD
REPEAT 10 [PROBLEM14]
END
```

▲ **B Problems**

16. Write a LOGO program to draw an equilateral triangle with each side measuring 100 units.

17. Write a LOGO program to draw a square with each side measuring 100 units.

Write a LOGO program to draw figures similar to those shown in Problems 18–27.

18.

19.

20.

21.

22.

23.

24.

25.

26.

27.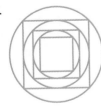

28. Write a LOGO program to draw a flower.

29. Write a LOGO program to draw a face.

30. Write a LOGO program to draw some recognizable animal.

31. Consider the following procedure, which Frederick S. Klotz from St. Patrick's College calls BUMPFD:

```
TO BUMPFD :D
FD :D/3 RT 90
FD :D/6 LT 90
FD :D/3 LT 90
FD :D/6 RT 90
FD :D/3
END
```

This LOGO program draws a line with a bump in the middle. After BUMPFD 100 the turtle ends up in exactly the same heading as FD 100 would have produced. Thus, BUMPFD behaves exactly like FD except in the path it draws.

In mathematics, the so-called "snowflake" curve was known before fractals, but is closely related to this new area of mathematics called fractal geometry. Consider the following LOGO programs along with some sample runs:

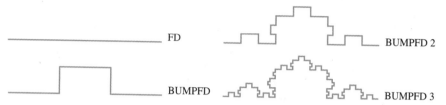

Experiment with these programs and the snowflake curve to design some patterns. Write a LOGO program that will draw the entire BUMPFD family for an input value of K where K is an integer.

32. You can draw coordinate axes and plot points using the following LOGO programs written by Will Watkins from Pan American University:

```
TO AXIS2
PU HOME PD
RT 90 FD 500 HOME
FD 500 HOME
END

TO PLOT2 :VECTOR
PU HOME
RT 90 FD FIRST :VECTOR
LT 90 FD LAST :VECTOR
PD FD 0.2
END
```

The first program draws a two-dimensional coordinate axis, and the second plots points. For example,

PLOT2 XC2 7 YC2 7*7

will plot a point that is 7 turtle steps to the right of the origin and 49 turtle steps above the x-axis. What will you obtain if you run the following LOGO programs?

```
TO FUNC :X
OUTPUT :X* :X/10
END

TO GRAPHFUNC :X :FINX
IF :X> :FINX STOP
PLOT2 EX2 :X YC2 FUNC :X
GRAPHFUNC :X+1 :FINX
END

DRAW AXIS2 GRAPHFUNC 30 35
```

Describe this program and show its output.

33. Here is a LOGO program that will generate golden rectangles.

```
TO GOLDEN.RECTANGLES :SIZE
IF :SIZE<1 STOP
SQUARE :SIZE
FD :SIZE RT 90 FD :SIZE
GOLDEN.RECTANGLES :SIZE*.61803
HIDETURTLE
END
```

An alternate procedure, called GOLDEN, is given below:

```
TO GOLDEN :SIZE
IF :SIZE<1 STOP
REPEAT 2 [FD :SIZE RT 90 FD :SIZE*1.61803 RT 90]
FD :SIZE RT 90 FD :SIZE
GOLDEN.RECTANGLES :SIZE*0.61803
HIDETURTLE
END
```

The first line names the procedure, the second line tells the procedure when to stop, and the third line draws the initial golden rectangle.

a. What does the fourth line do?

b. What does the fifth line do?

c. Draw the picture that would result if the fourth line were removed from the procedure.

▲ **Individual Research**

34. Find the total distance the turtle moves to the right of the line FD when drawing BUMPFD :D described in Problem 31.

    Reference Frederick S. Klotz, "Turtle Graphics and Mathematical Induction," *The Mathematics Teacher*, November 1987, pp. 636–639.

35. Write a paper on fractals using LOGO as a means of your investigation.

    Reference Jane F. Kern and Cherry C. Mauk, "Exploring Fractals — A Problem-solving Adventure Using Mathematics and LOGO," *The Mathematics Teacher*, March 1990, pp. 179–185.

"Today's society expects schools to ensure that all students have an opportunity to become mathematically literate . . . "

NCTM Standards

FEBRUARY

1

2 Jacques Binet (1786), matrix theory

3

4

5

6 Nicolaus Bernoulli (1695), probability

7 G. H. Hardy (1877), series, integration

8 D. Bernoulli (1700), probability, astronomy, physics

9 H. S. M. Coxeter

10

11 Josiah Gibbs (1839), chemistry, statistics

12

13 Gustav Lejeune Dirichlet (1805), complex numbers

14

15 Galileo Galilei (1564), classical physics, astronomy, p. 101
   Alfred North Whitehead (1861), logic, p. 80

16 Georges Rheticus (1514), trigonometry

17

18

19 Nicolaus Copernicus (1473), modern astronomy

20

21 Gerard Desargues (1591), projective geometry

22 Lambert Quetelet (1796), normal curve

23

24

25

26 Dominique Arago (1786), physics

27 Irving Fisher (1867), economics

28

29

Biographical Sketch

"Doing mathematics involves the same kind of excitement as exploring a wilderness."

H. S. M. Coxeter

Harold Scott Macdonald Coxeter was born in London, England, on February 9, 1907, and is a world-renowned geometer who is now a professor emeritus from the University of Toronto.

Professor Coxeter describes what he enjoys most about mathematics: "It enables one, for a while, to leave this real world, with its strife and sorrow, and to enter the ideal world of precisely but subtly related numbers, beautiful shapes, and logical deductions."

"Doing mathematics involves the same kind of excitement as exploring a wilderness. One often becomes entangled in the underbrush or finds the slope too steep, and so one needs to return to the beginning and rest till the next day. But when one reaches the top of a ridge and gazes over a wide expanse of glorious country, all the struggle has been worth while and one indulges in a special kind of euphoria."

Coxeter's interest in mathematics began early at about the age of 14. It grew out of his earlier interest in music, and the desire to compose music gradually transformed to the desire to compose mathematics.

Book Reports

Write a 500-word report on one of the following books:

- ▲ *Flatland, a Romance of Many Dimensions, by A Square,* Edwin A. Abbott (New York: Dover Publications, original edition printed in 1880).
- ▲ *Sphereland, a Fantasy About Curved Spaces and an Expanding Universe,* Dionys Burger (New York: Thomas Crowell Company, 1965).
- ▲ *Mathematical Recreations and Essays,* W. W. R. Ball and H. S. M. Coxeter (New York: Macmillan, 1962).

Important Terms

Acute angle [7.2]
Adjacent angles [7.2]
Adjacent side [7.4]
Alternate exterior angles [7.2]
Alternate interior angles [7.2]
Angle [7.2]
Arc [7.6]
Axiom [7.1]
Closed curve [7.7]
Closed network [7.6]
Compass [7.1]
Complementary angles [7.2]
Congruent [7.1]
Congruent angles [7.2]
Congruent triangles [7.3]
Construction [7.1]
Corresponding angles
    [7.2; 7.4]
Corresponding parts [7.3]
Corresponding sides [7.4]
Cosine [7.4]
Degree [7.2]
Edge [7.6]

Elliptic geometry [7.8]
Equilateral triangle [7.3]
Euclidean geometry [7.1]
Exterior angle [7.3]
Fractal geometry [7.7]
Genus [7.7]
Golden ratio [7.5]
Golden rectangle [7.5]
Great circle [7.8]
Half-line [7.2]
Horizontal lines [7.2]
Hyperbolic geometry [7.8]
Hypotenuse [7.4]
Isosceles triangle [7.3]
Jordan curve [7.7]
Line [7.1]
Line segment [7.1]
Line of symmetry [7.1]
LOGO [7.9]
Network [7.6]
Non-Euclidean geometry
    [7.8]
Obtuse angle [7.2]

Odd vertex [7.6]
Opposite side [7.4]
Parallel lines [7.1; 7.2]
Perpendicular lines [7.2]
Planar curve [7.7]
Plane [7.1]
Point [7.1]
Polygon [7.2]
Postulate [7.1]
Projective geometry [7.8]
Protractor [7.2]
Pseudosphere [7.8]
Ray [7.2]
Reflection [7.1]
Region [7.6]
Right angle [7.2]
Right triangle [7.3]
Saccheri quadrilateral [7.8]
Scalene triangle [7.3]
Similar figures [7.4]
Similar triangles [7.4]
Similarity [7.1]

Simple curve [7.7]
Sine [7.4]
Straight angle [7.2]
Straightedge [7.1]
Supplementary angles [7.2]
Surface [7.1]
Tangent [7.4]
Tessellation [7.7]
Theorem [7.1]
Topology [7.7]
Transformation [7.1]
Transformational geometry
    [7.1]
Transversal [7.2]
Traversable network [7.6]
Triangles [7.3]
Trigonometric ratios [7.4]
Undefined terms [7.1]
Vertex (pl. vertices) [7.2; 7.6]
Vertical angles [7.2]
Vertical lines [7.2]

Important Ideas

Euclidean Postulates [7.1]
Sum of the measures of angles in a triangle [7.3]
Exterior angles of a triangle [7.3]
Similar triangle theorem [7.4]
Pythagorean theorem [7.4]
What are a golden rectangle and the golden ratio? [7.5]

What is the Königsberg bridge problem? [7.6]
Euler's formula for networks [7.6]
Topologically equivalent figures [7.7]
What is the four-color problem? [7.7]
Lobachevskian postulate [7.8]
Comparison of major two-dimensional geometries [7.8]

Types of Problems

What are some of the categories of geometry? [7.1]
Construct line segments. [7.1]
Construct parallel lines. [7.1]
Find a line of symmetry for a given piece of art. [7.1]
Decide whether a given picture is symmetric. [7.1]
Visualize objects in three dimensions. [7.1]
Classify polygons with three to twelve sides. [7.2]
Construct an angle congruent to a given angle. [7.2]
Classify angles. [7.2]
Identify vertical, horizontal, intersecting, and parallel lines.
    [7.2]
Name the corresponding parts of congruent triangles. [7.3]
Find the measure of the third angle of a triangle. [7.3]

Find the measure of the exterior angles of a triangle. [7.3]
Construct a triangle congruent to a given triangle. [7.3]
Given a right triangle, find the length of a missing side. [7.4]
Given similar triangles, find the length of one of the sides.
    [7.4]
Evaluate a trigonometric ratio. [7.4]
Find the sine, cosine, and tangent for a given angle. [7.4]
Solve applied problems using triangles. [7.4]
Work applied problems involving the golden ratio. [7.5]
Decide whether a network is traversable. [7.6]
Work floor-plan problems. [7.6]
Solve applied problems involving traversable networks. [7.6]
Sort figures into topologically equivalent classes. [7.7]

Decide whether a given point is an interior or exterior point. [7.7]

Solve applied problems involving the four-color theorem. [7.7]

Decide whether a figure is a Saccheri quadrilateral. [7.8]

Work with tessellations. [7.8]

Know various LOGO commands. [7.9]

Be able to read LOGO programs. [7.9]

Be able to write LOGO programs. [7.9]

CHAPTER 7   Review Questions

1. If a 4-cm cube is painted green and then cut into 64 1-cm cubes, how many of those cubes will be painted on

   a. 4 sides?   **b.** 3 sides?   **c.** 2 sides?   **d.** 1 side?   **e.** 0 sides?

2. Which of the following illustrate at least one line of symmetry? If it does, show a line of symmetry. If it does not, tell why.

   a.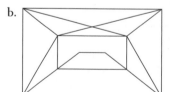

   United Kingdom (Great Britain)

   b.

   Nepal

   c.

   Safeway

   d.

   Vair

3. Suppose ℓ_1 and ℓ_2 are parallel lines. Classify the pairs of angles shown in the figure.

   a. $\angle 1$ and $\angle 5$ **b.** $\angle 5$ and $\angle 6$
   c. $\angle 2$ and $\angle 4$ **d.** $\angle 2$ and $\angle 8$
   e. $\angle 1$ and $\angle 7$ **f.** $\angle 1$ and $\angle 3$
   g. Can you classify any of the lines ℓ_1, ℓ_2, or ℓ_3 as perpendicular, horizontal, or vertical?

   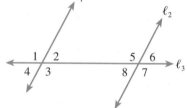

4. Find the value of the given trigonometric ratios. Round your answers to four decimal places.

   a. $\sin 59°$ **b.** $\tan 0°$ **c.** $\cos 18°$ **d.** $\tan 82°$

5. Indicate which of the networks or rooms are traversable. If the network is traversable, show how.

   a.   **b.**   **c.**

6. The world's most powerful lighthouse is on the coast of Brittany, France, and is about 160 ft tall. Suppose you are in a boat just off the coast, as shown in Figure

7.81. Determine your distance (to the nearest foot) from the base of the lighthouse if $\angle B = 12°$.

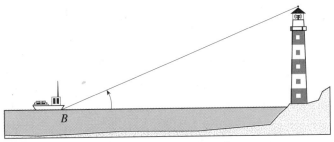

▲ **Figure 7.81** **Distance from ship to shore**

7. Draw a map with seven regions such that the indicated number of colors is required so that no two bordering regions have the same color.

   a. Two colors   **b.** Three colors   **c.** Four colors   **d.** Five colors

8. Take a strip of paper 11 in. by 1 in. and give it four half-twists; join the edges together. How many edges and sides does this band have? Cut the band down the center. What is the result?

9. What is a Saccheri quadrilateral? Discuss why this quadrilateral leads to different kinds of geometries.

10. Write a LOGO program to draw a regular pentagon with the length of each side 50 units.

Revisiting the Real World . . .

IN THE REAL WORLD "Carol, where have you been? I haven't seen you in over a year!" said Tom.

"I've been on an archeological dig," Carol responded. "Donald and I have been looking at the Garden Houses of Ostia. They were built in the 2nd century as part of the Roman Empire, and they were excavated in the first part of this century." "I've never heard of Ostia," Tom said with a swing of his arm. "Where or what is Ostia?" asked Tom.

"Ostia was a boom town in the second century," Carol said with enthusiasm. "Take a look at the map. The Tiber River was the lifeblood of the city Ostia, whose population reached 50,000 at its peak. They constructed an artificial harbor on the Tyrrhenian Sea, and the city became a major port of Rome, which is about 25 km away. Many apartment buildings were built, many four stories high, and they included not only living space, but also shops and gardens. We are trying to reconstruct life in the city."

"Those apartments look interesting," Tom commented. "It looks like there are 10 doors. With so many doors, I wonder if it is possible to pass through each door only once and stay either outside the apartment building or inside in the shaded hallway?"

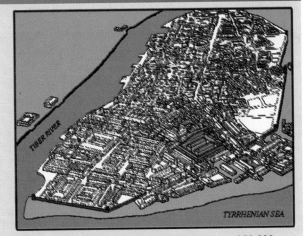

Ostia as it was in the 2nd century. The town of 50,000 required many apartment buildings (red).

Commentary: We use Polya's method.

Understand the Problem. Tom asked the question about entering and exiting the 10 doors of an apartment building. Here's a close-up of the floor plan.

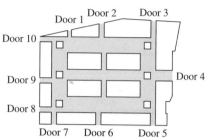

Devise a Plan. This problem looks very much like the floor-plan problem. We will form a conjecture for that problem, and then apply it to this apartment building of Ostia. When considering a room, there are two possible states: inside the room or outside the room. If there is an even number of doors between two rooms, then you can pass through all doors once. If there is an odd number of doors between two rooms, you begin in one room and end in the other. If there are more than two rooms, then it can be done if all the rooms have an even number of doors, or if there are no more than two rooms with an odd number of doors. For this problem, we can think of the diagram as two rooms: inside the building (room 1) and outside the building (room 2).

Carry Out the Plan. Since there are two rooms and 10 doors, we see that it is possible to go through all the doors exactly once.

Look Back. We can use a pencil to simulate a possible route. If we start inside the building, we will finish outside the building.

Group Research

Working in small groups is typical of most work environments, and being able to work with others to communicate specific ideas is an important skill to learn. Work with three or four other students to submit a single report based on each of the following questions.

1. In the figure at the right, there are eight square rooms making up a maze. Each square room has two walls that are mirrors and two walls that are open spaces. Identify the mirrored walls, and then solve the maze by showing how you can pass through all eight rooms consecutively without going through the same room twice. If that is not possible, tell why.

2. What exactly are fractals? Answer this question as a group project. Each group (three or four students is appropriate) will submit a single written paper answering this question.

    To get you started on your paper, we ask the following question that relates the ideas of series and fractals using the *snowflake curve.* Cut an equilateral triangle of side *a* out of paper, as shown in Figure 7.82a. Next, three equilateral triangles, each of side *a*/3, are cut out and placed in the middle of each side of the first triangle, as shown in Figure 7.82b. Then 12 equilateral triangles, each of side *a*/9, are placed halfway along each of the sides of this figure, as shown in Figure 7.82c. Figure 7.82d shows the result of adding 48 equilateral triangles, each of side *a*/27, to the previous figure. As part of the work on this paper, find the perimeter and the area of the snowflake curve formed if you continue this process indefinitely.

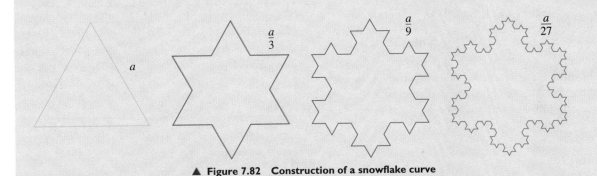

▲ **Figure 7.82** **Construction of a snowflake curve**

References
Anthony Barcellos, "The Fractal Geometry of Mandelbrot," *The College Mathematics Journal*, March 1984, pp. 98–114.
"Interview, Benoit B. Mandelbrot," *OMNI*, February 1984, pp. 65–66.
Benoit Mandelbrot, *Fractals: Form, Chance, and Dimension* (San Francisco: W. H. Freeman, 1977).
Benoit Mandelbrot, *The Fractal Geometry of Nature* (San Francisco: W. H. Freeman, 1982).

3. In Figures 7.1 and 7.2 we considered different views of a cube. Figure 7.83 shows a cube with a dot in the middle of each face. Draw a cube around the dots in Figure 7.84 so that each dot is in the middle of a face.

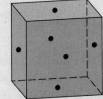

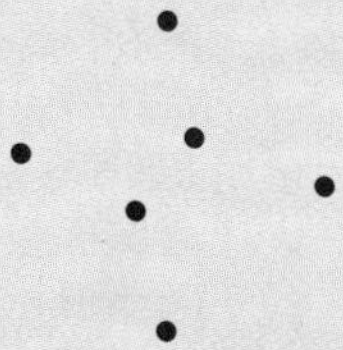

▲ **Figure 7.83** **Dots in a cube**

▲ **Figure 7.84** **Draw a cube so that each dot is in the center of a face of the cube.**

4. Place a dollar bill across the top of two glasses that are at least 3.5 in. apart. Now, describe how you can place a $.50 piece in the middle of the dollar bill without having it fall.

The Nature of Measurement

Neglect of mathematics works injury to all knowledge, since he who is ignorant of it cannot know the other sciences or the things of this world.

Roger Bacon

In the Real World . . .

"I can't wait until I move into my new home," said Susan. "I've been living out of boxes for longer than I want to remember."

"Have you picked out your carpet, tile, and made the other choices about color schemes yet?" asked Laurie.

"Heavens, no! They are just doing the framing and I have to build the decks, but I will get around to those selections soon," said Susan. "I hope I don't run out of money before I'm finished. My contractor gave me allowances for carpet, tile, light fixtures, and the like, but after looking around a bit I think those amounts are totally unrealistic. I think the goal was to come up with the lowest bid, and the dimensions of the house plan do not seem to match the amounts allowed. Can you help me make some estimates this Saturday?"

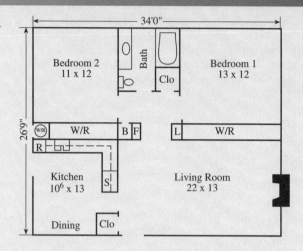

Contents

Perspective

Numbers are used to count and to measure. In counting, the numbers are considered exact unless the result has been rounded. In the previous chapter, we considered some measures related to angles and triangles. In this chapter, we study the topic of measurement in more detail.

Dimension refers to those properties called length, area, and volume. A figure having length only is said to be *one-dimensional*. A figure having area is said to be *two-dimensional*, and an object having volume is said to be *three-dimensional*. In more advanced mathematics, there is a much more complicated definition of dimension but, intuitively, if you can move an object back and forth only, it moves along a line and is one-dimensional. If you can move an object not only back and forth, but also right and left, then it moves in a plane and is two-dimensional. Finally, if you can move an object not only back and forth, right and left, but also up and down, then it is three-dimensional.

8.1 PRECISION, ACCURACY, AND ESTIMATION

What is your height? To answer that question, you must take a measurement.

Measure of Length

To measure an object is to assign a number to its size. The number representing its linear dimension, as measured from end to end, is called its **measure** or **length**.

Measurement is never exact, and you therefore need to decide how **precise** the measure should be. For example, the measurement might be to the nearest inch, nearest foot, or even nearest mile. The precision of a measurement depends not only on the instrument used but also on the purpose of your measurement. For example, if you are measuring the size of a room to lay carpet, the precision of your measurement might be different than if you are measuring the size of an airport hanger.

The **accuracy** refers to your answer. Suppose that you measure with an instrument that measures to the nearest tenth of a unit. You find one measurement to be 4.6 and another measurement to be 2.1. If, in the process of your work, you need to multiply these numbers, the result you obtain is

$$4.6 \times 2.1 = 9.66$$

This product is calculated to two decimal places, but it does not seem quite right that you obtain an answer that is more accurate (two decimal places) than the instrument you are using to do your measurements (one decimal place). In this book, we will require that the accuracy of your answers not exceed the precision of the measurement. This means that after the calculations are completed, the final answer should be rounded. The principle we will use is stated in the following box.

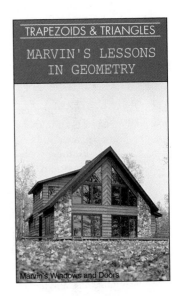

TRAPEZOIDS & TRIANGLES

MARVIN'S LESSONS IN GEOMETRY

Marvin's Windows and Doors

Accuracy of Measurements in This Book

All measurements are as precise as given in the text. If you are asked to make a measurement, the precision will be specified. Carry out all calculations without rounding. After you obtain a *final answer,* round this answer to be as accurate as *the least precise measurement.*

This means that, to avoid round-off error, you should round only once (at the end). This is particularly important if you are using a calculator, which will display 8, 10, 12, or even more decimal places.

You will also be asked to *estimate* the size of many objects in this chapter. As we introduce different units of measurement, you should remember some reference points so that you can make intelligent estimates. Many comparisons will be men-

Historical Note

← One Yard →

Early measurements were made in terms of the human body (digit, palm, cubit, span, and foot). Eventually, measurements were standardized in terms of the physical measurements of certain monarchs. King Henry I, for example, decreed that one yard was the distance from the tip of his nose to the end of his thumb. In 1790 the French Academy of Science was asked by the government to develop a logical system of measurement, and the original *metric system* came into being. By 1900 it had been adopted by more than 35 major countries. In 1906 there was a major effort to convert to the metric system in the United States, but it was opposed by big business and the attempt failed. In 1960 the metric system was revised and simplified to what is now known as the *SI system* (an abbreviation of *Système International d'Unités*). The term *metric* in this book will refer to the SI system. ▲

tioned in the text, but you need to remember only those that are meaningful for you to estimate other sizes or distances. You will also need to choose appropriate units of measurement. For example, you would not measure your height in yards or miles, or the distance from here to New York City in inches.

There have been numerous attempts to make the metric system mandatory in the United States. Figure 8.1 shows that the United States is the only major country not using the metric system. Today, big business is supporting the drive toward metric conversion, and it appears that the metric system will eventually come into use in the United States. In the meantime, it is important that we understand how to use both the United States and metric systems.

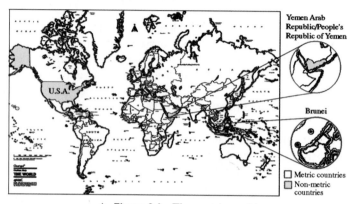

▲ **Figure 8.1   The metric world**

The most difficult problem in changing from the customary (U.S.) system to the metric system in the United States is not mathematical, but psychological. Many people fear that changing to the metric system will require complex multiplying and dividing and the use of confusing decimal points. For example, in a recent popular article, James Collier stated:

> For instance, if someone tells me it's 250 miles up to Lake George, or 400 out to Cleveland, I can pretty well figure out how long it's going to take and plan accordingly. Translating all of this into kilometers is going to be an awful headache. A kilometer is about 0.62 miles, so to convert miles into kilometers you divide by six and multiply by ten, and even that isn't accurate. Who can do that kind of thing when somebody is asking me are we almost there, the dog is beginning to drool and somebody else is telling you you're driving too fast?
>
> Of course, that won't matter, because you won't know how fast you're going anyway. I remember once driving in a rented car on a superhighway in France, and every time I looked down at the speedometer we were going 120. That kind of thing can give you the creeps. What's it going to be like when your wife keeps shouting, "Slow down, you're going almost 130"?
>
> But if you think kilometers will be hard to calculate . . .

The author of this article has missed the whole point. Why are kilometers hard to calculate? How does he know that it's 400 miles to Cleveland? He knows because the odometer on his car or a road sign told him. Won't it be just as easy to read an odometer calibrated to kilometers or a metric road sign telling him how far it is to Cleveland?

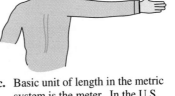

The real advantage of using the metric system is the ease of conversion from one unit of measurement to another. How many of you remember the difficulty you had in learning to change tablespoons to cups? Or pints to gallons?

In this book we will work with both the U.S. and the metric measurement systems. You should be familiar with both and be able to make estimates in both systems. The following box gives the standard units of length.

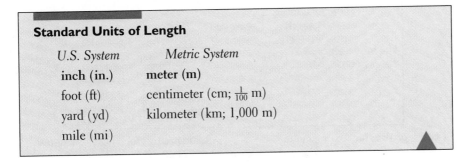

Standard Units of Length

U.S. System	Metric System
inch (in.)	**meter (m)**
foot (ft)	centimeter (cm; $\frac{1}{100}$ m)
yard (yd)	kilometer (km; 1,000 m)
mile (mi)	

To understand the size of any measurement, you need to see it, have experience with it, and take measurements using it as a standard unit. The basic unit of measurement for the U.S. system is the inch; it is shown in Figure 8.2. You can remember that an inch is about the distance from the joint of your thumb to the tip of your thumb. The basic unit of measurement for the metric system is the meter; it is also shown in Figure 8.2. You can remember that a meter is about the distance from your left ear to the tip of the fingers on the end of your outstretched right arm.

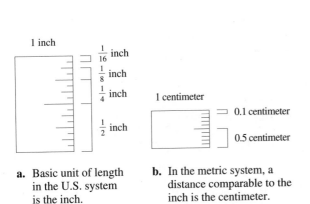

a. Basic unit of length in the U.S. system is the inch.

b. In the metric system, a distance comparable to the inch is the centimeter.

c. Basic unit of length in the metric system is the meter. In the U.S. system, a comparable unit is the yard.

▲ **Figure 8.2   Standard units of measurement for length**

For the larger distances of a mile and a kilometer, you will need to look at maps, or the odometer of your car. However, you should have some idea of these distances.*

It might help to have a visual image of certain prefixes as you progress through this chapter. Greek prefixes **kilo-, hecto-,** and **deka-** are used for measurements larger than the basic metric unit, and Latin prefixes **deci-, centi-,** and **milli-** are used for smaller quantities (see Figure 8.3). As you can see from Figure 8.3, a centimeter is $\frac{1}{100}$ of a meter; this means that 1 meter is equal to 100 centimeters.

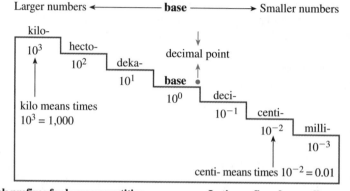

Greek prefixes for larger quantities   Latin prefixes for smaller quantities

▲ **Figure 8.3   Metric prefixes**

Now we will measure given line segments with different levels of precision. We will consider two different rulers, one marked to the nearest centimeter and another marked to the nearest $\frac{1}{10}$ centimeter.

◼◼◼ EXAMPLE 1

Measure the given segment

a. to the nearest centimeter. B ——————————————

b. to the nearest $\frac{1}{10}$ centimeter.

Solution

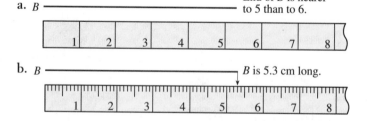

* We could tell you that a mile is 5,280 ft or that a kilometer is 1,000 m, but to do so does not give you any feeling for what these distances really are. You need to get into a car and watch the odometer to see how far you travel in going 1 mile. Most cars in the United States do not have odometers set to kilometers, and until they do it is difficult to measure in kilometers. You might, however, be familiar with a 10-kilometer race. It takes a good runner about 30 minutes to run 10 kilometers and an average runner about 45 minutes. You can walk a kilometer in about 6 minutes.

Perimeter

One application of both measurement and geometry involves finding the distance around a polygon. This distance is called the **perimeter** of the polygon.

> **Perimeter**
>
> The **perimeter** of a polygon is the sum of the lengths of the sides of that polygon.

Following are some formulas for finding the perimeters of the most common polygons.

An **equilateral triangle** is a triangle with all sides equal length.

$$\text{PERIMETER} = 3(\text{SIDE})$$
$$P = 3s$$

A **rectangle** is a quadrilateral with angles that are all right angles.

$$\text{PERIMETER} = 2(\text{LENGTH}) + 2(\text{WIDTH})$$
$$P = 2\ell + 2w$$

A **square** is a rectangle with all sides equal length.

$$\text{PERIMETER} = 4(\text{SIDE})$$
$$P = 4s$$

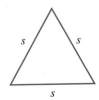

Equilateral triangle

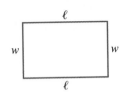

Rectangle

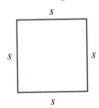

Square

■■■■ EXAMPLE 2

Find the perimeter of each polygon.

a.

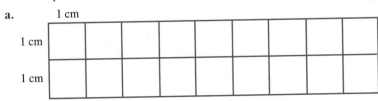

b.

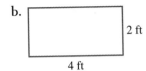

2 ft

4 ft

c.

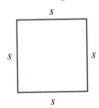

5 mi

5 mi

d.

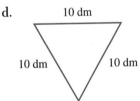

10 dm

10 dm 10 dm

e.

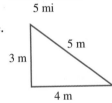

5 m

3 m

4 m

Solution

a. Rectangle is 2 cm by 9 cm, so
$$P = 2\ell + 2w$$
$$= 2(9) + 2(2)$$
$$= 18 + 4 = 22 \text{ cm}$$

b. Rectangle is 2 ft by 4 ft, so
$$P = 2\ell + 2w$$
$$= 2(4) + 2(2) = 8 + 4$$
$$= 12 \text{ ft}$$

c. Square, so
$$P = 4s = 4(5) = 20 \text{ mi}$$

d. Equilateral triangle, so
$$P = 3s = 3(10) = 30 \text{ dm}$$

e. Triangle (add lengths of sides), so
$$P = 3 + 4 + 5 = 12 \text{ m}$$

Circumference

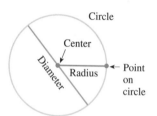

Circle

Center

Radius

Diameter

Point on circle

▲ **Figure 8.4   Circle**

Although a circle is not a polygon, sometimes we need to find the distance around a circle. This distance is called the **circumference**. For *any* circle, if you divide the circumference by the diameter, you will get the *same number* (see Figure 8.4). This number is given the name π (**pi**). The number π is an irrational number and is about 3.14 or $\frac{22}{7}$. We need this number π to state a formula for the circumference C:

$$C = d\pi \qquad \text{or} \quad C = 2\pi r$$
$$d = \text{DIAMETER} \qquad r = \text{RADIUS}$$

For a circle, notice that the radius is half the diameter. This means that, if you know the radius and want to find the diameter, you simply multiply by 2. If you know the diameter and want to find the radius, divide by 2.

EXAMPLE 3

Find the distance around each figure.

a.

4 ft

b.

28 cm

c.

5 m

d. $\frac{1}{2}$ of a circle

2 dm

e.
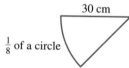
30 cm
$\frac{1}{8}$ of a circle

Solution

a. $C = 4\pi$

A decimal approximation (symbolized by $\approx$) is

$$C \approx 4(3.14) = 12.56$$

The circumference is 4π ft, which is about 13 ft.

b. $C = 28\pi$; this is about 88 cm.

c. $C = 2\pi(5) = 10\pi$; this is about 31 m.

d. This is half of a circle (called a **semicircle**); thus, the curved part is half of the circumference ($C = 2\pi \approx 6.28$), or about 3.14, which is added to the diameter:

$$3.14 + 2 = 5.14$$

The distance around the figure is about 5 dm.

e. This is one-eighth of a circle. The curved part is one-eighth of the circumference [$C = 2\pi(30) \approx 188.5$], or about 23.6, which is added here to the radius (on both sides): $23.6 + 30 + 30 \approx 84$ cm.

Many calculators have a single key marked π. If you press it, the display shows an approximation correct to several decimal places (the number of places depends on your calculator):

Press: $\boxed{\pi}$ Display: **3.141592654**

If your calculator doesn't have a key marked π, you may want to use this approximation to obtain the accuracy you want. In this book, the answers are found by using the π key on a calculator and then rounding the answer.

The ideas involving perimeter and circumference are sometimes needed to solve certain types of problems.

EXAMPLE 4

Suppose you have enough material for 70 ft of fence and want to build a rectangular pen 14 ft wide. What is the length of this pen?

Solution

PERIMETER = 2(LENGTH) + 2(WIDTH)   This is the formula for perimeter.

70   = 2(LENGTH) + 2(14)   Fill in the given information.

Let ℓ = LENGTH OF PEN

$70 = 2\ell + 28$

$42 = 2\ell$

$21 = \ell$

The pen will be 21 ft long.

PROBLEM SET 8.1

▲ **A Problems**

1. **IN YOUR OWN WORDS** Contrast precision and accuracy.

2. **IN YOUR OWN WORDS** What is the agreement about the accuracy of answers in this book?

3. **IN YOUR OWN WORDS** State the perimeter formulas for a square, rectangle, equilateral triangle, and regular pentagon.

4. **IN YOUR OWN WORDS** What is the formula for the circumference of a circle?

From memory, and without using any measuring devices, draw a line segment with approximate length as indicated in Problems 5–14.

5. 1 in.   6. 2 in.   7. 3 in.   8. $\frac{1}{2}$ in.   9. $\frac{1}{4}$ in.

10. 1 cm   11. 5 cm   12. 10 cm   13. 3 cm   14. 2 cm

Pick the best choices in Problems 15–39 by estimating. Do not measure. For metric measurements, do not attempt to convert to the U.S. system. The hardest part of the transition to the metric system is the transition to thinking in metrics.

15. The length of this math textbook is about
    A. 9 in.   B. 9 cm   C. 2 ft

16. The length of a car is about
    A. 1 m   B. 4 m   C. 10 m

17. The length of a dollar bill is about
    A. 3 in.   B. 6 in.   C. 9 in.

18. The width of a dollar bill is about
    A. 1.9 cm   B. 6.5 cm   C. 0.65 m

19. The perimeter of a dollar bill is
    A. 18 in.   B. 6 in.   C. 46 in.

20. The perimeter of a five-dollar bill is
    A. 19 cm   B. 6 cm   C. 46 cm

21. The length of a new pencil is
    A. 4 in.   B. 18 in.   C. 7 in.

22. The diameter of a new pencil is
    A. 0.25 cm   B. 0.25 in.   C. 0.25 ft

23. The circumference of an automobile tire is
    A. 60 cm   B. 60 in.   C. 1 m

24. The perimeter of this textbook is
    A. 34 in   B. 34 cm   C. 11 in.

25. The perimeter of a VISA credit card is
    A. 30 in.   B. 30 cm   C. 1 m

26. The perimeter of a sheet of notebook paper is
    A. 1 cm   B. 10 cm   C. 1 m

27. The perimeter of the screen on a console TV set is
    A. 30 in.   B. 100 cm   C. 100 in.

28. The perimeter of a classroom is
    A. 100 ft   B. 100 m   C. 100 yd

29. The distance from your home to the nearest grocery store is most likely to be
    A. 1 cm   B. 1 m   C. 1 km

30. The length of a 100-yard football field is
    A. 100 m   B. more than 100 m   C. less than 100 m

31. The distance from San Francisco to New York is about 3,000 miles. This distance is
    A. less than 3,000 km   B. more than 3,000 km   C. about 3,000 km

32. Suppose someone could run the 100-meter dash in 10 seconds flat. At the same rate, this person should be able to run the 100-yard dash in
    A. less than 10 sec   B. more than 10 sec   C. 10 sec

33. Your height is closest to
    A. 5 ft   B. 10 ft   C. 25 in.

34. An adult's height is most likely to be about
    A. 6 m   B. 50 cm   C. 170 cm

35. The distance around your waist is closest to
    A. 10 in.   B. 36 in.   C. 30 cm

36. The distance from floor to ceiling in a typical home is about
    A. 2.5 m   B. 0.5 m   C. 4.5 m

37. The length of a diagonal on a typical computer monitor is about
    A. 14 cm   B. 14 in.   C. 31 in.

38. The prefix *centi-* means
    A. one thousand   B. one-thousandth   C. one-hundredth

39. The prefix *kilo-* means
    A. one thousand   B. one-thousandth   C. one-hundredth

Measure the segments given in Problems 40–50 with the indicated precision.

40. To the nearest centimeter: ——————————

41. To the nearest centimeter: ————————————

42. To the nearest $\frac{1}{10}$ centimeter: ————————

43. To the nearest $\frac{1}{10}$ centimeter: ——————————

44. To the nearest $\frac{1}{10}$ centimeter: ——————————————

45. To the nearest inch: ——————————

46. To the nearest inch: ————————————

47. To the nearest inch: ——————————————

48. To the nearest eighth of an inch: ————————

49. To the nearest eighth of an inch: ——————————————

50. To the nearest eighth of an inch: ——————————————————

▲ **B Problems**

Find the perimeter, circumference, or distance around the figures given in Problems 51–69, by using the appropriate formula. Round approximate answers to two decimal places.

51. (rectangle, 5 in. by 4 in.)

52. (rectangle, 12 ft by 3 ft)

53. (square, 240 ft by 240 ft)

54.

6 yd   6 yd

6 yd

55.   3 dm

3 dm   3 dm

56.

50.00 ft

57.

2.40 m

58.

80.00 ft

59.

12.00 in.

60.   14 ft

9 ft   9 ft

14 ft

61.   18 in.

4 in.   4 in.

4 in.   4 in.

18 in.

62.   160 cm

40 cm   50 cm

280 cm

63.

50.00 ft

120.00 ft

64.

0.30 cm

0.30 cm

1.40 cm

1.00 cm

1.00 cm

1.00 cm

3.50 cm

|← 1.70 cm →|

65.

100.00 cm

66.

10.00 in.

67.

18.00 ft

68.

2.00 cm

|← 3.00 cm →|

69.

2.00 cm

|← 3.00 cm →|

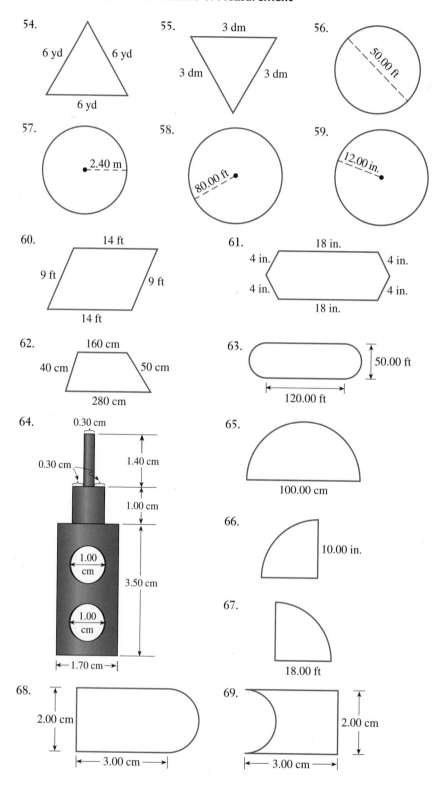

70. What is the width of a rectangular lot that has a perimeter of 410 ft and a length of 140 ft?

71. What is the length of a rectangular lot that has a perimeter of 750 m and a width of 75 m?

72. The perimeter of $\triangle ABC$ is 117 in. Find the lengths of the sides.

73. The perimeter of the pentagon is 280 cm. Find the lengths of the sides.

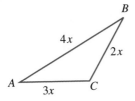

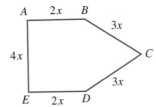

74. Find the dimensions of an equilateral triangle that has a perimeter measuring 198 dm.

75. Find the dimensions of a rectangle with a perimeter of 54 cm if the length is 5 less than three times the width.

76. One of the first recorded units of measurement is the **cubit.**

▲ **Figure 8.5   An ancient cubit, from the Louvre, Paris. The original measures 52.5 cm.**

   If we define a *cubit* as the distance from your elbow to your fingertips, then we may find that the cubit defined in your body is a little different from the Egyptian cubit on display at the Louvre (see Figure 8.5). How does your cubit compare with this Egyptian cubit?

Find your own metric measurements in Problems 77–82.

	Women		Men
77.	Height	77.	Height
78.	Bust	78.	Chest
79.	Waist	79.	Waist
80.	Hips	80.	Seat
81.	Distance from waist to hemline	81.	Neck
82.	Fist (measure around largest part of hand over knuckles while making a fist, excluding thumb)	82.	Length of shoe

▲ **Problem Solving**

83. In the sixth chapter of Genesis, the dimensions of Noah's ark are given as 300 cubits long, 50 cubits wide, and 30 cubits high. Use the Egyptian cubit in Figure 8.5 to convert these measurements to the nearest meter.

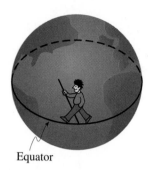

Equator

84. Suppose that we fit a band tightly around the earth at the equator. We wish to raise the band so that it is uniformly supported 6 ft above the earth at the equator.
    a. Guess how much extra length would have to be added to the band (not the supports) to do this.
    b. Calculate the amount of extra material that would be needed.

8.2   AREA

Suppose that you want to carpet your living room. The price of carpet is quoted as a price per square yard. A square yard is a measure of **area.** To measure the area of a plane figure, you fill it with **square units.** (See Figure 8.6.)

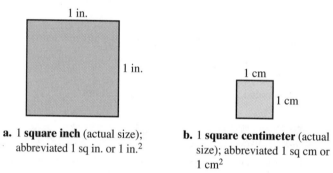

a. 1 **square inch** (actual size); abbreviated 1 sq in. or 1 in.^{2}

b. 1 **square centimeter** (actual size); abbreviated 1 sq cm or 1 cm^2

▲ **Figure 8.6   Common units of measurement for area**

EXAMPLE 1

What is the area of the shaded region?

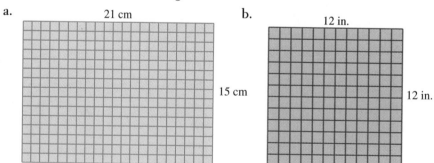

a. 21 cm   15 cm

b. 12 in.   12 in.

Solution

a. You can count the number of square centimeters in the shaded region; there are 315 squares. Also notice:

Across   Down
21 cm × 15 cm = 21 × 15 cm × cm = 315 cm^2

b. The shaded region is a **square foot.** You can count 144 square inches inside the region. Also notice:

Across   Down
12 in. × 12 in. = 144 in.²   —

As you can see from Example 1, the area of a rectangular or square region is the product of the distance across (length) and the distance down (width).

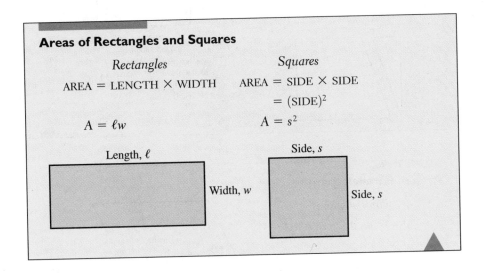

Areas of Rectangles and Squares

Rectangles   *Squares*

AREA = LENGTH × WIDTH   AREA = SIDE × SIDE
= (SIDE)²

$A = \ell w$ $A = s^2$

Length, ℓ Side, s

Width, w Side, s

■■■ EXAMPLE 2

How many square feet are there in a square yard?

Solution   Since 1 yd = 3 ft, we see from Figure 8.7 that

1 yd² = (3 ft)²   1 yd
= 9 ft²

1 yd

▲ **Figure 8.7   1 yd² = 9 ft²**   —

A **parallelogram** is a quadrilateral with two pairs of parallel sides, as shown in Figure 8.8. To find the area of a parallelogram, we can estimate the area by counting the number of square units inside the parallelogram (which may require estimation of partial square units), or we can show that the formula for the area of a parallelogram is the same as the formula for the area of a rectangle.

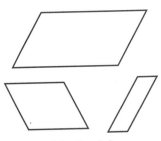

▲ **Figure 8.8   Parallelograms**

Geometric Justification of the Area Formula for a Parallelogram

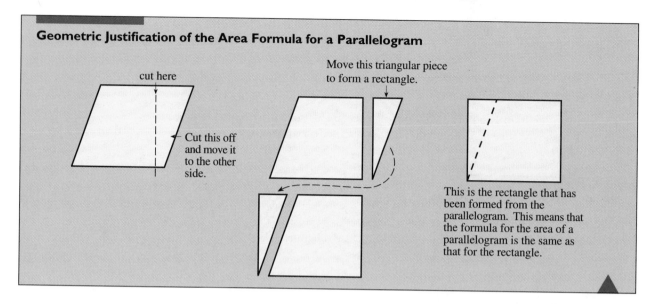

cut here

Cut this off and move it to the other side.

Move this triangular piece to form a rectangle.

This is the rectangle that has been formed from the parallelogram. This means that the formula for the area of a parallelogram is the same as that for the rectangle.

Area of Parallelogram

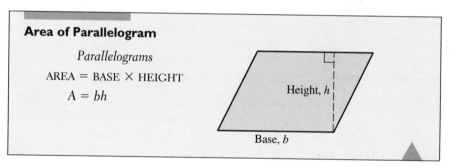

Parallelograms

AREA = BASE × HEIGHT

$A = bh$

Height, h

Base, b

▬▬ EXAMPLE 3

Find the area of each shaded region.

a.

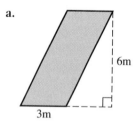

6m

3m

b.

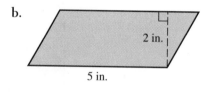

2 in.

5 in.

Solution

a. $A = 3 \text{ m} \times 6 \text{ m}$

$= 18 \text{ m}^2$

b. $A = 2 \text{ in.} \times 5 \text{ in.}$

$= 10 \text{ in.}^2$ ▬

You can find the area of a triangle by filling in and approximating the number of square units, by rearranging the parts, or by noticing that *every* triangle has an area that is exactly half that of a corresponding parallelogram.

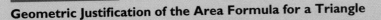

Geometric Justification of the Area Formula for a Triangle

These triangles have the same area.   These triangles have the same area.   These triangles have the same area.

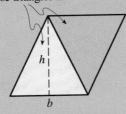

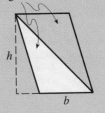

We can therefore state the following result.

Area of Triangles

Triangles

AREA = $\frac{1}{2}$ × BASE × HEIGHT

$A = \frac{1}{2}bh$

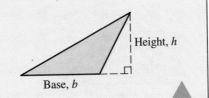

Height, h

Base, b

■ EXAMPLE 4

Find the area of each shaded region.

a.

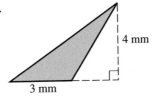

4 mm

3 mm

b.

3 km

5 km

Solution

a. $A = \frac{1}{2}$ × 3 mm × 4 mm   **b.** $A = \frac{1}{2}$ × 5 km × 3 km

$= 6$ mm² $= \frac{15}{2}$ km² or $7\frac{1}{2}$ km²   ▬

We can also find the area of a trapezoid by finding the area of triangles. A
trapezoid is a quadrilateral with two sides parallel. These sides are called the *bases*,
and the perpendicular distance between the bases is the height, as shown in Figure
8.9. The area formula can be found as the sum of the areas of triangle I and triangle
II:

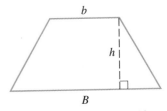

b

h

B

▲ **Figure 8.9   Trapezoid**

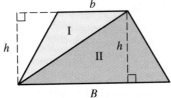

b

I

h h

II

B

Area of triangle I: $\frac{1}{2}bh$

Area of triangle II: $\frac{1}{2}Bh$

Total area: $\frac{1}{2}bh + \frac{1}{2}Bh$

If we use the distributive property, we obtain the area formula for a trapezoid, as shown in the following box.

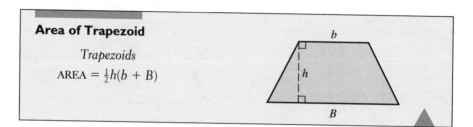

Area of Trapezoid

Trapezoids

$$\text{AREA} = \tfrac{1}{2}h(b + B)$$

▬▬ EXAMPLE 5

Find the area of each shaded region.

a.

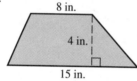

8 in.

4 in.

15 in.

b.

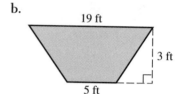

19 ft

3 ft

5 ft

Solution

a. $h = 4; b = 8; B = 15$

$A = \tfrac{1}{2}(4)(8 + 15)$

$\quad = 2(23)$

$\quad = 46$

The area is 46 in.²

b. $h = 3; b = 19; B = 5$

$A = \tfrac{1}{2}(3)(19 + 5)$

$\quad = \tfrac{3}{2}(24)$

$\quad = 36$

The area is 36 ft².   ▬

The last of our area formulas is for the area of a circle.

Area of Circle

Circles

$$A = \pi r^2 \quad \text{or} \quad A \approx 3.1416(\text{RADIUS})^2$$

Radius, r

Even though it is beyond the scope of this course to derive a formula for the area of a circle, we can give a geometric justification that may appeal to your intuition.

Geometric Justification of the Area Formula for a Circle

Consider a circle with radius r. Cut the circle in half:

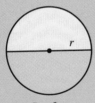

$C = 2\pi r$

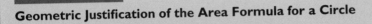

Half of the circumference is πr.

Cut along the dashed lines so that each half lays flat.

$r \times \pi$

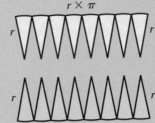

Fit these two pieces together:

$r \times \pi$

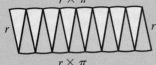

$r \times \pi$

$r \times \pi$

Therefore, it looks as if the area of a circle of radius r is about the same as the area of a rectangle of length πr and width r—that is, πr^2.

■■■■ EXAMPLE 6

Find the area of each shaded region to the nearest tenth unit.

a.

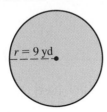

$r = 9$ yd

b.

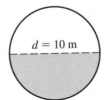

$d = 10$ m

Solution

a. On a calculator:

   Press: $\boxed{81}\ \boxed{\times}\ \boxed{\pi}\ \boxed{=}$

   To the nearest tenth, the area is 254.5 yd².

b. Notice that the shaded portion is only half the area of the circle.
   On a calculator:

   Press: $\boxed{25}\ \boxed{\times}\ \boxed{\pi}\ \boxed{\div}\ \boxed{2}\ \boxed{=}$

   semicircle

   Display: **39.26990817**

   To the nearest tenth, the area is 39.3 m².   ▬

Sometimes we measure area using one unit of measurement and then we want to convert the result to another.

■■■ EXAMPLE 7

Suppose your living room is 12 ft by 15 ft and you want to know how many square yards of carpet you need to cover this area.

Solution

Method I. $A = 12 \text{ ft} \times 15 \text{ ft}$

$= 180 \text{ ft}^2$

$= 180 \times 1 \text{ ft}^2$

$= 180 \times (\frac{1}{9} \text{ yd}^2)$ Since 1 yd = 3 ft,

$= 20 \text{ yd}^2$ 1 yd² = (1 yd) × (1 yd)

1 yd² = (3 ft) × (3 ft)

1 yd² = 9 ft²

$\frac{1}{9}$ yd² = 1 ft²

Method II.   Change feet to yards to begin the problem:

$12 \text{ ft} = 4 \text{ yd} \quad \text{and} \quad 15 \text{ ft} = 5 \text{ yd}$

$A = 4 \text{ yd} \times 5 \text{ yd}$

$= 20 \text{ yd}^2$ ▬

If the area is large, as with property, a larger unit is needed. This unit is called an *acre*.

Acre

An **acre** is 43,560 ft².

To convert from ft² to acres, divide by 43,560.

ESTIMATION HINT: Real estate brokers estimate an acre as 200 ft by 200 ft or 40,000 ft².

When working with acres, you usually need a calculator to convert square feet into acres, as shown in Example 8.

■■■ EXAMPLE 8

How many acres are there in a rectangular piece of property measuring 363 ft by 180 ft?

Solution

AREA $= 363 \text{ ft} \times 180 \text{ ft}$

$= 65,340 \text{ ft}^2$ ESTIMATE: $65,340 \approx 65,000$

To change ft² to acres, divide by 43,560. $65,000 \div 40,000 = 65 \div 40$

AREA $= 65,340 \text{ ft}^2$ $= 13 \div 8$

$= (65,340 \div 43,560) \text{ acres}$ $\approx 12 \div 8 = 1.5$

$= 1.5 \text{ acres}$ *Press:*   [65340] [÷] [43560] [=]   ▬

Surface Area

Suppose you wish to paint a box whose edges are each 3 ft, and you need to know how much paint to buy. To determine this you need to find the sum of the areas of all the faces — this is called the **surface area.** To find the surface area, add the area of each face.

▬▬ EXAMPLE 9

Find the amount of paint needed for a box with edges 3 ft.

Solution   A box (cube) has 6 faces of equal area.

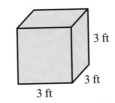

Number of faces of cube
↓
$$6 \times 9 \text{ ft}^2 = 54 \text{ ft}^2$$
Area of each face

3 ft
3 ft
3 ft

You need enough paint to cover 54 ft².

▬▬ EXAMPLE 10

Find the outside surface area.

a.

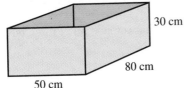

30 cm
80 cm
50 cm

b.

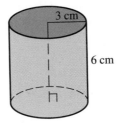

3 cm
6 cm

Solution

a. Notice that there is no top on the box. We find the sum of the areas of all the faces:

Front:	$30 \times 50 =$	1,500
Back:		1,500	Same as front
Side:	$80 \times 30 =$	2,400
Side:		2,400	Sides are the same size.
Bottom:	$80 \times 50 =$	4,000
Total:		11,800 cm²

b. Notice that the can has a bottom, but no lid. To find the surface area, find the area of a circle (the bottom) and think of the sides of the can as being "rolled out." The length of the resulting rectangle is the circumference of the can and

the width of the rectangle is the height of the can.

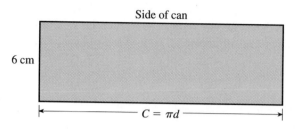 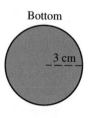

Side of can   Bottom

6 cm   3 cm

$C = \pi d$

Side: $A = \ell w = (\pi d)w = \pi(6)(6) = 36\pi$
Bottom: $A = \pi r^2 = \pi(3)^2 = 9\pi$
Surface area: $36\pi + 9\pi = 45\pi \approx 141.37167$

The surface area is about 141 cm².

EXAMPLE 11   Polya's Method

You want to paint 200 ft of a three-rail fence, and need to know how much paint to purchase.

Solution   We use Polya's problem-solving guidelines for this example.

Understand the Problem.   There is some additional information you need to gather before you can answer this question.

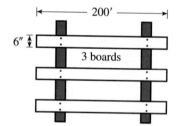

You want to paint 200 ft of a three-rail fence that is made up of three boards, each 6 inches wide. You want to know:

a. the number of square feet on one side of the fence.

b. the number of square feet to be painted if the posts and edges of the boards comprise 100 ft².

c. the number of gallons of paint to purchase if each gallon covers 325 ft².

Devise a Plan.   The paint coverage is found on the can (we assume it is 325 ft²). The next thing we need to know is the number of square feet to be painted. We estimate the number of square feet to be painted on the posts and the edges to be 100 ft². Now, calculate the number of square feet to be painted and divide by 325 (the number of square feet per gallon).

Carry Out the Plan.

a. We first calculate A, the number of square feet to be painted on one side of the fence:

$$A = (200 \text{ ft}) \times (6 \text{ in.}) \times 3$$
$$= (200 \text{ ft}) \times (\tfrac{1}{2} \text{ ft}) \times 3$$
$$= 300 \text{ ft}^2$$

b. AMOUNT TO BE PAINTED $= 2($AMOUNT ON ONE SIDE$) +$ EDGES AND POSTS

$$= 2(300 \text{ ft}^2) + 100 \text{ ft}^2$$

$$= 700 \text{ ft}^2$$

c. $\left(\begin{array}{c}\text{NUMBER OF SQUARE} \\ \text{FEET PAINTED}\end{array}\right) = \left(\begin{array}{c}\text{NUMBER OF SQUARE} \\ \text{FEET PER GALLON}\end{array}\right)\left(\begin{array}{c}\text{NUMBER OF} \\ \text{GALLONS}\end{array}\right)$

$$700 = 325 \left(\begin{array}{c}\text{NUMBER OF} \\ \text{GALLONS}\end{array}\right)$$

$$2.15 \approx \left(\begin{array}{c}\text{NUMBER OF} \\ \text{GALLONS}\end{array}\right) \qquad \text{Divide both sides by 325.}$$

Look Back.   If paint must be purchased by the gallon (as implied by the question), the amount to purchase is 3 gallons.   ▬

PROBLEM SET 8.2

▲ A Problems

1. **IN YOUR OWN WORDS** What do we mean by length?

2. **IN YOUR OWN WORDS** What do we mean by area?

3. **IN YOUR OWN WORDS** What do we mean by surface area?

4. **IN YOUR OWN WORDS** How do you find the area of a circle? Contrast with the procedure for finding the circumference of a circle.

Estimate the area of each figure in Problems 5–12 to the nearest square centimeter.

5.

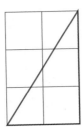

6.

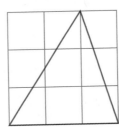

7.

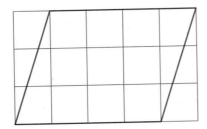

8.

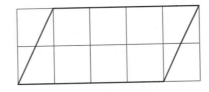

9.

10.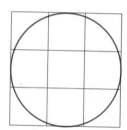

11.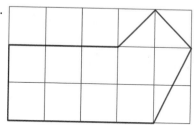

12.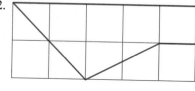

Pick the best choices in Problems 13–24 by estimating. Do not measure. For metric measurements, do not attempt to convert to the U.S. system.

13. The area of a dollar bill is
    A. 18 in.   B. 6 in.²   C. 18 in.²

14. The area of a five-dollar bill is
    A. 100 cm   B. 10 cm²   C. 100 cm²

15. The area of the front cover of this textbook is
    A. 70 in.²   B. 70 cm²   C. 70 in.

16. The area of a VISA credit card is
    A. 8 in.   B. 8 in.²   C. 8 cm²

17. The area of a sheet of notebook paper is
    A. 90 cm²   B. 10 in.²   C. 600 cm²

18. The area of a sheet of notebook paper is
    A. 90 in.²   B. 10 cm²   C. 600 in.²

19. The area of the screen of a console TV set is
    A. 4 ft²   B. 19 in.²   C. 100 in.²

20. The area of a classroom is
    A. 100 ft²   B. 1,000 ft²   C. 0.5 acre

21. The area of the floor space of the Superdome in New Orleans is
    A. 1 mi²   B. 1,000 m²   C. 10 acres

22. The state of California has an area of
    A. 5,000 acres   B. 150,000 mi²   C. 5,000 m²

23. The area of the bottom of your feet is
    A. 1 m²   B. 400 in.²   C. 400 cm²

24. It is known that your body's surface area is about 100 times the area that you will find if you trace your hand on a sheet of paper. Using this estimate, your body's surface area is
    A. 300 in.²   B. 3,000 in.²   C. 3,000 cm²

▲ **B Problems**

Find the area of each shaded region in Problems 25–46. (Assume that given measurements are exact, and round approximate answers to the nearest tenth of a square unit.)

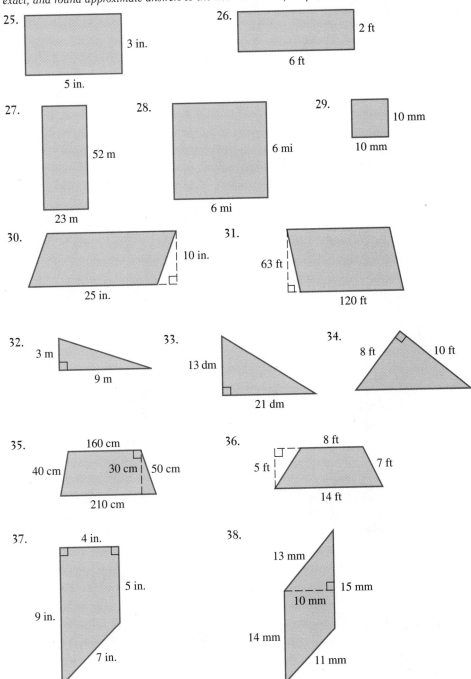

25.

3 in.

5 in.

26.

2 ft

6 ft

27.

52 m

23 m

28.

6 mi

6 mi

29.

10 mm

10 mm

30.

10 in.

25 in.

31.

63 ft

120 ft

32.

3 m

9 m

33.

13 dm

21 dm

34.

8 ft   10 ft

35.

160 cm

40 cm   30 cm   50 cm

210 cm

36.

8 ft

5 ft   7 ft

14 ft

37.

4 in.

5 in.

9 in.

7 in.

38.

13 mm

15 mm

10 mm

14 mm

11 mm

39.

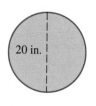

20 in.

40.

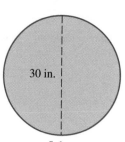

30 in.

41.

10 in.

42.

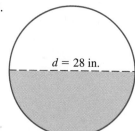

$d = 28$ in.

43. $r = 5$ dm

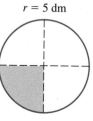

44. $r = 30$ cm

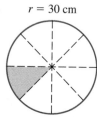

45.

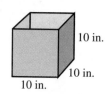

2 cm

3 cm

46.

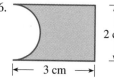

2 cm

3 cm

Find the outside surface area in Problems 47–58.

47.

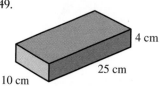

10 in.

10 in.

10 in.

48.

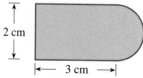

10 cm

10 cm   10 cm

49.

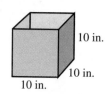

4 cm

25 cm

10 cm

50.

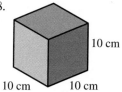

1 m

3 m

1 m

51.

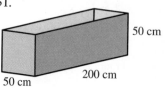

50 cm

200 cm

50 cm

52.

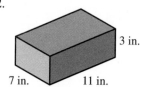

3 in.

7 in.   11 in.

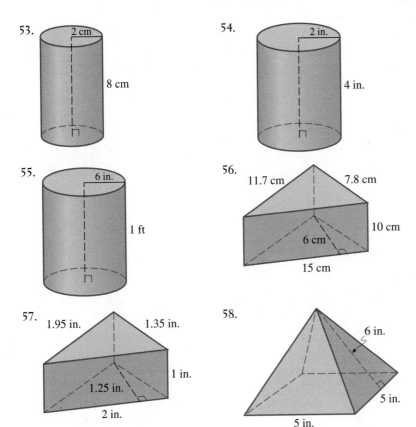

53. 2 cm 8 cm

54. 2 in. 4 in.

55. 6 in. 1 ft

56. 11.7 cm 7.8 cm 10 cm 6 cm 15 cm

57. 1.95 in. 1.35 in. 1 in. 1.25 in. 2 in.

58. 6 in. 5 in. 5 in.

59. What is the area of a television screen that measures 12 in. by 18 in.?

60. What is the area of a rectangular lot that measures 185 ft by 75 ft?

61. What is the area of a piece of $8\frac{1}{2}$-in. by 11-in. typing paper?

62. If a certain type of fabric comes in a bolt 3 feet wide, how long a piece must be purchased to have 24 square feet?

63. Find the cost of pouring a square concrete slab of a uniform thickness with sides of 25 ft if the cost (at the uniform thickness) is $5.75 per square foot.

64. What is the cost of seeding a rectangular lawn 100 ft by 30 ft if 1 pound of seed costs $5.85 and covers 150 square feet? Use estimation to decide whether your answer is reasonable.

65. If a mini-pizza has a 6-in. diameter, what is the number of square inches (to the nearest square inch)?

66. If a small pizza has a 10-in. diameter, what is the number of square inches (to the nearest square inch)?

67. If a medium pizza has a 12-in. diameter, what is the number of square inches (to the nearest square inch)?

68. If a large pizza has a 14-in. diameter, what is the number of square inches (to the nearest square inch)?

69. Which property costs less per square foot?

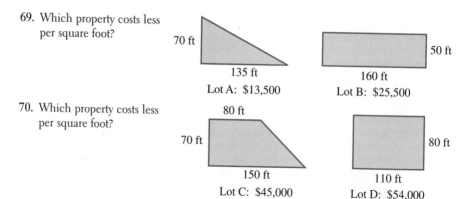

Lot A: $13,500   Lot B: $25,500

70. Which property costs less per square foot?

Lot C: $45,000   Lot D: $54,000

71. If a rectangular piece of property is 750 ft by 1,290 ft, what is the acreage (to the nearest tenth of an acre)?

72. How many square feet are there in $4\frac{1}{2}$ acres?

▲ **Problem Solving**

Find the area (to the nearest square inch) of the shaded region contained in the 10-in. squares in Problems 73–74. Assume that the arcs intersect the midpoints of the sides.

73.

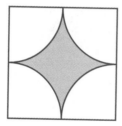

74.

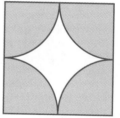

75. Figure 8.10 illustrates a strange and interesting relationship. The square in part **a** has an area of 64 cm² (8 cm by 8 cm). When *this same figure is cut and rearranged as* shown in part **b**, it appears to have an area of 65 cm². Where did this "extra" square centimeter come from?

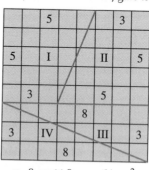

a. 8 cm × 8 cm = 64 cm²

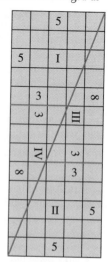

[*Hint:* Construct your own square 8 cm on a side, and then cut it into the four pieces as shown. Place the four pieces together as illustrated. Be sure to do your measuring and cutting very carefully. Satisfy yourself that this "extra" square centimeter has appeared. Can you explain this relationship?]

b. 13 cm × 5 cm = 65 cm²

▲ **Figure 8.10   Extra square cm?**

8.3   VOLUME AND CAPACITY

To measure area, we covered a region with square units and then found the area by using a mathematical formula. A similar procedure is used to find the amount of space inside a solid object, which is called its **volume.** We can imagine filling the space with **cubes.** A **cubic inch** and a **cubic centimeter** are shown in Figure 8.11.

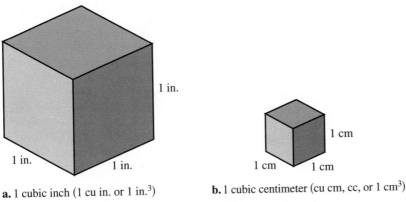

a. 1 cubic inch (1 cu in. or 1 in.³)   **b.** 1 cubic centimeter (cu cm, cc, or 1 cm³)

▲ **Figure 8.11   Cubic units used for measuring volume**

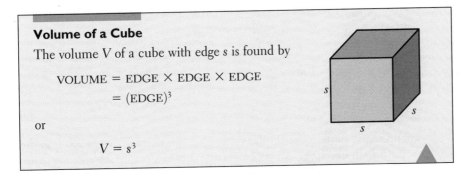

Volume of a Cube

The volume V of a cube with edge s is found by

$$\text{VOLUME} = \text{EDGE} \times \text{EDGE} \times \text{EDGE}$$
$$= (\text{EDGE})^3$$

or

$$V = s^3$$

If the solid is not a cube but is a box (called a **rectangular parallelepiped**) with edges of different lengths, the volume can be found similarly.

━━━ EXAMPLE 1

Find the volume of a box that measures 4 ft by 6 ft by 4 ft.

Solution   There are 24 cubic feet on the bottom layer of cubes. Do you see how many layers of cubes will fill the solid? (See Figure 8.12.) Since there are four layers with 24 cubes in each, the total is

$$4 \times 24 = 96$$

The volume is 96 ft³.

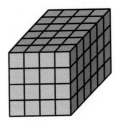

▲ **Figure 8.12   What is the volume?**

Historical Note

JOHANN BERNOULLI
(1667–1748)

The remarkable Bernoulli family of Switzerland produced at least eight noted mathematicians over three generations. Two brothers, Jakob (1654–1705) and Johann, were bitter rivals. In spite of their disagreements, they maintained continuous communication with each other and with the great mathematician Leibniz (see Historical Note on page 79). The brothers were extremely influential advocates of the newly born calculus. Johann was the most prolific of the clan, but was jealous and cantankerous; he tossed a son (Daniel) out of the house for winning an award he had expected to win himself. He was, however, the premier teacher of his time. Euler (see Historical Note on page 798) was his most famous student. Curiously, Johann and Jakob's father opposed his sons' study of mathematics and tried to force Jakob into theology. He chose the motto *Invito patre sidera verso*, which means "I study the stars against my father's will."

Volume of a Box (Parallelepiped)

The volume V of a box (parallelepiped) with edges ℓ, w, and h is

$$\text{VOLUME} = \text{LENGTH} \times \text{WIDTH} \times \text{HEIGHT}$$

or

$$V = \ell \times w \times h$$
$$= \ell w h$$

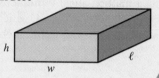

■ EXAMPLE 2

Find the volume of each solid.

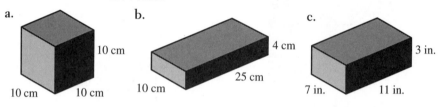

a. 10 cm 10 cm 10 cm

b. 4 cm 25 cm 10 cm

c. 3 in. 7 in. 11 in.

Solution

a. $V = s^3$

　$= (10 \text{ cm})^3$

　$= (10 \times 10 \times 10) \text{ cm}^3$

　$= 1{,}000 \text{ cm}^3$

b. $V = \ell w h$

　$= (25 \text{ cm})(10 \text{ cm})(4 \text{ cm})$

　$= (25 \times 10 \times 4) \text{ cm}^3$

　$= 1{,}000 \text{ cm}^3$

c. $V = \ell w h$

　$= 11 \text{ in.} \times 7 \text{ in.} \times 3 \text{ in.}$

　$= (11 \times 7 \times 3) \text{ in.}^3$

　$= 231 \text{ in.}^3$

Sometimes the dimensions for the volume we are finding are not given in the same units. In such cases, you must convert all units to a common unit. The common conversions are as follows:

1 ft = 12 in.	To convert feet to inches, multiply by 12. To convert inches to feet, divide by 12.
1 yd = 3 ft	To convert yards to feet, multiply by 3. To convert feet to yards, divide by 3.
1 yd = 36 in.	To convert yards to inches, multiply by 36. To convert inches to yards, divide by 36.

■ EXAMPLE 3

Suppose you are pouring a rectangular driveway with dimensions 24 ft by 65 ft. The depth of the driveway is 3 in. and concrete is ordered by the yard. By a "yard" of concrete, we mean a cubic yard. You cannot order part of a yard of concrete. How much concrete should you order?

Solution   There are three different measurements in this problem: inches, feet, and yards. Since we want the answer in cubic yards, we will convert all of these measurements to yards:

$$65 \text{ ft} = (65 \div 3) \text{ yd} = \tfrac{65}{3} \text{ yd} \qquad \text{This is the length, } \ell.$$
$$24 \text{ ft} = (24 \div 3) \text{ yd} = 8 \text{ yd} \qquad \text{This is the width, } w.$$
$$3 \text{ in.} = (3 \div 36) \text{ yd} = \tfrac{3}{36} \text{ yd} = \tfrac{1}{12} \text{ yd} \quad \text{This is the height (depth), } h.$$

$$
\begin{aligned}
V &= \ell wh \\
&= \tfrac{65}{3}(8)(\tfrac{1}{12}) \text{ yd}^3 \qquad \text{Think of 8 as } \tfrac{8}{1}. \\
&= \frac{65 \times 8 \times 1}{3 \times 12} \\
&= \frac{130}{9} \\
&= 14\tfrac{4}{9}
\end{aligned}
$$

You must order 15 cubic yards of concrete.

Capacity

One of the most common applications of volume involves measuring the amount of liquid a container holds, which we refer to as its **capacity.** For example, if a container is 2 ft by 2 ft by 12 ft, it is fairly easy to calculate the volume:

$$2 \times 2 \times 12 = 48 \text{ ft}^3$$

But this still doesn't tell us how much water the container holds. The capacities of a can of cola, a bottle of milk, an aquarium tank, the gas tank in your car, and a swimming pool can all be measured by the amount of fluid they can hold.

Standard Units of Capacity

U.S. System	Metric System
gallon (gal)	liter (L)
quart (qt; $\tfrac{1}{4}$ gal)	kiloliter (kl; 1,000 L)
ounce (oz; $\tfrac{1}{128}$ gal)	milliliter (ml; $\tfrac{1}{1,000}$ L)
cup (c; 8 oz)	

Most containers of liquid that you buy have capacities stated in both milliliters and ounces, or quarts and liters (see Figure 8.13). Some of these size statements are listed in Table 8.1 on page 582. The U.S. Bureau of Alcohol, Tobacco, and Firearms has made metric bottle sizes for liquor mandatory, so the half-pint, fifth, and quart have been replaced by 200-ml, 750-ml, and 1-L sizes. A typical dose of

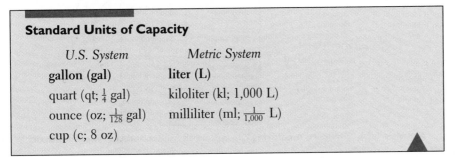

1.06 qt   0.95 ℓ

▲ **Figure 8.13   Standard capacities**

cough medicine is 5 ml, and 1 kl is 1,000 L, or about the amount of water one person would use for all purposes in two or three days.

⊘ You should remember some of these references for purposes of estimation. For example, remember that a can of Coke is 355 ml, and that a liter of milk is about the same as a quart of milk. A cup of coffee would be about 300 ml and a spoonful of medicine about 5 ml. ⊘

TABLE 8.1　Capacities of Common Grocery Items, as Shown on Labels

Item	U.S. capacity	Metric capacity
Milk	$\frac{1}{2}$ gal	1.89 L
Milk	1.06 qt	1 L
Budweiser	12 oz	355 ml
Coke	67.6 oz	2 L
Hawaiian Punch	1 qt	0.95 L
Del Monte pickles	1 pt 6 oz	651 ml

Since it is common practice to label capacities in both U.S. and metric measuring units, it will generally not be necessary for you to make conversions from one system to another. But if you do, it is easy to remember that a liter is just a little larger than a quart, just as a meter is a little larger than a yard.

To measure capacity, you use a measuring cup.

▬ EXAMPLE 4

Measure the amount of liquid in the measuring cup in Figure 8.14, in both the U.S. system and the metric system.

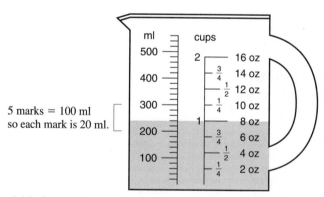

▲ Figure 8.14　Standard measuring cup with both metric and U.S. measurements

Solution

　　Metric:　240 ml　　U.S.:　About 1 c or 8 oz　　　　　▬

Some common relationships among volume and capacity measurements in the U.S. system are shown in Figure 8.15.

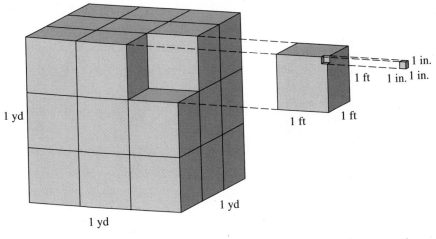

Volume:	$1 \text{ yd}^3 = 27 \text{ ft}^3$	$1 \text{ ft}^3 = 1{,}728 \text{ in.}^3$ $\quad 1 \text{ in.}^3$
Capacity:	about 200 gallons	7.48 gallons $\qquad 1 \text{ gal} = 231 \text{ in.}^3$

▲ **Figure 8.15   U.S. measurement relationship between volume and capacity**

In the U.S. system of measurement, the relationship between volume and capacity is not particularly convenient. One gallon of capacity occupies 231 in.^{3}. This means that, since the box in part **c** of Example 2 has a volume of 231 in.^{3}, we know that it will hold exactly 1 gallon of water.

To find the capacity of the 2-ft by 2-ft by 12-ft box mentioned earlier, we must change 48 ft^3 to cubic inches:

$$48 \text{ ft}^3 = 48 \times (1 \text{ ft}) \times (1 \text{ ft}) \times (1 \text{ ft})$$
$$= 48 \times 12 \text{ in.} \times 12 \text{ in.} \times 12 \text{ in.}$$
$$= 82{,}944 \text{ in.}^3 \quad \text{A calculator would help here.}$$

> ESTIMATE: Since 1 ft$^3 \approx$ 7.5 gal,
> 48 ft$^3 \approx$ 50 ft^3
> $\approx$ (50 × 7.5) gal
> = 375 gal

Since 1 gallon is 231 in.^{3}, the final step is to divide 82,944 by 231 to obtain approximately 359 gallons.

Calculator Comment

Converting from volume to capacity in the U.S. measurement system is best accomplished with a calculator. All of the steps for this problem can be combined in one sequence of calculator steps:

$\boxed{2} \ \boxed{\times} \ \boxed{2} \ \boxed{\times} \ \boxed{12} \qquad \boxed{\times} \ \boxed{12} \ \boxed{\times} \ \boxed{12} \ \boxed{\times} \ \boxed{12} \qquad \boxed{\div} \ \boxed{231} \ \boxed{=}$

This calculates the volume.　　This converts to cubic inches.　　This converts volume to capacity.

Display: **359.0649351**

The relationship between volume and capacity in the metric system is easier to remember. One cubic centimeter is one-thousandth of a liter. Notice that this is the same as a milliliter. For this reason, you will sometimes see cc used to mean cm³ or ml. These relationships are shown in Figure 8.16.

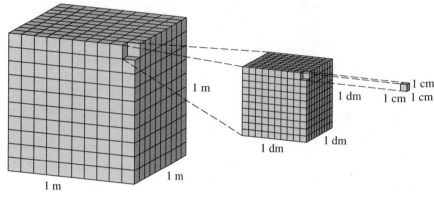

Volume: 1 m³ = 1,000 dm³   1 dm³ = 1,000 cm³   1 cm³ = 1 cc
Capacity: 1,000 L = 1 kl   1 L = 1,000 ml   1 ml = 1 cc

▲ **Figure 8.16   Metric measurement relationship between volume and capacity**

Relationship Between Volume and Capacity

1 liter = 1,000 cm³

1 gallon = 231 in.³;   1 ft³ ≈ 7.48 gal

████ EXAMPLE 5

How much water would each of the following containers hold?

a.   b.

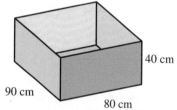   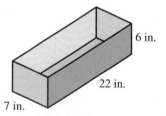

ESTIMATE for part b: Container is approximately $\frac{1}{2}$ ft × 2 ft × $\frac{1}{2}$ ft = $\frac{1}{2}$ ft³ ≈ ($\frac{1}{2}$ × 7.5) gal = 3.75 gal

Solution

a. $V = 90 \times 80 \times 40$ cm³

    $= 288{,}000$ cm³

Since each 1,000 cm³ is 1 liter,

$$\frac{288{,}000}{1{,}000} = 288$$

This container would hold 288 liters.

b. $V = 7 \times 22 \times 6$ in.

    $= 924$ in.³

Since each 231 in.³ is 1 gallon,

$$\frac{924}{231} = 4$$

This container would hold 4 gallons.

▬

▄ EXAMPLE 6

An economy swimming pool is advertised as being 20 ft × 25 ft × 5 ft. How many gallons will it hold?

Solution

$$V = 20 \text{ ft} \times 25 \text{ ft} \times 5 \text{ ft}$$
$$= 2,500 \text{ ft}^3$$

Since 1 ft³ ≈ 7.48 gal, the swimming pool contains approximately

$$2,500 \times 7.48 = 18,700 \text{ gallons}$$

PROBLEM SET 8.3

▲ A Problems

1. **IN YOUR OWN WORDS** Contrast finding length, area, and volume.

2. **IN YOUR OWN WORDS** Contrast volume and capacity.

3. **IN YOUR OWN WORDS** Compare the sizes of a cubic inch and a cubic centimeter.

4. **IN YOUR OWN WORDS** Compare the sizes of a quart and a liter.

5. **IN YOUR OWN WORDS** Compare a meter and a yard.

In Problems 6–8, find the volume of each solid by counting the number of cubic centimeters in each box.

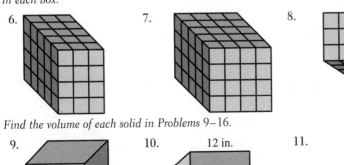

6. 7. 8.

Find the volume of each solid in Problems 9–16.

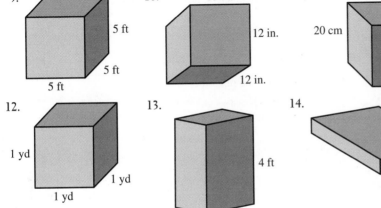

9. 5 ft, 5 ft, 5 ft

10. 12 in., 12 in., 12 in.

11. 20 cm, 20 cm, 20 cm

12. 1 yd, 1 yd, 1 yd

13. 4 ft, 2 ft, 3 ft

14. 20 mm, 2 mm, 10 mm

15.   16.

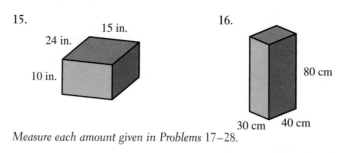

Measure each amount given in Problems 17–28.

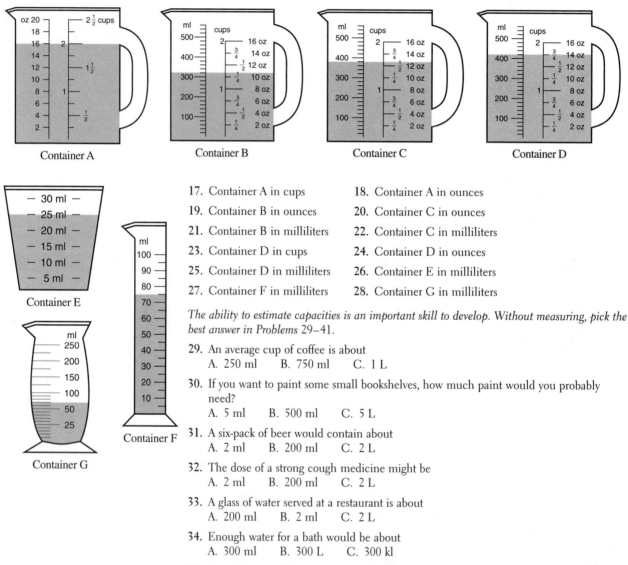

Container A   Container B   Container C   Container D

17. Container A in cups   18. Container A in ounces

19. Container B in ounces   20. Container C in ounces

21. Container B in milliliters   22. Container C in milliliters

23. Container D in cups   24. Container D in ounces

25. Container D in milliliters   26. Container E in milliliters

27. Container F in milliliters   28. Container G in milliliters

The ability to estimate capacities is an important skill to develop. Without measuring, pick the best answer in Problems 29–41.

29. An average cup of coffee is about
    A. 250 ml   B. 750 ml   C. 1 L

30. If you want to paint some small bookshelves, how much paint would you probably need?
    A. 5 ml   B. 500 ml   C. 5 L

31. A six-pack of beer would contain about
    A. 2 ml   B. 200 ml   C. 2 L

32. The dose of a strong cough medicine might be
    A. 2 ml   B. 200 ml   C. 2 L

33. A glass of water served at a restaurant is about
    A. 200 ml   B. 2 ml   C. 2 L

34. Enough water for a bath would be about
    A. 300 ml   B. 300 L   C. 300 kl

35. Enough gas to fill your car's empty tank would be about
    A. 15 L   B. 200 ml   C. 70 L

Container E

Container F

Container G

36. 50 kl of water would be about enough for
    A. taking a bath   B. taking a swim
    C. supplying the drinking water for a large city

37. Which measurement would be appropriate for administering some medication?
    A. ml   B. L   C. kl

38. You order some champagne for yourself and one companion. You would most likely order
    A. 2 ml   B. 700 ml   C. 20 L

39. The prefix *centi-* means
    A. one thousand   B. one-thousandth   C. one-hundredth

40. The prefix *milli-* means
    A. one thousand   B. one-thousandth   C. one-hundredth

41. The prefix *kilo-* means
    A. one thousand   B. one-thousandth   C. one-hundredth

▲ **B Problems**

What is the capacity for each of the containers in Problems 42–51? (Give answers in the U.S. system to the nearest tenth of a gallon or in metric to the nearest tenth of a liter.)

42.
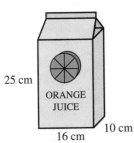
10 in.
10 in.
10 in.

43.

10 in.
10 in.
10 in.

44.
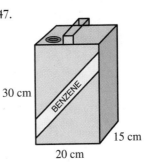
1 m
3 m
1 m

45.

25 cm
ORANGE JUICE
10 cm
16 cm

46.
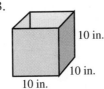
MILK
25 cm
5 cm
8 cm

47.
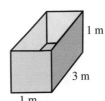
30 cm
BENZENE
15 cm
20 cm

48.
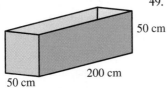
50 cm
200 cm
50 cm

49.
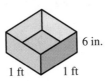
50 cm
6 in.
1 ft   1 ft

50.
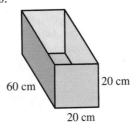
20 cm
60 cm
20 cm
20 cm

51.

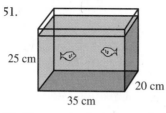

52. The exterior dimensions of a refrigerator are shown in the margin.
    a. How many cubic feet are contained within the refrigerator?
    b. If it is advertised as a 19-cu-ft refrigerator, how much space is taken up by the motor, insulation, and so on?

53. The exterior dimensions of a freezer are 48 inches by 36 inches by 24 inches, and it is advertised as being 27.0 cu ft. Is the advertised volume correctly stated?

54. Suppose that you must order concrete for a sidewalk 50 ft by 4 ft to a depth of 4 in. How much concrete is required? (Answer to the nearest $\frac{1}{2}$ cubic yard.)

▲ **Problem Solving**

*Use the plot plan shown in Figure 8.17 to answer Problems 55–58. Answer to the nearest cubic yard.** *

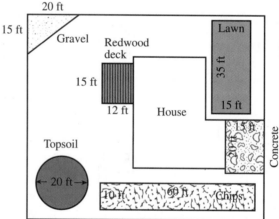

▲ **Figure 8.17   Plot plan**

55. How many cubic yards of sawdust are needed for preparation of the lawn area if it is to be spread to a depth of 6 inches?

56. How many cubic yards of chips are necessary if they are to be placed to a depth of 3 inches?

57. How much gravel is necessary if it is to be laid to a depth of 3 inches?

58. Suppose that you wish to pave the area labeled "Concrete." How much concrete is needed if it is to be poured to a depth of 4 inches?

59. How much water will a 7-m by 8-m by 2-m swimming pool contain (in kiloliters)?

---

* In practice, you would not round to the nearest yard, but rather would round up to ensure that you had enough material. However, for consistency in this book, we will round according to the rules developed in the first chapter.

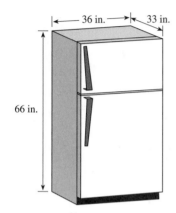

60. How much water will a 21-ft by 24-ft by 4-ft swimming pool contain (rounded to the nearest gallon)?

61. The total human population of the earth is about 5.0×10^9.
    a. If each person has the room of a prison cell (50 sq ft), and if there are about 2.8×10^7 sq ft in a square mile, how many people could fit into a square mile?
    b. How many square miles would be required to accommodate the entire human population of the earth?
    c. If the total land area of the earth is about 5.2×10^7 sq mi, and if all the land area were divided equally, how many acres of land would each person be allocated (1 sq mi = 640 acres)?

62. a. Guess what percentage of the world's population could be packed into a cubical box measuring $\frac{1}{2}$ mi on each side. [*Hint:* The volume of a typical person is about 2 cu ft.]
    b. Now calculate the answer to part **a**, using the earth's population as given in Problem 61.

8.4 MISCELLANEOUS MEASUREMENTS

In this chapter we have been discussing measurement.

Length

> If you are measuring length, you will use linear measures: ─────────
>
> in.; ft; yd; mi   cm; m; km

Area

> For area, you will use square measures: 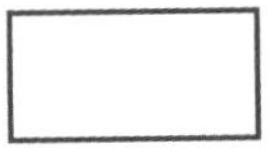
>
> in.²; ft²; yd²; mi²   cm²; m²; km²

Volume

> For volumes, you will use cubic measures:
>
> in.³; ft³   cm³ (or cc); m³; km³

In the previous section we found the volume and the capacity of boxes; now we can extend this to other solids. In Figure 8.18 we give some of the more common

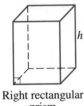

Right rectangular prism

$V = Bh$

Right circular cylinder

$V = Bh$

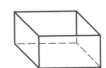

Pyramid

$V = \frac{1}{3}Bh$

Right circular cone

$V = \frac{1}{3}Bh$

Sphere

$V = \frac{4}{3}\pi r^3$

▲ **Figure 8.18   Common solids, with accompanying volume formulas** solids, along with the volume formulas. We will use B in each case to signify the area of the base and h for the height.

▬▬▬ **EXAMPLE 1**

Find the volume of the accompanying solid to the nearest cubic unit.

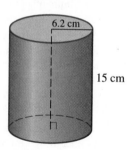

Solution   This is a right circular cylinder. Hence,

$$V = Bh$$

$$= \pi(6.2)^2(15)$$

$$\approx 1,811.4423$$

Notice that $B = \pi r^2$, where $r = 6.2$; the height of the cylinder is 15 ($h = 15$).
Press: $\boxed{\pi}$ $\boxed{\times}$ $\boxed{6.2}$ $\boxed{x^2}$ $\boxed{\times}$ $\boxed{15}$ $\boxed{=}$

The volume is $1,811$ cm³.

▬▬▬ **EXAMPLE 2**

Find the volume of the accompanying solid to the nearest cubic unit.

Solution   This is a right circular cone. Hence,

$$V = \frac{1}{3}Bh$$

$$= \frac{1}{3}\pi(5)^2 9$$ Notice that $B = \pi r^2$ and $r = 5$.

$$= 75\pi$$

$$\approx 235.6$$ You can use $\pi \approx 3.1416$ or the π key on your calculator.

The volume is 236 ft³.

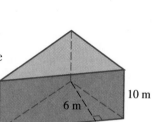

▬▬▬ **EXAMPLE 3**

Find the volume of the accompanying solid to the nearest cubic unit.

Solution   This is a right triangular prism. Hence,

$$V = Bh$$

$$= \frac{1}{2}(15)(6)(10)$$ Notice that $B = \frac{1}{2}ba$, where $a = 6$ (height of triangle), $b = 15$ (base of triangle), and $h = 10$ (height of prism).

$$= 450$$

The volume is 450 m³.

There once was a teacher named
  Streeter
Who taught stuff like meter and liter.
The kids moaned and groaned,
Until they were shown,
That meter and liter were neater!

▬▬▬ **EXAMPLE 4**

Find the volume of the accompanying solid to the nearest cubic unit.

Solution This is a sphere. Hence,

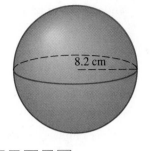

$$V = \frac{4}{3}\pi r^3$$

$$= \frac{4}{3}\pi(8.2)^3$$

$$\approx 2{,}309.564878$$ *Press:* [4] [×] [π] [×] [8.2] [^] [3] [÷] [3] [=]

The volume is 2,310 cm³. ▬

▬▬▬ **EXAMPLE 5**

Find the volume of the accompanying solid to the nearest cubic unit.

Solution This is a pyramid. Hence,

$$V = \frac{1}{3}Bh$$

$$= \frac{1}{3}(5)^2(3) \quad \text{Notice that the base is a square so } B = s^2 = 5^2.$$

$$= \frac{25}{3}(3)$$

$$= 25$$

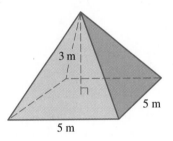

The volume is 25 m³. ▬

The measurements of length, area, volume, and capacity were discussed in the previous section. Two additional measurements we need to consider are mass and temperature.

Mass

The **mass** of an item is the amount of matter it comprises. The **weight** of an item is the heaviness of the matter.* The U.S. and metric units of measurement for mass or weight are given in the following box. Notice that, in the U.S. measurement system, an ounce is used as a weight measurement; this is not the same use of an ounce as a capacity measurement that we used earlier. The basic unit of measurement for mass in the metric system is the **gram,** which is defined as the mass of 1 cm³ of water, as shown in Figure 8.19.

1 cm³ of water   1 g

▲ **Figure 8.19 Mass of 1 g**

---

* Technically, weight is the force of gravity acting on a body. The mass of an object is the same on the moon as on Earth, whereas the weight of that same object would be different on the moon and on Earth, since they have different forces of gravity. For our purposes, you can use either word, *mass* or *weight*, because we are weighing things only on Earth.

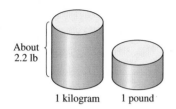

About 2.2 lb

1 kilogram   1 pound

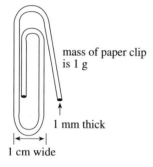

mass of paper clip is 1 g

1 mm thick

1 cm wide

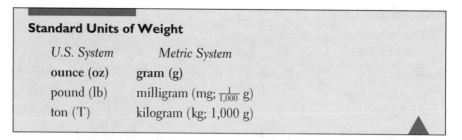

Standard Units of Weight

U.S. System	Metric System
ounce (oz)	**gram (g)**
pound (lb)	milligram (mg; $\frac{1}{1,000}$ g)
ton (T)	kilogram (kg; 1,000 g)

A paper clip weighs about 1 g, a cube of sugar about 3 g, and a nickel about 5 g. This book weighs about 1 kg, and an average-size person weighs from 50 to 100 kg. The weights of some common grocery items are shown in Table 8.2.

TABLE 8.2   Weights of Common Grocery Items, as Shown on Labels

Item	U.S. weight	Metric weight
Kraft cheese spread	5 oz	142 g
Del Monte tomato sauce	8 oz	227 g
Campbell cream of chicken soup	$10\frac{3}{4}$ oz	305 g
Kraft marshmallow cream	11 oz	312 g
Bag of sugar	5 lb	2.3 kg
Bag of sugar	22 lb	10 kg

We use a scale to measure weight. To weigh items, or ourselves, in metric units, we need only replace our U.S. weight scales with metric weight scales. As with other measures, we need to begin to think in terms of metric units, and to estimate the weights of various items. The multiple-choice questions in Problem Set 8.4 are designed to help you do this.

Temperature

The final quantity of measure that we'll consider in this chapter is temperature.

Standard Units of Temperature

The U.S. system has one common temperature unit:	The metric system has one common temperature unit:
Fahrenheit (°F)	**Celsius (°C)**

To work with temperatures, it is necessary to have some reference points.

U.S. Temperature		Metric Temperature	
Water freezes:	32°F	Water freezes:	0°C
Water boils:	212°F	Water boils:	100°C

We are usually interested in measuring temperature in three areas: atmospheric temperature (usually given in weather reports), body temperature (used to determine illness), and oven temperature (used in cooking). The same scales are used, of course, for measuring all of these temperatures. But notice the difference in the ranges of temperatures we're considering. The comparisons for Fahrenheit and Celsius are shown in Figure 8.20.

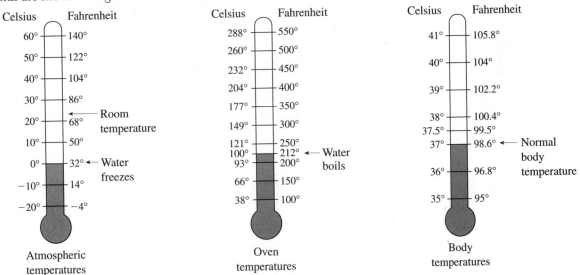

▲ **Figure 8.20 Temperature comparisons between Celsius and Fahrenheit**

Converting Units

As we have mentioned, one of the advantages (if not the chief advantage) of the metric system is the ease with which you can remember and convert units of measurement. The three basic metric units are related as shown in Figure 8.21.

Length: meter
Capacity: liter
Weight: gram

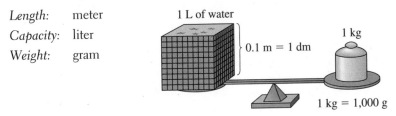

▲ **Figure 8.21 Relationship among meter, liter, and gram**

These units are combined with certain prefixes:

$$milli\text{- means } \frac{1}{1,000} \qquad centi\text{- means } \frac{1}{100} \qquad kilo\text{- means } 1,000$$

Other metric units are used less frequently, but they should be mentioned:

$$deci\text{- means } \frac{1}{10} \qquad deka\text{- means } 10 \qquad hecto\text{- means } 100$$

A listing of metric units (in order of size) is given in Table 8.3.

TABLE 8.3   Metric Measurements

Length	Capacity	Weight	Meaning	Memory aid
kilometer (km)	kiloliter (kl)	kilogram (kg)	1,000 units	Karl
hectometer (hm)	hectoliter (hl)	hectogram (hg)	100 units	Has
dekameter (dkm)	dekaliter (dkl)	dekagram (dkg)	10 units	Developed
meter (m)	liter (L)	gram (g)	1 unit	My
decimeter (dm)	deciliter (dl)	decigram (dg)	0.1 unit	Decimal
centimeter (cm)	centiliter (cl)	centigram (cg)	0.01 unit	Craving for
millimeter (mm)	milliliter (ml)	milligram (mg)	0.001 unit	Metrics

Basic Unit (meter, liter, gram)

To convert from one metric unit to another, you **simply move the decimal point.** For example, if the height of this book is 0.235 m, then it is also

Larger prefixes
$\begin{cases} 0.000235 \text{ km} & \leftarrow \text{Three decimal places} \\ 0.00235 \text{ hm} & \leftarrow \text{Two decimal places} \\ 0.0235 \text{ dkm} & \leftarrow \text{One decimal place} \end{cases}$

Smaller prefixes
$\begin{cases} 2.35 \text{ dm} & \leftarrow \text{One decimal place} \\ 23.5 \text{ cm} & \leftarrow \text{Two decimal places} \\ 235 \text{ mm} & \leftarrow \text{Three decimal places} \end{cases}$

■■■■ **EXAMPLE 6**

Write each metric measurement using each of the other prefixes.

a. 43 km   **b.** 60 L   **c.** 14.1 cg

Solution

a. All prefixes are smaller, so units become larger:

43 km	Given
430 hm	One place
4,300 dkm	Two places
43,000 m	Three places
430,000 dm	Four places
4,300,000 cm	Five places
43,000,000 mm	Six places

b. Larger prefixes, smaller numbers:

60 L	Given
6 dkl	One place
0.6 hl	Two places
0.06 kl	Three places

Smaller prefixes, larger numbers:

60 L	Given
600 dl	One place
6,000 cl	Two places
60,000 ml	Three places

c. Larger prefixes:

14.1 cg	Given
1.41 dg	One place
0.141 g	Two places
0.0141 dkg	Three places
0.00141 hg	Four places
0.000141 kg	Five places

Smaller prefixes:

14.1 cg	Given
141 mg	One place

To make a single conversion, use the pattern illustrated by Example 6 and simply count the number of decimal places.

■■■■ EXAMPLE 7

Make the indicated conversions.

a. 287 cm to km   **b.** 1.5 kl to L   **c.** 4.8 kg to g

Solution

a. From cm to km is five places; a larger prefix implies a smaller number, so move the decimal point to the left:

$$287 \text{ cm} = 0.00287 \text{ km}$$

b. From kl to L is three places; a smaller prefix implies a larger number, so move the decimal point to the right:

$$1.5 \text{ kl} = 1,500 \text{ L}$$

c. From kg to g is three places; a smaller prefix implies a larger number, so move the decimal point to the right:

$$4.8 \text{ kg} = 4,800 \text{ g}$$

PROBLEM SET 8.4

▲ A Problems

1. **IN YOUR OWN WORDS** Discuss the merits of the metric system as opposed to the U.S. measurement system. Why do you think the metric system has not yet been adopted in the United States?

2. **IN YOUR OWN WORDS** Discuss the relationships among measuring length, capacity, and weight.

3. **IN YOUR OWN WORDS** Explain how we change units within the metric system.

Name the metric unit you would use to measure each of the quantities in Problems 4–15.

4. The distance from New York to Chicago

5. The distance around your waist

6. Your height

7. The height of a building

8. The capacity of a wine bottle

9. The amount of gin in a martini

10. The capacity of a car's gas tank

11. The amount of water in a swimming pool

12. The weight of a pencil

13. The weight of an automobile

14. The outside temperature

15. The temperature needed to bake a cake

Without measuring, pick the best choice in Problems 16–35 by estimating.

16. A hamburger patty would weigh about
    A. 170 g   B. 240 mg   C. 2 kg

17. A can of carrots at the grocery store most likely weighs about
    A. 40 kg   B. 4 kg   C. 0.4 kg

18. A newborn baby would weigh about
    A. 490 mg   B. 4 kg   C. 140 kg

19. John tells you he weighs 150 kg. If John is an adult, he is
    A. underweight   B. about average   C. overweight

20. You have invited 15 people for Thanksgiving dinner. You should buy a turkey that weighs about
    A. 795 mg   B. 4 kg   C. 12 kg

21. Water boils at
    A. 0°C   B. 100°C   C. 212°C

22. If it is 32°C outside, you would most likely find people
    A. ice skating   B. water skiing

23. If the doctor says that your child's temperature is 37°C, your child's temperature is
    A. low   B. normal   C. high

24. You would most likely broil steaks at
    A. 120°C   B. 500°C   C. 290°C

25. A kilogram is __?__ a pound.
    A. more than   B. about the same as   C. less than

26. The prefix used to mean 1,000 is
    A. centi-   B. milli-   C. kilo-

27. The prefix used to mean $\frac{1}{1,000}$ is
    A. centi-   B. milli-   C. kilo-

28. The prefix used to mean $\frac{1}{100}$ is
    A. centi-   B. milli-   C. kilo-

News Clip

**Go Metric—
    Be a liter bug!**

Mike Keedy, Purdue University

29. 15 kg is a measure of
    A. length   B. capacity   C. weight   D. temperature

30. 28.5 m is a measure of
    A. length   B. capacity   C. weight   D. temperature

31. 6 L is a measure of
    A. length   B. capacity
    C. weight   D. temperature

32. 38°C is a measure of
    A. length   B. capacity
    C. weight   D. temperature

33. 7 ml is a measure of
    A. length   B. capacity
    C. weight   D. temperature

34. 68 km is a measure of
    A. length   B. capacity
    C. weight   D. temperature

35. 14.3 cm is a measure of
    A. length   B. capacity
    C. weight   D. temperature

Write each measurement given in Problems 36–43 using all of the metric prefixes.

	kilometer (km)	hectometer (hm)	dekameter (dkm)	meter (m)	decimeter (dm)	centimeter (cm)	millimeter (mm)
36.				9●			
38.						6●	
40.	4●						
42.					1	5	0●

	kiloliter (kl)	hectoliter (hl)	dekaliter (dkl)	liter (L)	deciliter (dl)	centiliter (cl)	milliliter (ml)
37.						6	3●
39.				3●	5		
41.			8●				
43.		3●	1				

▲ **B Problems**

In Problems 44–50 make the indicated conversions.

44. a. 1 cm = ＿＿ mm   b. 1 cm = ＿＿ m   c. 1 cm = ＿＿ km

45. a. 1 mm = ＿＿ cm   b. 1 mm = ＿＿ m   c. 1 mm = ＿＿ km

46. a. 1 m = ＿＿ mm   b. 1 m = ＿＿ cm   c. 1 m = ＿＿ km

47. a. 1 ml = ＿＿ dl   b. 1 ml = ＿＿ L   c. 1 ml = ＿＿ kl

48. a. 1 L = ＿＿ ml   b. 1 L = ＿＿ dl   c. 1 L = ＿＿ kl

49. a. 1 kl = ＿＿ ml   b. 1 kl = ＿＿ dl   c. 1 kl = ＿＿ L

50. a. 1 g = ＿＿ mg   b. 1 g = ＿＿ cg   c. 1 g = ＿＿ kg

Find the volumes of the solids in Problems 51–59 correct to the nearest cubic unit.

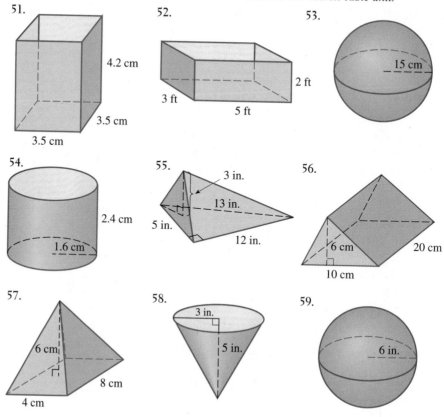

51.
4.2 cm
3.5 cm
3.5 cm

52.
2 ft
3 ft
5 ft

53.
15 cm

54.
2.4 cm
1.6 cm

55.
3 in.
13 in.
5 in.
12 in.

56.
6 cm
10 cm
20 cm

57.
6 cm
8 cm
4 cm

58.
3 in.
5 in.

59.
6 in.

▲ **Problem Solving**

60. Show how a spreadsheet could be used to print a table of temperature conversions between the Fahrenheit and Celsius scales. Your table should consist of a column showing Celsius temperatures from 0 to 40 in steps of 5 degrees, and a column to the right, with corresponding Fahrenheit temperatures.

61. A polyhedron is a simple closed surface in space whose boundary is composed of polygonal regions (see Figure 8.22). A rather surprising relationship exists among the numbers of vertices, edges, and sides of polyhedra. See if you can discover it by looking for patterns in the figures and filling in the blanks.

Triangular
pyramid

Quadrilateral
pyramid

Pentagonal
pyramid

Regular
tetrahedron

Regular
hexahedron (cube)

Regular
octahedron

Regular
dodecahedron

Regular
icosahedron

▲ **Figure 8.22 Some common polyhedra**

Figure	Number of:		
	Sides	Vertices	Edges
a. Triangular pyramid	4	4	6
b. Quadrilateral pyramid	5	___	8
c. Pentagonal pyramid	___	6	10
d. Regular tetrahedron	4	4	___
e. Cube	___	___	12
f. Regular octahedrom	___	6	___
g. Regular dodecahedron	___	___	30
h. Regular icosahedron	___	___	30

▲ **Individual Research**

62. **HISTORICAL QUESTION** In the Boston Museum of Fine Arts is a display of carefully made stone cubes found in the ruins of Mohenjo-Daro of the Indus. The stones are a set of weights that exhibit the binary pattern, 1, 2, 4, 8, 16, ... The fundamental unit displayed is just a bit lighter than the ounce in the U.S. measurement system. The old European standard of 16 oz for 1 pound may be a relic of the same idea. Write a paper showing how a set of such stones can successfully be used to measure any reasonable given weight of more than one unit.

63. Construct models for the regular polyhedra.

References
H. S. M. Coxeter, *Introduction to Geometry* (New York: Wiley, 1961).
Jean Pederson, "Plaited Platonic Puzzles," *Two-Year College Mathematics Journal*, Fall 1973, pp. 23–27.
Max Sobel and Evan Maletsky, *Teaching Mathematics: A Sourcebook* (Englewood Cliffs, N.J.: Prentice-Hall, 1975), pp. 173–184.
Charles W. Trigg, "Collapsible Models of the Regular Octahedron," *The Mathematics Teacher*, October 1972, pp. 530–533.

64. Which solids occur in nature? Find examples of each of the five regular solids. For example, the skeletons of marine animals called *radiolaria* show each of these forms.

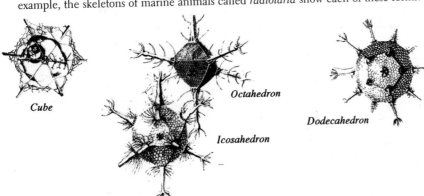

Cube

Octahedron

Icosahedron

Dodecahedron

Tetrahedron

References
David Bergamini, *Mathematics* (New York: Time, Inc., Life Science Library, 1963), Chapter 4.
Juithlynee Carson, "Fibonacci Numbers and Pineapple Phyllotaxy," *Two-Year College Mathematics Journal*, June 1978, pp. 132–136.

James Newman, *The World of Mathematics* (New York: Simon and Schuster, 1956). "Crystals and the Future of Physics," pp. 871–881; "On Being the Right Size," pp. 952–957; and "The Soap Bubble," pp. 891–900.

APPENDIX: U.S.–METRIC CONVERSIONS

This section may be considered as an appendix for reference use. The emphasis in this chapter has been on everyday use of the metric system. However, certain specialized applications require more precise conversions than we've considered. On the other hand, you don't want to become bogged down with arithmetic to the point where you say "nuts to the metric system."

The most difficult obstacle involving the change from the U.S. system to the metric system is not mathematical but psychological. However, if you have understood the chapter to this point, you realized in the last section that *working within the metric system is much easier than working within the U.S. system.*

With these ideas firmly in mind, and realizing that your everyday work with the metric system is discussed in the other sections of this chapter, we present an appendix of conversion factors between these measurement systems. Many calculators will make these conversions for you; check your owner's manual.

Length Conversions

U.S. to Metric			Metric to U.S.		
When you know	Multiply by	To find	When you know	Multiply by	To find
in.	2.54	cm	cm	0.39370	in.
ft	30.48	cm	m	39.37	in.
ft	0.3048	m	m	3.28084	ft
yd	0.9144	m	m	1.09361	yd
mi	1.60934	km	km	0.62137	mi

Capacity Conversions

U.S. to Metric			Metric to U.S.		
When you know	Multiply by	To find	When you know	Multiply by	To find
tsp	4.9289	ml	ml	0.20288	tsp
tbsp	14.7868	ml	ml	0.06763	tbsp
oz	29.5735	ml	ml	0.03381	oz
c	236.5882	ml	ml	0.00423	c
pt	473.1765	ml	ml	0.00211	pt
qt	946.353	ml	ml	0.00106	qt
qt	0.9464	L	L	1.05672	qt
gal	3.7854	L	L	0.26418	gal

Weight Conversions

When you know	U.S. to Metric Multiply by	To find	When you know	Metric to U.S. Multiply by	To find
oz	28.3495	g	g	0.0352739	oz
lb	453.59237	g	g	0.0022046	lb
lb	0.453592	kg	kg	2.2046226	lb
T	907.18474	kg			

Temperature Conversions

U.S. to Metric	Metric to U.S.
1. Subtract 32 from degrees Fahrenheit.	1. Multiply Celsius degrees by $\frac{9}{5}$.
2. Multiply this result by $\frac{5}{9}$ to get Celsius.	2. Add 32 to this result to get Fahrenheit.

"Many arts there are which beautify the mind of people; of all other none do more garnish and beautify it than those arts which are called mathematical."

H. Billingsley

Biographical Sketch

"Mathematics is a shared, exciting experience that can be enjoyed by everyone."

Wade Ellis, Jr.

Wade Ellis, Jr., was born in Ann Arbor, Michigan, on February 24, 1942, and is a master teacher at West Valley College in Saratoga, CA. He described a classroom experience to me that ended with the exclamation, "The class broke into wild applause and cheers." We all remember our good (and also probably some bad) teachers. When we are in a class that is taught by an instructor who instigates wild applause and cheers it can't help but be a memorable experience.

"Mathematics: a shared, exciting experience that can be enjoyed by everyone," is the way that Professor Ellis describes his subject. He says that what he enjoys most about mathematics is the pleasure of transferring beautiful, interesting, and difficult ideas from one mind to another.

Professor Ellis comes from a teaching family. His mother taught kindergarten and his father, Wade Ellis, Sr., was a professor of mathematics at Oberlin College and the University of Michigan. "They were my first and best teachers," says Wade.

Book Report

Write a 500-word report on the following book:

> ▲ *Ethnomathematics: A Multicultural View of Mathematical Ideas*, Marcia Ascher (Pacific Grove, CA: Brooks/Cole), 1991.

Important Terms

Accuracy [8.1]
Acre [8.2]
Area [8.2]
Capacity [8.3]
Celsius [8.4]
Centi- [8.1]
Circle [8.1]
Circumference [8.1]
Cone [8.4]
Cubic unit [8.3]
Cup [8.3]
Cylinder [8.4]
Deci- [8.1]
Deka- [8.1]

Diameter [8.1]
Equilateral triangle [8.1]
Fahrenheit [8.4]
Foot [8.1]
Gallon [8.3]
Gram [8.4]
Hecto- [8.1]
Inch [8.1]
Kilo- [8.1]
Length [8.1]
Liter [8.3]
Mass [8.4]
Measure [8.1]
Meter [8.1]

Metric system [8.1]
Mile [8.1]
Milli- [8.1]
Ounce [8.3; 8.4]
Parallelepiped [8.3]
Parallelogram [8.2]
Perimeter [8.1]
Pound [8.4]
Precision [8.1]
Prism [8.4]
Pyramid [8.4]
Quart [8.3]
Radius [8.1]

Rectangle [8.1]
Sphere [8.4]
Square [8.1]
Square unit [8.2]
Surface area [8.2]
Temperature [8.4]
Ton [8.4]
Trapezoid [8.2]
United States system [8.1]
Volume [8.3]
Weight [8.4]
Yard [8.1]

Important Ideas

Structure of the metric system; length (meter), capacity (liter), and mass (gram) [8.1–8.4]

Finding the distance around a closed figure (perimeter and circumference) [8.1]

From memory, approximate a given length in inches and in centimeters [8.1]

Know the area formulas for rectangles, squares, parallelograms, triangles, trapezoids, and circles [8.2]

Know the size of an acre [8.2]

Know the volume formulas for boxes (parallelepipeds), right rectangular prisms, right circular cylinders, pyramids, right circular cones, and spheres [8.3; 8.4]

Know the units for measuring capacity [8.3]

Know the relationship between volume and capacity [8.3]

Know the Fahrenheit and Celsius temperatures for both water freezing and water boiling [8.4]

Name a metric unit you would use to measure a given quantity [8.1–8.4]

Types of Problems

Distinguish the concepts of precision and accuracy [8.1]

Be able to measure length in both the United States and metric measurement systems [8.1]

Estimate lengths; choose an appropriate unit for measuring a given length [8.1]

Find the perimeter, circumference, or distance around a given figure [8.1]

Solve applied problems involving length [8.1]

Be able to measure area in both the United States and metric measurement systems [8.1]

Estimate areas; choose an appropriate unit for measuring a given area [8.2]

Find the area of a given figure [8.2]

Solve applied problems involving area [8.2]

Estimate capacities [8.3]

Find the capacity of a given container [8.3]

Solve applied problems involving capacity [8.3]

Measure the amount of a liquid [8.3]

Be able to measure mass (weight) in both the United States and metric measurement systems [8.4]

Estimate weights; choose an appropriate unit for measuring a given mass [8.4]

Be able to measure temperature in both the United States and metric measurement systems [8.4]

Be able to measure volume in both the United States and metric measurement systems [8.3; 8.4]

Estimate volumes; choose an appropriate unit for measuring a given volume [8.3; 8.4]

Find the volume of a given solid [8.3; 8.4]

Solve applied problems involving volume [8.3; 8.4]

Estimate temperatures [8.4]

Change units within the metric system [8.4]

Change units within the United States system [8.4]

Change units between the metric and United States systems [Appendix to Section 8.4]

CHAPTER 8   Review Questions

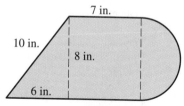

7 in.

10 in.

8 in.

6 in.

▲ **Figure 8.23**

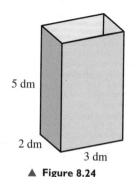

5 dm

2 dm

3 dm

▲ **Figure 8.24**

1. Without any measuring device, draw a segment approximately 10 cm long.

2. Measure this segment to the nearest tenth cm. _____

3. What is the width of your index finger in both metric and U.S. systems?

4. What is the distance (to the nearest inch) around the illustration in Figure 8.23?

5. Find the area (to the nearest square inch) of the illustration in Figure 8.23.

6. Find the volume of the box in Figure 8.24.

7. What is the capacity of the box in Figure 8.24?

8. What is the outside surface area of the box in Figure 8.24?

9. Find the volume (to the nearest tenth of a cubic foot) of a sphere with radius 1 ft.

10. What are the units of measurement in both the U.S. and metric measurement systems that you would use to measure the height of a building?

11. What does the measurement 8.6 ml measure?

12. A rectangular pool is 80 ft by 30 ft with a depth of 3 ft.
    a. What is the surface area of the top of the pool?
    b. What is the volume of the pool?
    c. What is the capacity of the pool (to the nearest gallon)?

13. Estimate the temperature in both the U.S. and metric measurement systems on a hot summer day.

14. What does the prefix *milli-* mean?

15. How many centimeters are equivalent to 10 km?

16. How many square yards of carpet are necessary to carpet an 11-ft by 16-ft room? (Assume that you cannot purchase part of a square yard.)

17. Consider the following advertisement from Round Table Pizza.

Pizzas
Made with fine natural cheeses, the choicest meats and the freshest vegetables, plus our own spicy sauce.

Original Style
Lavish toppings on a thin, crisp crust.

Mini Serves 1	**Small** Serves 1–2	**Medium** Serves 2–3	**Large** Serves 3–4
3.71	7.95	11.18	13.88

King Arthur's Supreme
Cheeses, pepperoni, sausage, salami, beef, linguica, mushrooms, green peppers, onions and black olives. Round Table's finest.
Shrimp and anchovies available upon request.

If you order a pizza, what size should you order if you want the best price per square inch? (That is, compare the price per square inch for the various pizza sizes; assume the diameters for the four sizes are 6-in., 10-in., 12-in., and 14-in.)

18. Suppose you were the owner of a pizza restaurant and were going to offer a 16-in. diameter pizza. What would you charge for the pizza if you wanted the price to be comparable with the prices of the other sizes at Round Table (see Problem 17)?

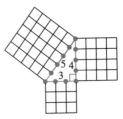

19. The Pythagorean theorem (Chapter 4) states that, for a right triangle with sides a and b and hypotenuse c, $a^2 + b^2 = c^2$. In this chapter, we saw that a^2 is the area of a square of length a. We can, therefore, illustrate the Pythagorean theorem for a triangle with sides 3, 4, and 5, as shown in Figure 8.25. Illustrate the Pythagorean theorem for a right triangle with sides 5, 12, and 13. On the other hand, if you draw a triangle with sides measuring 2, 3, and 4 units, you would find that $2^2 + 3^2 = 4 + 9 = 13$ and $4^2 = 16$ so $2^2 + 3^2 \neq 4^2$. Also notice that such a triangle is not a right triangle.

20. The distributive law (Chapter 4) states that
$$ab + ac = a(b + c)$$

In this chapter, we say ab and ac represent areas of two rectangles, one with sides a and b, and the other with sides a and c. Give a geometric justification for the distributive property.

▲ **Figure 8.25   Geometric interpretation of the Pythagorean theorem**

Revisiting the Real World . . .

IN THE REAL WORLD   "I can't wait until I move into my new home," said Susan. "I've been living out of boxes for longer than I want to remember."

"Have you picked out your carpet, tile, and made the other choices about color schemes yet?" asked Laurie.

"Heavens, no! They are just doing the framing and I have to build the decks, but I will get around to those selections soon," said Susan. "I hope I don't run out of money before I'm finished. My contractor gave me allowances for carpet, tile, light fixtures, and the like, but after looking around a bit I think those amounts are totally unrealistic. I think the goal was to come up with the lowest bid, and the dimensions on the house plan do not seem to match the amounts allowed. Can you help me make some estimates this Saturday?"

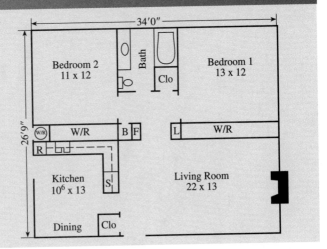

Commentary

We begin by making an estimate about the area of the house. We see that the house is 26 ft 9 in. by 34 ft so we find $30 \times 30 = 900$ as an estimate. To find the area of the house we multiply:

$$26\tfrac{9}{12} \times 34 = 26.75 \times 34 = 909.5$$

This is about 910 ft², so if we know that the approximate construction cost is $70 per square foot, the construction costs will be about $70 \times 910 = \$63,700$. Since the area of the living room is 22×13, we find its area to be 286 ft². If living room carpet and pad cost $52/yd², the cost for carpeting the living room will be

$$\underbrace{286 \div 9}_{\text{Change square feet to square yards.}} \times 52 = 1,652.444444$$

This assumes, however, that we can purchase part of a square yard. If we cannot purchase part of a yard, we must purchase 32 yd² because $286 \div 9 \approx 31.77$. Thus,

$32 \times 52 = 1,664$, and the carpet should cost $1,664. The area of Bedroom 1 is 13 ft $\times$ 12 ft $= 156$ ft^2 and the area of Bedroom 2 is 11 ft $\times$ 12 ft $= 132$ ft^2. If the bedrooms are carpeted with a carpet that costs $45 per square yard (including labor and pad), the total cost for carpeting the bedrooms can be found. The total area to be carpeted is

$$156 \text{ ft}^2 + 132 \text{ ft}^2 = 288 \text{ ft}^2$$

In square yards,

$$288 \div 9 = 32 \text{ yd}^2$$

The cost is $32 \times \$45 = \$1,440$. Finally, the kitchen/dining area is labeled $10^6 \times 13$. This is not scientific notation, but on architectural plans means 10 ft 6 in. by 13 ft, so the area is

$$10.5 \text{ ft} \times 13 \text{ ft} = 136.5 \text{ ft}^2$$

In these areas have hardwood floors at $9/ft^2, we would estimate the cost at $1,233 (that is, 137×9).

Group Research

Working in small groups is typical of most work environments, and being able to work with others to communicate specific ideas is an important skill to learn. Work with three or four other students to submit a single report based on each of the following questions.

1. Suppose the house whose floor plan is shown in the Commentary has an 8-ft ceiling in all rooms except the living room, which has a 10-ft cathedral ceiling. Approximately how many marbles would fit into this house?

2. Suppose you wish to build a spa on a wood deck. The deck is to be built 4 ft above level ground. It is to be 50 ft by 30 ft and is to contain a spa that is circular with a 14-ft diameter. The spa is 4 ft deep.
   a. How much water will the spa contain, and how much will it weigh? Assume that the spa itself weighs 550 lb.
   b. Draw plans for the wood deck.
   c. Draw up a materials list.
   d. Estimate the cost for this installation.

3. A Pythagorean triple is a set of three numbers that satisfy the Pythagorean theorem, $a^2 + b^2 = c^2$. Certain Pythagorean triples are well known, such as $\{3, 4, 5\}$ and $\{5, 12, 13\}$.
   a. Since $\{3, 4, 5\}$ is one Pythagorean triple, show that $\{3k, 4k, 5k\}$ is also a Pythagorean triple for any natural number k.
   b. Since $\{5, 12, 13\}$ is one Pythagorean triple, show that $\{5k, 12k, 13k\}$ is also a Pythagorean triple for any natural number k.
   c. Find five *different* sets of Pythagorean triples.

4. Here is a simple formula for finding Pythagorean triples. It was given to me in an elevator by my friend Bert Liberi (who is also a great mathematician). If m is any natural number greater than 1, then

$$\frac{1}{m} + \frac{1}{m + 2} = \frac{a}{b}$$

The reduced fraction $\dfrac{a}{b}$ will have the property that the set $\{a, b, c\}$ is a Pythagorean triple. For example, if $m = 2$, then

$$\frac{1}{2} + \frac{1}{2 + 2} = \frac{1}{2} + \frac{1}{4} = \frac{3}{4}$$

Thus, the first two numbers of the triple are 3 and 4. For the third number in the triple, we find

$$c = \sqrt{3^2 + 4^2} = \sqrt{9 + 16} = \sqrt{25} = 5$$

so the set is $\{3, 4, 5\}$. Find ten sets of Pythagorean triples not found in Problem 3.

The Nature of Probability

It is a truth very certain that, when it is not in our power to determine what is true, we ought to follow what is most probable.

— René Descartes

In the Real World . . .

Probability is the mathematics of uncertainty. "Wait a minute," interrupts Sammy, "I thought that mathematics was absolute and that there was no uncertainty about it!" Well, Sammy, suppose that you are playing a game of Monopoly. Do you know which property you'll land on in your next turn? "No, but neither do you!" That's right, but mathematics can tell us where you are *most likely* to land. There is nothing uncertain about probability; rather, probability is a way of describing uncertainty. For example, what is the probability of tossing a coin and obtaining heads? "That's easy; it's one-half," answers Sammy. Right, but *why* do you say it is one-half?

Contents

Perspective

We see examples of probability every day. Weather forecasts, stock market analyses, contests, children's games, political polls, game shows, and gambling all involve ideas of probability. Probability is the mathematics of uncertainty.

If you have ever played the lottery (hundreds of billions are wagered legally each year) or bought a life insurance policy, you have indirectly used probability. Animals and plants are bred to enhance certain traits to be passed through generations, and these breeding techniques are directed by the probability of genetics. Research in human inherited diseases such as cystic fibrosis, sickle-cell anemia, and Huntington's disease all involve probability.

In this chapter, we'll investigate the definition of probability and some of the procedures for dealing with probability.

9.1   INTRODUCTION TO PROBABILITY

Terminology

An **experiment** is an observation of any physical occurrence. The **sample space** of an experiment is the set of all its possible outcomes. An **event** is a subset of the sample space. If an event is the empty set, it is called the **impossible event**; and if it has only one element, it is called a **simple event**.

▬▬▬ EXAMPLE 1

a. What is the sample space for the experiment of tossing a coin and then rolling a die?

b. List the following events for the sample space in part **a**:

E = {rolling an even number on the die}

H = {tossing a head}

X = {rolling a six and tossing a tail}

Which of these (if any) are simple events?

Solution   A coin is considered *fair* if the outcomes of head and tail are **equally likely**. We also note that a fair die is one for which the outcomes from rolling it are *equally likely*. A die for which one outcome is more likely than the others is called a *loaded die*. In this book, we will assume fair dice and fair coins unless otherwise noted.

a. You can visualize the sample space as shown in Figure 9.1. If the sample space is very large, it is sometimes worthwhile to make a direct listing, as shown in Figure 9.1. Sometimes, it is helpful to build sample spaces by using what is called a **tree diagram**:

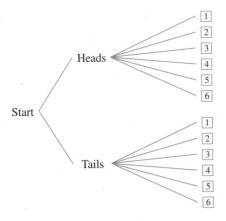

We see that S (the sample space) is

S = {H1, H2, H3, H4, H5, H6, T1, T2, T3, T4, T5, T6}

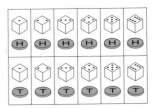

▲ **Figure 9.1   Sample space for tossing a coin and rolling a die**

b. An event must be a *subset* of the sample space. Notice that *E*, *H*, and X are all subsets of *S*. Therefore,

$$E = \{H2, H4, H6, T2, T4, T6\}$$
$$H = \{H1, H2, H3, H4, H5, H6\}$$
$$X = \{T6\}$$

X is a simple event, because it has only one element.   ▬

Two events *E* and *F* are said to be **mutually exclusive** if $E \cap F = \varnothing$; that is, they cannot both occur simultaneously.

▬▬▬ **EXAMPLE 2**

Suppose that you perform an experiment of rolling a die. Find the sample space, and then let $E = \{1, 3, 5\}$, $F = \{2, 4, 6\}$, $G = \{1, 3, 6\}$, and $H = \{2, 4\}$. Which of these are mutually exclusive?

Solution   The sample space for a single die, as shown in Figure 9.2, is

$$S = \{1, 2, 3, 4, 5, 6\} \quad \text{We look only at the spot on top.}$$

Evidently,

E and *F* are mutually exclusive, since $E \cap F = \varnothing$.
G and *H* are mutually exclusive, since $G \cap H = \varnothing$.
E and *H* are mutually exclusive, since $E \cap H = \varnothing$.

But

F and *H* are *not* mutually exclusive.
E and *G* are *not* mutually exclusive.
F and *G* are *not* mutually exclusive.   ▬

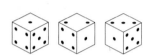

▲ **Figure 9.2   Sample space for a single die**

Probability

A **probabilistic model** deals with situations that are random in character and attempts to predict the outcomes of events with a certain stated or known degree of accuracy. For example, if we toss a coin, it is impossible to predict in advance whether the outcome will be a head or a tail. Our intuition tells us that it is equally likely to be a head or a tail, and somehow we sense that if we repeat the experiment of tossing a coin a large number of times, head will occur "about half the time." To check this out, I recently flipped a coin 1,000 times and obtained 460 heads and 540 tails. The percentage of heads is $\frac{460}{1,000} = 0.46 = 46\%$, which is called the *relative frequency*. If an experiment is repeated *n* times and an event occurs *m* times, then

$$\frac{m}{n} \text{ is called the } \textbf{relative frequency} \text{ of the event.}$$

Our task is to create a model that will assign a number *p*, called the *probability of an event*, which will predict the relative frequency. This means that for a *suffi-*

ciently large number of repetitions of an experiment,

$$p \approx \frac{m}{n}$$

Probabilities can be obtained in one of three ways:

1. **Empirical probabilities** (also called *a posteriori* models) are obtained from experimental data. For example, an assembly line producing brake assemblies for General Motors produces 1,500 items per day. The probability of a defective brake can be obtained by experimentation. Suppose the 1,500 brakes are tested and 3 are found to be defective. Then the empirical probability is the relative frequency of occurrence, namely

$$\frac{3}{1,500} = 0.002 \text{ or } 0.2\%$$

2. **Theoretical probabilities** (also called *a priori* models) are obtained by logical reasoning according to stated definitions. For example, the probability of rolling a die and obtaining a 3 is $\frac{1}{6}$, because there are six possible outcomes, each with an equal chance of occurring, so a 3 should appear $\frac{1}{6} \approx 17\%$ of the time.

3. **Subjective probabilities** are obtained by experience and indicate a measure of "certainty" on the part of the speaker. These probabilities are not necessarily arrived at through experimentation or theory. For example, a TV reporter studies the satellite maps and then issues a prediction about tomorrow's weather based on experience: 80% chance of rain tomorrow.

Our focus in this chapter is on theoretical probabilities, but we should keep in mind that our theoretical model should be predictive of the results obtained by experimentation (empirical probabilities); if they are not consistent, and we have been careful about our record-keeping in arriving at an empirical probability, we would conclude our theoretical model to be faulty.

The simplest probability model makes certain assumptions about the sample space. If the sample space can be divided into *mutually exclusive* and *equally likely* outcomes, we can define the probability of an event. Let's consider the experiment of tossing a single coin. A suitable sample space is

S = {heads, tails}

Suppose we wish to consider the event of obtaining heads; we'll call this event A. Then

A = {heads}

and this is a simple event.

We wish to define the probability of event A, which we denote by P(A). Notice that the outcomes in the sample space are mutually exclusive; that is, if one occurs, the other cannot occur. If we flip a coin, there are two possible outcomes, and *one and only one* outcome can occur on a toss. If each outcome in the sample space is

equally likely, we define the probability of A as

$$P(A) = \frac{\text{NUMBER OF SUCCESSFUL RESULTS}}{\text{NUMBER OF POSSIBLE RESULTS}}$$

A "successful" result is a result that corresponds to the event whose probability we are seeking — in this case, {heads}. Since we can obtain a head (success) in only one way, and the total number of possible outcomes is two, the probability of heads is given by this definition as

$$P(\text{heads}) = P(A) = \tfrac{1}{2}$$

This must correspond to the empirical results you would obtain if you repeated the experiment a large number of times. In Problem 47 you are asked to repeat this experiment 100 times.

Definition of Probability

If an experiment can result in any of n mutually exclusive and equally likely outcomes, and if s of these outcomes are considered favorable, then the **probability** of an event E, denoted by $P(E)$, is

$$P(E) = \frac{s}{n} = \frac{\text{NUMBER OF OUTCOMES FAVORABLE TO } E}{\text{NUMBER OF ALL POSSIBLE OUTCOMES}}$$

▰▰▰ **EXAMPLE 3**

Use the definition of probability to find first the probability of white and second the probability of black, using the spinner shown and assuming that the arrow will never lie on a border line.

Solution   Looking at the spinner, we note that it is divided into three areas of the same size. We assume that the spinner is equally likely to land in any of these three areas.

$P(\text{white}) = \dfrac{2}{3}$ ← Two sections are white.
        ← Three sections all together

$P(\text{black}) = \dfrac{1}{3}$ ← One section is black.
        ← Three sections all together

These are theoretical probabilities.

▰▰▰ **EXAMPLE 4**

Consider a jar that contains marbles as shown. Suppose that each marble has an equal chance of being picked from the jar. Find:

a. $P(\text{black})$ **b.** $P(\text{gray})$ **c.** $P(\text{white})$

Solution

$$P(\text{white}) = \frac{4}{12} \quad \leftarrow \text{4 black marbles in jar}$$
$$\phantom{P(\text{white})} \quad \leftarrow \text{12 marbles in jar}$$

$$\phantom{P(\text{white})} = \frac{1}{3} \quad \text{Reduce fractions.}$$

$$P(\text{gray}) = \frac{7}{12}$$

$$P(\text{white}) = \frac{1}{12}$$

This is a theoretical probability; also note that the probabilities of all the simple events sum to 1:

$$\frac{1}{3} + \frac{7}{12} + \frac{1}{12} = 1$$

Reduced fractions are used to state probabilities when the fractions are fairly simple. If, however, the fractions are not simple, and you have a calculator, it is acceptable to state the probabilities as decimals, as shown in Example 5.

■■■■ EXAMPLE 5

Suppose that, in a certain study, 46 out of 155 people showed a certain kind of behavior. Assign a probability to this behavior.

Solution

$$P(\text{behavior shown}) = \frac{46}{155} \approx 0.30$$

This is an empirical probability because it was arrived at through experimentation.

Calculator Comment

[46] [÷] [155] [=]

Display: *.2967741935*

■■■■ EXAMPLE 6

Consider two spinners as shown. You and an opponent are to spin a spinner simultaneously, and the one with the higher number wins. Which spinner should you choose, and why?

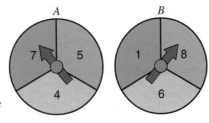

Solution We begin by listing the sample space:

B A	1	6	8	
4	(4, 1)	(4, 6)	(4, 8)	The times that A wins are enclosed in boxes.
5	(5, 1)	(5, 6)	(5, 8)	
7	(7, 1)	(7, 6)	(7, 8)	$P(A \text{ wins}) = \frac{4}{9}$; $P(B \text{ wins}) = \frac{5}{9}$

We would choose spinner *B* because it has a greater probability of winning.

■■■■ **EXAMPLE 7**

Suppose that a single card is selected from an ordinary deck of 52 cards. Find: **a.** $P(\text{ace})$ **b.** $P(\text{heart})$ **c.** $P(\text{face card})$

Solution   The sample space is shown in Figure 9.3.

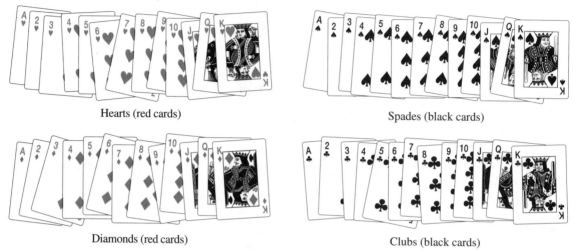

Hearts (red cards)   Spades (black cards)

Diamonds (red cards)   Clubs (black cards)

▲ **Figure 9.3   Sample space for a deck of cards**

a. An ace is a card with one spot. $P(\text{ace}) = \dfrac{4}{52} = \dfrac{1}{13}$

b. $P(\text{heart}) = \dfrac{13}{52} = \dfrac{1}{4}$

c. $P(\text{face card}) = \dfrac{12}{52}$ ← A face card is a jack, queen, or king.
         ← Number of cards in the sample space

$\qquad\qquad\quad = \dfrac{3}{13}$

■■■■ **EXAMPLE 8**

Suppose you are just beginning a game of Monopoly™. You roll a pair of dice. What is the probability that you land on a railroad on the first roll of the dice?

Solution   You need to know the locations of the railroads on a Monopoly board. There are four railroads, which are positioned so that only one can be reached on one roll of a pair of dice. The required number to roll is a 5. We begin by listing the sample space.

You might try {2, 3, 4, 5, 6, 7, 8, 9, 10, 11, 12}, but these possible outcomes are not equally likely, which you can see by considering a tree diagram. The roll of the first die has 6 possibilities, and *each* of these in turn can combine with any of 6 possibilities on the second die for a total of 36 possibilities. The sample space is summarized in Figure 9.4.

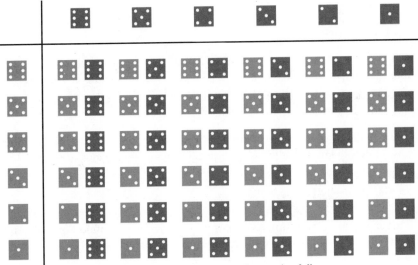

▲ Figure 9.4 Sample space for a pair of dice

Thus, $n = 36$ for the definition of probability. We need to look at Figure 9.4 to see how many possibilities there are for obtaining a 5. We find $(1, 4)$, $(2, 3)$, $(3, 2)$, and $(4, 1)$, so $s = 4$. Then

$$P(\text{five}) = \frac{4}{36} \quad \leftarrow \text{4 ways to obtain a 5}$$
$$\quad\quad\quad\quad \leftarrow \text{36 ways to roll a pair of dice}$$

$$= \frac{1}{9}$$

You should use Figure 9.4 when working probability problems dealing with rolling a pair of dice. You will be asked to perform an experiment (Problem 48) in which you roll a pair of dice 100 times and then compare the results you obtain with the probabilities you calculate by using Figure 9.4.

Probabilities of Unions and Intersections

The word *or* is translated as ∪ (union), and the word *and* is translated as ∩ (intersection). We will find the probabilities of combinations of events involving the words *or* and *and* by finding unions and intersections of events.

▬▬▬ EXAMPLE 9

Suppose that a single card is selected from an ordinary deck of cards.

a. What is the probability that it is a two or a king?

b. What is the probability that it is a two or a heart?

c. What is the probability that it is a two and a heart?

d. What is the probability that it is a two and a king?

Solution

a. $P(\text{two or a king}) = P(\text{two} \cup \text{king})$
Look at Figure 9.3.

two = {two of hearts, two of spades, two of diamonds, two of clubs}
king = {king of hearts, king of spades, king of diamonds, king of clubs}

two ∪ king = {two of hearts, two of spades, two of diamonds, two of clubs, king of hearts, king of spades, king of diamonds, king of clubs}

There are 8 possibilities for success. It is usually not necessary to list all of these possibilities to know that there are 8 possibilities — simply look at Figure 9.3.

$$P(\text{two} \cup \text{king}) = \frac{8}{52} = \frac{2}{13}$$

b. This seems to be very similar to part **a**, but there is one important difference. Look at the sample space and notice that although there are 4 twos and 13 hearts, the total number of successes is *not* 4 + 13 = 17, *but rather* 16.

two = { **two of hearts,** two of spades, two of diamonds, two of clubs}

heart = {ace of hearts, **two of hearts,** three of hearts, . . . , king of hearts}

{two ∪ heart = { **two of hearts,** two of spades, two of diamonds, two of clubs, ace of hearts, three of hearts, four of hearts, five of hearts, six of hearts, seven of hearts, eight of hearts, nine of hearts, ten of hearts, jack of hearts, queen of hearts, king of hearts}.

It is not necessary to list these possibilities. The purpose of doing so in this case was to reinforce the fact that there are *actually* 16 (not 17) possibilities.

$$P(\text{two} \cup \text{heart}) = \frac{16}{52} = \frac{4}{13}$$

c. two ∩ heart = {two of hearts} so there is one element in the intersection:

$$P(\text{two} \cap \text{heart}) = \frac{1}{52}$$

d. two ∩ king = ∅, so there are no elements in the intersection:

$$P(\text{two} \cap \text{king}) = \frac{0}{52} = 0$$

The probability of the empty set is 0, which means that the event cannot occur. In Problem Set 9.1, you will be asked to show that the probability of an event that *must* occur is 1. These are the two extremes. All other probabilities fall somewhere in between. The closer a probability is to 1, the more likely the event is to occur; the closer a probability is to 0, the less likely the event is to occur.

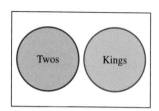

Mutually exclusive

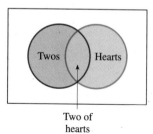

Two of
hearts

Not mutually exclusive

In Chapter 1, we found |two ∪ hearts|:
$$|T \cup H| = |T| + |H| - |T \cap H|$$
$$= 4 + 13 - 1 = 16$$

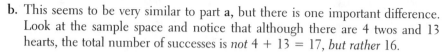

0   0.5   1

impossible even chance certain

$P(\emptyset) = 0$ $P(S) = 1$

Probability by Counting

If the probability you are finding has a sample space that is easy to count, and if all simple events in the sample space are equally likely, then there is a simple procedure that you can use to find probabilities.

1. Describe and identify the sample space, S. The number of elements in S is n.
2. Count the number of occurrences that satisfy the event of interest; call this the number of successes and denote it by s.
3. Compute the probability of the event using the formula:

$$P(E) = \frac{s}{n}$$

Keep in mind that this procedure will not always work. If it doesn't, you need a more complicated model, or else you must proceed experimentally.

PROBLEM SET 9.1

▲ **A Problems**

1. **IN YOUR OWN WORDS** What is the difference between empirical and theoretical probabilities?

2. **IN YOUR OWN WORDS** Define probability.

3. **IN YOUR OWN WORDS** Write a simple argument to convince someone why all probabilities must be between 0 and 1 (including 0 and 1).

For the spinners in Problems 4–7, assume that the pointer can never lie on a border line.

4. **a.** $P(A)$
   b. $P(B)$
   c. $P(C)$

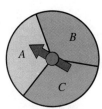

5. **a.** $P(D)$
   b. $P(E)$
   c. $P(F)$

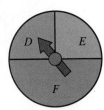

6. **a.** $P(G)$
   b. $P(H)$
   c. $P(I)$
   d. $P(J)$

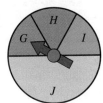

7. **a.** $P(\text{white})$
   b. $P(\text{black})$
   c. $P(\text{gray})$

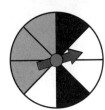

Give the probabilities in Problems 8–11 in decimal form (correct to two decimal places). A calculator may be helpful with these problems.

8. Last year, 1,485 calculators were returned to the manufacturer. If 85,000 were produced, assign a number to specify the probability that a particular calculator would be returned.

9. Last semester, a certain professor gave 13 A's out of 285 grades. If one of the 285 students is to be selected at random, what is the probability that his or her grade is an A?

10. Last year in Ferndale, CA, it rained on 75 days. What is the probability of rain on a day selected at random?

11. The campus vets club is having a raffle and is selling 1,500 tickets. If the people on your floor of the dorm bought 285 of those tickets, what is the probability that someone on your floor will hold the winning ticket?

Consider the jar containing marbles shown in Figure 9.5. Suppose each marble has an equal chance of being picked from the jar. Find the probabilities in Problems 12–14.

12. *P*(white)   13. *P*(black)   14. *P*(gray)

▲ **Figure 9.5   A jar of marbles**

Poker is a common game in which players are dealt five cards from a deck of cards. It can be shown that there are 2,598,960 different possible poker hands. The winning hands (from highest to lowest) are shown in Table 9.1. Find the requested probabilities in Problems 15–23. Use a calculator, and show your answers to whatever accuracy possible on your calculator.

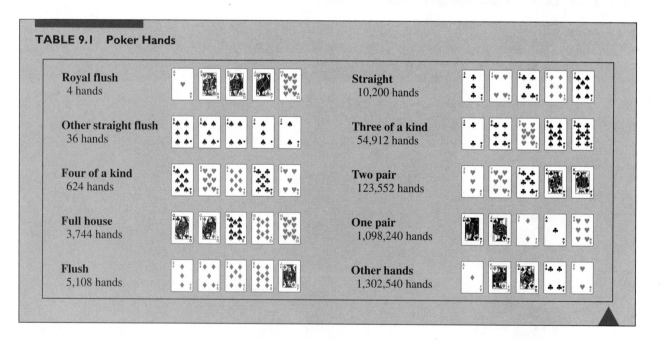

TABLE 9.1   Poker Hands

Royal flush 4 hands	**Straight** 10,200 hands
Other straight flush 36 hands	**Three of a kind** 54,912 hands
Four of a kind 624 hands	**Two pair** 123,552 hands
Full house 3,744 hands	**One pair** 1,098,240 hands
Flush 5,108 hands	**Other hands** 1,302,540 hands

15. *P*(royal flush)   16. *P*(straight flush)   17. *P*(four of a kind)

18. *P*(full house)   19. *P*(flush)   20. *P*(straight)

21. *P*(three of a kind)   22. *P*(two pair)   23. *P*(one pair)

▲ **B Problems**

A single card is selected from an ordinary deck of cards. The sample space is shown in Figure 9.3. Find the probabilities in Problems 24–33.

24. P(five of hearts) **25.** P(five) **26.** P(heart)

27. P(jack) **28.** P(diamond) **29.** P(even number)

30. P(five and a jack) **31.** P(five or a jack)

32. P(heart and a jack) **33.** P(heart or a jack)

Suppose that you toss a coin and roll a die in Problems 34–37. The sample space is shown in Figure 9.1.

34. What is the probability of obtaining:
  a. Tails *and* a five? **b.** Tails *or* a five? **c.** Heads *and* a two?

35. What is the probability of obtaining:
  a. Tails? **b.** One, two, three, *or* four? **c.** Heads *or* a two?

36. What is the probability of obtaining:
  a. Heads *and* an odd number? **b.** Heads *or* an odd number?

37. What is the probability of obtaining:
  a. Heads *and* a five? **b.** Heads *or* a five?

Use the sample space shown in Figure 9.4 to find the probabilities in Problems 38–46 for the experiment of rolling a pair of dice.

38. P(five) **39.** P(six) **40.** P(seven)

41. P(eight) **42.** P(nine) **43.** P(two)

44. P(four *or* five) **45.** P(even number) **46.** P(eight *or* ten)

Perform the experiments in Problems 47–50, tally your results, and calculate the probabilities (to the nearest hundredth).

47. Toss a coin 100 times. Make sure that, each time the coin is flipped, it rotates several times in the air and lands on a table or on the floor. Keep a record of the results of this experiment. Based on your experiment, what is P(heads)?

48. Roll a pair of dice 100 times. Keep a record of the results. Based on your experiment, find:
  a. P(two) **b.** P(three) **c.** P(four) **d.** P(five)
  e. P(six) **f.** P(seven) **g.** P(eight) **h.** P(nine)
  i. P(ten) **j.** P(eleven) **k.** P(twelve)

49. Flip three coins simultaneously 100 times, and note the results. The possible outcomes are:
  a. three heads **b.** two heads and one tail
  c. two tails and one head **d.** three tails

  Based on your experiment, find the probabilities of each of these events. Do these appear to be equally likely?

50. Simultaneously toss a coin and roll a die 100 times, and note the results. The possible outcomes are H1, H2, H3, H4, H5, H6, T1, T2, T3, T4, T5, and T6. Do these appear to be equally likely outcomes?

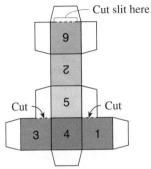

If you do not have any dice, you can use this pattern to construct your own.

51. Suppose it is certain that an earthquake will occur someday. What is the probability (to the nearest percent) that it will occur while you are at work? Assume you are at work 8 hours per day, 240 days per year.

52. Suppose it is certain that an earthquake will occur someday. What is the probability (to the nearest percent) that it will occur while you are at school? Assume you are at school 5 hours per day, 174 days per year.

53. Prepare three cards that are identical except for the color. One card is black on both sides, one is white on both sides, and one is black on one side and white on the other. One card is selected at random and placed flat on the table. You will see either a black or a white card; record the color of the face (the top). This is *not* the event with which we are concerned; rather, we are interested in finding the probability of the *other* side (the side you cannot see) being black or white. Record the underside, as shown in the following table. Repeat the experiment 50 times and find the probability of occurrence with respect to the known color. Do these appear to be equally likely outcomes?

Black on both sides

Black on one side, white on the other

White on both sides

Color of face	Frequency	Outcome (color of the underside)	Frequency	Probability
White		White		
		Black		
Black		White		
		Black		

54. Dice is a popular game in gambling casinos. Two dice are tossed, and various amounts are paid according to the outcome. If a seven or eleven occurs on the first roll, the player wins. What is the probability of winning on the first roll?

55. In dice, the player loses if the outcome of the first roll is a two, three, or twelve. What is the probability of losing on the first roll?

56. In dice, a pair of ones is called *snake eyes*. What is the probability of losing a dice game by rolling snake eyes?

B.C. By permission of Johnny Hart and Creators Syndicate, Inc.

57. Consider a die with only four sides, marked one, two, three, and four. Write out a sample space similar to the one in Figure 9.4 for rolling a pair of these dice.

58. Using the sample space you found in Problem 57, find the probability that the sum of the dice is the given number. Assume equally likely outcomes.
  a. P(two) **b.** P(three) **c.** P(four)

59. Using the sample space you found in Problem 57, find the probability that the sum of the dice is the given number. Assume equally likely outcomes.
  a. P(five) **b.** P(six) **c.** P(seven)

60. The game of Dungeons and Dragons uses nonstandard dice. Consider a die with eight sides marked one, two, three, four, five, six, seven, and eight. Write a sample space similar to the one in Figure 9.4 for rolling a pair of these dice.

Suppose you and an opponent each pick one of the spinners shown here. A "win" means spinning a higher number. Construct a sample space to answer each question, and tell which of the two spinners given in Problems 61–66 you would choose in each case.

61. A plays C 62. A plays B 63. B plays C
64. D plays E 65. E plays F 66. D plays F

A 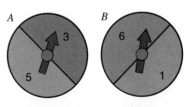 B

C D

E 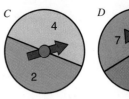 F

▲ **Problem Solving**

67. **The game of WIN.** Construct a set of nonstandard dice as shown in Figure 9.6.

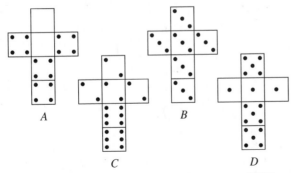

▲ **Figure 9.6 Faces on the dice for a game of WIN**

Suppose that one player picks die A and that the other picks die B. Then we can enumerate the sample space as shown here

B \ A	0	0	4	4	4	4
3	(3,0)	(3,0)	(3,4)	(3,4)	(3,4)	(3,4)
3	(3,0)	(3,0)	(3,4)	(3,4)	(3,4)	(3,4)
3	(3,0)	(3,0)	(3,4)	(3,4)	(3,4)	(3,4)
3	(3,0)	(3,0)	(3,4)	(3,4)	(3,4)	(3,4)
3	(3,0)	(3,0)	(3,4)	(3,4)	(3,4)	(3,4)
3	(3,0)	(3,0)	(3,4)	(3,4)	(3,4)	(3,4)

B wins | A wins

We see that A's probability of winning is $\frac{24}{36}$, or $\frac{2}{3}$. There are, of course, many other possible choices for the dice played. If you were to play the game of *WIN*, would you choose your die first or second? What is your probability of winning at *WIN*?

68. The question is to test your intuition. *Read* the following exercises (do *not* carry out the experiment).

   A. Open a phone book to any page in the white pages. Select 100 consecutive phone numbers and tally the number of times each of the digits 0, 1, 2, 3, 4, 5, 6, 7, 8, 9 occurs as a first digit.
   B. Repeat A for the third digit.
   C. Repeat A for the last digit.

   a. Do you think the occurrence of each digit will be about the same for each of these experiments?
   b. Do you think the results of all three experiments will be about the same?
   c. *Now,* after answering parts **a** and **b,** carry out the experiments described in A, B, and C.

▲ Individual Research

69. Michael Rossides has come up with a scheme for eliminating coins. This scheme involves probability and the fact that most cash registers today are computers. Suppose that every cash register could be programmed with a random number generator; that is, suppose that it were possible to pick a random number from 1 to 99. Rossides' system works as follows. Suppose you purchase items totaling $15.89. The computer would choose a number from 1 to 99 and then compare it with the cents portion of the purchase. In this case, if the random number is between 1 and 89 the price would be rounded up to $16; if it is between 90 and 99 the price would be rounded down to $15. For example, if you purchase a cup of coffee for $1.20, and the random number generator produces a random number from 1 to 20, the price is $2, but if it produces a number from 21 to 99 the price is $1. Write a paper commenting on this scheme.

70. Find the probability that the 13th day of a randomly chosen month will be a Friday. (*Hint:* It is not correct to say that the probability is $\frac{1}{7}$ because there are seven possible days in the week. In fact, it turns out that the 13th day of a month is more likely a Friday than any other day of the week.) Write a paper discussing this problem.

References

William Bailey, "Friday-the-Thirteenth," *The Mathematics Teacher*, Vol. LXII, No. 5 (1969), pp. 363–364.

C. V. Heuer, Solution to Problem E1541 in *American Mathematical Monthly*, Vol. 70, No. 7 (1963), p. 759.

G. L. Ritter et al., "An Aid to the Superstitious," *The Mathematics Teacher*, Vol. 70, No. 5 (1977), pp. 456–457.

9.2   MATHEMATICAL EXPECTATION

Expected Value

Smiles toothpaste is giving away $10,000. All you must do to have a chance to win is send a postcard with your name on it (the fine print says you do not need to buy a tube of toothpaste). Is it worthwhile to enter?

Suppose the contest receives 1 million postcards (a conservative estimate). We wish to compute the **expected value** (or your **expectation**) of winning this contest. We find the expectation for this contest by multiplying the amount to win by the probability of winning:

$$\text{EXPECTATION} = (\text{AMOUNT TO WIN}) \times (\text{PROBABILITY OF WINNING})$$

$$= \$10,000 \times \frac{1}{1,000,000}$$

$$= \$0.01$$

What does this expected value mean? It means that if you were to play this "game" a large number of times, you would expect your *average winnings per game* to be $0.01. A game is said to be **fair** if the expected value equals the cost of playing the game. If the expected value is positive, then the game is in your favor; and if the expected value is negative, then the game is not in your favor. Is this game fair? If the toothpaste company charges you 1¢ to play the game, then it is fair. But how much does the postcard cost? We see that this is not a fair game.

◼◼◼ EXAMPLE 1

Suppose that you draw a card from a deck of cards and are paid $10 if it is an ace. What is the expected value?

Solution

$$\text{EXPECTATION} = \$10 \times \frac{4}{52} \approx \$0.77$$

Since expected value involves a dollar amount, round your answer to the nearest cent.

Calculator Comment

[10] [×] [4] [÷] [52] [=]

Display: .7692307692

Sometimes there is more than one payoff, and we define the expected value (or expectation) as the sum of the expected values from each separate payoff.

◼◼◼ EXAMPLE 2

A recent contest offered one grand prize worth $10,000, two second prizes worth $5,000 each, and ten third prizes worth $1,000 each. What is the expected value if you assume that there are 1 million entries?

Solution

$$P(\text{1st prize}) = \frac{1}{1,000,000}; \quad P(\text{2nd prize}) = \frac{2}{1,000,000}; \quad P(\text{3rd prize}) = \frac{10}{1,000,000}$$

$$\text{EXPECTATION} = \underbrace{\$10,000}_{\text{Amount of 1st prize}} \times \underbrace{\frac{1}{1,000,000}}_{P(\text{1st prize})} + \$5,000 \times \overbrace{\frac{2}{1,000,000}}^{\text{2nd prize}} + \$1,000 \times \underbrace{\frac{10}{1,000,000}}_{\text{3rd prize}}$$

$$= \$0.01 + \$0.01 + \$0.01$$

$$= \$0.03$$

Calculator Comment

[10000] [÷] [1000000] [+]
[5000] [×] [2] [÷] [1000000]
[+] [1000] [×] [10] [÷]
[1000000] [=]

Display: .03

You can use mathematical expectation to help you make a decision, as suggested by the following example.

■■■ **EXAMPLE 3**

You are offered two games:

> *Game A:*   Two dice are rolled. You will be paid $3.60 if you roll two ones, and will not receive anything for any other outcome.
>
> *Game B:*   Two dice are rolled. You will be paid $36.00 if you roll any pair, but you must pay $3.60 for any other outcome.

Which game should you play?

Solution   You might say, "I'll play the first game because, if I play that game, I cannot lose anything." This strategy involves *minimizing your losses*. On the other hand, you can use a strategy that *maximizes your winnings*. In this book, we will base our decisions on maximizing the winnings. That is, we wish to select the game that provides the larger expectation.

> *Game A:*   EXPECTATION $= \$3.60 \times \dfrac{1}{36} = \$0.10$

> *Game B:*   When calculating the expected value with a charge (a loss), write that charge as a negative number (a negative payoff is a loss).

$$\text{EXPECTATION} = \$36.00 \times \frac{6}{36} + (-\$3.60) \times \frac{30}{36}$$
$$= \$6.00 + (-\$3.00)$$
$$= \$3.00$$

This means that, if you were to play each game 100 times, you would expect your winnings for Game A to be 100($0.10) or about $10 and those from playing Game B to be 100($3.00) or about $300. You should choose to play Game B. ■

Now we give a formal definition of expectation.

> **Mathematical Expectation**
>
> If an event E has several possible outcomes with probabilities p_1, p_2, p_3, . . . , and if for each of these outcomes the amount that can be won is a_1, a_2, a_3, . . . , respectively, then the **mathematical expectation** (or expected value) of E is
>
>   EXPECTATION $= a_1 p_1 + a_2 p_2 + a_3 p_3 + \cdots$

■■■ EXAMPLE 4

A contest offered the prizes shown here. What is the expected value for this contest?

WIN, WIN, WIN		
Prize	*Value*	*Probability*
Grand Prize Trip	$15,000	0.000008
Samsonite Luggage	$1,000	0.000016
Magic Chef Range	$625.00	0.000016
Murray Bicycle	$525.00	0.000016
Lawn Boy Mower	$390.00	0.000032
Weber Kettle	$250.00	0.000032

Solution

We note the following values:

$$a_1 = \$15,000; \; p_1 = 0.000008$$
$$a_2 = \$1,000; \; p_2 = 0.000016$$
$$a_3 = \$625; \; p_3 = 0.000016$$
$$a_4 = \$525; \; p_4 = 0.000016$$
$$a_5 = \$390; \; p_5 = 0.000032$$
$$a_6 = \$250; \; p_6 = 0.000032$$

EXPECTATION $= \$15{,}000(0.000008) + \$1{,}000(0.000016) + \$625(0.000016)$
$$+ \;\$525(0.000016) + \$390(0.000032) + \$250(0.000032)$$
$$\approx \$0.17$$

> **Calculator Comment**
>
> Look at EXPECTATION to see the obvious keys to press.
>
> *Display:* .17488 ▲

It is not necessary that the expectation or expected value be reported in terms of dollars and cents. Consider the following example.

■■■ EXAMPLE 5

Suppose a family has three children. What is the expected number of girls?

Solution The sample space for this problem can be found using a tree diagram: {GGG, GGB, GBG, GBB, BGG, BGB, BBG, BBB}. All possible probabilities are listed in the following table:

Number of Girls	Probability	Product
0	$\frac{1}{8}$	0
1	$\frac{3}{8}$	$\frac{3}{8}$
2	$\frac{3}{8}$	$\frac{6}{8}$
3	$\frac{1}{8}$	$\frac{3}{8}$

EXPECTED VALUE $= 0 + \frac{3}{8} + \frac{6}{8} + \frac{3}{8} = \frac{12}{8} = 1.5$

Notice from Example 5 that the expected value may be a number that can never occur; it is obvious that 1.5 girls will not occur. An expected value of 1.5 simply means that if we record the numbers of girls in a large number of different three-child families, the *average* number of girls for all these families will be 1.5. We will discuss averages in the next chapter.

Expectation with a Cost of Playing

Many games charge you a fee to play. If you must pay to play, this cost of playing should be taken into consideration when you calculate the expected value. Remember, if the expected value is 0, it is a fair game; if the expected value is positive, you should play, but if it is negative, you should not.

▬▬▬ EXAMPLE 6

Consider a game that consists of drawing a card from a deck of cards. If it is a face card, you win $20. Should you play the game if it costs $5 to play?

Solution

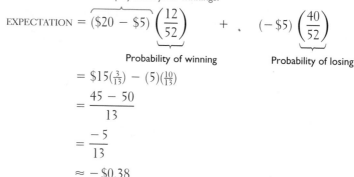

Subtract the cost of play from your winnings.

$$\text{EXPECTATION} = \overbrace{(\$20 - \$5)} \underbrace{\left(\frac{12}{52}\right)}_{\text{Probability of winning}} + \cdot \quad (-\$5)\underbrace{\left(\frac{40}{52}\right)}_{\text{Probability of losing}}$$

$$= \$15\left(\tfrac{3}{13}\right) - (5)\left(\tfrac{10}{13}\right)$$

$$= \frac{45 - 50}{13}$$

$$= \frac{-5}{13}$$

$$\approx -\$0.38$$

Calculator Comment

$15\ \boxed{\times}\ \boxed{12}\ \boxed{\div}\ \boxed{52}\ \boxed{+}$
$\boxed{5}\ \boxed{+/-}\ \boxed{\times}\ \boxed{(}\ \boxed{1}\ \boxed{-}$
$\boxed{12}\ \boxed{\div}\ \boxed{52}\ \boxed{)}\ \boxed{=}$
Display: $-.3846153846$

You should not play this game, because it has a negative expectation. ▬

You must understand the nature of the game to know whether to subtract the cost of playing, as we did in Example 6. If you surrender your money to play, then it must be subtracted, but if you "leave it on the table," then you do not subtract it. Consider the following example of a U.S. roulette game. In this game, your bet is placed on the table but is not collected until after the play of the game and it is determined that you lost.

▬▬▬ EXAMPLE 7

What is the expectation for playing roulette if you bet $1 on number 5?

Solution   A U.S. roulette wheel has 38 numbered slots (1–36, 0, and 00), as shown in Figure 9.7. Some of the more common bets and payoffs are shown. If the payoff is listed as 6 to 1, you would receive $6 for each $1 bet. In addition, you would keep the $1 you originally wagered. One play consists of having the dealer spin the wheel and a little ball in opposite directions. As the ball slows to a stop, it lands in one of the 38 numbered slots, which are colored black, red, or green. A single number bet has a payoff of 35 to 1. The $1 you bet is collected only if you lose.

Now, you can calculate the expected value:

$$\text{EXPECTATION} = \overbrace{35\left(\tfrac{1}{38}\right)}^{\text{Payoff for a win}} + \overbrace{(-1)\left(\tfrac{37}{38}\right)}^{\text{Payoff for a loss}}$$

Probability of winning   Probability of losing

$$\approx -\$0.05$$

The expected loss is about 5¢ per play.

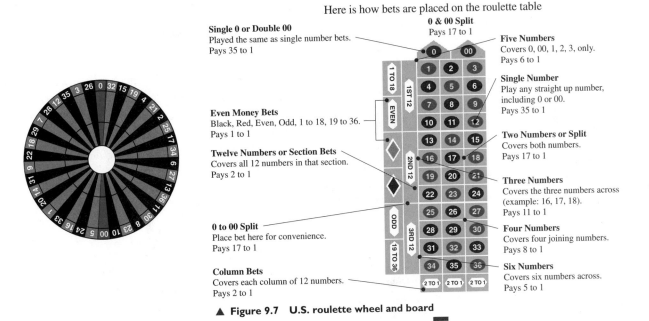

Here is how bets are placed on the roulette table

Single 0 or Double 00
Played the same as single number bets.
Pays 35 to 1

0 & 00 Split
Pays 17 to 1

Five Numbers
Covers 0, 00, 1, 2, 3, only.
Pays 6 to 1

Single Number
Play any straight up number,
including 0 or 00.
Pays 35 to 1

Even Money Bets
Black, Red, Even, Odd, 1 to 18, 19 to 36.
Pays 1 to 1

Two Numbers or Split
Covers both numbers.
Pays 17 to 1

Twelve Numbers or Section Bets
Covers all 12 numbers in that section.
Pays 2 to 1

Three Numbers
Covers the three numbers across
(example: 16, 17, 18).
Pays 11 to 1

0 to 00 Split
Place bet here for convenience.
Pays 17 to 1

Four Numbers
Covers four joining numbers.
Pays 8 to 1

Column Bets
Covers each column of 12 numbers.
Pays 2 to 1

Six Numbers
Covers six numbers across.
Pays 5 to 1

▲ **Figure 9.7   U.S. roulette wheel and board**

EXAMPLE 8

Eva, who is a realtor, knows that if she takes a listing to sell a house, it will cost her $1,000. However, if she sells the house, she will receive 6% of the selling price. If another realtor sells the house, Eva will receive 3% of the selling price. If the house remains unsold after 3 months, she will lose the listing and receive nothing. Suppose that the probabilities for selling a particular $200,000 house are as follows: The probability that Eva will sell the house is 0.4; the probability that another agent will sell the house is 0.2; and the probability that the house will remain unsold is 0.4. What is Eva's expectation if she takes this listing?

Solution   First, we must decide whether this is an "entrance fee" problem. Is Eva required to pay the $1,000 before the "game" of selling the house is played? The answer is yes, so the $1,000 must be subtracted from the payoffs. Now let's calculate those payoffs:

$$6\% \text{ of } \$200,000 = 0.06(\$200,000) = \$12,000$$
$$3\% \text{ of } \$200,000 = 0.03(\$200,000) = \$6,000$$

$$\overbrace{\text{Eva sells the house.}} \quad \overbrace{\text{Another agent sells.}} \quad \overbrace{\text{House doesn't sell.}}$$
$$\text{EXPECTATION} = (\$12,000 - \$1,000)(0.4) + (\$6,000 - \$1,000)(0.2) + (-\$1,000)(0.4)$$
$$= \$11,000(0.4) + \$5,000(0.2) - \$1,000(0.4)$$
$$= \$4,400 + \$1,000 - \$400$$
$$= \$5,000$$

Eva's expectation is $5,000.

PROBLEM SET 9.2

▲ **A Problems**

1. **IN YOUR OWN WORDS** *True or false?*   In roulette, if you bet on black, the probability of winning is $\frac{1}{2}$ because there are the same number of black and red spots. Explain.

2. **IN YOUR OWN WORDS** *True or false?*   An expected value of $5 means that you should expect to win $5 each time you play the game. Explain.

3. **IN YOUR OWN WORDS** *True or false?*   If the expected value of a game is positive, then it is a game you should play. Explain.

4. **IN YOUR OWN WORDS** If you were asked to choose between a sure $10,000, or an 80% chance of winning $15,000 and a 20% of winning nothing, which would you take?

   Game A: $E = \$10,000(1) = \$10,000$ This is a sure thing.
   Game B: $E = \$15,000(0.8) + \$0(0.2) = \$12,000$.

   True or false?   The better choice, according to expected value, is to take the 80% chance of winning $15,000. Explain.

5. **IN YOUR OWN WORDS** Suppose that you buy a lottery ticket for $1. The payoff is $50,000, with a probability of winning 1/1,000,000. Therefore, the expected value is

$$E = \$50,000 \left(\frac{1}{1,000,000} \right) = \$0.05$$

   Is this statement true or false? Explain.

Use estimation to select the best response in Problems 6–11. Do not calculate.

6. The expectation from playing a game in which you win $950 by correctly calling heads or tails when you flip a coin is about
   A. $500   B. $50   C. $950

7. The expectation from playing a game in which you win $950 by correctly calling heads or tails on each of five flips of a coin is about
   A. $500   B. $50   C. $950

8. If the expected value of playing a $1 game of blackjack is $0.04, then after playing the game 100 times you should have netted about
   A. $104   B. −$4   C. $4

9. If your expected value when playing a $1 game of roulette is −$0.05, then after playing the game 100 times you should have netted about
   A. −$105   B. −$5   C. $5

10. The probability of correctly guessing a telephone number is about
    A. 1 out of 100   B. 1 out of 1,000   C. 1 out of 1,000,000

11. Winning the grand prize in a state lottery is about as probable as
    A. having a car accident
    B. having an item fall out of the sky into your yard
    C. being a contestant on *Jeopardy*

12. A box contains one each of $1, $5, $10, $20, and $100 bills. You reach in and withdraw one bill. What is the expected value?

13. A box contains one each of $1, $5, $10, $20, and $100 bills. It costs $20 to reach in and withdraw one bill. What is the expected value?

14. Suppose that you roll two dice. You will be paid $5 if you roll a double. You will not receive anything for any other outcome. How much should you be willing to pay for the privilege of rolling the dice?

15. A magazine subscription service is having a contest in which the prize is $80,000. If the company receives 1 million entries, what is the expectation of the contest?

16. Suppose that you have 5 quarters, 5 dimes, 10 nickels, and 5 pennies in your pocket. You reach in and choose a coin at random so that you can tip your barber. What is the barber's expectation? What tip is the barber most likely to receive?

17. A game involves tossing two coins and receiving 50¢ if they are both heads. What is a fair price to pay for the privilege of playing?

18. Krinkles potato chips is having a "Lucky Seven Sweepstakes." The one grand prize is $70,000; 7 second prizes each pay $7,000; 77 third prizes each pay $700; and 777 fourth prizes each pay $70. What is the expectation of this contest, if there are 10 million entries?

19. A punch-out card contains 100 spaces. One space pays $100, five spaces pay $10, and the others pay nothing. How much should you pay to punch out one space?

▲ B Problems

20. In old gangster movies on TV, you often hear of "numbers runners" or the "numbers racket." This numbers game, which is still played today, involves betting $1 on the last three digits of the number of stocks sold on a particular day in the future as reported in *The Wall Street Journal*. If the payoff is $500, what is the expectation for this numbers game?

21. A realtor who takes the listing on a house to be sold knows that she will spend $800 trying to sell the house. If she sells it herself, she will earn 6% of the selling price. If another realtor sells a house from her list, the first realtor will earn only 3% of the price. If the house remains unsold after 6 months, she will lose the listing. Suppose that probabilities are as follows:

Event	Probability
Sell by herself	0.50
Sell by another realtor	0.30
Not sell in 6 months	0.20

What is the expected profit from listing an $85,000 house?

22. An oil-drilling company knows that it costs $25,000 to sink a test well. If oil is hit, the income for the drilling company will be $425,000. If only natural gas is hit, the income will be $125,000. If nothing is hit, there will be no income. If the probability of hitting oil is $\frac{1}{40}$ and if the probability of hitting gas is $\frac{1}{20}$, what is the expectation for the drilling company? Should the company sink the test well?

23. In Problem 22, suppose that the income for hitting oil is changed to $825,000 and the income for gas to $225,000. Now what is the expectation for the drilling company? Should the company sink the test well?

24. Suppose that you roll one die. You are paid $5 if you roll a one, and you pay $1 otherwise. What is the expectation?

25. A game involves drawing a single card from an ordinary deck. If an ace is drawn, you receive 50¢; if a face card is drawn, you receive 25¢; if the two of spades is drawn, you receive $1. If the cost of playing is 10¢, should you play?

26. Consider the following game in which a player rolls a single die. If a prime (2, 3, or 5) is rolled, the player wins $2. If a square (1 or 4) is rolled, the player wins $1. However, if the player rolls a perfect number (6), it costs the player $11. Is this a good deal for the player or not?

Consider the spinners in Problems 27–30. Determine which represent fair games. Assume that the cost to spin the wheel once is $5.00 and that you will receive the amount shown on the spinner after it stops.

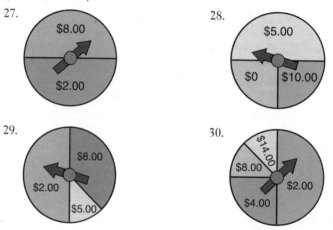

27.   28.   29.   30.

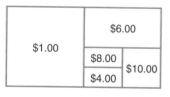

31. Assume that a dart is randomly thrown at the dart board shown in the margin and strikes the board every time. The payoffs are listed on the board. How much should you be willing to pay for the opportunity to play this game?

32. A company held a contest, and the following information was included in the fine print:

Prize	Number of Prizes	Probability of Winning Indicated Prize
$10,000	13	0.000005
$1,000	52	0.00002
$100	520	0.0002
$10	28,900	0.010886
TOTAL	29,485	0.011111

Read this information carefully, and calculate the expectation (to the nearest cent) for this contest.

33. A company held a bingo contest for which the following chances of winning were given:

Playing One Card, Your Chances of Winning Are at Least:			
	1 Time	7 Times	13 Times
$25 prize	1 in 21,252	1 in 3,036	1 in 1,635
$3 prize	1 in 2,125	1 in 304	1 in 163
$1 prize	1 in 886	1 in 127	1 in 68
Any prize	1 in 609	1 in 87	1 in 47

What is the expectation (to the nearest cent) from playing one card 13 times?

34. Calculate the expectation (to the nearest cent) for the *Reader's Digest* sweepstakes described. Assume there are 197,000,000 entries.

$10,500,000.00
SWEEPSTAKES ENTRY DOCUMENT

What is the expectation for $1 bets in Problems 35–44 on a U.S. roulette wheel? See Figure 9.7 on page 627.

35. Black

36. Odd

37. Single-number bet

38. Double-number bet

39. Three-number bet

40. Four-number bet

41. Five-number bet

42. Six-number bet

43. Twelve-number bet

44. Column bet

▲ **Problem Solving**

45. Consider a state lottery that has a weekly television show. On this show, a contestant receives the opportunity to win $1 million. The contestant picks from four hidden windows. Behind each is one of the following: $150,000, $200,000, $1 million, or a "stopper." Before beginning, the contestant is offered $100,000 to stop. Mathematically speaking, should the contestant take the $100,000?

46. Consider a state lottery that has a weekly television show. On this show, a contestant receives the opportunity to win $1 million. The contestant picks from four hidden windows. Behind each is one of the following: $150,000, $200,000, $1 million, or a "stopper." If the contestant picks the window containing $150,000 or $200,000, the contestant is asked whether they wish to quit or continue. Suppose a certain contestant has picked both the $150,000 and $200,000 windows. Mathematically speaking, should the contestant pick "one more time?"

47. **IN YOUR OWN WORDS** Read Problem 46. Discuss what you would do. Consider this situation: Suppose you owned a $350,000 home free and clear; would you gamble it on a 50–50 chance to win $1,000,000? According to the mathematical expectation, what should you do? Discuss.

48. **St. Petersburg paradox.** Suppose that you toss a coin and will win $1 if it comes up heads. If it comes up tails, you toss again. This time you will receive $2 if it comes up heads. If it comes up tails, toss again. This time you will receive $4 if it is heads and nothing if it comes up tails. What is the mathematical expectation for this game?

49. **St. Petersburg paradox.**   Suppose that you toss a coin and will win $1 if it comes up heads. If it comes up tails, you toss again. This time you will receive $2 if it comes up heads. If it comes up tails, toss again. This time you will receive $4 if it is heads. Continue in this fashion for a total of 10 flips of the coin, after which you receive nothing if it comes up tails. What is the mathematical expectation for this game?

50. **St. Petersburg paradox.**   Suppose that you toss a coin and will win $1 if it comes up heads. If it comes up tails, you toss again. This time you will receive $2 if it comes up heads. If it comes up tails, toss again. This time you will receive $4 if it is heads. Continue in this fashion for a total of 1,000 flips of the coin, after which you receive nothing if it comes up tails. What is the mathematical expectation for this game?

51. **IN YOUR OWN WORDS   St. Petersburg paradox.**   Suppose that you toss a coin and will win $1 if it comes up heads. If it comes up tails, you toss again. This time you will receive $2 if it comes up heads. If it comes up tails, toss again. This time you will receive $4 if it is heads. You continue in this fashion until you finally toss a head. Would you pay $100 for the privilege of playing this game? What is the mathematical expectation for this game?

52. **IN YOUR OWN WORDS** Suppose that you are in class and your instructor makes you the following legitimate offer. Take out a piece of paper and without communicating with your classmates write one of the following messages under your name:

    ☐ I share the wealth and will receive $1,000 if *everyone* in the class checks this box.

    ☐ I will not share the wealth and want a certain $100.

    If *everyone* checks the first box, then all will receive $1,000. If *one* person checks the second box, then only those who check the second box will receive $100. Which box would you check, and why?

53. **IN YOUR OWN WORDS** Repeat Problem 52 except change the stakes to $110 and $100, respectively.

54. **IN YOUR OWN WORDS** Repeat Problem 52 except change the stakes to $10,000 and $10, respectively.

55. **IN YOUR OWN WORDS** Repeat Problem 52, except change the stakes to be an A grade in the class if you *all* check the first box, but if anyone checks the second box, the following will occur: Those who check it will have a 50% chance of an A and a 50% chance of an F but those who do not check the second box will be given an F in this course.

▲ **Individual Research**

56. The questions in this problem are from a study by MacCrimmon, Stanbury, and Wehrung, "Real Money Lotteries: A Study of Ideal Risk, Context Effects, and Simple Processes," in *Cognitive Processes in Choice and Decision Behavior*, edited by Thomas Wallsten (Hillsdale, N.J.: Lawrence Erlbaum Associates, 1980, pp. 155–179).

    Question: You have five alternatives from which to choose. List your preferences for the alternatives from best to worst.

    1. sure win of $5 and no chance of loss

    2. 6.92% chance to win $20 and 93.08% chance to win $3.98

    3. 27.52% chance to win $20 and 72.48% chance to lose 69 cents

4. 61.85% chance to win $20 and 38.15% chance to lose $19.31

5. 90.46% chance to win $20 and 9.54% chance to lose $137.20

  a. Answer the question based on your own feelings.

  b. Answer the question using mathematical expectation as a basis for selecting your answer.

  c. Conduct a survey of at least 10 people and summarize your results.

  d. What are the conclusions of the study?

57. **a.** Answer each of the following questions* based on your own feelings.

  b. Answer questions (1) and (2) using mathematical expectation as a basis for selecting your answers.

  c. Conduct a survey of at least 10 people and summarize your results.

  1. Choose between A and B:

   A. A sure gain of $240

   B. 25% chance to gain $1,000 and a 75% chance to gain $0

  2. Choose between C and D:

   C. A sure loss of $700

   D. 75% chance to lose $1,000 and 25% chance to lose nothing

  3. Choose between E and F:

   E. Imagine that you have decided to see a concert and have paid the admission price of $10. As you enter the concert hall, you discover that you have lost your ticket. Would you pay $10 for another ticket?

   F. Image that you have decided to see a concert where the admission is $10. As you begin to enter the concert ticket line, you discover that you have lost one of your $10 bills. Would you still pay $10 for a ticket to the concert?

9.3 PROBABILITY MODELS

Engraving by Darcis: *Le Trente-et-un*

Historical Note

The engraving depicts gambling in 18th-century France. The mathematical theory of probability arose in France in the 17th century when a gambler, Chevalier de Méré, became interested in adjusting the stakes so that he could win more often than he lost. In 1654 he wrote to Blaise Pascal, who in turn sent his questions to Pierre de Fermat. Together they developed the first theory of probability. ▲

---

* These questions are from "A Bird in the Hand," by Carolyn Richbart and Lynn Richbart in *The Mathematics Teacher*, November 1996, pp. 674–676.

Complementary Probabilities

In Section 9.1 we looked at the probability of an event E that consists of n mutually exclusive and equally likely outcomes where s of these outcomes are considered favorable. Now we wish to expand our discussion. Let

s = NUMBER OF GOOD OUTCOMES (successes)

f = NUMBER OF BAD OUTCOMES (failures)

n = TOTAL NUMBER OF POSSIBLE OUTCOMES $(s + f = n)$

Then the probability that event E occurs is

$$P(E) = \frac{s}{n}$$

The probability that event E does not occur is

$$P(\overline{E}) = \frac{f}{n}$$

An important property is found by adding these probabilities:

$$P(E) + P(\overline{E}) = \frac{s}{n} + \frac{f}{n} = \frac{s+f}{n} = \frac{n}{n} = 1$$

Property of Complements

$$P(E) = 1 - P(\overline{E}) \quad \text{or} \quad P(\overline{E}) = 1 - P(E)$$

EXAMPLE 1

Use Table 9.1 (page 618) to find the probability of not obtaining one pair with a poker hand.

Solution   From Table 9.1 we see that $P(\text{pair}) = \dfrac{1,098,240}{2,598,960} \approx 0.42$, so that

$$P(\text{no pair}) = P(\overline{\text{pair}}) = 1 - P(\text{pair}) \approx 1 - 0.42 = 0.58$$

EXAMPLE 2

What is the probability of obtaining at least one head in three flips of a coin?

Solution   Let $F = \{$obtain at least one head in three flips of a coin$\}$.

Method I: Work directly; use a tree diagram to find the possibilities.

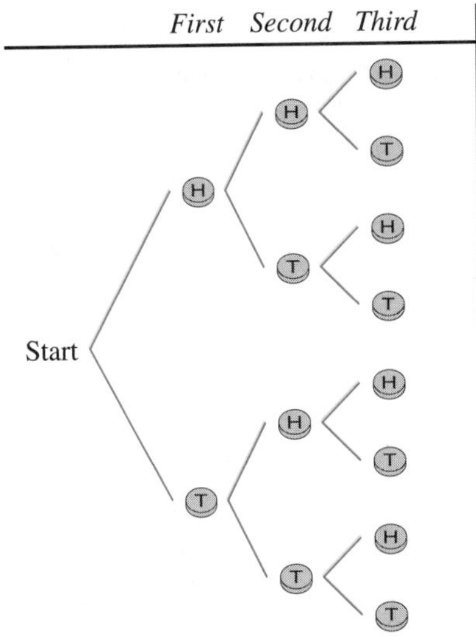

First	Second	Third		First	Second	Third	Success
		H		H	H	H	yes
	H						
		T		H	H	T	yes
H							
		H		H	T	H	yes
	T						
		T		H	T	T	yes
Start							
		H		T	H	H	yes
	H						
		T		T	H	T	yes
T							
		H		T	T	H	yes
	T						
		T		T	T	T	no

$P(F) = \frac{7}{8}$

Method II: For one coin, there are 2 outcomes (heads and tails); for two coins, there are 4 outcomes (HH, HT, TH, TT); and for three coins, there are 8 outcomes. We answer the question by finding the complement; $\overline{F}$ is the event of receiving no heads (that is, of obtaining all tails). *Without* drawing the tree diagram, we note that there is only one way of obtaining all tails (TTT). Thus

$P(F) = 1 - P(\overline{F}) = 1 - \frac{1}{8} = \frac{7}{8}$

Odds

Related to probability is the notion of odds. Instead of forming ratios

$$P(E) = \frac{s}{n} \quad \text{and} \quad P(\overline{E}) = \frac{f}{n}$$

we form the following ratios:

Odds in favor of an event E: $\quad \frac{s}{f}$ (ratio of good to bad)

Odds against an event E: $\quad \frac{f}{s}$ (ratio of bad to good)

Recall:

s = NUMBER OF SUCCESSES

f = NUMBER OF FAILURES

n = NUMBER OF POSSIBILITIES

The odds are
in on the
Academy Awards
LAS VEGAS
Neva...
The Odds Are
Against You...
Your chances of becoming
...h are only 1 out of
...h the book

GIROLAMO CARDANO
(1501–1576)

Cardano was a physician, mathematician, and astrologer. He investigated many interesting problems in probability and wrote a gambler's manual, probably because he was a compulsive gambler. He was the illegitimate son of a jurist and was at one time imprisoned for heresy — for publishing a controversial horoscope of Christ's life. He had a firm belief in astrology and predicted the date of his death. However, when the day came, he was alive and well — so he drank poison toward the end of the day to make his prediction come true!

■■■ EXAMPLE 3

If a jar has 2 quarters, 200 dimes, and 800 pennies, and a coin is to be chosen at random, what are the odds against picking a quarter?

Solution   We are interested in picking a quarter, so let $s = 2$; a failure is not obtaining a quarter so $f = 1,000$. (Find f by adding 200 and 800.)

$$\text{Odds in favor of obtaining a quarter:} \frac{s}{f} = \frac{2}{1,000} = \frac{1}{500}$$

$$\text{Odds against obtaining a quarter:} \frac{f}{s} = \frac{1,000}{2} = \frac{500}{1} \qquad \text{Do not write } \frac{500}{1} \text{ as 500.}$$

The odds against picking a quarter when a coin is selected at random are 500 to 1.

■■■ EXAMPLE 4   Polya's Method

The odds against winning a lottery are 50,000,000 to 1. Make up an example to help visualize these odds.

Solution   We use Polya's problem-solving guidelines for this example.

Understand the Problem.   It is difficult for the human mind to comprehend numbers like 50 million to 1, so the point of this problem is to help us "visualize" its magnitude.

Devise a Plan.   We will fill a house with 50 million ping pong balls and then paint 1 of them red.

Carry Out the Plan.   Imagine one red ping pong ball and 50,000,000 white ping pong balls. Winning a lottery is equivalent to reaching into a container containing these 50,000,001 ping pong balls and obtaining the red one. To visualize this, consider the size container you would need. Assume that a ping pong ball takes up a volume of 1 in.^{3}. A home of 1,200 ft^2 with an 8-ft ceiling has a volume of

$$1,200 \text{ ft}^2 \times 8 \text{ ft} = 9,600 \text{ ft}^3$$
$$= 9,600 \times (12 \text{ in.} \times 12 \text{ in.} \times 12 \text{ in.}) = 16,588,800 \text{ in.}^3$$

Look Back.   The number 16,588,800 is still not comprehensible. However, we can now say we would need to *fill* about 3 homes with ping pong balls and imagine that one ball is painted red. We do not even know which house contains the red ball, but we can imagine picking one of the three houses at random, and then reaching into that one house and choosing one ball. If the red one is chosen we win the lottery!

Sometimes you know the probability and want to find the odds, or you may know the odds and want to find the probability. These relationships are easy if you remember:

$$s + f = n$$

Finding the Odds Given the Probability

Suppose that you *know* $P(E)$.

   Odds in favor of an event E: $\dfrac{P(E)}{P(\overline{E})}$

   Odds against an event E: $\dfrac{P(\overline{E})}{P(E)}$

You can show these formulas to be true:

$$\frac{P(E)}{P(\overline{E})} = \frac{\dfrac{s}{n}}{\dfrac{f}{n}} = \frac{s}{n} \cdot \frac{n}{f} = \frac{s}{f} = \text{odds in favor}$$

The verification of the second formula is left for the problem set.

Finding the Probability Given Odds in Favor or Odds Against an Event

Suppose that you *know* the odds in favor of an event E:

   s to f

or the odds against an event E:

   f to s.

Then, $$P(E) = \frac{s}{s + f} \quad \text{and} \quad P(\overline{E}) = \frac{f}{s + f}$$

■■■■ **EXAMPLE 5**

If the probability of an event is 0.45, what are the odds in favor of the event?

Solution

$$P(E) = 0.45 = \frac{45}{100} = \frac{9}{20} \quad \text{This is given.}$$

$$P(\overline{E}) = 1 - \frac{9}{20} = \frac{11}{20}$$

Then the odds in favor of E are $\dfrac{P(E)}{P(\overline{E})} = \dfrac{\frac{9}{20}}{\frac{11}{20}} = \dfrac{9}{20} \cdot \dfrac{20}{11} = \dfrac{9}{11}.$

The odds in favor are 9 to 11.

■■■■ **EXAMPLE 6**

If the odds against you are 20 to 1, what is the probability of the event?

Solution   Odds against are f to s, so $f = 20$ and $s = 1$; thus,

$$P(E) = \frac{1}{20 + 1} = \frac{1}{21} \approx 0.048$$

■

■■■■ **EXAMPLE 7**

If the odds in favor of some event are 2 to 5, what is the probability of the event?

Solution   Odds in favor are s to f, so $s = 2$ and $f = 5$; thus,

$$P(E) = \frac{2}{2 + 5} = \frac{2}{7} \approx 0.286$$

■

Fundamental Counting Principle

In Example 2 we considered flipping a coin three times, and by drawing a tree diagram we found 8 possibilities. We also noted that for one coin there are 2 possibilities, and for 2 coins there are 4 possibilities. This application of a counting technique can be understood by looking at tree diagrams.

> **Fundamental Counting Principle**
>
> The **fundamental counting principle** gives the number of ways of performing two or more tasks. If task A can be performed in m ways, and if, after task A is performed, a second task, B, can be performed in n ways, then task A followed by task B can be performed in $m \cdot n$ ways.

■■■■ **EXAMPLE 8**

What is the probability that a two-child family will have children of opposite sex?

Solution   The fundamental counting principle tells us that, since there are 2 ways of having a child (B or G), for 2 children there is a total of

$$2 \cdot 2 = 4 \text{ ways}$$

First Child   Second Child

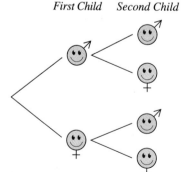

We verify this by looking at the tree diagram in the margin. There are 4 equally likely outcomes: BB, BG, GB, and GG. Thus, the probability of having a boy and a girl in a family of two children is

$$\frac{\text{NUMBER OF SUCCESSFUL OUTCOMES}}{\text{TOTAL NUMBER OF ALL POSSIBLE OUTCOMES}} = \frac{2}{4} = \frac{1}{2}$$

■

The fundamental counting principle can be used repeatedly for more than two possibilities.

■■■■ **EXAMPLE 9**

What is the probability that a four-child family will have two boys and two girls?

Solution   The fundamental counting principle tells us that the number of possibilities is

$$2 \times 2 \times 2 \times 2 = 16$$

Thus, $n = 16$. To find s we list the events that bring success:

{BBGG, BGBG, BGGB, GBGB, GBBG, GGBB}

Since there are 6 elements in this set, we see that $s = 6$. Thus, the desired probability is

$$\frac{6}{16} = \frac{3}{8}$$ ▬

▬▬ **EXAMPLE 10**

Consider the Dear Abby letter. What is a family's probability of having eight girls in a row? What are the odds?

Solution   First use the fundamental counting principle to find n:

$$2 \times 2 \times 2 \times 2 \times 2 \times 2 \times 2 \times 2 = 2^8$$

This is 256, so $P(\text{eight girls in a row}) = \dfrac{1}{256}$.

The odds in favor of 8 girls in a row are 1 to 255 ($s = 1$, $n = 256$, so $f = n - s = 256 - 1 = 255$). The odds against are 255 to 1. However, this does not answer the question in the letter. Since the couple already have 8 children, the probability of a boy for the 9th child is $\frac{1}{2}$. ▬

▬▬ **EXAMPLE 11**

What is the probability of getting a license plate that has a repeated letter or digit if you live in a state where the scheme is three letters followed by three numerals?

Solution   A license plate bears three letters followed by three digits. We begin by using the fundamental counting principle to find n (the total number of possibilities):

letters in the alphabet   number of digits
↓   ↓
$$\underbrace{26 \qquad \times 26 \times 26 \times}_{\text{three letters}} \quad \underbrace{10 \qquad \times 10 \times 10}_{\text{three digits}} = 17{,}576{,}000$$

To count the number of successes, you must understand the problem. To confirm this, which of the following plates would be considered a success?

ABC123	Failure
ABC122	Success; repeated digit
AAB456	Success; repeated letter
AAA111	Success; repeated letter and repeated numeral
XYZ890	Failure

Success is one repetition or two repetitions or three repetitions, ... Let R be the event that a repetition is received. It is difficult to count all of the possibilities for repetitions, but we can use the fundamental counting principle to count the number of license plates that *do not* have a repetition:

Calculator Comment

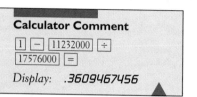

$\boxed{1}\ \boxed{-}\ \boxed{11232000}\ \boxed{\div}$
$\boxed{17576000}\ \boxed{=}$

Display: .3609467456

$$
\underset{\substack{\uparrow \\ \text{letters in the alphabet}}}{26} \quad \underset{\substack{\uparrow \\ \text{letters left after the first one (no repetitions)}}}{\times\ 25 \quad \times\ 24} \quad \times \quad \underset{\substack{\uparrow \\ \text{digits left}}}{\overset{\text{number of digits}}{10} \times 9 \times 8} = 11{,}232{,}000
$$

We complete the solution by using the property of complements:

$$P(R) = 1 - P(\overline{R}) = 1 - \frac{11{,}232{,}000}{17{,}576{,}000} \approx 0.36$$

This means that about 36% of all license plates in the state have a repeated letter or digit.

News Clip

The personalized license plates were done by Stephen Underwood, who not only assembled the plates to spell out the preamble of the U.S. Constitution, but also did it by using all 50 states *in alphabetical order!*

Conditional Probability

Frequently, we wish to compute the probability of an event but we have additional information that will alter the sample space. For example, suppose that a family has two children. What is the probability that the family has two boys?

$$P(2 \text{ boys}) = \frac{1}{4}$$ Sample space: BB, BG, GB, GG; 1 success out of 4 possibilities

Now, let's complicate the problem a little. Suppose that we know that the older child is a boy. We have altered the sample space as follows:

Original sample space: BB, BG, GB, GG; but we need to cross out the last two possibilities because we *know* that the older child is a boy:

Success
↓   These are crossed out.

Altered sample space: BB, BG, ~~GB~~, ~~GG~~; therefore,

Altered sample space has two elements.

$$P(2 \text{ boys given the oldest is a boy}) = \frac{1}{2}$$

This is a problem involving a **conditional probability** — namely, a *probability of an event* **given** *that another event F has occurred.* We denote this by

$$P(E|F)$$ *Read this as:*   "probability of E given F."

■■■■ EXAMPLE 12

Suppose that you toss two coins (or a single coin twice). What is the probability that two heads are obtained if you know that at least one head is obtained?

Solution   Consider an altered sample space: HH, HT, TH, ~~TT~~. The probability is $\frac{1}{3}$. ■

■■■■ EXAMPLE 13

Suppose that you draw two cards from a deck of cards. Find the following probabilities. Let $H = \{$the second card drawn is a heart$\}$.

a. $P(H)$
b. $P(H|$ a heart is drawn on the first draw$)$
c. $P(H|$ a heart is not drawn on the first draw$)$

Solution

a. $P(H) = \dfrac{13}{52} = \dfrac{1}{4} = 0.25$

Remember, it does not matter what happened on the first draw because we do not know what happened on that draw. The second card "does not remember" what is drawn on the first draw.

b. This time we know what happened on the first draw, so $n = 51$ and $s = 12$ (because a heart was drawn on the first draw):

$$P(H \mid \text{a heart is drawn on the first draw}) = \frac{12}{51} \approx 0.235$$

c. We still have $n = 51$, but this time $s = 13$:

$$P(H \mid \text{a heart is not drawn on the first draw}) = \frac{13}{51} \approx 0.255$$

■

■■■■■ **EXAMPLE 14**   **Polya's Method**

In a life science experiment it is necessary to examine fruit flies and to determine their sex and whether they have mutated after exposure to a certain dose of radiation. Suppose a single fruit fly is selected at random from the radiated fruit flies. Find the following probabilities (correct to the nearest hundredth).

a. It is male.

b. It is a normal male.

c. It is normal, given that it is a male.

d. It is a male, given that it is normal.

e. It is mutated, given that it is a male.

Solution   We use Polya's problem-solving guidelines for this example.

Understand the Problem.   We obviously need more information to find the desired probabilities. We need to know the number of fruit flies that are part of the experiment. Assume that the total number is 1,000. We also need some data. Upon examination, we find that for 1,000 fruit flies examined, there were 643 females and 357 males. Also, 403 of the females were normal and 240 were mutated; of the males, 190 were normal and 167 were mutated.

Devise a Plan.   We will display the data in table form to make it easy to calculate the desired probabilities.

Carry Out the Plan.

	Mutated	Normal	Total
Male	167	190	357
Female	240	403	643
Total	407	593	1,000

a. $P(\text{male}) = \dfrac{357}{1,000} \approx 0.36$ Round answers to the nearest hundredth.

b. $P(\text{normal male}) = \dfrac{190}{1{,}000} = 0.19$

c. $P(\text{normal}|\text{male}) = \dfrac{190}{357} \approx 0.53$ Note the altered sample space.

d. $P(\text{male}|\text{normal}) = \dfrac{190}{593} \approx 0.32$ Yet another altered sample space

e. $P(\text{mutated}|\text{male}) = \dfrac{167}{357} \approx 0.47$

Look Back.   All probabilities seem reasonable, and all are between 0 and 1.   ▬

▮▮▮▮ **EXAMPLE 15**

Two cards are drawn from a deck of cards. Find each of the following probabilities.

a. The first card drawn is a heart.

b. The first card drawn is not a heart.

c. The second card drawn is a heart if the first card drawn was a heart.

d. The second card drawn is not a heart if the first card drawn was a heart.

e. The second card drawn is a heart if the first card drawn was not a heart.

f. The second card drawn is not a heart if the first card drawn was not a heart.

g. The second card drawn is a heart.

h. Use a tree diagram to represent the indicated probabilities.

Solution   Let $H_1 = \{\text{first card drawn is a heart}\}$ and $H_2 = \{\text{second card drawn is a heart}\}$.

a. $P(H_1) = \dfrac{13}{52} = \dfrac{1}{4}$ There are 13 hearts in a deck of 52 cards.

b. $P(\overline{H}_1) = \dfrac{39}{52} = \dfrac{3}{4}$ There are 39 nonhearts in a deck of 52 cards.

c. $P(H_2|H_1) = \dfrac{12}{51} = \dfrac{4}{17}$ If a heart is obtained on the first draw, then for the second draw there are 12 hearts in the deck of 51 remaining cards.

d. $P(\overline{H}_2|H_1) = \dfrac{39}{51} = \dfrac{13}{17}$ If a heart is obtained on the first draw, then for the second draw there are 39 nonhearts in a deck of 51 remaining cards.

e. $P(H_2|\overline{H}_1) = \dfrac{13}{51}$ If a heart is not obtained on the first draw, then for the second draw there are 13 hearts in a deck of 51 remaining cards.

f. $P(\overline{H}_2|\overline{H}_1) = \dfrac{38}{51}$ If a nonheart is obtained on the first draw, then for the second draw there are 38 nonhearts in a deck of 51 cards.

g. $P(H_2) = \dfrac{1}{4}$ If we do not know what happened on the first draw, then this is not a conditional probability, so we consider this probability to be the same as $P(H_1)$.

h. We use a tree diagram to illustrate this situation.

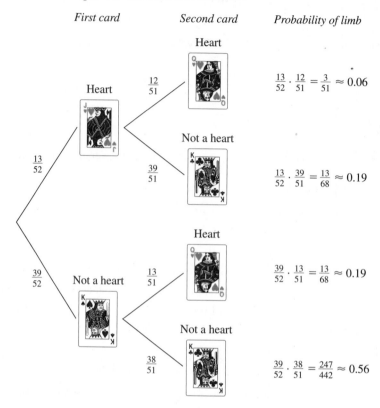

Note that we use the word *limb* to mean a sequence of branches that starts at the beginning. Also note that we have filled in the probabilities on each branch. To find the probabilities, we multiply as we move horizontally across a limb, and we add as we move vertically from limb to limb. Verify that the probabilities for the first card add to 1:

$$P(H_1) + P(\overline{H}_1) = \frac{13}{52} + \frac{39}{52} = \frac{52}{52} = 1$$

Also, for the second card, the probabilities for each branch add to 1. For $P(H_2)$, we can confirm our reasoning in part **g** by considering the first and third limbs of the tree diagram:

$$P(H_2) = \frac{13}{52} \cdot \frac{12}{51} + \frac{39}{52} \cdot \frac{13}{51} = \frac{663}{2,652} = \frac{1}{4}$$

Notice that conditional probabilities are found in the tree diagram by beginning at their condition.

Verify that all of the information in parts **a–g** can be found directly from the tree diagram. For this reason, we often use tree diagrams to assist us in finding probabilities and conditional probabilities.

Procedure for Using Tree Diagrams

Multiply when moving horizontally across a limb.

Add when moving vertically from limb to limb.

Conditional probabilities start at their condition; unconditional probabilities start at the beginning of the tree.

PROBLEM SET   9.3

▲ **A Problems**

1. **IN YOUR OWN WORDS** What is the property of complements?

2. **IN YOUR OWN WORDS** Compare and contrast odds and probability.

3. **IN YOUR OWN WORDS** Explain what we mean by "conditional probability."

4. **IN YOUR OWN WORDS** What is the fundamental counting principle?

5. **IN YOUR OWN WORDS** How many different sequences are there for the first five moves in a game of tick-tack-toe? Give reasons for your answer.

Find the requested probabilities in Problems 6–11.

6. $P(\overline{A})$ if $P(A) = 0.6$ 7. $P(\overline{B})$ if $P(B) = \frac{4}{5}$

8. $P(C)$ if $P(\overline{C}) = \frac{9}{13}$ 9. $P(D)$ if $P(\overline{D}) = 0.005$

10. Obtaining at least one head in four flips of a coin

11. Obtaining a sum of at least 4 in a roll of a pair of dice

List the outcomes considered a failure, and find the probability of failure, for the experiments named in Problems 12–17.

	Experiment	Success	P(success)
12.	Tossing a coin	head	$\frac{1}{2}$
13.	Rolling a die	four or six	$\frac{1}{3}$
14.	Guessing an answer on a 5-choice multiple-choice test	correct guess	$\frac{1}{5}$
15.	Card game	drawing a heart	0.18
16.	Baseball game	White Sox win	0.57
17.	Football game	your school wins	0.83

18. What are the odds in favor of drawing an ace from an ordinary deck of cards?

19. What are the odds in favor of drawing a heart from an ordinary deck of cards?

20. What are a four-child family's odds against having four boys?

21. Suppose the odds that a man will be bald by the time he is 60 are 9 to 1. State this as a probability.

22. Suppose the odds are 33 to 1 that someone will lie to you at least once in the next week. State this as a probability.

23. Racetracks quote the approximate odds for each race on a large display board called a *tote board*. Here's what it might say for a particular race:

Horse Number	Odds
1	2 to 1
2	15 to 1
3	3 to 2
4	7 to 5
5	1 to 1

    What would be the probability of winning for each of these horses?
    [*Note:*   The odds stated are for the horse's *losing*. Thus, $P(\text{horse 1 losing}) = \frac{2}{2+1} = \frac{2}{3}$, so $P(\text{horse 1 winning}) = 1 - \frac{2}{3} = \frac{1}{3}$.]

24. Three fair coins are tossed. What is the probability that at least one is a head?

25. A card is selected from an ordinary deck. What is the probability that it is not a face card?

26. Choose a natural number between 1 and 100, inclusive. What is the probability that the number chosen is not a multiple of 5?

27. Suppose that a family wants to have four children.
    a. What is the sample space?
    b. What is the probability of 4 girls? 4 boys?
    c. What is the probability of 1 girl and 3 boys? 1 boy and 3 girls?
    d. What is the probability of 2 boys and 2 girls?
    e. What is the sum of your answers in parts **b** through **d**?

28. What is a three-child family's probability of having exactly two boys, given that at least one of their children is a boy?

29. What is the probability of obtaining exactly three heads in four flips of a coin, given that at least two are heads?

30. Many states offer personalized license plates. The state of California, for example, allows personalized plates with seven spaces for numerals or letters, or one of the following four symbols:

    What is the total number of license plates possible using this counting scheme? (Assume that each available space is occupied by a numeral, letter, symbol, or space.)

▲ **B Problems**

A single card is drawn from a standard deck of cards. Find the probabilities if the given information is known about the chosen card in Problems 31–36. A face card is a jack, queen, or king.

31. $P(\text{face card} \mid \text{jack})$ 32. $P(\text{jack} \mid \text{face card})$ 33. $P(\text{heart} \mid \text{not a spade})$

34. $P(\text{two} \mid \text{not a face card})$ **35.** $P(\text{black} \mid \text{jack})$ **36.** $P(\text{jack} \mid \text{black})$

Two cards are drawn from a standard deck of cards, and one of the two cards is noted. Find the probabilities of the second card, given the information about the noted card provided in Problems 37–42.

37. $P(\text{ace} \mid \text{two})$ **38.** $P(\text{king} \mid \text{king})$ **39.** $P(\text{heart} \mid \text{heart})$

40. $P(\text{heart} \mid \text{spade})$ **41.** $P(\text{black} \mid \text{red})$ **42.** $P(\text{black} \mid \text{black})$

Suppose a single die is rolled. Find the probabilities requested in Problems 43–46.

43. 6, given that an odd number was rolled

44. 5, given that an odd number was rolled

45. odd, given that the rolled number was a 6

46. odd, given that the rolled number was a 5

Suppose a pair of dice are rolled. Consider the sum of the numbers on the top of the dice and find the probabilities requested in Problems 47–53.

47. 7, given that the sum is odd

48. 7, given that at least one die came up 2

49. 5, given that exactly one die came up 2

50. 3, given that exactly one die came up 2

51. 2, given that exactly one die came up 2

52. 8, given that a double was rolled

53. a double, given that an 8 was rolled

Use estimation to select the best response in Problems 54–59. Do not calculate.

54. Of these three means of travel, the safest is
   A. car   B. train   C. plane

55. Which of the following is most probable?
   A. Winning the grand prize in a state lottery
   B. Being struck by lightning
   C. Appearing on the *Tonight Show*

56. Which of the following is more probable?
   A. Flipping a coin 3 times and obtaining at least 2 heads
   B. Flipping a coin 4 times and obtaining at least 2 heads

57. Which of the following is more probable?
   A. Rolling a die 3 times and obtaining a six 2 times
   B. Rolling a die 3 times and obtaining a six at least 2 times

58. Which of the following is more probable?
   A. Correctly guessing all the answers on a 20-question true–false examination
   B. Flipping a coin 20 times and obtaining all heads

59. Which of the following is more probable?
   A. Correctly guessing all the answers on a 10-question 5-part multiple-choice test
   B. Your living room is filled with white ping pong balls. There is also one red ping

pong ball in the room. You reach in and select a ping pong ball at random and select the red ping pong ball.

60. Show that the odds against an event E can be found by computing $P(\overline{E})/P(E)$.

61. The Emory Harrison family of Tennessee had 13 boys.

    a. What is the probability of a 13-child family having 13 boys?
    b. What is the probability that the next child of the Harrison family will be a boy (giving them 14 boys in a row)?

▲ **Problem Solving**

62. One roulette system is to be $1 on black. If black comes up on the first spin of the wheel, you win $1 and the game is over. If black does not come up, double your bet ($2). If you win, the game is over and your net winnings for two spins is still $1 (show this). If black does not come up, double your bet again ($4). If you win, the game is over and your net winnings for three spins is still $1 (show this). Continue in this doubling procedure until you eventually win; in every case your net winnings amount to $1 (show this). What is the fallacy with this betting system?

63. "Last week I won a free Big Mac at McDonald's. I sure was lucky!" exclaimed Charlie. "Do you go there often?" asked Pat. "Only twenty or thirty times a month. And the odds of winning a Big Mac were only 20 to 1." "Don't you mean 1 to 20?" queried Pat. "Don't confuse me with details. I don't even understand odds at the racetrack, and I go all the time." Is Charlie or Pat correct about the odds?

9.4   COUNTING FORMULAS

We have used both tree diagrams and the fundamental counting principle to count the number of elements in the sample space and in the event we are considering. Tree diagrams work for all procedures where a selection (or choice) is to be made at each step of a process. The fundamental counting principle is a multistep procedure where the number of choices at each step is not affected by the choices made on a previous step. Even though these procedures are useful and important, there are two additional counting formulas used frequently in calculating probabilities.

The model for both of these counting formulas is that from a given set of elements, we are to pick, or choose, a certain number of elements; we can choose them so that the order of selection matters, or so that it does not matter. If the order of selection matters, we call it a *permutation*, and if the order is not important, it is called a *combination*.

Consider a set

$$A = \{a, b, c, d, e\}$$

Remember that when set symbols are used, the order in which the elements are listed is not important. Suppose now that you wish to select elements from A by taking them *in a certain order*. The selected elements are enclosed in parentheses to signify order and are called an **ordered pair** or **arrangement** of elements of A. For example, if the elements a and b are selected from A, then

there is one *subset* (order not important): $\{a, b\}$

there are two *arrangements* (order important): (a, b) and (b, a)

We also note that both arrangements and subsets are selected *without repetitions* (or sometimes we say *without replacement*).

"Five trillion, four hundred eighty billion, five hundred twenty-three million, two hundred ninety-seven thousand, one hundred and sixty-two . . ."

▬▬ EXAMPLE 1

Consider selecting the elements a, b, and c from set $A = \{a, b, c, d, e\}$. List all possible subsets of these elements, as well as all possible arrangements.

Solution

One subset: $\{a, b, c\}$

Six arrangements: $(a, b, c), (a, c, b), (b, a, c), (b, c, a), (c, a, b), (c, b, a)$

Sometimes it is difficult to list all the possible arrangements, so we often use a tree diagram to find them all. A tree diagram for the arrangements of this example is shown in the margin.   ▬

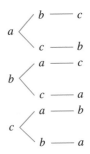

▬▬ EXAMPLE 2

Select two elements from set $A = \{a, b, c, d, e\}$. How many combinations (subsets) and how many permutations (arrangements) are possible?

Solution

Combinations (subsets)	$\{a, b\}$, $\{a, c\}$, $\{a, d\}$, $\{a, e\}$,
Permutations (arrangements)	$(a, b), (b, a), (a, c), (c, a), (a, d), (d, a), (a, e), (e, a),$
Combinations (subsets)	$\{b, c\}$, $\{b, d\}$, $\{b, e\}$,
Permutations (arrangements)	$(b, c), (c, b), (b, d), (d, b), (b, e), (e, b),$
Combinations (subsets)	$\{c, d\}$, $\{c, e\}$,
Permutations (arrangements)	$(c, d), (d, c), (c, e), (e, c),$
Combinations (subsets)	$\{d, e\}$,
Permutations (arrangements)	$(d, e), (e, d)$

There are 10 combinations of two elements from A and there are 20 permutations of two elements from A.   —

A name and some notation are needed for listing different subsets and arrangements of a given set.

If an arbitrary finite set S has n elements and r elements are selected from S (so that $r \leq n$), then

Combination

A **combination** is a subset obtained by selecting r elements from a set of n elements so that *order is not important*. The number of subsets or combinations is denoted by $_nC_r$ or $\binom{n}{r}$.

Permutation

A **permutation** is an arrangement obtained by selecting r elements from a set of n elements so that *order is important*. The number of arrangements or permutations is denoted by $_nP_r$.

From Example 2, we see that $_5C_2 = 10$ and $_5P_2 = 20$. We also note from the solution of Example 2 that $_5P_2 = 2 \cdot {_5C_2}$; that is, there are two arrangements for *each* subset (in this example).

■ EXAMPLE 3

Find $_5C_5$ and $_5P_5$.

Solution $_5C_5$ is the number of subsets of 5 elements taken from a set of 5 elements. For example, if $A = \{a, b, c, d, e\}$, then we see there is one subset consisting of 5 elements — namely, the set A itself. Thus, $_5C_5 = 1$. To find $_5P_5$, we can use a tree diagram; however, for this example it is easier to use the fundamental counting principle:

$$_5P_5 = 5 \cdot 4 \cdot 3 \cdot 2 \cdot 1 = 120$$ —

The pattern shown in Example 3 is common enough to warrant a special notation, which is called a *factorial*. This factorial notation uses an exclamation point such as $n!$, which is pronounced "n-factorial."

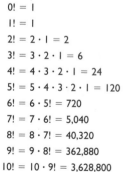

$0! = 1$
$1! = 1$
$2! = 2 \cdot 1 = 2$
$3! = 3 \cdot 2 \cdot 1 = 6$
$4! = 4 \cdot 3 \cdot 2 \cdot 1 = 24$
$5! = 5 \cdot 4 \cdot 3 \cdot 2 \cdot 1 = 120$
$6! = 6 \cdot 5! = 720$
$7! = 7 \cdot 6! = 5,040$
$8! = 8 \cdot 7! = 40,320$
$9! = 9 \cdot 8! = 362,880$
$10! = 10 \cdot 9! = 3,628,800$
$\vdots$

Factorial

For any counting number n, the **factorial of n** is defined by

$$n! = n(n-1)(n-2) \cdot \cdots \cdot 3 \cdot 2 \cdot 1$$

Also, $0! = 1$.

The calculations for small factorials are shown in the margin. You need not wonder why the notation "!" was chosen. For example,

$$52! \approx 8.065817517 \times 10^{67}$$

■ EXAMPLE 4

Simplify the given factorial expressions.

a. $5! - 4!$ **b.** $(5 - 4)!$ **c.** $\dfrac{7!}{5!}$ **d.** $\dfrac{200!}{198!}$ **e.** $\dfrac{8!}{3!(8 - 3)!}$

Solution

a. $5! = 5 \cdot 4 \cdot 3 \cdot 2 \cdot 1 = 120$ and $4! = 4 \cdot 3 \cdot 2 \cdot 1 = 24$;

$$5! - 4! = 120 - 24 = 96$$

b. $(5 - 4)! = 1! = 1$

c. $\dfrac{7!}{5!} = \dfrac{7 \cdot 6 \cdot \cancel{5} \cdot \cancel{4} \cdot \cancel{3} \cdot \cancel{2} \cdot \cancel{1}}{\cancel{5} \cdot \cancel{4} \cdot \cancel{3} \cdot \cancel{2} \cdot \cancel{1}} = 7 \cdot 6 = 42$

Instead of writing out all the factors and canceling, you can write

$$\frac{7!}{5!} = \frac{7 \cdot 6 \cdot 5!}{5!} = 42$$

d. $\dfrac{200!}{198!} = 200 \cdot 199 = 39{,}800$

e. $\dfrac{8!}{3!(8 - 3)!} = \dfrac{8!}{3!5!} = \dfrac{8 \cdot 7 \cdot 6 \cdot 5!}{3 \cdot 2 \cdot 5!} = 8 \cdot 7 = 56$

Example 4 illustrates a useful property of factorials.

> ### Multiplication Property of Factorials
> $$n! = n(n - 1)!$$

Now that we have factorial notation, we turn our attention to finding formulas for $_nC_r$ and $_nP_r$. For the combination formula, we turn to Pascal's triangle:

$$_0C_0 = 1$$
$$_1C_0 = 1 \qquad _1C_1 = 1$$
$$_2C_0 = 1 \qquad _2C_1 = 2 \qquad _2C_2 = 1$$
$$_3C_0 = 1 \qquad _3C_1 = 3 \qquad _3C_2 = 3 \qquad _3C_3 = 1$$
$$_4C_0 = 1 \qquad _4C_1 = 4 \qquad _4C_2 = 6 \qquad _4C_3 = 4 \qquad _4C_4 = 1$$
$$\vdots$$

That is, $_nC_r$ is the number in the nth row, rth diagonal of Pascal's triangle. We repeat the triangle here for easy reference.

Calculator Comment

Look for the factorial on your calculator: $\boxed{!}$ or $\boxed{x!}$

Historical Note

Factorial notation was first used by Christian Kramp in 1808.

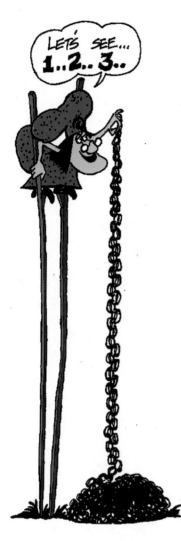

Pascal's Triangle (with Left Diagonals and Right Diagonals labeled 0–10):

Row																
0	1															
1	1	1														
2	1	2	1													
3	1	3	3	1												
4	1	4	6	4	1											
5	1	5	10	10	5	1										
6	1	6	15	20	15	6	1									
7	1	7	21	35	35	21	7	1								
8	1	8	28	56	70	56	28	8	1							
9	1	9	36	84	126	126	84	36	9	1						
10	1	10	45	120	210	252	210	120	45	10	1					
11	1	11	55	165	330	462	462	330	165	55	11	1				
12	1	12	66	220	495	792	924	792	495	220	66	12	1			
13	1	13	78	286	715	1287	1716	1716	1287	715	286	78	13	1		
14	1	14	91	364	1001	2002	3003	3432	3003	2002	1001	364	91	14	1	
15	1	15	105	455	1365	3003	5005	6435	6435	5005	3003	1365	455	105	15	1

EXAMPLE 5

Find **a.** $_8C_3$ **b.** $_6C_5$ **c.** $_{14}C_6$ **d.** $_{12}C_{10}$

Solution

a. $_8C_3 = 56$ row 8, diagonal 3

b. $_6C_5 = 6$ row 6, diagonal 5

c. $_{14}C_6 = 3{,}003$ row 14, diagonal 6

d. $_{12}C_{10} = 66$ row 12, diagonal 10

Permutations are related to combinations according to the following formula:

$$_nP_r = r! \, _nC_r$$

EXAMPLE 6

Find **a.** $_8P_3$ **b.** $_6P_5$ **c.** $_{14}P_6$ **d.** $_{12}P_{10}$

Solution

a. $_8P_3 = 3! \cdot {_8C_3} = 3! \cdot 56 = 336$

b. $_6P_5 = 5! \cdot {_6C_5} = 5! \cdot 6 = 720$

c. $_{14}P_6 = 6! \cdot 3{,}003 = 2{,}162{,}160$

d. $_{12}P_{10} = 10! \cdot 66 = 239{,}500{,}800$

EXAMPLE 7

Select three elements from set $A = \{a, b, c, d, e\}$. How many combinations (subsets) and how many permutations (arrangements) are possible?

Solution   We could solve the problem by direct enumeration (or a tree diagram), but instead we use Pascal's triangle:

$$_5C_3 = 10 \quad \text{and} \quad _5P_3 = 3! \cdot {_5C_3} = 6 \cdot 10 = 60$$

It is possible to derive formulas for both combinations and permutations (see Problems 54 and 55) using factorial notation.

Formulas for Combination and Permutation

If n and r are counting numbers so that $r \leq n$, then

$$_nC_r = \frac{n!}{r!(n-r)!} \quad \text{and} \quad _nP_r = \frac{n!}{(n-r)!}$$

▰▰▰ EXAMPLE 8

Evaluate the indicated numbers using the formulas. Check by using Pascal's triangle, if possible.

a. $_{10}C_6$ **b.** $_{13}P_3$ **c.** $_nC_2$ **d.** $_nP_{n-1}$ **e.** $_{52}C_5$ **f.** $_{52}P_5$

Solution

Calculator Comment

Check your calculator for combination and permutation keys. These functions are built into many calculators.

a. $_{10}C_6 = \dfrac{10!}{6!(10-6)!} = \dfrac{10 \cdot \overset{3}{\cancel{9}} \cdot \cancel{8} \cdot 7 \cdot \cancel{6!}}{\cancel{4} \cdot \cancel{3} \cdot \cancel{2} \cdot \cancel{6!}} = 210$

From Pascal's triangle, look at row 10, diagonal 6:   210.

b. $_{13}P_3 = \dfrac{13!}{(13-3)!} = \dfrac{13 \cdot 12 \cdot 11 \cdot 10!}{10!} = 1,716$

From Pascal's triangle, look at row 13, diagonal 3:   286.

$_{13}P_3 = 3! \cdot _{13}C_3 = 6 \cdot 286 = 1,716.$

c. $_nC_2 = \dfrac{n!}{2!(n-2)!} = \dfrac{n(n-1)(n-2)!}{2(n-2)!} = \dfrac{n(n-1)}{2}$

Cannot check this using Pascal's triangle.

d. $_nP_{n-1} = \dfrac{n!}{[n-(n-1)]!} = \dfrac{n!}{1!} = n!$

Cannot check this using Pascal's triangle.

e. $_{52}C_5 = \dfrac{52!}{5!(52-5)!} = \dfrac{52 \cdot 51 \cdot 50 \cdot 49 \cdot 48 \cdot 47!}{5 \cdot 4 \cdot 3 \cdot 2 \cdot 47!} = 2,598,960$

It is not practical to find this on Pascal's triangle.

f. $_{52}P_5 = \dfrac{52!}{(52-5)!} = \dfrac{52!}{47!} = 52 \cdot 51 \cdot 50 \cdot 49 \cdot 48 = 311,875,200$

You could also use the property that $_{52}P_5 = 5! \cdot _{52}C_5$ and the answer from part **e.** ▬

Classifying Combinations and Permutations

We have now looked at several counting schemes: tree diagrams, the fundamental counting principle, permutations, and combinations. In practice, you will generally not be told what type of counting problem you are dealing with — you will need to decide. Table 9.2 may help you with that decision.

TABLE 9.2   Comparison of Counting Techniques

Tree diagram	Fundamental counting principle	Combinations	Permutations
Use this to handle inconsistencies. It is the most general of the methods, but it is also the most tedious. You can use this method when the other methods fail.	Counts the total number of separate tasks	Number of ways of selecting r items out of n items	
	Repetitions are allowed.	Repetitions are not allowed.	
	If tasks 1, 2, 3, . . . , k can be performed in $n_1, n_2, n_3, \ldots, n_k$ ways, respectively, then the total number of ways the k tasks can be performed is $$n_1 \cdot n_2 \cdot n_3 \cdot \cdots \cdot n_k$$	Subsets	Arrangements
		Order does not matter.	Order matters.
		$${}_nC_r = \dfrac{n!}{r!(n-r)!}$$	$${}_nP_r = \dfrac{n!}{(n-r)!}$$

■■■■   EXAMPLE 9

Classify each of the given problems, and describe the process you would use to solve.

a. In how many ways can a diamond flush be drawn in poker? (*Hint:* A diamond flush is a hand of five diamonds.)

b. How many three-letter words can be formed?

c. Consider a club consisting of 12 members. In how many ways can the club elect a president, a vice-president, and a secretary?

d. Consider a club consisting of 12 members. In how many ways can the club appoint a committee consisting of 3 members?

e. How many ways can you select a salad dressing that comes in three brands, A, B, and C? Brand A sells Italian (8 and 12 oz), French (8, 12, and 16 oz), and Russian (8 and 12 oz). Brand B sells Italian (12 oz) and French (8 and 12 oz). Brand C sells Italian (12 oz), French (8, 12, and 16 oz), Russian (8 and 16 oz), and Ranch (8, 12, 16, and 32 oz).

f. It was recently reported that Wendy's spent nearly $1 million on thousands of taste tests to develop a better hamburger and came up with "The Big Classic." They tested 9 different buns, 40 sauces, 3 types of lettuce, 2 sizes of tomatoes, 4 different boxes, and 500 names. How many different hamburgers are possible if one choice is made from each category?

g. Find the number of bridge hands (13 cards) consisting of six hearts, four spades, and three diamonds.

h. A club with 42 members wants to elect a president, a vice-president, and a treasurer. From the other members, an advisory committee of five people is to be selected. In how many ways can this be done?

Solution

a. We are selecting 5 out of 13, and the order of selection does not matter, so this is a combination:

$$_{13}C_5 = 1,287 \qquad \text{See row 13, diagonal 5 of Pascal's triangle.}$$

b. We are selecting 3 out of 26 (we assume the English language with 26 letters), the order of selection does matter, and repetitions are allowed, so we use the fundamental counting principle:

$$26 \cdot 26 \cdot 26 = 17,576$$

c. We are selecting 3 out of 12, and the order of selection matters, so this is a permutation:

$$_{12}P_3 = 3! \cdot {}_{12}C_3 = 6 \cdot \underbrace{220}_{} = 1,320$$
$$\text{See row 12, diagonal 3.}$$

d. We are selecting 3 out of 12, and the order of selection for a committee does not matter, so this is a combination:

$$_{12}C_3 = 220 \qquad \text{See row 12, diagonal 3.}$$

e. We build a tree diagram:

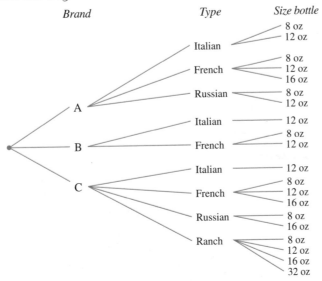

We simply count the number of limbs; there are 20 choices.

f. We are counting the number of separate tasks, so we use the fundamental counting principle:

$$9 \times 40 \times 3 \times 2 \times 4 \times 500 = 4,320,000$$

The news article didn't report the number of different hamburgers, but if the taste test was done at one location and each hamburger took 10 minutes to eat,

Historical Note

SOPHIE GERMAIN
(1776–1831)

Sophie Germain is one of the first women recorded in the history of mathematics. She did original mathematical research in number theory. In her time, women were not admitted to first-rate universities and were not, for the most part, taken seriously, so she wrote at first under the pseudonym LeBlanc. The situation is not too different from that portrayed by Barbra Streisand in the movie *Yentl*. Even though Germain's most important research was in number theory, she was awarded the prize of the French Academy for a paper entitled "Memoir on the Vibrations of Elastic Plates."

it would take over 80 years of nonstop eating to test all of the possibilities! Notice that Wendy's thought a different name and packaging changed the burger. If we don't count a name or packaging change as a different burger, then the number of different possible burgers is

$$9 \times 40 \times 3 \times 2 = 2,160$$

g. The order in which the cards are received is unimportant, so finding the number of bridge hands is a combination problem. However, the different categories use the fundamental counting principle:

Number of hearts   Spades   Diamonds

$$\overbrace{_{13}C_6} \quad \cdot \quad \overbrace{_{13}C_4} \quad \cdot \quad \overbrace{_{13}C_3} \quad = 1,716 \cdot 715 \cdot 286 \qquad \text{Pascal's triangle}$$
$$= 350,904,840$$

If, instead of Pascal's triangle, you use the formulas, then you would have

$$\frac{13!}{6!(13-6)!} \cdot \frac{13!}{4!(13-4)!} \cdot \frac{13!}{3!(13-3)!} = \frac{13! \cdot 13! \cdot 13!}{6! \cdot 7! \cdot 4! \cdot 9! \cdot 3! \cdot 10!}$$
$$= 350,904,840$$

h. This is both a permutation and a combination problem, with the final results calculated by using the fundamental counting principle:

Number of ways of selecting officers

$$\overbrace{_{42}P_3} \cdot \underbrace{_{39}C_5} = 42 \cdot 41 \cdot 40 \cdot \frac{39!}{5!(39-5)!} = 39,658,142,160$$

Number of ways of selecting committee

PROBLEM SET 9.4

▲ **A Problems**

1. **IN YOUR OWN WORDS** Contrast the fundamental counting principle and a permutation.

2. **IN YOUR OWN WORDS** Contrast a permutation and a combination.

3. **IN YOUR OWN WORDS** Describe a procedure for identifying the type of counting technique appropriate for a counting problem.

4. **IN YOUR OWN WORDS** Give two original examples of subsets that are combinations.

5. **IN YOUR OWN WORDS** Give two original examples of arrangements that are permutations.

Compute the results in Problems 6–16.

6. a. $7! - 5!$ b. $(7-5)!$ 7. a. $6! - 5!$ b. $(10-4)!$

8. a. $6!$ b. $6 \cdot 5!$ 9. a. $8! - 4!$ b. $(8-4)!$

10. a. $\dfrac{10!}{6!}$ b. $\dfrac{10!}{4! \cdot 6!}$ 11. a. $\dfrac{10!}{8!}$ b. $\dfrac{100!}{97!}$

12. a. $\dfrac{8!}{4!}$ b. $\left(\dfrac{8}{4}\right)!$ 13. a. $\dfrac{12!}{6!}$ b. $\left(\dfrac{12}{6}\right)!$

14. a. $\dfrac{8!}{5! \cdot 3!}$ b. $\dfrac{10!}{5!}$ 15. a. $\dfrac{9!}{6! \cdot 3!}$ b. $\dfrac{52!}{5! \cdot 47!}$

16. a. $\dfrac{52!}{3!(52 - 3)!}$ b. $\dfrac{12!}{3!(12 - 3)!}$

Evaluate each of the expressions in Problems 17–22.

17. a. $_7P_1$ b. $_7P_2$ c. $_7P_3$ d. $_7P_4$ e. $_7P_5$

18. a. $_6P_4$ b. $_{52}P_4$ c. $_6P_2$ d. $_9P_4$ e. $_{50}P_1$

19. a. $_9C_1$ b. $_9C_2$ c. $_9C_3$ d. $_9C_4$ e. $_9C_0$

20. a. $_5C_4$ b. $_5C_3$ c. $_7C_2$ d. $_4C_4$ e. $_{100}C_1$

21. a. $_7C_3$ b. $_{22}P_5$ c. $_5C_5$ d. $_5P_3$ e. $_{50}C_{48}$

22. a. $_8P_4$ b. $_{25}C_1$ c. $_{10}P_0$ d. $_gC_h$ e. $_sP_t$

▲ B Problems

Simplify the expressions in Problems 23–28.

23. $\dfrac{\frac{n!}{(n - r)!}}{r!}$ 24. $\dfrac{\frac{n!}{(n - 1)!}}{r!}$ 25. $\dfrac{n!}{(n - 2)!}$

26. $\dfrac{(n + 1)!}{n!}$ 27. $\dfrac{1}{n!} + \dfrac{1}{(n + 1)!}$ 28. $\dfrac{1}{n!} + \dfrac{1}{(n - 1)!}$

Classify Problems 29–41, and describe the process you would use to solve before answering the question.

29. In how many ways can a group of 7 choose a committee of 4?

30. In how many ways can a group of 7 elect a president, secretary, and treasurer?

31. How many subsets of size 3 can be formed from a set of 5 elements?

32. How many arrangements of size 3 can be formed from a set of 5 elements?

33. A shipment of 100 TV sets is received. Six sets are to be chosen at random and tested for defects. In how many ways can the six sets be chosen?

34. A night watchman visits 15 offices every night. To prevent others from knowing when he will be at a particular office, he varies the order of his visits. In how many ways can this be done?

35. A typical telephone number is 555-1212.
    a. What is the total number of possible phone numbers?
    b. What is the number of possible phone numbers if the first two digits cannot be ones or zeros?
    c. Answer the question in part **b**, but include area codes. (Assume all digits are available for area codes.)

36. A typical Social Security number is 575-38-4444.
    a. How many Social Security numbers are possible?
    b. How many Social Security numbers are possible if the first digit cannot be zero?
    c. How many Social Security numbers are possible if the first two digits cannot be zero?

37. A certain mathematics test consists of ten questions. Peppermint Patty wishes to answer the questions without reading them. In how many ways can she fill in the answer sheet if:
    a. the possible answers are true and false?
    b. the possible answers are true, false, and maybe?

38. Peppermint Patty is taking an examination and is asked to answer 10 out of 12 questions. In how many ways can she select the questions to be answered?

39. A certain manufacturing process calls for six chemicals to be mixed. One liquid is to be poured into the vat, and then the others are to be added in turn. All possibilities are tested to see which gives the best results. How many tests must be performed?

40. There are three boys and three girls at a party. In how many ways can they be seated in a row of four?

41. There are three boys and three girls at a party. In how many ways can they be seated in a row if they want to sit alternating boy, girl, boy, girl, boy, girl?

42. A bag contains 12 pieces of candy, 5 of which are lemon drops.
    a. If you reach in and select one piece of candy at random, what is the probability it is a lemon drop?
    b. In how many ways can 2 pieces of candy be selected?
    c. In how many ways can 2 lemon drops be selected?
    d. If you reach in and select two pieces of candy at random, what is the probability they are both lemon drops?

43. Suppose four cards are drawn from a deck of cards.
    a. In how many ways can four cards be drawn from the deck?
    b. In how many ways can four aces be drawn?
    c. What is the probability of drawing four aces?

44. Suppose five cards are drawn from a deck of cards.
    a. In how many ways can five cards be drawn from the deck?
    b. In how many ways can three aces be drawn?
    c. In how many ways can two kings be drawn?
    d. In how many ways can three aces and two kings be drawn?
    e. What is the probability of drawing a full house consisting of three aces and two kings?

▲ **Problem Solving**

45. The old lady jumping rope in the cartoon at the beginning of this section is counting one at a time. Assume that she jumps the rope 50 times per minute and that she jumps 8 hours a day, 5 days a week, 50 weeks a year. Estimate the length of time necessary for her to jump the rope the number of times indicated in the cartoon.

46. A marginal note in this section describes the longest paper-link chain. If there are 12 paper links per foot, and it takes 1 minute to construct each link, estimate the length of time necessary to build this paper-link chain.

47. These questions are based on the *Peanuts* cartoon.

PEANUTS reprinted by permission of UFS, Inc.

a. In how many ways can seven books be arranged on a bookshelf? (*Hint:*   The order in which books are arranged matters.)
b. In how many ways can three math books be arranged?
c. In how many ways can four science books be arranged?
d. Answer the question posed in the third frame of the cartoon.

48. The News Clip in the margin presents an argument proving that at least two New Yorkers must have *exactly* the same number of hairs on their heads! For this problem, just *suppose* you knew the number of hairs on your head, as well as the *exact* number of hairs on anyone else's head. Now multiply the number of hairs on your head by the number of hairs on your neighbor's head. Take this result and multiply by the number of hairs on the heads of each person in your town or city. Continue this process until you have done this for everyone in the entire world! Make a guess (you can use scientific notation, if you like) about the size of this answer. (*Hint:*   The author has worked this out and claims to know the exact answer.)

49. The advertisement reproduced here appeared in several national periodicals. It claims that the SEA graphic equalizer system can create 371,293 different sounds from the five possible zone levels on each of thirteen different controls. Can this claim be substantiated from the advertisement and the knowledge of this chapter? If so, explain how; if not, tell why this number is impossible.

News Clip

An example about reaching a conclusion about the number of objects in a set is given by M. Cohen and E. Nagel in *An Introduction to Logic*. They conclude that there are at least two people in New York City who have the same number of hairs on their heads. This conclusion was reached not through counting the hairs on the heads of 8 million inhabitants of the city, but through studies revealing that: (1) the maximum number of hairs on the human scalp could never be as many as 5,000 per square centimeter, and (2) the maximum area of the human scalp could never reach 1,000 square centimeters. We can now conclude that no head could contain even $5,000 \times 1,000 = 5,000,000$ hairs. Since this number is less than the population of New York City, it follows that at least two New Yorkers must have the same number of hairs on their heads!

The following television advertisement is the basis for Problems 50–53.

LITTLE CAESAR'S PIZZA PIZZA!!!

Customer: So what's this new deal?

Pizza chef: Two pizzas.

Customer [*Toward four-year-old boy*]: Two pizzas. Write that down.

Pizza chef: And on the two pizzas choose any toppings — up to five [*from a list of 11 toppings*].

Older boy: Do you . . .

Pizza chef: . . . have to pick the same toppings on each pizza? NO!

Four-year-old math whiz: Then the possibilities are endless.

Customer: What do you mean? Five plus five are ten.

Math whiz: Actually, there are 1,048,576 possibilities.

Customer: Ten was just a ballpark figure.

Old man: You got that right.

First aired on national television on November 8, 1993.

50. If we accept the facts of the advertisement and order one pizza per day, how long will we be able to order different pizzas before we are forced to order a pizza previously ordered?

51. How many different pizza orders are possible?

52. Jean Sherrod of Little Caesar's Enterprises, Inc. explains that an order where the first pizza is pepperoni and the second pizza is ham would be considered different from an order in which the first pizza is ham and the second pizza is pepperoni. Using this scheme, how many pizza orders are possible?

53. Double toppings are not included in the calculation in the advertisement. However, Little Caesar's does allow double toppings. If double toppings are possible, how many different pizzas orders are possible?

54. There are ten seats left aboard an airplane, of which six are aisle seats and four are window seats. If three people are assigned seats at random, what is the probability that they will all receive window seats?

55. What is the number of possible BINGO cards?

56. Use the fundamental counting principle to show that

$$_nP_r = \underbrace{n(n - 1)(n - 2) \cdot \cdots \cdot (n - r + 1)}_{r \text{ factors}}$$

57. Show that

$$_nP_r = \underbrace{n(n - 1)(n - 2) \cdot \cdots \cdot (n - r + 1)}_{r \text{ factors}} = \frac{n!}{(n - r)!}$$

58. Show that $_nC_r = \dfrac{n!}{r!(n - r)!}$

59. What numerical property is exhibited by the arrangement of billiard balls shown in Figure 9.8?

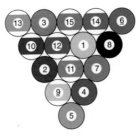

▲ **Figure 9.8** **Billiard balls**

▲ **Individual Research**

60. How can all the constructions of Euclidean geometry be done by folding paper? What assumptions are made when paper is folded to construct geometric figures? What is a hexaflexagon? What is origami?

References

Martin Gardner, *The Scientific American Book of Mathematical Puzzles and Diversions* (New York: Simon and Schuster, 1959), Chapter 1.

Donovan Johnson, *Paper Folding for the Mathematics Class* (Washington, D.C.: National Council of Teachers of Mathematics, 1957).

9.5 CALCULATED PROBABILITIES

In this section, we formulate some probability problems using tree diagrams, combinations, permutations, and the fundamental counting principle. We will also consider some additional models that will allow us to break up more complicated probability problems into some simpler types.

▬▬ EXAMPLE 1

A very popular lotto game in several states, as well as in most casinos, is a game called Keno. The game consists of a player trying to guess in advance which numbers will be selected from a pot containing 80 numbers. Twenty numbers are then selected at random. The player may choose from 1 to 15 spots and gets paid according to Table 9.3. Suppose a person picks one number and is paid $3.00 if the number picked is among the 20 chosen for the game. The cost for playing this game (which is collected in advance) is $1.00. Is this a fair game?

YOU CAN WIN $50,000									
KENO									
1	2	3	4	5	6	7	8	9	10
11	12	13	14	15	16	17	18	19	20
21	22	23	24	25	26	27	28	29	30
31	32	33	34	35	36	37	38	39	40
41	42	43	44	45	46	47	48	49	50
51	52	53	54	55	56	57	58	59	60
61	62	63	64	65	66	67	68	69	70
71	72	73	74	75	76	77	78	79	80

Solution In this problem we are picking 1 out of 80 and we will win if the one number we pick is among the 20 numbers chosen from the pot. Since the order in which the 20 numbers are picked from the pot does not matter, we see this is a combination:

$$P(\text{picking one number}) = \frac{\text{NUMBER OF WAYS OF CHOOSING 1 NO. FROM 20}}{\text{NUMBER OF WAYS OF CHOOSING 1 NO. FROM 80}}$$

$$= \frac{_{20}C_1}{_{80}C_1} = \frac{20}{80} = \frac{1}{4}$$

We now calculate the mathematical expectation:

$$\text{EXPECTATION} = \underbrace{\$2.00(\tfrac{1}{4})}_{\$3.00 \,-\, \$1.00 \text{ because you pay first}} + (-\$1.00)(\tfrac{3}{4}) = -\$0.25$$

🚫 It is common to work Example 1 incorrectly. Note that this is WRONG: $(\$3.00)(\tfrac{1}{4}) + (-\$1.00)(\tfrac{3}{4}) = 0$ or $\$3.00(\tfrac{1}{4}) = \$.75$ 🚫

No, it is not a fair game, since the expectation is negative.   ▬

The game of Keno becomes more complicated for picking more numbers.

▰▰▰ **EXAMPLE 2**

What is the mathematical expectation for playing a three-spot $1.00 Keno ticket?

Solution   There are four possibilities:

These Keno payoffs are found in Table 9.3 on page 663.

Pick 0 numbers, win $0.
Pick 1 number, win $0.   Picking 2 means picking exactly two numbers.
Pick 2 numbers, win $1.00.
Pick 3 numbers, win $40.00.

$$P(\text{pick 3}) = \frac{_{20}C_3}{_{80}C_3} = \frac{20 \cdot 19 \cdot 18}{80 \cdot 79 \cdot 78} \approx 0.0138753651$$

$$P(\text{pick 2}) = \frac{_{20}C_2 \cdot {_{60}C_1}}{_{80}C_3} = \frac{20 \cdot 19 \cdot 60 \cdot 3}{80 \cdot 79 \cdot 78} \approx 0.1387536514$$

$$P(\text{pick 1}) = \frac{_{20}C_1 \cdot {_{60}C_2}}{_{80}C_3} \approx 0.4308666018$$

$$P(\text{pick 0}) = \frac{_{60}C_3}{_{80}C_3} \approx 0.4165043817$$

We see that the expected value of one $1.00 play is

$$\begin{aligned}
\text{EXPECTED VALUE} = {}& \text{AMT. TO WIN FOR PICKING } 3 \cdot \text{PROB. OF PICKING } 3 \\
& + \text{AMT. TO WIN FOR PICKING } 2 \cdot \text{PROB. OF PICKING } 2 \\
& + \text{AMT. TO WIN FOR PICKING } 1 \cdot \text{PROB. OF PICKING } 1 \\
& + \text{AMT. TO WIN FOR PICKING } 0 \cdot \text{PROB. OF PICKING } 0 \\
\text{EXPECTED VALUE} \approx {}& (40 - 1)(0.0138753651) + (1 - 1)(0.1387536514) \\
& + (0 - 1)(0.4308666018) + (0 - 1)(0.4165043817) \\
\approx {}& -0.3062
\end{aligned}$$

This is a loss of about $0.31 for each play.   ▬

TABLE 9.3 Keno payoffs

MARK 1 SPOT

Catch	Play $1.00	Play $3.00	Play $5.00
Win	3.00	9.00	15.00

MARK 2 SPOTS

Catch	Play $1.00	Play $3.00	Play $5.00
2 Win	12.00	36.00	60.00

MARK 3 SPOTS

Catch	Play $1.00	Play $3.00	Play $5.00
2 Win	1.00	3.00	5.00
3 Win	40.00	120.00	200.00

MARK 4 SPOTS

Catch	Play $1.00	Play $3.00	Play $5.00
2 Win	1.00	3.00	5.00
3 Win	3.00	9.00	15.00
4 Win	113.00	339.00	565.00

MARK 5 SPOTS

Catch	Play $1.00	Play $3.00	Play $5.00
3 Win	1.00	3.00	5.00
4 Win	10.00	30.00	50.00
5 Win	750.00	2,250.00	3,750.00

MARK 6 SPOTS

Catch	Play $1.00	Play $3.00	Play $5.00
3 Win	1.00	3.00	5.00
4 Win	3.00	9.00	15.00
5 Win	90.00	270.00	450.00
6 Win	1,480.00	4,400.00	7,400.00

MARK 7 SPOTS

Catch	Play $1.00	Play $3.00	Play $5.00
4 Win	1.00	3.00	5.00
5 Win	18.00	54.00	90.00
6 Win	400.00	1,200.00	2,000.00
7 Win	8,000.00	24,000.00	40,000.00

MARK 8 SPOTS

Catch	Play $1.00	Play $3.00	Play $5.00
5 Win	8.00	24.00	40.00
6 Win	100.00	300.00	500.00
7 Win	1,480.00	4,440.00	7,400.00
8 Win	17,000.00	**50,000.00**	50,000.00

MARK 9 SPOTS

Catch	Play $1.00	Play $3.00	Play $5.00
5 Win	3.00	9.00	15.00
6 Win	42.00	126.00	210.00
7 Win	350.00	1,050.00	1,750.00
8 Win	4,200.00	12,600.00	21,000.00
9 Win	18,000.00	**50,000.00**	50,000.00

MARK 10 SPOTS

Catch	Play $1.00	Play $3.00	Play $5.00
5 Win	2.00	6.00	10.00
6 Win	20.00	60.00	100.00
7 Win	130.00	390.00	650.00
8 Win	900.00	2,700.00	4,500.00
9 Win	4,500.00	13,500.00	22,500.00
10 Win	20,000.00	**50,000.00**	50,000.00

MARK 11 SPOTS

Catch	Play $1.00	Play $3.00	Play $5.00
6 Win	9.00	27.00	45.00
7 Win	75.00	225.00	375.00
8 Win	380.00	1,140.00	1,900.00
9 Win	2,000.00	6,000.00	10,000.00
10 Win	12,500.00	37,500.00	**50,000.00**
11 Win	21,000.00	**50,000.00**	50,000.00

MARK 12 SPOTS

Catch	Play $1.00	Play $3.00	Play $5.00
6 Win	5.00	15.00	25.00
7 Win	28.00	84.00	140.00
8 Win	200.00	600.00	1,000.00
9 Win	850.00	2,550.00	4,250.00
10 Win	2,400.00	7,200.00	12,000.00
11 Win	13,000.00	39,000.00	**50,000.00**
12 Win	25,000.00	**50,000.00**	50,000.00

MARK 13 SPOTS
($2.00 minimum)

Catch	Play $2.00	Play $3.00	Play $5.00
6 Win	6.00	9.00	15.00
7 Win	24.00	36.00	60.00
8 Win	150.00	225.00	375.00
9 Win	1,400.00	2,100.00	3,500.00
10 Win	4,000.00	6,000.00	10,000.00
11 Win	18,000.00	27,000.00	45,000.00
12 Win	28,000.00	42,000.00	**50,000.00**
13 Win	**50,000.00**	50,000.00	50,000.00

MARK 14 SPOTS
($2.00 minimum)

Catch	Play $2.00	Play $3.00	Play $5.00
6 Win	4.00	6.00	10.00
7 Win	16.00	24.00	40.00
8 Win	64.00	96.00	160.00
9 Win	600.00	900.00	1,500.00
10 Win	1,600.00	2,400.00	4,000.00
11 Win	5,000.00	7,500.00	12,500.00
12 Win	24,000.00	36,000.00	**50,000.00**
13 Win	36,000.00	**50,000.00**	50,000.00
14 Win	**50,000.00**	50,000.00	50,000.00

MARK 15 SPOTS
($2.00 minimum)

Catch	Play $2.00	Play $3.00	Play $5.00
6 Win	2.00	3.00	5.00
7 Win	14.00	21.00	35.00
8 Win	42.00	63.00	105.00
9 Win	200.00	300.00	500.00
10 Win	800.00	1,200.00	2,000.00
11 Win	4,000.00	6,000.00	10,000.00
12 Win	16,000.00	24,000.00	40,000.00
13 Win	24,000.00	36,000.00	**50,000.00**
14 Win	**50,000.00**	50,000.00	50,000.00
15 Win	**50,000.00**	50,000.00	50,000.00

Independent Events

Consider the following problem dealing with four cards. Suppose four cards are taken from an ordinary deck of cards, and we form two stacks — one with an ace (one) and a deuce (two) and the other with an ace and a jack. If a card is drawn at random from each pile, what is the probability that a blackjack will occur (an ace and a jack)?

The tree diagram illustrates the possibilities. Notice that the probability of obtaining an ace from the first stack is 1/2 and the probability of obtaining a jack from the second stack is also 1/2.

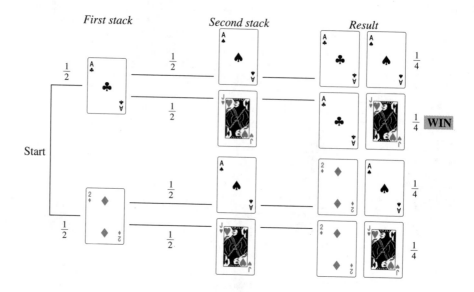

The probability of a blackjack is $\frac{1}{4}$. Notice *for this example* that

$$\text{PROBABILITY OF ACE FROM FIRST STACK} = \frac{1}{2}$$

$$\text{PROBABILITY OF JACK FROM SECOND STACK} = \frac{1}{2}$$

$$\text{PROBABILITY OF BLACKJACK} = \frac{1}{2} \times \frac{1}{2} = \frac{1}{4}$$

If one event (draw from the first stack) has no effect on the outcome of the second event (draw from the second stack), then we say that the events are **independent.**

To understand how independence is determined, consider the following alternative problem. Suppose that all four cards are put together into one pile. This is an entirely different situation. There are four possibilities for the first draw, and three possibilities for the second draw, as illustrated here.

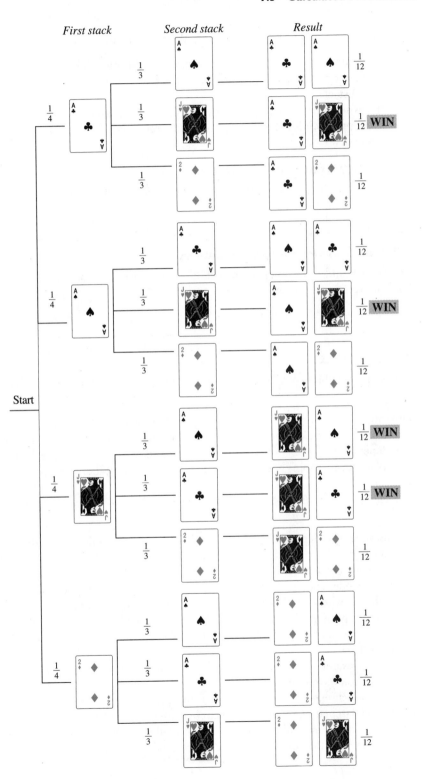

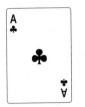 

Blackjack

The probability of a blackjack is $\frac{4}{12} = \frac{1}{3}$. We find this by looking at the sample space of equally likely possibilities. We *cannot* find this probability by multiplying the component parts for the first and second draw as we did when considering independent stacks.

For this second situation we see that the first and second draws are **dependent** because the probabilities of selecting a particular second card are influenced by the draw on the first card.

Probability of an Intersection

If two events are independent, then we can find the probability of an intersection by multiplication.

> **Multiplication Property of Probability**
>
> If events E and F are independent events, then we can find the probability of an intersection as follows:
>
> $$P(E \cap F) = P(E \text{ and } F) = P(E) \cdot P(F)$$

▬▬ EXAMPLE 3

What is the probability of black occurring on two successive plays on a U.S. roulette wheel?

Solution   Recall (Figure 9.7) that a U.S. roulette wheel has 38 compartments, 18 of which are black. Since the results of the spin of the wheel for one game are independent of the results of another spin, the plays are independent. Let $B_1 = \{$black on first spin$\}$ and $B_2 = \{$black on second spin$\}$; then

$$P(B_1 \cap B_2) = P(B_1) \cdot P(B_2) = \frac{18}{38} \cdot \frac{18}{38} \approx 0.2243767313$$

The idea of independence is a difficult idea to communicate outside of the classroom. On July 8, 1995, in a preliminary hearing in the O. J. Simpson murder trial, an objection was raised that an adequate foundation had not been laid to establish the independence (and consequently the use of the multiplication property of probability) when considering blood-type factors. It was a perfect place for the mathematical principle of independence to be introduced, but instead the response focused on how many previous times the multiplication principle had been allowed by the court.

Another example involving a misuse of the multiplication principle involves gamblers. If you have ever been at a casino and watched players bet on black or red on a roulette game, you have seen that if one color comes up several times in a row, players start betting on the other color because they think it is "due." The spins of a roulette wheel are independent and it is a fallacy to think that previous spins have any effect on any other spins. For example, let's continue with Example 3 and ask, what is the probability of black on the *next* or third spin:

$$P(B_3) = \frac{18}{38} \approx 0.4736842105 \quad \text{where } B_3 = \{\text{black on third spin}\}$$

But if we ask, what is the probability of obtaining black on the *next* three consecutive spins, it is

$$P(B_1 \cap B_2 \cap B_3) = P(B_1) \cdot P(B_2) \cdot P(B_3) = \left(\frac{18}{38}\right)^3 \approx 0.1062837148$$

You might also note that the multiplication property can be used for two *or more* independent events.

Independence and the multiplication property are often used with the idea of complementary events. We now consider the mathematics associated with an experiment called the **birthday problem.**

■■■■ EXAMPLE 4   Polya's Method

What is the probability that 4 unrelated persons share the same birthday? This example is considered in more detail as a Group Research Project at the end of this chapter.

Solution   We use Polya's problem-solving guidelines for this example.

Understand the Problem.   We assume that the birthdays of the individuals are independent, and we ignore leap years. We also are not considering the birthday year. We want to know, for example, if any 2 (or more) of the 4 persons have the same birthday — say, August 3.

Devise a Plan.   The first person's birthday can be any day (365 possibilities out of 365 days). We will find the probability that the second person's birthday is *different* from the first person's birthday. This is

$$1 - \tfrac{1}{365} = \tfrac{364}{365}$$

The probability that the third person's birthday is different from the first two birthdays is $\frac{363}{365}$ and the probability that the fourth person's birthday is different is $\frac{362}{365}$. Since these are independent events, we use the multiplication property of probability.

Carry Out the Plan.

$$P(\text{match}) = 1 - P(\text{no match}) = 1 - \frac{365}{365} \times \frac{364}{365} \times \frac{363}{365} \times \frac{362}{365} \approx 0.0163559125$$

Look Back.   There are about 2 chances out of 100 that there will be a birthday match with 4 persons. It is worth noting that if we consider this for 23 persons, we have

$$1 - \frac{_{365}P_{23}}{365^{23}} \approx 0.5072972343$$

This means that if you look at a group of more than 23 persons, it is more likely than not that there will be a birthday match. ▬

Probability of a Union

The multiplication property of probability is used to find the probability of an intersection (E and F); now let's turn to the probability of a union (E or F).

$$P(E \text{ or } F) = P(E \cup F) = \frac{|E \cup F|}{n}$$ Definition of probability

$$= \frac{|E| + |F| - |E \cap F|}{n}$$ Cardinality of a union

$$= \frac{|E|}{n} + \frac{|F|}{n} - \frac{|E \cap F|}{n}$$

$$= P(E) + P(F) - P(E \cap F)$$ Definition of probability

Addition Property of Probability

For any events E and F, the probability of their union can be found by

$$P(E \cup F) = P(E \text{ or } F) = P(E) + P(F) - P(E \cap F)$$

▬▬▬ **EXAMPLE 5**

Suppose a coin is tossed and a die is simultaneously rolled. What is the probability of tossing a tail or rolling a 4? Is this event more or less likely than flipping a coin and obtaining a head?

Solution   Let $T = \{\text{tail is tossed}\}$ and $F = \{4 \text{ is rolled}\}$.

$$P(T \cup F) = P(T) + P(F) - P(T \cap F)$$ Addition property

$$= P(T) + P(F) - P(T) \cdot P(F)$$ Multiplication property; T and F are independent

$$= \frac{1}{2} + \frac{1}{6} - \frac{1}{2} \cdot \frac{1}{6} = \frac{7}{12}$$

Since $\frac{7}{12} \approx 0.58$, we see that this event is more likely than flipping a coin and obtaining a head. ▬

Drawing With and Without Replacement

To highlight the difference between a combination model and a permutation model, we focus on the idea of *replacement*. Drawing **with replacement** means choosing the first item, noting the result, and then replacing the item back into the sample space before selecting the second item. Drawing **without replacement** means selecting the first item, noting the result, and then selecting a second item *without* replacing the first item.

■■■■ **EXAMPLE 6**

Find the probability of each event when drawing two cards from an ordinary deck of cards both with replacement and without replacement.

a. Drawing a spade on the first draw and a heart on the second draw with replacement

b. Drawing a spade on the first draw or a heart on the second draw with replacement

c. Drawing two hearts with replacement

d. Drawing a spade on the first draw and a heart on the second draw without replacement

e. Drawing a spade on the first draw or a heart on the second draw without replacement

f. Drawing two hearts without replacement

Rank the events from the most probable to the least probable.

Solution We state the decimal approximations for each probability for ease of comparison. Let $S_1 = \{$draw a spade on the first draw$\}$ and $H_2 = \{$draw a heart on the second draw$\}$.

With replacement

a. With replacement, the events are independent. Thus

$$P(S_1 \cap H_2) = P(S_1) \cdot P(H_2) = \tfrac{1}{4} \cdot \tfrac{1}{4} = \tfrac{1}{16} = 0.0625$$

b. $P(S_1 \cup H_2) = P(S_1) + P(H_2) - P(S_1 \cap H_2)$

$\qquad = \tfrac{1}{4} + \tfrac{1}{4} - \tfrac{1}{16}$ From part **a**

$\qquad = \tfrac{7}{16} = 0.4375$

c. With replacement, the events are independent. Then

$$P(\text{two hearts}) = \tfrac{1}{4} \cdot \tfrac{1}{4} = \tfrac{1}{16} = 0.0625$$

Without replacement

d. We use the permutation model since the order is important. The number of possibilities is $_{52}P_2 = 52 \cdot 51$; the number of successes (fundamental counting principle) is $13 \cdot 13$. Thus,

$$P(S_1 \cap H_2) = \frac{13 \cdot 13}{_{52}P_2} = \frac{13 \cdot 13}{52 \cdot 51} = \frac{13}{204} \approx 0.0637254902$$

e. $P(S_1 \cup H_2) = P(S_1) + P(H_2) - P(S_1 \cap H_2)$

$\qquad = \tfrac{1}{4} + \tfrac{1}{4} - \tfrac{13}{204}$ From part **d**

$\qquad = \tfrac{89}{204} \approx 0.4362745098$

f. Without replacement, draw both cards at once. We use combinations since the order is not important:

$$P(\text{two hearts}) = \frac{_{13}C_2}{_{52}C_2} = \frac{\dfrac{13 \cdot 12}{2}}{\dfrac{52 \cdot 51}{2}} = \frac{1}{17} \approx 0.0588235294$$

The ranking is: (1) part **b**, (2) part **e** (probabilities of parts **b** and **e** are about the same), (3) part **d**, (4) parts **a** and **c** (tie), (5) part **f**. Parts **a, c, d,** and **f** are about the same. ▬

Tree Diagrams

A powerful tool in handling probability problems is to use a device we have frequently used in this chapter — a tree diagram. If the events are independent, we find the probabilities using the multiplication property of probability.

▬▬ **EXAMPLE 7**

Consider a game consisting of at most three cuts with a deck of cards. You win and the game is over if a face card turns up on any of the cuts, but you lose if a face card does not turn up. (A face card is a jack, queen, or king.) If you stand to win or lose $1 on this game, should you play?

Solution

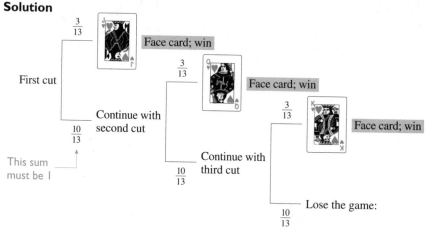

$$P(\text{lose the game}) = \tfrac{10}{13} \cdot \tfrac{10}{13} \cdot \tfrac{10}{13} \approx 0.455$$

$$P(\text{win}) = 1 - P(\text{lose}) \approx 1 - 0.455 = 0.545$$

$$\text{EXPECTATION} = \text{AMOUNT TO WIN} \cdot P(\text{win}) + \text{AMOUNT TO LOSE} \cdot P(\text{lose})$$
$$\approx \$1 \cdot (0.545) + (-\$1) \cdot (0.455)$$
$$= \$0.09$$

Since the mathematical expectation is positive, the recommendation is to play the game. If you played this game 1,000 times, you could expect to be ahead about 1,000($0.09) = $90.00. ▬

News Clip

Guest Essay: Extrasensory Perception (ESP)

Another presumed kind of extrasensory perception is the predictive dream. Everyone has an Aunt Matilda who had a vivid dream of a fiery car crash the night before Uncle Mortimer wrapped his Ford around a utility pole. I'm my own Aunt Matilda: when I was a kid I once dreamed of hitting a grand-slam home run and two days later I hit a bases-loaded triple. (Even believers in precognitive experiences don't expect an exact correspondence.) When one has such a dream and the predicted event happens, it's hard not to believe in precognition. But as the following derivation shows, such experiences are more rationally accounted for by coincidence.

Assume the probability to be one out of 10,000 that a particular dream matches a few vivid details of some sequence of events in real life. This is a pretty unlikely occurrence, and means that the chances of a nonpredictive dream are an overwhelming 9,999 out of 10,000. Also assume that whether or not a dream matches experience one day is independent of whether or not some other dream matches experience some other day. Thus, the probability of having two successive nonmatching dreams is, by the multiplication principle for probability, the product of 9,999/10,000 and 9,999/10,000. Likewise, the probability of having N straight nights of nonmatching dreams is $(9,999/10,000)^N$; for a year's worth of nonmatching or nonpredictive dreams, the probability is $(9,999/10,000)^{365}$.

Since $(9,999/10,000)^{365}$ is about .964, we can conclude that about 96.4 percent of the people who dream every night will have only nonmatching dreams during a one-year span. But that means that about 3.6 percent of the people who dream every night will have a predictive dream. 3.6 percent is not such a small fraction; it translates into millions of apparently precognitive dreams every year. Even if we change the probability to one in a million for such a predictive dream, we'll still get huge numbers of them by chance alone in a country the size of the United States. There's no need to invoke any special parapsychological abilities; the ordinariness of apparently predictive dreams does not need any explaining. What would need explaining would be the nonoccurrence of such dreams.

John Paulos, Innumeracy, *pp. 54–55. See the biography at the end of this chapter.*

PROBLEM SET 9.5

▲ A Problems

1. **IN YOUR OWN WORDS** What do we mean by independent events?

2. **IN YOUR OWN WORDS** What is the formula for the probability of an intersection?

3. **IN YOUR OWN WORDS** What is the formula for the probability of a union?

4. **IN YOUR OWN WORDS** What is the birthday problem?

5. **IN YOUR OWN WORDS** Comment on ESP as described in the guest essay by John Paulos.

6. **IN YOUR OWN WORDS** One system for betting in roulette is a variation of the one discussed in Problem 62, Section 9.3 (page 648). Instead of doubling your bet when you lose, begin with a $1 bet and double your bet each time you win until you win five times in a row. Comment on this betting "system."

Suppose events A, B, and C are independent and

$$P(A) = \tfrac{1}{2} \qquad P(B) = \tfrac{1}{3} \qquad P(C) = \tfrac{1}{6}$$

Find the probabilities in Problems 7–24.

7. $P(\overline{A})$ 8. $P(\overline{B})$ 9. $P(\overline{C})$ 10. $P(A \cap B)$

11. $P(A \cap C)$ **12.** $P(B \cap C)$ **13.** $P(A \cup B)$ **14.** $P(A \cup C)$

15. $P(B \cup C)$ **16.** $P(\overline{A \cap B})$ **17.** $P(\overline{A \cap C})$ **18.** $P(\overline{B \cap C})$

19. $P(\overline{A \cup B})$ **20.** $P(\overline{A \cup C})$ **21.** $P(\overline{B \cup C})$

22. $P(A \cap B \cap C)$ **23.** $P(\overline{A \cap B \cap C})$ **24.** $P[(A \cup B) \cap C]$

In Problems 25–36, suppose a die is rolled twice and let

$$A = \{first\ toss\ is\ a\ prime\} \quad B = \{first\ toss\ is\ a\ 3\}$$
$$C = \{second\ toss\ is\ a\ 2\} \quad D = \{second\ toss\ is\ a\ 3\}$$

Answer the questions or find the requested probabilities.

25. Are A and B independent?   **26.** Are A and C independent?

27. Are A and D independent?   **28.** Are B and C independent?

29. Are B and D independent?   **30.** Are C and D independent?

31. $P(A \cap B)$ **32.** $P(A \cap C)$

33. $P(A \cup B)$ **34.** $P(A \cup C)$

35. $P(B \cup D)$ **36.** $P(C \cup D)$

▲ B Problems

37. What is the expectation in playing Keno, of picking one number and paying $5.00 to play?

38. A "high rollers" Keno is offered in which you pay $749 to play a one-spot ticket. If you catch your number, you are paid $2,247. What is your expectation for this game?

39. A special "catch all" Keno ticket allows you to play a six-spot ticket which pays only if you pick all 6 numbers. It costs $5 to play and pays $27,777 if you win. What is your expectation for this game?

40. What is the expectation for playing Keno by picking two numbers and paying $1.00 to play?

41. What is the expectation for playing a three-spot Keno and paying $3.00 to play?

A certain slot machine has three identical independent wheels, each with 13 symbols as follows:

     1 bar, 2 lemons, 2 bells, 3 plums, 2 cherries, and 3 oranges.

Suppose you spin the wheels and one of the 13 symbols on each wheel is selected. Find the probabilities in Problems 42–46 and leave your answers in decimal form.

42. P(3 bars)   **43.** P(3 oranges)   **44.** P(3 plums)

45. P(cherries on the first wheel)

46. P(cherries on the first 2 wheels)

Suppose a slot machine has three independent wheels as shown in Figure 9.9. Find the probabilities in Problems 47–51.

47. P(3 bars)   **48.** P(3 bells)   **49.** P(3 cherries)

50. P(cherries on the first wheel)

51. P(cherries on the first 2 wheels)

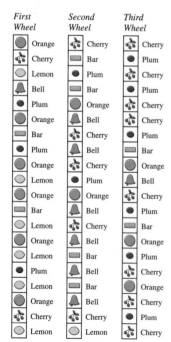

First Wheel	Second Wheel	Third Wheel
Orange	Cherry	Cherry
Cherry	Bar	Plum
Lemon	Plum	Cherry
Bell	Bar	Plum
Plum	Orange	Cherry
Orange	Bell	Cherry
Bar	Cherry	Plum
Plum	Bell	Bar
Orange	Cherry	Orange
Lemon	Plum	Bell
Orange	Orange	Cherry
Bar	Bell	Plum
Lemon	Cherry	Bar
Orange	Bell	Orange
Lemon	Bar	Plum
Plum	Bell	Cherry
Lemon	Bar	Orange
Orange	Bell	Cherry
Cherry	Cherry	Plum
Lemon	Lemon	Cherry

▲ **Figure 9.9**   **Slot machine wheels**

52. The payoffs for the slot machine shown in Figure 9.9 are as follows:

First one cherry	2 coins
First two cherries	5 coins
First two wheels are cherries and the third wheel a bar	10 coins
Three cherries	10 coins
Three oranges	14 coins
Three plums	14 coins
Three bells	20 coins
Three bars (jackpot)	50 coins

What is the mathematical expectation for playing the game with the coin being a quarter dollar? Assume the three wheels are independent.

53. What is the probability of obtaining five tails when a coin is flipped five times?

54. What is the probability of obtaining at least one tail when a coin is flipped five times?

55. A game consists of at most three cuts with a deck of 52 cards. You win $1 and the game is over if a heart turns up, but lose $1 otherwise. Should you play?

56. Repeat Problem 55 but remove the card that turns up on the cut.

57. A game consists of removing a card from a deck of 52 cards. If the card is a face card, you win $1 and the game is over. If it is not a face card, remove another card. If that one is a face card you win $1. Repeat the process again, and then again (for a total of 4 times). If you have not removed any face card, then you lose the game, and must pay $1. Should you play this game?

58. Repeat Problem 57 but replace the card before doing the following draw.

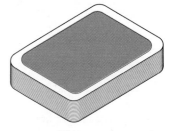

*Assume a jar has five red marbles and three black marbles. Draw out 2 marbles **a.** with replacement and **b.** without replacement. Find the requested probabilities in Problems 59–62.*

59. P(two red marbles)   60. P(two black marbles)

61. P(one red and one black marble)

62. P(red on the first draw and black on the second draw)

63. Suppose that in an assortment of 20 calculators there are 5 with defective switches.
    a. If one machine is selected at random, what is the probability it has a defective switch?
    b. If two machines are selected at random, what is the probability that both have defective switches?
    c. If three machines are selected at random, what is the probability that all three have defective switches?

64. Repeat Problem 63 but draw with replacement.

▲ Problem Solving

65. We know that in a game of U.S. roulette, the probability that the ball drops into any one slot is 1 out of 38. Suppose that there are two balls spinning at once, and that it is physically possible for two balls to fit into the same slot. What is the probability that *both* balls would drop into the same slot?

66. Chevalier de Méré used to bet that he could get at least one 6 in four rolls of a die. He also bet that, in 24 tosses of a pair of dice, he would get at least one 12. He found that he won more often than he lost with the first bet, but not with the second. He

did not know why, so he wrote to Pascal seeking the probabilities of these events. What are the probabilities for winning these two games?

67. In a book by John Fisher, *Never Give a Sucker an Even Break* (Pantheon Books, 1976), we find the following problem:

> Take a small opaque bottle and seven olives, two of which are green, five black. The green ones are considered the "unlucky" ones. Place all seven olives in the bottle, the neck of which should be of such a size that it will allow only one olive to pass through at a time. Ask the sucker to shake them and then wager that he will not be able to roll out three olives without getting an unlucky green one amongst them. If a green olive shows, he loses.

What is the probability that you win? Remember, the sucker bets that he can roll out three black olives without getting one of the green olives.

Many states conduct lotteries; a typical payoff ticket is shown at the right. Assume there are 20 *numbers chosen from a set of* 80 *possible numbers. Use this information for Problems* 68–72.

68. What is the expectation for a three-spot game?

69. What is the expectation for a four-spot game?

70. What is the expectation for a five-spot game?

71. What is the expectation for a six-spot game?

72. Which game illustrated on this Keno payoff ticket has the best expectation?

▲ **Individual Research**

73. Do some research on Keno probabilities. Write a paper on playing Keno.

   Reference Karl J. Smith, "Keno Expectation," *Two-Year College Mathematics Journal*, Vol. 3, No. 2, Fall 1972.

9.6   RUBIK'S CUBE AND INSTANT INSANITY*

Rubik's Cube

In the summer of 1974, a Hungarian architect invented a three-dimensional object that could rotate about *all three axes* (sounds impossible, doesn't it?). He wrote up the details of the cube and obtained a patent in 1975. The cube is now known

---

* The term "Rubik's Cube" is a trademark of Ideal Toy Corporation, Hollis, New York. The term "Rubik's Cube" as used in this book means the cube puzzle sold under any trademark. "Instant Insanity" is a trademark of Parker Brothers, Inc.

KENO PAYOUTS

NUMBER OF SPOTS MATCHED	PLAY $3.00...WIN:
6 SPOT GAME	
MATCH	PRIZE
6	$1,000
5	$25
4	$4
3	$1
Overall odds of winning a prize in this game — 1:6.2	
5 SPOT GAME	
MATCH	PRIZE
5	$250
4	$10
3	$2
Overall odds of winning a prize in this game — 1:10.3	
4 SPOT GAME	
MATCH	PRIZE
4	$50
3	$4
2	$1
Overall odds of winning a prize in this game — 1:3.9	
3 SPOT GAME	
MATCH	PRIZE
3	$20
2	$2
Overall odds of winning a prize in this game — 1:6.6	
2 SPOT GAME	
MATCH	PRIZE
2	$8
Overall odds of winning a prize in this game — 1:16.5	
1 SPOT GAME	
MATCH	PRIZE
1	$2
Overall odds of winning a prize in this game — 1:4.0	

worldwide as **Rubik's cube.** In case you have not seen one of these cubes, it is shown in Figure 9.10.

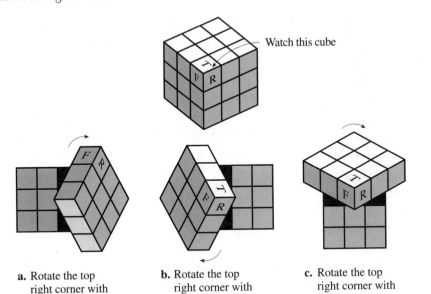

Watch this cube

a. Rotate the top
   right corner with
   the right face.

b. Rotate the top
   right corner with
   the front face.

c. Rotate the top
   right corner with
   the top face.

▲ **Figure 9.10   Rubik's cube**

Historical Note

The puzzle known as Rubik's Cube was designed by Ernö Rubik, an architect and teacher in Budapest, Hungary. It was also invented independently by Terutoshi Ishige, an engineer in Japan. Both applied for patents in the mid-1970s. The cubes were first manufactured in Hungary. In 1978, a Hungarian mathematics professor brought several with him to the International Congress of Mathematics in Helsinki, Finland. This formally introduced the cube to Europe and the rest of the world. They have been widely available in the United States since 1980. They are often used in mathematics classes to illustrate mathematical ideas, particularly in group theory.

When you purchase the cube it is arranged so that each face is showing a different color, but after a few turns it seems next to impossible to return to the start. In fact, the manufacturer claims there are 8.86×10^{22} possible arrangements, of which there is only one correct solution. This claim is incorrect; there are 2,048 possible solutions among the 8.86×10^{22} claimed arrangements (actually, $8.85801027 \times 10^{22}$). This means there is one solution for each 4.3×10^{19} (actually 43,252,003,274,489,856,000) arrangements.

■■■ EXAMPLE 1

If it took you one-half second for each arrangement, how long would it take you to move the cube into all arrangements?

Solution

4.3×10^{19}	Arrangements
2.15×10^{19}	Divide by 2 for the number of seconds.
3.58×10^{17}	Divide by 60 for number of minutes.
5.97×10^{15}	Divide by 60 for the number of hours.
2.49×10^{14}	Divide by 24 for the number of days.
6.81×10^{11}	Divide by 365.25 for the number of years.
6.81×10^{9}	Divide by 100 for the number of centuries.
6.81×10^{8}	Divide by 10 for the number of millennia.

Our problem ends here because it would require over

681,000,000 *millennia!*

If a high-speed computer listed 1,000,000 arrangements per minute, it would take over 80,000 millennia to print out the possibilities. Remember, it has only been 2 millennia since Christ! ▬

Let's consider what we call the **standard-position cube,** as shown in Figure 9.11. Label the faces Front (F), Right (R), Left (L), Back (B), Top (T), and Under (U), as shown. Hold the cube in your left hand with T up and F toward you so that L is against your left palm. Now describe the results of the moves in Example 2.

▬ **EXAMPLE 2**

a. Rotate the right face 90° clockwise; denote this move by R. Return the cube to standard position; we denote this move by R^{-1}.

b. Rotate the right face 180° clockwise; denote this by R^2. Return the cube to standard position by doing another R^2. Notice that $R^2R^2 = R^4$, which returns the cube to standard position.

c. Rotate the top 90° clockwise; call this T. Return the cube by doing T^{-1}.

d. TR means rotate the top face 90° clockwise, *then* rotate the right face 90° clockwise. Describe the steps necessary to return the cube to standard position.

Solution

a.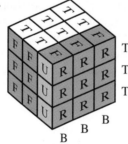

b.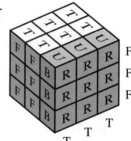

c.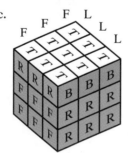

d.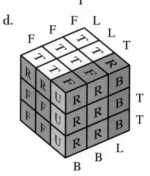

To return this cube to standard position you need $R^{-1} T^{-1}$. ▬

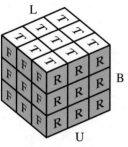

▲ **Figure 9.11 Standard-position Rubik's cube**

Now, a rearrangement of the 54 colored faces of a small cube is a *permutation*. We discussed permutations of a square in Problem 62, Section 4.6, at which time we called them *symmetries of a square*. The same type of algebraic structure can be used to find a solution of Rubik's cube. Although such a discussion is beyond the scope of this course, you can consult Question 6 of the Group Research Projects at the end of this chapter if you are interested.

Instant Insanity

An older puzzle, simpler than Rubik's cube, is called **Instant Insanity**. It provides four cubes colored red, white, blue, and green, as indicated in Figure 9.12. The puzzle is to assemble them into a $1 \times 1 \times 4$ block so that all four colors appear on each side of the block.

■■■■ EXAMPLE 3

In how many ways can this Instant Insanity puzzle be arranged?

Solution A common mistake is to assume that a cube can have 6 different arrangements (because of our experience with dice). In the case of Instant Insanity, we are interested not only in the top, but also with the other sides. There are 6 possible faces for the top and *then* 4 possible faces for the front. Thus, by the fundamental counting principle, there are

$6 \cdot 4 = 24$ arrangements for one cube

Now, for four cubes, again use the fundamental counting principle to find

$24 \cdot 24 \cdot 24 \cdot 24 = 331,776$

This is *not* the number of *different* possibilities. You are asked to find that number in the problem set. ▬

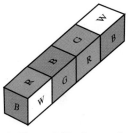

▲ **Figure 9.12 Instant Insanity**

PROBLEM SET 9.6

▲ A Problems

1. **IN YOUR OWN WORDS** Example 3 shows that Instant Insanity has 331,776 possibilities (not all different). If you could make one different move per second, how long do you think it would take you to go through all these arrangements? Estimate your answer first, then calculate the correct answer.

2. **IN YOUR OWN WORDS** Suppose a computer could print out 1,000 arrangements in Problem 1 every minute. How long would it take for this computer to print out all possibilities?

Show the result of the moves on Rubik's cube indicated in Problems 3–27. Remember that R, F, L, B, T, and U mean rotate 90° clockwise the right, front, left, back, top, and under faces, respectively. Use the standard Rubik's cube shown in Figure 9.11 as your starting point. Consider each clockwise rotation by first turning the cube so that the side you are rotating is facing you.

3. F 4. B 5. U 6. L 7. B^{-1}

8. F^{-1}	9. L^{-1}	10. T^{-1}	11. F^2	12. T^2
13. L^2	14. B^2	15. F^3	16. R^3	17. T^3
18. U^3	19. RL	20. TU	21. FB^{-1}	22. $F^{-1}B$
23. RT	24. FT	25. FT^{-1}	26. FR^{-1}	27. $F^{-1}T^{-1}$

▲ B Problems

Name the move or moves that will return the cube to standard position after making the move shown in Problems 28–37.

28. F	29. T^2	30. F^3	31. TU	32. FB^{-1}
33. B	34. L^{-1}	35. RT	36. FT^{-1}	37. $F^{-1}B^{-1}$

38. Are two consecutive moves on Rubik's cube commutative?

39. Are three consecutive moves on Rubik's cube associative?

40. We saw in Example 2 that $R^{-1}T^{-1}$ reversed the TR. Would $T^{-1}R^{-1}$ also reverse the move TR? Why or why not?

▲ Problem Solving

41. Find the number of *different* arrangements for the Instant Insanity blocks.

42. The most difficult part of understanding a solution to Rubik's cube is understanding the notation used by the author in stating the solution. In one solution, the Ledbetter–Nering algorithm (see the references in the Group Research Projects), the authors describe a sequence of moves that they call THE MAD DOG. This series of moves can be described using our notation by

$$(RTF)^5$$

Start with the standard-position cube and show the result after carrying out THE MAD DOG.

43. David Singmaster, in his book *Notes on Rubik's "Magic Cube"* (see the references in the Group Research Projects), describes a sequence of moves to put the upper corners in place (after certain other moves). Using our notation, this set of moves is

$$FU^2F^{-1}U^2R^{-1}UR$$

Start with the standard-position cube and show the result after carrying out this sequence of moves.

44. In *The Simple Solution to Rubik's Cube* (see the references in the Group Research Projects), James Nourse describes a process to orient the bottom corner cubes. Using our notation, this set of moves is $R^{-1}U^{-1}RU^{-1}R^{-1}U^2RU^2$. Start with the standard-position cube and show the result after carrying out this sequence of moves.

▲ Individual Research

45. This problem illustrates a numerical solution for the Instant Insanity puzzle. Let's associate numbers with the sides of the cubes of the Instant Insanity problem. Let

$$\text{White} = 1; \quad \text{Blue} = 2; \quad \text{Green} = 3; \quad \text{Red} = 5$$

Now, the product across the top must be 30, and the product across the bottom must also be 30 (why?). It follows that a solution must have a product of 900 for the faces on the top and bottom. Consider the four cubes:

Historical Note

There is money in writing a solution to Rubik's cube. The first to write such a book was David Singmaster. He was the one who brought the cube to England from the mathematics conference at which it was introduced. His book sold over 10,000 copies. James Nourse, a 33-year-old research associate at Stanford University, wrote *The Simple Solution to Rubik's Cube*. To date, 1,650,000 copies have been printed. Nourse says, "It's so difficult that it is approaching a real bona fide scientific problem," adding that the marriage of his scientific career and the cube's solution was a unique opportunity. "This is something that matches those abilities exactly," he said. "This is just a sideline that has turned into a gold mine."

```
     1              2              3              4
   ┌───┐          ┌───┐          ┌───┐          ┌───┐
   │ R │          │ B │          │ B │          │ B │
┌───┼───┼───┐  ┌───┼───┼───┐  ┌───┼───┼───┐  ┌───┼───┼───┐
│ R │ W │ G │  │ G │ B │ G │  │ W │ G │ R │  │ W │ R │ G │
└───┼───┼───┘  └───┼───┼───┘  └───┼───┼───┘  └───┼───┼───┘
   │ R │          │ W │          │ R │          │ W │
   ├───┤          ├───┤          ├───┤          ├───┤
   │ B │          │ R │          │ W │          │ G │
   └───┘          └───┘          └───┘          └───┘
```

Here are the products of top and bottom for the three possible arrangements of cube 1:

		Product
Top:	Red = 5	25
Bottom:	Red = 5	
Top:	Blue = 2	2
Bottom:	White = 1	
Top:	Red = 5	15
Bottom:	Green = 3	

By a clever and systematic analysis of the products of top and bottom for cubes 2, 3, and 4, you will find that there are several ways of solving the top and bottom for a product of 900. Only one of these gives a product of 900 for front and back. This is the solution to the Instant Insanity puzzle. Find the solution using this method.

> "The mathematics curriculum should include the continued study of probability so that all students can use experimental or theoretical probability, as appropriate, to represent and solve problems involving uncertainty."
>
> NCTM Standards

JULY

1 **Gottfried Wilhelm von Leibniz** (1646), cofounder of calculus, and one of history's greatest, p. 79

2 **William Burnside** (1852), probability

3

4 **John Paulos** (1945)

5

6

7

8

9

10 **Roger Cotes** (1682), numerical integration

11

12

13 **John Dee** (1527), educator

14

15 **George Green** (1793), calculus, electricity

16

17

18 **Jean Argand** (1768), complex numbers

19

20

21

22 **Friedrich Wilhelm Bessel** (1784), astronomy

23 **Thomas Bayes** (1761), inverse probability

24

25

26

27 **Johann Bernoulli** (1667), function concept, p. 580

28

29

30

31 **Gabriel Cramer** (1704), geometry, probability, systems

Biographical Sketch

"Doing mathematics depends on computational skills no more than writing novels does on typing skills."

John Paulos

John Paulos was born in Denver, Colorado, on July 4, 1945, and is the author of the entertaining book, *Innumeracy*. This book about mathematical illiteracy, which *Time* magazine describes as "an elegant little survival manual," seeks to explain why so many people are numerically inept.

Even though Dr. Paulos is now a widely respected professor of mathematics at Temple University, he didn't always like mathematics. "I hated math as a kid," Paulos admits. "It was dull, dreary, and I didn't like the quasi-militaristic attitude that seemed to prevail in class. I learned to love mathematics by browsing through books in the library . . . . When I was a freshman I wrote Bertrand Russell a letter, and to my surprise I received a response!"

"What I like most about mathematics is that it is fun. Too many people have been intimidated by officious and sometimes sexist teachers," says Paulos, himself a victim of inept instruction. "They feel that there are mathematical minds and nonmathematical minds. Misconceptions about risk threaten eventually to lead either to unfounded and crippling anxieties or to impossible and economically paralyzing demands for risk-free guarantees." With unusual frankness he proclaims, "I'm pained at the belief that mathematics is an esoteric discipline with little relation or connection to the 'real' world."

Book Reports

Write a 500-word report on one of the following books:

▲ *Innumeracy: Mathematical Illiteracy and Its Consequences.* John Allen Paulos (New York: Hill and Wang), 1988.
▲ *How to Take a Chance.* Darrell Huff and Irving Geis (New York: Norton), 1959.
▲ *Beat the Dealer.* Edward O. Thorp (New York: Vintage Books), 1966.

Important Terms

Arrangement [9.4]
Cards [9.1]
Combination [9.4]
Complementary probabilities [9.3]
Conditional probability [9.3]
Dependent events [9.5]
Dice [9.1]
Empirical probability [9.1]
Equally likely [9.1]

Event [9.1]
Expectation [9.2]
Expected value [9.2]
Experiment [9.1]
Factorial [9.4]
Fair game [9.2]
Fundamental counting principle [9.3]
Impossible event [9.1]
Independent events [9.5]

Instant Insanity [9.6]
Mathematical expectation [9.2]
Mutually exclusive [9.1]
Odds against [9.3]
Odds in favor [9.3]
Ordered pair [9.4]
Permutation [9.4]
Probabilistic model [9.1]
Probability [9.1]

Relative frequency [9.1]
Rubik's cube [9.6]
Sample space [9.1]
Simple event [9.1]
Subjective probability [9.1]
Theoretical probability [9.1]
Tree diagrams [9.1; 9.5]
With replacement [9.5]
Without replacement [9.5]

Important Ideas

Definition of probability [9.1]
Probability by counting [9.1]
Property of complements [9.3]
Know the difference between odds and probability [9.3]
Multiplication property of factorials [9.4]

Formula for combination [9.4]
Formula for permutation [9.4]
Classify and identify counting techniques [9.4]
Multiplication property of probability [9.5]
Addition property of probability [9.5]

Types of Problems

Find probabilities by looking at the sample space [9.1]
Decide which of two events is more probable by looking at a sample space [9.1]
Find the mathematical expectation for a simple event [9.2]
Find the mathematical expectation for a compound event [9.2]
Make decisions based on mathematical expectation [9.2]
Find expectations for roulette [9.2]
Find the probability of a complement [9.3]
Find the odds, given the probability [9.3]
Find the probability, given the odds [9.3]

Simplify expressions involving factorials [9.4]
Evaluate permutations and combinations [9.4]
Answer applied counting questions [9.4]
Calculate Keno probabilities [9.5]
Find the probabilities of unions, intersections, and complements [9.5]
Find slot machine probabilities [9.5]
Answer questions involving applied probabilities [9.5]
Use the notation involving Rubik's cube [9.6]
Explain the Instant Insanity problem [9.6]

CHAPTER 9   Review Questions

1. If a sample from an assembly line reveals four defective items out of 1,000 sampled items, what is the probability that any one item is defective?

2. What is the probability of obtaining a prime in a single roll of a die?

3. A pair of dice is rolled. What is the probability that the resulting sum is eight?

4. A card is selected from an ordinary deck of cards. What is the probability that it is a jack or better? (A jack or better is a jack, queen, king, or ace.)

5. For a roll of a pair of dice, find the following.
   a. P(5 on at least one of the dice)
   b. P(5 on one die or 4 on the other)
   c. P(5 on one die and 4 on the other)

6. If the probability of dropping an egg is 0.01, what is the probability of not dropping the egg?

7. If $P(E) = 0.9$, what are the odds in favor of E?

8. If the odds against you are 1,000 to 1, what is the probability of the event?

9. A single card is drawn from a standard deck of cards. What is the probability that it is an ace, if you know that one ace has already been removed from the deck?

10. A game consists of rolling a die and receiving \$12 if a one is rolled and nothing otherwise. What is the mathematical expectation?

11. Find the numerical value of each expression.

    a. $_5C_3$ **b.** $_8P_3$ **c.** $_{12}P_0$ **d.** $_{14}C_4$ $_{100}P_3$

12. If A and B are independent events with $P(A) = \frac{2}{3}$ and $P(B) = \frac{3}{5}$, what is $P(\overline{A \cup B})$?

13. In how many ways can a three-member committee be chosen from a group of 12 people?

14. In how many ways can five people line up at a bank teller's window?

15. Advertisements for Wendy's Old Fashioned Hamburgers claim that you can have your hamburgers 256 ways. If they offer catsup, onion, mustard, pickles, lettuce, tomato, mayonnaise, and relish, is their claim correct? Explain why or why not.

16. A box contains three orange balls and two purple balls, and two are selected at random. What is the probability of obtaining two orange balls if we know that the first draw was orange and we draw the second ball:

    a. after replacing the first ball?   **b.** without replacing the first ball?

17. A friend says she has two new baby cats to show you, but she doesn't know whether they're both male, both female, or a pair. You tell her that you want only a male, and she telephones the vet and asks, "Is at least one a male?" The vet answers, "Yes!" What is the probability that the other one is a male? Assume that a baby cat is equally likely to be female or male.

18. Consider the set of octahedral dice shown in Figure 9.13. Suppose your opponent chooses die D. Which die would you choose, and why, if the game consists of each player rolling their chosen die once and the higher number on top is the winner?

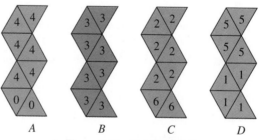

▲ **Figure 9.13   Octahedral dice**

19. The following problem appeared in an "Ask Marilyn" column, by Marilyn Vos Savant:

    There are five cars on display as prizes, and their five ignition keys are in a box. You get to pick one key out of the box and try it in the ignition of one car. If it fits, you win the car. What are your chances of winning a car?

    Answer this question.

20. Answer the question in Problem 19 assuming that you are in the audience and want the probability that the contestant will win the car you have selected.

Revisiting the Real World . . .

IN THE REAL WORLD   Probability is the mathematics of uncertainty. "Wait a minute," interrupts Sammy, "I thought that mathematics was absolute and that there was no uncertainty about it!" Well, Sammy, suppose that you are playing a game of Monopoly. Do you know which property you'll land on in your next turn? "No, but neither do you!" That's right, but mathematics can tell us where you are *most likely* to land. There is nothing uncertain about probability; rather, probability is a way of describing uncertainty. For example, what is the probability of tossing a coin and obtaining heads? "That's easy; it's one-half," answers Sammy. Right, but *why* do you say it is one-half?*

Park Place

Boardwalk

Jail

---

* It is assumed that you know the rules for playing a Monopoly game. MONOPOLY® is a registered trademark of Parker Brothers, Beverly, MA 01915.

Commentary

When you begin a Monopoly game, what is the most likely property on which you will land? Since Monopoly is played with a pair of dice, we look at Figure 9.4 to see that it is most likely that you will roll a 7 and thus land on CHANCE. As you move around the board, you are to purchase properties to form a monopoly (all the properties of a certain color). Is it possible to form a monopoly on one circuit around the board? According to Figure 9.4, it seems not to be possible, because it would be necessary to roll a 1, but this conclusion is not correct for two reasons. Look at Park Place and Boardwalk. On the first trip around the board you could land on Park Place and then roll a two. What is the probability that you would be able to purchase Boardwalk right after the opportunity to purchase Park Place? Once again, we turn to Figure 9.4, to see the $P(\text{rolling a two}) = \frac{1}{36} \approx 0.03$. This will occur about only 3 times out of 100. There are many ways of landing on both Park Place and Boardwalk on the first trip around the board. One possibility is to roll a 12, 12, 10, 3, and then 2 for a total of three turns (remember a double allows an extra roll). The probability of this happening is

$$P(\text{rolling } 12, 12, 10, 3, 2) = \frac{1}{36} \cdot \frac{1}{36} \cdot \frac{3}{36} \cdot \frac{2}{36} \cdot \frac{1}{36}$$

$$= \frac{6}{36^5} \approx 9.922903013 \times 10^{-8}$$

Is it possible to land on both of those properties *in two turns?* The answer is yes; roll 7, 7, 6, 6, 3, 4, 4 (purchase Park Place), and then 2 (purchase Boardwalk).

Notice that we didn't offer a possibility of rolling the same number three times in a row because that would land us in jail. Suppose we do land in jail, and decide not to pay the $50 to get out of jail. What is the probability of getting out of jail by waiting three turns to roll a double? To do ths we find

$$P(\text{not getting a double in 3 rolls}) = \left(\frac{5}{6}\right)^3 \approx 0.58;$$

$$P(\text{getting out of jail}) = 1 - \left(\frac{5}{6}\right)^3 \approx 0.42$$

Thus, it is more likely that you do not get out of jail.

Gene Zirkel wroke an article, "How to Win at Monopoly" (*MAYTC Journal,* Fall 1983, pp. 175–176) to ask the question, "What properties in Monopoly are the most profitable to own?" His conclusions: (1) light blue, (2) gold, (3) dark blue, (4) light purple, (5) yellow, (6) dark purple, (7) red, and (8) green. Thus, Park Place and Boardwalk (dark blue) come in third!

Group Research

Working in small groups is typical of most work environments, and being able to work with others to communicate specific ideas is an important skill to learn. Work with three or four other students to submit a single report based on each of the following questions.

1. **a.** In how many possible ways can you land on jail (just visiting) on your first turn when playing a Monopoly game?
   b. Is there another possible way (in addition to the ones mentioned in the Commentary) to make it from GO to Park Place on your first roll of the dice in a Monopoly game? If so, what is the probability not only that would happen, but also that you would obtain a 2 on your next roll to complete a set (a monopoly)?

2. **Birthday problem:**
   a. Experiment: Consider the birthdates of some famous mathematicians:

Abel	August 5, 1802
Cardano	September 21, 1576 (died)
Descartes	March 31, 1596
Euler	April 15, 1707
Fermat	August 20, 1601 (baptized)
Galois	October 25, 1811
Gauss	April 30, 1777
Newton	December 25, 1642
Pascal	June 19, 1623
Riemann	September 31, 1815

   Add to this list the birthdates of the members of your class. *But before you compile this list, guess the probability that at least two people in this group will have exactly the same birthday (not counting the year).* Be sure to make your guess *before* finding out the birthdates of your classmates. The answer, of course, depends on the number of people on the list. Ten mathematicians are listed and you may have 20 people in your class, giving 30 names on the list.

b. Find the probability of at least one birthday match among 3 randomly selected people. (See Example 4, Section 9.5.)

c. Find the probability of at least one birthday match among 23 randomly selected people. Have each person in your group pick 23 names at random from a biographical dictionary or a *Who's Who*, and verify empirically the probability you calculated.

d. Draw a graph showing the probability of a birthday match given a group of n people. How many people are necessary for the probability actually to reach 1?

e. In the previous parts of this problem, we interpreted two people having the same birthday as meaning at least 2 have the same birthday (see Example 4, Section 9.5). We now refine this idea. Find the following probabilities for a group of 5 randomly selected people:

  Exactly 2 of the 5 have the same birthday.

  Exactly 3 have the same birthday.

  Exactly 4 have the same birthday.

  All 5 have the same birthday.

  There are exactly two pairs sharing (a different) birthday.

  There is a full house of birthdays (that is, three share one birthday, and two share another).

  Show that the questions of this problem account for all the possibilities; that is, show that the sum of the probabilities for all of these possibilities is the same as for the original birthday problem involving 5 persons: What is the probability of a birthday match among 5 randomly selected people?

3. Prepare a strip of paper as shown in Figure 9.14a. Turn it over and mark the other side as shown in Figure 9.14b.

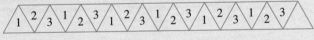

a.

b.

▲ **Figure 9.14   Strips for constructing a hexahexaflexagon. Make sure that each of the numbered triangles is equilateral.**

Starting from the left of Figure 9.14b, fold the 4 onto the 4, the 5 onto the 5, 6 onto 6, 4 onto 4, and so on until your paper looks like the one shown in Figures 9.14c and 9.14d.

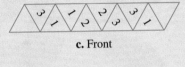

c. Front

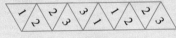

d. Back

Continue by folding 1 onto the 1 from the front, by folding the 1 onto the 1 from the back, and finally by bringing the 1 up from the bottom so that it rests on top of the 1 on the top. Your paper should look like the one shown in Figures 9.14e and 9.14f.

  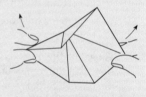

e. Front   **f.** Back   **g.** Folded
                                    hexahexaflexagon

Paste the blank onto the blank, and the result is called a **hexahexaflexagon,** as shown in Figure 9.14g. With a little practice you'll be able to "flex" your hexahexaflexagon (see Figure 9.15) so that you can obtain a side with all 1s, another with all 2s, . . . , and another with all 6s. After you have become fairly proficient at "flexing," count the number of flexes required to obtain all six "sides." What do you think is the fewest number of flexes necessary to obtain all six sides?

▲ **Figure 9.15   To "flex" your hexahexaflexagon, pinch together two of the triangles (left two figures). The inner edge may then be opened with the other hand (rightmost picture). If the hexahexaflexagon cannot be opened, an adjacent pair of triangles is pinched. If it opens, turn it inside out, finding a side that was not visible before. Be careful not to tear the hexahexaflexagon by forcing the flex.**

4. Consider the following classroom activity. Suppose the floor consists of square tiles 9 in. on each side. The players will toss a circular disk onto the floor. If the disk comes to rest on the edge of any tile, the player loses $1. Otherwise, the player wins $1. What is the probability of winning if the disk is:

   a. a dime   **b.** a quarter   **c.** a disk with a diameter of 4 in.

   d. Now, the real question: What size should the disk be so that the probability that the player wins is 0.45?

5. A puzzle sold under the name *The Avenger* is pictured in Figure 9.16.

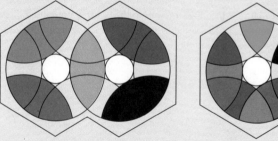 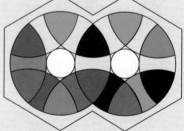

▲ **Figure 9.16   The Avenger Puzzle**

There are four problems posed in the article shown in the reference. Write a report on this article.

REFERENCE: "Group Theory, Rubik's Cube and The Avenger," *Games*, June/July 1987, pp. 44–45.

6. Consult one of the references and learn to solve Rubik's cube. Demonstrate your skill to the class. Nourse names the following categories:

20 minutes:	WHIZ
10 minutes:	SPEED DEMON
5 minutes:	EXPERT
3 minutes:	MASTER OF THE CUBE

a. Stage a contest in front of the class to see which members of your group can complete one face of a Rubik's cube.

b. Stage a contest to see which member of your group can solve the Rubik's cube puzzle the fastest. Report the results to the class.

REFERENCES:

Ledbetter and Nering, *The Solution to Rubik's Cube* (Rohnert Park, CA, Noah's Art Enterprises, 1980).

James G. Nourse, *The Simple Solution to Rubik's Cube* (New York: Bantam Books, 1981).

David Singmaster, *Notes on Rubik's "Magic Cube,"* 5th ed. (Hillside, N.J., Enslow Publishers, 1980).

The Nature of Statistics

There are three kinds of lies: lies, damned lies, and statistics.

G. B. Halsted

In the Real World . . .

"I signed up for Hunter in History 17," said Ben. "Why did you sign up for her?" asked Ted. "Don't you know she has given only three A's in the last 14 years?!" "That's just a rumor. Hunter grades on a curve," Ben re- plied. "Don't give me that 'on a curve' stuff," continued Ted, "I'll bet you don't even know what that means. Anyway, my sister-in-law had her last year and said it was so bad that . . . ."

Contents

Perspective

Undoubtedly, you have some idea about what is meant by the term *statistics*. For example, we hear about:

1. Statistics on population growth
2. Information about the depletion of the rain forests or the ozone layer
3. The latest statistics on the cost of living
4. The Gallup Poll's use of statistics to predict elec- tion outcomes
5. The Nielsen ratings, which indicate that one show has 30% more viewers than another
6. Baseball or sports statistics

We could go on and on, but you can see from these ex- amples that there are two main uses for the word **statistics**. First, we use the term to mean a mass of data, including charts and tables. This is the everyday, nontechnical use of the word. Second, the word refers to a methodology for collecting, analyzing, and interpreting data. In this chap- ter, we'll examine some of these statistical methods. We do not intend to present a statistics course, but rather to prepare you to use statistics in your everyday life, as well as possibly to prepare you to take a college level statistics course.

10.1 FREQUENCY DISTRIBUTIONS AND GRAPHS

Frequency Distributions

Computers, spreadsheets, and simulation programs have done a lot to help us deal with hundreds or thousands of pieces of information at the same time. We can deal with large batches of data by organizing them into groups, or **classes.** The difference between the lower limit of one class and the lower limit of the next class is called the **interval** of the class. After determining the number of values within a class, termed the **frequency,** you can use this information to summarize the data. The end result of this classification and tabulation is called a **frequency distribution.** For example, suppose that you roll a pair of dice 50 times and obtain these outcomes:

3, 2, 6, 5, 3, 8, 8, 7, 10, 9, 7, 5, 12, 9, 6, 11, 8,
11, 11, 8, 7, 7, 7, 10, 11, 6, 4, 8, 8, 7, 6, 4, 10, 7,
9, 7, 9, 6, 6, 9, 4, 4, 6, 3, 4, 10, 6, 9, 6, 11

We can organize these data in a convenient way by using a frequency distribution, as shown in Table 10.1.

TABLE 10.1 Frequency Distribution for 50 Rolls of a Pair of Dice

Outcome	Tally	Frequency
2	|	1
3	|||	3
4	⊞	5
5	||	2
6	⊞ ||||	9
7	⊞ |||	8
8	⊞ |	6
9	⊞ |	6
10	||||	4
11	⊞	5
12	|	1

▬ EXAMPLE 1

Make a frequency distribution for the information in Table 10.2.

TABLE 10.2 State Sales Tax Rates*

Alabama 4%	Hawaii 4%	Massachusetts 5%	New Mexico 5%	South Dakota 4%	
Alaska 0%	Idaho 5%	Michigan 6%	New York 4%	Tennessee 6%	
Arizona 5%	Illinois $6\frac{1}{4}$%	Minnesota $6\frac{1}{2}$%	North Carolina 6%	Texas 6.25%	
Arkansas 4.5%	Indiana 5%	Mississippi 7%	North Dakota 5%	Utah $4\frac{7}{8}$%	
California $7\frac{1}{4}$%	Iowa 5%	Missouri 5.725%	Ohio 5%	Vermont 5%	
Colorado 3%	Kansas 4.9%	Montana 0%	Oklahoma 4.5%	Virginia 4.5%	
Connecticut 6%	Kentucky 6%	Nebraska 5%	Oregon 0%	Washington $6\frac{1}{2}$%	
Delaware 0%	Louisiana 4%	Nevada 6.5%	Pennsylvania 6%	West Virginia 6%	
Florida 6%	Maine 6%	New Hampshire 0%	Rhode Island 7%	Wisconsin 5%	
Georgia 4%	Maryland 5%	New Jersey 6%	South Carolina 5%	Wyoming 4%	
		District of Columbia 6%			

* Does not include local sales taxes.

Solution Make three columns. First, list the sales tax categories; next, tally; and finally, count the tallies to determine the frequency of each (see margin). To ac-

Sales tax	Tally	Frequency
0%	ℍℍ	5
3%	│	1
4%	ℍℍ ‖	7
$4\frac{1}{2}$%	‖‖	3
$4\frac{7}{8}$%	│	1
4.9%	│	1
5%	ℍℍ ℍℍ ‖‖	13
5.725%	│	1
6%	ℍℍ ℍℍ │	11
$6\frac{1}{4}$%	‖	2
$6\frac{1}{2}$%	‖‖	3
7%	‖	2
$7\frac{1}{4}$%	│	1

count for irregularities (such as $4\frac{7}{8}$%), the categories are often grouped. For this example, the data are divided into 7 categories (or groups). The number of categories is usually arbitrary and chosen for convenience. This is called a **grouped frequency distribution**.

Grouped Frequency Distribution

Sales tax	Tally	Frequency
0–1%	ℍℍ	5
1⁺–2%		0
2⁺–3%	│	1
3⁺–4%	ℍℍ ‖	7
4⁺–5%	ℍℍ ℍℍ ℍℍ ‖‖	18
5⁺–6%	ℍℍ ℍℍ ‖	12
6⁺–7%	ℍℍ ‖	7
Over 7%	│	1

A procedure that is useful for organizing large sets of data is called a **stem-and-leaf plot**. Consider the data shown in Table 10.3* on page 691.

For these data we form stems representing decades of the ages of the actors. For example, Tom Hanks, who won best actor award in 1994 for his role in *Forrest Gump*, was 38 years of age when he won the award. We would say that the stem for this age is 3 (for 30), and the leaf is 8.

▰▰▰▰ EXAMPLE 2

Construct a stem-and-leaf plot for the best actor ages in Table 10.3.

Solution   *Stem-and-Leaf Plot of Ages of Best Actor, 1928–1996*

```
3 │ 0 1 1 1 2 2 3 3 4 4 5 5 5 5 7 7 7 8 8 8 8 8 8 9 9
4 │ 0 0 0 0 1 1 1 1 2 2 2 3 3 3 3 3 4 4 5 5 6 6 7 7 8 8 8 9 9 9
5 │ 1 1 2 2 3 5 5 6 6 6
6 │ 0 1 2
7 │ 6
```

Bar Graphs

A **bar graph** compares several related pieces of data using horizontal or vertical bars of uniform width. There must be some sort of scale or measurement on both the

---

* The idea for this example came from "Ages of Oscar-Winning Best Actors and Actresses" by Richard Brown and Gretchen Davis, *The Mathematics Teacher*, February 1990, pp. 96–102.

TABLE 10.3   Best Actors, 1928–1996

Year	Actor	Movie	Age	Year	Actor	Movie	Age
1928	Emil Jannings	The Way of All Flesh	44	1961	Maximilian Schell	Judgment at Nuremburg	31
1929	Warner Baxter	In Old Arizona	38	1962	Gregory Peck	To Kill a Mockingbird	46
1930	George Arliss	Disraeli	46	1963	Sidney Poitier	Lilies of the Field	39
1931	Lionel Barrymore	A Free Soul	53	1964	Rex Harrison	My Fair Lady	56
1932	Fredric March	Dr. Jekyll and Mr. Hyde	35	1965	Lee Marvin	Cat Ballou	41
1932	Wallace Berry	The Champ	47	1966	Paul Scofield	A Man for All Seasons	44
1933	Charles Laughton	The Private Life of Henry VIII	34	1967	Rod Steiger	In the Heat of the Night	42
1934	Clark Gable	It Happened One Night	33	1968	Cliff Robertson	Charly	43
1935	Victor McLaglen	The Informer	49	1969	John Wayne	True Grit	62
1936	Paul Muni	The Story of Louis Pasteur	41	1970	George C. Scott	Patton	43
1937	Spencer Tracy	Captains Courageous	37	1971	Gene Hackman	The French Connection	40
1938	Spencer Tracy	Boys' Town	38	1972	Marlon Brando	The Godfather	48
1939	Robert Donat	Goodbye Mr. Chips	34	1973	Jack Lemmon	Save the Tiger	48
1940	James Stewart	The Philadelphia Story	32	1974	Art Carney	Harry and Tonto	56
1941	Gary Cooper	Sergeant York	40	1975	Jack Nicholson	One Flew over the Cuckoo's Nest	38
1942	James Cagney	Yankee Doodle Dandy	43	1976	Peter Finch	Network	60
1943	Paul Lukas	On the Rhine	48	1977	Richard Dreyfuss	The Goodbye Girl	32
1944	Bing Crosby	Going My Way	43	1978	Jon Voight	Coming Home	40
1945	Ray Milland	The Lost Weekend	40	1979	Dustin Hoffman	Kramer vs. Kramer	42
1946	Fredric March	The Best Years of Our Lives	49	1980	Robert De Niro	Raging Bull	37
1947	Ronald Colman	A Double Life	56	1981	Henry Fonda	On Golden Pond	76
1948	Laurence Olivier	Hamlet	41	1982	Ben Kingsley	Gandhi	39
1949	Broderick Crawford	All the King's Men	38	1983	Robert Duvall	Tender Mercies	55
1950	Jose Ferrer	Cyrano de Bergerac	38	1984	F. Murray Abraham	Amadeus	45
1951	Humphrey Bogart	The African Queen	52	1985	William Hurt	Kiss of the Spider Woman	35
1952	Gary Cooper	High Noon	51	1986	Paul Newman	Color of Money	61
1953	William Holden	Stalag 17	35	1987	Michael Douglas	Wall Street	33
1954	Marlon Brando	On the Waterfront	30	1988	Dustin Hoffman	Rainman	51
1955	Ernest Borgnine	Marty	38	1989	Daniel Day-Lewis	My Left Foot	31
1956	Yul Brynner	The King and I	41	1990	Jeremy Irons	Reversal of Fortune	42
1957	Alec Guinness	The Bridge on the River Kwai	43	1991	Anthony Hopkins	Silence of the Lambs	55
1958	David Niven	Separate Tables	49	1992	Al Pacino	Scent of a Woman	52
1959	Charlton Heston	Ben Hur	35	1993	Tom Hanks	Philadelphia	37
1960	Burt Lancaster	Elmer Gantry	47	1994	Tom Hanks	Forrest Gump	38
				1995	Nicholas Cage	Leaving Las Vegas	31
				1996	Geoffrey Rush	Shine	45

Outcome	Tally	Frequency
2	I	1
3	III	3
4	HH	5
5	II	2
6	HH IIII	9
7	HH III	8
8	HH I	6
9	HH I	6
10	IIII	4
11	HH	5
12	I	1

TABLE 10.4 Frequency Distribution for 50 Rolls of a Pair of Dice

horizontal and vertical axes. An example of a bar graph is shown in Figure 10.1, which shows the data from Table 10.4.

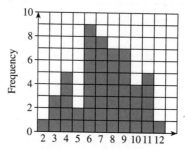

▲ **Figure 10.1   Outcomes of experiment of rolling a pair of dice**

EXAMPLE 3

Construct a bar graph for the data given in Example 1. Use the grouped categories.

Solution   To construct a bar graph, draw and label the horizontal and vertical axes, as shown in part **a** of Figure 10.2. It is helpful (although not necessary) to use graph paper. Next, draw marks indicating the frequency, as shown in part **b** of Figure 10.2. Finally, complete the bars and shade them, as shown in part **c**.

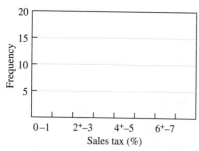

a. Drawing and labeling the axes.

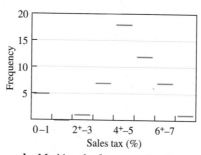

b. Marking the frequency levels.

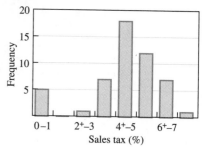

c. Completing and shading the bars.

▲ **Figure 10.2   Bar graph for sales tax data**

You will frequently need to look at and interpret bar graphs in which bars of different lengths are used for comparison purposes.

EXAMPLE 4

Refer to Figure 10.3 to answer the following questions.

a. What was the median price of a home in Sonoma County in August 1996?

b. In which month(s) did the median price of a home increase between 1995 and 1996?

c. In which month (if any) did the number of homes as well as the price of the homes increase between 1995 and 1996?

Solution

a. The median price of a home in August 1996 was $191,750.

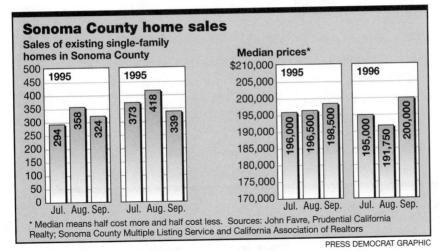

▲ **Figure 10.3   Sample bar graph**

b. The median price of home increased in September between 1995 and 1996.

c. In September, the number of homes increased from 324 to 339 and the median price increased from $198,500 to $200,000 between 1995 and 1996.   ▬

Line Graphs

A graph that uses a broken line to illustrate how one quantity changes with respect to another is called a **line graph.** A line graph is one of the most widely used kinds of graph.

▬ **EXAMPLE 5**

Draw a line graph for the data given in Example 1. Use the previously grouped categories.

Solution   The line graph uses points instead of bars to designate the locations of the frequencies. These points are then connected by line segments, as shown in Figure 10.4. To plot the points, use the frequency distribution to find the category (0–1%, for example); then plot a point showing the frequency (5, in this example). This step is shown in part **a** of Figure 10.4. The last step is to connect the dots with line segments, as shown in part **b.**

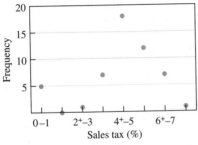

a. Plotting the points to represent frequency levels

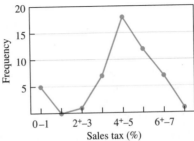

b. Connecting the dots with line segments

▲ **Figure 10.4   Constructing a line graph for sales tax data**   ▬

Just as with bar graphs, you need to be able to read and interpret line graphs.

▬▬ EXAMPLE 6

Refer to Figure 10.5 to answer the following questions.

a. In which year was the voter turnout in a presidential election the greatest?

b. During the period 1932–1996, were there more periods of economic growth or economic decline?

c. What was the voter turnout in 1996? Did a Republican or Democrat win that election? In which year was the voter turnout the closest to that in 1996? Did a Republican or a Democrat win in that election?

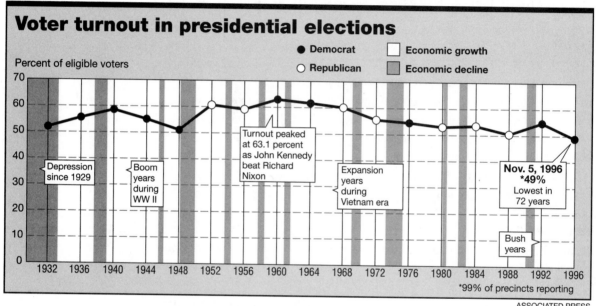

▲ **Figure 10.5   Voter turnout in presidential elections, 1932–1996**

Solution

a. Voter turnout was the greatest (63.1%) in 1960.

b. Growth is shown as a white region and decline as a shaded region. Since there are more white regions, we see that there have been more periods of economic growth than decline.

c. In 1996 the voter turnout was 49% and a Democrat won the election. Voter turnout had not been that low since 1949 (about 51%), when the winner was also Democrat.   ▬

Circle Graphs

Another type of commonly used graph is the **circle graph,** also known as a **pie chart.** This graph is particularly useful in illustrating how a whole quantity is divided into parts — for example, income or expenses in a budget.

To create a circle graph, first express the number in each category as a percentage of the total. Then convert this percentage to an angle in a circle. Remember that a circle is divided into 360°, so we multiply the percent by 360 to find the number of degrees for each category. You can use a protractor to construct a circle graph, as shown in Example 7.

■■■■ EXAMPLE 7

The 1998 expenses for Karlin Enterprises are shown in Figure 10.6.

KE KARLIN ENTERPRISES	EXPENSE REPORT FY 1998
Salaries	$ 72,000
Rents, taxes, insurance	$ 24,000
Utilities	$ 6,000
Advertising	$ 12,000
Shrinkage	$ 1,200
Materials and supplies	$ 1,200
Depreciation	$ 3,600
TOTAL	**$120,000**

▲ Figure 10.6 Karlin Enterprises expenses

Construct a circle graph showing the expenses for Karlin Enterprises.

Solution The first step in constructing a circle graph is to write the ratio of each entry to the total of the entries, as a percent. This is done by finding the total ($120,000) and then dividing each entry by that total:

Salaries: $\dfrac{72,000}{120,000} = 60\%$ Rents: $\dfrac{24,000}{120,000} = 20\%$

Utilities: $\dfrac{6,000}{120,000} = 5\%$ Advertising: $\dfrac{12,000}{120,000} = 10\%$

Shrinkage: $\dfrac{1,200}{120,000} = 1\%$ Materials/supplies: $\dfrac{1,200}{120,000} = 1\%$

Depreciation: $\dfrac{3,600}{120,000} = 3\%$

A circle has 360°, so the next step is to multiply each percent by 360°:

Salaries: $360° \times 0.60 = 216°$ Rents: $360° \times 0.20 = 72°$

Utilities: $360° \times 0.05 = 18°$ Advertising: $360° \times 0.10 = 36°$

Shrinkage: $360° \times 0.01 = 3.6°$ Depreciation: $360° \times 0.03 = 10.8°$

Materials/supplies: $360° \times 0.01 = 3.6°$

Finally, use a protractor to construct the circle graph as shown in Figure 10.7. ■

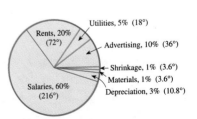

▲ Figure 10.7 Circle graph showing the expenses for Karlin Enterprises

Pictographs

A **pictograph** is a representation of data that uses pictures to show quantity. Consider the raw data shown in Table 10.5, and the bar graph of those data shown in Figure 10.8.

TABLE 10.5	Marital Status of Persons Age 65 and Older (in millions)*			
	Married	*Widowed*	*Divorced*	*Never Married*
Women	5.4	7.2	0.4	0.8
Men	7.5	1.4	0.3	0.6

* Figures are rounded to the nearest 100,000.

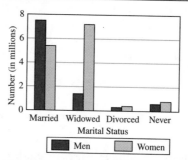

▲ **Figure 10.8   Bar graph showing the marital status of persons age 65 and older**

A pictograph uses a picture to illustrate data; it is normally used only in popular publications, rather than for scientific applications. For these data, suppose that we draw pictures of a woman and a man so that each picture represents 1 million persons, as shown in Figure 10.9.

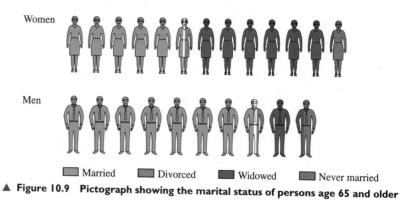

▲ **Figure 10.9   Pictograph showing the marital status of persons age 65 and older**

Misuses of Graphs

The scales on the axes of either bar or line graphs are frequently chosen to exaggerate or diminish real difference. Even worse, graphs are often presented with no

scale whatsoever. For instance, Figure 10.10b shows a graphical "comparison" between Anacin and "regular strength aspirin."

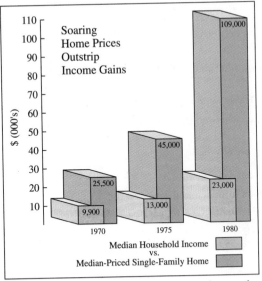

Soaring
Home Prices
Outstrip
Income Gains

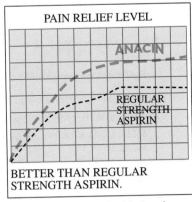

a. Heights of cubes (vertical scales) are used to graph
data, which is then represented as a volume.

b. No scale is shown in this Anacin
advertisement.

▲ **Figure 10.10** **Misuse of graphs**

The most misused type of graph is the pictograph. Consider the data from Table
10.5. Such data can be used to determine the height of a three-dimensional object,
as in part **a** of Figure 10.11. When an object (such as a person) is viewed as three-
dimensional, differences seem much larger than they actually are. Look at part **b**
of Figure 10.11 and notice that, as the height and width are doubled, the volume
is actually increased eightfold.

a. Pictograph showing the
number of widowed persons
of age 65 and older

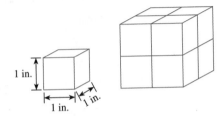

b. Change in volume as a result of
changes in length and width

▲ **Figure 10.11** **Examples of misuses in pictographs**

PROBLEM SET 10.1

▲ A Problems

1. **IN YOUR OWN WORDS** Distinguish between a frequency distribution and a stem-and-leaf plot.

2. **IN YOUR OWN WORDS** Write down what you think the following advertisement means:

   Nine out of ten dentists recommend Trident for their patients who chew gum.

3. **IN YOUR OWN WORDS** Write down what you think the following advertisement means:

   Eight out of ten owners said their cats prefer Whiskas.

Use the following information in Problems 4–7.

The heights of 30 students are as follows (rounded to the nearest inch):

   66, 68, 65, 70, 67, 67, 68, 64, 64, 66, 64, 70, 72, 71, 69, 64, 63, 70, 71, 63, 68, 67, 67, 65, 69, 65, 67, 66, 69, 69.

4. Prepare a frequency distribution.   5. Draw a stem-and-leaf plot.

6. Draw a bar graph.   7. Draw a line graph.

Use the following information in Problems 8–11.

The wages of employees of a small accounting firm are as follows:

   $20,000, $25,000, $25,000, $25,000, $30,000, $30,000, $35,000, $50,000, $60,000, $18,000, $18,000, $16,000, $14,000.

8. Prepare a frequency distribution.   9. Draw a stem-and leaf-plot.

10. Draw a bar graph.   11. Draw a line graph.

Use the following information in Problems 12–14.

The waiting times, in days, for a marriage license in the 50 states are as follows:

   Alabama, 0; Alaska, 3; Arizona, 0; Arkansas, 0; California, 0; Colorado, 0; Connecticut, 4; Delaware, 0; Florida, 3; Georgia, 3; Hawaii, 0; Idaho, 0; Illinois, 0; Indiana, 3; Iowa, 3; Kansas, 3; Kentucky, 0; Louisiana, 3; Maine, 3; Maryland, 2; Massachusetts, 3; Michigan, 3; Minnesota, 5; Mississippi, 3; Missouri, 0; Montana, 0; Nebraska, 0; Nevada, 0; New Hampshire, 3; New Jersey, 3; New Mexico, 0; New York, 0; North Carolina, 0; North Dakota, 0; Ohio, 5; Oklahoma, 0; Oregon, 3; Pennsylvania, 3; Rhode Island, 0; South Carolina, 1; South Dakota, 0; Tennessee, 3; Texas, 0; Utah, 0; Vermont, 3; Virginia, 0; Washington, 3; West Virginia, 3; Wisconsin, 5; Wyoming, 0.

12. Prepare a frequency distribution.

13. Draw a bar graph.   14. Draw a line graph.

Use the following information in Problems 15–16.

The purchasing power of the dollar (1967 = $1) is (to the nearest cent):

   1964, $1.07; 1966, $1.03; 1968, $0.96; 1970, $0.86; 1972, $0.80; 1974, $0.68; 1976, $0.59; 1978, $0.49; 1980, $0.38; 1982, $0.33; 1984, $0.31; 1986, $0.30; 1988, $0.29; 1990, $0.28; 1992, $0.27; 1994, $0.27

15. Draw a line graph.   16. Draw a bar graph.

Use the information in Table 10.6 in Problems 17–19.

17. Graph the minimum wage for the years 1956–1997 and on the same set of axes draw the 1995 value of the minimum wage for the years 1956–1997.

18. In what year will the lines in Problem 17 cross? Why do you think they cross at that point?

19. Draw a bar graph.

20. Figure 10.8 shows a bar graph for the information contained in Table 10.5. Draw a line graph to represent these data.

21. Figure 10.12 shows a line graph.
    a. During which month did Herb incur the most expenses?
    b. During which month did Lisa incur the least expenses?

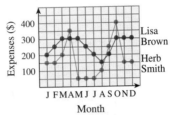

▲ **Figure 10.12   Expenses for two salespeople of the Leadwell Pencil Company**

	TABLE 10.6 U.S. Minimum Wage	
Year	Minimum wage	Value in 1995 dollars
1956	$1.00	$5.13
1961	$1.15	$5.38
1963	$1.25	$5.70
1967	$1.40	$5.86
1968	$1.60	$6.45
1974	$2.00	$5.85
1975	$2.10	$5.68
1976	$2.30	$5.88
1978	$2.65	$5.98
1979	$2.90	$5.95
1980	$3.10	$5.72
1981	$3.35	$5.65
1990	$3.80	$4.42
1991	$4.25	$4.74
1996	$4.75	$4.25
1997	$5.15	$4.10

Use the information in Figure 10.13 in Problems 22–25.

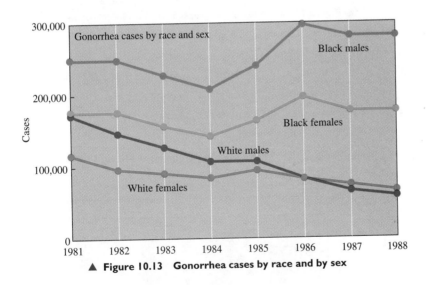

▲ **Figure 10.13   Gonorrhea cases by race and by sex**

22. What are the categories being graphed?

23. In what year were the numbers of cases in two of the categories the same?

24. During which period was the number of cases increasing for black females?

25. How many cases of gonorrhea were there among white males in 1985?

Use the bar graph in Figure 10.14 to answer the questions in Problems 26–31.

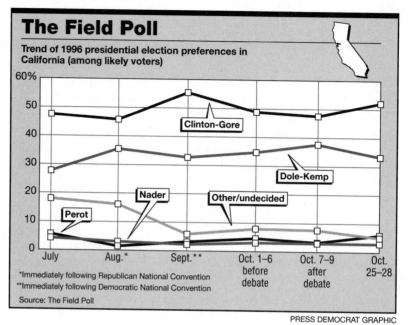

▲ **Figure 10.14   Political bar graph from the 1996 election**

26. How many candidates (not including other/undecided) are tracked?

27. In which month did Clinton–Gore have the most support?

28. What percent of the California voters favored Nader in August 1996?

29. According to this Field poll, which candidate won the 1996 presidential debate?

30. In which month was the undecided vote the greatest?

31. In which month was Perot's support the greatest?

Use the bar graph in Figure 10.15 to answer the questions in Problems 32–35.

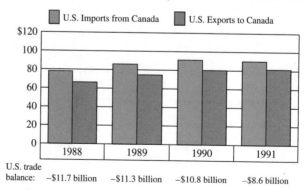

▲ **Figure 10.15   Bar graph showing U.S. trade with Canada**

32. What is being illustrated in this graph?

33. Did the United States have more imports or exports for the years 1988–1991?

34. What was the approximate balance of trade with Canada in 1991?

35. What was the approximate dollar amount of imports to Canada in 1991?

Use the bar graph in Figure 10.16 to answer the questions in Problems 36–39.

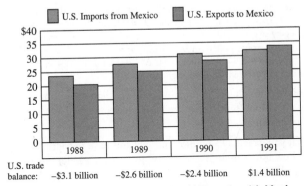

▲ **Figure 10.16   Bar graph showing U.S. trade with Mexico**

36. What is being illustrated in this graph?

37. Did the United States have more imports or exports for the years 1988–1991?

38. What was the approximate balance of trade with Mexico in 1991?

39. What was the approximate dollar amount of exports to Mexico in 1989?

▲ **B Problems**

When a person in California renews the registration for an automobile, the bar graph shown in Figure 10.17 is included with the bill.

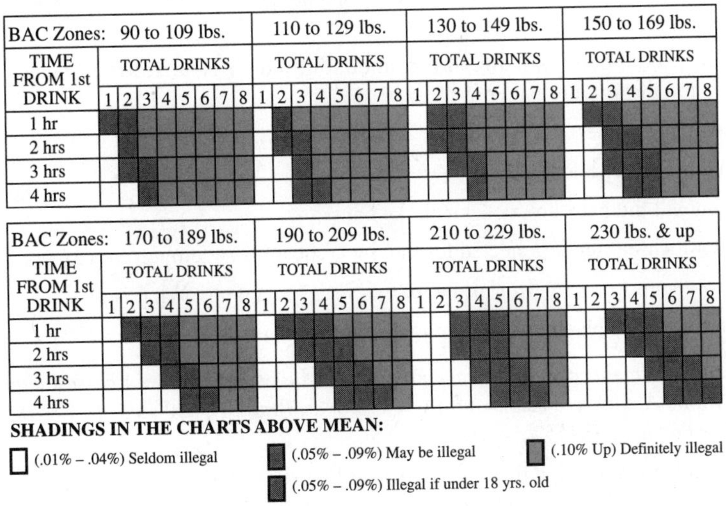

BAC Zones:	90 to 109 lbs.								110 to 129 lbs.								130 to 149 lbs.								150 to 169 lbs.							
TIME FROM 1st DRINK	TOTAL DRINKS								TOTAL DRINKS								TOTAL DRINKS								TOTAL DRINKS							
	1	2	3	4	5	6	7	8	1	2	3	4	5	6	7	8	1	2	3	4	5	6	7	8	1	2	3	4	5	6	7	8
1 hr																																
2 hrs																																
3 hrs																																
4 hrs																																

BAC Zones:	170 to 189 lbs.								190 to 209 lbs.								210 to 229 lbs.								230 lbs. & up							
TIME FROM 1st DRINK	TOTAL DRINKS								TOTAL DRINKS								TOTAL DRINKS								TOTAL DRINKS							
	1	2	3	4	5	6	7	8	1	2	3	4	5	6	7	8	1	2	3	4	5	6	7	8	1	2	3	4	5	6	7	8
1 hr																																
2 hrs																																
3 hrs																																
4 hrs																																

SHADINGS IN THE CHARTS ABOVE MEAN:

☐ (.01% – .04%) Seldom illegal   ■ (.05% – .09%) May be illegal   ■ (.10% Up) Definitely illegal

■ (.05% – .09%) Illegal if under 18 yrs. old

Prepared by the Department of Motor Vehicles in cooperation with the California Highway Patrol.

▲ **Figure 10.17   Blood Alcohol Concentration (BAC) charts**

Use Figure 10.17 to answer the questions in Problems 40–45. The following statement is included with the graph:

> There is no safe way to drive after drinking. These charts show that a few drinks can make you an unsafe driver. They show that drinking affects your BLOOD ALCOHOL CONCENTRATION (BAC). The BAC zones for various numbers of drinks and time periods are printed in white, gray, and red. HOW TO USE THESE CHARTS: First, find the chart that includes your weight. For example, if you weigh 160 lbs, use the "150 to 169" chart. Then look under "Total Drinks" at the "2" on this "150 to 169" chart. Now look below the "2" drinks, in the row for 1 hour. You'll see your BAC is in the gray shaded zone. This means that if you drive after 2 drinks in 1 hour, you could be arrested. In the gray zone, your chances of having an accident are 5 times higher than if you had no drinks. But if you had 4 drinks in 1 hour, your BAC would be in the red shaded area . . . and your chances of having an accident 25 times higher.

40. Suppose that you weigh 115 pounds and that you have two drinks in 2 hours. If you then drive, how much more likely are you to have an accident than if you had refrained from drinking?

41. Suppose that you weigh 115 pounds and that you have four drinks in 3 hours. If you then drive, how much more likely are you to have an accident than if you had refrained from drinking?

42. Suppose that you weigh 195 pounds and have two drinks in 2 hours. According to Figure 10.17, are you seldom illegal, maybe illegal, or definitely illegal?

43. Suppose that you weigh 195 pounds and that you have four drinks in 3 hours. According to Figure 10.17, are you seldom illegal, maybe illegal, or definitely illegal?

44. If you weigh 135 pounds, how many drinks in 3 hours would you need to be definitely illegal?

45. If you weigh 185 pounds and drink a six-pack of beer during a 3-hour baseball game, can you legally drive home?

A newspaper article discussing whether or not Social Security could be cut offered the four graphs shown at the top of page 703. Use the information in these graphs to answer the questions in Problems 46–52.

46. What is the difference between the amount of money collected from Social Security taxes and the amount paid out in 1992?

47. How many workers pay into Social Security for each person getting benefits in 1990?

48. What is a "baby boomer"?

49. When will the baby boomers break the bank?

50. How much money is projected for the Social Security system for the year 2000?

51. There seems to be an error in one of the three line graphs. What is it?

52. There seems to be an error in the pictograph. What is it?

The amount of time it takes for three leading pain relievers to reach your bloodstream is as follows: Brand A, 480 seconds; Brand B, 500 seconds; Brand C, 490 seconds. Use this information for Problems 53–55.

53. Draw a bar graph using the scale shown in Figure 10.18a.

54. Draw a bar graph using the scale shown in Figure 10.18b.

55. If you were an advertiser working on a promotion campaign for Brand A, which graph from Figure 10.18 would seem to give your product a more distinct advantage?

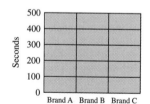

a. Scale from 0 to 500 seconds

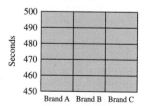

b. Scale from 450 to 500 seconds

▲ Figure 10.18   Time for pain reliever to reach your bloodstream

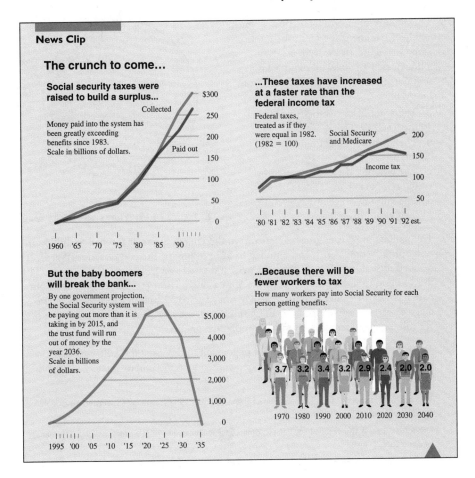

News Clip

The crunch to come...

Social security taxes were raised to build a surplus...

Money paid into the system has been greatly exceeding benefits since 1983.
Scale in billions of dollars.

Collected

Paid out

$300
250
200
150
100
50
0

1960 '65 '70 '75 '80 '85 '90

...These taxes have increased at a faster rate than the federal income tax

Federal taxes, treated as if they were equal in 1982.
(1982 = 100)

Social Security and Medicare

Income tax

200
150
100
50

'80 '81 '82 '83 '84 '85 '86 '87 '88 '89 '90 '91 '92 est.

But the baby boomers will break the bank...

By one government projection, the Social Security system will be paying out more than it is taking in by 2015, and the trust fund will run out of money by the year 2036.
Scale in billions of dollars.

$5,000
4,000
3,000
2,000
1,000
0

1995 '00 '05 '10 '15 '20 '25 '30 '35

...Because there will be fewer workers to tax

How many workers pay into Social Security for each person getting benefits.

3.7 3.2 3.4 3.2 2.9 2.4 2.0 2.0

1970 1980 1990 2000 2010 2020 2030 2040

56. The financial report from a small college shows the following income:

Tuition	$1,388,000
Development	90,000
Fund-raising	119,000
Athletic Dept.	44,000
Interest	75,000
Tax subsidy	29,000
TOTAL	$1,745,000

Draw a circle graph showing this income.

57. The financial report from a small college shows the following expenses:

Salaries	$1,400,000
Academic	68,000
Plant operations	196,000
Student activities	31,000
Athletic programs	50,000
TOTAL	$1,745,000

Draw a circle graph showing the expenses.

58. The amount of electricity used in a typical all-electric home is shown in the circle graph in Figure 10.19. If, in a certain month, a home used 1,100 kwh (kilowatt-hours), find the amounts of electricity used from the graph.

    a. The amount of electricity used by the water heater
    b. The amount of electricity used by the stove
    c. The amount of electricity used by the refrigerator
    d. The amount of electricity used by the clothes dryer

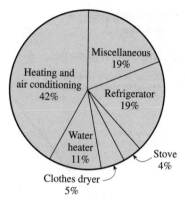

▲ **Figure 10.19   Home electricity usage**

59. The national debt is of growing concern in the United States. It reached $1 billion in 1916 during World War I, and climbed to $278 billion by the end of World War II. It reached its first trillion on October 1, 1981, and rose to $2 trillion on April 3, 1986. The third trillion milestone was reached on April 4, 1990. However, the magnitude of our national debt became headline news when it reached $4 trillion and became a campaign slogan for Ross Perot in 1992. On January 1, 1997, the national debt was just over 5.2 trillion. Represent this information on a line graph.

60. **Top women on Wall Street**   Draw a pictograph to represent the following data on women in big Wall Street firms. In 1996, women comprise approximately 11% of the top-tier executives.

    Regina Dolan is the chief financial officer of Paine Webber. She is one of 46 women among 465 managing directors.

    Robin Neustein is the chief of staff at Goldman Sachs. She is one of 9 women among 173 managing directors.

    Theresa Lang is the company treasurer for Merrill Lynch. She is one of 76 women among 694 managing directors.

What is wrong, if anything, with each of the statements in Problems 61–65? *Explain your reasoning.*

61. Consider the graph shown in Figure 10.20. Clearly, Anacin is better.

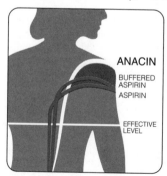

▲ **Figure 10.20   From an advertisement for a pain reliever**

62. Consider the graph shown in Figure 10.21. The potential commercial forest growth, as compared with the current commercial forest growth, is almost double (44.9 to 74.2).

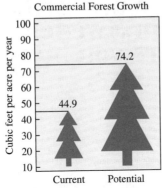

Commercial Forest Growth

▲ **Figure 10.21   From the U.S. Forest Service**

63. Consider the graph shown in Figure 10.22.

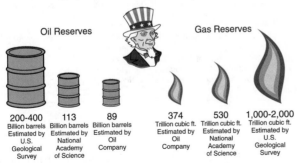

How much is "in the bank"?
Conflicting estimates of undiscovered oil and gas reserves—
which can be recovered and produced.

▲ **Figure 10.22   Pictograph showing oil and gas reserves**

64. The following graph was circulated throughout the mathematics department of a leading college.

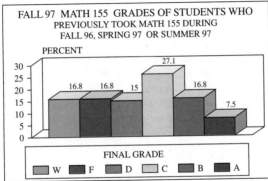

FALL 97 MATH 155 GRADES OF STUDENTS WHO
PREVIOUSLY TOOK MATH 155 DURING
FALL 96, SPRING 97 OR SUMMER 97

65. Consider the Saab advertisement.

There are two cars built in Sweden. Before you buy theirs, drive ours.

When people who know cars think about Swedish cars, they think of them as being strong and durable. And conquering some of the toughest driving conditions in the world.

But, unfortunately, when most people think about buying a Swedish car, the one they think about usually isn't ours. (Even though ours doesn't cost any more.)

Ours is the SAAB 99E. It's strong and durable. But it's a lot different from their car.

Our car has Front-Wheel Drive for better traction, stability and handling.

It has a 1.85 liter, fuel-injected, 4 cylinder, overhead cam engine as standard in every car. 4-speed transmission is standard too. Or you can get a 3-speed automatic (optional).

Our car has four-wheel disc brakes and dual-diagonal braking system so you can stop straight and fast every time.

It has a wide stance. (About 55 inches.) So it rides and handles like a sports car.

Outside, our car is smaller than a lot of "small" cars. 172″ overall length, 57″ overall width.

Inside, our car has bucket seats up front and a full five feet across in the back so you can easily accommodate five adults.

It has more headroom than a Rolls Royce and more room from the brake pedal to the back seat than a Mercedes 280. And it has factory air conditioning as an option.

There are a lot of other things that make our car different from their car. Like roll cage construction and a special "hot seat" for cold winter days.

So before you buy their car, stop by your nearest SAAB dealer and drive our car. The SAAB 99E. We think you'll buy it instead of theirs. **SAAB 99E**

▲ **Problem Solving**

66. The following advertisements from *Games* magazine show some vastly overpriced items. See if you can read between the lines and identify the item each ad is describing.

   a. Miniature, sparkling white crystals harvested in exotic Hawaii make perfect specimens for studying the wonders of science! Each educationally decorated packet contains hundreds of refined crystals. Completely safe and nontoxic. Only $3.95.

   b. It's like having U.S. history in your pocket! Miniature portraits of America's best-loved presidents, Washington sites, and more. Richly detailed copper and silver engraving. Set of four, just $4.50.

   c. Today multimillion-dollar computers need special programs to generate random numbers. But this GEOMETRICALLY DESIGNED device ingeniously bypasses computer technology to provide random numbers quickly and easily! Exciting? You bet! Batteries not required. Only $4.00.

   d. They said we couldn't do it, but we are! For a limited time, our LEADING HOLLYWOOD PROP HOUSE is dividing up and selling — yes, SELLING — parcels of this famous movie setting. You've seen it featured in *Lawrence of Arabia*, *Beach Blanket Bingo*, and *Ishtar!* Our warehouse must be cleared! Own a vial of movie history — a great conversation piece! Only $4.50.

e. As you command water — or any liquid — to flow upwards! Could you be tampering with the secrets of the Earth? Your friends will be amazed! Just $31.99.

▲ **Individual Research**

67. "You can clearly see that Bufferin is the most effective . . . "; "Penzoil is better suited . . . "; "Sylvania was preferred by . . . ." "How can anyone analyze the claims of the commercials we see and hear on a daily basis?" asked Betty. "I even subscribe to *Consumer Reports*, but so many of the claims seem to be unreasonable. I don't like to buy items by trial and error, and I really don't believe all the claims in advertisements." Collect examples of good statistical graphs and examples of misleading graphs. Use some of the leading newspapers and national magazines.

68. Carry out the following experiment: *A cat has two bowls of food; one bowl contains Whiskas and the other some other brand. The cat eats Whiskas and leaves the other untouched. Make a list of possible reasons why the cat ignored the second bowl. De-*scribe the circumstances under which you think the advertise could claim: *Eight out of ten owners said their cat preferred Whiskas.*

10.2   DESCRIPTIVE STATISTICS

Measures of Central Tendency

In Section 10.1, we organized data into a frequency distribution and then discussed their presentation in graphical form. However, some properties of data can help us interpret masses of information. We will use the following *Peanuts* cartoon to introduce the notion of *average*.

Do you suppose that Violet's dad bowled better on Monday nights (185 average) than on Thursday nights (170 average)? Don't be too hasty to say "yes" before you look at the scores that make up these averages:

	Monday Night	*Thursday Night*
Game 1	175	180
Game 2	150	130
Game 3	160	161
Game 4	180	185
Game 5	160	163
Game 6	183	185
Game 7	287	186
Totals	1,295	1,190

To find the averages used by Violet in the cartoon, we divide these totals by the number of games:

$$\begin{array}{cc} \textit{Monday Night} & \textit{Thursday Night} \\ \dfrac{1{,}295}{7} = 185 & \dfrac{1{,}190}{7} = 170 \end{array}$$

PEANUTS Reprinted by permission of UFS, Inc.

If we consider the averages, Violet's dad did better on Mondays; but if we consider the games separately, we see that he typically did better on Thursday (five out of seven games). Would any other properties of the bowling scores tell us this fact?

Since we must often add up a list of numbers in statistics, as we did above, we use the symbol Σx to mean *the sum of all the values that x can assume*. Similarly, Σx^2 means to square each value that x can assume, and then add the results; $(\Sigma x)^2$ means first to add the values and then square the result. The symbol Σ is the Greek capital letter sigma (which is chosen because S reminds us of "sum").

The average used by Violet is only one kind of statistical measure that can be used. It is the measure that most of us think of when we hear someone use the word *average*. It is called the *mean*. Other statistical measures, called **averages** or **measures of central tendency**, are defined in the following box.

The *mean* is the most sensitive average. It reflects the entire distribution and is the most common average.

Measures of Central Tendency: Mean, Median, Mode

1. **Mean.**   The number found by adding the data and then dividing by the number of data values. The mean is usually denoted by $\bar{x}$:

$$\bar{x} = \frac{\Sigma x}{n}$$

The *median* gives the middle value. It is useful when there are a few extraordinary values to distort the mean.

2. **Median.**   The middle number when the numbers in the data are arranged in order of size. If there are two middle numbers (in the case of an even number of data values), the median is the mean of these two middle numbers.

The *mode* is the average that measures "popularity." It is possible to have no mode or more than one mode.

3. **Mode.**   The value that occurs most frequently. If no number occurs more than once, there is no mode. It is possible to have more than one mode.

Consider these other measures of central tendency for Violet's dad's bowling scores.

a. Median.   Rearrange the data values from smallest to largest when finding the median:

Monday Night		*Thursday Night*
150		130
160		161
160		163
175	← Middle number →	180
180	is the median.	185
183		185
287		186

b. Mode.   Look for the number that occurs most frequently:

Monday Night	Thursday Night
150	130
160 ⎫ Most frequent	161
160 ⎭ is the mode.	163
175	180
180	185 ⎫ Most frequent
183	185 ⎭ is the mode.
287	186

If we compare the three measures of central tendency for the bowling scores, we find the following:

	Monday Night	Thursday Night
Mean	185	170
Median	175	180
Mode	160	185

We are no longer convinced that Violet's dad did better on Monday nights than on Thursday nights.

▬▬ EXAMPLE 1

Find the mean, median, and mode for the following sets of numbers.

a. 3, 5, 5, 8, 9   **b.** 4, 10, 9, 8, 9, 4, 5   **c.** 6, 5, 4, 7, 1, 9

Solution

a. *Mean:* $\dfrac{\text{SUM OF TERMS}}{\text{NUMBER OF TERMS}} = \dfrac{3 + 5 + 5 + 8 + 9}{5} = \dfrac{30}{5} = 6$

Median: Arrange in order: 3, 5, **5**, 8, 9. The middle term is the median: 5.

Mode: The most frequently occurring term is the mode: 5.

b. *Mean:* $\dfrac{4 + 10 + 9 + 8 + 9 + 4 + 5}{7} = \dfrac{49}{7} = 7$

Median: 4, 4, 5, **8**, 9, 9, 10; the median is 8.

Mode: The data have two modes: 4 and 9. If data have two modes, we say they are **bimodal.**

c. *Mean:* $\dfrac{6 + 5 + 4 + 7 + 1 + 9}{6.} = \dfrac{32}{6} \approx 5.33$

Median: 1, 4, $\underbrace{5, 6}$, 7, 9; the median is $\dfrac{11}{2} = 5.5$.

$$\dfrac{5 + 6}{2} = \dfrac{11}{2}$$

Mode: There is **no mode** because no term appears more than once.   ▬

Calculator Comment

Many calculators have built-in statistical functions. The method of inputting data varies from brand to brand, and you should check with your owner's manual. Once the data are entered, your calculator will find the mean, median, and sometimes the mode by pressing a single statistical button.   ▲

A rather nice physical model illustrates the idea of the mean. Consider a seesaw that consists of a plank and a movable support (called a *fulcrum*). We assume that the plank has no weight and is marked off into units as shown in Figure 10.23.

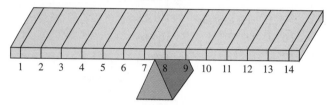

▲ **Figure 10.23   Fulcrum and plank model for mean**

Now let's place some 1-lb weights in the position of the numbers in some given distribution. The balance point for the plank is the mean.

For example, consider the data from part **a** of Example 1: 3, 5, 5, 8, 9. If weights are placed at these locations on the plank, the balance point is 6, as shown in Figure 10.24.

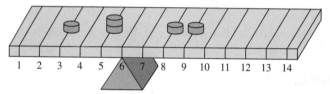

▲ **Figure 10.24   Balance point for the data set 3, 5, 5, 8, 9 is 6.**

■■■ EXAMPLE 2

Consider the number of days one must wait for a marriage license in the various states in the United States.

Days' Wait	Frequency
0	25
1	1
2	1
3	19
4	1
5	3
Total	50

What are the mean, the median, and the mode for these data?

Solution

Mean: To find the mean, we could, of course, add all 50 individual numbers, but instead, notice that

0 occurs 25 times, so write	0×25
1 occurs 1 time, so write	1×1
2 occurs 1 time, so write	2×1
3 occurs 19 times, so write	3×19
4 occurs 1 time, so write	4×1
5 occurs 3 times, so write	5×3

Thus, the mean is

$$\bar{x} = \frac{0 \times 25 + 1 \times 1 + 2 \times 1 + 3 \times 19 + 4 \times 1 + 5 \times 3}{50} = \frac{79}{50} = 1.58$$

Median: Since the median is the middle number, and since there are 50 values, the median is the mean of the 25th and 26th numbers (when they are arranged in order):

$$\left. \begin{array}{l} \text{25th term is 0} \\ \text{26th term is 1} \end{array} \right\} \quad \frac{0 + 1}{2} = \frac{1}{2}$$

Mode: The mode is the value that occurs most frequently, which is 0.   ▬

When finding the mean from a frequency distribution, you are finding what is called a *weighted mean*.

Weighted Mean

If a list of scores $x_1, x_2, x_3, \ldots, x_n$ occurs $w_1, w_2, \ldots, w_n$ times, respectively, then the **weighted mean** is

$$\bar{x} = \frac{\Sigma(w \cdot x)}{\Sigma w}$$

Measures of Position

The median divides the data into two equal parts; half the values are above the median and half are below the median. Sometimes we use benchmark positions that divide the data into more than two parts. **Quartiles,** denoted by Q_1 (first quartile), Q_2 (second quartile), and Q_3 (third quartile), divide the data into four equal parts. **Deciles** are nine values that divide the data into ten equal parts, and **percentiles** are 99 values that divide the data into 100 equal parts. For example, when you take the Scholastic Assessment Test (SAT), your score is recorded as a percentile score. If you scored in the 92nd percentile, it means that you scored better than approximately 92% of those who took the test.

Measures of Dispersion

The measures we've been discussing can help us interpret information, but they do not give the entire story. For example, consider these sets of data:

Set A: $\{8, 9, 9, 9, 10\}$ *Mean:* $\dfrac{8 + 9 + 9 + 9 + 10}{5} = 9$

Median: 9

Mode: 9

Set B: $\{2, 9, 9, 12, 13\}$ *Mean:* $\dfrac{2 + 9 + 9 + 12 + 13}{5} = 9$

Median: 9

Mode: 9

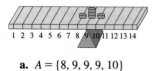

a. $A = \{8, 9, 9, 9, 10\}$

b. $B = \{2, 9, 9, 12, 13\}$

▲ **Figure 10.25   Visualization of dispersion of sets of data**

Notice that, for sets A and B, the measures of central tendency do not distinguish the data. However, if you look at the data placed on planks, as shown in Figure 10.25, you will see that the data in Set B are relatively widely dispersed along the plank, whereas the data in Set A are clumped around the mean.

We'll consider three **measures of dispersion:** the *range,* the *standard deviation,* and the *variance.*

> **Range**
>
> The **range** of a set of data is the difference between the largest and the smallest numbers in the set.

▬▬ EXAMPLE 3

Find the ranges for the following data sets:

a. Set $A = \{8, 9, 9, 9, 10\}$

b. Set $B = \{2, 9, 9, 12, 13\}$

Solution   Notice from Figure 10.25 that the mean for each of these sets of data is the same. The range is found by computing the difference between the largest and smallest values in the set.

a. $10 - 8 = 2$

b. $13 - 2 = 11$ ▬

Notice that the range is determined by only the largest and the smallest numbers in the set; it does not give us any information about the other numbers.

The range is used, along with quartiles, to construct a statistical tool called a *box plot.* For a given set of data, a **box plot** consists of a rectangular box positioned above a numerical scale, drawn from Q_1 (the first quartile) to Q_3 (the third quartile). The median (Q_2, or second quartile) is shown as a dashed line, and a segment is extended to the left to show the distance to the minimum value; another segment is extended to the right for the maximum value.

▬▬ EXAMPLE 4

Draw a box plot for the ages of the actors who received the best actor award at the Academy Awards, as reported in Table 10.3.

Solution We find the quartiles by first finding Q_2, the median. Since there are 70 entries, the median is the mean of the two middle items. For this example, the two middle items are 42 and 42, so

$$Q_2 = \frac{42 + 42}{2} = 42$$

Next, we find Q_1 (first quartile), which is the median of all items below Q_2:

$$Q_1 = 38$$

Finally, find Q_3 (third quartile), which is the median of all items above Q_2:

$$Q_3 = 48$$

The minimum age is 30 and the maximum age is 76, so we have a box plot as shown in Figure 10.26.

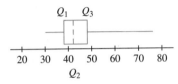

▲ **Figure 10.26   Box plot**

Sometimes a box plot is called a *box-and-whisker plot*. Its usefulness should be clear when you look at Figure 10.26. It shows:

1. the median (a measure of central tendency);
2. the location of the middle half of the data (represented by the extent of the box);
3. the range (a measure of dispersion);
4. the skewness (the nonsymmetry of both the box and the whiskers).

The **variance** and **standard deviation** are measures that use all the numbers in the data set to give information about the dispersion. When finding the variance, we must make a distinction between the variance of the entire population and the variance of a random sample from that population. When the variance is based on a set of sample scores, it is denoted by s^2; and when it is based on all scores in a population, it is denoted by σ^2 (σ is the lowercase Greek letter sigma). The variance for a sample is found by

$$s^2 = \frac{\Sigma(x - \bar{x})^2}{n - 1}$$

To understand this formula for the sample variance, we will consider an example before summarizing a procedure. Again, let's use the data sets we worked

Historical Note

FLORENCE NIGHTINGALE
(1820–1910)

Even though the origins of statistics are ancient, the development of modern statistical techniques began in the 16th century. Adolph Quetelet (1796–1874) was the first person to apply statistical methods to accumulated data. He correctly predicted (to his own surprise) the crime and mortality rates from year to year. Florence Nightingale was an early proponent of the use of statistics in her work. Since the advent of the computer, statistical methods have come within the reach of almost everyone. Today, statistical methods strongly influence agriculture, biology, business, chemistry, economics, education, electronics, medicine, physics, psychology, and sociology. If you decide to enter one of these disciplines for a career, you will no doubt be required to take at least one statistics course.

with in Example 3.

$$\text{Set } A = \{8, 9, 9, 9, 10\} \qquad \text{Set } B = \{2, 9, 9, 12, 13\}$$
$$\text{Mean is 9.} \qquad\qquad \text{Mean is 9.}$$

Find the deviations by subtracting the mean from each term:

$$8 - 9 = -1 \qquad 2 - 9 = -7$$
$$9 - 9 = 0 \qquad 9 - 9 = 0$$
$$9 - 9 = 0 \qquad 9 - 9 = 0$$
$$9 - 9 = 0 \qquad 12 - 9 = 3$$
$$10 - 9 = 1 \qquad 13 - 9 = 4$$
$$\uparrow \qquad\qquad \uparrow$$
$$\text{mean} \qquad\quad \text{mean}$$

If we sum these deviations (to obtain a measure of the total deviation), in each case we obtain 0, because the positive and negative differences "cancel each other out." Next we calculate the *square of each of these deviations*:

$$\text{Set } A = \{8, 9, 9, 9, 10\} \qquad \text{Set } B = \{2, 9, 9, 12, 13\}$$
$$(8 - 9)^2 = (-1)^2 = 1 \qquad (2 - 9)^2 = (-7)^2 = 49$$
$$(9 - 9)^2 = 0^2 = 0 \qquad (9 - 9)^2 = 0^2 = 0$$
$$(9 - 9)^2 = 0^2 = 0 \qquad (9 - 9)^2 = 0^2 = 0$$
$$(9 - 9)^2 = 0^2 = 0 \qquad (12 - 9)^2 = (3)^2 = 9$$
$$(10 - 9)^2 = (1)^2 = 1 \qquad (13 - 9)^2 = (4)^2 = 16$$

Finally, we find the sum of these squares and divide by one less than the number of items to obtain the variance:

$$\text{Set } A: \quad s^2 = \frac{1 + 0 + 0 + 0 + 1}{4} = \frac{2}{4} = 0.5$$

$$\text{Set } B: \quad s^2 = \frac{49 + 0 + 0 + 9 + 16}{4} = \frac{74}{4} = 18.5$$

The larger the variance, the more dispersion there is in the original data. However, we will continue in order to develop a true picture of the dispersion. Since we squared each difference (to eliminate the effect of positive and negative differences), it seems reasonable that we should find the square root of the variance as a more meaningful measure of dispersion. This number, called the **standard deviation,** is denoted by the lowercase Greek letter sigma (σ) when it is based on a population and by s when it is based on a sample. You will need a calculator to find square roots. For the data sets of Example 3:

$$\text{Set } A: \quad s = \sqrt{0.5} \qquad \text{Set } B: \quad s = \sqrt{18.5}$$
$$\approx 0.707 \qquad\qquad\qquad \approx 4.301$$

We summarize these steps in the following box.

> **Standard Deviation**
>
> The **standard deviation** of a sample, denoted by s, is the square root of the variance. To find it, carry out these steps:
>
> 1. Determine the mean of the set of numbers.
> 2. Subtract the mean from each number in the set.
> 3. Square each of these differences.
> 4. Find the sum of the squares of the differences.
> 5. Divide this sum by one less than the number of pieces of data. *This is the variance of the sample.*
> 6. Take the square root of the variance. *This is the standard deviation of the sample.*

■ EXAMPLE 5

Suppose that Missy received the following test scores in a math class: 92, 85, 65, 89, 96, and 71. What is the standard deviation for her test scores?

Solution First we calculate the mean, as in

Step 1: $\bar{x} = \dfrac{92 + 85 + 65 + 89 + 96 + 71}{6} = 83$ This is the mean.

We summarize steps 2 through 4 by using a table format:

Score	*(Deviation from the mean)²*
92	$(92 - 83)^2 = (9)^2 = 81$
85	$(85 - 83)^2 = (2)^2 = 4$
65	$(65 - 83)^2 = (-18)^2 = 324$
89	$(89 - 83)^2 = (6)^2 = 36$
96	$(96 - 83)^2 = (13)^2 = 169$
71	$(71 - 83)^2 = (-12)^2 = 144$

Step 5: Divide the sum by 5 (one less than the number of scores):

$$\frac{81 + 4 + 324 + 36 + 169 + 144}{6 - 1} = \frac{758}{5} = 151.6$$

We note that this number, 151.6, is called the variance. If you do not have access to a calculator, you can use the variance as a measure of dispersion. However, we assume you have a calculator and can find the standard deviation:

Step 6: $s = \sqrt{\dfrac{758}{5}} \approx 12.31$

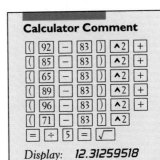

Calculator Comment

(92 − 83) ^2 +
(85 − 83) ^2 +
(65 − 83) ^2 +
(89 − 83) ^2 +
(96 − 83) ^2 +
(71 − 83) ^2
= ÷ 5 = √

Display: *12.31259518*

How can we use the standard deviation? We will begin the discussion in Section 10.3 with this question. For now, however, we will give one example.

Suppose that Missy obtained 65 on an examination for which the mean was 50 and the standard deviation was 15, whereas Shannon in another class scored 74 on an examination for which the mean was 80 and the standard deviation was 3. Did Shannon or Missy do better in her respective class? We see that Missy scored one standard deviation *above* the mean ($50 + 15 = 65$), whereas Shannon scored two standard deviations *below* the mean ($80 - 2 \times 3 = 74$); therefore Missy did better compared to her classmates than did Shannon.

PROBLEM SET   10.2

▲ A Problems

1. **IN YOUR OWN WORDS** What do we mean by average?

2. **IN YOUR OWN WORDS** What is a measure of central tendency?

3. **IN YOUR OWN WORDS** What is a measure of dispersion?

4. **IN YOUR OWN WORDS** Compare and contrast mean, median, and mode.

5. **IN YOUR OWN WORDS** In 1989, Andy Van Slyke's batting average was better than Dave Justice's; and in 1990 Andy once again beat Dave. Does it follows that Andy's combined 1989–1990 batting average is better than Dave's?

6. **IN YOUR OWN WORDS** Is the standard deviation always smaller than the variance?

7. **IN YOUR OWN WORDS** Suppose that a variance is zero. What can you say about the data?

8. **IN YOUR OWN WORDS** A professor gives five exams. Two students' scores have the same mean, although one student did better on all the tests except one. Give an example of such scores.

9. **IN YOUR OWN WORDS** A professor gives six exams. Two students' scores have the same mean, although one student's scores have a small standard deviation and the other student's scores have a large standard deviation. Give an example of such scores.

10. **IN YOUR OWN WORDS** Comment on the following quotation, which was printed in the May 28, 1989, issue of *The Community, Technical, and Junior College TIMES*: "Half the American students are below average."

11. **IN YOUR OWN WORDS** Comment on the following quotation: "I just received my SAT scores and I scored in the 78th percentile in English, and 50th percentile in math.

In Problems 12–23, find the three measures of central tendency (the mean, median, and mode).

	Andy	
	1989	*1990*
Hits	113	140
AB	476	493
Avg.	0.237	0.284

	Dave	
	1989	*1990*
Hits	12	124
AB	51	439
Avg.	0.235	0.282

12. 1, 2, 3, 4, 5
13. 17, 18, 19, 20, 21
14. 103, 104, 105, 106, 107
15. 765, 766, 767, 768, 769
16. 4, 7, 10, 7, 5, 2, 7
17. 15, 13, 10, 7, 6, 9, 10
18. 3, 5, 8, 13, 21
19. 1, 4, 9, 16, 25
20. 79, 90, 95, 95, 96
21. 70, 81, 95, 79, 85
22. 1, 2, 3, 3, 3, 4, 5
23. 0, 1, 1, 2, 3, 4, 16, 21

In Problems 24–35, find the range and the standard deviation (correct to two decimal places). If you do not have a calculator, find the range and the variance.

24. 1, 2, 3, 4, 5

25. 17, 18, 19, 20, 21

26. 103, 104, 105, 106, 107

27. 765, 766, 767, 768, 769

28. 4, 7, 10, 7, 5, 2, 7

29. 15, 13, 10, 7, 6, 9, 10

30. 3, 5, 8, 13, 21

31. 1, 4, 9, 16, 25

32. 79, 90, 95, 95, 96

33. 70, 81, 95, 79, 85

34. 1, 2, 3, 3, 3, 4, 5

35. 0, 1, 1, 2, 3, 4, 16, 21

36. Compare Problems 24–27. What do you notice about the range and standard deviation?

37. By looking at Problems 12–15 and 24–27, and discovering a pattern, find the mean and the standard deviation of the numbers 217,849, 217,850, 217,851, 217,852, and 217,853.

 B Problems

38. Find the mean, the median, and the mode of the following salaries of employees of the Moe D. Lawn Landscaping Company:

Salary	Frequency
$25,000	4
28,000	3
30,000	2
45,000	1

39. G. Thumb, the leading salesperson for the Moe D. Lawn Landscaping Company, turned in the following summary of sales for the week of October 23–28:

Date	Number of clients contacted by G. Thumb
Oct. 23	12
Oct. 24	9
Oct. 25	10
Oct. 26	16
Oct. 27	10
Oct. 28	21

   Find the mean, median, and mode.

40. Find the mean, the median, and the mode of the following scores:

Test Score	Frequency
90	1
80	3
70	10
60	5
50	2

41. A class obtained the following scores on a test:

Score	Frequency
90	1
80	6
70	10
60	4
50	3
40	1

Find the mean, the median, the mode, and the range for the class.

42. A class obtained the following test scores:

Score	Frequency
90	2
80	4
70	9
60	5
50	3
40	1
30	2
0	4

Find the mean, median, mode, and range for the class.

43. The county fair reported the following total attendance (in thousands).

Year:	1996	1995	1994	1993	1992	1991
Attendance:	366	391	358	373	346	364

Find the mean, median, and mode for the attendance figures (rounded to the nearest thousand).

44. The following salaries for the executives of a certain company are known:

Position	Salary
President	$170,000
1st VP	140,000
2nd VP	120,000
Supervising manager	54,000
Accounting manager	40,000
Personnel manager	40,000
Department manager	30,000
Department manager	30,000

Find the mean, the median, and the mode. Which measure seems to best describe the average executive salary for the company?

Use the nutritional information about candy bars given in Table 10.7 to answer the questions in Problems 45–50.

TABLE 10.7   Nutritional Information about Candy Bars

Company	Serving size (grams)	Total size (grams)	Calories from fat	Total fat (grams)
Almond Joy	36	180	90	10
Baby Ruth	60	280	110	12
Butterfinger	42	200	70	8
Clark Bar	50	240	90	10
5th Avenue	57	280	110	12
Heath Bar	40	210	110	13
Hershey choc. bar	43	230	120	13
Hershey choc. w/alm.	41	230	120	14
100 Grand	43	200	70	8
Kit Kat	42	220	110	12
Milky Way	61	280	100	11
Mr. Good Bar	49	280	160	18
Nestle milk choc.	41	220	110	13
Nut Rageous	45	250	140	15
Peppermint Patty	42	170	35	4
Reese's Cup	45	240	130	14
Rolo	54	230	110	12
Snickers	59	280	120	14
3 Musketeers	60	260	70	8
Twix	57	280	130	14

45. What is the mean calories from fat?

46. What is the median (in grams) of total fat?

47. What is the mode (in grams) of the serving size?

48. Divide the serving size into quartiles.

49. What are the mean and standard deviation for total size?

50. Draw a box plot for the calories from fat.

Find the standard deviation (rounded to the nearest unit) for the data indicated in Problems 51–56.

51. Problem 38   52. Problem 39   53. Problem 40

54. Problem 41   55. Problem 42   56. Problem 43

57. The number of miles driven on each of five tires was 17,000, 19,000, 19,000, 20,000, and 21,000. Find the mean, the range, and the standard deviation (rounded to the nearest unit) for these mileages.

58. Roll a single die until all six numbers occur at least once. Repeat the experiment 20 times. Find the mean, the median, the mode, and the range of the number of tosses.

59. Roll a pair of dice until all 11 numbers occur at least once. Repeat the experiment 20 times. Find the mean, the median, the mode, and the range of the number of tosses.

▲ **Problem Solving**

60. The example about averages of bowling scores at the beginning of this section is an instance of what is known as *Simpson's paradox*. The following example illustrates this paradox.

	Player A			Player B		
	At bat	Hits	Avg.	At bat	Hits	Avg.
Against right-handed pitchers	202	45	0.223	250	58	0.232
Against left-handed pitchers	250	71	0.284	108	32	0.296
Overall	452	116	0.257	358	90	0.251

Notice that Player A has a better overall batting average than Player B but yet is worse against both right-handed pitchers and left-handed pitchers. The following is an algebraic statement of Simpson's paradox:

Consider two populations for which the overall rate r of occurrence of some phenomenon in population A is greater than the corresponding rate R in population B. Suppose that each of the two populations is composed of the same two categories C_1 and C_2, and the rates of occurrence of the phenomenon for the two categories in population A are r_1 and r_2, and in population B are R_1 and R_2. If $r_1 < R_1$ and $r_2 < R_2$, despite the fact that $r > R$, then Simpson's paradox is said to have occurred.

Relate the variables in this statement to the numbers in the example.

61. An example of Simpson's paradox (see Problem 60) from real life is the fact that the overall federal income tax rate increased from 1974 to 1978, but decreased for each bracket. Make up a fictitious example to show how this might be possible.

62. An example of Simpson's paradox (see Problem 60) from real life is that the overall subscription renewal rate for *American History Illustrated* magazine increased from January to February but the rate decreased for each category of subscriber. Make up a fictitious example to show how this might be possible.

63. Let Q_1, Q_2, and Q_3 be the quartiles (from smallest to largest) for a large population of scores. *True or false* (give reasons):

$$Q_2 - Q_1 = Q_3 - Q_2$$

▲ **Individual Research**

64. If you roll a pair of dice 36 times, the expected number of times for rolling each of the numbers is given in the accompanying table. A graph of these data is shown in Figure 10.27.

a. Find the mean, the variance, and the standard deviation for this model.

Outcome	Expected Frequency
2	1
3	2
4	3
5	4
6	5
7	6
8	5
9	4
10	3
11	2
12	1

▲ Figure 10.27 **Distribution for rolling a pair of dice**

b. Roll a pair of dice 36 times. Construct a table and a graph similar to the ones shown in Figure 10.27. Find the mean, the variance, and the standard deviation for your experiment.

c. Compare the results of parts **a** and **b**. If this is a class problem, you might wish to pool data from the entire class before making the comparison.

65. Prepare a report or exhibit showing how statistics are used in baseball.

66. Prepare a report or exhibit showing how statistics are used in educational testing.

67. Prepare a report or exhibit showing how statistics are used in psychology.

68. Prepare a report or exhibit showing how statistics are used in business. Use a daily report of transactions on the New York Stock Exchange. What inferences can you make from the information reported?

69. Investigate the work of Adolph Quetelet, Francis Galton, Karl Pearson, R. A. Fisher, and Florence Nightingale. Prepare a report or an exhibit of their work in statistics.

70. "We need privacy and a consistent wind," said Wilbur. "Did you write to the Weather Bureau to find a suitable location?" "Well," replied Orville, "I received this list of possible locations and Kitty Hawk, North Carolina, looks like just what we want. Look at this . . . ." However, Orville and Wilbur spent many days waiting in frustration after they arrived in Kitty Hawk, because the winds weren't suitable. The Weather Bureau's information gave the averages, but the Wright brothers didn't realize that an acceptable average can be produced by unacceptable extremes. Write a paper explaining how it is possible to have an acceptable average produced by unacceptable extremes.

10.3 THE NORMAL CURVE

The cartoon in the margin suggests that most people do not like to think of themselves or their children as having "normal intelligence." But what do we mean by *normal* or *normal intelligence*?

News Clip

Baseball, as everyone knows, is played on a field with a bat and a ball. But baseball is also played on paper and computers with numbers and decimals. The name of this game-within-a-game is statistics, and to some fans, it is more engrossing and more real than the action on the field. Most sports keep statistics, but baseball statistics are in a league by themselves.

Wright Brothers' first flight at Kitty Hawk, North Carolina

Suppose we survey the results of 20 children's scores on an IQ test. The scores (rounded to the nearest 5 points) are 115, 90, 100, 95, 105, 95, 105, 105, 95, 125, 120, 110, 100, 100, 90, 110, 100, 115, 105, and 80. We can find $\bar{x} = 103$ and $s \approx 10.65$. A frequency graph of these data is shown in Figure 10.28a. If we consider 10,000 scores instead of only 20, we might obtain the frequency distribution shown in Figure 10.28b.

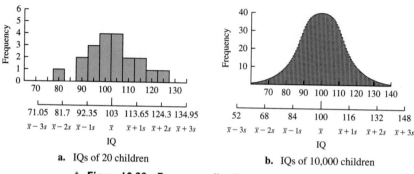

a. IQs of 20 children   b. IQs of 10,000 children

▲ Figure 10.28   Frequency distributions for IQ scores

The data illustrated in Figure 10.28 approximate a commonly used curve called a *normal frequency curve*, or simply a **normal curve**. (See Figure 10.29.)

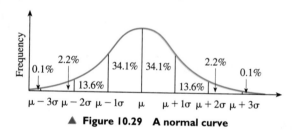

▲ Figure 10.29   A normal curve

If we obtain the frequency distribution of a large number of measurements (as with IQ), the corresponding graph tends to look normal, or **bell-shaped.** The normal curve has some interesting properties. In it, the mean, the median, and the mode all have the same value, and all occur exactly at the center of the distribution; we denote this value by the Greek letter mu (μ). The standard deviation for this distribution is σ (sigma). Roughly 68% of all values lie within the region from 1 standard deviation below to 1 standard deviation above the mean. About 95% lie within 2 standard deviations on either side of the mean, and virtually all (99.8%) values lie within 3 standard deviations on either side. These percentages are the same for all normal curves, regardless of the particular mean or standard deviation.

The normal distribution is a **continuous** (rather than a discrete) **distribution,** and it extends indefinitely in both directions, never touching the x-axis. It is symmetric about a vertical line drawn through the mean, μ. Graphs of this curve for several choices of σ are shown in Figure 10.30.

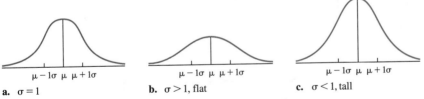

a. $\sigma = 1$ **b.** $\sigma > 1$, flat   **c.** $\sigma < 1$, tall

▲ **Figure 10.30 Variations of normal curves**

▰▰▰ EXAMPLE 1

Predict the distribution of IQ scores of 1,000 people if we assume that IQ scores are normally distributed, with a mean of 100 and a standard deviation of 15.

Solution   First, find the breaking points around the mean. For $\mu = 100$ and $\sigma = 15$:

$$\mu + \sigma = 100 + 15 = 115 \qquad \mu - \sigma = 100 - 15 = 85$$
$$\mu + 2\sigma = 100 + 2 \cdot 15 = 130 \qquad \mu - 2\sigma = 100 - 2 \cdot 15 = 70$$
$$\mu + 3\sigma = 100 + 3 \cdot 15 = 145 \qquad \mu - 3\sigma = 100 - 3 \cdot 15 = 55$$

We use Figure 10.29 to find that 34.1% of the scores will be between 100 and 115 (i.e., between μ and $\mu + 1\sigma$):

$$0.341 \times 1,000 = 341$$

About 13.6% will be between 115 and 130 (between $\mu + 1\sigma$ and $\mu + 2\sigma$):

$$0.136 \times 1,000 = 136$$

About 2.2% will be between 130 and 145 (between $\mu + 2\sigma$ and $\mu + 3\sigma$):

$$0.022 \times 1,000 = 22$$

About 0.1% will be above 145 (more than $\mu + 3\sigma$):

$$0.001 \times 1,000 = 1$$

Scores	%	Expected number
55 or below	0.1	1
55⁺–70	2.2	22
70⁺–85	13.6	136
85⁺–100	34.1	341
100⁺–115	34.1	341
115⁺–130	13.6	136
130⁺–145	2.2	22
Above 145	0.1	1
Totals	100.0	1,000

The distribution for intervals below the mean is identical, since the normal curve is the same to the left and to the right of the mean. The distribution is shown in the margin. ▰

▰▰▰ EXAMPLE 2

Suppose that an instructor "grades on a curve." Show the grading distribution on an examination of 45 students, if the scores are normally distributed with a mean of 73 and a standard deviation of 9.

Solution   Grading on a curve means determining students' grades according to the percentages shown in Figure 10.31.

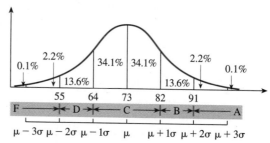

▲ **Figure 10.31   Grade distribution for a class "graded on a curve"**

We calculate these numbers as shown:

Calculation	Scores	Grade	Calculation	Number
Two or more standard deviations above the mean	91–100	A	$0.023 \times 45 = 1.04$	1
$\mu + 2\sigma = 73 + 2 \cdot 9 = 91$ $\mu + 1\sigma = 73 + 1 \cdot 9 = 82$	82–90	B	$0.136 \times 45 = 6.12$	6
Mean: $\mu = 73$ $\mu - 1\sigma = 73 - 1 \cdot 9 = 64$	64–81	C	$0.682 \times 45 = 30.69$	31
$\mu - 2\sigma = 73 - 2 \cdot 9 = 55$	55–63	D	same as B	6
Two or more standard deviations below the mean	0–54	F	same as A	1

Grading on a curve means that the person with the top score in this class of 45 receives an A; the next 6 ranked persons (from the top) receive B grades; the bottom score in the class receives an F; the next 6 ranked persons (from the bottom) receive D grades; and finally, the remaining 31 persons receive C grades. Notice that the majority of the class ($34.1\% + 34.1\% = 68.2\%$) will receive an "average" C grade.

Z-Scores

Sometimes we want to know the percent of occurrence for scores that do not happen to be 1, 2, or 3 standard deviations from the mean. For Example 2, we can see from Figure 10.31 that 34.1% of the scores are between the mean and 1 standard deviation above the mean. Suppose we wish to find the percent of scores that are between the mean and 1.2 standard deviations above the mean. To find this percent, we use the table given in Appendix C for this purpose. First, we introduce some terminology.

We use *z-scores* (sometimes called *standard scores*) to determine how far, in terms of standard deviations, that a given score is from the mean of the distribution. For example, if $z = 1$, we can use the table in Appendix C to find the percent of scores between the mean and the value that is 1 standard deviation above the mean. (From Figure 10.29, we know that the percent is 34.1% or 0.341.) In Appendix C, look in the row labeled (at the left) 1.0 and in the column headed 0.00: the entry is 0.3413, which is 34.13%. For $z = 1.2$, look at the entry in the row marked 1.2 and the 0.00 column: it is 0.3849.

Suppose we want to find $z = 1.68$; look at the row labeled 1.6 and the column headed 0.08 to find the entry 0.4535. This means that 45.35% of the values in a normal distribution are between the mean and 1.68 standard deviations above the mean.

We use the z-score to translate any normal curve into a *standard normal curve* (the particular normal curve with a mean of 0 and a standard deviation of 1) by using the definition.

---

z-Score

If x is a value from a normal distribution with mean μ and standard deviation σ, then its **z-score** is

$$z = \frac{x - \mu}{\sigma}$$

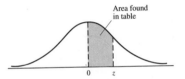

Area found in table

▲ **Figure 10.32 Percent of occurrence using a z-score**

The z-score is used with the table in Appendix C to find the percent of occurrences between the mean and the number of standard deviations above the mean specified by the z-score, as illustrated in Figure 10.32.

■■■■■ EXAMPLE 3

The Eureka Light Bulb Company tested a new line of light bulbs and found their lifetimes to be normally distributed, with a mean life of 98 hours and a standard deviation of 13 hours.

a. What percentage of bulbs will last less than 72 hours?

b. What percentage of bulbs will last less than 100 hours?

c. What is the probability that a bulb selected at random will last longer than 111 hours?

d. What is the probability that a bulb will last between 106 and 120 hours?

Solution Draw a normal curve with mean 98 and standard deviation 13, as shown in Figure 10.33.

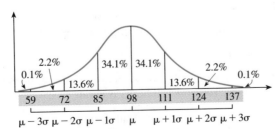

▲ **Figure 10.33 Light bulb lifetimes are normally distributed.**

For this example, we are given $\mu = 98$ and $\sigma = 13$.

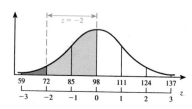

a. For $x = 72$, $z = \dfrac{72 - 98}{13} = -2$

This is 2 standard deviations below the mean (which is the same as the z-score). The percentage we seek is shown in blue:

About 2.3% (2.2% + 0.1% = 2.3%) will last less than 72 hours.

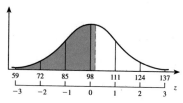

b. For $x = 100$, $z = \dfrac{100 - 98}{13} \approx 0.15$

This is 0.15 standard deviation above the mean (which is the same as the z-score). From the table in Appendix C, we find 0.0596 (this is shown in green). Since we want the percent of values less than 100, we must add 50% for the numbers below the mean (shown in blue). The percentage we seek is then $0.5000 + 0.0596 = 0.5596$ or about 56.0%.

c. For $x = 111$, $z = \dfrac{111 - 98}{13} = 1$

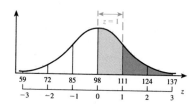

This is one standard deviation above the mean, which is the same as the z-score. The percentage we seek is shown in blue. We know that approximately 15.9% (13.6% + 2.2% + 0.1% = 15.9%) of the bulbs will last longer than 111 hours, so

$$P(\text{bulb life} > 111 \text{ hours}) \approx 0.159$$

We can also use the table in Appendix C. For $z = 1.00$, the table entry is 0.3413, and we are looking for the area to the right, so we compute

$$0.5000 - 0.3413 = 0.1587$$

d. We first find the z-scores using $\mu = 98$ and $\sigma = 13$. For $x = 106$,

$$z = \frac{106 - 98}{13} \approx 0.62$$

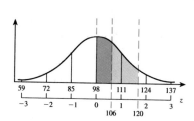

From the table in Appendix C, the area between this z-score and the mean is 0.2324 (shown in green). For $x = 120$,

$$z = \frac{120 - 98}{13} \approx 1.69$$

From the table in Appendix C, the area between this z-score and the mean is 0.4545. The desired answer (shown in yellow) is approximately

$$0.4545 - 0.2324 = 0.2221$$

Since percent and probability are the same, we see the probability that the life of the bulb is between 106 and 120 hours is about 22.2%.   ▬

Sometimes data do not fall into a normal distribution, but are **skewed,** which means their distribution has more tail on one side or the other. For example, Figure

10.34a shows that the 1941 scores on the SAT exam (when the test was first used) were normally distributed. However, by 1990, the scale had become skewed to the left, as shown in Figure 10.34b.

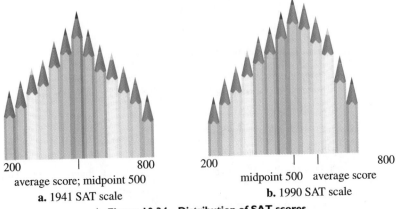

a. 1941 SAT scale

200   800
average score; midpoint 500

200   800
midpoint 500 average score

a. 1941 SAT scale   **b.** 1990 SAT scale

▲ **Figure 10.34 Distribution of SAT scores**

In a normal distribution, the mean, median, and mode all have the same value, but if the distribution is skewed, the relative positions of the mean, median, and mode would be as shown in Figure 10.35.

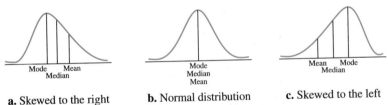

a. Skewed to the right   **b.** Normal distribution   **c.** Skewed to the left

▲ **Figure 10.35 Comparison of three distributions**

PROBLEM SET 10.3

▲ **A Problems**

1. **IN YOUR OWN WORDS** What does it mean to "grade on a curve"?

2. **IN YOUR OWN WORDS** What is a normal curve?

What percent of the total population is found between the mean and the z-scores given in Problems 3–14?

3. $z = 1.4$ 4. $z = 0.3$ 5. $z = 2.43$ 6. $z = 1.86$

7. $z = 3.25$ 8. $z = -0.6$ 9. $z = -2.33$ 10. $z = -0.50$

11. $z = -0.46$ 12. $z = -1.19$ 13. $z = -2.22$ 14. $z = -3.41$

In Problems 15–21, suppose that people's heights (in centimeters) are normally distributed, with a mean of 170 and a standard deviation of 5. We find the heights of 50 people.

15. How many would you expect to be between 165 and 175 cm tall?

16. How many would you expect to be between 170 and 180 cm?

17. How many would you expect to be taller than 168 cm?

18. How many would you expect to be taller than 176 cm?

19. How many would you expect to be shorter than 171 cm?

20. What is the probability that a person selected at random is taller than 163 cm?

21. What is the variance in heights for this experiment?

In Problems 22–26 suppose that, for a certain exam, a teacher grades on a curve. It is known that the mean is 50 and the standard deviation is 5. There are 45 students in the class.

22. How many students should receive a C?

23. How many students should receive an A?

24. What score would be necessary to obtain an A?

25. If an exam paper is selected at random, what is the probability that it will be a failing paper?

26. What is the variance in scores for this exam?

27. The graph shown in Figure 10.36 is from the February 1991 issue of *Scientific American*. If the curve in the middle is a standard normal curve, describe the upper curve (labeled **a**) and then describe the lower curve (labeled **b**).

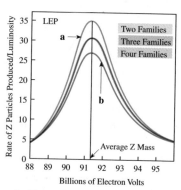

▲ **Figure 10.36   Normal curves**

▲ **B Problems**

28. **IN YOUR OWN WORDS** In a distribution that is skewed to the right, which has the greatest value — the mean, median, or mode? Explain why this is the case.

29. **IN YOUR OWN WORDS** In a distribution that is skewed to the left, which has the greatest value — the mean, median, or mode? Explain why this is the case.

30. Suppose that the breaking strength of a rope (in pounds) is normally distributed, with a mean of 100 pounds and a standard deviation of 16. What is the probability that a certain rope will break when subjected to a force of 130 pounds or less?

31. The diameter of an electric cable is normally distributed, with a mean of 0.9 inch and a standard deviation of 0.01 inch. What is the probability that the diameter will exceed 0.91 inch?

32. Suppose that the annual rainfall in Ferndale, California, is known to be normally distributed, with a mean of 35.5 inches and a standard deviation of 2.5 inches. About 2.3% of the time, the annual rainfall will exceed how many inches?

33. About what percent of the years will it rain more than 36 inches in Ferndale (see Problem 32)?

34. In Problem 32, what is the probability that the rainfall in a given year will exceed 30.5 inches in Ferndale?

35. The diameter of a pipe is normally distributed, with a mean of 0.4 inch and a variance of 0.0004. What is the probability that the diameter of a randomly selected pipe will exceed 0.44 inch?

36. The diameter of a pipe is normally distributed, with a mean of 0.4 inch and a variance of 0.0004. What is the probability that the diameter of a randomly selected pipe will exceed 0.41 inch?

37. The breaking strength (in pounds) of a certain new synthetic is normally distributed, with a mean of 165 and a variance of 9. The material is considered defective if the breaking strength is less than 159 pounds. What is the probability that a single, randomly selected piece of material will be defective?

38. Suppose the neck size of men is normally distributed, with a mean of 15.5 inches and a standard deviation of 0.5 inch. A shirt manufacturer is going to introduce a new line of shirts. Assume that if your neck size falls between two shirt sizes, you purchase the next larger shirt size. How many of each of the following sizes should be included in a batch of 1,000 shirts?

    a. 14   **b.** 14.5   c. 15   **d.** 15.5   **e.** 16   **f.** 16.5   **g.** 17

39. A package of Toys Galore Cereal is marked "Net Wt. 12 oz." The actual weight is normally distributed, with a mean of 12 oz and a variance of 0.04.

    a. What percent of the packages will weigh less than 12 oz?
    b. What weight will be exceeded by 2.3% of the packages?

40. Instant Dinner comes in packages with weights that are normally distributed, with a standard deviation of 0.3 oz. If 2.3% of the dinners weigh more than 13.5 oz, what is the mean weight?

▲ Individual Research

41. Select something that you think might be normally distributed (for example, the ring size of students at your college). Next, select 100 people and make the appropriate measurements (in this example, ring size). Calculate the mean and standard deviation. Illustrate your finding using a bar graph. Do your data appear to be normally distributed?

10.4 CORRELATION AND REGRESSION

In mathematical modeling, it is often necessary to deal with numerical data and to make assumptions regarding the relationship between two variables. For example, you may want to examine the relationship between

> IQ and salary
> Study time and grades
> Age and heart disease
> Runner's speed and runner's brand of shoe
> Math grades in the 8th grade and amount of TV viewing
> Teachers' salaries and beer consumption

All are attempts to relate two variables in some way or another. If it is established that there is a **correlation,** then the next step in the modeling process is to identify the nature of the relationship. This is called **regression analysis.** In this section we consider only linear relationships. It is assumed that you are familiar with the slope–intercept form of the equation of a line (namely, $y = mx + b$). If you need a review of graphing lines, you should look at Section 11.1.

We are interested in finding a *best-fitting line* by a technique called the **least squares method.** The derivations of the results and the formulas given in this section are, for the most part, based on calculus, and are therefore beyond the scope of this book. We will focus instead on how to use and interpret the formulas.

The first consideration is one of correlation. We want to know whether two variables are related. Let us call one variable x and the other y. These variables can be represented as ordered pairs (x, y) in a graph called a **scatter diagram.**

▬▬ EXAMPLE 1

A survey of 20 students compared the grade received on an examination with the length of time the student studied. Draw a scatter diagram to represent the data in the table.

Student number	1	2	3	4	5	6	7	8	9	10	11	12	13	14	15	16	17	18	19	20
Length of study time (nearest 5 min.)	30	40	30	35	45	15	15	50	30	0	20	10	25	25	25	30	40	35	20	15
Grade (100 possible)	72	85	75	78	89	58	71	94	78	10	75	43	68	60	70	68	82	75	65	62

Solution   Let x be the study time (in minutes) and let y be the grade (in points). The graph is shown in Figure 10.37.

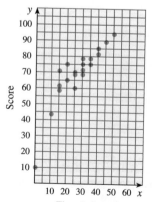

▲ **Figure 10.37   Correlation between study time and grade**

Correlation is a measure to determine whether there is a statistically significant linear relationship between two variables. Intuitively, it should assign a measure consistent with the scatter diagrams as shown in Figure 10.38.

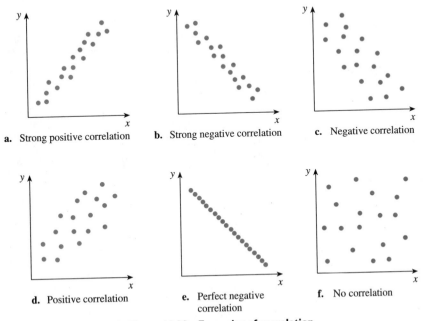

a. Strong positive correlation   **b.** Strong negative correlation   **c.** Negative correlation

d. Positive correlation   **e.** Perfect negative correlation   **f.** No correlation

▲ **Figure 10.38   Examples of correlation**

Such a measure, called the **linear correlation coefficient,** r, is defined so that it has the following properties:

1. r measures the correlation between x and y.
2. r is between -1 and 1.
3. If r is close to 0, it means there is little correlation.
4. If r is close to 1, it means there is a strong positive correlation.
5. If r is close to -1, it means there is a strong negative correlation.

To write a formula for r, let n denote the number of pairs of data present; and as before:

Σx	denotes the sum of the x-values.
Σx^2	means square the x-values and then sum.
$(\Sigma x)^2$	means sum the x-values and then square.
Σxy	means multiply each x-value by the corresponding y-value and then sum.
$n\Sigma xy$	means multiply n times Σxy.
$(\Sigma x)(\Sigma y)$	means multiply Σx times Σy.

> **Formula for Linear Correlation Coefficient**
>
> The linear correlation coefficient, r, is
>
> $$r = \frac{n\Sigma xy - (\Sigma x)(\Sigma y)}{\sqrt{n(\Sigma x^2) - (\Sigma x)^2} \sqrt{n(\Sigma y^2) - (\Sigma y)^2}}$$

■■■■ EXAMPLE 2

Find r for the following data (from Example 1):

Student number	1	2	3	4	5	6	7	8	9	10	11	12	13	14	15	16	17	18	19	20
Length of study time (nearest 5 min.)	30	40	30	35	45	15	15	50	30	0	20	10	25	25	25	30	40	35	20	15
Grade (100 possible)	72	85	75	78	89	58	71	94	78	10	75	43	68	60	70	68	82	75	65	62

Solution

Study Time, x	Score, y	xy	x^2	y^2
30	72	2,160	900	5,184
40	85	3,400	1,600	7,225
30	75	2,250	900	5,625
35	78	2,730	1,225	6,084
45	89	4,005	2,025	7,921
15	58	870	225	3,364
15	71	1,065	225	5,041
50	94	4,700	2,500	8,836
30	78	2,340	900	6,084
0	10	0	0	100
20	75	1,500	400	5,625
10	43	430	100	1,849
25	68	1,700	625	4,624
25	60	1,500	625	3,600
25	70	1,750	625	4,900
30	68	2,040	900	4,624
40	82	3,280	1,600	6,724
35	75	2,625	1,225	5,625
20	65	1,300	400	4,225
15	62	930	225	3,844
Total 535	1,378	40,575	17,225	101,104
↑	↑	↑	↑	↑
Σx	Σy	Σxy	Σx^2	Σy^2

$$r = \frac{n\Sigma xy - (\Sigma x)(\Sigma y)}{\sqrt{n(\Sigma x^2) - (\Sigma x)^2}\sqrt{n(\Sigma y^2) - (\Sigma y)^2}}$$

$$= \frac{20(40,575) - (535)(1,378)}{\sqrt{20(17,225) - (535)^2}\sqrt{20(101,104) - (1,378)^2}}$$

$$= \frac{74,270}{\sqrt{58,275}\sqrt{123,196}}$$

$$\approx 0.877$$

Example 2 shows a very strong positive correlation. But if r for this example had been 0.46, would we still have been able to conclude that there is a strong correlation? This question is a topic of major concern in statistics. The term **significance level** is used to denote the cutoff between results attributed to chance and results attributed to significant differences. Table 10.8 gives *critical values* for determining whether two variables are correlated. If $|r|$ is greater than the given table value, then you may assume a correlation exists between the variables. If you use the column labeled $\alpha = 0.05$, then we say the significance level is 5%. This means that the probability is 0.05 that you will say the variables are correlated when, in fact, the results should be attributed to chance, and similarly for a significance level of 1% ($\alpha = 0.01$). For Example 2, since $n = 20$, we see in Table 10.8 that $r = 0.877$ shows a significant linear correlation at both the 1% and 5% levels. On the other hand, if $r = 0.46$ and $n = 20$, then there is a significant linear correlation at a 5% level, but not at a 1% level.

▬▬ EXAMPLE 3

Find the critical value of the linear correlation coefficient for 13 pairs of data and a significance level of 0.05.

Solution From Table 10.8, the critical value is 0.553. For $n = 13$, any value greater than $r = 0.553$ or less than -0.553 is evidence of linear correlation. ▬

▬▬ EXAMPLE 4

If $r = -0.85$ and $n = 13$, are the variables correlated at a significance level of 1%?

Solution For $n = 13$ and $\alpha = 0.01$, the Table 10.8 table entry is 0.684. Since r is negative and $|r| > 0.684$, we see that there is a negative linear correlation. ▬

▬▬ EXAMPLE 5

The following table shows a sample of some past annual mean salaries for college professors, along with the annual per capita beer consumption (in gallons) for Americans. Find the correlation coefficient.

Year	1995	1993	1991	1990	1989	1988
Mean teacher salary	$58,400	$57,400	$55,800	$53,200	$50,100	$47,200
Per capita beer consumption	22.6	22.8	23.1	24.0	25.4	24.7

TABLE 10.8
Correlation coefficient, r

n	$\alpha = 0.05$	$\alpha = 0.01$
4	0.950	0.999
5	0.878	0.959
6	0.811	0.917
7	0.754	0.875
8	0.707	0.834
9	0.666	0.798
10	0.632	0.765
11	0.602	0.735
12	0.576	0.708
13	0.553	0.684
14	0.532	0.661
15	0.514	0.641
16	0.497	0.623
17	0.482	0.606
18	0.468	0.590
19	0.456	0.575
20	0.444	0.561
25	0.396	0.505
30	0.361	0.463
35	0.335	0.430
40	0.312	0.402
45	0.294	0.378
50	0.279	0.361
60	0.254	0.330
70	0.236	0.305
80	0.220	0.286
90	0.207	0.269
100	0.196	0.256

The derivation of this table is beyond the scope of this course. It shows the critical values of the *Pearson correlation coefficient*.

Calculator Comment

Once again, we remind you to use a calculator to carry out calculations such as the ones shown in Example 5. You should check the work shown in this example on your own calculator. Keep in mind that many calculators have built-in function keys for finding the correlation. ▲

Solution

$$n = 6 \qquad\qquad \Sigma x^2 = 1.738705 \times 10^{10}$$
$$\Sigma x = 322{,}100 \qquad \Sigma y^2 = 3{,}395.46$$
$$\Sigma y = 142.6 \qquad\quad (\Sigma x)^2 = 1.0374841 \times 10^{11}$$
$$\Sigma xy = 7{,}632{,}720 \qquad (\Sigma y)^2 = 20{,}334.76$$

$$r = \frac{n\Sigma xy - (\Sigma x)(\Sigma y)}{\sqrt{n(\Sigma x^2) - (\Sigma x)^2}\,\sqrt{n(\Sigma y^2) - (\Sigma y)^2}}$$

$$= \frac{6(7{,}632{,}720) - (322{,}100)(142.6)}{\sqrt{6(1.738705 \times 10^{10}) - 1.0374841 \times 10^{11}}\,\sqrt{6(3{,}395.46) - 20{,}334.76}}$$

$$\approx -0.9151$$

The number $r \approx -0.9151$ is statistically significant at the 5% level, but not at the 1% level. This means that there is a negative correlation between these two phenomena. ▬

The significance of the negative correlation in Example 5 implies that professors' salaries and beer drinking are related, but be careful! In any event, the techniques in this chapter can be used only to establish a *statistical* linear relationship. *We cannot establish the existence or absence of any inherent cause-and-effect relationship on the basis of a correlation analysis.*

Best-Fitting Line

The final step in our discussion of correlation is to find the best-fitting line. That is, we want to find a line $y' = mx + b$ so that the sum of the distances of the data points from this line will be as small as possible. (We use y' instead of y to distinguish between the actual second component, y, and the predicted y-value, y'.) Since some of these distances may be positive and some negative, and since we do not want large opposites to "cancel each other out," we minimize the sum of the *squares* of these distances. Therefore, the regression line is sometimes called the *least squares line.*

Least Squares Line

The **least squares** (or **regression**) **line** is $y' = mx + b$, where

$$m = \frac{n(\Sigma xy) - (\Sigma x)(\Sigma y)}{n(\Sigma x^2) - (\Sigma x)^2} \qquad b = \frac{\Sigma y - m(\Sigma x)}{n}$$

This is the *line of best fit.*

▬▬▬ **EXAMPLE 6**

Find the best-fitting line for the data in Example 1:

Student number	1	2	3	4	5	6	7	8	9	10	11	12	13	14	15	16	17	18	19	20
Length of study time (nearest 5 min.)	30	40	30	35	45	15	15	50	30	0	20	10	25	25	25	30	40	35	20	15
Grade (100 possible)	72	85	75	78	89	58	71	94	78	10	75	43	68	60	70	68	82	75	65	62

Solution   For Example 1, $n = 20$ and

535	1,378	40,575	17,225	101,104
↑	↑	↑	↑	↑
Σx	Σy	Σxy	Σx^2	Σy^2

$$m = \frac{n(\Sigma xy) - (\Sigma x)(\Sigma y)}{n(\Sigma x^2) - (\Sigma x)^2} = \frac{20(40,575) - (535)(1,378)}{20(17,225) - (535)^2} \approx 1.27447$$

$$b = \frac{\Sigma y - m(\Sigma x)}{n} = \frac{1,378 - 1.27447(535)}{20} \approx 34.8078$$

Then we approximate the best-fitting line as the line with equation $y' = 1.3x + 35$. This line (along with the data points) is shown in Figure 10.39.

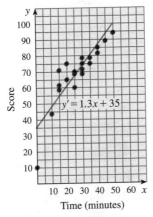

▲ **Figure 10.39   Regression line**

━━━ **EXAMPLE 7**

Use the regression line in Example 6 to predict the score of a person who studied $\frac{1}{2}$ hour.

Solution $x = 30$ minutes, so $y' = 1.3x + 35 = 1.3(30) + 35 = 74$.   ▬

A Final Word of Caution.   You should use the regression line only if r indicates that there is a significant linear correlation, as shown in Table 10.8.

PROBLEM SET   10.4

▲ **A Problems**

1. **IN YOUR OWN WORDS** What do we mean by correlation?

2. **IN YOUR OWN WORDS** How do you find a linear correlation coefficient?

3. **IN YOUR OWN WORDS** How do you determine whether there is a linear correlation between two variables x and y?

4. **IN YOUR OWN WORDS** What is a least squares line?

5. **IN YOUR OWN WORDS** Discuss the correlation shown by the following chart taken from the November 1987 issue of *Scientific American*.

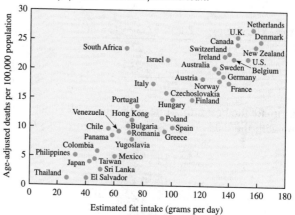

6. **IN YOUR OWN WORDS** Discuss the correlation shown by the following chart taken from the April 1991 issue of *Scientific American*.

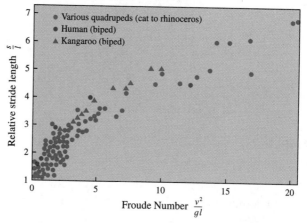

In Problems 7–18 a sample of paired data gives a linear correlation coefficient r. In each case, use Table 10.8 to determine whether there is a significant linear correlation.

7. $n = 10, r = 0.7$; 1% level

8. $n = 30, r = 0.4$; 1% level

9. $n = 30, r = 0.4$; 5% level

10. $n = 15, r = -0.732$; 5% level

11. $n = 35, r = -0.413$; 1% level

12. $n = 50, r = -0.3416$; 1% level

13. $n = 25, r = 0.521$; 1% level

14. $n = 100, r = -0.4109$; 1% level

15. $n = 40, r = -0.416$; 5% level

16. $n = 20, r = -0.214$; 5% level

17. $n = 10, r = -0.56$; 1% level

18. $n = 10, r = 0.7$; 5% level

Draw a scatter diagram and find r for the data shown in each table in Problems 19–24.

19. x	y		20. x	y		21. x	y		22. x	y		23. x	y		24. x	y
4	0		1	1		1	30		0	25		85	80		10	20
5	−10		2	5		3	22		1	19		90	40		20	48
10	−10		3	8		3	19		2	16		100	30		30	60
10	−20		4	13		5	15		3	12		102	28		30	58
						8	10		4	10		105	25		50	70
														60	75

Find the regression line for the data points given in Problems 25–30.

25. x	y		26. x	y		27. x	y		28. x	y		29. x	y		30. x	y
4	0		1	1		1	30		0	25		85	80		10	20
5	−10		2	5		3	22		1	19		90	40		20	48
10	−10		3	8		3	19		2	16		100	30		30	60
10	−20		4	13		5	15		3	12		102	28		30	58
						8	10		4	10		105	25		50	70
														60	75

▲ B Problems

31. U.S. wine consumption over the last few years (in gallons per person per year) is shown in the following table.

Year	1988	1989	1990	1991	1993	1995
Wine consumption	2.24	2.09	2.05	1.85	1.74	1.80

Compare these numbers with the U.S. beer consumption (found in Example 5, page 733). Are beer drinking and wine drinking correlated? What is the correlation coefficient?

32. A group of ten people were selected and given a standard IQ test. These scores were then compared with their high school grades.

IQ	117	105	111	96	135	81	103	99	107	109
Grade (GPA)	3.1	2.8	2.5	2.8	3.4	1.9	2.1	3.2	2.9	2.3

Find r and determine whether it is statistically significant at the 1% level.

33. A new computer circuit was tested and the times (in nanoseconds) required to carry out different subroutines were recorded.

Difficulty	1	2	2	3	4	5	5	5
Time	10	11	13	8	15	18	21	19

Find r and determine whether it is statistically significant at the 1% level.

34. Find the regression line for the data in Problem 32. Assume x is the IQ and y is the GPA.

35. Find the regression line for the data in Problem 33. Assume x is the difficulty level and y is the time.

36. The following data are measurements of temperature ($x =$ °F) and chirping frequency ($y =$ chirps per second) for the striped ground cricket. Is there a correlation between

temperature and chirping frequency, and if so, is it significant at the 5% or the 1% level?

| Temperature | 31.4 | 22.0 | 34.1 | 29.1 | 27.0 | 24.0 | 20.9 | 27.8 | 20.8 | 28.5 | 26.4 | 28.1 | 27.0 | 28.6 | 24.6 |
| Frequency | 20.0 | 16.0 | 19.8 | 18.4 | 17.1 | 15.5 | 14.7 | 17.1 | 15.4 | 16.2 | 15.0 | 17.2 | 16.0 | 17.0 | 14.4 |

37. The following data are the number of years of full-time education (x) and the annual salary in thousands of dollars (y) for 15 persons. Is there a correlation between education and salary, and if so, is it significant at the 5% or the 1% level?

| Education | 20 | 27 | 28 | 18 | 13 | 18 | 9 | 16 | 16 | 12 | 12 | 19 | 16 | 14 | 13 |
| Salary | 35.2 | 24.6 | 23.7 | 33.3 | 24.4 | 33.4 | 11.2 | 32.3 | 25.1 | 22.1 | 18.9 | 37.8 | 25.9 | 28.4 | 29.6 |

38. A researcher chooses and interviews a group of 15 male workers in an automobile plant. The researcher then gives a score (x) ranging from 1 to 20 based on a scale of patriotism — the higher the score, the more patriotic the person appears to be. Each person is then given a written test and is scored (y) on their patriotism. Is there a correlation between the researcher score and the test score, and if so, is it significant at the 5% or the 1% level?

| Researcher | 10 | 14 | 15 | 17 | 17 | 18 | 18 | 19 | 16 | 18 | 20 | 12 | 14 | 9 | 17 |
| Test | 15 | 12 | 19 | 8 | 9 | 16 | 17 | 6 | 11 | 14 | 12 | 11 | 10 | 12 | 6 |

39. A bank records the number of mortgage applications and its own prevailing interest rate (at the first of the month) for each of 16 consecutive months. Is there a correlation between the interest rate (x) and the number of applicants (y), and if so, it is significant at the 5% or the 1% level?

| Interest | 9.5 | 9.9 | 10.0 | 10.5 | 11.0 | 11.5 | 11.0 | 12.0 | 12.0 | 12.5 | 13.0 | 13.5 | 13.0 | 12.5 | 11.5 | 11.5 |
| Number | 27 | 29 | 25 | 25 | 19 | 20 | 17 | 13 | 15 | 10 | 10 | 6 | 5 | 5 | 11 | 14 |

40. Find the best-fitting line for the data in Problem 36.

41. Find the best-fitting line for the data in Problem 37.

42. Find the best-fitting line for the data in Problem 38.

43. Find the best-fitting line for the data in Problem 39.

10.5   SAMPLING

The first three sections of this chapter dealt with what is called **descriptive statistics,** which is concerned with the accumulation of data, measures of central tendency, and dispersion. A second branch of statistics is **inferential statistics,** which is concerned with making generalizations or predictions about a population based on a sample from that population.

A **sample** is a group of items chosen to represent a larger group. The larger group is called a **population.** Thus, the sample is a proper subset of the population. The sample is analyzed; then based on this analysis, some conclusion about the entire population is made. Sampling necessarily involves some error, because the sample and the population are not identical. A great deal of effort in statistics is devoted to the methodology of sampling. Here are some common sampling procedures:

Simple random sampling: The sample is obtained in a way that allows every member of the population to have the same chance of being chosen. This is the type of sampling we assumed so far in Chapters 9–10.

Systematic sampling: The sample is obtained by drawing every kth item on a list or production line. The first item is determined by using a random number.

Cluster sampling: This procedure is applied on a geographical basis. The result is sometimes known as an **area sample.**

Stratified sampling: The entire population is divided into parts, called *strata*, according to some factor (such as sex, age, or income). When a population has varied characteristics, it is desirable to separate the population into homogeneous strata, and then take a random sample from each stratum.

The inference drawn from a poll can, of course, be wrong, so statistics is also concerned with estimating the error involved in predictions based on samples. In 1936 the *Literary Digest* predicted that Alfred Landon would defeat Franklin D. Roosevelt — who was subsequently reelected president by a landslide. (The magazine ceased publication the following year.) In 1948 the *Chicago Daily Tribune* drew an incorrect conclusion from its polls and declared in a headline that Thomas Dewey had just been elected president over Harry S. Truman. And in 1976, the *Milwaukee Sentinel* printed the erroneous headline shown here about the Wisconsin Democratic primary, again based on the result of its polls.

In an attempt to minimize error in their predictions, statisticians follow very careful procedures:

Step 1. Propose some hypothesis about a population.

Step 2. Gather a sample from the population.

Step 3. Analyze the data.

Step 4. Accept or reject the hypothesis.

Suppose that you want to decide whether a certain coin is a "fair" coin. You decide to test the hypothesis, "This is a fair coin," by flipping the coin 100 times. This provides a *sample*. Suppose the result is

Heads: 55   Tails: 45

Do you accept or reject the hypothesis that "This coin is fair"? The expected number of heads is 50, but certainly a fair coin might well produce the results obtained.

As you can readily see, two types of errors are possible:

Type I: Rejection of the hypothesis when it is true

Type II: Acceptance of the hypothesis when it is false

How can we minimize the possibility of making either error? Let's carry this example further, and repeat the experiment of flipping the coin 100 times:

Trial Number:	1	2	3	4	5	6	. . .
Number of Heads:	55	52	54	57	59	55	. . .

If the coin is fair and we repeat the experiment a large number of times, it can be shown mathematically that the distribution should be normal, with a mean of 50 and a standard deviation of 5, as shown in Figure 10.40. The question is whether or not to accept the *unknown* test coin as fair.

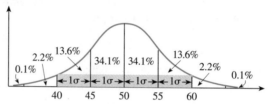

▲ **Figure 10.40   Given normal distribution for the number of heads upon flipping a fair coin 100 times**

Suppose that you are willing to accept the coin as fair only if the number of heads falls between 45 and 55 (that is, within 1 standard deviation of the expected number). If you adopt this standard, you know you will be correct 68% of the time if the coin is fair. How do you know this? Look at Figure 10.40, and note that 34.1% of the results are within $+1\sigma$ and 34.1% are within -1σ of the mean; the total is $34.1\% + 34.1\% = 68.2\%$.

But a friend says, "Yes, you will be correct 68% of the time, but you will also be rejecting a lot of fair coins!" You respond, "But suppose that a coin really is a bad coin (it really favors heads), with a mean number of heads of 60 and a standard deviation of 5. If I adopted the same standard ($\pm 1\sigma$), I'd be accepting all the coins in the shaded region of Figure 10.41."

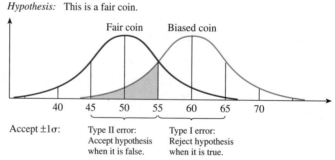

▲ **Figure 10.41   Comparison of Type I and Type II errors**

As you can see, a decrease in Type I error probability increases the Type II error probability, and vice versa. Deciding which type of error to minimize depends on the stakes involved and on some statistical calculations that go beyond the scope of this course.

Consider a company that produces two types of valves. The first type is used in jet aircraft, and the failure of this valve might cause many deaths. A sample of the valve is taken and tested, and the company must accept or reject the entire shipment on the basis of these test results. Under these circumstances, the company would rather reject many good valves than accept a bad one. On the other hand, the second valve is used in toy airplanes; the failure of this valve would merely cause the crash of the model. In this case, the company wouldn't want to reject too many good valves, so they would minimize the probability of rejecting the good valves.

Many times we read of a poll in the newspaper or hear of it on the evening news, and a percentage is given. For example, "The candidate was favored by 48% of those sampled." What is not often stated (or is stated only in small print) is that there is a margin of error and a confidence level. For example, "The margin of error is 4 percent at a confidence level of 95%." This means that we are 95% confident that the actual percent of people who favor the candidate is between 44% and 52%.

PROBLEM SET 10.5

▲ A Problems

1. **IN YOUR OWN WORDS** When you hear the word *statistics*, what comes to mind? Look up the word in a dictionary, and see if you wish to supplement or change any part of your answer.

2. **IN YOUR OWN WORDS** Explain the difference between descriptive and inferential statistics.

▲ Individual Research

3. Five identical containers (shoe boxes, paper cups, etc.) must be prepared for this problem, with contents as follows: There are five boxes containing red and white items (such as marbles, poker chips, or colored slips of paper).

Box	Contents
#1	15 red and 15 white
#2	30 red and 0 white
#3	25 red and 5 white
#4	20 red and 10 white
#5	10 red and 20 white

Select one of the boxes at random so that you don't know its contents.

Step 1. Shake the box.

Step 2. Select one marker, note the result, and return it to the box.

Step 3. Repeat the first two steps 20 times with the same box.

a. What do you think is inside the box you have sampled?

    b. Could you have guessed the contents by repeating the experiment five times? Ten times? Do you think you should have more than 20 observations per experiment? Discuss.

4. Conduct a survey asking the following questions.

    a. Do you doodle?

    b. If you doodle, which of the following do you doodle?

        (1) circles, curves, or spirals
        (2) squares, rectangles, or other straight-line designs
        (3) people or human features
        (4) symbols, such as stars, arrows, or other objects
        (5) words, letters, or numerals
        (6) other

    c. When do you most often doodle?

        (1) while at a meeting or class
        (2) while on the phone at home
        (3) when solving a problem
        (4) when you have nothing else to do
        (5) when writing a letter
        (6) at meals

5. Conduct a survey to determine the major worry of college students.

6. Conduct a survey to determine whether there is a significant correlation between math scores in the 8th grade and amount of TV viewing.

7. Suppose that John hands you a coin to flip and wants to bet on the outcome. Now, John has tried this sort of thing before, and you suspect that the coin is "rigged." You decide to test this hypothesis by taking a sample. You flip the coin twice, and it is heads both times. You say, "Aha, I knew it was rigged!" John replies, "Don't be silly. Any coin can come up heads twice in a row." The following scheme was devised by mathematician John von Neumann to allow fair results even if the coin is somewhat biased. The coin is flipped twice. If it comes up heads both times or tails both times, it is flipped twice again. If it comes up heads–tails, this will decide the outcome in favor of the first party; and if it comes up tails–heads, this will decide the outcome in favor of the second party. Show that this will result in a fair toss even if the coins are biased.

"If you want to understand nature, you must be conversant with the language in which nature speaks to us."
Richard Feynman, *Everybody Counts: A Report to the Nation*

SEPTEMBER

1

2 **Maurice Fréchet** (1878), statistics, topology

3 **James Sylvester** (1814), determinants, p. 839

4

5 **Girolamo Saccheri** (1667), non-Euclidean geometry

6 **François Viète** (1540), trigonometry, algebra, geometry

7 **Compte de Buffon** (1707), probability

8 **Marin Mersenne** (1588), number theory, prolific writer

9

10 **Charles Peirce** (1839), four-color problem

11 **Franz Neumann** (1798), electromagnetic induction

12

13

14

15

16

17 **Georg Bernhard Riemann** (1826), integral calculus, p. 530

18 **Adrien-Marie Legendre** (1752), function theory

19

20

21

22

23

24 **Girolamo Cardano** (1501), theory of equations, imaginary numbers, p. 636

25

26

27

28

29

30

BIRTHDAYS

Biographical Sketch

"Statistics is . . . closely related to our daily life."

Ann Watkins

Ann Watkins was born in Los Angeles, California, on January 10, 1949. She fell in love with statistics as a graduate student, not independently of the fact that a professor at UCLA told her that statistics could make her fortune, as it had his. While waiting, she is a professor of mathematics at California State University, Northridge, and past editor of *The College Mathematics Journal*.

Professor Watkins likes teaching statistics. "When I am teaching statistics, I always feel that the students are learning something that will be valuable to them. After arithmetic, statistics is the branch of mathematics that is most closely related to our daily life. Every day I can get some example out of the newspaper to use in class. Did you ever wonder what it means when an election poll says there is a margin of error of plus or minus 3%?"

Professor Watkins believes that it's possible for a person to understand statistics who does not have good skills in algebra. "Learning statistics depends more on a person's verbal and reasoning skills than on his or her ability to manipulate symbols."

Book Reports

Write a 500-word report on one of the following books:

▲ *The Mathematical Experience*, Philip J. Davis and Reuben Hersh (Boston: Houghton Mifflin, 1981).

▲ *How to Lie with Statistics*, Darrell Huff (New York: Norton, 1954).

Important Terms

Average [10.2]
Bar graph [10.1]
Box plot [10.1]
Circle graph [10.1]
Classes [10.1]
Continuous distribution [10.3]
Correlation [10.4]
Decile [10.2]
Frequency [10.1]
Frequency distribution [10.1]
Graph [10.1]

Grouped frequency distribution [10.1]
Interval [10.1]
Least-squares line [10.4]
Least squares method [10.4]
Line graph [10.1]
Linear correlation coefficient [10.4]
Mean [10.2]
Measures of central tendency [10.2]
Measures of dispersion [10.2]

Measures of position [10.2]
Median [10.2]
Mode [10.2]
Normal curve [10.3]
Percentile [10.2]
Pictograph [10.1]
Pie chart [10.1]
Population [10.5]
Quartile [10.2]
Range [10.2]
Regression analysis [10.4]
Sample [10.5]

Scatter diagram [10.4]
Significance level [10.4]
Skewed distribution [10.3]
Standard deviation [10.2]
Stem-and-leaf plot [10.1]
Type I error [10.5]
Type II error [10.5]
Variance [10.2]
Weighted mean [10.2]
z-score [10.3]

Important Ideas

Different uses of statistics [10.1]
Recognize and describe misuses of graphs [10.1]
Decide on an appropriate measure of central tendency [10.2]
Formula for weighted mean [10.2]
Formula for standard deviation [10.2]
Percentages for the standard deviations on a normal curve [10.3]

Formula for the linear correlation coefficient [10.4]
Formulas for the slope and y-intercept of the least-squares (or regression) line [10.4]
Difference between a sample and a population [10.5]
Type I and Type II sampling errors [10.5]
Difference between descriptive and inferential statistics [10.5]

Types of Problems

Prepare a frequency distribution [10.1]
Draw a bar graph [10.1]
Draw a line graph [10.1]
Draw a stem-and-leaf plot [10.1]
Draw a circle graph [10.1]
Draw a pictograph [10.1]
Read and interpret bar graphs, line graphs, circle graphs, and pictographs [10.1]
Find the mean, median, and mode for a set of data [10.2]
Find the range, standard deviation, and variance for a set of data [10.2]
Find the expected numbers for ranges of a normally distributed set of data [10.3]

Determine the probability of falling within a certain range of a normally distributed set of data [10.3]
Decide whether there is a significant linear correlation between two given variables [10.4]
Discuss the type of correlation for a given data set [10.4]
Determine whether there is a significant linear correlation, given the nuimber of items and the correlation coefficient [10.4]
Find the correlation coefficient for a given set of data [10.4]
Make an inference about a population by taking a sample [10.5]

CHAPTER 10   Review Questions

1. **a.** Make a frequency table for the following results of tossing a coin 40 times:

   HTTTT HHTHH TTHHT THTHT HHTTH THHTH HHTHT TTTTT

   b. Draw a bar graph for the number of heads and tails given in part **a**.

2. The table in the margin shows the U.S. government's expenditures for social welfare.
   a. Draw a bar graph to represent these data.
   b. Suppose that you had the following viewpoint: *Social welfare expenditures rose only $0.3 trillion from 1989 to 1992.* Draw a graph that shows very little increase in the expenditures.
   c. Suppose you had the following viewpoint: *Social welfare expenditures rose $308,000,000,000 from 1989 to 1992.* Draw a graph that shows a tremendous increase in the expenditures.
   d. In view of parts **b** and **c**, discuss the possibilities of using statistics to mislead or support different views.

Year	Expenditures (billions)
1989	$ 957
1990	$1,105
1991	$1,162
1992	$1,265

3. A small grocery store stocked several sizes of Copycat cola last year. The sales figures are shown in the table. If the store manager decides to cut back the variety and stock only one size, which measure of central tendency will be most useful in making this decision?

Size	No. of cases sold
6 oz	5
10 oz	21
12 oz	24

4. A student's scores in a certain math class are 72, 73, 74, 85, and 91. Find the mean, the median, and the mode. Which measure of central tendency is most representative of the student's scores?

5. The 1993–1994 school superintendents' salaries in Sonoma County are shown:

Rohnert Park, $90,000
Analy, $84,522
Rincon Valley, $81,000
Gravenstein, $73,378
Old Adobe, $71,000
Wright, $67,691
Harmony, $64,009
Alexander Valley, $54,528
West Side, $44,474
Montgomery, $39,000
Santa Rosa, $86,958
Winsor, $83,500
Petaluma, $80,814

Bellevue, $72,057
Guerneville, $69,965
Healdsburg, $67,562
Waugh, $60,000
Cinnabar, $54,500
Oak Grove, $43,625
Kenwood, $28,550
Sonoma Valley, $86,004
SCOE, $83,473
Forestville, $76,944
Sebastopol, $69,123
Piner-Olivet, $67,300
Two Rock, $57,760

Monte Rio, $49,016
Fort Ross, $42,515
Dunham, $16,950
Cloverdale, $84,665
Mark West, $82,543
Roseland, $74,370
Twin Hills, $71,564
Geyersville, $69,000
Bennett Valley, $65,600
Wilmar, $57,000
Liberty, $47,412
Horicon, $41,800

Find the mean, the median, and the mode. Which measure of central tendency is most representative of the superintendents' salaries?

6. The 1993–1994 NBA ticket prices are shown:

New York, $39.66
Orlando, $31.28
Portland, $27.81
Washington, $24.83
LA Clippers, $22.45
Minnesota, $20.74
Chicago, $36.45
San Antonio, $29.93
Seattle, $27.78

Miami, $24.62
Charlotte, $22.43
Indiana, $20.71
Phoenix, $36.06
New Jersey, $29.62
Houston, $27.36
Atlanta, $24.26
Dallas, $22.30
LA Lakers, $32.84

Detroit, $29.41
Cleveland, $26.89
Sacramento, $22.89
Milwaukee, $21.59
Boston, $31.45
Utah, $28.61
Golden State, $26.74
Philadelphia, $22.70
Denver, $21.14

Find the mean, the median, and the mode. Which measure of central tendency is most representative of the NBA ticket prices?

7. Consider the line graph showing the median family income in 1990 dollars.

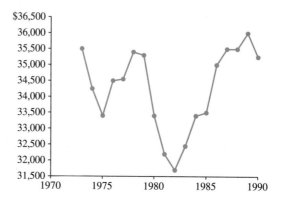

a. In which year (1973–1990) was the median family income the least?
b. In which year (1973–1990) was the median family income the greatest?
c. What was the approximate median family income in 1990?

The following table compares age with blood pressure. Use this table in Problems 8–9.

Age (in years)	20	25	30	40	50	35	68	55
Blood pressure	85	91	84	93	100	86	94	92

8. Draw a scatter diagram and find the regression line for this set of data.

9. Find the linear correlation coefficient and determine whether the variables are significantly correlated at either the 1% or 5% level.

10. Read the *Dear Abby* column in the margin. Abby's answer was consoling and gracious, but not very statistical. If pregnancy durations have a normal distribution with a mean of 266 days and a standard deviation of 16 days, what is the probability of having a 314-day pregnancy?

Revisiting the Real World . . .

IN THE REAL WORLD "I signed up for Hunter in History 17," said Ben. "Why did you sign up for her?" asked Ted. "Don't you know she has given only three A's in the last 14 years?!" "That's just a rumor. Hunter grades on a curve," Ben replied. "Don't give me that 'on a curve' stuff," continued Ted, "I'll bet you don't even know what that means. Anyway, my sister-in-law had her last year and said it was so bad that . . . ."

Commentary

Assume that Hunter does, indeed, grade on a curve. Show the grading distribution if 200 students take an examination, with a mean of 73 and a standard deviation of 9. What are the grades Hunter would give?

Score	Grade	Number	Percentage of Class
91 or above	A	5	2.3%
82–90	B	27	13.6%
64–81	C	136	68.2%
55–63	D	27	13.6%
Below 55	F	5	2.3%

As students we often think of the grades in a class from a selfish standpoint (that is, what grade did I get) rather than what it *means* to *grade on a curve* or how class grades are determined from the raw data. Consider the following test scores: 96, 92, 92, 89, 88, 87, 87, 87, 87, 80, 79, 79, 78, 76, 76, 76, 74, 73, 72, 72, 71, 71, 70, 66, 66, 60, 53, 20. Let's find the mean (rounded to the nearest unit):

$$\bar{x} = \frac{\Sigma(\text{test scores})}{n} = \frac{2{,}117}{28} = 75.60714286 \approx 76$$

The median is the mean of the 14th and 15th items (in order from highest to lowest): 76. The mode is the most frequently occurring item, 87. Which measure of central tendency is most appropriate for these test scores? Under ordinary circumstances, the most appropriate is the mean. The range for the scores is 76 points. We will now calculate the standard deviation for this class of 28 scores.

Score	(Deviation from the mean)^2
96	$(96 - 76)^2 = 20^2 = 400$
92	$(92 - 76)^2 = 16^2 = 256$ times 2 (occurs 2 times)
89	$(89 - 76)^2 = 13^2 = 169$
88	$(88 - 76)^2 = 12^2 = 144$
87	$(87 - 76)^2 = 11^2 = 121$ times 4 (occurs 4 times)
80	$(80 - 76)^2 = 4^2 = 16$
79	$(79 - 76)^2 = 3^2 = 9$ times 2
78	$(78 - 76)^2 = 2^2 = 4$
76	$(76 - 76)^2 = 0^2 = 0$ times 3
74	$(74 - 76)^2 = (-2)^2 = 4$
73	$(73 - 76)^2 = (-3)^2 = 9$
72	$(72 - 76)^2 = (-4)^2 = 16$ times 2
71	$(71 - 76)^2 = (-5)^2 = 25$ times 2
70	$(70 - 76)^2 = (-6)^2 = 36$
66	$(66 - 76)^2 = (-10)^2 = 100$ times 2
60	$(60 - 76)^2 = (-16)^2 = 256$
53	$(53 - 76)^2 = (-23)^2 = 529$
20	$(20 - 76)^2 = (-56)^2 = 3{,}136$

NOTE: We work with the rounded mean of 76 for this illustration; if you use a calculator, work with calculator accuracy, namely, $\bar{x} = 75.60714286$; in that case the standard deviation will be slightly different.

Sum of squared deviations: 5,999; standard deviation: $\sqrt{\dfrac{5{,}999}{28 - 1}} \approx 14.91$.

If you use a calculator for this standard deviation, the result is 14.90050779; the discrepancy is because of round-off error in the above list.

Working in small groups is typical of most work environments, and being able to work with others to communicate specific ideas is an important skill to learn. Work with three or four other students to submit a single report based on each of the following questions.

1. a. Make a grading scale for the 28 scores given in the Commentary.
   b. Find the mean, median, mode, range, and standard deviation of the ages of the actors winning the best actor award at the Academy Awards. See Table 10.3 on page 691.

2. Toss a toothpick onto a hardwood floor 1,000 times as described in Figure 10.42, or toss 1,000 toothpicks, one at a time, onto the floor. Let ℓ be the length of the toothpick and d be the distance between the parallel lines determined by the floorboards.

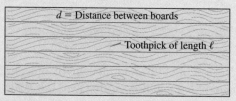

▲ Figure 10.42   Buffon's Needle Problem. Equipment needed: A box of toothpicks (of uniform length) and a large sheet of paper with equidistant parallel lines. A hardwood floor works very well instead of using a sheet of paper. The length of a toothpick should be less than the perpendicular distance between the parallel lines.

   a. Guess the probability p that a toothpick will cross a line. *Do this before you begin the experiment.* The members of your group should reach a consensus before continuing.
   b. Perform the experiment and find p empirically. That is, to find p, divide the number of toothpicks crossing a line by the number of toothpicks tossed (1,000 in this case).
   c. By direct measurement, find ℓ and d.
   d. Calculate 2ℓ and pd, and $\dfrac{2\ell}{pd}$.
   e. Formulate a conclusion. This is an experiment known as Buffon's needle problem.

3. You are interested in knowing the number and ages of children (0–18 years) in a part (or all) of your community. You will need to sample 50 families, finding the number of children in each family and the age of each child. It is important that you select the 50 families at random. How to do this is a subject of a course in statistics. For this problem, however, follow these steps:

   Step 1. Determine the geographic boundaries of the area with which you are concerned.

   Step 2. Consider various methods for selecting the families at random. For example, could you:
      (i) select the first 50 homes at which someone is at home when you call?
      (ii) select 50 numbers from a phone book that covers the same geographic boundaries as those described in step 1?
      Using (i) or (ii) could result in a biased sample. Can you guess why this might

be true? In a statistics course you might explore other ways of selecting the homes. For this problem, use one of these methods.

Step 3. Consider different ways of asking the question. Can the way the family is approached affect the response?

Step 4. Gather your data.

Step 5. Organize your data. Construct a frequency distribution for the children, with integral values from 0 to 18.

Step 6. Find out the number of families who actually live in the area you've selected. If you can't do this, assume that the area has 1,000 families.

  a. What is the average number of children per family?

  b. What percent of the children are in the first grade (age 6)?

  c. If all the children ages 12–15 are in junior high, how many are in junior high for the geographic area you are considering?

  d. See if you can actually find out the answers to parts **b** and **c**, and compare these answers with your projections.

  e. What other inferences can you make from your data?

11 The Nature of Graphs and Functions

> Algebra is but written geometry and geometry is but figured algebra.
>
> G. B. Halsted

In the Real World . . .

"Well, Billy, can I make it?" asked Evel. "If you can accelerate to the proper speed, and if the wind is not blowing too much, I think you can." answered Billy. "It will be one huge money-maker, but I want some assurance that it can be done!" retorted Evel. Suppose a daredevil wants to attempt to skycycle ride across the Snake River. What factors are necessary to determine whether he will make it?

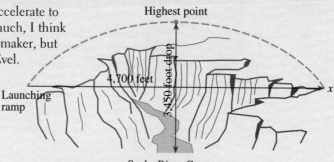

Snake River Canyon

Contents

Perspective

One of the most revolutionary ideas in the history of human thought was the joining together of algebra and geometry. It was basically a simple idea (as are most revolutionary ideas) — namely, representing an ordered pair from algebra as a point on a flat surface, called a plane. This was summarized by the great mathematician Lagrange, who said, "As long as algebra and geometry proceeded along separate paths, their advance was slow and their application limited. But when these sciences joined company, they drew from each other fresh vitality and thenceforth marched on at rapid pace toward perfection."

In this chapter, we move from elementary mathematics, to presenting an overview of the underpinnings of college mathematics. Some of the basic building blocks for advanced mathematics are functions, functional notation, and limits. These ideas are, in turn, the building blocks for a discussion of the nature of calculus. It is not the intent of this chapter to teach calculus, but to discuss the *nature* of calculus in a context that will enable you to see why its invention was not only necessary, but inevitable.

11.1   GRAPHING LINES

Solving Equations with Two Variables

Let's consider an equation with two variables — say, x and y. If there are two values x and y that make an equation true, then we say that the ordered pair (x, y) **satisfies** the equation and that it is a **solution** of the equation.

▬ EXAMPLE 1

Tell whether each ordered pair satisfies the equation $3x - 2y = 7$.

a. $(1, -2)$ **b.** $(-2, 1)$ **c.** $(-1, -5)$

Solution You should substitute each pair of values of x and y into the given equation to see whether the equation is true or false. If it is true, the ordered pair is a solution.

a. $(1, -2)$ means $x = 1$ and $y = -2$:

$$3x - 2y = 3(1) - 2(-2)$$
$$= 3 + 4$$
$$= 7 \quad \text{Since } 7 = 7, \text{ we see that } (1, -2) \text{ is a solution.}$$

b. $(-2, 1)$ means $x = -2$ and $y = 1$ (compare with part **a** and notice that the order is important in determining which variable takes which value):

$$3x - 2y = 3(-2) - 2(1)$$
$$= -6 - 2$$
$$= -8$$

Since $-8 \neq 7$, we see that $(-2, 1)$ is not a solution.

c. $(-1, -5)$:

$$3x - 2y = 3(-1) - 2(-5)$$
$$= -3 + 10$$
$$= 7 \quad \text{The ordered pair } (-1, -5) \text{ is a solution.}$$ ▬

Cartesian Coordinate System

Have we found all the ordered pairs that satisfy the equation $3x - 2y = 7$? Can you find others? Let's represent this information by *drawing a graph*. Recall from elementary algebra that a **Cartesian coordinate system** is formed by drawing two perpendicular number lines, called **axes**. The point of intersection is called the **origin**, and the plane is divided into four parts called **quadrants**, which are labeled as shown in Figure 11.1.

Point $P(a, b)$ in Figure 11.1 is found in the plane by counting a units from the origin in a horizontal direction (either positive or negative), and then b units (either positive or negative) in a vertical direction. The number a is sometimes called the **abscissa** and b the **ordinate** of the point P. Together, a and b are called the **coordinates** of the point P.

News Clip

As long as algebra and geometry proceeded along separate paths, their advance was slow and their application limited. But when these sciences joined company, they drew from each other fresh vitality and thenceforth marched on at rapid pace toward perfection.

Lagrange

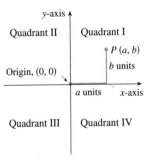

▲ **Figure 11.1   Cartesian coordinate system**

Historical Note

RENÉ DESCARTES
(1596–1650)

René Descartes, the person after whom we name the coordinate system, was a person of frail health. He had a lifelong habit of lying in bed until late in the morning or even the early afternoon. It is said that these hours in bed were probably his most productive. During one of these periods, it is presumed, Descartes made the discovery of the coordinate system. ▲

Suppose we plot the ordered pairs in Example 1a and 1c along with four others, as shown in Figure 11.2a.

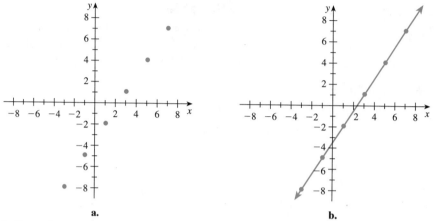

a.

b.

▲ **Figure 11.2** **(a) Points satisfying $3x - 2y = 7$ and (b) the line passing through the points**

Graphing a Line

The process of graphing a line requires that you find ordered pairs that make an equation true. To do this, *you* must choose convenient values for x and then solve the resulting equation to find a corresponding value for y.

▬▬▬ EXAMPLE 2

Find three ordered pairs that satisfy the equation $y = -2x + 3$.

Solution   *You* choose any x-value — say, $x = 1$. Substitute this value into the given equation to find a corresponding y-value:

$$y = -2x + 3 \qquad \text{Given equation}$$
$$= -2(1) + 3 \qquad \text{Substitute chosen value.}$$
$$= -2 + 3$$
$$= 1$$

You choose this value.
↓

The first ordered pair is $(1, 1)$.
↑

You find this value by substitution into the equation.

Choose a second value — say, $x = 2$. Then,

$$y = -2x + 3 \qquad \text{Start with given equation.}$$
$$= -2(2) + 3 \qquad \text{Substitute.}$$
$$= -1 \qquad \text{Simplify.}$$

The second ordered pair is $(2, -1)$.

Choose a third value — say, $x = -1$. Then,

$y = -2x + 3$

$\quad = -2(-1) + 3$

$\quad = 5$

The third ordered pair is $(-1, 5)$.

Have we found *all* the ordered pairs that satisfy the equation $y = -2x + 3$? Can you find others? Suppose we plot the ordered pairs in Example 2, along with three additional points, as shown in Figure 11.3. Do you notice anything about the arrangement of these points in the plane? Suppose that we draw a line passing through these points, as shown in Figure 11.4. This line represents the set of *all* ordered pairs that satisfy the equation.

Graphing a Line by Plotting Points

If we carry out the process of finding three ordered pairs that satisfy an equation of a line, and then we draw the line through those points, we say that we are **graphing** the line, and the final set of points we have drawn represents the **graph** of the line.

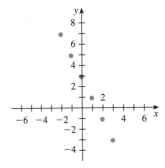

▲ **Figure 11.3 Points that satisfy $y = -2x + 3$**

Procedure for Graphing a Line

To graph a line, find two ordered pairs that lie on the line. These two points determine the line. Find a third point as a check. Draw the line (using a straightedge) passing through these three points.

If the three points don't lie on a straight line, then you have made an error.

■■■■ EXAMPLE 3

Graph $y = 2x + 2$.

Solution It is generally easier to pick x and find y.

If $x = 0$: $y = 2x + 2$ Given equation

$\qquad\quad = 2(0) + 2$ Substitute.

$\qquad\quad = 2$ Simplify.   Plot the point $(0, 2)$.

If $x = 1$: $y = 2x + 2$

$\qquad\quad = 2(1) + 2$

$\qquad\quad = 4$ Plot $(1, 4)$.

If $x = 2$: $y = 2x + 2$

$\qquad\quad = 2(2) + 2$

$\qquad\quad = 6$ Plot $(2, 6)$.

Draw the line passing through the three plotted points, as shown in Figure 11.5.

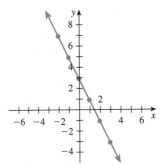

▲ **Figure 11.4 Representation of all points that satisfy $y = -2x + 3$**

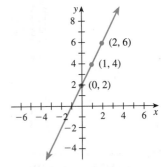

▲ **Figure 11.5 Graph of $y = 2x + 2$**

Graphing a Line by Slope–Intercept

Let's introduce some terminology.

> ### Standard Form
>
> The **standard form** of the equation of a line is
>
> $$Ax + By + C = 0$$
>
> where A, B, and C are real numbers (A and B not both 0) and (x, y) is any point on the line. This equation is called:
>
> a **first-degree** equation in two variables, or
>
> a **linear** equation in two variables.

Graphing a line by plotting points is generally not the most efficient way. Consider the standard form of the equation of a line:

$$Ax + By + C = 0$$

If $B \neq 0$, we can solve for y:

$$By = -Ax - C$$

$$y = -\frac{A}{B}x - \frac{C}{B}$$

Notice that if $m = -\dfrac{A}{B}$ and $b = -\dfrac{C}{B}$, then the first-degree equation in two variables x and y (with $B \neq 0$) can be written in the form

$$y = mx + b$$

This is a very useful way of writing the equation of a line. Note that lowercase b in this form is not the same as capital B in the standard form. This form is easy to use when you *know* a value for x and want to find a value for y. We call x, the first component, the **independent variable** and calculate the value of y, the second component, called the **dependent variable.**

The points where a graph crosses the coordinate axes are usually easy to find, and they are often used to help sketch the curve. A **y-intercept** is a point where a graph crosses the y-axis and, consequently, it is a point with a first component of 0. When we say b is a y-intercept, we mean the curve crosses the y-axis at the point $(0, b)$. An **x-intercept** is a point where a graph crosses the x-axis, and it has a second component of 0. If we speak of an x-intercept of a, we mean the graph crosses the x-axis at the point $(a, 0)$.

The **slope** of a line passing through the points (x_1, y_1) and (x_2, y_2) is the steepness of that line, and is measured by rise divided by run:

$$m = \frac{\text{RISE}}{\text{RUN}} = \frac{\text{CHANGE IN THE VERTICAL DISTANCE}}{\text{CHANGE IN THE HORIZONTAL DISTANCE}} = \frac{y_2 - y_1}{x_2 - x_1}$$

Figure 11.6 shows how slope can be used to measure the steepness of a roof.

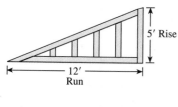

▲ **Figure 11.6   Cross-section of roof gables is an example of slope; it is called the *pitch* of a roof.**

> **Slope–Intercept Form**
>
> The first-degree equation
>
> $$y = mx + b$$
>
> where b is the y-intercept and m is the slope, is called the **slope–intercept form** of the equation of a line.

■ EXAMPLE 4

Graph the line $x + 2y = 6$.

Solution   Solve y to put the equation into slope–intercept form:

$$x + 2y = 6$$
$$2y = -x + 6 \qquad \text{Subtract } x \text{ from both sides.}$$
$$y = -\tfrac{1}{2}x + 3 \qquad \text{Divide both sides (all terms) by 2.}$$

The y-intercept is $(0, 3)$; plot this point.

The slope is $m = -\tfrac{1}{2}$; *start* at the y-intercept and count down 1 unit and over (rightward) 2 units:

$$m = \frac{\text{RISE}}{\text{RUN}} = -\frac{1}{2} = \frac{-1}{2}$$

Plot this point, which we call a **slope point,** meaning that it is found by counting out the slope (rise and run) from a given point. Using this slope point, as well as the y-intercept, we now draw the line as shown in Figure 11.7.

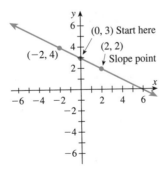

▲ **Figure 11.7   Graph of $x + 2y = 6$**

A general procedure can now be stated for graphing a line using the slope–intercept method. This procedure is shown in Figure 11.8.

■ EXAMPLE 5

Sketch the graph of $2x - 3y + 6 = 0$.

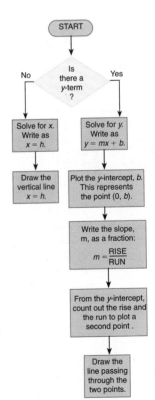

▲ **Figure 11.8   General procedure for graphing lines**

Calculator Comment

Many calculators available today do a variety of types of graphing problems. The cost is still somewhat prohibitive ($50–$70), but it is worthwhile to show how easily you can graph lines and other curves by using a calculator.

There is generally a button labeled $\boxed{Y=}$. Press this button and input the equation you wish to graph. For Example 5, we wish to graph

$$y = \tfrac{2}{3}x + 2$$

After inputting the equation, we usually obtain the graph by pressing a button labeled $\boxed{\text{GRAPH}}$. The graph on a TI-82 graphing calculator is shown.

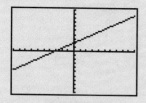

```
Y₁⊟(2/3)X+2
Xmin=-10 Ymin=-10
Xmax=10  Ymax=10
Xscl=1   Yscl=1
```

To see the coordinates of points on a graph, we can press $\boxed{\text{TRACE}}$, which displays the coordinates of the point on the curve corresponding to the location of the cursor. ▲

Solution *For most graphs*, the most efficient method is to solve the equation for y, and then find the y-intercept and the slope by inspection. This is the general procedure illustrated in Figure 11.8.

$$2x - 3y + 6 = 0$$
$$2x + 6 = 3y$$
$$y = \tfrac{2}{3}x + 2$$

Thus, $b = 2$; plot the y-intercept $(0, 2)$ as shown in Figure 11.9.

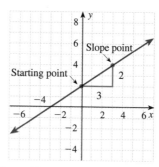

▲ **Figure 11.9   Graph of** $2x - 3y + 6 = 0$

Since $m = \tfrac{2}{3}$, we start at the y-intercept and move up 2 and over (rightward) 3; plot another point (which we have called a slope point). Draw the line passing through the two plotted points. ▬

Even though the slope–intercept form for graphing a line is the preferred method, sometimes it is not convenient to plot the y-intercept.

EXAMPLE 6

Sketch the graph of $7x + y + 28 = 0$.

Solution   Since $y = -7x - 28$, we have $m = -7$ and $b = -28$. The y-intercept is a little awkward to plot, so find some other point to use.

If $x = -4$, then

$$y = -7x - 28$$
$$= -7(-4) - 28 = 28 - 28 = 0$$

and $(-4, 0)$ is a point on the line. Write the slope as $m = -7 = \tfrac{-7}{1}$ and find a slope point by counting down 7 and over 1. Note that another slope point can be found by using $m = -7 = \tfrac{7}{-1}$. The graph is shown in Figure 11.10.

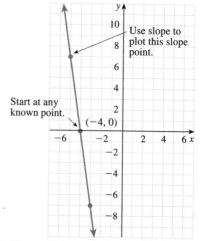

▲ **Figure 11.10   Graph of** $7x + y + 28 = 0$ ▬

In Example 6 note that $m = -7 = \frac{7}{-1}$ means "up 7 and back (leftward) 1," and $m = -7 = \frac{-7}{1}$ means "down 7 and over 1." A *positive slope* (as in Example 5) indicates that the line *increases* (goes uphill) from left to right, and a *negative slope* (as in Example 6) indicates that the line *decreases* (goes downhill) from left to right. A *zero slope* (as in the following example) indicates that the line is horizontal. For a vertical line, we say that the *slope does not exist*.

▬ **EXAMPLE 7**

Graph **a.** $y = 4$ **b.** $x = 3$

Solution

a. Since the only requirement is that y (the second component) equal 4, we see that there is no restriction on the choice for x. Thus, $(0, 4)$, $(1, 4)$, and $(-2, 4)$ all satisfy the equation that $y = 4$. If you plot and connect these points, you will see that the line formed is a **horizontal line**. A horizontal line has slope equal to zero.

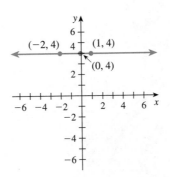

b. Since the only requirement is that x (the first component) equal 3, we see that there is no restriction on the choice for y. Thus, $(3, 2)$, $(3, -1)$, and $(3, 0)$ all satisfy the condition that $x = 3$. If you plot and connect these points, you will see that the line formed is a **vertical line**. A vertical line does not have any slope. Note that a line with no slope is vertical, but a line with zero slope is horizontal.

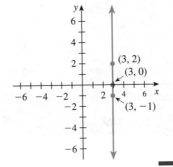

PROBLEM SET **11.1**

▲ **A Problems**

1. **IN YOUR OWN WORDS** What does it mean to solve a linear equation in two variables?

2. **IN YOUR OWN WORDS** What is a y-intercept? What is the slope?

3. **IN YOUR OWN WORDS** Outline a procedure for graphing a line by plotting points.

4. **IN YOUR OWN WORDS** Outline a procedure for graphing a line using the slope–intercept method.

5. **IN YOUR OWN WORDS** Contrast horizontal and vertical lines and their slopes.

Use the map of Venus shown in Figure 11.11 to name the landmarks at the locations specified in Problems 6–10.

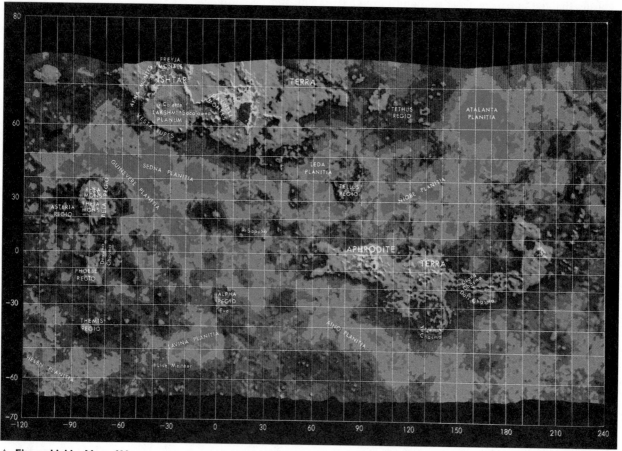

▲ **Figure 11.11   Map of Venus**

6. $(90, 0)$ 7. $(0, 65)$ 8. $(165, 65)$ 9. $(-80, 32)$ 10. $(-30, -50)$

For each equation in Problems 11–22, find three ordered pairs that satisfy the equation.

11. $y = x + 5$ 12. $y = 2x - 1$ 13. $y = 2x + 5$ 14. $y = x - 4$

15. $y = x - 1$ 16. $y = x + 1$ 17. $y = -2x + 1$ 18. $y = -3x + 1$

19. $y = 2x + 1$ 20. $y + 3x = 1$ 21. $3x + 4y = 8$ 22. $x + 2y = 4$

Graph the lines in Problems 23–34. These are the same equations as those in Problems 11–22.

23. $y = x + 5$ 24. $y = 2x - 1$ 25. $y = 2x + 5$ 26. $y = x - 4$

27. $y = x - 1$ 28. $y = x + 1$ 29. $y = -2x + 1$ 30. $y = -3x + 1$

31. $y = 2x + 1$ 32. $y + 3x = 1$ 33. $3x + 4y = 8$ 34. $x + 2y = 4$

Graph the lines through the given points and with the given slopes as indicated in Problems 35–46.

35. $(2, 5)$; $m = \frac{1}{3}$ 36. $(7, -3)$; $m = \frac{1}{5}$ 37. $(-4, -3)$; $m = \frac{3}{4}$

38. $(5, 0)$; $m = -\frac{2}{3}$ **39.** $(1, -1)$; $m = -\frac{1}{7}$ **40.** $(6, 3)$; $m = -\frac{2}{5}$

41. $(-1, 3)$; $m = 1$ **42.** $(1, -1)$; $m = 2$ **43.** $(1, 3)$; $m = 3$

44. $(2, 3)$; $m = 0$ **45.** $(2, 3)$; no slope **46.** $(2, 3)$; $m = 1.5$

▲ B Problems

Graph the lines in Problems 47–64.

47. $x + y + 4 = 0$ **48.** $3x + y + 2 = 0$ **49.** $4x - y + 5 = 0$

50. $2x - y - 3 = 0$ **51.** $2x + 3y + 6 = 0$ **52.** $x + 3y - 2 = 0$

53. $3x + 2y - 5 = 0$ **54.** $3x + 4y - 5 = 0$ **55.** $5x - 4y - 8 = 0$

56. $x = 5$ **57.** $y = 5$ **58.** $y = -2$

59. $x = -2$ **60.** $x + y - 100 = 0$ **61.** $5x - 3y = 27$

62. $y = 50x$ **63.** $y = -100x$ **64.** $y = -0.001x$

▲ Problem Solving

65. The Brazilian Institute of Geography and Statistics reported the following fertility rates for Brazil, measured as the average number of children born per woman of childbearing age:

$$1970: 5.8 \qquad 1980: 4.4 \qquad 1991: 2.7 \qquad 1994: 2.2$$

Do you think the fertility rate is declining linearly? (That is, if you plot these points, is the graph linear?) Support your answer.

66. Pick two data points from Problem 65 and find the equation of the line between these points. Predict the fertility rate in the year 2000. Will this answer vary depending on the selected data points? Support your answer.

67. Find the best-fitting line for the data given in Problem 65. Use this best-fitting line to predict the fertility rate in the year 2000. [*Hint:* Refer to Section 10.4 for the method of finding the best-fitting line.]

68. If a present value P is invested at the simple interest rate r for t years, then the *future value* after t years is given by the formula $A = P(1 + rt)$. Suppose you invest \$10,000 at 8% simple interest per year.
  a. Graph the amount you will have in t years.
  b. What is the slope of the graph?
  c. What is the A-intercept of the graph?

69. A business purchasing an item for business purposes may use *straight-line depreciation* to obtain a tax deduction. The formula for the present value, P, after t years is

$$P = C - \left(\frac{C - s}{L}\right)t$$

where C is the cost and s is the scrap value after L years. The number L is called the *useful life* of the item.
  a. If a certain piece of equipment costs \$20,000 and has a scrap value of \$2,000 after 8 years, write an equation to represent the present value after t years.
  b. Graph the amount you will have in t years.
  c. What is the slope of the graph?

70. Suppose the *profit P* (in dollars) of a certain item is given by

$$P = 1.25x - 850$$

where x is the number of items sold.
   a. Graph this profit relationship.
   b. Interpret the value of P when $x = 0$.
   c. The slope of this graph is called the *marginal profit*. Find the marginal profit, and give an interpretation.

71. Suppose the *cost C* (in dollars) of x items is given by

$$C = 2.25x + 550$$

   a. Graph this cost relationship.
   b. Interpret the value of C when $x = 0$.
   c. The slope of the graph is called the *marginal cost*. Find the marginal cost and give an interpretation.

72. The population of Texas was 14.2 million in 1980 and 17.0 million in 1990. Let x be the year (let 1980 be the base year; that is, $x = 0$ represents the year 1980 and $x = 10$ represents 1990) and y be the population. Use this information to write a linear equation, and then use this equation to predict the population of Texas in the year 2000.

73. Show that the equation of a line passing through (h, k) with slope m is

$$y - k = m(x - h)$$

   This is called the **point–slope form.**

74. Begin with the point–slope form in Problem 73 and derive the **two-point form.** If the line passes through (x_1, y_1) and (x_2, y_2), then

$$y - y_1 = \left(\frac{y_2 - y_1}{x_2 - x_1}\right)(x - x_1)$$

75. Show that the equation of the line with x-intercept a and y-intercept b is

$$\frac{x}{a} + \frac{y}{b} = 1 \qquad (a \neq 0, b \neq 0)$$

   This is called the **intercept form.**

11.2   GRAPHING HALF-PLANES

Once you know how to graph a line, you can also graph linear inequalities. We begin by noting that every line divides a plane into three parts, as shown in Figure 11.12. Two parts are labeled I and II; these are called **half-planes.** The third part is called the **boundary** and is the line separating the half-planes. The solution of a first-degree inequality in two unknowns is the set of all ordered pairs that satisfy the given inequality. This solution set is a half-plane. The following table offers some examples of first-degree inequalities with two unknowns, along with some associated terminology.

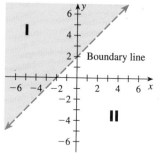

▲ **Figure 11.12   Half-planes**

Example	Inequality Symbol	Boundary Included	Term
$3x - y > 5$	$>$	no	**open half-plane**
$3x - y < 5$	$<$	no	open half-plane
$3x - y \geq 5$	$\geq$	yes	**closed half-plane**
$3x - y \leq 5$	$\leq$	yes	closed half-plane

We can now summarize the procedure for graphing a first-degree inequality in two unknowns.

Procedure for Graphing a Linear Inequality

Step 1. Graph the boundary.
Replace the inequality symbol with an equality symbol and draw the resulting line. This is the boundary line.
Use a solid line when the boundary is included ($\leq$ or $\geq$).
Use a dashed line when the boundary is not included ($<$ or $>$).

Step 2. Test a point.
Choose any point in the plane that is not on the boundary line; the point $(0, 0)$ is usually the simplest choice.
If this **test point** makes the *inequality* true, shade in the half-plane that contains the test point. That is, the shaded plane is the solution set.*
If the test point makes the *inequality* false, shade in the other half-plane for the solution.

This process sounds complicated, but if you know how to draw lines from equations, you will not find this difficult.

■■■ EXAMPLE I

Graph $3x - y \geq 5$.

Solution   Note that the inequality symbol is $\geq$, so the boundary is included.

   Step 1.   Graph the boundary; draw the (solid) line corresponding to

$$3x - y = 5 \qquad \text{Replace the inequality symbol with an equality symbol.}$$
$$y = 3x - 5$$

   The y-intercept is -5 and the slope is 3; the boundary line is shown in Figure 11.13.

   Step 2.   Choose a test point; we choose $(0, 0)$. Plot $(0, 0)$ in Figure 11.13 and note that it lies in one of the half-planes determined by the boundary

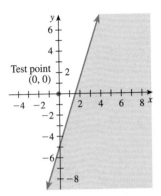

▲ **Figure 11.13   Graph of** $3x - y \geq 5$

---

* A highlighter pen does a nice job of shading in your work.

line. We now check this test point with the given *inequality*:

$$3x - y \geq 5$$

$$3(0) - (0) \geq 5$$

$$0 \geq 5 \quad \text{This is false.}$$

You can usually test this in your head.

Therefore, shade the half-plane that does *not* contain $(0, 0)$, as shown in Figure 11.13.

▬▬ EXAMPLE 2

Graph $y < x$.

Solution   Note that the inequality symbol is $<$, so the boundary line is not included.

Step 1.   Draw the (dashed) boundary line, $y = x$, as shown in Figure 11.14.

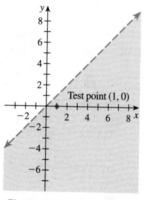

▲ **Figure 11.14   Graph of** $y < x$

Step 2.   Choose a test point. We can't pick $(0, 0)$, because $(0, 0)$ is on the boundary line. Since we must choose some point not on the boundary, we choose $(1, 0)$:

$$y < x$$

$$0 < 1 \quad \text{This is true.}$$

Therefore, shade the half-plane that contains the test point, as shown in Figure 11.14.

PROBLEM SET   11.2

▬▬▬▬▬▬▬▬▬▬▬▬▬▬▬▬▬▬▬▬▬▬▬▬▬▬▬▬▬▬▬▬▬▬▬

▲ A Problems

1. **IN YOUR OWN WORDS** What is a first-degree inequality in two variables?

2. **IN YOUR OWN WORDS** Outline a procedure for graphing first-degree inequalities in two variables.

What is wrong, if anything, with each of the statements in Problems 3–10? Explain your reasoning.

3. The linear inequality $2x + 5y < 2$ does not have a boundary line, because the inequality symbol is $<$.

4. A good test point for the linear inequality $y \geq x$ is $(0, 0)$.

5. The test point $(0, 0)$ satisfies the inequality $2y - 3x < 2$.

6. The test point $(0, 0)$ satisfies the inequality $3x - 2y \geq -1$.

7. The test point $(0, 0)$ satisfies the inequality $3x > 2y$.

8. The test point $(-2, 4)$ satisfies the inequality $y > 2x - 1$.

9. The test point $(-2, 4)$ satisfies the inequality $5x + 2y \leq 9$.

10. The test point $(-2, 4)$ satisfies the inequality $4x < 3y$.

▲ **B Problems**

Graph the first-degree inequalities in two unknowns in Problems 11–25.

11. $y \leq 2x + 1$ 12. $y \geq 5x - 3$ 13. $y > 5x - 3$ 14. $y \geq -2x + 3$

15. $y > 3x - 3$ 16. $y < -2x + 5$ 17. $3x \leq 2y$ 18. $2x < 3y$

19. $x \geq y$ 20. $y > x$ 21. $y \geq 0$ 22. $x \leq 0$

23. $x - 3y \geq 9$ 24. $6x - 2y < 1$ 25. $3x + 2y > 1$

▲ **Problem Solving**

26. Will a baseball player's batting average always change when the player goes up to bat? That is, under what conditions will the batting average stay the same? *Hint:* Let a be the number of times at bat, and let h be the number of hits. Then the batting average is $\dfrac{h}{a}$. If the player gets a hit, then the new batting average is

$$\frac{h + 1}{a + 1}$$

and if the player does not get a hit, then the new batting average is

$$\frac{h}{a + 1}$$

▲ **Individual Research**

27. This problem is a continuation of Problem 26. A player's batting average really isn't simply the ratio $\dfrac{h}{a}$. It is the value of $\dfrac{h}{a}$ rounded to the nearest thousandth. It is possible that a batting average could be raised or lowered, but the reported batting average might remain the same when rounded. Write a paper on this topic.

Reference James M. Sconyers, "Serendipity: Batting Averages to Greatest Integers," *The Mathematics Teacher*, April 1980, pp. 278–280.

11.3 GRAPHING CURVES

Curves by Plotting Points

When cannons were introduced in the 13th century, their primary use was to demoralize the enemy. It was much later that they were used for strategic purposes.

In fact, cannons existed nearly three centuries before enough was known about the behavior of projectiles to use them with any accuracy. The cannonball does not travel in a straight line, it was discovered, because of an unseen force that today we know as *gravity*. Consider Figure 11.15. It is a scale drawing (graph) of the path of a cannonball fired in a particular way.

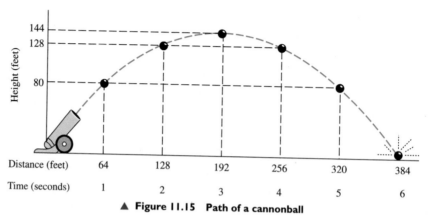

▲ **Figure 11.15   Path of a cannonball**

The path described by a projectile is called a **parabola.** Any projectile — a ball, an arrow, a bullet, a rock from a slingshot, even water from the nozzle of a hose or sprinkler — will travel a parabolic path. Note that this parabolic curve has a maximum height and is symmetric about a vertical line through that height. In other words, the ascent and descent paths are symmetric.

■■■ EXAMPLE 1

Graph $y = x^2$.

Solution   We will choose x-values and find corresponding y-values:

Let $x = 0$: $y = 0^2 = 0$; plot $(0, 0)$.
Let $x = 1$: $y = 1^2 = 1$; plot $(1, 1)$.
Let $x = -1$: $y = (-1)^2 = 1$; plot $(-1, 1)$.

Notice in Figure 11.16 that these points do not fall in a straight line. If we find two more points, we can see the shape of the graph:

Let $x = 2$: $y = 2^2 = 4$; plot $(2, 4)$.
Let $x = -2$: $y = (-2)^2 = 4$; plot $(-2, 4)$.

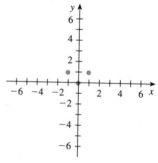

▲ **Figure 11.16   Points satisfying the equation $y = x^2$ do not lie on a straight line.**

Connect the points to form a smooth curve, as shown in Figure 11.17.

▲ **Figure 11.17   Graph of $y = x^2$**

The curve shown in Figure 11.17 is a parabola that is said to *open upward*. The lowest point, $(0, 0)$ in Example 1, is called the **vertex**. The following example is a parabola that *opens downward*. Notice that the number a in $y = ax^2$ is negative.

EXAMPLE 2

Sketch $y = -\frac{1}{2}x^2$.

Solution

Let $x = 0$: $y = -\frac{1}{2}(0)^2 = 0$; plot $(0, 0)$.

Let $x = 1$: $y = -\frac{1}{2}(1)^2 = -\frac{1}{2}$; plot $(1, -\frac{1}{2})$.

Let $x = -1$: $y = -\frac{1}{2}(-1)^2 = -\frac{1}{2}$; plot $(-1, -\frac{1}{2})$.

Let $x = 2$: $y = -\frac{1}{2}(2)^2 = -2$; plot $(2, -2)$.

Let $x = -2$: $y = -\frac{1}{2}(-2)^2 = -2$; plot $(-2, -2)$.

Let $x = 4$: $y = -\frac{1}{2}(4)^2 = -8$; plot $(4, -8)$.

Connect these points to form a smooth curve, as shown in Figure 11.18.

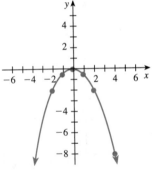

▲ Figure 11.18 Graph of $y = -\frac{1}{2}x^2$

Notice from Example 2 that you can use symmetry to plot points on one side of the vertex.

You can sketch many different curves by plotting points. The procedure is to decide whether you should pick x-values and find the corresponding y-values, or pick y-values and find the corresponding x-values. Find enough ordered pairs so that you can connect those points with a smooth graph. Many graphs in mathematics are not smooth, but we will not consider those in this course.

EXAMPLE 3

Sketch the graph of the curve $x = y^2 - 6y + 4$.

Solution   We use the same procedure to find points on the graph, except that in this equation we see that it will be easier to choose y-values and find corresponding x-values.

Let $y = 0$: $x = 0^2 - 6 \cdot 0 + 4 = 4$; plot $(4, 0)$.

Let $y = 1$: $x = 1^2 - 6 \cdot 1 + 4 = 1 - 6 + 4 = -1$; plot $(-1, 1)$.

Let $y = 2$: $x = 2^2 - 6 \cdot 2 + 4 = 4 - 12 + 4 = -4$; plot $(-4, 2)$.

Let $y = 3$: $x = 3^2 - 6 \cdot 3 + 4 = 9 - 18 + 4 = -5$; plot $(-5, 3)$.

Let $y = 4$: $x = 4^2 - 6 \cdot 4 + 4 = 16 - 24 + 4 = -4$; plot $(-4, 4)$.

Let $y = 5$: $x = 5^2 - 6 \cdot 5 + 4 = 25 - 30 + 4 = -1$; plot $(-1, 5)$.

Let $y = 6$: $x = 6^2 - 6 \cdot 6 + 4 = 36 - 36 + 4 = 4$; plot $(4, 6)$.

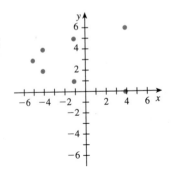

How many points should we find? For lines, we needed to find two points, with a third point as a check. For curves that are not lines, we do not have a particular number of points to check. We must plot as many points as are necessary to enable us to draw a smooth curve. We connect the points we have plotted to draw the curve shown in Figure 11.19.

Exponential Curves

An **exponential equation** is one in which a variable appears as an exponent. Consider the equation $y = 2^x$. This equation represents a doubling process. The graph of such an equation is called an **exponential curve** and is graphed by plotting points.

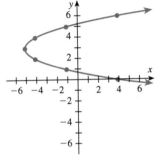

▲ Figure 11.19 Graph of $x = y^2 - 6y + 4$

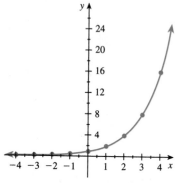

▲ **Figure 11.20   Graph of** $y = 2^x$

EXAMPLE 4

Sketch the graph of $y = 2^x$.

Solution   Choose x-values and find corresponding y-values. These values form ordered pairs (x, y). Plot enough ordered pairs so that you can see the general shape of the curve, and then connect the points with a smooth curve.

Let $x = 0$: $y = 2^0 = 1$; plot $(0, 1)$.

Let $x = 1$: $y = 2^1 = 2$; plot $(1, 2)$.

Let $x = 2$: $y = 2^2 = 4$; plot $(2, 4)$.

Let $x = 3$: $y = 2^3 = 8$; plot $(3, 8)$.

Let $x = -1$: $y = 2^{-1} = 1/2$; plot $(-1, \frac{1}{2})$.

Let $x = -2$: $y = 2^{-2} = 1/4$; plot $(-2, \frac{1}{4})$.

Let $x = -3$: $y = 2^{-3} = 1/8$; plot $(-3, \frac{1}{8})$.

Connect the points with a smooth curve, as shown in Figure 11.20.   ▬

Population Growth

Population growth is described by an exponential equation. Natural growth is dependent on an irrational number known as e. Many calculators have an e^x key that can be used to find a decimal approximation for this number: $e \approx 2.718281828$.

Population Growth Model

The population, P, at some future date can be predicted if you know the population P_0 at some time, and the annual growth rate, r. The predicted population t years after the given time is

$$P = P_0 e^{rt}$$

EXAMPLE 5   Polya's Method

Marsha and Tony have owned a parcel of property within the city limits of their small town for a number of years. Now a developer is threatening to build a large housing project on the land next to their property, but to do so the developer is seeking a zoning change from the City Planning Commission. The hearing is next week, and Tony believes that the population projections presented by the developer are vague and unfounded. Tony needs a mathematical method for projecting population and needs to communicate his findings to the Planning Commission.

Solution   We use Polya's problem-solving guidelines for this example.

Understand the Problem.   Tony calls his local Chamber of Commerce and finds that the growth rate of his town is now 5%. Also, according to the 1990 census, the population was 2,500.

Devise a Plan. Tony decides to draw a graph showing the population between the years 1990 and 2010. The formula for population growth is $P = P_0 e^{rt}$.

Carry Out the Plan. For this problem,

$P_0 = 2,500$ and $r = 5\% = 0.05$ Remember, to change a percent to a decimal, move the decimal point two places to the left.

The equation of the graph is $P_0 e^{rt} \approx 2,500 e^{0.05t}$.

If $t = 0$: $P = 2,500 e^0 = 2,500$.   Plot the point (0, 2500). This is the 1990 population; 1990 is called the base year. It is also called the "present time" (even if it is now 1999). Thus, if $t = 5$, the population corresponds to the year 1995. If $t = 10$, the population is for the year 2000.

If $t = 10$: $P = 2,500 e^{0.05(10)} \approx 4,120$.

Press: $\boxed{2500}$ $\boxed{\times}$ $\boxed{e}$ $\boxed{\wedge}$ $\boxed{(}$ $\boxed{.05}$ $\boxed{\times}$ $\boxed{10}$ $\boxed{)}$ $\boxed{=}$

Display: **4121.803177**

This means that the predicted population in the year 2000 is 4,122. Plot the point (10, 4120).

If $t = 20$: $P = 2,500 e^{0.05(20)} \approx 6,800$.

Press: $\boxed{2500}$ $\boxed{\times}$ $\boxed{e}$ $\boxed{\wedge}$ $\boxed{(}$ $\boxed{.05}$ $\boxed{\times}$ $\boxed{20}$ $\boxed{)}$ $\boxed{=}$

Display: **6795.704571**   Plot the point (20, 6800).

This means that the predicted population for the year 2010 is 6,800. We have plotted these points in Figure 11.21.

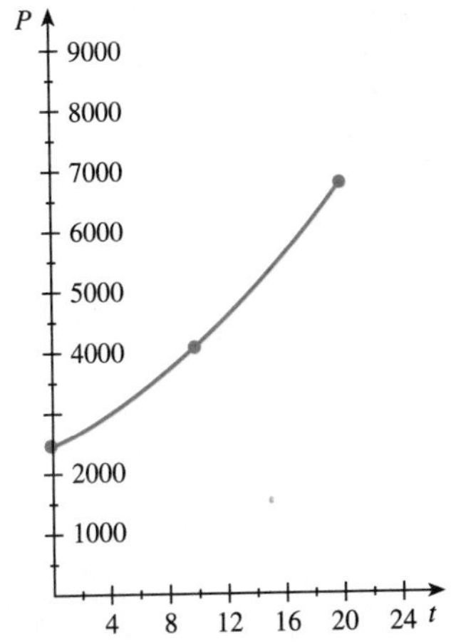

Look Back. To do a good job with persuasion, you need facts and data to back up your argument. The visual impact of a graph can be quite impressive. ▬

▲ **Figure 11.21 Graph showing population P from 1990 to 2010 (base year, t = 0, is 1990)**

Growth of Money

One of the most important concepts for intelligent functioning in today's world is interest and how interest works. We can now return to this idea and look at graphs showing the growth of money as the result of interest accumulation. Simple interest is represented by a line, whereas compound interest requires an exponential graph. This means that, for short periods of time, there is very little difference between simple and compound interest, but as time increases, the differences become significant.

▬ **EXAMPLE 6**

Suppose that you deposit $100 at 10% interest. Graph the total amounts you will have if you invest your money at simple interest and if you invest your money at compound interest. Recall the appropriate formulas:

Simple interest: $A = P(1 + rt)$
Compound interest: $A = P(1 + r)^t$

Solution   For this example, $P = 100$ and $r = 0.10$; the variables are t and A. We begin by finding some ordered pairs. A calculator is necessary for calculating compound interest.

Year	Simple Interest	Compound Interest
$t = 0$	$A = 100(1 + 0.1 \cdot 0) = 100$ Point (0, 100)	$A = 100(1 + 0.1)^0 = 100$ Point (0, 100)
$t = 1$	$A = 100(1 + 0.1 \cdot 1) = 110$ Point (1, 110)	$A = 100(1 + 0.1)^1 = 110$ Point (1, 110)
$t = 10$	$A = 100(1 + 0.1 \cdot 10) = 200$ Point (10, 200)	$A = 100(1 + 0.1)^{10} = 259$ Point (10, 259)
$t = 20$	$A = 100(1 + 0.1 \cdot 20) = 300$ Point (20, 300)	$A = 100(1 + 0.1)^{20} = 673$ Point (20, 673)

The graphs through these points are shown in Figure 11.22.

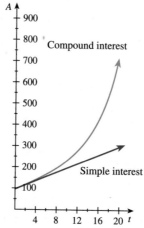

▲ **Figure 11.22   Comparison of simple interest and compound interest**   ▬

PROBLEM SET   11.3

▲ **A Problems**

1. **IN YOUR OWN WORDS** What is a parabola?

2. **IN YOUR OWN WORDS** What is e?

3. **IN YOUR OWN WORDS** Explain how you might make a prediction of population growth.

Sketch the graph of each equation in Problems 4–30.

4. $y = 3x^2$ 5. $y = 2x^2$ 6. $y = 10x^2$

7. $y = -x^2$ 8. $y = -2x^2$ 9. $y = -3x^2$

10. $y = -5x^2$ 11. $y = 5x^2$ 12. $y = \frac{1}{2}x^2$

13. $x = 2y^2$ 14. $x = -3y^2$ 15. $x = \frac{1}{2}y^2$

16. $y = 2x^3$ 17. $y = -x^3$ 18. $x = -y^3$

19. $y = \frac{1}{3}x^2$ 20. $y = -\frac{2}{3}x^2$ 21. $y = \frac{1}{10}x^2$

22. $y = -\frac{1}{10}x^2$ 23. $y = \frac{1}{1,000}x^2$ 24. $y = x^2 - 4$

25. $y = x^2 + 4$ 26. $y = 9 - x^2$ 27. $y = -3x^2 + 4$

28. $y = 2x^2 - 3$ 29. $y = x^2 - 2x + 1$ 30. $y = x^2 + 4x + 4$

▲ B Problems

Sketch the graphs of the equations in Problems 31–42.

31. $y = 3^x$ 32. $y = 4^x$ 33. $y = 5^x$ 34. $y = -2^x$

35. $y = -6^x$ 36. $y = -7^x$ 37. $y = 10^x$ 38. $y = 100 - 2^x$

39. $y = (\frac{1}{2})^x$ 40. $y = (\frac{1}{3})^x$ 41. $y = (\frac{1}{10})^x$ 42. $y = -(\frac{1}{10})^x$

43. **a.** Graph $h = -16t^2 + 96t$.
    b. Does the parabola in part **a** open upward or downward?
    c. If we relate the graph in part **a** to the cannonball in Figure 11.15, can t be negative? Draw the graph of the parabola in part **a** for $0 \le t \le 6$.

44. **a.** Graph $d = 16t^2$.
    b. Does the parabola in part **a** open upward or downward?
    c. Graph the parabola in part **a** for $0 \le t \le 16$.

Draw the graphs in Problems 45–50.

45. $y = -2x^2 + 4x - 2$ 46. $y = \frac{1}{4}x^2 - \frac{1}{2}x + \frac{1}{4}$ 47. $y = \frac{1}{2}x^2 + x + \frac{1}{2}$

48. $y = x^2 - 2x + 3$ 49. $x = 3y^2 + 12y + 14$ 50. $x = 4y - y^2 - 4$

51. Change the growth rate in Example 5 to 6%, and graph the population curve.

52. Change the growth rate in Example 5 to 1.5%, and graph the population curve.

53. Rework Example 6 for a 12% interest rate.

54. Rework Example 6 for a 5% interest rate.

▲ Problem Solving

55. In archeology, carbon-14 dating is a standard method of determining the age of certain artifacts. Decay rate can be measured by half-life, which is the time required for one-half of a substance to decompose. Carbon-14 has a half-life of approximately 5,600 years. If 100 grams of carbon were present originally, then the amount A of carbon present today is given by the formula.

$$A = 100(\tfrac{1}{2})^{t/5,600}$$

where t is the time since the artifact was alive until today. Graph this relationship by letting $t = 5,600$, $t = 11,200$, $t = 16,800$ . . . , and adjust the scale accordingly.

56. The healing law for skin wounds states that $A = \dfrac{A_0}{e^{0.1t}}$, where A is the number of square centimeters of unhealed skin after t days when the original area of the wound was A_0. Graph the healing curve for a wound of 50 cm^2.

57. A learning curve describes the rate at which a person learns certain specific tasks. If N is the number of words per minute typed by a student, then, for an average student

$$N = 80(1 - e^{-0.016n})$$

where n is the number of days of instruction. Graph this formula.

58. The equation for the standard normal curve (the normal curve with mean 0 and standard deviation 1) graphs as an exponential curve. Graph this curve, whose equation is

$$y = \frac{e^{-x^2/2}}{\sqrt{2\pi}}$$

by using a calculator or by plotting points.

59. Graph the curve $y = 2^{-x^2}$ and compare with the standard normal curve (Problem 58). List some similarities and some differences.

▲ **Individual Research**

60. The population in California was 31,910,000 in January 1995, and 32,231,000 in January 1996. Predict California's population in the year 2000. Check the Internet or an almanac to verify the 1997 population using the information of this problem.

61. The population in Sebastopol, California, was 7,475 in January 1995, and 7,525 in January 1996. Predict Sebastopol's population in the year 2000.

62. Predict the population of your city or state for the year 2000.

11.4 FUNCTIONS

The idea of looking at two sets of variables at the same time was introduced in the previous section. Sets of ordered pairs provide a very compact and useful way to represent relationships between various sets of numbers. To consider this idea, let's look at the following cartoon.

B.C. By permission of Johnny Hart and Creators Syndicate, Inc.

The distance an object will fall depends on (among other things) the length of time it falls. If we let the variable d be the distance the object has fallen (in feet) and the variable t be the time it has fallen (in seconds), and if we disregard air resistance, the formula is

$$d = 16t^2$$

Therefore, in the *B.C.* cartoon, if the well is 16 seconds deep (and we neglect the time it takes for the sound to come back up), we know that the depth of the well (in feet) is

$$d = 16(16)^2$$
$$= 16(256)$$
$$= 4,096$$

The formula $d = 16t^2$ gives rise to a set of data:

Time (in seconds)	0	1	2	3	4	. . .	15	16
Distance (in ft)	0	16	64	144	256	. . .	3,600	4,096

For every nonnegative value of t, there is a corresponding value of d. We can represent the data in the table as a set of ordered pairs in which the first component represents a value for t and the second component represents a corresponding value for d. For this example, we have $(0, 0), (1, 16), (2, 64), (3, 144), (4, 256), . . . ,$ $(15, 3600), (16, 4096)$.

Whenever we have a situation comparable to the one illustrated by this example — namely, whenever the first component of an ordered pair is associated with exactly one second component — we call the set of ordered pairs a *function*.

First component (values for t)
↓
(x, y)
↑
Second component (values for d)

> **Function**
>
> A **function** is a set of ordered pairs in which the first component is associated with exactly one second component. ▲

Not all sets of ordered pairs are functions, as we can see from the following examples.

▬ EXAMPLE 1

Which of the following sets of ordered pairs are functions?

a. $\{(0, 0), (1, 2), (2, 4), (3, 9), (4, 16)\}$
b. $\{(0, 0), (1, 1), (1, -1), (4, 2), (3, -2)\}$
c. $\{(1, 3), (2, 3), (3, 3), (4, 3)\}$
d. $\{(3, 1), (3, 2), (3, 3), (3, 4)\}$

Solution

a. We see that

$0 \rightarrow 0$
$1 \rightarrow 2$
$2 \rightarrow 4$
$3 \rightarrow 9$
$4 \rightarrow 16$

Sometimes it is helpful to think of the first component as the "picker" and the second component as the "pickee." Given an ordered pair (x, y), we find that each replacement for x "picks" a partner, or a second value. We can symbolize this by $x \rightarrow y$.

Since each first component is associated with exactly one second component, the set is a function.

b. For this set,

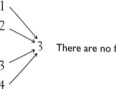

$$0 \longrightarrow 0$$

Since 1 picks two values as a partner or second component, we call the number 1 a "fickle picker." But a function is a set of ordered pairs for which there are no fickle pickers.

$$4 \longrightarrow 2$$
$$3 \longrightarrow -2$$

Since the first component can be associated with more than one second component, the set is not a function.

c. For this set,

There are no fickle pickers, so it is a function.

This is an example of a function.

d. Finally,

The number 3 is a fickle picker, so it is not a function.

Since the first component is associated with several second components, the set is not a function. ■

Another way to consider functions is with the idea of a **function machine**, as shown in Figure 11.23. Think of this machine as having an input where items are entered and an output where results are obtained, much like a vending machine. If a number 2 is dropped into the input, a function machine will output a single value. If the name of the function machine is f, then the output value is called "f of 2" and is written as $f(2)$. This is called **functional notation**.

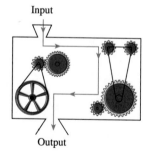

Input

Output

▲ **Figure 11.23　Function machine**

■ EXAMPLE 2

If you input each of the given values into a function machine named g, what is the resulting output value?

a. 4　　**b.** -3　　**c.** π　　**d.** x

Solution

a. Input is 4; output is $g(4)$, pronounced "gee of four." To calculate $g(4)$, first add 2, then multiply by 5:

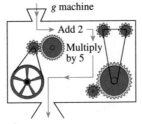

g machine
Add 2
Multiply by 5

Input value · Output value
$$\downarrow$$
$$(4 + 2) \times 5 = 6 \times 5 = \overbrace{30}$$

We write $g(4) = 30$.

b. Input is -3; output is $g(-3)$, pronounced "gee of negative three":

$$g(-3) = (-3 + 2) \times 5 = -1 \times 5 = -5$$

c. Input is π; output is $g(\pi)$, pronounced "gee of pi":

$$g(\pi) = (\pi + 2) \times 5 = 5(\pi + 2)$$

d. Input is x; output is $g(x)$, pronounced "gee of ex":

$$g(x) = (x + 2) \times 5 = 5(x + 2)$$

▬ EXAMPLE 3

If a function machine f squares the input value, we write $f(x) = x^2$, where x represents the input value. We usually define functions by simply saying "Let $f(x) = x^2$." Identify the value of f for the given value.

a. $f(2)$ b. $f(8)$ c. $f(-3)$ d. $f(t)$

Solution

a. $f(2) = 2^2 = 4$ b. $f(8) = 8^2 = 64$

c. $f(-3) = (-3)^2 = 9$ d. $f(t) = t^2$

PROBLEM SET 11.4

▲ A Problems

1. **IN YOUR OWN WORDS** What is a function?

2. **IN YOUR OWN WORDS** Explain the notation $f(x)$.

Which of the sets in Problems 3–14 are functions?

3. $\{(1, 4), (2, 5), (4, 7), (9, 12)\}$

4. $\{(4, 1), (5, 2), (7, 4), (12, 9)\}$

5. $\{(1, 1), (2, 1), (3, 4), (4, 4), (5, 9), (6, 9)\}$

6. $\{(1, 1), (1, 2), (4, 3), (4, 4), (9, 5), (9, 6)\}$

7. $\{(4, 3), (17, 29), (18, 52), (4, 19)\}$

8. $\{(13, 4), (29, 4), (5, 4), (9, 4)\}$

9. $\{(19, 4), (52, 18), (29, 17), (3, 4)\}$ **10.** $\{(4, 9), (4, 4), (4, 29), (4, 19)\}$

11. $\{(5, 0)\}$ **12.** $\{(0, 0)\}$ **13.** $\{1, 2, 3, 4, 5\}$ **14.** $\{69, 82, 44, 37\}$

15. The velocity v (in feet per second) of the rock dropped into the well in the B.C. cartoon at the beginning of this section is also related to time t (in seconds) by the formula

$$v = 32t$$

Complete the table showing the time and velocity of the rock.

Time (in sec)	0	1	2	3	4 . . .	8 . . .	16
Velocity (in ft per sec)	0	32	a.	b.	c.	d.	e.

16. Using the table of values in Problem 15, find the velocity of the rock at the instant it was released, after 8 seconds, and when it hit the bottom of the well. Write your answers in the form (t, v).

17. An independent distributor bought a new vending machine for $2,000. It had a probable scrap value of $100 at the end of its expected 10-year life. The value V at the end of n years is given by

$$V = 2,000 - 190n$$

Complete the table showing the year and the value of the machine.

Year	0	1	3	5	7	9	10
Value	2,000	1,810	a.	b.	c.	d.	e.

18. Using the table of values in Problem 17, find the value of the machine when it is purchased, when it is 5 years old, and when it is scrapped. Write your answers in the form (t, V).

Tell what the output value is for each of the function machines in Problems 19–24 for **(a)** 4, **(b)** 6, **(c)** -8, **(d)** $\frac{1}{2}$, **(e)** x.

19.   **20.**   **21.**

22.   **23.**   **24.**

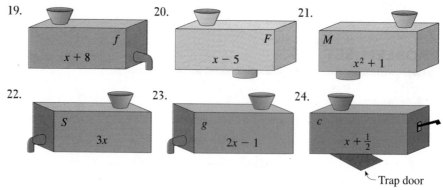

Trap door

▲ **B Problems**

25. If $f(x) = x - 7$, find
    a. $f(15)$ **b.** $f(-9)$ **c.** $f(p)$

26. If $g(x) = 2x$, find
    a. $g(100)$ **b.** $g(-25)$ **c.** $g(m)$

27. If $h(x) = 3x - 1$, find
    a. $h(0)$ **b.** $h(-10)$ **c.** $h(a)$

28. If $f(x) = x^2 + 1$, find
    a. $f(-3)$ **b.** $f(\frac{1}{2})$ **c.** $f(b)$

29. If $g(x) = \dfrac{x}{2}$, find

    a. $g(10)$ **b.** $g(-4)$ **c.** $g(3)$

30. If $h(x) = 0.6x$, find

    a. $h(4.1)$ **b.** $h(2.3)$ **c.** $h(\frac{5}{2})$

▲ **Problem Solving**

Find $\dfrac{f(x + h) - f(x)}{h}$ for the functions given in Problems 31–36.

31. $f(x) = 3x - 5$ 32. $f(x) = 3x^2 - 5$ 33. $f(x) = x^3$

34. $f(x) = 5x^3$ 35. $f(x) = \dfrac{1}{x}$ 36. $f(x) = \dfrac{1}{x^2}$

37. Let A be the set of the following cities:

    A = {Arm, MI; Bone, ID; Cheek, TX; Doublehead, AL; Elbow Lake, MN}

    Suppose that these cities are connected by direct service as shown in Figure 11.24. We define a set of ordered pairs (x, y) if and only if x and y are connected by direct service. Find the elements of this set of ordered pairs. Is this set a function?

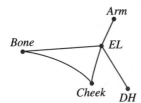

▲ **Figure 11.24 City network**

38. Suppose that we modify the network of Problem 37 by making the lines one-way, as shown by the arrows in Figure 11.25. We define a set of all pairs (x, y) for which there is one-way service from x to y. Find the elements of this set. Is this set a function?

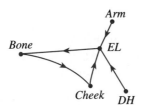

▲ **Figure 11.25 Directed network**

39. From a square whose side has length x, create a new square whose side is 5 in. longer. Find an expression for the difference between the areas of the two squares as a function of x. Graph this expression.

40. "Look out! That old abandoned well is very dangerous, Huck. I think that it should be capped so that nobody could fall in and kill his self." "Ah, shucks, Tom, let's climb down and see how deep it is. I'll bet it is a mile down to the bottom." If we assume that it is not possible for Tom or Huck to climb down into the well, how can they find out the depth of the well? Assume that the well is a mile deep and explain how Tom and Huck might determine this fact.

1. Slope of a line

2. Tangent line to a circle

3. Area of a region bounded by line segments

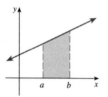

4. Average position and velocity

5. Average of a finite collection of numbers

▲ **Figure 11.26　Topics from elementary mathematics**

11.5　WHAT IS CALCULUS?

If there is an event that marked the coming of age of mathematics in Western culture, it must surely be the essentially simultaneous development of the calculus by Newton and Leibniz in the seventeenth century. Before this remarkable synthesis, mathematics had often been viewed as merely a strange but harmless pursuit, indulged in by those with an excess of leisure time. After the calculus, mathematics became virtually the only acceptable language for describing the physical universe. This view of mathematics and its association with the scientific method has come to dominate the Western view of how the world ought to be explained.

What distinguishes calculus from algebra, geometry, and trigonometry is the transition from static or discrete applications (see Figure 11.26) to those that are dynamic or continuous (see Figure 11.27). For example, in elementary mathematics we consider the slope of a line, but in calculus we define the (nonconstant) slope of a nonlinear curve. In elementary mathematics we find average values of position and velocity, but in calculus we can find instantaneous changes of velocity and acceleration. In elementary mathematics we find the average of a finite collection of numbers, but in calculus we can find the average value of a function with infinitely many values over an interval.

Calculus is the mathematics of motion and change, which is why calculus is a prerequisite for many courses. Whenever we move from the static to the dynamic, we would consider using calculus.

The development of calculus in the 17th century by Newton and Leibniz was the result of their attempt to answer some fundamental questions about the world and the way things work. These investigations led to two fundamental concepts of calculus — namely, the idea of a *derivative* and that of an *integral*. The breakthrough in the development of these concepts was the formulation of a mathematical tool called a *limit*.

1. **Limit:** The limit is a mathematical tool for studying the *tendency* of a function as its variable *approaches* some value. Calculus is based on the concept of limit. We shall introduce the limit of a function informally in the next section.

2. **Derivative:** The derivative is defined as a certain type of limit, and it is used initially to compute rates of change and slopes of tangent lines to curves. The study of derivatives is called *differential calculus*. Derivatives can be used in sketching graphs and in finding the extreme (largest and smallest) values of functions.

3. **Integral:** The integral is found by taking a special limit of a sum of terms, and the study of this process is called *integral calculus*. Area, volume, arc length, work, and hydrostatic force are a few of the many quantities that can be expressed as integrals.

Let us take an intuitive look at each of these three essential ideas of calculus.

The Limit: Zeno's Paradox

In Chapter 5 we first introduced Zeno's paradox. Zeno (ca. 500 B.C.) was a Greek philosopher known primarily for his famous paradoxes. One of those concerns a race between Achilles, a legendary Greek hero, and a tortoise. When the race begins, the (slower) tortoise is given a head start, as shown in Figure 11.28.

▲ **Figure 11.28   Achilles and the tortoise**

Is it possible for Achilles to overtake the tortoise? Zeno pointed out that by the time Achilles reaches the tortoise's starting point, $a_1 = t_0$, the tortoise will have moved ahead to a new point t_1. When Achilles gets to this next point, a_2, the tortoise will be at a new point, t_2. The tortoise, even though much slower than Achilles, keeps moving forward. Although the distance between Achilles and the tortoise is getting smaller and smaller, the tortoise will apparently always be ahead.

Of course, common sense tells us that Achilles will overtake the slow tortoise, but where is the error in reasoning? The error is in the assumption that an infinite amount of time is required to cover a distance divided into an infinite number of segments. This discussion is getting at an essential idea in calculus — namely, the notion of a limit.

Consider the successive positions for both Achilles and the tortoise:

Starting position

↓

Achilles: $a_0, \quad a_1, \quad a_2, \quad a_3, \quad a_4, \quad \dots$

Tortoise: $t_0, \quad t_1, \quad t_2, \quad t_3, \quad t_4, \quad \dots$

After the start, the positions for Achilles, as well as those for the tortoise, form sets of positions that are ordered with positive integers. Such ordered listings are called *sequences* (see Section 6.3).

For Achilles and the tortoise we have two sequences, $\{a_1, a_2, a_3, \dots, a_n, \dots\}$ and $\{t_1, t_2, t_3, \dots, t_n, \dots\}$, where $a_n < t_n$ for all values of n. Both the sequence for Achilles' position and the sequence for the tortoise's position have limits, and it is precisely at that limit point that Achilles overtakes the tortoise. The idea of limit will be discussed in the next section and it is this limit idea that allows us to define the other two basic concepts of calculus: the derivative and the integral. Even if the solution to Zeno's paradox using limits seems unnatural at first, do not be discouraged. It took over 2,000 years to refine the ideas of Zeno and provide conclusive answers to those questions about limits. The following example will provide an intuitive preview of a limit.

CALCULUS

1. Slope of a curve

2. Tangent line to a general curve

3. Area of a region bounded by curves

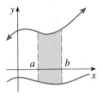

4. Instantaneous changes in position and velocity

5. Average of an infinite collection of numbers

▲ **Figure 11.27   Topics from calculus**

▰▰▰ **EXAMPLE 1**

The sequence

$$\tfrac{1}{2}, \quad \tfrac{2}{3}, \quad \tfrac{3}{4}, \quad \tfrac{4}{5}, \; \ldots$$

can be described by writing a *general term*: $\dfrac{n}{n+1}$ where $n = 1, 2, 3, 4, \ldots$. Can you guess the limit, L, of this sequence? We will say that L is the number that the sequence with general term $\dfrac{n}{n+1}$ tends toward as n becomes large without bound. We will define a notation to summarize this idea:

$$L = \lim_{n \to \infty} \frac{n}{n+1}$$

Solution   As you consider larger and larger values for n, you find a sequence of fractions:

$$\tfrac{1}{2}, \tfrac{2}{3}, \tfrac{3}{4}, \; \ldots, \; \tfrac{1{,}000}{1{,}001}, \tfrac{1{,}001}{1{,}002}, \; \ldots, \; \tfrac{9{,}999{,}999}{10{,}000{,}000}, \; \ldots$$

It is reasonable to guess that the sequence of fractions is approaching the number 1.   ▰▰▰

The Derivative: The Tangent Problem

A **tangent line** (or, if the context is clear, simply say "tangent") to a circle at a given point P is a line that intersects the circle at P and only at P. (See Figure 11.29a.) This characterization does not apply for curves in general, as you can see by looking at Figure 11.29b.

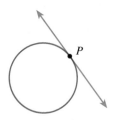

a. At each point P on a circle, there is one line that intersects the circle exactly once.

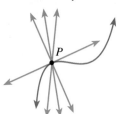

b. At a point P on a curve, there may be several lines that intersect that curve only once.

▲ **Figure 11.29   Tangent line**

To find a tangent line, begin by considering a line that passes through two points on the curve, as shown in Figure 11.30a. This line is called a **secant line.**

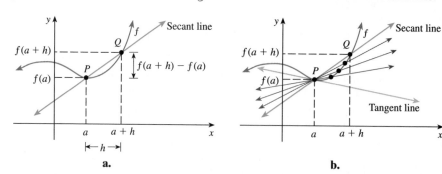

▲ **Figure 11.30   Secant line**

The coordinates of the two points P and Q are $P(a, f(a))$ and $Q(a + h, f(a + h))$. The slope of the secant line is

$$m = \frac{\text{RISE}}{\text{RUN}} = \frac{f(a+h) - f(a)}{h}$$

Now imagine that Q moves along the curve toward P, as shown in Figure 11.30b. You can see that the secant line approaches a limiting position as h approaches zero. We define this limiting position to be the tangent line. The slope of the tangent line is defined as a limit of the sequence of slopes of a set of secant lines. Once again, we can use limit notation to summarize this idea: We say that the slopes of the secant lines, as h becomes small, tend toward a number that we call the slope of the tangent line. We will define the following notation to summarize this idea:

$$\lim_{h \to 0} \frac{f(a + h) - f(a)}{h}$$

This limit forms the definition of derivative and forms the foundation for what is called **differential calculus.**

The Integral: The Area Problem

You probably know the formula for the area of a circle with radius r:

$$A = \pi r^2$$

The Egyptians were the first to use this formula over 5,000 years ago, but the Greek Archimedes (ca. 300 B.C.) showed how to derive the formula for the area of a circle by using a limiting process. Consider the areas of inscribed polygons, as shown in Figure 11.31.

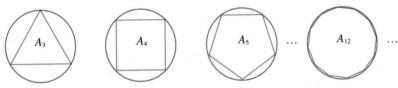

▲ **Figure 11.31   Approximating the area of a circle**

Even though Archimedes did not use the following notation, here is the essence of what he did, using a method called "exhaustion":

Let A_3 be the area of the inscribed equilateral triangle;

A_4 be the area of the inscribed square;

A_5 be the area of the inscribed regular pentagon.

How can we find the area of this circle? As you can see from Figure 11.31, if we consider the area of A_3, then A_4, then A_5, . . . , we should have a sequence of areas such that each successive area more closely approximates that of the circle. We write this idea as a limit statement:

$$A = \lim_{n \to \infty} A_n$$

In this course we will use limits in yet a different way to find the areas of regions enclosed by curves. For example, consider the area shown in blue in Figure 11.32. We can approximate the area by using rectangles. If A_n is the area of the nth rectangle, then the total area can be approximated by finding the sum

$$A_1 + A_2 + A_3 + \cdots + A_{n-1} + A_n$$

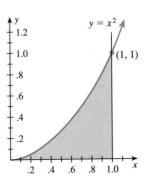

▲ **Figure 11.32   Area under a curve**

News Clip

HOW GLOBAL CLIMATE
IS MODELED
We find a good example of mathematical modeling by looking
at the work being done with
weather prediction. In theory, if
the correct assumptions could be
programmed into a computer,
along with appropriate mathematical statements of the ways
global climate conditions operate, we would have a model to
predict the weather throughout
the world. In the global climate
model, a system of equations calculates time-dependent changes
in wind as well as temperature
and moisture changes in the atmosphere and on the land. The
model may also predict alterations in the temperature of the
ocean's surface. At the National
Center for Atmospheric Research, a CRAY supercomputer
is used to do this modeling.

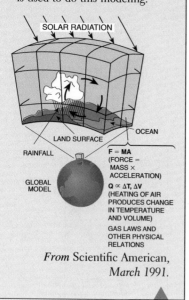

From Scientific American,
March 1991.

This process is shown in Figure 11.33.

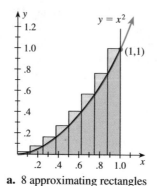 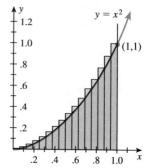

a. 8 approximating rectangles   **b.** 16 approximating rectangles

▲ **Figure 11.33   Approximating the area using rectangles**

The area problem leads to a process called *integration,* and the study of integration forms what is called **integral calculus.** Similar reasoning allows us to calculate such quantities as volume, the length of a curve, the average value, or amount
of work required for a particular task.

Mathematical Modeling

A real-life situation is usually far too complicated to be precisely and mathematically
defined. When confronted with a problem in the real world, therefore, it is usually
necessary to develop a mathematical framework based on certain assumptions about
the real world. This framework can then be used to find a solution to the real-world
problem. The process of developing this body of mathematics is referred to as **mathematical modeling.** Most mathematical models are dynamic (not static), but are
continually being revised (modified) as additional relevant information becomes
known.

Some mathematical models are quite accurate, particularly those used in the
physical sciences. For example, one of the first models we will consider in calculus
is a model for the path of a projectile. Other rather precise models predict such
things as the time of sunrise and sunset, or the speed at which an object falls in a
vacuum. Some mathematical models, however, are less accurate, especially those
that involve examples from the life sciences and social sciences. Only recently has
modeling in these disciplines become precise enough to be expressed in terms of
calculus.

What, precisely, is a mathematical model? Sometimes, mathematical modeling
can mean nothing more than a textbook word problem. But mathematical modeling
can also mean choosing appropriate mathematics to solve a problem that has previously been unsolved. In this book, we use the term *mathematical modeling* to
mean something between these two extremes. That is, it is a process we will apply
to some real-life problem that does not have an obvious solution. It usually cannot
be solved by applying a single formula.

The first step of what we call mathematical modeling involves *abstraction.*

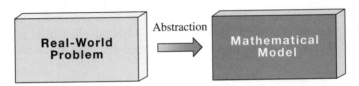

With the method of abstraction, certain assumptions about the real world are made, variables are defined, and appropriate mathematics is developed. The next step is to simplify the mathematics or derive related mathematical facts from the mathematical model.

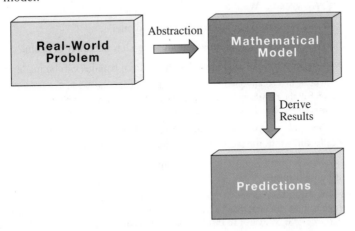

The results derived from the mathematical model should lead us to some predictions about the real world. The next step is to gather data from the situation being modeled, and then to compare those data with the predictions. If the two do not agree, then the gathered data are used to modify the assumptions used in the model.

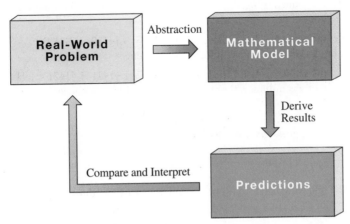

Mathematical modeling is an ongoing process. As long as the predictions match the real world, the assumptions made about the real world are regarded as correct, as are the defined variables. On the other hand, as discrepancies are noticed, it is necessary to construct a closer and a more dependable mathematical model. You might wish to read the article in *Scientific American* described in the News Clip.

PROBLEM SET   11.5

▲ **A Problems**

1. **IN YOUR OWN WORDS** What are the three main topics of calculus?

2. **IN YOUR OWN WORDS** What is a mathematical model?

3. **IN YOUR OWN WORDS** Why are mathematical models necessary or useful?

4. **IN YOUR OWN WORDS** An analogy to Zeno's tortoise paradox can be made as follows:

   A woman standing in a room cannot walk to a wall. To do so, she would first have to go half the distance, then half the remaining distance, and then again half of what still remains. This process can always be continued and can never be ended.

   Draw an appropriate figure for this problem and then present an argument using sequences to show that the woman will, indeed, reach the wall.

5. **IN YOUR OWN WORDS** Zeno's paradoxes remind us of an argument that might lead to an absurd conclusion:

   Suppose I am playing baseball and decide to steal second base. To run from first to second base, I must first go half the distance, then half the remaining distance, and then again half of what remains. This process is continued so that I never reach second base. Therefore it is pointless to steal base.

   Draw an appropriate figure for this problem and then present a mathematical argument using sequences to show that the conclusion is absurd.

6. Consider the sequence 0.3, 0.33, 0.333, 0.3333, ... What do you think is the appropriate limit of this sequence?

7. Consider the sequence 0.2, 0.27, 0.272, 0.2727, ... What do you think is the appropriate limit of this sequence?

8. Consider the sequence 3, 3.1, 3.14, 3.141, 3.1415, 3.14159, 3.141592, ... What do you think is the appropriate limit of this sequence?

9. Consider the sequence 1, 1.4, 1.41, 1.414, 1.4142, 1.41421, 1.414213, 1.4142135, ... What do you think is the appropriate limit of this sequence?

▲ **B Problems**

In Problems 10–13, **guess** *the requested limits.*

10. $\lim\limits_{n\to\infty} \dfrac{2n}{n+4}$ 11. $\lim\limits_{n\to\infty} \dfrac{2n}{3n+1}$ 12. $\lim\limits_{n\to\infty} \dfrac{3n}{n^2+2}$ 13. $\lim\limits_{n\to\infty} \dfrac{3n^2+1}{2n^2-1}$

14. Copy the following figures on your paper. Draw what you think is an appropriate tangent line for each curve at the point P.

   a.   b.

c.   P

d.   P

▲ **Problem Solving**

15. Calculate the sum of the areas of the rectangles shown in Figure 11.33a.

16. Calculate the sum of the areas of the rectangles shown in Figure 11.33b.

17. Use the results of Problems 15–16 to make a guess about the blue shaded area under the curve.

▲ **Individual Research**

18. The definition of probability we gave in Chapter 9 is often cited as an elementary example of mathematical modeling. Recast our definition of theoretical probability in terms of mathematical modeling as described in this section.

11.6 LIMITS

The making of a motion picture is a complex process, and editing all the film into a movie requires that all the frames of the action be labeled in chronological order. For example, R21-435 might signify the 435th frame of the 21st reel. A mathematician might refer to the movie editor's labeling procedure by saying the frames are arranged in a *sequence*. We introduced sequences in Chapter 6, but let's briefly review that idea. A **sequence** is a succession of numbers that are listed according to a given prescription or rule. Specifically, if n is a positive integer, the sequence whose nth term is the number a_n can be written as

$$a_1, \quad a_2, \quad \ldots, \quad a_n, \ldots$$

or more simply

$$\{a_n\}$$

The number a_n is called the **general term** of the sequence. We will deal only with infinite sequences, so each term a_n has a **successor** a_{n+1} and for $n > 1$, a **predecessor** a_{n-1}. For example, by associating each positive integer n with its reciprocal $\frac{1}{n}$, we obtain the sequence denoted by

$$\left\{\frac{1}{n}\right\}, \quad \text{which represents the succession of numbers } 1, \tfrac{1}{2}, \tfrac{1}{3}, \ldots, \tfrac{1}{n}, \ldots$$

The general term is denoted by $a_n = \dfrac{1}{n}$.

The following examples illustrate the notation and terminology used in connection with sequences.

These consecutive frames from a 16-mm movie film show a racehorse. If this film were projected at a rate of 24 frames per second, the viewer would have the illusion of seeing the racehorse in action as he makes his way to the finish line.

EXAMPLE 1

Find the 1st, 2nd, and 15th terms of the sequence $\{a_n\}$ where the general term is

$$a_n = \left(\frac{1}{2}\right)^{n-1}$$

Solution   If $n = 1$, then $a_1 = (\frac{1}{2})^{1-1} = 1$. Similarly,

$$a_2 = (\tfrac{1}{2})^{2-1} = \tfrac{1}{2}$$
$$a_{15} = (\tfrac{1}{2})^{15-1} = (\tfrac{1}{2})^{14} = 2^{-14}$$

The Limit of a Sequence

It is often desirable to examine the behavior of a given sequence $\{a_n\}$ as n gets arbitrarily large. For example, consider the sequence

$$a_n = \frac{n}{n+1}$$

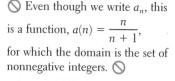 Even though we write a_n, this is a function, $a(n) = \dfrac{n}{n+1}$, for which the domain is the set of nonnegative integers.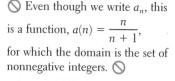

Because $a_1 = \frac{1}{2}$, $a_2 = \frac{2}{3}$, $a_3 = \frac{3}{4}$, . . . , we can plot the terms of this sequence on a number line as shown in Figure 11.34a, or the sequence can be plotted in two dimensions as shown in Figure 11.34b.

Even though we know these tracks to be parallel (they do not actually approach each other), they appear to meet at the horizon. This concept of convergence at a point is used in calculus to give specific values for otherwise unmeasurable quantities.

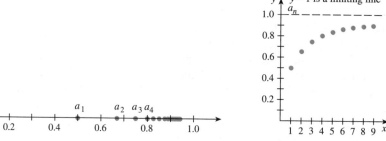

a. Graphing a sequence in one dimension

b. Graphing a sequence in two dimensions

▲ **Figure 11.34   Graphing the sequence $a_n = \dfrac{n}{n+1}$**

By looking at either graph in Figure 11.34, we can see that the terms of the sequence are approaching 1. We say that the limit of this sequence is $L = 1$.

In general, we say that the sequence **converges to the limit L** and write

$$\lim_{n \to \infty} a_n = L$$

if L is a *unique* number that the value of the expression a_n is moving ever and ever closer to as $n \to \infty$. The value L does not need to exist. If a limit does not exist, we say the sequence **diverges.**

EXAMPLE 2

Find the limit L, if it exists.

a. $\displaystyle\lim_{n \to \infty} \frac{n}{n+1}$ **b.** $\displaystyle\lim_{n \to \infty} n$ **c.** $\displaystyle\lim_{n \to \infty} \frac{10,000}{n}$ **d.** $\displaystyle\lim_{n \to \infty} \frac{2n-1}{n}$

Solution

a. $\lim\limits_{n\to\infty} \dfrac{n}{n+1}$ This sequence is shown in Figure 11.34. We see

$$\lim\limits_{n\to\infty} \dfrac{n}{n+1} = 1$$

b. $\lim\limits_{n\to\infty} n$ This sequence is 1, 2, 3, 4, . . . , and we see that the values are not

approaching any limit, so we say this limit does not exist.

c. $\lim\limits_{n\to\infty} \dfrac{10{,}000}{n}$

This sequence is 10,000; 5,000; 3,333.33; . . . ; 2,500; 2,000; . . . . For large n we see that the value of the sequence is close to 0:

$$n = 10{,}000: \quad \dfrac{10{,}000}{n} = \dfrac{10{,}000}{10{,}000} = 1$$

$$\vdots$$

$$n = 10{,}000{,}000: \quad \dfrac{10{,}000}{n} = \dfrac{10{,}000}{10{,}000{,}000} = 0.001$$

$$\vdots$$

It appears that $\lim\limits_{n\to\infty} \dfrac{10{,}000}{n} = 0$

d. $\lim\limits_{n\to\infty} \dfrac{2n-1}{n}$

This sequence can be shown with the following table for selected values of n:

n	1	10	100	1,000	10,000	100,000
$\dfrac{2n-1}{n}$	1	1.9	1.99	1.999	1.9999	1.99999

In this case it appears that $\lim\limits_{n\to\infty} \dfrac{2n-1}{n} = 2.$ ▬

Example 2 leads to some conclusions.

Limits to Infinity

$$\lim\limits_{n\to\infty} \dfrac{1}{n} = 0$$

Furthermore, for any A, and k a positive integer,

$$\lim\limits_{n\to\infty} \dfrac{A}{n^k} = 0$$

■■■■ **EXAMPLE 3**

Find the limit of each of these convergent sequences:

a. $\left\{\dfrac{100}{n}\right\}$ b. $\left\{\dfrac{2n^2 + 5n - 7}{n^3}\right\}$ c. $\left\{\dfrac{3n^4 + n - 1}{5n^4 + 2n^2 + 1}\right\}$

Solution

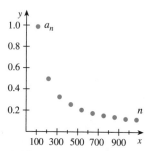

▲ **Figure 11.35   Graphical representation of $a_n = 100/n$**

a. As n grows arbitrarily large, $\dfrac{100}{n}$ get smaller and smaller. Thus,

$$\lim_{n \to \infty} \frac{100}{n} = 0$$

A graphical representation is shown in Figure 11.35.

b. We begin by algebraically rewriting the expression:

$$\frac{2n^2 + 5n - 7}{n^3} = \frac{2}{n} + \frac{5}{n^2} - \frac{7}{n^3}$$

We now use the limit-to-infinity property:

$$\lim_{n \to \infty} \frac{2n^2 + 5n - 7}{n^3} = \lim_{n \to \infty}\left[\frac{2n^2}{n^3} + \frac{5n}{n^3} + \frac{-7}{n^3}\right]$$

$$= \lim_{n \to \infty}\left[\frac{2}{n} + \frac{5}{n^2} + \frac{-7}{n^3}\right]$$

If we assume that the limit of a sum is the sum of the limits, then we have

$$= 0 + 0 + 0$$
$$= 0$$

A graph is shown in Figure 11.36.

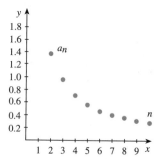

▲ **Figure 11.36   Graph of $a_n = \dfrac{2n^2 + 5n - 7}{n^3}$**

c. Divide the numerator and denominator by n^4 to obtain

$$\lim_{n \to \infty} \frac{3n^4 + n - 1}{5n^4 + 2n^2 + 1} = \lim_{n \to \infty} \frac{\dfrac{3n^4}{n^4} + \dfrac{n}{n^4} - \dfrac{1}{n^4}}{\dfrac{5n^4}{n^4} + \dfrac{2n^2}{n^4} + \dfrac{1}{n^4}} \qquad \text{Divide each term by } n^4.$$

$$= \lim_{n \to \infty} \frac{3 + \dfrac{1}{n^3} - \dfrac{1}{n^4}}{5 + \dfrac{2}{n^2} + \dfrac{1}{n^4}} = \frac{3}{5}$$

▬▬ **EXAMPLE 4**

Show that the following sequences diverge:

a. $\{(-1)^n\}$ **b.** $\left\{ \dfrac{n^5 + n^3 + 2}{7n^4 + n^2 + 3} \right\}$

Solution

a. The sequence defined by $\{(-1)^n\}$ is $-1, 1, -1, 1, \ldots$, and this sequence diverges by oscillation because the nth term is always either 1 or -1. Thus a_n cannot approach one specific number L as n grows large. The graph is shown in Figure 11.37a.

b. $\lim_{n \to \infty} \dfrac{n^5 + n^3 + 2}{7n^4 + n^2 + 3} = \dfrac{1 + \dfrac{1}{n^2} + \dfrac{2}{n^5}}{\dfrac{7}{n} + \dfrac{1}{n^3} + \dfrac{3}{n^5}}$

The numerator tends toward 1 as $n \to \infty$, and the denominator approaches 0. Hence the quotient gets ever larger and larger, passing by all numbers, so it cannot approach a specific number L; thus the sequence must diverge. The graph is shown in Figure 11.37b.

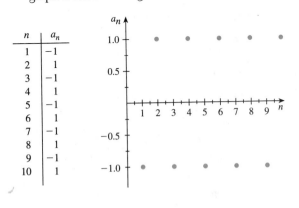

n	a_n
1	-1
2	1
3	-1
4	1
5	-1
6	1
7	-1
8	1
9	-1
10	1

a. $a_n = (-1)^n$

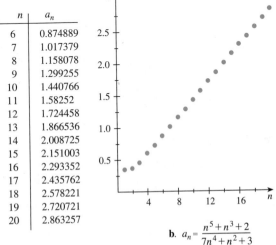

n	a_n
6	0.874889
7	1.017379
8	1.158078
9	1.299255
10	1.440766
11	1.58252
12	1.724458
13	1.866536
14	2.008725
15	2.151003
16	2.293352
17	2.435762
18	2.578221
19	2.720721
20	2.863257

b. $a_n = \dfrac{n^5 + n^3 + 2}{7n^4 + n^2 + 3}$

▲ **Figure 11.37**   **Graphs and values for two divergent sequences**

If $\lim\limits_{n \to \infty} a_n$ does not exist because the numbers a_n become arbitrarily large as $n \to \infty$, we write $\lim\limits_{n \to \infty} a_n = \infty$. We summarize this more precisely in the following box.

Limit Notation

$\lim\limits_{n \to \infty} a_n = \infty$ means that for any real number A, we have $a_n > A$ for all sufficiently large n.

$\lim\limits_{n \to \infty} b_n = -\infty$ means that for any real number B, we have $b_n < B$ for all sufficiently large n.

The answer to Example 4b can be rewritten in this notation as

$$\lim_{n \to \infty} \frac{n^5 + n^3 + 2}{7n^4 + n^2 + 3} = \infty$$

Also notice that $\lim\limits_{n \to \infty} (-5n) = -\infty$ (that is, exceeds all bounds), whereas $\lim\limits_{n \to \infty} (-1)^n$ does not exist. Thus, the answer to Example 4a is *not* ∞ or $-\infty$.

PROBLEM SET 11.6

▲ **A Problems**

1. **IN YOUR OWN WORDS** What do we mean by the limit of a sequence?

2. **IN YOUR OWN WORDS** Outline a procedure for finding the limit of a sequence.

Write out the first five terms (beginning with $n = 1$) of the sequences given in Problems 3–9.

3. $\{1 + (-1)^n\}$ 4. $\left\{ \left(\frac{-1}{2} \right)^{n+2} \right\}$ 5. $\left\{ \frac{3n + 1}{n + 2} \right\}$ 6. $\left\{ \frac{n^2 - n}{n^2 + n} \right\}$

7. $\{a_n\}$ where $a_1 = 256$ and $a_n = \sqrt{a_{n-1}}$ for $n \geq 2$.

8. $\{a_n\}$ where $a_1 = -1$ and $a_n = n + a_{n-1}$ for $n \geq 2$.

9. $\{a_n\}$ where $a_1 = 1$ and $a_n = (a_{n-1})^2 + a_{n-1} + 1$ for $n \geq 2$.

Compute the limit of the convergent sequences in Problems 10–17.

10. $\left\{ \frac{5n + 8}{n} \right\}$ 11. $\left\{ \frac{5n}{n + 7} \right\}$ 12. $\left\{ \frac{2n + 1}{3n - 4} \right\}$ 13. $\left\{ \frac{4 - 7n}{8 + n} \right\}$

14. $\left\{ \frac{8n^2 + 800n + 5,000}{2n^2 - 1,000n + 2} \right\}$ 15. $\left\{ \frac{100n + 7,000}{n^2 - n - 1} \right\}$

16. $\left\{ \frac{8n^2 + 6n + 4,000}{n^3 + 1} \right\}$ 17. $\left\{ \frac{n^3 - 6n^2 + 85}{2n^3 - 5n + 170} \right\}$

▲ **B Problems**

Find the limit (if it exists) as $n \to \infty$ *for each of the sequences in Problems* 18–29.

18. $\frac{1}{3}, \frac{1}{6}, \frac{1}{9}, \frac{1}{12}, \ldots, \frac{1}{3n}, \ldots$

19. $0.69, 0.699, 0.6999, 0.69999, \ldots$

20. $5, 5\frac{1}{2}, 5\frac{2}{3}, 5\frac{3}{4}, 5\frac{4}{5}, \ldots$

21. $-6, -6.5, -6.25, -6.125, -6.0625, \ldots$

22. $\lim\limits_{n \to \infty} \dfrac{3n^2 - 7n + 2}{5n^4 + 9n^2}$

23. $\lim\limits_{n \to \infty} \dfrac{4n^4 + 10n - 1}{9n^3 - 2n^2 - 7n + 3}$

24. $\lim\limits_{n \to \infty} \dfrac{2n^4 + 5n^2 - 6}{3n + 8}$

25. $\lim\limits_{n \to \infty} \dfrac{10n^3 + 13}{7n^2 - n + 2}$

26. $\lim\limits_{n \to \infty} \dfrac{15 + 9n - 2n^2}{6n^2 - 4n + 1}$

27. $\lim\limits_{n \to \infty} \dfrac{n^4 + 5n^3 + 8n^2 - 4n + 12}{2n^6 - 7n^4 + 3n}$

28. $\lim\limits_{n \to \infty} \dfrac{7n^5 - 21n^3 + 52}{-8n^3 + n^2 + 20n - 9}$

29. $\lim\limits_{n \to \infty} \dfrac{12n^5 + 7n^4 - 3n^3 + 2n}{n^5 + 8n^3 + 14}$

▲ **Problem Solving**

30. Twenty-four milligrams of a drug is administered into the body. At the end of each hour, the amount of drug present is half what it was at the end of the previous hour. What amount of the drug is present at the end of 5 hours? At the end of n hours?

31. The Fibonacci Rabbit Problem The general term in a sequence can be defined in a number of ways. In this problem, we consider a sequence whose nth term is defined by a *recursion formula* — that is, a formula in which the nth term is given in terms of previous terms in the sequence. The problem was originally examined by Leonardo Pisano (also called Fibonacci) in the 13th century, and was first introduced in Chapter 6. Recall that we suppose rabbits breed in such a way that each pair of adult rabbits produces a pair of baby rabbits each month.

Number of Months	Number of Pairs	Pairs of Rabbits (the pairs shown in color are ready to reproduce in the next month)
Start	1	
1	1	
2	2	
3	3	
4	5	
5	8	
⋮	⋮	Same pair (rabbits never die)

Further assume that young rabbits become adults after two months and produce another pair of offspring at that time. A rabbit breeder begins with one adult pair. Let a_n
denote the number of adult pairs of rabbits in this "colony" at the end of n months.

a. Explain why $a_1 = 1$, $a_2 = 1$, $a_3 = 2$, $a_4 = 3$, and in general, $a_{n+1} = a_{n-1} + a_n$ for
$n = 2, 3\ 4, \ldots$.

b. The *growth rate* of the colony during the $(n + 1)$st month is

$$r_n = \frac{a_{n+1}}{a_n}$$

Compute r_n for $n = 1, 2, 3, \ldots, 10$.

c. Assume that the growth rate sequence $\{r_n\}$ defined in part **b** converges, and let
$L = \lim_{n \to \infty} r_n$. Use the recursion formula in part **a** to show that

$$\frac{a_{n+1}}{a_n} = 1 + \frac{a_{n-1}}{a_n}$$

and conclude that L must satisfy the equation

$$L = 1 + \frac{1}{L}$$

Use this information to compute L.

"The development of the calculus represents one of the great
intellectual accomplishments in human history."

NCTM *STANDARDS*

JUNE

1

2

3

4

5

6 **Johann Müller (Regiomontanus) (1436),** applied algebra to geometry

7

8

9

10

11 **Robert Hamilton (1743),** abstract algebra

12

13

14

15

16 **Julius Plücker (1801),** analytic geometry

17

18 **Pythagoras** (ca. 580 B.C.), geometry, Pythagorean theorem

19 **Blaise Pascal (1623),** first calculating machine, analytic geometry, philosophy, p. 11

20

21 **Siméon Poisson (1781),** probability, statistics

22

23 **Alan Turing (1912),** computer science, p. 169

24

25

26

27 **Augustus De Morgan (1806),** logic; one of the most influential teachers in history

28 **Henri Lebesgue (1875),** calculus

29

30

Biographical Sketch

"I enjoy being able to deduce the answer, rather than simply having it memorized."

Marcia Sward

Marcia Sward was born in Maywood, Illinois, on February 1, 1939. She is the Executive Director of the Mathematical Association of America and previously was the Executive Director of the Mathematical Sciences Education Board of the National Research Council.

Dr. Sward was most influenced in mathematics by two teachers, Martha Hildebrandt in high school, and Winnifred Asprey at Vassar. "These teachers were excited about mathematics, and encouraged me in mathematics," Marcia reminisced. "At the time Vassar was an all women's college and it did not seem unusual to be a mathematics major, but when I transferred to the University of Illinois I found that I was one of 6 women in over 300 TAs and graduate assistants in mathematics. My high school teacher, Martha Hildebrandt, advised me to 'Shoot for the stars.'"

"I'm not very good at remembering things, so what I enjoy most about mathematics is its deductive nature," Dr. Sward confessed. "I enjoy being able to deduce the answer, rather than simply having it memorized. I think this is why I decided to go into mathematics. . . . When I'm not at the office, I love white-water rafting, hiking, yoga, and playing the piano. I think it is important that students see that mathematicians do things other than teach and administer."

Book Report

Write a 500-word report on the book:

▲ *To Infinity and Beyond: A Cultural History of the Infinite.* Eli Maor (Boston: Birkhäuser, 1987).

Important Terms

Abscissa [11.1]
Axes [11.1]
Boundary [11.2]
Calculus [11.5]
Cartesian coordinate system [11.1]
Closed half-plane [11.2]
Converge [11.6]
Coordinates [11.1]
Dependent variable [11.1]
Derivative [11.5]
Differential calculus [11.5]
Diverge [11.6]
Exponential curve [11.3]

Exponential equation [11.3]
First-degree equation [11.1]
Function [11.4]
Function machine [11.4]
Functional notation [11.4]
General term [11.6]
Half-plane [11.2]
Horizontal line [11.1]
Independent variable [11.1]
Integral [11.5]
Integral calculus [11.5]
Limit [11.5]
Limit of a sequence [11.6]
Line [11.1]

Linear equation [11.1]
Mathematical modeling [11.5]
Open half-plane [11.2]
Ordinate [11.1]
Origin [11.1]
Parabola [11.3]
Point–slope form [11.1]
Population growth [11.3]
Predecessor [11.6]
Quadrant [11.1]
Satisfy [11.1]
Secant line [11.5]
Sequence [11.6]

Slope [11.1]
Slope–intercept form [11.1]
Slope point [11.1]
Solution [11.1]
Standard form [11.1]
Successor [11.6]
Tangent line [11.5]
Test point [11.2]
Two-point form [11.1]
Vertex [11.3]
Vertical line [11.1]
x-intercept [11.1]
y-intercept [11.1]

Important Ideas

Graphing a line by plotting points [11.1]
Graphing a line by slope–intercept [11.1]
Distinguish vertical lines and horizontal lines [11.1]
Graphing half-planes [11.2]
Graphing equations by plotting points [11.3]
Graphing parabolas [11.3]
Graphing exponential curves [11.3]

Population growth model [11.3]
Money growth model [11.3]
Using functional notation [11.4]
Discuss the nature of calculus [11.5]
Mathematical modeling [11.5]
Limit of a sequence [11.6]
Limits to infinity [11.6]

Types of Problems

Graph lines by plotting points and by using the slope–intercept [11.1]
Draw a line when given a point and the slope [11.1]
Solve applied problems involving lines, including future value, depreciation, profit, marginal profit, cost, and marginal cost [11.1]
Graph first-degree inequalities with two unknowns [11.2]
Graph parabolas and exponential curves by plotting points [11.3]
Solve applied problems involving exponential and parabolic models [11.3]
Decide whether a given set is a function [11.4]
Determine the output value for a function [11.4]
Evaluate functions [11.4]

Find $\dfrac{f(x + h) - f(x)}{h}$ for a given function f [11.4]
Describe the limit process, including Zeno's paradox [11.5]
Describe the derivative, including the tangent problem [11.5]
Describe the integral, including the area problem [11.5]
Be able to guess the limit of a sequence [11.5]
Be able to draw the line tangent to a curve at a specified point [11.5]
Approximate an area by using rectangles [11.5]
Write out the first 5 terms of a sequence when given a general term [11.6]
Find the limit of a sequence [11.6]
Solve applied problems involving limits of sequences [11.6]

CHAPTER 11 Review Questions

Graph the lines, curves, or half-planes in Problems 1–8.

1. $5x - y = 15$

2. $y = -\frac{4}{5}x - 3$

3. $2x + 3y = 15$

4. $x = -\frac{2}{3}y + 1$

5. $x = 150$

6. $x < 3y$

7. $y = 1 - x^2$

8. $y = -2^x$

9. Is $\{(4, 3), (5, -2), (6, 3)\}$ a function? Tell why or why not.

Find the value of each function in Problems 10–13.

10. $f(x) = 3x + 2$; find $f(6)$.

11. $g(x) = x^2 - 3$; find $g(0)$.

12. $F(x) = 5x + 25$; find $F(10)$.

13. $m(x) = 5$; find $m(10)$.

Evaluate the limits in Problems 14–16.

14. $\lim\limits_{n \to \infty} \dfrac{1}{n}$

15. $\lim\limits_{n \to \infty} \dfrac{3n^4 + 20}{7n^4}$

16. $\lim\limits_{n \to \infty} (2n + 3)$

17. What are the main ideas of calculus? Briefly describe each of these main ideas.

18. An amount of money A results from investing a sum P at a simple interest rate of 7% for 10 years, and is specified by the formula

$$A = 1.7P$$

Graph this equation, where P is the independent variable.

19. If a cannonball is fired upward with an initial velocity of 128 feet per second, its height can be calculated according to the formula

$$y = 128t - 16t^2$$

where t is the length of time (in seconds) after the cannonball is fired. Sketch this equation by letting $t = 0, 1, 2, \ldots, 7, 8$. Connect these points with part of a parabola.

20. Draw a population curve for a city whose growth rate is 1.3% and whose present population is 53,000. The equation is

$$P = P_0 e^{rt}$$

Let $t = 0, 10, \ldots, 50$ to help you find points for graphing this curve.

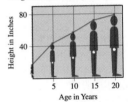

Height Measured Every Five Years

A rough estimate of a man's growth from childhood to maturity is obtained by measuring his height every five years. The straight lines between heads show the rate of growth.

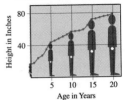

Height Measured Every Year

A more accurate approximation of growth is obtained by taking yearly measurements. The straight lines connecting the heads now begin to blend into a continuous curve.

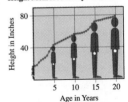

Height Measured Every Six Months

Measurements taken every six months give a still more accurate rate-of-growth line. With increasingly smaller intervals, the lines merge into an ever-smoother curve.

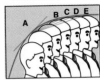

Growth Rate at an Instant

The lines AE, AD, AC and AB above show average growth rates for successively smaller periods of time. But for the Instant A, the growth rate is shown by the tangent at A.

Revisiting the Real World . . .

IN THE REAL WORLD   "Well, Billy, can I make it?" asked Evel. "If you can accelerate to the proper speed, and if the wind is not blowing too much, I think you can," answered Billy. "It will be one huge money-maker, but I want some assurance that it can be done!" retorted Evel. Suppose a daredevil wants to attempt a skycycle ride across the Snake River. What factors are necessary to determine whether he will make it?

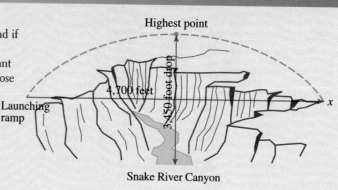

Commentary: We use Polya's method.

Understand the Problem   We need to know whether the skycycle ride across the Snake River Canyon has any chance of succeeding. There are many factors to consider: the speed at launch, the angle of the launch ramp, the speed of the skycycle, and the competence of Evel, the driver. We wish to consider all possibilities and conclude that the trip is possible, or that the trip is impossible.

Devise a Plan.   We will hire an engineering company to consider all the factors dealing with the mathematics and the physics of the launch. Suppose it is concluded that the path of Evel Knievel's skycycle is a parabola that can be described by the equation

$$y = -0.0005x^2 + 2.39x$$

This is assuming that the ramp is at the origin and x is the horizontal distance traveled. Graph this relationship.

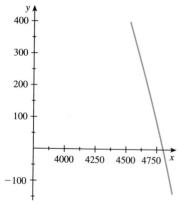

Using your graph, answer Evel's question: Will he make it? Assume that the actual distance the skycycle must travel is 4,700 ft.

Carry Out the Plan.   We can draw the graph representing the equation provided by the consulting firm. We seek to answer the question: Will the point of intersection of this graph and the x-axis be shorter or longer than 4,700 (the width of the Snake River Canyon)? We look at a detail of the first graph, shown in the margin.

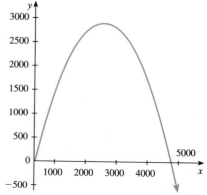

Look Back.   Since the x-intercept is greater than 4,700, we conclude that Evel will make it.

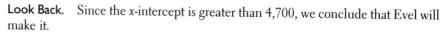

Group Research

Working in small groups is typical of most work environments, and being able to work with others to communicate specific ideas is an important skill to learn. Work with three or four other students to submit a single report based on each of the following questions.

1. If the path of a baseball is parabolic and is 200 ft wide at the base and 50 ft high at the vertex, write an equation that specifies the path of the baseball if the origin is the point of departure for the ball, and the form of the equation is $y - k = a(x - h)^2$, where (h, k) are the coordinates of the highest point of the baseball.

2. According to the Centers for Disease Control, the number of AIDS-related cases is shown in Table 11.1.
   a. Use the formula $A = A_0 e^{rt}$ to predict the number of cases in 1996.
   b. Plot the data points and approximate by using a normal curve with mean at 1988 and standard deviation $\sqrt{5}$. Predict the number of cases in 1996.
   c. Research the number of AIDS cases in 1996 and decide whether the model in part a or part b makes a better prediction.

3. Prepare a list of women mathematicians from the history of mathematics. Answer the question, "Why Were So Few Mathematicians Female?"

   References
   Teri Perl, *Math Equals: Biographies of Women Mathematicians plus Related Activities* (Reading, MA: Addison-Wesley Publishing Co., 1978).
   Loretta Kelley, "Why Were So Few Mathematicians Female?" *The Mathematics Teacher*, October 1996.
   Barbara Sicherman and Carol H. Green, eds. *Notable American Women: The Modern Period*. A Biographical Dictionary. (Cambridge, MA: Belknap Press, Harvard University Press, 1980).
   Outstanding Women in Mathematics and Science (National Women's History Project, Windsor, CA 95492, 1991).

4. Prepare a list of black mathematicians from the history of mathematics.

   Reference Virginia Newell et al., eds. *Black Mathematicians and Their Works* (Ardmore, PA: Dorrance & Company, 1980).

5. Prepare a list of mathematicians with the first name of Karl.

6. Write a news article about a historical mathematician as if you were a contemporary of the person you are writing about. Put it in newspaper style and include other newsworthy items from the period.

TABLE 11.1   Number of AIDS-Related Cases

Year	No. of new cases
1982	920
1983	2,573
1984	5,237
1985	9,328
1986	14,705
1987	19,333
1988	21,978

The Nature of Mathematical Systems

Now, when these studies reach the point of intercommunion and connection with one another, and come to be considered in their mutual affinities, then, I think, but not till then, will the pursuit of them have a value for our objects; otherwise there is no profit in them.

Plato

In the Real World . . .

"I think I'm going crazy, Bill," said George. "I can't figure out a good rating system for the state league. I've got over a hundred teams and I need to come up with a statewide rating system to rank all of those teams. Any ideas?"

"As a matter of fact, yes!" said Bill enthusiastically. "Have you ever heard of the Harbin Football Team rating system? I think they use it in Ohio. They use it to determine the high school football teams that are eligi-

ble to compete in post-season playoffs. Each team earns points for games it wins and for games that a defeated opponent wins. Here's how it works: Level 1 points are awarded for each game a team wins. A level-2 point is awarded for each game a defeated opponent wins."

"I hear ya, but I still don't get it. How do you put all that together?" asked George. "Well," said Bill, "it's very easy . . . . We simply use a matrix . . ."

Contents

Perspective

Solving systems of equations is a procedure that arises throughout mathematics. One of the fundamental building blocks of this book has been problem solving, and in most real-life situations, solving a problem requires (1) understanding the problem, (2) devising a method of solution, (3) carrying out the method, and (4) checking. Devising a method of solution may involve many quantities that are unknown, as well as many relationships between and among those quantities. Systems of equations and inequalities may facilitate the solution of a particular problem. This chapter begins by reviewing those methods, which you have probably studied in a previous course, and then proceeds to two methods that are particularly suitable for calculator or computer solutions — namely, Gauss–Jordan elimination and inverse matrix methods. The chapter concludes by considering some applied problems that involve finding a maximum or minimum value subject to a set of constraints.

12.1   SYSTEMS OF LINEAR EQUATIONS

Two or more equations that are to be solved at the same time make up a **system of equations.** The **simultaneous solution** of a system of equations is the intersection of the solution sets of the individual equations. We use a brace to show that we are looking for a simultaneous solution. If all the equations in a system are linear, it is called a **linear system.**

We begin by reviewing those methods of solving linear systems that you first encountered in your previous algebra courses, and then we generalize first to *non-linear systems*, and later in this chapter to more complicated linear systems.

Graphing Method

The graph of each equation in a system of linear equations in two variables is a line; in the Cartesian plane, two lines must be related to each other in one of three ways:

1. They intersect at a single point.
2. The graphs are parallel lines. In this case, the solution set is empty, and the system is called *inconsistent.* In general, any system that has an empty solution set is referred to as an **inconsistent system.**
3. The graphs are the same line. In this case, there are infinitely many points in the solution set, and any solution of one equation is also a solution of the other. Such a system is called a **dependent system.**

In other words: Graph the equations and check for the intersection point. If the graphs do not intersect, then the equations represent an inconsistent system. If they coincide, then they represent a dependent system.

■■■■ EXAMPLE 1

Solve the given systems by graphing:

a. $\begin{cases} 2x - 3y = -8 \\ x + y = 6 \end{cases}$ **b.** $\begin{cases} 2x - 3y = -8 \\ 4x - 6y = 0 \end{cases}$ **c.** $\begin{cases} 2x - 3y = -8 \\ y = \frac{2}{3}x + \frac{8}{3} \end{cases}$

Solution

a. Graph the line $2x - 3y = -8$:

$$3y = 2x + 8$$
$$y = \tfrac{2}{3}x + \tfrac{8}{3}$$

Graph the line $x + y = 6$:

$$y = -x + 6$$

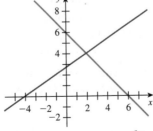

▲ **Figure 12.1   Graph of** $\begin{cases} 2x - 3y = -8 \\ x + y = 6 \end{cases}$

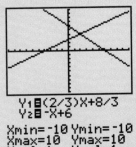
Look at the point(s) of intersection, as shown in Figure 12.1. The solution appears to be (2, 4), which can be verified by direct substitution into both the given equations.

b. The graphs of the given lines are shown in Figure 12.2.

Notice that these lines are parallel; you can show this analytically by noting that the slopes of the lines are the same. Since they are distinct parallel lines, there is not a point of intersection. This is an *inconsistent system*.

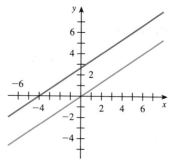

▲ **Figure 12.2   Graph of** $\begin{cases} 2x - 3y = -8 \\ 4x - 6y = 0 \end{cases}$

c. The graphs of the given lines are shown in Figure 12.3.

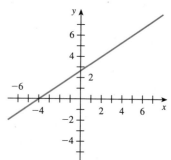

▲ **Figure 12.3   Graph of** $\begin{cases} 2x - 3y = -8 \\ y = \frac{2}{3}x + \frac{8}{3} \end{cases}$

The equations represent the same line. This is a *dependent system*.   ▬

Substitution Method

The graphing method can give solutions as accurate as the graphs you can draw, and consequently it is adequate for many applications. However, there is often a need for more exact methods.

In general, given a system, the procedure for solving is to write a simpler equivalent system. Two systems are said to be **equivalent** if they have the same solution set. In this book we limit ourselves to finding only real roots. There are several ways to go about writing equivalent systems. The first nongraphical method we consider comes from the substitution property of real numbers and leads to a **substitution method** for solving systems.

> **Substitution Method for Solving Systems of Equations with Two Equations and Two Unknowns**
>
> 1. *Solve* one of the equations for one of the variables.
> 2. *Substitute* the expression that you obtain into the other equation.
> 3. *Solve* the resulting equation.
> 4. *Substitute* that solution into either of the original equations to find the value of the other variable.
> 5. *State* the solution.

▬▬ EXAMPLE 2

Solve: $\begin{cases} 2p + 3q = 5 \\ q = -2p + 7 \end{cases}$

Solution Since $q = -2p + 7$, substitute $-2p + 7$ for q in the other equation:

$$2p + 3q = 5$$
$$2p + 3(-2p + 7) = 5$$
$$2p - 6p + 21 = 5$$
$$-4p = -16$$
$$p = 4$$

Now, substitute 4 for p in either of the given equations:

$$q = -2p + 7$$
$$= -2(4) + 7$$
$$= -1$$

The solution is $(p, q) = (4, -1)$. If the variables are not x and y, then you must also show the variables along with the ordered pair. This establishes which variable is associated with which number — namely, $p = 4$ and $q = -1$. ▬

Calculator Comment

If you are graphing with variables other than x and y, you will need to reassign them. For this example, we let $x = p$ and $y = q$:

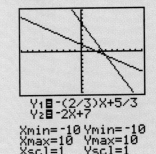

Linear Combination (Addition) Method

A third method for solving systems is called the **linear combination method** (or, as it is often called, the **addition method**). It involves substitution and the idea that if equal quantities are added to equal quantities, the resulting equation is equivalent to the original system. In general, addition will not simplify the system unless the numerical coefficients of one or more terms are opposites. However, you can often force them to be opposites by multiplying one or both of the given equations by nonzero constants.

Linear Combination Method for Solving Systems of Equations

1. *Multiply* one or both of the equations by a constant or constants so that the coefficients of one of the variables become opposites.
2. *Add* corresponding members of the equations to obtain a new equation in a single variable.
3. *Solve* the derived equation for that variable.
4. *Substitute* the value of the found variable into either of the original equations, and solve for the second variable.
5. *State* the solution.

In other words: Multiply one or both of the equations by a constant or constants so that the coefficients of one of the variables are opposites, and then add the equations to eliminate the variable.

▬▬▬ **EXAMPLE 3**

Solve the given system by the linear combination method.

$$\begin{cases} 3x + 5y = -2 \\ 2x + 3y = 0 \end{cases}$$

Solution   Multiply **both sides** of the first equation by 2, and **both sides** of the second equation by -3. This procedure, denoted as shown below, forces the coefficients of x to be opposites:

$$\begin{matrix} 2 \\ -3 \end{matrix} \begin{cases} 3x + 5y = -2 \\ 2x + 3y = 0 \end{cases}$$

This means you should add the equations.

$$\begin{matrix} \downarrow \\ + \end{matrix} \begin{cases} 6x + 10y = -4 \\ -6x - 9y = 0 \end{cases}$$

$$\overline{\qquad\qquad y = -4} \quad \text{Mentally, add the equations.}$$

If $y = -4$, then $2x + 3y = 0$ means $2x + 3(-4) = 0$, which implies that $x = 6$. The solution is $(6, -4)$.   ▬

PROBLEM SET   12.1

▲ **A Problems**

1. **IN YOUR OWN WORDS** What is a system of linear equations?

2. **IN YOUR OWN WORDS** Distinguish inconsistent and dependent systems.

3. **IN YOUR OWN WORDS** Describe the graphing method for solving a system of equations.

4. **IN YOUR OWN WORDS** Describe the substitution method for solving a system of equations.

5. **IN YOUR OWN WORDS** Describe the addition method for solving a system of equations.

Solve the systems in Problems 6–13 by graphing.

6. $\begin{cases} x - y = 2 \\ 2x + 3y = 9 \end{cases}$
   7. $\begin{cases} 3x - 4y = 16 \\ -x + 2y = -6 \end{cases}$
   8. $\begin{cases} y = 3x + 1 \\ x - 2y = 8 \end{cases}$

9. $\begin{cases} 2x - 3y = 12 \\ -4x + 6y = 18 \end{cases}$
   10. $\begin{cases} x - 6 = y \\ 4x + y = 9 \end{cases}$
   11. $\begin{cases} 6x + y = -5 \\ x + 3y = 2 \end{cases}$

12. $\begin{cases} 4x - 3y = -1 \\ -2x + 3y = -1 \end{cases}$
   13. $\begin{cases} 3x + 2y = 5 \\ 4x - 3y = 1 \end{cases}$

Solve the systems in Problems 14–23 by the substitution method.

14. $\begin{cases} y = 3 - 2x \\ 3x + 2y = -17 \end{cases}$
   15. $\begin{cases} 5x - 2y = -19 \\ x = 3y + 4 \end{cases}$
   16. $\begin{cases} 2x - 3y = 15 \\ y = \frac{2}{3}x - 8 \end{cases}$

17. $\begin{cases} x + y = 12 \\ 0.6y = 0.5(12) \end{cases}$

18. $\begin{cases} 4y + 5x = 2 \\ y = \frac{5}{4}x + 2 \end{cases}$

19. $\begin{cases} \dfrac{x}{3} - y = 7 \\ x + \dfrac{y}{2} = 7 \end{cases}$

20. $\begin{cases} 3t_1 + 5t_2 = 1{,}541 \\ t_2 = 2t_1 + 160 \end{cases}$

21. $\begin{cases} x = -7y - 3 \\ 2x + 5y = 3 \end{cases}$

22. $\begin{cases} x = 3y - 4 \\ 5x - 4y = -9 \end{cases}$

23. $\begin{cases} x + 3y = 0 \\ x = 5y + 16 \end{cases}$

Solve the systems in Problems 24–33 by the addition method.

24. $\begin{cases} x + y = 16 \\ x - y = 10 \end{cases}$

25. $\begin{cases} x + y = 560 \\ x - y = 490 \end{cases}$

26. $\begin{cases} 6r - 4s = 10 \\ 2s = 3r - 15 \end{cases}$

27. $\begin{cases} 3u + 2v = 5 \\ 4v = -7 - 6u \end{cases}$

28. $\begin{cases} 3a_1 + 4a_2 = -9 \\ 5a_1 + 7a_2 = -14 \end{cases}$

29. $\begin{cases} 5s_1 + 2s_2 = 23 \\ 2s_1 + 7s_2 = 34 \end{cases}$

30. $\begin{cases} s + t = 12 \\ s - 2t = -4 \end{cases}$

31. $\begin{cases} 2u - 3v = 16 \\ 5u + 2v = 21 \end{cases}$

32. $\begin{cases} 2x + 5y = 7 \\ 2x + 6y = 14 \end{cases}$

33. $\begin{cases} 5x + 4y = 5 \\ 15x - 2y = 8 \end{cases}$

▲ B Problems

Solve the systems in Problems 34–49 for all real solutions, using any suitable method.

34. $\begin{cases} x + y = 7 \\ x - y = -1 \end{cases}$

35. $\begin{cases} x - y = 8 \\ x + y = 2 \end{cases}$

36. $\begin{cases} -x + 2y = 2 \\ 4x - 7y = -5 \end{cases}$

37. $\begin{cases} x - 6y = -3 \\ 2x + 3y = 9 \end{cases}$

38. $\begin{cases} y = 3x + 1 \\ x - 2y = 8 \end{cases}$

39. $\begin{cases} 2x + 3y = 9 \\ x = 5y - 2 \end{cases}$

40. $\begin{cases} x - y = 2 \\ 2x + 3y = 9 \end{cases}$

41. $\begin{cases} 3x - 4y = 16 \\ -x + 2y = -6 \end{cases}$

42. $\begin{cases} 2x - y = 6 \\ 4x + y = 3 \end{cases}$

43. $\begin{cases} 6x + 9y = -4 \\ 9x + 3y = 1 \end{cases}$

44. $\begin{cases} 100x - y = 0 \\ 50x + y = 300 \end{cases}$

45. $\begin{cases} x = \frac{3}{4}y - 2 \\ 3y - 4x = 5 \end{cases}$

46. $\begin{cases} q + d = 147 \\ 0.25q + 0.10d = 24.15 \end{cases}$

47. $\begin{cases} x + y = 10 \\ 0.4x + 0.9y = 0.5(10) \end{cases}$

48. $\begin{cases} 12x - 5y = -39 \\ y = 2x + 9 \end{cases}$

49. $\begin{cases} y = 2x - 1 \\ y = -3x - 9 \end{cases}$

▲ Problem Solving

50. **JOURNAL PROBLEM** (From "When Does a Dog Become Older Than Its Owner?" by Anne Larson Quinn and Karen R. Larson, *The Mathematics Teacher*, December 1996, pp. 734–737.) The premise of this problem is that since dogs supposedly age seven times as quickly as humans, at some point the dog will become "older" than its owner. Karen wanted to determine exactly on which day this milestone would occur for her and her dog, Sydney, so that they could celebrate the occasion. Here are the basic facts. Karen was born on December 7, 1940, and Sydney was born on April 18, 1992. For every year that Karen aged, Sydney aged seven equivalent people-years. On what date are Karen and Sydney "the same age"?

51. Assume that Sydney in Problem 50 is a cat instead of a dog, and assume that cats age four times as quickly as humans. Using this assumption, when will Karen and Sydney be "the same age"?

52. Assume that you obtain a dog that was born today. When will this dog be older (in dog-years) than you are?

53. **HISTORICAL QUESTION** The Louvre Tablet from the Babylonian civilization is dated about 1500 B.C. It shows a system equivalent to

$$\begin{cases} xy = 1 \\ x + y = a \end{cases}$$

Solve this system for x and y in terms of a. This is not a linear system, and you will need to use the quadratic formula after substituting.

54. **HISTORICAL QUESTION** The following problem was written by Leonhard Euler:

Two persons owe conjointly 29 pistoles;* they both have money, but neither of them enough to enable him, singly, to discharge this common debt. The first debtor says therefore to the second, "If you give me $\frac{2}{3}$ of your money, I can immediately pay the debt." The second answers that he also could discharge the debt, if the other would give him $\frac{3}{4}$ of his money. Required, how many pistoles each had?

12.2   PROBLEM SOLVING WITH SYSTEMS

Problem solving has been a major theme in this book since it was first introduced in Chapter 1. The procedure that we described as "evolving to a single variable" is equivalent to solving a system by substitution. However, as we see in this chapter, substitution is only one of the techniques for solving systems. In this chapter, we look at systems that can be solved by a variety of techniques. This section introduces some of the usual types of word problems, and as we develop the rest of this chapter, we will introduce some of the more unusual types of problems that use systems of equations.

Coin Problems

As we have seen, substitution is one of the primary tools used in solving word problems. Many word problems are given in a form that indicates more than two variables. There are at least three ways you can reduce the number of variables in an applied problem:

1. Substitute numbers for variables. Use this type of substitution when the value of a variable is known.
2. Substitute variables for other variables by using a known formula.
3. Substitute variables for other variables when relationships between those variables are given in the problem.

The first type of problem we consider in this section is coin problems. We begin here because coin problems vividly illustrate the need to be careful in defining

---

* A *pistole* is a unit of money.

the variables you use. In money problems, you must distinguish between the *number of coins* and the *value of the coins* because, for example, the number of quarters you have is not the same as the value of those quarters. The formulas you will need for money problems are:

$$\text{VALUE OF QUARTERS} = 25(\text{NUMBER OF QUARTERS})$$
$$\text{VALUE OF DIMES} = 10(\text{NUMBER OF DIMES})$$
$$\text{VALUE OF NICKELS} = 5(\text{NUMBER OF NICKELS})$$
$$\text{VALUE OF PENNIES} = \text{NUMBER OF PENNIES}$$

Notice that the values in the preceding formulas are in terms of cents and not dollars.

■■■ EXAMPLE 1

A box of coins has dimes and nickels totaling $4.20. If there are three more nickels than dimes, how many of each type of coin is in the box?

Solution   Restate the problem in equation form. It does not matter if you use more than two variables.

$$\begin{cases} \text{VALUE OF DIMES} + \text{VALUE OF NICKELS} = \text{TOTAL VALUE} \\ \text{NUMBER OF NICKELS} = \text{NUMBER OF DIMES} + 3 \end{cases}$$

We used five different unknowns in this statement.

Substitute known numbers:

Substitute: **TOTAL VALUE = 420**
                                                    ↓

$$\begin{cases} \text{VALUE OF DIMES} + \text{VALUE OF NICKELS} = \text{TOTAL VALUE} \\ \text{NUMBER OF NICKELS} = \text{NUMBER OF DIMES} + 3 \end{cases}$$

Substitute using a formula:

   10(NUMBER OF DIMES)   **5(NUMBER OF NICKELS)**   Substitute.
              ↓   ↓

$$\begin{cases} \text{VALUE OF DIMES} \quad + \quad \text{VALUE OF NICKELS} \quad = 420 \\ \text{NUMBER OF NICKELS} \quad = \text{NUMBER OF DIMES} + 3 \end{cases}$$

$$\begin{cases} 10(\text{NUMBER OF DIMES}) + 5(\text{NUMBER OF NICKELS}) = 420 \\ \text{NUMBER OF NICKELS} = \text{NUMBER OF DIMES} + 3 \end{cases}$$

Now that we have only two unknowns, we choose variables. Let $d =$ NUMBER OF DIMES and $n =$ NUMBER OF NICKELS.

$$\begin{cases} 10d + 5n = 420 \\ n = d + 3 \end{cases}$$

Solve *this* system by substitution:

$$10d + 5n = 420$$
$$10d + 5(d + 3) = 420 \quad \text{Substitute } n = d + 3.$$
$$10d + 5d + 15 = 420$$
$$15d = 405$$
$$d = 27$$

If $d = 27$, then $n = d + 3 = 27 + 3 = 30$. There are 27 dimes and 30 nickels.
 Check: Three more nickels than dimes and the value is
$27(\$0.10) + 30(\$0.05) = \$4.20$.   ▬

Combining Rates

On a recent visit to an airport, I was walking from the terminal to the gate and there was a moving sidewalk. This caused me to think (of course) of a variety of algebra problems. If I walk at a rate of 100 feet per minute and the sidewalk travels at 80 ft per minute, then my rate is:

100 feet per minute if I walk to the terminal without using the moving sidewalk;

80 feet per minute if I stand on the moving sidewalk;

$100 + 80 = 180$ feet per minute if I walk on the moving sidewalk;

$100 - 80 = 20$ feet per minute if I walk against the movement of the moving sidewalk.

This illustrates a general principle. When you move in air, in water, on a treadmill, or on some other medium that is also moving, you combine rates. *If you move in the same direction, your rate is added to the rate of the medium. If you move against the movement of the medium, the rates are subtracted.* Consider the following examples.

▬ EXAMPLE 2

A deep-sea fishing boat travels at 20 mph going out with the tide. Sometimes it comes in against the tidal current and is able to make only 15 mph. What is the boat's rate without the current, and what is the rate at which the tide moves the water?

Solution   We are given:

RATE WITH THE TIDE is 20 mph.

RATE AGAINST THE TIDE is 15 mph.

This gives us the following system of equations:

$$\begin{cases} (\text{RATE OF BOAT}) + (\text{RATE OF TIDE}) = 20 & \text{Same direction: Add rates.} \\ (\text{RATE OF BOAT}) - (\text{RATE OF TIDE}) = 15 & \text{Opposite direction: Subtract rates.} \end{cases}$$

We assume that the rate of the boat is greater so that we obtain a positive distance.

There are two unknowns; let $b = $ RATE OF BOAT and $t = $ RATE OF TIDE.

$$+\begin{cases} b + t = 20 \\ b - t = 15 \end{cases}$$
$$\overline{\quad 2b = 35}$$
$$b = 17.5$$

Substitute to find t:

$$17.5 + t = 20$$
$$t = 2.5$$

The boat's speed is 17.5 mph, and the tide moves at 2.5 mph.

▬▬ EXAMPLE 3

On a certain day a large bird was clocked at 80 mph flying with the wind, but could fly only 10 mph against the wind. What is the speed of the bird in still air?

Solution We are given:

RATE WITH THE WIND is 80 mph.

RATE AGAINST THE WIND is 10 mph.

This gives us the following system:

$$\begin{cases} \text{BIRD'S RATE } + \text{ WIND'S RATE } = 80 \\ \text{BIRD'S RATE } - \text{ WIND'S RATE } = 10 \end{cases}$$

Let $b = $ BIRD'S RATE and $w = $ WIND'S RATE:

$$+\begin{cases} b + w = 80 \\ b - w = 10 \end{cases} \quad \text{Assume the bird's rate is greater than the wind's rate.}$$
$$\overline{\quad 2b = 90}$$
$$b = 45$$

The bird's rate in still air is 45 mph.

⊘ Remember to answer the question that was asked. In Example 3, we were not asked to find the rate of the wind. ⊘

Supply and Demand

We have solved word problems with linear combinations and with substitution. Sometimes it is useful to solve word problems by graphing. One such application has to do with supply and demand. If supply greatly exceeds demand, then money will be lost because of unsold items. On the other hand, if demand greatly exceeds supply, then money will be lost because of insufficient inventory. The most desirable situation is when the supply and demand are equal. The point for which the supply and demand are equal is called the **equilibrium point.** In this section we assume that supply and demand functions are linear.

▬▬ EXAMPLE 4   **Polya's Method**

Suppose you have a small product that is marketable to the students on your campus. You want to know what price to charge for this product to maximize your profit.

Solution   We use Polya's problem-solving guidelines for this example.

Understand the Problem.   There is not sufficient information given to answer this question. *A little market research shows that only* 200 *people would buy the product if it were priced at* $10, *but* 2,000 *would buy it at* $1. This information represents the **demand.** Let p = PRICE and n = NUMBER OF ITEMS. Then, because the demand, n, is determined by the price, let the price, p, be the independent variable. That is, let the ordered pairs be (p, n). From the given information, we see that the demand curve passes through the points $(10, 200)$ and $(1, 2000)$. If we assume that demand is linear (a straight line), we label the line passing through these points as the "demand" line.

    There is still insufficient information. *We call a local shop and find that they can make the product during slack time and could supply* 300 *items at a price that allows you to sell them for* $2. *To supply more, the shop must use overtime. If the shop supplies* 1,500 *items, you will have to charge* $6 *for your product.* This information represents the **supply.** Use (p, n) as defined for demand. From the given information, the supply curve passes through $(2, 300)$ and $(6, 1500)$. If we assume that supply is linear, we draw the line through these points and label it the "supply" line, as shown in Figure 12.4.

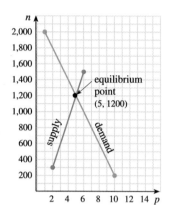

▲ **Figure 12.4   Supply and demand**

Devise a Plan.   We now have sufficient information to solve the problem. Profit is maximized at the equilibrium point. This is the point of intersection of the supply and demand lines. We will look for the intersection point of the supply and demand lines.

Carry Out the Plan.   We see from Figure 12.4 that the equilibrium point is $(5, 1200)$. This means that the price charged should be $5. It also says that you should expect to sell 1,200 items.

Look Back.   As a real modeling problem, you would need to research the parts shown in italic, but for our purposes in this book, you will be supplied this information. ▬

Age Problems

If you are comparing birthdates of individuals, then the *younger* person has the *larger* birthdate (year). For example, if you were born in 1975, and your sister is 6 years younger, then her birthdate is

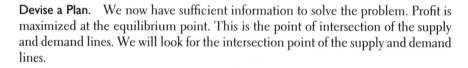

▬ **EXAMPLE 5**

Ron Howard is 9 years older than actor Tom Cruise. If the sum of their birth years is 3,915, in what year was Tom Cruise born?

Solution Let

x = Ron Howard's birth year

y = Tom Cruise's birth year

Since Ron Howard is 9 years older, you add 9 to his birth year to obtain Cruise's birth year: $x + 9 = y$. Thus, we have the following system:

$$\begin{cases} x + 9 = y \\ x + y = 3,915 \end{cases}$$

We solve this system by substituting $x + 9$ for y in the second equation (this is easy since the first equation is already solved for y):

$$x + (x + 9) = 3,915$$
$$2x + 9 = 3,915$$
$$2x = 3,906$$
$$x = 1953$$

We are looking for Tom Cruise's birth year, so $y = x + 9 = 1953 + 9 = 1962$.

⊘ The solution to the equation may not be the answer to the question asked. ⊘

Mixture Problems

Another common textbook application problem is the so-called **mixture problem.** The process involves combining two (or more) ingredients to obtain a mixture. Each ingredient has a certain quantity, and the quantity of the mixture is the sum of the quantities of the added ingredients. Quantities are measured in some appropriate unit; for example, consider the following spreadsheet:

	INGREDIENT I	+	INGREDIENT II	=	MIXTURE
a.	5 lb	+	10 lb	=	15 lb
b.	x lb	+	y lb	=	$(x + y)$ lb
c.	18 oz	+	20 oz	=	38 oz
d.	s oz	+	t oz	=	$(s + t)$ oz
e.	8 L	+	p L	=	$(8 + p)$ L

This idea seems easy enough, but with mixture problems we have additional considerations. We begin with a simple example of mixing peanuts and cashews.

■■■ EXAMPLE 6

a. If you mix 16 lb of peanuts with 4 lb of cashews, how many pounds are in this mixture (called mixture I)?

b. If you mix 2 lb of peanuts with 8 lb of cashews, how many pounds are in this mixture (called mixture II)?

c. If you mix mixtures I and II, what are the percentage of peanuts and the percentage of cashews?

Solution

	PEANUTS	+	CASHEWS	=	MIXTURE	
a.	16 lb +		4 lb =		20 lb	(mixture I)
b.	2 lb +		8 lb =		10 lb	(mixture II)
c.	(16 + 2) lb +		(4 + 8) lb =		(20 + 10) lb	(final mixture)

Notice the double check; columns add down and rows add across to determine the total amount in the final mixture. As percentages these are:

⊘ The percentages of the parts must total 100%. ⊘

$$\frac{18}{30} = 0.6 \text{ or } 60\% \qquad \frac{12}{30} = 0.4 \text{ or } 40\%$$

Mixture problems are often stated using percentages. The following example is a restatement of Example 6.

■■■ EXAMPLE 7

If you mix together 20 lb of a peanut/cashew mixture consisting of 80% peanuts with 10 lb of a peanut/cashew mixture consisting of 20% peanuts to obtain 30 lb of a peanut/cashew mixture consisting of 60% peanuts, how many pounds of peanuts and cashews are present in mixture I, mixture II, and the final mixture?

Solution   We must mix together

	MIXTURE I	+	MIXTURE II	=	FINAL MIXTURE
Given:	20 lb	+	10 lb	=	30 lb

We now use percentages to find the component parts of the final mixture (and to answer the question):

	MIXTURE I	+	MIXTURE II	=	FINAL MIXTURE
PEANUTS	[0.8(20) = 16 lb]	+	[0.2(10) = 2 lb] =		18 lb
CASHEWS	[0.2(20) = 4 lb]	+	[0.8(10) = 8 lb] =		12 lb
TOTAL	[(16 + 4) lb]	+	[(2 + 8) lb]	=	(18 + 12) lb

Check: Final mixture should be 60% peanuts: 0.60(30) = 18 lb.

The last of this related trilogy of introductory examples repeats the information in the previous two examples in the form of a typical mixture problem. If you have difficulty with this example, you can look back at the previous two examples.

■■■ EXAMPLE 8

How much of a peanut/cashew mixture consisting of 80% peanuts must be mixed with a peanut/cashew mixture consisting of 20% peanuts to obtain 30 lb of a peanut/cashew mixture consisting of 60% peanuts?

Solution   We must mix together

$$\text{MIXTURE I} + \text{MIXTURE II} = \text{FINAL MIXTURE}$$

Given: 30 lb

Let x = number of pounds of mixture I

y = number of pounds of mixture II

$$x \quad + \quad y \quad = \quad 30$$

These mixtures are made up of peanuts and cashews:

PEANUTS $0.8x \quad + \quad 0.2y \quad = 0.6(30) = 18$ lb

CASHEWS $0.2x \quad + \quad 0.8y \quad = 0.4(30) = 12$ lb

Notice that the percentages add to 100% (or as decimals add to 1). This gives rise to the following system:

$$\begin{cases} x + y = 30 \\ 0.8x + 0.2y = 18 \\ 0.2x + 0.8y = 12 \end{cases}$$

You notice that we have three equations with two unknowns. We will discuss such systems in general in the following section, but for a problem like this one we can use substitution. From the first equation, we have $y = 30 - x$, which we substitute into either of the other equations:

$$0.8x + 0.2y = 18$$
$$0.8x + 0.2(30 - x) = 18$$
$$0.8x + 6 - 0.2x = 18$$
$$0.6x + 6 = 18$$
$$0.6x = 12$$
$$x = 20$$

If $x = 20$, then $y = 10$; combine 20 pounds of Mixture I with 10 pounds of Mixture II.

Check: $\begin{cases} x + y = 30 \\ 0.8x + 0.2y = 18 \\ 0.2x + 0.8y = 12 \end{cases}$ $\begin{array}{l} 20 + 10 = 30 \quad ✔ \\ 0.8(20) + 0.2(10) = 18 \quad ✔ \\ 0.2(20) + 0.8(10) = 12 \quad ✔ \end{array}$ ▬

▬▬▬ EXAMPLE 9

How many liters of water must be added to 3 liters of an 80% acid solution to obtain a 30% acid solution? (By an 80% acid solution we mean a solution that is 80% acid and 20% water; remember that the sum of the percentages of all ingredients must be 100%.)

Solution   The basic relationship for this mixture is

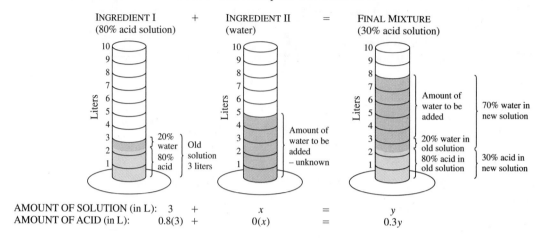

INGREDIENT I (80% acid solution)	+	INGREDIENT II (water)	=	FINAL MIXTURE (30% acid solution)

AMOUNT OF SOLUTION (in L):   3   + x = y
AMOUNT OF ACID (in L):   0.8(3)   +   0(x)   =   0.3y

where x = AMOUNT OF WATER TO BE ADDED

y = AMOUNT AFTER MIXING

This leads to the system

$$\begin{cases} 3 + x = y \\ 2.4 + 0 = 0.3y \end{cases}$$

Solve the second equation for y to obtain $y = 8$, and substitute this into the first equation to obtain

$$3 + x = 8$$
$$x = 5$$

The amount of water to be added is 5 liters.   ▬

▬▬ EXAMPLE 10

Milk containing 10% butterfat and cream containing 80% butterfat are mixed to produce half-and-half, which is 50% butterfat. How many gallons of each must be mixed to make 140 gallons of half-and-half?

Solution

	BEFORE MIXING		AFTER MIXING
AMT OF SOLUTION:	MILK + CREAM	=	HALF-AND-HALF
AMT OF BUTTERFAT:	0.10(MILK) + 0.80(CREAM)	=	0.50(HALF-AND-HALF)

We see that there are two unknowns; let

m = AMOUNT OF MILK

c = AMOUNT OF CREAM

140 = AMOUNT OF HALF-AND-HALF

We have the system of equations (with variables):

AMOUNT OF SOLUTION: $\begin{cases} m + c = 140 \\ 0.10m + 0.80c = 0.50(140) \end{cases}$
AMOUNT OF BUTTERFAT:

Multiply both sides of the second equation by -10:

$$+\begin{cases} m + c = 140 \\ -m - 8c = -700 \end{cases}$$
$$\overline{-7c = -560}$$
$$c = 80$$

If $c = 80$, then $m + 80 = 140$ so that $m = 60$. You must mix 60 gallons of milk and 80 gallons of cream.

PROBLEM SET   12.2

▲ A Problems

Carefully interpret each problem, restate the information and relationships until you are able to write the equations, solve the equations, and state an answer.

Problems 1–3 are patterned after Example 1.

1. A box contains twenty-five more dimes than nickels. How many of each type of coin is there if the total value is $7.15?

2. A box contains only dimes and quarters. The number of dimes is three less than twice the number of quarters. If the total value of the coins is $22.65, how many dimes are in the box?

3. Forty-two coins have a total value of $6.90. If the coins are all nickels and quarters, how many are nickels?

Problems 4–6 are patterned after Examples 2–3.

4. A boat travels 25 mph relative to the riverbank while going downstream and only 15 mph returning upstream. What is the boat's speed in still water?

5. A plane travels 200 mph relative to the ground while flying with a strong wind and only 150 mph returning against it. What is the plane's speed in still air?

6. An airliner flies at 510 mph in the jet stream, but only 270 mph returning against it. What is the speed of the plane in still air?

Problems 7–9 are patterned after Example 4.

7. The demand for a product varies from 150,000 units at $110 per unit to 300,000 at $20 per unit. Also, 300,000 could be supplied at $90 per unit, whereas only 200,000 could be supplied for $10 each. Find the equilibrium point for the system.

8. California Instruments, a manufacturer of calculators, finds by test marketing its calculators at UCLA that 180 calculators could be sold if they were priced at $10, but only 20 calculators could be sold if they were priced at $40. On the other hand, they find that 20 calculators can be supplied at $10 each. If they were supplied at $40 each,

overtime shifts could be used to raise the supply to 180 calculators. What is the optimum price for the calculators?

9. A manufacturer of lapel buttons test marketed a new item at the University of California, Davis. It was found that 900 items could be sold if they were priced at $1, but only 300 items could be sold if the price were raised to $7. On the other hand, they find that 600 items can be supplied at $1 each. If they were supplied at $9 each, overtime shifts could be used to raise the supply to 1,000 items. What is the optimum price for the items?

Problems 10–12 are patterned after Example 5.

10. Mel Brooks is seven years older than fellow funnyman Dom DeLuise. If the sum of their birth years is 3,859, when was DeLuise born?

11. The sum of the birth years of actresses Debra Winger and Meryl Streep is 3,904. If Debra is six years younger, in what year was she born?

12. Ray Charles is fifteen years older than fellow musician José Feliciano. If the sum of their years of birth is 3,875, then in what year was Feliciano born?

Problems 13–18 are patterned after Examples 6–7. The information is based on the spreadsheet shown on page 807. Suppose Ingredient I is made up of 80% micoden and 20% water, Ingredient II is made up of 30% micoden, 50% bixon, and 20% water, and these ingredients are mixed together.

	INGREDIENT I	INGREDIENT II	MIXTURE
a.	5 lb	10 lb	15 lb
b.	x lb	y lb	$(x + y)$ lb
c.	8 L	p L	$(8 + p)$ L

13. How much micoden is in the mixture **a**?

14. How much water is in the mixture **b**?

15. How much bixon is in the mixture **c**?

16. What is the percentage of water in mixture **a**?

17. What is the percentage of bixon in mixture **a**?

18. What is the percentage of micoden in mixture **a**?

Problems 19–21 are patterned after Examples 8–10.

19. How many ounces of a base metal (no silver) must be alloyed with 100 ounces of 21% silver alloy to obtain an alloy that is 15% silver?

20. An after-shave lotion is 50% alcohol. If you have 6 fluid ounces of the lotion, how much water must be added to reduce the mixture to 20% alcohol?

21. Milk containing 20% butterfat is mixed with cream containing 60% butterfat to produce half-and-half, which is 50% butterfat. How many gallons of each must be mixed to make 180 gallons of half-and-half?

▲ B Problems

Problems 22–58 provide a variety of word problems. Answer each question.

22. Suppose a car rental agency gives the following choices:

    Option A: $30 per day plus 40¢ per mile
    Option B: Flat $50 per day (unlimited miles)

    At what mileage are both rates the same if you rent the car for three days?

23. Suppose a car rental agency gives the following options:

    Option A: $40 per day plus 50¢ per mile
    Option B: Flat $60 per day with unlimited mileage

    At what mileage are both rates the same if you rent the car for four days?

24. The supply for a certain commodity is linear and determined to be $p = 0.005n + 12$, whereas the demand is linear with $p = 150 - 0.01n$, where p is the price and n is the number of items. What is the equilibrium point (p, n)?

25. A certain item has a linear supply curve $p = 0.0005n - 3$ and a linear demand $p = 8 - 0.0006n$, where p is the price and n is the number of items. What is the equilibrium point (p, n)?

26. There are six more dimes than quarters in a container. How many of each coin is there if the total value is $3.75?

27. A box contains $8.40 in quarters and dimes. The number of quarters is twice the number of dimes. How many of each type of coin is in the box?

28. A canoeist rows downstream in $1\frac{1}{2}$ hr and back upstream in 3 hr. What is the rate of the current if the canoeist rows 9 miles in each direction?

29. A plane makes a 660-mile flight with the wind in $2\frac{1}{2}$ hours. Returning against the wind takes 3 hours. Find the wind speed.

30. Charles Bronson was born nine years before another movie hardguy, Clint Eastwood. If the sum of their years of birth is 3,853, in what year was Eastwood born?

31. Kristy McNichol is just a year older than fellow actress Tatum O'Neal. The sum of their years of birth is 3,935. In what year was Kristy born?

32. You have a 24% silver alloy and some pure silver. How much of each must be mixed to obtain 100 oz of 43% silver?

33. How much water must be added to a gallon of 80% antifreeze to obtain a 60% mixture?

34. How much antifreeze must be added to a gallon of 60% antifreeze to obtain an 80% mixture?

35. The combined area of New York and California is 204,192 square miles. The area of California is 108,530 square miles more than that of New York. Find the land area of each state.

36. The area of Texas is 208,044 square miles greater than that of Florida. Their combined area is 316,224 square miles. What is the area of each state?

37. Forty-two coins have a total value of $9.50. If the coins are all nickels and quarters, how many are quarters?

38. A collection of coins has a value of $4.76. There is the same number of nickels and dimes but four fewer pennies than nickels or dimes. How many pennies are in the collection if there are 86 coins?

39. A bunch of change contains nickels, dimes, and quarters. There are the same number of dimes and quarters, and there are eight more nickels than either dimes or quarters. How many dimes are there if the value of the 98 coins is $12.40?

40. A box contains $8.40 in nickels, dimes, and pennies. How many of each type of coin is in the box if the number of dimes is six less than twice the number of pennies, and there is an equal number of dimes and nickels in the box?

41. Sherlock Holmes was called in as a consultant to solve the Great Bank Robbery. He was told that the thief had made away with a bag of money containing $5, $10, and $20 bills totaling $1,390. When the bankers were checking serial numbers to see how many of each denomination were taken, Holmes said, "It is elementary, since there were five times as many $10 bills as $5 bills and three more than twice as many $20 bills as $5 bills." How many of each denomination were taken?

42. A plane makes an 870-mile flight in $3\frac{1}{3}$ hours against a strong head wind, but returns in 50 minutes less with the wind. What is the plane's speed without the wind?

43. A plane with a tail wind makes its 945-mile flight in 3 hours. The return flight against the wind takes a half hour longer. What is the wind speed?

44. A chemist has two solutions of sulfuric acid. One is a 50% solution, and the other is a 75% solution. How many liters of each does the chemist mix to get 10 liters of a 60% solution?

45. A dairy has cream containing 23% butterfat and milk that is 3% butterfat. How much of each must be mixed to obtain 30 gallons of a richer milk containing 4% butterfat?

46. A winery has a large amount of a wine labeled "Lot I," which is a mixture of 92% Merlot wine and 8% Cabernet Sauvignon wine. It also has a second wine labeled "Lot II," which is a mixture of 88% Cabernet Sauvignon and 12% Merlot. The winemaster decides to mix together these two lots to obtain 500 gallons of a mixture that is 50% Merlot and 50% Cabernet Sauvignon. How much of each should be used to obtain this blend?

47. A pain remedy contains 12% aspirin, and a stronger formula has 25% aspirin, but is otherwise the same. A chemist mixes some of each to obtain 100 mg of a mixture with 20% aspirin. How much of each is used?

48. The combined height of the World Trade Center and the Empire State Building is 2,600 ft. The Trade Center is 100 ft taller. What is the height of each of the New York City skyscrapers?

49. The Standard Oil building and the Sears Tower have a combined height of 2,590 ft. The Sears Tower is 318 ft taller. What is the height of each of these Chicago towers?

50. The combined length of the Golden Gate and San Francisco Bay bridges is 6,510 ft. If the Golden Gate is 1,890 ft longer, what is its length?

51. End to end, the Verrazano–Narrows and the George Washington bridges would span 7,760 ft. If the Verrazano–Narrows is the longer of the two New York structures by 760 ft, how long is it?

52. Noxin Electronics has investigated the feasiblity of introducing a new line of magnetic tape. The study shows that both supply and demand are linear. The supply can increase from 1,000 items at $2 each to 5,000 units at $4 each. The demand ranges from 1,000 items at $4 to 7,000 at $3. What is the equilibrium point of this supply-and-demand system?

53. Noxin Electronics is considering producing a small cassette line. Research shows linear demand to be from 40,000 cassettes at $2 to 100,000 at $1. Similarly, the supply goes from 20,000 cassettes at 50¢ to 80,000 at $5. What is the equilibrium point?

54. Sterling silver contains 92.5% silver. How many grams of pure silver and sterling silver must be mixed to get 100 grams of a 94% alloy?

55. The radiator of a car holds 17 quarts of liquid. If it now contains 15% antifreeze, how many quarts must be replaced by antifreeze to give the car a 60% solution in its radiator?

56. How many gallons of 24% butterfat cream must be mixed with 500 gallons of 3% butterfat milk to obtain a 4% butterfat milk?

57. The supply curve for a new software product is given by $n = 2.5p - 500$, and the demand curve for the same product is $n = 200 - 0.5p$, where n is the number of items and p is the number of dollars.

    a. At $250 for the product, how many items would be supplied? How many would be demanded?
    b. At what price would no items be supplied?
    c. At what price would no items be demanded?
    d. What is the equilibrium price for this product?
    e. How many units will be produced at the equilibrium price?

58. The supply curve for a certain commodity is given by $n = 2,500p - 500$, and the demand curve for the same product is $n = 31,500 - 1,500p$, where n is the number of items and p is the number of dollars.

    a. At $15 per unit of the commodity, how many items would be supplied? How many would be demanded?
    b. At what price would no items be supplied?
    c. At what price would no items be demanded?
    d. What is the equilibrium price for this product?
    e. How many units will be produced at the equilibrium price?

▲ **Problem Solving**

What is wrong, if anything, with each of the statements in Problems 59–60? Explain your reasoning.

59. A bottle and a cork cost $1.10. What is the cost of the cork alone, if the bottle costs a dollar more? *Answer:* The cork costs 10¢ and the bottle costs $1.

60. You have identical cups, one containing coffee and one containing cream. One teaspoon of the cream is added to the coffee and stirred in. Now a teaspoon of the coffee/cream mixture is added back to the cup of cream and stirred in. Both cups contain identical amounts of liquid, but is there more cream in the coffee or more coffee in the cream? *Answer:*   There is obviously more cream in the coffee since it was added before any coffee was removed.

12.3   MATRIX SOLUTION OF A SYSTEM OF EQUATIONS

One of the most common types of problems to which we can apply mathematics in a variety of different disciplines is to the solution of systems of equations. In fact, common real-world problems require the simultaneous solution of systems involving 3, 4, 5, or even 20 or 100 unknowns. The methods of the previous section will not suffice and, in practice, techniques that will allow computer or calculator help in solving systems are common. In this section, we introduce a way of solving large systems of equations in a general way so that we can handle the solution of a system of m equations with n unknowns.

Definition of a Matrix

You are already familiar with matrices from everyday experiences. For example, the following table shows a rental chart in the form of a matrix.

Country	Fiat	Opel	Renault
Austria	$149	$219	$289
Belgium	$130	$222	$273
Denmark	$164	$269	$408
France	$189	$206	$215
G. Britain	$160	$208	$225
Holland	$156	$213	$299
Italy	$179	$259	$353
Spain	$156	$210	$247
Sweden	$178	$246	$281

Considering the numerical entries in this chart as a 9 by 3 matrix, we would say that the price of the Renault in Belgium is found in row 2, column 3 (namely, $273). In contrast, the entry in row 3, column 2 ($269), is the cost of the Opel in Denmark.

Consider a system with unknowns x_1 and x_2. We use **subscripts** 1 and 2 to denote the unknowns, instead of using variables x and y, because we want to be able to handle n unknowns, which we can easily denote as $x_1, x_2, x_3, \ldots, x_n$; if we continued by using $x, y, z, \ldots$ for systems in general, we would soon run out of letters. Here is the way we will write a general system of two equations with two unknowns:

$$\begin{cases} a_{11}x_1 + a_{12}x_2 = b_1 \\ a_{21}x_1 + a_{22}x_2 = b_2 \end{cases}$$

The coefficients of the unknowns use **double subscripts** to denote their position in the system; a_{11} is used to denote the numerical coefficient of the first variable in the first row; a_{12} denotes the numerical coefficient of the second variable in the first row; and so on. The constants are denoted by b_1 and b_2.

We now separate the parts of this system of equations into rectangular **arrays** of numbers. An **array** of numbers is called a **matrix**. A matrix is denoted by enclosing the array in large brackets.

Let [A] be the matrix (array) of coefficients: $\begin{bmatrix} a_{11} & a_{12} \\ a_{21} & a_{22} \end{bmatrix}$

Let [X] be the matrix of unknowns: $\begin{bmatrix} x_1 \\ x_2 \end{bmatrix}$

Let [B] be the matrix of constants: $\begin{bmatrix} b_1 \\ b_2 \end{bmatrix}$

We will write the system of equations as a **matrix equation**, $[A][X] = [B]$, but before we do this, we will do some preliminary work with matrices.

Matrices are classified by the number of (horizontal) **rows** and (vertical) **columns.** The numbers of rows and columns of a matrix need not be the same; but if they are, the matrix is called a **square matrix.**

The **order** or **dimension** of a matrix is given by an expression **$m \times n$** (pronounced "m by n"), where m is the number of rows and n is the number of columns. For example, [A] (shown above) is a matrix of order 2×2, and matrices (plural for matrix) [X] and [B] have order 2×1.

Matrix Form of a System of Equations

We write a system of equations in the form of an **augmented matrix**. The *matrix* refers to the matrix of coefficients, and we *augment* (add to, or affix) this matrix by writing the constant terms at the right of the matrix (separated by a dashed line):

$$\begin{cases} a_{11}x_1 + a_{12}x_2 = b_1 \\ a_{21}x_1 + a_{22}x_2 = b_2 \end{cases} \text{ in matrix form is } \begin{bmatrix} a_{11} & a_{12} & \vdots & b_1 \\ a_{21} & a_{22} & \vdots & b_2 \end{bmatrix}$$

■ EXAMPLE I

Give the order of each system, and then write it in augmented matrix form.

a. $\begin{cases} 2x + y = 3 \\ 3x - y = 2 \\ 4x + 3y = 7 \end{cases}$
b. $\begin{cases} 5x - 3y + z = -3 \\ 2x + 5z = 14 \end{cases}$
c. $\begin{cases} x_1 - 3x_3 + x_5 = -3 \\ x_2 + x_4 = -1 \\ x_3 + x_5 = 7 \\ x_1 + x_2 - x_3 + 4x_4 = -8 \\ x_1 + x_2 + x_3 + x_4 + x_5 = 8 \end{cases}$

Solution   Note that some coefficients are negative and some are zero.

a. $\begin{bmatrix} 2 & 1 & \vdots & 3 \\ 3 & -1 & \vdots & 2 \\ 4 & 3 & \vdots & 7 \end{bmatrix}$
   Order: 3×3

b. $\begin{bmatrix} 5 & -3 & 1 & \vdots & -3 \\ 2 & 0 & 5 & \vdots & 14 \end{bmatrix}$
   Order: 2×4

c. $\begin{bmatrix} 1 & 0 & -3 & 0 & 1 & \vdots & -3 \\ 0 & 1 & 0 & 1 & 0 & \vdots & -1 \\ 0 & 0 & 1 & 0 & 1 & \vdots & 7 \\ 1 & 1 & -1 & 4 & 0 & \vdots & -8 \\ 1 & 1 & 1 & 1 & 1 & \vdots & 8 \end{bmatrix}$ Order: 5×6

■ EXAMPLE 2

Write a system of equations (use $x_1, x_2, x_3, \ldots$) that has the given augmented matrix.

a. $\begin{bmatrix} 2 & 1 & -1 & \vdots & -3 \\ 3 & -2 & 1 & \vdots & 9 \\ 1 & -4 & 3 & \vdots & 17 \end{bmatrix}$
b. $\begin{bmatrix} 1 & 0 & 0 & \vdots & 3 \\ 0 & 1 & 0 & \vdots & -2 \\ 0 & 0 & 1 & \vdots & 21 \end{bmatrix}$

c. $\begin{bmatrix} 1 & 0 & 0 & \vdots & 5 \\ 0 & 1 & 0 & \vdots & 12 \\ 0 & 0 & 1 & \vdots & -3 \\ 0 & 0 & 0 & \vdots & 4 \end{bmatrix}$
d. $\begin{bmatrix} 1 & 0 & 0 & \vdots & -7 \\ 0 & 1 & 0 & \vdots & 3 \\ 0 & 0 & 1 & \vdots & -1 \\ 0 & 0 & 0 & \vdots & 0 \end{bmatrix}$

Solution

a. $\begin{cases} 2x_1 + x_2 - x_3 = -3 \\ 3x_1 - 2x_2 + x_3 = 9 \\ x_1 - 4x_2 + 3x_3 = 17 \end{cases}$
b. $\begin{cases} x_1 = 3 \\ x_2 = -2 \\ x_3 = 21 \end{cases}$

c. $\begin{cases} x_1 = 5 \\ x_2 = 12 \\ x_3 = -3 \\ 0 = 4 \end{cases}$
d. $\begin{cases} x_1 = -7 \\ x_2 = 3 \\ x_3 = -1 \\ 0 = 0 \end{cases}$

The goal of this section is to solve a system of m equations with n unknowns. We have already looked at systems of two equations with two unknowns. In previous courses you may have solved three equations with three unknowns. Now, however, we want to be able to solve problems with two equations and five unknowns, or three equations and two unknowns, or systems with any number of linear equations and unknowns. The procedure for this section — **Gauss–Jordan elimination** — is a general method for solving all these types of systems. We write the system in augmented matrix form (as in Example 1), then carry out a process that transforms the matrix until the solution is obvious. Look back at Example 2 — the solution to part **b** is obvious. Part **c** shows $0 = 4$ in the last equation, so this system has no solution (0 cannot equal 4), and part **d** shows $0 = 0$ (which is true for all replacements of the variable), which means that the solution is found by looking at the other equations (namely, $x_1 = -7$, $x_2 = 3$, and $x_3 = -1$). The terms with nonzero coefficients in these examples (that is, parts **b**, **c**, and **d**) are arranged on a diagonal, and such a system is said to be in **diagonal form.**

Elementary Row Operations and Pivoting

What process will allow us to transform a matrix into diagonal form? We begin with some steps called **elementary row operations.** Elementary row operations change the *form* of a matrix, but the new form represents an equivalent system. Matrices that represent equivalent systems are called **equivalent matrices;** we now introduce the elementary row operations, which allow us to write equivalent matrices. Let us work with a system consisting of three equations and three unknowns (any size will work the same way).

$$
\begin{array}{ll}
\textit{System Format} & \textit{Matrix Format} \\[4pt]
\begin{cases} 2x - 2y + 4z = 14 \\ x - y - 2z = -9 \\ 3x + 2y + z = 16 \end{cases} \quad [A] = &
\left[\begin{array}{ccc:c} 2 & -2 & 4 & 14 \\ 1 & -1 & -2 & -9 \\ 3 & 2 & 1 & 16 \end{array} \right]
\end{array}
$$

For the discussion, we call this matrix A, and denote it by [A].

Elementary Row Operation 1:   RowSwap

Interchanging two equations is equivalent to interchanging two rows in the matrix format, and certainly, if we do this, the solution to the system will be the same:

$$
\begin{array}{ll}
\textit{System Format} & \textit{Matrix Format} \\[4pt]
\begin{cases} x - y - 2z = -9 \\ 2x - 2y + 4z = 14 \\ 3x + 2y + z = 16 \end{cases} &
\left[\begin{array}{ccc:c} 1 & -1 & -2 & -9 \\ 2 & -2 & 4 & 14 \\ 3 & 2 & 1 & 16 \end{array} \right]
\end{array}
$$

In this example we interchanged the first and the second rows of the matrix. If we denote the original matrix as [A], then we indicate the operation of interchanging the first and second rows of matrix A by RowSwap([A],1,2).

Elementary Row Operation 2:   Row+

Since adding the entries of one equation to the corresponding entries (similar terms) of another equation will not change the solution to a system of equations, the second elementary row operation is called row addition. (This is the step called linear combinations in the previous section.) In terms of matrices, we see that this operation corresponds to adding one row to another:

$$
\text{System Format} \qquad\qquad \text{Matrix Format}
$$

$$
\begin{cases} 2x - 2y + 4z = 14 \\ x - y - 2z = -9 \\ 3x + 2y + z = 16 \end{cases} \quad [A] = \begin{bmatrix} 2 & -2 & 4 & \vdots & 14 \\ 1 & -1 & -2 & \vdots & -9 \\ 3 & 2 & 1 & \vdots & 16 \end{bmatrix}
$$

Add row 1 to row 3:

$$
\text{System Format} \qquad\qquad \text{Matrix Format}
$$

$$
\begin{cases} 2x - 2y + 4z = 14 \\ x - y - 2z = -9 \\ 5x + 5z = 30 \end{cases} \qquad \begin{bmatrix} 2 & -2 & 4 & \vdots & 14 \\ 1 & -1 & -2 & \vdots & -9 \\ 5 & 0 & 5 & \vdots & 30 \end{bmatrix}
$$

Notice that only row 3 changes; we call the row being added to (that is, the row that is being changed) the **target row.** We indicate this operation by Row+([A],1,3).

$$
\underset{\text{target row}}{\uparrow}
$$

Elementary Row Operation 3: *Row

Multiplying or dividing both sides of an equation by any nonzero number does not change the simultaneous solution; then, in matrix format, the solution will not be changed if any row is multiplied or divided by a nonzero constant. In this context we call the constant a **scalar.** For example, we can multiply both sides of the first equation of the original system by $\frac{1}{2}$. (Note that dividing both sides of an equation by 2 can be considered as multiplying both sides by $\frac{1}{2}$.)

$$
\text{System Format} \qquad\qquad \text{Matrix Format}
$$

$$
\begin{cases} 2x - 2y + 4z = 14 \\ x - y - 2z = -9 \\ 3x + 2y + z = 16 \end{cases} \quad [A] = \begin{bmatrix} 2 & -2 & 4 & \vdots & 14 \\ 1 & -1 & -2 & \vdots & -9 \\ 3 & 2 & 1 & \vdots & 16 \end{bmatrix}
$$

Multiply the first row by $\frac{1}{2}$ to obtain:

$$
\text{System Format} \qquad\qquad \text{Matrix Format}
$$

$$
\begin{cases} x - y + 2z = 7 \\ x - y - 2z = -9 \\ 3x + 2y + z = 16 \end{cases} \qquad \begin{bmatrix} 1 & -1 & 2 & \vdots & 7 \\ 1 & -1 & -2 & \vdots & -9 \\ 3 & 2 & 1 & \vdots & 16 \end{bmatrix}
$$

$$
\underset{\downarrow}{\text{scalar}}
$$

We indicate this operation by *Row($\frac{1}{2}$,[A],1)

$$
\underset{\text{target row}}{\uparrow}
$$

Elementary Row Operation 4: *Row+

When solving systems, more often than not we need to multiply both sides of an equation by a scalar before adding to make the coefficients opposites. Elementary

---

Calculator Comment

Check your owner's manual for the matrix operations. Most will have a [MATRIX] key and you will need to name the matrix. Most will allow for three matrices (called [A], [B], and [C]). Then you will need to input the order (most will handle up to order 6×6). After inputting a matrix, it is a good idea to recall it to make sure it is input correctly. For this example, after a row swap of rows 1 and 2 the display shows:

$$
\begin{array}{cccc} [1 & -1 & -2 & -9] \\ [2 & -2 & 4 & 14] \\ [3 & 2 & 1 & 16] \end{array}
$$

row operation 4 combines row operations 2 and 3 so that this can be accomplished in one step. Let us return to the original system:

$$\text{System Format} \qquad\qquad \text{Matrix Format}$$

$$\begin{cases} 2x - 2y + 4z = 14 \\ x - y - 2z = -9 \\ 3x + 2y + z = 16 \end{cases} \qquad [A] = \begin{bmatrix} 2 & -2 & 4 & \vdots & 14 \\ 1 & -1 & -2 & \vdots & -9 \\ 3 & 2 & 1 & \vdots & 16 \end{bmatrix}$$

We can change this system by multiplying the second equation by -2 and adding the result to the first equation. In matrix terminology we say that we multiply the second row by -2 and add it to the first row. Denote this by

$$\underset{\uparrow}{*\text{Row}+(-2,[A],2,1)}$$

scalar ↓

target row

$$\text{System Format} \qquad\qquad \text{Matrix Format}$$

$$\begin{cases} 8z = 32 \\ x - y - 2z = -9 \\ 3x + 2y + z = 16 \end{cases} \qquad \begin{bmatrix} 0 & 0 & 8 & \vdots & 32 \\ 1 & -1 & -2 & \vdots & -9 \\ 3 & 2 & 1 & \vdots & 16 \end{bmatrix}$$

Once again, multiply the second row, this time by -3, and add it to the third row. Wait! Why -3? Where did that come from? The idea is the same one we used in the linear combination method — we use a number that will give a zero coefficient to the x in the third equation.

$$\text{System Format} \qquad\qquad \text{Matrix Format}$$

$$\begin{cases} 8z = 32 \\ x - y - 2z = -9 \\ 5y + 7z = 43 \end{cases} \qquad \begin{bmatrix} 0 & 0 & 8 & \vdots & 32 \\ 1 & -1 & -2 & \vdots & -9 \\ 0 & 5 & 7 & \vdots & 43 \end{bmatrix}$$

Note that the multiplied row is not changed; instead, the changed row is the one to which the multiplied row is added. We call the original row the **pivot row.** Note also that we did not work with the original matrix [A] but rather with the previous answer, so we indicate this by

scalar   pivot row
↓   ↓

$$*\text{Row}+(-3,[\text{Ans}],2,3)$$

↑

target row

There you have it! You can carry out these four elementary operations until you have a system for which the solution is obvious, as illustrated by the following example.

■■■■ **EXAMPLE 3**

Solve using both system format and matrix format: $\begin{cases} 2x - 5y = 5 \\ x - 2y = 1 \end{cases}$

Solution

System Format	Matrix Format	Operation Performed
$\begin{cases} 2x - 5y = 5 \\ x - 2y = 1 \end{cases}$ $[A] = \begin{bmatrix} 2 & -5 & \vdots & 5 \\ 1 & -2 & \vdots & 1 \end{bmatrix}$		RowSwap$([A],1,2)$

$\begin{cases} x - 2y = 1 \\ 2x - 5y = 5 \end{cases}$ $\begin{bmatrix} 1 & -2 & \vdots & 1 \\ 2 & -5 & \vdots & 5 \end{bmatrix}$

This matrix is called [Ans] because it is the result of the previous matrix operation.

$*\mathrm{Row}+(-2,[\mathrm{Ans}],1,2)$

$\begin{cases} x - 2y = 1 \\ -y = 3 \end{cases}$ $\begin{bmatrix} 1 & -2 & \vdots & 1 \\ 0 & -1 & \vdots & 3 \end{bmatrix}$

This matrix is now referred to as [Ans].

$*\mathrm{Row}(-1,[\mathrm{Ans}],2)$

$\begin{cases} x - 2y = 1 \\ y = -3 \end{cases}$ $\begin{bmatrix} 1 & -2 & \vdots & 1 \\ 0 & 1 & \vdots & -3 \end{bmatrix}$ $*\mathrm{Row}+(2,[\mathrm{Ans}],2,1)$

$\begin{cases} x = -5 \\ y = -3 \end{cases}$ $\begin{bmatrix} 1 & 0 & \vdots & -5 \\ 0 & 1 & \vdots & -3 \end{bmatrix}$

The solution $(-5, -3)$ is now obvious.   ▬

As you study Example 3, first look at how the elementary row operations led to a system equivalent to the first — but one for which the solution is obvious. Next, try to decide *why* a particular row operation was chosen when it was. Many students quickly learn the elementary row operations, but then use a series of (almost random) steps until the obvious solution results. This often works, but is not very efficient. The steps chosen in Example 3 illustrate a very efficient method of using the elementary row operations to determine a system whose solution is obvious. Let us restate the elementary row operations and the operation called **pivoting.**

Pivoting

There are four **elementary row operations** for producing equivalent matrices:

1. **RowSwap**   Interchange any two rows.
2. **Row+**   Row addition — add a row to any other row.
3. ***Row**   Scalar multiplication — multiply (or divide) all the elements of a row by the same nonzero real number.
4. ***Row+**   Multiply all the entries of a row (**pivot row**) by a nonzero real number and add each resulting product to the corresponding entry of another specified row (**target row**). ⊘ This operation changes only the target row. ⊘

These elementary row operations are used together in a process called **pivoting:**

1. Divide all entries in the row in which the pivot appears (called the **pivot row**) by the nonzero pivot element so that the pivot entry becomes a 1. This uses elementary row operation 3.
2. Obtain zeros above and below the pivot element by using elementary row operation 4.

KARL FRIEDRICH GAUSS
(1777–1855)

Karl Gauss has been profiled in a previous Historical Note, but we cannot ignore him here with our discussion of this procedure for solving a system of equations. The process for solving equations described in this section was, for years, attributed solely to Gauss, and for that reason is often referred to as *Gaussian elimination*. Camille Jordan is known for his work in algebra and a branch of mathematics called Galois theory. In 1878, however, Jordan worked with forms of systems of equations that we study in this section.

Gauss–Jordan Elimination

You are now ready to see the method worked out by Gauss and Jordan. It efficiently uses the elementary row operations to diagonalize the matrix. That is, the first pivot is the first entry in the first row, first column; the second is the entry in the second row, second column; and so on until the solution is obvious. A **pivot** element is an element that is used to eliminate elements above and below it in a given column by applying elementary row operations.

Gauss–Jordan Elimination

Step 1. Select as the first pivot the element in the first row, first column; pivot.

Step 2. The next pivot is the element in the second row, second column; pivot.

Step 3. Repeat the process until you arrive at the last row, or until the pivot element is a zero. If it is a zero and you can interchange that row with a row below it, so that the pivot element is no longer a zero, do so and continue. If it is zero and you cannot interchange rows so that it is not a zero, continue with the next row. The final matrix is called the **row-reduced form.**

EXAMPLE 4

Solve $\begin{cases} x + 2y - z = 0 \\ 2x + 3y - 2z = 3 \\ -x - 4y + 3z = -2 \end{cases}$

Solution   We solve this system by choosing the steps according to the Gauss–Jordan method.

$$[A] = \begin{bmatrix} 1 & 2 & -1 & \vdots & 0 \\ 2 & 3 & -2 & \vdots & 3 \\ -1 & -4 & 3 & \vdots & -2 \end{bmatrix}$$

First pivot (row 1, col. 1)

$$\rightarrow \begin{bmatrix} 1 & 2 & -1 & \vdots & 0 \\ 0 & -1 & 0 & \vdots & 3 \\ 0 & -2 & 2 & \vdots & -2 \end{bmatrix}$$

*Row+(−2,[A],1,2)
*Row+(1,[Ans],1,3)

$$\rightarrow \begin{bmatrix} 1 & 2 & -1 & \vdots & 0 \\ 0 & 1 & 0 & \vdots & -3 \\ 0 & -2 & 2 & \vdots & -2 \end{bmatrix}$$

*Row(−1,[Ans],2)

Second pivot (row 2, col. 2)

$$\rightarrow \begin{bmatrix} 1 & 0 & -1 & \vdots & 6 \\ 0 & 1 & 0 & \vdots & -3 \\ 0 & 0 & 2 & \vdots & -8 \end{bmatrix}$$

*Row+(−2,[Ans],2,1)
*Row+(2,[Ans],2,3)

$$\rightarrow \begin{bmatrix} 1 & 0 & -1 & \vdots & 6 \\ 0 & 1 & 0 & \vdots & -3 \\ 0 & 0 & 1 & \vdots & -4 \end{bmatrix}$$

*Row(0.5,[Ans],3)

Third pivot (row 3, col. 3)

$$\rightarrow \begin{bmatrix} 1 & 0 & 0 & \vdots & 2 \\ 0 & 1 & 0 & \vdots & -3 \\ 0 & 0 & 1 & \vdots & -4 \end{bmatrix}$$

*Row+(1,[Ans],3,1)

The solution $(2, -3, -4)$ is found by inspection since this matrix represents the system

$$\begin{cases} 1x + 0y + 0z = 2 \\ 0x + 1y + 0z = -3 \\ 0x + 0y + 1z = -4 \end{cases}$$

Calculator Comment

Pay special attention to the interpretation of what you see with a calculator, and the correct answer.

▬▬ EXAMPLE 5

A rancher has to mix three types of feed for her cattle. The following analysis shows the amounts per bag (100 lb) of grain:

Grain	Protein	Carbohydrates	Sodium
A	7 lb	88 lb	1 lb
B	6 lb	90 lb	1 lb
C	10 lb	70 lb	2 lb

How many bags of each type of grain should she mix to provide 71 lb of protein, 854 lb of carbohydrates, and 12 lb of sodium?

Solution Let a, b, and c be the number of bags of grains A, B, and C, respectively, that are needed for the mixture. Then

Grain	Protein	Carbohydrates	Sodium
A	$7a$ lb	$88a$ lb	a lb
B	$6b$ lb	$90b$ lb	b lb
C	$10c$ lb	$70c$ lb	$2c$ lb
Total needed:	71 lb	854 lb	12 lb

Thus, $\begin{cases} 7a + 6b + 10c = 71 \\ 88a + 90b + 70c = 854. \\ a + b + 2c = 12 \end{cases}$

Let $[A] = \begin{bmatrix} 7 & 6 & 10 & \vdots & 71 \\ 88 & 90 & 70 & \vdots & 854 \\ 1 & 1 & 2 & \vdots & 12 \end{bmatrix}$.

We show the steps in Gauss–Jordan elimination:

$$\begin{bmatrix} 7 & 6 & 10 & \vdots & 71 \\ 88 & 90 & 70 & \vdots & 854 \\ 1 & 1 & 2 & \vdots & 12 \end{bmatrix} \rightarrow \begin{bmatrix} 1 & 1 & 2 & \vdots & 12 \\ 88 & 90 & 70 & \vdots & 854 \\ 7 & 6 & 10 & \vdots & 71 \end{bmatrix}$$

$$\text{RowSwap}([A],1,3)$$

$$\rightarrow \begin{bmatrix} 1 & 1 & 2 & \vdots & 12 \\ 0 & 2 & -106 & \vdots & -202 \\ 0 & -1 & -4 & \vdots & -13 \end{bmatrix} \rightarrow \begin{bmatrix} 1 & 1 & 2 & \vdots & 12 \\ 0 & 1 & -53 & \vdots & -101 \\ 0 & -1 & -4 & \vdots & -13 \end{bmatrix}$$

$$*\text{Row}+(-88,[\text{Ans}],1,2)$$
$$*\text{Row}+(-7,[\text{Ans}],1,3)$$
$$\qquad\qquad *\text{Row}(\tfrac{1}{2},[\text{Ans}],2)$$

$$\rightarrow \begin{bmatrix} 1 & 0 & 55 & \vdots & 113 \\ 0 & 1 & -53 & \vdots & -101 \\ 0 & 0 & -57 & \vdots & -114 \end{bmatrix} \rightarrow \begin{bmatrix} 1 & 0 & 55 & \vdots & 113 \\ 0 & 1 & -53 & \vdots & -101 \\ 0 & 0 & 1 & \vdots & 2 \end{bmatrix}$$

$$*\text{Row}+(-1,[\text{Ans}],2,1)$$
$$*\text{Row}+(1,[\text{Ans}],2,3)$$
$$\qquad\qquad *\text{Row}(-\tfrac{1}{57},[\text{Ans}],3)$$

$$\rightarrow \begin{bmatrix} 1 & 0 & 0 & \vdots & 3 \\ 0 & 1 & 0 & \vdots & 5 \\ 0 & 0 & 1 & \vdots & 2 \end{bmatrix}$$

$$*\text{Row}+(-55,[\text{Ans}],3,1)$$
$$*\text{Row}+(53,[\text{Ans}],3,2)$$

Mix three bags of grain A, five bags of grain B, and two bags of grain C.   ▬

PROBLEM SET   12.3

▲ **A Problems**

1. **IN YOUR OWN WORDS** What is a matrix?

2. **IN YOUR OWN WORDS** List the elementary row operations and briefly describe each.

3. **IN YOUR OWN WORDS** What is Gauss–Jordan elimination?

In Problems 4–12, decide whether the statement is true or false. If it is false, tell what is wrong.

4. Entry a_{34} is in row 3, column 4.

5. The matrix for the system $\begin{cases} x_1 = 1 \\ x_2 = 2 \\ x_3 = 0 \\ x_4 = 5 \end{cases}$ has order 4×4.

6. In $*$Row+ notation, the first number listed is the target row.

7. In $*$Row and $*$Row+ notation, the target row is the last number listed.

8. If row 3 of a matrix [A] is added to row 5 of [A], then the correct notation is $*$Row+([A],3,5).

9. If row 7 of a matrix [B] is multiplied by -2 and then added to row 6 of [B], then the correct notation is *Row+(-2,[B],7,6).

10. The notation *Row+(3^{-1},[C],4,2) means multiply row 2 of matrix [C] by $\frac{1}{3}$ and add the corresponding entries to the entries in row 4.

11. In the notation *Row+(3,[A],4,5), the target row is 3.

12. For the matrix $[A] = \begin{bmatrix} 0 & 7 & 8 & \vdots & 3 \\ 1 & 2 & 3 & \vdots & 4 \\ 0 & 1 & 3 & \vdots & 4 \end{bmatrix}$

the first step in Gauss–Jordan elimination is RowSwap([A],1,3).

13. Write each system in augmented matrix form.

a. $\begin{cases} 4x + 5y = -16 \\ 3x + 2y = 5 \end{cases}$
b. $\begin{cases} x + y + z = 4 \\ 3x + 2y + z = 7 \\ x - 3y + 2z = 0 \end{cases}$
c. $\begin{cases} x_1 + 2x_2 - 5x_3 + x_4 = 5 \\ x_1 - 3x_3 + 6x_4 = 0 \\ x_3 - 3x_4 = -15 \\ x_2 - 5x_3 + 5x_4 = 2 \end{cases}$

14. Write a system of equations that has the given augmented matrix.

a. $\begin{bmatrix} 6 & 7 & 8 & \vdots & 3 \\ 1 & 2 & 3 & \vdots & 4 \\ 0 & 1 & 3 & \vdots & 4 \end{bmatrix}$
b. $\begin{bmatrix} 1 & 0 & 0 & \vdots & 3 \\ 0 & 1 & 2 & \vdots & 4 \end{bmatrix}$
c. $\begin{bmatrix} 1 & 0 & 0 & \vdots & 32 \\ 0 & 1 & 0 & \vdots & 27 \\ 0 & 0 & 1 & \vdots & -5 \\ 0 & 0 & 0 & \vdots & 3 \end{bmatrix}$

Given the matrices in Problems 15–18, perform elementary row operations to obtain a 1 in the row 1, column 1 position. Answers for Problems 15–18 may vary.

15. $[A] = \begin{bmatrix} 3 & 1 & 2 & \vdots & 1 \\ 0 & 2 & 4 & \vdots & 5 \\ 1 & 3 & -4 & \vdots & 9 \end{bmatrix}$

16. $[B] = \begin{bmatrix} -2 & 3 & 5 & \vdots & 9 \\ 1 & 0 & 2 & \vdots & -8 \\ 0 & 1 & 0 & \vdots & 5 \end{bmatrix}$

17. $[C] = \begin{bmatrix} 2 & 4 & 10 & \vdots & -12 \\ 6 & 3 & 4 & \vdots & 6 \\ 10 & -1 & 0 & \vdots & 1 \end{bmatrix}$

18. $[A] = \begin{bmatrix} 5 & 20 & 15 & \vdots & 6 \\ 7 & -5 & 3 & \vdots & 2 \\ 12 & 0 & 1 & \vdots & 4 \end{bmatrix}$

Given the matrices in Problems 19–22, perform elementary row operations to obtain zeros under the 1 in the first column.

19. $[A] = \begin{bmatrix} 1 & 2 & -3 & \vdots & 0 \\ 0 & 3 & 1 & \vdots & 4 \\ 2 & 5 & 1 & \vdots & 6 \end{bmatrix}$

20. $[B] = \begin{bmatrix} 1 & 3 & -5 & \vdots & 6 \\ -3 & 4 & 1 & \vdots & 2 \\ 0 & 5 & 1 & \vdots & 3 \end{bmatrix}$

21. $[C] = \begin{bmatrix} 1 & 2 & 4 & \vdots & 1 \\ -2 & 5 & 0 & \vdots & 2 \\ -4 & 5 & 1 & \vdots & 3 \end{bmatrix}$

22. $[A] = \begin{bmatrix} 1 & 5 & 3 & \vdots & 2 \\ 2 & 3 & -1 & \vdots & 4 \\ 3 & 2 & 1 & \vdots & 0 \end{bmatrix}$

Given the matrices in Problems 23–26, perform elementary row operations to obtain a 1 in the second row, second column without changing the entries in the first column.

23. $[A] = \begin{bmatrix} 1 & 3 & 5 & \vdots & 2 \\ 0 & 2 & 6 & \vdots & -8 \\ 0 & 3 & 4 & \vdots & 1 \end{bmatrix}$

24. $[B] = \begin{bmatrix} 1 & 5 & -3 & \vdots & 5 \\ 0 & 3 & 9 & \vdots & -15 \\ 0 & 2 & 1 & \vdots & 5 \end{bmatrix}$

25. $[C] = \begin{bmatrix} 1 & 4 & -1 & \vdots & 6 \\ 0 & 5 & 1 & \vdots & 3 \\ 0 & 4 & 6 & \vdots & 5 \end{bmatrix}$

26. $[A] = \begin{bmatrix} 1 & 3 & -2 & \vdots & 0 \\ 0 & 4 & 2 & \vdots & 9 \\ 0 & 3 & 6 & \vdots & 1 \end{bmatrix}$

Given the matrices in Problems 27–28, perform elementary row operations to obtain a zero (or zeros) above and below the 1 in the second column without changing the entries in the first column.

27. $[A] = \begin{bmatrix} 1 & 5 & -3 & | & 2 \\ 0 & 1 & 4 & | & 5 \\ 0 & 3 & 4 & | & 2 \end{bmatrix}$

28. $[B] = \begin{bmatrix} 1 & 6 & -3 & 4 & | & 1 \\ 0 & 1 & 7 & 3 & | & 0 \\ 0 & 3 & 4 & 0 & | & -2 \\ 0 & -2 & 3 & 1 & | & 0 \end{bmatrix}$

Given the matrices in Problems 29–30, perform elementary row operations to obtain a 1 in the third row, third column without changing the entries in the first two columns.

29. $[A] = \begin{bmatrix} 1 & 3 & 4 & | & 5 \\ 0 & 1 & -3 & | & 6 \\ 0 & 0 & 5 & | & 10 \end{bmatrix}$

30. $[B] = \begin{bmatrix} 1 & -3 & 4 & | & -5 \\ 0 & 1 & 3 & | & 6 \\ 0 & 0 & 8 & | & 12 \end{bmatrix}$

Given the matrices in Problems 31–32, perform elementary row operations to obtain zeros above the 1 in the third column without changing the entries in the first or second columns.

31. $[A] = \begin{bmatrix} 1 & 3 & -1 & | & 5 \\ 0 & 1 & 2 & | & 6 \\ 0 & 0 & 1 & | & 4 \end{bmatrix}$

32. $[B] = \begin{bmatrix} 1 & 6 & -3 & | & -2 \\ 0 & 1 & 4 & | & 5 \\ 0 & 0 & 1 & | & 3 \end{bmatrix}$

▲ B Problems

Solve the systems in Problems 33–50 by the Gauss–Jordan method.

33. $\begin{cases} x + y = 7 \\ x - y = -1 \end{cases}$

34. $\begin{cases} x - y = 8 \\ x + y = 2 \end{cases}$

35. $\begin{cases} -x + 2y = 2 \\ 4x - 7y = -5 \end{cases}$

36. $\begin{cases} x - 6y = -3 \\ 2x + 3y = 9 \end{cases}$

37. $\begin{cases} 4x - 3y = 1 \\ 5x + 2y = 7 \end{cases}$

38. $\begin{cases} 3x + 7y = 5 \\ 4x + 9y = 7 \end{cases}$

39. $\begin{cases} x - y = 2 \\ 2x + 3y = 9 \end{cases}$

40. $\begin{cases} 3x - 4y = 16 \\ -x + 2y = -6 \end{cases}$

41. $\begin{cases} x - y = 1 \\ x + z = 1 \\ y - z = 1 \end{cases}$

42. $\begin{cases} x + y = 2 \\ x - z = 1 \\ -y + z = 1 \end{cases}$

43. $\begin{cases} x + 5z = 9 \\ y + 2z = 2 \\ 2x + 3z = 4 \end{cases}$

44. $\begin{cases} x + 2z = 13 \\ 2x + y = 8 \\ -2y + 9z = 41 \end{cases}$

45. $\begin{cases} 4x + y = -2 \\ 3x + 2z = -9 \\ 2y + 3z = -5 \end{cases}$

46. $\begin{cases} 5x + z = 9 \\ x - 5z = 7 \\ x + y - z = 0 \end{cases}$

47. $\begin{cases} x + y = -2 \\ y + z = 2 \\ x - y - z = -1 \end{cases}$

48. $\begin{cases} x + y = -1 \\ y + z = -1 \\ x + y + z = 1 \end{cases}$ 49. $\begin{cases} x + 2z = 9 \\ 2x + y = 13 \\ 2y + z = 8 \end{cases}$ 50. $\begin{cases} x + 2z = 0 \\ 3x - y + 2z = 0 \\ 4x + y = 6 \end{cases}$

▲ **Problem Solving**

51. To control a certain type of crop disease, it is necessary to use 23 gal of chemical A and 34 gal of chemical B. The dealer can order commercial spray I, each container of which holds 5 gal of chemical A and 2 gal of chemical B, and commercial spray II, each container of which holds 2 gal of chemical A and 7 gal of chemical B. How many containers of each type of commercial spray should be used to obtain exactly the right proportion of chemicals needed?

52. A candy maker mixes chocolate, milk, and mint extract to produce three kinds of candy (I, II, and III) with the following proportions:

   I: 7 lb chocolate, 5 gal milk, 1 oz mint extract
   II: 3 lb chocolate, 2 gal milk, 2 oz mint extract
   III: 4 lb chocolate, 3 gal milk, 3 oz mint extract

   If 67 lb of chocolate, 48 gal of milk, and 32 oz of mint extract are available, how much of each kind of candy can be produced?

53. Using the data from Problem 52, how much of each type of candy can be produced with 62 lb of chocolate, 44 gal of milk, and 32 oz of mint extract?

12.4 INVERSE MATRICES

The availability of computer software and calculators that can carry out matrix operations has considerably increased the importance of a matrix solution to systems of equations, which involves the inverse of a matrix. If we let [A] be the matrix of coefficients of a system of equations, [X] the matrix of unknowns, and [B] the matrix of constants, we can then represent the system of equations by the **matrix equation**

$$[A][X] = [B]$$

If we can define an inverse matrix, denoted by $[A]^{-1}$, we should be able to solve the *system* by finding

$$[X] = [A]^{-1}[B]$$

To understand this simple process, we need to understand what it means for matrices to be inverses and develop a basic algebra for matrices. Even though these matrix operations are difficult with a pencil and paper, they are easy with the aid of computer software or a calculator that does matrix operations.

⊘ Note that $[A]^{-1}$ does not mean

$$\frac{1}{[A]},$$

but rather means the inverse of matrix [A]. ⊘

Matrix Operations

The next box gives a definition of matrix equality along with the fundamental matrix operations.

Matrix Operations

Equality

[M] = [N] if and only if matrices [M] and [N] are the same order and the corresponding entries are the same.

Addition

[M] + [N] = [S] if and only if [M] and [N] are the same order and the entries of [S] are found by adding the corresponding entries of [M] and [N].

Multiplication by a scalar

c[M] = [M]c is the matrix in which each entry of [M] is multiplied by the scalar (real number) c.

Subtraction

[M] − [N] = [D] if and only if [M] and [N] are the same order and the entries of [D] are found by subtracting the entries of [N] from the corresponding entries of [M].

Multiplication

Let [M] be an $m \times r$ matrix and [N] an $r \times n$ matrix. The product matrix [M][N] = [P] is an $m \times n$ matrix. The entry in the ith row and jth column of [M][N] is *the sum of the products formed by multiplying each entry of the ith row of [M] by the corresponding element in the jth column of [N].*

All of these definitions, except multiplication, are straightforward, so we will consider multiplication separately after Example 1. If an addition or multiplication cannot be performed because of the order of the given matrices, the matrices are said to be **nonconformable**.

▬▬ EXAMPLE 1

Let $[A] = [5 \quad 2 \quad 1]$, $[B] = [4 \quad 8 \quad -5]$, $[C] = \begin{bmatrix} 7 & 3 & 2 \\ 5 & -4 & -3 \end{bmatrix}$,

$[D] = \begin{bmatrix} 4 & -2 & 1 \\ -3 & 3 & -1 \\ 2 & 4 & -1 \end{bmatrix}$, $[E] = \begin{bmatrix} 3 & 4 & -1 \\ 2 & 0 & 5 \\ -4 & 2 & 3 \end{bmatrix}$. Find:

a. [A] + [B]   **b.** [A] + [C]   **c.** [D] + [E]   **d.** (−5)[C]   **e.** 2[A] − 3[B]

Solution

a. $[A] + [B] = [5 \quad 2 \quad 1] + [4 \quad 8 \quad -5]$
$$= [5 + 4 \quad 2 + 8 \quad 1 + (-5)]$$
$$= [9 \quad 10 \quad -4]$$

b. [A] + [C] is not defined because [A] and [C] are nonconformable.

c. $[D] + [E] = \begin{bmatrix} 4 & -2 & 1 \\ -3 & 3 & -1 \\ 2 & 4 & -1 \end{bmatrix} + \begin{bmatrix} 3 & 4 & -1 \\ 2 & 0 & 5 \\ -4 & 2 & 3 \end{bmatrix} = \begin{bmatrix} 7 & 2 & 0 \\ -1 & 3 & 4 \\ -2 & 6 & 2 \end{bmatrix}$

Add entry by entry.

Calculator Comment

Matrix operations are particularly easy using a calculator that handles matrices. For this example:

[A] + [B] = [9 10 −4]
[A] + [C] error
[E] + [D] =
[7 2 0]
[−1 3 4]
[−2 6 2]
−5*[C] =
[−35 −15 −10]
[−25   20   15]
2[A] − 3[B] =
[−2 −20 17]

d. $(-5)[C] = (-5)\begin{bmatrix} 7 & 3 & 2 \\ 5 & -4 & -3 \end{bmatrix} = \begin{bmatrix} -35 & -15 & -10 \\ -25 & 20 & 15 \end{bmatrix}$

Multiply each entry by -5.

e. $2[A] - 3[B] = 2[5 \quad 2 \quad 1] + (-3)[4 \quad 8 \quad -5]$

$= [10 \quad 4 \quad 2] + [-12 \quad -24 \quad 15]$

$= [-2 \quad -20 \quad 17]$

■ EXAMPLE 2

We illustrate matrix multiplication for several different examples.

Solution

a. $\underbrace{[2 \quad 3 \quad 4 \quad 5]}_{1 \times 4 \text{ matrix}} \underbrace{\begin{bmatrix} a \\ b \\ c \\ d \end{bmatrix}}_{4 \times 1 \text{ matrix}} = \underbrace{[2a + 3b + 4c + 5d]}_{1 \times 1 \text{ matrix answer}}$

To be conformable, these numbers must be the same

b. $\underbrace{[2 \quad 3 \quad 4 \quad 5]}_{1 \times 4 \text{ matrix}} \underbrace{\begin{bmatrix} 1 \\ -3 \\ 0 \\ 2 \end{bmatrix}}_{4 \times 1 \text{ matrix}} = \underbrace{[2(1) + 3(-3) + 4(0) + 5(2)]}_{1 \times 1 \text{ matrix}} = [3]$

Same

c. $\underbrace{[5 \quad 3 \quad 2]}_{1 \times 3 \text{ matrix}} \underbrace{\begin{bmatrix} 1 \\ 2 \\ 3 \\ 4 \end{bmatrix}}_{4 \times 1 \text{ matrix}}$ not defined

Not the same

d. $\underbrace{[2 \quad 3 \quad 4 \quad 5]}_{1 \times 4 \text{ matrix}} \underbrace{\begin{bmatrix} 1 & 2 \\ -3 & 1 \\ 0 & -2 \\ 2 & 3 \end{bmatrix}}_{4 \times 2 \text{ matrix}} = \underbrace{[2(1) + 3(-3) + 4(0) + 5(2)}_{\text{(row 1, column 1) entry}}$

Same: Answer is a 1×2 matrix.

$[2 \quad 3 \quad 4 \quad 5]\begin{bmatrix} 1 & 2 \\ -3 & 1 \\ 0 & -2 \\ 2 & 3 \end{bmatrix} = [3 \quad \underbrace{2(2) + 3(1) + 4(-2) + 5(3)}_{\text{(row 1, column 2) entry}}]$

$= [3 \quad 14]$

e.

$$[2 \quad -1 \quad 2] \begin{bmatrix} 4 & 2 & -1 \\ 1 & 0 & 2 \\ 3 & -1 & 3 \end{bmatrix} = [\underline{2(4) + (-1)(1) + 2(3)} \qquad \qquad]$$

$$[2 \quad -1 \quad 2] \begin{bmatrix} 4 & 2 & -1 \\ 1 & 0 & 2 \\ 3 & -1 & 3 \end{bmatrix} = [13 \quad \underline{2(2) + (-1)(0) + 2(-1)} \quad .]$$

$$[2 \quad -1 \quad 2] \begin{bmatrix} 4 & 2 & -1 \\ 1 & 0 & 2 \\ 3 & -1 & 3 \end{bmatrix} = [13 \quad 2 \quad \underline{2(-1) + (-1)(2) + 2(3)}]$$

$$= [13 \quad 2 \quad 2]$$

This example is written out to demonstrate clearly what is happening. Your work, however, should look like this:

$$[2 \quad -1 \quad 2] \begin{bmatrix} 4 & 2 & -1 \\ 1 & 0 & 2 \\ 3 & -1 & 3 \end{bmatrix} = [8 - 1 + 6 \quad 4 + 0 - 2 \quad -2 - 2 + 6]$$

$$= [13 \quad 2 \quad 2]$$

Calculator Comment

One of the principal advantages of using matrix multiplication to solve systems of equations is the ease with which matrix multiplication can be done using a matrix calculator. You should check the following calculations using a calculator.

EXAMPLE 3

Find $[A][B]$ and $[B][A]$.

a. Let $[A] = \begin{bmatrix} 1 & 2 & 3 & 4 \\ 5 & 6 & 7 & 8 \end{bmatrix}$ and $[B] = \begin{bmatrix} -3 & 1 & -2 \\ 0 & -1 & 5 \\ -4 & 3 & -1 \\ 2 & 3 & -2 \end{bmatrix}$.

b. Let $[A] = \begin{bmatrix} 3 & -1 & 4 \\ 2 & 1 & 0 \\ -1 & 3 & 2 \end{bmatrix}$ and $[B] = \begin{bmatrix} 5 & 1 & -1 \\ 2 & 3 & -2 \\ 0 & 3 & 4 \end{bmatrix}$.

Solution

a. $[A][B] = \begin{bmatrix} 1 & 2 & 3 & 4 \\ 5 & 6 & 7 & 8 \end{bmatrix} \begin{bmatrix} -3 & 1 & -2 \\ 0 & -1 & 5 \\ -4 & 3 & -1 \\ 2 & 3 & -2 \end{bmatrix}$

$$= \begin{bmatrix} 1(-3) + 2(0) + 3(-4) + 4(2) & 1(1) + 2(-1) + 3(3) + 4(3) & 1(-2) + 2(5) + 3(-1) + 4(-2) \\ 5(-3) + 6(0) + 7(-4) + 8(2) & 5(1) + 6(-1) + 7(3) + 8(3) & 5(-2) + 6(5) + 7(-1) + 8(-2) \end{bmatrix}$$

$$= \begin{bmatrix} -7 & 20 & -3 \\ -27 & 44 & -3 \end{bmatrix}$$

$$[B][A] = \begin{bmatrix} -3 & 1 & -2 \\ 0 & -1 & 5 \\ -4 & 3 & -1 \\ 2 & 3 & -2 \end{bmatrix} \begin{bmatrix} 1 & 2 & 3 & 4 \\ 5 & 6 & 7 & 8 \end{bmatrix}$$

[B] and [A] are not conformable. Note that $[A][B] \neq [B][A]$.

b. $[A][B] = \begin{bmatrix} 3 & -1 & 4 \\ 2 & 1 & 0 \\ -1 & 3 & 2 \end{bmatrix} \begin{bmatrix} 5 & 1 & -1 \\ 2 & 3 & -2 \\ 0 & 3 & 4 \end{bmatrix}$

$$= \begin{bmatrix} 3(5) + (-1)(2) + 4(0) & 3(1) + (-1)(3) + 4(3) & 3(-1) + (-1)(-2) + 4(4) \\ 2(5) + (1)(2) + 0(0) & 2(1) + (1)(3) + 0(3) & 2(-1) + (1)(-2) + 0(4) \\ (-1)(5) + 3(2) + 2(0) & (-1)(1) + 3(3) + 2(3) & (-1)(-1) + 3(-2) + 2(4) \end{bmatrix}$$

$$= \begin{bmatrix} 13 & 12 & 15 \\ 12 & 5 & -4 \\ 1 & 14 & 3 \end{bmatrix}$$

$$[B][A] = \begin{bmatrix} 5 & 1 & -1 \\ 2 & 3 & -2 \\ 0 & 3 & 4 \end{bmatrix} \begin{bmatrix} 3 & -1 & 4 \\ 2 & 1 & 0 \\ -1 & 3 & 2 \end{bmatrix} = \begin{bmatrix} 18 & -7 & 18 \\ 14 & -5 & 4 \\ 2 & 15 & 8 \end{bmatrix}$$

Once again, note that $[A][B] \neq [B][A]$. ▬

▬▬ **EXAMPLE 4**

Let $[A] = \begin{bmatrix} 1 & 2 & 3 \\ 4 & -1 & 5 \\ 3 & 2 & -1 \end{bmatrix}$, $[X] = \begin{bmatrix} x \\ y \\ z \end{bmatrix}$, and $[B] = \begin{bmatrix} 3 \\ 16 \\ 5 \end{bmatrix}$.

What is $[A][X] = [B]$?

Solution $[A][X] = \begin{bmatrix} x + 2y + 3z \\ 4x - y + 5z \\ 3x + 2y - z \end{bmatrix}$ so $[A][X] = [B]$ is a matrix equation rep-

resenting the system $\begin{cases} x + 2y + 3z = 3 \\ 4x - y + 5z = 16 \\ 3x + 2y - z = 5 \end{cases}$ ▬

Communication Matrices

A **communication matrix** is a square matrix in which the entries symbolize the occurrence of some facet or event with a 1 and the nonoccurrence with a 0. Consider the following example.

■■■ EXAMPLE 5   Polya's Method

Use matrix notation to summarize the following information.

> The United States has diplomatic relations with Russia and with Mexico, but not with Cuba. Mexico has diplomatic relations with the United States and Russia, but not with Cuba. Russia has diplomatic relations with the United States, Mexico, and Cuba. Finally, Cuba has diplomatic relations with Russia, but not with the United States and not with Mexico. Note that a country is not considered to have diplomatic relations with itself.

a. Write a matrix showing direct communication possibilities.

b. Write a matrix showing the channels of communication that are open to the various countries if they are willing to speak through an intermediary.

Solution   We use Polya's problem-solving guidelines for this example.

Understand the Problem.   Part **a** is fairly easy to understand, so we will do this first:

$$[A] = \begin{array}{c} \\ \text{U.S.} \\ \text{Russia} \\ \text{Cuba} \\ \text{Mexico} \end{array} \begin{array}{cccc} \text{U.S.} & \text{Russia} & \text{Cuba} & \text{Mexico} \\ \left[\begin{array}{cccc} 0 & 1 & 0 & 1 \\ 1 & 0 & 1 & 1 \\ 0 & 1 & 0 & 0 \\ 1 & 1 & 0 & 0 \end{array}\right] \end{array}$$

For part **b,** we want to find a matrix that will tell us the number of ways the countries can communicate through an intermediary. For example, the United States can talk to Cuba by talking through Russia. A country can also communicate with itself (to test security, perhaps) if it does so through an intermediary.

Devise a Plan.   The plan could list all possibilities by considering one possibility at a time, but instead we will use matrix multiplication to find $[A]^2$; we will then verify that this is the desired matrix.

Carry Out the Plan.

$$[A]^2 = \begin{bmatrix} 0 & 1 & 0 & 1 \\ 1 & 0 & 1 & 1 \\ 0 & 1 & 0 & 0 \\ 1 & 1 & 0 & 0 \end{bmatrix} \begin{bmatrix} 0 & 1 & 0 & 1 \\ 1 & 0 & 1 & 1 \\ 0 & 1 & 0 & 0 \\ 1 & 1 & 0 & 0 \end{bmatrix}$$

$$= \begin{bmatrix} 0+1+0+1 & 0+0+0+1 & 0+1+0+0 & 0+1+0+0 \\ 0+0+0+1 & 1+0+1+1 & 0+0+0+0 & 1+0+0+0 \\ 0+1+0+0 & 0+0+0+0 & 0+1+0+0 & 0+1+0+0 \\ 0+1+0+0 & 1+0+0+0 & 0+1+0+0 & 1+1+0+0 \end{bmatrix}$$

$$= \begin{bmatrix} 2 & 1 & 1 & 1 \\ 1 & 3 & 0 & 1 \\ 1 & 0 & 1 & 1 \\ 1 & 1 & 1 & 2 \end{bmatrix}$$

Look Back. We see that the United States can communicate with Cuba (entry in the first row, third column) through an intermediary. The zeros indicate that Russia and Cuba cannot communicate through intermediaries (they can communicate only directly). Matrix $[A]^3$ tells us how many ways the countries can communicate if they use two intermediaries.

Algebraic Properties of Matrices

Properties for an algebra of matrices can also be developed. The $m \times n$ **zero matrix,** denoted by $[0]$, is the matrix with m rows and n columns in which each entry is 0. The **identity matrix for multiplication,** denoted by $[I_n]$, is the square matrix with n rows and n columns consisting of a 1 in each position on the **main diagonal** (entries $m_{11}, m_{22}, m_{33}, \ldots, m_{nn}$) and zeros elsewhere:

$$[I_2] = \begin{bmatrix} 1 & 0 \\ 0 & 1 \end{bmatrix}, \quad [I_3] = \begin{bmatrix} 1 & 0 & 0 \\ 0 & 1 & 0 \\ 0 & 0 & 1 \end{bmatrix}, \quad [I_4] = \begin{bmatrix} 1 & 0 & 0 & 0 \\ 0 & 1 & 0 & 0 \\ 0 & 0 & 1 & 0 \\ 0 & 0 & 0 & 1 \end{bmatrix}$$

The **additive inverse** of a matrix $[M]$ is denoted by $[-M]$ and is defined by $(-1)[M]$; the **multiplicative inverse** of a matrix $[M]$ is denoted by $[M]^{-1}$ if it exists. Table 12.1 summarizes the properties of matrices. Assume that $[M]$, $[N]$, and $[P]$ all have order $n \times n$, which forces them to be conformable for the given operations. If the context makes the order of the identity obvious, we sometimes write $[I]$ to denote the identity matrix; that is, we assume that $[I]$ means $[I_n]$.

$\oslash \ [M]^{-1} \neq \dfrac{1}{[M]} \ \oslash$

TABLE 12.1 Properties of Matrices

Property	Addition	Multiplication
Commutative	$[M] + [N] = [N] + [M]$	$[M][N] \neq [N][M]$
Associative	$([M] + [N]) + [P] = [M] + ([N] + [P])$	$([M][N])[P] = [M]([N][P])$
Identity	$[M] + [0] = [0] + [M]$	$[I][M] = [M][I] = [M]$
Inverse	$[M] + [-M] = [-M] + [M] = [0]$	$[M][M]^{-1} = [M]^{-1}[M] = [I]$
Distributive	$[M]([N] + [P]) = [M][N] + [M][P]$ and $([N] + [P])[M] = [N][M] + [P][M]$	

Inverse Property

The property from this list that is particularly important for us in solving systems of equations is the inverse property. There are two unanswered questions about the inverse property. Given a *square* matrix $[M]$, when does $[M]^{-1}$ exist? And if it exists, how do you find it?

> **Inverse of a Matrix**
>
> If $[A]$ is a square matrix and if there exists a matrix $[A]^{-1}$ such that
>
> $$[A]^{-1}[A] = [A][A]^{-1} = [I]$$
>
> where $[I]$ is the identity matrix for multiplication, then $[A]^{-1}$ is called the **inverse** of $[A]$ for multiplication.

Usually, in the context of matrices, when we talk simply of the inverse of $[A]$ we mean the inverse of $[A]$ for multiplication, and when we talk of the identity matrix we mean the identity matrix for multiplication.

■■■■ EXAMPLE 6

Show that $[A]$ and $[B]$ are inverses, given:

a. $[A] = \begin{bmatrix} 2 & 1 \\ 3 & 2 \end{bmatrix}$; $[B] = \begin{bmatrix} 2 & -1 \\ -3 & 2 \end{bmatrix}$

b. $[A] = \begin{bmatrix} 0 & 1 & 2 \\ -1 & 1 & 2 \\ 1 & -2 & -5 \end{bmatrix}$; $[B] = \begin{bmatrix} 1 & -1 & 0 \\ 3 & 2 & 2 \\ -1 & -1 & -1 \end{bmatrix}$

Solution   **a.** We must show that $[A][B] = [I]$ and $[B][A] = [I]$.

$$[A][B] = \begin{bmatrix} 2 & 1 \\ 3 & 2 \end{bmatrix} \begin{bmatrix} 2 & -1 \\ -3 & 2 \end{bmatrix} = \begin{bmatrix} 4-3 & -2+2 \\ 6-6 & -3+4 \end{bmatrix} = \begin{bmatrix} 1 & 0 \\ 0 & 1 \end{bmatrix} = [I]$$

$$[B][A] = \begin{bmatrix} 2 & -1 \\ -3 & 2 \end{bmatrix} \begin{bmatrix} 2 & 1 \\ 3 & 2 \end{bmatrix} = \begin{bmatrix} 4-3 & 2-2 \\ -6+6 & -3+4 \end{bmatrix} = \begin{bmatrix} 1 & 0 \\ 0 & 1 \end{bmatrix} = [I]$$

Since $[A][B] = [B][A] = [I]$, we see that $[B] = [A]^{-1}$.

$$\textbf{b.} \ \ [A][B] = \begin{bmatrix} 0 & 1 & 2 \\ -1 & 1 & 2 \\ 1 & -2 & -5 \end{bmatrix} \begin{bmatrix} 1 & -1 & 0 \\ 3 & 2 & 2 \\ -1 & -1 & -1 \end{bmatrix}$$

$$= \begin{bmatrix} 0+3-2 & 0+2-2 & 0+2-2 \\ -1+3-2 & 1+2-2 & 0+2-2 \\ 1-6+5 & -1-4+5 & 0-4+5 \end{bmatrix} = \begin{bmatrix} 1 & 0 & 0 \\ 0 & 1 & 0 \\ 0 & 0 & 1 \end{bmatrix} = [I]$$

$$[B][A] = \begin{bmatrix} 1 & -1 & 0 \\ 3 & 2 & 2 \\ -1 & -1 & -1 \end{bmatrix} \begin{bmatrix} 0 & 1 & 2 \\ -1 & 1 & 2 \\ 1 & -2 & -5 \end{bmatrix}$$

$$= \begin{bmatrix} 0+1+0 & 1-1+0 & 2-2+0 \\ 0-2+2 & 3+2-4 & 6+4-10 \\ 0+1-1 & -1-1+2 & -2-2+5 \end{bmatrix} = \begin{bmatrix} 1 & 0 & 0 \\ 0 & 1 & 0 \\ 0 & 0 & 1 \end{bmatrix} = [I]$$

Since $[A][B] = [I] = [B][A]$, then $[B] = [A]^{-1}$.   ■

If a given matrix has an inverse, we say it is **nonsingular**. The unanswered question, however, is how to *find* an inverse matrix.

Calculator Comment

If you have a calculator that does matrix operations, use a $\boxed{\text{MATRIX}}$ function and then press $[A]^{-1}$ to find the inverse of a matrix $[A]$. We will show this step in the margin as we find the inverse matrix in the next two examples.

▬▬ EXAMPLE 7

Find the inverse of $[A] = \begin{bmatrix} 1 & 2 \\ 1 & 4 \end{bmatrix}$.

Solution   Find a matrix $[B]$, if it exists, so that $[A][B] = [I]$; since we do not know $[B]$, let its entries be variables:

$$[B] = \begin{bmatrix} x_1 & x_2 \\ y_1 & y_2 \end{bmatrix}$$

$$\text{Then, } [A][B] = \begin{bmatrix} 1 & 2 \\ 1 & 4 \end{bmatrix} \begin{bmatrix} x_1 & x_2 \\ y_1 & y_2 \end{bmatrix}$$

$$= \begin{bmatrix} x_1 + 2y_1 & x_2 + 2y_2 \\ x_1 + 4y_1 & x_2 + 4y_2 \end{bmatrix}$$

$$= \begin{bmatrix} 1 & 0 \\ 0 & 1 \end{bmatrix}$$

Calculator Comment

If you input $\boxed{\text{MATRIX}}$ $[A]$ and then press $[A]^{-1}$, the output will look like:

$$\begin{bmatrix} 2 & -1 \\ -.5 & .5 \end{bmatrix}$$

By definition of equality of matrices, we see that

$$\begin{cases} x_1 + 2y_1 = 1 \\ x_1 + 4y_1 = 0 \end{cases} \text{ and } \begin{cases} x_2 + 2y_2 = 0 \\ x_2 + 4y_2 = 1 \end{cases}$$

Solve each of these systems simultaneously to find $x_1 = 2$, $x_2 = -1$, $y_1 = \frac{1}{2}$, $y_2 = \frac{1}{2}$. Thus, the inverse is

$$[B] = \begin{bmatrix} x_1 & x_2 \\ y_1 & y_2 \end{bmatrix} = \begin{bmatrix} 2 & -1 \\ -\frac{1}{2} & \frac{1}{2} \end{bmatrix}$$

▬▬ EXAMPLE 8

Find the inverse for $[A] = \begin{bmatrix} 1 & -1 & 0 \\ 3 & 2 & 2 \\ -1 & -1 & -1 \end{bmatrix}$.

Solution   We need to find a matrix $\begin{bmatrix} x_1 & x_2 & x_3 \\ y_1 & y_2 & y_3 \\ z_1 & z_2 & z_3 \end{bmatrix}$ so that

$$\begin{bmatrix} 1 & -1 & 0 \\ 3 & 2 & 2 \\ -1 & -1 & -1 \end{bmatrix} \begin{bmatrix} x_1 & x_2 & x_3 \\ y_1 & y_2 & y_3 \\ z_1 & z_2 & z_3 \end{bmatrix} = \begin{bmatrix} 1 & 0 & 0 \\ 0 & 1 & 0 \\ 0 & 0 & 1 \end{bmatrix}$$

Calculator Comment

Use the $\boxed{\text{MATRIX}}$ function and input [A]. Then press $[A]^{-1}$ to find:

$$\begin{bmatrix} 0 & 1 & 2 \\ -1 & 1 & 2 \\ 1 & -2 & -5 \end{bmatrix}$$

The definition of equality of matrices gives rise to three systems of equations:

$$\begin{cases} x_1 - y_1 + 0z_1 = 1 \\ 3x_1 + 2y_1 + 2z_1 = 0 \\ -x_1 - y_1 - z_1 = 0 \end{cases} \quad \begin{cases} x_2 - y_2 + 0z_2 = 0 \\ 3x_2 + 2y_2 + 2z_2 = 1 \\ -x_2 - y_2 - z_2 = 0 \end{cases} \quad \begin{cases} x_3 - y_3 + 0z_3 = 0 \\ 3x_3 + 2y_3 + 2z_3 = 0 \\ -x_3 - y_3 - z_3 = 1 \end{cases}$$

We could solve these as three separate systems using Gauss–Jordan elimination; however, all the steps would be identical since the coefficients are the same in each system. Therefore, suppose we augment the matrix of the coefficients by the *three* columns of constants and do all three at once. Write the augmented matrix as $[A\,|\,I]$:

$$\begin{bmatrix} 1 & -1 & 0 & \vdots & 1 & 0 & 0 \\ 3 & 2 & 2 & \vdots & 0 & 1 & 0 \\ -1 & -1 & -1 & \vdots & 0 & 0 & 1 \end{bmatrix} \rightarrow \begin{bmatrix} 1 & -1 & 0 & \vdots & 1 & 0 & 0 \\ 0 & 5 & 2 & \vdots & -3 & 1 & 0 \\ 0 & -2 & -1 & \vdots & 1 & 0 & 1 \end{bmatrix}$$

$$*\text{Row}+(-3,[A],1,2)$$
$$*\text{Row}+(1,[\text{Ans}],1,3)$$

$$\rightarrow \begin{bmatrix} 1 & -1 & 0 & \vdots & 1 & 0 & 0 \\ 0 & 1 & 0 & \vdots & -1 & 1 & 2 \\ 0 & -2 & -1 & \vdots & 1 & 0 & 1 \end{bmatrix} \rightarrow \begin{bmatrix} 1 & 0 & 0 & \vdots & 0 & 1 & 2 \\ 0 & 1 & 0 & \vdots & -1 & 1 & 2 \\ 0 & 0 & -1 & \vdots & -1 & 2 & 5 \end{bmatrix}$$

$$*\text{Row}+(2,[\text{Ans}],3,2) \qquad\qquad *\text{Row}+(1,[\text{Ans}],2,1)$$
$$*\text{Row}+(2,[\text{Ans}],2,3)$$

$$\rightarrow \begin{bmatrix} 1 & 0 & 0 & \vdots & 0 & 1 & 2 \\ 0 & 1 & 0 & \vdots & -1 & 1 & 2 \\ 0 & 0 & 1 & \vdots & 1 & -2 & -5 \end{bmatrix}$$

$$\text{Row}*(-1,[\text{Ans}],3)$$

Now, if we relate this to the original three systems, we see that the inverse matrix is

$$\begin{bmatrix} 0 & 1 & 2 \\ -1 & 1 & 2 \\ 1 & -2 & -5 \end{bmatrix}$$

By studying Example 8, we are led to a procedure for finding the inverse of a nonsingular matrix.

Procedure for Finding the Inverse of a Matrix

To find the inverse of a *square* matrix A:

1. Augment [A] with [I]; that is, write $[A\,|\,I]$, where [I] is the identity matrix of the same order as [A].
2. Perform elementary row operations using Gauss–Jordan elimination to change the matrix A into the identity matrix (if possible).
3. If at any time you obtain all zeros in a row or column to the left of the dashed line, then there will be no inverse.
4. If steps 1 and 2 can be performed, the result in the augmented part is the inverse of [A].

■■■ EXAMPLE 9

Find the inverse, if possible, of $[A] = \begin{bmatrix} 1 & 2 \\ 0 & 0 \end{bmatrix}$

Solution   Write the augmented matrix $[A \vdots I]$:

$$\begin{bmatrix} 1 & 2 & \vdots & 1 & 0 \\ 0 & 0 & \vdots & 0 & 1 \end{bmatrix}$$

We want to make the left-hand side look like the corresponding identity matrix. This is impossible since there are no elementary row operations that will put it into the required form. Thus, there is no inverse.   ■

■■■ EXAMPLE 10

Find the inverse, if possible, of $[A] = \begin{bmatrix} 1 & 2 \\ 1 & 4 \end{bmatrix}$

Solution   Enter the augmented matrix $[A \vdots I]$:

$$\begin{bmatrix} 1 & 2 & \vdots & 1 & 0 \\ 1 & 4 & \vdots & 0 & 1 \end{bmatrix} \rightarrow \begin{bmatrix} 1 & 2 & \vdots & 1 & 0 \\ 0 & 2 & \vdots & -1 & 1 \end{bmatrix} \rightarrow \begin{bmatrix} 1 & 2 & \vdots & 1 & 0 \\ 0 & 1 & \vdots & -\frac{1}{2} & \frac{1}{2} \end{bmatrix}$$

$\qquad$ *Row+(−1,[A],1,2) $\qquad$ *Row(.5,[Ans],2)

$$\rightarrow \begin{bmatrix} 1 & 0 & \vdots & 2 & -1 \\ 0 & 1 & \vdots & -\frac{1}{2} & \frac{1}{2} \end{bmatrix}$$

*Row+(−2,[Ans],2,1)

The inverse is on the right of the dashed line: $\begin{bmatrix} 2 & -1 \\ -\frac{1}{2} & \frac{1}{2} \end{bmatrix}$.

When a matrix has fractional entries, it is often rewritten with a fractional coefficient and integer entries to simplify the arithmetic. For example, we could rewrite this inverse matrix as

$$[A]^{-1} = \frac{1}{2} \begin{bmatrix} 4 & -2 \\ -1 & 1 \end{bmatrix}$$ ■

■■■ EXAMPLE 11

Find the inverse, if possible, of $[A] = \begin{bmatrix} 0 & 1 & 2 \\ 2 & -1 & 1 \\ -1 & 1 & 0 \end{bmatrix}$

Solution   Write the augmented matrix $[A \vdots I]$ and make the left-hand side look like the corresponding identity matrix (if possible):

$$\begin{bmatrix} 0 & 1 & 2 & \vdots & 1 & 0 & 0 \\ 2 & -1 & 1 & \vdots & 0 & 1 & 0 \\ -1 & 1 & 0 & \vdots & 0 & 0 & 1 \end{bmatrix} \rightarrow \begin{bmatrix} -1 & 1 & 0 & \vdots & 0 & 0 & 1 \\ 2 & -1 & 1 & \vdots & 0 & 1 & 0 \\ 0 & 1 & 2 & \vdots & 1 & 0 & 0 \end{bmatrix}$$

$\qquad\qquad\qquad$ RowSwap([A],1,3)

$$\rightarrow \begin{bmatrix} 1 & -1 & 0 & \vdots & 0 & 0 & -1 \\ 2 & -1 & 1 & \vdots & 0 & 1 & 0 \\ 0 & 1 & 2 & \vdots & 1 & 0 & 0 \end{bmatrix} \rightarrow \begin{bmatrix} 1 & -1 & 0 & \vdots & 0 & 0 & -1 \\ 0 & 1 & 1 & \vdots & 0 & 1 & 2 \\ 0 & 1 & 2 & \vdots & 1 & 0 & 0 \end{bmatrix}$$

*Row(−1,[Ans],1)   *Row+(−2,[Ans],1,2)

$$\rightarrow \begin{bmatrix} 1 & 0 & 1 & \vdots & 0 & 1 & 1 \\ 0 & 1 & 1 & \vdots & 0 & 1 & 2 \\ 0 & 0 & 1 & \vdots & 1 & -1 & -2 \end{bmatrix} \rightarrow \begin{bmatrix} 1 & 0 & 0 & \vdots & -1 & 2 & 3 \\ 0 & 1 & 0 & \vdots & -1 & 2 & 4 \\ 0 & 0 & 1 & \vdots & 1 & -1 & -2 \end{bmatrix}$$

*Row+(1,[Ans],2,1)   *Row+(−1,[Ans],3,1)
*Row+(−1,[Ans],2,3)   *Row+(−1,[Ans],3,2)

Thus $[A]^{-1} = \begin{bmatrix} -1 & 2 & 3 \\ -1 & 2 & 4 \\ 1 & -1 & -2 \end{bmatrix}$ ▬

Calculator Comment

The answer to Example 11 was "nice," but most matrices will have "ugly" entries. If you are using a calculator, those fractions will be shown in decimal form. For example, the inverse of

$$[A] = \begin{bmatrix} 3 & -1 & 4 \\ 2 & 1 & 0 \\ -1 & 3 & 2 \end{bmatrix}$$

is found by calculator to be (approximately)

$$[A]^{-1} = \begin{bmatrix} .0526315789 & .3684210526 & -.1052631579 \\ -.1052631579 & .2631578947 & .2105263158 \\ .1842105263 & -.2105263158 & .1315789474 \end{bmatrix}$$

Most real-life applications will involve inverses such as the one shown here.

Systems of Equations

In Example 4 we saw how a system of equations can be written in matrix form. We can now see how to solve a system of linear equations by using the inverse. Consider a system of n linear equations with n unknowns whose matrix of coefficients $[A]$ has an inverse $[A]^{-1}$:

$[A][X] = [B]$	Given system
$[A]^{-1}[A][X] = [A]^{-1}[B]$	Multiply both sides by $[A]^{-1}$.
$([A]^{-1}[A])[X] = [A]^{-1}[B]$	Associative property
$[I][X] = [A]^{-1}[B]$	Inverse property
$[X] = [A]^{-1}[B]$	Identity property

In other words: To solve a system of equations, look to see whether the number of equations is the same as the number of unknowns. If so, find the inverse of the matrix of the coefficients (if it exists) and multiply it (on the right) by the matrix of the constants.

■ EXAMPLE 12

Solve $\begin{cases} y + 2z = 0 \\ 2x - y + z = -1 \\ y - x = 1 \end{cases}$

Solution Write in matrix form:

$$[A] = \begin{bmatrix} 0 & 1 & 2 \\ 2 & -1 & 1 \\ -1 & 1 & 0 \end{bmatrix}, \quad [X] = \begin{bmatrix} x \\ y \\ z \end{bmatrix}, \quad [B] = \begin{bmatrix} 0 \\ -1 \\ 1 \end{bmatrix}$$

From Example 11, $[A]^{-1} = \begin{bmatrix} -1 & 2 & 3 \\ -1 & 2 & 4 \\ 1 & -1 & -2 \end{bmatrix}$. Thus,

$$[X] = [A]^{-1}[B] = \begin{bmatrix} -1 & 2 & 3 \\ -1 & 2 & 4 \\ 1 & -1 & -2 \end{bmatrix} \begin{bmatrix} 0 \\ -1 \\ 1 \end{bmatrix} = \begin{bmatrix} 1 \\ 2 \\ -1 \end{bmatrix}$$

Therefore, the solution to the system is $(x, y, z) = (1, 2, -1)$. ■

Calculator Comment

The calculator, of course, is what really makes this inverse method worthwhile. It makes the solution of any system *that has the same number of equations and variables* almost trivial. For Example 12, simply enter [A] and [B] and then press

$[A]^{-1}[B]$

The output is: $\begin{bmatrix} 1 \\ 2 \\ -1 \end{bmatrix}$

That is all there is! Does this make the method worthwhile?

Historical Note

JAMES SYLVESTER
(1814–1897)

A second mathematician responsible for much of what we study in this section is James Sylvester. He was a contemporary and good friend of Cayley (see Historical Note on page 829). Cayley would try out many of his ideas on Sylvester, and Sylvester would react and comment on Cayley's work. Sylvester originated many of the ideas that Cayley used, and even suggested the term *matrix* to Cayley. It should also be noted that he founded the *American Journal of Mathematics* while he taught in the United States at Johns Hopkins University (from 1877 to 1883).

The method of solving a system by using the inverse matrix is very efficient if you know the inverse. Unfortunately, *finding* the inverse for one system is usually more work than using another method to solve the system. However, there are certain applications that yield the same system over and over, and the only thing to change is the constants. In this case you should think about the inverse method. And, finally, computers and calculators can find approximations for inverse matrices quite easily, so this method becomes the method of choice if you have access to this technology.

PROBLEM SET   12.4

▲ **A Problems**

1. **IN YOUR OWN WORDS** When are two matrices equal?

2. **IN YOUR OWN WORDS** Describe the process for adding matrices.

3. **IN YOUR OWN WORDS** Describe the process for multiplying matrices.

4. **IN YOUR OWN WORDS** Make up an original example showing that matrix multiplication is not commutative.

5. **IN YOUR OWN WORDS** What is the inverse of a matrix?

6. **IN YOUR OWN WORDS** What is a nonsingular matrix?

7. **IN YOUR OWN WORDS** Describe a procedure for finding the inverse of a square matrix.

8. **IN YOUR OWN WORDS** Describe a procedure for using the inverse of a matrix to solve a system of equations.

In Problems 9–26 find the indicated matrices if possible.

$$[A] = \begin{bmatrix} 1 & 2 \\ 4 & 0 \\ -1 & 3 \\ 2 & 1 \end{bmatrix} \qquad [B] = \begin{bmatrix} 4 & 2 \\ -1 & 3 \end{bmatrix} \qquad [C] = \begin{bmatrix} 1 & 0 & 0 & 0 \\ 0 & 1 & 0 & 0 \\ 0 & 0 & 1 & 0 \\ 0 & 0 & 0 & 1 \end{bmatrix}$$

$$[D] = \begin{bmatrix} 4 & 1 & 3 & 6 \\ -1 & 0 & -2 & 3 \end{bmatrix} \qquad [E] = \begin{bmatrix} 1 & 0 & 2 \\ 3 & -1 & 2 \\ 4 & 1 & 0 \end{bmatrix}$$

$$[F] = \begin{bmatrix} 1 & 4 & 0 \\ 3 & -1 & 2 \\ -2 & 1 & 5 \end{bmatrix} \qquad [G] = \begin{bmatrix} 8 & 1 & 6 \\ 3 & 5 & 7 \\ 4 & 9 & 2 \end{bmatrix}$$

9. $[E] + [F]$

10. $[E][F]$

11. $[E][G]$

12. $[E][F] + [E][G]$

13. $[E]([F] + [G])$

14. $2[E] - [G]$

15. $[F][G]$

16. $[G][F]$

17. $([E][F])[G]$

18. $[E]([F][G])$

19. $[A][B]$

20. $[B][D]$

21. $[B]^2$

22. $[C][A]$

23. $([B] + [C])[A]$

24. $[B][A] + [C][A]$

25. $[C]^3$

26. $[C][D]$

▲ **B Problems**

Find the inverse of each matrix in Problems 27–34, if it exists.

27. $\begin{bmatrix} 4 & -7 \\ -1 & 2 \end{bmatrix}$

28. $\begin{bmatrix} 8 & 6 \\ -2 & 4 \end{bmatrix}$

29. $\begin{bmatrix} 1 & 3 \\ 2 & 0 \end{bmatrix}$

30. $\begin{bmatrix} 1 & 0 & 2 \\ 2 & 1 & 0 \\ 0 & -2 & 9 \end{bmatrix}$

31. $\begin{bmatrix} 6 & 1 & 20 \\ 1 & -1 & 0 \\ 0 & 1 & 3 \end{bmatrix}$

32. $\begin{bmatrix} 4 & 1 & 0 \\ 2 & -1 & 4 \\ -3 & 2 & 1 \end{bmatrix}$

33. $\begin{bmatrix} 1 & 0 & 0 & 1 \\ 0 & 2 & 0 & 0 \\ 0 & 0 & 0 & 1 \\ 2 & 0 & 1 & 0 \end{bmatrix}$ 34. $\begin{bmatrix} 0 & 1 & 2 & 0 \\ 0 & 0 & 0 & 1 \\ 1 & 1 & 3 & 0 \\ 2 & 4 & 0 & 0 \end{bmatrix}$

Solve the systems in Problems 35–61 by solving the corresponding matrix equation with an inverse, if possible.

Problems 35–40 use the inverse found in Problem 27.

35. $\begin{cases} 4x - 7y = -2 \\ -x + 2y = 1 \end{cases}$ 36. $\begin{cases} 4x - 7y = -65 \\ -x + 2y = 18 \end{cases}$ 37. $\begin{cases} 4x - 7y = 48 \\ -x + 2y = -13 \end{cases}$

38. $\begin{cases} 4x - 7y = 2 \\ -x + 2y = 3 \end{cases}$ 39. $\begin{cases} 4x - 7y = 5 \\ -x + 2y = 4 \end{cases}$ 40. $\begin{cases} 4x - 7y = -3 \\ -x + 2y = 8 \end{cases}$

Problems 41–46 use the inverse found in Problem 28.

41. $\begin{cases} 8x + 6y = 12 \\ -2x + 4y = -14 \end{cases}$ 42. $\begin{cases} 8x + 6y = 16 \\ -2x + 4y = 18 \end{cases}$ 43. $\begin{cases} 8x + 6y = -6 \\ -2x + 4y = -26 \end{cases}$

44. $\begin{cases} 8x + 6y = -28 \\ -2x + 4y = 18 \end{cases}$ 45. $\begin{cases} 8x + 6y = -26 \\ -2x + 4y = 12 \end{cases}$ 46. $\begin{cases} 8x + 6y = -36 \\ -2x + 4y = -2 \end{cases}$

Problems 47–52 all use the same inverse.

47. $\begin{cases} 2x + 3y = -9 \\ x - 6y = -3 \end{cases}$ 48. $\begin{cases} 2x + 3y = 2 \\ x - 6y = 16 \end{cases}$ 49. $\begin{cases} 2x + 3y = 2 \\ x - 6y = -14 \end{cases}$

50. $\begin{cases} 2x + 3y = 9 \\ x - 6y = 42 \end{cases}$ 51. $\begin{cases} 2x + 3y = -22 \\ x - 6y = 49 \end{cases}$ 52. $\begin{cases} 2x + 3y = 12 \\ x - 6y = -24 \end{cases}$

Problems 53–58 use the inverse found in Problem 30.

53. $\begin{cases} x + 2z = 7 \\ 2x + y = 16 \\ -2y + 9z = -3 \end{cases}$ 54. $\begin{cases} x + 2z = 4 \\ 2x + y = 0 \\ -2y + 9z = 19 \end{cases}$ 55. $\begin{cases} x + 2z = 4 \\ 2x + y = 0 \\ -2y + 9z = 31 \end{cases}$

56. $\begin{cases} x + 2z = 7 \\ 2x + y = 1 \\ -2y + 9z = 28 \end{cases}$ 57. $\begin{cases} x + 2z = 12 \\ 2x + y = 0 \\ -2y + 9z = 10 \end{cases}$ 58. $\begin{cases} x + 2z = 5 \\ 2x + y = 8 \\ -2y + 9z = 9 \end{cases}$

Problems 59–61 use the inverse found in Problem 31.

59. $\begin{cases} 6x + y + 20z = 27 \\ x - y = 0 \\ y + 3z = 4 \end{cases}$ 60. $\begin{cases} 6x + y + 20z = 14 \\ x - y = 1 \\ y + 3z = 1 \end{cases}$ 61. $\begin{cases} 6x + y + 20z = 11 \\ x - y = 5 \\ y + 3z = -3 \end{cases}$

▲ **Problem Solving**

62. The Seedy Vin Company produces Riesling, Charbono, and Rosé wines. There are three procedures for producing each wine (the procedures for production affect the cost of the final product). One procedure allows an outside company to bottle the wine; a second allows the wine to be produced and bottled at the winery; a third allows the wine to be estate bottled. The amount of wine produced by the Seedy Vin

Company by each method is shown by the matrix:

$$[W] = \begin{array}{c} \\ \\ \\ \end{array} \begin{bmatrix} \overset{\text{Riesling}}{2} & \overset{\text{Charbono}}{1} & \overset{\text{Rosé}}{3} \\ 4 & 3 & 6 \\ 1 & 2 & 4 \end{bmatrix} \begin{array}{l} \text{outside bottling} \\ \text{produced and bottled at winery} \\ \text{estate bottled} \end{array}$$

Suppose the cost for each method of production is given by the matrix

$$[C] = [\text{outside} \quad \text{winery} \quad \text{estate}] = [1 \quad 4 \quad 6]$$

Suppose the production cost of a unit of each type of wine is given by the matrix:

$$[D] = \begin{bmatrix} 40 \\ 60 \\ 30 \end{bmatrix} \begin{array}{l} \text{Riesling} \\ \text{Charbono} \\ \text{Rosé} \end{array}$$

a. Find the cost of producing each of the three types of wine. *Hint:* This is [C][W].
b. Find the dollar amount for producing each unit of the different types of wine for these methods of production. *Hint:* This is [W][D].
c. Find ([C][W])[D] and [C]([W][D]). Give a verbal interpretation for these products.

63. Sociologists often study the dominance of one group over another. Suppose in a certain society there are four classes, which we will call abigweel, aweel, upancomer, and pon. *Hint:* These are pronounced "A big wheel," "A wheel," "Up and comer," and "peon."

Abigweel dominate aweel, upancomer, and pon.
Aweel dominate upancomer and pon.
Upancomer dominate pon.

a. Write a communication matrix, [D], representing these dominance relationships.
b. We say that an individual has two-stage dominance over another if 1 appears in the matrix $[D]^2$ for those individuals. Write the two-stage dominance matrix.
c. We define the **power** of an individual as the sum of the entries in the appropriate row of the matrix

$$[P] = [D] + [D]^2$$

Determine the power of abigweel, aweel, upancomer, and pon.
d. Rank the abigweel, aweel, upancomer, and pon.

12.5   SYSTEMS OF INEQUALITIES

In previous sections, we have discussed the simultaneous solution of a system of equations. In this section, we discuss the graphical solution of a simultaneous **system of inequalities.** The solution of a *system of inequalities* refers to the intersection of the solutions of the individual inequalities in the system. The procedure for graphing the solution of a system of inequalities is to graph the solution of the individual inequalities and then shade in the intersection of all the graphs.

Systems of Linear Inequalities

■■■■ EXAMPLE 1

Solve: $\begin{cases} y > -20x + 110 \\ x \geq 0 \\ y \geq 0 \\ x \leq 8 \end{cases}$

Solution   Graph each half-plane, but instead of shading, mark the appropriate regions with arrows. For this example, the two inequalities

$$x \geq 0$$
$$y \geq 0$$

give the first quadrant, which is marked with arrows as shown in the margin. Next, graph $y = -20x + 110$ and use a test point, say, $(0, 0)$, to determine the half-plane. Finally, draw $x = 8$ and mark it with arrows to show that $x \leq 8$, as shown in the margin.

The last step is to shade the intersection, as shown in Figure 12.5.

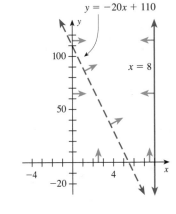

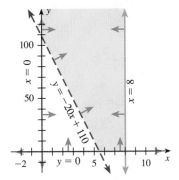

▲ **Figure 12.5   Graph of a system of inequalities**

■

■■■■ EXAMPLE 2

Graph the solution of the system: $\begin{cases} 2x + y \leq 3 \\ x - y > 5 \\ x \geq 0 \\ y \geq -10 \end{cases}$

Solution   The graph of the individual inequalities and their intersection is shown in Figure 12.6. Note the use of arrows to show the solutions of the individual inequalities. This device replaces the use of a lot of shading, which can be confusing if there are many inequalities in the system. In this book we show the intersection (solution) in color. You can show the intersection in your work in a variety of different ways, but a color highlighter works well.   ■

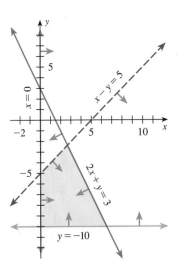

▲ **Figure 12.6   System of inequalities**

PROBLEM SET 12.5

▲ **A Problems**

1. **IN YOUR OWN WORDS** What is a system of linear inequalities?

2. **IN YOUR OWN WORDS** Describe a procedure for solving a system of linear inequalities.

Graph the solution of each system given in Problems 3–38.

3. $\begin{cases} x \geq 0 \\ y \leq 0 \end{cases}$

4. $\begin{cases} x \geq 0 \\ y \geq 0 \end{cases}$

5. $\begin{cases} x \leq 0 \\ y \leq 0 \end{cases}$

6. $\begin{cases} y \geq 0 \\ x < 8 \\ y < 5 \end{cases}$

7. $\begin{cases} x \geq 0 \\ y \geq 0 \\ x < 5 \\ y < 6 \end{cases}$

8. $\begin{cases} x \geq 0 \\ y \geq 0 \\ x < 500 \\ y < 1{,}000 \end{cases}$

9. $\begin{cases} -10 < x < 6 \\ 2 \leq y \leq 5 \end{cases}$

10. $\begin{cases} -4 \leq x \leq -2 \\ -5 \leq y \leq 9 \end{cases}$

11. $\begin{cases} -3 \leq x \leq 3 \\ 10 \leq 3y \leq 15 \end{cases}$

12. $\begin{cases} -7 \leq 3x \leq 6 \\ -10 \leq 3y \leq 16 \end{cases}$

13. $\begin{cases} 2y \geq -x + 4 \\ x \leq 3 \\ y \leq 3 \\ x \geq 0 \end{cases}$

14. $\begin{cases} 2x + 3y \geq 12 \\ y \leq 6 \\ 0 \leq x \leq 6 \end{cases}$

15. $\begin{cases} y \leq -\frac{3}{4}x + 3 \\ x \geq 0 \\ y \geq 0 \end{cases}$

16. $\begin{cases} 2x + y > 3 \\ 3x - y < 2 \end{cases}$

17. $\begin{cases} y \leq 3x - 4 \\ y \geq -2x + 5 \end{cases}$

18. $\begin{cases} 3x - 2y \geq 6 \\ 2x + 3y \leq 6 \end{cases}$

19. $\begin{cases} y - 5 \leq 0 \\ y \geq 0 \end{cases}$

20. $\begin{cases} x - 10 \leq 0 \\ x \geq 0 \end{cases}$

21. $\begin{cases} y - 25 \leq 0 \\ y \geq 0 \end{cases}$

22. $\begin{cases} x - y \geq 0 \\ y \leq 0 \end{cases}$

23. $\begin{cases} x + 5 \geq 0 \\ x \leq 0 \end{cases}$

▲ **B Problems**

24. $\begin{cases} -10 \leq x \\ x \leq 6 \\ -3 < y \\ y < 8 \end{cases}$

25. $\begin{cases} -5 < x \\ 3 \geq x \\ 5 > y \\ 2 \leq y \end{cases}$

26. $\begin{cases} -5 < x \\ x \leq 2 \\ -4 \leq y \\ y < 9 \end{cases}$

27. $\begin{cases} y \geq \frac{3}{4}x - 4 \\ y \leq -\frac{3}{4}x + 11 \\ x \geq 6 \end{cases}$

28. $\begin{cases} y \geq \frac{3}{2}x + 3 \\ y \leq \frac{3}{2}x + 6 \\ 3 \leq y \leq 6 \end{cases}$

29. $\begin{cases} 5x + 2y \leq 30 \\ 5x + 2y \geq 20 \\ x \geq 0 \\ y \geq 0 \end{cases}$

30. $\begin{cases} 5x - 2y + 30 \geq 0 \\ 5x - 2y + 20 \leq 0 \\ x \leq 0 \\ y \geq 0 \end{cases}$

31. $\begin{cases} 8x + 3y \leq 9 \\ y - 4 \geq -\frac{8}{3}(x + 2) \\ -5 \leq y \leq 3 \end{cases}$

32. $\begin{cases} x + y - 9 \leq 0 \\ x + y + 3 \geq 0 \\ x - y \leq 7 \\ y - x \leq 5 \end{cases}$

33. $\begin{cases} x \geq 0 \\ y \geq 0 \\ x + y \leq 9 \\ 2x - 3y \geq -6 \\ x - y \leq 3 \end{cases}$

34. $\begin{cases} x \geq 0 \\ y \geq 0 \\ x + y \leq 8 \\ y \leq 4 \\ x \leq 6 \end{cases}$

35. $\begin{cases} 2x + y \leq 8 \\ y \leq 5 \\ x - y \leq 2 \\ 3x - y \geq 5 \end{cases}$

36. $\begin{cases} 2x + 3y \leq 30 \\ 3x + 2y \geq 20 \\ x \geq 0 \\ y \geq 0 \end{cases}$

37. $\begin{cases} 2x - 3y + 30 \geq 0 \\ 3x - 2y + 20 \leq 0 \\ x \leq 0 \\ y \geq 0 \end{cases}$

38. $\begin{cases} x + y - 10 \leq 0 \\ x + y + 4 \geq 0 \\ x - y \leq 6 \\ y - x \leq 4 \end{cases}$

12.6   MODELING WITH LINEAR PROGRAMMING

This section gives us the opportunity to use mathematical modeling to solve an important type of real-life problem. The problem of interest is to maximize or minimize a linear expression subject to a set of limitations. This type of problem was first analyzed during World War II and developed into a modeling procedure known as **linear programming.** It is used to maximize profits, minimize costs, find efficient shipping schedules, minimize waste, secure the proper mix of ingredients, control inventories, and efficiently assign tasks to personnel.

We begin with some terminology. A **convex set** is a set that contains the line segment joining any two of its points. The expression to be maximized or minimized is called the **objective function,** and the limitations are called **constraints.** In linear programming these constraints are specified by a system of linear inequalities whose solution forms a convex set S. Each point in the set S is called a **feasible solution,** and a point at which the objective function takes on a maximum or a minimum value is called an **optimum solution.**

Linear Programming Theorem

A linear expression in two variables

$$c_1 x + c_2 y$$

defined over a convex set S whose sides are line segments takes on its maximum value at a corner point of S and its minimum value at a corner point of S. If S is unbounded, there may or may not be an optimum value, but if there is, then it must occur at a corner point.

Procedure for Solving a Linear Programming Problem

To solve a linear programming problem:

Step 1. Find the objective function (the quantity to be maximized or minimized).
Step 2. Graph the constraints defined by a system of linear inequalities; this graph is called the set S.
Step 3. Find the corners of S; for each corner, this may require the solution of a system of two equations with two unknowns.
Step 4. Find the value of the objective function for the coordinates of each corner point. The largest value is the maximum; the smallest value is the minimum.

■ EXAMPLE I

Maximize $T = 4x + 5y$ subject to:
$$\begin{cases} 2x + 5y \le 25 \\ 6x + 5y \le 45 \\ x \ge 0 \\ y \ge 0 \end{cases}$$

Solution The objective function is the function to be maximized or minimized; for this example, it is $T = 4x + 5y$.

Next, graph the set of constraints. This is the feasible region and is shown in Figure 12.7. The goal of this linear programming problem is to find that point in the feasible set that gives the largest possible value of T. Consider the following list of possible solutions:

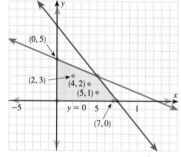

▲ **Figure 12.7 Graph of feasible solutions**

A Point in the Feasible Set	$T = 4x + 5y$ Value of Objective Function
(2, 3)	$T = 4(2) + 5(3) = 23$
(4, 2)	$T = 4(4) + 5(2) = 26$
(5, 1)	$T = 4(5) + 5(1) = 25$
(7, 0)	$T = 4(7) + 5(0) = 28$
(0, 5)	$T = 4(0) + 5(5) = 25$

The point from the table that makes the objective function the largest is $(7, 0)$. But is this the largest for all feasible solutions? How about $(6, 1)$ or $(5, 3)$? The linear programming theorem tells us that the maximum value occurs at a corner point. The corner points are labeled A, B, C, and D on Figure 12.7. Some corner points can be found by inspection — for example, $D = (0, 0)$ and $C = (0, 5)$. Other corner points may require some work with boundary lines.

⊘ Use the equations, not the inequalities, for the boundaries. ⊘

Point A: Solve the system $\begin{cases} y = 0 \\ 6x + 5y = 45 \end{cases}$

Solve by substitution to find $x = \frac{15}{2}$ and $y = 0$.

Point B: Solve the system $\begin{cases} 2x + 5y = 25 \\ 6x + 5y = 45 \end{cases}$

Solve by adding to find $x = 5$ and $y = 3$.

We now have the coordinates of all the corner points. This is the *entire* list of all the points we need to check from the feasible set.

$$T = 4x + 5y$$

Corner Point	Objective Function
A: $(\frac{15}{2}, 0)$	$T = 4(\frac{15}{2}) + 5(0) = 30$
B: $(5, 3)$	$T = 4(5) + 5(3) = 35$
C: $(0, 5)$	$T = 4(0) + 5(5) = 25$
D: $(0, 0)$	$T = 4(0) + 5(0) = 0$

Look for the maximum value of T on *this* list; it is 35 so the maximum value of T is 35. ■

◼ EXAMPLE 2   Polya's Method

A farmer has 100 acres on which to plant two crops, corn and wheat, and the problem is to maximize the profit. There are several considerations; the first is the expense:

	Cost per Acre	
	Corn	Wheat
seed	$ 12	$ 40
fertilizer	$ 58	$ 80
planting/care/harvesting	$ 50	$ 90
TOTAL	$120	$210

After the harvest, the farmer must store the crops while awaiting proper market conditions. Each acre yields an average of 110 bushels of corn or 30 bushels of wheat. The limitations of resources are as follows:

Available capital: $15,000
Available storage facilities: 4,000 bushels

If the net profit (after all expenses have been subtracted) per bushel of corn is $1.30 and for wheat is $2.00, how should the farmer plant the 100 acres to maximize the profits?

Solution   We use Polya's problem-solving guidelines for this example.

Understand the Problem.   First, you might try to solve this problem by using your intuition.

Plant 100 acres in wheat

Production:	100 acres @ 30 bu/acre = 3,000 bushels
Net profit:	3,000 bu × \$2.00/bu = \$6,000
Costs:	100 acres @ \$210/acre = \$21,000
	These costs are more than the \$15,000 available, so the farmer cannot plant 100 acres in wheat.

Plant 100 acres in corn

Production:	100 acres @ 110 bu/acre = 11,000 bushels
Net profit:	11,000 bu × \$1.30/bu = \$14,300
Costs:	100 acres @ \$120/acre = \$12,000
	These costs can be met with the available \$15,000, but the production of 11,000 bushels exceeds the available storage capacity of 4,000 bu.

Clearly, some mix of wheat and corn is necessary. Let us build a mathematical model.

Devise a Plan.

Let x = number of acres to be planted in corn;

      y = number of acres to be planted in wheat.

With a mathematical model, we must make certain assumptions. For this example, we assume:

⊘ These first two assumptions (constraints) will apply in almost every linear programming model. ⊘

$x \geq 0$	The number of acres of corn cannot be negative.
$y \geq 0$	The number of acres of wheat cannot be negative.
$x + y \leq 100$	The amount of available land is 100 acres. We do not assume that $x + y = 100$, because it might be more profitable to leave some land unplanted.

EXPENSES $\leq$ 15,000

      The total expenses cannot exceed \$15,000. We also know

EXPENSES = EXPENSES FOR CORN + EXPENSE FOR WHEAT

      = $120x + 210y$

      Thus, this constraint (in terms of x and y) is
      $120x + 210y \leq 15,000$

TOTAL YIELD $\leq$ 4,000

      The total yield cannot exceed the storage capacity of 4,000 bushels. We also know

TOTAL YIELD = YIELD FOR CORN + YIELD FOR WHEAT

      = $110x + 30y$

      Thus, this constraint (in terms of x and y) is
      $110x + 30y \leq 4,000$

The farmer wants to maximize the profit. Let P = TOTAL PROFIT.

TOTAL PROFIT = $\quad$ PROFIT FROM CORN $\quad + \quad$ PROFIT FROM WHEAT

$$= \text{CORN VALUE} \times \text{CORN AMOUNT} + \text{WHEAT VALUE} \times \text{WHEAT AMOUNT}$$

$$= \quad\quad 1.30 \times 110x \quad\quad + \quad\quad 2.00 \times 30y$$

$$= 143x + 60y$$

We have formulated the following linear programming problem:

Maximize: $\quad P = 143x + 60y$

Subject to: $\quad \begin{cases} x \geq 0 \\ y \geq 0 \\ x + y \leq 100 \\ 120x + 210y \leq 15{,}000 \\ 110x + 30y \leq 4{,}000 \end{cases}$

Carry Out the Plan. To solve this problem, we must graph the constraints (see Figure 12.8). Note that the corner points are labeled A, B, C, and D. Find the coordinates of these points:

$A(0, 0) \quad\quad$ By inspection

$B(0, \frac{500}{7}) \quad\quad$ Solve the system $\begin{cases} 120x + 210y = 15{,}000 \\ x = 0 \end{cases}$

$C(\frac{400}{11}, 0) \quad\quad$ Solve the system $\begin{cases} 110x + 30y = 4{,}000 \\ y = 0 \end{cases}$

$D(20, 60) \quad\quad$ Solve the system $\begin{cases} 110x + 30y = 4{,}000 \\ 120x + 210y = 15{,}000 \end{cases}$

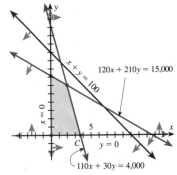

▲ **Figure 12.8 Farmer problem**

Use the linear programming theorem to check these corner points:

Corner Point	Objective Function, $P = 143x + 60y$
$A(0, 0)$	$P = 143(0) + 60(0) = 0$
$B(0, \frac{500}{7})$	$P = 143(0) + 60(\frac{500}{7}) \approx 4{,}286$
$C(\frac{400}{11}, 0)$	$P = 143(\frac{400}{11}) + 60(0) = 5{,}200$
$D(20, 60)$	$P = 143(20) + 60(60) = 6{,}460$

Look Back. The maximum value of P is at $(20, 60)$. This means that, to maximize the profit subject to the constraints, the farmer should plant 20 acres in corn, plant 60 acres in wheat, and leave 20 acres unplanted. ▬

Notice from the graph in Example 2 that some of the constraints could be eliminated from the problem and everything else would remain unchanged. For example, the boundary $x + y = 100$ was not necessary in finding the maximum value of P. Such a condition is said to be a **superfluous constraint**. It is not uncommon to have superfluous constraints in a linear programming problem. Suppose, however, that the farmer in Example 2 contracted to have the grains stored at a neighboring farm and now the contract calls for *at least* 4,000 bushels to be stored.

This change from $110x + 30y \leq 4,000$ to $110x + 30y \geq 4,000$ *now* makes the condition $x + y \leq 100$ important to the solution of the problem (see Problem 39). Therefore, you must be careful about superfluous constraints even though they do not affect the solution at the present time.

The next example is solved more succinctly to show you the way your work will probably look.

▬▬▬ EXAMPLE 3   Polya's Method

The Sticky Widget Company makes two types of widgets: regular and deluxe. Each widget is produced at a station consisting of a machine and a person who finishes the widgets by hand. The regular widget requires 2 hr of machine time and 1 hr of finishing time. The deluxe widget requires 3 hr of machine time and 5 hr of finishing time. The profit on the regular widget is \$25; on the deluxe widget it is \$30. If the workday is 8 hr, how many of each type of widget should be produced at each station to maximize the profit?*

Solution   We use Polya's problem-solving guidelines for this example.

Understand the Problem.   We are trying to find the number of regular and deluxe widgets that should be produced to maximize the profits.

Devise a Plan.

Let $x =$ NUMBER OF REGULAR WIDGETS PRODUCED
       $y =$ NUMBER OF DELUXE WIDGETS PRODUCED

Maximize profit: $P = 25x + 30y$

Subject to: $\begin{cases} x \geq 0 \\ y \geq 0 \\ 2x + 3y \leq 8 \\ x + 5y \leq 8 \end{cases}$ The workday is no more than 8 hr.

The set of feasible solutions is found by graphing this system of inequalities, as shown in Figure 12.9. The corner points are found by considering the intersection of the boundary lines.

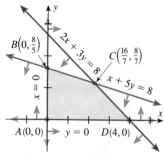

▲ **Figure 12.9   Maximizing profit**

---

* It is difficult to set up textbook problems to model real-life problems, but our goal in this book is to teach you a *process* of problem solving. Remember that, in real life, you cannot be sure of what information to use and what information is not necessary. This problem, for example, would be stated in real life as: Maximize the profit. All of the other information would be part of the assumptions you would gather for the solution of the problem. Deciding on the assumptions often requires a great deal of work when solving *real-life* problems.

Carry Out the Plan.   The linear programming theorem requires the values summarized with the following table.

Corner point	System to solve	Solution	Objective function $25x + 30y$	Comment
A	$\begin{cases} x = 0 \\ y = 0 \end{cases}$	$(0, 0)$	$25(0) + 30(0) = 0$	**minimum**
B	$\begin{cases} x = 0 \\ x + 5y = 8 \end{cases}$	$(0, \frac{8}{5})$	$25(0) + 30(\frac{8}{5}) = 48$	
C	$\begin{cases} 2x + 3y = 8 \\ x + 5y = 8 \end{cases}$	$(\frac{16}{7}, \frac{8}{7})$	$25(\frac{16}{7}) + 30(\frac{8}{7}) \approx 91.43$	
D	$\begin{cases} y = 0 \\ 2x + 3y = 8 \end{cases}$	$(4, 0)$	$25(4) + 30(0) = 100$	**maximum**

Look Back.   Profits are maximized if only regular widgets are produced. The company should produce four regular widgets per day at each station.   ▬

The method discussed here can be generalized to higher dimensions by means of more sophisticated methods for solving these linear programming problems. The general method for solving a linear programming problem is called the *simplex method* and is usually discussed in a course called linear programming or finite mathematics.

PROBLEM SET   12.6

▲ **A Problems**

1. **IN YOUR OWN WORDS** What is a linear programming problem?
2. **IN YOUR OWN WORDS** What is an objective function?
3. **IN YOUR OWN WORDS** What do we mean by constraints?

For Problems 4–11, decide whether the given point is a feasible solution for the constraints

$$\begin{cases} x \geq 0 \\ y \geq 0 \\ 3x + 2y \leq 10 \\ 2x + 4y \leq 8 \end{cases}$$

4. $(1, 3)$ **5.** $(2, 1)$ **6.** $(1, 2)$ **7.** $(-1, 4)$ **8.** $(2, 2)$ **9.** $(0, 4)$

For Problems 10–15, decide whether the given point is a corner point for the constraints

$$\begin{cases} x \geq 0 \\ y \geq 0 \\ 2x + 3y \geq 120 \\ 2x + y \geq 80 \end{cases}$$

10. $(0, 0)$ **11.** $(0, 80)$ **12.** $(80, 0)$ **13.** $(60, 0)$ **14.** $(30, 20)$ **15.** $(20, 30)$

▲ B Problems

Find the corner points for the set of feasible solutions for the constraints given in Problems 16–27.

16. $\begin{cases} x \geq 0 \\ y \geq 0 \\ 2x + y \leq 12 \\ x + 2y \leq 9 \end{cases}$

17. $\begin{cases} x \geq 0 \\ y \geq 0 \\ 2x + 5y \leq 20 \\ 2x + y \leq 12 \end{cases}$

18. $\begin{cases} x \geq 0 \\ y \geq 0 \\ 3x + 2y \leq 12 \\ x + 2y \leq 8 \end{cases}$

19. $\begin{cases} x \geq 0 \\ y \geq 0 \\ x \leq 10 \\ y \leq 8 \\ 3x + 2y \geq 12 \end{cases}$

20. $\begin{cases} x \geq 0 \\ y \geq 0 \\ x + y \leq 8 \\ y \leq 4 \\ x \leq 6 \end{cases}$

21. $\begin{cases} x \geq 0 \\ y \geq 0 \\ x + y \geq 6 \\ -2x + y \geq -16 \\ y \leq 9 \end{cases}$

22. $\begin{cases} x \geq 0 \\ y \geq 0 \\ 3x + 2y \leq 8 \\ x + 5y \leq 8 \end{cases}$

23. $\begin{cases} x \geq 0 \\ y \geq 0 \\ x \leq 8 \\ y \geq 2 \\ x + y \leq 10 \\ x \leq 3y \end{cases}$

24. $\begin{cases} x \geq 0 \\ y \geq 0 \\ 10x + 5y \geq 200 \\ 2x + 5y \geq 100 \\ 3x + 4y \geq 120 \end{cases}$

25. $\begin{cases} x \geq 0 \\ y \geq 0 \\ x + y \leq 9 \\ 2x - 3y \geq -6 \\ x - y \leq 3 \end{cases}$

26. $\begin{cases} x \geq 0 \\ y \geq 0 \\ 2x + y \geq 8 \\ y \leq 5 \\ x - y \leq 2 \\ 3x - 2y \geq 5 \end{cases}$

27. $\begin{cases} x \geq 0 \\ y \geq 0 \\ 2x + y \geq 8 \\ x - 2y \leq 7 \\ x - y \geq -3 \\ x \leq 9 \end{cases}$

Find the optimum value for each objective function given in Problems 28–33.

28. Maximize $W = 30x + 20y$ subject to the constraints of Problem 16.

29. Maximize $T = 100x + 10y$ subject to the constraints of Problem 17.

30. Maximize $P = 100x + 100y$ subject to the constraints of Problem 18.

31. Minimize $K = 140x + 250y$ subject to the constraints of Problem 23.

32. Minimize $A = 2x - 3y$ subject to the constraints of Problem 20.

33. Minimize $K = 6x + 18y$ subject to the constraints of Problem 19.

▲ Problem Solving

Write a linear programming model, including the objective function and the set of constraints, for Problems 34–37. DO NOT SOLVE, but be sure to define all your variables.

34. The Wadsworth Widget Company manufactures two types of widgets: regular and deluxe. Each widget is produced at a station consisting of a machine and a person who finishes each widget by hand. The regular widget requires 3 hr of machine time and 2 hr of finishing time. The deluxe widget requires 2 hr of machine time and 4 hr of finishing time. The profit on the regular widget is $25; on the deluxe widget it is $30. If the workday is 8 hours, how many of each type of widget should be produced at each station per day to maximize the profit?

35. A convalescent hospital wishes to provide, at a minimum cost, a diet that has a minimum of 200 g of carbohydrates, 100 g of protein, and 20 g of fat per day. These requirements can be met with two foods, A and B:

Food	Carbohydrates	Protein	Fats
A	10 g	2 g	3 g
B	5 g	5 g	4 g

If food A costs $0.29 per gram and food B costs $0.15 per gram, how many grams of each food should be purchased for each patient per day to meet the minimum requirements at the lowest cost?

36. Karlin Enterprises manufactures two games. Standing orders require that at least 24,000 space-battle games and 5,000 football games be produced per month. The Gainesville plant can produce 600 space-battle games and 100 football games per day; the Sacramento plant can produce 300 space-battle games and 100 football games per day. If the Gainesville plant costs $20,000 per day to operate and the Sacramento factory costs $15,000 per day, find the number of days per month each factory should operate to minimize the cost. (Assume each month is 30 days.)

37. Brown Bros., Inc. is an investment company doing an analysis of the pension fund for a certain company. The fund has a maximum of $10 million to invest in two places: no more than $8 million in stocks yielding 12%, and at least $2 million in long-term bonds yielding 8%. The stock-to-bond investment ratio cannot be more than 3 to 1. How should Brown Bros. advise its client so that the investments yield the maximum yearly return?

Solve the linear programming problems in Problems 38–42.

38. Suppose the net profit per bushel of corn in Example 2 increased to $2.00 and the net profit per bushel of wheat dropped to $1.50. Maximize the profit if the other conditions in the example remain the same.

39. Suppose the farmer in Example 2 contracted to have the grain stored at a neighboring farm and the contract calls for at least 4,000 bushels to be stored. How many acres should be planted in corn and how many in wheat to maximize profit if the other conditions in Example 2 remain the same?

40. The Thompson Company manufactures two industrial products, standard ($45 profit per item) and economy ($30 profit per item). These items are built using machine time and manual labor. The standard product requires 3 hr of machine time and 2 hr of manual labor. The economy model requires 3 hr of machine time and no manual labor. If the week's supply of manual labor is limited to 800 hr and machine time to 15,000 hr, how much of each type of product should be produced each week to maximize the profit?

41. The following carbohydrate information is given on the side of the respective cereal boxes (for 1 oz of cereal with $\frac{1}{2}$ cup of whole milk):

	Starch and related carbohydrates	Sucrose and other sugars
Kellogg's Corn Flakes	23 g	7 g
Post Honeycombs	14 g	17 g

What is the minimum cost to receive at least 322 g starch and 119 g sucrose by consuming these two cereals if Corn Flakes cost $0.07 per ounce and Honeycombs cost $0.19 per ounce?

42. Your broker tells you of two investments she thinks are worthwhile. She advises a new issue of Pertec stock, which should yield 20% over the next year, and then to balance your account she advises Campbell Municipal Bonds with a 10% yearly yield. The stock-to-bond ratio should be no less than 3 to 1. If you have no more than $100,000 to invest and do not want to invest more than $70,000 in Pertec or less than $20,000 in bonds, how much should be invested in each to maximize your return?

"How can students compete in a mathematical society when they leave school knowing so little mathematics?"

Lester Thurow, in *Everybody Counts*

Biographical Sketch

"Explaining is an important part of solving problems — to make your thoughts and reasons understood by others."

Lynn Arthur Steen

Lynn Arthur Steen was born in Chicago, Illinois, on January 1, 1941, and is Professor of Mathematics at St. Olaf College. He is very active in national professional organizations and is the author of the 1989 report on mathematics education in the United States, *Everybody Counts*, published by the prestigious National Academy of Sciences.

When asked to describe mathematics, he said, "Mathematics is a superb 'liberal arts' discipline, since it opens doors to an enormous number of other fields. Math is not just for scientists and engineers. If you want to be a doctor or a lawyer, mathematics provides excellent preparation — both for technical skills and for logical reasoning. Journalists and politicians, homeowners and citizens all benefit from the rigorous quantitative thinking that one experiences in mathematics courses."

Professor Steen is a dynamic lecturer and writer. He explains, "Mathematics has the power to explain, but only if it is put in words that people can understand. Otherwise it is like the noise made in the forest by a tree falling where no one can hear it."

Book Report

Write a 500-word report on the following book:

▲ *On the Shoulders of Giants: New Approaches to Numeracy*, Lynn Arthur Steen, Editor (Washington, D.C.: National Academy Press, 1990).

Important Terms

Addition method [12.1]
Array [12.3]
Augmented matrix [12.3]
Communication matrix [12.4]
Constraint [12.6]
Convex set [12.6]
Dependent system [12.1]
Dimension [12.3]
Double subscripts [12.3]
Elementary row operations [12.3]

Equivalent matrices [12.3]
Equivalent systems [12.1]
Feasible solution [12.6]
Gauss–Jordan elimination [12.3]
Graphing method [12.1]
Identity matrix [12.4]
Inconsistent system [12.1]
Inverse matrix [12.4]
Inverse property [12.4]
Linear combination method [12.1]

Linear programming [12.6]
Linear system [12.1]
Matrix [12.3]
Matrix equation [12.3; 12.4]
Nonconformable matrices [12.4]
Objective function [12.6]
Optimum solution [12.6]
Order [12.3]
Pivot [12.3]
Row+ [12.3]
Row-reduced form [12.3]

RowSwap [12.3]
Simultaneous solution [12.1]
Square matrix [12.3]
Subscript [12.3]
Substitution method [12.1]
Superfluous constraint [12.6]
Supply and demand [12.2]
System [12.1]
System of inequalities [12.5]
*Row [12.3]
*Row+ [12.3]
Zero matrix [12.4]

Important Ideas

Solving systems of equations by graphing, substitution, and addition [12.1]
Four elementary row operations [12.3]
Process of pivoting [12.3]
Relationship between a system of equations and a corresponding matrix [12.3]
Know the matrix operations of addition, multiplication by a scalar, subtraction, and multiplication [12.4]

Know the properties of matrices, including commutative, associative, identity, inverse, and distributive properties [12.4]
Procedure for finding the inverse of a matrix [12.4]
Linear programming theorem [12.6]
Procedure for solving a linear programming problem [12.6]

Types of Problems

Solve systems of equations by selecting the most appropriate method [12.1]
Solve applied problems, including coin problems, combining rates, supply and demand, and mixture problems [12.1]
Perform elementary row operations on a given matrix [12.3]
Solve systems of equations by the Gauss–Jordan method [12.3]
Carry out matrix operations, including finding the inverse of a given matrix [12.4]
Solve a system of equations using the inverse matrix method [12.4]

Solve a system of inequalities [12.5]
Decide whether a given point is a feasible solution for a set of constraints [12.6]
Decide whether a given point is a corner point for a set of constraints [12.6]
Find the corner points for a set of feasible solutions [12.6]
Maximize or minimize an objective function subject to a set of constraints [12.6]
Solve applied problems using a linear programming model [12.6]

CHAPTER 12 Review Questions

In Problems 1–4, perform the indicated matrix operations, if possible, where

$$[A] = \begin{bmatrix} 1 & 0 \\ 2 & -1 \end{bmatrix} \quad [B] = \begin{bmatrix} 2 & -1 & 0 \\ 1 & 0 & 1 \end{bmatrix} \quad [C] = \begin{bmatrix} 2 & 0 \\ 1 & 2 \\ -1 & 1 \end{bmatrix} \quad [D] = \begin{bmatrix} 0 & 1 & 0 \\ -1 & 0 & 0 \\ 0 & 1 & -1 \end{bmatrix}$$

1. $[C][B] - 3[D]$ **2.** $[A][B][C]$ **3.** $[B][A][D]$ **4.** $[C][A][B]$

Find the inverse of each matrix in Problems 5–6.

5. $\begin{bmatrix} 2 & 1 \\ -\frac{3}{2} & -\frac{1}{2} \end{bmatrix}$ 6. $\begin{bmatrix} 1 & 3 & 3 \\ 1 & 4 & 3 \\ 1 & 3 & 4 \end{bmatrix}$

Solve the systems in Problems 7–11 by the indicated method.

7. By graphing: $\begin{cases} 2x - y = 2 \\ 3x - 2y = 1 \end{cases}$

8. By addition: $\begin{cases} x + 3y = 3 \\ 4x - 6y = -6 \end{cases}$

9. By substitution: $\begin{cases} y = 1 - 2x \\ 5x + 2y = 1 \end{cases}$

10. By Gauss–Jordan: $\begin{cases} x + y + z = 2 \\ x + 2y - 2z = 1 \\ x + y + 3z = 4 \end{cases}$

11. By using an inverse matrix: $\begin{cases} 2x + y = 13 \\ -\frac{3}{2}x - \frac{1}{2}y = 10 \end{cases}$

12. Graph the solution of the system: $\begin{cases} 2x - y + 2 \leq 0 \\ 2x - y + 12 \geq 0 \\ 3x + 2y + 10 \geq 0 \\ 3x + 2y - 18 \leq 0 \end{cases}$

13. To manufacture a certain alloy, it is necessary to use 33 oz of metal A and 56 oz of metal B. It is cheaper for the manufacturer if she buys and mixes two products that come as metal bars: Product I, each bar of which contains 3 oz of metal A and 5 oz of metal B; Product II, each bar of which contains 4 oz of metal A and 7 oz of metal B. How many bars of each product should she use to produce the desired alloy?

14. A manufacturer of auto accessories uses three basic parts, A, B, and C, in its three products, in the following proportions:

	A	B	C
Product I	2	1	1
Product II	2	2	1
Product III	3	2	2

The inventory shows 1,250 of part A, 900 of part B, and 750 of part C on hand. How many of each product may be manufactured using all the inventory on hand?

15. A farmer has 500 acres on which to plant two crops: corn and wheat. To produce these crops, there are certain expenses:

	Cost per Acre	
	Corn	Wheat
Seed	$ 12	$10
Fertilizer	$ 58	$20
Planting/care/harvesting	$ 50	$30
TOTAL	$120	$60

After the harvest, the farmer must store the crops while awaiting proper market conditions. Each acre yields an average of 100 bushels of corn or 40 bushels of wheat. The farmer has available capital of $24,000 and has contracted to store at least 18,000 bushels. If the net profit (after all expenses have been subtracted) per bushel of corn is $2.10 and for wheat is $2.50, how should the farmer plant the 500 acres to maximize the profit, and what is the maximum profit?

Revisiting the Real World . . .

IN THE REAL WORLD "I think I'm going crazy, Bill," said George. "I can't figure out a good rating system for the state league. I've got over a hundred teams and I need to come up with a statewide rating system to rank all of those teams. Any ideas?"

"As a matter of fact, yes!" said Bill enthusiastically. "Have you ever heard of the Harbin Football Team rating system? I think they use it in Ohio. They use it to determine the high school football teams that are eligible to compete in post-season playoffs. Each team earns points for games it wins and for games that a defeated opponent wins. Here's how it works: Level 1 points are awarded for each game a team wins. A level-2 point is awarded for each game a defeated opponent wins."

"I hear ya, but I still don't get it. How do you put all that together?" asked George. "Well," said Bill, "it's very easy . . . . We simply use a matrix . . ."

Commentary

To analyze this problem, we will use a communication matrix. Team A beats teams B and D; team B beats teams C and D; team C beats team B; and team D beats team C. Represent this in matrix form showing level-1 points:

$$[M] = \begin{array}{c} \\ A \\ B \\ C \\ D \end{array}\begin{array}{c} A\ B\ C\ D \\ \begin{bmatrix} 0 & 1 & 0 & 1 \\ 0 & 0 & 1 & 1 \\ 0 & 1 & 0 & 0 \\ 0 & 0 & 1 & 0 \end{bmatrix} \end{array} \qquad [M][T] = \begin{bmatrix} 0 & 1 & 0 & 1 \\ 0 & 0 & 1 & 1 \\ 0 & 1 & 0 & 0 \\ 0 & 0 & 1 & 0 \end{bmatrix}\begin{bmatrix} 1 \\ 1 \\ 1 \\ 1 \end{bmatrix} = \begin{bmatrix} 2 \\ 2 \\ 1 \\ 1 \end{bmatrix}$$

Now, the number of level-1 points awarded each team is the sum of the row representing that team. We can use matrix multiplication to show this by letting [T] be a column matrix consisting of all ones. The number of rows of [T] is the same as the number of teams. Thus, the number of level-1 points is shown as the column matrix at the right. A level-2 point is awarded for each game a defeated opponent wins. This means, for example, if team A beats team B (level-1 point) but team B beats teams C and D, then team A receives one level-2 point for each of these outcomes. It is easy to verify that the level-2 points are represented by $[M]^2$:

$$[M]^2 = \begin{bmatrix} 0 & 1 & 0 & 1 \\ 0 & 0 & 1 & 1 \\ 0 & 1 & 0 & 0 \\ 0 & 0 & 1 & 0 \end{bmatrix}\begin{bmatrix} 0 & 1 & 0 & 1 \\ 0 & 0 & 1 & 1 \\ 0 & 1 & 0 & 0 \\ 0 & 0 & 1 & 0 \end{bmatrix} = \begin{bmatrix} 0 & 0 & 2 & 1 \\ 0 & 1 & 1 & 0 \\ 0 & 0 & 1 & 1 \\ 0 & 1 & 0 & 0 \end{bmatrix}$$

As before, we can find the total number of level-2 points by adding the row entries, or by multiplying $[M]^2[T]$. The total number of points, [P], awarded is $[P] = ([M] + [M]^2)[T]$:

$$\left(\begin{bmatrix} 0 & 1 & 0 & 1 \\ 0 & 0 & 1 & 1 \\ 0 & 1 & 0 & 0 \\ 0 & 0 & 1 & 0 \end{bmatrix} + \begin{bmatrix} 0 & 0 & 2 & 1 \\ 0 & 1 & 1 & 0 \\ 0 & 0 & 1 & 1 \\ 0 & 1 & 0 & 0 \end{bmatrix}\right)\begin{bmatrix} 1 \\ 1 \\ 1 \\ 1 \end{bmatrix} = \begin{bmatrix} 0 & 1 & 2 & 2 \\ 0 & 1 & 2 & 1 \\ 0 & 1 & 1 & 1 \\ 0 & 1 & 1 & 0 \end{bmatrix}\begin{bmatrix} 1 \\ 1 \\ 1 \\ 1 \end{bmatrix} = \begin{bmatrix} 5 \\ 4 \\ 3 \\ 2 \end{bmatrix}$$

The ranking of the teams would be A, B, C, and then D.

Group Research

Working in small groups is typical of most work environments, and being able to work with others to communicate specific ideas is an important skill to learn. Work with three or four other students to submit a single report based on each of the following questions.

1. If two teams tie, enter 0.5 in the communication matrix instead of 1. Team A beats F, and ties C; team B beats A, C, and F; team C beats E and F; team D ties A and beats F; team E beats A and F; team F beats C and ties D. Rank these teams.

2. Suppose your group conducts an experiment at a local department store. You walk up a rising escalator and you take one step per second to reach the top in 20 seconds. Next, you walk up the same rising escalator at the rate of two steps per second and this time it takes 32 steps. How many steps would be required to reach the top on a stopped escalator?

3. Two ranchers sold a herd of cattle and received as many dollars for each animal as there were cattle in the herd. With the money they bought a flock of sheep at $10 a head and then a lamb with the rest of the money (less than $10). Finally, they divided the animals between them, with one rancher obtaining an extra sheep and the other the lamb. The rancher who got the lamb was given his friend's new watch as compensation. What is the value of the watch?

4. Suppose your group has just been hired by a company called Alco. You are asked to analyze its operations and make some recommendations about how it can comply at a minimum cost with recent orders of the Environmental Protection Agency (EPA).

    To prepare your report you study the operation and obtain the following information:

    Alco Cement Company produces cement.

    The EPA has ordered Alco to reduce the amount of emissions released into the atmosphere during production.

    Alco wants to comply, but wants to do so at the least possible cost.

    Present production is 2.5 million barrels of cement, and 2 pounds of dust are emitted for every barrel of cement produced.

    The cement is produced in kilns that are presently equipped with mechanical collectors.

    To reduce the emissions to the required level, the mechanical collectors must be replaced either by four-field electrostatic precipitators, which would reduce emission to 0.5 pound of dust per barrel of cement, or by five-field precipitators, which would reduce emission to 0.2 pound per barrel.

    The capital and operating costs for the four-field precipitator are 14¢ per barrel of cement produced; for the five-field precipitator, costs are 18¢ per barrel.

    To comply with the EPA, Alco must reduce particulate emission by at least 4.2 million pounds.*

    Use mathematical modeling to write your paper. Mathematical modeling involves creating equations and procedures to make predictions about the real world. Typical textbook problems focus on limited, specific skills, but in the real world you need to sift through the given information to decide what information you need and what information you do not need. You may need to do some research to gather data not provided.

---

* This research project is adapted from R. E. Kohn, "A Mathematical Programming Model for Air Pollution Control," *Science and Mathematics*, June 1969, pp. 487–499.

APPENDIX A   Problem Solving

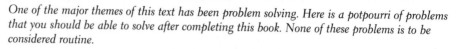

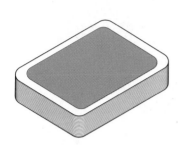

One of the major themes of this text has been problem solving. Here is a potpourri of problems that you should be able to solve after completing this book. None of these problems is to be considered routine.

1. How many cards must you draw from a deck of 52 playing cards to be sure that at least two are from the same suit?

2. How many people must be in a room to be sure that at least four of them have the same birthday (not necessarily the same year)?

3. Find the units digit of $3^{1998} - 2^{1998}$?

4. If a year had two consecutive months with a Friday the thirteenth, which months must they be?

5. How many months have 28 days?

6. What is the largest number that is a divisor of both 210 and 330?

7. If Ann is 3 years older than Brittany, Brittany is 2 years older than Chelsea, Deidre is 5 years older than Elysse, and Elysse is a year older than Fawn, then what can be said of Deidre's age as compared to Chelsea's age?

8. The News Clip shows a letter printed in the "Ask Marilyn" column of *Parade* magazine (Sept. 27, 1992). How would you answer it? *Hint:* We won't give you the answer, but will quote one line from Marilyn's answer: "So the question should be not why the smaller one yields that much, but why it yields that little."

9. If a megamile is one million miles and a kilomile is one thousand miles, how many kilomiles are there in 2.376 megamiles?

10. A long straight fence having a pole every 8 feet is 1,440 feet long. How many fence poles are used in the fence?

11. If $(a, b) = a \times b + a + b$, what is the value of $((1, 2), (3, 4))$?

12. If it is known that all Angelenos are Venusians and all Venusians are Los Angeles residents, then what must necessarily be the conclusion?

13. If 1 is the first odd number, what is the 473rd odd number?

14. If $1 + 2 + 3 + \cdots + n = \dfrac{n(n + 1)}{2}$, what is the sum of the first 100,000 counting numbers beginning with 1?

15. A four-inch cube is painted red on all sides. It is then cut into one-inch cubes. What fraction of all the one-inch cubes are painted on one side only?

16. If slot machines had two arms and people had one arm, then it is probable that our number system would be based on the digits 0, 1, 2, 3, and 4 only. How would the number we know as 18 be written in such a number system?

17. If $M(a, b)$ stands for the larger number in the parentheses, and $m(a, b)$ stands for the lesser number in the parentheses, what is the value of $M(m(1, 2), m(2, 3))$?

18. If a group of 50 persons consists of 20 males, 12 children, and 25 women, how many men are in the group?

19. There are only five regular polyhedra and we have shown the patterns that give those polyhedra. Name the polyhedron obtained from each of the patterns shown below.

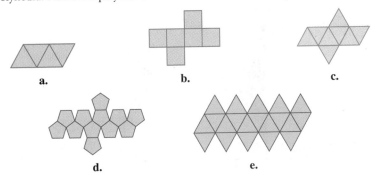

    a.   **b.**   **c.**

        d.   **e.**

20. Jack and Jill decided to exercise together by walking around a lake. Jack walks around the lake in 16 minutes and Jill jogs around the lake in 10 minutes. If Jack and Jill start at the same time and at the same place, and continue to exercise until they return to the starting point at the same time, how long will they be exercising?

21. What is the 1,000th positive integer that is not divisible by 3?

22. A frugal man allows himself a glass of wine before dinner on every third day, an after-dinner chocolate every fifth day, and a steak dinner once a week. If it happens that he enjoys all three luxuries on March 31, what will be the date of the next steak dinner that is preceded by wine and followed by an after-dinner chocolate?

23. How many trees must be cut to make a trillion one-dollar bills? To answer this question you need to make some assumptions. Assume that a pound of paper is equal to a pound of wood, and also assume that a dollar bill weighs about one gram. This implies that a pound of wood yields about 450 dollar bills. Furthermore, estimate that an average tree has a height of 50 ft and a diameter of 12 inches. Finally, assume that wood yields about 50 lb/ft³.

24. Estimate the volume of beer in the six-pack shown in the photograph.

25. Given a square with sides equal to 8 inches, with two inscribed semicircles of radius 4. What is the area of the shaded region?

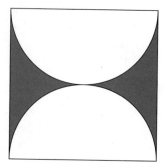

26. Critique the statement given in the News Clip.

27. The two small circles have radii of 2 and 3. Find the ratio of the area of the smallest circle to the area of the shaded region.

28. A large container filled with water is to be drained, and you would like to drain the container as quickly as possible. You can drain the container with either one 1-in. diameter hose or two $\frac{1}{2}$-in. hoses. Which do you think would be faster (one 1-in. drain or two $\frac{1}{2}$-in. drains), and why?

29. A gambler went to the horse races two days in a row. On the first day she doubled her money and spent $30. The second day she tripled her money and spent $20, after which she had what she started with the first day. How much did she start with?

30. What conclusions can you draw from the graphs that are reprinted here from the July 19, 1993, issue of *Newsweek*?

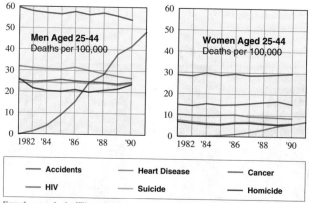

Figures for causes other than HIV unavailable for 1991. Source: Centers for Disease Control and Prevention

31. A charter flight has signed up 100 travelers. They are told that if they can sign up an additional 25 persons, they can save $78 each. What is the cost per person if 100 persons make the trip?

32. A hospital wishes to provide for its patients a diet that has a minimum of 100 g of carbohydrates, 60 g of protein, and 40 g of fats per day. These requirements can be met with two foods:

Food	Carbohydrates	Protein	Fats
A	6 g	3 g	1 g
B	2 g	2 g	2 g

It is also important to minimize costs; food A costs $0.14 per gram and food B costs $0.06 per gram. How many grams of each food should be bought for each patient per day to meet the minimum daily requirements at the lowest cost?

33. There is a map of a small village in the margin. To walk from A to B, Sarah obviously must walk at least 7 blocks (all the blocks are the same length). What is the number of shortest paths from A to B?

34. What is the smallest number of operations needed to build up to the number 100 if you start at 0 and use only two operations: doubling or increasing by 1. *Challenge:* Answer the same question for the positive integer n.

35. On January 1, 1997, the U.S. national debt was $5.2 trillion and on that date there were 266.5 million people. How long would it take to pay off this debt if *every* person pays $1 per day?

36. Supply the missing number in the following sequence: 10, 11, 12, 13, 14, 15, 16, 17, 20, 22, 24, ___, 100, 121, 10000.

37. Write a BASIC program to print out the first 20 Fibonacci numbers.

38. Answer the question asked in the News Clip from the "Ask Marilyn" column of *Parade* magazine (July 16, 1995).

39. How many different configurations can you see in the following figure?

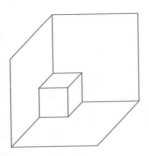

40. Five cards are drawn at random from a pack of cards that have been numbered consecutively from 1 to 104, and have been thoroughly shuffled. What is the probability that the numbers on the cards as drawn are in increasing order of magnitude?

TABLE Trigonometric Ratios

Degrees	sin x	cos x	tan x	Degrees	sin x	cos x	tan x
1	0.0175	0.9998	0.0175	46	0.7193	0.6947	1.0355
2	0.0349	0.9994	0.0349	47	0.7314	0.6820	1.0724
3	0.0523	0.9986	0.0524	48	0.7431	0.6691	1.1106
4	0.0698	0.9976	0.0699	49	0.7547	0.6561	1.1504
5	0.0872	0.9962	0.0875	50	0.7660	0.6428	1.1918
6	0.1045	0.9945	0.1051	51	0.7771	0.6293	1.2349
7	0.1219	0.9925	0.1228	52	0.7880	0.6157	1.2799
8	0.1392	0.9903	0.1405	53	0.7986	0.6018	1.3270
9	0.1564	0.9877	0.1584	54	0.8090	0.5878	1.3764
10	0.1736	0.9848	0.1763	55	0.8192	0.5736	1.4281
11	0.1908	0.9816	0.1944	56	0.8290	0.5592	1.4826
12	0.2079	0.9781	0.2126	57	0.8387	0.5446	1.5399
13	0.2250	0.9744	0.2309	58	0.8480	0.5299	1.6003
14	0.2419	0.9703	0.2493	59	0.8572	0.5150	1.6643
15	0.2588	0.9659	0.2679	60	0.8660	0.5000	1.7321
16	0.2756	0.9613	0.2867	61	0.8746	0.4848	1.8040
17	0.2924	0.9563	0.3057	62	0.8829	0.4695	1.8807
18	0.3090	0.9511	0.3249	63	0.8910	0.4540	1.9626
19	0.3256	0.9455	0.3443	64	0.8988	0.4384	2.0503
20	0.3420	0.9397	0.3640	65	0.9063	0.4226	2.1445
21	0.3584	0.9336	0.3839	66	0.9135	0.4067	2.2460
22	0.3746	0.9272	0.4040	67	0.9205	0.3907	2.3559
23	0.3907	0.9205	0.4245	68	0.9272	0.3746	2.4751
24	0.4067	0.9135	0.4452	69	0.9336	0.3584	2.6051
25	0.4226	0.9063	0.4663	70	0.9397	0.3420	2.7475
26	0.4384	0.8988	0.4877	71	0.9455	0.3256	2.9042
27	0.4540	0.8910	0.5095	72	0.9511	0.3090	3.0777
28	0.4695	0.8829	0.5317	73	0.9563	0.2924	3.2709
29	0.4848	0.8746	0.5543	74	0.9613	0.2756	3.4874
30	0.5000	0.8660	0.5774	75	0.9659	0.2588	3.7321
31	0.5150	0.8572	0.6009	76	0.9703	0.2419	4.0108
32	0.5299	0.8480	0.6249	77	0.9744	0.2250	4.3315
33	0.5446	0.8387	0.6494	78	0.9781	0.2079	4.7046
34	0.5592	0.8290	0.6745	79	0.9816	0.1908	5.1446
35	0.5736	0.8192	0.7002	80	0.9848	0.1736	5.6713
36	0.5878	0.8090	0.7265	81	0.9877	0.1564	6.3138
37	0.6018	0.7986	0.7536	82	0.9903	0.1392	7.1154
38	0.6157	0.7880	0.7813	83	0.9925	0.1219	8.1444
39	0.6293	0.7771	0.8098	84	0.9945	0.1045	9.5144
40	0.6428	0.7660	0.8391	85	0.9962	0.0872	11.4300
41	0.6561	0.7547	0.8693	86	0.9976	0.0698	14.3007
42	0.6691	0.7431	0.9004	87	0.9986	0.0523	19.0812
43	0.6820	0.7314	0.9325	88	0.9994	0.0349	28.6362
44	0.6947	0.7193	0.9657	89	0.9998	0.0175	57.2898
45	0.7071	0.7071	1.0000	90	1.0000	0.0000	undefined

TABLE   Standard Normal Distribution: z-scores

For a particular value, this table gives the percent of scores between the mean and the z-value of a normally distributed random variable.

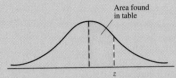

Area found in table

z	0.00	0.01	0.02	0.03	0.04	0.05	0.06	0.07	0.08	0.09
0.0	0.0000	0.0040	0.0080	0.0120	0.0160	0.0199	0.0239	0.0279	0.0319	0.0359
0.1	0.0398	0.0438	0.0478	0.0517	0.0557	0.0596	0.0636	0.0675	0.0714	0.0753
0.2	0.0793	0.0832	0.0871	0.0910	0.0948	0.0987	0.1026	0.1064	0.1103	0.1141
0.3	0.1179	0.1217	0.1255	0.1293	0.1331	0.1368	0.1406	0.1443	0.1480	0.1517
0.4	0.1554	0.1591	0.1628	0.1664	0.1700	0.1736	0.1772	0.1808	0.1844	0.1879
0.5	0.1915	0.1950	0.1985	0.2019	0.2054	0.2088	0.2123	0.2157	0.2190	0.2224
0.6	0.2257	0.2291	0.2324	0.2357	0.2389	0.2422	0.2454	0.2486	0.2517	0.2549
0.7	0.2580	0.2611	0.2642	0.2673	0.2704	0.2734	0.2764	0.2794	0.2823	0.2852
0.8	0.2881	0.2910	0.2939	0.2967	0.2995	0.3023	0.3051	0.3078	0.3106	0.3133
0.9	0.3159	0.3186	0.3212	0.3238	0.3264	0.3289	0.3315	0.3340	0.3365	0.3389
1.0	0.3413	0.3438	0.3461	0.3485	0.3508	0.3531	0.3554	0.3577	0.3599	0.3621
1.1	0.3643	0.3665	0.3686	0.3708	0.3729	0.3749	0.3770	0.3790	0.3810	0.3830
1.2	0.3849	0.3869	0.3888	0.3907	0.3925	0.3944	0.3962	0.3980	0.3997	0.4015
1.3	0.4032	0.4049	0.4066	0.4082	0.4099	0.4115	0.4131	0.4147	0.4162	0.4177
1.4	0.4192	0.4207	0.4222	0.4236	0.4251	0.4265	0.4279	0.4292	0.4306	0.4319
1.5	0.4332	0.4345	0.4357	0.4370	0.4382	0.4394	0.4406	0.4418	0.4429	0.4441
1.6	0.4452	0.4463	0.4474	0.4484	0.4495	0.4505	0.4515	0.4525	0.4535	0.4545
1.7	0.4554	0.4564	0.4573	0.4582	0.4591	0.4599	0.4608	0.4616	0.4625	0.4633
1.8	0.4641	0.4649	0.4656	0.4664	0.4671	0.4678	0.4686	0.4693	0.4699	0.4706
1.9	0.4713	0.4719	0.4726	0.4732	0.4738	0.4744	0.4750	0.4756	0.4761	0.4767
2.0	0.4772	0.4778	0.4783	0.4788	0.4793	0.4798	0.4803	0.4808	0.4812	0.4817
2.1	0.4821	0.4826	0.4830	0.4834	0.4838	0.4842	0.4846	0.4850	0.4854	0.4857
2.2	0.4861	0.4864	0.4868	0.4871	0.4875	0.4878	0.4881	0.4884	0.4887	0.4890
2.3	0.4893	0.4896	0.4898	0.4901	0.4904	0.4906	0.4909	0.4911	0.4913	0.4916
2.4	0.4918	0.4920	0.4922	0.4925	0.4927	0.4929	0.4931	0.4932	0.4934	0.4936
2.5	0.4938	0.4940	0.4941	0.4943	0.4945	0.4946	0.4948	0.4949	0.4951	0.4952
2.6	0.4953	0.4955	0.4956	0.4957	0.4959	0.4960	0.4961	0.4962	0.4963	0.4964
2.7	0.4965	0.4966	0.4967	0.4968	0.4969	0.4970	0.4971	0.4972	0.4973	0.4974
2.8	0.4974	0.4975	0.4976	0.4977	0.4977	0.4978	0.4979	0.4979	0.4980	0.4981
2.9	0.4981	0.4982	0.4982	0.4983	0.4984	0.4984	0.4985	0.4985	0.4986	0.4986
3.0	0.4987	0.4987	0.4987	0.4988	0.4988	0.4989	0.4989	0.4989	0.4990	0.4990

NOTE: For values of z above 3.09, use 0.4999.

APPENDIX D   Glossary

Abelian group [4.6] A group that is also commutative.

Abscissa [11.1] The horizontal coordinate in a two-dimensional system of rectangular coordinates, usually denoted by x.

Absolute value [4.3] The absolute value of a number is the distance of that number from the origin. Symbolically,

$$|n| = \begin{cases} n & \text{if } n \geq 0 \\ -n & \text{if } n < 0 \end{cases}$$

Accuracy [8.1] One speaks of an *accurate statement* in the sense that it is true and correct, or of an accurate computation in the sense that it contains no numerical error. Accurate to a certain decimal place means that all digits preceding and including the given one are correct.

Acre [8.2] A unit commonly used in the United States system for measuring land. It contains 43,560 ft².

Acute angle [7.2] An angle whose measure is smaller than a right angle.

Acute triangle [7.3] A triangle with three acute angles.

Addition law of exponents [1.4] To multiply two numbers with like bases, add the exponents; that is, $b^m \cdot b^n = b^{m+n}$.

Addition method [12.1] The method of solution of a system of equations in which the coefficients of one of the variables are opposites so that when the equations are added, one of the variables is eliminated.

Addition of integers [4.3] If the integers to be added have the same sign, the answer will also have that same sign and will have a magnitude equal to the sum of the absolute values of the given integers. If the integers to be added have opposite signs, the answer will have the sign of the integer with the larger absolute value, and will have a magnitude equal to the difference of the absolute values. Finally, if one or both of the given integers is 0, use the property that $n + 0 = n$ for any integer n.

Addition of matrices [12.4] $[M] + [N] = [S]$ if and only if $[M]$ and $[N]$ are the same order and the entries of $[S]$ are found by adding the corresponding entries of $[M]$ and $[N]$.

Addition of rational numbers [4.4]

$$\frac{a}{b} + \frac{c}{d} = \frac{ad}{bd} + \frac{bc}{bd} = \frac{ad + bc}{bd}$$

Addition property of equations [5.4] The solution of an equation is unchanged by adding the same number to both sides of the equation.

Addition property of inequality [5.5] The solution of an inequality is unchanged if you add the same number to both sides of the inequality.

Additive identity [4.6] The number 0, which has the property that $a + 0 = a$ for any number a.

Additive inverse [4.6; 12.4] See *Opposite*. The additive inverse of a matrix $[M]$ is denoted by $[-M]$ and is defined by $(-1)[M]$.

Add-on interest [6.2] It is a method of calculating interest and installments on a loan. The amount of interest is calculated according to the formula $I = Prt$ and is then added to the amount of the loan. This sum, divided by the number of payments, is the amount of monthly payment.

Address Designation of the location of data within internal memory or on a magnetic disk or tape.

Adjacent angles [7.2] Two angles are adjacent if they share a common side.

Adjacent side [7.4] In a right triangle, an acute angle is made up of two sides; one of those sides is the hypotenuse, and the other side is called the adjacent side.

Adjusted balance method [6.2] A method of calculating credit card interest using the formula $I = Prt$ in which P is the balance owed after the current payment is subtracted.

Algebra [4.7] A generalization of arithmetic. Letters called variables are used to denote numbers, which are related by laws that hold (or are assumed) for any of the numbers in the set.

Algebraic expression Any meaningful combination of numbers, variables, and signs of operation.

Alternate exterior angles [7.2] Two *alternate angles* are angles on opposite sides of a transversal cutting two parallel lines, each having one of the lines for one of its sides. They are *alternate exterior angles* if neither lies between the two lines cut by the transversal.

Alternate interior angles [7.2] Two *alternate angles* are angles on opposite sides of a transversal cutting two parallel lines, each having one of the lines for one of its sides. They are *alternate interior angles* if both lie between the two lines cut by the transversal.

Amortization [6.6] The process of paying off a debt by systematically making partial payments until the debt (principal) and interest are repaid.

Amortization schedule [6.6] A table showing the schedule of payments of a loan detailing the amount of each payment that goes to repay the principal and how much goes to pay interest.

Amortized loan [6.6] A loan that is fully paid off with the last periodic payment.

And [1.2; 2.1] See *Conjunction*. In everyday usage, it is used to join together elements that are connected in two sets simultaneously.

AND-gate [2.6] An electrical circuit that simulates conjunction; that is, the circuit is on when two switches are on.

Angle [7.2] Two rays or segments with a common endpoint.

Annual compounding [6.1] In the compound interest formula, it is when $n = 1$.

Annual percentage rate [6.2] The percentage rate charged on a loan based on the actual amount owed and the actual time it is owed. The approximation formula for annual percentage rate (APR) is $\text{APR} = 2Nr/(N + 1)$.

Annuity [6.5] A sequence of payments into or out of an interest-bearing account. If the payments are made into an interest-bearing account at the end of each time period, and if the frequency of payments is the same as the frequency of compounding, the annuity is called an *ordinary annuity*.

Antecedent [2.2] See *Conditional*.

Apportionment problem [4.5] The problem of apportioning a legislative body.

APR [6.2] Abbreviation for annual percentage rate. See *Compound interest formula*.

Arc [7.6] Part of the circumference of a circle.

Area [8.2] A number describing the two-dimensional content of a set. Specifically, it is the number of square units enclosed in a plane figure.

Area formulas [8.2] Square, s^2; rectangle, ℓw; parallelogram, bh; triangle, $\frac{1}{2}bh$; circle, πr^2; trapezoid, $\frac{1}{2}h(b_1 + b_2)$.

Argument [2.1] Statements and conclusion as a form of logical reasoning.

Arithmetic sequence [6.3] A sequence, each term of which is equal to the sum of the preceding term and a constant, written a_1, $a_2 = a_1 + d$, $a_3 = a_1 + 2d$, . . . ; the nth term of an arithmetic sequence is $a_1 + (n - 1)d$, where a_1 is the first term and d is the *common difference*. Also called an *arithmetic progression*. See also *Sequence*.

Arithmetic series [6.4] The indicated sum of the terms of an arithmetic sequence. The sum of n terms is denoted by A_n and

$$A_n = \frac{n}{2}(a_1 + a_n) \text{ or } A_n = \frac{n}{2}[2a_1 + (n - 1)d]$$

Arrangement [9.4] Same as *permutation*.

Array [12.3] An arrangement of items into rows and columns. See *Matrix*.

Artificial intelligence [3.4] A field of study devoted to computer simulation of human intelligence.

ASCII code [3.5] A standard computer code used to facilitate the interchange of information among various types of computer equipment.

Assignment [5.9] A computer term for setting the value of one variable to match the value of another.

Associative property [4.1] A property of grouping that applies to certain operations (addition and multiplication, for example, but not to subtraction or division): If a, b, and c are real numbers, then

$$(a + b) + c = a + (b + c) \text{ and } (ab)c = a(bc)$$

Assuming the antecedent [2.4] Same as *direct reasoning*.

Assuming the consequent [2.4] A logical fallacy; same as *fallacy of the converse*.

Augmented matrix [12.3] A matrix that results after affixing an additional column to an existing matrix.

Average [10.2] A single number that is used to typify or represent a set of numbers. In this book, it refers to the *mean*, *median*, or *mode*.

Average daily balance method [6.2] A method of calculating credit card interest using the formula $I = Prt$ in which P is the average daily balance owed for a current month, and t is the number of days in the month divided by 365.

Axes [11.1] The intersecting lines of a Cartesian coordinate system. The horizontal axis is called the x-axis, and the vertical axis is called the y-axis. The axes divide the plane into four parts called *quadrants*.

Axiom [1.3; 7.1] A statement that is accepted without proof.

Axis of symmetry A curve is symmetric with respect to a line, called the axis of symmetry, if for any point P on the curve, there is a point Q also on the curve such that the axis of symmetry is the perpendicular bisector of the line segment PQ.

Balloon payment A single larger payment made at the end of the time period of an installment loan that is not amortized.

Bar graph [10.1] See *Graph*.

Base [1.4] See *Exponent*.

Base of an exponential In $y = b^x$, the *base* is b ($b \neq 1$).

BASIC [5.9] Beginner's All-purpose Symbolic Instruction Code; it is a higher-level computer language common to most microcomputers.

Because [2.3] A logical operator for p because q which is defined to mean

$$(p \wedge q) \wedge (q \to p)$$

Bell-shaped curve [10.3] See *Normal curve*.

Belong to a set [1.2] To be an element of a set.

Biconditional [2.3] A logical operator for simple statements p and q which is true when p and q have the same truth values, and is false when p and q have different truth values.

Billion [1.4] A name for $10^9 = 1,000,000,000$.

Binary numeration system [3.5] A numeration system with two numerals, 0 and 1.

Binomial [5.1] A polynomial with exactly two terms.

Binomial theorem [5.1] For any positive integer n,

$$(a + b)^n = \sum_{k=0}^{n}\binom{n}{k}a^{n-k}b^k$$

Bisect To divide into two equal or congruent parts.

Bit [3.5] Binary digit, the smallest unit of data storage with a value of either 0 or 1, thereby representing whether a circuit is open or closed.

Bond An interest-bearing certificate issued by a government or business, promising to pay the holder a specified amount (usually $1,000) on a certain date.

Boot To start a computer. To do this the computer must have access to an operating system on a floppy disk or hard disk.

Boundary [11.2] See *Half-plane*.

Box plot [10.1] A rectangular box positioned above a numerical scale associated with a given set of data. It shows the maximum value, the minimum value, and the quartiles for the data.

Braces See *Grouping symbols*.

Brackets See *Grouping symbols*.

Branch [5.9] In BASIC programming, a branch is a means of transferring from one point in a program to another under certain conditions. In particular, it refers to the IF-THEN and IF-ELSE commands.

Bug An error in the design or makeup of a computer program (software bug) or a hardware component of the system (hardware bug).

Bulletin boards [3.4] An online means of communicating with others on a particular topic.

Byte The fundamental block of data that can be processed by a computer. In most microcomputers, a byte is a group of eight adjacent bits and the rough equivalent of one alphanumeric symbol.

Calculus [11.5] The field of mathematics that deals with differentiation and integration of functions, and related concepts and applications.

Canceling The process of reducing a fraction by dividing the same number into both the numerator and the denominator.

Canonical form [4.2] When a given number is written as a product of prime factors in ascending order, it is said to be in canonical form.

Capacity [8.3] A measurement for the amount of liquid a container holds.

Cardinal number A number that designates the manyness of a set; the number of units, but not the order in which they are arranged.

Cardinality [1.2] The number of elements in a set.

Cards [9.1] A deck of 52 matching objects that are identical on one side and on the other side are divided into four suits (hearts, diamonds, spades, and clubs). The objects, called cards, are labeled 2, 3, . . . 9, 10, J, Q, K, and A in each suit.

Cartesian coordinate system [11.1] Two intersecting lines, called *axes*, used to locate points in a plane called a *Cartesian plane*. If the intersecting lines are perpendicular, the system is called a *rectangular coordinate system*.

Cartesian plane [11.1] See *Cartesian coordinate system*.

CD-ROM A form of mass storage. It is a cheap read-only device, which means that you can only use the data stored on it when it was created, but it can store a massive amount of material, such as an entire encyclopedia.

Cell [5.3] A specific location on a spreadsheet. It is designated using a letter (column heading) followed by a numeral (row heading). A cell can contain a letter, word, sentence, number, or formula.

Celsius [8.4] A metric measurement for temperature for which the freezing point of water is 0° and the boiling point of water is 100°.

Center See *Circle*.

Centi- [8.1] A prefix that means 1/100.

Centigram [8.4] One hundredth of a gram.

Centimeter [8.1] One hundredth of a meter.

Circle [8.1] The set of points in a plane that are a given distance from a given point. The given point is called the *center*, and the given distance is called the *radius*. The diameter is twice the radius. The *unit circle* is the circle with center at (0, 0) and $r = 1$.

Circle graph [10.1] See *Graph*.

Circular definition A definition that relies on the use of the word being defined, or other words that rely on the word being defined.

Circumference [8.1] The distance around a circle. The formula for finding the circumference is $C = \pi D$ or $C = 2\pi r$.

Classes [10.1] One of the groupings when organizing data. The difference between the lower limit of one class and the lower limit of the next class is called the *interval* of the class. The number of values within a class is called the *frequency*.

Closed See *Closure property*.

Closed curve [7.7] A curve that has no endpoints.

Closed-ended loan [6.2] An installment loan.

Closed half-plane [11.2] See *Half-plane*.

Closed network [7.6] A network that connects each point.

Closed set [4.1] A set that satisfies the closure property for some operation.

Closing The process of settlement on a real estate loan.

Closing costs Costs paid at the closing of a real estate loan.

Closure property [4.1] A set S is *closed* for an operation ○ if $a \circ b$ is an element of S for all elements a and b in S. This property is called the *closure property*.

Code The lines of a program are referred to as *code*. One refers to part of the program as "a block of code," or to the program in programming language as "source code," or finally to the compiled program as "object code."

Coefficient Any factor of a term is said to be the coefficient of the remaining factors. Generally, the word *coefficient* is taken to be the numerical coefficient of the variable factors.

Column [5.3; 12.3] A vertical arrangement of numbers or entries of a matrix. It is denoted by letters A, B, C, . . . on a spreadsheet.

Combination [9.4] It is a selection of objects from a given set without regard to the order in which they are selected. Sometimes it refers to the number of ways this selection can be done and is denoted by $_nC_r$ or $\binom{n}{r}$ and is pronounced "*n* choose *r*." The formula for finding it is

$$\binom{n}{r} = \frac{n!}{r!(n-r)!}$$

Command [5.9] A single instruction to prompt a computer to perform a specific predefined operation.

Common denominator For two or more fractions, a common multiple of the denominators.

Common difference [6.3] The difference between successive terms of an arithmetic sequence.

Common factor [5.2] A factor that two or more terms of a polynomial have in common.

Common fraction Franctions written in the form of one integer divided by a whole number are common fractions. For example, 1/10 is common fraction representation and 0.1 is the decimal representation of the same number.

Common ratio [6.3] The ratio between successive terms of a geometric sequence.

Communication matrix [12.4] A square matrix in which the entries symbolize the occurrence of some facet or event with a 1 and the nonoccurrence with a 0.

Communications package [3.4] A program that allows one computer to communicate with another computer.

Commutative group [4.6] A group that also satisfies the property that

$$a \circ b = b \circ a$$

for some operation ○ and elements a and b in the set.

Commutative property [4.1] A property of order that applies to certain operations (addition and multiplication, for example, but not to subtraction and division). If a and b are real numbers, then

$$a + b = b + a \text{ and } ab = ba$$

Comparison property [5.5] For any two numbers x and y, exactly one of the following is true: (1) $x = y$; x is equal to y (the same as) (2) $x > y$; x is greater than y (bigger than) (3) $x < y$; x is less than y (smaller than). This is sometimes known as the *trichotomy property*.

Comparison rate for home loans A formula for comparing terms of a home loan. The formula is

$$\text{APR} = 0.125 \left(\text{POINTS} + \frac{\text{ORIGINATION FEE}}{\text{AMOUNT OF LOAN}} \right)$$

Compass [7.1] An instrument for scribing circles or for measuring distances between two points.

Compiler An operating system program that converts an entire program written in a higher-level language into machine language before the program is executed.

Complement [1.2] (1) Two numbers less than 1 are called complements if their sum is 1. (2) The complement of a set is everything not in the set relative to a given universe.

Complementary angles [7.2] Two angles are complementary if the sum of their measures is 90°.

Complementary probabilities [9.3]

$$P(E) = 1 - P(\overline{E})$$

Two probabilities are *complementary* if

$$P(E) + P(\overline{E}) = 1$$

Complex decimal A form that mixes decimal and fractional form, such as $0.12\frac{1}{2}$.

Complex fraction A rational expression a/b where a or b (or both) have fractional form.

Components See *Ordered pair.*

Composite number [4.2] Sometimes simply referred to as a *composite*; it is a positive integer that has more than two divisors.

Compound interest [6.1] A method of calculating interest by adding the interest to the principal at the end of the compounding period so that this sum is used in the interest calculation for the next period.

Compound interest formula [6.1] $A = P(1 + i)^N$, where A = future value; P = present value (or principal); r = annual interest rate (APR); t = number of years; n = number of times compounded per year;

$$i = \frac{r}{n}; \text{ and } N = nt.$$

Compound statement [2.1] A statement formed by combining simple statements with one or more operators.

Compounding [6.1] The process of adding interest to the principal so that in the next time period the interest is calculated on this sum.

Computer [3.4] A device which, under the direction of a program, can process data, alter its own program instructions, and perform computations and logical operations without human intervention.

Computer abuse [3.4] A misuse of a computer.

Conclusion [2.1] The statement that follows (or is to be proved to follow) as a consequence of the hypothesis of the theorem.

Conditional [2.2] The statement "if p, then q," symbolized by $p \rightarrow q$. The statement p is called the *antecedent* and q is called the *consequent.*

Conditional equation [5.4] See *Equation.*

Conditional inequality [5.5] See *Inequality.*

Conditional probability [9.3] A probability that is found on the condition that a certain event has occurred. The notation $P(E|F)$ is the probability of event E *on the condition* that event F has occurred.

Cone [8.4] A solid with a circle for its base and a curved surface tapering evenly to an apex so that any point on this surface is in a straight line between the circumference of the base and the apex.

Congruent [7.1] Of the same size and shape; if one is placed on top of the other, the two figures will coincide exactly in all their parts.

Congruent angles [7.2] Two angles that have the same measure.

Congruent modulo m [4.7] Two real numbers a and b are congruent modulo m, written $a = b \pmod{m}$, if a and b differ by a multiple of m.

Congruent triangles [7.3] Two triangles that have the same size and shape.

Conjecture [1.3] A guess or prediction based on incomplete or uncertain evidence.

Conjunction [2.1] The conjunction of two simple statements p and q is true whenever both p and q are true, and is false otherwise. The common translation of conjunction is "and."

Consecutive numbers Counting numbers that differ by 1.

Consequent [2.2] See *Conditional.*

Consistent system [12.1] If a system of equations has at least one solution, it is said to be consistent; otherwise it is said to be *inconsistent.*

Constant Symbol with exactly one possible value.

Constant function A function of the form $f(x) = c$.

Constraint [12.6] A limitation placed on an objective function. See *Linear programming.*

Construction [7.1] The process of drawing a figure that will satisfy certain given conditions.

Contained in a set [1.2] An element is contained in a set if it is a member of the set.

Continuous distribution [10.3] A probability distribution that includes all x-values (as opposed to a discrete distribution, which allows a finite number of x-values).

Contradiction [5.4] An open equation for which the solution set is empty.

Contrapositive [2.2] For the implication $p \rightarrow q$, the contrapositive is $\sim q \rightarrow \sim p$.

Converge [11.6] To draw near to. A series is said to converge when the sum of the first n terms approaches a limit as n increases without bound. We say that the sequence converges to a limit L if the values of the successive terms of the sequence get closer and closer to the number L as $n \rightarrow \infty$.

Converse [2.2] For the implication $p \rightarrow q$, the converse is $q \rightarrow p$.

Convex set [12.6] A set that contains the line segment joining any two of its points.

Coordinates [11.1] A numerical description for a point. Also see *Ordered pair.*

Coordinate plane See *Cartesian coordinate system.*

Correlation [10.4] The interdependence between two sets of numbers. It is a relationship between two quantities, such that when one changes the other does (simultaneous increasing or decreasing is called *positive correlation*; and one increasing, the other decreasing, *negative correlation*).

Corresponding angles [7.2; 7.4] Angles in different triangles that are similarly related to the rest of the triangle.

Corresponding parts [7.3] Points, angles, lines, etc., in different figures, similarly related to the rest of the figures.

Corresponding sides [7.4] Sides of different triangles that are similarly related to the rest of the triangle.

Cosine [7.4] In a right triangle ABC with right angle C,

$$\cos A = \frac{\text{adjacent side of } A}{\text{hypotenuse}}$$

Counterclockwise In the direction of rotation opposite to that in which the hands move around the dial of a clock.

Counterexample [2.4] An example that is used to disprove a proposition.

Counting numbers [4.1] See *Natural numbers*.

CPU <u>C</u>entral <u>P</u>rocessing <u>U</u>nit, the primary section of the computer that contains the memory, logic, and arithmetic procedures necessary to process data and perform computations. The CPU also controls the functions performed by the input, output, and memory devices.

Credit card [6.2] A card signifying that the person or business issued the card has been approved open-ended credit. It can be used at certain restaurants, airlines, and stores accepting that card.

Cryptography [4.8] The writing or deciphering of messages in code.

Cube (1) A solid with six equal square sides. (2) In an expression such as x^3, which is pronounced "*x* cubed," it means *xxx*.

Cube root See *Root of a number*.

Cubed See *Cube*.

Cubic unit [8.3] A three-dimensional unit. It is the result of cubing a unit of measurement.

Cup [8.3] A unit of measurement in the United States measurement system that is equivalent to 8 fluid ounces.

Cursor [5.9] Indicator (often flashing) on a computer or calculator display to designate where the next character input will be placed.

Cylinder [8.4] Suppose we are given two parallel planes and two simple closed curves C_1 and C_2 in these planes for which lines joining corresponding points of C_1 and C_2 are parallel to a given line L. A cylinder is a closed surface consisting of two bases that are plane regions bounded by such curves C_1 and C_2 and a lateral surface that is the union of all line segments joining corresponding points of C_1 and C_2.

Cybertext [4.8] A secret or coded message.

Daily compounding [6.1] In the compound interest formula, it is when $n = 365$ (exact interest) or when $n = 360$ (ordinary interest). In this book, use ordinary interest unless otherwise indicated.

Data processing [3.4] The recording and handling of information by means of mechanical or electronic equipment.

Database A collection of information. A database manager is a program that is in charge of the information stored in a database.

Database manager [3.4] A computer program that allows a user to interface with a database.

Dealer's cost [6.2] The actual amount that a dealer pays for the goods sold.

Debug The organized process of testing for, locating, and correcting errors within a program.

Decagon [7.2] A polygon having ten sides.

Deci- [8.1] A prefix that means 1/10.

Deciles [10.1] Nine values that divide a data set into ten equal parts.

Decimal [3.2] Any number written in decimal notation. The digits represent powers of ten with whole numbers and fractions being separated by a period, called a *decimal point*. Sometimes called a Hindu–Arabic numeral.

Decimal fraction A number in decimal notation that has fractional parts, such as 23.25. If a common fraction p/q is written as a decimal fraction, the result will either be a *terminating decimal* as with $\frac{1}{4} = 0.25$ or a *repeating decimal* as with $\frac{2}{3} = 0.6666 \ldots .$

Decimal point [3.2; 4.6] See *Decimal*.

Decoding key [4.8] A key that allows one to unscramble a coded message.

Deductive reasoning [1.3; 2.1] A formal structure based on a set of axioms and a set of undefined terms. New terms are defined in terms of the given undefined terms and new statements, or *theorems*, are derived from the axioms by proof.

Degree [5.1; 7.2] (1) The degree of a term in one variable is the exponent of the variable, or it is the sum of the exponents of the variables if there are more than one. The degree of a polynomial is the degree of its highest-degree term. (2) A unit of measurement of an angle that is equal to 1/360 of a revolution.

Deka- [8.1] A prefix that means 10.

Deleted point A single point that is excluded from the domain.

De Morgan's laws [1.2; 2.3] For sets X and Y,
$$\overline{X \cup Y} = \overline{X} \cap \overline{Y} \qquad \text{and} \qquad \overline{X \cap Y} = \overline{X} \cup \overline{Y}$$

Demand [12.2] The number of items that can be sold at a given price.

Denominator [4.4] See *Rational number*.

Dense set [4.6] A set of numbers with the property that between any two points of the set, there exists another point in the set that is between the two given points.

Denying the antecedent [2.4] A logical fallacy; same as the *fallacy of the inverse*.

Denying the consequent [2.4] Same as *indirect reasoning*.

Dependent events [9.5] Two events are dependent if the occurrence of one influences the occurrence of the other.

Dependent system [12.1] If every ordered pair satisfying one equation in a system of equations also satisfies every other equation of the given system, then we describe the system as dependent.

Dependent variable [11.1] The variable associated with the second component of an ordered pair.

Derivative [11.5] One of the fundamental operations of calculus; it is the instantaneous rate of change of a function with respect to the variable.

Description method A method of defining a set by describing the set (as opposed to listing its elements).

Diameter [8.1] See *Circle*.

Dice [9.1] Plural for the word *die*, which is a small, marked cube used in games of chance.

Die [9.1] See *Dice*.

Difference The result of a subtraction.

Difference of squares [5.2] A mathematical expression in the form $a^2 - b^2$.

Differential calculus [11.5] That branch of calculus concerned with applications of the derivative.

Dimension [12.3] A configuration having length only is said to be of one dimension; area and not volume, two dimensions; volume, three dimensions. In reference to matrices, it is the numbers of rows and columns.

Direct reasoning [2.4] One of the principal forms of logical reasoning. It is an argument of the form $[(p \rightarrow q) \wedge p] \rightarrow q$.

Discount A reduction from a usual or list price.

Disjoint sets [1.2] Sets that have no elements in common.

Disjunction [2.1] The disjunction of two simple statements p and q is false whenever both p and q are false, and is true otherwise. The common translation of conjunction is "or."

Disk drive A mechanical device that uses the rotating surface of a magnetic disk for the high-speed transfer and storage of data.

Distributive law for exponents [1.4]

$$(1)\ (ab)^m = a^m b^m; \quad (2)\ \left(\frac{a}{b}\right)^m = \frac{a^m}{b^m}$$

Distributive property (for multiplication over addition) [4.1] If a, b, and c are real numbers, then $a(b + c) = ab + ac$ and $(a + b)c = ac + bc$ for the basic operations. That is, the number outside the parentheses indicating a sum or difference is distributed to each of the numbers inside the parentheses.

Diverge [11.6] A sequence that does not converge is said to diverge.

Dividend The number or quantity to be divided. In a/b the dividend is a.

Divisibility [4.2] If m and d are counting numbers, and if there is a counting number k so that $m = d \cdot k$, we say that d is a divisor of m, d is a factor of m, d divides m, and m is a multiple of d.

Division [4.3] $a/b = x$ is $a \div b = x$ and means $a = bx$.

Division by zero [4.3] In the definition of division, $b \neq 0$, because if $b = 0$, then $bx = 0$, regardless of the value of x. If $a \neq 0$, then there is no such number. On the other hand, if $a = 0$, then $0/0 = 1$ checks from the definition, and so also does $0/0 = 2$, which means that $1 = 2$, another contradiction. Thus, division by 0 is excluded.

Division of integers [4.3] The quotient of two integers is the quotient of the absolute values, and is positive if the given integers have the same sign, and negative if the given numbers have opposite signs. Furthermore, division by zero is not possible and division into 0 gives the answer 0.

Division of rational numbers [4.4]

$$\frac{a}{b} \div \frac{c}{d} = \frac{ad}{bc} \quad (c \neq 0)$$

Division property of equations [5.4] The solution of an equation is unchanged by dividing both sides of the equation by the same nonzero number.

Divisor The quantity by which the dividend is to be divided. In a/b, b is the divisor.

Dodecagon [7.2] A polygon with 12 sides.

DOS <u>D</u>isk <u>O</u>perating <u>S</u>ystem, the set of programs that allows both the user and the computer to communicate with a disk drive. The manufacturer is Microsoft, so sometimes it is referred to as MS DOS.

Dot matrix printer A type of printer that forms letters using tiny dots, usually 9 or 24 (for better quality) dots per letter.

Double subscripts [12.3] Two subscripts on a variable, as in a_{12}.

Download [3.4] To copy a program or data from some outside source onto your own computer.

Domain The *domain* of a variable is the set of replacements for the variable. The *domain* of a graph of an equation with two variables x and y is the set of permissible real-number replacements for x.

Double negative $-(-a) = a$

Down payment An amount paid at the time a product is financed. The purchase price minus the down payment is equal to the amount financed.

Edge [7.6] A line or a line segment that is the intersection of two plane faces of a geometric figure, or that is in the boundary of a plane figure.

Either . . . or [2.1; 2.3] A logical operator for "either p or q," which is defined to mean $(p \vee q) \wedge \sim(q \wedge p)$.

Element [1.2] One of the individual objects that belong to a set.

Elementary operations Refers to the operations of addition, subtraction, multiplication, and division.

Elementary row operations [12.3] There are four elementary row operations for producing equivalent matrices: (1) *RowSwap*: Interchange any two rows. (2) *Row+*: Row addition — add a row to any other row. (3) **Row*: Scalar multiplication — multiply (or divide) all the elements of a row by the same nonzero real number. (4) **Row+*: Multiply all the entries of a row (*pivot row*) by a nonzero real number and add each resulting product to the corresponding entry of another specified row (*target row*).

Elliptic geometry [7.8] A non-Euclidean geometry is which a Saccheri quadrilateral is constructed with summit angles obtuse.

e-mail [3.4] Electronic mail sent from one computer to another.

Empirical probability [9.1] A probability obtained empirically by experimentation.

Empty set [1.2] See *Set*.

Encoding key [4.8] A key that allows one to scramble, or encode a message.

Encrypt [4.8] To scramble a message so that it cannot be read by an unwanted person.

Equal angles Two angles that have the same measure.

Equal sets Sets that are the same.

Equal to Two numbers are equal if they represent the same quantity or are identical. In mathematics, a relationship that satisfies the axioms of equality.

Equality, axioms of For $a, b, c \in \mathbb{R}$,

reflexive: $a = a$
symmetric: If $a = b$, then $b = a$.
transitive: If $a = b$ and $b = c$, then $a = c$.
substitution: If $a = b$, then a may be replaced throughout by b (or b by a) in any statement without changing the truth or falsity of the statement

Equally likely outcomes [9.1] Outcomes whose probabilities of occurring are the same.

Equation [5.4] A statement of equality. If always true, an equation is called an *identity*; if always false, it is called a *contradiction*. If it is sometimes true and sometimes false, it is called a *conditional equation*. Values that make an equation true are said to *satisfy* the equation and are called *solutions* or *roots* of the equation. Equations with the same solutions are called *equivalent equations*.

Equation of a graph Every point on the graph has coordinates that satisfy the equation, and every ordered pair that satisfies the equation has coordinates that lie on the graph.

Equation properties [5.4] There are four equation properties: (1)

Addition property: Adding the same number to both sides of an equation results in an equivalent equation. (2) *Subtraction property*: Subtracting the same number from both sides of an equation results in an equivalent equation. (3) *Multiplication property*: Multiplying both sides of a given equation by the same nonzero number results in an equivalent equation. (4) *Division property*: Dividing both sides of a given equation by the same nonzero number results in an equivalent equation.

Equilateral triangle [7.3; 8.1] A triangle whose three sides all have the same length.

Equilibrium point [12.2] A point for which the supply and demand are equal.

Equivalent equations [5.4] See *Equation*.

Equivalent matrices [12.3] Matrices that represent equivalent systems.

Equivalent sets [8.1] Sets that have the same cardinality.

Equivalent systems [12.1] Systems that have the same solution set.

Estimation [1.4] An approximation (usually mental) of size or value used to form an opinion.

Euclidean geometry [7.1] The study of geometry based on the assumptions of Euclid. These basic assumptions are called Euclid's postulates.

Euclid's postulates [7.1] 1. A straight line can be drawn from any point to any other point. 2. A straight line extends infinitely far in either direction. 3. A circle can be described with any point as center and with a radius equal to any finite straight line drawn from the center. 4. All right angles are equal to each other. 5. Given a straight line and any point not on this line, there is one and only one line through that point that is parallel to the given line.

Euler circles [1.2] The representation of sets using interlocking circles.

Evaluate [5.3; 6.4] To *evaluate* an expression means to replace the variables by given numerical values and then simplify the resulting numerical expression. To *evaluate* a circle function means to find its approximate numerical value. To *evaluate* a summation means to find its value.

Event [9.1] A subset of a sample space.

Exact interest [6.1] The calculation of interest assuming that there are 365 days in a year.

Exclusive or [2.1] A translation of *p or q* which includes *p* or *q*, but not both. In this book we translate the exclusive or as "either *p* or *q*."

Expand a summation [6.4] To write out a summation notation showing the individual terms, without a sigma.

Expanded notation [3.2] A way of writing a number which lists the meaning of each grouping symbol and the number of items in that group. For example, 382.5, written in expanded notation, is
$$3 \times 10^2 + 8 \times 10^1 + 2 \times 10^0 + 5 \times 10^{-1}$$

Expectation [9.2] See *Mathematical expectation*.

Expected value [9.2] See *Mathematical expectation*.

Experiment [9.1] An observation of any physical occurrence.

Exponent [1.4] Where *b* is any nonzero real number and *n* is any natural number, exponent is defined as follows:
$$b^n = \underbrace{b \cdot b \cdots b}_{n \text{ factors}} \quad b^0 = 1 \quad b^{-n} = \frac{1}{b^n}$$

b is called the *base*, *n* is called the *exponent*, and b^n is called a *power*.

Exponential curve [11.3] The graph of an exponential equation. It indicates an increasingly steep rise, and passes through the point (0, 1).

Exponential equation [11.3] An equation of the form $y = b^x$ where *b* is positive and not equal to 1.

Exponential notation [1.4] A notation involving exponents.

Expression Numbers, variables, functions, and their arguments that can be evaluated to obtain a single result.

Exterior angle [7.3] An exterior angle of a triangle is the angle on the other side of an extension on one side of the triangle.

Extraneous root A number obtained in the process of solving an equation that is not a root of the equation to be solved.

Extremes [5.7] See *Proportion*.

Factor [5.2] (noun) Each of the numbers multiplied to form a product is called a factor of the product. (verb) To write a given number as a product.

Factorial [9.4] For a natural number *n*, the product of all the positive integers less than or equal to *n*. It is denoted by *n*! and is defined by
$$n! = n(n - 1)(n - 2) \cdots \cdot 4 \cdot 3 \cdot 2 \cdot 1$$
Also, 0! = 1.

Factoring [4.2] The process of determining the factors of a product.

Factorization The result of factoring a number or an expression.

Fahrenheit [8.4] A unit of measurement in the United States system for measuring temperature based on system where the freezing point of water is 32° and the boiling point of water is 212°.

Fair coin [9.2] A coin for which heads and tails are equally likely.

Fair game [9.2] A game for which the mathematical expectation is zero.

Fallacy [2.4] An invalid form of reasoning.

Fallacy of the converse [2.4] An invalid form of reasoning that has the form $[(p \rightarrow q) \wedge q]$ and reaches the incorrect conclusion *p*.

Fallacy of the inverse [2.4] An invalid form of reasoning that has the form $[(p \rightarrow q) \wedge (\sim p)]$ and reaches the incorrect conclusion $\sim q$.

False chain pattern [2.4] An invalid form of reasoning that has the form $[(p \rightarrow q) \wedge (p \rightarrow r)]$ and reaches the incorrect conclusion $q \rightarrow r$.

Feasible solution [12.6] A set of values that satisfies the set of constraints in a linear programming problem.

Fibonacci sequence [6.3] The sequence 1, 1, 2, 3, 5, 8, 13, 21, ... The general term is $s_n = s_{n-1} + s_{n-2}$, for any given s_1 and s_2.

Field [4.6] A set with two operations satisfying the closure, commutative, associative, identity, and inverse properties for both operations. A field also satisfies a distributive property combining both operations.

Finance charge [6.2] A charge made for the use of someone else's money.

Finite series [6.4] A series with *n* terms, where *n* is a counting number.

Finite set See *Set.*

First component See *Ordered pair.*

First-degree equation [11.1] With one variable, an equation of the form $ax + b = 0$; with two variables, an equation of the form $y = mx + b$.

Five-percent offer [6.2] An offer made that is 105% of the price paid by the dealer. That is, it is an offer that is 5% over the cost.

Fixed-point form [1.4] The usual decimal representation of a number. It is usually used in the context of writing numbers in scientific notation or in floating-point form. See *Floating-point form.*

Floating-point form [1.4] It is a calculator or computer variation of scientific notation in which a number is written as a number between one and ten times a power of ten where the power of ten is understood. For example, 2.678×10^{11} is scientific notation and 2.67800000000 E11 and 2.67800000000 +11 are floating-point representations. The fixed-point representation is the usual decimal representation of 267,800,000,000.

Floppy disk Storage medium, which is a flexible platter ($3\frac{1}{2}$ or $5\frac{1}{4}$ inches in diameter) of mylar plastic coated with a magnetic material. Data are represented on the disk by electrical impulses.

FOIL [5.1] A method for multiplying binomials that requires First terms, Outer terms + Inner terms, Last terms:

$$(a + b)(c + d) = ac + (ad + bc) + bd$$

Foot [8.1] A unit of linear measure in the United States system that is equal to 12 inches.

Foreclose If the scheduled payments are not made, the lender takes the right to redeem the mortgage and keeps the collateral property.

Formula A general answer, rule, or principle stated in mathematical notation.

FOR-NEXT [5.9] A BASIC computer language command that sets up a loop to repeat a given set of steps a specific number of times.

Fractal [7.7] A family of shapes involving chance whose irregularities are statistical in nature. They are shapes used, for example, to model coastlines, growth, and boundaries of clouds. Fractals model curves as well as surfaces. The term *fractal set* is also used in place of the word *fractal.*

Fraction [4.4] See *Rational number.*

Frequency [10.1] See *Classes.*

Frequency distribution [10.1] For a collection of data, the tabulation of the number of elements in each class.

Function [11.4] A rule that assigns to each element in the domain a single (unique) element.

Function machine [11.4] A device used to help us understand the nature of functions. It is the representation of a function as a machine in which some number is input and then after being "processed" through the machine, outputs a single value.

Functional notation [11.4] The representation of a function f using the notation $f(x)$.

Fundamental counting principle [9.3] If one task can be performed in m ways and a second task can be performed in n ways, then the number of ways that the tasks can be performed one after the other is mn.

Fundamental operators [2.2] In symbolic logic, the fundamental operators are the connectives *and, or,* and *not.*

Fundamental property of equations If P and Q are algebraic expressions, and k is a real number, then each of the following is equivalent to $P = Q$:

Addition	$P + k = Q + k$
Subtraction	$P - k = Q - k$
Nonzero multiplication	$kP = kQ, k \neq 0$
Nonzero division	$\dfrac{P}{k} = \dfrac{Q}{k}, k \neq 0$

Fundamental property of fractions [4.4] If both the numerator and denominator are multiplied by the same nonzero number, the resulting fraction will be the same. That is, $\dfrac{PK}{QK} = \dfrac{P}{Q}$, $(Q, K \neq 0)$.

Fundamental property of inequalities If P and Q are algebraic expressions, and k is a real number, then each of the following is equivalent to $P < Q$:

Addition	$P + k < Q + k$
Subtraction	$P - k < Q - k$
Positive multiplication	$kP < kQ, k > 0$
Positive division	$\dfrac{P}{K} < \dfrac{Q}{k}, k > 0$
Negative multiplication	$kP > kQ, k < 0$
Negative division	$\dfrac{P}{k} > \dfrac{Q}{k}, k < 0$

This property also applies for $\leq$, $>$, and $\geq$.

Fundamental theorem of arithmetic [4.2] Every counting number greater than 1 is either a prime or a product of primes, and the prime factorization is unique (except for the order in which the factors appear).

Future value [6.1] See *Compound interest formula.*

Future value formula [6.1] For simple interest: $A = P(1 + rt)$; for compound interest: $A = P(1 + i)^N$.

Fuzzy logic [2.1] A relatively new branch of logic used in computer programming that does not use the law of the excluded middle.

Gallon [8.3] A measure of capacity in the United States system that is equal to 4 quarts.

Gates [2.6] In circuit logic, it is a symbolic representation of a particular circuit.

Gauss–Jordan elimination [12.3] A method for solving a system of equations that uses the following steps. *Step 1:* Select as the first pivot the element in the first row, first column, and pivot. *Step 2:* The next pivot is the element in the second row, second column; pivot. *Step 3:* Repeat the process until you arrive at the last row, or until the pivot element is a zero. If it is a zero and you can interchange that row with a row below it, so that the pivot element is no longer a zero, do so and continue. If it is zero and you cannot interchange rows so that it is not a zero, continue with the next row. The final matrix is called the **row-reduced form.**

GCF [4.2] An abbreviation for *greatest common factor.*

General term [11.6] The nth term of a sequence or series.

Genus [7.7] The number of cuts that can be made without cutting a figure into two pieces. The genus is equivalent to the number of holes in the object.

Geometric sequence [6.3] A sequence for which the ratio of each term to the preceding term is a constant, written $g_1, g_2, g_3, \ldots$. The nth term of a geometric sequence is $g_n = g_1 r^{n-1}$, where g_1 is the first term and r is the *common ratio*. It is also called a *geometric progression*.

Geometric series [6.4] The indicated sum of the terms of a geometric sequence. The sum of n terms is denoted by G_n and

$$G_n = \frac{g_1(1 - r^n)}{1 - r}, r \neq 1$$

If $|r| < 1$, then $G = \frac{g_1}{1 - r}$, where G is the sum of the infinite geometric series. If $|r| \geq 1$, the infinite geometric series has no sum.

Geometry The branch of mathematics that treats the shape and size of things. Technically, it is the study of invariant properties of given elements under specified groups of transformations.

GIGO Garbage In, Garbage Out, an old axiom regarding the use of computers.

Golden ratio [7.5] The division of a line segment AB by an interior point P so that

$$\frac{AB}{AP} = \frac{AP}{PB}$$

It follows that this ratio is a root of the equation $x^2 - x - 1 = 0$, or $x = \frac{1}{2}(1 + \sqrt{5})$. This ratio is called the golden ratio and is said to be considered pleasing to the eye.

Golden rectangle [7.5] A rectangle R with the property that it can be divided into a square and a rectangle similar to R; a rectangle whose sides form a golden ratio.

Googol [1.4] The number with 1 followed by 100 zeros — that is, 10,000,000,000,000,000,000,000,000,000,000,000,000,000,-000,000,000,000,000,000,000,000,000,000,000,000,000,000,-000,000

Grace period [6.2] A period of time between when an item is purchased and when it is paid during which no interest is charged.

Gram [8.4] A unit of weight in the metric system. It is equal to the weight of one cubic centimeter of water at 4°C.

Graph [10.1; 11.1] (1) In statistics, it is a drawing that shows the relation between certain sets of numbers. Common forms are bar graphs, line graphs, pictographs, and pie charts (circle graphs). (2) A drawing that shows the relation between certain sets of numbers. It may be one-dimensional ($\mathbb{R}$), two-dimensional ($\mathbb{R}^2$), or three-dimensional ($\mathbb{R}^3$).

Graph of an equation See *Equation of a graph*.

Graphing method [12.1] A method of solving a system of equations that finds the solution by looking at the intersection of the individual graphs. It is an approximate method of solving a system of equations, and depends on the accuracy of the graph that is drawn.

Great circle [7.8] A circle on a sphere that has its diameter equal to that of the sphere.

Greater than If a lies to the right of b on a number line, then a is greater than b, $a > b$. Formally, $a > b$ if and only if $a - b$ is positive.

Greater than or equal to Written $a \geq b$, means $a > b$ or $a = b$.

Greatest common factor [4.2] The largest divisor common to a given set of numbers.

Group [4.6] A set with one defined operation that satisfies the closure, associative, identity, and inverse properties.

Grouped frequency distribution [10.1] If the data are grouped before they are tallied, then the resulting distribution is called a *grouped frequency distribution*.

Grouping symbols Parentheses ( ), brackets [ ], and braces { } indicate the order of operations and are also sometimes used to indicate multiplication, as in $(2)(3) = 6$. Also called *symbols of inclusion*.

Growth formula [11.3] $P = P_0 e^{rt}$

Half-line [7.2] A ray, with or without its endpoint. The half-line is said to be closed if it includes the endpoint, and open if it does not include the endpoint.

Half-plane [11.2] The part of a plane that lies on one side of a line in the plane. It is a *closed* half-plane if the line is included. It is an *open* half-plane if the line is not included. The line is the *boundary* of the half-plane in either case.

Hard copy The output produced on paper by a printer.

Hard disk A secondary storage device often installed inside the main component. A hard disk is used to store the programs one usually uses, as well as any necessary data.

Hard drive [3.5] That part of a computer on which data can be stored by the operator.

Hardware [3.4] The physical components (mechanical, magnetic, electronic) of a computer system.

Hecto- [8.1] A prefix meaning 100.

Heptagon [7.2] A polygon having seven sides.

Hexagon [7.2] A polygon havng six sides.

Higher-level language A computer programming language (e.g., BASIC, PASCAL, LOGO) that approaches the syntax of English and is easier both to use and to learn than machine language. It is also not system-dependent.

Hindu-Arabic numerals [3.2] Same as the usual decimal numeration system that is in everyday use.

Horizontal line [7.2; 11.1] A line with zero slope. Its equation has the form $y = $ constant.

Hyperbolic geometry [7.8] A non-Euclidean geometry in which a Saccheri quadrilateral is constructed with summit angles acute.

Hypotenuse [4.5; 7.4] The longest side in a right triangle.

Hypothesis [2.1] An assumed proposition used as a premise in proving something else.

Identity [4.6] A statement of equality that is true for all values of the variable. It also refers to a number I so that for some operation $\circ$, $I \circ a = a \circ I = a$ for every number a in a given set.

Identity matrix [12.4] A matrix satisfying the identity property. It is a square matrix consisting of ones along the main diagonal and zeros elsewhere.

If–then [2.2; 5.9] In symbolic logic, it is a connective also called *implication*. In BASIC computer language, IF-THEN is a branching command that will conditionally transfer from one place in a program to another place.

Implication [2.3] A statement that follows from other statements. It is also a proposition formed from two given propositions by connecting them with an "if . . . , then . . ." form. It is symbolized by $p \rightarrow q$.

Impossible event [9.1] An event for which the probability is zero — that is, an event that cannot happen.

Improper fraction [4.4] A fraction for which the numerator is greater than the denominator.

Improper subset [1.2] See *Subset*.

Inch [8.1] A linear measurement in the United States system equal in length to the following segment: ——————————

Inclusive or [2.1] This is the same as *disjunction*. The compound statement "p or q" is called the inclusive or.

Inconsistent system [12.1] A system for which no replacements of the variable make the equations true simultaneously.

Independent events [9.5] Events E and F are *independent* if the occurrence of one in no way affects the occurrence of the other.

Independent system A system of equations such that no one of them is necessarily satisfied by a set of values of the variables that satisfy all the others.

Independent variable [11.1] The variable associated with the first component of an ordered pair.

Indirect reasoning [2.4] One of the principal forms of logical reasoning. It is an argument of the form $[(p \rightarrow q) \wedge \sim q] \rightarrow p$.

Inductive reasoning [1.3] A type of reasoning accomplished by first observing patterns and then predicting answers for more complicated similar problems.

Inequality [5.5] A statement of order. If always true, an inequality is called an *absolute inequality*; if always false, an inequality is called a *contradiction*. If sometimes true and sometimes false, it is called a *conditional inequality*. Values that make the statement true are said to *satisfy* the inequality. A *string of inequalities* may be used to show the order of three or more quantities.

Inequality symbols [5.5] The symbols $>$, $\geq$, $<$, and $\leq$. Also called *order symbols*.

Infinite series [6.4] The indicated sum of an infinite sequence.

Infinite set See *Set*.

Infinity symbol ∞

Inflation [6.1] An increase in the amount of money in circulation, resulting in a fall in its value and a rise in prices. In this book, we assume annual compounding with the future value formula; that is, use $A = P(1 + r)^n$, where r is projected annual inflation rate, n is the number of years, and P is the present value.

Information retrieval [3.4] The locating and displaying of specific material from a description of its content.

Input [5.9] A method of putting information into a computer. It includes downloading a program, typing on a keyboard, pressing on a pressure-sensitive screen. In BASIC programming, it refers to a command (INPUT) that stops a program to wait for input material.

Input device [3.5] Component of a system that allows the entry of data or a program into a computer's memory.

Installment loan [6.2] A financial problem in which an item is paid for over a period of time. It is calculated using add-on interest or compound interest.

Installments [6.2] Part of a debt paid at regular intervals over a period of time.

Instant Insanity [9.6] A puzzle game consisting of four blocks with different colors on the faces. The object of the game is to arrange the four blocks in a row so that no color is repeated on one side as the four blocks are rotated through 360°.

Integers [4.3] $\mathbb{Z} = \{. . . , -3, -2, -1, 0, 1, 2, 3, . . .\}$, composed of the natural numbers, their opposites, and 0.

Integral [11.5] A fundamental concept of calculus that involves finding the area bounded by a curve, the x-axis, and two vertical lines.

Integral calculus [11.5] That branch of calculus that involves applications of the integral.

Integrated circuit The plastic or ceramic body that contains a chip and the leads connecting it to other components.

Interactive Software that allows continuous two-way communication between the user and the program.

Intercept form [11.1] The form

$$\frac{x}{a} + \frac{y}{b} = 1$$

of a linear equation where the x-intercept is a and the y-intercept is b.

Intercepts The point or points where a line or a curve crosses a coordinate axis. The x-intercepts are sometimes called the *zeros* of the equation.

Interest [6.1] An amount of money paid for the use of another's money. See *Compound interest*.

Interest-only loan A loan in which periodic payments are for interest only so that the principal amount of the loan remains the same.

Interest rate [6.1] The percentage rate paid on financial problems. In this book it is denoted by r and is assumed to be an annual rate unless otherwise stated.

Interface The electronics necessary for a computer to communicate with a peripheral.

Internet [3.4] A network of computers from all over the world that are connected together. It can be accessed through institutions or servers such as America Online, Compuserve, or Prodigy.

Intersection [1.2] The *intersection* of sets A and B, denoted by $A \cap B$, is the set consisting of elements in *both* A and B.

Interval [10.1] See *Classes*.

Inverse [2.2; 4.6; 12.4] (1) In symbolic logic, for an implication $p \rightarrow q$, the inverse is the statement $\sim p \rightarrow q$. (2) For addition, see *Opposites*. For multiplication, see *Reciprocal*. (3) For matrices, if $[A]$ is a square matrix, and if there exists a matrix $[A]^{-1}$ such that

$$[A]^{-1}[A] = [A][A]^{-1} = [I]$$

where $[I]$ is the identity matrix for multiplication, then $[A]^{-1}$ is called the inverse of $[A]$ for multiplication.

Invert In relation to the fraction a/b, it means to interchange the numerator and the denominator to obtain the fraction b/a.

Irrational number [4.5] A number that can be expressed as a non-repeating, nonterminating decimal; the set of irrational numbers is denoted by $\mathbb{Q}'$.

Isosceles triangle [7.3] A triangle with two sides the same length.

Jordan curve [7.7] Also called a *simple closed curve*. For example, a curve such as a circle or an ellipse or a rectangle that is closed and does not intersect itself.

Juxtaposition When two variables, a number and a variable, or a symbol and a parenthesis are written next to each other with no operation symbol, as in xy, $2x$, or $3(x + y)$. Juxtaposition is used to indicate multiplication.

K The symbol represents 1,024 (or 2^{10}). For example, 48K bytes of memory is the same as $48 \times 1,024$ or 49,152 bytes. It is sometimes used as an approximation for 1,000.

Keyboard [3.5] Typewriter-like device that allows the user to input data and commands into a computer.

Kilo- [8.1] A prefix that means 1,000.

Kilogram [8.1] 1,000 grams.

Kiloliter [8.3] 1,000 liters.

Kilometer [8.1] 1,000 meters.

Laptop [3.4] A small portable computer.

Laser printer A type of high-quality printer that produces almost typeset-quality output.

Law of contraposition [2.2] A conditional may always be replaced by its contrapositive without having its truth value affected.

Law of detachment [2.4] Same as *direct reasoning*.

Law of double negation [2.2] $\sim(\sim p) \Leftrightarrow p$

Law of square roots [4.5] There are 4 laws of square roots.

(1) $\sqrt{0} = 0$ (2) $\sqrt{a^2} = a$

(3) $\sqrt{ab} = \sqrt{a}\sqrt{b}$ (4) $\sqrt{\dfrac{a}{b}} = \dfrac{\sqrt{a}}{\sqrt{b}}$

Law of the excluded middle [2.1] Every simple statement is either true or false.

Laws of exponents [1.4] There are 5 laws of exponents.

Addition law	$b^m \cdot b^n = b^{m+n}$
Multiplication law	$(b^n)^m = b^{mn}$
Subtraction law	$\dfrac{b^m}{b^n} = b^{m-n}$
Distributive laws	$(ab)^m = a^m b^m$
	$\left(\dfrac{a}{b}\right)^m = \dfrac{a^m}{b^m}$

LCD An abbreviation for least common denominator.

LCM [4.2] An abbreviation for least common multiple.

Least common denominator (LCD) [4.4] The smallest number that is exactly divisible by each of the given numbers.

Least common multiple (LCM) [4.2] The smallest number that each of a given set of numbers divides into.

Least-squares line [10.4] A line $y = mx + b$ so that the sum of the squares of the vertical distances of the data points from this line will be as small as possible.

Least-squares method [10.4] A method based on the principle that the best prediction of a quantity that can be deduced from a set of measurements or observations is that for which the sum of the squares of the deviations of the observed values from the predictions is a minimum.

Leg [4.5] One of the two sides of a right triangle that are not the hypotenuse.

Length [8.1] A measurement of an object from end to end.

Less than If a is to the left of b on a number line, then a is less than b, $a < b$. Formally, $a < b$ if and only if $b > a$.

Less than or equal to Written $a \leq b$, means $a < b$ or $a = b$.

LET [5.9] A BASIC command that assigns a value to a variable.

Like terms [5.1] Terms that differ only in their numerical coefficients. Also called *similar terms*.

Limit [11.5] The formal definition of a limit is beyond the scope of this course. Intuitively, it is the tendency of a function to approach some value as its variable approaches a given value.

Limit of a sequence [11.6] The formal definition of a limit of a sequence is beyond the scope of this course. Intuitively, it is an accumulation point such that there are an infinite number of terms of the sequence arbitrarily close to the accumulation point.

Line [7.1; 11.1] In mathematics, it is an undefined term. It is a curve that is straight, so it is sometimes referred to as a *straight line*. It extends in both directions, is considered one-dimensional, so it has no thickness.

Line graph [10.1] See *Graph*.

Line of credit [6.2; 7.3] A preapproved credit limit on a credit account. The maximum amount of credit to be extended to a borrower. That is, it is a promise by a lender to extend credit up to some predetermined amount.

Line of symmetry [7.1] A line with the property that for a given curve, any point P on the curve has a corresponding point Q (called the reflection point of P) so that the perpendicular bisector of PQ is on the line of symmetry.

Line segment [7.1] A part of a line between two points on the line.

Linear [5.1] Pertaining to a line. In two variables, a set of points satisfying the equation $Ax + By + C = 0$.

Linear combination method [12.1] A method of solving a system of equations by obtaining (by multiplication) equations with opposite numerical coefficients for a variable and then adding the equations to eliminate that variable.

Linear correlation coefficient [10.4] A measure to determine whether there is a statistically significant linear relationship between two variables.

Linear equation [5.4; 11.1] A first-degree equation with one or two variables. For example, $x + 5 = 0$ and $x + y + 5 = 0$ are linear. An equation is linear in a certain variable if it is first-degree in that variable. For example, $x + y^2 = 0$ is linear in x, but not y.

Linear function A function whose equation can be written in the form $f(x) = mx + b$.

Linear inequality A first-degree inequality with one or two variables.

Linear polynomial A first-degree polynomial.

Linear programming [12.6] A type of problem that seeks to maximize or minimize a function called the *objective function* subject to a set of *restrictions* (linear inequalities) called *constraints*.

Linear programming theorem [12.6] A linear expression in two variables, $c_1x + c_2y$, defined over a convex set S whose sides are line segments takes on its maximum value at a corner point of S and its minimum value at a corner point of S. If S is unbounded, there may or may not be an optimum value, but if there is, then it must occur at a corner point.

Linear system [12.1] A system of equations, each of which is first-degree.

Liter [8.3] The basic unit of capacity in the metric system. It is the capacity of 1 cubic decimeter.

Literal equation An equation with more than one variable.

LOAD [5.9] A BASIC program command that copies a program into a usable portion of a computer.

Logic [2.1] The science of correct reasoning.

Logical conclusion [2.1] The statement that follows logically as a consequence of the hypotheses of a theorem.

Logical equivalence [2.3] Two statements are *logically equivalent* if they have the same truth values.

LOGO [7.9] A computer language used in drawing geometrical designs. It is also useful in learning a programming language.

Loop [5.9] A principle of computer programming that directs a computer to perform the same operation over and over a specified number of times.

Lowest common denominator [4.4] For two or more fractions, the smallest common multiple of the denominators.

Lump-sum problem [6.5] A financial problem that deals with a single sum of money, called a *lump sum*. Contrast with a *periodic payment* problem.

Main diagonal The entries a_{11}, a_{22}, a_{33}, . . . in a matrix.

Mass [8.4] In this course, it is the amount of matter an object comprises. Formally, it is a measure of the tendency of a body to oppose changes in its velocity.

Mathematical expectation [9.2] A calculation defined as the product of an amount to be won and the probability that it is won. If there is more than one amount to be won, it is the sum of the expectations of all the prizes. It is also called the *expected value* or *expectation*.

Mathematical modeling [11.5] An iterative procedure that makes assumptions about real-world problems to formulate the problem in mathematical terms. After the mathematical problem is solved, it is tested for accuracy in the real world, and revised for the next step in the iterative process.

Mathematical system A set with at least one defined operation and some developed properties.

Matrix [12.3] A rectangular array of terms called *elements*.

Matrix equation [12.3; 12.4] An equation whose elements are matrices.

Maximum loan [6.6] In this book, it refers to the maximum amount of loan that can be obtained for a home with a given amount of income and a given amount of debt.

Mean [10.2] The number found by adding the data and dividing by the number of values in the data set. The sample mean is usually denoted by $\bar{x}$.

Means [5.7] See *Proportion*.

Measure [8.1] Comparison to some unit recognized as standard.

Measures of central tendency [10.2] Refers to the averages of mean, median, and mode.

Measures of dispersion [10.2] Refers to the measures of range, standard deviation, and variance.

Measures of position [10.2] Measures that divide a data set by position, including median, quartiles, deciles, and percentiles.

Median [10.2] The middle number when the numbers in the data are arranged in order of size. If there are two middle numbers (in the case of an even number of data values), the median is the mean of these two middle numbers.

Member [1.2] See *Set*.

Meter [8.1] The basic unit for measuring length in the metric system.

Metric system [8.1] A decimal system of weights and measures in which the gram, the meter, and the liter are the basic units of mass, length, and capacity, respectively. One gram is the mass of one cm^3 of water and one liter is the same as $1,000 \ cm^3$. In this book, the metric system refers to SI metric system as revised in 1960.

Mile [8.1] A unit of linear measurement in the United States system that is equal to 5,280 ft.

Milli- [8.1] A prefix that means 1/1,000.

Milligram [8.4] 1/1,000 of a gram.

Milliliter [8.3] 1/1,000 of a liter.

Millimeter [8.1] 1/1,000 of a meter.

Million [1.4] A name for $10^6 = 1,000,000$.

Minicomputer [3.4] An everyday name for a personal computer.

Minus Refers to the operation of subtraction. The symbol "$-$" means minus only when it appears between two numbers, two variables, or between numbers and variables.

Mixed number A number that has both a counting number part and a proper fraction part: for example, $3\frac{1}{2}$.

Mode [10.4] The value in a data set that occurs most frequently. If no number occurs more than once, there is no mode. It is possible to have more than one mode.

Modem [3.4] A device connected to a computer that allows the computer to communicate with other computers using electric cables or phone lines.

Modular code [4.8] A code based on modular arithmetic.

Modulo 5 [4.7] A mathematical system consisting of five elements having the property that every number is equivalent to one of these five elements if they have the same remainder when divided by 5.

Modulo n A mathematical system consisting of n elements having the property that every number is equivalent to one of these n elements if they have the same remainder when divided by n.

Modus ponens [2.4] Same as *direct reasoning*.

Modus tollens [2.4] Same as *indirect reasoning*.

Monitor [3.5] An output device for communicating with a computer. It is similar to a television screen.

Monomial [5.1] A polynomial with one and only one term.

Monthly compounding [6.1] In the compound interest formula, it is when $n = 12$.

Monthly payment [6.2] In an installment loan, it is a periodic payment that is made once every month.

Mortgage An agreement, or loan contract, in which a borrower pledges a home or other real estate as security.

Mouse [3.5] A small plastic "box" usually with two buttons on top and a ball on the bottom so that it can be rolled around on a pad. It is attached to the computer by a long cord and is used to take over some of the keyboard functions.

Multiplication [4.1] For $b \neq 0$, $a \times b$ means

$$\underbrace{b + b + b + \cdots + b}_{a \text{ addends}}$$

If $a = 0$, then $0 \times b = 0$.

Multiplication law of equality [5.4] If $a = b$, then $ac = bc$. Also called the *multiplication property of equality* or a *fundamental property of equations*.

Multiplication law of exponents [1.4] To raise a power to a power, multiply the exponents. That is, $(b^n)^m = b^{mn}$.

Multiplication law of inequality [5.5]
1. If $a > b$ and $c > 0$, then $ac > bc$.
2. If $a > b$ and $c < 0$. then $ac < bc$.

Multiplication of integers [4.3] If the integers to be multiplied both have the same sign, the result is positive and the magnitude of the answer is the product of the absolute values of the integers. If the integers to be multiplied have opposite signs, the product is negative and has magnitude equal to the product of the absolute values of the given integers. Finally, if one or both of the given integers is 0, the product is 0.

Multiplication of matrices [12.4] Let [M] be an $m \times r$ matrix and [N] an $r \times n$ matrix. The product matrix $[M][N] = [P]$ is an $m \times n$ matrix. The entry in the ith row and the jth column of $[M][N]$ is the sum of the products formed by multiplying each entry of the ith row of [M] by the corresponding element in the jth column of [N].

Multiplication of rational numbers [4.4]

$$\frac{a}{b} \times \frac{c}{d} = \frac{ac}{bd}$$

Multiplication principle [3.1] In a numeration system, multiplication of the value of a symbol by some number. Also, see *Fundamental counting principle*.

Multiplication property of equations [5.4] Both sides of an equation may be multiplied or divided by any nonzero number to obtain an equivalent equation.

Multiplication property of inequality [5.5] Both sides of an inequality may be multiplied or divided by a positive number, and the order of the inequality will remain unchanged. The order is reversed if both sides are multiplied or divided by a negative number. That is, if $a < b$ then $ac < bc$ if $c > 0$ and $ac > bc$ if $c < 0$. This also applies to $\leq$, $>$, and $\geq$.

Multiplicative identity [4.6] The number 1, with the property that $1 \cdot a = a$ for any real number a.

Multiplicative inverse [4.6] See *Reciprocal*.

Multiplicity [5.4] If a root for an equation appears more than once, it is called a *root of multiplicity*.

For example,

$$(x - 1)(x - 1)(x - 1)(x - 2)(x - 2)(x - 3) = 0$$

has roots 1, 2, and 3. The root 1 has multiplicity three and root 2 has multiplicity two.

Mutually exclusive [9.1] Events are *mutually exclusive* if their intersection is empty.

Natural numbers [4.1] $\mathbb{N} = \{1, 2, 3, 4, 5, \ldots\}$, the positive integers, also called the *counting numbers*.

Negation [2.1] A logical connective that changes the truth value of a given statement. The negation of p is symbolized by $\sim p$.

Negative number [4.3] A number less than zero.

Negative sign The symbol "$-$" when used in front of a number, as in -5. Do not confuse with the same symbol used for subtraction.

Neither . . . nor [2.3] A logical operator for "neither p nor q," which is defined to mean $\sim(p \vee q)$.

Network [3.4; 7.6] (1) A linking together of computers. (2) A set of points connected by arcs or by line segments.

n-gon [7.2] A polygon with n sides.

No p is q [2.3] A logical operator for "no p is q," which is defined to mean $p \rightarrow \sim q$.

Nonagon [7.2] A polygon with 9 sides.

Nonconformable matrices [12.4] Matrices that cannot be added or multiplied because their dimensions are not compatible.

Non-Euclidean geometry [7.8] A geometry that results when Euclid's fifth postulate is not accepted.

Nonrepeating decimal [4.6] A decimal representation of a number that does not repeat.

Nonterminating decimal [4.6] A decimal representation of a number that does not terminate.

Normal curve [10.3] A graphical representation of a normal distribution. Its high point occurs at the mean, it is symmetric with respect to this mean, and on each side of the mean has an area that includes 34.1% of the population within one standard deviation, 13.6% from one to two standard deviations, and about 2.3% of the population more than two standard deviations from the mean.

Not [2.1] A common translation for the connective of negation.

NOT-gate [2.6] A logical gate that changes the truth value of a given statement.

Null set See *Set*.

Number [3.1] A *number* represents a given quantity, as opposed to a *numeral*, which is the symbol for the number. In mathematics, it generally refers to a specific set of numbers — for example, counting numbers, whole numbers, integers, rationals, or real numbers. If the set is not specified, the assumed usage is to the set of real numbers.

Number line [4.6] A line used to display a set of numbers graphically (the axis for a one-dimensional graph).

Numeral [3.1] Symbol used to denote a number.

Numeration system [3.1] A system of symbols with rules of combination for representing all numbers.

Numerator [4.4] See *Rational number*.

Numerical coefficient See *Coefficient*.

Objective function [12.6] The function to be maximized or minimized in a linear programming problem.

Obtuse angle [7.2] An angle that is greater than a right angle and smaller than a straight angle.

Obtuse triangle [7.3] A triangle with one obtuse angle.

Octagon [7.2] A polygon with 8 sides.

Odd vertex [7.6] In a network, a vertex with an odd number of arcs or line segments connected at that vertex.

Odds [9.3] If $s + f = n$, where s is the number of outcomes considered favorable to an event E and n is the total number of possibilities, then the *odds in favor of* E is s/f and the *odds against* E is f/s.

One-dimensional coordinate system A real-number line.

One-to-one correspondence Between two sets A and B, this means each element of A can be matched with exactly one element of B and also each element of B can be matched with exactly one element of A.

Online [3.4] To be connected to a computer network.

Open-end loan [6.2] A preapproved line of credit that the borrower can access as long as timely payments are made and the credit line is not exceeded. It is usually known as a credit card loan.

Open half-plane [11.2] See *Half-plane*.

Opposite side [7.4] In a right triangle, an acute angle is made up of two sides. The opposite side of the angle refers to the third side that is not used to make up the sides of the angle.

Opposites [4.6] Opposites x and $-x$ are the same distance from 0 on the number line but in opposite directions; $-x$ is also called the *additive inverse* of x. Do not confuse the symbol "$-$" meaning opposite with the same symbol as used to mean subtraction or negative.

Optimum solution [12.6] The maximum or minimum value in a linear programming problem.

Or [1.2; 2.1] A common translation for the connective of disjunction.

OR-gate [2.6] An electrical circuit that simulates disjunction. That is, the circuit is on when either of two switches is on.

Order [5.5; 12.3] Refers to an order symbol. In reference to matrices, it refers to the number of rows and columns in a matrix. When used in relation to a matrix, it is the same as the *dimension* of the matrix.

Order of an inequality [5.5] Refers to a $>$, $\geq$, $<$, or $\leq$ relationship.

Order of operations [1.3] If no grouping symbols are used in a numerical expression, first perform all multiplication and division from left to right, and then perform all addition and subtraction from left to right.

Order symbols [5.5] Refers to $>$, $\geq$, $<$, $\leq$ in an inequality. Also called inequality symbols.

Ordered pair [9.4; 11.1] A pair of numbers, written (x, y), in which the order of naming is important. The numbers x and y are sometimes called the *first* and *second components* of the pair and are called the *coordinates* of the point designated by (x, y).

Ordinary annuity [6.5] See *Annuity*.

Ordinary interest [6.1] The calculation of interest assuming a year has 360 days. In this book, we assume ordinary interest unless otherwise stated.

Ordinate [11.1] The vertical coordinate in a two-dimensional system of rectangular coordinates, usually denoted by y.

Origin [11.1] The point designating 0 on a number line. In two dimensions, the point of intersection of the coordinate axes; the coordinates are $(0, 0)$.

Origination fee A fee paid to obtain a real estate loan.

Ounce [8.3; 8.4] (1) A unit of capacity in the United States system that is equal to 1/128 of a gallon. (2) A unit of mass in the United States system that is equal to 1/16 of a pound.

Output device [3.6] Component of a system that allows the output of data. The most common output device is a printer.

Overlapping sets Sets whose intersection is not empty.

Parabola [11.3] A set of points in the plane equidistant from a given point (called the *focus*) and a given line (called the *directrix*). It is the path of a projectile. The *axis of symmetry* is the axis of the parabola. The point where the axis cuts the parabola is the *vertex*.

Parallel circuit [2.6] Two switches connected together so that if either of the two switches is turned on, the circuit is on.

Parallel lines [7.1; 7.2] Two nonintersecting straight lines in the same plane.

Parallelepiped [8.3] A polyhedron, all of whose faces are parallelograms.

Parallelogram [7.3; 8.2] A quadrilateral with its opposite sides parallel.

Parentheses See *Grouping symbols*.

Partial sum [6.4] If $s_1, s_2, s_3, \ldots$ is a sequence, then the partial sums are $S_1 = s_1$, $S_2 = s_1 + s_2$, $S_3 = s_1 + s_2 + s_3, \ldots$.

Pascal's triangle [1.1; 9.4] A triangular array of numbers that is bordered by ones and the sum of two adjacent numbers in one row is equal to the number in the next row between the two numbers.

$$
\begin{array}{ccccccccccc}
 & & & & & 1 & & & & & \\
 & & & & 1 & & 1 & & & & \\
 & & & 1 & & 2 & & 1 & & & \\
 & & 1 & & 3 & & 3 & & 1 & & \\
 & 1 & & 4 & & 6 & & 4 & & 1 & \\
1 & & 5 & & 10 & & 10 & & 5 & & 1 \\
 & & & & & \vdots & & & & &
\end{array}
$$

Password [3.4] A word or set of symbols that allows access to a computer account.

Pattern recognition [3.4] A computer function that entails the automatic identification and classification of shapes, forms, or relationships.

Pentagon [7.2] A polygon with 5 sides.

Percent The ratio of a given number to 100; hundredths; denoted by %; that is, 5% means 5/100.

Percent markdown The percent of an original price used to find the amount of discount.

Percent problem A is $P\%$ of W is formulated as a proportion

$$\frac{P}{100} = \frac{A}{W}$$

Percentage The total amount in a percentage problem.

Percentiles [10.1] Ninety-nine values that divide a data set into one hundred equal parts.

Perfect number [4.2] An integer that is equal to the sum of all of its factors except the number itself. For example, 28 is a perfect number since $28 = 1 + 2 + 4 + 7 + 14$.

Perfect square [4.5] $1^2 = 1$, $2^2 = 4$, $3^2 = 9$, . . . , so the perfect squares are 1, 4, 9, 16, 25, 36, 49, ...

Perimeter [8.1] The distance around a polygon.

Periodic payment problem [6.5] A financial problem that involves monthly or other periodic payments.

Peripheral [3.4] A device, such as a printer, that is connected to and operated by a computer.

Permutation [9.4] A selection of objects from a given set with regard to the order in which they are selected. Sometimes it refers to the number of ways this selection can be done and is denoted by $_nP_r$. The formula for finding it is

$$_nP_r = \frac{n!}{(n - r)!}$$

Perpendicular lines [7.2] Two lines are perpendicular if they meet at right angles.

Pi (π) A number that is defined as the ratio of the circumference to the diameter of a circle. It cannot be represented exactly as a decimal, but it is a number between 3.1415 and 3.1416.

Pictograph [10.1] See *Graph*.

Pie chart [10.1] See *Graph*.

Pirating Stealing software by copying it illegally for the use of someone other than the person who paid for it.

Pivot [12.3] A process that uses elementary row operations to carry out the following steps: (1) Divide all entries in the row in which the pivot appears (called the *pivot row*) by the nonzero pivot element so that the pivot entry becomes a 1. This uses elementary row operation 3. (2) Obtain zeros above and below the pivot element by using elementary row operation 4.

Pivot row [12.3] In an elementary row operation, it is the row that is multiplied by a constant. See *Elementary row operations*.

Pivoting [12.3] See *Pivot*.

Pixel [3.5] Any of the thousands (or millions) of tiny dots that make up a computer or calculator image.

Place-value names Trillions, hundred billions, ten billions, billions, hundred millions, ten millions, millions, hundred thousands, ten thousands, thousands, hundreds, tens, units, tenths, hundredths, thousandths, ten-thousandths, hundred-thousandths, and millionths (from large to small).

Planar curve [7.7] A curve completely contained in a plane.

Plane [7.1] In mathematics, it is an undefined term. It is flat and level and extends infinitely in horizontal and vertical directions. It is considered two-dimensional.

Plot a point To mark the position of a point.

Point [4.6; 6.6; 7.1] (1) In the decimal representation of a number, it is a mark that divides the whole number part of a number from its fractional part. (2) In geometry, it is an undefined word that signifies a position, but that has no dimension or size. (3) In relation to a home loan, it represents 1% of the value of a loan, so that 3 points would be a fee paid to a lender equal to 3% of the amount of the loan.

Point–slope form [11.1] An algebraic form of an equation of a line that is given in terms of a point (x_1, y_1) and slope m of a given line: $y - y_1 = m(x - x_1)$.

Polygon [7.2] A geometric figure that has three or more straight sides that all lie in a plane so that the starting point and the ending point are the same.

Polynomial [5.1] An algebraic expression that may be written as a sum (or difference) of terms. Each *term* of a polynomial contains multiplication only.

Population [10.5] The total set of items (actual or potential) defined by some characteristic of the items.

Population growth [11.3] The population, P, at some future time can be predicted if you know the population P_0 at some time, and the annual growth rate, r. The predicted population t years after the given time is $P = P_0e^{rt}$.

Positional system [3.1] A numeration system in which the position of a symbol in the representation of a number determines the meaning of that symbol.

Positive number [4.3] A number greater than 0.

Positive sign The symbol "+" when used in front of a number or an expression.

Postulate [7.1] A statement that is accepted without proof.

Pound [8.4] A unit of measurement for mass in the United States system. It is equal to 16 oz.

Power [1.4] See *Exponent*.

Precision [8.1] The accuracy of the measurement; for example, a measurement is taken to the nearest inch, nearest foot, or nearest mile. It is not to be confused with accuracy that applies to the calculation.

Predecessor [11.6] In a sequence, the predecessor of an element a_n is the preceding element, a_{n-1}.

Premise [1.3; 2.1] A previous statement or assertion that serves as the basis for an argument.

Present value [6.1] See *Compound interest formula*.

Present value of an annuity [6.6] A financial formula that seeks the present value from periodic payments over a period of time. The formula is

$$P = m\left[\frac{1 - (1 + i)^{-N}}{i}\right]$$

Previous balance method [6.2] A method of calculating credit card interest using the formula $I = Prt$ in which P is the balance owed before the current payment is subtracted.

Prime factorization [4.2] The factorization of a number so that all of the factors are primes and so that their product is equal to the given number.

Prime numbers [4.2] $P = \{2, 3, 5, 7, 11, 13, 17, 19, 23, . . .\}$; a number with exactly two factors: 1 and the number itself.

Principal [6.1] See *Compound interest formula*.

PRINT [5.9] A BASIC command that causes a computer to output a line of type, an answer to a calculation, or a line space.

Printer [3.5] An output device for a computer.

Prism [8.4] In this book, it refers to a right prism, which is also called a parallelepiped or more commonly a box.

Probabilistic model [9.1] A model that deals with situations that are random in character and attempts to predict the outcomes of events with a certain stated or known degree of accuracy.

Probability [9.1] If an experiment can result in any of n ($n \geq 1$) mutually exclusive and equally likely outcomes, and if s of these are considered favorable to event E, then $P(E) = s/n$.

Problem solving procedure [1.1] 1. *Read the problem.* Note what it is all about. Focus on processes rather than numbers. You can't work a problem you don't understand. 2. *Restate the problem.* Write a verbal description of the problem using operation signs and an equal sign. Look for equality. If you can't find equal quantities, you will never formulate an equation. 3. *Choose a variable.* If there is a single unknown, choose a variable. 4. *Substitute.* Replace the verbal phrases by known numbers and by the variable. 5. *Solve the equation.* This is the easy step. Be sure your answer makes sense by checking it with the original question in the problem. Use estimation to eliminate unreasonable answers. 6. *State the answer.* There were no variables defined when you started, so $x = 3$ is not an answer. Pay attention to units of measure and other details of the problem. Remember to answer the question that was asked.

Product The result of multiplication.

Profit formula $P = S - C$, where P represents the profit, S represents the selling price (or revenue), and C the cost (or overhead).

Program [3.4; 5.9] A set of step-by-step instructions that instruct a computer what to do in a specified situation.

Progression See *Sequence.*

Projective geometry [7.8] The study of those properties of geometric configurations that are invariant under projection. It was developed to satisfy the need for depth in works of art.

Prompt [5.9] In a computer program, a prompt is a direction that causes the program to print some message to help the user understand what is happening at a particular time.

Proper divisor [4.2] A divisor of a number that is less than the number itself.

Proper fraction [4.4] A fraction for which the numerator is less than the denominator.

Proper subset [1.2] See *Subset.*

Proof [2.4] A logical argument that establishes the truth of a statement.

Property of complements [9.3] See *Complementary* probabilities.

Property of proportions [5.7] If the product of the means equals the product of the extremes, then the ratios form a proportion. Also, if the ratios form a proportion, then the product of the means equals the product of the extremes.

Property of rational expressions [4.4] Let $P, Q, R, S,$ and K be any polynomials such that all values of the variable that cause division by zero are excluded from the domain.

Equality $\dfrac{P}{Q} = \dfrac{R}{S}$ if and only if $PS = QR$.

Fundamental property $\dfrac{PK}{QK} = \dfrac{P}{Q}$

Addition $\dfrac{P}{Q} + \dfrac{R}{S} = \dfrac{PS + QR}{QS}$

Subtraction $\dfrac{P}{Q} - \dfrac{R}{S} = \dfrac{PS - QR}{QS}$

Multiplication $\dfrac{P}{Q} \cdot \dfrac{R}{S} = \dfrac{PR}{QS}$

Division $\dfrac{P}{Q} \div \dfrac{R}{S} = \dfrac{PS}{QR}$

Property of zero [5.4] $AB = 0$ if and only if $A = 0$ or $B = 0$ (or both). Also called the *zero-product rule.*

Proportion [5.7] A statement of equality between two ratios. For example,

$$\frac{a}{b} = \frac{c}{d}$$

For this proportion, a and d are called the *extremes*; b and c are called the *means.*

Protractor [7.2] A device used to measure angles.

Pseudosphere [7.8] The surface of revolution of a tractrix about its asymptote. It is sometimes called a "four-dimensional sphere."

Pull-down windows [5.9] In many computer programs, you obtain additional information or directions on a menu that is hidden until it is ready to be used, at which time it is "pulled down."

Pyramid [8.4] A solid figure having a polygon as a base, the sides of which form the bases of triangular surfaces meeting at a common vertex.

Pythagorean theorem [4.5] If a triangle with legs a and b and hypotenuse c is a right triangle, then $a^2 + b^2 = c^2$.

Quadrant [11.1] See *Axes.*

Quadratic [5.1] A second-degree polynomial.

Quadratic equation [5.4] An equation of the form

$$ax^2 + bx + c = 0, a \neq 0.$$

Quadratic formula [5.4] If $ax^2 + bx + c = 0$ and $a \neq 0$, then

$$x = \frac{-b \pm \sqrt{b^2 - 4ac}}{2a}$$

The radicand $b^2 - 4ac$ is called the *discriminant* of the quadratic.

Quadrilateral [7.2] A polygon having four sides.

Quart [8.3] A measure of capacity in the United States system equal to 1/4 of a gallon.

Quarterly compounding [6.1] In the compound interest formula, it is when $n = 12$.

Quartiles [10.1] Three values that divide a data set into four equal parts.

Quotient The result of a division.

Radical form [4.5] The $\sqrt{\ }$ symbol in an expression such as $\sqrt{2}$. The number 2 is called the *radicand* and an expression involving a radical is called a *radical expression.*

Radicand [4.5] See *Radical.*

Radius [8.1] The distance of a point on a circle from the center of the same circle.

RAM [3.5] Random-Access Memory, or memory where each location is uniformly accessible, often used for the storage of a program and the data being processed.

Range [10.2] In statistics, it is the difference between the largest and the smallest numbers in the data set.

Rate (1) In percent problems, it is the percent. (2) In tax problems, it is the level of taxation, written as a percent. (3) In financial problems, it refers to the APR.

Ratio [5.7] The quotient of two numbers or expressions.

Rational equation An equation that has at least one variable in the denominator.

Rational number [4.4] A number belonging to the set $\mathbb{Q}$ defined by

$$\mathbb{Q} = \left\{ \frac{a}{b} \middle| \ a \text{ is an integer}, b \text{ is a nonzero integer} \right\}$$

a is called the *numerator* and b is called the *denominator*. A rational number is also called a *fraction*.

Ray [7.2] If P is a point on a line, then a ray from the point P is all points on the line on one side of P.

Real number line [4.6] A line on which points are associated with real numbers in a one-to-one fashion.

Real numbers [4.6] The set of all rational and irrational numbers, denoted by $\mathbb{R}$.

Reciprocal [4.6] The reciprocal of n is $\frac{1}{n}$, also called the *multiplicative inverse of n*.

Rectangle [7.3; 8.1] A quadrilateral whose angles are all right angles.

Rectangular coordinate system [11.1] See *Cartesian coordinate system*.

Rectangular coordinates See *Ordered pair*.

Rectangular parallelepiped In this book, it refers to a box all of whose angles are right angles.

Reduced fraction [4.4] A fraction so that the numerator and denominator have no common divisors (other than 1).

Reducing fractions [4.4] The process by which we make sure that there are no common factors (other than 1) for the numerator and denominator of a fraction.

Reflection [7.1] Given a line L and a point P, we call the point P' the *reflection* about the line L if PP' is perpendicular to L and is also bisected by L.

Region [7.6] In a network, a separate part of the plane.

Regression analysis [10.4] The analysis used to determine the relationship or a correlation between two variables.

Relation A set of ordered pairs.

Relative frequency [9.1] If an experiment is repeated n times and an event occurs m times, then the relative frequency is the ratio m/n.

Relatively prime [4.2] Two integers are relatively prime if they have no common factors other than ± 1; two polynomials are relatively prime if they have no common factors except constants.

Remainder When an integer m is divided by a positive integer n, and a quotient q is obtained for which $m = nq + r$ with $0 \le r < n$, then r is the remainder.

Repeating decimal [4.6] See *Decimal fraction*.

Replication [5.3] On a spreadsheet, the operation of copying a formula from one place to another.

Resolution [3.5] The number of dots (or pixels) determines the clarity, or resolution, of the image on the monitor.

Revolving credit [6.2] It is the same as open-end or credit card credit.

Rhombus [7.3] A parallelogram with adjacent sides equal.

Right angle [7.2] An angle of 90°.

Right circular cone [8.4] A cone with a circular base for which the base is perpendicular to its axis.

Right circular cylinder [8.4] A cylinder with a circular base for which the base is perpendicular to its axis.

Right prism [8.4] A prism whose base is perpendicular to the lateral edges.

Right triangle [7.3] A triangle with one right angle.

Rise [11.1] See *Slope*.

ROM [3.5] Read-Only Memory, or memory that cannot be altered either by the user or a loss of power. In microcomputers, the ROM usually contains the operating system and system programs.

Root of a number [4.5] An nth root (n is a natural number) of a number b is a only if $a^n = b$. If $n = 2$, then the root is called a *square root*; if $n = 3$, it is called a *cube root*.

Root of an equation [5.4] See *Solution*.

Rounding a number Dropping decimals after a certain significant place. The procedure for rounding is: 1. Locate the rounding place digit. 2. Determine the rounding place digit: It stays the same if the first digit to its right is a 0, 1, 2, 3, or 4; it increases by 1 if the digit to the right is a 5, 6, 7, 8, or 9. 3. Change digits: All digits to the left of the rounding digit remain the same (unless there is a carry) and all digits to the right of the rounding digit are changed to zeros. 4. Drop zeros: If the rounding place digit is to the left of the decimal point, drop all trailing zeros; if the rounding place digit is to the right of the decimal point, drop all trailing zeros to the right of the rounding place digit.

Row [5.3] A horizontal arrangement of numbers or entries of a matrix. It is denoted by numerals 1, 2, 3, . . . on a spreadsheet.

Row+ [12.3] An elementary row transformation that causes one row of a matrix (called the *pivot row*) to be added to another row (called the *target row*). The answer to this addition replaces the entries in the target row, entry-by-entry.

Row-reduced form [12.3] The final matrix after the process of Gauss–Jordan elimination.

RowSwap [12.3] An elementary row operation that causes two rows of a matrix to be switched, entry-by-entry.

Rubik's cube [9.6] A three-dimensional cube that can rotate about all three axes. It is a puzzle that has the object of returning the faces to a single color position.

Rules of divisibility [4.2] A number N is divisible by:

  1
  2 if the last digit is divisible by 2.
  3 if the sum of the digits is divisible by 3.
  4 if the number formed by the last two digits is divisible by 4.
  5 if the last digit is 0 or 5.
  6 if the number is divisible by 2 and by 3.
  8 if the number formed by the last three digits is divisible by 8.
  9 if the sum of the digits is divisible by 9.
  10 if the last digit is 0.
  12 if the number is divisible by 3 and by 4.

Run [11.1] See *Slope*.

Saccheri quadrilateral [7.8] A rectangle with base angles A and B right angles, and with sides $\overline{AC}$ and $\overline{BD}$ the same length.

Sale price A reduced price usually offered to stimulate sales. It can be found by subtracting the discount from the original price, or by multiplying the original price by the complement of the markdown.

Sales tax A tax levied by government bodies that is based on the sale price of an item.

Sample [10.5] A finite portion of a population.

Sample space [9.1] The set of possible outcomes for an experiment.

Satisfy [5.4; 11.1] See *Equation* or *Inequality*.

SAVE [5.9] A BASIC command that causes a program to be saved for later use.

Scalar A real number.

Scalene triangle [7.3] A triangle with no sides having the same length.

Scatter diagram [10.4] A diagram showing the frequencies with which joint values of variables are observed. One variable is indicated along the x-axis and the other along the y-axis.

Scientific notation [1.4] Writing a number as the product of a number between 1 and 10 and a power of 10: For any real number n, $n = m \cdot 10^c$, $1 \leq m < 10$, and c is an integer. Calculators often switch to scientific notation to represent large or small numbers. The usual notation is 8.234 05, where the space separates the number from the power; thus 8.234 05 means 8.234×10^5.

Secant line [11.5] A line passing through two points of a given curve.

Second component [9.4] See *Ordered pair*.

Semiannual compounding [6.1] In the compound interest formula, it is when $n = 2$.

Semicircle Half a circle.

Sequence [6.3; 11.6] An *infinite sequence* is a function whose domain is the set of counting numbers. It is sometimes called a *progression*. A *finite sequence* with n terms is a function whose domain is the set of numbers $\{1, 2, 3, \ldots, n\}$.

Series [6.4] The indicated sum of a finite or an infinite sequence of terms.

Series circuit [2.6] Two switches connected together so that the circuit is on only if both switches are on.

Set [1.2] A collection of particular things, called the *members* or *elements* of the set. A set with no elements is called the *null set* or *empty set* and is denoted by the symbol $\emptyset$. All elements of a *finite set* may be listed, whereas the elements of an *infinite set* continue without end.

SI system See *Metric system*.

Sieve of Eratosthenes [4.2] A method for determining a set of primes less than some counting number n. Write out the consecutive numbers from 1 to n. Cross out 1, since it is not classified as a prime number. Draw a circle around 2, the smallest prime number. Then cross out every following multiple of 2, since each is divisible by 2 and thus is not prime. Draw a circle around 3, the next prime number. Then cross out each succeeding multiple of 3. Some of these numbers, such as 6 and 12, will already have been crossed out because they are also multiples of 2. Circle the next open prime, 5, and cross out all subsequent multiples of 5. The next prime number is 7; circle 7 and cross out multiples of 7. Continue this process until you have crossed out the primes up to $\sqrt{n}$. All of the remaining numbers on the list are prime.

Sigma notation [6.4] Sigma, the Greek letter corresponding to S, is written Σ. It is used to indicate the process of summing the first to the nth terms of a set of numbers $s_1, s_2, s_3, \ldots, s_n$, which is written as

$$\sum_{k=1}^{n} s_k$$

This notation is also called *summation notation*.

Signed number An integer.

Significance level [10.4] Deviations between hypothesis and observations that are so improbable under the hypothesis as not to be due merely to sampling errors or fluctuations are said to be *statistically significant*. The significance level is set at an acceptable level for a deviation to be statistically significant.

Similar figures [7.4] Two geometric figures are similar if they have the same shape, but not necessarily the same size.

Similar terms [5.1] Terms that differ only in their numerical coefficients.

Similar triangles [7.4] Triangles that have the same shape.

Similarity [7.1] Two geometric figures are *similar* if they have the same shape.

Simple curve [7.7] A curve that does not intersect itself.

Simple event [9.1] An event for which the sample space has only one element.

Simple grouping system [3.1] A numeration system is a grouping system if the position of the symbols is not important, and each symbol larger than 1 represents a group of another symbols.

Simple interest formula [6.1] $I = Prt$.

Simple statement [2.1] A statement that does not contain a connective.

Simplify [4.4; 4.5] (1) A *polynomial*: Combine similar terms and write terms in order of descending degree. (2) A *fraction* (a *rational expression*): Simplify numerator and denominator, factor if possible, and eliminate all common factors. (3) A *square root*: The *radicand* (the number under the radical sign) has no factor with an exponent larger than 1 when it is written in factored form; the radicand is not written as a fraction or by using negative exponents; and there are no square root symbols in the denominators of fractions.

Simulation [3.4] Use of a computer program to simulate some real-world situation.

Simultaneous solution [12.1] The solution of a simultaneous system of equations.

Sine [7.4] In a right triangle ABC with right angle C,

$$\sin A = \frac{\text{opposite side of } A}{\text{hypotenuse}}$$

Sinking fund [6.5] A financial problem in which the monthly payment must be found to obtain a known future value. The formula is

$$m = \frac{Ai}{(1 + i)^N - 1}$$

Skewed distribution [10.3] A statistical distribution that is not symmetric, but favors the occurrence on one side of the mean or the other.

Slope [11.1] The slope of a line passing through (x_1, y_1) and (x_2, y_2) is denoted by m, and is found by

$$m = \frac{y_2 - y_1}{x_2 - x_1} = \frac{\text{VERTICAL CHANGE}}{\text{HORIZONTAL CHANGE}} = \frac{\text{RISE}}{\text{RUN}}$$

Slope–intercept form [11.1] $y = mx + b$

Slope point [11.1] A point that is found after counting out the rise and the run from the y-intercept.

Software [3.4] The routines, programs, and associated documentation in a computer system.

Software packages [3.4] A commercially available computer program that is written to carry out a specific purpose, for example, a database program or a word-processing program.

Solution [5.4; 11.1] The values or ordered pairs of values for which an equation, a system of equations, inequality, or system of inequalities is true. Also called *roots*.

Solution set The set of all solutions to an equation.

Solve To find the values of the variable that satisfy the equation.

Solve a proportion [5.7] To find the missing term of a proportion. Procedure: First, find the product of the means or the product of the extremes, whichever does not contain the unknown term; next, divide this product by the number that is opposite the unknown term.

Some [2.1] A word used to mean *at least one*.

Sphere [8.4] The set of all points in space that are a given distance from a given point.

Spreadsheet [3.4; 5.3] A rectangular grid used to collect and perform calculations on data. *Rows* are horizontal and are labeled with numbers and *columns* are vertical and are labeled with letters to designate *cells* such as A4, P604. Each cell can contain text, numbers, or formulas.

Square [7.3; 8.1] A quadrilateral with all sides the same length and all angles right angles.

Square matrix [12.3] A matrix with the same number of rows and columns.

Square numbers [4.5] Numbers that are squares of the counting numbers: 1, 4, 9, 16, 25, 36, 49, 64, 81, 100, 121, 144, 169, ...

Square root [4.5] See *Root of a number*.

Square unit [8.2] A two-dimensional unit. It is the result of squaring a unit of measurement.

Standard deviation [10.2] It is a measure of the variation from a trend. In particular, it is the square root of the mean of the squares of the deviations from the mean.

Standard form [11.1] The standard form of the equation of a line is $Ax + By + C = 0$.

Statement [2.1] A declarative sentence that is either true or false, but not both true and false.

Statistics [10.1] Methods of obtaining and analyzing data.

Stem-and-leaf plot [10.1] A procedure for organizing data that can be divided into two categories. The first category is listed at the left, and the second category at the right.

Sticker price [6.2] In this book, it refers to the manufacturer's total price of a new automobile as listed on the window of the car.

Straight angle [7.2] An angle whose rays point in opposite directions; an angle whose measure is 180°.

Straightedge [7.1] A device used as an aid in drawing a straight line segment.

Street problem [1.1] A problem that asks the number of possible routes from one location to another along some city's streets. The assumptions are that we always move in the correct direction and that we do not cut through the middle of a block, but rather stay on the streets or alleys.

Subjective probability [9.1] A probability obtained by experience and used to indicate a measure of "certainty" on the part of the speaker. These probabilities are not necessarily arrived at through experimentation or theory.

Subscript [12.3] A small number or letter written below and to the right or left of a letter as a mark of distinction.

Subset [1.2] A set contained within a set. There are 2^n subsets of a set with n distinct elements. A subset is *improper* if it is equivalent to the given set; otherwise it is *proper*.

Substitution property [1.3] The property that allows us to replace one quantity or unknown by another equal quantity. That is, if $a = b$, then a may be substituted for b in any mathematical statement without affecting the truth or falsity of the mathematical statement.

Substitution method [12.1] The method of solution of a system of equations in which one of the equations is solved for one of the variables and substituted into another equation.

Subtraction [4.1] The operation of subtraction is defined by:

$$a - b = x \quad \text{means} \quad a = b + x$$

Subtraction law of exponents [1.4] To divide two numbers with the same base, subtract the exponents. That is,

$$\frac{b^m}{b^n} = b^{m-n}$$

Subtraction of integers [4.3]

$$a - b = a + (-b)$$

Subtraction of rational numbers [4.4]

$$\frac{a}{b} - \frac{c}{d} = \frac{ad}{bd} - \frac{bc}{bd} = \frac{ad - bc}{bd}$$

Subtraction principle [3.1] In reference to numeration systems, it is subtracting the value of some symbol from the value of the other symbols. For example, in the Roman numeration system, IX uses the subtraction principle because the position of the I in front of the X indicates that the value of I (which is 1) is to be subtracted from the value of X (which is 10): IX = 9.

Subtraction property of equations [5.4] The solution of an equation is unchanged by subtracting the same number from both sides of the equation.

Subtraction property of inequality [5.5] See *Addition property of inequality*.

Successor [11.6] In a sequence, the successor of an element a_n is the following element, a_{n+1}.

Sum The result of an addition.

Summation notation [6.4] See *Sigma notation*.

Superfluous constraint [12.6] In a linear programming problem, a constraint that does not change the outcome if it is deleted.

Supplementary angles [7.2] Two angles whose sum is 180°.

Supply [12.2] The number of items that can be supplied at a given price.

Surface [7.1] In mathematics, it is an undefined term. It is the outer face or exterior of an object; it has an extent or magnitude having length and breadth, but no thickness.

Surface area [8.2] The area of the outside faces of a solid.

Syllogism [2.4] A logical argument that involves three propositions, usually two premises and a conclusion, the conclusion necessarily being true if the premises are true.

Symbols of inclusion See *Grouping symbols.*

Symmetric property of equality [5.4] If $a = b$, then $b = a$.

Syntax error The breaking of a rule governing the structure of the programming language being used.

System (of equations) [12.1] A set of equations that are to be solved *simultaneously.* A brace symbol is used to show the equations belonging to the system.

System of inequalities [12.5] A set of inequalities that are to be solved simultaneously. The solution is the set of all ordered pairs (x, y) that satisfy all the given inequalities. It is found by finding the intersection of the half-planes defined by each inequality.

Tangent [7.4] In a right triangle ABC with right angle C,

$$\tan A = \frac{\text{opposite side of } A}{\text{adjacent side of } A}$$

Tangent line [11.5] A tangent line to a circle is a line that contains exactly one point of the circle. The tangent line to a curve at a point P is the limiting position, if this exists, of the secant line through a fixed point P on the curve and a variable point P' on the curve so that P' approaches P along the curve.

Target row [12.3] In an elementary row operation, it is the row that is changed. See *Elementary row operation.*

Tautology [2.3] A compound statement is a tautology if all values on its truth table are true.

Temperature [8.4] The degree of hotness or coldness.

Term [5.1; 6.3] A number, a variable, or a product of numbers and variables. See *Polynomial.* A *term of a sequence* is one of the elements of that sequence.

Terminating decimal [4.6] See *Decimal fraction.*

Tessellation [7.7] A mosaic, repetitive pattern.

Test point [11.2] A point that is chosen to find the appropriate half-plane when graphing a linear inequality in two variables.

Theorem [1.3; 2.4; 7.1] A statement that has been proved. See *Deductive reasoning.*

Theoretical probability [9.1] A probability obtained by logical reasoning according to stated definitions.

Time [6.1] In a financial problem, the length of time (in years) from the present value to the future value.

*Row [12.3] An elementary row operation that multiplies each entry of a row of a matrix (called the *target row*) by some number, called a *scalar.* The elements of the row are replaced term-by-term by the products. It is denoted by *Row.

*Row+ [12.3] An elementary row operation that multiplies each entry of a row of a matrix (called the *pivot row*) by some number (called a *scalar*), and then adds that product, term-by-term, to the numbers in another row (called the *target row*). The results replace the entries in the target row, term-by-term. It is denoted by *Row+.

Ton [8.4] A measurement of mass in the United States system; it is equal to 2,000 lb.

Topology [7.7] That branch of geometry that deals with the *topological properties* of figures. If one figure can be transformed into another by stretching or contracting, then the figures are said to be *topologically equivalent.*

Trailing zeros Sometimes zeros are placed after the decimal point or after the last digit to the right of the decimal point, and if these zeros do not change the value of the number, they are called *trailing zeros.*

Transformation [7.1] A passage from one figure or expression to another, such as a reflection, translation, rotation, contraction, or dilation.

Transformational geometry [7.1] The geometry that studies transformations.

Transitive reasoning [2.4] If $a = b$ and $b = c$, then $a = c$.

Translating The process of writing an English sentence in mathematical symbols.

Transversal [7.2] A line that intersects two parallel lines.

Trapezoid [7.3; 8.2] A quadrilateral that has two parallel sides.

Traversable network [7.6] A network is said to be *traversable* if it can be traced in one sweep without lifting the pencil from the paper and without tracing the same edge more than once. Vertices may be passed through more than once.

Tree diagram [9.1] A device used to list all the possibilities for an experiment.

Triangle [7.2; 7.3] A polygon with three sides.

Trichotomy [5.5] Exactly one of the following is true, for any real numbers a and b: $a < b$, $a > b$, or $a = b$.

Trigonometric ratios [7.4] The sine, cosine, and tangent ratios are known as the *trigonometric ratios.*

Trillion [1.4] A name for $10^{12} = 1,000,000,000,000$.

Trinomial [5.1] A polynomial with exactly three terms.

Truth set [2.3] The set of values that makes a given statement true.

Truth table [2.2] A table that shows the truth values of all possibilities for compound statements.

Truth value [2.1] The truth value of a simple statement is true or false. The truth value of a compound statement is true or false and depends only on the truth values of its simple component parts. It is determined by using the rules for connecting those parts with well-defined operators.

Two-point form [11.1] The equation of a line passing through (x_1, y_1) and (x_2, y_2) is

$$y - y_1 = \left(\frac{y_2 - y_1}{x_2 - x_1} \right)(x - x_1)$$

Type I error [10.5] Rejection of a hypothesis based on sampling when, in fact, the hypothesis is true.

Type II error [10.5] Acceptance of a hypothesis based on sampling when, in fact, it is false.

Undefined terms [7.1] To avoid circular definitions, it is necessary to include certain terms without specific mathematical definition.

Union [1.2] The union of sets A and B, denoted by $A \cup B$, is the set consisting of elements in A or in B or in both A and B.

Unit circle A circle with radius 1 centered at the origin.

Unit scale The distance between the points marked 0 and 1 on a number line.

United States system [8.1] The measurement system used in the United States.

Universal set [1.2] The set that contains all of the elements under consideration for a problem or a set of problems.

Unless [2.3] A logical operator for "p unless q," which is defined to mean $\sim q \rightarrow p$.

Upload [3.4] The process of copying a program from your computer to the network.

User-friendly A term used to describe software that is easy to use. It includes built-in safeguards to keep the user from changing important parts of a program.

Valid argument [2.1] In logic, refers to a correctly inferred logical argument.

Variable A symbol that represents unspecified elements of a given set. On a calculator, it refers to the name given to a location in the memory that can be assigned a value.

Variable expression An expression that contains at least one variable.

Variance [10.2] The square of the standard deviation.

Venn diagram [1.2] A diagram used to illustrate relationships among sets.

Vertex [7.2; 7.6; 11.3] (1) A *vertex* of a polygon is a corner point, or a point of intersection of two sides. (2) A *vertex* of a parabola is the lowest point for a parabola that opens upward; the highest point for one that opens downward; the leftmost point for one that opens to the right; and the rightmost point for one that opens to the left.

Vertical angles [7.2] Two angles such that each side of one is a prolongation through the vertex of a side of the other.

Vertical line [7.2; 11.1] A line with undefined slope. Its equation has the form $x = $ constant.

Volume [8.3] A number describing three-dimensional content of a set. Specifically, it is the number of cubic units enclosed in a solid figure.

Weight [8.4] In everyday usage, the heaviness of an object. In scientific usage, the gravitational pull of a body.

Weighted mean [10.2] If the scores $x_1, x_2, x_3, \ldots, x_n$ occur $w_1, w_2, \ldots w_n$ times, respectively, then the *weighted mean* is

$$\bar{x} = \frac{\Sigma(w \cdot x)}{\Sigma w}$$

Well-defined set A set for which there is no doubt about whether a particular element is included in the given set.

Whole numbers [4.3; 4.4] The positive integers and zero; $\mathbb{W} = \{0, 1, 2, 3, \ldots\}$.

Windows A graphical environment for IBM format computers.

With replacement [9.5] If there is more than one step for an experiment, to perform the experiment with replacement means that the object chosen on the first step is replaced before the next steps are completed.

Without replacement [9.5] If there is more than one step for an experiment, to perform the experiment without replacement means that the object chosen on the first step is not replaced before the next steps are completed.

Word processing [3.4] The process of creating, modifying, deleting, and formatting textual materials.

WorldWide Web [3.4] A network that connects computers from all over the world. It is abbreviated by www.

WYSIWYG: An acronym for $\underline{W}$hat $\underline{Y}$ou $\underline{S}$ee $\underline{I}$s $\underline{W}$hat $\underline{Y}$ou $\underline{G}$et.

x-axis The horizontal axis in a Cartesian coordinate system.

x-intercept [11.1] The place where a graph passes through the x-axis.

y-axis The vertical axis in a Cartesian coordinate system.

y-intercept [11.1] The place where a graph passes through the y-axis. For a line $y = mx + b$, it is the point b.

Yard [8.1] A linear measure in the United States system; it has the same length as 3 ft.

Z-score [10.3] A measure to determine the distance (in terms of standard deviations) that a given score is from the mean of that distribution.

Zero [4.3; 4.6] The number that divides the positive and negative numbers; it is also called the *identity element* for addition; that is, it satisfies the property that $x + 0 = 0 + x = x$ for all numbers x.

Zero matrix [12.4] A matrix with all entries equal to 0.

Zero multiplication theorem If a is any real number, then $a \cdot 0 = 0 \cdot a = 0$.

Zero-product rule [5.4] If $a \cdot b = 0$, then either $a = 0$ or $b = 0$.

APPENDIX E   Selected Answers

CHAPTER 1 THE NATURE OF PROBLEM SOLVING

1.1   Problem Solving, page 15

7. first diagonal   9. Answers vary; yes, it is the same.
11. 20   13. 56   15. 2 up, 3 over; 10 paths
17. 4 up, 6 over; 210 paths   19. 4 down, 3 over; 35 paths
21. 2 up, 4 over; 15 paths   23. Answers vary.
25. There is a total of 27 boxes.
27. **a.** 2   **b.** 4   **c.** 8   **d.** 16   **29.** 37 paths
31. 11 paths   37. The fly flies 20 miles.
39. 8 miles or 12 miles
41. 1,089; does not work for palindromes
43. 6, 8, 10, 14, and 15   45. They are the same.
47. 7 cards   49. They don't bury the living.   51. triplets
53. 11   57. 200 miles
59. Ahmes should tie one end of the rope onto the ankh at the
outer edge of the funnel, then walk around the outer perimeter
of the funnel until he returns to the ankh. The rope will have
wrapped around the ankh at the center so that if he ties off the
rope he will have a rope bridge to the center.

1.2   Problem Solving with Sets, page 31

3. $\overline{X \cup Y} = \overline{X} \cap \overline{Y}$ and $\overline{X \cap Y} = \overline{X} \cup \overline{Y}$ 7. {2, 6, 8, 10}
9. {3, 4, 5}   11. {2, 3, 5, 6, 8, 9}   13. {1, 3, 4, 5, 6, 7, 10}
15. {1, 2, 3, 4, 5, 6}   17. {5}   19. {5, 6, 7}

21.   23.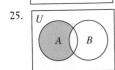

25. 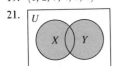   27.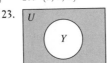

29.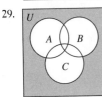

31. 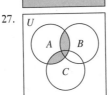   33.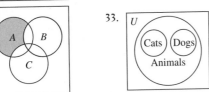

35.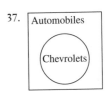

37.

39. Let U = {SRJC students},
F = {females}, and A = {students
over 25}.
$|U|$ = 29,000;
$|F|$ = 0.58(29,000) = 16,820;
$|\overline{F}|$ = 29,000 − 16,820 = 12,180;
$|A|$ = 0.62(29,000) = 17,980;
$|\overline{F} \cap A|$ = 0.40(17,980) = 7,192;
$|F \cap A|$ = 0.60(17,980) = 10,788.
Thus, I: 4,988; II: 7,192; III: 10,788;
IV: 6,032.

41. $\overline{A \cup B}$ 43. $A \cap (B \cup C)$

45.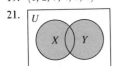
$A \cup B$ $\overline{A} \cup \overline{B}$

47.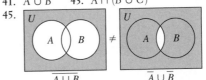
$(A \cup B) \cup C$ $A \cup (B \cup C)$

49.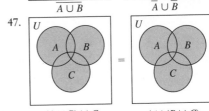
$A \cap (B \cup C)$ $(A \cap B) \cap (A \cap C)$

51. Steffi Graf, II; Michael Stich,
V; Martina Navratilova, III;
Stefan Edberg, V; Chris Evert
Lloyd, VI; Boris Becker, V;
Evonne Goolagong, II; Pat
Cash, V; Virginia Wade, II;
John McEnroe, IV

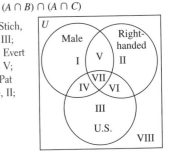

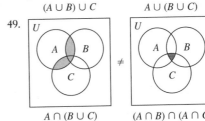

53. a. 50   **b.** 15   **55.** There are 34 people playing.

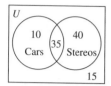
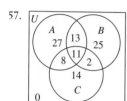

57. 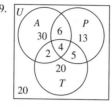   **59.**

A = {use shampoo A}
B = {use shampoo B}
C = {use shampoo C}

A = {drive alone}
P = {carpool}
T = {public transportation}
20 used none of the above
means of transportation.

61.

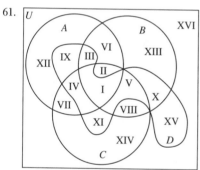

63. a. region 32   **b.** region 2
65. A = 3, B = 6, C = 1, D = 5, E = 8, F = 4, G = 9, H = 2, I = 7

1.3 Inductive and Deductive Reasoning, page 43

7. a. 32   **b.** 12   **9. a.** 29   **b.** 35   **11. a.** $\frac{11}{2}$ **b.** 8
13. a. 14   **b.** 6   **15. a.** 50   **b.** 2   **17.** 12   **19.** 6
21. 4, 8, 3, 7, 2, 6, 1, 5, 9, 4, 8, 3, . . .
23. 6, 3, 9, 6, 3, 9, 6, 3, 9, . . .   **25.** $50^2 = 2,500$
27. a. $3 + 2 \times 4$ **b.** $3(2 + 4)$
29. a. $8 \times 9 + 10$ **b.** $8(9 + 10)$
31. a. $3^2 + 2^3$ **b.** $3^3 - 2^2$ **33. a.** $4^2 + 9^2$ **b.** $(4 + 9)^2$
35. a. $3(n + 4) = 16$ **b.** $5(n + 1) = 5n + 5$
37. a. $n^2 + 8n = 6$ **b.** $n^2 + 8n = n(8 + n)$ **39.** $A = \frac{1}{2}bh$
41. $A = \frac{1}{2}h(a + b)$ **43.** $V = \ell wh$ **45.** $V = \pi r^2 h$
47. $9 \times 54,321 - 1$; 488,888   **49.** 98,888,888,888
51. 7,777,777,707
53. The figure has: 36 1-by-1 squares; 25 2-by-2 squares; 16 3-by-3 squares; 9 4-by-4 squares; 4 5-by-5 squares; and one 6-by-6 square for a total of 91 squares.

55. There are $4 \times 4 \times 4 = 64$ cubes; there are 6 faces and each face has 4 cubes painted on one side, so there are $6 \times 4 = 24$ cubes with one painted face. The fraction is $\frac{24}{64} = \frac{3}{8}$.
57. Balance 3 with 3; if it balances, the heavier one is in the three not weighed; if it does not balance, then the pan balance will show which set of 3 contains the heavier one. In any case, the first weighing narrows it down to a set of 3 coins, one of which is the heavy coin. Take 2 from this set of heavy 3 and balance them. If they balance, the heavy coin is the one not weighed; if they do not balance, then the pan balance will show the heavier one. In any case, it takes two weighings.

59. Arrange the cards to match $\begin{bmatrix} 2 & 7 & 6 \\ 9 & 5 & 1 \\ 4 & 3 & 8 \end{bmatrix}$.

61. a. One diagonal sum is 38,307 and the other sums are 21,609.
   b. One diagonal sum is 13,546,875 and the other sums are 20,966,014. If you find a 3 × 3 magic square with nine distinct square numbers, send me your solution and if you are correct I'll send you $100.

1.4 Scientific Notation and Estimation, page 62

7. a. 3.2×10^3; 3.2   03   **b.** 4×10^{-4}; 4. -04
   c. 6.4×10^{10}; 6.4   10
9. a. 6.3×10^7; 6.3   07   **b.** 5.629×10^3; 5.629   03
   c. 3.4×10^{-13}; 3.4 -13
11. a. 49   **b.** 72,000,000,000   **c.** 4,560
13. a. 216   **b.** 0.00000 041   **c.** 0.00000 048
17. 3.0×10^{10} **19.** 2.2×10^8; 2.5×10^{-6}
21. 3,600,000   **23.** 0.00000 003
25. Answers vary; 35 miles
27. Estimate $20 \times 400 = 8,000$ hr; $24 \times 360 = 8,640$
29. Estimate $4 \times 4,000 = 16,000$; $4 \times \$3,900 = \$15,600$
31. Estimate $1,500 \times 12 = 1,500 \times 10 + 1,500 \times 2$
   $= 15,000 + 3,000 = 18,000$; $\$1,543 \times 12 = \$18,516$
33. $36,000 \div 12 = 3,000$ per month $\approx 2,800 \div 4$
   $= 700$ per week $\approx 800 \div 40 = 20$ per hour; $\$31,200 \div 52$
   $= \$600$ per week; $\$600 \div 40 = \$15$ per hour.
35. Estimate $0.5 + 2.5 + 2 + 1 + 2 + 2 + 1 + 1 + 2 + 2 + 1$
   $+ 1 + 2 = 20$; $\$.59 + \$2.50 + \$1.89 + \$.99 + \$1.97 + \$1.99$
   $+ \$.99 + \$1.09 + \$1.79 + \$1.89 + \$.79 + \$1.39 + \$1.61$
   $= \$19.48$
37. a. 1.2×10^9 **b.** 3×10^2
39. a. 1.2×10^{-3} **b.** 2×10^7
41. a. 4×10^{-8} **b.** 1.2
43. Estimate $10/\text{cm}^2$, so about 120 oranges
45. about 200 chairs   **47.** $72 \times 10^{12} \times 10^9 = 7.2 \times 10^{22}$
49. approximately 1.86×10^{10} oranges
51. A million pennies would be about a mile high.
53. a. Answers vary.   **b.** 9^{9^9} **c.** This number is too large for your calculator; $9^{9^9} = 9^{387,420,489}$; the largest power of 9 most calculators will handle is $9^{104} \approx 1.7 \times 10^{99}$.
55. $3.36271772 \times 10^{23}$
57. It is your favorite counting number repeated six times. If n is your choice, then $n \times 259 \times 429 = 111,111n$.

59. a. Number the rings 1, 2, 3, 4 (in order of smallest to largest):

Start	1st	2nd	3rd	4th	5th	6th	7th	8th	9th	10th
1							1	1			
2	2						1	1			
3	3	3		3 1	1 1		1 2	2 2	2 1	1	
4	4 1	4 1 2	4 2	4 3 2	4 3 2	4 3	4 3	3 4	3 4	2 3 4	

11th	12th	13th	14th	15th
				1
		2	2	
1	1 3	3	3	3
2 3 4 2	4 2 1 4	1 4	4	

b. 5 rings; $2^5 - 1 = 31$ moves

61. The units column will contain — 1,000 digits
The tens column will contain — 1,000 − 9 digits
The hundreds column will contain — 1,000 − 99 digits
The thousands column will contain — 1,000 − 999 digits
Sum — 4,000 − 1,107 = 2,893 digits

Thus, the time required is 2,893 seconds or 2,893 ÷ 60 minutes = 48 minutes, 13 seconds.

63. Look for a pattern; consider each column separately.
The units column will contain — 100 zeros
The tens column will contain — 100 − 9 zeros
Hundreds column — 100 − 99
Sum — 300 − 108 = 192 zeros

65. Answers vary; a brick is about 8 in. long, $3\frac{1}{2}$ in. wide, and 2 in. high. The wall would require about 27,000 bricks.

67. Answers vary; the church contains about 6,000,000 bricks.

69. a. 5.6×10^5 people per square mile   **b.** about 10,000 square miles   **c.** 6.05 acres

---

Chapter 1 Review Questions, page 70

1. Understand the problem, devise a plan, carry out the plan, and then look back.

2. Use Pascal's triangle; look at 5 blocks down and 4 blocks over to find 126.

3. Turn the chessboard so that it forms a triangle with the rook at the top. To get the appropriate square, look at 7 blocks down and 4 blocks over using Pascal's triangle to find 330 paths.

4. By patterns: $1 \times 1 = 1$; $11 \times 11 = 121$; $111 \times 111 = 12,321$; . . . ; $111,111,111^2 = 12,345,678,987,654,321$

5. Order of operations: (1) First, perform any operations enclosed in parentheses. (2) Next, perform multiplications and divisions as they occur by working from left to right. (3) Finally, perform additions and subtractions as they occur by working from left to right.

6. a. $A \cup B$
$= \{1, 3, 5, 7, 9\} \cup \{2, 4, 6, 9, 10\}$
$= \{1, 2, 3, 4, 5, 6, 7, 9, 10\}$
b. $A \cap B = \{1, 3, 5, 7, 9\} \cap \{2, 4, 6, 9, 10\} = \{9\}$
c. $\overline{B} = \overline{\{2, 4, 6, 9, 10\}} = \{1, 3, 5, 7, 8\}$ **You need to look at the universal set, U, in obtaining this result.**
d. $\overline{A} \cap (B \cup A)$
$= \{2, 4, 6, 8, 10\} \cap \{1, 2, 3, 4, 5, 6, 7, 9, 10\}$
$= \{2, 4, 6, 10\}$

7. a.

b.

c.
d.

8. a. $A \cup (B \cap C) \neq (A \cup B) \cap C$; disproved
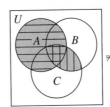 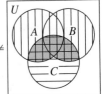

b. $(A \cup B) \cap C = (A \cap C) \cup (B \cap C)$; proved
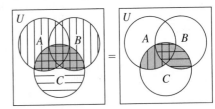

9. Inductive reasoning; the answer was found by looking at the pattern of questions.

10. Answers vary. **a.** $\boxed{2}\ \boxed{y^x}\ \boxed{63}\ \boxed{=}$ or $\boxed{2}\ \boxed{\wedge}\ \boxed{63}$
$\boxed{\text{ENTER}}$ *9.223372037E18*
b. $\boxed{9.22}\ \boxed{\text{EE}}\ \boxed{18}\ \boxed{÷}\ \boxed{6.34}\ \boxed{\text{EE}}\ \boxed{6}\ \boxed{=}$
1.454258675E12

11. A. $50/person times 263 million is about $13 billion; not even close.
B. $1 × 60 sec × 60 min × 24 hr × 365.25 days × 1,000 years is about $3.1 billion; not even close.
C. 1 million people × $80,000 + 1 billion people × 200 ≈ 280 billion; choice C comes closest.

12. $C = \dfrac{\ell w}{15}$, where C is the capacity of the boat, ℓ and w are the length and width of the boat.

13. $\frac{3}{4}$ hr × 365 = 273.75 hr. This is approximately $2\frac{3}{4}$ hr/book. Could not possibly be a complete transcription of each book.

14. Draw a Venn diagram; begin with the innermost part first: 15 have all three. (See diagram on the next page.)

Next, use the information for two overlapping sets to fill in 2, 20, and 10.

Fill in the regions in each circle not yet completed to fill in 5, 5, and 3.

Finally, total all the numbers to see that 10 of the students have none of these items.

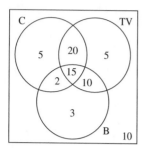

15. The scientific notation for a number is that number written as a power of 10 times another number x, such that x is between 1 and 10; that is $1 \leq x < 10$.

16. An ice cube is about 1 in.^{3} and a cubic foot has 12^3 in.^{3}. A classroom 30 ft $\times$ 50 ft $\times$ 10 ft would hold about 2.592×10^7 ice cubes. $\dfrac{7 \times 10^{16}}{2.592 \times 10^7} \approx 2.7 \times 10^9$ classrooms; this is not meaningful, but is shown here because it is typical of first attempts. It is important to look for a meaningful comparison. Let's try again. The world's largest dam is in Arizona and holds back a volume of 274,015 yd^3 (Source: *1996 Information Please Almanac*). This is

$$274{,}015 \times (3 \times 12)^3 \approx 12{,}784{,}443{,}840 \text{ in.}^3 \approx 1.3 \times 10^{10} \text{ in.}^3$$

The iceberg would fill 5,475,404 such dams. There are 35 dams listed in the almanac, so it is safe to say the size of the iceberg is larger than the capacity of all the dams in the world.

17. 52×10^{12}; $(5.2 \times 10^{12}) \div (2.6665 \times 10^8) \approx 1.9501 \times 10^4$; each person's share is about $19,500.

18. Classroom volume is 20 ft $\times$ 30 ft $\times$ 10 ft = 6,000 ft^3 = 10,368,000 in.^{3}.

    A dollar bill is 6 in. $\times$ 2.5 in. $\times \dfrac{1}{233}$ in. $\approx$ 0.0643776824 in.^{3}

    $$\dfrac{10{,}368{,}000}{0.0643776824} \approx \$161{,}049{,}600/\text{classroom}$$

    The national debt is $5,200,000,000,000 so

    $$\dfrac{5.2 \times 10^{12}}{161{,}049{,}600} \approx 32{,}288 \text{ classrooms}$$

    This number of classrooms is still hard to comprehend, so we will fill them with $100 bills instead. Now we can estimate it will take 322 classrooms. How many classrooms are in your city? Imagine them *filled* with $100 bills.

19. Answers vary; 20 is unhappy because $2^2 + 0^2 = 4$, which is unhappy. 100 is happy because $1^2 + 0^2 + 0^2 = 1$, which is happy.

20. By patterns: $7^1 = 7$; $7^2 = 49$ ends in 9; $7^3 = 343$ ends in 3; $7^4 = 2{,}401$ ends in 1; 7^5 ends in 7; 7^6 ends in 9; 7^7 ends in 3; 7^8 ends in a 1, ... Looking ahead, we see that $7^{1{,}000}$ must end in 1.

CHAPTER 2 THE NATURE OF LOGIC

2.1 Deductive Reasoning, page 84

7. a, b, and d are statements. 9. a and b are statements.
11. Some dogs do not have fleas.
13. All people pay taxes. 15. Some triangles are squares.
17. Some counting numbers are not divisible by 1.
19. All integers are odd.
21. **a.** T **b.** F **c.** T **d.** F
23. **a.** Prices or taxes will rise. **b.** Prices will not rise and taxes will rise. **c.** Prices will rise or taxes will fall. **d.** Prices will not rise or taxes will not rise.
25. **a.** F **b.** T **c.** T **d.** T
27. *Answers may vary.* **a.** Paul is peculiar and likes to read math texts. **b.** Paul is not peculiar and likes to read math texts. **c.** Paul is peculiar or does not like to read math texts. **d.** Paul is not peculiar or does not like to read math texts.
29. **a.** T **b.** F **c.** T **d.** F 31. **a.** T **b.** F
33. **a.** F **b.** T 35. **a.** T **b.** T 37. **a.** T **b.** F
39. deductive reasoning; answers vary
41. If p is true, then $\sim(\sim p)$ is $\sim(\sim T)$ or $\sim F$, which is T; if p is false, then $\sim(\sim p)$ is $\sim(\sim F)$ or $\sim T$, which is F. In either case, p has the same truth value as $\sim(\sim p)$.
43. $\sim s \wedge \sim a$ where s: Sam will seek the nomination;
    a: Sam will accept the nomination.
45. $\ell \wedge \sim e$ where ℓ: Fat Albert lives to eat;
    e: Fat Albert eats to live.
47. $a \vee p$ where a: The successful applicant for the job has a B.A. in liberal arts;
    p: The successful applicant for the job has a B.A. in psychology.
49. Let p: Dinner includes soup; d: Dinner includes salad; v: Dinner includes the vegetable of the day.
    a. $(p \wedge d) \vee v$ **b.** $p \wedge (d \vee v)$
51. F 53. F 55. T 57. T
59. Melissa did change her mind.

2.2 Truth Tables and the Conditional, page 94

5.
p	q	$\sim p$	$\sim p \vee q$
T	T	F	T
T	F	F	F
F	T	T	T
F	F	T	T

7.
p	q	$p \wedge q$	$\sim(p \wedge q)$
T	T	T	F
T	F	F	T
F	T	F	T
F	F	F	T

9.
r	$\sim r$	$\sim(\sim r)$
T	F	T
F	T	F

11.

p	q	~q	p ∧ ~q
T	T	F	F
T	F	T	T
F	T	F	F
F	F	T	F

13.

p	q	~p	~p ∧ q	~q	(~p ∧ q) ∨ ~q
T	T	F	F	F	F
T	F	F	F	T	T
F	T	T	T	F	T
F	F	T	F	T	T

15.

p	q	p → q	p ∨ (p → q)
T	T	T	T
T	F	F	T
F	T	T	T
F	F	T	T

17.

p	q	~q	p → ~q	~p	q → ~p	(p → ~q) → (q → ~p)
T	T	F	F	F	F	T
T	F	T	T	F	T	T
F	T	F	T	T	T	T
F	F	T	T	T	T	T

19.

p	q	p ∨ q	p ∧ (p ∨ q)	[p ∧ (p ∨ q)] → p
T	T	T	T	T
T	F	T	T	T
F	T	T	F	T
F	F	F	F	T

21.

p	q	r	p ∨ q	(p ∨ q) ∨ r
T	T	T	T	T
T	T	F	T	T
T	F	T	T	T
T	F	F	T	T
F	T	T	T	T
F	T	F	T	T
F	F	T	F	T
F	F	F	F	F

23.

p	q	r	p ∨ q	~r	(p ∨ q) ∧ (~r)	[(p ∨ q) ∧ ~ r] ∧ r
T	T	T	T	F	F	F
T	T	F	T	T	T	F
T	F	T	T	F	F	F
T	F	F	T	T	T	F
F	T	T	T	F	F	F
F	T	F	T	T	T	F
F	F	T	F	F	F	F
F	F	F	F	T	F	F

25. Statement: ~p → ~q.   Converse: ~q → ~p.
Inverse: p → q.   Contrapositive: q → p.

27. Statement: ~t → ~s.   Converse: ~s → ~t.
Inverse: t → s.   Contrapositive: s → t.

29. Statement: If I get paid, then I will go Saturday.
Converse: If I go Saturday, then I will get paid.
Inverse: If I do not get paid, then I will not go Saturday.
Contrapositive: If I do not go Saturday, then I do not get paid.

31. If it is a triangle, then it is a polygon.

33. If you are a good person, then you will go to heaven.

35. If we make a proper use of those means which the God of Nature has placed in our power, then we are not weak.

37. If it is work, then it is noble.   39. F → F is true.

41. F → F is true.

43. a. a → (e ∨ f)   b. (a ∧ e) → q   c. (a ∧ f) → q

45. [(m ∨ t ∨ w ∨ h) ∧ s] → q   47. t → (m ∧ s ∧ p)

49. Assume d, c, and b are true, and w is false:
(~T ∧ T) → (F ∧ T) is true.

51. (q ∧ ~d) → n where q: The qualifying person is a child; d: This child is your dependent; n: You enter your child's name.

53. (b ∨ c) → n where b: The amount on line 32 is $86,025; c: The amount on line 32 is less than $86,025; n: You multiply the number of exemptions by $2,500.

55. (m ∨ d) → s where m: You are a student; d: You are a disabled person; s: You see line 6 of instructions.

57. p is T and q is T.   a. T   b. F   c. F

59. p is F and q is T.   a. F   b. T   c. T

61. This is not a statement, since it gives rise to a paradox and is neither true nor false.

2.3 Operators and Laws of Logic, page 102

5. yes   7. no   9. no   11. no

13.

p	q	p ∨ q	p ∧ q	~(p ∧ q)	(p ∧ q) ∧ ~ (p ∧ q)
T	T	T	T	F	F
T	F	T	F	T	T
F	T	T	F	T	T
F	F	F	F	T	F

15.

p	q	~q	~q → p
T	T	F	T
T	F	T	T
F	T	F	T
F	F	T	F

17. Let s: Smoking is good for your health; d: Drinking is good for your health. Neither s nor d: ~(s ∨ d).

19. Let ℓ: I obtain the loan, i: I have an income of $85,000 per year. If ℓ, then i: ℓ → i.

21. Let g: I can go with you; e: I have a previous engagement. Not g because e: (~g ∧ e) ∧ (e → ~g).

23. Let i: I will invest my money in stocks; s: I will put my money in a savings account. Either i or s: $(i \vee s) \wedge \sim(i \wedge s)$.

25. Let h: One has once heartily laughed; w: One has once wholly laughed; b: One is altogether irreclaimably bad. No $(h \wedge w)$ is b: $(h \wedge w) \to \sim b$.

27. I did not go or I paid \$100.

29. If we visit New York, then we will visit the Statue of Liberty.

31. If the sun is not shining, then I will not go to the park.

33.

p	q	$\sim p$	$\sim q$	$p \vee q$	$\sim p \wedge \sim q$	$\sim(p \vee q)$
T	T	F	F	T	F	F
T	F	F	T	T	F	F
F	T	T	F	T	F	F
F	F	T	T	F	T	T

same

35.

p	q	$p \to q$	$\sim q$	$\sim p$	$\sim q \to \sim p$	$(p \to q) \leftrightarrow (\sim q \to \sim p)$
T	T	T	F	F	T	T
T	F	F	T	F	F	T
F	T	T	F	T	T	T
F	F	T	T	T	T	T

Thus, $(p \to q) \Leftrightarrow (\sim q \to \sim p)$.

37.

p	q	$p \to q$	$\sim(p \to q)$	$\sim q$	$p \wedge \sim q$	$\sim(p \to q) \leftrightarrow (p \wedge \sim q)$
T	T	T	F	F	F	T
T	F	F	T	T	T	T
F	T	T	F	F	F	T
F	F	T	F	T	F	T

Thus, $\sim(p \to q) \Leftrightarrow (p \wedge \sim q)$.

39. $p \wedge \sim q$ 41. $\sim p \wedge \sim q$

43. John did not go to Macy's and he did not go to Sears.

45. Tim is here or he is at home.

47. I can't go with you, and I will not go with Bill.

49. $x + 2 = 5$ and $x \neq 3$.   51. $x = -5$ and $x^2 \neq 25$.

53. Let s: The person is single (assume $\bar{s}$: The person is married); a: The person has assets of at least \$50,000; c: The person has a gross income of \$72,000; m: The person is married; d: The person has a combined income of \$100,000; q: The person qualifies for a loan of \$200,000. Symbolic statement: $[(s \wedge a \wedge c) \vee (\sim s \wedge a \wedge d)] \to q$: Liz does not qualify.

55. Let a: An alteration is made to the building; r: The building is redecorated; t: Tacks are inserted onto the building; n: Nails are hammered into the building; p: Permission is obtained. Symbolic statement: $\sim(a \vee r \vee t \vee m)$ unless p. This means: $\sim p \to \sim(a \vee r \vee t \vee m)$.

57. Every person would say that he or she is a member of the Veracious party. Thus the second person's assertion was true, and the third person was lying. Hence the third was a member of the Deceit party.

59. Question: Is it true that number 1 is the hot seat if and only if you are telling the truth? If the answer is yes, then number 1 is the hot seat. If the answer is no, then number 2 is the hot seat.

2.4   The Nature of Proof, page 112

7. We show this is a fallacy by constructing a truth table:

p	q	r	$p \to q$	$p \to r$	$q \to r$	$(p \to q) \wedge (p \to r)$	$[(p \to q) \wedge (p \to r)] \to (q \to r)$
T	T	T	T	T	T	T	T
T	T	F	T	F	F	F	T
T	F	T	F	T	T	F	T
T	F	F	F	F	T	F	T
F	T	T	T	T	T	T	T
F	T	F	T	T	F	T	F
F	F	T	T	T	T	T	T
F	F	F	T	T	T	T	T

Not all Ts

Since the result does not show all Ts, it is a fallacy. This is the false chain pattern.

9.

Problem-solving language	*Detective language*
Understand the problem. What is the unknown? *Devise a plan.* Do you know a related problem? Can you simplify the problem?	*Understand the case.* What are you looking for? *Investigate the case.* Have you solved a similar case? What are the facts?
Carry out the plan. What information is important? What information is not important? What pieces of information fit together logically? Which information is consistent with the given information?	*Analyze the facts/data.* What information is important? What information is not important? What pieces of information do not seem to fit together logically? Which data are inconsistent with the given information?
Look back. Examine the solution obtained. Does it make sense?	*Reexamine the facts.* Do the facts support the solution? Can we obtain a conviction?

11. invalid; fallacy of the converse

13. invalid; fallacy of the inverse

15. valid; by law of the excluded middle

17. valid; direct

19. valid; direct

21. valid; transitive
23. valid; direct
25. valid; direct
27. invalid; false chain
29. valid; indirect
31. valid; indirect
33. invalid; fallacy of the converse
35. valid; transitive
37. valid; transitive
39. valid; indirect
41. valid; contrapositive and transitive (twice)
43. I become lazy. (direct)
45. If you climb the highest mountain, then you are happy. (transitive)
47. $b = 0$ (excluded middle)
49. We do not interfere with the publication of false information. (indirect)
51. I will not eat that piece of pie. (indirect)
53. We will go to Europe.
55. If it is a baby, then it cannot manage a crocodile. Or, babies cannot manage crocodiles.
57. None of my poultry are officers. Here is another possibility: My poultry are not officers.
59. Airsecond Aircraft Company suffers financial setbacks.
61. The janitor could not have taken the elevator because the building fuses were blown.
63. No kitten with green eyes plays with a gorilla.
65. No dream of mine ever fails to come true.

2.5   Problem Solving Using Logic, page 122

1. Moe sees two hands and two black hats and concludes his hat must be black. If Moe's hat were white, then Harry and Larry would each have a solution because they would each see one white hat and one black hat with two hands raised. Thus Moe knows that his hat must be black.
3. Gary was 17 and had 15 marbles; Harry was 10 and had 12 marbles; Iggy was 3 and had 12 marbles; Jack was 18 and had 9 marbles. They shot in the order of Gary, Jack, Harry, and then Iggy.
5. There can be only one yellow flower, so there must be 49 red flowers.
7. Curly committed the murder.
9. Only one question is necessary: Ask Connie (who falsely claimed to have the mixed bag) to pull out one fruit. Suppose she pulls out a peach; this means she has the bag containing two peaches. Then Alice, who falsely claimed two peaches, must have two plums. This leaves Betty with the mixed bag. Suppose she pulls out a plum; this means she has the bag containing two plums. Then Betty, who falsely claimed two plums, must have two peaches. This leaves Alice with the mixed bag.
11. Pitcher: Jones; catcher: Smith; first base: Brown; second base: White; third base: Adams; shortstop: Miller; left field: Green; center field: Hunter; right field: Knight

2.6   Logic Circuits, page 127

5.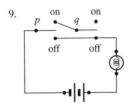

7.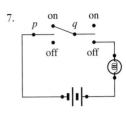

9.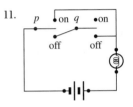

11.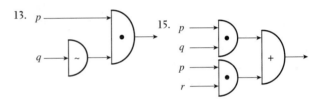

13.   15.

17.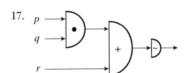

19.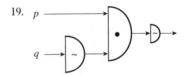

21. Notice $\sim p \rightarrow q \Leftrightarrow p \vee q$

23. $\sim q \rightarrow \sim p \Leftrightarrow \sim p \vee q$

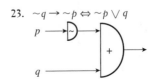

25. Let the committee members be a, b, and c, respectively. Light *on* represents a majority. The circuit is shown at the right.

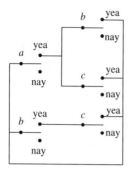

Chapter 2 Review Questions, page 131

1. A definition is accepted without proof, but a theorem requires proof.

2.

p	q	$\sim p$	$p \wedge q$	$p \vee q$	$p \rightarrow q$	$p \leftrightarrow q$
T	T	F	T	T	T	T
T	F	F	F	T	F	F
F	T	T	F	T	T	F
F	F	T	F	F	T	T

3.

p	q	$p \wedge q$	$\sim(p \wedge q)$
T	T	T	F
T	F	F	T
F	T	F	T
F	F	F	T

4.

p	q	$\sim q$	$p \vee \sim q$	$\sim p$	$(p \vee \sim q)$ $\wedge \sim p$	$[(p \vee \sim q) \wedge \sim p]$ $\rightarrow \sim q$
T	T	F	T	F	F	T
T	F	T	T	F	F	T
F	T	F	F	T	F	T
F	F	T	T	T	T	T

5.

p	q	r	$p \wedge q$	$(p \wedge q) \wedge r$	$[(p \wedge q) \wedge r] \rightarrow p$
T	T	T	T	T	T
T	T	F	T	F	T
T	F	T	F	F	T
T	F	F	F	F	T
F	T	T	F	F	T
F	T	F	F	F	T
F	F	T	F	F	T
F	F	F	F	F	T

6.

p	q	$\sim p$	$\sim q$	$p \wedge q$
T	T	F	F	T
T	F	F	T	F
F	T	T	F	F
F	F	T	T	F

(continued)

$\sim(p \wedge q)$	$\sim p \vee \sim q$	$\sim(p \wedge q) \leftrightarrow (\sim p \vee \sim q)$
F	F	T
T	T	T
T	T	T
T	T	T

7.

$p \rightarrow q$
p
$\therefore q$

p	q	$p \rightarrow q$	$(p \rightarrow q) \wedge p$	$[(p \rightarrow q) \wedge p] \rightarrow q$
T	T	T	T	T
T	F	F	F	T
F	T	T	F	T
F	F	T	F	T

8. Answers vary. If you eat too much, then you will be fat. You are not fat. Therefore, you do not eat too much.

9. Answers vary; fallacy of the converse, fallacy of the inverse, or false chain pattern

10. Yes; it is indirect reasoning.

11. **a.** Some birds do not have feathers. **b.** No apples are rotten. **c.** Some cars have two wheels. **d.** All smart people attend college. **e.** You go on Tuesday and you can win the lottery.

12. **a.** T $(F \rightarrow F)$ **b.** T $(F \rightarrow T)$ **c.** T $(T \rightarrow T)$ **d.** T $(F \rightarrow F)$ **e.** T $(F \rightarrow F)$

13. **a.** If P is a prime number, then $P + 2$ is a prime number. **b.** Either P or $P + 2$ is a prime number.

14. **a.** Let p: This machine is a computer; q: This machine is capable of self-direction. $p \rightarrow \sim q$ **b.** Contrapositive: $\sim(\sim q) \rightarrow \sim p$; $q \rightarrow \sim p$. If this machine is capable of self-direction, then it is not a computer.

15. Let p: There are a finite number of primes. q: There is some natural number, greater than 1, that is not divisible by any prime. Then the argument in symbolic form is:

$p \rightarrow q$
$\sim q$
$\therefore \sim p$

Conclusion: There are infinitely many primes (assuming that "not a finite number" is the same as "infinitely many").

16. **a.**

b.

17. Let d: I attend to my duties; r: I am rewarded; ℓ: I am lazy. Symbolic argument:

(1) $d \rightarrow r$ (2) $\ell \rightarrow \sim r$ (1) $d \rightarrow r$
(2) $\ell \rightarrow \sim r$ (3) ℓ (4) $\sim r$

(3) ℓ (4) $\therefore \sim r$ $\therefore \sim d$
           Direct reasoning   Indirect reasoning

Conclusion: I do not attend to my duties.

18. Let o: This is organic food; h: This is healthy food; s: This is an artificial sweetener; p: This is a prune. Symbolic argument:

(1) $o \rightarrow h$ (3) $p \rightarrow o$
(2) $s \rightarrow \sim h$ (1) $o \rightarrow h$

(3) $p \rightarrow o$ (4) $p \rightarrow h$ Transitive
            (2') $h \rightarrow \sim s$ Law of contraposition

           $p \rightarrow \sim s$ Transitive

Conclusion: No prune is an artificial sweetener.

19. Let s: This is a square; r: This is a rectangle; q: This is a quadrilateral; p: This is a polygon. Symbolic argument:

(1) $s \rightarrow r$
(2) $r \rightarrow q$

$\therefore s \rightarrow q$ Transitive
(3) $q \rightarrow p$

$\therefore s \rightarrow p$ Transitive
Conclusion: All squares are polygons.

20.

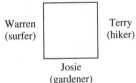

Maureen (baker)

Warren (surfer)   Terry (hiker)

Josie (gardener)

CHAPTER 3 THE NATURE OF CALCULATION

3.1   Early Numeration Systems, page 143

7. **a.** Egyptian **b.** Roman, Babylonian
   c. Egyptian, Roman, Babylonian
   d. Egyptian, Roman, Babylonian
   e. Roman, Babylonian **f.** Roman

9. $\frac{1}{200}$ **11.** 100,010 **13.** $\frac{1}{12}$ **15.** $\frac{1}{100}$ **17.** 1,000,001

19. 1,997 **21.** 261 **23.** 671 **25.** 28 **27.** 25

29. 709 **31.** 2,001 **33.** 400,000 **35.** 9,712

37. One; the only one known to be going to St. Ives is myself.

39. **a.** ∩∩∩∩∩∩∩ ||||| **b.** LXXV
   c. ▼ ◁ ▼▼▼▼▼

41. **a.** 99999∩∩| **b.** DXXI
   c. ▼▼▼▼▼▼▼▼◁◁◁◁ ▼

43. **a.** 𐤀999999999∩∩∩∩∩∩∩∩∩ ||||||||||
   b. MCMXCVIII **c.** ◁◁◁▼▼▼◁▼▼▼▼▼▼▼

45. 133

47. 9∩∩∩∩∩∩∩∩ ||||||

49. ∩∩∩ |||||| || **51.** ▼▼▼▼▼▼▼▼

53. ◁◁▼▼▼▼▼   **55.** ◁◁◁▼▼▼▼▼▼▼▼▼

57. The dimensions are 20 by 30.

3.2   Hindu–Arabic Numeration System, page 149

1.

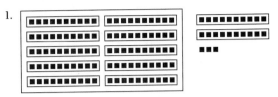

3. Let $\boxed{\text{X}}$ represent ■■■■■■■■■ of the larger groups.

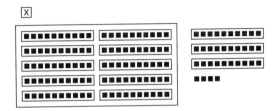

5. $b^n = b \cdot b \cdot \cdots \cdot b$, where there are n factors of b (for any counting number n)

7. 5 units **9.** 5 thousandths **11.a.** 100,000 **b.** 1,000

13. **a.** 0.0001 **b.** 0.001 **15. a.** 5,000 **b.** 500

17. **a.** 0.06 **b.** 0.00009 **19.** 10,234 **21.** 521,658

23. 7,000,000.03   500,457.34 **27.** 20,600.40769

29. 7×10^5 ¬ $+ 8 \times 10^3 + 4 \times 10^2 + 7 \times 10^0$

31. $2 \times 10^1 +$ $- 5 \times 10^{-1} + 7 \times 10^{-2} + 2 \times 10^{-3}$

33. $5 \times 10^2 +$ 1×10^0

35. $2 \times 10^6 +$ $5 \times 10^3 + 6 \times 10^2 + 8 \times 10^1 + 1 \times 10^0$

37. $5 \times 10^3 + ?$ $4 \times 10^1 + 5 \times 10^0 + 5 \times 10^{-1}$

39. $1 \times 10^5 + 1$ **41.** $8 \times 10^0 + 5 \times 10^{-5}$

43. $5 \times 10^4 + 7$ $\times 10^2 + 8 \times 10^1 + 5 \times 10^0$
   $+ 9 \times 10^{-1} +$ $+ 6 \times 10^{-3} + 1 \times 10^{-4}$

45. 3,201 **47.**   **49.** 8,009,026

51.   53.

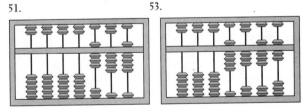

55.   57.

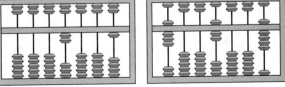

59. 22, 23, 24, 30, 31, 32, 33, 34, 40, ...

3.3 Different Numeration Systems, page 156

7. a. 13 b. 23_{five} c. $10_{thirteen}$ d. 15_{eight} e. 1101_{two}
   f. 11_{twelve}
9. a. $6 \times 8^2 + 4 \times 8^1 + 3 \times 8^0$
   b. $5 \times 12^3 + 3 \times 12^2 + 8 \times 12^1 + 7 \times 12^0 + 9 \times 12^{-1}$
11. a. $6 \times 8^7 + 4 \times 8^6 + 2 \times 8^5 + 5 \times 8^1 + 1 \times 8^0$
    b. $1 \times 3^3 + 2 \times 3^1 + 1 \times 3^0 + 2 \times 3^{-1} + 2 \times 3^{-2}$
    $+ 1 \times 3^{-3}$
13. a. $3 \times 5^0 + 4 \times 5^{-1} + 2 \times 5^{-3} + 3 \times 5^{-4} + 1 \times 5^{-5}$
    b. $2 \times 4^3 + 3 \times 4^1 + 3 \times 4^0 + 1 \times 4^{-1}$
15. 751   17. 4,307   19. 53   21. 1,862   23. 2,250
25. 582   27. 2,001   29. 21310_{four} 31. $2E7_{twelve}$
33. 1147_{eight} 35. 1001100111_{two} 37. 2214_{seven}
39. $28T3_{twelve}$ 41. 1030_{four} 43. 7 weeks, 3 days
45. 4 ft, 7 in.   47. 3 gross, 5 doz, 8 units
49. $84 = 314_{five}$ so you would need 8 coins
51. $954_{twelve} = 1,360$ 53. $44 = 62_{seven}$; 6 weeks and 2 days
55. $29 = 15_{twenty-four}$; 1 day and 5 hours

3.4 History of Calculating Devices, page 167

9. a. 45 b. 72 c. 315 d. 423 e. 612

Answers to Problems 11–21 vary.

11. yes; speed, complicated computations   13. yes; repetition
15. yes; repetition   17. yes; ability to make corrections easily
19. yes (but not completely); it can help with some of the technical
    aspects
21. yes; speed, complicated computations, repetition   23. E
25. D

3.5 Computers and the Binary Numeration System, page 175

1. The CPU is the microprocessor the computer uses; for example,
   INTEL Pentium Pro.
3. ROM is read-only memory, and RAM is random-access
   memory.
5. 39   7. 167   9. 13   11. 11   13. 29   15. 27
17. 99   19. 184   21. 1101_{two} 23. 100011_{two}
25. 110011_{two} 27. 1000000_{two} 29. 10000000_{two}
31. 1100011011_{two} 33. 68 79   35. 69 78 68   37. HAVE
39. STUDY   41. 101_{two} 43. 1101_{two} 45. 101_{two}
47. Let p: There is peace; q: There is war. Consider the truth table:

p	$\sim p$	$p \vee \sim p$
T	F	T
F	T	T
		↑

tautology, so the computer must be correct

49. a. 101_{two} b. 110_{two}
51. a. $111\,000\,100\,101_{two}$ b. $11\,000\,110\,010_{two}$
53. a. 5_{eight} b. 4_{eight} 55. 007750_{eight} 57. 773521_{eight}

59. The trick works because card 1 represents each number from 1
    to 31 (inclusive) having a one in the units column of its binary
    representation; card 2, each number having a one in the twos
    column of its binary representation; cards 3, 4, and 5 similarly.
    The final result leaves only one number showing through the
    remaining hole; this number is the chosen number.

Chapter 3 Review Questions, page 180

1. Answers vary. The position in which the individual digits are
   listed is relevant; examples will vary.
2. Answers vary. Addition is easier in a simple grouping system;
   examples will vary.
3. Answers vary. It uses ten symbols; it is positional; it has a place-
   holder symbol (0); and it uses 10 as its basic unit for grouping.
4. Answers vary. Should include finger calculating, Napier's rods,
   Pascal's calculator, Leibniz' reckoning machine, Babbage's
   difference and analytic engines, ENIAC, UNIVAC, Atanasoff's
   and Eckert and Mauchly's computers, and the dispute they had
   in proving their position in the history of computers. Should
   also include the role and impact of the Apple and Macintosh
   computers, as well as the supercomputers (such as the Cray).
5. Answers vary. Should include illegal (breaking into another's
   computer; adding, modifying, or destroying information;
   copying programs without authorization or permission) and
   ignorance (assuming that output information is correct, or not
   using software for purposes for which it was intended).
6. Answers vary.   a. the physical components (mechanical,
   magnetic, electronic) of a computer system   b. the routine
   programs and associated documentation in a computer system
   c. the process of creating, modifying, deleting, and formatting
   text and materials   d. a device connected to a computer that
   allows the computer to communicate with other computers
   using electronic cables or phone lines   e. electronic mail —
   that is, messages sent along computer modems   f. Random-
   Access Memory or memory where each location is uniformly
   accessible, often used for the storage of a program and data
   being processed   g. an electronic place to exchange
   information with others, usually on a particular topic
7. 10^9
8. $4 \times 10^2 + 3 \times 10^1 + 6 \times 10^0 + 2 \times 10^{-1} + 1 \times 10^{-5}$
9. $5 \times 8^2 + 2 \times 8^1 + 3$
10. $1 \times 2^6 + 0 \times 2^5 + 0 \times 2^4 + 1 \times 2^3 + 1 \times 2^2 + 1 \times 2^1$
    $+ 0 \times 2^0$
11. 4,020,005.62
12. $1 \times 2^4 + 1 \times 2^3 + 1 \times 2^2 + 0 \times 2^1 + 1 \times 2^0 = 29$
13. $1 \times 2^6 + 1 \times 2^5 + 1 \times 2^4 + 1 \times 2^3 + 0 \times 2^2$
    $+ 1 \times 2^1 + 1 \times 2^0 = 123$
14. $1 \times 3^2 + 2 \times 3^1 + 2 \times 3^0 = 17$
15. $8 \times 12^2 + 2 \times 12^1 + 1 \times 12^0 = 1,177$
16.
$$\begin{array}{l} 0 \ \text{r. 1} \\ 2\overline{)1} \ \text{r. 1} \\ 2\overline{)3} \ \text{r. 0} \\ 2\overline{)6} \ \text{r. 0} \\ 2\overline{)12} \qquad 12 = 1100_{two} \end{array}$$

17.
```
      0   r. 1
   2) 1   r. 1
   2) 3   r. 0
   2) 6   r. 1
   2)13   r. 0
   2)26   r. 0
   2)52
       52 = 110100_two
```

18.
```
       0    r. 1
   2)  1    r. 1
   2)  3    r. 1
   2)  7    r. 1
   2)  15   r. 1
   2)  31   r. 0
   2)  62   r. 0
   2)  124  r. 1
   2)  249  r. 1
   2)  499  r. 0
   2)  998  r. 1
   2)1997
       1,997 = 11111001101_two
```

19. $11110100001001000000_{two}$

20. **a.**
```
        0    r. 9
   12)  9    r. 2
   12)  110  r. E
   12)1,331      1,331 = 92E_twelve
```
   b.
```
       0   r. 4
   5)  4   r. 0
   5)  20  r. 0
   5)100        100 = 400_five
```

CHAPTER 4 THE NATURE OF NUMBERS

4.1   Natural Numbers, page 190

1. $\mathbb{N} = \{1, 2, 3, 4, 5, \ldots\}$
3. Subtraction is defined in terms of addition: $m - n = x$ means $m = n + x$.
5. For an operation $\circ$ and elements a and b in a set S, $a \circ b = b \circ a$.
7. For operations $\circ$ and $\otimes$ and elements a, b, and c in S, $a \otimes (b \circ c) = (a \otimes b) \circ (a \otimes c)$
9. **a.** $4 + 4 + 4$ **b.** $3 + 3 + 3 + 3$
11. **a.** $1,845 + 1,845$ **b.** $2 + 2 + \cdots + 2$ (a total of 1,845 terms)
13. **a.** $b + b + b + \cdots + b$ (a total of a terms)
    b. $a + a + a + \cdots + a$ (a total of b terms)
15. commutative   17. associative   19. commutative
21. commutative   23. commutative   25. commutative
27. None are associative.
29. **a.** 4   **b.** 4   **c.** 1   **d.** 7   **e.** 4
31. **a.** yes   **b.** yes   **c.** yes
33. **a.** no; $b \star c \neq c \star b$ **b.** yes; $b \star (b \star c) = (b \star b) \star c$
35. **a.** $\square$ **b.** $\circ$ **c.** yes; no, $\circ \bullet \triangle \neq \triangle \bullet \circ$ **d.** yes
37. **a.** 492   **b.** 328   **c.** 343   **d.** 495   **e.** 352
39. yes; $X \cap Y = Y \cap X$; use Venn diagrams
41. yes; even + even = even   43. yes; odd × odd = odd

45. It is not commutative or associative: $2 \oplus 3 = 7$ but $3 \oplus 2 = 8$; also $(2 \oplus 3) \oplus 4 \neq 2 \oplus (3 \oplus 4)$.

47.
$\star$	ℓ	r	a	f
ℓ	a	f	r	ℓ
r	f	a	ℓ	r
a	r	ℓ	f	a
f	ℓ	r	a	f

49. Answers vary; $0 + 1 + 2 - 3 - 4 + 5 - 6 + 7 + 8 - 9 = 1$

4.2   Prime Numbers, page 205

9. **a.** not prime; $3^2 \cdot 7$ **b.** prime; use sieve
    c. prime; use sieve   **d.** prime; check primes under 45
11. **a.** not prime; $7 \cdot 13$ **b.** not prime; $3 \cdot 29$
    c. not prime; $3 \cdot 37$ **d.** not prime; $3 \cdot 23 \cdot 29$
13. **a.** T   **b.** F   **c.** F   **d.** F
15. **a.** F   **b.** T   **c.** F   **d.** F
17. **a.** 13   **b.** 19   **c.** 31   **d.** 997
19. **a.** $2^2 \cdot 7$ **b.** $2^2 \cdot 19$ **c.** $5 \cdot 43$ **d.** 5^3
21. **a.** $2^3 \cdot 3 \cdot 5$ **b.** $2 \cdot 3^2 \cdot 5$ **c.** $3 \cdot 5^2$ **d.** $3 \cdot 5^2 \cdot 13$
23. prime   25. prime   27. $13 \cdot 29$ 29. $3 \cdot 5 \cdot 7$
31. prime   33. $3^2 \cdot 5 \cdot 7$ 35. $3^4 \cdot 7$ 37. $19 \cdot 151$
39. 12; 360   41. 3; 2,052   43. 4; 180   45. 1; 252
47. 15; 1,800
49. The least common multiple of 75 and 90 is 450; 2:30 A.M. (450 minutes later)
51. Answers vary; all primes except 2 and 3 are one more or one less than a multiple of 6.
53. Answers vary; see answer to Problem 16 for possible pairs.
55. Answers vary; $10 = 5 + 5$; $12 = 7 + 5$; $14 = 11 + 3$; $16 = 13 + 3$; $18 = 13 + 5$; $20 = 17 + 3$; $40 = 37 + 3$; $80 = 73 + 7$; $100 = 97 + 3$
57. Suppose 23 is the largest prime. Consider
    $M = 2 \cdot 3 \cdot 5 \cdot 7 \cdot 11 \cdot 13 \cdot 17 \cdot 19 \cdot 23 + 1$
    which is either prime or composite. If M is prime, then 23 is not the largest prime. If M is composite, then it has a prime divisor. Try all prime divisors: 2, 3, 5, 7, 11, 13, 17, 19, and 23. None of these divide into M, so M must have a prime divisor that is larger than 23. In either case, 23 is not the largest prime.
59. $333,333,331 = 17 \cdot 19,607,843$ (See *International Mathematics Magazine*, February 1962, for a method of solution that does not involve calculators; otherwise, use a calculator to divide by consecutive primes until you reach 17.)
61. Conjecture: None are prime.
63. Answers vary; some possibilities are: 2, 5, 17, 37, 101, 197, 257, 401, 577, 677, 1297, 1601, 2917, 3137, 4357, 5477, 7057, 8101, 8837. A computer could also be used to answer this question.
65. Answers vary; if $n = 1$, $\sqrt{1 + 24(1)} = 5$, a prime; if $n = 2$, $\sqrt{1 + 24(2)} = 7$, a prime; if $n = 3$, $\sqrt{1 + 24(3)} = \sqrt{73}$, not a prime. If you read the problem as "all primes are contained in the list," then the formula is also a fraud because "n" is also such a formula.
67. The last digit is 5.

4.3   Integers, page 217

1. If $x = 0$, then $x + 0 = 0 + x = x$. To add nonzero integers x and y, look at the signs of x and y:

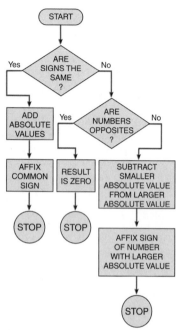

3. If $x = 0$, then $x \cdot 0 = 0 \cdot x = 0$. To multiply nonzero integers x and y, look at the signs of x and y:

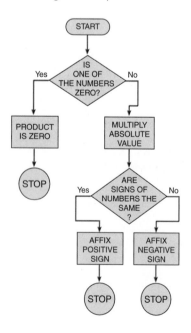

5. $0 \div 5$ means $\frac{0}{5}$, which is equal to 0; on the other hand, $5 \div 0$ is not defined.

7. **a.** You moved to the *right* 7 units.   **b.** 6 (or +6)

9. **a.** 8   **b.** -2 11. **a.** 10   **b.** -4

13. **a.** -7 **b.** 12   15. **a.** 3   **b.** -3

17. **a.** -18 **b.** -20 19. **a.** -70 **b.** 70

21. **a.** -3 **b.** 7   23. **a.** 132   **b.** -1

25. **a.** -56 **b.** -75 27. **a.** -4 **b.** 4

29. **a.** 150   **b.** -19 31. **a.** 0   **b.** 10

33. **a.** 11   **b.** 174   35. **a.** -35 **b.** -4

37. **a.** 0   **b.** -8 39. **a.** 1   **b.** 22

41. **a.** 15   **b.** -15 43. **a.** -26 **b.** 2

45. **a.** 8   **b.** -2 47. **a.** 2   **b.** 18

49. **a.** 11   **b.** -43 51. **a.** $-3{,}628{,}800$ **b.** -5

53. **a.** 4   **b.** -8 **c.** 16   **d.** -32 **e.** negative

55. **a.** empty intersection   **b.** no, since it does not include zero

57. **a.** An operation $\star$ is associative for a set S if $(a \star b) \star c = a \star (b \star c)$ for all elements a, b, and c in S.   **b.** Yes, addition is associative for the integers.   **c.** not associative for subtraction since $(2 - 3) - 5 \neq 2 - (3 - 5)$ **d.** Yes, multiplication is associative for the integers.   **e.** not associative for division since $20 \div (10 \div 5) \neq (20 \div 10) \div 5$

59. Answers vary; the set $\{-1, 0, 1\}$ is closed for multiplication.

61. Answers are not unique.
$$4 = 4 + (4 - 4) \div 4; \qquad 5 = (4 \cdot 4 + 4) \div 4;$$
$$6 = 4 + (4 + 4) \div 4; \qquad 7 = (44 \div 4) - 4;$$
$$8 = 4 + 4 + 4 - 4; \qquad 9 = 4 + 4 + (4 \div 4);$$
$$10 = (44 - 4) \div 4$$

4.4   Rational Numbers, page 229

7. **a.** $\frac{1}{5}$ **b.** $\frac{1}{4}$ 9. **a.** 2   **b.** 2   11. **a.** 3   **b.** $\frac{2}{3}$

13. **a.** $\frac{1}{8}$ **b.** $\frac{1}{3}$ 15. **a.** $\frac{6}{35}$ **b.** $\frac{3}{20}$ 17. **a.** $\frac{17}{21}$ **b.** $\frac{7}{8}$

19. **a.** $\frac{10}{3}$ **b.** $\frac{5}{2}$ **c.** $\frac{5}{6}$ 21. **a.** $\frac{1}{9}$ **b.** $\frac{71}{63}$ **c.** $\frac{1}{15}$

23. **a.** $\frac{2}{3}$ **b.** 21   **c.** $\frac{141}{26}$ 25. **a.** 1   **b.** -1 **c.** -1

27. **a.** $\frac{-1}{25}$ **b.** $\frac{18}{35}$ **c.** $\frac{4}{45}$

29. **a.** $\frac{4}{5}$ (*Note:* Use distributive property.)   **b.** $\frac{4}{5}$

31. **a.** $\frac{80}{27}$ **b.** $\frac{-15}{56}$ 33. **a.** 20   **b.** 6

35. **a.** $-\frac{5}{43}$ **b.** $\frac{28}{33}$

37. **a.** $\frac{2{,}137}{10{,}800}$ 39. $\frac{971}{3{,}060}$ 41. $\frac{10{,}573}{13{,}020}$

43.

Region	Number	Ratio	Adams	Jefferson	Webster
North	8,700	3.625	4	3	4
South	5,600	2.333	3	2	2
East	7,200	3	3	3	3
West	3,500	1.458	2	1	1
Total:	25,000		12	9	10

Region	Hamilton		
North	3.828;	3 + 1 =	4
South	2.464;	2 =	2
East	3.168;	3 =	3
West	1.54;	1 + 1 =	2
Total:			11

45.

Region	Number	Ratio	Adams	Jefferson	Webster
North	18,200	3.64	4	3	4
South	12,900	2.58	3	2	3
East	17,600	3.52	4	3	4
West	13,300	2.66	3	2	3
Total:	62,000		14	10	14

Region	Hamilton		
North	3.522; 3 + 1	=	4
South	2.497; 2	=	2
East	3.406; 3	=	3
West	2.574; 2 + 1	=	3
Total:			12

47. Given any two rationals $\frac{a}{b}$ and $\frac{c}{d}$; to show that $\frac{a}{b} - \frac{c}{d}$ is rational.

Now, $\frac{a}{b} - \frac{c}{d} = \frac{ad - bc}{bd}$ by the definition of subtraction. Also, ad and bc are integers and bd is a nonzero integer since a, c are integers and b, d are nonzero integers and the set of integers is closed for multiplication. Finally, $ad - bc$ is an integer because the integers are closed for subtraction. Therefore, $\frac{ad - bc}{bd}$ is a rational by the definition of a rational number.

49. The set of integers, $\mathbb{Z}$, is closed for addition, subtraction, and multiplication, but not for division since $4 \div 5$ is not an integer.

51. Yes; answers vary.

53. Answers vary; suppose a wise person gives an old nag horse to the estate so there are 18 horses. Then $\frac{1}{2}(18) = 9$ horses; $\frac{1}{3}(18) = 6$ horses; $\frac{1}{9}(18) = 2$ horses. Since the eldest picked 9 horses, and the middle son 6 horses, and the youngest 2 horses, they picked a total of 17 horses. Certainly, none picked the old nag horse, which was now returned to the wise person.

55. The average number for 1, 2, 3, 4, . . . , N is $\frac{N + 1}{2}$, so for $N = 17$, the average ranking is 9. The youngest one must receive 2 horses whose ranks add up to 18 ($9 \times 2 = 18$). The middle one must receive 6 horses whose ranks add up to 54 ($9 \times 6 = 54$), and the eldest must receive 9 horses whose ranks add up to 81 ($9 \times 9 = 81$). The following table shows one possible solution.

Eldest: 2, 3, 4, 7, 11, 12, 13, 14, 15   Rank: 81, average rank is 9

Middle   5, 6, 8, 9, 10, 16   Rank: 54, average rank is 9

Youngest: 1, 17   Rank: 18, average rank is 9

57. Answers vary; $\frac{47}{60} = \frac{1}{3} + \frac{1}{5} + \frac{1}{4}$ or $\frac{1}{2} + \frac{1}{4} + \frac{1}{30}$

59. Answers vary; $\frac{7}{17} = \frac{1}{3} + \frac{1}{15} + \frac{1}{85}$ or $\frac{1}{3} + \frac{1}{17} + \frac{1}{51}$

61. There is only one solution: $2 + 2 = 2^2$ and $2 - 2 = 0^2$.

4.5   Irrational Numbers, page 239

9. a. 14   **b.** 30   **c.** 36   **11. a.** 2.5   **b.** 2.4   **c.** 0.25

13. a. irrational; 3.162   **b.** irrational; 5.477

15. a. rational; 13   **b.** rational; 20

17. a. rational; 32   **b.** rational; 44

19. a. $10\sqrt{10}$ **b.** $20\sqrt{7}$ **c.** $8\sqrt{35}$ **d.** $21\sqrt{10}$

21. a. $\frac{1}{2}\sqrt{2}$ **b.** $\frac{1}{3}\sqrt{3}$ **c.** $\frac{1}{5}\sqrt{15}$ **d.** $\frac{1}{7}\sqrt{21}$

23. a. $\frac{1}{2}\sqrt{2}$ **b.** $-\frac{1}{3}\sqrt{3}$ **c.** $\frac{2}{5}\sqrt{5}$ **d.** $\frac{1}{2}\sqrt{10}$

25. a. $\sqrt{69}$ **b.** $2\sqrt{17}$ **27. a.** $2\sqrt{15}$ **b.** $2\sqrt{2}$

29. a. $\frac{4 - \sqrt{2}}{2}$ **b.** $\frac{3 - \sqrt{5}}{2}$

31. a. $1 - 3\sqrt{x}$ **b.** $-3 - \sqrt{x}$

33. a. $\frac{2x}{5y}\sqrt{y}$ **b.** $\frac{1}{4x}\sqrt{5xy}$

35. $\frac{2 - \sqrt{76}}{12} = \frac{2 - 2\sqrt{19}}{12} = \frac{1 - \sqrt{19}}{6}$

37. $\frac{12 + \sqrt{148}}{2} = \frac{12 + 2\sqrt{37}}{2} = 6 + \sqrt{37}$

39. 24 ft   **41.** 10 ft   **43.** $\sqrt{18}$ ft or $3\sqrt{2}$ ft

45. $\sqrt{325}$ or $5\sqrt{13}$; 18 ft; 55 ft   **47.** 15 ft

49. Answers vary; 1.2323323332. . .

51. Answers vary; 0.0919919991. . .

53. Solution comes from Pythagorean theorem.   **a.** 2-in. square   **b.** same   **c.** 7-in. square   **d.** 9-in. square   **e.** 15-in. square

55.

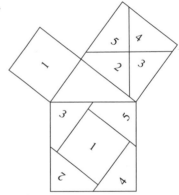

57. He can put it in a box along the diagonal since $3^2 + 4^2 = 5^2$.

59. Answers vary; any square number can be written as the sum of two consecutive triangular numbers.

61. Entries in the second diagonal: 1, 3, 6, 10, 15, ...

4.6   Real Numbers, page 253

5. Let $\mathbb{S}$ be any set, let $\circ$ be any operation, and let a, b, and c be any elements of $\mathbb{S}$. We say that $\mathbb{S}$ is a **group** for the operation of $\circ$ if the following properties are satisfied: 1. The set $\mathbb{S}$ is *closed* for $\circ$: $(a \circ b) \in \mathbb{S}$. 2. The set $\mathbb{S}$ is *associative* for $\circ$: $(a \circ b) \circ c = a \circ (b \circ c)$. 3. The set $\mathbb{S}$ satisfies the *identity* for $\circ$: There exists a number $I \in \mathbb{S}$ so that $x \circ I = I \circ x = x$ for *every* $x \in \mathbb{S}$. 4. The set $\mathbb{S}$ satisfies the *inverse* for $\circ$: For *each* $x \in \mathbb{S}$, there exists a corresponding $x^{-1} \in \mathbb{S}$ so that $x \circ x^{-1} = x^{-1} \circ x = I$, where I is the identity element in $\mathbb{S}$.

7. a. $\mathbb{N}, \mathbb{Z}, \mathbb{Q}, \mathbb{R}$ **b.** $\mathbb{Q}, \mathbb{R}$ **c.** $\mathbb{Q}', \mathbb{R}$ **d.** $\mathbb{Q}', \mathbb{R}$ **e.** $\mathbb{Q}, \mathbb{R}$ **f.** $\mathbb{Q}, \mathbb{R}$ **g.** $\mathbb{Q}, \mathbb{R}$ **h.** $\mathbb{N}, \mathbb{Z}, \mathbb{Q}, \mathbb{R}$

9. a. 1.5   b. 0.7   11. a. 0.4   b. 0.47
13. a. $0.8\overline{3}$ b. $0.\overline{285714}$ 15. a. 0.12   b. $2.1\overline{6}$
17. a. $0.\overline{6}$ b. $2.\overline{153846}$ 19. a. -0.8 b. $-0.\overline{6}$
21. a. $\frac{1}{2}$ b. $\frac{4}{5}$ 23. a. $\frac{2}{25}$ b. $\frac{3}{25}$ 25. a. $\frac{9}{20}$ b. $\frac{117}{500}$
27. a. $\frac{987}{10}$ b. $\frac{63}{100}$ 29. a. $\frac{153}{10}$ b. $\frac{139}{20}$ 31. commutative
33. commutative   35. associative   37. commutative
39. inverse
41. $\mathbb{N}$ is not a group for $+$ since there is no identity.
43. $\mathbb{W}$ is not a group for $+$ since the inverse property is not satisfied.
45. $\mathbb{Z}$ is a group for $+$.   47. $\mathbb{Q}$ is a group for $+$.
49. $\mathbb{Q}$ is a group for $\times$.
51. a.

$\times$	1	2	3	4
1	1	2	3	4
2	2	4	6	8
3	3	6	9	12
4	4	8	12	16

b.

*	1	2	3	4
1	2	2	2	2
2	4	4	4	4
3	6	6	6	6
4	8	8	8	8

c. Operation $\times$: Not closed; associative; identity is 1; no inverse property; commutative. Operation *: Not closed; not associative; no identity; no inverse property; not commutative
53. $a \circ b = a + b$ 55. $a \circ b = a - b$ 57. $a \circ b = ab + 1$
59. $a \circ b = 2a - b$
61. Answers vary. It will terminate if the only factors of the denominator are 2, 5, or powers of those numbers.

4.7   A Finite Algebra, page 264

5. a. 3   b. 10   7. a. 5   b. 10   9. a. 12   b. 8
11. a. 8   b. 6   13. a. 8   b. 1   15. a. T   b. F
17. a. T   b. T   19. a. T   b. T
21. a. 0, (mod 5)   b. 8, (mod 12)
23. a. 3, (mod 4)   b. 3, (mod 5)
25. a. 2, (mod 5)   b. 2, (mod 8)
27. a. 3, (mod 5)   b. 10, (mod 11)
29. a. 4, (mod 7)   b. 4, (mod 5)
31. a. 5, (mod 6)   b. 3, (mod 7)
33. a. 1, 3, (mod 4)   b. 2, (mod 9)
35. a. 0, (mod 4)   b. 2, (mod 4)   37. all x's, (mod 2)
39. a. 1, (mod 7); Sunday   b. 5, (mod 7); Thursday
    c. 4, (mod 7); Wednesday   d. 2, (mod 7); Monday
41. Let $x =$ the number of miles to aunt's house. Then the total mileage for the six round trips is $12x \equiv 8$, (mod 10). Solving, $x \equiv 4, 9$, (mod 10). The possible distances are 4, 14, 24, 34, . . . or 9, 19, 29, 39, ...
43. She lives 14 miles from your house.
45. Answers vary; it is a group for addition modulo 7.

47.

$+$	0	1	2	3	4	5	6	7	8	9	10
0	0	1	2	3	4	5	6	7	8	9	10
1	1	2	3	4	5	6	7	8	9	10	0
2	2	3	4	5	6	7	8	9	10	0	1
3	3	4	5	6	7	8	9	10	0	1	2
4	4	5	6	7	8	9	10	0	1	2	3
5	5	6	7	8	9	10	0	1	2	3	4
6	6	7	8	9	10	0	1	2	3	4	5
7	7	8	9	10	0	1	2	3	4	5	6
8	8	9	10	0	1	2	3	4	5	6	7
9	9	10	0	1	2	3	4	5	6	7	8
10	10	0	1	2	3	4	5	6	7	8	9

$\times$	0	1	2	3	4	5	6	7	8	9	10
0	0	0	0	0	0	0	0	0	0	0	0
1	0	1	2	3	4	5	6	7	8	9	10
2	0	2	4	6	8	10	1	3	5	7	9
3	0	3	6	9	1	4	7	10	2	5	8
4	0	4	8	1	5	9	2	6	10	3	7
5	0	5	10	4	9	3	8	2	7	1	6
6	0	6	1	7	2	8	3	9	4	10	5
7	0	7	3	10	6	2	9	5	1	8	4
8	0	8	5	2	10	7	4	1	9	6	3
9	0	9	7	5	3	1	10	8	6	4	2
10	0	10	9	8	7	6	5	4	3	2	1

49. yes   51. yes   53. 8   55. 1
57. 0, 1, 2, 3, 4, 5, 6, 7, 8, 9; the check digit is 10.

4.8   Cryptography, page 271

1. 14-5-22-5-18-29-19-1-25-29-14-5-22-5-18-28
3. 25-15-21-29-2-5-20-29-25-15-21-18-29-12-9-6-5-28
5. ARE WE HAVING FUN YET
7. FAILURE TEACHES SUCCESS
9. Divide by 8.
11. Multiply by 6.
13. Divide by 2, then subtract 2, then divide by 4.
15. MOHOY .CQ MOHOYZ
17. ZFREIOPEZFRLE WQOC
19. HUMPTY DUMPTY IS A FALL GUY.
21. MIDAS HAD A GILT COMPLEX.
23. D UPEOTEHEQAUXUJGEWAUOEHEQHGHEXAUNTDD-CRZEDN UULEHRQECEPCLLED UPEVUJEHERCRNUOXJGTAB
25. ANYONE WHO SLAPS CATSUP, MUSTARD, AND RELISH ON HIS HOT DOG IS TRULY A MAN FOR ALL SEASONINGS.
27. CRYSTAL-CLEAR AIR OF ROCKY MOUNTAINS PROVIDES IDEAL ENVIRONMENT FOR WEATHER STATION.
29. $3,915 + 15 + 4,826 = 8,756$

Chapter 4   Review Questions, page 276

1. $-4 + 5(-3) = -4 - 15 = -19$

2. $\dfrac{4}{7} \cdot \dfrac{9}{9} + \dfrac{5}{9} \cdot \dfrac{7}{7} = \dfrac{36}{63} + \dfrac{35}{63} = \dfrac{71}{63}$

3. $30 = 2^1 \cdot 3^1 \cdot 5^1 \cdot 7^0 \cdot 11^0$

   $42 = 2^1 \cdot 3^1 \cdot 5^0 \cdot 7^1 \cdot 11^0$

   $99 = 2^0 \cdot 3^2 \cdot 5^0 \cdot 7^0 \cdot 11^1$

   l.c.m. $= 2^1 \cdot 3^2 \cdot 5^1 \cdot 7^1 \cdot 11 = 6{,}930$

   $\dfrac{7}{2 \cdot 3 \cdot 5} \cdot \dfrac{3 \cdot 7 \cdot 11}{3 \cdot 7 \cdot 11} = \dfrac{1{,}617}{2 \cdot 3^2 \cdot 5 \cdot 7 \cdot 11}$

   $\dfrac{5}{2 \cdot 3 \cdot 7} \cdot \dfrac{3 \cdot 5 \cdot 11}{3 \cdot 5 \cdot 11} = \dfrac{825}{2 \cdot 3^2 \cdot 5 \cdot 7 \cdot 11}$

   $\dfrac{5}{3^2 \cdot 11} \cdot \dfrac{2 \cdot 5 \cdot 7}{2 \cdot 5 \cdot 7} = \dfrac{350}{2 \cdot 3^2 \cdot 5 \cdot 7 \cdot 11}$

   $\dfrac{2{,}792}{2 \cdot 3^2 \cdot 5 \cdot 7 \cdot 11} = \dfrac{1{,}396}{3^2 \cdot 5 \cdot 7 \cdot 11} = \dfrac{1{,}396}{3{,}465}$

4. $-\sqrt{10} \cdot \sqrt{10} = -10$

5. $\left(\dfrac{11}{12} + 2\right) = \dfrac{-11}{12} = 2 + \left(\dfrac{11}{12} + \dfrac{-11}{12}\right) = 2$

6. $\dfrac{3^{-1} + 4^{-1}}{6} = \dfrac{\frac{1}{3} + \frac{1}{4}}{6} = \dfrac{\frac{7}{12}}{6} = \dfrac{7}{12} \times \dfrac{1}{6} = \dfrac{7}{72}$

7. $\dfrac{-7}{9} \cdot \dfrac{99}{174} + \dfrac{-7}{9} \cdot \dfrac{75}{174} = \dfrac{-7}{9}\left(\dfrac{99}{174} + \dfrac{75}{174}\right) = \dfrac{-7}{9}(1) = -\dfrac{7}{9}$

8. $\dfrac{-3 + \sqrt{3^2 + 4(2)(3)}}{2(2)} = \dfrac{-3 + \sqrt{33}}{4}$

9. $\dfrac{8}{3}$ (Note: $2\frac{2}{3}$ is mixed-number form, but $\frac{8}{3}$ is reduced.)

10. $\dfrac{16}{18} = \dfrac{2 \cdot 8}{2 \cdot 9} = \dfrac{8}{9}$ 11. $\dfrac{100}{825} = \dfrac{25 \cdot 4}{25 \cdot 33} = \dfrac{4}{33}$

12. $\dfrac{184}{207} = \dfrac{23 \cdot 8}{23 \cdot 9} = \dfrac{8}{9}$ 13. $\dfrac{1{,}209}{2{,}821} = \dfrac{3 \cdot 13 \cdot 31}{7 \cdot 13 \cdot 31} = \dfrac{3}{7}$

14. 0.375, rational   15. $2.\overline{3}$, rational   16. $0.\overline{428571}$, rational

17. 6.25, rational   18. $0.\overline{230769}$, rational   19. 89 is prime

20. 101 is prime   21. 349 is prime (check primes up to 17)

22. $1{,}001 = 7 \cdot 11 \cdot 13$ (use a factor tree)

23. $6{,}825 = 3 \cdot 5^2 \cdot 7 \cdot 13$ (use a factor tree)

24. $\dfrac{x}{5} \equiv 2,$ (mod 8) means $x = 5 \cdot 2 = 10 \equiv 2,$ (mod 8)

25. $2x \equiv 3,$ (mod 7); consider the set $\{0, 1, 2, 3, 4, 5, 6\}$ and try each, one at a time, to find $2 \cdot 5 = 10 \equiv 3,$ (mod 7), so $x \equiv 5,$ (mod 7).

26. $2x^2 + 7x + 1 \equiv 0,$ (mod 2); consider the set $\{0, 1\}$ and try each: $x \equiv 1,$ (mod 2)

27. $49 = 7^2$
   $1{,}001 = 7^1 \cdot 11^1 \cdot 13$
   $2{,}401 = 7^4$
   g.c.f. $= 7^1 = 7$

28. l.c.m. $= 7^4 \cdot 11^1 \cdot 13 = 343{,}343$

29. $1 \odot 2 = 1 \times 2 + 1 + 2 = 2 + 1 + 2 = 5;$
   $3 \odot 4 = 3 \times 4 + 3 + 4 = 12 + 3 + 4 = 19;$
   $5 \odot 19 = 5 \times 19 + 5 + 19 = 95 + 5 + 19 = 119$

30. $(1 \downarrow 2) \uparrow (2 \downarrow 3) = 1 \uparrow 2 = 2$

31. For $b \neq 0$, multiplication is defined as: $a \times b$ means $\underbrace{b + b + b + \cdots + b}_{a \text{ addends}}$. If $a = 0$, then $0 \times b = 0$.

32. $a - b = a + (-b)$

33. $\dfrac{a}{b} = m$ means $a = bm$ where $b \neq 0$.

34. If we use the definition of division $\dfrac{x}{0} = m$, then $0 \cdot m = x$. If $x \neq 0$, then there is no solution because $0 \cdot m = 0$ for all numbers m. Also, if $\dfrac{0}{0} = m$, then $0 = 0 \cdot m$, which is true for every number m. Thus, $\dfrac{0}{0} = 0$ checks, and $\dfrac{0}{0} = 1$ checks, and since two numbers equal to the same number must also be equal, we obtain the statement $0 = 1$.

35. Answers vary; $34.1011011101111011 \ldots$

36. A **field** is a set $\mathbb{R}$, with two operations $+$ and $\times$ satisfying the following properties for any elements $a, b, c \in \mathbb{R}$:

	Addition, $+$	Multiplication, $\times$
Closure:	1. $(a + b) \in \mathbb{R}$	2. $ab \in \mathbb{R}$
Associative:	3. $(a + b) + c$ $= a + (b + c)$	4. $(a \times b) \times c$ $= a \times (b \times c)$
Identity:	5. There exists $0 \in \mathbb{R}$ so that $0 + a = a + 0$ $= a$ for every element a in $\mathbb{R}$.	6. There exists $1 \in \mathbb{R}$ so that $1 \times a = a \times 1$ $= a$ for every element a in $\mathbb{R}$.
Inverse:	7. For each $a \in \mathbb{R}$, there is a unique number $(-a) \in \mathbb{R}$ so that $a + (-a)$ $= (-a) + a = 0$	8. For each $a \in \mathbb{R}$, $a \neq 0$, there is a unique number $\dfrac{1}{a} \in \mathbb{R}$ so that $a \times \dfrac{1}{a} = \dfrac{1}{a} \times a$ $= 1$
Commutative:	9. $a + b = b + a$	10. $ab = ba$

Distributive for multiplication over addition:
   11. $a \times (b + c) = a \times b + a \times c$

37. Let a, b, and c be any elements in $\mathbb{N}$.

Closure: $a \not b \in \mathbb{N}$ because the g.c.f. is the product of factors and $\mathbb{N}$ is closed for multiplication.

Associative: $(a \not b) \not c = a \not (b \not c)$ because the g.c.f. is the product of factors and $\mathbb{N}$ is associative for multiplication.

Identity: Look for I so that $a \not I = I \not a = a$, but there is no such number.

Inverse: Since there is no identity, there can be no inverse property.

$\mathbb{N}$ is not a group for $\emptyset$, and therefore cannot be a commutative group.

38. You can use a sieve or prime factorizations. A door will be opened or closed by a tenant only if the door number can be divided evenly by the number of that tenant. For example, door 9 will be touched (opened or closed) by tenants 1, 3, and 9; door 10 by tenants 1, 2, 5, and 10. Thus, the only doors left open are those with an odd number of divisors. The open doors are the perfect squares: 1, 4, 9, 16, 25, 36, . . . , 841, 900, 961. There are 31 doors left open.

39. 8 ft = 96 in. and $\dfrac{96}{8}$ = 12, so 12 stairs are necessary. The total length of the segments is $12 + 8 = 20$ ft, and the length of the diagonal is $\sqrt{12^2 + 8^2} = \sqrt{208} = 4\sqrt{13}$.

40.

Region	Number	Ratio	Adams	Jefferson
N	2,400,000	9.6	10	9
E	2,700,000	10.8	11	10
S	2,800,000	11.2	12	11
W	4,100,000	16.4	17	16
NE	4,000,000	16	16	16
SE	2,500,000	10	10	10
SW	3,500,000	14	14	14
NW	2,600,000	10.4	11	10
Total:	24,600,000		101	96

Region	Webster	Hamilton
N	10	9.756098; 9 + 1 = 10
E	11	10.97561; 10 + 1 = 11
S	11	11.38211; 11 11
W	16	16.66667; 16 + 1 = 17
NE	16	16.26016; 16 16
SE	10	10.16260; 10 10
SW	14	14.22764; 14 14
NW	10	10.56911; 10 + 1 = 11
Total:	98	100

CHAPTER 5 THE NATURE OF ALGEBRA

5.1 Polynomials, page 288

9. $-x - 8$ **11.** $3x - 6y - 4z$ **13.** $5x^2 - 7x - 7$
15. $-x^2 - 5x + 3$ **17.** $x - 31$ **19.** $16x^2 + 9x - 3$
21. $-3x^2 + 15x - 17$
23. a. $x^2 + 5x + 6$ **b.** $y^2 + 6y + 5$ **c.** $z^2 + 4z - 12$
    d. $s^2 + s - 20$
25. a. $c^2 - 6c - 7$ **b.** $z^2 + 2z - 15$ **c.** $2x^2 - x - 1$
    d. $2x^2 - 5x + 3$
27. a. $x^2 + 2xy + y^2$ **b.** $x^2 - 2xy + y^2$ **c.** $x^2 - y^2$
    d. $a^2 - b^2$
29. a. $x^2 + 8x + 16$ **b.** $y^2 - 6y + 9$ **c.** $s^2 + 2st + t^2$
    d. $u^2 - 2uv + v^2$

31. $6x^3 + x^2 - 12x + 5$ **33.** $5x^4 - 9x^3 + 13x^2 + 3x$
35. $-4x^3 + 25x^2 - 19x + 22$ **37.** $7x^2$
39.

$x + 1$ | x^2 | x | x | x | x ; x | 1 1 1 1 (with $x + 4$)

$(x + 1)(x + 4) = x^2 + 5x + 4$

41.

$(x + 3)(x + 4) = x^2 + 7x + 12$

43.

$(2x + 1)(2x + 3) = 4x^2 + 8x + 3$

45. $(x - 1)^3 = x^3 - 3x^2 + 3x - 1$
47. $(x + y)^6 = x^6 + 6x^5y + 15x^4y^2 + 20x^3y^3 + 15x^2y^4 + 6xy^5 + y^6$
49. $(x - y)^8 = x^8 - 8x^7y + 28x^6y^2 - 56x^5y^3 + 70x^4y^4$
    $- 56x^3y^5 + 28x^2y^6 - 8xy^7 + y^8$
51. $16x^4 - 96x^3y + 216x^2y^2 - 216xy^3 + 81y^4$
53. $91x^2y^{12} + 14xy^{13} + y^{14}$
55. $(6x + 2)(51x - 7) = 306x^2 + 60x - 14$
57. $(6b + 15)(10 - 2b) = 150 + 30b - 12b^2$
59. $(10x + y)(10x + z) = 100x^2 + 10xz + 10xy + yz$
    $= 100x^2 + 10x(z + y) + yz$
    $= 100x^2 + 10x(10) + yz$
    $= 100x^2 + 100x + yz$
    $= 100(x^2 + x) + yz$
    $= 100[x(x + 1)] + yz$

5.2 Factoring, page 295

3. $2x(5y - 3)$ **5.** $2x(4y - 3)$
7. $(x - 3)(x - 1)$ **9.** $(x - 3)(x - 2)$
11. $(x - 4)(x - 3)$ **13.** $(x - 6)(x + 5)$
15. $(x - 7)(x + 5)$ **17.** $(3x + 10)(x - 1)$
19. $(2x - 1)(x - 3)$ **21.** $(3x + 1)(x - 2)$
23. $(2x + 1)(x + 4)$
25. $(3x - 2)(x + 1)$
27. $x(5x - 3)(x + 2)$
29. $x^2(7x + 3)(x - 2)$
31. $(x - 8)(x + 8)$
33. $25(x^2 + 2)$
35. $(x - 1)(x + 1)(x^2 + 1)$

37.

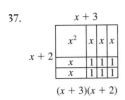

$(x + 3)(x + 2)$

39.

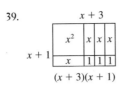

$(x + 3)(x + 1)$

41.

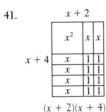

$(x + 2)(x + 4)$

43.

Move this unshaded
piece to form area:
$(x - 1)(x + 1)$

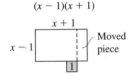

Moved piece

45.

Move this unshaded
piece to form area:
$(x - 1)(x + 2)$

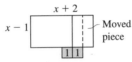

Moved piece

47.

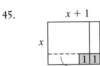

Move this unshaded
piece to form area:
$(x - 2)(x + 1)$

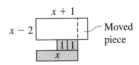

Moved piece

49. The dimensions are $x - 13$ feet by $x + 11$ feet.
51. The time is $2x - 1$ hours.
53. $(3x + 2)(2x + 1)$ **55.** $x^2(x - 3)(x + 3)(x - 2)(x + 2)$
57. $x^2(4y + 5z)(5y - 2z)$
59. Let $x - 1$, x, and $x + 1$ be the three integers. Then $(x - 1)(x + 1) = x^2 - 1$, so the square of the middle integer is one more than the product of the first and third.

5.3   Evaluation, Applications, and Spreadsheets, page 305

5. a. $+(2/3)*A1\verb|^|2$ **b.** $5*A1\verb|^|2 - 6*A2\verb|^|2$
7. a. $12*(A1\verb|^|2 + 4)$ **b.** $+(15*A1 + 7)/2$

9. a. $+(5 - A1)*(A1 + 3)\verb|^|2$ **b.** $6*(A1 + 3)*(2*A1 - 7)\verb|^|2$
11. a. $+(1/4)*A1\verb|^|2 - (1/2)*A1 + 12$
b. $+(2/3)*A1\verb|^|2 + (1/3)*A1 - 17$
13. a. $4x + 3$ **b.** $5x^2 - 3x + 4$
15. a. $\frac{5}{4}x + 14^2$ **b.** $(\frac{5}{4}x + 14)^2$ **17. a.** $\frac{x}{y}(z)$ **b.** $\frac{x}{yz}$
19. $A = 13$ **21.** $C = 8$ **23.** $E = 21$ **25.** $G = 19$
27. $I = 2$ **29.** $K = 4$ **31.** $M = 9$ **33.** $P = 20$
35. $R = 11$ **37.** $T = 17$ **39.** $V = 48$ **41.** $X = 100$
45. Genotype: black, 42.25%; black (recessive brown), 45.5%; brown, 12.25%. Phenotype: black, 87.75%; brown, 12.25%
47. Genotypes and phenotypes are the same: red, 4%; pink, 32%; white, 64%
49. 0   **51.** 8   **53.** 9

	A	B	C	D	E	F	G
55.	1	3	−2	5	−7	12	−19
57.	1	3	4	5	6	7	8
59.	1	3	3	9	9	27	27

61.

	A	B	C
1	−3	−10	−5
2	−8	−6	−4
3	−7	−2	−9

63.

	A	B	C
1	107	100	105
2	102	104	106
3	103	108	101

65. $(B + b)^2 = B^2 + 2Bb + b^2$; $B = 50\%$ and $b = 50\%$
67. Answers vary. For the general population the result is FF is 49%, Ff is 42%, and ff is 9%; free hanging is 91% and attached is 9%.

69.

	A	B	C	D	E
1	NAME	SALES	COST	PROFIT	COMMISSION
2				$+B2 - C2$	$.08*D2$
3	Replicate Row 2 for Rows 3 to 21.				
4					
5					

71. Place 1 (or the starting number) in Cell A1, 1 (or the second number) in Cell A2. Cell A3 should have the formula $+A1 +A2$ with cells below A3 in column A replicated. Cell B1 should have the heading QUOTIENTS with cell B2 containing the formula $+A2/A1$. The cells below B2 in column B are replicated.

5.4   Equations, page 315

3. a. 15   **b.** 22   **c.** 8   **5. a.** −4   **b.** 3   **c.** −8
7. a. −56   **b.** 40   **c.** −108
9. a. 0   **b.** 0   **c.** 0
11. a. $C = 10$ **b.** $D = 4$ **c.** $E = \frac{1}{3}$
13. a. $I = 7$ **b.** $J = 15$ **c.** $K = 9$
15. a. $P = -8$ **b.** $Q = -2$ **c.** $R = \frac{11}{5}$
17. $T = -3$ **19.** $V = -4$ **21.** $Y = -28$ **25.** 0, 14
27. $\frac{2}{5}, -\frac{2}{5}$ **29.** 0, 1, −1   **31.** $-\frac{2}{3}, \frac{2}{3}$ **33.** $\frac{3 \pm \sqrt{5}}{2}$

35. 3   **37.** $3 \pm \sqrt{3}$ **39.** no solution   **41.** $\frac{3}{2}, \frac{2}{3}$
43. $-\frac{1}{6}, 3$ **45.** $0, \frac{2}{9}$ **47.** 1.73, 5.20   **49.** $1.46, -0.34$
51. $1.73, -1.67$ **53.** $0.02, -4.11$
55. a. 3.3 mg   **b.** 75 mg   **57.** 444

5.5   Inequalities, page 322

3.
$-6 \, -4 \, -2 \quad 0 \quad 2 \quad 4 \quad 6 \quad 8$

5.
$-6 \, -4 \, -2 \quad 0 \quad 2 \quad 4 \quad 6 \quad 8$

7.
$-6 \, -4 \, -2 \quad 0 \quad 2 \quad 4 \quad 6 \quad 8$

9.
$-6 \, -4 \, -2 \quad 0 \quad 2 \quad 4 \quad 6 \quad 8$

11.
$-6 \, -4 \, -2 \quad 0 \quad 2 \quad 4 \quad 6 \quad 8$

13.
$-8 \, -4 \quad 0 \quad 4 \quad 8$

15.
$-50 \, -25 \quad 0 \quad 25 \quad 50 \quad 75$

17.
$-100 \, -50 \quad 0 \quad 50 \quad 100$

19.
$-50 \, -25 \quad 0 \quad 25 \quad 50 \quad 75$

21. $x \geq -4$ **23.** $x \geq -2$ **25.** $y > -6$ **27.** $s > -2$
29. $m < 5$ **31.** $x > -1$ **33.** $x \leq 1$ **35.** $s < -6$
37. $a \geq 2$ **39.** $s > 2$ **41.** $u \leq 2$ **43.** $w < -1$
45. $A > 0$ **47.** $C > \frac{19}{2}$ **49.** $E < -1$ **51.** $G > 7$
53. $I > \frac{5}{4}$ **55.** $K \geq -\frac{5}{7}$ **57.** any number greater than -5
59. any number less than -4 **61.** The number is 2.
63. The number is less than 2.
65. The height and width must each be 12 in. or less and the length 48 in. or less.

5.6   Algebra in Problem Solving, page 332

3. The number is 5.   **5.** The number is -4.
7. The number is -6.
9. The integers are 58 and 59.
11. The integers are 17, 18, and 19.
13. The integers are 17, 18, 19, and 20.
15. The smaller integer is 23.
17. The house is worth $177,000.
19. The daughter correctly solved 10 problems.
21. $\dfrac{6 + \sqrt{302}}{2} \approx 11.7, \dfrac{-6 + \sqrt{302}}{2} \approx 5.7$, and 13
23. $2\sqrt{65}$, or approximately 16 ft.   **25.** 9.75 miles
27. 250 mi   **29.** 12.5 sec
31. 6 min (or 0.1 hour)
33. 4.5 mi each
35. 7.75 and 9.25 mph
37. There are 120 lilies.
39. The second monkey jumped 50 cubits into the air.

5.7   Ratios, Proportions, and Problem Solving, page 343

5. 20 to 1   **7.** 53 to 50   **9.** 14 to 1   **11.** 18 to 1
13. $\frac{11}{25}$ **15.** $\frac{17}{7}$ **17. a.** yes   **b.** yes   **c.** yes
19. a. no   **b.** yes   **c.** yes   **21. a.** no   **b.** no   **c.** yes
23. a. <   **b.** <   **c.** <   **25. a.** <   **b.** >   **c.** >
27. a. <   **b.** >   **c.** >   **29.** $B = 36$ **31.** $D = 56$
33. $F = 25$ **35.** $H = 20$ **37.** $J = \frac{15}{4}$ or 3.75
39. $L = 15$ **41.** $N = 2$ **43.** $Q = \frac{63}{2}$ or 31.5
45. $S = 32$ **47.** $U = 1$ **49.** $X = \frac{4}{5}$ or 0.8
51. $Z = \frac{5}{6}$ or $0.8\overline{3}$ **53.** $0.91   **55.** 286 miles
57. 24 gallons   **59.** 38 minutes   **61.** 159.25   **63.** $\frac{2}{3}$
65. $3\frac{1}{3}$ ft or 3 ft 4 in.   **67.** $780
69. Since she started with 32 gallons and ended with 1 pt (of pure soft drink) the amount of soft drink served was 31 gal, 3 qt, and 1 pt.

5.8   Modeling Uncategorized Problems, page 354

3. The numbers are 20 and 22.   **5.** The integer is 2.
7. There are two numbers, 1 and 2.
9. Yes, there are 73 people.
11. Standard Oil: 1,136 ft; Sears: 1,454 ft
13. 10,100   **15.** 198 hours   **17.** 70
19. $x = 2\ell - \frac{1}{30}c\ell - 4$
21. The selling price of the house was $224,000.
23. Cut off 2 in.; the box is $8 \times 9 \times 12$ in.
25. The wind speed is 21 mph.
27. The wind's speed is 22.5 mph.
29. There are 120 different meals.
31. The reading would not change.
33. a. $2, 5, 8, 10, 13, \ldots$ **b.** No; $2 \cdot 8 = 16$ and 16 is not sacred.

5.9   Introduction to Computer BASIC, page 368

5.
```
6 + 5 = 11
6 - 5 = 1
6*5 = 30    6/3 = 2
```

7.
```
9
HELLO, I LIKE YOU!
4 + 5 = 9
```

9.
```
THE MISSISSIPPI IS WET.
```

11.

```
10 PRINT "THE MISSISSIPPI IS WET."
20 END
```

13.

```
Calculate the perimeter and area of a rectangle.
What is the length? [input 5 here]
What is the width? [input 4 here]
The perimeter is 18 AREA is 20
```

15.

```
10 PRINT 12850*(1 + .09/12)^360
20 END
```

17.

```
10 PRINT (2*5.8 - 8.9^2)/5
20 END
```

19.

```
10 PRINT "WHAT IS X";
20 INPUT X
30 PRINT "12.8X^2 + 14.76X + 6 1/3 = "; 12.8*X^2 + 14.76*X + (6 +1/3)
40 END
```

21.

```
10 PRINT "WHAT IS P";
20 INPUT P
30 PRINT "WHAT IS R";
40 INPUT R
50 PRINT "P(1 + R/12)^360 = "; P*(1 + R/12)^360
60 END
```

23.

```
10 LET R=2
20 LET PI=3.1416
30 PRINT "A = "; PI*R^2
40 END
```

25.

```
10 PRINT "What are your sales for this week";
20 INPUT S
30 PRINT "Your salary for this week is "; .05*S + 100
40 END
```

27.
1	5
2	7
3	9
4	11
5	13
6	15
7	17
8	19
9	21
10	23

29.
Line	A	K	Output
10	1		
20		1	
30	1		
40		2	
30	2		
40		3	
30	6		
40		4	
30	24		
40		5	
30	120		
40		6	
30	720		
50			720

31.
NUMBER	NUMBER SQUARED
1	1
2	4
3	9
4	16
5	25
6	36
7	49
8	64
9	81
10	100

33.
EVALUATE F
1
2
3
4
5
0
13
28
45
64
Values checked

35.
```
10 LET SUM = 0
20 FOR J = 1 TO 100
30    LET SUM = SUM + J
40 NEXT J
50 PRINT SUM
60 END
```

37.
```
10 LET SUM = 0
20 FOR J = 1 TO 100
30    LET SUM = SUM + J^2
40 NEXT J
50 PRINT SUM
60 END
```

39.
```
10 FOR K = 0 TO 99
20    PRINT 100 - K
30 NEXT K
40 END
```

41.
```
10  PRINT "How many scores to average";
20  INPUT N
30  PRINT "What is first score";
40  INPUT S
50  FOR K = 1 TO N - 1
60     PRINT "What is next score";
70     INPUT T
80     LET S = S + T
90  NEXT K
100 IF N > 0 then
110     PRINT "The average is"; S/N
120 ELSE
130 PRINT "Error dividing by 0."
140 END
```

43.

```
10 PRINT "I will solve any linear equation, ax + b = 0."
20 PRINT "What is a";
30 INPUT A
40 PRINT "What is b";
50 INPUT B
60 IF A=0 THEN
70    PRINT "a cannot be 0; try again."
80 END IF
90 PRINT "The root is"; -B/A
100 END
```

45. **a.** Let x = number;

$x + 9$;

$(x + 9)^2 = x^2 + 18x + 81$;

$x^2 + 18x + 81 - x^2 = 18x + 81$;

$18x + 81 - 61 = 18x + 20$;

$2(18x + 20) = 36x + 40$;

$36x + 40 + 24 = 36x + 64$;

$36x + 64 - 36x = 64$;

$\sqrt{64} = 8$;

Thus, the answer is always 8.

b.

```
10 PRINT "THINK OF A NUMBER N. WHAT IS THIS NUMBER";
20 INPUT N
30 LET X = N + 9
40 PRINT X
50 LET X = X^2
60 PRINT X
70 LET X = X - N^2
80 PRINT X
90 LET X = X - 61
100 PRINT X
110 LET X = 2*X
120 PRINT X
130 LET X = X + 24
140 PRINT X
150 LET X = X - 36*N
160 PRINT X
170 LET X = SQR(X)
180 PRINT X
190 END
```

Chapter 5   Review Questions, page 375

1. $(x - 1)(x^2 + 2x + 8) = (x - 1)(x^2) + (x - 1)(2x) + (x - 1)(8)$
$$= x^3 - x^2 + 2x^2 - 2x + 8x - 8$$
$$= x^3 + x^2 + 6x - 8$$

2. $x^2(x^2 - y) - xy(x^2 - 1) = x^4 - x^2y - x^3y + xy$

3. $a^2b + ab^2 = ab(a + b)$ 4. $x^2 - 5x - 6 = (x - 6)(x + 1)$

5. $3x^2 - 27 = 3(x^2 - 9) = 3(x - 3)(x + 3)$

6. $x^2 + 5x + 6 = (x + 3)(x + 2)$

7. $2x + 5 = 13$ **8.** $3x + 1 = 7x$
        $2x = 8$ $1 = 4x$
        $x = 4$ $\frac{1}{4} = x$

9. $\frac{2x}{3} = 6$ **10.** $2x - 7 = 5x$
        $2x = 18$ $-7 = 3x$
        $x = 9$ $-\frac{7}{3} = x$

11. $\frac{P}{100} = \frac{3}{20}$
        $P = 15$ $\boxed{3}$ $\boxed{\times}$ $\boxed{100}$ $\boxed{\div}$ $\boxed{20}$ $\boxed{=}$

12. $\frac{25}{W} = \frac{80}{12}$
        $W = 3.75$ $\boxed{25}$ $\boxed{\times}$ $\boxed{12}$ $\boxed{\div}$ $\boxed{80}$ $\boxed{=}$

13. $x^2 = 4x + 5$
        $x^2 - 4x - 5 = 0$
        $(x - 5)(x + 1) = 0$
        $x = 5, -1$

14. $4x^2 + 1 = 6x$
    $4x^2 - 6x + 1 = 0$
        $x = \dfrac{6 \pm \sqrt{36 - 4(4)(1)}}{2(4)}$
    $x = \dfrac{6 \pm \sqrt{20}}{8} = \dfrac{6 \pm 2\sqrt{5}}{8} = \dfrac{3 \pm \sqrt{5}}{4}$

15. $3 < -x$ **16.** $2 - x \geq 4$
        $x < -3$ $-x \geq 2$
                                $x \leq -2$

17. $3x + 2 \leq x + 6$ **18.** $14 > 5x - 1$
        $2x \leq 4$ $15 > 5x$
        $x \leq 2$ $3 > x$
                                    $x < 3$

19. a. $\frac{2}{3}$ ___ $\frac{67}{100}$ **b.** $\frac{3}{4}$ ___ $\frac{75}{100}$ **c.** $\frac{23}{27}$ ___ $\frac{92}{107}$
    $2(100)$? $3(67)$ $3(100)$? $4(75)$ $23(107)$? $27(92)$
    $200 < 201$ $300 = 300$ $2,461 < 2,484$

    d. 0.05 ___ 0.1 **e.** 0.99 ___ 0.909
        0.05 ? 0.10 0.990 ? 0.909
            $<$ $>$

20. 1 is to 5 as 2.5 is to how much? $\frac{1}{5} = \frac{2.5}{x}$; this means

    $x = 5(2.5) = 12.5$; $12\frac{1}{2}$ cups of flour.

21. Let $x, x + 2, x + 4$, and $x + 6$ be the four consecutive even integers. Then
    $x + (x + 2) + (x + 4) + (x + 6) = 100$
                    $4x + 12 = 100$
                        $4x = 88$
                        $x = 22$
    The integers are 22, 24, 26, and 28.

22. $(T + t)^2 = T^2 + 2Tt + t^2$
    Genotypes: 27% tall: $T^2 = (0.52)^2 = 0.2704$;
    50% tall (recessive short): $2Tt = 2(0.52)(0.48) = 0.4992$;
    23% short: $t^2 = (0.48)^2 = 0.2304$.
    Phenotypes: 77% tall: $0.2704 + 0.4992 = 0.7696$;
    23% short: 0.2304

23.

(NEW YORK TO CHICAGO) + (CHICAGO TO SF) + (SF TO HONOLULU) = 4,980
            ↑   ↑
    (CHICAGO TO SF − 1,140)   (CHICAGO TO SF + 540)
    Let x = DISTANCE FROM CHICAGO TO SF
                    $(x - 1,140) + x + (x + 540) = 4,980$
                                $3x - 600 = 4,980$
The distance from Chicago to San Francisco $3x = 5,580$
is 1,860 mi, from New York to Chicago is 720 mi, $x = 1,860$
and from San Francisco to Honolulu is 2,400 mi.

24. 10 PRINT "What is x";
    20 INPUT X
    30 PRINT "18x^2 + 10 = "; 18*X^2 + 10
    40 END

25.   A   B   C
        1                    $18*A1^2 + 10$
        2
    The value of x is placed in cell A1.

CHAPTER 6 THE NATURE OF SEQUENCES, SERIES, AND FINANCIAL MANAGEMENT

6.1   Interest, page 389

5. B; interest rate is not stated, but you should still recognize a reasonable answer.
7. C; price is not stated, but you should still recognize a reasonable answer.
9. C is the most reasonable.   **11.** D   **13.** C
15. \$1,400; \$1,469.33; \$69.33 more
17. \$2,720; \$2,809.86; \$89.86 more
19. \$17,000; \$48,231.47; \$31,231.47 more   **21.** \$903.46
23. \$1,028.25   **25.** \$755.59   **27.** \$16,536.79   **29.** \$50.73
31. \$572,177.99   **33.** \$3,019,988.95   **35.** \$1,548.13
37. \$24,264.50   **39.** \$13,276.75
41. a. \$1.44   **b.** \$2.31   **c.** \$3.00   **d.** \$2.08
43. a. \$139.76   **b.** \$201.88   **c.** \$1,149.16   **d.** \$1,863.51
45. \$1,220.19   **47.** \$1,960.07   **49.** \$10,310
51. \$540,541.20   **53.** \$2,181,175.28   **55.** \$8,705.51
57. \$15,415.93   **59.** \$2,266.50
61. We want to find Y so that the compounded amount is equal to the amount with the simple interest formula; that is:

$$P\left(1 + \frac{r}{n}\right)^{nt} = P(1 + Y)$$

$$P\left(1 + \frac{r}{n}\right)^{n} = P(1 + Y) \quad t = 1 \text{ (one year)}$$

$$\left(1 + \frac{r}{n}\right)^{n} = 1 + Y \quad \text{Divide both sides by } P.$$

$$\left(1 + \frac{r}{n}\right)^{n} - 1 = Y \quad \text{Subtract 1 from both sides.}$$

63. $800,000   **65.** $200,000

6.2   Installment Buying, page 400

7. B   **9.** B   **11.** B   **13.** B   **15.** B   **17.** B
19. B   **21.** A   **23.** C   **25.** 16%   **27.** 10%
29. 11%   **31.** $3.75   **33.** $37.50   **35.** $16.44
37. $13,033   **39.** $31,808   **41.** $53,804
43. $650 interest; $131.25 per month; 25.0% APR
45. $720 interest; $130 per month; 28.8% APR
47. $240 interest; $51.67 per month; 23.0% APR
49. $882 interest; $82.83 per month; 27.2% APR
51. $6,191.50 interest; $814.86 per month; 5.7% APR
53. $5,098.32   **55.** $19,608.56   **57.** 9.0% (0.09)
59. 15.8% (0.1584181185)   **61.** 8.7% (0.08702)
63. 17.8% (0.1781059459)   **65.** 11.8% (0.1175569754)
67. 8% add-on rate; APR is about 15.7%
69. 11% add-on rate; APR is about 21.6%
71. previous balance method, $45; adjusted balance method, $40.50; average daily balance method, $43.35
73. $67.58

6.3   Sequences, page 415

7. a. geometric   **b.** $r = 2$ **c.** 32
9. a. arithmetic   **b.** $d = 10$ **c.** 35
11. a. Fibonacci-type   **b.** $s_1 = 5, s_2 = 15$ **c.** 35
13. a. geometric   **b.** $r = \frac{1}{5}$ **c.** $\frac{1}{5}$
15. a. geometric   **b.** $r = 3$ **c.** 27
17. a. none of the classified types   **b.** Differences are $-2, 1,$ $-2, 1, -2, \ldots$ **c.** 5
19. a. geometric   **b.** $r = 2$ **c.** 96
21. a. both arithmetic and geometric   **b.** $d = 0$ or $r = 1$ **c.** 10
23. a. Fibonacci-type   **b.** $s_1 = 3, s_2 = 6$ **c.** 24
25. a. geometric   **b.** $r = \frac{3}{2}$ **c.** $\frac{81}{2}$
27. a. geometric   **b.** $r = 4^{-1}$ or $\frac{1}{4}$ **c.** 4
29. a. arithmetic   **b.** $d = \frac{1}{10}$ **c.** $\frac{3}{5}$
31. a. arithmetic   **b.** $d = \frac{1}{12}$ **c.** $\frac{11}{12}$
33. a. 0, 3, 6   **b.** arithmetic; $d = 3$
35. a. 1, 0, -1 **b.** arithmetic; $d = -1$
37. a. 0, $-10, -20$ **b.** arithmetic; $d = -10$
39. a. $0, \frac{1}{2}, \frac{2}{3}$ **b.** neither   **41. a.** 1, 3, 6   **b.** neither
43. a. 1, 9, 36   **b.** neither
45. a. 1, -1, 1   **b.** geometric, $r = -1$
47. a. $\frac{2}{3}, \frac{2}{3}, \frac{2}{3}$ **b.** both; $d = 0$ or $r = 1$
49. a. 1, -2, 3   **b.** neither
51. -200 **53.** 625   **55.** $3, 1, \frac{1}{3}, \frac{1}{9}, \frac{1}{27}$ **57.** 1, 2, 3, 5, 8
59. It is Fibonacci.
61. 1st number: x
2nd number: y
3rd number: $x + y$
4th number: $x + 2y$
5th number: $2x + 3y$

6th number: $3x + 5y$
7th number: $5x + 8y$
8th number: $8x + 13y$
9th number: $13x + 21y$
10th number: $21x + 34y$
SUM: $55x + 88y = 11(5x + 8y)$

6.4   Series, page 427

5. 20   **7.** 12   **9.** 90   **11.** 12   **13.** 10
15. 33   **17.** 2   **19.** 4   **21.** 2,000   **23.** $-\frac{200}{3}$
25. 25   **27.** 15   **29.** 110   **31.** 10,000   **33.** 5,050
35. $n(n + 1)$ **37.** 11,500   **39.** 2,030   **41.** 4,048
43. 36; 84   **45.** 465   **47.** 406 blocks   **49.** 5,050 blocks
51. $P = 1,000,000 \cdot 2^d$ **53.** $G_3 = 13$ games
55. $G = 200$ cm   **57.** $G = 1,500$ revolutions   **59.** 190
61. a. 55   **b.** 42,925 blocks

6.5   Annuities, page 437

7. $i = 0.06/12; N = 36; $1,966.81$
9. $i = 0.01; N = 36; $2,153.84$
11. $i = 0.06/12; N = 360; $50,225.75$
13. $i = 0.01; N = 360; $174,748.21$
15. $i = 0.06/12; N = 120; $16,387.93$
17. $i = 0.01; N = 120; $23,003.87$
19. $i = 0.06/12; N = 420; $213,706.54$
21. $i = 0.01; N = 420; $964,643.92$
23. $i = 0.06/12; N = 360; $652,934.78$
25. $39,529.09   **27.** $13,211.40   **29.** $6,629.90
31. $112,885.15   **33.** $60,030.54   **35.** $136,340.66
37. $4,014.26   **39.** $1,219.65   **41.** $55,281.31
43. $2,153.09   **45.** $441.06   **47.** $98.36
49. $15,793.29   **51.** $5,426.39   **53.** $255,310
55. $3,530.70   **57.** $1,023,118

6.6   Amortization, page 444

3. m is the amount of a periodic payment (usually a monthly payment); n is the number of payments made each year; t is the number of years; r is the annual interest rate; $i = r/n$; $N = nt$; A is the future value; and P is the present value
5. $i = 0.06/12; N = 60; $2,586.28$
7. $i = 0.05/12; N = 360; $27,942.24$
9. $i = 0.08/12; N = 360; $20,442.52$
11. $i = 0.06/12; N = 360; $175,131.20$
13. $264.20   **15.** $363.17   **17.** $1,206.93   **19.** $2,674.39
21. $1,907.79   **23.** $6,882.42   **25.** $10,827.33
27. $5,429.91   **29.** $12,885.18   **31.** $7,407.76
33. $9,299.39   **35.** $6.38   **37.** $148.31   **39.** $430.73
41. $73.35   **43.** $780.54   **45.** $10,186.47   **47.** $117.239
49. present value of an annuity; $367,695.71
51. Present value of an annuity is $83,358.72, which is worth considerably more than the lump sum of $50,000. The $750 per month is the better choice.

53. Present value of an annuity ($m = \$1,000,000, r = 0.05, n = 1$, and $t = 20$) is \$12,462,210.34. This would be a fair price to receive for the \$20,000,000 lottery prize.
55. \$206,029.43
57. \$1,510.92 (30 yr); \$543,931.20 total for interest and principal (30 yr); \$1,838.25 per mo for 15 years with total payments of \$330,885.00; the savings is \$213,046.20
59. price range of \$175,322.31 to \$204,386.77
61. **a.** increases **b.** decreases
63. **a.** decreases **b.** increases

6.7 Building Financial Power, page 448

7. present value   9. annuity   11. amortization
13. future value; \$1,404.93   15. present value; \$5,674.27
17. **a.** future value **b.** \$1,347.85
19. **a.** present value **b.** \$5,536.76
21. **a.** sinking fund **b.** \$1,670.92
23. **a.** future value **b.** \$27,081.62
25. **a.** present value of an annuity **b.** \$5,756.94
27. **a.** amortization **b.** \$1,028.61
29. **a.** ordinary annuity **b.** \$23,193.91
31. **a.** present value **b.** \$7,215.46
33. **a.** ordinary annuity **b.** \$21,867.63
35. **a.** sinking fund **b.** \$437.06
37. **a.** present value **b.** \$29,927.57
39. **a.** ordinary annuity **b.** \$175,610.81
41. **a.** future value **b.** \$224,628.30
43. **a.** future value **b.** \$19,898.24
45. **a.** future value **b.** \$39,960.19
47. **a.** present value **b.** \$12,878.55
49. **a.** present value **b.** \$2,219
51. **a.** present value of an annuity **b.** \$3,433,081
53. amortization; \$1,328.94
55. amortization schedule required (not shown here)
57. amortization; \$3,180.90
59. amortization schedule required (not shown here)

Chapter 6 Review Questions, page 455

1. A sequence is a list of numbers having a first term, a second term, and so on; a series is the indicated sum of the terms of a sequence. An arithmetic sequence is one that has a common difference; a geometric sequence is one that has a common ratio; and a Fibonacci-type sequence is one that, given the first two terms, is found by adding the previous two terms. The sum of an arithmetic sequence is $A_n = n\left(\dfrac{a_1 + a_n}{2}\right)$ or $A_n = \dfrac{n}{2}[2a_1 + (n-1)d]$, and the sum of a geometric sequence is $G_n = \dfrac{g_1(1 - r^n)}{1 - r}$.

2. Answers vary; a good procedure is to ask a series of questions. **Is it a lump-sum problem?** If it is, then what is the unknown? If FUTURE VALUE is the unknown, then it is a *future value* problem. If PRESENT VALUE is the unknown, then it is a *present value* problem. **Is it a periodic-payment problem?** If it is, then is the periodic payment known? If the PERIODIC PAYMENT IS KNOWN and you want to find the future value, then it is an *ordinary annuity* problem. If the PERIODIC PAYMENT IS KNOWN and you want to find the present value, then it is a *present value of an annuity* problem. If the PERIODIC PAYMENT IS UNKNOWN and you know the future value, then it is a *sinking fund* problem. If the PERIODIC PAYMENT IS UNKNOWN and you know the present value, then it is an *amortization* problem.

3. **a.** arithmetic; $a_n = 5n$ **b.** geometric; $g_n = 5 \cdot 2^{n-1}$
   c. Fibonacci type; $s_1 = 5, s_2 = 10, s_n = s_{n-1} + s_{n-2}, n \geq 3$
   d. none of these (add 5, 10, 15, 20, . . .); 55, 80
   e. geometric; $g_n = 5 \cdot 10^{n-1}$ **f.** none of these (alternate terms); 5, 50

4. **a.** $\displaystyle\sum_{k=1}^{3} (k^2 - 2k + 1) = (1^2 - 2(1) + 1) + (2^2 - 2(2) + 1)$
$$+ (3^2 - 3(2) + 1) = 5$$

   b. $\displaystyle\sum_{k=1}^{4} \frac{k-1}{k+1} = \frac{0}{2} + \frac{1}{3} + \frac{2}{4} + \frac{3}{5}$
$$= \frac{43}{30} \text{ or } 1.4\overline{3}$$

5. parents, grandparents, great-grandparents, . . . ; that is, 2, 4, 8, 16, ... Find G_{10} where $g_1 = 2$ and $r = 2$;
$$G_{10} = \frac{2(1 - 2^{10})}{1 - 2} = 2,046.$$ Thus, there are a minimum of 2,046 people.

6. There will be 72 divisions in 24 hours; $g_{73} = 2^{10} \cdot 2^{72} = 2^{82}$

7. $18,579(1 + 0.05) = \$19,507.95$. You should offer \$19,500 for the car.

8. $I = Prt = 13,500(0.029)(2) = 783$;
   $A = P + I = 13,500 + 783 = 14,283$;
   monthly payment is $\$14,283 \div 24 = \$595.13$.
   The total interest is \$783, and the monthly payment is \$595.13.

9. $\text{APR} = \dfrac{2Nr}{N+1} = \dfrac{2(24)(0.029)}{25} = 0.05568$; the APR is 5.568%.

10. $A = 48(353.04) = 16,945.92$;
    $I = A - P = 16,945.92 - 11,450.00 = 5,495.92$; also,

$$I = Prt$$

$5,495.92 = 11,450(r)(4)$	**48 months is 4 years, so $t = 4$.**
$1,373.98 = 11,450r$	**Divide both sides by 4.**
$0.12 \approx r$	**Divide both sides by 11,450.**

Finally, $\text{APR} = \dfrac{2Nr}{N+1} = \dfrac{2(48)r}{49} \approx 0.235$. The APR is about 23.5%.

11. The adjusted balance method is most advantageous to the consumer.

PREVIOUS BALANCE METHOD

$I = Prt$

$= 525(0.09)(\frac{1}{12})$

$= 3.9375$

The finance charge is $3.94.

ADJUSTED BALANCE METHOD

$I = Prt$

$= (525 - 100)(0.09)(\frac{1}{12})$

$= 3.185$

The finance charge is $3.19.

AVERAGE DAILY BALANCE

$I = Prt$

$= \dfrac{525 \times 7 + 425 \times 24}{31}(0.09)(\frac{31}{365})$

$= 3.421232877$

The finance charge is $3.42.

12. For inflation, use future value where $n = 1$; for this problem, $P = 185,000$, $t = 30$, and $r = 0.08$. Then $A = P(1 + i)^N = 185,000(1.08)^{30} \approx 1,861,591.52$.

13. $A = \$1,000,000$; $t = 50$; $r = 0.09$; $n = 12$;

$N = nt = 50(12) = 600$; $i = \dfrac{r}{n} = \dfrac{0.09}{12}$.

Thus, $P = A(1 + i)^{-N}$

$= 1,000,000\left(1 + \dfrac{0.09}{12}\right)^{-600} \approx 11,297.10$.

Deposit $11,297.10 to have a million dollars in 50 years.

14. Amortization, where $P = 154,000$, $n = 12$, $r = 0.08$,

$i = \dfrac{r}{n} = \dfrac{0.08}{12}$, $N = nt = 12(20) = 240$.

Thus, $m = \dfrac{Pi}{1 - (1 + i)^{-N}}$

$= \dfrac{154,000\left(\dfrac{0.08}{12}\right)}{1 - \left(1 + \dfrac{0.08}{12}\right)^{-240}}$

$\approx 1,288.117706$ **(by calculator)**

The monthly payments are $1,288.12.

15. Present value, where $A = \$100,000$, $r = 0.064$; $t = 3\frac{4}{12} = \frac{40}{12}$; $n = 12$;

$N = nt = 12(\frac{40}{12}) = 40$ and $i = \dfrac{r}{n} = \dfrac{0.064}{12}$. Thus,

$P = A(1 + i)^{-N} = 100,000\left(1 + \dfrac{0.064}{12}\right)^{-40} \approx 80,834.49$.

CHAPTER 7 THE NATURE OF GEOMETRY

7.1   Geometry, page 467

Constructions in Problems 9–15 can be verified by comparison with the art in text.

17. symmetric   19. symmetric   21. symmetric

23. not symmetric (look at eagle's head)   25. symmetric
27. not symmetric   29. not symmetric

Answers for Problems 30–35 may vary.

31.   33.   35.

37. C   39. C
41. (1) letters with no symmetry; (2) letters with horizontal line symmetry; (3) letters with vertical line symmetry; (4) letters with symmetry around both vertical and horizontal lines

7.2   Polygons and Angles, page 478

9. pentagon   11. decagon   13. hexagon
15. dodecagon   17. heptagon   19. nonagon

Constructions in Problems 21–25 can be verified by comparison with the art in text.

27. right   29. acute   31. obtuse   33. obtuse
35. vertical   37. adjacent or supplementary
39. corresponding   41. alternate interior angles
43. alternate exterior angles   45. F   47. T   49. F
51. $m\angle 1 = m\angle 3 = m\angle 5 = m\angle 7 = 115°$;
    $m\angle 2 = m\angle 4 = m\angle 6 = m\angle 8 = 65°$
53. $m\angle 1 = m\angle 3 = m\angle 5 = m\angle 7 = 153°$;
    $m\angle 2 = m\angle 4 = m\angle 6 = m\angle 8 = 27°$
55. $m\angle 1 = m\angle 3 = m\angle 5 = m\angle 7 = 163°$;
    $m\angle 2 = m\angle 4 = m\angle 6 = m\angle 8 = 17°$
57. $x = 132.5°$; $y = 47.5°$

7.3   Triangles, page 485

5. $\overline{GH} \simeq \overline{G'H'}$; $\overline{GI} \simeq \overline{G'I'}$; $\overline{HI} \simeq \overline{H'I'}$; $\angle G \simeq \angle G'$;
   $\angle H \simeq \angle H'$; $\angle I \simeq \angle I'$
7. $\overline{WX} \simeq \overline{YZ}$; $\overline{WZ} \simeq \overline{YX}$; $\overline{XZ} \simeq \overline{ZX}$; $\angle W \simeq \angle Y$; $\angle WXZ \simeq \angle YZX$;
   $\angle WZX \simeq \angle YXZ$
9. 62°   11. 35°   13. 49°   15. 95°
17. 95°   19. 115°

Constructions in Problems 21–25 can be verified by comparison with the art in text.

27. 10°   29. 12.5°   31. 5°   33. 25°; 65°; 90°

Constructions in Problems 35–37 may be verified by comparison with the art in text.

39. a. F   b. T   c. T   d. F   e. T   43. 1,080°

7.4   Similar Triangles, page 494

7. a 9. b 11. c 13. $\dfrac{b}{c}$ 15. $\dfrac{a}{c}$ 17. similar

19. not similar   **21.** similar   **23.** $\sqrt{80}$ or $4\sqrt{5}$ **25.** 4
27. $\frac{16}{3}$ **29.** $\frac{8}{3}$ **31.** 0.8290   **33.** 0.8746
35. 0.5878   **37.** 0   **39.** 0.4452   **41.** 3.7321
43. $\sin A = \frac{5}{13}$; $\cos A = \frac{12}{13}$; $\tan A = \frac{5}{12}$
45. $\sin A = \frac{3}{5}$; $\cos A = \frac{4}{5}$; $\tan A = \frac{3}{4}$
47. $\sin A = \dfrac{1}{6} \approx 0.1667$; $\cos A = \dfrac{\sqrt{35}}{6} \approx 0.9860$;

   $\tan A = \dfrac{1}{\sqrt{35}} \approx 0.1690$

49. $\sin A = \dfrac{\sqrt{3}}{2} \approx 0.8660$; $\cos A = \dfrac{1}{2} = 0.5000$;

   $\tan A = \sqrt{3} \approx 1.7321$

51. $\sin A = \dfrac{6}{\sqrt{37}} \approx 0.9864$; $\cos A = \dfrac{1}{\sqrt{37}} \approx 0.1644$;

   $\tan A = 6 = 6.000$
53. 125 ft   **55.** 29 ft   **57.** 109 ft
59. $w = 12$ in.   **61.** 63,400,000 miles

7.5   Golden Rectangles, page 503

7. 1, 3, 4, 7, 11, 18, 29, 47, 76, 123; the ratios are 3, 1.33, 1.75, 1.57, 1.64, 1.61, 1.62, 1.62, 1.62; it is τ
9. Ratio of s to $\frac{1}{2}b$ is 1.62; b to h is 1.57; these are both about the same as τ.
11. Ratio is 1.875; this is close to τ.
13. **a.** $5'' \times 3''$ **b.** 1.67 (about the same)
15. **a.** $9\frac{1}{4}$ in. by 8 in.   **b.** 1.16 (not too close)   **23.** 8
25. 1, 2, 1.50, 1.67, 1.60, 1.63, 1.62, 1.62, 1.62, 1.62
27. 4 reflections, 8 paths

29. $\frac{2}{1} = 2$; $\frac{3}{2} = 1.5$; $\frac{5}{3} \approx 1.67$; $\frac{8}{5} = 1.6$; $\frac{13}{8} \approx 1.63$; $\frac{21}{13} \approx 1.62$
31. $\dfrac{-5 + 5\sqrt{5}}{2} \approx 3.1$ ft   **33.** 15 cm   **35.** 1.41421

7.6   Königsberg Bridge Problem, page 511

5. traversable   **7.** not traversable   **9.** traversable
11. not traversable   **13.** not traversable   **15.** not traversable
17. not traversable   **19.** not traversable   **21.** traversable
23. Transform the problem into a network; it is not traversable since there are more than two odd vertices.
25. It is possible; Bank and South Kensington are the only odd vertices.
27. No; there are now four odd vertices.
29. A tetrahedron has four triangular faces; the sum of the measures of the angles on each face is 180°, so the sum of the measures of the face angles of a tetrahedron is 720°.

31.

	V	E	R	V + R − 2
(4)	6	6	2	6
(5)	6	7	3	7
(6)	7	10	5	10
(7)	4	6	4	6
(8)	5	10	7	10
(9)	4	8	6	8
(10)	5	12	9	12
(11)	12	16	6	16
(12)	8	10	4	10
(13)	8	12	6	12
(14)	7	12	7	12
(15)	8	11	5	11

$$V + R - 2 = E$$

33. Look for a pattern; the number of paths to Room n is the $(n + 1)$th Fibonacci number.
35. one edge   **37.** two interlocking pieces   **39.** yes

7.7   Topology and Fractals, page 522

3. A and F; B and G; C and D; as well as H and I are topologically equivalent.
5. A and F are simple closed curves.
7. A, B, C, D, E, and F are the same class; G is different.
9. A and C are inside; B is outside.
11. A and C are inside; B is outside.
13. A and B are inside; C is outside.   **15.** Four colors
17. C, G, I, J, L, M, N, S, U, V, W, and Z are topologically equivalent; also D and O; E, F, T, and Y; K and X; as well as Q and R.
19. The images are topologically equivalent.
21. Construction varies.

7.8   Projective and Non-Euclidean Geometries, page 532

5. It is not a Saccheri quadrilateral.
7. It is not a Saccheri quadrilateral.
9. Their sum is greater than 180°.   **11.** North Pole

7.9   Introduction to Computer LOGO, page 541

5.   **7.**   **9.**

11.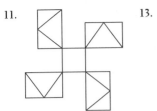

13. [pentagon figure]

15. ||||||||||

17. TO: SQUARER 100

19.
```
TO PROBLEM19
RT 30
TRIANGLER 10
TRIANGLER 20
TRIANGLER 30
TRIANGLER 40
TRIANGLER 50
END
```

21.
```
TO PROBLEM21
REPEAT 8[PROBLEM20 RT 45]
END
```

23.
```
TO PROBLEM23
REPEAT 4[PROBLEM22 LT 90]
END
```

25.
```
TO PROBLEM25
LT 45
REPEAT 4[PETAL 30 RT 90]
RT 45
REPEAT 4 PETAL 20 RT 90
END
```

27.
```
TO PROBLEM27
CIRCLE 12
PROBLEM 26T 13
CIRCLE 25
PROBLEM 26T 30
CIRCLE 40
PROBLEM 26T 53
END
```

31.
```
TO BUMPFD :K :D
1IF :K = 0 FD :D STOP
BUMPFD :K-1 :D/3 RT 90
BUMPFD :K-1 :D/6 LT 90
BUMPFD :K-1 :D/3 LT 90
BUMPFD :K-1 :D/6 RT 90
BUMPFD :K-1 :D/3
END
```

33. **a.** The fourth line positions the turtle to draw the next golden rectangle.   **b.** The fifth line draws the second golden rectangle inside the first.   **c.** Answers vary.

Chapter 7 Review Questions, page 548

1. **a.** 0
   b. 8 (corners of cube)
   c. 24 (2 such pieces on each of the 12 edges)
   d. 24 (4 on each face)
   e. 8 (interior unpainted pieces)

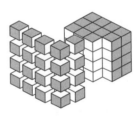

2. **a.** Flag is symmetric; picture is not (if you include the flagpole).   **b.** no   **c.** yes   **d.** yes
3. **a.** corresponding angles
   b. adjacent angles or supplementary angles
   c. vertical angles
   d. alternate interior angles
   e. alternate exterior angles
   f. vertical angles
   g. ℓ_1 and ℓ_2 are parallel (given); ℓ_3 is horizontal.
4. **a.** sin 59° = 0.8572   **b.** tan 0° = 0   **c.** cos 18° = 0.9511
   d. tan 82° = 7.1154
5. **a.** Classify the vertices; they are all odd (size 3) so there are more than 2 odd vertices. Thus it is not traversable.
   b. Classify the vertices; the outside corners and the corners at the bottom of the inner rectangle are even; the intersecting lines at the top show an even vertex, and the two vertices at the top of the inner rectangle are odd (size 5). Start at either of these odd vertices and end at the other to show that the network is traversable.
   c. There are 7 rooms, including the exterior. There are two rooms with an odd number of doors, so begin in either of those rooms and end in the other — it is possible.

6. Use the definition of the tangent ratio:

$$\tan 12° = \frac{160}{d}$$ Let d be the distance to the lighthouse.

$$d \tan 12° = 160$$ Multiply both sides by d.

$$d = \frac{160}{\tan 12°}$$ By calculator: $\boxed{160}$ $\boxed{÷}$ $\boxed{12}$ $\boxed{\text{TAN}}$ $\boxed{=}$

             Display: 752.7408175

$$d \approx 753$$ By table: Use .2126 for tan 12°.

7. Answers vary.

a.

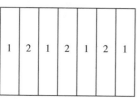

b.

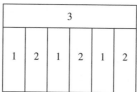

c.

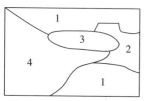

d. No such map exists.

8. There are two edges and two sides; the result after cutting is two interlocking loops.

9. A Saccheri quadrilateral is a quadrilateral *ABCD* with base angles *A* and *B* right angles and with sides *AC* and *BD* with equal lengths. If the summit angles *C* and *D* are right angles, then the result is Euclidean geometry. If they are acute, then the result is hyperbolic geometry. If they are obtuse, the result is elliptic geometry.

10.

```
TO PENTAGON :SIDE
FD :SIDE
RT 72
FD :SIDE
RT 72
FD :SIDE
RT 72
FD :SIDE
RT 72
FD :SIDE
RT 72
END
```

TO PENTAGON: 50

CHAPTER 8 THE NATURE OF MEASUREMENT

8.1 Precision, Accuracy, and Estimation, page 559

5. —————

7. ————————————

9. —— 11. ————————

13. ——————

15. A   17. B   19. A   21. C   23. B   25. B
27. C   29. C   31. B   33. A   35. B   37. B
39. A   41. 3 cm   43. 3.2 cm   45. 1 in.   47. 2 in.
49. $1\frac{3}{8}$ in.   51. 18 in.   53. 960 ft   55. 9 dm
57. 15.08 m   59. 75.40 in.   61. 52 in.   63. 397.08 ft
65. 257.08 cm   67. 64.27 ft   69. 11.14 cm   71. 300 m
73. The sides are 40 cm, 60 cm, 60 cm, 40 cm, and 80 cm.
75. 8 cm by 19 cm
83. The approximate measurements of the ark are 158 m long, 26 m wide, and 16 m high.

8.2 Area, page 573

5. 3 cm²   7. 12 cm²   9. 3 cm²   11. 10 or 11 cm²
13. C   15. A   17. C   19. A   21. C   23. C
25. 15 in.²   27. 1,196 m²   29. 100 mm²   31. 7,560 ft²
33. 136.5 dm²   35. 5,550 cm²   37. 28 in.²   39. 314.2 in.²
41. 78.5 in.²   43. 19.6 dm²   45. 7.6 cm²   47. 3,125 cm²
49. 780 cm²   51. 35,000 cm²   53. 36π cm² $\approx$ 113.1 cm²
55. 1.25π ft² $\approx$ 3.9 ft²   57. 7.80 in.²   59. 216 in.²
61. $93\frac{1}{2}$ in.²   63. $3,593.75   65. 28 in.²   67. 113 in.²
69. A; $2.86/ft² for Lot A and $3.19/ft² for Lot B.
71. 22.2 acres   73. 21 in.²
75. The edges of the two triangles and two trapezoids do not really form a diagonal in the new rectangle.

8.3 Volume and Capacity, page 585

7. 80 cm³   9. 125 ft³   11. 8,000 cm³   13. 24 ft³
15. 3,600 in.³   17. 2 c   19. 11 oz   21. 320 ml
23. $1\frac{3}{4}$ c   25. 420 ml   27. 75 ml   29. A   31. C
33. A   35. C   37. A   39. C   41. A   43. 15.6 L
45. 4 L   47. 9 L   49. 3.7 gal   51. 17.5 L
53. No; exterior dimensions are 24 ft³, so it cannot be 27 ft³ inside.
55. 10 yd³   57. 1 yd³   59. 112 kl
61. a. 560,000 people/mi²   b. 8,929 mi²
     c. about 6.656 acres per person!

8.4 Miscellaneous Measurements, page 595

5. centimeter   7. meter   9. milliliter   11. kiloliter
13. kilogram   15. Celsius   17. C   19. C   21. B
23. B   25. A   27. B   29. C   31. B   33. B

35. A
37. 0.000063 kiloliter; 0.00063 hectoliter; 0.0063 dekaliter; 0.063 liter; 0.63 deciliter; 6.3 centiliter; **63 milliliter**
39. 0.0035 kiloliter; 0.035 hectoliter; 0.35 dekaliter; **3.5 liter;** 35 deciliter; 350 centiliter; 3,500 milliliter
41. 0.08 kiloliter; 0.8 hectoliter; **8 dekaliter;** 80 liter; 800 deciliter; 8,000 centiliter; 80,000 milliliter
43. 0.31 kiloliter; **3.1 hectoliter;** 31 dekaliter; 310 liter; 3,100 deciliter; 31,000 centiliter; 310,000 milliliter
45. **a.** 0.1   **b.** 0.001   **c.** 0.000001
47. **a.** 0.01   **b.** 0.001   **c.** 0.000001
49. **a.** 1,000,000   **b.** 10,000   **c.** 1,000   **51.** 51 cm³
53. 14,137 cm³   **55.** 30 in.³   **57.** 64 cm³   **59.** 905 in.³
61. **a.** given   **b.** 5   **c.** 6   **d.** 6   **e.** 6, 8   **f.** 8, 12
    g. 12, 20   **h.** 20, 12; *Pattern:*
    NUMBER OF SIDES + NUMBER OF VERTICES
    =NUMBER OF EDGES + 2

Chapter 8 Review Questions, page 604

1. ———————————————————————
2. 2.1 cm   **3.** Answer vary; $\frac{1}{2}$ in.; 1.5 cm
4. First find the distance around the semicircle:
   $C = \frac{1}{2}\pi d = 4\pi \approx 12.56637061$. The distance around is
   $10 + 7 + 4\pi + 7 + 6$. To the nearest inch, the perimeter is 43 in.
5. First find the area of the semicircle:
   $A = \frac{1}{2}\pi(4)^2 = 8\pi \approx 25.13274123$. The area of the trapezoid is
   $A = \frac{1}{2}(8)(7 + 13) = 80$. The area of the entire figure is
   $8\pi + 80 \approx 105.1327412$. To the nearest square inch, the area
   is 105 in.².
6. $V = \ell wh$
   $= 2(3)(5)$
   $= 30$
   The volume is 30 dm³.
7. Since 1 L = 1 dm³, we see from Problem 6 that the box holds 30 liters.
8. 56 dm²
9. $V = \frac{4}{3}\pi r^3 = \frac{4}{3}\pi(1)^3 = \frac{4}{3}\pi \approx 4.188790205$; the volume is 4.2 ft³.
10. feet and meters   **11.** capacity
12. **a.** 2,400 ft²   **b.** 7,200 ft³   **c.** 53,856 gal
13. Answers vary; 40°C; 100°F
14. one-thousandth
15. 10 km = 1,000,000 cm
16. 11 ft × 16 ft = $3\frac{2}{3}$ yd × $5\frac{1}{3}$ yd
    $= \frac{11}{3} \times \frac{16}{3}$ yd²
    $= \frac{176}{9}$ yd²
    ≈ 19.6 yd²
    You need to purchase 20 square yards.
17. The 6-in. pizza is about $0.13/in.²; the 10-in. and 12-in. pizzas are about $0.10/in.², and the 14-in. pizza is about $0.09/in.². The best value is the large size.
18. Answers vary; the size is 201 in.²; price for a large should be about $0.09 per inch; I would price it at $17.95 ($16.95 to $18.25 is acceptable; calculate $201 \times 0.09 \approx 18.09$).

19.

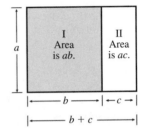

$5^2 + 12^2 = 25 + 144 = 169 = 13^2$

20. Use the diagram to give a geometric justification of the distributive law.

	I Area is ab.	II Area is ac.

Area of large rectangle is $a(b + c)$, and is the same as

Area of rectangle I: ab
Area of rectangle II: ac
Area of large rectangle is
$ab + ac$.

CHAPTER 9 THE NATURE OF PROBABILITY

9.1   Introduction to Probability, page 617

5. **a.** $\frac{1}{4}$ **b.** $\frac{1}{4}$ **c.** $\frac{1}{2}$ **7. a.** $\frac{3}{8}$ **b.** $\frac{1}{4}$ **c.** $\frac{3}{8}$
9. about 0.05   **11.** 0.19   **13.** $\frac{4}{18} = \frac{2}{9}$
15. $P(\text{royal flush}) \approx 0.000001539077169$
17. $P(\text{four of a kind}) \approx 0.0002400960384$
19. $P(\text{flush}) \approx 0.00196540155$
21. $P(\text{three of a kind}) \approx 0.02112845138$
23. $P(\text{one pair}) \approx 0.42256902761$ **25.** $P(\text{five}) = \frac{1}{13}$
27. $P(\text{jack}) = \frac{1}{13}$ **29.** $P(\text{even number}) = \frac{5}{13}$
31. $P(\text{five or a jack}) = \frac{2}{13}$ **33.** $P(\text{heart or a jack}) = \frac{4}{13}$
35. **a.** $\frac{1}{2}$ **b.** $\frac{2}{3}$ **c.** $\frac{7}{12}$ **37. a.** $\frac{1}{12}$ **b.** $\frac{7}{12}$
39. $P(\text{six}) = \frac{5}{36}$ **41.** $P(\text{eight}) = \frac{5}{36}$ **43.** $P(\text{two}) = \frac{1}{36}$
45. $P(\text{even number}) = \frac{1}{2}$ **47.** Answers vary; $P(H) = \frac{1}{2}$ **55.** $\frac{1}{9}$
57.

	1	2	3	4
1	(1, 1)	(1, 2)	(1, 3)	(1, 4)
2	(2, 1)	(2, 2)	(2, 3)	(2, 4)
3	(3, 1)	(3, 2)	(3, 3)	(3, 4)
4	(4, 1)	(4, 2)	(4, 3)	(4, 4)

59. **a.** $P(\text{five}) = \frac{1}{4}$ **b.** $P(\text{six}) = \frac{3}{16}$ **c.** $P(\text{seven}) = \frac{1}{8}$
61. Pick A: $P(A \text{ winning}) = \frac{3}{4}$
63. Each spinner has a probability of 0.5 of winning, so pick either.
65. Pick E: $P(E) = \frac{2}{3}$

9.2 Mathematical Expectation, page 628

7. B **9.** B **11.** B **13.** $E = \$7.20$ **15.** $E = \$0.08$
17. $E = 0.125$; A fair price would be to pay $0.25 for two plays of the game.
19. $1.50 **21.** $E = 2{,}515$
23. $E = 6{,}875$; they should dig the well because the expectation is positive.
25. $E \approx 0.02$; yes, you should play the game.
27. $E = 0$; fair game **29.** $E = -0.5$; not a fair game; don't play
31. $E = 4$; pay up to $4.00 to play this game
33. $E = 0.0484013102$; the expectation is about $0.05.
35. -0.05 **37.** -0.05 **39.** -0.05 **41.** -0.08
43. -0.05
45. No, the player should not stop for $100,000 since the mathematical expectation is $337,500.

9.3 Probability Models, page 645

7. $\frac{1}{5}$ **9.** 0.995 **11.** $\frac{33}{36} = \frac{11}{12}$
13. one, two, three, or five; $\frac{2}{3}$
15. drawing a spade, club, or diamond; 0.82
17. your school loses; 0.17 **19.** $\frac{13}{39} = \frac{1}{3}$; odds of 1 to 3
21. $\frac{9}{1}$; $P(\text{bald}) = \frac{9}{9+1} = \frac{9}{10}$
23. $P(\#1) = \frac{1}{3}$; $P(\#2) = \frac{1}{16}$; $P(\#3) = \frac{2}{5}$; $P(\#4) = \frac{5}{12}$; $P(\#5) = \frac{1}{2}$. *Note:* At a horse race, the odds are determined by track betting and the sum of the probabilities is not necessarily 1.
25. $P(\text{not a face card}) = 1 - P(\text{face card}) = 1 - \frac{12}{52} = \frac{10}{13}$
27. **a.** BBBB; BBBG; BBGB; BBGG; BGBB; BGBG; BGGB; BGGG; GBBB; GBBG; GBGB; GBGG; GGBB; GGBG; GGGB; GGGG **b.** $P(4 \text{ girls}) = P(4 \text{ boys}) = \frac{1}{16}$
c. $P(1 \text{ girl and } 3 \text{ boys}) = P(3 \text{ girls and } 1 \text{ boy}) = \frac{1}{4}$
d. $P(2 \text{ boys and } 2 \text{ girls}) = \frac{6}{16} = \frac{3}{8}$ **e.** 1
29. $\frac{4}{11}$ **31.** 1 **33.** $\frac{1}{3}$ **35.** $\frac{1}{2}$ **37.** $\frac{4}{51}$ **39.** $\frac{12}{51} = \frac{4}{17}$
41. $\frac{26}{51}$ **43.** 0 **45.** 0 **47.** $\frac{1}{3}$ **49.** $\frac{1}{5}$ **51.** 0
53. $\frac{1}{5}$ **55.** C **57.** B **59.** B
61. **a.** 0.000122 **b.** $\frac{1}{2}$ **63.** Pat is correct.

9.4 Counting Formulas, page 656

7. **a.** 600 **b.** 720 **9.** **a.** 40,296 **b.** 24
11. **a.** 90 **b.** 970,200 **13.** **a.** 665,280 **b.** 2
15. **a.** 84 **b.** 2,598,960
17. **a.** 7 **b.** 42 **c.** 210 **d.** 840 **e.** 2,520
19. **a.** 9 **b.** 36 **c.** 84 **d.** 126 **e.** 1
21. **a.** 35 **b.** 3,160,080 **c.** 1 **d.** 60 **e.** 1,225
23. $\dfrac{\frac{n!}{(n-r)!}}{r!} = \dfrac{n!}{(n-r)!} \cdot \dfrac{1}{r!} = \dfrac{n!}{r!(n-r)!}$

25. $\dfrac{n!}{(n-2)!} = n(n-1)$

27. $\dfrac{1}{n!} + \dfrac{1}{(n+1)!} = \dfrac{(n+1)! + n!}{n!(n+1)!}$
$= \dfrac{(n+1)n! + n!}{n!(n+1)!}$
$= \dfrac{n!(n+1+1)}{n!(n+1)!}$
$= \dfrac{n+2}{(n+1)!}$

29. combination; $_7C_4 = 35$ **31.** combination: $_5C_3 = 10$
33. combination; $_{100}C_6 = 1{,}192{,}052{,}400$
35. fundamental counting principle (FCP) **a.** $10^7 = 10{,}000{,}000$
b. $8^2 \cdot 10^5 = 6{,}400{,}000$ **c.** $8^2 \cdot 10^8 = 6{,}400{,}000{,}000$
37. FCP **a.** $2^{10} = 1{,}024$ **b.** $3^{10} = 59{,}049$
39. permutation; $_6P_6 = 720$ **41.** FCP; $3 \cdot 3 \cdot 2 \cdot 2 \cdot 1 \cdot 1 = 36$
43. **a.** $_{52}C_4 = 270{,}725$ **b.** $_4C_4 = 1$ **c.** $P(4 \text{ aces}) = \frac{1}{270{,}725}$
45. 9,134.2 centuries
47. **a.** $_7P_7 = 7! = 5{,}040$ **b.** $_3P_3 = 6$ **c.** $_4P_4 = 24$
d. 720 ways
49. There are 5 dials and each has 13 positions, so by the fundamental counting principle, there are 13^5 different positions. Since $13^5 = 371{,}293$, the claim is substantiated.
51. 524,800 **53.** 9,541,896 **55.** $5.524464741 \times 10^{26}$

9.5 Calculated Probabilities, page 671

7. $\frac{1}{2}$ **9.** $\frac{5}{6}$ **11.** $\frac{1}{12}$ **13.** $\frac{2}{3}$ **15.** $\frac{4}{9}$ **17.** $\frac{11}{12}$
19. $\frac{1}{3}$ **21.** $\frac{5}{9}$ **23.** $\frac{35}{36}$ **25.** no **27.** yes **29.** yes
31. $\frac{1}{6}$ **33.** $\frac{1}{2}$ **35.** $\frac{11}{36}$ **37.** $-\$1.25$ **39.** $-\$1.42$
41. $-\$0.92$ **43.** $P(3 \text{ oranges}) = \frac{27}{2{,}197} \approx 0.01229$
45. $P(\text{1st cherry}) = \frac{2}{13} \approx 0.1538$
47. $P(3 \text{ bars}) = P(\text{bar}) \cdot P(\text{bar}) \cdot P(\text{bar}) = \frac{2}{20} \cdot \frac{4}{20} \cdot \frac{2}{20} = 0.002$
49. $P(3 \text{ cherries}) = \frac{2}{20} \cdot \frac{5}{20} \cdot \frac{8}{20} = 0.01$
51. $P(\text{cherries on first two wheels}) = \frac{2}{20} \cdot \frac{5}{20} = 0.025$
53. $P(\text{five tails}) = (\frac{1}{2})^5 = \frac{1}{32}$ **55.** $E = 0.15625$; yes, play
57. $E = 0.32484994$; yes, play **59.** **a.** $\frac{25}{64}$ **b.** $\frac{5}{14}$
61. **a.** $\frac{15}{32}$ **b.** $\frac{15}{28}$
63. **a.** $\frac{1}{4}$ **b.** $\frac{1}{19} \approx 0.053$ **c.** $\frac{1}{114} \approx 0.009$ **65.** $\frac{1}{38}$
67. 0.71 **69.** $-\$0.46$
71. $-\$0.55$

9.6 Rubik's Cube and Instant Insanity, page 677

3.

5.

7.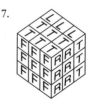

9.

11.

13.

15.

17.

19.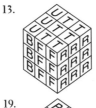

21.

23.

25.

27.

29. T^2 or $(T^{-1})^2$
31. $U^{-1}T^{-1}$
33. B^{-1}
35. $T^{-1}R^{-1}$
37. BF
39. no
41. 41,472
43.

Chapter 9   Review Questions, page 681

1. $P(\text{defective}) = \frac{4}{1,000} = 0.004$
2. The possible primes are 2, 3, and 5 so $P(\text{prime}) = \frac{3}{6} = \frac{1}{2}$.
3. Look at Figure 9.4. $P(\text{eight}) = \frac{5}{36}$
4. $P(\text{jack or better}) = \frac{16}{52} = \frac{4}{13}$
5. Look at Figure 9.4. **a.** $P(5 \text{ on at least one of the dice}) = \frac{11}{36}$
   b. $P(5 \text{ on one die or 4 on the other}) = \frac{20}{36} = \frac{5}{9}$
   c. $P(5 \text{ on one die and 4 on the other}) = \frac{2}{36} = \frac{1}{18}$
6. $P(E) = 0.01; P(\bar{E}) = 1 - P(E) = 1 - 0.01 = 0.99$

7. $P(E) = \frac{9}{10}; P(\bar{E}) = 1 - \frac{9}{10} = \frac{1}{10};$
   odds in favor $= \dfrac{P(E)}{P(\bar{E})} = \dfrac{\frac{9}{10}}{\frac{1}{10}} = \frac{9}{10} \cdot \frac{10}{1} = \frac{9}{1}$ or 9 to 1
8. Let E be the event. Given $f = 1,000$ and $s = 1$,
$$P(E) = \frac{s}{s + f} = \frac{1}{1,000 + 1} = \frac{1}{1,001}$$
9. $P(\text{ace}) = \frac{3}{51} = \frac{1}{17}$
10. $E = P(\text{one}) \cdot \$12 = \frac{1}{6}(12) = 2$; the expected value is \$2.
11. Answers for **a–e** are found using Pascal's triangle:
    a. $_5C_3 = 10$
    b. $_8P_3 = 3! \cdot {_8C_3} = 6 \cdot 56 = 336$
    c. $_{12}P_0 = 1 \cdot 1 = 1$
    d. $_{14}C_4 = 1,001$
    e. $_{100}P_3 = \dfrac{100!}{97!} = 100 \cdot 99 \cdot 98 = 970,200$
12. $P(\overline{A \cup B}) = 1 - P(A \cup B)$
    $= 1 - [P(A) + P(B) - P(A \cap B)]$
    $= 1 - P(A) - P(B) + P(A \cap B)$
    $= 1 - P(A) - P(B) + P(A)P(B)$
    $= 1 - \frac{2}{3} - \frac{3}{5} + \frac{2}{3} \cdot \frac{3}{5} = \frac{2}{15}$
13. $_{12}C_3 = 220$; they can form 220 different committees.
14. $_5P_5 = 5! = 120$; they can form 120 different lineups at the teller's window.
15. Fundamental counting principle; each item can be included or not included, and there are 8 items, so the number of possibilities is $2^8 = 256$. The claim is correct.
16. **a.** $P(\text{orange} \mid \text{orange}) = \frac{3}{5}$
    b. $P(\text{orange} \mid \text{orange}) = \frac{2}{4} = \frac{1}{2}$
17. List the sample space: MM $\Big\}$ one success out of 3 possibilities;
    MF
    FM $\Big\}$ $P(\text{other a male}) = \frac{1}{3}$
    FF ← Vet rules this out, so delete this from the sample space.
18. Write out the possible sample spaces. The best possibility is with die C:

		5	5	5	5	1	1	1	1
	2	5	5	5	5	2	2	2	2
	2	5	5	5	5	2	2	2	2
	2	5	5	5	5	2	2	2	2
C	2	5	5	5	5	2	2	2	2
	2	5	5	5	5	2	2	2	2
	2	5	5	5	5	2	2	2	2
	6	6	6	6	6	6	6	6	6
	6	6	6	6	6	6	6	6	6

(Header column labeled *D*)

We see that there are 64 possibilities and that die C wins 40 of those times; thus,
$$P(C \text{ wins}) = \frac{40}{64} = \frac{5}{8}$$
19. $P(\text{winning}) = \frac{1}{5}$. It is tempting to say 1 out of 25, but after you pick a key, one of five cars will fit that key.
20. $P(\text{winning a particular car}) = \frac{1}{5} \cdot \frac{1}{5} = \frac{1}{25}$

CHAPTER 10 THE NATURE OF STATISTICS

10.1   Frequency Distributions and Graphs, page 698

5.

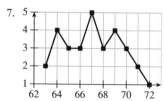

| 6 | 3 | 3 | 4 | 4 | 4 | 4 | 5 | 5 | 5 | 6 | 6 | 6 | 7 | 7 | 7 | 7 | 7 | 8 | 8 | 8 | 9 | 9 | 9 | 9 |
| 7 | 0 | 0 | 0 | 1 | 1 | 2 |

7.

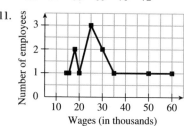

9. (in thousands)

1	4	6	8	8
2	0	5	5	5
3	0	0	5
4	
5	0
6	0

11.

13.

15.

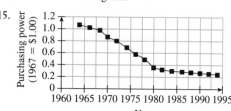

17.

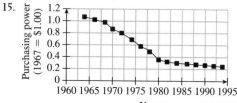

19.

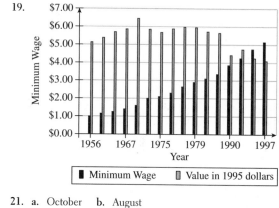

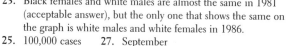

21. a. October   **b.** August

23. Black females and white males are almost the same in 1981 (acceptable answer), but the only one that shows the same on the graph is white males and white females in 1986.

25. 100,000 cases   **27.** September

29. Dole won the debate because his support went up after the debate, and Clinton's went down.

31. July (October is also acceptable because they are very close)

33. more imports   **35.** $90 billion   **37.** more imports

39. $25 billion   **41.** 25 times   **43.** likely DUI   **45.** no

47. 3.4 persons   **49.** 2036

51. The line graph at the right begins at 50 rather than 0.

53.

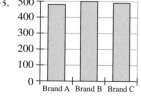

55. the graph in Figure 10.18b

57.

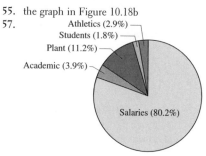

59.

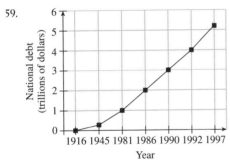

61. Answers vary; graph is meaningless without a scale.
63. The graph is appropriate and accurate.

10.2   Descriptive Statistics, page 716

13. mean = 19; median = 19; no mode
15. mean = 767; median = 767; no mode
17. mean = 10; median = 10; mode = 10
19. mean = 11; median = 9; no mode
21. mean = 82; median = 81; no mode
23. mean = 6; median = 2.5; mode = 1
25. range = 4; var = 2.5; $s \approx 1.58$
27. range = 4; var = 2.5; $s \approx 1.58$
29. range = 9; var = 10; $s \approx 3.16$
31. range = 24; var = 93.5; $s \approx 9.67$
33. range = 25; var = 83; $s = 9.11$
35. range = 21; var = 62.86; $s \approx 7.93$
37. mean = 217,851; $s \approx 1.58$
39. mean = 13; median = 11; mode = 10
41. mean = 68; median = 70; mode = 70; range = 50
43. mean = 366; median = 365; no mode
45. 105.25   47. 42
49. mean = 238; standard deviation ≈ 35
51. $s = 6,008$ (6,008.3)   53. $s \approx 10$ (9.8)
55. $s \approx 27$ (26.7)
57. mean = 19,200; range = 4,000; $s \approx 1,483$

10.3   The Normal Curve, page 727

3. 0.4192   5. 0.4925   7. 0.4999   9. 0.4901
11. 0.1772   13. 0.4868   15. 34 people   17. 33 people
19. 29 people   21. 25   23. 1   25. .0228
27. Answers vary; **a.** less variance **b.** more variance
29. Answers vary; mode (see Figure 10.35)   31. 0.1587
33. 0.4207   35. 0.0228   37. 0.0228
39. **a.** 50% **b.** 12.4 oz

10.4   Correlation and Regression, page 735

7. no   9. yes   11. no
13. yes   15. yes   17. no

19. $r = -0.765$;
    not significant

21. $r = -0.954$,
    significant at 5%

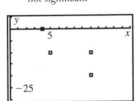

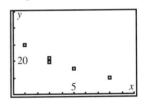

23. $r = -0.890$,
    significant at 5%

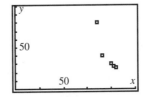

25. $y' = -1.95x + 4.146$
27. $y' = -2.7143x + 30.0571$
29. $y' = -2.380x + 270.00$
31. $r = 0.8830$, significant at 5%
33. $r = 0.8421$, significant at 1%
35. $y' = 2.4545x + 6.0909$

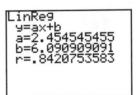

```
LinReg
 y=ax+b
 a=2.454545455
 b=6.090909091
 r=.8420753583
```

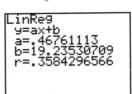

37. $r = 0.358$,
    no significant correlation

39. $r = -0.936$,
    significant at 1%

```
LinReg
 y=ax+b
 a=.46761113
 b=19.23530709
 r=.3584296566
```

```
LinReg
 y=ax+b
 a=-6.186388358
 b=87.17895046
 r=-.9356261053
```

41. $y' = 0.468x + 19.235$ 43. $y' = -6.186x + 87.179$

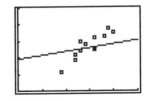

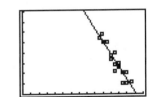

Chapter 10 Review Questions, page 744

1. **a.** Heads:  (18); Tails ┼┼┼┼ ┼┼┼┼ ┼┼┼┼ ┼┼┼┼ || (22)

   b.

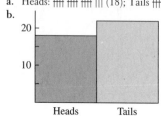

2. **a.**   **b.**

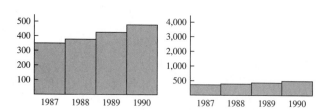

   c.

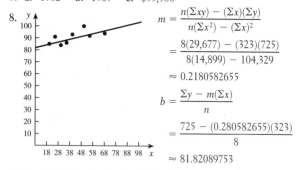

   d. Answers vary; impressions can be greatly influenced by using faulty or inappropriate scale (or even worse, no scale at all).
3. mean = 10.56; median = 10; mode = 12; mode is the most appropriate measure
4. mean = 79; median = 74; no mode; mean is the most appropriate measure
5. mean = $64,916.31; median = $69,000; no mode; the median is the most appropriate measure
6. mean = $27.13; median = $26.89; no mode; the median is the most appropriate measure
7. **a.** 1982 **b.** 1989 **c.** $35,500
8.

$$m = \frac{n(\Sigma xy) - (\Sigma x)(\Sigma y)}{n(\Sigma x^2) - (\Sigma x)^2}$$

$$= \frac{8(29,677) - (323)(725)}{8(14,899) - 104,329}$$

$$\approx 0.2180582655$$

$$b = \frac{\Sigma y - m(\Sigma x)}{n}$$

$$= \frac{725 - (0.280582655)(323)}{8}$$

$$\approx 81.82089753$$

The best-fitting line is $y' = 0.22x + 81.8$.

9. $n = 8$ $\Sigma x^2 = 14,899$
   $\Sigma x = 323$ $\Sigma y^2 = 65,907$
   $\Sigma y = 725$ $(\Sigma x)^2 = 104,329$
   $\Sigma xy = 29,677$ $(\Sigma y)^2 = 525,625$

$$r = \frac{n\Sigma xy - (\Sigma x)(\Sigma y)}{\sqrt{n(\Sigma x^2) - (\Sigma x)^2}\sqrt{n(\Sigma y^2) - (\Sigma y)^2}}$$

$$= \frac{8(29,677) - (323)(725)}{\sqrt{8(14,899) - 104,329}\sqrt{8(65,907) - 525,625}}$$

$$\approx 0.6582620404$$

$r \approx 0.658$; not significant at 1% level or 5% level

10. $314 - 266 = 48$; since the standard deviation is 16 days we note this is 3 standard deviations greater than the mean. From the standard normal curve, this should happen about 0.001 (0.1%) of the time. This means that approximately 1 in 1,000 pregnancies will have a 314-day duration.

CHAPTER 11 THE NATURE OF GRAPHS AND FUNCTIONS

11.1 Graphing Lines, page 757

7. Maxwell Montes 9. Rhea Mons 11.–21. Answers vary.

23.   25.

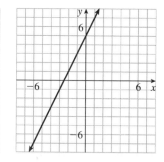

27.   29.

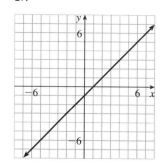
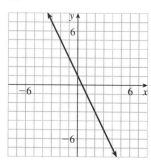

31.
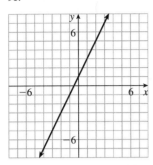

33.
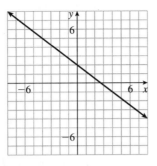

47.
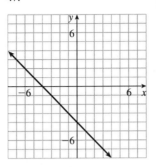

49.
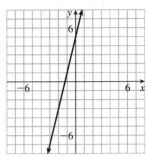

35.
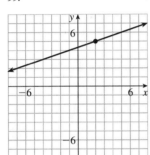

37.
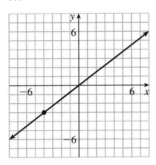

51.
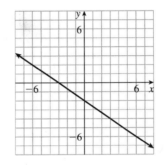

53.
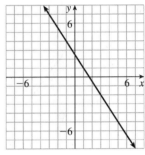

39.
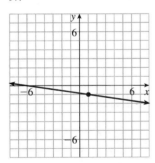

41.
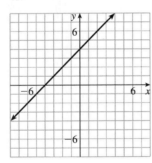

55.
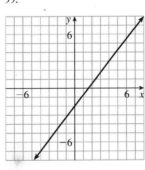

57.
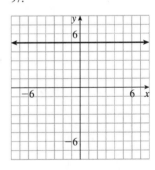

43.
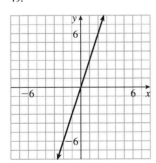

45.
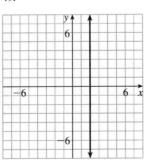

59.
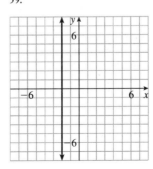

61.
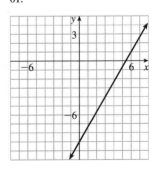

63.

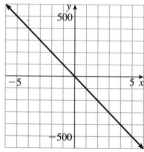

65. It is declining at a rate that is nearly linear, but you can use slopes to prove that it is not linear. (Looking at a graph does not provide sufficient information to make a decision.)

67. The estimated rate in the year 2000 is 1.4.

69. **a.** $P = -2,250t + 20,000$
    b.

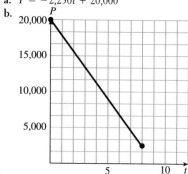

   c. $-2,250$

71. **a.**

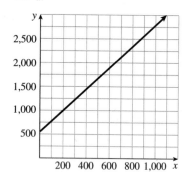

   b. When $x = 0$, $C = 550$; the fixed costs are $550.
   c. 2.25; it is the increase in cost corresponding to each unit increase in number of items

73. $y = mx + b$ passes through (h, k) means $k = mh + b$ and $b = k - mh$; substitute back into the original equation:
$$y = mx + (k - mh)$$
$$y - k = mx - mh$$
$$y - k = m(x - h)$$

75. Since the line passes through $(0, b)$ and $(a, 0)$, $a \neq 0$, $b \neq 0$, we have from Problem 74:
$$y - 0 = \left(\frac{0 - b}{a - 0} \right)(x - a)$$
$$y = \frac{-b}{a}(x - a)$$
$$ay = -bx + ab$$
$$bx + ay = ab$$
$$\frac{x}{a} + \frac{y}{b} = 1 \qquad \text{Divide by } ab \ (a \neq 0, b \neq 0).$$

11.2 Graphing Half-Planes, page 762

3. F; the boundary is $2x + 5y = 2$ 5. T

7. F; (0, 0) is not a test point because it lies on the boundary.

9. T

11.

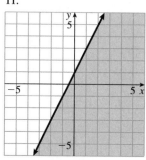

13.

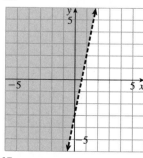

15.

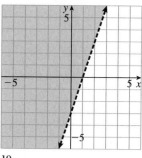

17.

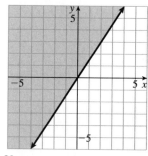

19.

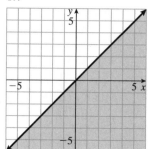

21.

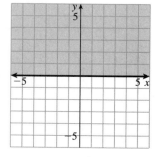

23.
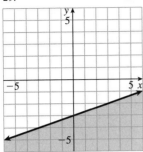

25.
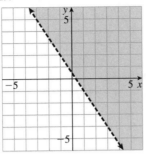

17.
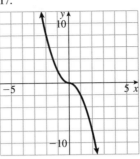

19.
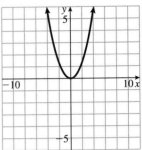

11.3   Graphing Curves, page 768

5.
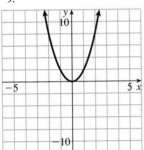

7.
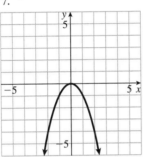

21.
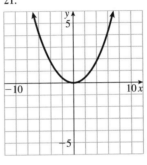

23.
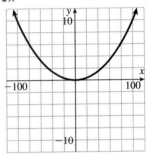

9.
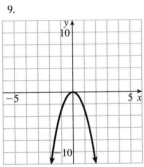

11.
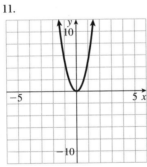

25.
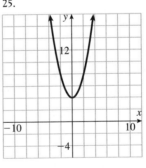

27.
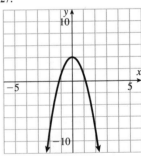

13.
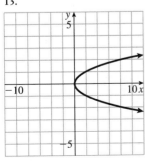

15.
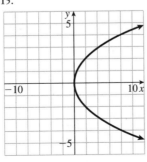

29.
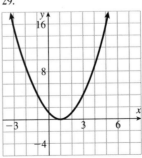

31.
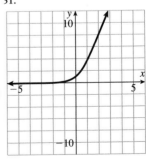

33.

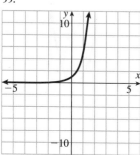

35.

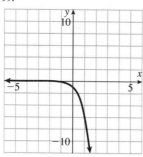

45.

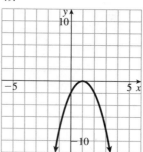

47.

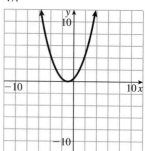

37.

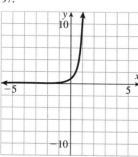

39.

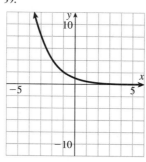

49.

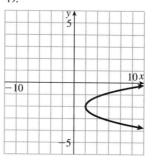

41.

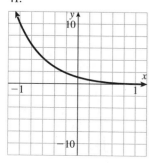

43. a.

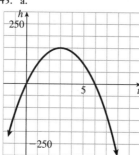

51.

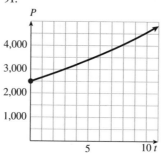

b. downward

c.

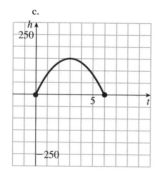

53.

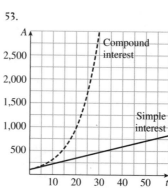

55.

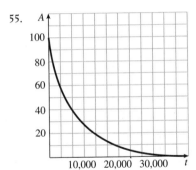

57.

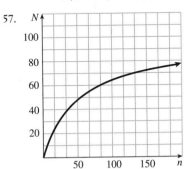

59.

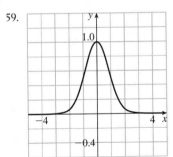

The shape of the two curves is the same. The y-intercepts are $(0, 0.4)$, and $(0, 1)$, respectively. Look at both graphs on the same axis:

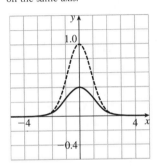

61. 7,728; answers vary

11.4   Functions, page 773

3. function   **5.** function
7. not a function   **9.** function
11. function   **13.** not a function
15. **a.** 64   **b.** 96   **c.** 128   **d.** 256   **e.** 512
17. **a.** 1,430   **b.** 1,050   **c.** 670   **d.** 290   **e.** 100
19. **a.** $f(4) = 12$ **b.** $f(6) = 14$ **c.** $f(-8) = 0$
   d. $f(\frac{1}{2}) = 8\frac{1}{2}$ **e.** $f(x) = x + 8$
21. **a.** $M(4) = 17$ **b.** $M(6) = 37$ **c.** $M(-8) = 65$
   d. $M(\frac{1}{2}) = 1\frac{1}{4}$ **e.** $M(x) = x^2 + 1$
23. **a.** $g(4) = 7$ **b.** $g(6) = 11$ **c.** $g(-8) = -17$
   d. $g(\frac{1}{2}) = 0$ **e.** $g(x) = 2x - 1$
25. **a.** 8   **b.** -16 **c.** $p - 7$
27. **a.** -1 **b.** -31 **c.** $3a - 1$
29. **a.** 5   **b.** -2 **c.** $\frac{3}{2}$
31. 3   **33.** $3x^2 + 3xh + h^2$

35. $\dfrac{-1}{x(x + h)}$

37. $\{(a, e), (e, a), (b, e), (e, b), (b, c), (c, b), (c, e), (e, c), (e, d),$
   $(d, e)\}$; it is not a function
39. $f(x) = 10x + 25$

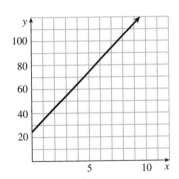

11.5   What Is Calculus?, page 782

7. Answers vary; $\frac{3}{11}$ **9.** Answers vary; $\sqrt{2}$
11. Answers vary; $\frac{2}{3}$ **13.** Answers vary; $\frac{3}{2}$
15. 0.3984375   **17.** Answers vary; $\frac{1}{3}$

11.6   Limits, page 788

3. 0, 2, 0, 2, 0   **5.** $\frac{4}{3}, \frac{7}{4}, 2, \frac{13}{6}, \frac{16}{7}$
7. 256, 16, 4, 2, $\sqrt{2}$ **9.** 1, 3, 13, 183, 33673
11. 5   **13.** -7 **15.** 0
17. $\frac{1}{2}$ **19.** 0.7   **21.** -6
23. no limit   **25.** no limit
27. 0   **29.** 12

Chapter 11   Review Questions, page 793

1. Solve for y: $5x - y = 15$
                 $5x - 15 = y$
The y-intercept is -15; the slope is 5; rise of 5 and run of 1.

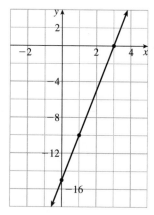

2. $y = -\frac{4}{5}x - 3$
The y-intercept is -3 and the slope is $-\frac{4}{5}$, rise of -4, and run of 5.

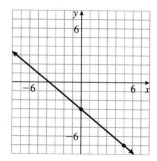

3. Solve for y: $2x + 3y = 15$
                    $3y = -2x + 15$
                    $y = -\frac{2}{3}x + 5$
The y-intercept is 5; the slope is $-\frac{2}{3}$; rise of -2 and run of 3.

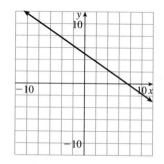

4. Solve for y: $x = -\frac{2}{3}y + 1$
                    $3x = -2y + 3$
                    $2y = -3x + 3$
                    $y = -\frac{3}{2}x + \frac{3}{2}$
y-intercept is $\frac{3}{2}$; the slope is $-\frac{3}{2}$; rise of -3 and run of 2.

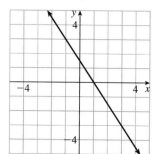

5. $x = 150$; recognize this as a vertical line; watch the scale.

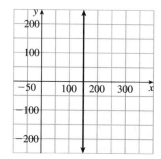

6. $x < 3y$
Graph the boundary $x = 3y$.

Test point: $(0, 3)$
    $x < 3y$
    $0 < 3(3)$ is true

Shade the half-plane on the same side as the test point.

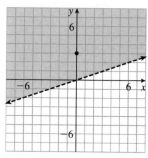

7. $y = 1 - x^2$

x	y
1	0
-1	0
2	-3
3	-8

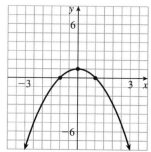

8. $y = -2^x$

x	y
0	-1
1	-2
2	-4
3	-8
-1	$-\frac{1}{2}$
-2	$-\frac{1}{4}$
-3	$-\frac{1}{8}$

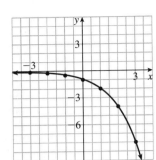

9. Yes, it is a function because each x value (namely, 4, 5, and 6) is associated with exactly one y-value.

10. $f(6) = 3(6) + 2 = 20$

11. $g(0) = 0^2 - 3 = -3$

12. $F(10) = 5(10) + 25 = 75$

13. $m(10) = 5$

14. $\lim\limits_{n \to \infty} \dfrac{1}{n} = 0$

15. $\lim\limits_{n \to \infty} \dfrac{3n^4 + 20}{7n^4} = \lim\limits_{n \to \infty} \dfrac{\frac{3n^4}{n^4} + \frac{20}{n^4}}{\frac{7n^4}{n^4}} = \lim\limits_{n \to \infty} \dfrac{3 + \frac{20}{n^4}}{7} = \dfrac{3}{7}$

16. $\lim\limits_{n \to \infty} (2n + 3)$ This limit does not exist.

17. The main ideas of calculus are limits, derivatives, and integrals. $\lim\limits_{n \to \infty} a_n = L$ means that the sequence a_n becomes closer and closer to the number L as n becomes larger and larger. The derivative illustrates the idea of a tangent line. Take a look at the series of illustrations showing the instantaneous growth rate. Every point on the curve has a tangent, which indicates a rate of growth at that particular instant in time. Finally, the integral is used to find the area under a curve.

Comparing Several Rates of Growth

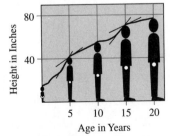

Each point on the curve has a tangent which indicates rate of growth at that point. The steepest tangent shown here occurs at about age three: the boy's growth was most rapid then.

18. Label the horizontal axis P and the vertical axis A. The intercept is 0 and the slope is 1.7 — that is, a rise of 1.7 and a run of 1 (or a rise of 17 and a run of 10). Note that since P represents an amount of money, it is nonnegative.

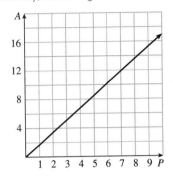

19. Form a table of values:

t	0	1	2	3	4	5	6	7	8
y	0	112	192	240	256	240	192	112	0

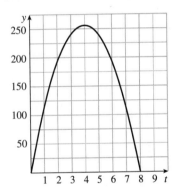

20. Form a table of values for $P = 53{,}000e^{0.013t}$, label the horizontal axis t and the vertical axis P (in units of thousands).

t	0	10	20	30	40	50
P	53,000	60,358	68,737	78,280	89,147	101,524

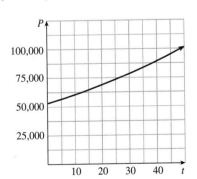

CHAPTER 12 THE NATURE OF MATHEMATICAL SYSTEMS

12.1 Systems of Linear Equations, page 800

7. $(4, -1)$

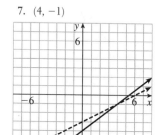

9. no solutions; inconsistent

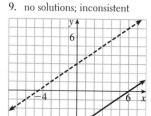

11. $(-1, 1)$

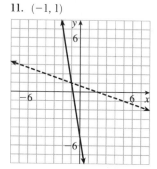

13. $(1, 1)$

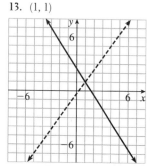

15. $(-5, -3)$ 17. $(2, 10)$ 19. $(9, -4)$ 21. $(4, -1)$
23. $(6, -2)$ 25. $(525, 35)$ 27. inconsistent
29. $(s_1, s_2) = (3, 4)$ 31. $(u, v) = (5, -2)$ 33. $(\frac{3}{5}, \frac{1}{2})$
35. $(5, -3)$ 37. $(3, 1)$ 39. $(3, 1)$ 41. $(4, -1)$
43. $(\frac{1}{3}, -\frac{2}{3})$ 45. inconsistent   47. $(8, 2)$ 49. $(-\frac{8}{5}, -\frac{21}{5})$
51. 2,601 days after the cat's birth — namely, June 3, 1999
53. $\left(\dfrac{a + \sqrt{a^2 - 4}}{2}, \dfrac{a - \sqrt{a^2 - 4}}{2} \right),$
$\left(\dfrac{a - \sqrt{a^2 - 4}}{2}, \dfrac{a + \sqrt{a^2 - 4}}{2} \right)$

12.2 Problem Solving with Systems, page 811

1. The box has 31 nickels and 56 dimes.
3. There are 18 nickels.
5. The plane's speed in still air is 175 mph.
7. The equilibrium point for the system is (50, 250 000).
9. The optimum price for the items is $3.
11. Debra Winger was born in 1955.
13. Mixture **a** has 7 lb of micoden.
15. Mixture **c** has $0.5p$ L of bixon.
17. Mixture **a** contains $33\frac{1}{3}\%$ bixon.

19. There are 40 oz of the base metal.
21. Mix 45 gal of milk with 135 gal of cream.
23. Both rates are the same if you drive 160 miles.
25. The equilibrium point is (2, 10 000).
27. There are 14 dimes and 28 quarters.
29. The wind speed is 22 mph.
31. Kristy McNichol was born in 1967.   33. Add $\frac{1}{3}$ gal water.
35. NY is 47,831 sq mi and CA is 156,361 sq mi.
37. There are 37 quarters.   39. There are 30 dimes.
41. The robbery included 14 $5 bills, 70 $10 bills, and 31 $20 bills.
43. The wind speed is 22.5 mph.
45. Mix together 28.5 gal of milk with 1.5 gal of cream.
47. Mix together 38.46 mg of 12% aspirin with 61.54 mg of 25% aspirin.
49. The Sears Tower is 1,454 ft and the Standard Oil building is 1,136 ft.
51. The Verrazano–Narrows bridge is 4,260 ft.
53. The equilibrium point is $2 for 40,000 items.
55. Replace 9 qt of the liquid with antifreeze.
57. **a.** 125 items would be supplied; 75 would be demanded
b. No items would be supplied at $200.   **c.** No items would be demanded at $400.   **d.** The equilibrium price is $\frac{700}{3} \approx \$233.33$.   **e.** The number of items produced at the equilibrium price is $\frac{250}{3} \approx 83$.
59. False; the cork costs 5¢ and the bottle $1.05.

12.3 Matrix Solution of a System of Equations, page 824

5. false; it is 4×5 7. true   9. true
11. false; the target row is 5.

13. **a.** $\begin{bmatrix} 4 & 5 & \vdots & -16 \\ 3 & 2 & \vdots & 5 \end{bmatrix}$ **b.** $\begin{bmatrix} 1 & 1 & 1 & \vdots & 4 \\ 3 & 2 & 1 & \vdots & 7 \\ 1 & -3 & 2 & \vdots & 0 \end{bmatrix}$

c. $\begin{bmatrix} 1 & 2 & -5 & 1 & \vdots & 5 \\ 1 & 0 & -3 & 6 & \vdots & 0 \\ 0 & 0 & 1 & -3 & \vdots & -15 \\ 0 & 1 & -5 & 5 & \vdots & 2 \end{bmatrix}$

15. $\begin{bmatrix} 1 & 3 & -4 & \vdots & 9 \\ 0 & 2 & 4 & \vdots & 5 \\ 3 & 1 & 2 & \vdots & 1 \end{bmatrix}$ 17. $\begin{bmatrix} 1 & 2 & 5 & \vdots & -6 \\ 6 & 3 & 4 & \vdots & 6 \\ 10 & -1 & 0 & \vdots & 1 \end{bmatrix}$

RowSwap([A], 1, 3)   *Row($\frac{1}{2}$, [C], 1)

19. $\begin{bmatrix} 1 & 2 & -3 & \vdots & 0 \\ 0 & 3 & 1 & \vdots & 4 \\ 0 & 1 & 7 & \vdots & 6 \end{bmatrix}$ 21. $\begin{bmatrix} 1 & 2 & 4 & \vdots & 1 \\ 0 & 9 & 8 & \vdots & 4 \\ 0 & 13 & 17 & \vdots & 7 \end{bmatrix}$

*Row+(-2, [A], 1, 3)   *Row+(2, [C], 1, 2)
*Row+(4, [Ans], 1, 3)

23. $\begin{bmatrix} 1 & 3 & 5 & \vdots & 2 \\ 0 & 1 & 3 & \vdots & -4 \\ 0 & 3 & 4 & \vdots & 1 \end{bmatrix}$ 25. $\begin{bmatrix} 1 & 4 & -1 & \vdots & 6 \\ 0 & 1 & -5 & \vdots & -2 \\ 0 & 4 & 6 & \vdots & 5 \end{bmatrix}$

*Row($\frac{1}{2}$, [A], 2)   *Row+(-1, [C], 3, 2)

27. $\begin{bmatrix} 1 & 0 & -23 &|& -23 \\ 0 & 1 & 4 &|& 5 \\ 0 & 0 & -8 &|& -13 \end{bmatrix}$ 29. $\begin{bmatrix} 1 & 3 & 4 &|& 5 \\ 0 & 1 & -3 &|& 6 \\ 0 & 0 & 1 &|& 2 \end{bmatrix}$

   *Row+(−5, [A], 2, 1)   *Row($\frac{1}{5}$, [A], 3)
   *Row+(−3, [Ans], 2, 3)

31. $\begin{bmatrix} 1 & 3 & 0 &|& 9 \\ 0 & 1 & 0 &|& -2 \\ 0 & 0 & 1 &|& 4 \end{bmatrix}$ 33. (3, 4)   35. (4, 3)

   *Row+(−2, [A], 3, 2)
   Row+([Ans], 3, 1)
37. (1, 1)   39. (3, 1)   41. $(\frac{3}{2}, \frac{1}{2}, -\frac{1}{2})$ 43. (−1, −2, 2)
45. (−1, 2, −3)   47. (1, −3, 5)   49. (5, 3, 2)
51. Mix 3 containers of Spray 1 with 4 containers of Spray II.
53. Mix 3 units of Candy I, 7 units of Candy II, and 5 units of Candy III.

12.4   Inverse Matrices, page 840

9. $\begin{bmatrix} 2 & 4 & 2 \\ 6 & -2 & 4 \\ 2 & 2 & 5 \end{bmatrix}$ 11. $\begin{bmatrix} 16 & 19 & 10 \\ 29 & 16 & 15 \\ 35 & 9 & 31 \end{bmatrix}$

13. $\begin{bmatrix} 13 & 25 & 20 \\ 25 & 31 & 23 \\ 42 & 24 & 33 \end{bmatrix}$ 15. $\begin{bmatrix} 20 & 21 & 34 \\ 29 & 16 & 15 \\ 7 & 48 & 5 \end{bmatrix}$

17. $\begin{bmatrix} 34 & 117 & 44 \\ 45 & 143 & 97 \\ 109 & 100 & 151 \end{bmatrix}$ 19. $\begin{bmatrix} 2 & 8 \\ 16 & 8 \\ -7 & 7 \\ 7 & 7 \end{bmatrix}$

21. $\begin{bmatrix} 14 & 14 \\ -7 & 7 \end{bmatrix}$ 23. not conformable

25. $\begin{bmatrix} 1 & 0 & 0 & 0 \\ 0 & 1 & 0 & 0 \\ 0 & 0 & 1 & 0 \\ 0 & 0 & 0 & 1 \end{bmatrix}$ 27. $\begin{bmatrix} 2 & 7 \\ 1 & 4 \end{bmatrix}$ 29. $\begin{bmatrix} 0 & \frac{1}{2} \\ \frac{1}{3} & -\frac{1}{6} \end{bmatrix}$

31. $\begin{bmatrix} 3 & -17 & -20 \\ 3 & -18 & -20 \\ -1 & 6 & 7 \end{bmatrix}$ 33. $\begin{bmatrix} 1 & 0 & -1 & 0 \\ 0 & \frac{1}{2} & 0 & 0 \\ -2 & 0 & 2 & 1 \\ 0 & 0 & 1 & 0 \end{bmatrix}$

35. (3, 2)   37. (5, −4)   39. (38, 21)   41. (3, −2)
43. (3, −5)   45. (−4, 1)   47. (3, 1)   49. (−2, 2)
51. (1, −8)   53. (5, 6, 1)   55. (−26, 52, 15)
57. (88, −176, −38)   59. (1, 1, 1)   61. (8, 3, −2)
63. a.

	Abigweel	Aweel	Upancomer	Pon
Abigweel	0	1	1	1
Aweel	0	0	1	1
Upancomer	0	0	0	1
Pon	0	0	0	0

b. $\begin{bmatrix} 0 & 0 & 1 & 2 \\ 0 & 0 & 0 & 1 \\ 0 & 0 & 0 & 0 \\ 0 & 0 & 0 & 0 \end{bmatrix}$ c. $\begin{bmatrix} 0 & 1 & 2 & 3 \\ 0 & 0 & 1 & 2 \\ 0 & 0 & 0 & 1 \\ 0 & 0 & 0 & 0 \end{bmatrix}$

   The power of Abigweel is 6; Aweel, 3; Upancomer, 1; and Pon, 0.

d. Ranking: Abigweel, Aweel, Upancomer, and Pon

12.5   Systems of Inequalities, page 844

3.
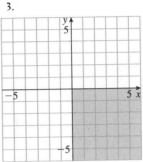

5.
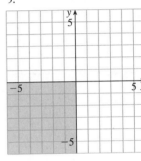

7.
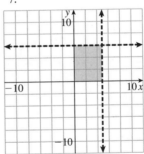

9.
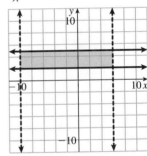

11.
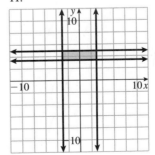

13.
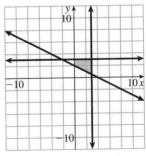

15.

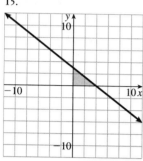

17.

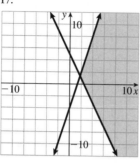

31.

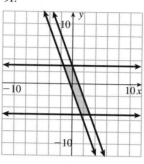

33.

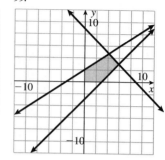

19.

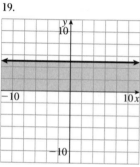

21.

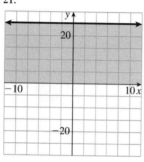

35.

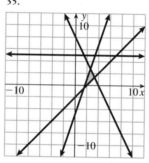

37.

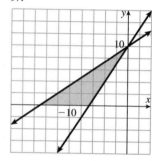

23.

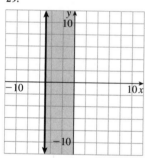

25.

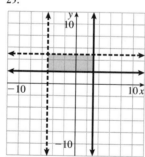

27.

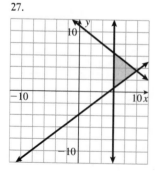

29.

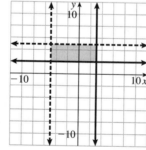

12.6   Modeling with Linear Programming, page 851

5. yes   **7.** no   **9.** no   **11.** yes   **13.** yes   **15.** no

17. $(0, 0)$, $(0, 4)$, $(5, 2)$, $(6, 0)$

19. $(0, 6)$, $(0, 8)$, $(10, 8)$, $(10, 0)$, $(4, 0)$

21. $(0, 6)$, $(0, 9)$, $(\frac{25}{2}, 9)$, $(8, 0)$, $(6, 0)$

23. $(0, 2)$, $(0, 10)$, $(\frac{15}{2}, \frac{5}{2})$, $(6, 2)$

25. $(0, 0)$, $(0, 2)$, $(\frac{21}{5}, \frac{24}{5})$, $(6, 3)$, $(3, 0)$

27. $(4, 0)$, $(\frac{5}{3}, \frac{14}{3})$, $(9, 12)$, $(9, 1)$, $(7, 0)$

29. maximum $T = 600$ at $(6, 0)$

31. minimum $K = 500$ at $(0, 2)$

33. minimum $K = 24$ at $(4, 0)$

35. Let x = number of grams of food A
   y = number of grams of food B
   Minimize $C = 0.29x + 0.15y$
   Subject to $\begin{cases} x \geq 0, y \geq 0 \\ 10x + 5y \geq 200 \\ 2x + 5y \geq 100 \\ 3x + 4y \geq 20 \end{cases}$

37. Let x = amount invested in stock
   y = amount invested in bonds
   Maximize $T = 0.12x + 0.08y$
   Subject to $\begin{cases} x \geq 0, y \geq 0 \\ x \leq 8, y \geq 2 \\ x + y \leq 10 \\ x \leq 3y \end{cases}$

39. The maximum profit $P = \$14{,}300$ is achieved with all 100 acres planted in corn.

41. The minimum cost of $\$1.19$ is obtained with 17 oz of Corn Flakes and no Honeycombs.

Chapter 12   Review Questions, page 856

1. $[C][B] - 3[D] = \begin{bmatrix} 2 & 0 \\ 1 & 2 \\ -1 & 1 \end{bmatrix} \begin{bmatrix} 2 & -1 & 0 \\ 1 & 0 & 1 \end{bmatrix} - 3 \begin{bmatrix} 0 & 1 & 0 \\ -1 & 0 & 0 \\ 0 & 1 & -1 \end{bmatrix}$

$= \begin{bmatrix} 4 & -2 & 0 \\ 4 & -1 & 2 \\ -1 & 1 & 1 \end{bmatrix} - \begin{bmatrix} 0 & 3 & 0 \\ -3 & 0 & 0 \\ 0 & 3 & -3 \end{bmatrix}$

$= \begin{bmatrix} 4 & -5 & 0 \\ 7 & -1 & 2 \\ -1 & -2 & 4 \end{bmatrix}$

2. $[A][B][C] = \begin{bmatrix} 1 & 0 \\ 2 & -1 \end{bmatrix} \begin{bmatrix} 2 & -1 & 0 \\ 1 & 0 & 1 \end{bmatrix} \begin{bmatrix} 2 & 0 \\ 1 & 2 \\ -1 & 1 \end{bmatrix}$

$= \begin{bmatrix} 2 & -1 & 0 \\ 3 & -2 & -1 \end{bmatrix} \begin{bmatrix} 2 & 0 \\ 1 & 2 \\ -1 & 1 \end{bmatrix} = \begin{bmatrix} 3 & -2 \\ 5 & -5 \end{bmatrix}$

3. $[B]$ is a 2×3 matrix and $[A]$ is 2×2 so these matrices are not conformable for multiplication.

4. $[C][A][B] = \begin{bmatrix} 2 & 0 \\ 1 & 2 \\ -1 & 1 \end{bmatrix} \begin{bmatrix} 1 & 0 \\ 2 & -1 \end{bmatrix} \begin{bmatrix} 2 & -1 & 0 \\ 1 & 0 & 1 \end{bmatrix}$

$= \begin{bmatrix} 2 & 0 \\ 5 & -2 \\ 1 & -1 \end{bmatrix} \begin{bmatrix} 2 & -1 & 0 \\ 1 & 0 & 1 \end{bmatrix} = \begin{bmatrix} 4 & -2 & 0 \\ 8 & -5 & -2 \\ 1 & -1 & -1 \end{bmatrix}$

5. $\begin{bmatrix} 2 & 1 & \vdots & 1 & 0 \\ -\frac{3}{2} & -\frac{1}{2} & \vdots & 0 & 1 \end{bmatrix} \rightarrow \begin{bmatrix} 1 & \frac{1}{2} & \vdots & \frac{1}{2} & 0 \\ -\frac{3}{2} & -\frac{1}{2} & \vdots & 0 & 1 \end{bmatrix}$

$\rightarrow \begin{bmatrix} 1 & \frac{1}{2} & \vdots & \frac{1}{2} & 0 \\ 0 & \frac{1}{4} & \vdots & \frac{3}{4} & 1 \end{bmatrix} \rightarrow \begin{bmatrix} 1 & \frac{1}{2} & \vdots & \frac{1}{2} & 0 \\ 0 & 1 & \vdots & 3 & 4 \end{bmatrix}$

$\rightarrow \begin{bmatrix} 1 & 0 & \vdots & -1 & -2 \\ 0 & 1 & \vdots & 3 & 4 \end{bmatrix}$; Inverse is $\begin{bmatrix} -1 & -2 \\ 3 & 4 \end{bmatrix}$.

6. $\begin{bmatrix} 1 & 3 & 3 & \vdots & 1 & 0 & 0 \\ 1 & 4 & 3 & \vdots & 0 & 1 & 0 \\ 1 & 3 & 4 & \vdots & 0 & 0 & 1 \end{bmatrix} \rightarrow \begin{bmatrix} 1 & 3 & 3 & \vdots & 1 & 0 & 0 \\ 0 & 1 & 0 & \vdots & -1 & 1 & 0 \\ 0 & 0 & 1 & \vdots & -1 & 0 & 1 \end{bmatrix}$

$\rightarrow \begin{bmatrix} 1 & 0 & 3 & \vdots & 4 & -3 & 0 \\ 0 & 1 & 0 & \vdots & -1 & 1 & 0 \\ 0 & 0 & 1 & \vdots & -1 & 0 & 1 \end{bmatrix}$

$\rightarrow \begin{bmatrix} 1 & 0 & 0 & \vdots & 7 & -3 & -3 \\ 0 & 1 & 0 & \vdots & -1 & 1 & 0 \\ 0 & 0 & 1 & \vdots & -1 & 0 & 1 \end{bmatrix}$; Inverse is $\begin{bmatrix} 7 & -3 & -3 \\ -1 & 1 & 0 \\ -1 & 0 & 1 \end{bmatrix}$.

7. Graph $2x - y = 2$ by writing $y = 2x - 2$; y-intercept is -2 and slope is $2 = \frac{2}{1}$. Graph $3x - 2y = 1$ by writing $y = \frac{3}{2}x - \frac{1}{2}$; y-intercept is $-\frac{1}{2}$ and slope is $\frac{3}{2}$. The intersection point appears to be $(3, 4)$. Check:

$$2(3) - 4 = 6 - 4 = 2$$
$$3(3) - 2(4) = 9 - 8 = 1$$

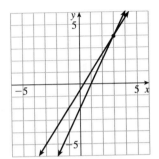

8. $2 \begin{cases} x + 3y = 3 \\ 4x - 6y = -6 \end{cases} + \begin{cases} 2x + 6y = 6 \\ 4x - 6y = -6 \end{cases}$

$ 6x = 0$
$ x = 0$

If $x = 0$ then $0 + 3y = 3$ or $y = 1$. The solution is $(0, 1)$.

9. Substitute the first equation into the second:

$$5x + 2(1 - 2x) = 1$$
$$5x + 2 - 4x = 1$$
$$x = -1$$

If $x = -1$, then $y = 1 - 2(-1) = 1 + 2 = 3$; the solution is $(-1, 3)$.

10. $\begin{bmatrix} 1 & 1 & 1 & \vdots & 2 \\ 1 & 2 & -2 & \vdots & 1 \\ 1 & 1 & 3 & \vdots & 4 \end{bmatrix} \rightarrow \begin{bmatrix} 1 & 1 & 1 & \vdots & 2 \\ 0 & 1 & -3 & \vdots & -1 \\ 0 & 0 & 2 & \vdots & 2 \end{bmatrix}$

$\rightarrow \begin{bmatrix} 1 & 0 & 4 & \vdots & 3 \\ 0 & 1 & -3 & \vdots & -1 \\ 0 & 0 & 2 & \vdots & 2 \end{bmatrix} \rightarrow \begin{bmatrix} 1 & 0 & 4 & \vdots & 3 \\ 0 & 1 & -3 & \vdots & -1 \\ 0 & 0 & 1 & \vdots & 1 \end{bmatrix}$

$\rightarrow \begin{bmatrix} 1 & 0 & 0 & \vdots & -1 \\ 0 & 1 & 0 & \vdots & 2 \\ 0 & 0 & 1 & \vdots & 1 \end{bmatrix}$; The solution is $(-1, 2, 1)$.

11. Let $[A] = \begin{bmatrix} 2 & 1 \\ -\frac{3}{2} & -\frac{1}{2} \end{bmatrix}$, $[X] = \begin{bmatrix} x \\ y \end{bmatrix}$, $[B] = \begin{bmatrix} 13 \\ 10 \end{bmatrix}$; the solution is $[X] = [A]^{-1}[B]$. We found the inverse in Problem 5; then

$[X] = \begin{bmatrix} -1 & -2 \\ 3 & 4 \end{bmatrix} \begin{bmatrix} 13 \\ 10 \end{bmatrix} = \begin{bmatrix} -33 \\ 79 \end{bmatrix}$. The solution is $(-33, 79)$.

12.

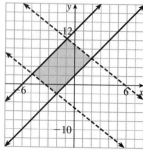

13. Let x = number of bars of Product I.
    y = number of bars of Product II.

	A	B
Product I:	3x	5x
Product II:	4y	7y
Total:	33	56

Thus,

$$\begin{cases} 3x + 4y = 33 \\ 5x + 7y = 56 \end{cases}$$

The solution to this system is (7, 3). This means she uses 7 bars of Product I and 3 bars of Product II.

14. Let x, y, and z be the number of products I, II, and III that can be manufactured. This leads to the following system:

$$\begin{cases} 2x + 2y + 3z = 1{,}250 \\ x + 2y + 2z = 900 \\ x + y + 2z = 750 \end{cases}$$

The solution to this system is (100, 150, 250). This means the number of product I to be manufactured is 100; product II, 150; and product III, 250.

15. Let x = number of bushels of corn; y = number of bushels of wheat. Maximize profit, $P = 210x + 100y$. The constraints are:

$$\begin{cases} x \geq 0 \\ y \geq 0 \\ x + y \leq 500 \\ 100x + 40y \leq 18{,}000 \\ 120x + 60y \leq 24{,}000 \end{cases}$$

The corner points are (100, 200), (0, 400), and (180, 0).

Corner Point	Value of Objective Function
(100, 200)	41,000
(0, 400)	40,000
(180, 0)	37,800

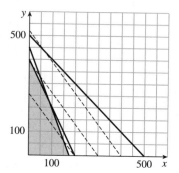

This means the farmer will maximize the profit if 100 acres of corn and 200 acres of wheat are planted (100 acres left unplanted).

APPENDIX A   Problem Solving, page 860

1. Must draw at least five cards.   **3.** 5   **5.** 12 months
7. Nothing can be said about Deidre's age as compared to
   Chelsea's age.
9. 2,376 kilomiles   **11.** 119   **13.** 945   **15.** $\frac{3}{8}$ **17.** 2
19. **a.** tetrahedron   **b.** cube (hexahedron)   **c.** octahedron
    d. icosahedron   **e.** dodecahedron
21. 1,499
23. More than a million trees must be cut down to print a trillion
    one-dollar bills.
25. The area is approximately 13.7 in.^{2}.   **27.** $\frac{1}{3}$
29. She had $22 at the start of the first day.
31. The cost is $390 if 100 make the trip.
33. There are 33 paths.   **35.** It will take until May 2050.
37.

```
10 REM FIBONACCI SEQUENCE
20 PRINT "This program calculates the first 20 numbers in a Fibonacci sequence."
30 PRINT
40 PRINT "NUMBER", "FIBONACCI SEQUENCE"
45 REM SET THE VALUES FOR THE FIRST TWO NUMBERS
50 LET FIRST = 1
60 LET SECOND = 1
70 PRINT "1", FIRST
80 PRINT "2", SECOND
95 REM RESET VALUES TO USE IN THE FORMULA
100 LET NEXTLAST = FIRST
110 LET LAST = SECOND
120 FOR J = 3 TO 20
125    REM CALCULATE THE NEXT FIBONACCI NUMBER IN THE SEQUENCE
130    NEW = LAST + NEXTLAST
140    PRINT J, NEW
145    REM RESET VALUES FOR THE NEXT STEP
150    LET NEXTLAST = LAST
160    LET LAST = NEW
170 NEXT J
180 END
```

39. 3

Credits

CHAPTER 1

13 ■ Photo by Freeman/Grishaber. Courtesy of Photo Edit.

19 ■ From "The Thoth Maneuver," by Clifford A. Pickover, *Discover*, March 1996, p. 108. Clifford Pickover/© 1996. Reprinted with permission of *Discover* Magazine. Nenad Jakesevic and Sonja Lamut/© 1996. Reprinted with permission of *Discover* Magazine.

26 ■ Photo © David Weintraub.

30 ■ Frontispiece of a book written by Augustus De Morgan in 1838.

31 ■ Problem 6 from the 1987 Examination of the California Assessment Program (along with analysis from 1989) in *A Question of Thinking: A First Look at Students' Performance on Open-ended Questions in Mathematics.*

33 ■ Photo © Robert E. Daemmrich/Tony Stone Images.

35 ■ Circle intersection puzzle from *Math Puzzles and Logic Problems* © 1995 by Dell Magazines, a division of Penny Marketing Limited Partnership, reprinted by permission of Dell Magazines.

39 ■ *B. C.* cartoon reprinted by permission of Johnny Hart and Creators Syndicate.

41 ■ Cartoon used with permission from *The Saturday Evening Post* © 1960. ■ Quotation in News Clip from *Everybody Counts*, A Report to the Nation on the Future of Mathematics Education, National Research Council, Washington, D.C., 1989, p. 3.

47 ■ Photo courtesy of Corbis-Bettmann.

50 ■ *B. C.* cartoon reprinted by permission of Johnny Hart and Creators Syndicate.

51 ■ Photo by NASA. Courtesy of Stock Boston.

53 ■ Photos courtesy of Texas Instruments, Inc.

54 ■ Quotation in News Clip from *Everybody Counts*, p. 61.

59 ■ DENNIS THE MENACE® used by permission of Hank Ketcham and © by North American Syndicate.

60 ■ Photo courtesy of Anaheim Public Information.

61 ■ *Graffiti* reprinted by permission of Newspaper Enterprise Association, Inc.

62 ■ Quotations in News Clip from *Teacher's Conference*, 1703, *Principal's Association*, 1815, *National Association of Teachers*, 1907, *The Rural American Teacher*, 1929, *PTA Gazette*, 1941, and *Federal Teacher*, 1950.

65 ■ Photo of oranges © Fredrik Bodin/Stock Boston. ■ Photo of crowd © John Coletti/Stock Boston. ■ Photo of chairs by David Gilkey. Courtesy of the *Daily Camera*, Boulder, Colorado.

67 ■ Photos courtesy of Grundtvig Church.

68 ■ Photo © Rafael Macia/Photo Researchers.

CHAPTER 2

81 ■ Photo courtesy Vermont Agency of Transportation.

86 ■ *B. C.* cartoon reprinted by permission of Johnny Hart and Creators Syndicate.

88 ■ Cartoon by Robert Mankoff from Saturday Review, June 11, 1977. Copyright © 1977 by Robert Mankoff.

90 ■ PIXIES reprinted by permission of United Feature Syndicate, Inc.

105 ■ Copyright © 1977 by Sidney Harris— American Scientist magazine. Reprinted by permission.

115 ■ Beetle Bailey cartoon © 1974 by King Features Syndicate, Inc.

134 ■ From "Ask Marilyn," by Marilyn vos Savant, *Parade Magazine*, October 31, 1993. Reprinted with permission from Parade, copyright © 1993.

CHAPTER 3

138 ■ Drawing by Alan Dunn, 1952, appeared in *The New Yorker Magazine*.

141 ■ Photo by Erich Lessing/Photo Edit.

146 ■ From *Margarita Philosophica Nova*, 1512. Courtesy of Museum of the History of Science, University of Oxford.

153 ■ Photo courtesy of NASA.

159 ■ Cartoon reprinted by permission of Datamation Magazine. Copyright © by Technical Publishing Company, A Division of Dun-Donnelley Publishing Corporation, 1969. All rights reserved.

160 ■ Photo courtesy of International Business Machines Corporation.

161 ■ Photo of slide rule courtesy of Pickett Industries. ■ Photo of Pascal's calculator courtesy of International Business Machines Corporation. ■ Photo of Leibniz' reckoning machine courtesy of International Business Machines Corporation. ■ Photos of Babbage's calculating machine: *left*, by David Parker/Science Museum/SPL/Photo

935

648 ■ Photo courtesy AP/Wide World Photos.

649 ■ Cartoon by Doug. Reprinted by permission from *The Saturday Evening Post* © 1976.

659 ■ PEANUTS reprinted by permission of United Feature Syndicate, Inc. ■ Advertisement courtesy of U.S. JVC Corp.

660 ■ Text of Little Caesar's Pizza Pizza television advertisement.

671 ■ Essay from *Innumeracy: Mathematical Illiteracy and Its Consequences*, by John Paulos, pp. 54–55. Copyright © 1988 John Paulos.

674 ■ Keno Payouts courtesy of The California Lottery.

675 ■ "The Thinker" cartoon courtesy of The Bettmann Archive.

682 ■ Advertisement courtesy of Wendy's International, Inc.

CHAPTER 10

693 ■ Graph based on data from the *Press Democrat*, Santa Rosa, CA.

694 ■ Graph based on data from the *Press Democrat*, Santa Rosa, CA.

697 ■ Anacin advertisement from *Product Movers* Advertising supplement. Home price graph based on data from the *Press Democrat*, Santa Rosa, CA, January 29, 1981.

700 ■ Graph (*top*) based on data from the *Press Democrat*, Santa Rosa, CA.

701 ■ Illustration of Blood Alcohol Concentration chart courtesy of TECH•arts based on a chart prepared by the Department of Motor Vehicles in cooperation with the California Highway Patrol.

704 ■ Advertisement courtesy of Anacin.

705 ■ Graph (*top*) from the U.S. Forest Service.

706 ■ Illustration courtesy of TECH•arts, from an advertisement © 1971 by SAAB-Scania of America, Inc.

707 ■ Problem 67 adapted from "Caveat Emptor," by Robert Leighton in *Games*, January 1988, p. 38. ■ PEANUTS reprinted by permission of United Feature Syndicate, Inc.

720 ■ Problem 60 is from "Instances of Simpson's Paradox," by Thomas R. Knapp in the *College Mathematics Journal*, July 1985, pp. 209–211.

721 ■ Photo courtesy of the U.S. Air Force Museum, Dayton, Ohio.

736 ■ Illustration (Problem 5) by Andrew Christie from "Diet and Cancer," by Leonard A. Cohen, *Scientific American*, November 1987, p. 44. Copyright © 1987 by Scientific American, Inc. All rights reserved. ■ Illustration (Problem 6) by Patricia J. Wynne from "How Dinosaurs Ran," by R. McNeill Alexander, *Scientific American*, April 1991, p. 132. Copyright © 1991 by Scientific American, Inc. All rights reserved.

738 ■ Photo courtesy Corbis-Bettmann.

746 ■ News Clip taken from a DEAR ABBY column by Abigail Van Buren. © 1973. Dist. by UNIVERSAL PRESS SYNDICATE. Reprinted with permission. All rights reserved.

CHAPTER 11

758 ■ Map of Venus courtesy of NASA; artwork by TECH•arts.

770 ■ *B. C.* cartoon reprinted by permission of Johnny Hart and Creators Syndicate.

774 ■ *B. C.* cartoon reprinted by permission of Johnny Hart and Creators Syndicate.

780 ■ Illustration in News Clip by Hank Iken from "Plateau Uplift and Climatic Change," by William F. Ruddman and John E. Kutzbach, *Scientific American*, March 1991, p. 68. Copyright © 1991 by Scientific American, Inc. All rights reserved.

784 ■ Photo © David Young-Wolff/Photo Edit.

CHAPTER 12

822 ■ Cartoon courtesy of Patrick J. Boyle.

847 ■ Photo property of AT&T Archives. Reprinted with permission of AT&T.

APPENDIX A

860 ■ News Clip from "Ask Marilyn," by Marilyn vos Savant, *Parade Magazine*, September 27, 1992. Reprinted with permission from Parade, copyright © 1992.

863 ■ News Clip from "Ask Marilyn," by Marilyn vos Savant, *Parade Magazine*, July 16, 1995. Reprinted with permission from Parade, copyright © 1995.

FRONT ENDPAPER

A Worldwide Timeline, The Global Nature of Mathematics ■ Illustration of Aztec stone calendar. *Source*: National Museum of Anthropology and History, Mexico City. After M. Coe, in ARCHAEO-ASTRONOMY IN PRE-COLUMBIAN AMERICA edited by Anthony F. Aveni, Copyright © 1975. By Permission of the University of Texas Press.

Mathematical

1202	Fibonacci: arithmetic, algebra, geometry, sequences, *Liber Abaci*
1250	Sacrobosco: Hindu-Arabic numerals
1260	Campanus translates Euclid
1267	Roger Bacon: *Opus*
1280	Geometry used as the basis of painting
1281	Li Yeh introduces notation for negative numbers
1303	Chu Shi-Kie: algebra, solutions of equations, Pascal's triangle
1325	Thomas Bradwardine: arithmetic, geometry, star polygons
1360	Nicole Oresme: coordinates, fractional exponents
1370	Chinese version of Pascal's triangle
1400	In Florence, commercial activity results in several books on mercantile arithmetic
1425	Use of perspective gives depth to Renaissance painting
1435	Ulugh Beg: trig tables
1460	Georg von Peurbach: arithmetic, table of sines
1464	Regiomontanus: establishes trigonometry
1470	First printed arithmetic book

Timeline: 1200 1210 1220 1230 1240 1250 1260 1270 1280 1290 1300 1310 1320 1330 1340 1350 1360 1370 1380 1390 1400 1410 1420 1430 1440 1450 1460 1470

Cultural

1206	Genghis Khan becomes chief prince of the Mongols
1209	Francis of Assisi initiates brotherhood
1211	Genghis Khan invades China
1215	Magna Carta signed
1233	Start of the Papal inquisition
1240	Amiens Cathedral rebuilt
1260	Chartres Cathedral completed
1273	Thomas Aquinas: *Summa Theologicae*
1275	Moses de Leon: *Zohar*, major source for the cabala
1299	Florentine bankers are forbidden to use Hindu numerals
1307–21	Dante: *Divine Comedy*
1321	Chaucer
1322	The pope forbids the use of counterpoint in church music
1324	Mansa Musa's pilgrimage to Mecca
1325	Aztecs establish Tenochtitlán
1347–51	Approximately 75 million die of the Black Death
1378	Beginning of the Great Schism
1390	Chaucer: *Canterbury Tales*
1413	First recorded use of perspective, by Brunelleschi
1417	End of Great Schism
1429	Joan of Arc raises siege of Orleans
1435	Rogier Van der Weyden
1436	Fra Angelico begins frescoes at San Marco
1450	Florence is center of Renaissance
1454	Gutenberg prints Bible
1465	First printed music
1470	First illustrated books

1480–1599

Mathematics	Decade	History & Culture
1482 First printing of Euclid's *Elements*	**1480**	1484 Botticelli: *Birth of Venus*
1489 Johann Widmann: first use of + and − signs	**1490**	1492 Columbus discovers America
1492 Pellos: use of decimal point		1497 Vasco da Gama rounds Cape of Good Hope
1505 Leonardo da Vinci: geometry, art, optics	**1500**	1504 Michelangelo: *David*
1506 Scipione dal Ferro: cubic equations		1507 Leonardo da Vinci begins *Leda and the Swan*
1510 Albrecht Dürer: perspective, polyhedra, curves	**1510**	1509 First slaves brought to America
1514 Dürer's *Melancholia* contains magic square		1513 Machiavelli: *The Prince*
1525 Stifel: number mysticism; Rudolff: algebra, decimals	**1520**	1517 Luther launches Reformation
1527 Petrus Apianus; Pascal's triangle		1520 Magellan discovers the straits; Luther excommunicated
1530 Copernicus: astronomy, trigonometry	**1530**	1534 Henry VIII becomes head of the Church of England
1540 Gemma Frisius: arithmetic	**1540**	1543 Publication of Copernicus' work
1545 Tartaglia: cubic equations, arithmetic		
Ferrari: quartic equations		
1550 Cardano: *Ars Magna*	**1550**	1558 Elizabeth crowned Queen of England
Scheubel: algebra		
1557 Adam Riese: originator of the radical sign	**1560**	1567 Bothwell abducts Mary, Queen of Scots
Robert Recorde: arithmetic, algebra, first use of = sign		1569 Tycho Brahe begins construction of a 19-foot quadrant
1564 Galileo Galilei born		
1572 Bombelli: algebra, cubic equations	**1570**	1573 Francis Drake sees Pacific Ocean
1579 Viète: advocated use of decimal notation	**1580**	1583 Pope Gregory XIII creates new calendar
1580 Viète: algebra, geometry, much modern notation		1588 England defeats Spanish Armada
1583 Clavius: arithmetic, algebra, geometry	**1590**	
1593 Adrianus Romanus: value of π	**1599**	1596 Discovery of the Marquesas; invention of the flush toilet

Century of Enlightenment—1600 to 1699

Mathematics	Decade	History & Culture
1600 Galileo: physics, astronomy, projectiles; earth revolves around sun	**1600**	1600 Shakespeare: *Hamlet*
1601 Pierre de Fermat born		1605 Cervantes: *Don Quixote*
		1609 Pocahontas saves John Smith
1610 Kepler: astronomy, continuity	**1610**	1611 King James Bible published
1617 Napier: logarithms, Napier's rods		1618 Beginning of the Thirty Years War
1621 Diophantus: *Arithmetica* published	**1620**	1620 Pilgrims land at Plymouth
1623 Blaise Pascal: Pascal's triangle		
1630 Mersenne: number theory	**1630**	1636 Harvard College founded, the first American college
1631 Oughtred: first table of natural logs		
1635 Cavalieri: number theory		
1637 Descartes: *Discourse on Method*, analytic geometry	**1640**	1643 Moliere founds *Theatre de la Comedie Francaise*
Fermat's Last Theorem stated in the margin of a book		1646 Building of the Taj Mahal
1640 Desargues: projective geometry		1649 Cromwell abolishes English monarchy
1642 Pascal: conics, probability, computing machines;	**1650**	1654 Coronation of Louis XIV of France
John Wallis: algebra, imaginary numbers		1656 First pendulum clock
1654 Pascal-Fermat correspondence begins study of probability		1659 Birth of Alessandro Scarlatti
1658 Huygens invents the pendulum clock—theory of curves	**1660**	1665 Great Plague in London kills 75,000
1670 Sir Christopher Wren: architecture, imaginary numbers	**1670**	Newton's experiments on gravitation
1678 Ceva: nature of concurrency	**1680**	1668 La Fontaine: *Fables*
1680 Sir Isaac Newton: calculus, gravitation, series, hydrodynamics		1677 Spinoza: *Ethics*
1682 Gottfried Leibniz: calculus, determinants, symbolic logic, notation, computing machines		1680 Stradivari makes the first cello
		1683 First public museum
1690 Nicolaus Bernoulli: probability curves	**1690**	1685 J. S. Bach and Handel born
	1699	1689 Peter the Great becomes Czar of Russia

Mathematical

1700	Jacob and Johann Bernoulli: applied calculus, probability
1715	Brook Taylor: series, geometry, calculus of finite differences
1720	Abraham de Moivre: probability, calculus, complex numbers
1733	Saccheri: beginnings of analytic geometry
1735	Emilie de Breteuil: Newtonian studies
1740	Colin Maclaurin: series, physics, higher plane curves
1742	Goldbach's conjecture
1748	Maria Agnesi: analytic geometry
1750	Leonhard Euler: number theory, applied mathematics
1760	Compte de Buffon: connection between probability and π
1761	Johann Peter: population statistics
1770	Johann Lambert: irrationality of π, non-Euclidean geometry, map projections
1780	Lagrange: calculus, number theory
1790	Metric system invented
1796	Karl Gauss: Num $= \Delta + \Delta + \Delta$
1797	Caroline Herschel: astronomy
1799	Metric system adopted in France
1805	Laplace: probability, differential equations, method of least squares, integrals; punched cards to operate Jacquard loom
1815	George Boole born
1820	Sophie Germain: theory of numbers
1822	Feuerbach: geometry of the triangle
1824	Abel: elliptic functions, equations, series, calculus
1825	Bolyai and Lobachevski: non-Euclidean geometry
1830	Cauchy: calculus, complex variables
1831	Galois dies in a duel
1832	Babbage: calculating machines; Galois: groups, theory of equations
1837	Trisection of an angle and duplication of the cube proved impossible
1843	Hamilton: quaternions
1849	De Morgan: probability, logic
1850	Cayley: invariants, hyperspace, matrices and determinants
1852	Byron: first programming in weaving industry
1854	Riemann: calculus; Boole: logic, Laws of Thought
1855	Dirichlet: number theory

Decade markers: 1700 | 1710 | 1720 | 1730 | 1740 | 1750 | 1760 | 1770 | 1780 | 1790 | 1800 | 1810 | 1820 | 1830 | 1840 | 1850

Cultural

1705	Haley predicts return of 1682 comet
1706	Benjamin Franklin born
1710	Leibniz: *Théodicée*
1712	Last execution for witchcraft in England
1722	Bach creates a standard musical scale
1726	Jonathan Swift: *Gulliver's Travels*
1738	Bach: *Mass in B Minor*
1740	Accession of Fredrick the Great; Israel Baal Shem Toh founds Hasidism
1742	Handel: *The Messiah*
1756–63	The Seven Years War
1759	Voltaire: *Candide*
1762	Rousseau: *Social Contract*
1764	Paris Pantheon started
1767	Isaac Watts: steam engine; Industrial Revolution begins
1776	American Declaration of Independence
1779	Mozart: *Don Giovanni*
1783	Czarina Catherine
1789	French Revolution
1796	Smallpox vaccine
1799	Napoleon rules France; Rosetta Stone found
1803	Robert Fulton: first steamboat
1804	Beethoven: *Eroica* symphony; Haiti independence
1808	Goethe: *Faust, Part I*
1810	Goya: *The Disasters of War*
1811	First mechanical press
1812	Canned food
1815	Battle of Waterloo
1821	Rosetta Stone deciphered
1826	First photograph
1828	Alexander Dumas: *The Three Musketeers*
1830	Simon Bolivar liberates South America
1834	Refrigeration
1836	The first telegraph
1842	Henry Wadsworth Longfellow: *The Village Blacksmith*
1846	Anesthesia used in surgery
1848	Karl Marx: *Communist Manifesto*
1851	Herman Melville: *Moby Dick*
1855	Walt Whitman: *Leaves of Grass*